SECTION GUIDE

MANUAL OF STEEL CONSTRUCTION

LOAD & RESISTANCE FACTOR DESIGN

FIRST EDITION

Copyright © 1986

by

American Institute of Steel Construction, Inc.

Printed in the United States of America
Third Printing — 3/90

FOREWORD

The American Institute of Steel Construction, founded in 1921, is the non-profit technical specifying and trade organization for the fabricated structural steel industry in the United States. Executive and engineering headquarters of AISC are maintained in Chicago.

The Institute is supported by three classes of membership: Active Members totaling 400 companies engaged in the fabrication and erection of structural steel, Associate Members who are allied product manufacturers and Professional Members who are individuals or firms engaged in the practice of architecture or engineering. Professional Members also include architectural and engineering educators. The continuing financial support and active participation of Active Members in the engineering, research and development activities of the Institute make possible the publishing of this *Load and Resistance Factor Design Manual of Steel Construction*.

The Institute's objectives are to improve and advance the use of fabricated structural steel through research and engineering studies to develop the most efficient and economical design of structures. It also conducts programs to improve product quality.

To accomplish these objectives the Institute publishes manuals, textbooks, specifications and technical booklets. Best known and most widely used is the *Manual of Steel Construction* which holds a highly respected position in engineering literature. Outstanding among AISC standards are the *Specification for the Design, Fabrication and Erection of Structural Steel for Buildings* and the *Code of Standard Practice for Steel Buildings and Bridges*.

The Institute also assists designers, contractors, educators and others by publishing technical information and timely articles on structural applications through two quarterly publications, *Engineering Journal* and *Modern Steel Construction*. In addition, public appreciation of aesthetically designed structures is encouraged through its annual award programs: Prize Bridges, Architectural Awards of Excellence and student Fellowship Awards.

The First Edition of the *Load and Resistance Factor Design Manual* was produced under the guidance of the AISC Manual and Textbook Committee, made up of experienced and knowledgeable engineers from industry and member fabricator companies. The valuable assistance furnished by the American Iron and Steel Institute and the Welded Steel Tube Institute in assembling data and in generating many of the tables by electronic computers is gratefully acknowledged.

PROP.

PART 1
Dimensions and Properties

AMERICAN INSTITUTE OF STEEL CONSTRUCTION

STRUCTURAL STEELS

PRODUCT AVAILABILITY

Section A3.1 of the AISC Load and Resistance Factor Design *Specification for Structural Steel Buildings* lists fourteen ASTM specifications for structural steel approved for use in building construction.

Six of these steels are available in hot-rolled structural shapes, plates, and bars. One steel, ASTM A514, is only available in plates. Table 1 shows five groups of shapes and eleven ranges of thicknesses of plates and bars available in the various minimum yield stress* and tensile strength levels afforded by the seven steels. For complete information on each steel, reference should be made to the appropriate ASTM specification. A listing of the shape sizes included in each of the five groups follows in Table 2, corresponding to the groupings given in Table A of ASTM Specification A6.

Seven additional grades of steel, other than those covering hot-rolled shapes, plates, and bars, are listed in Part 6, Sect. A3.1. These steels cover pipe, cold- and hot-formed tubing, and cold- and hot-rolled sheet and strip.

The principal producers of shapes listed in Part 1 of this Manual are shown in Table 3. Availability and the principal producers of structural tubing are shown in Tables 4 and 5. For additional information on availability and classification of structural steel plates and bars, refer to the separate discussion beginning on pg. 1-121.

Space does not permit inclusion in Table 3, or in the listing of shapes and plates in Part 1 of this Manual, of all rolled shapes, or plates of greater thickness that are occasionally used in construction. For such products, reference should be made to the various producers' catalogs.

SELECTION OF THE APPROPRIATE STRUCTURAL STEEL

ASTM A36 is the all-purpose carbon grade steel widely used in building and bridge construction. ASTM A529 structural carbon steel, ASTM A441 and A572 high-strength low-alloy structural steels, ASTM A242 and A588 atmospheric corrosion-resistant high-strength low-alloy structural steels, and ASTM A514 quenched and tempered alloy structural steel plate may each have certain advantages over ASTM A36 structural carbon steel, depending on the application. These high-strength steels have proven to be economical choices where lighter members, resulting from use of higher design strengths, are not penalized because of instability, local buckling, deflection, or other similar reasons. They are frequently used in tension members, beams in continuous and composite construction where deflections can be minimized, and columns having low slenderness ratios. The reduction of dead load, and associated savings in shipping costs, can be significant factors. However, higher strength steels are not to be used indiscriminately. Effective use of all steels depends on thorough cost and engineering analysis.

With suitable procedures and precautions, all steels listed in the AISC Specification are suitable for welded fabrication.

ASTM A242 and A588 atmospheric corrosion-resistant high-strength low-alloy steels can be used in the bare (uncoated) condition in most atmospheres. Where boldly exposed under such conditions, exposure to the normal atmosphere causes a tightly adherent oxide to form on the surface which protects the steel from further atmospheric corrosion. To achieve the benefits of the enhanced atmospheric corrosion resistance of these bare steels, it is necessary that design, detailing, fabrication,

---

*As used in the AISC LRFD Specification, "yield stress" denotes either the specified minimum yield point (for those steels that have a yield point) or specified minimum yield strength (for those steels that do not have a yield point).

erection and maintenance practices proper for such steels be observed. Designers should consult with the steel producers on the atmospheric corrosion-resistant properties and limitations of these steels prior to use in the bare condition. When either A242 or A588 steel is used in the coated condition, the coating life is typically longer than with other steels. Although A242 and A588 steels are more expensive than other high-strength, low-alloy steels, the reduction in maintenance resulting from the use of these steels usually offsets their higher initial cost.

BRITTLE FRACTURE CONSIDERATIONS IN STRUCTURAL DESIGN
General Considerations

As the temperature decreases, an increase is generally noted in the yield stress, tensile strength, modulus of elasticity, and fatigue strength of the structural steels. In contrast, the ductility of these steels, as measured by reduction in area or by elongation, decreases with decreasing temperatures. Furthermore, there is a temperature below which a structural steel subjected to tensile stresses may fracture by cleavage,* with little or no plastic deformation, rather than by shear,* which is usually preceded by a considerable amount of plastic deformation or yielding.

Fracture that occurs by cleavage at a nominal tensile stress below the yield stress is commonly referred to as brittle fracture. Generally, a brittle fracture can occur in a structural steel when there is a sufficiently adverse combination of tensile stress, temperature, strain rate and geometrical discontinuity (notch) present. Other design and fabrication factors may also have an important influence. Because of the interrelation of these effects, the exact combination of stress, temperature, notch and other conditions that will cause brittle fracture in a given structure cannot be readily calculated. Consequently, designing against brittle fracture often consists mainly of (1) avoiding conditions that tend to cause brittle fracture, and (2) selecting a steel appropriate for the application. A discussion of these factors is given in the following sections. References 1 through 5 cover the subject in much more detail.

Conditions Causing Brittle Fracture

It has been established that plastic deformation can occur only in the presence of shear stresses. Shear stresses are always present in a uniaxial or in biaxial state-of-stress. However, in a triaxial state-of-stress, the maximum shear stress approaches zero as the principal stresses approach a common value, and thus, under equal triaxial tensile stresses failure occurs by cleavage rather than by shear. Consequently, triaxial tensile stresses tend to cause brittle fracture and should be avoided. A triaxial state-of- stress can result from a uniaxial loading when notches or geometrical discontinuities are present.

Increased strain rates tend to increase the possibility of brittle behavior. Thus, structures that are loaded at fast rates are more susceptible to brittle fracture. However, a rapid strain rate or impact load is not a required condition for a brittle fracture.

Cold work, and the strain aging that normally follows, generally increases the likelihood of brittle fracture. This behavior is usually attributed to the previously mentioned reduction in ductility. The effect of cold work that occurs in cold forming operations can be minimized by selecting a generous forming radius and, thus, limiting

---

*Shear and cleavage are used in the metallurgical sense (macroscopically) to denote different fracture mechanisms. Reference 2, as well as most elementary textbooks on metallurgy, discusses these mechanisms.

the amount of strain. The amount of strain that can be tolerated depends on both the steel and the application.

When tensile residual stresses are present, such as those resulting from welding, they add to any applied tensile stress and thus, the actual tensile stress in the member will be greater than the applied stress. Consequently, the likelihood of brittle fracture in a structure that contains high residual stresses may be minimized by a postweld heat treatment. The decision to use a postweld heat treatment should be made with assurance that the anticipated benefits are needed and will be realized, and that possible harmful effects can be tolerated. Many modern steels for welded construction are designed to be used in the less costly as-welded condition when possible. The soundness and mechanical properties of welded joints in some steels may be adversely affected by a postweld heat treatment.

Welding may also contribute to the problem of brittle fracture by introducing notches and flaws into a structure and by causing an unfavorable change in microstructure of the base metal. However, by properly designing welds, including care in selecting their location and by using good welding practice, such detrimental effects can be minimized. The proper electrode must be selected so that the weld metal will be as resistant to brittle fracture as the base metal.

Selecting a Steel

The best guide in selecting a steel that is appropriate for a given application is experience with existing and past structures. The A36 steel has been successfully used in a great number of applications, such as buildings, transmission towers, transportation equipment, and bridges, even at the lowest atmospheric temperatures encountered in the U.S. Therefore, it appears that any of the structural steels, when designed and fabricated in an appropriate manner, could be used for similar applications with little likelihood of brittle fracture. Consequently, brittle fracture is not usually experienced in such structures unless unusual temperature, notch, and stress conditions are present. Nevertheless, it is always desirable to avoid or minimize the previously cited adverse conditions that increase the susceptibility to brittle fracture.

A requirement that steels must absorb a certain amount of energy, 15 ft-lb or higher (Charpy V-notch test), at the lowest expected operating temperature is used in some applications where notch toughness is considered very important. Although this may be a suitable requirement for some severe applications, it should be emphasized that a large number of structures have performed successfully, even though the steels used did not meet a 15 ft-lb requirement at the lowest operating temperature. Therefore, this requirement should not be applied indiscriminately to all structures.

LAMELLAR TEARING

The information on strength and ductility presented in the previous sections generally pertains to loadings applied in the planar direction (longitudinal or transverse orientation) of the steel plate or shape. It should be noted that elongation and area reduction values may well be significantly lower in the through-thickness direction than in the planar direction. This inherent directionality is of small consequence in many applications, but does become important in the design and fabrication of structures containing massive members with highly restrained welded joints.

With the increasing trend toward heavy welded-plate construction, there has been a broader recognition of occurrences of lamellar tearing in some highly restrained joints of welded structures, especially those using thick plates and heavy structural shapes. The restraint induced by some joint designs in resisting weld deposit shrinkage can impose tensile strain sufficiently high to cause separation or tearing on planes parallel to the rolled surface of the structural member being joined. The incidence of this phenomenon can be reduced or eliminated through greater understanding by designers, detailers and fabricators of (1) the inherent directionality of constructional forms of steel; (2) the high restraint developed in certain types of connections, and (3) the need to adopt appropriate weld details and welding procedures with proper weld metal for through-thickness connections. Further, steels can be specified to be produced by special practices and/or processes to enhance through-thickness ductility and thus assist in reducing the incidence of lamellar tearing. Steels produced by such practices are available from several producers. However, unless precautions are taken in both design and fabrication, lamellar tearing may still occur in thick plates and heavy shapes of such steels at restrained through-thickness connections. Some guidelines in minimizing potential problems have been developed.[6]

REFERENCES

1. *Brockenbrough, R. L. and B. G. Johnston*   USS Steel Design Manual   *US Steel, 1981.*
2. *Parker, E. R.*   Brittle Behavior of Engineering Structures   *John Wiley & Sons, New York, N.Y., 1957.*
3. *Welding Research Council*   Control of Steel Construction to Avoid Brittle Failure,   *1957.*
4. *Lightner, M. W. and R. W. Vanderbeck*   Factors Involved in Brittle Fracture   *Regional Technical Meetings, American Iron and Steel Institute, 1956.*
5. *Rolfe, S. T. and J. M. Barsom*   Fracture and Fatigue Control in Structures—Applications of Fracture Mechanics   *Prentice-Hall, Inc., Englewood Cliffs, N.J., 1977.*
6. *American Institute of Steel Construction, Inc.*   Commentary on Highly Restrained Welded Connections   *AISC Engineering Journal, 3rd Qtr., 1973, Chicago, Ill.*

STRUCTURAL SHAPES—
DESIGNATIONS, DIMENSIONS AND PROPERTIES

The hot-rolled shapes shown in Part 1 of this Manual are published in ASTM Specification A6/A6M-84c, *Standard Specification for General Requirements for Rolled Steel Plates, Shapes, Sheet Piling and Bars for Structural Use.*

W shapes have essentially parallel flange surfaces. The profile of a W shape of a given nominal depth and weight available from different producers is essentially the same except for the size of fillets between the web and flange.

HP bearing pile shapes have essentially parallel flange surfaces and equal web and flange thicknesses. The profile of an HP shape of a given nominal depth and weight available from different producers is essentially the same.

American Standard beams (S) and American Standard channels (C) have a slope of approximately 16⅔% (2 in 12 inches) on their inner flange surfaces. The profiles of S and C shapes of a given nominal depth and weight available from different producers are essentially the same.

The letter M designates shapes that cannot be classified as W, HP, or S shapes. Similarly, MC designates channels that cannot be classified as C shapes. Because many of the M and MC shapes are only available from a limited number of producers, or are infrequently rolled, their availability should be checked prior to specifying these shapes. They have various slopes on their inner flange surfaces, dimensions for which may be obtained from the respective producing mills.

The flange thickness given in the tables for S, M, C and MC shapes is the *average* flange thickness.

In calculating the theoretical weights, properties and dimensions of the rolled shapes listed in Part 1 of this Manual, fillets and roundings have been included for all shapes except angles. The properties of these rolled shapes are based on the *smallest* theoretical size fillets produced; dimensions for detailing are based on the *largest* theoretical size fillets produced. These properties and dimensions are either exact or slightly conservative for all producers who offer them, except as noted in the footnotes to Table 3.

Equal leg and unequal leg angle (L) shapes of the same nominal size available from different producers have profiles which are essentially the same, except for the size of fillet between the legs and the shape of the ends of the legs. The k distance given in the tables for each angle is based on the largest theoretical size fillet available. Availability of certain angles is subject to rolling accumulation and geographical location, and should be checked with material suppliers.

TABLE 1
Availability of Shapes, Plates and Bars According to ASTM Structural Steel Specifications

Steel Type	ASTM Designation		F_y Minimum Yield Stress (ksi)	F_u Tensile Stress[a] (ksi)	Shapes Group per ASTM A6					Plates and Bars										
					[b]1	2	3	4	5	To ½" Incl.	Over ½" to ¾" Incl.	Over ¾" to 1¼" Incl.	Over 1¼" to 1½" Incl.	Over 1½" to 2" Incl.	Over 2" to 2½" Incl.	Over 2½" to 4" Incl.	Over 4" to 5" Incl.	Over 5" to 6" Incl.	Over 6" to 8" Incl.	Over 8"
Carbon	A36		32	58–80															■	
	A36		36	58–80[c]	■	■	■	■	■	■	■	■	■	■	■	■	■	■	■	
	A529		42	60–85	■	■				■	■									
High-Strength Low-Alloy	A441		40	60										■	■					
	A441		42	63			■						■							
	A441		46	67		■					■									
	A441		50	70	■					■										
	A572 Grade	42	42	60	■	■	■	■	■	■	■	■	■	■						
		50	50	65	■	■	■	■		■	■	■	■							
		60	60	75	■	■				■	■									
		65	65	80	■	■				■	■									
Corrosion-Resistant High-Strength Low-Alloy	A242		42	63			■						■							
	A242		46	67		■					■									
	A242		50	70	■					■										
	A588		42	63													■		■	
	A588		46	67											■					
	A588		50	70	■	■	■	■	■	■	■	■	■	■						
Quenched & Tempered Alloy	A514[d]		90	100–130												■				
	A514[d]		100	110–130						■	■	■	■	■	■					

[a]Minimum unless a range is shown.
[b]Includes bar-size shapes.
[c]For shapes over 426 lbs./ft, minimum of 58 ksi only applies.
[d]Plates only

■ Available.
☐ Not available.

TABLE 2
Structural Shape Size Groupings for Tensile Property Classification

Structural Shape	Group 1	Group 2	Group 3	Group 4	Group 5
W Shapes	W 24x55,62 W 21x44 to 57 incl W 18x35 to 71 incl W 16x26 to 57 incl W 14x22 to 53 incl W 12x14 to 58 incl W 10x12 to 45 incl W 8x10 to 48 incl W 6x9 to 25 incl W 5x16,19 W 4x13	W 40x149 to 268 incl W 36x135 to 256 incl W 33x118 to 152 incl W 30x90 to 211 incl W 27x84 to 194 incl W 24x68 to 176 incl W 21x62 to 166 incl W 18x76 to 143 incl W 16x67 to 100 incl W 14x61 to 132 incl W 12x65 to 106 incl W 10x49 to 112 incl W 8x58,67	W 40x277 to 328 incl W 36x230 to 359 incl W 33x201 to 354 incl W 30x261 W 27x235 to 258 incl W 24x192 to 229 incl W 21x182 to 223 incl W 18x158 to 192 incl W 14x145 to 211 incl W 12x120 to 190 incl	W 40x362 to 655 incl W 36x393 to 798 incl W 33x387 to 619 incl W 30x292 to 581 incl W 27x281 to 539 incl W 24x250 to 492 incl W 21x248 to 402 incl W 18x211 to 311 incl W 14x233 to 550 incl W 12x210 to 336 incl	W 36x848 W 14x605 to 730 incl
M Shapes	to 20 lb/ft incl				
S Shapes	to 35 lb/ft incl	over 35 lb/ft			
HP Shapes		to 102 lb/ft incl	Over 102 lb/ft		
American Standard Channels (C)	to 20.7 lb/ft incl	over 20.7 lb/ft			
Miscellaneous Channels (MC)	to 28.5 lb/ft incl	over 28.5 lb/ft			
Angles (L). Structural Bar-Size	to ½ in. incl	over ½ to ¾ in. incl	over ¾ in.		

Notes: Structural tees from W, M and S shapes fall in the same group as the structural shape from which they are cut. Group 4 and Group 5 shapes and Group 3 shapes with flange thickness greater than 1-½ in. are generally contemplated for application as columns or compression components. When used in other applications (e.g. trusses) and when thermal cutting or welding is required, the material specification and fabrication procedures should be carefully reviewed to minimize the possibility of cracking. Low toughness properties at the web mid-thickness and at the web-flange intersection may increase the possibility of cracking, especially at regions of stress concentration. The use of killed steel and special metallurgical requirements may be helpful to improve properties. The effects of tensile stresses resulting from welding and handling during fabrication and erection, as well as from design loads, should be considered. When connections are used in tension, bolting should be considered as an alternative to welding.

AMERICAN INSTITUTE OF STEEL CONSTRUCTION

TABLE 3
Principal Producers of Structural Shapes

B. Bethlehem Steel Corp.
E. Roanoke Electric Steel Corp.

F. Florida Steel Corp.
H. Chaparral Steel
I. Inland Steel Co.
L. LTV Steel Co.

M. SMI Steel, Inc.
N. Northwestern Steel & Wire
O. Ohio River Steel
R. Nucor Corporation

S. North Star Steel Co.
U. United States Steel Corp.
Y. Bayou Steel Corp.

Section, Weight per Ft	Producer Code	Section, Weight per Ft	Producer Code
W 40—all	*	W 10x22–30	B, I, N, U
W 36x328–848	*	W 10x12–19	B, I, N, H
W 36x135–359	B^6, U	W 8x31–67	B, I, N^5, U
W 33x263–619	*	W 8x24–28	B, I, N, U
W 33x118–354	B^6, U	W 8x18–21	B, I, N, H
W 30x235–581	*	W 8x10–15	B, I, N, H
W 30x99–235	B^6, U	W 6x15–25	B, I, N, H
W 30x90	*	W 6x9–16	B, I, N, H
W 27x194–539	*	W 5x16–19	B, I
W 27x84–217	B^6, U	W 4x13	B, H, Y
W 24x176–492	*		
W 24x55–176	B^6, I^6, U	S 24—all	B
W 21x166–402	*	S 20—all	B
W 21x44–166	B^6, I^6, U	S 18—all	B, U
W 18x130–311	*	S 15—all	B, U
W 18x76–143	B^6, I^6, U	S 12—all	B, U
W 18x50–71	B, I, N^1, U	S 10—all	B, I, U, O
W 18x35–46	B, I, N, U	S 8—all	B, I, H, O
W 16x67–100	B, I, U		
W 16x36–57	B, I, N^2, U	S 7x15.3	B, I
W 16x26–31	B, I, N, U	S 6—all	B, I, Y, H, O
W 14x145–730	B, U		
W 14x90–132	B, U	S 5x10	I, Y
W 14x61–82	B, I, U	S 4x9.5	I, Y, H, O
W 14x43–53	B, I, U	S 4x7.7	I, Y, H, O
W 14x30–38	B, I, N, U	S 3x7.5	I, Y, H
W 14x22–26	B, I, N, U	S 3x5.7	I, Y, H
W 12x65–336	B^3, I, U^3		
W 12x53–58	B, I, U	M 12x11.8	L, H
W 12x40–50	B, I, N, U	M 10x9	L, H
W 12x26–35	B, I, N, U	M 8x6.5	L, H
W 12x14–22	B, I, N, U, H	M 6x4.4	L, Y
W 10x49–112	B, I, N^4, U	M 5x18.9	B
W 10x33–45	B, I, N, U	M 4x13	I

^{1}W 18x65–71 excluded.
^{2}W 16x57 excluded.
^{3}W 12x210–336 excluded.
^{4}W 10x60–112 excluded.
^{5}W 8x48–67 excluded.
^{6}Check producers for size availability in heavier shapes.

*Not available from domestic producers.

Note: Maximum lengths of shapes obtainable vary widely with producers, but a conservative range for all mills is from 60 to 75 ft. Some mills will accept orders for lengths up to 120 ft, but only for certain shapes and subject to special arrangement. Consult the producers for unusual length requirements.

TABLE 3
Principal Producers of Structural Shapes (cont'd)

B. Bethlehem Steel Corp.   F. Florida Steel Corp.   M. SMI Steel, Inc.   S. North Star Steel Co.
E. Roanoke Electric Steel Corp. H. Chaparral Steel   N. Northwestern Steel & Wire U. United States Steel Corp.
                              I. Inland Steel Co.   O. Ohio River Steel   Y. Bayou Steel Corp.
                              L. LTV Steel Co.   R. Nucor Corporation

Section, Weight per Ft	Producer Code	Section, Weight per Ft	Producer Code
HP 14—all	B, U	MC 18—all	B, U
HP 13—all	I	MC 13—all	B, I, U
HP 12—all	B^6, I, U^7	MC 12x31–50	B, I, U
HP 10—all	B, I, N, U		
HP 8x36	B, I, N, U	MC 12x10.6	L, O
		MC 10x28.5—41.1	B, U
C 15—all	B, I, N^8, U	MC 10x22 & 25	B, I, U
C 12x30	B, N, U	MC 10x8.4	L, O
C 12x20.7–25	B, N, U, H, O	MC 10x6.5	L
C 10x30	B, I, N	MC 9—all	B
C 10x25	B, I, N	MC 8x21.4–22.8	B
C 10x15.3–20	B, I, N, H, O	MC 8x18.7–20	B, O
		MC 8x8.5	L
C 9x13.4–15	B, I, H, O	MC 7x19.1–22.7	B, U
C 8x18.75	B, I, N, O	MC 7x17.6	U
C 8x13.75	B, I, N, H, O	MC 6x18	B, I
C 8x11.5	B, I, N, H, O, M	MC 6x15.3	B, I, O
		MC 6x15.1–16.3	B, O
C 7x9.8–12.25	B, I, N, H, O	MC 6x12	B, O
C 6x13	I, N, Y, S, O		
C 6x10.5	I, N, Y, H, S, O		
C 6x8.2	I, N, Y, H, S, O, M		
C 5x9	I, N, Y, H, S, F		
C 5x6.7	I, N, Y, H, S, M, F		
C 4x7.25	I, J, N, Y, H, S, F		
C 4x5.4	I, J, N, Y, H, S, M, F		
C 3x6	N, Y, S, F		
C 3x4.1–5	I, N, Y, H, S, E, M, F		

[6]HP 12x84 excluded.   [8]C 15x40-50 excluded.
[7]HP 12x63 and HP 12x84 excluded.

Note: Maximum lengths of shapes obtainable vary widely with producers, but a conservative range for all mills is from 60 to 75 ft. Some mills will accept orders for lengths up to 120 ft, but only for certain shapes and subject to special arrangement. Consult the producers for unusual length requirements.

TABLE 3
Principal Producers of Structural Shapes (cont'd)

B. Bethlehem Steel Corp.
E. Roanoke Electric Steel Corp.

F. Florida Steel Corp.
H. Chaparral Steel
I. Inland Steel Co.
L. LTV Steel Co.

M. SMI Steel, Inc.
N. Northwestern Steel & Wire
O. Ohio River Steel
R. Nucor Corporation

S. North Star Steel Co.
U. United States Steel Corp.
Y. Bayou Steel Corp.

Section by Leg Length, Thickness	Producer Code	Section by Leg Length, Thickness	Producer Code
L 8 x8 x1⅛	B, I, U	L 3 x3 x ½	I, N, Y, R, M, F
x1	B, I, U	x ⁷⁄₁₆	F
x ⅞	B, I, U	x ⅜	I, N, U, Y, H, S, R, E, M, F
x ¾	B, I, U	x ⁵⁄₁₆	I, N, U, Y, H, S, R, E, M, F
x ⅝	B, I, U	x ¼	I, N, U, Y, H, S, R, E, M, F
x ⁹⁄₁₆	B, I, U	x ³⁄₁₆	I, N, Y, S, R, E, M, F
x ½	B, I, U		
		L 2½x2½x ½	N, Y, R, M, F
L 6 x6 x1	B	x ⅜	N, Y, H, S, R, E, M, F
x ⅞	B, Y	x ⁵⁄₁₆	N, Y, H, S, R, E, M, F
x ¾	B, I, Y, R	x ¼	N, Y, H, S, R, E, M, F
x ⅝	B, I, Y, R, M	x ³⁄₁₆	N, H, S, R, E, M, F
x ⁹⁄₁₆	B, I, R, M		
x ½	B, I, Y, H, R, O, M	L 2 x2 x ⅜	N, Y, H, S, R, E, M, F
x ⁷⁄₁₆	B, I, M	x ⁵⁄₁₆	N, Y, H, S, R, E, F
x ⅜	B, I, Y, H, R, O, M	x ¼	N, Y, H, S, R, E, M, F
x ⁵⁄₁₆	H, R, M	x ³⁄₁₆	N, H, S, R, E, M, F
		x ⅛	S, R, E, F
L 5 x5 x ⅞	B, I, Y		
x ¾	B, I, Y		
x ⅝	I, Y		
x ½	B, I, N, Y, H, R, M	L 8 x6 x1	B, I, U
x ⁷⁄₁₆	B, I, R	x ⅞	B
x ⅜	B, I, N, Y, H, R, M	x ¾	B, I, U
x ⁵⁄₁₆	B, I, N, Y, H, R, M	x ⅝	
		x ⁹⁄₁₆	B, I, U
L 4 x4 x ¾	I, Y	x ½	B, I, U
x ⅝	I, Y	x ⁷⁄₁₆	B, I
x ½	I, N, Y, H, S, R, M, F		
x ⁷⁄₁₆	I, R, F		
x ⅜	I, N, Y, H, S, R, O, M, F	L 8 x4 x1	B, U
x ⁵⁄₁₆	I, N, Y, H, S, R, O, M, F	x ¾	B, U
x ¼	I, N, Y, H, S, R, O, M, F	x ⁹⁄₁₆	B, U
		x ½	B, U
L 3½x3½x ½	I, N, Y, R, M, F		
x ⁷⁄₁₆	F		
x ⅜	I, N, Y, H, S, R, M, F		
x ⁵⁄₁₆	I, N, Y, H, S, R, M, F		
x ¼	I, N, Y, H, S, R, M, F		

Note: Maximum lengths of shapes obtainable vary widely with producers, but a conservative range for all mills is from 60 to 75 ft. Some mills will accept orders for lengths up to 120 ft, but only for certain shapes and subject to special arrangement. Consult the producers for unusual length requirements.

TABLE 3
Principal Producers of Structural Shapes (cont'd)

B. Bethlehem Steel Corp.
E. Roanoke Electric Steel Corp.
F. Florida Steel Corp.
H. Chaparral Steel
I. Inland Steel Co.
L. LTV Steel Co.
M. SMI Steel, Inc.
N. Northwestern Steel & Wire
O. Ohio River Steel
R. Nucor Corporation
S. North Star Steel Co.
U. United States Steel Corp.
Y. Bayou Steel Corp.

Section by Leg Length, Thickness	Producer Code	Section by Leg Length, Thickness	Producer Code
L 7x4 $\times^3/_4$	B, I, Y	L 4 x3 $\times^5/_8$	I
$\times^5/_8$	B, I	$\times^1/_2$	I, N, Y, R, F
$\times^1/_2$	B, I, Y, O	$\times^7/_{16}$	F
$\times^3/_8$	B, I, Y, O	$\times^3/_8$	I, N, Y, H, S, R, M, F
		$\times^5/_{16}$	I, N, Y, H, S, R, M, F
		$\times^1/_4$	I, N, Y, H, S, R, M, F
L 6x4 $\times^3/_4$	B, I, N, Y		
$\times^5/_8$	B, I, N, Y	L 3½x3 $\times^1/_2$	N, R, F
$\times^9/_{16}$	B	$\times^7/_{16}$	F
$\times^1/_2$	B, I, N, Y, R, O, M	$\times^3/_8$	I, N, S, R, M, F
$\times^7/_{16}$	B, I, Y	$\times^5/_{16}$	I, N, S, R, M, F
$\times^3/_8$	B, I, N, Y, R, O, M	$\times^1/_4$	I, N, S, R, M, F
$\times^5/_{16}$	B, I, N, Y, R, M		
		L 3½x2½x½	I, R, F
L 6x3½x½	I, N, Y, R, M	$\times^7/_{16}$	F
$\times^3/_8$	B, I, N, Y, R, M	$\times^3/_8$	I, N, R, F
$\times^5/_{16}$	B, I, N, Y, R, M	$\times^5/_{16}$	I, N, R, F
		$\times^1/_4$	I, N, R, F
L 5x3½x$^3/_4$	I, Y		
$\times^5/_8$	I, Y	L 3 x2½x½	I, F
$\times^1/_2$	I, N, Y, S, R, M	$\times^7/_{16}$	F
$\times^7/_{16}$	I	$\times^3/_8$	I, N, H, S, R, M, F
$\times^3/_8$	I, N, Y, S, R, M	$\times^5/_{16}$	I, N, H, S, R, E, M, F
$\times^5/_{16}$	I, N, Y, S, R, M	$\times^1/_4$	I, N, H, S, R, E, M, F
$\times^1/_4$	N, Y, S, R, M	$\times^3/_{16}$	I, R, E, M, F
L 5x3 $\times^1/_2$	I, N, Y, S, R, F	L 3 x2 $\times^1/_2$	I, M, F
$\times^7/_{16}$	F	$\times^7/_{16}$	F
$\times^3/_8$	I, N, Y, S, R, M, F	$\times^3/_8$	I, N, H, S, R, M, F
$\times^5/_{16}$	I, N, Y, S, R, M, F	$\times^5/_{16}$	I, N, H, S, R, E, M, F
$\times^1/_4$	Y, S, R, M, F	$\times^1/_4$	I, N, H, S, R, E, M, F
		$\times^3/_{16}$	I, N, R, E, M, F
L 4x3½x$^5/_8$	I		
$\times^1/_2$	I, N, R, F	L 2½x2 $\times^3/_8$	I, N, H, S, R, E, M, F
$\times^7/_{16}$	F	$\times^5/_{16}$	N, H, S, R, E, M, F
$\times^3/_8$	I, N, S, R, M, F	$\times^1/_4$	I, N, H, S, R, E, M, F
$\times^5/_{16}$	I, N, S, R, M, F	$\times^3/_{16}$	N, H, S, R, E, M, F
$\times^1/_4$	I, N, S, R, M, F		

Note: Maximum lengths of shapes obtainable vary widely with producers, but a conservative range for all mills is from 60 to 75 ft. Some mills will accept orders for lengths up to 120 ft, but only for certain shapes and subject to special arrangement. Consult the producers for unusual length requirements.

AMERICAN INSTITUTE OF STEEL CONSTRUCTION

TABLE 4
Availability of Steel Pipe and Structural Tubing According to ASTM Material Specifications

Steel	ASTM Specification	Grade	F_y Minimum Yield Stress (ksi)	F_u Minimum Tensile Stress (ksi)	Shape		Availability
					Round	Square & Rectangular	
Electric-Resistance Welded	A53 Type E	B	35	60	■		Note 3
Seamless	Type S	B	35	60	■		Note 3
Cold Formed	A500	A	33	45	■		Note 1
		B	42	58	■		Note 1
		C	46	62	■		Note 1
		A	39	45		■	Note 1
		B	46	58		■	Note 2
		C	50	62		■	Note 1
Hot Formed	A501	—	36	58			Note 1
High-Strength Low-Alloy	A618	I	50	70	■		Note 1
		II	50	70	■		Note 1
		III	50	65	■		Note 1

Notes:
1. Available in mill quantities only; consult with producers.
2. Normally stocked in local steel service centers.
3. Normally stocked by local pipe distributors.
▨ Available.
☐ Not available.

TABLE 5
Producers of Structural Tubing

A. Harris Tube Company
Ac. Acme Roll Forming Company
B. Bock Industries
C. Copperweld Corporation

Dm. Delta Metalforming Company
E. Eugene Welding Company
Ex. Ex-L Tube, Inc.
H. Hanna Steel Corporation
I. Independence Tube Corporation

J. James Steel & Tube Company
K. Kaiser Steel Tubing, Inc.
U. UNR Leavitt
W. Welded Tube Company of America

Nominal Size and Thickness	Producer Code	Nominal Size and Thickness	Producer Code
16x16—All	W	10x4x$\frac{1}{2}$	A, B, C, U, W
14x14—All	W	10x4x$\frac{3}{8}$, $\frac{5}{16}$, $\frac{1}{4}$, $\frac{3}{16}$	A, B, C, Dm, U, W
12x12—All	A, B, W	10x2x$\frac{3}{8}$, $\frac{5}{16}$	A, B, W
10x10x$\frac{5}{8}$	B, C	10x2x$\frac{1}{4}$, $\frac{3}{16}$	A, B, Dm, U, W
10x10x$\frac{1}{2}$, $\frac{3}{8}$, $\frac{5}{16}$, $\frac{1}{4}$	A, B, C, W	8x6x$\frac{1}{2}$	A, B, C, U, W
8x8x$\frac{5}{8}$	B, C	8x6x$\frac{3}{8}$, $\frac{5}{16}$, $\frac{1}{4}$, $\frac{3}{16}$	A, B, C, Dm, U, W
8x8x$\frac{1}{2}$	A, B, C, U, W	8x4x$\frac{1}{2}$	A, B, C, U, W
8x8x$\frac{3}{8}$, $\frac{5}{16}$, $\frac{1}{4}$	A, B, C, Dm, U, W	8x4x$\frac{3}{8}$, $\frac{5}{16}$	A, B, C, Dm, I, U, W
8x8x$\frac{3}{16}$	A, B, C, Dm, W	8x4x$\frac{1}{4}$, $\frac{3}{16}$	A, Ac, B, C, Dm, I, U, W
7x7x$\frac{1}{2}$	A, B, C, U, W	8x3x$\frac{3}{8}$, $\frac{5}{16}$	A, B, C, Dm, I, U, W
7x7x$\frac{3}{8}$, $\frac{5}{16}$, $\frac{1}{4}$, $\frac{3}{16}$	A, B, C, Dm, U, W	8x3x$\frac{1}{4}$	A, Ac, B, C, Dm, I, U, W
6x6x$\frac{1}{2}$	A, B, C, U, W	8x3x$\frac{3}{16}$	A, B, C, Dm, I, U, W
6x6x$\frac{3}{8}$, $\frac{5}{16}$	A, B, C, Dm, I, U, W	8x2x$\frac{3}{8}$	A, B, W
6x6x$\frac{1}{4}$, $\frac{3}{16}$	A, Ac, B, C, Dm, I, U, W	8x2x$\frac{5}{16}$	B, I, W
5x5x$\frac{1}{2}$	A, B, C, U, W	8x2x$\frac{1}{4}$	A, Ac, B, C, Dm, I U, W
5x5x$\frac{3}{8}$, $\frac{5}{16}$	A, B, C, Dm, I, U, W	8x2x$\frac{3}{16}$	A, B, C, Dm, I, U, W
5x5x$\frac{1}{4}$, $\frac{3}{16}$	A, Ac, B, C, Dm, I, J, U, W	7x5x$\frac{1}{2}$	A, B, C, U, W
4x4x$\frac{1}{2}$	A, B, C, U, W	7x5x$\frac{3}{8}$, $\frac{5}{16}$	A, B, C, Dm, I, U, W
4x4x$\frac{3}{8}$, $\frac{5}{16}$	A, B, C, Dm, I, U, W	7x5x$\frac{1}{4}$	A, Ac, B, C, Dm, I, U, W
4x4x$\frac{1}{4}$, $\frac{3}{16}$	All Producers	7x5x$\frac{3}{16}$	A, B, C, Dm, I, U, W
3$\frac{1}{2}$x3$\frac{1}{2}$x$\frac{5}{16}$	A, Dm, I, W	7x4x$\frac{3}{8}$, $\frac{5}{16}$	A, C, Dm, I, U, W
3$\frac{1}{2}$x3$\frac{1}{2}$x$\frac{1}{4}$, $\frac{3}{16}$	A, Ac, B, C, Dm, E, Ex, I, J, K, U, W	7x4x$\frac{1}{4}$	A, Ac, C, Dm, I, U, W
3x3x$\frac{5}{16}$	A, Dm, I, W	7x4x$\frac{3}{16}$	A, C, Dm, I, U, W
3x3x$\frac{1}{4}$	A, Ac, B, C, Dm, E, Ex, I, J, K, U, W	7x3x$\frac{3}{8}$, $\frac{5}{16}$	A, B, C, Dm, I, U, W
3x3x$\frac{3}{16}$	All Producers	7x3x$\frac{1}{4}$, $\frac{3}{16}$	A, Ac, B, C, Dm, I, U, W
2$\frac{1}{2}$x2$\frac{1}{2}$x$\frac{1}{4}$, $\frac{3}{16}$	All Producers	6x4x$\frac{1}{2}$	A, B, C, U, W
2x2x$\frac{1}{4}$	A, Ac, B, C, Dm, Ex, I, J, K, U, W	6x4x$\frac{3}{8}$, $\frac{5}{16}$	A, B, C, Dm, I, U, W
2x2x$\frac{3}{16}$	All Producers	6x4x$\frac{1}{4}$	A, Ac, B, C, Dm, I, J, U, W
20x12x$\frac{1}{2}$, $\frac{3}{8}$, $\frac{5}{16}$	W	6x4x$\frac{3}{16}$	B, C, Dm, I, J, U, W
20x8x$\frac{1}{2}$, $\frac{3}{8}$, $\frac{5}{16}$	W	6x3x$\frac{3}{8}$, $\frac{5}{16}$	A, Dm, I, U, W
20x4x$\frac{1}{2}$, $\frac{3}{8}$, $\frac{5}{16}$	W	6x3x$\frac{1}{4}$, $\frac{3}{16}$	A, Ac, B, C, Dm, I, U, W
18x6x$\frac{1}{2}$, $\frac{3}{8}$, $\frac{5}{16}$	W	6x2x$\frac{3}{8}$	A, B, W
16x12x$\frac{1}{2}$, $\frac{3}{8}$, $\frac{5}{16}$	W	6x2x$\frac{5}{16}$	A, B, I, W
16x8x$\frac{1}{2}$, $\frac{3}{8}$, $\frac{5}{16}$	B, W	6x2x$\frac{1}{4}$, $\frac{3}{16}$	A, Ac, B, C, Dm, E, Ex, I, J, K, U, W
16x4x$\frac{1}{2}$, $\frac{3}{8}$, $\frac{5}{16}$	B, W	5x4x$\frac{3}{8}$, $\frac{5}{16}$	A, Dm, I, U, W
14x10x$\frac{1}{2}$, $\frac{3}{8}$, $\frac{5}{16}$	B, W	5x4x$\frac{1}{4}$	A, Ac, B, C, Dm, I, U, W
14x6x$\frac{1}{2}$, $\frac{3}{8}$, $\frac{5}{16}$, $\frac{1}{4}$	B, W	5x4x$\frac{3}{16}$	A, B, C, Dm, I, U, W
14x4x$\frac{1}{2}$, $\frac{3}{8}$, $\frac{5}{16}$, $\frac{1}{4}$	B, W	5x3x$\frac{1}{2}$	A, B, C, U
12x8x$\frac{5}{8}$	B, C	5x3x$\frac{3}{8}$, $\frac{5}{16}$	A, B, C, Dm, I, U, W
12x8x$\frac{1}{2}$	A, B, C, W	5x3x$\frac{1}{4}$, $\frac{3}{16}$	A, Ac, B, C, Dm, E, Ex, I, J, K, U, W
12x8x$\frac{3}{8}$, $\frac{5}{16}$, $\frac{1}{4}$	A, B, C, W	5x2x$\frac{5}{16}$	I, W
12x6x$\frac{5}{8}$	B, C	5x2x$\frac{1}{4}$, $\frac{3}{16}$	Ac, B, C, Dm, E, Ex, I, J, K, U, W
12x6x$\frac{1}{2}$, $\frac{3}{8}$, $\frac{5}{16}$, $\frac{1}{4}$	A, B, C, U, W	4x3x$\frac{5}{16}$	Dm, I, W
12x6x$\frac{3}{16}$	B, C, W	4x3x$\frac{1}{4}$, $\frac{3}{16}$	Ac, B, C, Dm, E, Ex, I, J, K, U, W
12x4x$\frac{1}{2}$, $\frac{3}{8}$, $\frac{5}{16}$, $\frac{1}{4}$	A, B, C, U, W	4x2x$\frac{5}{16}$	A, I, W
12x4x$\frac{3}{16}$	A, B, C, W	4x2x$\frac{1}{4}$	A, Ac, B, C, Dm, E, Ex, I, J, K, U, W
12x2x$\frac{1}{4}$, $\frac{3}{16}$	B, U	4x2x$\frac{3}{16}$	All Producers
10x6x$\frac{5}{8}$	B, C	3x2x$\frac{1}{4}$	A, Ac, B, C, Dm, E, Ex, I, J, K, U, W
10x6x$\frac{1}{2}$	A, B, C, U, W	3x2x$\frac{3}{16}$	All Producers
10x6x$\frac{3}{8}$, $\frac{5}{16}$, $\frac{1}{4}$	A, B, C, Dm, U, W		
10x6x$\frac{3}{16}$	A, B, C, Dm, W		

<u>Notes</u>

DIMENSIONS AND PROPERTIES
W Shapes
M Shapes
S Shapes
HP Shapes
American Standard Channels (C)
Miscellaneous Channels (MC)
Angles (L)

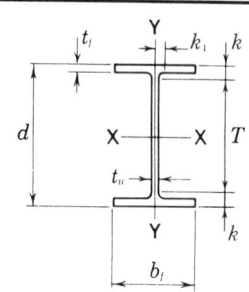

W SHAPES
Dimensions

Desig-nation	Area A	Depth d		Web Thickness t_w		$\dfrac{t_w}{2}$	Flange Width b_f		Flange Thickness t_f		T	k	k_1
	In.²	In.		In.		In.	In.		In.		In.	In.	In.
W 40x328	96.4	40.00	40	0.910	15/16	1/2	17.910	17⅞	1.730	1¾	33¾	3⅛	1 11/16
x298	87.6	39.69	39¾	0.830	13/16	7/16	17.830	17⅞	1.575	1 9/16	33¾	3	1⅝
x268	78.8	39.37	39⅜	0.750	¾	⅜	17.750	17¾	1.415	1 7/16	33¾	2 13/16	1 9/16
x244	71.7	39.06	39	0.710	11/16	⅜	17.710	17¾	1.260	1¼	33¾	2⅝	1 9/16
x221	64.8	38.67	38⅝	0.710	11/16	⅜	17.710	17¾	1.065	1 1/16	33¾	2 7/16	1 9/16
x192	56.5	38.20	38¼	0.710	11/16	⅜	17.710	17¾	0.830	13/16	33¾	2¼	1 9/16
W 40x655[a]	192.0	43.62	43⅝	1.970	2	1	16.870	16⅞	3.540	3 9/16	33¾	4 15/16	2¼
x593[a]	174.0	42.99	43	1.790	1 13/16	1	16.690	16¾	3.230	3¼	33¾	4⅝	2⅛
x531[a]	156.0	42.34	42⅜	1.610	1⅝	13/16	16.510	16½	2.910	2 15/16	33¾	4 5/16	2
x480[a]	140.0	41.81	41¾	1.460	1 7/16	¾	16.360	16⅜	2.640	2⅝	33¾	4	2
x436[a]	128.0	41.34	41⅜	1.340	1 5/16	11/16	16.240	16¼	2.400	2⅜	33¾	3 13/16	1 15/16
x397[a]	116.0	40.95	41	1.220	1¼	⅝	16.120	16⅛	2.200	2 3/16	33¾	3⅝	1⅞
x362[a]	106.0	40.55	40½	1.120	1⅛	9/16	16.020	16	2.010	2	33¾	3⅜	1 13/16
x324	95.3	40.16	40⅛	1.000	1	½	15.905	15⅞	1.810	1 13/16	33¾	3 3/16	1¾
x297	87.4	39.84	39⅞	0.930	15/16	½	15.825	15⅞	1.650	1⅝	33¾	3 1/16	1 11/16
x277	81.3	39.69	39¾	0.830	13/16	7/16	15.830	15⅞	1.575	1 9/16	33¾	3	1⅝
x249	73.3	39.38	39⅜	0.750	¾	⅜	15.750	15¾	1.420	1 7/16	33¾	2 13/16	1 9/16
x215	63.3	38.98	39	0.650	⅝	5/16	15.750	15¾	1.220	1¼	33¾	2⅝	1 9/16
x199	58.4	38.67	38⅝	0.650	⅝	5/16	15.750	15¾	1.065	1 1/16	33¾	2 7/16	1 9/16
W 40x183[b]	53.7	38.98	39	0.650	⅝	5/16	11.810	11¾	1.220	1¼	33¾	2⅝	1 9/16
x167	49.1	38.59	38⅝	0.650	⅝	5/16	11.810	11¾	1.025	1	33¾	2 7/16	1 9/16
x149	43.8	38.20	38¼	0.630	⅝	5/16	11.810	11¾	0.830	13/16	33¾	2¼	1½
W 36x848[a]	249.0	42.45	42½	2.520	2½	1¼	18.130	18⅛	4.530	4½	31⅛	5 11/16	2¼
x798[a]	234.0	41.97	42	2.380	2⅜	1 3/16	17.990	18	4.290	4 5/16	31⅛	5 7/16	2 3/16
x720[a]	211.0	41.19	41¼	2.165	2 3/16	1⅛	17.775	17¾	3.900	3⅞	31⅛	5 1/16	2 1/16
x650[a]	190.0	40.47	40½	1.970	2	1	17.575	17⅝	3.540	3 9/16	31⅛	4 11/16	2
x588[a]	172.0	39.84	39⅞	1.790	1 13/16	1	17.400	17⅜	3.230	3¼	31⅛	4⅜	1⅞
x527[a]	154.0	39.21	39¼	1.610	1⅝	13/16	17.220	17¼	2.910	2 15/16	31⅛	4 1/16	1¾
x485[a]	142.0	38.74	38¾	1.500	1½	¾	17.105	17⅛	2.680	2 11/16	31⅛	3 13/16	1¾
x439[a]	128.0	38.26	38¼	1.360	1⅜	11/16	16.965	17	2.440	2 7/16	31⅛	3 9/16	1⅝
x393[a]	115.0	37.80	37¾	1.220	1¼	⅝	16.830	16⅞	2.200	2 3/16	31⅛	3 5/16	1⅝
x359[a]	105.0	37.40	37⅜	1.120	1⅛	9/16	16.730	16¾	2.010	2	31⅛	3⅛	1 9/16
x328[a]	96.4	37.09	37⅛	1.020	1	½	16.630	16⅝	1.850	1⅞	31⅛	3	1½
x300[a]	88.3	36.74	36¾	0.945	15/16	½	16.655	16⅝	1.680	1 11/16	31⅛	2 13/16	1½
x280[a]	82.4	36.52	36½	0.885	⅞	7/16	16.595	16⅝	1.570	1 9/16	31⅛	2 11/16	1½
x260	76.5	36.26	36¼	0.840	13/16	7/16	16.550	16½	1.440	1 7/16	31⅛	2 9/16	1½
x245	72.1	36.08	36⅛	0.800	13/16	7/16	16.510	16½	1.350	1⅜	31⅛	2½	1 7/16
x230	67.6	35.90	35⅞	0.760	¾	⅜	16.470	16½	1.260	1¼	31⅛	2⅜	1 7/16
W 36x256[a]	75.4	37.43	37⅜	0.960	1	½	12.215	12¼	1.730	1¾	32⅛	2⅝	1 5/16
x232[a]	68.1	37.12	37⅛	0.870	⅞	7/16	12.120	12⅛	1.570	1 9/16	32⅛	2½	1¼
x210	61.8	36.69	36¾	0.830	13/16	7/16	12.180	12⅛	1.360	1⅜	32⅛	2 5/16	1¼
x194	57.0	36.49	36½	0.765	¾	⅜	12.115	12⅛	1.260	1¼	32⅛	2 3/16	1 3/16
x182	53.6	36.33	36⅜	0.725	¾	⅜	12.075	12⅛	1.180	1 3/16	32⅛	2⅛	1 3/16

[a]For application refer to Notes in Table 2.
[b]Heavier shapes in this series are available from some producers.

W SHAPES
Properties

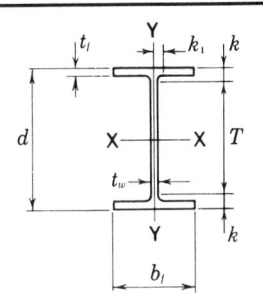

Nominal Wt. per Ft Lb.	Compact Section Criteria $\frac{b_f}{2t_f}$	Compact Section Criteria $\frac{h_c}{t_w}$	F_y''' Ksi	X_1 Ksi	$X_2 \times 10^6$ (1/Ksi)^2	Elastic Properties Axis X-X I In.^{4}	Axis X-X S In.^{3}	Axis X-X r In.	Axis Y-Y I In.^{4}	Axis Y-Y S In.^{3}	Axis Y-Y r In.	Plastic Modulus Z_x In.^{3}	Z_y In.^{3}
328	5.2	37.6	45	2530	3800	26800	1340	16.7	1660	185	4.15	1510	286
298	5.7	41.2	38	2300	5430	24200	1220	16.6	1490	167	4.12	1370	257
268	6.3	45.6	31	2090	8070	21500	1090	16.5	1320	149	4.09	1220	229
244	7.0	48.1	28	1900	11900	19200	983	16.4	1170	132	4.04	1100	203
221	8.3	48.1	28	1730	18400	16600	858	16.0	988	112	3.90	967	172
192	10.7	48.1	28	1570	29600	13500	708	15.5	770	87	3.69	807	135
655	2.4	17.4	—	5230	240	56500	2590	17.2	2860	339	3.86	3060	541
593	2.6	19.1	—	4790	337	50400	2340	17.0	2520	302	3.81	2750	481
531	2.8	21.2	—	4340	496	44300	2090	16.9	2200	266	3.75	2450	422
480	3.1	23.4	—	3920	723	39500	1890	16.8	1940	237	3.72	2180	374
436	3.4	25.5	—	3610	1020	35400	1710	16.6	1720	212	3.67	1980	334
397	3.7	28.0	—	3290	1440	32000	1560	16.6	1540	191	3.65	1790	300
362	4.0	30.5	—	3030	2000	28900	1420	16.5	1380	173	3.61	1630	270
324	4.4	34.2	55	2720	3030	25600	1280	16.4	1220	153	3.57	1460	239
297	4.8	36.8	47	2500	4240	23200	1170	16.3	1090	138	3.54	1330	215
277	5.0	41.2	38	2350	5370	21900	1100	16.4	1040	132	3.58	1250	204
249	5.5	45.6	31	2120	7940	19500	992	16.3	926	118	3.56	1120	182
215	6.5	52.6	23	1830	14000	16700	858	16.2	796	101	3.54	963	156
199	7.4	52.6	23	1690	20300	14900	769	16.0	695	88.2	3.45	868	137
183	4.8	52.6	23	1900	13700	13300	682	15.7	336	56.9	2.50	781	89.6
167	5.8	52.6	23	1750	20500	11600	599	15.3	283	47.9	2.40	692	76.0
149	7.1	54.3	22	1610	31400	9780	512	14.9	229	38.8	2.29	597	62.2
848	2.0	12.5	—	7100	71	67400	3170	16.4	4550	501	4.27	3830	799
798	2.1	13.2	—	6720	87	62600	2980	16.4	4200	467	4.24	3570	743
720	2.3	14.5	—	6130	123	55300	2690	16.2	3680	414	4.18	3190	656
650	2.5	16.0	—	5590	175	48900	2420	16.0	3230	367	4.12	2840	580
588	2.7	17.6	—	5130	246	43500	2180	15.9	2850	328	4.07	2550	517
527	3.0	19.6	—	4630	365	38300	1950	15.8	2490	289	4.02	2270	454
485	3.2	21.0	—	4300	488	34700	1790	15.6	2250	263	3.98	2070	412
439	3.5	23.1	—	3900	704	31000	1620	15.6	1990	235	3.95	1860	367
393	3.8	25.8	—	3540	1040	27500	1450	15.5	1750	208	3.90	1660	325
359	4.2	28.1	—	3240	1470	24800	1320	15.4	1570	188	3.87	1510	292
328	4.5	30.9	—	2980	2040	22500	1210	15.3	1420	171	3.84	1380	265
300	5.0	33.3	58	2720	2930	20300	1110	15.2	1300	156	3.83	1260	241
280	5.3	35.6	51	2560	3730	18900	1030	15.1	1200	144	3.81	1170	223
260	5.7	37.5	46	2370	5100	17300	953	15.0	1090	132	3.78	1080	204
245	6.1	39.4	41	2230	6460	16100	895	15.0	1010	123	3.75	1010	190
230	6.5	41.4	37	2100	8190	15000	837	14.9	940	114	3.73	943	176
256	3.5	33.8	56	2840	2870	16800	895	14.9	528	86.5	2.65	1040	137
232	3.9	37.3	46	2580	4160	15000	809	14.8	468	77.2	2.62	936	122
210	4.5	39.1	42	2320	6560	13200	719	14.6	411	67.5	2.58	833	107
194	4.8	42.4	36	2140	8820	12100	664	14.6	375	61.9	2.56	767	97.7
182	5.1	44.8	32	2020	11300	11300	623	14.5	347	57.6	2.55	718	90.7

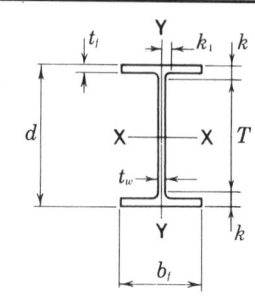

W SHAPES
Dimensions

Desig-nation	Area A	Depth d		Web Thickness t_w		$\frac{t_w}{2}$	Flange Width b_f		Flange Thickness t_f		T	k	k_1
	In.²	In.		In.		In.	In.		In.		In.	In.	In.
W 36x170[b]	50.0	36.17	36⅛	0.680	11/16	3/8	12.030	12	1.100	1⅛	32⅛	2	1 3/16
x160	47.0	36.01	36	0.650	5/8	5/16	12.000	12	1.020	1	32⅛	1 15/16	1⅛
x150	44.2	35.85	35⅞	0.625	5/8	5/16	11.975	12	0.940	15/16	32⅛	1⅞	1⅛
x135	39.7	35.55	35½	0.600	5/8	5/16	11.950	12	0.790	13/16	32⅛	1 11/16	1⅛
W 33x619[a]	181.0	38.47	38½	1.970	2	1	16.910	16⅞	3.540	3 9/16	29¾	4⅜	1¾
x567[a]	166.0	37.91	37⅞	1.810	1 13/16	1	16.750	16¾	3.270	3¼	29¾	4 1/16	1 11/16
x515[a]	151.0	37.36	37⅜	1.650	1 5/8	13/16	16.590	16⅝	2.990	3	29¾	3 13/16	1⅝
x468[a]	137.0	36.81	36¾	1.520	1½	¾	16.455	16½	2.720	2¾	29¾	3½	1 9/16
x424[a]	124.0	36.34	36⅜	1.380	1⅜	11/16	16.315	16⅜	2.480	2½	29¾	3 5/16	1 7/16
x387[a]	113.0	35.95	36	1.260	1¼	5/8	16.200	16¼	2.280	2¼	29¾	3⅛	1⅜
x354[a]	104.0	35.55	35½	1.160	1 3/16	5/8	16.100	16⅛	2.090	2 1/16	29¾	2⅞	1⅜
x318[a]	93.5	35.16	35⅛	1.040	1 1/16	9/16	15.985	16	1.890	1⅞	29¾	2 11/16	1 5/16
x291[a]	85.6	34.84	34⅞	0.960	1	½	15.905	15⅞	1.730	1¾	29¾	2 9/16	1¼
x263[a]	77.4	34.53	34½	0.870	⅞	7/16	15.805	15¾	1.570	1 9/16	29¾	2⅜	1 3/16
x241	70.9	34.18	34⅛	0.830	13/16	7/16	15.860	15⅞	1.400	1⅜	29¾	2 3/16	1 3/16
x221	65.0	33.93	33⅞	0.775	¾	3/8	15.805	15¾	1.275	1¼	29¾	2 1/16	1 3/16
x201	59.1	33.68	33⅝	0.715	11/16	3/8	15.745	15¾	1.150	1⅛	29¾	1 15/16	1⅛
W 33x169[b]	49.5	33.82	33⅞	0.670	11/16	3/8	11.500	11½	1.220	1¼	29¾	2 1/16	1⅛
x152	44.7	33.49	33½	0.635	5/8	5/16	11.565	11⅝	1.055	1 1/16	29¾	1⅞	1⅛
x141	41.6	33.30	33¼	0.605	5/8	5/16	11.535	11½	0.960	15/16	29¾	1¾	1 1/16
x130	38.3	33.09	33⅛	0.580	9/16	5/16	11.510	11½	0.855	⅞	29¾	1 11/16	1 1/16
x118	34.7	32.86	32⅞	0.550	9/16	5/16	11.480	11½	0.740	¾	29¾	1 9/16	1 1/16
W 30x581[a]	170.0	35.39	35⅜	1.970	2	1	16.200	16¼	3.540	3 9/16	26¾	4 5/16	1 11/16
x526[a]	154.0	34.76	34¾	1.790	1 13/16	1	16.020	16	3.230	3¼	26¾	4	1⅝
x477[a]	140.0	34.21	34¼	1.630	1⅝	13/16	15.865	15⅞	2.950	3	26¾	3¾	1 9/16
x433[a]	127.0	33.66	33⅝	1.500	1½	¾	15.725	15¾	2.680	2 11/16	26¾	3 7/16	1½
x391[a]	114.0	33.19	33¼	1.360	1⅜	11/16	15.590	15⅝	2.440	2 7/16	26¾	3¼	1 7/16
x357[a]	104.0	32.80	32¾	1.240	1¼	5/8	15.470	15½	2.240	2¼	26¾	3	1⅜
x326[a]	95.7	32.40	32⅜	1.140	1⅛	9/16	15.370	15⅜	2.050	2 1/16	26¾	2 13/16	1 5/16
x292[a]	85.7	32.01	32	1.020	1	½	15.255	15¼	1.850	1⅞	26¾	2⅝	1¼
x261	76.7	31.61	31⅝	0.930	15/16	½	15.155	15⅛	1.650	1⅝	26¾	2 7/16	1 3/16
x235	69.0	31.30	31¼	0.830	13/16	7/16	15.055	15	1.500	1½	26¾	2¼	1⅛
x211	62.0	30.94	31	0.775	¾	3/8	15.105	15⅛	1.315	1 5/16	26¾	2⅛	1⅛
x191	56.1	30.68	30⅝	0.710	11/16	3/8	15.040	15	1.185	1 3/16	26¾	1 15/16	1 1/16
x173	50.8	30.44	30½	0.655	5/8	5/16	14.985	15	1.065	1 1/16	26¾	1⅞	1 1/16
W 30x148[b]	43.5	30.67	30⅝	0.650	5/8	5/16	10.480	10½	1.180	1 3/16	26¾	2	1
x132	38.9	30.31	30¼	0.615	5/8	5/16	10.545	10½	1.000	1	26¾	1¾	1 1/16
x124	36.5	30.17	30⅛	0.585	9/16	5/16	10.515	10½	0.930	15/16	26¾	1 11/16	1
x116	34.2	30.01	30	0.565	9/16	5/16	10.495	10½	0.850	⅞	26¾	1⅝	1
x108	31.7	29.83	29⅞	0.545	9/16	5/16	10.475	10½	0.760	¾	26¾	1 9/16	1
x 99	29.1	29.65	29⅝	0.520	½	¼	10.450	10½	0.670	11/16	26¾	1 7/16	1
x 90	26.4	29.53	29½	0.470	½	¼	10.400	10⅜	0.610	9/16	26¾	1 5/16	1

[a]For application refer to Notes in Table 2.
[b]Heavier shapes in this series are available from some producers.

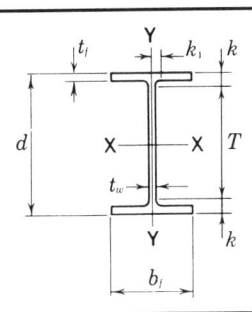

W SHAPES
Properties

Nominal Wt. per Ft	Compact Section Criteria			X_1	$X_2 \times 10^6$	Elastic Properties						Plastic Modulus	
						Axis X-X			Axis Y-Y				
	$\dfrac{b_f}{2t_f}$	$\dfrac{h_c}{t_w}$	F_y'''			I	S	r	I	S	r	Z_x	Z_y
Lb.			Ksi	Ksi	$(1/\text{Ksi})^2$	In.^{4}	In.^{3}	In.	In.^{4}	In.^{3}	In.	In.^{3}	In.^{3}
170	5.5	47.8	28	1900	14500	10500	580	14.5	320	53.2	2.53	668	83.8
160	5.9	50.0	26	1780	18600	9750	542	14.4	295	49.1	2.50	624	77.3
150	6.4	52.0	24	1680	24200	9040	504	14.3	270	45.1	2.47	581	70.9
135	7.6	54.1	22	1520	38000	7800	439	14.0	225	37.7	2.38	509	59.7
619	2.4	15.2	—	5910	142	41800	2170	15.2	2870	340	3.98	2560	537
567	2.6	16.6	—	5460	191	37700	1990	15.1	2580	308	3.94	2330	485
515	2.8	18.2	—	5000	268	33700	1810	14.9	2290	276	3.89	2110	433
468	3.0	19.7	—	4600	374	30100	1630	14.8	2030	247	3.85	1890	387
424	3.3	21.7	—	4180	535	26900	1480	14.7	1800	221	3.81	1700	345
387	3.6	23.8	—	3850	740	24300	1350	14.7	1620	200	3.79	1550	312
354	3.8	25.8	—	3540	1030	21900	1230	14.5	1460	181	3.74	1420	282
318	4.2	28.8	—	3200	1530	19500	1110	14.4	1290	161	3.71	1270	250
291	4.6	31.2	—	2940	2130	17700	1010	14.4	1160	146	3.69	1150	226
263	5.0	34.5	54	2670	3100	15800	917	14.3	1030	131	3.66	1040	202
241	5.7	36.1	49	2430	4590	14200	829	14.1	932	118	3.63	939	182
221	6.2	38.7	43	2240	6440	12800	757	14.1	840	106	3.59	855	164
201	6.8	41.9	36	2040	9390	11500	684	14.0	749	95.2	3.56	772	147
169	4.7	44.7	32	2160	8150	9290	549	13.7	310	53.9	2.50	629	84.4
152	5.5	47.2	29	1940	12900	8160	487	13.5	273	47.2	2.47	559	73.9
141	6.0	49.6	26	1800	17800	7450	448	13.4	246	42.7	2.43	514	66.9
130	6.7	51.7	24	1660	25100	6710	406	13.2	218	37.9	2.39	467	59.5
118	7.8	54.5	22	1510	37700	5900	359	13.0	187	32.6	2.32	415	51.3
581	2.3	13.7	—	6470	97	33000	1870	13.9	2530	312	3.86	2210	492
526	2.5	15.1	—	5950	135	29300	1680	13.8	2230	278	3.80	1990	438
477	2.7	16.6	—	5420	193	26100	1530	13.7	1970	249	3.75	1790	390
433	2.9	18.0	—	4970	271	23200	1380	13.5	1750	222	3.71	1610	348
391	3.2	19.9	—	4510	386	20700	1250	13.5	1550	198	3.68	1430	310
357	3.5	21.8	—	4150	536	18600	1140	13.4	1390	179	3.65	1300	279
326	3.7	23.7	—	3860	735	16800	1030	13.2	1240	162	3.61	1190	252
292	4.1	26.5	—	3460	1110	14900	928	13.2	1100	144	3.58	1060	223
261	4.6	29.0	—	3110	1690	13100	827	13.1	959	127	3.54	941	196
235	5.0	32.5	61	2820	2450	11700	746	13.0	855	114	3.52	845	175
211	5.7	34.9	53	2510	3950	10300	663	12.9	757	100	3.49	749	154
191	6.3	38.0	44	2280	5840	9170	598	12.8	673	89.5	3.46	673	138
173	7.0	41.2	38	2070	8540	8200	539	12.7	598	79.8	3.43	605	123
148	4.4	41.5	37	2310	6180	6680	436	12.4	227	43.3	2.28	500	68.0
132	5.3	43.9	33	2050	10500	5770	380	12.2	196	37.2	2.25	437	58.4
124	5.7	46.2	30	1930	13500	5360	355	12.1	181	34.4	2.23	408	54.0
116	6.2	49.6	28	1800	17700	4930	329	12.0	164	31.3	2.19	378	49.2
108	6.9	45.6	26	1680	24200	4470	299	11.9	146	27.9	2.15	346	43.9
99	7.8	51.9	24	1560	34100	3990	269	11.7	128	24.5	2.10	312	38.6
90	8.5	57.5	19	1430	47000	3620	245	11.7	115	22.1	2.09	283	34.7

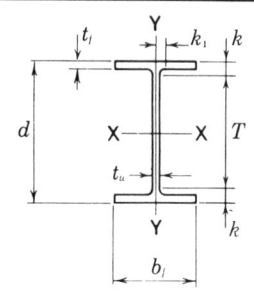

W SHAPES
Dimensions

Desig-nation	Area A	Depth d	Web		Flange		Distance		
			Thickness t_w	$\frac{t_w}{2}$	Width b_f	Thickness t_f	T	k	k_1
	In.²	In.	In.	In.	In.	In.	In.	In.	In.
W 27x539[a]	158.0	32.52 32½	1.970 2	1	15.255 15¼	3.540 3⁹⁄₁₆	24	4¼	1⅝
x494[a]	145.0	31.97 32	1.810 1¹³⁄₁₆	1	15.095 15⅛	3.270 3¼	24	4	1⁹⁄₁₆
x448[a]	131.0	31.42 31⅜	1.650 1⅝	¹³⁄₁₆	14.940 15	2.990 3	24	3¹¹⁄₁₆	1½
x407[a]	119.0	30.87 30⅞	1.520 1½	¾	14.800 14¾	2.720 2¾	24	3⁷⁄₁₆	1⁷⁄₁₆
x368[a]	108.0	30.39 30⅜	1.380 1⅜	¹¹⁄₁₆	14.665 14⅝	2.480 2½	24	3³⁄₁₆	1⁵⁄₁₆
x336[a]	98.7	30.00 30	1.260 1¼	⅝	14.545 14½	2.280 2¼	24	3	1⁵⁄₁₆
x307[a]	90.2	29.61 29⅝	1.160 1³⁄₁₆	⅝	14.445 14½	2.090 2¹⁄₁₆	24	2¹³⁄₁₆	1¼
x281[a]	82.6	29.29 29¼	1.060 1¹⁄₁₆	⁹⁄₁₆	14.350 14⅜	1.930 1¹⁵⁄₁₆	24	2⅝	1³⁄₁₆
x258	75.7	28.98 29	0.980 1	½	14.270 14¼	1.770 1¾	24	2½	1⅛
x235	69.1	28.66 28⅝	0.910 ¹⁵⁄₁₆	½	14.190 14¼	1.610 1⅝	24	2⁵⁄₁₆	1⅛
x217	63.8	28.43 28⅜	0.830 ¹³⁄₁₆	⁷⁄₁₆	14.115 14⅛	1.500 1½	24	2³⁄₁₆	1¹⁄₁₆
x194	57.0	28.11 28⅛	0.750 ¾	⅜	14.035 14	1.340 1⁵⁄₁₆	24	2¹⁄₁₆	1
x178	52.3	27.81 27¾	0.725 ¾	⅜	14.085 14⅛	1.190 1³⁄₁₆	24	1⅞	1¹⁄₁₆
x161	47.4	27.59 27⅝	0.660 ¹¹⁄₁₆	⅜	14.020 14	1.080 1¹⁄₁₆	24	1¹³⁄₁₆	1
x146	42.9	27.38 27⅜	0.605 ⅝	⁵⁄₁₆	13.965 14	0.975 1	24	1¹¹⁄₁₆	1
W 27x129[b]	37.8	27.63 27⅝	0.610 ⅝	⁵⁄₁₆	10.010 10	1.100 1⅛	24	1¹³⁄₁₆	¹⁵⁄₁₆
x114	33.5	27.29 27¼	0.570 ⁹⁄₁₆	⁵⁄₁₆	10.070 10⅛	0.930 ¹⁵⁄₁₆	24	1⅝	¹⁵⁄₁₆
x102	30.0	27.09 27⅛	0.515 ½	¼	10.015 10	0.830 ¹³⁄₁₆	24	1⁹⁄₁₆	¹⁵⁄₁₆
x 94	27.7	26.92 26⅞	0.490 ½	¼	9.990 10	0.745 ¾	24	1⁷⁄₁₆	¹⁵⁄₁₆
x 84	24.8	26.71 26¾	0.460 ⁷⁄₁₆	¼	9.960 10	0.640 ⅝	24	1⅜	¹⁵⁄₁₆
W 24x492[a]	144.0	29.65 29⅝	1.970 2	1	14.115 14⅛	3.540 3⁹⁄₁₆	21	4⁵⁄₁₆	1⁹⁄₁₆
x450[a]	132.0	29.09 29⅛	1.810 1¹³⁄₁₆	1	13.955 14	3.270 3¼	21	4¹⁄₁₆	1½
x408[a]	119.0	28.54 28½	1.650 1⅝	¹³⁄₁₆	13.800 13¾	2.990 3	21	3¾	1⅜
x370[a]	108.0	27.99 28	1.520 1½	¾	13.660 13⅝	2.720 2¾	21	3½	1⁵⁄₁₆
x335[a]	98.4	27.52 27½	1.380 1⅜	¹¹⁄₁₆	13.520 13½	2.480 2½	21	3¼	1¼
x306[a]	89.8	27.13 27⅛	1.260 1¼	⅝	13.405 13⅜	2.280 2¼	21	3¹⁄₁₆	1³⁄₁₆
x279[a]	82.0	26.73 26¾	1.160 1³⁄₁₆	⅝	13.305 13¼	2.090 2¹⁄₁₆	21	2⅞	1⅛
x250[a]	73.5	26.34 26⅜	1.040 1¹⁄₁₆	⁹⁄₁₆	13.185 13⅛	1.890 1⅞	21	2¹¹⁄₁₆	1⅛
x229	67.2	26.02 26	0.960 1	½	13.110 13⅛	1.730 1¾	21	2½	1
x207	60.7	25.71 25¾	0.870 ⅞	⁷⁄₁₆	13.010 13	1.570 1⁹⁄₁₆	21	2⅜	1
x192	56.3	25.47 25½	0.810 ¹³⁄₁₆	⁷⁄₁₆	12.950 13	1.460 1⁷⁄₁₆	21	2¼	1
x176	51.7	25.24 25¼	0.750 ¾	⅜	12.890 12⅞	1.340 1⁵⁄₁₆	21	2⅛	¹⁵⁄₁₆
x162	47.7	25.00 25	0.705 ¹¹⁄₁₆	⅜	12.955 13	1.220 1¼	21	2	1¹⁄₁₆
x146	43.0	24.74 24¾	0.650 ⅝	⁵⁄₁₆	12.900 12⅞	1.090 1¹⁄₁₆	21	1⅞	1¹⁄₁₆
x131	38.5	24.48 24½	0.605 ⅝	⁵⁄₁₆	12.855 12⅞	0.960 ¹⁵⁄₁₆	21	1¾	1¹⁄₁₆

[a]For application refer to Notes in Table 2.
[b]Heavier shapes in this series are available from some producers.

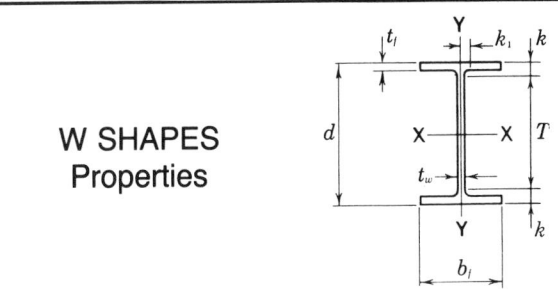

W SHAPES
Properties

Nom-inal Wt. per Ft Lb.	Compact Section Criteria $\frac{b_f}{2t_f}$	$\frac{h_c}{t_w}$	F_y''' Ksi	X_1 Ksi	$X_2 \times 10^6$ $(1/\text{Ksi})^2$	Axis X-X I In.^{4}	S In.^{3}	r In.	Axis Y-Y I In.^{4}	S In.^{3}	r In.	Z_x In.^{3}	Z_y In.^{3}
539	2.2	12.3	—	7160	66	25500	1570	12.7	2110	277	3.66	1880	437
494	2.3	13.4	—	6620	88	22900	1440	12.6	1890	250	3.61	1710	394
448	2.5	14.7	—	6070	123	20400	1300	12.5	1670	224	3.57	1530	351
407	2.7	15.9	—	5600	170	18100	1170	12.3	1480	200	3.52	1380	313
368	3.0	17.6	—	5100	243	16100	1060	12.2	1310	179	3.48	1240	279
336	3.2	19.2	—	4690	336	14500	970	12.1	1170	161	3.45	1130	252
307	3.5	20.9	—	4320	463	13100	884	12.0	1050	146	3.42	1020	227
281	3.7	22.9	—	3980	631	11900	811	12.0	953	133	3.40	933	206
258	4.0	24.7	—	3670	873	10800	742	11.9	859	120	3.37	850	187
235	4.4	26.6	—	3360	1230	9660	674	11.8	768	108	3.33	769	168
217	4.7	29.2	—	3120	1640	8870	624	11.8	704	99.8	3.32	708	154
194	5.2	32.3	61	2800	2520	7820	556	11.7	618	88.1	3.29	628	136
178	5.9	33.4	57	2550	3740	6990	502	11.6	555	78.8	3.26	567	122
161	6.5	36.7	47	2320	5370	6280	455	11.5	497	70.9	3.24	512	109
146	7.2	40.0	40	2110	7900	5630	411	11.4	443	63.5	3.21	461	97.5
129	4.5	39.7	41	2390	5340	4760	345	11.2	184	36.8	2.21	395	57.6
114	5.4	42.5	35	2100	9220	4090	299	11.0	159	31.5	2.18	343	49.3
102	6.0	47.0	29	1890	14000	3620	267	11.0	139	27.8	2.15	305	43.4
94	6.7	49.4	26	1740	19900	3270	243	10.9	124	24.8	2.12	278	38.8
84	7.8	52.7	23	1570	31100	2850	213	10.7	106	21.2	2.07	244	33.2
492	2.0	10.9	—	7950	43	19100	1290	11.5	1670	237	3.41	1550	375
450	2.1	11.9	—	7430	57	17100	1170	11.4	1490	214	3.36	1410	337
408	2.3	13.1	—	6780	79	15100	1060	11.3	1320	191	3.33	1250	300
370	2.5	14.2	—	6220	110	13400	957	11.1	1160	170	3.28	1120	267
335	2.7	15.6	—	5700	156	11900	864	11.0	1030	152	3.23	1020	238
306	2.9	17.1	—	5250	215	10700	789	10.9	919	137	3.20	922	214
279	3.2	18.6	—	4840	297	9600	718	10.8	823	124	3.17	835	193
250	3.5	20.7	—	4370	436	8490	644	10.7	724	110	3.14	744	171
229	3.8	22.5	—	4020	605	7650	588	10.7	651	99.4	3.11	676	154
207	4.1	24.8	—	3650	876	6820	531	10.6	578	88.8	3.08	606	137
192	4.4	26.6	—	3410	1150	6260	491	10.5	530	81.8	3.07	559	126
176	4.8	28.7	—	3140	1590	5680	450	10.5	479	74.3	3.04	511	115
162	5.3	30.6	—	2870	2260	5170	414	10.4	443	68.4	3.05	468	105
146	5.9	33.2	58	2590	3420	4580	371	10.3	391	60.5	3.01	418	93.2
131	6.7	35.6	50	2330	5290	4020	329	10.2	340	53.0	2.97	370	81.5

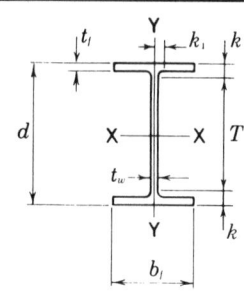

W SHAPES
Dimensions

Desig-nation	Area A	Depth d		Web Thickness t_w		$\frac{t_w}{2}$	Flange Width b_f		Thickness t_f		T	k	k_1
	In.²	In.		In.		In.	In.		In.		In.	In.	In.
W 24x117	34.4	24.26	24¼	0.550	⁹⁄₁₆	⁵⁄₁₆	12.800	12¾	0.850	⅞	21	1⅝	1
x104	30.6	24.06	24	0.500	½	¼	12.750	12¾	0.750	¾	21	1½	1
W 24x103[b]	30.3	24.53	24½	0.550	⁹⁄₁₆	⁵⁄₁₆	9.000	9	0.980	1	21	1¾	¹³⁄₁₆
x 94	27.7	24.31	24¼	0.515	½	¼	9.065	9⅛	0.875	⅞	21	1⅝	1
x 84	24.7	24.10	24⅛	0.470	½	¼	9.020	9	0.770	¾	21	1⁹⁄₁₆	¹⁵⁄₁₆
x 76	22.4	23.92	23⅞	0.440	⁷⁄₁₆	¼	8.990	9	0.680	¹¹⁄₁₆	21	1⁷⁄₁₆	¹⁵⁄₁₆
x 68	20.1	23.73	23¾	0.415	⁷⁄₁₆	¼	8.965	9	0.585	⁹⁄₁₆	21	1⅜	¹⁵⁄₁₆
W 24x 62	18.2	23.74	23¾	0.430	⁷⁄₁₆	¼	7.040	7	0.590	⁹⁄₁₆	21	1⅜	¹⁵⁄₁₆
x 55	16.2	23.57	23⅝	0.395	⅜	³⁄₁₆	7.005	7	0.505	½	21	1⁵⁄₁₆	¹⁵⁄₁₆
W 21x402[a]	118.0	26.02	26	1.730	1¾	⅞	13.405	13⅜	3.130	3⅛	18¼	3⅞	1⁷⁄₁₆
x364[a]	107.0	25.47	25½	1.590	1⁹⁄₁₆	¹³⁄₁₆	13.265	13¼	2.850	2⅞	18¼	3⅝	1⅜
x333[a]	97.9	25.00	25	1.460	1⁷⁄₁₆	¾	13.130	13⅛	2.620	2⅝	18¼	3⅜	1⁵⁄₁₆
x300[a]	88.2	24.53	24½	1.320	1⁵⁄₁₆	¹¹⁄₁₆	12.990	13	2.380	2⅜	18¼	3⅛	1¼
x275[a]	80.8	24.13	24⅛	1.220	1¼	⅝	12.890	12⅞	2.190	2³⁄₁₆	18¼	3	1³⁄₁₆
x248[a]	72.8	23.74	23¾	1.100	1⅛	⁹⁄₁₆	12.775	12¾	1.990	2	18¼	2¾	1⅛
x223	65.4	23.35	23⅜	1.000	1	½	12.675	12⅝	1.790	1¹³⁄₁₆	18¼	2⁹⁄₁₆	1¹⁄₁₆
x201	59.2	23.03	23	0.910	¹⁵⁄₁₆	½	12.575	12⅝	1.630	1⅝	18¼	2⅜	1
x182	53.6	22.72	22¾	0.830	¹³⁄₁₆	⁷⁄₁₆	12.500	12½	1.480	1½	18¼	2¼	1
x166	48.8	22.48	22½	0.750	¾	⅜	12.420	12⅜	1.360	1⅜	18¼	2⅛	¹⁵⁄₁₆
x147	43.2	22.06	22	0.720	¾	⅜	12.510	12½	1.150	1⅛	18¼	1⅞	1¹⁄₁₆
x132	38.8	21.83	21⅞	0.650	⅝	⁵⁄₁₆	12.440	12½	1.035	1¹⁄₁₆	18¼	1¹³⁄₁₆	1
x122	35.9	21.68	21⅝	0.600	⅝	⁵⁄₁₆	12.390	12⅜	0.960	¹⁵⁄₁₆	18¼	1¹¹⁄₁₆	1
x111	32.7	21.51	21½	0.550	⁹⁄₁₆	⁵⁄₁₆	12.340	12⅜	0.875	⅞	18¼	1⅝	¹⁵⁄₁₆
x101	29.8	21.36	21⅜	0.500	½	¼	12.290	12¼	0.800	¹³⁄₁₆	18¼	1⁹⁄₁₆	¹⁵⁄₁₆
W 21x 93	27.3	21.62	21⅝	0.580	⁹⁄₁₆	⁵⁄₁₆	8.420	8⅜	0.930	¹⁵⁄₁₆	18¼	1¹¹⁄₁₆	1
x 83	24.3	21.43	21⅜	0.515	½	¼	8.355	8⅜	0.835	¹³⁄₁₆	18¼	1⁹⁄₁₆	¹⁵⁄₁₆
x 73	21.5	21.24	21¼	0.455	⁷⁄₁₆	¼	8.295	8¼	0.740	¾	18¼	1½	¹⁵⁄₁₆
x 68	20.0	21.13	21⅛	0.430	⁷⁄₁₆	¼	8.270	8¼	0.685	¹¹⁄₁₆	18¼	1⁷⁄₁₆	⅞
x 62	18.3	20.99	21	0.400	⅜	³⁄₁₆	8.240	8¼	0.615	⅝	18¼	1⅜	⅞
W 21x 57	16.7	21.06	21	0.405	⅜	³⁄₁₆	6.555	6½	0.650	⅝	18¼	1⅜	⅞
x 50	14.7	20.83	20⅞	0.380	⅜	³⁄₁₆	6.530	6½	0.535	⁹⁄₁₆	18¼	1⁵⁄₁₆	⅞
x 44	13.0	20.66	20⅝	0.350	⅜	³⁄₁₆	6.500	6½	0.450	⁷⁄₁₆	18¼	1³⁄₁₆	⅞

[a]For application refer to Notes in Table 2.
[b]Heavier shapes in this series are available from some producers.

W SHAPES
Properties

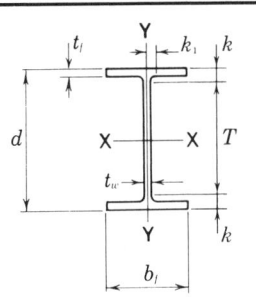

Nominal Wt. per Ft	Compact Section Criteria			X_1	$X_2 \times 10^6$	Elastic Properties						Plastic Modulus	
						Axis X-X			Axis Y-Y				
	$\frac{b_f}{2t_f}$	$\frac{h_c}{t_w}$	F_y'''			I	S	r	I	S	r	Z_x	Z_y
Lb.			Ksi	Ksi	$(1/Ksi)^2$	In.^{4}	In.^{3}	In.	In.^{4}	In.^{3}	In.	In.^{3}	In.^{3}
117	7.5	39.2	42	2090	8190	3540	291	10.1	297	46.5	2.94	327	71.4
104	8.5	43.1	34	1860	12900	3100	258	10.1	259	40.7	2.91	289	62.4
103	4.6	39.2	42	2400	5280	3000	245	9.96	119	26.5	1.99	280	41.5
94	5.2	41.9	37	2180	7800	2700	222	9.87	109	24.0	1.98	254	37.5
84	5.9	45.9	30	1950	12200	2370	196	9.79	94.4	20.9	1.95	224	32.6
76	6.6	49.0	27	1760	18600	2100	176	9.69	82.5	18.4	1.92	200	28.6
68	7.7	52.0	24	1590	29000	1830	154	9.55	70.4	15.7	1.87	177	24.5
62	6.0	50.1	25	1700	25100	1550	131	9.23	34.5	9.80	1.38	153	15.7
55	6.9	54.6	21	1540	39600	1350	114	9.11	29.1	8.30	1.34	134	13.3
402	2.1	10.8	—	8000	41	12200	937	10.2	1270	189	3.27	1130	296
364	2.3	11.8	—	7340	57	10800	846	10.0	1120	168	3.23	1010	263
333	2.5	12.8	—	6790	78	9610	769	9.91	994	151	3.19	915	237
300	2.7	14.2	—	6200	111	8480	692	9.81	873	134	3.15	816	210
275	2.9	15.4	—	5720	150	7620	632	9.71	785	122	3.12	741	189
248	3.2	17.1	—	5210	215	6760	569	9.63	694	109	3.09	663	169
223	3.5	18.8	—	4700	319	5950	510	9.54	609	96.1	3.05	589	149
201	3.9	20.6	—	4290	453	5310	461	9.47	542	86.1	3.02	530	133
182	4.2	22.6	—	3910	649	4730	417	9.40	483	77.2	3.00	476	119
166	4.6	25.0	—	3590	904	4280	380	9.36	435	70.1	2.98	432	108
147	5.4	26.1	—	3140	1590	3630	329	9.17	376	60.1	2.95	373	92.6
132	6.0	28.9	—	2840	2350	3220	295	9.12	333	53.5	2.93	333	82.3
122	6.5	31.3	—	2630	3160	2960	273	9.09	305	49.2	2.92	307	75.6
111	7.1	34.1	55	2400	4510	2670	249	9.05	274	44.5	2.90	279	68.2
101	7.7	37.5	45	2200	6400	2420	227	9.02	248	40.3	2.89	253	61.7
93	4.5	32.3	61	2680	3460	2070	192	8.70	92.9	22.1	1.84	221	34.7
83	5.0	36.4	48	2400	5250	1830	171	8.67	81.4	19.5	1.83	196	30.5
73	5.6	41.2	38	2140	8380	1600	151	8.64	70.6	17.0	1.81	172	26.6
68	6.0	43.6	34	2000	10900	1480	140	8.60	64.7	15.7	1.80	160	24.4
62	6.7	46.9	29	1820	15900	1330	127	8.54	57.5	13.9	1.77	144	21.7
57	5.0	46.3	30	1960	13100	1170	111	8.36	30.6	9.35	1.35	129	14.8
50	6.1	49.4	26	1730	22600	984	94.5	8.18	24.9	7.64	1.30	110	12.2
44	7.2	53.6	22	1550	36600	843	81.6	8.06	20.7	6.36	1.26	95.4	10.2

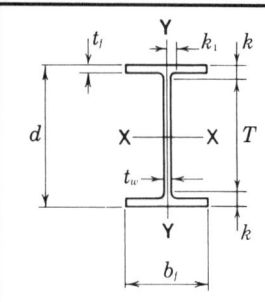

W SHAPES
Dimensions

Designation	Area A	Depth d		Web Thickness tw		tw/2	Flange Width bf		Thickness tf		T	k	k1
	In.²	In.		In.		In.	In.		In.		In.	In.	In.
W 18x311[a]	91.5	22.32	22 3/8	1.520	1 1/2	3/4	12.005	12	2.740	2 3/4	15 1/2	3 7/16	1 3/16
x283[a]	83.2	21.85	21 7/8	1.400	1 3/8	11/16	11.890	11 7/8	2.500	2 1/2	15 1/2	3 3/16	1 3/16
x258[a]	75.9	21.46	21 1/2	1.280	1 1/4	5/8	11.770	11 3/4	2.300	2 5/16	15 1/2	3	1 1/8
x234[a]	68.8	21.06	21	1.160	1 3/16	5/8	11.650	11 5/8	2.110	2 1/8	15 1/2	2 3/4	1
x211[a]	62.1	20.67	20 5/8	1.060	1 1/16	9/16	11.555	11 1/2	1.910	1 15/16	15 1/2	2 9/16	1
x192	56.4	20.35	20 3/8	0.960	1	1/2	11.455	11 1/2	1.750	1 3/4	15 1/2	2 7/16	15/16
x175	51.3	20.04	20	0.890	7/8	7/16	11.375	11 3/8	1.590	1 9/16	15 1/2	2 1/4	7/8
x158	46.3	19.72	19 3/4	0.810	13/16	7/16	11.300	11 1/4	1.440	1 7/16	15 1/2	2 1/8	7/8
x143	42.1	19.49	19 1/2	0.730	3/4	3/8	11.220	11 1/4	1.320	1 5/16	15 1/2	2	13/16
x130	38.2	19.25	19 1/4	0.670	11/16	3/8	11.160	11 1/8	1.200	1 3/16	15 1/2	1 7/8	13/16
x119	35.1	18.97	19	0.655	5/8	5/16	11.265	11 1/4	1.060	1 1/16	15 1/2	1 3/4	15/16
x106	31.1	18.73	18 3/4	0.590	9/16	5/16	11.200	11 1/4	0.940	15/16	15 1/2	1 5/8	15/16
x 97	28.5	18.59	18 5/8	0.535	9/16	5/16	11.145	11 1/8	0.870	7/8	15 1/2	1 9/16	7/8
x 86	25.3	18.39	18 3/8	0.480	1/2	1/4	11.090	11 1/8	0.770	3/4	15 1/2	1 7/16	7/8
x 76	22.3	18.21	18 1/4	0.425	7/16	1/4	11.035	11	0.680	11/16	15 1/2	1 3/8	13/16
W 18x 71	20.8	18.47	18 1/2	0.495	1/2	1/4	7.635	7 5/8	0.810	13/16	15 1/2	1 1/2	7/8
x 65	19.1	18.35	18 3/8	0.450	7/16	1/4	7.590	7 5/8	0.750	3/4	15 1/2	1 7/16	7/8
x 60	17.6	18.24	18 1/4	0.415	7/16	1/4	7.555	7 1/2	0.695	11/16	15 1/2	1 3/8	13/16
x 55	16.2	18.11	18 1/8	0.390	3/8	3/16	7.530	7 1/2	0.630	5/8	15 1/2	1 5/16	13/16
x 50	14.7	17.99	18	0.355	3/8	3/16	7.495	7 1/2	0.570	9/16	15 1/2	1 1/4	13/16
W 18x 46	13.5	18.06	18	0.360	3/8	3/16	6.060	6	0.605	5/8	15 1/2	1 1/4	13/16
x 40	11.8	17.90	17 7/8	0.315	5/16	3/16	6.015	6	0.525	1/2	15 1/2	1 3/16	13/16
x 35	10.3	17.70	17 3/4	0.300	5/16	3/16	6.000	6	0.425	7/16	15 1/2	1 1/8	3/4
W 16x100	29.4	16.97	17	0.585	9/16	5/16	10.425	10 3/8	0.985	1	13 5/8	1 11/16	15/16
x 89	26.2	16.75	16 3/4	0.525	1/2	1/4	10.365	10 3/8	0.875	7/8	13 5/8	1 9/16	7/8
x 77	22.6	16.52	16 1/2	0.455	7/16	1/4	10.295	10 1/4	0.760	3/4	13 5/8	1 7/16	7/8
x 67	19.7	16.33	16 3/8	0.395	3/8	3/16	10.235	10 1/4	0.665	11/16	13 5/8	1 3/8	13/16
W 16x 57	16.8	16.43	16 3/8	0.430	7/16	1/4	7.120	7 1/8	0.715	11/16	13 5/8	1 3/8	7/8
x 50	14.7	16.26	16 1/4	0.380	3/8	3/16	7.070	7 1/8	0.630	5/8	13 5/8	1 5/16	13/16
x 45	13.3	16.13	16 1/8	0.345	3/8	3/16	7.035	7	0.565	9/16	13 5/8	1 1/4	13/16
x 40	11.8	16.01	16	0.305	5/16	3/16	6.995	7	0.505	1/2	13 5/8	1 3/16	13/16
x 36	10.6	15.86	15 7/8	0.295	5/16	3/16	6.985	7	0.430	7/16	13 5/8	1 1/8	3/4
W 16x 31	9.12	15.88	15 7/8	0.275	1/4	1/8	5.525	5 1/2	0.440	7/16	13 5/8	1 1/8	3/4
x 26	7.68	15.69	15 3/4	0.250	1/4	1/8	5.500	5 1/2	0.345	3/8	13 5/8	1 1/16	3/4

[a]For application refer to Notes in Table 2.

W SHAPES
Properties

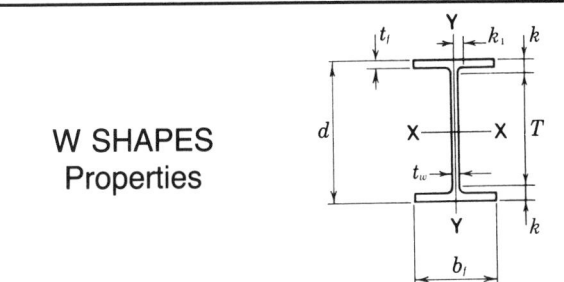

Nom-inal Wt. per Ft	Compact Section Criteria					Elastic Properties						Plastic Modulus	
						Axis X-X			Axis Y-Y				
	$\frac{b_f}{2t_f}$	$\frac{h_c}{t_w}$	F_y'''	X_1	$X_2 \times 10^6$	I	S	r	I	S	r	Z_x	Z_y
Lb.			Ksi	Ksi	(1/Ksi)^2	In.^{4}	In.^{3}	In.	In.^{4}	In.^{3}	In.	In.^{3}	In.^{3}
311	2.2	10.6	—	8160	38	6960	624	8.72	795	132	2.95	753	207
283	2.4	11.5	—	7520	52	6160	564	8.61	704	118	2.91	676	185
258	2.6	12.5	—	6920	71	5510	514	8.53	628	107	2.88	611	166
234	2.8	13.8	—	6360	97	4900	466	8.44	558	95.8	2.85	549	149
211	3.0	15.1	—	5800	140	4330	419	8.35	493	85.3	2.82	490	132
192	3.3	16.7	—	5320	194	3870	380	8.28	440	76.8	2.79	442	119
175	3.6	18.0	—	4870	274	3450	344	8.20	391	68.8	2.76	398	106
158	3.9	19.8	—	4430	396	3060	310	8.12	347	61.4	2.74	356	94.8
143	4.2	21.9	—	4060	557	2750	282	8.09	311	55.9	2.72	322	85.4
130	4.6	23.9	—	3710	789	2460	256	8.03	278	49.9	2.70	291	76.7
119	5.3	24.5	—	3340	1210	2190	231	7.90	253	44.9	2.69	261	69.1
106	6.0	27.2	—	2990	1880	1910	204	7.84	220	39.4	2.66	230	60.5
97	6.4	30.0	—	2750	2580	1750	188	7.82	201	36.1	2.65	211	55.3
86	7.2	33.4	57	2460	4060	1530	166	7.77	175	31.6	2.63	186	48.4
76	8.1	37.8	45	2180	6520	1330	146	7.73	152	27.6	2.61	163	42.2
71	4.7	32.4	61	2680	3310	1170	127	7.50	60.3	15.8	1.70	145	24.7
65	5.1	35.7	50	2470	4540	1070	117	7.49	54.8	14.4	1.69	133	22.5
60	5.4	38.7	43	2290	6080	984	108	7.47	50.1	13.3	1.69	123	20.6
55	6.0	41.2	38	2110	8540	890	98.3	7.41	44.9	11.9	1.67	112	18.5
50	6.6	45.2	31	1920	12400	800	88.9	7.38	40.1	10.7	1.65	101	16.6
46	5.0	44.6	32	2060	10100	712	78.8	7.25	22.5	7.43	1.29	90.7	11.7
40	5.7	51.0	25	1810	17200	612	68.4	7.21	19.1	6.35	1.27	78.4	9.95
35	7.1	53.5	22	1590	30300	510	57.6	7.04	15.3	5.12	1.22	66.5	8.06
100	5.3	24.3	—	3450	1040	1490	175	7.10	186	35.7	2.51	198	54.9
89	5.9	27.0	—	3090	1630	1300	155	7.05	163	31.4	2.49	175	48.1
77	6.8	31.2	—	2680	2790	1110	134	7.00	138	26.9	2.47	150	41.1
67	7.7	35.9	50	2350	4690	954	117	6.96	119	23.2	2.46	130	35.5
57	5.0	33.0	59	2650	3400	758	92.2	6.72	43.1	12.1	1.60	105	18.9
50	5.6	37.4	46	2340	5530	659	81.0	6.68	37.2	10.5	1.59	92.0	16.3
45	6.2	41.2	38	2120	8280	586	72.7	6.65	32.8	9.34	1.57	82.3	14.5
40	6.9	46.6	30	1890	12900	518	64.7	6.63	28.9	8.25	1.57	72.9	12.7
36	8.1	48.1	28	1700	20800	448	56.5	6.51	24.5	7.00	1.52	64.0	10.8
31	6.3	51.6	24	1740	20000	375	47.2	6.41	12.4	4.49	1.17	54.0	7.03
26	8.0	56.8	20	1470	40900	301	38.4	6.26	9.59	3.49	1.12	44.2	5.48

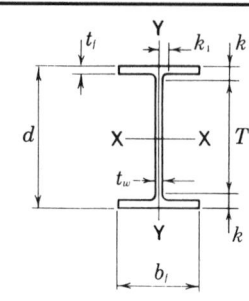

W SHAPES
Dimensions

Desig-nation	Area A	Depth d		Web Thickness t_w		$\dfrac{t_w}{2}$	Flange Width b_f		Thickness t_f		Distance T	k	k_1
	In.²	In.		In.		In.	In.		In.		In.	In.	In.
W 14x730[a]	215.0	22.42	22⅜	3.070	3¹⁄₁₆	1⁹⁄₁₆	17.890	17⅞	4.910	4¹⁵⁄₁₆	11¼	5⁹⁄₁₆	2³⁄₁₆
x665[a]	196.0	21.64	21⅝	2.830	2¹³⁄₁₆	1⁷⁄₁₆	17.650	17⅝	4.520	4½	11¼	5³⁄₁₆	2¹⁄₁₆
x605[a]	178.0	20.92	20⅞	2.595	2⅝	1⁵⁄₁₆	17.415	17⅜	4.160	4³⁄₁₆	11¼	4¹³⁄₁₆	1¹⁵⁄₁₆
x550[a]	162.0	20.24	20¼	2.380	2⅜	1³⁄₁₆	17.200	17¼	3.820	3¹³⁄₁₆	11¼	4½	1¹³⁄₁₆
x500[a]	147.0	19.60	19⅝	2.190	2³⁄₁₆	1⅛	17.010	17	3.500	3½	11¼	4³⁄₁₆	1¾
x455[a]	134.0	19.02	19	2.015	2	1	16.835	16⅞	3.210	3³⁄₁₆	11¼	3⅞	1⅝
W 14x426[a]	125.0	18.67	18⅝	1.875	1⅞	¹⁵⁄₁₆	16.695	16¾	3.035	3¹⁄₁₆	11¼	3¹¹⁄₁₆	1⁹⁄₁₆
x398[a]	117.0	18.29	18¼	1.770	1¾	⅞	16.590	16⅝	2.845	2⅞	11¼	3½	1½
x370[a]	109.0	17.92	17⅞	1.655	1⅝	¹³⁄₁₆	16.475	16½	2.660	2¹¹⁄₁₆	11¼	3⁵⁄₁₆	1⁷⁄₁₆
x342[a]	101.0	17.54	17½	1.540	1⁹⁄₁₆	¹³⁄₁₆	16.360	16⅜	2.470	2½	11¼	3⅛	1⅜
x311[a]	91.4	17.12	17⅛	1.410	1⁷⁄₁₆	¾	16.230	16¼	2.260	2¼	11¼	2¹⁵⁄₁₆	1⁵⁄₁₆
x283[a]	83.3	16.74	16¾	1.290	1⁵⁄₁₆	¹¹⁄₁₆	16.110	16⅛	2.070	2¹⁄₁₆	11¼	2¾	1¼
x257[a]	75.6	16.38	16⅜	1.175	1³⁄₁₆	⅝	15.995	16	1.890	1⅞	11¼	2⁹⁄₁₆	1³⁄₁₆
x233[a]	68.5	16.04	16	1.070	1¹⁄₁₆	⁹⁄₁₆	15.890	15⅞	1.720	1¾	11¼	2⅜	1³⁄₁₆
x211[a]	62.0	15.72	15¾	0.980	1	½	15.800	15¾	1.560	1⁹⁄₁₆	11¼	2¼	1⅛
x193	56.8	15.48	15½	0.890	⅞	⁷⁄₁₆	15.710	15¾	1.440	1⁷⁄₁₆	11¼	2⅛	1¹⁄₁₆
x176	51.8	15.22	15¼	0.830	¹³⁄₁₆	⁷⁄₁₆	15.650	15⅝	1.310	1⁵⁄₁₆	11¼	2	1¹⁄₁₆
x159	46.7	14.98	15	0.745	¾	⅜	15.565	15⅝	1.190	1³⁄₁₆	11¼	1⅞	1
x145	42.7	14.78	14¾	0.680	¹¹⁄₁₆	⅜	15.500	15½	1.090	1¹⁄₁₆	11¼	1¾	1

[a]For application refer to Notes in Table 2.

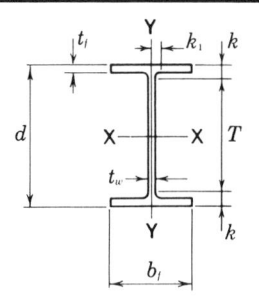

W SHAPES
Properties

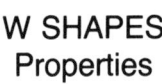

Nom-inal Wt. per Ft	Compact Section Criteria					Elastic Properties						Plastic Modulus	
	$\dfrac{b_f}{2t_f}$	$\dfrac{h_c}{t_w}$	F_y'''	X_1	$X_2 \times 10^6$	Axis X-X			Axis Y-Y			Z_x	Z_y
						I	S	r	I	S	r		
Lb.			Ksi	Ksi	$(1/\text{Ksi})^2$	In.^{4}	In.^{3}	In.	In.^{4}	In.^{3}	In.	In.^{3}	In.^{3}
730	1.8	3.7	—	17500	1.90	14300	1280	8.17	4720	527	4.69	1660	816
665	2.0	4.0	—	16300	2.46	12400	1150	7.98	4170	472	4.62	1480	730
605	2.1	4.4	—	15100	3.20	10800	1040	7.80	3680	423	4.55	1320	652
550	2.3	4.8	—	14200	4.15	9430	931	7.63	3250	378	4.49	1180	583
500	2.4	5.2	—	13100	5.49	8210	838	7.48	2880	339	4.43	1050	522
455	2.6	5.7	—	12200	7.30	7190	756	7.33	2560	304	4.38	936	468
426	2.8	6.1	—	11500	8.89	6600	707	7.26	2360	283	4.34	869	434
398	2.9	6.4	—	10900	11.0	6000	656	7.16	2170	262	4.31	801	402
370	3.1	6.9	—	10300	13.9	5440	607	7.07	1990	241	4.27	736	370
342	3.3	7.4	—	9600	17.9	4900	559	6.98	1810	221	4.24	672	338
311	3.6	8.1	—	8820	24.4	4330	506	6.88	1610	199	4.20	603	304
283	3.9	8.8	—	8120	33.4	3840	459	6.79	1440	179	4.17	542	274
257	4.2	9.7	—	7460	46.1	3400	415	6.71	1290	161	4.13	487	246
233	4.6	10.7	—	6820	64.9	3010	375	6.63	1150	145	4.10	436	221
211	5.1	11.6	—	6230	91.8	2660	338	6.55	1030	130	4.07	390	198
193	5.5	12.8	—	5740	125	2400	310	6.50	931	119	4.05	355	180
176	6.0	13.7	—	5280	173	2140	281	6.43	838	107	4.02	320	163
159	6.5	15.3	—	4790	249	1900	254	6.38	748	96.2	4.00	287	146
145	7.1	16.8	—	4400	348	1710	232	6.33	677	87.3	3.98	260	133

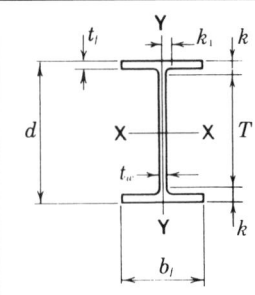

W SHAPES
Dimensions

Desig- nation	Area A	Depth d	Web		Flange			Distance		
			Thickness t_w	$\frac{t_w}{2}$	Width b_f	Thickness t_f		T	k	k_1
	In.^{2}	In.	In.	In.	In.	In.		In.	In.	In.
W 14x132	38.8	14.66	14⅝	0.645	⅝	⁵⁄₁₆	14.725	14¾	1.030	1	11¼	1¹¹⁄₁₆	¹⁵⁄₁₆
x120	35.3	14.48	14½	0.590	⁹⁄₁₆	⁵⁄₁₆	14.670	14⅝	0.940	¹⁵⁄₁₆	11¼	1⅝	¹⁵⁄₁₆
x109	32.0	14.32	14⅜	0.525	½	¼	14.605	14⅝	0.860	⅞	11¼	1⁹⁄₁₆	⅞
x 99	29.1	14.16	14⅛	0.485	½	¼	14.565	14⅝	0.780	¾	11¼	1⁷⁄₁₆	⅞
x 90	26.5	14.02	14	0.440	⁷⁄₁₆	¼	14.520	14½	0.710	¹¹⁄₁₆	11¼	1⅜	⅞
W 14x 82	24.1	14.31	14¼	0.510	½	¼	10.130	10⅛	0.855	⅞	11	1⅝	1
x 74	21.8	14.17	14⅛	0.450	⁷⁄₁₆	¼	10.070	10⅛	0.785	¹³⁄₁₆	11	1⁹⁄₁₆	¹⁵⁄₁₆
x 68	20.0	14.04	14	0.415	⁷⁄₁₆	¼	10.035	10	0.720	¾	11	1½	¹⁵⁄₁₆
x 61	17.9	13.89	13⅞	0.375	⅜	³⁄₁₆	9.995	10	0.645	⅝	11	1⁷⁄₁₆	¹⁵⁄₁₆
W 14x 53	15.6	13.92	13⅞	0.370	⅜	³⁄₁₆	8.060	8	0.660	¹¹⁄₁₆	11	1⁷⁄₁₆	¹⁵⁄₁₆
x 48	14.1	13.79	13¾	0.340	⁵⁄₁₆	³⁄₁₆	8.030	8	0.595	⅝	11	1⅜	⅞
x 43	12.6	13.66	13⅝	0.305	⁵⁄₁₆	³⁄₁₆	7.995	8	0.530	½	11	1⁵⁄₁₆	⅞
W 14x 38	11.2	14.10	14⅛	0.310	⁵⁄₁₆	³⁄₁₆	6.770	6¾	0.515	½	12	1¹⁄₁₆	⅝
x 34	10.0	13.98	14	0.285	⁵⁄₁₆	³⁄₁₆	6.745	6¾	0.455	⁷⁄₁₆	12	1	⅝
x 30	8.85	13.84	13⅞	0.270	¼	⅛	6.730	6¾	0.385	⅜	12	¹⁵⁄₁₆	⅝
W 14x 26	7.69	13.91	13⅞	0.255	¼	⅛	5.025	5	0.420	⁷⁄₁₆	12	¹⁵⁄₁₆	⁹⁄₁₆
x 22	6.49	13.74	13¾	0.230	¼	⅛	5.000	5	0.335	⁵⁄₁₆	12	⅞	⁹⁄₁₆

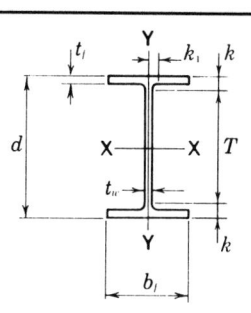

W SHAPES
Properties

Nom-inal Wt. per Ft	Compact Section Criteria		F_y'''	X_1	$X_2 \times 10^6$	Elastic Properties						Plastic Modulus	
	$\dfrac{b_f}{2t_f}$	$\dfrac{h_c}{t_w}$				Axis X-X			Axis Y-Y			Z_x	Z_y
						I	S	r	I	S	r		
Lb.			Ksi	Ksi	$(1/\text{Ksi})^2$	In.⁴	In.³	In.	In.⁴	In.³	In.	In.³	In.³
132	7.1	17.7	—	4180	428	1530	209	6.28	548	74.5	3.76	234	113
120	7.8	19.3	—	3830	601	1380	190	6.24	495	67.5	3.74	212	102
109	8.5	21.7	—	3490	853	1240	173	6.22	447	61.2	3.73	192	92.7
99	9.3	23.5	—	3190	1220	1110	157	6.17	402	55.2	3.71	173	83.6
90	10.2	25.9	—	2900	1750	999	143	6.14	362	49.9	3.70	157	75.6
82	5.9	22.4	—	3600	846	882	123	6.05	148	29.3	2.48	139	44.8
74	6.4	25.3	—	3290	1190	796	112	6.04	134	26.6	2.48	126	40.6
68	7.0	27.5	—	3020	1650	723	103	6.01	121	24.2	2.46	115	36.9
61	7.7	30.4	—	2720	2460	640	92.2	5.98	107	21.5	2.45	102	32.8
53	6.1	30.8	—	2830	2250	541	77.8	5.89	57.7	14.3	1.92	87.1	22.0
48	6.7	33.5	57	2580	3220	485	70.3	5.85	51.4	12.8	1.91	78.4	19.6
43	7.5	37.4	46	2320	4900	428	62.7	5.82	45.2	11.3	1.89	69.6	17.3
38	6.6	39.6	41	2190	6850	385	54.6	5.87	26.7	7.88	1.55	61.5	12.1
34	7.4	43.1	35	1970	10600	340	48.6	5.83	23.3	6.91	1.53	54.6	10.6
30	8.7	45.4	31	1750	17600	291	42.0	5.73	19.6	5.82	1.49	47.3	8.99
26	6.0	48.1	28	1890	13900	245	35.3	5.65	8.91	3.54	1.08	40.2	5.54
22	7.5	53.5	22	1610	27300	199	29.0	5.54	7.00	2.80	1.04	33.2	4.39

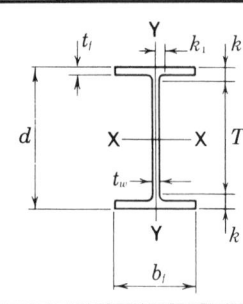

W SHAPES
Dimensions

Desig-nation	Area A	Depth d		Web Thickness t_w		$\dfrac{t_w}{2}$	Flange Width b_f		Thickness t_f		Distance T	k	k_1
	In.²	In.		In.		In.	In.		In.		In.	In.	In.
W 12x336ᵃ	98.8	16.82	16⅞	1.775	1¾	⅞	13.385	13⅜	2.955	2¹⁵⁄₁₆	9½	3¹¹⁄₁₆	1½
x305ᵃ	89.6	16.32	16⅜	1.625	1⅝	¹³⁄₁₆	13.235	13¼	2.705	2¹¹⁄₁₆	9½	3⁷⁄₁₆	1⁷⁄₁₆
x279ᵃ	81.9	15.85	15⅞	1.530	1½	¾	13.140	13⅛	2.470	2½	9½	3³⁄₁₆	1⅜
x252ᵃ	74.1	15.41	15⅜	1.395	1⅜	¹¹⁄₁₆	13.005	13	2.250	2¼	9½	2¹⁵⁄₁₆	1⁵⁄₁₆
x230ᵃ	67.7	15.05	15	1.285	1⁵⁄₁₆	¹¹⁄₁₆	12.895	12⅞	2.070	2¹⁄₁₆	9½	2¾	1¼
x210ᵃ	61.8	14.71	14¾	1.180	1³⁄₁₆	⅝	12.790	12¾	1.900	1⅞	9½	2⅝	1¼
x190ᵃ	55.8	14.38	14⅜	1.060	1¹⁄₁₆	⁹⁄₁₆	12.670	12⅝	1.735	1¾	9½	2⁷⁄₁₆	1³⁄₁₆
x170ᵃ	50.0	14.03	14	0.960	¹⁵⁄₁₆	½	12.570	12⅝	1.560	1⁹⁄₁₆	9½	2¼	1⅛
x152	44.7	13.71	13¾	0.870	⅞	⁷⁄₁₆	12.480	12½	1.400	1⅜	9½	2⅛	1¹⁄₁₆
x136	39.9	13.41	13⅜	0.790	¹³⁄₁₆	⁷⁄₁₆	12.400	12⅜	1.250	1¼	9½	1¹⁵⁄₁₆	1
x120	35.3	13.12	13⅛	0.710	¹¹⁄₁₆	⅜	12.320	12⅜	1.105	1⅛	9½	1¹³⁄₁₆	1
x106	31.2	12.89	12⅞	0.610	⅝	⁵⁄₁₆	12.220	12¼	0.990	1	9½	1¹¹⁄₁₆	¹⁵⁄₁₆
x 96	28.2	12.71	12¾	0.550	⁹⁄₁₆	⁵⁄₁₆	12.160	12⅛	0.900	⅞	9½	1⅝	⅞
x 87	25.6	12.53	12½	0.515	½	¼	12.125	12⅛	0.810	¹³⁄₁₆	9½	1½	⅞
x 79	23.2	12.38	12⅜	0.470	½	¼	12.080	12⅛	0.735	¾	9½	1⁷⁄₁₆	⅞
x 72	21.1	12.25	12¼	0.430	⁷⁄₁₆	¼	12.040	12	0.670	¹¹⁄₁₆	9½	1⅜	⅞
x 65	19.1	12.12	12⅛	0.390	⅜	³⁄₁₆	12.000	12	0.605	⅝	9½	1⁵⁄₁₆	¹³⁄₁₆
W 12x 58	17.0	12.19	12¼	0.360	⅜	³⁄₁₆	10.010	10	0.640	⅝	9½	1⅜	¹³⁄₁₆
x 53	15.6	12.06	12	0.345	⅜	³⁄₁₆	9.995	10	0.575	⁹⁄₁₆	9½	1¼	¹³⁄₁₆
W 12x 50	14.7	12.19	12¼	0.370	⅜	³⁄₁₆	8.080	8⅛	0.640	⅝	9½	1⅜	¹³⁄₁₆
x 45	13.2	12.06	12	0.335	⁵⁄₁₆	³⁄₁₆	8.045	8	0.575	⁹⁄₁₆	9½	1¼	¹³⁄₁₆
x 40	11.8	11.94	12	0.295	⁵⁄₁₆	³⁄₁₆	8.005	8	0.515	½	9½	1¼	¾
W 12x 35	10.3	12.50	12½	0.300	⁵⁄₁₆	³⁄₁₆	6.560	6½	0.520	½	10½	1	⁹⁄₁₆
x 30	8.79	12.34	12⅜	0.260	¼	⅛	6.520	6½	0.440	⁷⁄₁₆	10½	¹⁵⁄₁₆	½
x 26	7.65	12.22	12¼	0.230	¼	⅛	6.490	6½	0.380	⅜	10½	⅞	½
W 12x 22	6.48	12.31	12¼	0.260	¼	⅛	4.030	4	0.425	⁷⁄₁₆	10½	⅞	½
x 19	5.57	12.16	12⅛	0.235	¼	⅛	4.005	4	0.350	⅜	10½	¹³⁄₁₆	½
x 16	4.71	11.99	12	0.220	¼	⅛	3.990	4	0.265	¼	10½	¾	½
x 14	4.16	11.91	11⅞	0.200	³⁄₁₆	⅛	3.970	4	0.225	¼	10½	¹¹⁄₁₆	½

ᵃFor application refer to Notes in Table 2.

W SHAPES
Properties

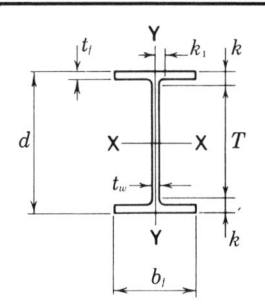

Nominal Wt. per Ft	Compact Section Criteria $\frac{b_f}{2t_f}$	$\frac{h_c}{t_w}$	F_y'''	X_1	$X_2 \times 10^6$	Elastic Properties Axis X-X I	S	r	Axis Y-Y I	S	r	Plastic Modulus Z_x	Z_y
Lb.			Ksi	Ksi	$(1/\text{Ksi})^2$	In.^{4}	In.^{3}	In.	In.^{4}	In.^{3}	In.	In.^{3}	In.^{3}
336	2.3	5.5	—	12800	6.05	4060	483	6.41	1190	177	3.47	603	274
305	2.4	6.0	—	11800	8.17	3550	435	6.29	1050	159	3.42	537	244
279	2.7	6.3	—	11000	10.8	3110	393	6.16	937	143	3.38	481	220
252	2.9	7.0	—	10100	14.7	2720	353	6.06	828	127	3.34	428	196
230	3.1	7.6	—	9390	19.7	2420	321	5.97	742	115	3.31	386	177
210	3.4	8.2	—	8670	26.6	2140	292	5.89	664	104	3.28	348	159
190	3.7	9.2	—	7940	37.0	1890	263	5.82	589	93.0	3.25	311	143
170	4.0	10.1	—	7190	54.0	1650	235	5.74	517	82.3	3.22	275	126
152	4.5	11.2	—	6510	79.3	1430	209	5.66	454	72.8	3.19	243	111
136	5.0	12.3	—	5850	119	1240	186	5.58	398	64.2	3.16	214	98.0
120	5.6	13.7	—	5240	184	1070	163	5.51	345	56.0	3.13	186	85.4
106	6.2	15.9	—	4660	285	933	145	5.47	301	49.3	3.11	164	75.1
96	6.8	17.7	—	4250	405	833	131	5.44	270	44.4	3.09	147	67.5
87	7.5	18.9	—	3880	586	740	118	5.38	241	39.7	3.07	132	60.4
79	8.2	20.7	—	3530	839	662	107	5.34	216	35.8	3.05	119	54.3
72	9.0	22.6	—	3230	1180	597	97.4	5.31	195	32.4	3.04	108	49.2
65	9.9	24.9	—	2940	1720	533	87.9	5.28	174	29.1	3.02	96.8	44.1
58	7.8	27.0	—	3070	1470	475	78.0	5.28	107	21.4	2.51	86.4	32.5
53	8.7	28.1	—	2820	2100	425	70.6	5.23	95.8	19.2	2.48	77.9	29.1
50	6.3	26.2	—	3170	1410	394	64.7	5.18	56.3	13.9	1.96	72.4	21.4
45	7.0	29.0	—	2870	2070	350	58.1	5.15	50.0	12.4	1.94	64.7	19.0
40	7.8	32.9	59	2580	3110	310	51.9	5.13	44.1	11.0	1.93	57.5	16.8
35	6.3	36.2	49	2420	4340	285	45.6	5.25	24.5	7.47	1.54	51.2	11.5
30	7.4	41.8	37	2090	7950	238	38.6	5.21	20.3	6.24	1.52	43.1	9.56
26	8.5	47.2	29	1820	13900	204	33.4	5.17	17.3	5.34	1.51	37.2	8.17
22	4.7	41.8	37	2160	8640	156	25.4	4.91	4.66	2.31	0.847	29.3	3.66
19	5.7	46.2	30	1880	15600	130	21.3	4.82	3.76	1.88	0.822	24.7	2.98
16	7.5	49.4	26	1610	32000	103	17.1	4.67	2.82	1.41	0.773	20.1	2.26
14	8.8	54.3	22	1450	49300	88.6	14.9	4.62	2.36	1.19	0.753	17.4	1.90

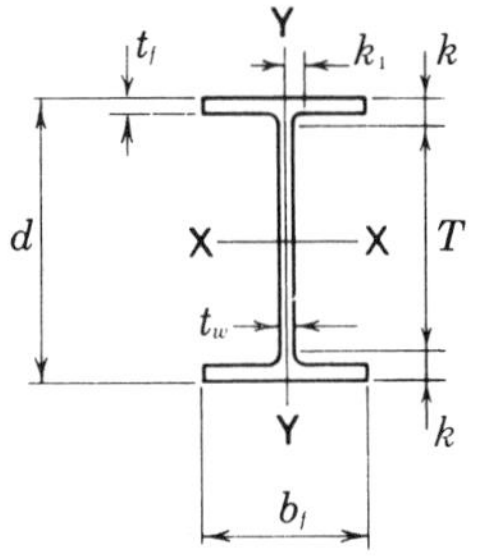

W SHAPES
Dimensions

Designation	Area A	Depth d		Web Thickness t_w		Web $\frac{t_w}{2}$	Flange Width b_f		Flange Thickness t_f		Distance T	Distance k	Distance k_1
	In.^{2}	In.		In.		In.	In.		In.		In.	In.	In.
W 10x112	32.9	11.36	11 3/8	0.755	3/4	3/8	10.415	10 3/8	1.250	1 1/4	7 5/8	1 7/8	15/16
x100	29.4	11.10	11 1/8	0.680	11/16	3/8	10.340	10 3/8	1.120	1 1/8	7 5/8	1 3/4	7/8
x 88	25.9	10.84	10 7/8	0.605	5/8	5/16	10.265	10 1/4	0.990	1	7 5/8	1 5/8	13/16
x 77	22.6	10.60	10 5/8	0.530	1/2	1/4	10.190	10 1/4	0.870	7/8	7 5/8	1 1/2	13/16
x 68	20.0	10.40	10 3/8	0.470	3/8	1/4	10.130	10 1/8	0.770	3/4	7 5/8	1 3/8	3/4
x 60	17.6	10.22	10 1/4	0.420	7/16	1/4	10.080	10 1/8	0.680	11/16	7 5/8	1 5/16	3/4
x 54	15.8	10.09	10 1/8	0.370	3/8	3/16	10.030	10	0.615	5/8	7 5/8	1 1/4	11/16
x 49	14.4	9.98	10	0.340	5/16	3/16	10.000	10	0.560	9/16	7 5/8	1 3/16	11/16
W 10x 45	13.3	10.10	10 1/8	0.350	3/8	3/16	8.020	8	0.620	5/8	7 5/8	1 1/4	11/16
x 39	11.5	9.92	9 7/8	0.315	5/16	3/16	7.985	8	0.530	1/2	7 5/8	1 1/8	11/16
x 33	9.71	9.73	9 3/4	0.290	5/16	3/16	7.960	8	0.435	7/16	7 5/8	1 1/16	11/16
W 10x 30	8.84	10.47	10 1/2	0.300	5/16	3/16	5.810	5 3/4	0.510	1/2	8 5/8	15/16	1/2
x 26	7.61	10.33	10 3/8	0.260	1/4	1/8	5.770	5 3/4	0.440	7/16	8 5/8	7/8	1/2
x 22	6.49	10.17	10 1/8	0.240	1/4	1/8	5.750	5 3/4	0.360	3/8	8 5/8	3/4	1/2
W 10x 19	5.62	10.24	10 1/4	0.250	1/4	1/8	4.020	4	0.395	3/8	8 5/8	13/16	1/2
x 17	4.99	10.11	10 1/8	0.240	1/4	1/8	4.010	4	0.330	5/16	8 5/8	3/4	1/2
x 15	4.41	9.99	10	0.230	1/4	1/8	4.000	4	0.270	1/4	8 5/8	11/16	7/16
x 12	3.54	9.87	9 7/8	0.190	3/16	1/8	3.960	4	0.210	3/16	8 5/8	5/8	7/16

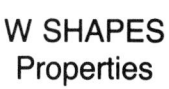
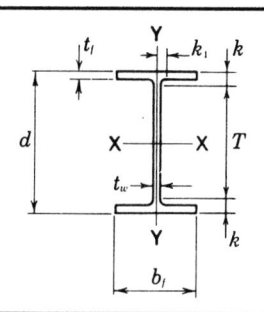

W SHAPES
Properties

Nominal Wt. per Ft	Compact Section Criteria			X_1	$X_2 \times 10^6$	Elastic Properties						Plastic Modulus	
	$\frac{b_f}{2t_f}$	$\frac{h_c}{t_w}$	F_y'''			Axis X-X			Axis Y-Y			Z_x	Z_y
						I	S	r	I	S	r		
Lb.			Ksi	Ksi	$(1/\text{Ksi})^2$	In.⁴	In.³	In.	In.⁴	In.³	In.	In.³	In.³
112	4.2	10.4	—	7080	56.7	716	126	4.66	236	45.3	2.68	147	69.2
100	4.6	11.6	—	6400	83.8	623	112	4.60	207	40.0	2.65	130	61.0
88	5.2	13.0	—	5680	132	534	98.5	4.54	179	34.8	2.63	113	53.1
77	5.9	14.8	—	5010	213	455	85.9	4.49	154	30.1	2.60	97.6	45.9
68	6.6	16.7	—	4460	334	394	75.7	4.44	134	26.4	2.59	85.3	40.1
60	7.4	18.7	—	3970	525	341	66.7	4.39	116	23.0	2.57	74.6	35.0
54	8.2	21.2	—	3580	778	303	60.0	4.37	103	20.6	2.56	66.6	31.3
49	8.9	23.1	—	3280	1090	272	54.6	4.35	93.4	18.7	2.54	60.4	28.3
45	6.5	22.5	—	3650	758	248	49.1	4.32	53.4	13.3	2.01	54.9	20.3
39	7.5	25.0	—	3190	1300	209	42.1	4.27	45.0	11.3	1.98	46.8	17.2
33	9.1	27.1	—	2710	2510	170	35.0	4.19	36.6	9.20	1.94	38.8	14.0
30	5.7	29.5	—	2890	2160	170	32.4	4.38	16.7	5.75	1.37	36.6	8.84
26	6.6	34.0	55	2500	3790	144	27.9	4.35	14.1	4.89	1.36	31.3	7.50
22	8.0	36.9	47	2150	7170	118	23.2	4.27	11.4	3.97	1.33	26.0	6.10
19	5.1	35.4	51	2420	5160	96.3	18.8	4.14	4.29	2.14	0.874	21.6	3.35
17	6.1	36.9	47	2210	7820	81.9	16.2	4.05	3.56	1.78	0.844	18.7	2.80
15	7.4	38.5	43	1930	14300	68.9	13.8	3.95	2.89	1.45	0.810	16.0	2.30
12	9.4	46.6	30	1550	35400	53.8	10.9	3.90	2.18	1.10	0.785	12.6	1.74

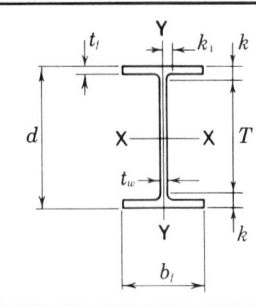

W SHAPES
Dimensions

Desig-nation	Area A	Depth d		Web Thickness t_w		$\dfrac{t_w}{2}$	Flange Width b_f		Thickness t_f		Distance T	k	k_1
	In.²	In.		In.		In.	In.		In.		In.	In.	In.
W 8x 67	19.7	9.00	9	0.570	9/16	5/16	8.280	8¼	0.935	15/16	6⅛	1 7/16	11/16
x 58	17.1	8.75	8¾	0.510	½	¼	8.220	8¼	0.810	13/16	6⅛	1 5/16	11/16
x 48	14.1	8.50	8½	0.400	⅜	3/16	8.110	8⅛	0.685	11/16	6⅛	1 3/16	⅝
x 40	11.7	8.25	8¼	0.360	⅜	3/16	8.070	8⅛	0.560	9/16	6⅛	1 1/16	⅝
x 35	10.3	8.12	8⅛	0.310	5/16	3/16	8.020	8	0.495	½	6⅛	1	9/16
x 31	9.13	8.00	8	0.285	5/16	3/16	7.995	8	0.435	7/16	6⅛	15/16	9/16
W 8x 28	8.25	8.06	8	0.285	5/16	3/16	6.535	6½	0.465	7/16	6⅛	15/16	9/16
x 24	7.08	7.93	7⅞	0.245	¼	⅛	6.495	6½	0.400	⅜	6⅛	⅞	9/16
W 8x 21	6.16	8.28	8¼	0.250	¼	⅛	5.270	5¼	0.400	⅜	6⅝	13/16	½
x 18	5.26	8.14	8⅛	0.230	¼	⅛	5.250	5¼	0.330	5/16	6⅝	¾	7/16
W 8 x15	4.44	8.11	8⅛	0.245	¼	⅛	4.015	4	0.315	5/16	6⅝	¾	½
x 13	3.84	7.99	8	0.230	¼	⅛	4.000	4	0.255	¼	6⅝	11/16	7/16
x 10	2.96	7.89	7⅞	0.170	3/16	⅛	3.940	4	0.205	3/16	6⅝	⅝	7/16
W 6x 25	7.34	6.38	6⅜	0.320	5/16	3/16	6.080	6⅛	0.455	7/16	4¾	13/16	7/16
x 20	5.87	6.20	6¼	0.260	¼	⅛	6.020	6	0.365	⅜	4¾	¾	7/16
x 15	4.43	5.99	6	0.230	¼	⅛	5.990	6	0.260	¼	4¾	⅝	⅜
W 6x 16	4.74	6.28	6¼	0.260	¼	⅛	4.030	4	0.405	⅜	4¾	¾	7/16
x 12	3.55	6.03	6	0.230	¼	⅛	4.000	4	0.280	¼	4¾	⅝	⅜
x 9	2.68	5.90	5⅞	0.170	3/16	⅛	3.940	4	0.215	3/16	4¾	9/16	⅜
W 5x 19	5.54	5.15	5⅛	0.270	¼	⅛	5.030	5	0.430	7/16	3½	13/16	7/16
x 16	4.68	5.01	5	0.240	¼	⅛	5.000	5	0.360	⅜	3½	¾	7/16
W 4x 13	3.83	4.16	4⅛	0.280	¼	⅛	4.060	4	0.345	⅜	2¾	11/16	7/16

W SHAPES
Properties

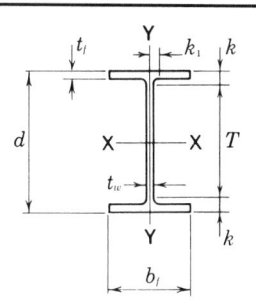

Nom-inal Wt. per Ft	Compact Section Criteria					Elastic Properties						Plastic Modulus	
	$\frac{b_f}{2t_f}$	$\frac{h_c}{t_w}$	F_y'''	X_1	$X_2 \times 10^6$	Axis X-X			Axis Y-Y			Z_x	Z_y
						I	S	r	I	S	r		
Lb.			Ksi	Ksi	$(1/\text{Ksi})^2$	In.^{4}	In.^{3}	In.	In.^{4}	In.^{3}	In.	In.^{3}	In.^{3}
67	4.4	11.1	—	6620	73.9	272	60.4	3.72	88.6	21.4	2.12	70.2	32.7
58	5.1	12.4	—	5820	122	228	52.0	3.65	75.1	18.3	2.10	59.8	27.9
48	5.9	15.8	—	4860	238	184	43.3	3.61	60.9	15.0	2.08	49.0	22.9
40	7.2	17.6	—	4080	474	146	35.5	3.53	49.1	12.2	2.04	39.8	18.5
35	8.1	20.4	—	3610	761	127	31.2	3.51	42.6	10.6	2.03	34.7	16.1
31	9.2	22.2	—	3230	1180	110	27.5	3.47	37.1	9.27	2.02	30.4	14.1
28	7.0	22.2	—	3480	931	98.0	24.3	3.45	21.7	6.63	1.62	27.2	10.1
24	8.1	25.8	—	3020	1610	82.8	20.9	3.42	18.3	5.63	1.61	23.2	8.57
21	6.6	27.5	—	2890	2090	75.3	18.2	3.49	9.77	3.71	1.26	20.4	5.69
18	8.0	29.9	—	2490	3890	61.9	15.2	3.43	7.97	3.04	1.23	17.0	4.66
15	6.4	28.1	—	2670	3440	48.0	11.8	3.29	3.41	1.70	0.876	13.6	2.67
13	7.8	29.9	—	2370	5780	39.6	9.91	3.21	2.73	1.37	0.843	11.4	2.15
10	9.6	40.5	39	1760	17900	30.8	7.81	3.22	2.09	1.06	0.841	8.87	1.66
25	6.7	15.5	—	4410	369	53.4	16.7	2.70	17.1	5.61	1.52	18.9	8.56
20	8.2	19.1	—	3550	846	41.4	13.4	2.66	13.3	4.41	1.50	14.9	6.72
15	11.5	21.6	—	2740	2470	29.1	9.72	2.56	9.32	3.11	1.46	10.8	4.75
16	5.0	19.1	—	4010	591	32.1	10.2	2.60	4.43	2.20	0.966	11.7	3.39
12	7.1	21.6	—	3100	1740	22.1	7.31	2.49	2.99	1.50	0.918	8.30	2.32
9	9.2	29.2	—	2360	4980	16.4	5.56	2.47	2.19	1.11	0.905	6.23	1.72
19	5.8	14.0	—	5140	192	26.2	10.2	2.17	9.13	3.63	1.28	11.6	5.53
16	6.9	15.8	—	4440	346	21.3	8.51	2.13	7.51	3.00	1.27	9.59	4.57
13	5.9	10.6	—	5560	154	11.3	5.46	1.72	3.86	1.90	1.00	6.28	2.92

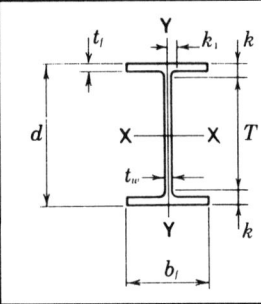

M SHAPES
Dimensions

Desig-nation	Area A	Depth d		Web Thickness t_w		$\frac{t_w}{2}$	Flange Width b_f		Thickness t_f		Distance T	k	Grip	Max. Flge. Fastener
	In.²	In.		In.		In.	In.		In.		In.	In.	In.	In.
M 14x18	5.10	14.00	14	0.215	3/16	1/8	4.000	4	0.270	1/4	12¾	5/8	1/4	3/4
M 12x11.8	3.47	12.00	12	0.177	3/16	1/8	3.065	3⅛	0.225	1/4	10⅞	9/16	1/4	—
M 10x9	2.65	10.00	10	0.157	3/16	1/8	2.690	2¾	0.206	3/16	8⅞	9/16	3/16	—
M 8x6.5	1.92	8.00	8	0.135	1/8	1/16	2.281	2¼	0.189	3/16	7	1/2	3/16	—
M 6x20	5.89	6.00	6	0.250	1/4	1/8	5.938	6	0.379	3/8	4¼	7/8	3/8	7/8
M 6x4.4	1.29	6.00	6	0.114	1/8	1/16	1.844	1⅞	0.171	3/16	5⅛	7/16	3/16	—
M 5x18.9	5.55	5.00	5	0.316	5/16	3/16	5.003	5	0.416	7/16	3¼	7/8	7/16	7/8
M 4x13	3.81	4.00	4	0.254	1/4	1/8	3.940	4	0.371	3/8	2⅜	13/16	3/8	3/4

M SHAPES
Properties

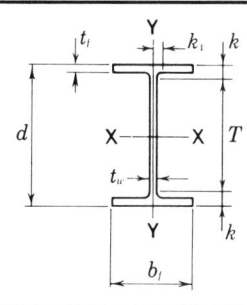

Nom-inal Wt. per Ft	Compact Section Criteria $\frac{b_f}{2t_f}$	$\frac{h_c}{t_w}$	F_y'''	X_1	$X_2 \times 10^6$	Elastic Properties Axis X-X I	S	r	Axis Y-Y I	S	r	Plastic Modulus Z_x	Z_y
Lb.			Ksi	Ksi	(1/Ksi)²	In.⁴	In.³	In.	In.⁴	In.³	In.	In.³	In.³
18	7.4	60.3	18	1420	55300	148	21.1	5.38	2.64	1.32	0.719	24.9	2.20
11.8	6.8	62.5	16	1390	63700	71.9	12.0	4.55	0.980	0.639	0.532	14.3	1.09
9	6.5	58.4	19	1450	51200	38.8	7.76	3.83	0.609	0.453	0.480	9.19	0.765
6.5	6.0	53.8	22	1700	26000	18.5	4.62	3.10	0.343	0.301	0.423	5.42	0.502
20	7.8	18.5	—	4090	473	39.0	13.0	2.57	11.6	3.90	1.40	14.5	6.25
4.4	5.4	47.0	29	1890	15600	7.20	2.40	2.36	0.165	0.179	0.358	2.80	0.296
18.9	6.0	11.2	—	5710	134	24.1	9.63	2.08	7.86	3.14	1.19	11.0	5.02
13	5.3	10.4	—	6500	79.9	10.5	5.24	1.66	3.36	1.71	0.939	6.05	2.74

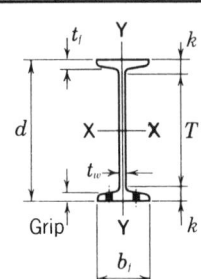

S SHAPES
Dimensions

Desig-nation	Area A	Depth d		Web Thickness t_w	$\frac{t_w}{2}$	Flange Width b_f		Thickness t_f	Distance T	k	Grip	Max. Flge. Fastener
	In.²	In.	In.	In.	In.	In.		In.	In.	In.	In.	In.
S 24x121	35.6	24.50	24½	0.800	13/16	7/16	8.050	8	1.090	1 1/16	20½	2	1 1/8	1
x106	31.2	24.50	24½	0.620	5/8	5/16	7.870	7 7/8	1.090	1 1/16	20½	2	1 1/8	1
S 24x100	29.3	24.00	24	0.745	3/4	3/8	7.245	7 1/4	0.870	7/8	20½	1 3/4	7/8	1
x90	26.5	24.00	24	0.625	5/8	5/16	7.125	7 1/8	0.870	7/8	20½	1 3/4	7/8	1
x80	23.5	24.00	24	0.500	1/2	1/4	7.000	7	0.870	7/8	20½	1 3/4	7/8	1
S 20x96	28.2	20.30	20¼	0.800	13/16	7/16	7.200	7 1/4	0.920	15/16	16¾	1 3/4	15/16	1
x86	25.3	20.30	20¼	0.660	11/16	3/8	7.060	7	0.920	15/16	16¾	1 3/4	15/16	1
S 20x75	22.0	20.00	20	0.635	5/8	5/16	6.385	6 3/8	0.795	13/16	16¾	1 5/8	13/16	7/8
x66	19.4	20.00	20	0.505	1/2	1/4	6.255	6 1/4	0.795	13/16	16¾	1 5/8	13/16	7/8
S 18x70	20.6	18.00	18	0.711	11/16	3/8	6.251	6 1/4	0.691	11/16	15	1½	11/16	7/8
x54.7	16.1	18.00	18	0.461	7/16	1/4	6.001	6	0.691	11/16	15	1½	11/16	7/8
S 15x50	14.7	15.00	15	0.550	9/16	5/16	5.640	5 5/8	0.622	5/8	12¼	1 3/8	9/16	3/4
x42.9	12.6	15.00	15	0.411	7/16	1/4	5.501	5½	0.622	5/8	12¼	1 3/8	9/16	3/4
S 12x50	14.7	12.00	12	0.687	11/16	3/8	5.477	5½	0.659	11/16	9 1/8	1 7/16	11/16	3/4
x40.8	12.0	12.00	12	0.462	7/16	1/4	5.252	5 1/4	0.659	11/16	9 1/8	1 7/16	5/8	3/4
S 12x35	10.3	12.00	12	0.428	7/16	1/4	5.078	5 1/8	0.544	9/16	9 5/8	1 3/16	1/2	3/4
x31.8	9.35	12.00	12	0.350	3/8	3/16	5.000	5	0.544	9/16	9 5/8	1 3/16	1/2	3/4
S 10x35	10.3	10.00	10	0.594	5/8	5/16	4.944	5	0.491	1/2	7¾	1 1/8	1/2	3/4
x25.4	7.46	10.00	10	0.311	5/16	3/16	4.661	4 5/8	0.491	1/2	7¾	1 1/8	1/2	3/4
S 8x23	6.77	8.00	8	0.441	7/16	1/4	4.171	4 1/8	0.426	7/16	6	1	7/16	3/4
x18.4	5.41	8.00	8	0.271	1/4	1/8	4.001	4	0.426	7/16	6	1	7/16	3/4
S 7x20	5.88	7.00	7	0.450	7/16	1/4	3.860	3 7/8	0.392	3/8	5 1/8	15/16	3/8	5/8
x15.3	4.50	7.00	7	0.252	1/4	1/8	3.662	3 5/8	0.392	3/8	5 1/8	15/16	3/8	5/8
S 6x17.25	5.07	6.00	6	0.465	7/16	1/4	3.565	3 5/8	0.359	3/8	4¼	7/8	3/8	5/8
x12.5	3.67	6.00	6	0.232	1/4	1/8	3.332	3 3/8	0.359	3/8	4¼	7/8	3/8	—
S 5x14.75	4.34	5.00	5	0.494	1/2	1/4	3.284	3 1/4	0.326	5/16	3 3/8	13/16	5/16	—
x10	2.94	5.00	5	0.214	3/16	1/8	3.004	3	0.326	5/16	3 3/8	13/16	5/16	—
S 4x 9.5	2.79	4.00	4	0.326	5/16	3/16	2.796	2 3/4	0.293	5/16	2½	3/4	5/16	—
x 7.7	2.26	4.00	4	0.193	3/16	1/8	2.663	2 5/8	0.293	5/16	2½	3/4	5/16	—
S 3x 7.5	2.21	3.00	3	0.349	3/8	3/16	2.509	2½	0.260	1/4	1 5/8	11/16	1/4	—
x 5.7	1.67	3.00	3	0.170	3/16	1/8	2.330	2 3/8	0.260	1/4	1 5/8	11/16	1/4	—

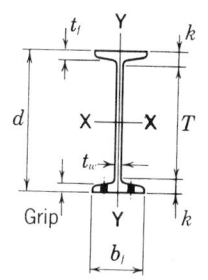

S SHAPES
Properties

Nom-inal Wt. per Ft	Compact Section Criteria					Elastic Properties						Plastic Modulus	
	$\frac{b_f}{2t_f}$	$\frac{h_c}{t_w}$	F_y'''	X_1	$X_2 \times 10^6$	Axis X-X			Axis Y-Y			Z_x	Z_y
						I	S	r	I	S	r		
Lb.			Ksi	Ksi	$(1/Ksi)^2$	In.^{4}	In.^{3}	In.	In.^{4}	In.^{3}	In.	In.^{3}	In.^{3}
121	3.7	26.4	—	3310	1770	3160	258	9.43	83.3	20.7	1.53	306	36.2
106	3.6	34.1	55	2960	2470	2940	240	9.71	77.1	19.6	1.57	279	33.2
100	4.2	28.3	—	3000	2940	2390	199	9.02	47.7	13.2	1.27	240	23.9
90	4.1	33.7	56	2710	4090	2250	187	9.21	44.9	12.6	1.30	222	22.3
80	4.0	42.1	36	2450	5480	2100	175	9.47	42.2	12.1	1.34	204	20.7
96	3.9	21.6	—	3730	1160	1670	165	7.71	50.2	13.9	1.33	198	24.9
86	3.8	26.2	—	3350	1630	1580	155	7.89	46.8	13.3	1.36	183	23.0
75	4.0	27.1	—	3140	2290	1280	128	7.62	29.8	9.32	1.16	153	16.7
66	3.9	34.1	55	2800	3250	1190	119	7.83	27.7	8.85	1.19	140	15.3
70	4.5	21.8	—	3590	1470	926	103	6.71	24.1	7.72	1.08	125	14.4
54.7	4.3	33.6	57	2770	3400	804	89.4	7.07	20.8	6.94	1.14	105	12.1
50	4.5	23.2	—	3450	1540	486	64.8	5.75	15.7	5.57	1.03	77.1	9.97
42.9	4.4	31.0	—	2960	2470	447	59.6	5.95	14.4	5.23	1.07	69.3	9.02
50	4.2	13.9	—	5070	333	305	50.8	4.55	15.7	5.74	1.03	61.2	10.3
40.8	4.0	20.7	—	4050	682	272	45.4	4.77	13.6	5.16	1.06	53.1	8.85
35	4.7	23.4	—	3500	1310	229	38.2	4.72	9.87	3.89	0.980	44.8	6.79
31.8	4.6	28.6	—	3190	1710	218	36.4	4.83	9.36	3.74	1.00	42.0	6.40
35	5.0	13.8	—	4960	374	147	29.4	3.78	8.36	3.38	0.901	35.4	6.22
25.4	4.7	26.4	—	3430	1220	124	24.7	4.07	6.79	2.91	0.954	28.4	4.96
23	4.9	14.5	—	4770	397	64.9	16.2	3.10	4.31	2.07	0.798	19.3	3.68
18.4	4.7	23.7	—	3770	821	57.6	14.4	3.26	3.73	1.86	0.831	16.5	3.16
20	4.9	12.3	—	5380	252	42.4	12.1	2.69	3.17	1.64	0.734	14.5	2.96
15.3	4.7	21.9	—	3960	666	36.7	10.5	2.86	2.64	1.44	0.766	12.1	2.44
17.25	5.0	9.9	—	6250	143	26.3	8.77	2.28	2.31	1.30	0.675	10.6	2.36
12.5	4.6	19.9	—	4290	477	22.1	7.37	2.45	1.82	1.09	0.705	8.47	1.85
14.75	5.0	7.5	—	7750	63.1	15.2	6.09	1.87	1.67	1.01	0.620	7.42	1.88
10	4.6	17.4	—	4630	348	12.3	4.92	2.05	1.22	0.809	0.643	5.67	1.37
9.5	4.8	8.7	—	6830	87.4	6.79	3.39	1.56	0.903	0.646	0.569	4.04	1.13
7.7	4.5	14.7	—	5240	207	6.08	3.04	1.64	0.764	0.574	0.581	3.51	0.964
7.5	4.8	5.6	—	9160	28	2.93	1.95	1.15	0.586	0.468	0.516	2.36	0.826
5.7	4.5	11.4	—	6160	106	2.52	1.68	1.23	0.455	0.390	0.522	1.95	0.653

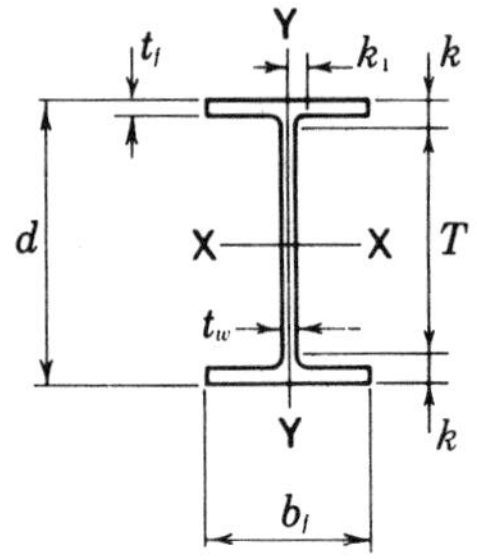

HP SHAPES
Dimensions

Designation	Area A	Depth d		Web Thickness t_w		Web $\frac{t_w}{2}$	Flange Width b_f		Flange Thickness t_f		Distance T	Distance k	Distance k_1
	In.^{2}	In.		In.		In.	In.		In.		In.	In.	In.
HP 14x117	34.4	14.21	14¼	0.805	13/16	7/16	14.885	14⅞	0.805	13/16	11¼	1½	1 1/16
x102	30.0	14.01	14	0.705	11/16	⅜	14.785	14¾	0.705	11/16	11¼	1⅜	1
x 89	26.1	13.83	13⅞	0.615	⅝	5/16	14.695	14¾	0.615	⅝	11¼	1 5/16	15/16
x 73	21.4	13.61	13⅝	0.505	½	¼	14.585	14⅝	0.505	½	11¼	1 3/16	⅞
HP 13x100	29.4	13.15	13⅛	0.765	¾	⅜	13.205	13¼	0.765	¾	10¼	1 7/16	1
x 87	25.5	12.95	13	0.665	11/16	⅜	13.105	13⅛	0.665	11/16	10¼	1⅜	15/16
x 73	21.6	12.75	12¾	0.565	9/16	5/16	13.005	13	0.565	9/16	10¼	1¼	15/16
x 60	17.5	12.54	12½	0.460	7/16	¼	12.900	12⅞	0.460	7/16	10¼	1⅛	⅞
HP 12x 84	24.6	12.28	12¼	0.685	11/16	⅜	12.295	12¼	0.685	11/16	9½	1⅜	1
x 74	21.8	12.13	12⅛	0.605	⅝	5/16	12.215	12¼	0.610	⅝	9½	1 5/16	15/16
x 63	18.4	11.94	12	0.515	½	¼	12.125	12⅛	0.515	½	9½	1¼	⅞
x 53	15.5	11.78	11¾	0.435	7/16	¼	12.045	12	0.435	7/16	9½	1⅛	⅞
HP 10x 57	16.8	9.99	10	0.565	9/16	5/16	10.225	10¼	0.565	9/16	7⅝	1 3/16	13/16
x 42	12.4	9.70	9¾	0.415	7/16	¼	10.075	10⅛	0.420	7/16	7⅝	1 1/16	¾
HP 8x 36	10.6	8.02	8	0.445	7/16	¼	8.155	8⅛	0.445	7/16	6⅛	15/16	⅝

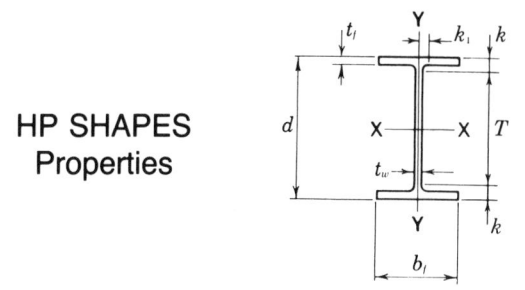

HP SHAPES
Properties

Nominal Wt. per Ft	Compact Section Criteria $\frac{b_f}{2t_f}$	Compact Section Criteria $\frac{h_c}{t_w}$	F_y'''	X_1	$X_2 \times 10^6$	Elastic Properties Axis X-X I	Axis X-X S	Axis X-X r	Axis Y-Y I	Axis Y-Y S	Axis Y-Y r	Plastic Modulus Z_x	Plastic Modulus Z_y
Lb.			Ksi	Ksi	$(1/\text{Ksi})^2$	In.^{4}	In.^{3}	In.	In.^{4}	In.^{3}	In.	In.^{3}	In.^{3}
117	9.2	14.2	—	3870	659	1220	172	5.96	443	59.5	3.59	194	91.4
102	10.5	16.2	—	3400	1090	1050	150	5.92	380	51.4	3.56	169	78.8
89	11.9	18.5	—	2960	1840	904	131	5.88	326	44.3	3.53	146	67.7
73	14.4	22.6	—	2450	3880	729	107	5.84	261	35.8	3.49	118	54.6
100	8.6	13.6	—	4020	571	886	135	5.49	294	44.5	3.16	153	68.6
87	9.9	15.7	—	3510	970	755	117	5.45	250	38.1	3.13	131	58.5
73	11.5	18.4	—	3000	1790	630	98.8	5.40	207	31.9	3.10	110	48.8
60	14.0	22.7	—	2460	3880	503	80.3	5.36	165	25.5	3.07	89.0	39.0
84	9.0	14.2	—	3860	670	650	106	5.14	213	34.6	2.94	120	53.2
74	10.0	16.0	—	3440	1050	569	93.8	5.11	186	30.4	2.92	105	46.6
63	11.8	18.9	—	2940	1940	472	79.1	5.06	153	25.3	2.88	88.3	38.7
53	13.8	22.3	—	2500	3650	393	66.8	5.03	127	21.1	2.86	74.0	32.2
57	9.0	13.9	—	3920	631	294	58.8	4.18	101	19.7	2.45	66.5	30.3
42	12.0	18.9	—	2920	1970	210	43.4	4.13	71.7	14.2	2.41	48.3	21.8
36	9.2	14.2	—	3840	685	119	29.8	3.36	40.3	9.88	1.95	33.6	15.2

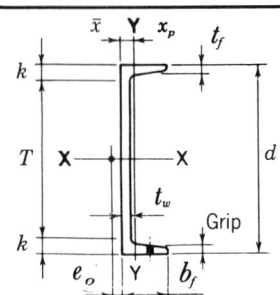

CHANNELS
AMERICAN STANDARD
Dimensions

Desig-nation	Area A	Depth d	Web Thickness t_w		$\frac{t_w}{2}$	Flange Width b_f		Average thickness t_f	Distance T	k	Grip	Max. Flge. Fastener
	In.^{2}	In.	In.		In.	In.		In.	In.	In.	In.	In.
C 15x50	14.7	15.00	0.716	$^{11}/_{16}$	$^3/_8$	3.716	$3^3/_4$	0.650	$^5/_8$	$12^1/_8$	$1^7/_{16}$	$^5/_8$	1
x40	11.8	15.00	0.520	$^1/_2$	$^1/_4$	3.520	$3^1/_2$	0.650	$^5/_8$	$12^1/_8$	$1^7/_{16}$	$^5/_8$	1
x33.9	9.96	15.00	0.400	$^3/_8$	$^3/_{16}$	3.400	$3^3/_8$	0.650	$^5/_8$	$12^1/_8$	$1^7/_{16}$	$^5/_8$	1
C 12x30	8.82	12.00	0.510	$^1/_2$	$^1/_4$	3.170	$3^1/_8$	0.501	$^1/_2$	$9^3/_4$	$1^1/_8$	$^1/_2$	$^7/_8$
x25	7.35	12.00	0.387	$^3/_8$	$^3/_{16}$	3.047	3	0.501	$^1/_2$	$9^3/_4$	$1^1/_8$	$^1/_2$	$^7/_8$
x20.7	6.09	12.00	0.282	$^5/_{16}$	$^1/_8$	2.942	3	0.501	$^1/_2$	$9^3/_4$	$1^1/_8$	$^1/_2$	$^7/_8$
C 10x30	8.82	10.00	0.673	$^{11}/_{16}$	$^5/_{16}$	3.033	3	0.436	$^7/_{16}$	8	1	$^7/_{16}$	$^3/_4$
x25	7.35	10.00	0.526	$^1/_2$	$^1/_4$	2.886	$2^7/_8$	0.436	$^7/_{16}$	8	1	$^7/_{16}$	$^3/_4$
x20	5.88	10.00	0.379	$^3/_8$	$^3/_{16}$	2.739	$2^3/_4$	0.436	$^7/_{16}$	8	1	$^7/_{16}$	$^3/_4$
x15.3	4.49	10.00	0.240	$^1/_4$	$^1/_8$	2.600	$2^5/_8$	0.436	$^7/_{16}$	8	1	$^7/_{16}$	$^3/_4$
C 9x20	5.88	9.00	0.448	$^7/_{16}$	$^1/_4$	2.648	$2^5/_8$	0.413	$^7/_{16}$	$7^1/_8$	$^{15}/_{16}$	$^7/_{16}$	$^3/_4$
x15	4.41	9.00	0.285	$^5/_{16}$	$^1/_8$	2.485	$2^1/_2$	0.413	$^7/_{16}$	$7^1/_8$	$^{15}/_{16}$	$^7/_{16}$	$^3/_4$
x13.4	3.94	9.00	0.233	$^1/_4$	$^1/_8$	2.433	$2^3/_8$	0.413	$^7/_{16}$	$7^1/_8$	$^{15}/_{16}$	$^7/_{16}$	$^3/_4$
C 8x18.75	5.51	8.00	0.487	$^1/_2$	$^1/_4$	2.527	$2^1/_2$	0.390	$^3/_8$	$6^1/_8$	$^{15}/_{16}$	$^3/_8$	$^3/_4$
x13.75	4.04	8.00	0.303	$^5/_{16}$	$^1/_8$	2.343	$2^3/_8$	0.390	$^3/_8$	$6^1/_8$	$^{15}/_{16}$	$^3/_8$	$^3/_4$
x11.5	3.38	8.00	0.220	$^1/_4$	$^1/_8$	2.260	$2^1/_4$	0.390	$^3/_8$	$6^1/_8$	$^{15}/_{16}$	$^3/_8$	$^3/_4$
C 7x14.75	4.33	7.00	0.419	$^7/_{16}$	$^3/_{16}$	2.299	$2^1/_4$	0.366	$^3/_8$	$5^1/_4$	$^7/_8$	$^3/_8$	$^5/_8$
x12.25	3.60	7.00	0.314	$^5/_{16}$	$^3/_{16}$	2.194	$2^1/_4$	0.366	$^3/_8$	$5^1/_4$	$^7/_8$	$^3/_8$	$^5/_8$
x 9.8	2.87	7.00	0.210	$^3/_{16}$	$^1/_8$	2.090	$2^1/_8$	0.366	$^3/_8$	$5^1/_4$	$^7/_8$	$^3/_8$	$^5/_8$
C 6x13	3.83	6.00	0.437	$^7/_{16}$	$^3/_{16}$	2.157	$2^1/_8$	0.343	$^5/_{16}$	$4^3/_8$	$^{13}/_{16}$	$^5/_{16}$	$^5/_8$
x10.5	3.09	6.00	0.314	$^5/_{16}$	$^3/_{16}$	2.034	2	0.343	$^5/_{16}$	$4^3/_8$	$^{13}/_{16}$	$^3/_8$	$^5/_8$
x 8.2	2.40	6.00	0.200	$^3/_{16}$	$^1/_8$	1.920	$1^7/_8$	0.343	$^5/_{16}$	$4^3/_8$	$^{13}/_{16}$	$^5/_{16}$	$^5/_8$
C 5x 9	2.64	5.00	0.325	$^5/_{16}$	$^3/_{16}$	1.885	$1^7/_8$	0.320	$^5/_{16}$	$3^1/_2$	$^3/_4$	$^5/_{16}$	$^5/_8$
x 6.7	1.97	5.00	0.190	$^3/_{16}$	$^1/_8$	1.750	$1^3/_4$	0.320	$^5/_{16}$	$3^1/_2$	$^3/_4$	—	—
C 4x 7.25	2.13	4.00	0.321	$^5/_{16}$	$^3/_{16}$	1.721	$1^3/_4$	0.296	$^5/_{16}$	$2^5/_8$	$^{11}/_{16}$	$^5/_{16}$	$^5/_8$
x 5.4	1.59	4.00	0.184	$^3/_{16}$	$^1/_{16}$	1.584	$1^5/_8$	0.296	$^5/_{16}$	$2^5/_8$	$^{11}/_{16}$	—	—
C 3x 6	1.76	3.00	0.356	$^3/_8$	$^3/_{16}$	1.596	$1^5/_8$	0.273	$^1/_4$	$1^5/_8$	$^{11}/_{16}$	—	—
x 5	1.47	3.00	0.258	$^1/_4$	$^1/_8$	1.498	$1^1/_2$	0.273	$^1/_4$	$1^5/_8$	$^{11}/_{16}$	—	—
x 4.1	1.21	3.00	0.170	$^3/_{16}$	$^1/_{16}$	1.410	$1^3/_8$	0.273	$^1/_4$	$1^5/_8$	$^{11}/_{16}$	—	—

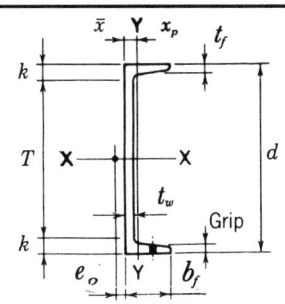

CHANNELS
AMERICAN STANDARD
Properties

Nom-inal Wt. per Ft	x̄	Shear Center Location e_o	PNA Location x_p	Axis X-X				Axis Y-Y			
				I	Z	S	r	I	Z	S	r
Lb.	In.	In.	In.	In.⁴	In.³	In.³	In.	In.⁴	In.³	In.³	In.
50	0.798	0.583	0.488	404	68.2	53.8	5.24	11.0	8.17	3.78	0.867
40	0.777	0.767	0.390	349	57.2	46.5	5.44	9.23	6.87	3.37	0.886
33.9	0.787	0.896	0.330	315	50.4	42.0	5.62	8.13	6.23	3.11	0.904
30	0.674	0.618	0.366	162	33.6	27.0	4.29	5.14	4.33	2.06	0.763
25	0.674	0.746	0.305	144	29.2	24.1	4.43	4.47	3.84	1.88	0.780
20.7	0.698	0.870	0.252	129	25.4	21.5	4.61	3.88	3.49	1.73	0.799
30	0.649	0.369	0.439	103	26.6	20.7	3.42	3.94	3.78	1.65	0.669
25	0.617	0.494	0.366	91.2	23.0	18.2	3.52	3.36	3.19	1.48	0.676
20	0.606	0.637	0.292	78.9	19.3	15.8	3.66	2.81	2.71	1.32	0.692
15.3	0.634	0.796	0.223	67.4	15.8	13.5	3.87	2.28	2.35	1.16	0.713
20	0.583	0.515	0.325	60.9	16.8	13.5	3.22	2.42	2.47	1.17	0.642
15	0.586	0.682	0.243	51.0	13.5	11.3	3.40	1.93	2.05	1.01	0.661
13.4	0.601	0.743	0.217	47.9	12.5	10.6	3.48	1.76	1.95	0.962	0.669
18.75	0.565	0.431	0.343	44.0	13.8	11.0	2.82	1.98	2.17	1.01	0.599
13.75	0.553	0.604	0.251	36.1	10.9	9.03	2.99	1.53	1.73	0.854	0.615
11.5	0.571	0.697	0.209	32.6	9.55	8.14	3.11	1.32	1.58	0.781	0.625
14.75	0.532	0.441	0.308	27.2	9.68	7.78	2.51	1.38	1.64	0.779	0.564
12.25	0.525	0.538	0.255	24.2	8.40	6.93	2.60	1.17	1.43	0.703	0.571
9.8	0.540	0.647	0.203	21.3	7.12	6.08	2.72	0.968	1.26	0.625	0.581
13	0.514	0.380	0.317	17.4	7.26	5.80	2.13	1.05	1.36	0.642	0.525
10.5	0.499	0.486	0.255	15.2	6.15	5.06	2.22	0.866	1.15	0.564	0.529
8.2	0.511	0.599	0.198	13.1	5.13	4.38	2.34	0.693	0.993	0.492	0.537
9	0.478	0.427	0.262	8.96	4.36	3.56	1.83	0.632	0.918	0.450	0.489
6.7	0.484	0.552	0.217	7.49	3.51	3.00	1.95	0.479	0.763	0.378	0.493
7.25	0.459	0.386	0.264	4.59	2.81	2.29	1.47	0.433	0.697	0.343	0.450
5.4	0.457	0.502	0.241	3.85	2.26	1.93	1.56	0.319	0.569	0.283	0.449
6	0.455	0.322	0.291	2.07	1.72	1.38	1.08	0.305	0.544	0.268	0.416
5	0.438	0.392	0.242	1.85	1.50	1.24	1.12	0.247	0.466	0.233	0.410
4.1	0.436	0.461	0.284	1.66	1.30	1.10	1.17	0.197	0.401	0.202	0.404

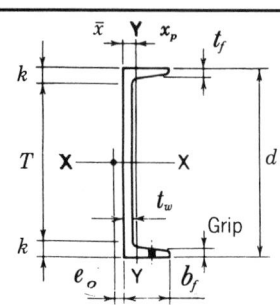

CHANNELS
MISCELLANEOUS
Dimensions

Desig-nation	Area A	Depth d	Web Thickness t_w		$\dfrac{t_w}{2}$	Flange Width b_f		Average thickness t_f		T	k	Grip	Max. Flge. Fastener
	In.²	In.	In.		In.	In.		In.		In.	In.	In.	In.
MC 18x58	17.1	18.00	0.700	¹¹⁄₁₆	³⁄₈	4.200	4¼	0.625	⁵⁄₈	15¼	1³⁄₈	⁵⁄₈	1
x51.9	15.3	18.00	0.600	⁵⁄₈	⁵⁄₁₆	4.100	4⅛	0.625	⁵⁄₈	15¼	1³⁄₈	⁵⁄₈	1
x45.8	13.5	18.00	0.500	½	¼	4.000	4	0.625	⁵⁄₈	15¼	1³⁄₈	⁵⁄₈	1
x42.7	12.6	18.00	0.450	⁷⁄₁₆	¼	3.950	4	0.625	⁵⁄₈	15¼	1³⁄₈	⁵⁄₈	1
MC 13x50	14.7	13.00	0.787	³⁄₁₆	³⁄₈	4.412	4⅜	0.610	⁵⁄₈	10¼	1³⁄₈	⁵⁄₈	1
x40	11.8	13.00	0.560	⁹⁄₁₆	¼	4.185	4⅛	0.610	⁵⁄₈	10¼	1³⁄₈	⁹⁄₁₆	1
x35	10.3	13.00	0.447	⁷⁄₁₆	¼	4.072	4⅛	0.610	⁵⁄₈	10¼	1³⁄₈	⁹⁄₁₆	1
x31.8	9.35	13.00	0.375	³⁄₈	³⁄₁₆	4.000	4	0.610	⁵⁄₈	10¼	1³⁄₈	⁹⁄₁₆	1
MC 12x50	14.7	12.00	0.835	¹³⁄₁₆	⁷⁄₁₆	4.135	4⅛	0.700	¹¹⁄₁₆	9⅜	1⁵⁄₁₆	¹¹⁄₁₆	1
x45	13.2	12.00	0.712	¹¹⁄₁₆	³⁄₈	4.012	4	0.700	¹¹⁄₁₆	9⅜	1⁵⁄₁₆	¹¹⁄₁₆	1
x40	11.8	12.00	0.590	⁹⁄₁₆	⁵⁄₁₆	3.890	3⅞	0.700	¹¹⁄₁₆	9⅜	1⁵⁄₁₆	¹¹⁄₁₆	1
x35	10.3	12.00	0.467	⁷⁄₁₆	¼	3.767	3¾	0.700	¹¹⁄₁₆	9⅜	1⁵⁄₁₆	¹¹⁄₁₆	1
x31	9.12	12.00	0.370	³⁄₈	³⁄₁₆	3.670	3⅝	0.700	¹¹⁄₁₆	9⅜	1⁵⁄₁₆	¹¹⁄₁₆	1
MC 12x10.6	3.10	12.00	0.190	³⁄₁₆	⅛	1.500	1½	0.309	⁵⁄₁₆	10⅝	¹¹⁄₁₆	—	—
MC 10x41.1	12.1	10.00	0.796	¹³⁄₁₆	³⁄₈	4.321	4⅜	0.575	⁹⁄₁₆	7½	1¼	⁹⁄₁₆	⅞
x33.6	9.87	10.00	0.575	⁹⁄₁₆	⁵⁄₁₆	4.100	4⅛	0.575	⁹⁄₁₆	7½	1¼	⁹⁄₁₆	⅞
x28.5	8.37	10.00	0.425	⁷⁄₁₆	³⁄₁₆	3.950	4	0.575	⁹⁄₁₆	7½	1¼	⁹⁄₁₆	⅞
MC 10x25	7.35	10.00	0.380	³⁄₈	³⁄₁₆	3.405	3⅜	0.575	⁹⁄₁₆	7½	1¼	⁹⁄₁₆	⅞
x22	6.45	10.00	0.290	⁵⁄₁₆	⅛	3.315	3⅜	0.575	⁹⁄₁₆	7½	1¼	⁹⁄₁₆	⅞
MC 10x 8.4	2.46	10.00	0.170	³⁄₁₆	¹⁄₁₆	1.500	1½	0.280	¼	8⅝	¹¹⁄₁₆	—	—
MC 10x 6.5	1.91	10.00	0.152	⅛	¹⁄₁₆	1.127	1⅛	0.202	³⁄₁₆	9⅛	⁷⁄₁₆	—	—

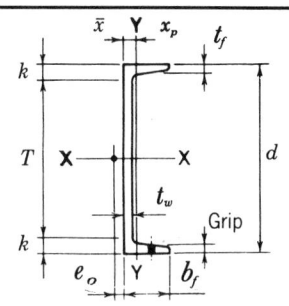

CHANNELS
MISCELLANEOUS
Properties

Nom-inal Wt. per Ft	$\bar{x}$	Shear Center Location e_o	PNA Location x_p	Axis X-X				Axis Y-Y			
				I	Z	S	r	I	Z	S	r
Lb.	In.	In.	In.	In.^{4}	In.^{3}	In.^{3}	In.	In.^{4}	In.^{3}	In.^{3}	In.
58	0.862	0.695	0.472	676	94.6	75.1	6.29	17.8	9.94	5.32	1.02
51.9	0.858	0.797	0.422	627	86.5	69.7	6.41	16.4	9.13	5.07	1.04
45.8	0.866	0.909	0.372	578	78.4	64.3	6.56	15.1	8.42	4.82	1.06
42.7	0.877	0.969	0.347	554	74.4	61.6	6.64	14.4	8.10	4.69	1.07
50	0.974	0.815	0.564	314	60.5	48.4	4.62	16.5	10.1	4.79	1.06
40	0.963	1.03	0.450	273	50.9	42.0	4.82	13.7	8.57	4.26	1.08
35	0.980	1.16	0.394	252	46.2	38.8	4.95	12.3	7.95	3.99	1.10
31.8	1.00	1.24	0.358	239	43.1	36.8	5.06	11.4	7.60	3.81	1.11
50	1.05	0.741	0.610	269	56.1	44.9	4.28	17.4	10.2	5.65	1.09
45	1.04	0.844	0.549	252	51.7	42.0	4.36	15.8	9.35	5.33	1.09
40	1.04	0.952	0.488	234	47.3	39.0	4.46	14.3	8.59	5.00	1.10
35	1.05	1.07	0.426	216	42.8	36.1	4.59	12.7	7.91	4.67	1.11
31	1.08	1.18	0.416	203	39.3	33.8	4.71	11.3	7.44	4.39	1.12
10.6	0.269	0.284	0.129	55.4	11.6	9.23	4.22	0.382	0.639	0.310	0.351
41.1	1.09	0.864	0.601	158	38.9	31.5	3.61	15.8	8.71	4.88	1.14
33.6	1.08	1.06	0.490	139	33.4	27.8	3.75	13.2	7.51	4.38	1.16
28.5	1.12	1.21	0.415	127	29.6	25.3	3.89	11.4	6.83	4.02	1.17
25	0.953	1.03	0.364	110	25.8	22.0	3.87	7.35	5.21	3.00	1.00
22	0.990	1.13	0.468	103	23.6	20.5	3.99	6.50	4.86	2.80	1.00
8.4	0.284	0.332	0.122	32.0	7.86	6.40	3.61	0.328	0.552	0.270	0.365
6.5	0.180	0.167	0.096	22.1	5.73	4.42	3.40	0.112	0.250	0.118	0.242

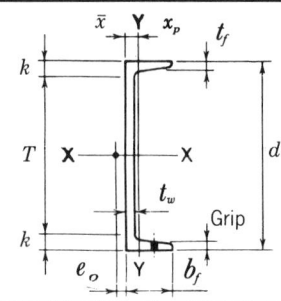

CHANNELS
MISCELLANEOUS
Dimensions

Desig-nation	Area A	Depth d	Web		Flange			Distance		Grip	Max. Flge. Fastener
			Thickness t_w	$\frac{t_w}{2}$	Width b_f		Average thickness t_f	T	k		
	In.²	In.	In.	In.	In.		In.	In.	In.	In.	In.
MC 9x25.4	7.47	9.00	0.450	7/16	1/4	3.500	3½	0.550	9/16	6⅝	1 3/16	9/16	7/8
x23.9	7.02	9.00	0.400	3/8	3/16	3.450	3½	0.550	9/16	6⅝	1 3/16	9/16	7/8
MC 8x22.8	6.70	8.00	0.427	7/16	3/16	3.502	3½	0.525	½	5⅝	1 3/16	½	7/8
x21.4	6.28	8.00	0.375	3/8	3/16	3.450	3½	0.525	½	5⅝	1 3/16	½	7/8
MC 8x20	5.88	8.00	0.400	3/8	3/16	3.025	3	0.500	½	5¾	1⅛	½	7/8
x18.7	5.50	8.00	0.353	3/8	3/16	2.978	3	0.500	½	5¾	1⅛	½	7/8
MC 8x 8.5	2.50	8.00	0.179	3/16	1/16	1.874	1⅞	0.311	5/16	6½	¾	5/16	5/8
MC 7x22.7	6.67	7.00	0.503	½	1/4	3.603	3⅝	0.500	½	4¾	1⅛	½	7/8
x19.1	5.61	7.00	0.352	3/8	3/16	3.452	3½	0.500	½	4¾	1⅛	½	7/8
MC 7x17.6	5.17	7.00	0.375	3/8	3/16	3.000	3	0.475	½	4⅞	1 1/16	½	¾
MC 6x18	5.29	6.00	0.379	3/8	3/16	3.504	3½	0.475	½	3⅞	1 1/16	½	7/8
x15.3	4.50	6.00	0.340	5/16	3/16	3.500	3½	0.385	3/8	4¼	7/8	3/8	7/8
MC 6x16.3	4.79	6.00	0.375	3/8	3/16	3.000	3	0.475	½	3⅞	1 1/16	½	¾
x15.1	4.44	6.00	0.316	5/16	3/16	2.941	3	0.475	½	3⅞	1 1/16	½	¾
MC 6x12	3.53	6.00	0.310	5/16	⅛	2.497	2½	0.375	3/8	4⅜	13/16	3/8	5/8

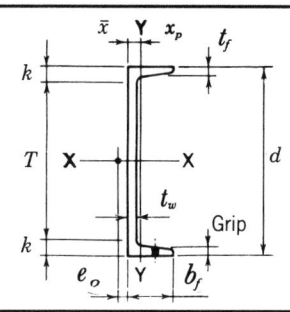

CHANNELS
MISCELLANEOUS
Properties

Nom-inal Wt. per Ft	$\bar{x}$	Shear Center Loca-tion e_o	PNA Loca-tion x_p	Axis X-X				Axis Y-Y			
				I	Z	S	r	I	Z	S	r
Lb.	In.	In.	In.	In.^{4}	In.^{3}	In.^{3}	In.	In.^{4}	In.^{3}	In.^{3}	In.
25.4	0.970	0.986	0.411	88.0	23.2	19.6	3.43	7.65	5.23	3.02	1.01
23.9	0.981	1.04	0.386	85.0	22.2	18.9	3.48	7.22	5.05	2.93	1.01
22.8	1.01	1.04	0.415	63.8	18.8	16.0	3.09	7.07	4.88	2.84	1.03
21.4	1.02	1.09	0.449	61.6	18.0	15.4	3.13	6.64	4.71	2.74	1.03
20	0.840	0.843	0.364	54.5	16.2	13.6	3.05	4.47	3.57	2.05	0.872
18.7	0.849	0.889	0.341	52.5	15.4	13.1	3.09	4.20	3.44	1.97	0.874
8.5	0.428	0.542	0.155	23.3	6.91	5.83	3.05	0.628	0.882	0.434	0.501
22.7	1.04	1.01	0.473	47.5	16.2	13.6	2.67	7.29	4.86	2.85	1.05
19.1	1.08	1.15	0.567	43.2	14.3	12.3	2.77	6.11	4.34	2.57	1.04
17.6	0.873	0.890	0.366	37.6	12.7	10.8	2.70	4.01	3.26	1.89	0.881
18	1.12	1.17	0.622	29.7	11.5	9.91	2.37	5.93	4.14	2.48	1.06
15.3	1.05	1.16	0.495	25.4	9.82	8.47	2.38	4.97	3.28	2.03	1.05
16.3	0.927	0.930	0.464	26.0	10.2	8.68	2.33	3.82	3.18	1.84	0.892
15.1	0.940	0.982	0.537	25.0	9.69	8.32	2.37	3.51	3.00	1.75	0.889
12	0.704	0.725	0.292	18.7	7.38	6.24	2.30	1.87	1.79	1.04	0.728

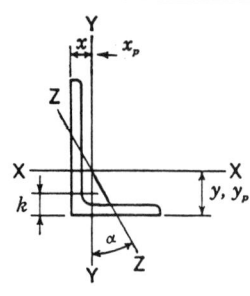

ANGLES
Equal legs and unequal legs
Properties for designing

Size and Thickness	k	Weight per Ft	Area	Axis X-X					
				I	S	r	y	Z	y_p
In.	In.	Lb.	In.²	In.⁴	In.³	In.	In.	In.³	In.
L 9x4x ⅝	1⅛	26.3	7.73	64.9	11.5	2.90	3.36	19.7	2.81
⁹⁄₁₆	1¹⁄₁₆	23.8	7.00	59.1	10.4	2.91	3.33	17.9	2.78
½	1	21.3	6.25	53.2	9.34	2.92	3.31	16.0	2.75
L 8x8x1⅛	1¾	56.9	16.7	98.0	17.5	2.42	2.41	31.6	1.05
1	1⅝	51.0	15.0	89.0	15.8	2.44	2.37	28.5	0.938
⅞	1½	45.0	13.2	79.6	14.0	2.45	2.32	25.3	0.827
¾	1⅜	38.9	11.4	69.7	12.2	2.47	2.28	22.0	0.715
⅝	1¼	32.7	9.61	59.4	10.3	2.49	2.23	18.6	0.601
⁹⁄₁₆	1³⁄₁₆	29.6	8.68	54.1	9.34	2.50	2.21	16.8	0.543
½	1⅛	26.4	7.75	48.6	8.36	2.50	2.19	15.1	0.484
L 8x6x1	1½	44.2	13.0	80.8	15.1	2.49	2.65	27.3	1.50
⅞	1⅜	39.1	11.5	72.3	13.4	2.51	2.61	24.2	1.44
¾	1¼	33.8	9.94	63.4	11.7	2.53	2.56	21.1	1.38
⅝	1⅛	28.5	8.36	54.1	9.87	2.54	2.52	17.9	1.31
⁹⁄₁₆	1¹⁄₁₆	25.7	7.56	49.3	8.95	2.55	2.50	16.2	1.28
½	1	23.0	6.75	44.3	8.02	2.56	2.47	14.5	1.25
⁷⁄₁₆	¹⁵⁄₁₆	20.2	5.93	39.2	7.07	2.57	2.45	12.8	1.22
L 8x4x1	1½	37.4	11.0	69.6	14.1	2.52	3.05	24.3	2.50
¾	1¼	28.7	8.44	54.9	10.9	2.55	2.95	18.9	2.38
⁹⁄₁₆	1¹⁄₁₆	21.9	6.43	42.8	8.35	2.58	2.88	14.5	2.28
½	1	19.6	5.75	38.5	7.49	2.59	2.86	13.0	2.25
L 7x4x ¾	1¼	26.2	7.69	37.8	8.42	2.22	2.51	14.8	1.88
⅝	1⅛	22.1	6.48	32.4	7.14	2.24	2.46	12.6	1.81
½	1	17.9	5.25	26.7	5.81	2.25	2.42	10.3	1.75
⅜	⅞	13.6	3.98	20.6	4.44	2.27	2.37	7.87	1.69

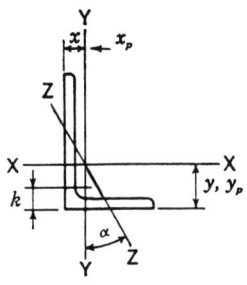

ANGLES
Equal legs and unequal legs
Properties for designing

Size and Thickness	Axis Y-Y						Axis Z-Z	
	I	S	r	x	Z	x_p	r	Tan
In.	In.⁴	In.³	In.	In.	In.³	In.	In.	α
L 9x4x ⅝	8.32	2.65	1.04	0.858	4.97	0.430	.847	0.216
⁹⁄₁₆	7.63	2.41	1.04	0.834	4.48	0.389	.850	0.218
½	6.92	2.17	1.05	0.810	3.98	0.347	.854	0.220
L 8x8x1⅛	98.0	17.5	2.42	2.41	31.6	1.05	1.56	1.000
1	89.0	15.8	2.44	2.37	28.5	0.938	1.56	1.000
⅞	79.6	14.0	2.45	2.32	25.3	0.827	1.57	1.000
¾	69.7	12.2	2.47	2.28	22.0	0.715	1.58	1.000
⅝	59.4	10.3	2.49	2.23	18.6	0.601	1.58	1.000
⁹⁄₁₆	54.1	9.34	2.50	2.21	16.8	0.543	1.59	1.000
½	48.6	8.36	2.50	2.19	15.1	0.484	1.59	1.000
L 8x6x1	38.8	8.92	1.73	1.65	16.2	0.813	1.28	0.543
⅞	34.9	7.94	1.74	1.61	14.4	0.718	1.28	0.547
¾	30.7	6.92	1.76	1.56	12.5	0.621	1.29	0.551
⅝	26.3	5.88	1.77	1.52	10.5	0.522	1.29	0.554
⁹⁄₁₆	24.0	5.34	1.78	1.50	9.52	0.472	1.30	0.556
½	21.7	4.79	1.79	1.47	8.51	0.422	1.30	0.558
⁷⁄₁₆	19.3	4.23	1.80	1.45	7.50	0.371	1.31	0.560
L 8x4x1	11.6	3.94	1.03	1.05	7.72	0.688	0.846	0.247
¾	9.36	3.07	1.05	0.953	5.81	0.527	0.852	0.258
⁹⁄₁₆	7.43	2.38	1.07	0.882	4.38	0.402	0.861	0.265
½	6.74	2.15	1.08	0.859	3.90	0.359	0.865	0.267
L 7x4x ¾	9.05	3.03	1.09	1.01	5.65	0.549	0.860	0.324
⅝	7.84	2.58	1.10	0.963	4.74	0.463	0.865	0.329
½	6.53	2.12	1.11	0.917	3.83	0.375	0.872	0.335
⅜	5.10	1.63	1.13	0.870	2.90	0.285	0.880	0.340

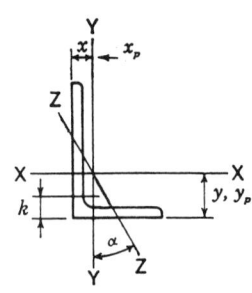

ANGLES
Equal legs and unequal legs
Properties for designing

Size and Thickness	k	Weight per Ft	Area	Axis X-X					
				I	S	r	y	Z	y_p
In.	In.	Lb.	In.^{2}	In.^{4}	In.^{3}	In.	In.	In.^{3}	In.
L 6x6 x1	1½	37.4	11.0	35.5	8.57	1.80	1.86	15.5	0.917
⅞	1⅜	33.1	9.73	31.9	7.63	1.81	1.82	13.8	0.811
¾	1¼	28.7	8.44	28.2	6.66	1.83	1.78	12.0	0.703
⅝	1⅛	24.2	7.11	24.2	5.66	1.84	1.73	10.2	0.592
⁹⁄₁₆	1¹⁄₁₆	21.9	6.43	22.1	5.14	1.85	1.71	9.26	0.536
½	1	19.6	5.75	19.9	4.61	1.86	1.68	8.31	0.479
⁷⁄₁₆	¹⁵⁄₁₆	17.2	5.06	17.7	4.08	1.87	1.66	7.34	0.422
⅜	⅞	14.9	4.36	15.4	3.53	1.88	1.64	6.35	0.363
⁵⁄₁₆	¹³⁄₁₆	12.4	3.65	13.0	2.97	1.89	1.62	5.35	0.304
L 6x4 x ⅞	1⅜	27.2	7.98	27.7	7.15	1.86	2.12	12.7	1.44
¾	1¼	23.6	6.94	24.5	6.25	1.88	2.08	11.2	1.38
⅝	1⅛	20.0	5.86	21.1	5.31	1.90	2.03	9.51	1.31
⁹⁄₁₆	1¹⁄₁₆	18.1	5.31	19.3	4,83	1.90	2.01	8.66	1.28
½	1	16.2	4.75	17.4	4.33	1.91	1.99	7.78	1.25
⁷⁄₁₆	¹⁵⁄₁₆	14.3	4.18	15.5	3.83	1.92	1.96	6.88	1.22
⅜	⅞	12.3	3.61	13.5	3.32	1.93	1.94	5.97	1.19
⁵⁄₁₆	¹³⁄₁₆	10.3	3.03	11.4	2.79	1.94	1.92	5.03	1.16
L 6x3½x ½	1	15.3	4.50	16.6	4.24	1.92	2.08	7.50	1.50
⅜	⅞	11.7	3.42	12.9	3.24	1.94	2.04	5.76	1.44
⁵⁄₁₆	¹³⁄₁₆	9.8	2.87	10.9	2.73	1.95	2.01	4.85	1.41
L 5x5 x ⅞	1⅜	27.2	7.98	17.8	5.17	1.49	1.57	9.33	0.798
¾	1¼	23.6	6.94	15.7	4.53	1.51	1.52	8.16	0.694
⅝	1⅛	20.0	5.86	13.6	3.86	1.52	1.48	6.95	0.586
½	1	16.2	4.75	11.3	3.16	1.54	1.43	5.68	0.475
⁷⁄₁₆	¹⁵⁄₁₆	14.3	4.18	10.0	2.79	1.55	1.41	5.03	0.418
⅜	⅞	12.3	3.61	8.74	2.42	1.56	1.39	4.36	0.361
⁵⁄₁₆	¹³⁄₁₆	10.3	3.03	7.42	2.04	1.57	1.37	3.68	0.303

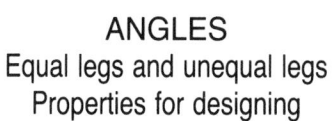

ANGLES
Equal legs and unequal legs
Properties for designing

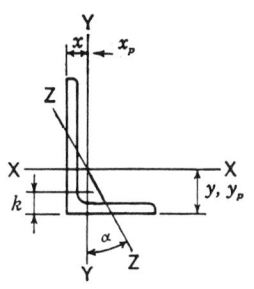

Size and Thickness	Axis Y-Y						Axis Z-Z	
	I	S	r	x	Z	x_p	r	Tan
In.	In.^{4}	In.^{3}	In.	In.	In.^{3}	In.	In.	α
L 6x6　x1	35.5	8.57	1.80	1.86	15.5	0.917	1.17	1.000
⅞	31.9	7.63	1.81	1.82	13.8	0.811	1.17	1.000
¾	28.2	6.66	1.83	1.78	12.0	0.703	1.17	1.000
⅝	24.2	5.66	1.84	1.73	10.2	0.592	1.18	1.000
⁹⁄₁₆	22.1	5.14	1.85	1.71	9.26	0.536	1.18	1.000
½	19.9	4.61	1.86	1.68	8.31	0.479	1.18	1.000
⁷⁄₁₆	17.7	4.08	1.87	1.66	7.34	0.422	1.19	1.000
⅜	15.4	3.53	1.88	1.64	6.35	0.363	1.19	1.000
⁵⁄₁₆	13.0	2.97	1.89	1.62	5.35	0.304	1.20	1.000
L 6x4　x ⅞	9.75	3.39	1.11	1.12	6.31	0.665	0.857	0.421
¾	8.68	2.97	1.12	1.08	5.47	0.578	0.860	0.428
⅝	7.52	2.54	1.13	1.03	4.62	0.488	0.864	0.435
⁹⁄₁₆	6.91	2.31	1.14	1.01	4.19	0.442	0.866	0.438
½	6.27	2.08	1.15	0.987	3.75	0.396	0.870	0.440
⁷⁄₁₆	5.60	1.85	1.16	0.964	3.30	0.349	0.873	0.443
⅜	4.90	1.60	1.17	0.941	2.85	0.301	0.877	0.446
⁵⁄₁₆	4.18	1.35	1.17	0.918	2.40	0.252	0.882	0.448
L 6x3½x ½	4.25	1.59	0.972	0.833	2.91	0.375	0.759	0.344
⅜	3.34	1.23	0.988	0.787	2.20	0.285	0.767	0.350
⁵⁄₁₆	2.85	1.04	0.996	0.763	1.85	0.239	0.772	0.352
L 5x5　x ⅞	17.8	5.17	1.49	1.57	9.33	0.798	0.973	1.000
¾	15.7	4.53	1.51	1.52	8.16	0.694	0.975	1.000
⅝	13.6	3.86	1.52	1.48	6.95	0.586	0.978	1.000
½	11.3	3.16	1.54	1.43	5.68	0.475	0.983	1.000
⁷⁄₁₆	10.0	2.79	1.55	1.41	5.03	0.418	0.986	1.000
⅜	8.74	2.42	1.56	1.39	4.36	0.361	0.990	1.000
⁵⁄₁₆	7.42	2.04	1.57	1.37	3.68	0.303	0.994	1.000

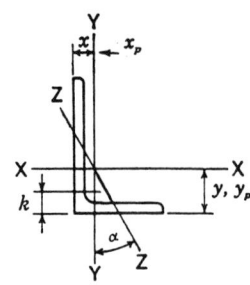

ANGLES
Equal legs and unequal legs
Properties for designing

Size and Thickness	k	Weight per Ft	Area	Axis X-X					
				I	S	r	y	Z	y_p
In.	In.	Lb.	In.²	In.⁴	In.³	In.	In.	In.³	In.
L 5x3½x¾	1¼	19.8	5.81	13.9	4.28	1.55	1.75	7.65	1.13
⅝	1⅛	16.8	4.92	12.0	3.65	1.56	1.70	6.55	1.06
½	1	13.6	4.00	9.99	2.99	1.58	1.66	5.38	1.00
⁷⁄₁₆	¹⁵⁄₁₆	12.0	3.53	8.90	2.64	1.59	1.63	4.77	0.969
⅜	⅞	10.4	3.05	7.78	2.29	1.60	1.61	4.14	0.938
⁵⁄₁₆	¹³⁄₁₆	8.7	2.56	6.60	1.94	1.61	1.59	3.49	0.906
¼	¾	7.0	2.06	5.39	1.57	1.62	1.56	2.83	0.875
L 5x3 x⅝	1	15.7	4.61	11.4	3.55	1.57	1.80	6.27	1.31
½	1	12.8	3.75	9.45	2.91	1.59	1.75	5.16	1.25
⁷⁄₁₆	¹⁵⁄₁₆	11.3	3.31	8.43	2.58	1.60	1.73	4.57	1.22
⅜	⅞	9.8	2.86	7.37	2.24	1.61	1.70	3.97	1.19
⁵⁄₁₆	¹³⁄₁₆	8.2	2.40	6.26	1.89	1.61	1.68	3.36	1.16
¼	¾	6.6	1.94	5.11	1.53	1.62	1.66	2.72	1.13
L 4x4 x¾	1⅛	18.5	5.44	7.67	2.81	1.19	1.27	5.07	0.680
⅝	1	15.7	4.61	6.66	2.40	1.20	1.23	4.33	0.576
½	⅞	12.8	3.75	5.56	1.97	1.22	1.18	3.56	0.469
⁷⁄₁₆	¹³⁄₁₆	11.3	3.31	4.97	1.75	1.23	1.16	3.16	0.414
⅜	¾	9.8	2.86	4.36	1.52	1.23	1.14	2.74	0.357
⁵⁄₁₆	¹¹⁄₁₆	8.2	2.40	3.71	1.29	1.24	1.12	2.32	0.300
¼	⅝	6.6	1.94	3.04	1.05	1.25	1.09	1.88	0.242
L 4x3½x⅝	1¹⁄₁₆	14.7	4.30	6.37	2.35	1.22	1.29	4.24	0.614
½	¹⁵⁄₁₆	11.9	3.50	5.32	1.94	1.23	1.25	3.50	0.500
⁷⁄₁₆	⅞	10.6	3.09	4.76	1.72	1.24	1.23	3.11	0.469
⅜	¹³⁄₁₆	9.1	2.67	4.18	1.49	1.25	1.21	2.71	0.438
⁵⁄₁₆	¾	7.7	2.25	3.56	1.26	1.26	1.18	2.29	0.406
¼	¹¹⁄₁₆	6.2	1.81	2.91	1.03	1.27	1.16	1.86	0.375

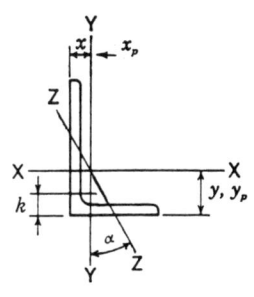

ANGLES
Equal legs and unequal legs
Properties for designing

Size and Thickness	Axis Y-Y						Axis Z-Z	
	I	S	r	x	Z	x_p	r	Tan
In.	In.^{4}	In.^{3}	In.	In.	In.^{3}	In.	In.	α
L 5x3½x¾	5.55	2.22	0.977	0.996	4.10	0.581	0.748	0.464
⅝	4.83	1.90	0.991	0.951	3.47	0.492	0.751	0.472
½	4.05	1.56	1.01	0.906	2.83	0.400	0.755	0.479
⁷⁄₁₆	3.63	1.39	1.01	0.883	2.49	0.353	0.758	0.482
⅜	3.18	1.21	1.02	0.861	2.16	0.305	0.762	0.486
⁵⁄₁₆	2.72	1.02	1.03	0.838	1.82	0.256	0.766	0.489
¼	2.23	0.830	1.04	0.814	1.47	0.206	0.770	0.492
L 5x3 x⅝	3.06	1.39	0.815	0.796	2.61	0.461	0.644	0.349
½	2.58	1.15	0.829	0.750	2.11	0.375	0.648	0.357
⁷⁄₁₆	2.32	1.02	0.837	0.727	1.86	0.331	0.651	0.361
⅜	2.04	0.888	0.845	0.704	1.60	0.286	0.654	0.364
⁵⁄₁₆	1.75	0.753	0.853	0.681	1.35	0.240	0.658	0.368
¼	1.44	0.614	0.861	0.657	1.09	0.194	0.663	0.371
L 4x4 x¾	7.67	2.81	1.19	1.27	5.07	0.680	0.778	1.000
⅝	6.66	2.40	1.20	1.23	4.33	0.576	0.779	1.000
½	5.56	1.97	1.22	1.18	3.56	0.469	0.782	1.000
⁷⁄₁₆	4.97	1.75	1.23	1.16	3.16	0.414	0.785	1.000
⅜	4.36	1.52	1.23	1.14	2.74	0.357	0.788	1.000
⁵⁄₁₆	3.71	1.29	1.24	1.12	2.32	0.300	0.791	1.000
¼	3.04	1.05	1.25	1.09	1.88	0.242	0.795	1.000
L 4x3½x⅝	4.52	1.84	1.03	1.04	3.33	0.537	0.719	0.745
½	3.79	1.52	1.04	1.00	2.73	0.438	0.722	0.750
⁷⁄₁₆	3.40	1.35	1.05	0.978	2.42	0.386	0.724	0.753
⅜	2.95	1.17	1.06	0.955	2.11	0.334	0.727	0.755
⁵⁄₁₆	2.55	0.994	1.07	0.932	1.78	0.281	0.730	0.757
¼	2.09	0.808	1.07	0.909	1.44	0.227	0.734	0.759

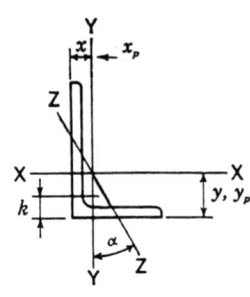

ANGLES
Equal legs and unequal legs
Properties for designing

Size and Thickness	k	Weight per Ft	Area	Axis X-X					
				I	S	r	y	Z	y_p
In.	In.	Lb.	In.^{2}	In.^{4}	In.^{3}	In.	In.	In.^{3}	In.
L 4 x3 x⅝	1¹⁄₁₆	13.6	3.98	6.03	2.30	1.23	1.37	4.12	0.813
½	¹⁵⁄₁₆	11.1	3.25	5.05	1.89	1.25	1.33	3.41	0.750
⁷⁄₁₆	⅞	9.8	2.87	4.52	1.68	1.25	1.30	3.03	0.719
⅜	¹³⁄₁₆	8.5	2.48	3.96	1.46	1.26	1.28	2.64	0.688
⁵⁄₁₆	¾	7.2	2.09	3.38	1.23	1.27	1.26	2.23	0.656
¼	¹¹⁄₁₆	5.8	1.69	2.77	1.00	1.28	1.24	1.82	0.625
L 3½x3½x½	⅞	11.1	3.25	3.64	1.49	1.06	1.06	2.68	0.464
⁷⁄₁₆	¹³⁄₁₆	9.8	2.87	3.26	1.32	1.07	1.04	2.38	0.410
⅜	¾	8.5	2.48	2.87	1.15	1.07	1.01	2.08	0.355
⁵⁄₁₆	¹¹⁄₁₆	7.2	2.09	2.45	0.976	1.08	0.990	1.76	0.299
¼	⅝	5.8	1.69	2.01	0.794	1.09	0.968	1.43	0.241
L 3½x3 x½	¹⁵⁄₁₆	10.2	3.00	3.45	1.45	1.07	1.13	2.63	0.500
⁷⁄₁₆	⅞	9.1	2.65	3.10	1.29	1.08	1.10	2.34	0.469
⅜	¹³⁄₁₆	7.9	2.30	2.72	1.13	1.09	1.08	2.04	0.438
⁵⁄₁₆	¾	6.6	1.93	2.33	0.954	1.10	1.06	1.73	0.406
¼	¹¹⁄₁₆	5.4	1.56	1.91	0.776	1.11	1.04	1.41	0.375
L 3½x2½x½	¹⁵⁄₁₆	9.4	2.75	3.24	1.41	1.09	1.20	2.53	0.750
⁷⁄₁₆	⅞	8.3	2.43	2.91	1.26	1.09	1.18	2.26	0.719
⅜	¹³⁄₁₆	7.2	2.11	2.56	1.09	1.10	1.16	1.97	0.688
⁵⁄₁₆	¾	6.1	1.78	2.19	0.927	1.11	1.14	1.67	0.656
¼	¹¹⁄₁₆	4.9	1.44	1.80	0.755	1.12	1.11	1.36	0.625
L 3 x3 x½	¹³⁄₁₆	9.4	2.75	2.22	1.07	0.898	0.932	1.93	0.458
⁷⁄₁₆	¾	8.3	2.43	1.99	0.954	0.905	0.910	1.72	0.406
⅜	¹¹⁄₁₆	7.2	2.11	1.76	0.833	0.913	0.888	1.50	0.352
⁵⁄₁₆	⅝	6.1	1.78	1.51	0.707	0.922	0.865	1.27	0.296
¼	⁹⁄₁₆	4.9	1.44	1.24	0.577	0.930	0.842	1.04	0.240
³⁄₁₆	½	3.71	1.09	0.962	0.441	0.939	0.820	0.794	0.182

ANGLES
Equal legs and unequal legs
Properties for designing

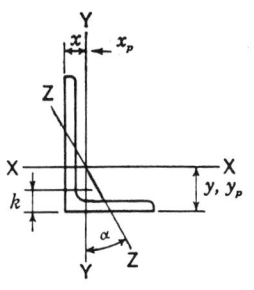

Size and Thickness	Axis Y-Y						Axis Z-Z	
	I	S	r	x	Z	x_p	r	Tan
In.	In.^{4}	In.^{3}	In.	In.	In.^{3}	In.	In.	α
L 4 x3 x⅝	2.87	1.35	0.849	0.871	2.48	0.498	0.637	0.534
½	2.42	1.12	0.864	0.827	2.03	0.406	0.639	0.543
⁷⁄₁₆	2.18	0.992	0.871	0.804	1.79	0.359	0.641	0.547
⅜	1.92	0.866	0.879	0.782	1.56	0.311	0.644	0.551
⁵⁄₁₆	1.65	0.734	0.887	0.759	1.31	0.261	0.647	0.554
¼	1.36	0.599	0.896	0.736	1.06	0.211	0.651	0.558
L 3½x3½x½	3.64	1.49	1.06	1.06	2.68	0.464	0.683	1.000
⁷⁄₁₆	3.26	1.32	1.07	1.04	2.38	0.410	0.684	1.000
⅜	2.87	1.15	1.07	1.01	2.08	0.355	0.687	1.000
⁵⁄₁₆	2.45	0.976	1.08	0.990	1.76	0.299	0.690	1.000
¼	2.01	0.794	1.09	0.968	1.43	0.241	0.694	1.000
L 3½x3 x½	2.33	1.10	0.881	0.875	1.98	0.429	0.621	0.714
⁷⁄₁₆	2.09	0.975	0.889	0.853	1.76	0.379	0.622	0.718
⅜	1.85	0.851	0.897	0.830	1.53	0.328	0.625	0.721
⁵⁄₁₆	1.58	0.722	0.905	0.808	1.30	0.276	0.627	0.724
¼	1.30	0.589	0.914	0.785	1.05	0.223	0.631	0.727
L 3½x2½x½	1.36	0.760	0.704	0.705	1.40	0.393	0.534	0.486
⁷⁄₁₆	1.23	0.677	0.711	0.682	1.24	0.348	0.535	0.491
⅜	1.09	0.592	0.719	0.660	1.07	0.301	0.537	0.496
⁵⁄₁₆	0.939	0.504	0.727	0.637	0.907	0.254	0.540	0.501
¼	0.777	0.412	0.735	0.614	0.735	0.205	0.544	0.506
L 3 x3 x½	2.22	1.07	0.898	0.932	1.93	0.458	0.584	1.000
⁷⁄₁₆	1.99	0.954	0.905	0.910	1.72	0.406	0.585	1.000
⅜	1.76	0.833	0.913	0.888	1.50	0.352	0.587	1.000
⁵⁄₁₆	1.51	0.707	0.922	0.865	1.27	0.296	0.589	1.000
¼	1.24	0.577	0.930	0.842	1.04	0.240	0.592	1.000
³⁄₁₆	0.962	0.441	0.939	0.820	0.794	0.182	0.596	1.000

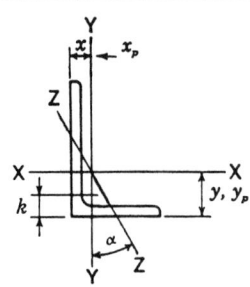

ANGLES
Equal legs and unequal legs
Properties for designing

Size and Thickness		k	Weight per Ft	Area	Axis X-X					
					I	S	r	y	Z	y_p
In.		In.	Lb.	In.^{2}	In.^{4}	In.^{3}	In.	In.	In.^{3}	In.
L 3	x2½x½	⅞	8.5	2.50	2.08	1.04	0.913	1.00	1.88	0.500
	⁷⁄₁₆	¹³⁄₁₆	7.6	2.21	1.88	**0.928**	0.920	0.978	1.68	0.469
	⅜	¾	6.6	1.92	1.66	0.810	0.928	0.956	1.47	0.438
	⁵⁄₁₆	¹¹⁄₁₆	5.6	1.62	1.42	0.688	0.937	0.933	1.25	0.406
	¼	⅝	4.5	1.31	1.17	0.561	0.945	0.911	1.02	0.375
	³⁄₁₆	⁹⁄₁₆	3.39	0.996	0.907	0.430	0.954	0.888	0.781	0.344
L 3	x2 x½	¹³⁄₁₆	7.7	2.25	1.92	1.00	0.924	1.08	1.78	0.750
	⁷⁄₁₆	¾	6.8	2.00	1.73	0.894	0.932	1.06	1.59	0.719
	⅜	¹¹⁄₁₆	5.9	1.73	1.53	0.781	0.940	1.04	1.40	0.688
	⁵⁄₁₆	⅝	5.0	1.46	1.32	0.664	0.948	1.02	1.19	0.656
	¼	⁹⁄₁₆	4.1	1.19	1.09	0.542	0.957	0.993	0.973	0.625
	³⁄₁₆	½	3.07	0.902	0.842	0.415	0.966	0.970	0.746	0.594
L 2½x2½x½		¹³⁄₁₆	7.7	2.25	1.23	0.724	0.739	0.806	1.31	0.450
	⅜	¹¹⁄₁₆	5.9	1.73	0.984	0.566	0.753	0.762	1.02	0.347
	⁵⁄₁₆	⅝	5.0	1.46	0.849	0.482	0.761	0.740	0.869	0.293
	¼	⁹⁄₁₆	4.1	1.19	0.703	0.394	0.769	0.717	0.711	0.238
	³⁄₁₆	½	3.07	0.902	0.547	0.303	0.778	0.694	0.545	0.180
L 2½x2 x⅜		¹¹⁄₁₆	5.3	1.55	0.912	0.547	0.768	0.831	0.986	0.438
	⁵⁄₁₆	⅝	4.5	1.31	0.788	0.466	0.776	0.809	0.843	0.406
	¼	⁹⁄₁₆	3.62	1.06	0.654	0.381	0.784	0.787	0.691	0.375
	³⁄₁₆	½	2.75	0.809	0.509	0.293	0.793	0.764	0.532	0.344
L 2 x2 x⅜		¹¹⁄₁₆	4.7	1.36	0.479	0.351	0.594	0.636	0.633	0.340
	⁵⁄₁₆	⅝	3.92	1.15	0.416	0.300	0.601	0.614	0.541	0.288
	¼	⁹⁄₁₆	3.19	0.938	0.348	0.247	0.609	0.592	0.445	0.234
	³⁄₁₆	½	2.44	0.715	0.272	0.190	0.617	0.569	0.343	0.179
	⅛	⁷⁄₁₆	1.65	0.484	0.190	0.131	0.626	0.546	0.235	0.121

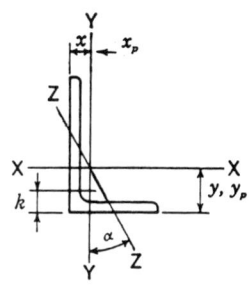

ANGLES
Equal legs and unequal legs
Properties for designing

Size and Thickness	Axis Y-Y						Axis Z-Z	
	I	S	r	x	Z	x_p	r	Tan
In.	In.^{4}	In.^{3}	In.	In.	In.^{3}	In.	In.	α
L 3 x2½x½	1.30	0.744	0.722	0.750	1.35	0.417	0.520	0.667
7/16	1.18	0.664	0.729	0.728	1.20	0.369	0.521	0.672
3/8	1.04	0.581	0.736	0.706	1.05	0.320	0.522	0.676
5/16	0.898	0.494	0.744	0.683	0.889	0.270	0.525	0.680
¼	0.743	0.404	0.753	0.661	0.724	0.219	0.528	0.684
3/16	0.577	0.310	0.761	0.638	0.553	0.166	0.533	0.688
L 3 x2 x½	0.672	0.474	0.546	0.583	0.891	0.375	0.428	0.414
7/16	0.609	0.424	0.553	0.561	0.788	0.333	0.429	0.421
3/8	0.543	0.371	0.559	0.539	0.684	0.289	0.430	0.428
5/16	0.470	0.317	0.567	0.516	0.577	0.244	0.432	0.435
¼	0.392	0.260	0.574	0.493	0.468	0.198	0.435	0.440
3/16	0.307	0.200	0.583	0.470	0.357	0.150	0.439	0.446
L 2½x2½x½	1.23	0.724	0.739	0.806	1.31	0.450	0.487	1.000
3/8	0.984	0.566	0.753	0.762	1.02	0.347	0.487	1.000
5/16	0.849	0.482	0.761	0.740	0.869	0.293	0.489	1.000
¼	0.703	0.394	0.769	0.717	0.711	0.238	0.491	1.000
3/16	0.547	0.303	0.778	0.694	0.545	0.180	0.495	1.000
L 2½x2 x3/8	0.514	0.363	0.577	0.581	0.660	0.309	0.420	0.614
5/16	0.446	0.310	0.584	0.559	0.561	0.262	0.422	0.620
¼	0.372	0.254	0.592	0.537	0.457	0.213	0.424	0.626
3/16	0.291	0.196	0.600	0.514	0.350	0.162	0.427	0.631
L 2 x2 x3/8	0.479	0.351	0.594	0.636	0.633	0.340	0.389	1.000
5/16	0.416	0.300	0.601	0.614	0.541	0.288	0.390	1.000
¼	0.348	0.247	0.609	0.592	0.445	0.234	0.391	1.000
3/16	0.272	0.190	0.617	0.569	0.343	0.179	0.394	1.000
1/8	0.190	0.131	0.626	0.546	0.235	0.121	0.398	1.000

Notes

STRUCTURAL TEES

Dimensions and Properties

Structural tees are obtained by splitting the webs of various beams, generally with the aid of rotary shears, and straightening to meet established permissible variations listed in *Standard Mill Practice* in Part 1 of this Manual.

Although structural tees may be obtained by off-center splitting, or by splitting on two lines, as specified on order, the Dimensions and Properties for Designing are based on a depth of tee equal to ½ the published beam depth. The table shows properties and dimensions for these full depth tees. Values of Q_s are given for $F_y = 36$ ksi and $F_y = 50$ ksi, for those tees having stems which exceed the limiting width-thickness ratio λ_r of AISC LRFD Specification Sect. B5. Since the cross section is comprised entirely of unstiffened elements, $Q_a = 1.0$ and $Q = Q_s$ for all tee sections. The Flexural-Torsional Properties Table also lists the dimensional values (r_o and H_o) and cross-section constants (J and C_w) needed for checking torsional and flexural-torsional buckling.

USE OF TABLE

The table may be used as follows for checking the limit states of (1) flexural buckling and (2) torsional or flexural-torsional buckling. The lower of the two limit states must be used for design.

(1) Flexural Buckling

Where no value of Q_s is shown, the design compressive strength for this limit state is given by AISC LRFD Specification Sect. E2. Where a value of Q_s is shown, the strength must be reduced in accordance with Appendix B5 if $\lambda_c \sqrt{Q} \leq \sqrt{2}$.

(2) Torsional or Flexural-Torsional Buckling

The design compressive strength for this limit state is given by AISC LRFD Specification Appendix E3. This involves calculations with J and C_w.

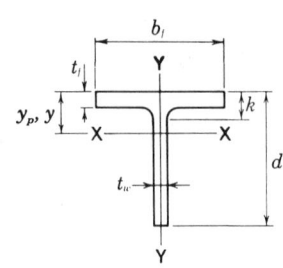

STRUCTURAL TEES
Cut from W shapes
Dimensions

Designation	Area	Depth of Tee d	Stem Thickness t_w	$\dfrac{t_w}{2}$	Area of Stem	Flange Width b_f	Flange Thickness t_f	Distance k
	In.²	In.	In.	In.	In.²	In.	In.	In.
WT 18 x179.5	52.7	18.70 $18\frac{11}{16}$	1.120 $1\frac{1}{8}$	$\frac{9}{16}$	20.9	16.730 $16\frac{3}{4}$	2.010 2	$3\frac{1}{8}$
x164	48.2	18.54 $18\frac{9}{16}$	1.020 1	$\frac{1}{2}$	18.9	16.630 $16\frac{5}{8}$	1.850 $1\frac{7}{8}$	3
x150	44.1	18.370 $18\frac{3}{8}$	0.945 $\frac{15}{16}$	$\frac{1}{2}$	17.4	16.655 $16\frac{5}{8}$	1.680 $1\frac{11}{16}$	$2\frac{13}{16}$
x140	41.2	18.260 $18\frac{1}{4}$	0.885 $\frac{7}{8}$	$\frac{7}{16}$	16.2	16.595 $16\frac{5}{8}$	1.570 $1\frac{9}{16}$	$2\frac{11}{16}$
x130	38.2	18.130 $18\frac{1}{8}$	0.840 $\frac{13}{16}$	$\frac{7}{16}$	15.2	16.550 $16\frac{1}{2}$	1.440 $1\frac{7}{16}$	$2\frac{9}{16}$
x122.5	36.0	18.040 18	0.800 $\frac{13}{16}$	$\frac{7}{16}$	14.4	16.510 $16\frac{1}{2}$	1.350 $1\frac{3}{8}$	$2\frac{1}{2}$
x115	33.8	17.950 18	0.760 $\frac{3}{4}$	$\frac{3}{8}$	13.6	16.470 $16\frac{1}{2}$	1.260 $1\frac{1}{4}$	$2\frac{3}{8}$
WT 18 x128	37.7	18.710 $18\frac{11}{16}$	0.960 1	$\frac{1}{2}$	18.0	12.215 $12\frac{1}{4}$	1.730 $1\frac{3}{4}$	$2\frac{5}{8}$
x116	34.1	18.560 $18\frac{9}{16}$	0.870 $\frac{7}{8}$	$\frac{7}{16}$	16.1	12.120 $12\frac{1}{8}$	1.570 $1\frac{9}{16}$	$2\frac{1}{2}$
x105	30.9	18.345 $18\frac{3}{8}$	0.830 $\frac{13}{16}$	$\frac{7}{16}$	15.2	12.180 $12\frac{1}{8}$	1.360 $1\frac{3}{8}$	$2\frac{5}{16}$
x 97	28.5	18.245 $18\frac{1}{4}$	0.765 $\frac{3}{4}$	$\frac{3}{8}$	14.0	12.115 $12\frac{1}{8}$	1.260 $1\frac{1}{4}$	$2\frac{3}{16}$
x 91	26.8	18.165 $18\frac{1}{8}$	0.725 $\frac{3}{4}$	$\frac{3}{8}$	13.2	12.075 $12\frac{1}{8}$	1.180 $1\frac{3}{16}$	$2\frac{1}{8}$
x 85	25.0	18.085 $18\frac{1}{8}$	0.680 $\frac{11}{16}$	$\frac{3}{8}$	12.3	12.030 12	1.100 $1\frac{1}{8}$	2
x 80	23.5	18.005 18	0.650 $\frac{5}{8}$	$\frac{5}{16}$	11.7	12.000 12	1.020 1	$1\frac{15}{16}$
x 75	22.1	17.925 $17\frac{7}{8}$	0.625 $\frac{5}{8}$	$\frac{5}{16}$	11.2	11.975 12	0.940 $\frac{15}{16}$	$1\frac{7}{8}$
x 67.5	19.9	17.775 $17\frac{3}{4}$	0.600 $\frac{5}{8}$	$\frac{5}{16}$	10.7	11.950 12	0.790 $\frac{13}{16}$	$1\frac{11}{16}$
WT 16.5x177	52.1	17.775 $17\frac{3}{4}$	1.160 $1\frac{3}{16}$	$\frac{5}{8}$	20.6	16.100 $16\frac{1}{8}$	2.090 $2\frac{1}{16}$	$2\frac{7}{8}$
x159	46.7	17.580 $17\frac{9}{16}$	1.040 $1\frac{1}{16}$	$\frac{9}{16}$	18.3	15.985 16	1.890 $1\frac{7}{8}$	$2\frac{11}{16}$
x145.5	42.8	17.420 $17\frac{7}{16}$	0.960 1	$\frac{1}{2}$	16.7	15.905 $15\frac{7}{8}$	1.730 $1\frac{3}{4}$	$2\frac{9}{16}$
x131.5	38.7	17.265 $17\frac{1}{4}$	0.870 $\frac{7}{8}$	$\frac{7}{16}$	15.0	15.805 $15\frac{3}{4}$	1.570 $1\frac{9}{16}$	$2\frac{3}{8}$
x120.5	35.4	17.090 $17\frac{1}{8}$	0.830 $\frac{13}{16}$	$\frac{7}{16}$	14.2	15.860 $15\frac{7}{8}$	1.400 $1\frac{3}{8}$	$2\frac{3}{16}$
x110.5	32.5	16.965 17	0.775 $\frac{3}{4}$	$\frac{3}{8}$	13.1	15.805 $15\frac{3}{4}$	1.275 $1\frac{1}{4}$	$2\frac{1}{16}$
x100.5	29.5	16.840 $16\frac{7}{8}$	0.715 $\frac{11}{16}$	$\frac{3}{8}$	12.0	15.745 $15\frac{3}{4}$	1.150 $1\frac{1}{8}$	$1\frac{15}{16}$
WT 16.5x 84.5	24.8	16.910 $16\frac{15}{16}$	0.670 $\frac{11}{16}$	$\frac{3}{8}$	11.3	11.500 $11\frac{1}{2}$	1.220 $1\frac{1}{4}$	$2\frac{1}{16}$
x 76	22.4	16.745 $16\frac{3}{4}$	0.635 $\frac{5}{8}$	$\frac{5}{16}$	10.6	11.565 $11\frac{5}{8}$	1.055 $1\frac{1}{16}$	$1\frac{7}{8}$
x 70.5	20.8	16.650 $16\frac{5}{8}$	0.605 $\frac{5}{8}$	$\frac{5}{16}$	10.1	11.535 $11\frac{1}{2}$	0.960 $\frac{15}{16}$	$1\frac{3}{4}$
x 65	19.2	16.545 $16\frac{1}{2}$	0.580 $\frac{9}{16}$	$\frac{5}{16}$	9.60	11.510 $11\frac{1}{2}$	0.855 $\frac{7}{8}$	$1\frac{11}{16}$
x 59	17.3	16.430 $16\frac{3}{8}$	0.550 $\frac{9}{16}$	$\frac{5}{16}$	9.04	11.480 $11\frac{1}{2}$	0.740 $\frac{3}{4}$	$1\frac{9}{16}$
WT 15 x117.5	34.5	15.650 $15\frac{5}{8}$	0.830 $\frac{13}{16}$	$\frac{7}{16}$	13.0	15.055 15	1.500 $1\frac{1}{2}$	$2\frac{1}{4}$
x105.5	31.0	15.470 $15\frac{1}{2}$	0.775 $\frac{3}{4}$	$\frac{3}{8}$	12.0	15.105 $15\frac{1}{8}$	1.315 $1\frac{5}{16}$	$2\frac{1}{8}$
x 95.5	28.1	15.340 $15\frac{3}{8}$	0.710 $\frac{11}{16}$	$\frac{3}{8}$	10.9	15.040 15	1.185 $1\frac{3}{16}$	$1\frac{15}{16}$
x 86.5	25.4	15.220 $15\frac{1}{4}$	0.655 $\frac{5}{8}$	$\frac{5}{16}$	9.97	14.985 15	1.065 $1\frac{1}{16}$	$1\frac{7}{8}$
WT 15 x 74	21.7	15.335 $15\frac{5}{16}$	0.650 $\frac{5}{8}$	$\frac{5}{16}$	9.96	10.480 $10\frac{1}{2}$	1.180 $1\frac{3}{16}$	2
x 66	19.4	15.155 $15\frac{1}{8}$	0.615 $\frac{5}{8}$	$\frac{5}{16}$	9.32	10.545 $10\frac{1}{2}$	1.000 1	$1\frac{3}{4}$
x 62	18.2	15.085 $15\frac{1}{8}$	0.585 $\frac{9}{16}$	$\frac{5}{16}$	8.82	10.515 $10\frac{1}{2}$	0.930 $\frac{15}{16}$	$1\frac{11}{16}$
x 58	17.1	15.005 15	0.565 $\frac{9}{16}$	$\frac{5}{16}$	8.48	10.495 $10\frac{1}{2}$	0.850 $\frac{7}{8}$	$1\frac{5}{8}$
x 54	15.9	14.915 $14\frac{7}{8}$	0.545 $\frac{9}{16}$	$\frac{5}{16}$	8.13	10.475 $10\frac{1}{2}$	0.760 $\frac{3}{4}$	$1\frac{9}{16}$
x 49.5	14.5	14.825 $14\frac{7}{8}$	0.520 $\frac{1}{2}$	$\frac{1}{4}$	7.71	10.450 $10\frac{1}{2}$	0.670 $\frac{11}{16}$	$1\frac{7}{16}$

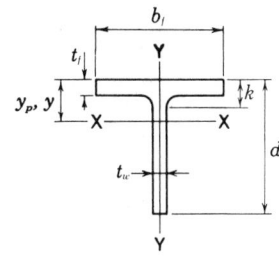

STRUCTURAL TEES
Cut from W shapes
Properties

Nominal Wt. per Ft	$\frac{h_c}{t_w}$	Axis X-X						Axis Y-Y				Q_s^*	
		I	S	r	y	Z	y_p	I	S	r	Z	F_y, ksi	
Lb.		In.^{4}	In.^{3}	In.	In.	In.^{3}	In.	In.^{4}	In.^{3}	In.	In.^{3}	36	50
179.5	14.1	1500	104.0	5.33	4.33	189	1.58	786	94.0	3.86	146	—	—
164	15.4	1350	94.1	5.29	4.21	168	1.45	711	85.5	3.84	132	—	—
150	16.7	1230	86.1	5.27	4.13	153	1.33	648	77.8	3.83	120	—	0.927
140	17.8	1140	80.0	5.25	4.07	142	1.24	599	72.2	3.81	112	—	0.867
130	18.7	1060	75.1	5.26	4.05	133	1.16	545	65.9	3.78	102	0.981	0.816
122.5	19.7	995	71.0	5.26	4.03	125	1.09	507	61.4	3.75	94.9	0.943	0.770
115.	20.7	934	67.0	5.25	4.01	118	1.03	470	57.1	3.73	88.1	0.896	0.715
128	16.9	1200	87.4	5.66	4.92	156	1.54	264	43.2	2.65	68.6	—	0.927
116	18.7	1080	78.5	5.63	4.82	140	1.40	234	38.6	2.62	61.0	0.994	0.831
105	19.6	985	73.1	5.65	4.87	131	1.27	206	33.8	2.58	53.5	0.960	0.791
97	21.2	901	67.0	5.62	4.80	120	1.18	187	30.9	2.56	48.9	0.887	0.705
91	22.4	845	63.1	5.62	4.77	113	1.11	174	28.8	2.55	45.4	0.831	0.635
85	23.9	786	58.9	5.61	4.73	105	1.04	160	26.6	2.53	41.9	0.767	0.565
80	25.0	740	55.8	5.61	4.74	100	0.980	147	24.6	2.50	38.6	0.720	0.521
75	26.0	698	53.1	5.62	4.78	95.5	0.923	135	22.5	2.47	35.5	0.677	0.486
67.5	27.1	636	49.7	5.66	4.96	94.3	1.24	113	18.9	2.38	29.8	0.634	0.457
177	12.9	1320	96.8	5.03	4.16	174	1.62	729	90.6	3.74	141	—	—
159	14.4	1160	85.8	4.99	4.02	154	1.46	645	80.7	3.71	125	—	—
145.5	15.6	1050	78.3	4.97	3.94	140	1.34	581	73.1	3.69	113	—	0.993
131.5	17.2	943	70.2	4.94	3.84	125	1.22	517	65.5	3.66	101	—	0.907
120.5	18.1	871	65.8	4.96	3.85	116	1.12	466	58.8	3.63	90.9	—	0.867
110.5	19.3	799	60.8	4.96	3.81	107	1.03	420	53.2	3.59	82.1	0.968	0.801
100.5	21.0	725	55.5	4.95	3.78	97.7	0.938	375	47.6	3.56	73.4	0.896	0.715
84.5	22.4	649	51.1	5.12	4.21	90.8	1.08	155	27.0	2.50	42.2	0.827	0.630
76	23.6	592	47.4	5.14	4.26	84.5	0.967	136	23.6	2.47	37.0	0.775	0.574
70.5	24.8	552	44.7	5.15	4.29	79.8	0.901	123	21.3	2.43	33.5	0.728	0.529
65	25.8	513	42.1	5.18	4.36	75.6	0.832	109	18.9	2.39	29.7	0.685	0.492
59	27.3	469	39.2	5.20	4.47	74.8	0.862	93.6	16.3	2.32	25.7	0.621	0.447
117.5	16.2	674	55.1	4.42	3.42	98.2	1.15	427	56.8	3.52	87.5	—	0.952
105.5	17.4	610	50.5	4.43	3.40	89.5	1.03	378	50.1	3.49	77.2	—	0.897
95.5	19.0	549	45.7	4.42	3.35	80.8	0.933	336	44.7	3.46	68.9	0.981	0.816
86.5	20.6	497	41.7	4.42	3.31	73.4	0.848	299	39.9	3.43	61.4	0.913	0.735
74	20.8	466	40.6	4.63	3.84	72.2	1.04	113	21.7	2.28	34.0	0.896	0.715
66	22.0	421	37.4	4.66	3.90	66.8	0.921	98.0	18.6	2.25	29.2	0.853	0.664
62	23.1	396	35.3	4.66	3.90	63.1	0.867	90.4	17.2	2.23	27.0	0.801	0.601
58	23.9	373	33.7	4.67	3.94	60.4	0.815	82.1	15.7	2.19	24.6	0.767	0.565
54	24.8	349	32.0	4.69	4.01	57.7	0.757	73.0	13.9	2.15	22.0	0.733	0.533
49.5	26.0	322	30.0	4.71	4.09	57.4	0.912	63.9	12.2	2.10	19.3	0.685	0.492

*Where no value of Q_s is shown, the Tee complies with LRFD Specification Sect. E2.

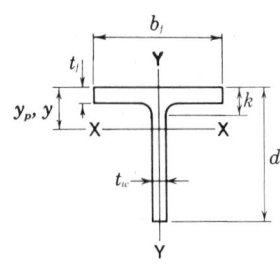

STRUCTURAL TEES
Cut from W shapes
Dimensions

Designation	Area	Depth of Tee d		Stem			Area of Stem	Flange					Dis-tance k
				Thickness t_w		$\frac{t_w}{2}$		Width b_f		Thickness t_f			
	In.²	In.		In.		In.	In.²	In.		In.			In.
WT 13.5x108.5	31.9	14.215	14³⁄₁₆	0.830	¹³⁄₁₆	⁷⁄₁₆	11.8	14.115	14⅛	1.500	1½		2³⁄₁₆
x 97	28.5	14.055	14¹⁄₁₆	0.750	¾	⅜	10.5	14.035	14	1.340	1⁵⁄₁₆		2¹⁄₁₆
x 89	26.1	13.905	13⅞	0.725	¾	⅜	10.1	14.085	14⅛	1.190	1³⁄₁₆		1⅞
x 80.5	23.7	13.795	13¾	0.660	¹¹⁄₁₆	⅜	9.10	14.020	14	1.080	1¹⁄₁₆		1¹³⁄₁₆
x 73	21.5	13.690	13¾	0.605	⅝	⁵⁄₁₆	8.28	13.965	14	0.975	1		1¹¹⁄₁₆
WT 13.5x 64.5	18.9	13.815	13¹³⁄₁₆	0.610	⅝	⁵⁄₁₆	8.42	10.010	10	1.100	1⅛		1¹³⁄₁₆
x 57	16.8	13.645	13⅝	0.570	⁹⁄₁₆	⁵⁄₁₆	7.78	10.070	10⅛	0.930	¹⁵⁄₁₆		1⅝
x 51	15.0	13.545	13½	0.515	½	¼	6.98	10.015	10	0.830	¹³⁄₁₆		1⁹⁄₁₆
x 47	13.8	13.460	13½	0.490	½	¼	6.60	9.990	10	0.745	¾		1⁷⁄₁₆
x 42	12.4	13.355	13⅜	0.460	⁷⁄₁₆	¼	6.14	9.960	10	0.640	⅝		1⅜
WT 12 x 88	25.8	12.625	12⅝	0.750	¾	⅜	9.46	12.890	12⅞	1.340	1⁵⁄₁₆		2⅛
x 81	23.9	12.500	12½	0.705	¹¹⁄₁₆	⅜	8.81	12.955	13	1.220	1¼		2
x 73	21.5	12.370	12⅜	0.650	⅝	⁵⁄₁₆	8.04	12.900	12⅞	1.090	1¹⁄₁₆		1⅞
x 65.5	19.3	12.240	12¼	0.605	⅝	⁵⁄₁₆	7.41	12.855	12⅞	0.960	¹⁵⁄₁₆		1¾
x 58.5	17.2	12.130	12⅛	0.550	⁹⁄₁₆	⁵⁄₁₆	6.67	12.800	12¾	0.850	⅞		1⅝
x 52	15.3	12.030	12	0.500	½	¼	6.01	12.750	12¾	0.750	¾		1½
WT 12 x 51.5	15.1	12.26	12¼	0.550	⁹⁄₁₆	⁵⁄₁₆	6.74	9.000	9	0.980	1		1¾
x 47	13.8	12.155	12⅛	0.515	½	¼	6.26	9.065	9⅛	0.875	⅞		1⅝
x 42	12.4	12.050	12	0.470	½	¼	5.66	9.020	9	0.770	¾		1⁹⁄₁₆
x 38	11.2	11.960	12	0.440	⁷⁄₁₆	¼	5.26	8.990	9	0.680	¹¹⁄₁₆		1⁷⁄₁₆
x 34	10.0	11.865	11⅞	0.415	⁷⁄₁₆	¼	4.92	8.965	9	0.585	⁹⁄₁₆		1⅜
WT 12 x 31	9.11	11.870	11⅞	0.430	⁷⁄₁₆	¼	5.10	7.040	7	0.590	⁹⁄₁₆		1⅜
x 27.5	8.10	11.785	11¾	0.395	⅜	³⁄₁₆	4.66	7.005	7	0.505	½		1⁵⁄₁₆
WT 10.5x 83	24.4	11.240	11¼	0.750	¾	⅜	8.43	12.420	12⅜	1.360	1⅜		2⅛
x 73.5	21.6	11.030	11	0.720	¾	⅜	7.94	12.510	12½	1.150	1⅛		1⅞
x 66	19.4	10.915	10⅞	0.650	⅝	⁵⁄₁₆	7.09	12.440	12½	1.035	1¹⁄₁₆		1¹³⁄₁₆
x 61	17.9	10.840	10⅞	0.600	⅝	⁵⁄₁₆	6.50	12.390	12⅜	0.960	¹⁵⁄₁₆		1¹¹⁄₁₆
x 55.5	16.3	10.755	10¾	0.550	⁹⁄₁₆	⁵⁄₁₆	5.92	12.340	12⅜	0.875	⅞		1⅝
x 50.5	14.9	10.680	10⅝	0.500	½	¼	5.34	12.290	12¼	0.800	¹³⁄₁₆		1⁹⁄₁₆
WT 10.5x 46.5	13.7	10.810	10¾	0.580	⁹⁄₁₆	⁵⁄₁₆	6.27	8.420	8⅜	0.930	¹⁵⁄₁₆		1¹¹⁄₁₆
x 41.5	12.2	10.715	10¾	0.515	½	¼	5.52	8.355	8⅜	0.835	¹³⁄₁₆		1⁹⁄₁₆
x 36.5	10.7	10.620	10⅝	0.455	⁷⁄₁₆	¼	4.83	8.295	8¼	0.740	¾		1½
x 34	10.0	10.565	10⅝	0.430	⁷⁄₁₆	¼	4.54	8.270	8¼	0.685	¹¹⁄₁₆		1⁷⁄₁₆
x 31	9.13	10.495	10½	0.400	⅜	³⁄₁₆	4.20	8.240	8¼	0.615	⅝		1⅜
WT 10.5x 28.5	8.37	10.530	10½	0.405	⅜	³⁄₁₆	4.26	6.555	6½	0.650	⅝		1⅜
x 25	7.36	10.415	10⅜	0.380	⅜	³⁄₁₆	3.96	6.530	6½	0.535	⁹⁄₁₆		1⁵⁄₁₆
x 22	6.49	10.330	10⅜	0.350	⅜	³⁄₁₆	3.62	6.500	6½	0.450	⁷⁄₁₆		1³⁄₁₆

STRUCTURAL TEES
Cut from W shapes
Properties

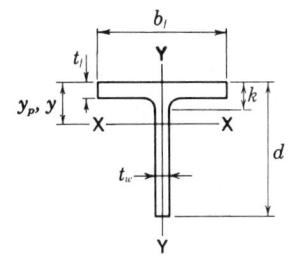

Nominal Wt. per Ft	h_c/t_w	Axis X-X						Axis Y-Y				Q_s*	
		I	S	r	y	Z	y_p	I	S	r	Z	F_y, ksi	
Lb.		In.⁴	In.³	In.	In.	In.³	In.	In.⁴	In.³	In.	In.³	36	50
108.5	14.6	502	45.2	3.97	3.11	81.1	1.13	352	49.9	3.32	77.0	—	—
97	16.2	444	40.3	3.95	3.03	71.8	1.02	309	44.1	3.29	67.9	—	0.963
89	16.7	414	38.2	3.98	3.05	67.6	0.928	278	39.4	3.26	60.8	—	0.937
80.5	18.4	372	34.4	3.96	2.99	60.8	0.845	248	35.4	3.24	54.5	—	0.851
73	20.0	336	31.2	3.95	2.95	55.0	0.768	222	31.7	3.21	48.8	0.938	0.765
64.5	19.9	323	31.0	4.13	3.39	55.1	0.945	92.2	18.4	2.21	28.8	0.938	0.765
57	21.3	289	28.3	4.15	3.42	50.4	0.833	79.4	15.8	2.18	24.7	0.883	0.700
51	23.5	258	25.3	4.14	3.37	45.0	0.750	69.6	13.9	2.15	21.7	0.780	0.578
47	24.7	239	23.8	4.16	3.41	42.4	0.692	62.0	12.4	2.12	19.4	0.728	0.529
42	26.3	216	21.9	4.18	3.48	39.2	0.621	52.8	10.6	2.07	16.6	0.664	0.476
88	14.4	319	32.2	3.51	2.74	57.8	1.00	240	37.2	3.04	57.3	—	—
81	15.3	293	29.9	3.50	2.70	53.3	0.921	221	34.2	3.05	52.6	—	—
73	16.6	264	27.2	3.50	2.66	48.2	0.833	195	30.3	3.01	46.6	—	0.947
65.5	17.8	238	24.8	3.52	2.65	43.9	0.750	170	26.5	2.97	40.7	—	0.887
58.5	19.6	212	22.3	3.51	2.62	39.2	0.672	149	23.2	2.94	35.7	0.960	0.791
52	21.6	189	20.0	3.51	2.59	35.1	0.600	130	20.3	2.91	31.2	0.874	0.690
51.5	19.6	204	22.0	3.67	3.01	39.2	0.841	59.7	13.3	1.99	20.7	0.951	0.781
47	20.9	186	20.3	3.67	2.99	36.1	0.764	54.5	12.0	1.98	18.8	0.896	0.715
42	22.9	166	18.3	3.67	2.97	32.5	0.685	47.2	10.5	1.95	16.3	0.810	0.610
38	24.5	151	16.9	3.68	3.00	30.1	0.622	41.3	9.18	1.92	14.3	0.741	0.541
34	26.0	137	15.6	3.70	3.06	27.9	0.560	35.2	7.85	1.87	12.3	0.681	0.489
31	25.1	131	15.6	3.79	3.46	28.4	1.28	17.2	4.90	1.38	7.87	0.724	0.525
27.5	27.3	117	14.1	3.80	3.50	25.6	1.53	14.5	4.15	1.34	6.67	0.626	0.450
83	12.4	226	25.5	3.04	2.39	46.3	0.984	217	35.0	2.98	53.9	—	—
73.5	13.0	204	23.7	3.08	2.39	42.4	0.864	188	30.0	2.95	46.3	—	—
66	14.4	181	21.1	3.06	2.33	37.6	0.780	166	26.7	2.93	41.1	—	—
61	15.6	166	19.3	3.04	2.28	34.3	0.724	152	24.6	2.92	37.8	—	0.993
55.5	17.1	150	17.5	3.03	2.23	31.0	0.662	137	22.2	2.90	34.1	—	0.917
50.5	18.8	135	15.8	3.01	2.18	27.9	0.605	124	20.2	2.89	30.9	0.990	0.826
46.5	16.2	144	17.9	3.25	2.74	31.8	0.812	46.4	11.0	1.84	17.4	—	0.968
41.5	18.2	127	15.7	3.22	2.66	28.0	0.728	40.7	9.75	1.83	15.3	—	0.856
36.5	20.6	110	13.8	3.21	2.60	24.4	0.647	35.3	8.51	1.81	13.3	0.908	0.730
34	21.8	103	12.9	3.20	2.59	22.9	0.606	32.4	7.83	1.80	12.2	0.853	0.664
31	23.5	93.8	11.9	3.21	2.58	21.1	0.554	28.7	6.97	1.77	10.9	0.784	0.583
28.5	23.2	90.4	11.8	3.29	2.85	21.2	0.638	15.3	4.67	1.35	7.42	0.793	0.592
25	24.7	80.3	10.7	3.30	2.93	20.8	0.771	12.5	3.82	1.30	6.09	0.733	0.533
22	26.8	71.1	9.68	3.31	2.98	17.6	1.06	10.3	3.18	1.26	5.09	0.638	0.460

*Where no value of Q_s is shown, the Tee complies with LRFD Specification Sect. E2.

AMERICAN INSTITUTE OF STEEL CONSTRUCTION

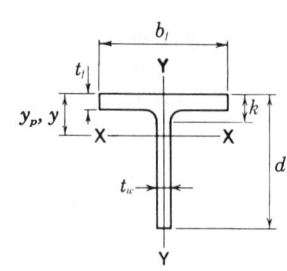

STRUCTURAL TEES
Cut from W shapes
Dimensions

Designation	Area	Depth of Tee d		Stem Thickness t_w		$\frac{t_w}{2}$	Area of Stem	Flange Width b_f		Flange Thickness t_f		Distance k
	In.²	In.		In.		In.	In.²	In.		In.		In.
WT 9x71.5	21.0	9.745	9¾	0.730	¾	⅜	7.11	11.220	11¼	1.320	1⁵⁄₁₆	2
x65	19.1	9.625	9⅝	0.670	¹¹⁄₁₆	⅜	6.45	11.160	11⅛	1.200	1³⁄₁₆	1⅞
x59.5	17.5	9.485	9½	0.655	⅝	⁵⁄₁₆	6.21	11.265	11¼	1.060	1¹⁄₁₆	1¾
x53	15.6	9.365	9⅜	0.590	⁹⁄₁₆	⁵⁄₁₆	5.53	11.200	11¼	0.940	¹⁵⁄₁₆	1⅝
x48.5	14.3	9.295	9¼	0.535	⁹⁄₁₆	⁵⁄₁₆	4.97	11.145	11⅛	0.870	⅞	1⁹⁄₁₆
x43	12.7	9.195	9¼	0.480	½	¼	4.41	11.090	11⅛	0.770	¾	1⁷⁄₁₆
x38	11.2	9.105	9⅛	0.425	⁷⁄₁₆	¼	3.87	11.035	11	0.680	¹¹⁄₁₆	1⅜
WT 9x35.5	10.4	9.235	9¼	0.495	½	¼	4.57	7.635	7⅝	0.810	¹³⁄₁₆	1½
x32.5	9.55	9.175	9⅛	0.450	⁷⁄₁₆	¼	4.13	7.590	7⅝	0.750	¾	1⁷⁄₁₆
x30	8.82	9.120	9⅛	0.415	⁷⁄₁₆	¼	3.78	7.555	7½	0.695	¹¹⁄₁₆	1⅜
x27.5	8.10	9.055	9	0.390	⅜	³⁄₁₆	3.53	7.530	7½	0.630	⅝	1⁵⁄₁₆
x25	7.33	8.995	9	0.355	⅜	³⁄₁₆	3.19	7.495	7½	0.570	⁹⁄₁₆	1¼
WT 9x23	6.77	9.030	9	0.360	⅜	³⁄₁₆	3.25	6.060	6	0.605	⅝	1¼
x20	5.88	8.950	9	0.315	⁵⁄₁₆	³⁄₁₆	2.82	6.015	6	0.525	½	1³⁄₁₆
x17.5	5.15	8.850	8⅞	0.300	⁵⁄₁₆	³⁄₁₆	2.65	6.000	6	0.425	⁷⁄₁₆	1⅛
WT 8x50	14.7	8.485	8½	0.585	⁹⁄₁₆	⁵⁄₁₆	4.96	10.425	10⅜	0.985	1	1¹¹⁄₁₆
x44.5	13.1	8.375	8⅜	0.525	½	¼	4.40	10.365	10⅜	0.875	⅞	1⁹⁄₁₆
x38.5	11.3	8.260	8¼	0.455	⁷⁄₁₆	¼	3.76	10.295	10¼	0.760	¾	1⁷⁄₁₆
x33.5	9.84	8.165	8⅛	0.395	⅜	³⁄₁₆	3.23	10.235	10¼	0.665	¹¹⁄₁₆	1⅜
WT 8x28.5	8.38	8.215	8¼	0.430	⁷⁄₁₆	¼	3.53	7.120	7⅛	0.715	¹¹⁄₁₆	1⅜
x25	7.37	8.130	8⅛	0.380	⅜	³⁄₁₆	3.09	7.070	7⅛	0.630	⅝	1⁵⁄₁₆
x22.5	6.63	8.065	8⅛	0.345	⅜	³⁄₁₆	2.78	7.035	7	0.565	⁹⁄₁₆	1¼
x20	5.89	8.005	8	0.305	⁵⁄₁₆	³⁄₁₆	2.44	6.995	7	0.505	½	1³⁄₁₆
x18	5.28	7.930	7⅞	0.295	⁵⁄₁₆	³⁄₁₆	2.34	6.985	7	0.430	⁷⁄₁₆	1⅛
WT 8x15.5	4.56	7.940	8	0.275	¼	⅛	2.18	5.525	5½	0.440	⁷⁄₁₆	1⅛
x13	3.84	7.845	7⅞	0.250	¼	⅛	1.96	5.500	5½	0.345	⅜	1¹⁄₁₆

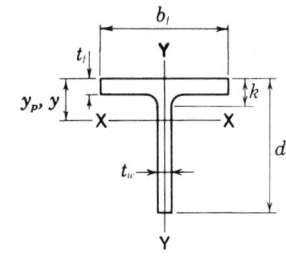

STRUCTURAL TEES
Cut from W shapes
Properties

Nom-inal Wt. per Ft	$\frac{h_c}{t_w}$	Axis X-X						Axis Y-Y				$Q_s{}^*$	
		I	S	r	y	Z	y_p	I	S	r	Z	F_y, ksi	
Lb.		In.^{4}	In.^{3}	In.	In.	In.^{3}	In.	In.^{4}	In.^{3}	In.	In.^{3}	36	50
71.5	11.0	142	18.5	2.60	2.09	34.0	0.938	156	27.7	2.72	42.7	—	—
65	11.9	127	16.7	2.58	2.02	30.5	0.856	139	24.9	2.70	38.3	—	—
59.5	12.3	119	15.9	2.60	2.03	28.7	0.778	126	22.5	2.69	34.6	—	—
53	13.6	104	14.1	2.59	1.97	25.2	0.695	110	19.7	2.66	30.2	—	—
48.5	15.0	93.8	12.7	2.56	1.91	22.6	0.640	100	18.0	2.65	27.6	—	—
43	16.7	82.4	11.2	2.55	1.86	19.9	0.570	87.6	15.8	2.63	24.2	—	0.937
38	18.9	71.8	9.83	2.54	1.80	17.3	0.505	76.2	13.8	2.61	21.1	0.990	0.826
35.5	16.2	78.2	11.2	2.74	2.26	20.0	0.683	30.1	7.89	1.70	12.3	—	0.963
32.5	17.8	70.7	10.1	2.72	2.20	18.0	0.629	27.4	7.22	1.69	11.2	—	0.877
30	19.3	64.7	9.29	2.71	2.16	16.5	0.583	25.0	6.63	1.69	10.3	0.964	0.796
27.5	20.6	59.5	8.63	2.71	2.16	15.3	0.538	22.5	5.97	1.67	9.27	0.913	0.735
25	22.6	53.5	7.79	2.70	2.12	13.8	0.489	20.0	5.35	1.65	8.29	0.823	0.625
23	22.3	52.1	7.77	2.77	2.33	13.9	0.558	11.3	3.72	1.29	5.85	0.831	0.635
20	25.5	44.8	6.73	2.76	2.29	12.0	0.489	9.55	3.17	1.27	4.97	0.690	0.496
17.5	26.8	40.1	6.21	2.79	2.39	12.0	0.450	7.67	2.56	1.22	4.03	0.638	0.460
50	12.1	76.8	11.4	2.28	1.76	20.7	0.706	93.1	17.9	2.51	27.4	—	—
44.5	13.5	67.2	10.1	2.27	1.70	18.1	0.631	81.3	15.7	2.49	24.0	—	—
38.5	15.6	56.9	8.59	2.24	1.63	15.3	0.549	69.2	13.4	2.47	20.5	—	0.988
33.5	18.0	48.6	7.36	2.22	1.56	13.0	0.481	59.5	11.6	2.46	17.7	—	0.861
28.5	16.5	48.7	7.77	2.41	1.94	13.8	0.589	21.6	6.06	1.60	9.43	—	0.942
25	18.7	42.3	6.78	2.40	1.89	12.0	0.521	18.6	5.26	1.59	8.16	0.990	0.826
22.5	20.6	37.8	6.10	2.39	1.86	10.8	0.471	16.4	4.67	1.57	7.23	0.904	0.725
20	23.3	33.1	5.35	2.37	1.81	9.43	0.421	14.4	4.12	1.57	6.37	0.784	0.583
18	24.1	30.6	5.05	2.41	1.88	8.93	0.378	12.2	3.50	1.52	5.42	0.754	0.553
15.5	25.8	27.4	4.64	2.45	2.02	8.27	0.413	6.20	2.24	1.17	3.52	0.668	0.479
13	28.4	23.5	4.09	2.47	2.09	8.12	0.372	4.80	1.74	1.12	2.74	0.563	0.406

*Where no value of Q_s is shown, the Tee complies with LRFD Specification Sect. E2.

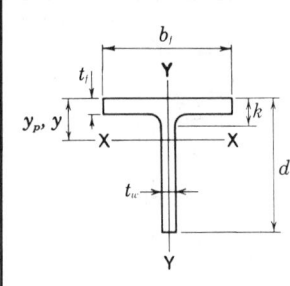

STRUCTURAL TEES
Cut from W shapes
Dimensions

Designation	Area	Depth of Tee d		Stem Thickness t_w		$\dfrac{t_w}{2}$	Area of Stem	Flange Width b_f		Flange Thickness t_f		Distance k
	In.²	In.		In.		In.	In.²	In.		In.		In.
WT 7x365	107	11.210	11¼	3.070	3¹/₁₆	1⁹/₁₆	34.4	17.890	17⅞	4.910	4¹⁵/₁₆	5⁹/₁₆
x332.5	97.8	10.820	10⅞	2.830	2¹³/₁₆	1⁷/₁₆	30.6	17.650	17⅝	4.520	4½	5³/₁₆
x302.5	88.9	10.460	10½	2.595	2⅝	1⁵/₁₆	27.1	17.415	17⅜	4.160	4³/₁₆	4¹³/₁₆
x275	80.9	10.120	10⅛	2.380	2⅜	1³/₁₆	24.1	17.200	17¼	3.820	3¹³/₁₆	4½
x250	73.5	9.800	9¾	2.190	2³/₁₆	1⅛	21.5	17.010	17	3.500	3½	4³/₁₆
x227.5	66.9	9.510	9½	2.015	2	1	19.2	16.835	16⅞	3.210	3³/₁₆	3⅞
x213	62.6	9.335	9⅜	1.875	1⅞	¹⁵/₁₆	17.5	16.695	16¾	3.035	3¹/₁₆	3¹¹/₁₆
x199	58.5	9.145	9⅛	1.770	1¾	⅞	16.2	16.590	16⅝	2.845	2⅞	3½
x185	54.4	8.960	9	1.655	1⅝	¹³/₁₆	14.8	16.475	16½	2.660	2¹¹/₁₆	3⁵/₁₆
x171	50.3	8.770	8¾	1.540	1⁹/₁₆	¹³/₁₆	13.5	16.360	16⅜	2.470	2½	3⅛
x155.5	45.7	8.560	8½	1.410	1⁷/₁₆	¾	12.1	16.230	16¼	2.260	2¼	2¹⁵/₁₆
x141.5	41.6	8.370	8⅜	1.290	1⁵/₁₆	¹¹/₁₆	10.8	16.110	16⅛	2.070	2¹/₁₆	2¾
x128.5	37.8	8.190	8¼	1.175	1³/₁₆	⅝	9.62	15.995	16	1.890	1⅞	2⁹/₁₆
x116.5	34.2	8.020	8	1.070	1¹/₁₆	⁹/₁₆	8.58	15.890	15⅞	1.720	1¾	2⅜
x105.5	31.0	7.860	7⅞	0.980	1	½	7.70	15.800	15¾	1.560	1⁹/₁₆	2¼
x 96.5	28.4	7.740	7¾	0.890	⅞	⁷/₁₆	6.89	15.710	15¾	1.440	1⁷/₁₆	2⅛
x 88	25.9	7.610	7⅝	0.830	¹³/₁₆	⁷/₁₆	6.32	15.650	15⅝	1.310	1⁵/₁₆	2
x 79.5	23.4	7.490	7½	0.745	¾	⅜	5.58	15.565	15⅝	1.190	1³/₁₆	1⅞
x 72.5	21.3	7.390	7⅜	0.680	¹¹/₁₆	⅜	5.03	15.500	15½	1.090	1¹/₁₆	1¾

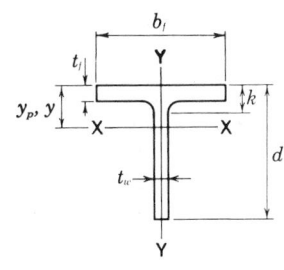

STRUCTURAL TEES
Cut from W shapes
Properties

Nom-inal Wt. per Ft	$\dfrac{h_c}{t_w}$	Axis X-X						Axis Y-Y				Q_s*	
		I	S	r	y	Z	y_p	I	S	r	Z	F_y, ksi	
Lb.		In.^{4}	In.^{3}	In.	In.	In.^{3}	In.	In.^{4}	In.^{3}	In.	In.^{3}	36	50
365	1.9	739	95.4	2.62	3.47	211	3.00	2360	264	4.69	408	—	—
332.5	2.0	622	82.1	2.52	3.25	182	2.77	2080	236	4.62	365	—	—
302.5	2.2	524	70.6	2.43	3.05	157	2.55	1840	211	4.55	326	—	—
275	2.4	442	60.9	2.34	2.85	136	2.35	1630	189	4.49	292	—	—
250	2.6	375	52.7	2.26	2.67	117	2.16	1440	169	4.43	261	—	—
227.5	2.8	321	45.9	2.19	2.51	102	1.99	1280	152	4.38	234	—	—
213	3.0	287	41.4	2.14	2.40	91.7	1.88	1180	141	4.34	217	—	—
199	3.2	257	37.6	2.10	2.30	82.9	1.76	1090	131	4.31	201	—	—
185	3.4	229	33.9	2.05	2.19	74.4	1.65	994	121	4.27	185	—	—
171	3.7	203	30.4	2.01	2.09	66.2	1.54	903	110	4.24	169	—	—
155.5	4.0	176	26.7	1.96	1.97	57.7	1.41	807	99.4	4.20	152	—	—
141.5	4.4	153	23.5	1.92	1.86	50.4	1.29	722	89.7	4.17	137	—	—
128.5	4.9	133	20.7	1.88	1.75	43.9	1.18	645	80.7	4.13	123	—	—
116.5	5.3	116	18.2	1.84	1.65	38.2	1.08	576	72.5	4.10	110	—	—
105.5	5.8	102	16.2	1.81	1.57	33.4	0.980	513	65.0	4.07	99.0	—	—
96.5	6.4	89.8	14.4	1.78	1.49	29.4	0.903	466	59.3	4.05	90.2	—	—
88	6.9	80.5	13.0	1.76	1.43	26.3	0.827	419	53.5	4.02	81.4	—	—
79.5	7.7	70.2	11.4	1.73	1.35	22.8	0.751	374	48.1	4.00	73.0	—	—
72.5	8.4	62.5	10.2	1.71	1.29	20.2	0.688	338	43.7	3.98	66.3	—	—

*Where no value of Q_s is shown, the Tee complies with LRFD Specification Sect. E2.

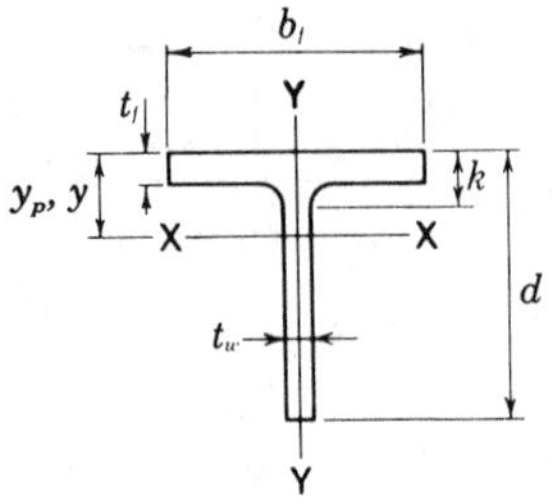

STRUCTURAL TEES
Cut from W shapes
Dimensions

Designation	Area	Depth of Tee d		Stem Thickness t_w		$\frac{t_w}{2}$	Area of Stem	Flange Width b_f		Flange Thickness t_f		Distance k
	In.²	In.		In.		In.	In.²	In.		In.		In.
WT 7x66	19.4	7.330	7⅜	0.645	⅝	5/16	4.73	14.725	14¾	1.030	1	1 11/16
x60	17.7	7.240	7¼	0.590	9/16	5/16	4.27	14.670	14⅝	0.940	15/16	1⅝
x54.5	16.0	7.160	7⅛	0.525	½	¼	3.76	14.605	14⅝	0.860	⅞	1 9/16
x49.5	14.6	7.080	7⅛	0.485	⅝	¼	3.43	14.565	14⅝	0.780	¾	1 7/16
x45	13.2	7.010	7	0.440	7/16	¼	3.08	14.520	14½	0.710	11/16	1⅜
WT 7x41	12.0	7.155	7⅛	0.510	½	¼	3.65	10.130	10⅛	0.855	⅞	1⅝
x37	10.9	7.085	7⅛	0.450	7/16	¼	3.19	10.070	10⅛	0.785	13/16	1 9/16
x34	9.99	7.020	7	0.415	7/16	¼	2.91	10.035	10	0.720	¾	1½
x30.5	8.96	6.945	7	0.375	⅜	3/16	2.60	9.995	10	0.645	⅝	1 7/16
WT 7x26.5	7.81	6.960	7	0.370	⅜	3/16	2.58	8.060	8	0.660	11/16	1 7/16
x24	7.07	6.895	6⅞	0.340	5/16	3/16	2.34	8.030	8	0.595	⅝	1⅜
x21.5	6.31	6.830	6⅞	0.305	5/16	3/16	2.08	7.995	8	0.530	½	1 5/16
WT 7x19	5.58	7.050	7	0.310	5/16	3/16	2.19	6.770	6¾	0.515	½	1 1/16
x17	5.00	6.990	7	0.285	5/16	3/16	1.99	6.745	6¾	0.455	7/16	1/16
x15	4.42	6.920	6⅞	0.270	¼	⅛	1.87	6.730	6¾	0.385	⅜	15/16
WT 7x13	3.85	6.955	7	0.255	¼	⅛	1.77	5.025	5	0.420	7/16	15/16
x11	3.25	6.870	6⅞	0.230	¼	⅛	1.58	5.000	5	0.335	5/16	⅞

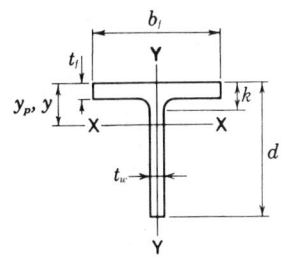

STRUCTURAL TEES
Cut from W shapes
Properties

Nom-inal Wt. per Ft	$\frac{h_c}{t_w}$	Axis X-X						Axis Y-Y				$Q_s{}^*$
		I	S	r	y	Z	y_p	I	S	r	Z	F_y, ksi
Lb.		In.^{4}	In.^{3}	In.	In.	In.^{3}	In.	In.^{4}	In.^{3}	In.	In.^{3}	36	50
66	8.8	57.8	9.57	1.73	1.29	18.6	0.658	274	37.2	3.76	56.6	—	—
60	9.7	51.7	8.61	1.71	1.24	16.5	0.602	247	33.7	3.74	51.2	—	—
54.5	10.9	45.3	7.56	1.68	1.17	14.4	0.549	223	30.6	3.73	46.4	—	—
49.5	11.8	40.9	6.88	1.67	1.14	12.9	0.500	201	27.6	3.71	41.8	—	—
45	13.0	36.4	6.16	1.66	1.09	11.5	0.456	181	25.0	3.70	37.8	—	—
41	11.2	41.2	7.14	1.85	1.39	13.2	0.594	74.2	14.6	2.48	22.4	—	—
37	12.7	36.0	6.25	1.82	1.32	11.5	0.541	66.9	13.3	2.48	20.3	—	—
34	13.7	32.6	5.69	1.81	1.29	10.4	0.498	60.7	12.1	2.46	18.5	—	—
30.5	15.2	28.9	5.07	1.80	1.25	9.16	0.448	53.7	10.7	2.45	16.4	—	0.973
26.5	15.4	27.6	4.94	1.88	1.38	8.87	0.484	28.8	7.16	1.92	11.0	—	0.958
24	16.8	24.9	4.48	1.87	1.35	8.00	0.440	25.7	6.40	1.91	9.82	—	0.882
21.5	18.7	21.9	3.98	1.86	1.31	7.05	0.395	22.6	5.65	1.89	8.66	0.947	0.775
19	19.8	23.3	4.22	2.04	1.54	7.45	0.412	13.3	3.94	1.55	6.07	0.934	0.760
17	21.5	20.9	3.83	2.04	1.53	6.74	0.371	11.7	3.45	1.53	5.32	0.857	0.669
15	22.7	19.0	3.55	2.07	1.58	6.25	0.329	9.79	2.91	1.49	4.49	0.810	0.610
13	24.1	17.3	3.31	2.12	1.72	5.89	0.383	4.45	1.77	1.08	2.77	0.737	0.537
11	26.7	14.8	2.91	2.14	1.76	5.20	0.325	3.50	1.40	1.04	2.19	0.621	0.447

*Where no value of Q_s is shown, the Tee complies with LRFD Specification Sect. E2.

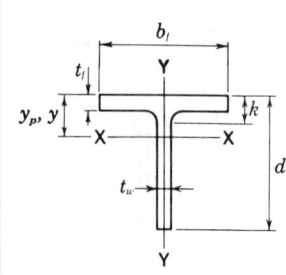

STRUCTURAL TEES
Cut from W shapes
Dimensions

Designation	Area	Depth of Tee d		Stem Thickness t_w		$\frac{t_w}{2}$	Area of Stem	Flange Width b_f		Flange Thickness t_f		Distance k
	In.²	In.		In.		In.	In.²	In.		In.		In.
WT 6x168	49.4	8.410	8⅜	1.775	1¾	⅞	14.9	13.385	13⅜	2.955	2¹⁵⁄₁₆	3¹¹⁄₁₆
x152.5	44.8	8.160	8⅛	1.625	1⅝	¹³⁄₁₆	13.3	13.235	13¼	2.705	2¹¹⁄₁₆	3⁷⁄₁₆
x139.5	41.0	7.925	7⅞	1.530	1½	¾	12.1	13.140	13⅛	2.470	2½	3³⁄₁₆
x126	37.0	7.705	7¾	1.395	1⅜	¹¹⁄₁₆	10.7	13.005	13	2.250	2¼	2¹⁵⁄₁₆
x115	33.9	7.525	7½	1.285	1⁵⁄₁₆	¹¹⁄₁₆	9.67	12.895	12⅞	2.070	2¹⁄₁₆	2¾
x105	30.9	7.355	7⅜	1.180	1³⁄₁₆	⅝	8.68	12.790	12¾	1.900	1⅞	2⅝
x 95	27.9	7.190	7¼	1.060	1¹⁄₁₆	⁹⁄₁₆	7.62	12.670	12⅝	1.735	1¾	2⁷⁄₁₆
x 85	25.0	7.015	7	0.960	¹⁵⁄₁₆	½	6.73	12.570	12⅝	1.560	1⁹⁄₁₆	2¼
x 76	22.4	6.855	6⅞	0.870	⅞	⁷⁄₁₆	5.96	12.480	12½	1.400	1⅜	2⅛
x 68	20.0	6.705	6¾	0.790	¹³⁄₁₆	⁷⁄₁₆	5.30	12.400	12⅜	1.250	1¼	1¹⁵⁄₁₆
x 60	17.6	6.560	6½	0.710	¹¹⁄₁₆	⅜	4.66	12.320	12⅜	1.105	1⅛	1¹³⁄₁₆
x 53	15.6	6.445	6½	0.610	⅝	⁵⁄₁₆	3.93	12.220	12¼	0.990	1	1¹¹⁄₁₆
x 48	14.1	6.355	6⅜	0.550	⁹⁄₁₆	⁵⁄₁₆	3.50	12.160	12⅛	0.900	⅞	1⅝
x 43.5	12.8	6.265	6¼	0.515	½	¼	3.23	12.125	12⅛	0.810	¹³⁄₁₆	1½
x 39.5	11.6	6.190	6¼	0.470	½	¼	2.91	12.080	12⅛	0.735	¾	1⁷⁄₁₆
x 36	10.6	6.125	6⅛	0.430	⁷⁄₁₆	¼	2.63	12.040	12	0.670	¹¹⁄₁₆	1⅜
x 32.5	9.54	6.060	6	0.390	⅜	³⁄₁₆	2.36	12.000	12	0.605	⅝	1⁵⁄₁₆
WT 6x 29	8.52	6.095	6⅛	0.360	⅜	³⁄₁₆	2.19	10.010	10	0.640	⅝	1⅜
x 26.5	7.78	6.030	6	0.345	⅜	³⁄₁₆	2.08	9.995	10	0.575	⁹⁄₁₆	1¼
WT 6x 25	7.34	6.095	6⅛	0.370	⅜	³⁄₁₆	2.26	8.080	8⅛	0.640	⅝	1⅜
x 22.5	6.61	6.030	6	0.335	⁵⁄₁₆	³⁄₁₆	2.02	8.045	8	0.575	⁹⁄₁₆	1¼
x 20	5.89	5.970	6	0.295	⁵⁄₁₆	³⁄₁₆	1.76	8.005	8	0.515	½	1¼
WT 6x 17.5	5.17	6.250	6¼	0.300	⁵⁄₁₆	³⁄₁₆	1.88	6.560	6½	0.520	½	1
x 15	4.40	6.170	6⅛	0.260	¼	⅛	1.60	6.520	6½	0.440	⁷⁄₁₆	¹⁵⁄₁₆
x 13	3.82	6.110	6⅛	0.230	¼	⅛	1.41	6.490	6½	0.380	⅜	⅞
WT 6x 11	3.24	6.155	6⅛	0.260	¼	⅛	1.60	4.030	4	0.425	⁷⁄₁₆	⅞
x 9.5	2.79	6.080	6⅛	0.235	¼	⅛	1.43	4.005	4	0.350	⅜	¹³⁄₁₆
x 8	2.36	5.995	6	0.220	¼	⅛	1.32	3.990	4	0.265	¼	¾
x 7	2.08	5.955	6	0.200	³⁄₁₆	⅛	1.19	3.970	4	0.225	¼	¹¹⁄₁₆

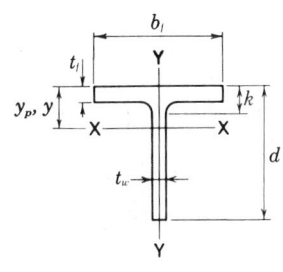

STRUCTURAL TEES
Cut from W shapes
Properties

Nominal Wt. per Ft	$\frac{h_c}{t_w}$	Axis X-X						Axis Y-Y				Q_s^*	
		I	S	r	y	Z	y_p	I	S	r	Z	F_y, ksi	
Lb.		In.^{4}	In.^{3}	In.	In.	In.^{3}	In.	In.^{4}	In.^{3}	In.	In.^{3}	36	50
168	2.7	190	31.2	1.96	2.31	68.4	1.84	593	88.6	3.47	137	—	—
152.5	3.0	162	27.0	1.90	2.16	59.1	1.69	525	79.3	3.42	122	—	—
139.5	3.2	141	24.1	1.86	2.05	51.9	1.56	469	71.3	3.38	110	—	—
126	3.5	121	20.9	1.81	1.92	44.8	1.42	414	63.6	3.34	97.9	—	—
115	3.8	106	18.5	1.77	1.82	39.4	1.31	371	57.5	3.31	88.4	—	—
105	4.1	92.1	16.4	1.73	1.72	34.5	1.21	332	51.9	3.28	79.7	—	—
95	4.6	79.0	14.2	1.68	1.62	29.8	1.10	295	46.5	3.25	71.3	—	—
85	5.1	67.8	12.3	1.65	1.52	25.6	0.994	259	41.2	3.22	63.0	—	—
76	5.6	58.5	10.8	1.62	1.43	22.0	0.896	227	36.4	3.19	55.6	—	—
68	6.1	50.6	9.46	1.59	1.35	19.0	0.805	199	32.1	3.16	49.0	—	—
60	6.8	43.4	8.22	1.57	1.28	16.2	0.716	172	28.0	3.13	42.7	—	—
53	8.0	36.3	6.91	1.53	1.19	13.6	0.637	151	24.7	3.11	37.5	—	—
48	8.8	32.0	6.12	1.51	1.13	11.9	0.580	135	22.2	3.09	33.7	—	—
43.5	9.4	28.9	5.60	1.50	1.10	10.7	0.527	120	19.9	3.07	30.2	—	—
39.5	10.3	25.8	5.03	1.49	1.06	9.49	0.480	108	17.9	3.05	27.2	—	—
36	11.3	23.2	4.54	1.48	1.02	8.48	0.439	97.5	16.2	3.04	24.6	—	—
32.5	12.4	20.6	4.06	1.47	0.985	7.50	0.398	87.2	14.5	3.02	22.0	—	—
29	13.5	19.1	3.76	1.50	1.03	6.97	0.426	53.5	10.7	2.51	16.3	—	—
26.5	14.1	17.7	3.54	1.51	1.02	6.46	0.389	47.9	9.58	2.48	14.6	—	—
25	13.1	18.7	3.79	1.60	1.17	6.90	0.454	28.2	6.97	1.96	10.7	—	—
22.5	14.5	16.6	3.39	1.58	1.13	6.12	0.411	25.0	6.21	1.94	9.50	—	0.998
20	16.5	14.4	2.95	1.57	1.08	5.30	0.368	22.0	5.51	1.93	8.41	—	0.887
17.5	18.1	16.0	3.23	1.76	1.30	5.71	0.394	12.2	3.73	1.54	5.73	—	0.856
15	20.9	13.5	2.75	1.75	1.27	4.83	0.337	10.2	3.12	1.52	4.78	0.891	0.710
13	23.6	11.7	2.40	1.75	1.25	4.20	0.295	8.66	2.67	1.51	4.08	0.767	0.565
11	20.9	11.7	2.59	1.90	1.63	4.63	0.402	2.33	1.16	0.847	1.83	0.891	0.710
9.5	23.1	10.1	2.28	1.90	1.65	4.11	0.348	1.88	0.939	0.822	1.49	0.797	0.596
8	24.7	8.70	2.04	1.92	1.74	3.72	0.639	1.41	0.706	0.773	1.13	0.741	0.541
7	27.2	7.67	1.83	1.92	1.76	3.32	0.760	1.18	0.594	0.753	0.950	0.626	0.450

*Where no value of Q_s is shown, the Tee complies with LRFD Specification Sect. E2.

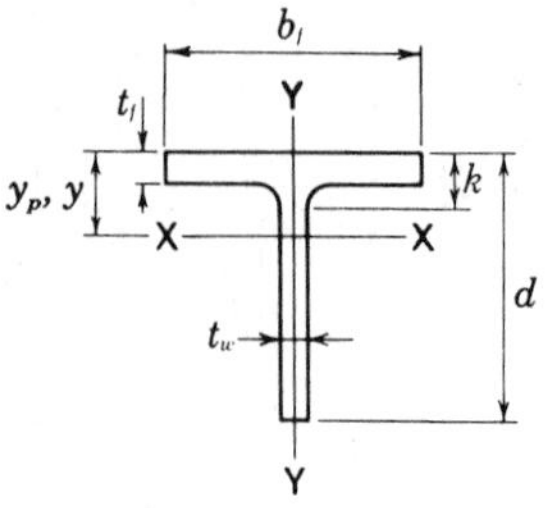

STRUCTURAL TEES
Cut from W shapes
Dimensions

Designation	Area	Depth of Tee d		Stem Thickness t_w		$\frac{t_w}{2}$	Area of Stem	Flange Width b_f		Flange Thickness t_f		Distance k
	In.^{2}	In.		In.		In.	In.^{2}	In.		In.		In.
WT 5x56	16.5	5.680	5 5/8	0.755	3/4	3/8	4.29	10.415	10 3/8	1.250	1 1/4	1 7/8
x50	14.7	5.550	5 1/2	0.680	11/16	3/8	3.77	10.340	10 3/8	1.120	1 1/8	1 3/4
x44	12.9	5.420	5 3/8	0.605	5/8	5/16	3.28	10.265	10 1/4	0.990	1	1 5/8
x38.5	11.3	5.300	5 1/4	0.530	1/2	1/4	2.81	10.190	10 1/4	0.870	7/8	1 1/2
x34	9.99	5.200	5 1/4	0.470	1/2	1/4	2.44	10.130	10 1/8	0.770	3/4	1 3/8
x30	8.82	5.110	5 1/8	0.420	7/16	1/4	2.15	10.080	10 1/8	0.680	11/16	1 5/16
x27	7.91	5.045	5	0.370	3/8	3/16	1.87	10.030	10	0.615	5/8	1 1/4
x24.5	7.21	4.990	5	0.340	5/16	3/16	1.70	10.000	10	0.560	9/16	1 3/16
WT 5x22.5	6.63	5.050	5	0.350	3/8	3/16	1.77	8.020	8	0.620	5/8	1 1/4
x19.5	5.73	4.960	5	0.315	5/16	3/16	1.56	7.985	8	0.530	1/2	1 1/8
x16.5	4.85	4.865	4 7/8	0.290	5/16	3/16	1.41	7.960	8	0.435	7/16	1 1/16
WT 5x15	4.42	5.235	5 1/4	0.300	5/16	3/16	1.57	5.810	5 3/4	0.510	1/2	15/16
x13	3.81	5.165	5 1/8	0.260	1/4	1/8	1.34	5.770	5 3/4	0.440	7/16	7/8
x11	3.24	5.085	5 1/8	0.240	1/4	1/8	1.22	5.750	5 3/4	0.360	3/8	3/4
WT 5x 9.5	2.81	5.120	5 1/8	0.250	1/4	1/8	1.28	4.020	4	0.395	3/8	13/16
x 8.5	2.50	5.055	5	0.240	1/4	1/8	1.21	4.010	4	0.330	5/16	3/4
x 7.5	2.21	4.995	5	0.230	1/4	1/8	1.15	4.000	4	0.270	1/4	11/16
x 6	1.77	4.935	4 7/8	0.190	3/16	1/8	0.938	3.960	4	0.210	3/16	5/8

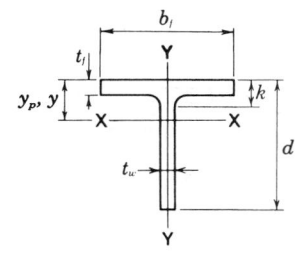

STRUCTURAL TEES
Cut from W shapes
Properties

Nom-inal Wt. per Ft	$\frac{h_c}{t_w}$	Axis X-X						Axis Y-Y				Q_s^*	
		I	S	r	y	Z	y_p	I	S	r	Z	F_y, ksi	
Lb.		In.⁴	In.³	In.	In.	In.³	In.	In.⁴	In.³	In.	In.³	36	50
56	5.2	28.6	6.40	1.32	1.21	13.4	0.791	118	22.6	2.68	34.6	—	—
50	5.8	24.5	5.56	1.29	1.13	11.4	0.711	103	20.0	2.65	30.5	—	—
44	6.5	20.8	4.77	1.27	1.06	9.65	0.631	89.3	17.4	2.63	26.5	—	—
38.5	7.4	17.4	4.04	1.24	0.990	8.06	0.555	76.8	15.1	2.60	22.9	—	—
34	8.4	14.9	3.49	1.22	0.932	6.85	0.493	66.8	13.2	2.59	20.0	—	—
30	9.4	12.9	3.04	1.21	0.884	5.87	0.438	58.1	11.5	2.57	17.5	—	—
27	10.6	11.1	2.64	1.19	0.836	5.05	0.395	51.7	10.3	2.56	15.7	—	—
24.5	11.6	10.0	2.39	1.18	0.807	4.52	0.361	46.7	9.34	2.54	14.2	—	—
22.5	11.2	10.2	2.47	1.24	0.907	4.65	0.413	26.7	6.65	2.01	10.1	—	—
19.5	12.5	8.84	2.16	1.24	0.876	3.99	0.359	22.5	5.64	1.98	8.59	—	—
16.5	13.6	7.71	1.93	1.26	0.869	3.48	0.305	18.3	4.60	1.94	7.01	—	—
15	14.8	9.28	2.24	1.45	1.10	4.01	0.380	8.35	2.87	1.37	4.42	—	—
13	17.0	7.86	1.91	1.44	1.06	3.39	0.330	7.05	2.44	1.36	3.75	—	0.902
11	18.4	6.88	1.72	1.46	1.07	3.02	0.282	5.71	1.99	1.33	3.05	0.999	0.836
9.5	17.7	6.68	1.74	1.54	1.28	3.10	0.349	2.15	1.07	0.874	1.68	—	0.872
8.5	18.4	6.06	1.62	1.56	1.32	2.90	0.311	1.78	0.888	0.844	1.40	—	0.841
7.5	19.2	5.45	1.50	1.57	1.37	3.03	0.306	1.45	0.723	0.810	1.15	0.977	0.811
6	23.3	4.35	1.22	1.57	1.36	2.50	0.323	1.09	0.551	0.785	0.872	0.793	0.592

*Where no value of Q_s is shown, the Tee complies with LRFD Specification Sect. E2.

AMERICAN INSTITUTE OF STEEL CONSTRUCTION

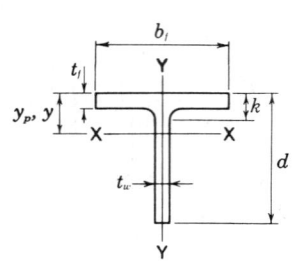

STRUCTURAL TEES
Cut from W shapes
Dimensions

Designation	Area	Depth of Tee d		Stem Thickness t_w		$\dfrac{t_w}{2}$	Area of Stem	Flange Width b_f		Flange Thickness t_f		Distance k
	In.²	In.		In.		In.	In.²	In.		In.		In.
WT 4 x33.5	9.84	4.500	4½	0.570	9/16	5/16	2.56	8.280	8¼	0.935	15/16	1 7/16
x29	8.55	4.375	4⅜	0.510	½	¼	2.23	8.220	8¼	0.810	13/16	1 5/16
x24	7.05	4.250	4¼	0.400	⅜	3/16	1.70	8.110	8⅛	0.685	11/16	1 3/16
x20	5.87	4.125	4⅛	0.360	⅜	3/16	1.48	8.070	8⅛	0.560	9/16	1 1/16
x17.5	5.14	4.060	4	0.310	5/16	3/16	1.26	8.020	8	0.495	½	1
x15.5	4.56	4.000	4	0.285	5/16	3/16	1.14	7.995	8	0.435	7/16	15/16
WT 4 x14	4.12	4.030	4	0.285	5/16	3/16	1.15	6.535	6½	0.465	7/16	15/16
x12	3.54	3.965	4	0.245	¼	⅛	0.971	6.495	6½	0.400	⅜	⅞
WT 4 x10.5	3.08	4.140	4⅛	0.250	¼	⅛	1.03	5.270	5¼	0.400	⅜	13/16
x 9	2.63	4.070	4⅛	0.230	¼	⅛	0.936	5.250	5¼	0.330	5/16	¾
WT 4 x 7.5	2.22	4.055	4	0.245	¼	⅛	0.993	4.015	4	0.315	5/16	¾
x 6.5	1.92	3.995	4	0.230	¼	⅛	0.919	4.000	4	0.255	¼	11/16
x 5	1.48	3.945	4	0.170	3/16	⅛	0.671	3.940	4	0.205	3/16	⅝
WT 3 x12.5	3.67	3.190	3¼	0.320	5/16	3/16	1.02	6.080	6⅛	0.455	7/16	13/16
x10	2.94	3.100	3⅛	0.260	¼	⅛	0.806	6.020	6	0.365	⅜	¾
x 7.5	2.21	2.995	3	0.230	¼	⅛	0.689	5.990	6	0.260	¼	⅝
WT 3 x 8	2.37	3.140	3⅛	0.260	¼	⅛	0.816	4.030	4	0.405	⅜	¾
x 6	1.78	3.015	3	0.230	¼	⅛	0.693	4.000	4	0.280	¼	⅝
x 4.5	1.34	2.950	3	0.170	3/16	⅛	0.502	3.940	4	0.215	3/16	9/16
WT 2.5x 9.5	2.77	2.575	2⅝	0.270	¼	⅛	0.695	5.030	5	0.430	7/16	13/16
x 8	2.34	2.505	2½	0.140	¼	⅛	0.601	5.000	5	0.360	⅜	¾
WT 2 x 6.5	1.91	2.080	2⅛	0.280	¼	⅛	0.582	4.060	4	0.345	⅜	11/16

STRUCTURAL TEES
Cut from W shapes
Properties

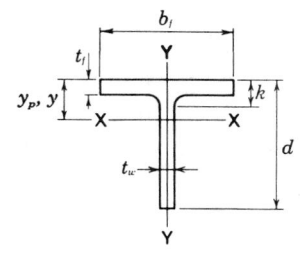

Nom-inal Wt. per Ft	$\frac{h_c}{t_w}$	Axis X-X						Axis Y-Y				Q_s*	
		I	S	r	y	Z	y_p	I	S	r	Z	F_y, ksi	
Lb.		In.^{4}	In.^{3}	In.	In.	In.^{3}	In.	In.^{4}	In.^{3}	In.	In.^{3}	36	50
33.5	5.6	10.9	3.05	1.05	0.936	6.29	0.594	44.3	10.7	2.12	16.3	—	—
29	6.2	9.12	2.61	1.03	0.874	5.25	0.520	37.5	9.13	2.10	13.9	—	—
24	7.9	6.85	1.97	0.986	0.777	3.94	0.435	30.5	7.52	2.08	11.4	—	—
20	8.8	5.73	1.69	0.988	0.735	3.25	0.364	24.5	6.08	2.04	9.25	—	—
17.5	10.2	4.81	1.43	0.967	0.688	2.71	0.321	21.3	5.31	2.03	8.06	—	—
15.5	11.1	4.28	1.28	0.968	0.667	2.39	0.285	18.5	4.64	2.02	7.04	—	—
14	11.1	4.22	1.28	1.01	0.734	2.38	0.315	10.8	3.31	1.62	5.05	—	—
12	12.9	3.53	1.08	0.999	0.695	1.98	0.273	9.14	2.81	1.61	4.29	—	—
10.5	13.8	3.90	1.18	1.12	0.831	2.11	0.292	4.89	1.85	1.26	2.84	—	—
9	15.0	3.41	1.05	1.14	0.834	1.86	0.251	3.98	1.52	1.23	2.33	—	—
7.5	14.0	3.28	1.07	1.22	0.998	1.91	0.276	1.70	0.849	0.876	1.33	—	—
6.5	15.0	2.89	0.974	1.23	1.03	1.74	0.240	1.37	0.683	0.843	1.08	—	—
5	20.2	2.15	0.717	1.20	0.953	1.27	0.188	1.05	0.532	0.841	0.828	0.913	0.735
12.5	7.8	2.28	0.886	0.789	0.610	1.68	0.302	8.53	2.81	1.52	4.28	—	—
10	9.6	1.76	0.693	0.774	0.560	1.29	0.244	6.64	2.21	1.50	3.36	—	—
7.5	10.8	1.41	0.577	0.797	0.558	1.03	0.185	4.66	1.56	1.45	2.37	—	—
8	9.6	1.69	0.685	0.844	0.676	1.25	0.294	2.21	1.10	0.966	1.70	—	—
6	10.8	1.32	0.564	0.861	0.677	1.01	0.222	1.50	0.748	0.918	1.16	—	—
4.5	14.6	0.950	0.408	0.842	0.623	0.720	0.170	1.10	0.557	0.905	0.858	—	—
9.5	7.0	1.01	0.485	0.605	0.487	0.967	0.275	4.56	1.82	1.28	2.76	—	—
8	7.9	0.845	0.413	0.601	0.458	0.798	0.234	3.75	1.50	1.27	2.29	—	—
6.5	5.3	0.526	0.321	0.524	0.440	0.616	0.236	1.93	0.950	1.00	1.46	—	—

*Where no value of Q_s is shown, the Tee complies with LRFD Specification Sect. E2.

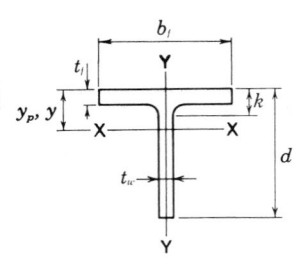

STRUCTURAL TEES
Cut from M shapes
Dimensions

Designation	Area	Depth of Tee d	Stem Thickness t_w		$\frac{t_w}{2}$	Area of Stem	Flange Width b_f		Flange Thickness t_f		Distance k	Grip	Max. Flge. Fastener
	In.²	In.	In.		In.	In.²	In.		In.		In.	In.	In.
MT 7 x 9	2.55	7.000	7	0.215	³⁄₁₆	⅛	1.50	4.000	4	0.270	¼	⅝	¼	¾
MT 6 x 5.9	1.73	6.000	6	0.177	³⁄₁₆	⅛	1.06	3.065	3⅛	0.225	¼	⁹⁄₁₆	¼	—
MT 5 x 4.5	1.32	5.000	5	0.157	³⁄₁₆	⅛	0.785	2.690	2¾	0.206	³⁄₁₆	⁹⁄₁₆	³⁄₁₆	—
MT 4 x 3.25	0.958	4.000	4	0.135	⅛	¹⁄₁₆	0.540	2.281	2¼	0.189	³⁄₁₆	½	³⁄₁₆	—
MT 3 x10	2.94	3.000	3	0.250	¼	⅛	0.750	5.938	6	0.379	⅜	⅞	⅜	⅞
MT 3 x 2.2	0.646	3.000	3	0.114	⅛	¹⁄₁₆	0.342	1.844	1⅞	0.171	³⁄₁₆	⁷⁄₁₆	³⁄₁₆	—
MT 2.5x 9.45	2.78	2.500	2½	0.316	⁵⁄₁₆	³⁄₁₆	0.790	5.003	5	0.416	⁷⁄₁₆	⅞	⁷⁄₁₆	⅞
MT 2 x 6.5	1.90	2.000	2	0.254	¼	⅛	0.508	3.940	4	0.371	⅜	¹³⁄₁₆	⅜	¾

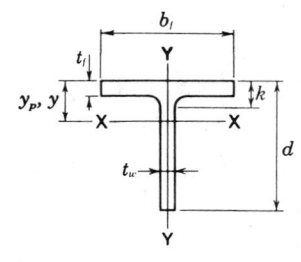

STRUCTURAL TEES
Cut from M shapes
Properties

Nom-inal Wt. per Ft	$\frac{h_c}{t_w}$	Axis X-X						Axis Y-Y				$Q_s{}^*$	
		I	S	r	y	Z	y_p	I	S	r	Z	F_y, ksi	
Lb.		In.⁴	In.³	In.	In.	In.³	In.	In.⁴	In.³	In.	In.³	36	50
9	30.1	13.1	2.69	2.27	2.12	4.88	1.06	1.32	0.660	0.719	1.16	0.523	0.376
5.9	31.3	6.60	1.60	1.95	1.89	2.89	1.09	0.490	0.320	0.532	0.577	0.483	0.348
4.5	29.2	3.46	0.997	1.62	1.53	1.81	0.778	0.305	0.227	0.480	0.405	0.549	0.396
3.25	26.9	1.57	0.556	1.28	1.17	1.01	0.446	0.172	0.150	0.423	0.265	0.634	0.457
10	9.2	1.54	0.624	0.724	0.531	1.19	0.248	5.80	1.95	1.40	3.39	—	—
2.2	23.5	0.577	0.267	0.945	0.836	0.522	0.200	0.083	0.090	0.358	0.155	0.780	0.578
9.45	5.6	1.05	0.527	0.615	0.511	1.03	0.278	3.93	1.57	1.19	2.66	—	—
6.5	5.2	0.431	0.271	0.476	0.410	0.547	0.243	1.68	0.853	0.939	1.47	—	—

*Where no value of Q_s is shown, the Tee complies with LRFD Specification Sect. E2.

AMERICAN INSTITUTE OF STEEL CONSTRUCTION

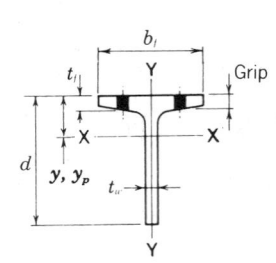

STRUCTURAL TEES
Cut from S shapes
Dimensions

Designation	Area	Depth of Tee d		Stem Thickness t_w		$\frac{t_w}{2}$	Area of Stem	Flange Width b_f		Flange Thickness t_f		Distance k	Grip	Max. Flge. Fastener
	In.²	In.		In.		In.	In.²	In.		In.		In.	In.	In.
ST 12 x60.5	17.8	12.250	12¼	0.800	¹³⁄₁₆	⁷⁄₁₆	9.80	8.050	8	1.090	1¹⁄₁₆	2	1⅛	1
x53	15.6	12.250	12¼	0.620	⅝	⁵⁄₁₆	7.59	7.870	7⅞	1.090	1¹⁄₁₆	2	1⅛	1
ST 12 x50	14.7	12.000	12	0.745	¾	⅜	8.94	7.245	7¼	0.870	⅞	1¾	⅞	1
x45	13.2	12.000	12	0.625	⅝	⁵⁄₁₆	7.50	7.125	7⅛	0.870	⅞	1¾	⅞	1
x40	11.7	12.000	12	0.500	½	¼	6.00	7.000	7	0.870	⅞	1¾	⅞	1
ST 10 x48	14.1	10.150	10⅛	0.800	¹³⁄₁₆	⁷⁄₁₆	8.12	7.200	7¼	0.920	¹⁵⁄₁₆	1¾	¹⁵⁄₁₆	1
x43	12.7	10.150	10⅛	0.660	¹¹⁄₁₆	⅜	6.70	7.060	7	0.920	¹⁵⁄₁₆	1¾	¹⁵⁄₁₆	1
ST 10 x37.5	11.0	10.000	10	0.635	⅝	⁵⁄₁₆	6.35	6.385	6⅜	0.795	¹³⁄₁₆	1⅝	¹³⁄₁₆	⅞
x33	9.70	10.000	10	0.505	½	¼	5.05	6.225	6¼	0.795	¹³⁄₁₆	1⅝	¹³⁄₁₆	⅞
ST 9 x35	10.3	9.000	9	0.711	¹¹⁄₁₆	⅜	6.40	6.251	6¼	0.691	¹¹⁄₁₆	1½	¹¹⁄₁₆	⅞
x27.35	8.04	9.000	9	0.461	⁷⁄₁₆	¼	4.15	6.001	6	0.691	¹¹⁄₁₆	1½	¹¹⁄₁₆	⅞
ST 7.5x25	7.35	7.500	7½	0.550	⁹⁄₁₆	⁵⁄₁₆	4.13	5.640	5⅝	0.622	⅝	1⅜	⁹⁄₁₆	¾
x21.45	6.31	7.500	7½	0.411	⁷⁄₁₆	¼	3.08	5.501	5½	0.622	⅝	1⅜	⁹⁄₁₆	¾
ST 6 x25	7.35	6.000	6	0.687	¹¹⁄₁₆	⅜	4.12	5.477	5½	0.659	¹¹⁄₁₆	1⁷⁄₁₆	¹¹⁄₁₆	¾
x20.4	6.00	6.000	6	0.462	⁷⁄₁₆	¼	2.77	5.252	5¼	0.659	¹¹⁄₁₆	1⁷⁄₁₆	⅝	¾
ST 6 x17.5	5.15	6.000	6	0.428	⁷⁄₁₆	¼	2.57	5.078	5⅛	0.545	⁹⁄₁₆	1³⁄₁₆	½	¾
x15.9	4.68	6.000	6	0.350	⅜	³⁄₁₆	2.10	5.000	5	0.544	⁹⁄₁₆	1³⁄₁₆	½	¾
ST 5 x17.5	5.15	5.000	5	0.594	⅝	⁵⁄₁₆	2.97	4.944	5	0.491	½	1⅛	½	¾
x12.7	3.73	5.000	5	0.311	⁵⁄₁₆	³⁄₁₆	1.55	4.661	4⅝	0.491	½	1⅛	½	¾
ST 4 x11.5	3.38	4.000	4	0.441	⁷⁄₁₆	¼	1.76	4.171	4⅛	0.425	⁷⁄₁₆	1	⁷⁄₁₆	¾
x 9.2	2.70	4.000	4	0.271	¼	⅛	1.08	4.001	4	0.425	⁷⁄₁₆	1	⁷⁄₁₆	¾
ST 3.5x10	2.94	3.500	3½	0.450	⁷⁄₁₆	¼	1.57	3.860	3⅞	0.392	⅜	¹⁵⁄₁₆	⅜	⅝
x 7.65	2.25	3.500	3½	0.252	¼	⅛	0.882	3.662	3⅝	0.392	⅜	¹⁵⁄₁₆	⅜	⅝
ST 3 x 8.625	2.53	3.000	3	0.465	⁷⁄₁₆	¼	1.39	3.565	3⅝	0.359	⅜	⅞	⅜	⅝
x 6.25	1.83	3.000	3	0.232	¼	⅛	0.696	3.332	3⅜	0.359	⅜	⅞	⅜	—
ST 2.5x 7.375	2.17	2.500	2½	0.494	¼	¼	1.23	3.284	3¼	0.326	⁵⁄₁₆	¹³⁄₁₆	⁵⁄₁₆	—
x 5	1.47	2.500	2½	0.214	³⁄₁₆	⅛	0.535	3.004	3	0.326	⁵⁄₁₆	¹³⁄₁₆	⁵⁄₁₆	—
ST 2 x 4.75	1.40	2.000	2	0.326	⁵⁄₁₆	³⁄₁₆	0.652	2.796	2¾	0.293	⁵⁄₁₆	¾	⁵⁄₁₆	—
x 3.85	1.13	2.000	2	0.193	³⁄₁₆	⅛	0.386	2.663	2⅝	0.293	⁵⁄₁₆	¾	⁵⁄₁₆	—
ST 1.5x 3.75	1.10	1.500	1½	0.349	⅜	³⁄₁₆	0.523	2.509	2½	0.260	¼	¹¹⁄₁₆	¼	—
x 2.85	0.835	1.500	1½	0.170	³⁄₁₆	⅛	0.255	2.330	2⅜	0.260	¼	¹¹⁄₁₆	¼	—

STRUCTURAL TEES
Cut from S shapes
Properties

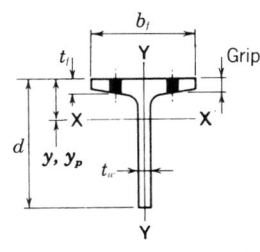

Nom-inal Wt. per Ft	$\dfrac{h_c}{t_w}$	Axis X-X						Axis Y-Y				Q_s*	
		I	S	r	y	Z	y_p	I	S	r	Z	F_y, ksi	
Lb.		In.^{4}	In.^{3}	In.	In.	In.^{3}	In.	In.^{4}	In.^{3}	In.	In.^{3}	36	50
60.5	13.2	259	30.1	3.82	3.63	54.5	1.28	41.7	10.4	1.53	18.1	—	—
53	17.0	216	24.1	3.72	3.28	43.3	1.03	38.5	9.80	1.57	16.6	—	0.907
50	14.1	215	26.3	3.83	3.84	47.5	2.20	23.8	6.58	1.27	12.0	—	—
45	16.8	190	22.6	3.79	3.60	41.1	1.48	22.5	6.31	1.30	11.2	—	0.937
40	21.1	162	18.7	3.72	3.29	33.6	0.922	21.1	6.04	1.34	10.4	0.878	0.695
48	10.8	143	20.3	3.18	3.13	36.9	1.40	25.1	6.97	1.33	12.5	—	—
43	13.1	125	17.2	3.14	2.91	31.1	0.985	23.4	6.63	1.36	11.6	—	—
37.5	13.6	109	15.8	3.15	3.07	28.6	1.40	14.9	4.66	1.16	8.37	—	—
33	17.0	93.1	12.9	3.10	2.81	23.4	0.855	13.8	4.43	1.19	7.70	—	0.907
35	10.9	84.7	14.0	2.87	2.94	25.1	1.81	12.1	3.86	1.08	7.21	—	—
27.35	16.8	62.4	9.61	2.79	2.50	17.3	0.747	10.4	3.47	1.14	6.07	—	0.922
25	11.6	40.6	7.73	2.35	2.25	14.0	0.872	7.85	2.78	1.03	5.01	—	—
21.45	15.5	33.0	6.00	2.29	2.01	10.8	0.613	7.19	2.61	1.07	4.54	—	0.988
25	7.0	25.2	6.05	1.85	1.84	11.0	0.770	7.85	2.87	1.03	5.19	—	—
20.4	10.3	18.9	4.28	1.78	1.58	7.71	0.581	6.78	2.58	1.06	4.45	—	—
17.5	11.7	17.2	3.95	1.83	1.64	7.12	0.548	4.94	1.95	0.980	3.41	—	—
15.9	14.3	14.9	3.31	1.78	1.51	5.94	0.485	4.68	1.87	1.00	3.22	—	—
17.5	6.9	12.5	3.63	1.56	1.56	6.58	0.702	4.18	1.69	0.901	3.11	—	—
12.7	13.2	7.83	2.06	1.45	1.20	3.70	0.408	3.39	1.46	0.954	2.49	—	—
11.5	7.3	5.03	1.77	1.22	1.15	3.19	0.447	2.15	1.03	0.798	1.84	—	—
9.2	11.8	3.51	1.15	1.14	0.941	2.07	0.341	1.86	0.932	0.831	1.59	—	—
10	6.1	3.36	1.36	1.07	1.04	2.47	0.425	1.59	0.821	0.734	1.48	—	—
7.65	10.9	2.19	0.816	0.987	0.817	1.48	0.309	1.32	0.720	0.766	1.23	—	—
8.625	5.0	2.13	1.02	0.917	0.914	1.85	0.401	1.15	0.648	0.675	1.18	—	—
6.25	10.0	1.27	0.552	0.833	0.691	1.01	0.275	0.911	0.547	0.705	0.929	—	—
7.375	3.8	1.27	0.740	0.764	0.789	1.34	0.379	0.833	0.507	0.620	0.936	—	—
5	8.7	0.681	0.353	0.681	0.569	0.650	0.243	0.608	0.405	0.643	0.685	—	—
4.75	4.3	0.470	0.325	0.580	0.553	0.592	0.255	0.451	0.323	0.569	0.566	—	—
3.85	7.3	0.316	0.203	0.528	0.448	0.381	0.209	0.382	0.287	0.581	0.483	—	—
3.75	2.8	0.204	0.191	0.430	0.432	0.351	0.223	0.293	0.234	0.516	0.412	—	—
2.85	5.7	0.118	0.101	0.376	0.329	0.196	0.175	0.227	0.195	0.522	0.327	—	—

*Where no value of Q_s is shown, the Tee complies with LRFD Specification Sect. E2.

AMERICAN INSTITUTE OF STEEL CONSTRUCTION

<u>Notes</u>

DOUBLE ANGLES
Properties of Sections

Properties of double angles in contact and separated are listed in the following tables. Each table shows properties of double angles in contact, and the radius of gyration about the Y-Y axis when the legs of the angles are separated. Values of Q_s are given for $F_y = 36$ ksi and $F_y = 50$ ksi for those angles exceeding the width-thickness ratio λ_r of AISC LRFD Specification B5. Since the cross section is comprised entirely of unstiffened elements, $Q_a = 1.0$ and $Q = Q_s$, for all angle sections. The torsion properties table also lists the dimensional values (r_o and H) needed for checking torsional and flexural-torsional buckling.

USE OF TABLE

The table may be used as follows for checking the limit states of (1) flexural buckling and (2) torsional or flexural-torsional buckling. The lower of the two limit states must be used for design.

(1) Flexural Buckling

While no value of Q_s is shown, the design compressive strength for this limit state is given by AISC LRFD Specification Section E2. Where a value of Q_s is shown, the strength must be reduced in accordance with Appendix B5 if $\lambda_c \sqrt{Q} \le \sqrt{2}$.

(2) Torsional or Flexural-Torsional Buckling

The design compressive strength for this limit state is given by AISC LRFD Specification Appendix E3. This involves calculations with J and C_w. These torsional constants are obtained by summing the respective values for single angles listed with the angle properties in Part 1 of this Manual.

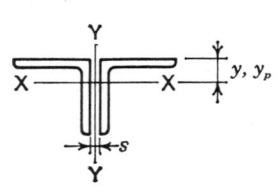

DOUBLE ANGLES
Two equal leg angles
Properties of sections

Designation	Wt. per Ft 2 Angles	Area of 2 Angles	Axis X-X					
			I	S	r	y	Z	y_p
	Lb.	In.^{2}	In.^{4}	In.^{3}	In.	In.	In.^{3}	In.
L 8x8x1⅛	113.8	33.5	195.0	35.1	2.42	2.41	63.2	1.05
1	102.0	30.0	177.0	31.6	2.44	2.37	56.9	0.938
⅞	90.0	26.5	159.0	28.0	2.45	2.32	50.5	0.827
¾	77.8	22.9	139.0	24.4	2.47	2.28	43.9	0.715
⅝	65.4	19.2	118.0	20.6	2.49	2.23	37.1	0.601
½	52.8	15.5	97.3	16.7	2.50	2.19	30.1	0.484
L 6x6x1	74.8	22.0	70.9	17.1	1.80	1.86	30.9	0.917
⅞	66.2	19.5	63.8	15.3	1.81	1.82	27.5	0.811
¾	57.4	16.9	56.3	13.3	1.83	1.78	24.0	0.703
⅝	48.4	14.2	48.3	11.3	1.84	1.73	20.4	0.592
½	39.2	11.5	39.8	9.23	1.86	1.68	16.6	0.479
⅜	29.8	8.72	30.8	7.06	1.88	1.64	12.7	0.363
L 5x5x ⅞	54.4	16.0	35.5	10.3	1.49	1.57	18.7	0.798
¾	47.2	13.9	31.5	9.06	1.51	1.52	16.3	0.694
½	32.4	9.50	22.5	6.31	1.54	1.43	11.4	0.475
⅜	24.6	7.22	17.5	4.84	1.56	1.39	8.72	0.361
⁵⁄₁₆	20.6	6.05	14.8	4.08	1.57	1.37	7.35	0.303
L 4x4x ¾	37.0	10.9	15.3	5.62	1.19	1.27	10.1	0.680
⅝	31.4	9.22	13.3	4.80	1.20	1.23	8.66	0.576
½	25.6	7.50	11.1	3.95	1.22	1.18	7.12	0.469
⅜	19.6	5.72	8.72	3.05	1.23	1.14	5.49	0.357
⁵⁄₁₆	16.4	4.80	7.43	2.58	1.24	1.12	4.64	0.300
¼	13.2	3.88	6.08	2.09	1.25	1.09	3.77	0.242

DOUBLE ANGLES
Two equal leg angles
Properties of sections

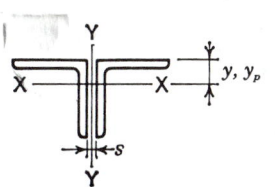

Designation	Axis Y-Y			Q_s*			
	Radii of Gyration			Angles in Contact		Angles Separated	
	Back to Back of Angles, In.			$F_y =$ 36 ksi	$F_y =$ 50 ksi	$F_y =$ 36 ksi	$F_y =$ 50 ksi
	0	⅜	¾				
L 8x8x1⅛	3.42	3.55	3.69	—	—	—	—
1	3.40	3.53	3.67	—	—	—	—
⅞	3.38	3.51	3.64	—	—	—	—
¾	3.36	3.49	3.62	—	—	—	—
⅝	3.34	3.47	3.60	—	—	.997	.935
½	3.32	3.45	3.58	.995	.921	.911	.834
L 6x6x1	2.59	2.73	2.87	—	—	—	—
⅞	2.57	2.70	2.85	—	—	—	—
¾	2.55	2.68	2.82	—	—	—	—
⅝	2.53	2.66	2.80	—	—	—	—
½	2.51	2.64	2.78	—	—	—	.961
⅜	2.49	2.62	2.75	.995	.921	.911	.834
L 5x5x ⅞	2.16	2.30	2.45	—	—	—	—
¾	2.14	2.28	2.42	—	—	—	—
½	2.10	2.24	2.38	—	—	—	—
⅜	2.09	2.22	2.35	—	—	.982	.919
⁵⁄₁₆	2.08	2.21	2.34	.995	.921	.911	.834
L 4x4x ¾	1.74	1.88	2.03	—	—	—	—
⅝	1.72	1.86	2.00	—	—	—	—
½	1.70	1.83	1.98	—	—	—	—
⅜	1.68	1.81	1.95	—	—	—	—
⁵⁄₁₆	1.67	1.80	1.94	—	—	.997	.935
¼	1.66	1.79	1.93	.995	.921	.911	.834

*Where no value of Q_s is shown, the angles comply with LRFD Specification Sect. E2.

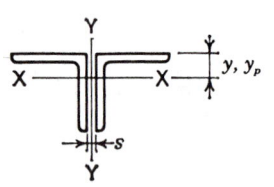

DOUBLE ANGLES
Two equal leg angles
Properties of sections

Designation	Wt. per Ft 2 Angles	Area of 2 Angles	Axis X-X					
			I	S	r	y	Z	y_p
	Lb.	In.^{2}	In.^{4}	In.^{3}	In.	In.	In.^{3}	In.
L 3½x3½x⅜	17.0	4.97	5.73	2.30	1.07	1.01	4.15	0.355
5/16	14.4	4.18	4.90	1.95	1.08	.990	3.52	0.299
¼	11.6	3.38	4.02	1.59	1.09	.968	2.86	0.241
L 3 x3 x½	18.8	5.50	4.43	2.14	.898	.932	3.87	0.458
⅜	14.4	4.22	3.52	1.67	.913	.888	3.00	0.352
5/16	12.2	3.55	3.02	1.41	.922	.865	2.55	0.296
¼	9.8	2.88	2.49	1.15	.930	.842	2.08	0.240
3/16	7.4	2.18	1.92	.882	.939	.820	1.59	0.182
L 2½x2½x⅜	11.8	3.47	1.97	1.13	.753	.762	2.04	0.347
5/16	10.0	2.93	1.70	.964	.761	.740	1.74	0.293
¼	8.2	2.38	1.41	.789	.769	.717	1.42	0.238
3/16	6.1	1.80	1.09	.606	.778	.694	1.09	0.180
L 2 x2 x⅜	9.4	2.72	.958	.702	.594	.636	1.27	0.340
5/16	7.8	2.30	.832	.681	.601	.614	1.08	0.288
¼	6.4	1.88	.695	.494	.609	.592	0.890	0.234
3/16	4.9	1.43	.545	.381	.617	.569	0.686	0.179
⅛	3.30	.960	.380	.261	.626	.546	0.471	0.121

DOUBLE ANGLES
Two equal leg angles
Properties of sections

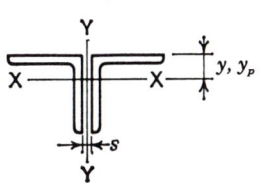

Designation	Axis Y-Y			Q_s*			
	Radii of Gyration			Angles in Contact		Angles Separated	
	Back to Back of Angles, In.			$F_y =$ 36 ksi	$F_y =$ 50 ksi	$F_y =$ 36 ksi	$F_y =$ 50 ksi
	0	3/8	3/4				
L 3½x3½x3/8	1.48	1.61	1.75	—	—	—	—
5/16	1.47	1.60	1.74	—	—	—	.986
1/4	1.46	1.59	1.73	—	.982	.965	.897
L 3 x3 x1/2	1.29	1.43	1.59	—	—	—	—
3/8	1.27	1.41	1.56	—	—	—	—
5/16	1.26	1.40	1.55	—	—	—	—
1/4	1.26	1.39	1.53	—	—	—	.961
3/16	1.25	1.38	1.52	.995	.921	.911	.834
L 2½x2½x3/8	1.07	1.21	1.36	—	—	—	—
5/16	1.06	1.20	1.35	—	—	—	—
1/4	1.05	1.19	1.34	—	—	—	—
3/16	1.04	1.18	1.32	—	—	.982	.919
L 2 x2 x3/8	.870	1.01	1.17	—	—	—	—
5/16	.859	1.00	1.16	—	—	—	—
1/4	.849	.989	1.14	—	—	—	—
3/16	.840	.977	1.13	—	—	—	—
1/8	.831	.965	1.11	.995	.921	.911	.834

Where no value of Q_s is shown, the angles comply with LRFD Specification Sect. E2.

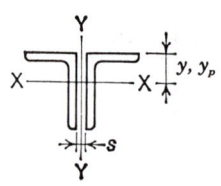

DOUBLE ANGLES
Two unequal leg angles
Properties of sections
Long legs back to back

Designation	Wt. per Ft 2 Angles	Area of 2 Angles	Axis X-X					
			I	S	r	y	Z	y_p
	Lb.	In.^{2}	In.^{4}	In.^{3}	In.	In.	In.^{3}	In.
L 8x6 x1	88.4	26.0	161.0	30.2	2.49	2.65	54.5	1.50
3/4	67.6	19.9	126.0	23.3	2.53	2.56	42.2	1.38
1/2	46.0	13.5	88.6	16.0	2.56	2.47	29.1	1.25
L 8x4 x1	74.8	22.0	139.0	28.1	2.52	3.05	48.5	2.50
3/4	57.4	16.9	109.0	21.8	2.55	2.95	37.7	2.38
1/2	39.2	11.5	77.0	15.0	2.59	2.86	26.1	2.25
L 7x4 x 3/4	52.4	15.4	75.6	16.8	2.22	2.51	29.6	1.88
1/2	35.8	10.5	53.3	11.6	2.25	2.42	20.6	1.75
3/8	27.2	7.97	41.1	8.88	2.27	2.37	15.7	1.69
L 6x4 x 3/4	47.2	13.9	49.0	12.5	1.88	2.08	22.3	1.38
5/8	40.0	11.7	42.1	10.6	1.90	2.03	19.0	1.31
1/2	32.4	9.50	34.8	8.67	1.91	1.99	15.6	1.25
3/8	24.6	7.22	26.9	6.64	1.93	1.94	11.9	1.19
L 6x3½x 3/8	23.4	6.84	25.7	6.49	1.94	2.04	11.5	1.44
5/16	19.6	5.74	21.8	5.47	1.95	2.01	9.70	1.41
L 5x3½x 3/4	39.6	11.6	27.8	8.55	1.55	1.75	15.3	1.13
1/2	27.2	8.00	20.0	5.97	1.58	1.66	10.8	1.00
3/8	20.8	6.09	15.6	4.59	1.60	1.61	8.28	0.938
5/16	17.4	5.12	13.2	3.87	1.61	1.59	6.99	0.906
L 5x3 x 1/2	25.6	7.50	18.9	5.82	1.59	1.75	10.3	1.25
3/8	19.6	5.72	14.7	4.47	1.61	1.70	7.95	1.19
5/16	16.4	4.80	12.5	3.77	1.61	1.68	6.71	1.16
1/4	13.2	3.88	10.2	3.06	1.62	1.66	5.45	1.13

DOUBLE ANGLES
Two unequal leg angles
Properties of sections
Long legs back to back

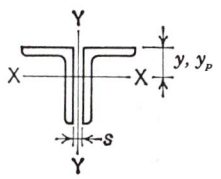

Designation	Axis Y-Y			Q_s*			
	Radii of Gyration Back to Back of Angles, In.			Angles in Contact		Angles Separated	
				$F_y =$ 36 ksi	$F_y =$ 50 ksi	$F_y =$ 36 ksi	$F_y =$ 50 ksi
	0	⅜	¾				
L 8x6 x1	2.39	2.52	2.66	—	—	—	—
¾	2.35	2.48	2.62	—	—	—	—
½	2.32	2.44	2.57	—	—	.911	.834
L 8x4 x1	1.47	1.61	1.75	—	—	—	—
¾	1.42	1.55	1.69	—	—	—	—
½	1.38	1.51	1.64	—	—	.911	.834
L 7x4 x ¾	1.48	1.62	1.76	—	—	—	—
½	1.44	1.57	1.71	—	—	.965	.897
⅜	1.43	1.55	1.68	—	—	.839	.750
L 6x4 x ¾	1.55	1.69	1.83	—	—	—	—
⅝	1.53	1.67	1.81	—	—	—	—
½	1.51	1.64	1.78	—	—	—	.961
⅜	1.50	1.62	1.76	—	—	.911	.834
L 6x3½x ⅜	1.26	1.39	1.53	—	—	.911	.834
⁵⁄₁₆	1.26	1.38	1.51	—	—	.825	.733
L 5x3½x ¾	1.40	1.53	1.68	—	—	—	—
½	1.35	1.49	1.63	—	—	—	—
⅜	1.34	1.46	1.60	—	—	.982	.919
⁵⁄₁₆	1.33	1.45	1.59	—	—	.911	.834
L 5x3 x ½	1.12	1.25	1.40	—	—	—	—
⅜	1.10	1.23	1.37	—	—	.982	.919
⁵⁄₁₆	1.09	1.22	1.36	—	—	.911	.834
¼	1.08	1.21	1.34	—	—	.804	.708

*Where no value of Q_s is shown, the angles comply with LRFD Specification Sect. E2.

AMERICAN INSTITUTE OF STEEL CONSTRUCTION

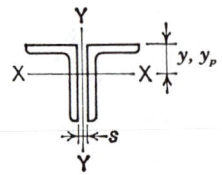

DOUBLE ANGLES
Two unequal leg angles
Properties of sections
Long legs back to back

| Designation | Wt. per Ft 2 Angles | Area of 2 Angles | Axis X-X | | | | | |
			I	S	r	y	Z	y_p
	Lb.	In.^{2}	In.^{4}	In.^{3}	In.	In.	In.^{3}	In.
L 4 x3½x½	23.8	7.00	10.6	3.87	1.23	1.25	7.00	0.500
⅜	18.2	5.34	8.35	2.99	1.25	1.21	5.42	0.438
⁵⁄₁₆	15.4	4.49	7.12	2.53	1.26	1.18	4.59	0.406
¼	12.4	3.63	5.83	2.05	1.27	1.16	3.73	0.375
L 4 x3 x½	22.2	6.50	10.1	3.78	1.25	1.33	6.81	0.750
⅜	17.0	4.97	7.93	2.92	1.26	1.28	5.28	0.688
⁵⁄₁₆	14.4	4.18	6.76	2.47	1.27	1.26	4.47	0.656
¼	11.6	3.38	5.54	2.00	1.28	1.24	3.63	0.625
L 3½x3 x⅜	15.8	4.59	5.45	2.25	1.09	1.08	4.08	0.438
⁵⁄₁₆	13.2	3.87	4.66	1.91	1.10	1.06	3.46	0.406
¼	10.8	3.13	3.83	1.55	1.11	1.04	2.82	0.375
L 3½x2½x⅜	14.4	4.22	5.12	2.19	1.10	1.16	3.94	0.688
⁵⁄₁₆	12.2	3.55	4.38	1.85	1.11	1.14	3.35	0.656
¼	9.8	2.88	3.60	1.51	1.12	1.11	2.73	0.625
L 3 x2½x⅜	13.2	3.84	3.31	1.62	.928	.956	2.93	0.438
¼	9.0	2.63	2.35	1.12	.945	.911	2.04	0.375
³⁄₁₆	6.8	1.99	1.81	.859	.954	.888	1.56	0.344
L 3 x2 x⅜	11.8	3.47	3.06	1.56	.940	1.04	2.79	0.688
⁵⁄₁₆	10.0	2.93	2.63	1.33	.948	1.02	2.38	0.656
¼	8.2	2.38	2.17	1.08	.957	.993	1.95	0.625
³⁄₁₆	6.1	1.80	1.68	.830	.966	.970	1.49	0.594
L 2½x2 x⅜	10.6	3.09	1.82	1.09	.768	.831	1.97	0.438
⁵⁄₁₆	9.0	2.62	1.58	.932	.776	.809	1.69	0.406
¼	7.2	2.13	1.31	.763	.784	.787	1.38	0.375
³⁄₁₆	5.5	1.62	1.02	.586	.793	.764	1.06	0.344

DOUBLE ANGLES
Two unequal leg angles
Properties of sections
Long legs back to back

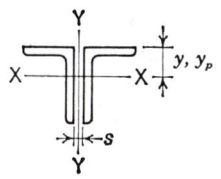

Designation	Axis Y-Y			Q_s*			
	Radii of Gyration Back to Back of Angles, In.			Angles in Contact		Angles Separated	
				$F_y =$ 36 ksi	$F_y =$ 50 ksi	$F_y =$ 36 ksi	$F_y =$ 50 ksi
	0	⅜	¾				
L 4 x3½x½	1.44	1.58	1.72	—	—	—	—
⅜	1.42	1.56	1.70	—	—	—	—
⁵⁄₁₆	1.42	1.55	1.69	—	—	.997	.935
¼	1.41	1.54	1.67	—	.982	.911	.834
L 4 x3 x½	1.20	1.33	1.48	—	—	—	—
⅜	1.18	1.31	1.45	—	—	—	—
⁵⁄₁₆	1.17	1.30	1.44	—	—	.997	.935
¼	1.16	1.29	1.43	—	—	.911	.834
L 3½x3 x⅜	1.22	1.36	1.50	—	—	—	—
⁵⁄₁₆	1.21	1.35	1.49	—	—	—	.986
¼	1.20	1.33	1.48	—	—	.965	.897
L 3½x2½x⅜	0.976	1.11	1.26	—	—	—	—
⁵⁄₁₆	0.966	1.10	1.25	—	—	—	.986
¼	0.958	1.09	1.23	—	—	.965	.897
L 3 x2½x⅜	1.02	1.16	1.31	—	—	—	—
¼	1.00	1.13	1.28	—	—	—	.961
³⁄₁₆	.993	1.12	1.27	—	—	.911	.834
L 3 x2 x⅜	.777	.917	1.07	—	—	—	—
⁵⁄₁₆	.767	.903	1.06	—	—	—	—
¼	.757	.891	1.04	—	—	—	.961
³⁄₁₆	.749	.879	1.03	—	—	.911	.834
L 2½x2 x⅜	.819	.961	1.12	—	—	—	—
⁵⁄₁₆	.809	.948	1.10	—	—	—	—
¼	.799	.935	1.09	—	—	—	—
³⁄₁₆	.790	.923	1.07	—	—	.982	.919

*Where no value of Q_s is shown, the angles comply with LRFD Specification Sect. E2.

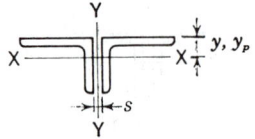

DOUBLE ANGLES
Two unequal leg angles
Properties of sections
Short legs back to back

Designation	Wt. per Ft 2 Angles	Area of 2 Angles	Axis X-X					
			I	S	r	y	Z	y_p
	Lb.	In.²	In.⁴	In.³	In.	In.	In.³	In.
L 8x6 x1	88.4	26.0	77.6	17.8	1.73	1.65	32.4	0.813
¾	67.6	19.9	61.4	13.8	1.76	1.56	24.9	0.621
½	46.0	13.5	43.4	9.58	1.79	1.47	17.0	0.422
L 8x4 x1	74.8	22.0	23.3	7.88	1.03	1.05	15.4	0.688
¾	57.4	16.9	18.7	6.14	1.05	.953	11.6	0.527
½	39.2	11.5	13.5	4.29	1.08	.859	7.81	0.359
L 7x4 x ¾	52.4	15.4	18.1	6.05	1.09	1.01	11.3	0.549
½	35.8	10.5	13.1	4.23	1.11	.917	7.66	0.375
⅜	27.2	7.97	10.2	3.26	1.13	.870	5.80	0.285
L 6x4 x ¾	47.2	13.9	17.4	5.94	1.12	1.08	10.9	0.578
⅝	40.0	11.7	15.0	5.07	1.13	1.03	9.24	0.488
½	32.4	9.50	12.5	4.16	1.15	.987	7.50	0.396
⅜	24.6	7.22	9.81	3.21	1.17	.941	5.71	0.301
L 6x3½x ⅜	23.4	6.84	6.68	2.46	.988	.787	4.41	0.285
⁵⁄₁₆	19.6	5.74	5.70	2.08	.996	.763	3.70	0.239
L 5x3½x ¾	39.6	11.6	11.1	4.43	.977	.996	8.20	0.581
½	27.2	8.00	8.10	3.12	1.01	.906	5.65	0.400
⅜	20.8	6.09	6.37	2.41	1.02	.861	4.32	0.305
⁵⁄₁₆	17.4	5.12	5.44	2.04	1.03	.838	3.63	0.256
L 5x3 x ½	25.6	7.50	5.16	2.29	.829	.750	4.22	0.375
⅜	19.6	5.72	4.08	1.78	.845	.704	3.21	0.286
⁵⁄₁₆	16.4	4.80	3.49	1.51	.853	.681	2.69	0.240
¼	13.2	3.88	2.88	1.23	.861	.657	2.17	0.194

DOUBLE ANGLES
Two unequal leg angles
Properties of sections
Short legs back to back

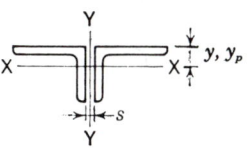

Designation	Axis Y-Y			Q_s*			
	Radii of Gyration			Angles in Contact		Angles Separated	
	Back to Back of Angles, In.			$F_y =$ 36 ksi	$F_y =$ 50 ksi	$F_y =$ 36 ksi	$F_y =$ 50 ksi
	0	⅜	¾				
L 8x6 x1	3.64	3.78	3.92	—	—	—	—
¾	3.60	3.74	3.88	—	—	—	—
½	3.56	3.69	3.83	.995	.921	.911	.834
L 8x4 x1	3.95	4.10	4.25	—	—	—	—
¾	3.90	4.05	4.19	—	—	—	—
½	3.86	4.00	4.14	.995	.921	.911	.834
L 7x4 x ¾	3.35	3.49	3.64	—	—	—	—
½	3.30	3.44	3.59	—	.982	.965	.897
⅜	3.28	3.42	3.56	.926	.838	.839	.750
L 6x4 x ¾	2.80	2.94	3.09	—	—	—	—
⅝	2.78	2.92	3.06	—	—	—	—
½	2.76	2.90	3.04	—	—	—	.961
⅜	2.74	2.87	3.02	.995	.921	.911	.834
L 6x3½x ⅜	2.81	2.95	3.09	.995	.921	.911	.834
⁵⁄₁₆	2.80	2.94	3.08	.912	.822	.825	.733
L 5x3½x ¾	2.33	2.48	2.63	—	—	—	—
½	2.29	2.43	2.57	—	—	—	—
⅜	2.27	2.41	2.55	—	—	.982	.919
⁵⁄₁₆	2.26	2.39	2.54	.995	.921	.911	.834
L 5x3 x ½	2.36	2.50	2.65	—	—	—	—
⅜	2.34	2.48	2.63	—	—	.982	.919
⁵⁄₁₆	2.33	2.47	2.61	.995	.921	.911	.834
¼	2.32	2.46	2.60	.891	.797	.804	.708

*Where no value of Q_s is shown, the angles comply with LRFD Specification Sect. E2.

AMERICAN INSTITUTE OF STEEL CONSTRUCTION

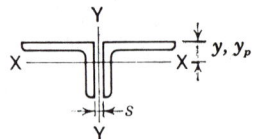

DOUBLE ANGLES
Two unequal leg angles
Properties of sections
Short legs back to back

Designation	Wt. per Ft 2 Angles	Area of 2 Angles	Axis X-X					
			I	S	r	y	Z	y_p
	Lb.	In.²	In.⁴	In.³	In.	In.	In.³	In.
L 4 x3½x½	23.8	7.00	7.58	3.03	1.04	1.00	5.47	0.438
⅜	18.2	5.34	5.97	2.35	1.06	.955	4.21	0.334
5⁄16	15.4	4.49	5.10	1.99	1.07	.932	3.56	0.281
¼	12.4	3.63	4.19	1.62	1.07	.909	2.89	0.227
L 4 x3 x½	22.2	6.50	4.85	2.23	.864	.827	4.06	0.406
⅜	17.0	4.97	3.84	1.73	.879	.782	3.11	0.311
5⁄16	14.4	4.18	3.29	1.47	.887	.759	2.63	0.261
¼	11.6	3.38	2.71	1.20	.896	.736	2.13	0.211
L 3½x3 x⅜	15.8	4.59	3.69	1.70	.897	.830	3.06	0.328
5⁄16	13.2	3.87	3.17	1.44	.905	.808	2.59	0.276
¼	10.8	3.13	2.61	1.18	.914	.785	2.10	0.223
L 3½x2½x⅜	14.4	4.22	2.18	1.18	.719	.660	2.15	0.301
5⁄16	12.2	3.55	1.88	1.01	.727	.637	1.81	0.254
¼	9.8	2.88	1.55	.824	.735	.614	1.47	0.205
L 3 x2½x⅜	13.2	3.84	2.08	1.16	.736	.706	2.10	0.320
¼	9.0	2.63	1.49	.808	.753	.661	1.45	0.219
3⁄16	6.8	1.99	1.15	.620	.761	.638	1.11	0.166
L 3 x2 x⅜	11.8	3.47	1.09	.743	.559	.539	1.37	0.289
5⁄16	10.0	2.93	.941	.634	.567	.516	1.16	0.244
¼	8.2	2.38	.784	.520	.574	.493	0.937	0.198
3⁄16	6.1	1.80	.613	.401	.583	.470	0.713	0.150
L 2½x2 x⅜	10.6	3.09	1.03	.725	.577	.581	1.32	0.309
5⁄16	9.0	2.62	.893	.620	.584	.559	1.12	0.262
¼	7.2	2.13	.745	.509	.592	.537	0.915	0.213
3⁄16	5.5	1.62	.583	.392	.600	.514	0.701	0.162

DOUBLE ANGLES
Two unequal leg angles
Properties of sections
Short legs back to back

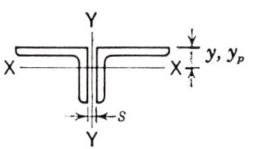

Designation	Axis Y-Y			Q_s*			
	Radii of Gyration Back to Back of Angles, In.			Angles in Contact		Angles Separated	
	0	⅜	¾	$F_y=$ 36 ksi	$F_y=$ 50 ksi	$F_y=$ 36 ksi	$F_y=$ 50 ksi
L 4 x3½x½	1.76	1.89	2.04	—	—	—	—
⅜	1.74	1.87	2.01	—	—	—	—
⁵⁄₁₆	1.73	1.86	2.00	—	—	.997	.935
¼	1.72	1.85	1.99	.995	.921	.911	.834
L 4 x3 x½	1.82	1.96	2.11	—	—	—	—
⅜	1.80	1.94	2.08	—	—	—	—
⁵⁄₁₆	1.79	1.93	2.07	—	—	.997	.935
¼	1.78	1.92	2.06	.995	.921	.911	.834
L 3½x3 x⅜	1.53	1.67	1.82	—	—	—	—
⁵⁄₁₆	1.52	1.66	1.80	—	—	—	.986
¼	1.52	1.65	1.79	—	.982	.965	.897
L 3½x2½x⅜	1.60	1.74	1.89	—	—	—	—
⁵⁄₁₆	1.59	1.73	1.88	—	—	—	.986
¼	1.58	1.72	1.86	—	.982	.965	.897
L 3 x2½x⅜	1.33	1.47	1.62	—	—	—	—
¼	1.31	1.45	1.60	—	—	—	.961
³⁄₁₆	1.30	1.44	1.58	.995	.921	.911	.834
L 3 x2 x⅜	1.40	1.55	1.70	—	—	—	—
⁵⁄₁₆	1.39	1.53	1.68	—	—	—	—
¼	1.38	1.52	1.67	—	—	—	.961
³⁄₁₆	1.37	1.51	1.66	.995	.921	.911	.834
L 2½x2 x⅜	1.13	1.28	1.43	—	—	—	—
⁵⁄₁₆	1.12	1.26	1.42	—	—	—	—
¼	1.11	1.25	1.40	—	—	—	—
³⁄₁₆	1.10	1.24	1.39	—	—	.982	.919

*Where no value of Q_s is shown, the angles comply with LRFD Specification Sect. E2.

Notes

COMBINATION SECTIONS

Standard rolled shapes are frequently combined to produce efficient and economical structural members for special applications. Experience has established a demand for certain combinations. When properly sized and connected to satisfy the design and specification criteria, these members may be used as struts, lintels, eave struts and light crane and trolley runways. The W section with channel attached to the web is not recommended for use as a trolley or crane runway member.

Properties of several combined sections are tabulated for those combinations that experience has proven to be in popular demand.

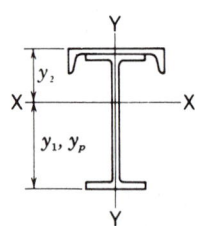

COMBINATION SECTIONS
W shapes and channels
Properties of sections

Beam	Channel	Total Weight per Ft	Total Area	Axis X-X			
				I	$S_1 = I/y_1$	$S_2 = I/y_2$	r
		Lb.	In.^{2}	In.^{4}	In.^{3}	In.^{3}	In.
W 12x 26	C 10x15.3	41.3	12.14	299	36.3	70.5	4.96
x 26	C 12x20.7	46.7	13.74	318	36.8	82.2	4.81
W 14x 30	C 10x15.3	45.3	13.34	420	46.1	84.6	5.61
x 30	C 12x20.7	50.7	14.94	448	46.8	98.3	5.47
W 16x 36	C 12x20.7	56.7	16.69	670	62.8	123	6.34
x 36	C 15x33.9	69.9	20.56	748	64.6	160	6.03
W 18x 50	C 12x20.7	70.7	20.79	1120	97.4	166	7.34
x 50	C 15x33.9	83.9	24.66	1250	100	211	7.11
W 21x 62	C 12x20.7	82.7	24.39	1800	138	218	8.59
x 62	C 15x33.9	95.9	28.26	2000	142	272	8.41
x 68	C 12x20.7	88.7	26.09	1960	152	232	8.68
x 68	C 15x33.9	101.9	29.96	2180	156	287	8.52
W 24x 68	C 12x20.7	88.7	26.19	2450	168	258	9.67
x 68	C 15x33.9	101.9	30.06	2720	173	321	9.50
x 84	C 12x20.7	104.7	30.79	3040	212	303	9.93
x 84	C 15x33.9	117.9	34.66	3340	217	368	9.82
W 27x 84	C 15x33.9	117.9	34.76	4050	237	404	10.8
x 94	C 15x33.9	127.9	37.66	4530	268	436	11.0
W 30x 99	C 15x33.9	132.9	39.06	5540	300	480	11.9
x 99	C 18x42.7	141.7	41.70	5830	304	533	11.8
x116	C 15x33.9	149.9	44.16	6590	360	544	12.2
x116	C 18x42.7	158.7	46.80	6900	365	599	12.1
W 33x118	C 15x33.9	151.9	44.66	7900	395	596	13.3
x118	C 18x42.7	160.7	47.30	8280	400	656	13.2
x141	C 15x33.9	174.9	51.56	9580	484	689	13.6
x141	C 18x42.7	183.7	54.20	10000	490	751	13.6
W 36x150	C 15x33.9	183.9	54.16	11500	546	765	14.6
x150	C 18x42.7	192.7	56.80	12100	553	832	14.6

COMBINATION SECTIONS
W shapes and channels
Properties of sections

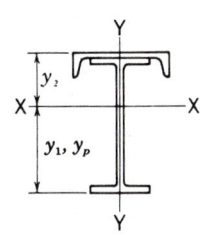

Beam	Channel	Axis X-X			Axis Y-Y			
		y_1	Z	y_p	I	S	r	Z
		In.	In.^{3}	In.	In.^{4}	In.^{3}	In.	In.^{3}
W 12x 26	C 10x15.3	8.22	47.0	11.30	84.7	16.9	2.64	24.0
x 26	C 12x20.7	8.63	48.8	11.55	146	24.4	3.26	33.6
W 14x 30	C 10x15.3	9.12	60.5	12.56	87.0	17.4	2.55	24.8
x 30	C 12x20.7	9.57	62.3	12.87	149	24.8	3.15	34.4
W 16x 36	C 12x20.7	10.67	83.6	14.56	154	25.6	3.03	36.3
x 36	C 15x33.9	11.58	88.6	15.21	340	45.3	4.06	61.3
W 18x 50	C 12x20.7	11.51	128	16.08	169	28.2	2.85	42.0
x 50	C 15x33.9	12.47	134	16.90	355	47.3	3.79	67.0
W 21x 62	C 12x20.7	13.01	182	18.06	187	31.1	2.77	47.2
x 62	C 15x33.9	14.06	190	19.36	373	49.7	3.63	72.2
x 68	C 12x20.7	12.93	200	17.60	194	32.3	2.72	49.8
x 68	C 15x33.9	13.95	208	19.32	380	50.6	3.56	74.8
W 24x 68	C 12x20.7	14.53	224	19.15	199	33.2	2.76	50.0
x 68	C 15x33.9	15.67	234	21.66	385	51.4	3.58	75.0
x 84	C 12x20.7	14.35	275	18.49	223	37.2	2.69	58.1
x 84	C 15x33.9	15.40	288	21.61	409	54.6	3.44	83.1
W 27x 84	C 15x33.9	17.07	320	23.86	421	56.1	3.48	83.6
x 94	C 15x33.9	16.92	357	23.56	439	58.5	3.41	89.2
W 30x 99	C 15x33.9	18.51	408	24.34	443	59.1	3.37	89.1
x 99	C 18x42.7	19.18	418	26.43	682	75.8	4.04	113
x116	C 15x33.9	18.30	480	23.77	479	63.9	3.29	99.6
x116	C 18x42.7	18.93	492	26.04	718	79.8	3.92	124
W 33x118	C 15x33.9	20.01	529	25.43	502	66.9	3.35	102
x118	C 18x42.7	20.69	544	27.77	741	82.3	3.96	126
x141	C 15x33.9	19.79	634	24.83	561	74.8	3.30	117
x141	C 18x42.7	20.42	652	26.96	800	88.9	3.84	141
W 36x150	C 15x33.9	21.15	716	25.84	585	78.0	3.29	121
x150	C 18x42.7	21.81	738	27.91	824	91.6	3.81	145

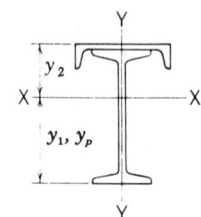

COMBINATION SECTIONS
S shapes and channels
Properties of sections

Beam	Channel	Total Weight per Ft	Total Area	Axis X-X			
				I	$S_1 = I/y_1$	$S_2 = I/y_2$	r
		Lb.	In.^{2}	In.^{4}	In.^{3}	In.^{3}	In.
S 10x25.4	C 8x11.5	36.9	10.84	176	27.2	46.6	4.02
	C 10x15.3	40.7	11.95	186	27.6	52.9	3.94
S 12x31.8	C 8x11.5	43.3	12.73	299	39.8	63.2	4.84
	C 10x15.3	47.1	13.84	316	40.4	71.4	4.78
S 15x42.9	C 8x11.5	54.4	15.98	585	64.9	94.2	6.05
	C 10x15.3	58.2	17.09	616	65.8	105	6.01
S 20x66	C 10x15.3	81.3	23.89	1530	130	181	8.00
	C 12x20.7	86.7	25.49	1620	132	203	7.97
S 24x80	C 10x15.3	95.3	27.99	2610	188	252	9.66
	C 12x20.7	100.7	29.59	2750	191	278	9.65

COMBINATION SECTIONS
S shapes and channels
Properties of sections

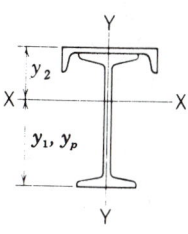

Beam	Channel	Axis X-X			Axis Y-Y			
		y_1	Z	y_p	I	S	r	Z
		In.	In.³	In.	In.⁴	In.³	In.	In.³
S 10x25.4	C 8x11.5	6.45	35.7	8.81	39.4	9.8	1.91	14.5
	C 10x15.3	6.73	36.9	9.02	74.2	14.8	2.49	20.8
S 12x31.8	C 8x11.5	7.50	52.6	10.30	42.0	10.5	1.82	16.0
	C 10x15.3	7.82	53.9	10.61	76.8	15.4	2.36	22.2
S 15x42.9	C 8x11.5	9.01	85.7	11.58	47.0	11.8	1.71	18.6
	C 10x15.3	9.37	88.2	12.77	81.8	16.4	2.19	24.9
S 20x66	C 10x15.3	11.81	171	14.41	95.1	19.0	2.00	31.2
	C 12x20.7	12.29	178	15.99	157	26.1	2.48	40.8
S 24x80	C 10x15.3	13.86	244	16.46	110	21.9	1.98	36.6
	C 12x20.7	14.38	254	18.05	171	28.5	2.41	46.2

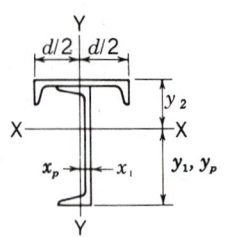

COMBINATION SECTIONS
Two channels
Properties of sections

Vertical Channel	Horizontal Channel	Total Weight per Ft	Total Area	Axis X-X						
				I	$S_1 = I/y_1$	$S_2 = I/y_2$	r	y_1	Z	y_p
		Lb.	In.^{2}	In.^{4}	In.^{3}	In.^{3}	In.	In.	In.^{3}	In.
C 3x 4.1	C 4x 5.4	9.5	2.80	3.0	1.4	3.0	1.04	2.20	2.16	2.67
C 4x 5.4	C 4x 5.4	10.8	3.18	6.5	2.3	4.9	1.43	2.86	3.39	3.56
	C 5x 6.7	12.1	3.56	6.9	2.3	5.5	1.39	2.94	3.62	3.61
C 5x 6.7	C 5x 6.7	13.4	3.94	12.8	3.5	8.0	1.80	3.60	5.23	4.50
	C 6x 8.2	14.9	4.37	13.4	3.6	8.9	1.75	3.70	5.50	4.57
	C 7x 9.8	16.5	4.84	14.0	3.7	9.8	1.70	3.79	5.81	4.62
C 6x 8.2	C 5x 6.7	14.9	4.37	21.5	5.1	10.9	2.22	4.22	7.31	5.37
	C 6x 8.2	16.4	4.80	22.5	5.2	12.1	2.16	4.34	7.61	5.45
	C 7x 9.8	18.0	5.27	23.4	5.2	13.3	2.11	4.45	7.93	5.53
	C 8x11.5	19.7	5.78	24.3	5.3	14.5	2.05	4.55	8.30	5.58
	C 9x13.4	21.6	6.34	25.2	5.4	15.8	1.99	4.64	8.72	5.63
	C 10x15.3	23.5	6.89	26.0	5.5	16.9	1.94	4.70	9.16	5.65
C 7x 9.8	C 6x 8.2	18.0	5.27	35.3	7.1	15.7	2.59	4.95	10.2	6.32
	C 7x 9.8	19.6	5.74	36.7	7.2	17.3	2.53	5.08	10.6	6.40
	C 8x11.5	21.3	6.25	38.0	7.3	18.8	2.47	5.20	10.9	6.48
	C 9x13.4	23.2	6.81	39.3	7.4	20.5	2.40	5.31	11.4	6.54
	C 10x15.3	25.1	7.36	40.5	7.5	21.9	2.34	5.39	11.8	6.58
C 8x11.5	C 6x 8.2	19.7	5.78	52.4	9.5	19.6	3.01	5.53	13.4	7.18
	C 7x 9.8	21.3	6.25	54.5	9.6	21.6	2.95	5.68	13.8	7.27
	C 8x11.5	23.0	6.76	56.4	9.7	23.6	2.89	5.82	14.2	7.35
	C 9x13.4	24.9	7.32	58.4	9.8	25.6	2.82	5.95	14.6	7.44
	C 10x15.3	26.8	7.87	60.0	9.9	27.5	2.76	6.06	15.1	7.49
	C 12x20.7	32.2	9.47	64.4	10.2	32.6	2.61	6.30	16.4	7.62

COMBINATION SECTIONS
Channels and angles
Properties of sections
Long leg of angle turned out

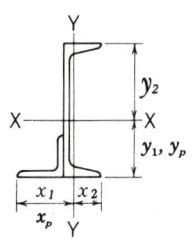

Channel	Angle	Axis Y-Y						
		I	$S_1 = I/x_1$	$S_2 = I/x_2$	r	x_1	Z	x_p
		In.^{4}	In.^{3}	In.^{3}	In.	In.	In.^{3}	In.
C 6x 8.2	L 2½x2½x¼	2.6	1.0	1.4	0.85	2.60	2.02	2.60
	L 3 x2½x¼	3.6	1.2	1.9	0.98	3.01	2.38	3.09
	L 3½x3 x¼	4.9	1.4	2.4	1.11	3.40	2.82	3.57
	x⁵⁄₁₆	5.7	1.7	2.7	1.14	3.31	3.27	3.54
	L 4 x3 x¼	6.5	1.7	3.1	1.26	3.79	3.30	4.06
C 7x 9.8	L 2½x2½x¼	3.0	1.1	1.6	0.86	2.67	2.31	2.62
	L 3 x2½x¼	4.0	1.3	2.0	0.98	3.09	2.66	3.11
	L 3½x3 x¼	5.4	1.6	2.6	1.10	3.48	3.11	3.59
	x⁵⁄₁₆	6.3	1.8	2.9	1.14	3.40	3.57	3.57
	L 4 x3 x¼	7.1	1.8	3.2	1.25	3.88	3.59	4.08
	x⁵⁄₁₆	8.3	2.2	3.6	1.29	3.78	4.16	4.05
C 8x11.5	L 3 x2½x¼	4.6	1.4	2.2	0.99	3.16	3.00	3.13
	L 3½x3 x¼	6.0	1.7	2.7	1.10	3.56	3.45	3.61
	x⁵⁄₁₆	6.9	2.0	3.0	1.14	3.48	3.91	3.59
	L 4 x3 x¼	7.8	2.0	3.4	1.24	3.97	3.93	4.10
	x⁵⁄₁₆	9.0	2.3	3.8	1.28	3.87	4.51	4.08
	L 5 x3½x⁵⁄₁₆	14.7	3.2	5.6	1.57	4.64	5.97	5.05
C 9x13.4	L 3 x2½x¼	5.2	1.6	2.3	0.99	3.22	3.38	3.14
	L 3½x3 x¼	6.7	1.8	2.9	1.10	3.64	3.83	3.63
	x⁵⁄₁₆	7.7	2.2	3.2	1.14	3.55	4.31	3.61
	L 4 x3 x¼	8.5	2.1	3.6	1.23	4.05	4.32	4.12
	x⁵⁄₁₆	9.9	2.5	4.0	1.28	3.96	4.91	4.10
	L 5 x3½x⁵⁄₁₆	15.8	3.3	5.9	1.56	4.74	6.38	5.08
C 10x15.3	L 3½x3 x¼	7.4	2.0	3.1	1.11	3.70	4.25	3.64
	x⁵⁄₁₆	8.5	2.3	3.4	1.15	3.62	4.73	3.63
	L 4 x3 x¼	9.4	2.3	3.8	1.23	4.12	4.74	4.14
	x⁵⁄₁₆	10.8	2.7	4.2	1.28	4.03	5.34	4.12
	L 5 x3½x⁵⁄₁₆	16.9	3.5	6.1	1.55	4.83	6.82	5.09
	x³⁄₈	19.2	4.1	6.7	1.60	4.73	7.70	5.07

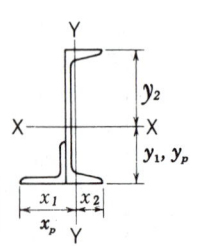

COMBINATION SECTIONS
Channels and angles
Properties of sections
Long leg of angle turned out

Channel	Angle	Total Weight per Ft	Total Area	Axis X-X						
				I	$S_1 = I/y_1$	$S_2 = I/y_2$	r	y_1	Z	y_p
		Lb.	In.^{2}	In.^{4}	In.^{3}	In.^{3}	In.	In.	In.^{3}	In.
C 12x20.7	L 3½x3 x¼	26.1	7.65	164	33.2	23.2	4.63	4.94	31.4	3.23
	x⁵⁄₁₆	27.3	8.02	170	35.8	23.5	4.61	4.75	32.2	2.80
	L 4 x3 x¼	26.5	7.78	167	34.4	23.4	4.63	4.86	31.8	3.01
	x⁵⁄₁₆	27.9	8.18	173	37.2	23.6	4.60	4.66	32.6	2.67
	L 5 x3½x⁵⁄₁₆	29.4	8.65	180	40.2	23.9	4.56	4.47	33.5	2.53
	x⅜	31.1	9.14	186	43.4	24.1	4.51	4.29	34.2	2.25
	L 6 x4 x⅜	33.0	9.70	192	46.6	24.3	4.45	4.12	35.3	2.11
	x½	36.9	10.84	202	53.2	24.7	4.32	3.80	36.7	1.68
C 12x25	L 3½x3 x¼	30.4	8.91	180	35.4	26.1	4.50	5.09	35.8	3.98
	x⁵⁄₁₆	31.6	9.28	187	38.0	26.4	4.49	4.92	36.8	3.50
	L 4 x3 x¼	30.8	9.04	183	36.6	26.3	4.50	5.02	36.3	3.82
	x⁵⁄₁₆	32.2	9.44	190	39.3	26.6	4.49	4.84	37.3	3.30
	L 5 x3½x⁵⁄₁₆	33.7	9.91	197	42.3	26.9	4.46	4.67	38.3	3.05
	x⅜	35.4	10.40	204	45.4	27.2	4.43	4.49	39.3	2.77
	L 6 x4 x⅜	37.3	10.96	211	48.7	27.5	4.39	4.33	40.4	2.65
	x½	41.2	12.10	223	55.3	28.0	4.29	4.03	42.2	2.20
C 15x33.9	L 4 x3 x¼	39.7	11.65	383	58.7	45.1	5.73	6.52	60.1	5.39
	x⁵⁄₁₆	41.1	12.05	395	62.4	45.6	5.73	6.33	61.8	4.89
	L 5 x3½x⁵⁄₁₆	42.6	12.52	408	66.5	46.1	5.71	6.14	63.4	4.30
	x⅜	44.3	13.01	421	70.8	46.5	5.69	5.94	64.8	3.69
	L 6 x4 x⅜	46.2	13.57	434	75.4	46.9	5.65	5.76	66.2	3.48
	x½	50.1	14.71	458	84.8	47.7	5.58	5.40	68.6	2.92

COMBINATION SECTIONS
Channels and angles
Properties of sections
Long leg of angle turned out

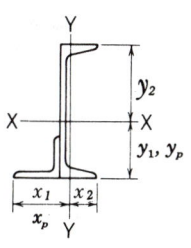

Channel	Angle		Axis Y-Y						
		I	$S_1 = I/x_1$	$S_2 = I/x_2$	r	x_1	Z	x_p	
		In.⁴	In.³	In.³	In.	In.	In.³	In.	
C 12x20.7	L 3½x3 x¼	9.5	2.5	3.7	1.12	3.84	5.45	3.69	
	x⁵⁄₁₆	10.7	2.8	4.0	1.16	3.77	5.94	3.67	
	L 4 x3 x¼	11.6	2.7	4.4	1.22	4.28	5.94	4.18	
	x⁵⁄₁₆	13.2	3.2	4.8	1.27	4.20	6.56	4.16	
	L 5 x3½x⁵⁄₁₆	19.9	4.0	6.8	1.52	5.02	8.06	5.15	
	x⅜	22.5	4.6	7.5	1.57	4.93	8.97	5.13	
	L 6 x4 x⅜	33.2	5.8	10.3	1.85	5.72	11.1	6.10	
	x½	40.6	7.3	11.9	1.93	5.52	13.7	6.05	
C 12x25	L 3½x3 x¼	10.2	2.6	3.8	1.07	3.87	5.88	3.74	
	x⁵⁄₁₆	11.4	3.0	4.2	1.11	3.81	6.40	3.72	
	L 4 x3 x¼	12.3	2.8	4.5	1.17	4.32	6.38	4.23	
	x⁵⁄₁₆	13.9	3.3	5.0	1.22	4.25	7.02	4.22	
	L 5 x3½x⁵⁄₁₆	20.8	4.1	7.0	1.45	5.09	8.54	5.20	
	x⅜	23.5	4.7	7.7	1.50	5.00	9.48	5.18	
	L 6 x4 x⅜	34.5	5.9	10.7	1.77	5.81	11.7	6.15	
	x½	42.3	7.5	12.4	1.87	5.63	14.3	6.11	
C 15x33.9	L 4 x3 x¼	16.8	3.7	5.8	1.20	4.49	8.82	4.27	
	x⁵⁄₁₆	18.7	4.2	6.3	1.25	4.43	9.47	4.26	
	L 5 x3½x⁵⁄₁₆	26.2	4.9	8.5	1.45	5.30	11.0	5.24	
	x⅜	29.3	5.6	9.2	1.50	5.23	12.0	5.23	
	L 6 x4 x⅜	41.3	6.8	12.4	1.75	6.06	14.2	6.21	
	x½	50.3	8.5	14.3	1.85	5.89	16.9	6.17	

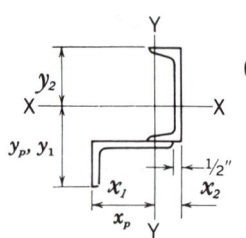

COMBINATION SECTIONS
Channels and angles
Properties of sections
Short leg of angle turned down

Channel	Angle	Total Weight per Ft	Total Area	Axis X-X					
				$S_1 = I/y_1$	$S_2 = I/y_2$	r	y_1	Z	y_p
		Lb.	In.^{2}	In.^{3}	In.^{3}	In.	In.	In.^{3}	In.
C 6x 8.2	L 3x2½x¼	12.7	3.71	6.0	5.9	2.61	4.21	7.86	2.79
	L 3x3 x¼	13.1	3.84	6.1	6.2	2.68	4.56	8.23	3.25
	L 4x3 x¼	14.0	4.09	6.4	6.2	2.63	4.46	8.32	3.18
	L 5x3 x⁵⁄₁₆	16.4	4.80	7.5	6.4	2.55	4.16	8.77	3.00
	L 6x3½x⁵⁄₁₆	18.0	5.27	7.7	6.8	2.56	4.45	9.32	3.46
C 7x 9.8	L 3x2½x¼	14.3	4.18	8.0	7.8	3.00	4.70	10.5	2.95
	L 3x3 x¼	14.7	4.31	8.0	8.2	3.07	5.05	10.9	3.38
	L 4x3 x¼	15.6	4.56	8.5	8.2	3.03	4.93	11.0	3.29
	L 5x3 x⁵⁄₁₆	18.0	5.27	10.0	8.5	2.95	4.60	11.6	3.11
	L 6x3½x⁵⁄₁₆	19.6	5.74	10.3	8.9	2.96	4.87	12.2	3.50
C 8x11.5	L 3x2½x¼	16.0	4.69	10.4	10.2	3.39	5.20	13.7	3.52
	L 3x3 x¼	16.4	4.82	10.4	10.6	3.45	5.55	14.2	3.73
	L 4x3 x¼	17.3	5.07	10.9	10.6	3.42	5.42	14.3	3.42
	L 5x3 x⁵⁄₁₆	19.7	5.78	12.9	11.0	3.36	5.06	14.9	3.21
	L 6x3½x⁵⁄₁₆	21.3	6.25	13.3	11.4	3.36	5.31	15.6	3.61
C 9x13.4	L 3x2½x¼	17.9	5.25	13.1	12.9	3.78	5.71	17.4	4.18
	L 3x3 x¼	18.3	5.38	13.1	13.4	3.84	6.07	18.0	4.42
	L 4x3 x¼	19.2	5.63	13.8	13.5	3.81	5.93	18.3	3.88
	L 5x3 x⁵⁄₁₆	21.6	6.34	16.2	13.9	3.76	5.54	19.0	3.32
	L 6x3½x⁵⁄₁₆	23.2	6.81	16.7	14.4	3.77	5.78	19.7	3.71
C 10x15.3	L 3x2½x¼	19.8	5.80	16.2	16.0	4.17	6.22	21.4	4.77
	L 3x3 x¼	20.2	5.93	16.1	16.5	4.22	6.58	22.1	5.01
	L 4x3 x¼	21.1	6.18	17.0	16.6	4.20	6.43	22.5	4.48
	L 5x3 x⁵⁄₁₆	23.5	6.89	19.9	17.1	4.17	6.02	23.5	3.44
	L 6x3½x⁵⁄₁₆	25.1	7.36	20.5	17.7	4.18	6.25	24.2	3.81
C 12x20.7	L 3x2½x¼	25.2	7.40	24.3	24.7	4.90	7.32	32.6	6.17
	L 3x3 x¼	25.6	7.53	24.0	25.3	4.95	7.69	33.4	6.45
	L 4x3 x¼	26.5	7.78	25.3	25.5	4.95	7.54	34.3	6.01
	L 5x3 x⁵⁄₁₆	28.9	8.49	29.2	26.3	4.94	7.11	36.4	4.74
	L 6x3½x⁵⁄₁₆	30.5	8.96	30.1	27.1	4.97	7.33	37.5	4.41
C 15x33.9	L 3x3 x¼	38.8	11.40	42.7	47.2	5.95	9.45	61.1	8.70
	L 4x3 x¼	39.7	11.65	44.5	47.7	5.96	9.31	62.5	8.39
	L 5x3 x⁵⁄₁₆	42.1	12.36	50.1	49.1	6.01	8.91	66.5	7.50
	L 6x3½x⁵⁄₁₆	43.7	12.83	51.4	50.3	6.05	9.15	69.0	7.41

COMBINATION SECTIONS
Channels and angles
Properties of sections
Short leg of angle turned down

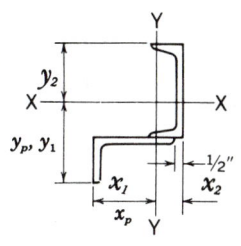

Channel	Angle	Axis Y-Y					
		$S_1 = I/x_1$	$S_2 = I/x_2$	r	x_1	Z	x_p
		In.³	In.³	In.	In.	In.³	In.
C 6x 8.2	L 3x2½x¼	2.4	4.4	1.22	2.25	3.70	2.65
	L 3x3 x¼	2.8	4.6	1.26	2.18	4.01	2.59
	L 4x3 x¼	3.8	6.6	1.64	2.85	5.46	3.46
	L 5x3 x⁵⁄₁₆	6.0	9.3	2.05	3.33	8.29	4.12
	L 6x3½x⁵⁄₁₆	8.4	12.1	2.47	3.82	11.3	4.84
C 7x 9.8	L 3x2½x¼	2.6	5.0	1.19	2.32	4.03	2.72
	L 3x3 x¼	2.9	5.2	1.23	2.25	4.35	2.67
	L 4x3 x¼	3.9	7.5	1.60	2.95	5.82	3.55
	L 5x3 x⁵⁄₁₆	6.1	10.5	2.01	3.47	8.71	4.24
	L 6x3½x⁵⁄₁₆	8.6	13.6	2.44	3.98	11.8	4.99
C 8x11.5	L 3x2½x¼	2.7	5.6	1.16	2.37	4.40	2.78
	L 3x3 x¼	3.0	5.8	1.20	2.31	4.73	2.73
	L 4x3 x¼	4.0	8.3	1.55	3.03	6.21	3.62
	L 5x3 x⁵⁄₁₆	6.3	11.7	1.97	3.58	9.16	4.34
	L 6x3½x⁵⁄₁₆	8.7	15.2	2.40	4.13	12.3	5.12
C 9x13.4	L 3x2½x¼	2.8	6.2	1.14	2.40	4.81	2.84
	L 3x3 x¼	3.2	6.5	1.18	2.35	5.15	2.79
	L 4x3 x¼	4.2	9.2	1.51	3.10	6.65	3.70
	L 5x3 x⁵⁄₁₆	6.4	12.9	1.92	3.68	9.66	4.43
	L 6x3½x⁵⁄₁₆	8.9	16.9	2.36	4.26	12.8	5.23
C 10x15.3	L 3x2½x¼	3.0	6.8	1.12	2.42	5.25	2.88
	L 3x3 x¼	3.4	7.1	1.16	2.37	5.59	2.84
	L 4x3 x¼	4.3	10.0	1.48	3.15	7.10	3.74
	L 5x3 x⁵⁄₁₆	6.5	14.0	1.88	3.76	10.1	4.49
	L 6x3½x⁵⁄₁₆	9.0	18.3	2.31	4.36	13.3	5.31
C 12x20.7	L 3x2½x¼	3.6	8.6	1.10	2.47	6.47	3.01
	L 3x3 x¼	4.0	8.9	1.13	2.43	6.82	2.97
	L 4x3 x¼	4.7	12.2	1.40	3.25	8.37	3.89
	L 5x3 x⁵⁄₁₆	6.9	17.0	1.78	3.92	11.5	4.67
	L 6x3½x⁵⁄₁₆	9.3	22.4	2.19	4.59	14.8	5.52
C 15x33.9	L 3x3 x¼	5.6	13.5	1.10	2.48	9.54	3.12
	L 4x3 x¼	5.9	17.2	1.30	3.35	11.1	4.11
	L 5x3 x⁵⁄₁₆	7.8	23.4	1.61	4.12	14.5	5.02
	L 6x3½x⁵⁄₁₆	10.2	30.7	1.97	4.88	18.0	5.90

STEEL PIPE AND STRUCTURAL TUBING
Dimensions and properties

GENERAL

When designing and specifying steel pipe or tubing as compression members, refer to comments in the notes for Columns, Steel Pipe and Structural Tubing, p. 2-33. For standard mill practices and tolerances, refer to p. 1-175. For material specifications and availability, see Table 4 (Part 1).

STEEL PIPE

The tables of dimensions and properties of steel pipe (unfilled) list a selected range of sizes of Standard, Extra Strong, and Double-Extra Strong pipe. For a complete range of sizes manufactured, refer to manufacturers' catalogs.

STRUCTURAL TUBING

The tables of dimensions and properties of square and rectangular structural tubing (unfilled) list a selected range of frequently used sizes. For dimensions and properties of other sizes, refer to manufacturers' catalogs.

The tables are based on an outside corner radius equal to two times the specified wall thickness. Material specifications stipulate that the outside corner radius may vary up to three times the specified wall thickness. This variation should be considered in those details where a close match or fit is important.

PIPE
Dimensions and properties

Dimensions				Weight per Ft Lbs. Plain Ends	Properties					
Nominal Diameter In.	Outside Diameter In.	Inside Diameter In.	Wall Thickness In.		Area In.^{2}	I In.^{4}	S In.^{3}	r In.	J In.^{4}	Z In.^{3}
Standard Weight										
½	.840	.622	.109	.85	.250	.017	.041	.261	.034	.059
¾	1.050	.824	.113	1.13	.333	.037	.071	.334	.074	.100
1	1.315	1.049	.133	1.68	.494	.087	.133	.421	.175	.187
1¼	1.660	1.380	.140	2.27	.669	.195	.235	.540	.389	.324
1½	1.900	1.610	.145	2.72	.799	.310	.326	.623	.620	.448
2	2.375	2.067	.154	3.65	1.07	.666	.561	.787	1.33	.761
2½	2.875	2.469	.203	5.79	1.70	1.53	1.06	.947	3.06	1.45
3	3.500	3.068	.216	7.58	2.23	3.02	1.72	1.16	6.03	2.33
3½	4.000	3.548	.226	9.11	2.68	4.79	2.39	1.34	9.58	3.22
4	4.500	4.026	.237	10.79	3.17	7.23	3.21	1.51	14.5	4.31
5	5.563	5.047	.258	14.62	4.30	15.2	5.45	1.88	30.3	7.27
6	6.625	6.065	.280	18.97	5.58	28.1	8.50	2.25	56.3	11.2
8	8.625	7.981	.322	28.55	8.40	72.5	16.8	2.94	145	22.2
10	10.750	10.020	.365	40.48	11.9	161	29.9	3.67	321	39.4
12	12.750	12.000	.375	49.56	14.6	279	43.8	4.38	559	57.4
Extra Strong										
½	.840	.546	.147	1.09	.320	.020	.048	.250	.040	.072
¾	1.050	.742	.154	1.47	.433	.045	.085	.321	.090	.125
1	1.315	.957	.179	2.17	.639	.106	.161	.407	.211	.233
1¼	1.660	1.278	.191	3.00	.881	.242	.291	.524	.484	.414
1½	1.900	1.500	.200	3.63	1.07	.391	.412	.605	.782	.581
2	2.375	1.939	.218	5.02	1.48	.868	.731	.766	1.74	1.02
2½	2.875	2.323	.276	7.66	2.25	1.92	1.34	.924	3.85	1.87
3	3.500	2.900	.300	10.25	3.02	3.89	2.23	1.14	8.13	3.08
3½	4.000	3.364	.318	12.50	3.68	6.28	3.14	1.31	12.6	4.32
4	4.500	3.826	.337	14.98	4.41	9.61	4.27	1.48	19.2	5.85
5	5.563	4.813	.375	20.78	6.11	20.7	7.43	1.84	41.3	10.1
6	6.625	5.761	.432	28.57	8.40	40.5	12.2	2.19	81.0	16.6
8	8.625	7.625	.500	43.39	12.8	106	24.5	2.88	211	33.0
10	10.750	9.750	.500	54.74	16.1	212	39.4	3.63	424	52.6
12	12.750	11.750	.500	65.42	19.2	362	56.7	4.33	723	75.1
Double-Extra Strong										
2	2.375	1.503	.436	9.03	2.66	1.31	1.10	.703	2.62	1.67
2½	2.875	1.771	.552	13.69	4.03	2.87	2.00	.844	5.74	3.04
3	3.500	2.300	.600	18.58	5.47	5.99	3.42	1.05	12.0	5.12
4	4.500	3.152	.674	27.54	8.10	15.3	6.79	1.37	30.6	9.97
5	5.563	4.063	.750	38.16	11.3	33.6	12.1	1.72	67.3	17.5
6	6.625	4.897	.864	53.16	15.6	66.3	20.0	2.06	133	28.9
8	8.625	6.875	.875	72.42	21.3	162	37.6	2.76	324	52.8

The listed sections are available in conformance with ASTM Specification A53 Gr. B or A501. Other sections are made to these specifications. Consult with pipe manufacturers or distributors for availability.

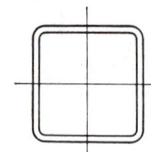

STRUCTURAL TUBING
Square
Dimensions and properties

	Dimensions			Properties**					
Nominal* Size	Wall Thickness		Weight per Ft	Area	I	S	r	J	Z
In.	In.		Lb.	In.²	In.⁴	In.³	In.	In.⁴	In.³
16 x 16	.5000	½	103.30	30.4	1200	150	6.29	1890	175
	.3750	⅜	78.52	23.1	931	116	6.35	1450	134
	.3125	⁵⁄₁₆	65.87	19.4	789	98.6	6.38	1220	113
14 x 14	.5000	½	89.68	26.4	791	113	5.48	1250	132
	.3750	⅜	68.31	20.1	615	87.9	5.54	963	102
	.3125	⁵⁄₁₆	57.36	16.9	522	74.6	5.57	812	86.1
12 x 12	.5000	½	76.07	22.4	485	80.9	4.66	777	95.4
	.3750	⅜	58.10	17.1	380	63.4	4.72	599	73.9
	.3125	⁵⁄₁₆	48.86	14.4	324	54.0	4.75	506	62.6
	.2500	¼	39.43	11.6	265	44.1	4.78	410	50.8
10 x 10	.6250	⅝	76.33	22.4	321	64.2	3.78	529	77.6
	.5000	½	62.46	18.4	271	54.2	3.84	439	64.6
	.3750	⅜	47.90	14.1	214	42.9	3.90	341	50.4
	.3125	⁵⁄₁₆	40.35	11.9	183	36.7	3.93	289	42.8
	.2500	¼	32.63	9.59	151	30.1	3.96	235	34.9
8 x 8	.6250	⅝	59.32	17.4	153	38.3	2.96	258	47.2
	.5000	½	48.85	14.4	131	32.9	3.03	217	39.7
	.3750	⅜	37.69	11.1	106	26.4	3.09	170	31.3
	.3125	⁵⁄₁₆	31.84	9.36	90.9	22.7	3.12	145	26.7
	.2500	¼	25.82	7.59	75.1	18.8	3.15	118	21.9
	.1875	³⁄₁₆	19.63	5.77	58.2	14.6	3.18	90.6	16.8
7 x 7	.5000	½	42.05	12.4	84.6	24.2	2.62	141	29.6
	.3750	⅜	32.58	9.58	68.7	19.6	2.68	112	23.5
	.3125	⁵⁄₁₆	27.59	8.11	59.5	17.0	2.71	95.6	20.1
	.2500	¼	22.42	6.59	49.4	14.1	2.74	78.3	16.5
	.1875	³⁄₁₆	17.08	5.02	38.5	11.0	2.77	60.2	12.7

*Outside dimensions across flat sides.
**Properties are based upon a nominal outside corner radius equal to two times the wall thickness.

AMERICAN INSTITUTE OF STEEL CONSTRUCTION

STRUCTURAL TUBING
Square
Dimensions and properties

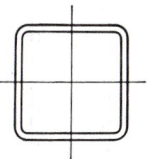

	Dimensions			Properties**					
Nominal* Size	Wall Thickness		Weight per Ft	Area	I	S	r	J	Z
In.	In.		Lb.	In.²	In.⁴	In.³	In.	In.⁴	In.³
6 x 6	.5000	½	35.24	10.4	50.5	16.8	2.21	85.6	20.9
	.3750	⅜	27.48	8.08	41.6	13.9	2.27	68.5	16.8
	.3125	5⁄16	23.34	6.86	36.3	12.1	2.30	58.9	14.4
	.2500	¼	19.02	5.59	30.3	10.1	2.33	48.5	11.9
	.1875	3⁄16	14.53	4.27	23.8	7.93	2.36	37.5	9.24
5 x 5	.5000	½	28.43	8.36	27.0	10.8	1.80	46.8	13.7
	.3750	⅜	22.37	6.58	22.8	9.11	1.86	38.2	11.2
	.3125	5⁄16	19.08	5.61	20.1	8.02	1.89	33.1	9.70
	.2500	¼	15.62	4.59	16.9	6.78	1.92	27.4	8.07
	.1875	3⁄16	11.97	3.52	13.4	5.36	1.95	21.3	6.29
4 x 4	.5000	½	21.63	6.36	12.3	6.13	1.39	21.8	8.02
	.3750	⅜	17.27	5.08	10.7	5.35	1.45	18.4	6.72
	.3125	5⁄16	14.83	4.36	9.58	4.79	1.48	16.1	5.90
	.2500	¼	12.21	3.59	8.22	4.11	1.51	13.5	4.97
	.1875	3⁄16	9.42	2.77	6.59	3.30	1.54	10.6	3.91
3.5 x 3.5	.3125	5⁄16	12.70	3.73	6.09	3.48	1.28	10.4	4.35
	.2500	¼	10.51	3.09	5.29	3.02	1.31	8.82	3.70
	.1875	3⁄16	8.15	2.39	4.29	2.45	1.34	6.99	2.93
3 x 3	.3125	5⁄16	10.58	3.11	3.58	2.39	1.07	6.22	3.04
	.2500	¼	8.81	2.59	3.16	2.10	1.10	5.35	2.61
	.1875	3⁄16	6.87	2.02	2.60	1.73	1.13	4.28	2.10
2.5 x 2.5	.2500	¼	7.11	2.09	1.69	1.35	.899	2.92	1.71
	.1875	3⁄16	5.59	1.64	1.42	1.14	.930	2.38	1.40
2 x 2	.2500	¼	5.41	1.59	.766	.766	.694	1.36	1.00
	.1875	3⁄16	4.32	1.27	.668	.668	.726	1.15	0.84

*Outside dimensions across flat sides.
**Properties are based upon a nominal outside corner radius equal to two times the wall thickness.

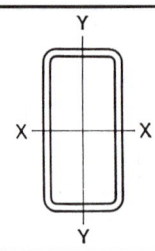

STRUCTURAL TUBING
Rectangular
Dimensions and properties

Dimensions			Properties**									
				X-X Axis				Y-Y Axis				
Nominal* Size	Wall Thickness	Weight per Ft	Area	I_x	S_x	Z_x	r_x	I_y	S_y	Z_y	r_y	J
In.	In.	Lb.	In.^{2}	In.^{4}	In.^{3}	In.^{3}	In.	In.^{4}	In.^{3}	In.^{3}	In.	In.^{4}
20x12	.5000 ½	103.30	30.4	1650	165	201	7.37	750	125	141	4.97	1650
	.3750 ⅜	78.52	23.1	1280	128	154	7.44	583	97.2	109	5.03	1270
	.3125 ⁵⁄₁₆	65.87	19.4	1080	108	130	7.47	495	82.5	91.8	5.06	1070
20x 8	.5000 ½	89.68	26.4	1270	127	162	6.94	300	75.1	84.7	3.38	806
	.3750 ⅜	68.31	20.1	988	98.8	125	7.02	236	59.1	65.6	3.43	625
	.3125 ⁵⁄₁₆	57.36	16.9	838	83.8	105	7.05	202	50.4	55.6	3.46	529
20x 4	.5000 ½	76.07	22.4	889	88.9	123	6.31	61.6	30.8	36.0	1.66	205
	.3750 ⅜	58.10	17.1	699	69.9	95.3	6.40	50.3	25.1	28.5	1.72	165
	.3125 ⁵⁄₁₆	48.86	14.4	596	59.6	80.8	6.44	43.7	21.8	24.3	1.74	143
18x 6	.5000 ½	76.07	22.4	818	90.9	119	6.05	141	47.2	53.9	2.52	410
	.3750 ⅜	58.10	17.1	641	71.3	92.2	6.13	113	37.6	42.1	2.57	322
	.3125 ⁵⁄₁₆	48.86	14.4	546	60.7	78.1	6.17	97.0	32.3	35.8	2.60	274
16x12	.5000 ½	89.68	26.4	962	120	144	6.04	618	103	118	4.84	1200
	.3750 ⅜	68.31	20.1	748	93.5	111	6.11	482	80.3	91.3	4.90	922
	.3125 ⁵⁄₁₆	57.36	16.9	635	79.4	93.8	6.14	409	68.2	77.2	4.93	777
16x 8	.5000 ½	76.07	22.4	722	90.2	113	5.68	244	61.0	69.7	3.30	599
	.3750 ⅜	58.10	17.1	565	70.6	87.6	5.75	193	48.2	54.2	3.36	465
	.3125 ⁵⁄₁₆	48.86	14.4	481	60.1	74.2	5.79	165	41.2	45.9	3.39	394
16x 4	.5000 ½	62.46	18.4	481	60.2	82.2	5.12	49.3	24.6	29.0	1.64	157
	.3750 ⅜	47.90	14.1	382	47.8	64.2	5.21	40.4	20.2	23.0	1.69	127
	.3125 ⁵⁄₁₆	40.35	11.9	327	40.9	54.5	5.25	35.1	17.6	19.7	1.72	110
14x10	.5000 ½	76.07	22.4	608	86.9	105	5.22	361	72.3	83.6	4.02	730
	.3750 ⅜	58.10	17.1	476	68.0	81.5	5.28	284	56.8	64.8	4.08	564
	.3125 ⁵⁄₁₆	48.86	14.4	405	57.9	69.0	5.31	242	48.4	54.9	4.11	477

*Outside dimensions across flat sides.
**Properties are based upon a nominal outside corner radius equal to two times the wall thickness.

STRUCTURAL TUBING
Rectangular
Dimensions and properties

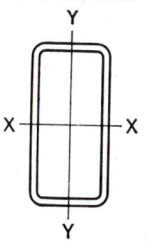

Dimensions			Properties**									
				X-X Axis				Y-Y Axis				
Nominal* Size	Wall Thickness	Weight per Ft	Area	I_x	S_x	Z_x	r_x	I_y	S_y	Z_y	r_y	J
In.	In.	Lb.	In.^{2}	In.^{4}	In.^{3}	In.^{3}	In.	In.^{4}	In.^{3}	In.^{3}	In.	In.^{4}
14x6	.5000 ½	62.46	18.4	426	60.8	78.3	4.82	111	37.1	42.9	2.46	296
	.3750 ⅜	47.90	14.1	337	48.1	61.1	4.89	89.1	29.7	33.6	2.52	233
	.3125 5/16	40.35	11.9	288	41.2	51.9	4.93	76.7	25.6	28.7	2.54	199
	.2500 ¼	32.63	9.59	237	33.8	42.3	4.97	63.4	21.1	23.4	2.57	162
14x4	.5000 ½	55.66	16.4	335	47.8	64.8	4.52	43.1	21.5	25.5	1.62	134
	.3750 ⅜	42.79	12.6	267	38.2	50.8	4.61	35.4	17.7	20.3	1.68	108
	.3125 5/16	36.10	10.6	230	32.8	43.3	4.65	30.9	15.4	17.4	1.71	93.1
	.2500 ¼	29.23	8.59	189	27.0	35.4	4.69	25.8	12.9	14.3	1.73	77.0
12x8	.6250 ⅝	76.33	22.4	418	69.7	87.1	4.32	222	55.3	65.6	3.14	481
	.5000 ½	62.46	18.4	353	58.9	72.4	4.39	188	46.9	54.7	3.20	401
	.3750 ⅜	47.90	14.1	279	46.5	56.5	4.45	149	37.3	42.7	3.26	312
	.3125 5/16	40.35	11.9	239	39.8	47.9	4.49	128	32.0	36.3	3.28	265
	.2500 ¼	32.63	9.59	196	32.6	39.1	4.52	105	26.3	29.6	3.31	216
12x6	.6250 ⅝	67.82	19.9	337	56.2	72.9	4.11	112	37.2	44.5	2.37	286
	.5000 ½	55.66	16.4	287	47.8	60.9	4.19	96.0	32.0	37.4	2.42	241
	.3750 ⅜	42.79	12.6	228	38.1	47.7	4.26	77.2	25.7	29.4	2.48	190
	.3125 5/16	36.10	10.6	196	32.6	40.6	4.30	66.6	22.2	25.1	2.51	162
	.2500 ¼	29.23	8.59	161	26.9	33.2	4.33	55.2	18.4	20.6	2.53	132
	.1875 3/16	22.18	6.52	124	20.7	25.4	4.37	42.8	14.3	15.8	2.56	101
12x4	.5000 ½	48.85	14.4	221	36.8	49.4	3.92	36.9	18.5	22.0	1.60	110
	.3750 ⅜	37.69	11.1	178	29.6	39.0	4.01	30.5	15.2	17.6	1.66	89.0
	.3125 5/16	31.84	9.36	153	25.5	33.3	4.05	26.6	13.3	15.1	1.69	76.9
	.2500 ¼	25.82	7.59	127	21.1	27.3	4.09	22.3	11.1	12.5	1.71	63.6
	.1875 3/16	19.63	5.77	98.2	16.4	21.0	4.13	17.5	8.75	9.63	1.74	49.3
12x2	.2500 ¼	22.42	6.59	92.2	15.4	21.4	3.74	4.62	4.62	5.38	.837	15.9
	.1875 3/16	17.08	5.02	72.0	12.0	16.6	3.79	3.76	3.76	4.24	.865	12.8

*Outside dimensions across flat sides.
**Properties are based upon a nominal outside corner radius equal to two times the wall thickness.

AMERICAN INSTITUTE OF STEEL CONSTRUCTION

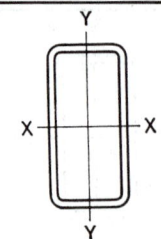

STRUCTURAL TUBING
Rectangular
Dimensions and properties

Dimensions			Properties**										
				X-X Axis				Y-Y Axis				J
Nominal* Size	Wall Thickness	Weight per Ft	Area	I_x	S_x	Z_x	r_x	I_y	S_y	Z_y	r_y	
In.	In.	Lb.	In.²	In.⁴	In.³	In.³	In.	In.⁴	In.³	In.³	In.	In.⁴
10x6	.6250 ⅝	59.32	17.4	211	42.2	54.2	3.48	93.5	31.2	37.7	2.32	221
	.5000 ½	48.85	14.4	181	36.2	45.6	3.55	80.8	26.9	31.9	2.37	187
	.3750 ⅜	37.69	11.1	145	29.0	35.9	3.62	65.4	21.8	25.2	2.43	147
	.3125 5⁄16	31.84	9.36	125	25.0	30.7	3.65	56.5	18.8	21.5	2.46	126
	.2500 ¼	25.82	7.59	103	20.6	25.1	3.69	46.9	15.6	17.7	2.49	103
	.1875 3⁄16	19.63	5.77	79.8	16.0	19.3	3.72	36.5	12.2	13.6	2.51	79.1
10x4	.5000 ½	42.05	12.4	136	27.1	36.1	3.31	30.8	15.4	18.5	1.58	86.9
	.3750 ⅜	32.58	9.58	110	22.0	28.7	3.39	25.5	12.8	14.9	1.63	70.4
	.3125 5⁄16	27.59	8.11	95.5	19.1	24.6	3.43	22.4	11.2	12.8	1.66	60.8
	.2500 ¼	22.42	6.59	79.3	15.9	20.2	3.47	18.8	9.39	10.6	1.69	50.4
	.1875 3⁄16	17.08	5.02	61.7	12.3	15.6	3.51	14.8	7.39	8.20	1.72	39.1
10x2	.3750 ⅜	27.48	8.08	75.4	15.1	21.5	3.06	4.85	4.85	6.05	.775	16.5
	.3125 5⁄16	23.34	6.86	66.1	13.2	18.5	3.10	4.42	4.42	5.33	.802	14.9
	.2500 ¼	19.02	5.59	55.5	11.1	15.4	3.15	3.85	3.85	4.50	.830	12.8
	.1875 3⁄16	14.53	4.27	43.7	8.74	11.9	3.20	3.14	3.14	3.56	.858	10.3
8x6	.5000 ½	42.05	12.4	103	25.8	32.2	2.89	65.7	21.9	26.4	2.31	135
	.3750 ⅜	32.58	9.58	83.7	20.9	25.6	2.96	53.5	17.8	21.0	2.36	107
	.3125 5⁄16	27.59	8.11	72.4	18.1	21.9	2.99	46.4	15.5	18.0	2.39	91.3
	.2500 ¼	22.42	6.59	60.1	15.0	18.0	3.02	38.6	12.9	14.8	2.42	74.9
	.1875 3⁄16	17.08	5.02	46.8	11.7	13.9	3.05	30.1	10.0	11.4	2.45	57.6
8x4	.5000 ½	35.24	10.4	75.1	18.8	24.7	2.69	24.6	12.3	15.0	1.54	64.1
	.3750 ⅜	27.48	8.08	61.9	15.5	19.9	2.77	20.6	10.3	12.2	1.60	52.2
	.3125 5⁄16	23.34	6.86	53.9	13.5	17.1	2.80	18.1	9.05	10.5	1.62	45.2
	.2500 ¼	19.02	5.59	45.1	11.3	14.1	2.84	15.3	7.63	8.72	1.65	37.5
	.1875 3⁄16	14.53	4.27	35.3	8.83	11.0	2.88	12.0	6.02	6.77	1.68	29.1

*Outside dimensions across flat sides.
**Properties are based upon a nominal outside corner radius equal to two times the wall thickness.

STRUCTURAL TUBING
Rectangular
Dimensions and properties

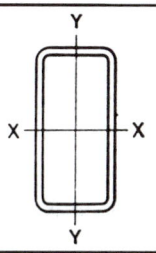

Nominal* Size	Wall Thickness		Weight per Ft	Area	X-X Axis				Y-Y Axis				J
					I_x	S_x	Z_x	r_x	I_y	S_y	Z_y	r_y	
In.	In.		Lb.	In.²	In.⁴	In.³	In.³	In.	In.⁴	In.³	In.³	In.	In.⁴
8x3	.3750	⅜	24.93	7.33	51.0	12.7	17.0	2.64	10.4	6.92	8.31	1.19	29.9
	.3125	⁵⁄₁₆	21.21	6.23	44.7	11.2	14.7	2.68	9.25	6.16	7.24	1.22	26.3
	.2500	¼	17.32	5.09	37.6	9.40	12.2	2.72	7.90	5.26	6.05	1.25	22.1
	.1875	³⁄₁₆	13.25	3.89	29.6	7.40	9.49	2.76	6.31	4.21	4.73	1.27	17.3
8x2	.3750	⅜	22.37	6.58	40.1	10.0	14.2	2.47	3.85	3.85	4.83	.765	12.6
	.3125	⁵⁄₁₆	19.08	5.61	35.5	8.87	12.3	2.51	3.52	3.52	4.28	.792	11.4
	.2500	¼	15.62	4.59	30.1	7.52	10.3	2.56	3.08	3.08	3.63	.819	9.84
	.1875	³⁄₁₆	11.97	3.52	23.9	5.97	8.02	2.60	2.52	2.52	2.88	.847	7.94
7x5	.5000	½	35.24	10.4	63.5	18.1	23.1	2.48	37.2	14.9	18.2	1.90	79.9
	.3750	⅜	27.48	8.08	52.2	14.9	18.5	2.54	30.8	12.3	14.6	1.95	64.2
	.3125	⁵⁄₁₆	23.34	6.86	45.5	13.0	15.9	2.58	26.9	10.8	12.6	1.98	55.3
	.2500	¼	19.02	5.59	38.0	10.9	13.2	2.61	22.6	9.04	10.4	2.01	45.6
	.1875	³⁄₁₆	14.53	4.27	29.8	8.50	10.2	2.64	17.7	7.10	8.10	2.04	35.3
7x4	.3750	⅜	24.93	7.33	44.0	12.6	16.0	2.45	18.1	9.06	10.8	1.57	43.3
	.3125	⁵⁄₁₆	21.21	6.23	38.5	11.0	13.8	2.49	16.0	7.98	9.36	1.60	37.5
	.2500	¼	17.32	5.09	32.3	9.23	11.5	2.52	13.5	6.75	7.78	1.63	31.2
	.1875	³⁄₁₆	13.25	3.89	25.4	7.26	8.91	2.55	10.7	5.34	6.06	1.66	24.2
7x3	.3750	⅜	22.37	6.58	35.7	10.2	13.5	2.33	9.08	6.05	7.32	1.18	25.1
	.3125	⁵⁄₁₆	19.08	5.61	31.5	9.00	11.8	2.37	8.11	5.41	6.40	1.20	22.0
	.2500	¼	15.62	4.59	26.6	7.61	9.79	2.41	6.95	4.63	5.36	1.23	18.5
	.1875	³⁄₁₆	11.97	3.52	21.1	6.02	7.63	2.45	5.57	3.71	4.21	1.26	14.6
6x4	.5000	½	28.43	8.36	35.3	11.8	15.4	2.06	18.4	9.21	11.5	1.48	42.1
	.3750	⅜	22.37	6.58	29.7	9.90	12.5	2.13	15.6	7.82	9.44	1.54	34.6
	.3125	⁵⁄₁₆	19.08	5.61	26.2	8.72	10.9	2.16	13.8	6.92	8.21	1.57	30.1
	.2500	¼	15.62	4.59	22.1	7.36	9.06	2.19	11.7	5.87	6.84	1.60	25.0
	.1875	³⁄₁₆	11.97	3.52	17.4	5.81	7.06	2.23	9.32	4.66	5.34	1.63	19.5

*Outside dimensions across flat sides.
**Properties are based upon a nominal outside corner radius equal to two times the wall thickness.

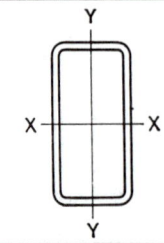

STRUCTURAL TUBING
Rectangular
Dimensions and properties

	Dimensions			Properties**									
				X-X Axis				Y-Y Axis					
Nominal* Size	Wall Thickness		Weight per Ft	Area	I_x	S_x	Z_x	r_x	I_y	S_y	Z_y	r_y	J
In.	In.		Lb.	In.²	In.⁴	In.³	In.³	In.	In.⁴	In.³	In.³	In.	In.⁴
6x3	.3750	⅜	19.82	5.83	23.8	7.92	10.4	2.02	7.78	5.19	6.34	1.16	20.3
	.3125	⁵⁄₁₆	16.96	4.98	21.1	7.03	9.11	2.06	6.98	4.65	5.56	1.18	17.9
	.2500	¼	13.91	4.09	17.9	5.98	7.62	2.09	6.00	4.00	4.67	1.21	15.1
	.1875	³⁄₁₆	10.70	3.14	14.3	4.76	5.97	2.13	4.83	3.22	3.68	1.24	11.9
6x2	.3750	⅜	17.27	5.08	17.8	5.94	8.33	1.87	2.84	2.84	3.61	.748	8.72
	.3125	⁵⁄₁₆	14.83	4.36	16.0	5.34	7.33	1.92	2.62	2.62	3.22	.775	7.94
	.2500	¼	12.21	3.59	13.8	4.60	6.18	1.96	2.31	2.31	2.75	.802	6.88
	.1875	³⁄₁₆	9.42	2.77	11.1	3.70	4.88	2.00	1.90	1.90	2.20	.829	5.56
5x4	.3750	⅜	19.82	5.83	18.7	7.50	9.44	1.79	13.2	6.58	8.08	1.50	26.3
	.3125	⁵⁄₁₆	16.96	4.98	16.6	6.65	8.24	1.83	11.7	5.85	7.05	1.53	22.9
	.2500	¼	13.91	4.09	14.1	5.65	6.89	1.86	9.98	4.99	5.90	1.56	19.1
	.1875	³⁄₁₆	10.70	3.14	11.2	4.49	5.39	1.89	7.96	3.98	4.63	1.59	14.9
5x3	.5000	½	21.63	6.36	16.9	6.75	9.20	1.63	7.33	4.88	6.35	1.07	18.2
	.3750	⅜	17.27	5.08	14.7	5.89	7.71	1.70	6.48	4.32	5.35	1.13	15.6
	.3125	⁵⁄₁₆	14.83	4.36	13.2	5.27	6.77	1.74	5.85	3.90	4.72	1.16	13.8
	.2500	¼	12.21	3.59	11.3	4.52	5.70	1.77	5.05	3.37	3.98	1.19	11.7
	.1875	³⁄₁₆	9.42	2.77	9.06	3.62	4.49	1.81	4.08	2.72	3.15	1.21	9.21
5x2	.3125	⁵⁄₁₆	12.70	3.73	9.74	3.90	5.31	1.62	2.16	2.16	2.70	.762	6.24
	.2500	¼	10.51	3.09	8.48	3.39	4.51	1.66	1.92	1.92	2.32	.789	5.43
	.1875	³⁄₁₆	8.15	2.39	6.89	2.75	3.59	1.70	1.60	1.60	1.86	.816	4.40
4x3	.3125	⁵⁄₁₆	12.70	3.73	7.45	3.72	4.75	1.41	4.71	3.14	3.88	1.12	9.89
	.2500	¼	10.51	3.09	6.45	3.23	4.03	1.45	4.10	2.74	3.30	1.15	8.41
	.1875	³⁄₁₆	8.15	2.39	5.23	2.62	3.20	1.48	3.34	2.23	2.62	1.18	6.67
4x2	.3125	⁵⁄₁₆	10.58	3.11	5.32	2.66	3.60	1.31	1.71	1.71	2.17	.743	4.58
	.2500	¼	8.81	2.59	4.69	2.35	3.09	1.35	1.54	1.54	1.88	.770	4.01
	.1875	³⁄₁₆	6.87	2.02	3.87	1.93	2.48	1.38	1.29	1.29	1.52	.798	3.26
3x2	.2500	¼	7.11	2.09	2.21	1.47	1.92	1.03	1.15	1.15	1.44	.742	2.63
	.1875	³⁄₁₆	5.59	1.64	1.86	1.24	1.57	1.06	.977	.977	1.18	.771	2.16

*Outside dimensions across flat sides.
**Properties are based upon a nominal outside corner radius equal to two times the wall thickness.

BARS AND PLATES
Product availability

Plates and bars are readily available in seven of the structural steel specifications listed in Sect. A3.1 of the AISC LRFD Specification. These are: ASTM A36, A242, A441, A529, A572, A588, and A514. Bars are available in all of these steels except A514. Table 1, pg. 1–8, shows the availability of each steel in terms of plate thickness.

The Manual user is referred to the discussion on Selection of the Appropriate Structural Steel, pg. 1–3, for guidance in selection of both plate and structural shapes. For additional information designers should consult the steel producers.

CLASSIFICATION

Bars and plates are generally classified as follows:

Bars: 6 in. or less in width, .203 in. and over in thickness.
         Over 6 in. to 8 in. in width, .230 in. and over in thickness.
Plates: Over 8 in. to 48 in. in width, .230 in. and over in thickness.
         Over 48 in. in width, .180 in. and over in thickness.

BARS

Bars are available in various widths, thicknesses, diameters and lengths. The preferred practice is to specify widths in ¼-in. increments and thickness and diameter in ⅛-in. increments.

PLATES

Defined according to rolling procedure:

Sheared plates are rolled between horizontal rolls and trimmed (sheared or gas cut) on all edges.
Universal (UM) plates are rolled between horizontal and vertical rolls and trimmed (sheared or gas cut) on ends only.
Stripped plates are furnished to required widths by shearing or gas cutting from wider sheared plates.

Sizes

Plate mills are located in various districts, but the sizes of plates produced differ greatly and the catalogs of individual mills should be consulted for detail data. The extreme width of UM plates currently rolled is 60 in. and for sheared plates it is 200 in., but their availability together with limiting thickness and lengths should be checked with the mills before specifying. The preferred increments for width and thickness are:

Widths: Various. The catalogs of individual mills should be consulted to determine the most economical widths.
Thickness: 1/32 in. increments up to ½ in.
           1/16 in. increments over ½ to 1 in.
           ⅛ in. increments over 1 in. to 3 in.
           ¼ in. increments over 3 in.

Ordering

Plate thickness may be specified in inches or by weight per square foot, but no decimal edge thickness can be assured by the latter method. Separate tolerance tables apply to each method.

"*Sketch*" plates (i.e., plates whose dimensions and cuts are detailed), exclusive of those with re-entrant cuts, can be supplied by most mills by shearing or gas cutting, depending on thickness.

"*Full circles*" are also available, either by shearing up to 1 in. thickness, or by gas cutting for heavier gages.

Invoicing

Standard practice is to invoice plates to the fabricator at theoretical weight at point of shipment. Permissible variations in weight are in accordance with the tables of ASTM Specification A6.

All sketch plates, including circles, are invoiced at theoretical weight and, except as noted, are subject to the same weight variations as apply to rectangular plates. Odd shapes in most instances require gas cutting, for which gas cutting extras are applicable.

All plates ordered gas cut for whatever reason, or beyond published shearing limits, take extras for gas cutting in addition to all other extras. Rolled steel bearing plates are often gas cut to prevent distortion due to shearing but would also take the regular extra for the thickness involved.

Extras for thickness, width, length, cutting, quality and quantity, etc., which are added to the base price of plates, are subject to revision, and should be obtained by inquiry to the producer. The foregoing general statements are made as a guide toward economy in design.

FLOOR PLATES

Floor plates having raised patterns are available from several mills, each offering their own style of surface projections and in a variety of widths, thicknesses, and lengths. A maximum width of 96 in. and a maximum thickness of 1 in. are available, but availability of matching widths, thicknesses, and lengths should be checked with the producer. Floor plates are generally not specified to chemical composition limits or mechanical property requirements; a commercial grade of carbon steel is furnished. However, when strength or corrosion resistance is a consideration, raised pattern floor plates are procurable in any of the regular steel specifications. As in the case of plain plates, the individual manufacturers should be consulted for precise information. The nominal or ordered thickness is that of the flat plate, exclusive of the height of raised pattern. The usual weights are as follows:

THEORETICAL WEIGHTS OF ROLLED FLOOR PLATES

Gauge No.	Theoretical Weight per Sq. Ft, Lb.	Nominal Thickness, In.	Theoretical Weight per Sq. Ft, Lb.	Nominal Thickness, In.	Theoretical Weight per Sq. Ft, Lb.
18	2.40	1/8	6.16	1/2	21.47
16	3.00	3/16	8.71	9/16	24.02
14	3.75	1/4	11.26	5/8	26.58
13	4.50	5/16	13.81	3/4	31.68
12	5.25	3/8	16.37	7/8	36.78
		7/16	18.92	1	41.89

Note: Thickness is measured near the edge of the plate, exclusive of raised pattern.

SQUARE AND ROUND BARS
Weight and area

Size In.	Weight Lb. per Ft ■	Weight Lb. per Ft ●	Area Sq. In. ▨	Area Sq. In. ◯	Size In.	Weight Lb. per Ft ■	Weight Lb. per Ft ●	Area Sq. In. ▨	Area Sq. In. ◯
0					3	30.63	24.05	9.000	7.069
1/16	0.013	0.010	0.0039	0.0031	1/16	31.91	25.07	9.379	7.366
1/8	0.053	0.042	0.0156	0.0123	1/8	33.23	26.10	9.766	7.670
3/16	0.120	0.094	0.0352	0.0276	3/16	34.57	27.15	10.160	7.980
1/4	0.213	0.167	0.0625	0.0491	1/4	35.94	28.23	10.563	8.296
5/16	0.332	0.261	0.0977	0.0767	5/16	37.34	29.32	10.973	8.618
3/8	0.479	0.376	0.1406	0.1105	3/8	38.76	30.44	11.391	8.946
7/16	0.651	0.512	0.1914	0.1503	7/16	40.21	31.58	11.816	9.281
1/2	0.851	0.668	0.2500	0.1963	1/2	41.68	32.74	12.250	9.621
9/16	1.077	0.846	0.3164	0.2485	9/16	43.19	33.92	12.691	9.968
5/8	1.329	1.044	0.3906	0.3068	5/8	44.71	35.12	13.141	10.321
11/16	1.608	1.263	0.4727	0.3712	11/16	46.27	36.34	13.598	10.680
3/4	1.914	1.503	0.5625	0.4418	3/4	47.85	37.58	14.063	11.045
13/16	2.246	1.764	0.6602	0.5185	13/16	49.46	38.85	14.535	11.416
7/8	2.605	2.046	0.7656	0.6013	7/8	51.09	40.13	15.016	11.793
15/16	2.991	2.349	0.8789	0.6903	15/16	52.76	41.43	15.504	12.177
1	3.403	2.673	1.0000	0.7854	4	54.44	42.76	16.000	12.566
1/16	3.841	3.017	1.1289	0.8866	1/16	56.16	44.11	16.504	12.962
1/8	4.307	3.382	1.2656	0.9940	1/8	57.90	45.47	17.016	13.364
3/16	4.798	3.769	1.4102	1.1075	3/16	59.67	46.86	17.535	13.772
1/4	5.317	4.176	1.5625	1.2272	1/4	61.46	48.27	18.063	14.186
5/16	5.862	4.604	1.7227	1.3530	5/16	63.28	49.70	18.598	14.607
3/8	6.433	5.053	1.8906	1.4849	3/8	65.13	51.15	19.141	15.033
7/16	7.032	5.523	2.0664	1.6230	7/16	67.01	52.63	19.691	15.466
1/2	7.656	6.013	2.2500	1.7671	1/2	68.91	54.12	20.250	15.904
9/16	8.308	6.525	2.4414	1.9175	9/16	70.83	55.63	20.816	16.349
5/8	8.985	7.057	2.6406	2.0739	5/8	72.79	57.17	21.391	16.800
11/16	9.690	7.610	2.8477	2.2365	11/16	74.77	58.72	21.973	17.257
3/4	10.421	8.185	3.0625	2.4053	3/4	76.78	60.30	22.563	17.721
13/16	11.179	8.780	3.2852	2.5802	13/16	78.81	61.90	23.160	18.190
7/8	11.963	9.396	3.5156	2.7612	7/8	80.87	63.51	23.766	18.665
15/16	12.774	10.032	3.7539	2.9483	15/16	82.96	65.15	24.379	19.147
2	13.611	10.690	4.0000	3.1416	5	85.07	66.81	25.000	19.635
1/16	14.475	11.369	4.2539	3.3410	1/16	87.21	68.49	25.629	20.129
1/8	15.366	12.068	4.5156	3.5466	1/8	89.38	70.20	26.266	20.629
3/16	16.283	12.788	4.7852	3.7583	3/16	91.57	71.92	26.910	21.135
1/4	17.227	13.530	5.0625	3.9761	1/4	93.79	73.66	27.563	21.648
5/16	18.197	14.292	5.3477	4.2000	5/16	96.04	75.43	28.223	22.166
3/8	19.194	15.075	5.6406	4.4301	3/8	98.31	77.21	28.891	22.691
7/16	20.217	15.879	5.9414	4.6664	7/16	100.61	79.02	29.566	23.221
1/2	21.267	16.703	6.2500	4.9087	1/2	102.93	80.84	30.250	23.758
9/16	22.344	17.549	6.5664	5.1572	9/16	105.29	82.69	30.941	24.301
5/8	23.447	18.415	6.8906	5.4119	5/8	107.67	84.56	31.641	24.850
11/16	24.577	19.303	7.2227	5.6727	11/16	110.07	86.45	32.348	25.406
3/4	25.734	20.211	7.5625	5.9396	3/4	112.50	88.36	33.063	25.967
13/16	26.917	21.140	7.9102	6.2126	13/16	114.96	90.29	33.785	26.535
7/8	28.126	22.090	8.2656	6.4918	7/8	117.45	92.24	34.516	27.109
15/16	29.362	23.061	8.6289	6.7771	15/16	119.96	94.22	35.254	27.688
3									

SQUARE AND ROUND BARS
Weight and area

Size In.	Weight Lb. per Ft ■	Weight Lb. per Ft ●	Area Sq. In. ▨	Area Sq. In. ◎	Size In.	Weight Lb. per Ft ■	Weight Lb. per Ft ●	Area Sq. In. ▨	Area Sq. In. ◎
6	122.50	96.21	36.000	28.274	9	275.63	216.48	81.000	63.617
1/16	125.07	98.23	36.754	28.866	1/16	279.47	219.49	82.129	64.504
1/8	127.66	100.26	37.516	29.465	1/8	283.33	222.53	83.266	65.397
3/16	130.28	102.32	38.285	30.069	3/16	287.23	225.59	84.410	66.296
1/4	132.92	104.40	39.063	30.680	1/4	291.15	228.67	85.563	67.201
5/16	135.59	106.49	39.848	31.296	5/16	295.10	231.77	86.723	68.112
3/8	138.29	108.61	40.641	31.919	3/8	299.07	234.89	87.891	69.029
7/16	141.02	110.75	41.441	32.548	7/16	303.07	238.03	89.066	69.953
1/2	143.77	112.91	42.250	33.183	1/2	307.10	241.20	90.250	70.882
9/16	146.55	115.10	43.066	33.824	9/16	311.15	244.38	91.441	71.818
5/8	149.35	117.30	43.891	34.472	5/8	315.24	247.59	92.641	72.760
11/16	152.18	119.52	44.723	35.125	11/16	319.34	250.81	93.848	73.708
3/4	155.04	121.77	45.563	35.785	3/4	323.48	254.06	95.063	74.662
13/16	157.92	124.03	46.410	36.450	13/16	327.64	257.33	96.285	75.622
7/8	160.83	126.32	47.266	37.122	7/8	331.82	260.61	97.516	76.589
15/16	163.77	128.63	48.129	37.800	15/16	336.04	263.92	98.754	77.561
7	166.74	130.95	49.000	38.485	10	340.28	267.25	100.000	78.540
1/16	169.73	133.30	49.879	39.175	1/16	344.54	270.60	101.254	79.525
1/8	172.74	135.67	50.766	39.871	1/8	348.84	273.98	102.516	80.516
3/16	175.79	138.06	51.660	40.574	3/16	353.16	277.37	103.785	81.513
1/4	178.86	140.48	52.563	41.282	1/4	357.50	280.78	105.063	82.516
5/16	181.96	142.91	53.473	41.997	5/16	361.88	284.22	106.348	83.525
3/8	185.08	145.36	54.391	42.718	3/8	366.28	287.67	107.641	84.541
7/16	188.23	147.84	55.316	43.445	7/16	370.70	291.15	108.941	85.562
1/2	191.41	150.33	56.250	44.179	1/2	375.16	294.65	110.250	86.590
9/16	194.61	152.85	57.191	44.918	9/16	379.64	298.17	111.566	87.624
5/8	197.84	155.38	58.141	45.664	5/8	384.14	301.70	112.891	88.664
11/16	201.10	157.94	59.098	46.415	11/16	388.67	305.26	114.223	89.710
3/4	204.38	160.52	60.063	47.173	3/4	393.23	308.84	115.563	90.763
13/16	207.69	163.12	61.035	47.937	13/16	397.82	312.45	116.910	91.821
7/8	211.03	165.74	62.016	48.707	7/8	402.43	316.07	118.266	92.886
15/16	214.39	168.38	63.004	49.483	15/16	407.07	319.71	119.629	93.956
8	217.78	171.04	64.000	50.265	11	411.74	323.38	121.000	95.033
1/16	221.19	173.73	65.004	51.054	1/16	416.43	327.06	122.379	96.116
1/8	224.64	176.43	66.016	51.849	1/8	421.15	330.77	123.766	97.205
3/16	228.11	179.15	67.035	52.649	3/16	425.89	334.49	125.160	98.301
1/4	231.60	181.90	68.063	53.456	1/4	430.66	338.24	126.563	99.402
5/16	235.12	184.67	69.098	54.269	5/16	435.46	342.01	127.973	100.510
3/8	238.67	187.45	70.141	55.088	3/8	440.29	345.80	129.391	101.623
7/16	242.25	190.26	71.191	55.914	7/16	445.14	349.61	130.816	102.743
1/2	245.85	193.09	72.250	56.745	1/2	450.02	353.44	132.250	103.869
9/16	249.48	195.94	73.316	57.583	9/16	454.92	357.30	133.691	105.001
5/8	253.13	198.81	74.391	58.426	5/8	459.85	361.17	135.141	106.139
11/16	256.82	201.70	75.473	59.276	11/16	464.81	365.06	136.598	107.284
3/4	260.53	204.62	76.563	60.132	3/4	469.80	368.98	138.063	108.434
13/16	264.26	207.55	77.660	60.994	13/16	474.81	372.91	139.535	109.591
7/8	268.02	210.50	78.766	61.862	7/8	479.84	376.87	141.016	110.753
15/16	271.81	213.48	79.879	62.737	15/16	484.91	380.85	142.504	111.922
					12	490.00	384.85	144.000	113.097

AREA OF RECTANGULAR SECTIONS
Square inches

Width In.	Thickness, Inches													
	3/16	1/4	5/16	3/8	7/16	1/2	9/16	5/8	11/16	3/4	13/16	7/8	15/16	1
1/4	0.047	0.063	0.078	0.094	0.109	0.125	0.141	0.156	0.172	0.188	0.203	0.219	0.234	0.250
1/2	0.094	0.125	0.156	0.188	0.219	0.250	0.281	0.313	0.344	0.375	0.406	0.438	0.469	0.500
3/4	0.141	0.188	0.234	0.281	0.328	0.375	0.422	0.469	0.516	0.563	0.609	0.656	0.703	0.750
1	0.188	0.250	0.313	0.375	0.438	0.500	0.563	0.625	0.688	0.750	0.813	0.875	0.938	1.00
1 1/4	0.234	0.313	0.391	0.469	0.547	0.625	0.703	0.781	0.859	0.938	1.02	1.09	1.17	1.25
1 1/2	0.281	0.375	0.469	0.563	0.656	0.750	0.844	0.938	1.03	1.13	1.22	1.31	1.41	1.50
1 3/4	0.328	0.438	0.547	0.656	0.766	0.875	0.984	1.09	1.20	1.31	1.42	1.53	1.64	1.75
2	0.375	0.500	0.625	0.750	0.875	1.00	1.13	1.25	1.38	1.50	1.63	1.75	1.88	2.00
2 1/4	0.422	0.563	0.703	0.844	0.984	1.13	1.27	1.41	1.55	1.69	1.83	1.97	2.11	2.25
2 1/2	0.469	0.625	0.781	0.938	1.09	1.25	1.41	1.56	1.72	1.88	2.03	2.19	2.34	2.50
2 3/4	0.516	0.688	0.859	1.03	1.20	1.38	1.55	1.72	1.89	2.06	2.23	2.41	2.58	2.75
3	0.563	0.750	0.938	1.13	1.31	1.50	1.69	1.88	2.06	2.25	2.44	2.63	2.81	3.00
3 1/4	0.609	0.813	1.02	1.22	1.42	1.63	1.83	2.03	2.23	2.44	2.64	2.84	3.05	3.25
3 1/2	0.656	0.875	1.09	1.31	1.53	1.75	1.97	2.19	2.41	2.63	2.84	3.06	3.28	3.50
3 3/4	0.703	0.938	1.17	1.41	1.64	1.88	2.11	2.34	2.58	2.81	3.05	3.28	3.52	3.75
4	0.750	1.00	1.25	1.50	1.75	2.00	2.25	2.50	2.75	3.00	3.25	3.50	3.75	4.00
4 1/4	0.797	1.06	1.33	1.59	1.86	2.13	2.39	2.66	2.92	3.19	3.45	3.72	3.98	4.25
4 1/2	0.844	1.13	1.41	1.69	1.97	2.25	2.53	2.81	3.09	3.38	3.66	3.94	4.22	4.50
4 3/4	0.891	1.19	1.48	1.78	2.09	2.38	2.67	2.97	3.27	3.56	3.86	4.16	4.45	4.75
5	0.938	1.25	1.56	1.88	2.19	2.50	2.81	3.13	3.44	3.75	4.06	4.38	4.69	5.00
5 1/4	0.984	1.31	1.64	1.97	2.30	2.63	2.95	3.28	3.61	3.94	4.27	4.59	4.92	5.25
5 1/2	1.03	1.38	1.72	2.06	2.41	2.75	3.09	3.44	3.78	4.13	4.47	4.81	5.16	5.50
5 3/4	1.08	1.44	1.80	2.16	2.52	2.88	3.23	3.59	3.95	4.31	4.67	5.03	5.39	5.75
6	1.13	1.50	1.88	2.25	2.63	3.00	3.38	3.75	4.13	4.50	4.88	5.25	5.63	6.00
6 1/4	1.17	1.56	1.95	2.34	2.73	3.13	3.52	3.91	4.30	4.69	5.08	5.47	5.86	6.25
6 1/2	1.22	1.63	2.03	2.44	2.84	3.25	3.66	4.06	4.47	4.88	5.28	5.69	6.09	6.50
6 3/4	1.27	1.69	2.10	2.53	2.95	3.38	3.80	4.22	4.64	5.06	5.48	5.91	6.33	6.75
7	1.31	1.75	2.19	2.63	3.06	3.50	3.94	4.38	4.81	5.25	5.69	6.13	6.56	7.00
7 1/4	1.36	1.81	2.27	2.72	3.17	3.63	4.08	4.53	4.98	5.44	5.89	6.34	6.80	7.25
7 1/2	1.41	1.88	2.34	2.81	3.28	3.75	4.22	4.69	5.16	5.63	6.09	6.56	7.03	7.50
7 3/4	1.45	1.94	2.42	2.91	3.39	3.88	4.36	4.84	5.33	5.81	6.30	6.78	7.27	7.75
8	1.50	2.00	2.50	3.00	3.50	4.00	4.50	5.00	5.50	6.00	6.50	7.00	7.50	8.00
8 1/2	1.59	2.13	2.66	3.19	3.72	4.25	4.78	5.31	5.84	6.38	6.91	7.44	7.97	8.50
9	1.69	2.25	2.81	3.38	3.94	4.50	5.06	5.63	6.19	6.75	7.31	7.88	8.44	9.00
9 1/2	1.78	2.38	2.97	3.56	4.16	4.75	5.34	5.94	6.53	7.13	7.72	8.31	8.91	9.50
10	1.88	2.50	3.13	3.75	4.38	5.00	5.63	6.25	6.88	7.50	8.13	8.75	9.38	10.00
10 1/2	1.97	2.63	3.28	3.94	4.59	5.25	5.91	6.56	7.22	7.88	8.53	9.19	9.84	10.50
11	2.06	2.75	3.44	4.13	4.81	5.50	6.19	6.88	7.56	8.25	8.94	9.63	10.31	11.00
11 1/2	2.16	2.88	3.59	4.31	5.03	5.75	6.47	7.19	7.91	8.63	9.34	10.06	10.78	11.50
12	2.25	3.00	3.75	4.50	5.25	6.00	6.75	7.50	8.25	9.00	9.75	10.50	11.25	12.00

WEIGHT OF RECTANGULAR SECTIONS
Pounds per linear foot

Width In.	Thickness, Inches													
	3/16	1/4	5/16	3/8	7/16	1/2	9/16	5/8	11/16	3/4	13/16	7/8	15/16	1
1/4	0.16	0.21	0.27	0.32	0.37	0.43	0.48	0.53	0.58	0.64	0.69	0.74	0.80	0.85
1/2	0.32	0.43	0.53	0.64	0.74	0.85	0.96	1.06	1.17	1.28	1.38	1.49	1.60	1.70
3/4	0.48	0.64	0.80	0.96	1.12	1.28	1.44	1.60	1.75	1.91	2.07	2.23	2.39	2.55
1	0.64	0.85	1.06	1.28	1.49	1.70	1.91	2.13	2.34	2.55	2.76	2.98	3.19	3.40
1 1/4	0.80	1.06	1.33	1.60	1.86	2.13	2.39	2.66	2.92	3.19	3.46	3.72	3.99	4.25
1 1/2	0.96	1.28	1.60	1.91	2.23	2.56	2.87	3.19	3.51	3.83	4.15	4.47	4.79	5.10
1 3/4	1.12	1.49	1.86	2.23	2.61	2.98	3.35	3.72	4.09	4.47	4.84	5.21	5.58	5.95
2	1.28	1.70	2.13	2.55	2.98	3.40	3.83	4.25	4.68	5.10	5.53	5.95	6.38	6.81
2 1/4	1.44	1.91	2.39	2.87	3.35	3.83	4.31	4.79	5.26	5.74	6.22	6.70	7.18	7.66
2 1/2	1.60	2.13	2.66	3.19	3.72	4.25	4.79	5.32	5.85	6.38	6.91	7.44	7.98	8.51
2 3/4	1.75	2.34	2.92	3.51	4.09	4.68	5.26	5.85	6.43	7.02	7.60	8.19	8.77	9.36
3	1.91	2.55	3.19	3.83	4.47	5.10	5.74	6.38	7.02	7.66	8.29	8.93	9.57	10.2
3 1/4	2.07	2.76	3.46	4.15	4.84	5.53	6.22	6.91	7.60	8.29	8.99	9.68	10.4	11.1
3 1/2	2.23	2.98	3.72	4.47	5.21	5.95	6.70	7.44	8.19	8.93	9.68	10.4	11.2	11.9
3 3/4	2.39	3.19	3.99	4.79	5.58	6.38	7.18	7.98	8.77	9.57	10.4	11.2	12.0	12.8
4	2.55	3.40	4.25	5.10	5.95	6.81	7.66	8.51	9.36	10.2	11.1	11.9	12.8	13.6
4 1/4	2.71	3.62	4.52	5.42	6.33	7.23	8.13	9.04	9.94	10.8	11.8	12.7	13.6	14.5
4 1/2	2.87	3.83	4.79	5.74	6.70	7.66	8.61	9.57	10.5	11.5	12.4	13.4	14.4	15.3
4 3/4	3.03	4.04	5.05	6.06	7.07	8.08	9.09	10.1	11.1	12.1	13.1	14.1	15.2	16.2
5	3.19	4.25	5.32	6.38	7.44	8.51	9.57	10.6	11.7	12.8	13.8	14.9	16.0	17.0
5 1/4	3.35	4.47	5.58	6.70	7.82	8.93	10.0	11.2	12.3	13.4	14.5	15.6	16.7	17.9
5 1/2	3.51	4.68	5.85	7.02	8.19	9.36	10.5	11.7	12.9	14.0	15.2	16.4	17.5	18.7
5 3/4	3.67	4.89	6.11	7.34	8.56	9.78	11.0	12.2	13.5	14.7	15.9	17.1	18.3	19.6
6	3.83	5.10	6.38	7.66	8.93	10.2	11.5	12.8	14.0	15.3	16.6	17.9	19.1	20.4
6 1/4	3.99	5.32	6.65	7.98	9.30	10.6	12.0	13.3	14.6	16.0	17.3	18.6	19.9	21.3
6 1/2	4.15	5.53	6.91	8.29	9.68	11.1	12.4	13.8	15.2	16.6	18.0	19.4	20.7	22.1
6 3/4	4.31	5.74	7.18	8.61	10.0	11.5	12.9	14.4	15.8	17.2	18.7	20.1	21.5	23.0
7	4.47	5.95	7.44	8.93	10.4	11.9	13.4	14.9	16.4	17.9	19.4	20.8	22.3	23.8
7 1/4	4.63	6.17	7.71	9.25	10.8	12.3	13.9	15.4	17.0	18.5	20.0	21.6	23.1	24.7
7 1/2	4.79	6.38	7.98	9.57	11.2	12.8	14.4	16.0	17.5	19.1	20.7	22.3	23.9	25.5
7 3/4	4.94	6.59	8.24	9.89	11.5	13.2	14.8	16.5	18.1	19.8	21.4	23.1	24.7	26.4
8	5.10	6.81	8.51	10.2	11.9	13.6	15.3	17.0	18.7	20.4	22.1	23.8	25.5	27.2
8 1/2	5.42	7.23	9.04	10.8	12.7	14.5	16.3	18.1	19.9	21.7	23.5	25.3	27.1	28.9
9	5.74	7.66	9.57	11.5	13.4	15.3	17.2	19.1	21.1	23.0	24.9	26.8	28.7	30.6
9 1/2	6.06	8.08	10.1	12.1	14.1	16.2	18.2	20.2	22.2	24.2	26.3	28.3	30.3	32.3
10	6.38	8.51	10.6	12.8	14.9	17.0	19.1	21.3	23.4	25.5	27.6	29.8	31.9	34.0
10 1/2	6.70	8.93	11.2	13.4	15.6	17.9	20.1	22.3	24.6	26.8	29.0	31.3	33.5	35.7
11	7.02	9.36	11.7	14.0	16.4	18.7	21.1	23.4	25.7	28.1	30.4	32.8	35.1	37.4
11 1/2	7.34	9.78	12.2	14.7	17.1	19.6	22.0	24.5	26.9	29.3	31.8	34.2	36.7	39.1
12	7.66	10.2	12.8	15.3	17.9	20.4	23.0	25.5	28.1	30.6	33.2	35.7	38.3	40.8

CRANE RAILS

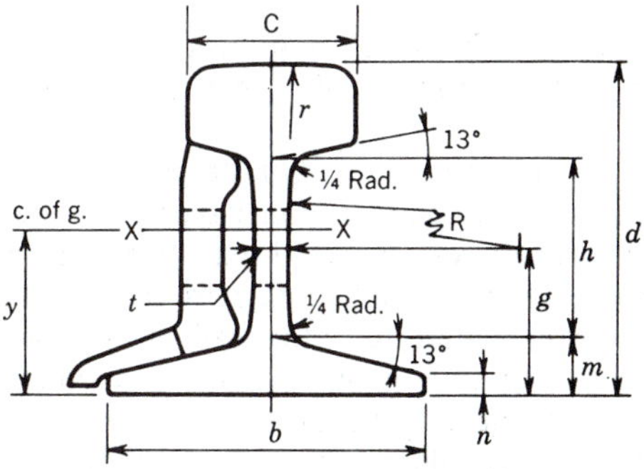

A.S.C.E. 40, 60 & 85 lb.

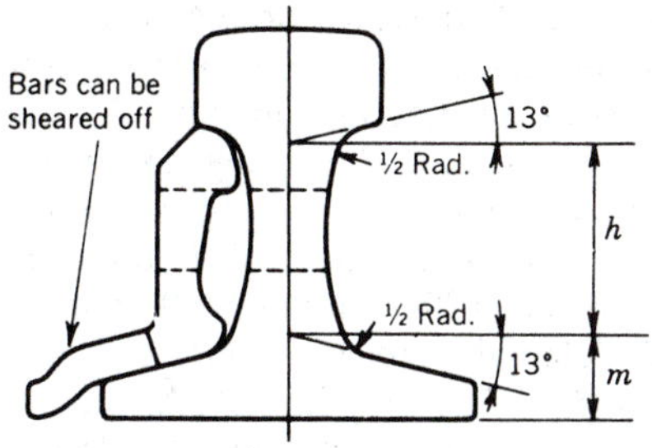

C

r

13°

¼ Rad.

c. of g.

X — X

R

¼ Rad.

13°

t

y

d

h

g

m

b

n

Bethlehem 104 lb.

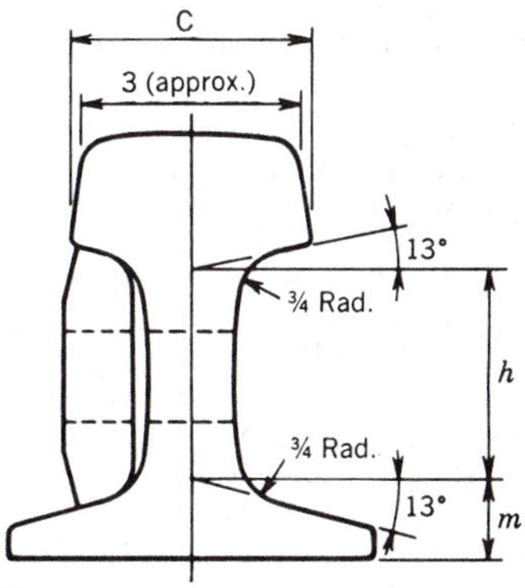

Bars can be sheared off

13°

½ Rad.

½ Rad.

13°

h

m

Bethlehem 135 lb.

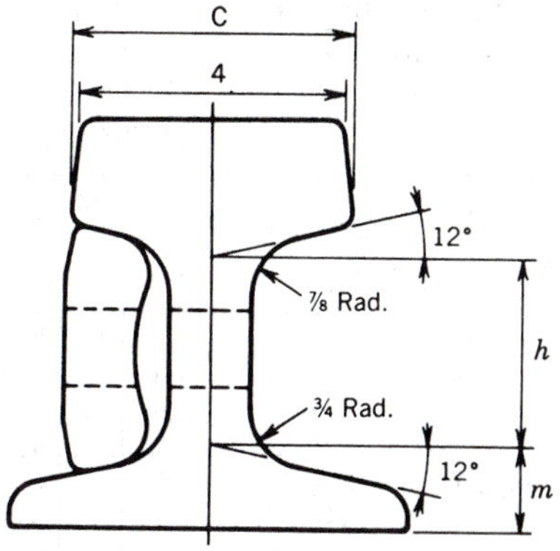

C

3 (approx.)

13°

¾ Rad.

¾ Rad.

13°

h

m

Bethlehem 171 lb.

C

4

12°

⅞ Rad.

¾ Rad.

12°

h

m

Bethlehem 175 lb.

C

4 1/32 (approx.)

12°

2 Rad.

1⅛ Rad.

12°

h

$2^{53}/_{64}$

m

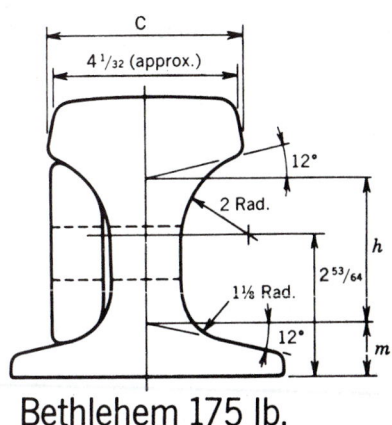

Nomenclature of sketch for A.S.C.E. rails also applies to the other sections.

AMERICAN INSTITUTE OF STEEL CONSTRUCTION

CRANE RAILS

GENERAL NOTES

The ASCE rails and the 104- to 175-lb. crane rails listed below are recommended for crane runway use. For completes details and for profiles and properties of rails not listed, consulted manufacturers' catalogs.

Rails should be arranged so that joints on opposite sides of the crane runway will be staggered with respect to each other and with due consideration to the wheelbase of the crane. Rail joints should not occur at crane girder splices. Light 40-lb. rails are available in 30 ft lengths, 60-lb. rails in 30-, 33- or 39-ft lengths, standard rails in 33- or 39- ft lengths and crane rails up to 60 ft. Consult manufacturer for availability of other lengths. Odd lengths, which must be included to complete a run or obtain the necessary stagger, should not be less than 10 ft long. For crane rail service, 40-lb. rails are furnished to manufacturers' specifications and tolerances. 60 and 85-lb. rails are furnished to manufacturers' specifications and tolerances, or to ASTM A1. Crane rails are furnished to ASTM A759. Rails will be furnished with standard drilling (see pg. 1-130) in both standard and odd lengths unless stipulated otherwise on order. For controlled cooling, heat treatment and rail end preparation, see manufacturers' catalogs. Purchase orders for crane rails should be noted "For crane service."

DIMENSIONS AND PROPERTIES

Type	Classification	Nominal Wt. per yd.	d	Gage g	Base			Head		Web			Properties—Axis X-X				
					b	m	n	c	r	t	h	R	Area	I	S		y
															Hd.	Bse.	
		Lb.	In.	In.	In.	In.	In.	In.	In.	In.	In.	In.	In.^{2}	In.^{4}	In.^{3}	In.^{3}	In.
ASCE	Light	40	$3\frac{1}{2}$	$1\frac{71}{128}$	$3\frac{1}{2}$	$\frac{5}{8}$	$\frac{7}{32}$	$1\frac{7}{8}$	12	$\frac{25}{64}$	$1\frac{55}{64}$	12	3.94	6.54	3.59	3.89	1.68
ASCE	Light	60	$4\frac{1}{4}$	$1\frac{115}{128}$	$4\frac{1}{4}$	$\frac{49}{64}$	$\frac{9}{32}$	$2\frac{3}{8}$	12	$\frac{31}{64}$	$2\frac{17}{64}$	12	5.93	14.6	6.64	7.12	2.05
ASCE	Std.	85	$5\frac{3}{16}$	$2\frac{17}{64}$	$5\frac{3}{16}$	$\frac{57}{64}$	$\frac{19}{64}$	$2\frac{9}{16}$	12	$\frac{9}{16}$	$2\frac{3}{4}$	12	8.33	30.1	11.1	12.2	2.47
Bethlehem	Crane	104	5	$2\frac{7}{16}$	5	$1\frac{1}{16}$	$\frac{1}{2}$	$2\frac{1}{2}$	12	1	$2\frac{7}{16}$	$3\frac{1}{2}$	10.3	29.8	10.7	13.5	2.21
Bethlehem	Crane	135	$5\frac{3}{4}$	$2\frac{15}{32}$	$5\frac{3}{16}$	$1\frac{1}{16}$	$\frac{15}{32}$	$3\frac{7}{16}$	14	$1\frac{1}{4}$	$2\frac{13}{16}$	12	13.3	50.8	17.3	18.1	2.81
Bethlehem	Crane	171	6	$2\frac{5}{8}$	6	$1\frac{1}{4}$	$\frac{5}{8}$	4.3	Flat	$1\frac{1}{4}$	$2\frac{3}{4}$	Vert.	16.8	73.4	24.5	24.4	3.01
Bethlehem	Crane	175	6	$2\frac{21}{32}$	6	$1\frac{9}{64}$	$\frac{1}{2}$	$4\frac{1}{4}$	18	$1\frac{1}{2}$	$3\frac{7}{64}$	Vert.	17.1	70.5	23.4	23.6	2.98

For maximum wheel loadings see manufacturers' catalogs.

CRANE RAILS
Splices

WELDED SPLICES

When welded splices are specified, consult the manufacturer for recommended rail end preparation, welding procedure and method of ordering. Although joint continuity, made possible by this method of splicing, is desirable, it should be cautioned that the careful control required in all stages of the welding operation may be difficult to meet during crane rail installation.

In any event, rails should not be spliced by welding straps in the webs, nor should they be attached to structural supports by welding. Rails with holes for joint bar bolts should not be used in making welded splices.

BOLTED SPLICES

It is often more desirable to use properly installed and maintained bolted splice bars in making up rail joints for crane service.

Standard rail drilling and joint bar punching, as furnished by manufacturers of light standard rails for track work, include round holes in rail ends and slotted holes in joint bars to receive standard oval neck track bolts. Holes in rails are oversize and punching in joint bars is spaced to allow ⅟₁₆ to ⅛ in. clearance between rail ends (see manufacturers' catalogs for spacing and dimensions of holes and slots). Although this construction is satisfactory for track and light crane service, its use in general crane service may lead to joint failure.

For best service in bolted splices, it is recommended that tight joints be stipulated for all rails for crane service. This will require rail ends to be finished by milling or grinding, and the special rail drilling and joint bar punching tabulated below. Special rail drilling is accepted by some mills, or rails may be ordered blank for shop drilling. End finishing of standard rails can be done at the mill; light rails must be end finished in the fabricating shop or ground at the site prior to erection. In the crane rail range, from 104 to 175 lbs. per yard, rails and joint bars are manufactured to obtain a tight fit and no further special end finishing, drilling or punching is required. Because of cumulative tolerance variations in holes, bolt diameters and rail ends, a slight gap may sometimes occur in the so-called tight joints. Conversely, it may sometimes be necessary to ream holes through joint bar and rail to permit entry of bolts.

Joint bars for crane service are provided in various sections to match the rails. Joint bars for light and standard rails may be purchased blank for special shop punching to obtain tight joints. See Bethlehem Steel Corp. Booklet 3351 for dimensions, material specifications and the identification necessary to match the crane rail section.

Joint bar bolts, as distinguished from oval neck track bolts, have straight shanks to the head and are manufactured to ASTM A449 specifications. Nuts are manufactured to ASTM A563 Gr. B specifications. ASTM A325 bolts and nuts may be used. Bolt assembly includes an alloy steel spring washer, furnished to AREA specification.

After installation, bolts should be retightened within 30 days and every three months thereafter.

CRANE RAILS

Splices for tight joints

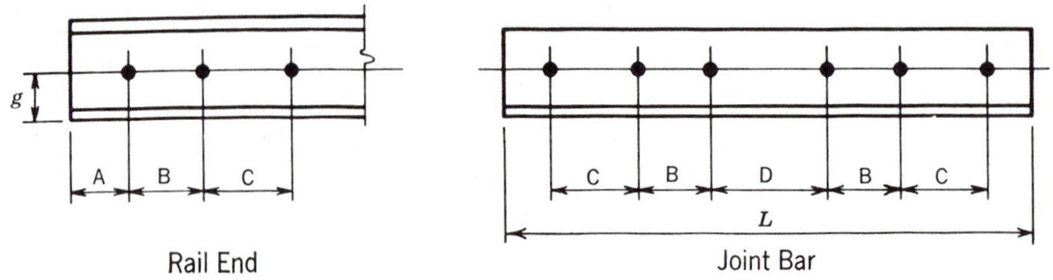
g
A
B
C
Rail End
C
B
D
B
C
L
Joint Bar

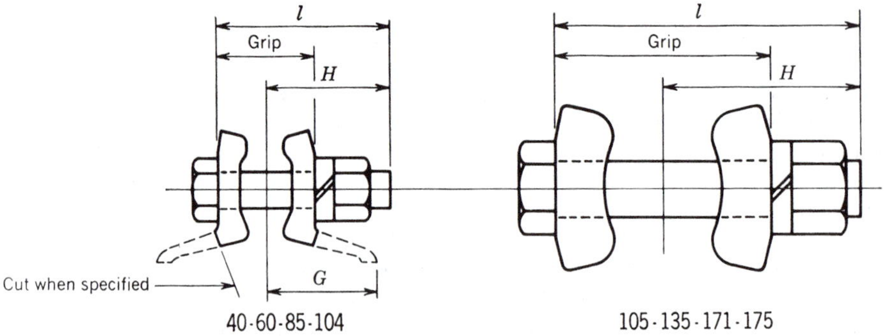
l
Grip
H
G
Cut when specified
40-60-85-104
l
Grip
H
105-135-171-175

Rail						Joint Bar						Bolt				Washer		Wt. 2 Bars Bolts, Nuts Washers	
Wt. per Yard	Drilling					Punching				L	G	Dia.	Grip	l	H	Inside Dia.	Thickness and Width	With Fig.	Less Fig.
	g	Hole Dia.	A	B	C	Hole Dia.	D	B	C										
Lb.	In.	In.	In.	In.	In.	In.	In.	In.	In.	In.	In.	In.	In.	In.	In.	In.	In.	Lb.	Lb.
40	1 71/128	13/16*	2 1/2	5	...	13/16*	4 15/16*	5	...	20	2 3/16	3/4	1 15/16	3 1/2	2 1/2	13/16	7/16 x 3/8	20.0	16.5
60	1 115/128	13/16*	2 1/2	5	...	13/16*	4 15/16*	5	...	24	2 11/16	3/4	2 19/32	4	2 11/16	13/16	7/16 x 3/8	36.5	29.6
85	2 17/64	15/16*	2 1/2	5	...	15/16*	4 15/16*	5	...	24	3 11/32	7/8	3 5/32	4 3/4	3 3/16	15/16	7/16 x 3/8	56.6	45.3
104	2 7/16	1 1/16	4	5	6	1 1/16	7 15/16	5	6	34	3 1/2	1	3 1/2	5 1/4	3 1/2	1 1/16	7/16 x 1/2	73.5	55.4
135	2 15/32	1 3/16	4	5	6	1 3/16	7 15/16	5	6	34	...	1 1/8	3 5/8	5 1/2	3 11/16	1 3/16	7/16 x 1/2	...	75.3
171	2 5/8	1 3/16	4	5	6	1 3/16	7 15/16	5	6	34	...	1 1/8	4 7/16	6 1/4	4 1/16	1 3/16	7/16 x 1/2	...	90.8
175	2 21/32	1 3/16	4	5	6	1 3/16	7 15/16	5	6	34	...	1 1/8	4 1/8	6 1/4	3 15/16	1 3/16	7/16 x 1/2	...	87.7

*Special rail drilling and joint bar punching.

CRANE RAILS
Fastenings

HOOK BOLTS

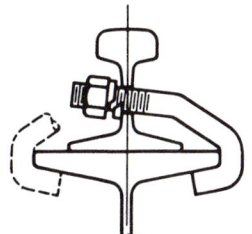

Hook bolts are used primarily with light rails when attached to beams with flanges too narrow for clamps. Rail adjustment up to ± ½ inch is inherent in the threaded shank. Hook bolts are paired alternately 3 to 4 inches apart, spaced at about 24 inch centers. The special rail drilling required must be done at the fabricator's shop.

RAIL CLAMPS

Although a variety of satisfactory rail clamps are available from track accessory manufacturers, two, frequently recommended for crane runway use, are the fixed and floating types illustrated below. These are available in forgings or pressed steel, either for single bolts or for double bolts as shown. The fixed type features adjustment through eccentric punching of fillers and positive attachment of rail to support. The floating type permits longitudinal and controlled transverse movement through clamp clearances and filler adjustment, useful in allowing for thermal expansion and contraction of rails, and possible misalignment of supports. Both types should be spaced 3 ft or less apart.

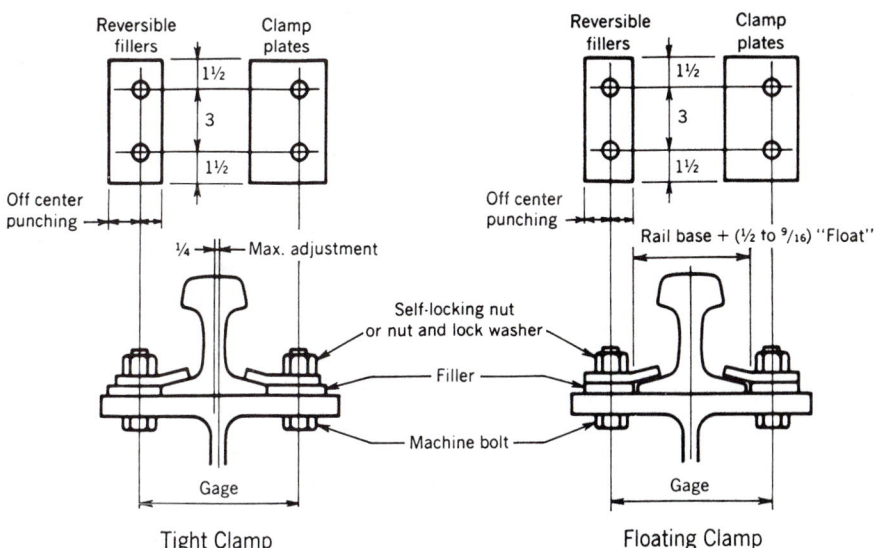

Tight Clamp   Floating Clamp

Dimensions shown above are suggested. See manufacturers' catalogs for recommended gages, bolt sizes and detail dimensions not shown.

Notes

TORSION PROPERTIES
W shapes

Torsional analysis is not required for the routine design of most structural steel members. Where torsional analysis is required, the table of Torsion Properties for W Shapes will be of assistance in utilizing current analysis methods.

NOMENCLATURE

C_w Warping constant for a section, in.^{6}
E Modulus of Elasticity of steel (29,000 ksi)
G Shear modulus of elasticity of steel (11,200 ksi)
J Torsional Constant for a section, in.^{4}
Q_f Statical moment for a point in the flange directly above the vertical edge of the web, in.^{3}
Q_w Statical moment at mid-depth of the section, in.^{3}
S_w Warping statical moment at a point in the section, in.^{4}
W_{no} Normalized warping function at a point at the flange edge, in.^{2}

TORSION PROPERTIES
W shapes

Designation	Torsional Constant J	Warping Constant C_w	$\sqrt{\dfrac{EC_w}{GJ}}$	Normalized Warping Constant W_{no}	Warping Statical Moment S_w	Statical Moment Q_f	Statical Moment Q_w
	In.^{4}	In.^{6}	In.	In.^{2}	In.^{4}	In.^{3}	In.^{3}
W 40x328	74.2	607000	145	171	1330	287	755
x298	56.3	540000	158	170	1190	261	684
x268	41.1	475000	173	168	1060	234	612
x244	30.4	417000	188	167	934	208	550
x221	21.2	349000	206	166	785	176	483
x192	13.7	268000	225	165	608	137	404
W 40x655	596	1140000	70.4	169	2520	534	1530
x593	451	989000	75.4	166	2240	484	1380
x531	329	848000	81.7	163	1950	433	1230
x480	245	739000	88.4	160	1730	391	1090
x436	186	649000	95.1	158	1540	354	992
x397	142	577000	103	156	1380	323	894
x362	109	511000	110	154	1240	294	813
x324	79.4	446000	121	152	1100	264	730
x297	61.2	397000	130	151	986	240	665
x277	51.1	378000	138	151	940	230	624
x249	37.7	333000	151	149	836	208	560
x215	24.4	283000	173	149	714	179	481
x199	18.1	245000	187	148	621	157	434
W 40x183	19.6	119000	126	111	402	134	391
x167	14.0	99300	136	111	336	113	346
x149	9.62	79600	146	110	270	92	299
W 36x848	1270	1620000	57.5	172	3530	674	1910
x798	1070	1480000	59.8	169	3270	634	1790
x720	804	1270000	64.0	166	2870	571	1590
x650	600	1090000	68.6	162	2520	513	1420
x588	453	950000	73.7	159	2240	465	1280
x527	330	816000	80.0	156	1960	415	1130
x485	260	727000	85.2	154	1770	380	1040
x439	195	637000	91.9	152	1570	344	928
x393	143	554000	100	150	1390	309	830
x359	109	493000	108	148	1240	281	757
x328	84.5	441000	117	146	1130	258	691
x300	64.2	398000	127	146	1020	235	628
x280	52.6	366000	134	145	944	219	585
x260	41.5	330000	144	144	858	200	538
x245	34.6	306000	151	143	799	187	505
x230	28.6	282000	160	143	740	175	472
W 36x256	53.3	168000	90.6	109	576	176	520
x232	39.8	148000	98.3	108	512	159	468
x210	28.0	128000	109	108	446	138	416
x194	22.2	116000	116	107	407	128	383
x182	18.4	107000	123	106	378	120	359
x170	15.1	98500	130	105	349	111	334
x160	12.4	90200	137	105	321	103	312
x150	10.1	82200	145	105	294	95.1	291
x135	6.99	68100	159	104	245	79.9	255
W 33x619	567	870000	63.0	148	2210	463	1280
x567	444	768000	66.9	145	1990	425	1170
x515	338	672000	71.8	143	1770	385	1060
x468	256	587000	77.1	140	1570	348	947
x424	193	514000	83.1	138	1400	315	852
x387	149	458000	89.2	136	1260	288	773
x354	115	408000	96.1	135	1130	263	709
x318	84.4	357000	105	133	1000	237	634
x291	65.0	319000	113	132	906	216	577
x263	48.5	281000	123	130	808	195	519
x241	35.8	250000	135	130	721	174	469
x221	27.5	224000	145	129	650	158	428
x201	20.5	198000	158	128	580	142	386

TORSION PROPERTIES
W shapes

Designation	Torsional Constant J	Warping Constant C_w	$\sqrt{\dfrac{EC_w}{GJ}}$	Normalized Warping Constant W_{no}	Warping Statical Moment S_w	Statical Moment	
						Q_f	Q_w
	In.^{4}	In.^{6}	In.	In.^{2}	In.^{4}	In.^{3}	In.^{3}
W 33x169	17.7	82400	110	93.7	329	109	314
x152	12.4	71700	123	93.8	286	95.1	279
x141	9.70	64400	131	93.3	258	86.5	257
x130	7.37	56600	141	92.8	228	76.9	233
x118	5.30	48300	154	92.2	196	66.6	207
W 30x581	537	636000	55.4	129	1850	402	1110
x526	405	550000	59.3	126	1630	364	993
x477	307	480000	63.6	124	1450	329	896
x433	231	417000	68.3	122	1280	297	805
x391	174	364000	73.6	120	1140	268	716
x357	134	323000	79.1	118	1020	245	650
x326	103	286000	85.0	117	919	223	595
x292	74.9	249000	92.8	115	812	200	530
x261	53.8	215000	102	114	710	177	470
x235	40.0	190000	111	112	633	160	422
x211	27.9	166000	124	112	556	141	374
x191	20.6	146000	135	111	494	126	337
x173	15.3	129000	148	110	439	113	303
x148	14.6	49400	93.9	77.3	239	86.8	250
x132	9.72	42100	106	77.3	204	74.0	219
x124	7.99	38600	112	76.9	188	68.8	204
x116	6.43	34900	119	76.5	171	62.8	189
x108	4.99	30900	127	76.1	152	56.1	173
x 99	3.77	26800	136	75.7	133	49.5	156
x 90	2.94	24000	146	75.2	119	45.1	142
W 27x539	499	440000	47.8	111	1490	342	940
x494	391	386000	50.6	108	1340	313	856
x448	297	336000	54.1	106	1190	283	766
x407	225	291000	57.9	104	1050	255	688
x368	169	254000	62.3	102	930	231	620
x336	131	225000	66.8	101	836	211	564
x307	101	199000	71.5	99.4	750	192	511
x281	78.8	178000	76.5	98.2	680	176	466
x258	61.0	159000	82.2	97.1	613	161	424
x235	46.3	140000	88.4	96.0	548	146	384
x217	37.0	128000	94.7	95.0	503	135	354
x194	26.5	111000	104	93.9	442	120	314
x178	19.5	98300	114	93.7	393	107	284
x161	14.7	87300	124	92.9	352	96.6	256
x146	10.9	77200	135	92.2	314	87.0	231
x129	11.2	32500	86.9	66.4	183	69.5	197
x114	7.33	27600	98.7	66.4	155	59.2	171
x102	5.29	24000	108	65.7	137	52.7	153
x 94	4.03	21300	117	65.4	122	47.3	139
x 84	2.81	17900	129	64.9	103	40.6	122
W 24x492	456	283000	40.1	92.1	1150	281	774
x450	357	247000	42.4	90.1	1030	257	703
x408	271	214000	45.2	88.1	909	233	626
x370	205	184000	48.2	86.3	802	209	562
x335	154	160000	51.8	84.6	709	189	509
x306	119	141000	55.4	83.3	636	173	461
x279	91.7	125000	59.4	82.0	570	157	418
x250	67.3	108000	64.4	80.6	502	141	372
x229	51.8	95800	69.2	79.6	451	128	338
x207	38.6	83900	75.0	78.5	401	116	303
x192	31.0	76200	79.7	77.7	367	107	280
x176	24.1	68400	85.9	77.0	333	97.8	255
x162	18.5	62600	93.6	77.0	304	89.4	234
x146	13.4	54600	103	76.3	268	79.5	209
x131	9.50	47100	113	75.6	233	69.7	185
x117	6.72	40800	125	74.9	204	61.5	164
x104	4.72	35200	139	74.3	178	54.1	144

TORSION PROPERTIES
W shapes

Designation	Torsional Constant J	Warping Constant C_w	$\sqrt{\dfrac{EC_w}{GJ}}$	Normalized Warping Constant W_{no}	Warping Statical Moment S_w	Statical Moment	
						Q_f	Q_w
	In.^{4}	In.^{6}	In.	In.^{2}	In.^{4}	In.^{3}	In.^{3}
W 24x103	7.11	16600	77.8	53.0	117	49.4	140
x 94	5.26	15000	85.8	53.1	105	44.4	127
x 84	3.70	12800	94.8	52.6	91.3	39.0	112
x 76	2.68	11100	104	52.2	79.8	34.4	100
x 68	1.87	9430	114	51.9	68.0	29.5	88.3
W 24x 62	1.71	4620	83.7	40.7	42.3	23.2	76.6
x 55	1.18	3870	92.0	40.4	35.7	19.8	67.1
W 21x402	297	165000	37.9	76.7	805	210	564
x364	225	142000	40.4	75.0	709	189	505
x333	174	124000	42.9	73.5	632	172	457
x300	130	107000	46.2	71.9	556	154	408
x275	101	94100	49.1	70.7	499	141	370
x248	75.2	81800	53.0	69.5	441	127	331
x223	54.9	70600	57.6	68.3	388	113	295
x201	41.3	61800	62.1	67.3	345	102	265
x182	31.0	54300	67.2	66.4	307	92.3	238
x166	23.9	48500	72.6	65.6	277	84.4	216
x147	15.4	41100	83.0	65.4	235	71.4	187
x132	11.3	36000	90.9	64.7	208	64.0	167
x122	8.98	32700	97.1	64.2	191	59.2	154
x111	6.83	29200	105	63.7	172	53.7	139
x101	5.21	26200	114	63.2	155	49.0	127
W 21x 93	6.03	9940	65.3	43.6	85.3	38.2	110
x 83	4.34	8630	71.8	43.0	75.0	34.2	98.0
x 73	3.02	7410	79.7	42.5	65.2	30.3	86.2
x 68	2.45	6760	84.6	42.3	59.9	28.0	79.9
x 62	1.83	5960	91.8	42.0	53.2	25.1	72.2
W 21x 57	1.77	3190	68.3	33.4	35.6	20.9	64.3
x 50	1.14	2570	76.2	33.1	28.9	17.2	55.0
x 44	0.77	2110	84.3	32.8	24.0	14.5	47.7
W 18x311	177	75700	33.2	58.8	483	141	376
x283	135	65600	35.4	57.5	427	127	338
x258	104	57400	37.7	56.4	382	116	306
x234	79.7	49900	40.2	55.2	339	105	274
x211	59.3	43200	43.4	54.2	299	94.3	245
x192	45.2	37900	46.5	53.3	267	85.7	221
x175	34.2	33200	50.1	52.5	237	77.2	199
x158	25.4	28900	54.2	51.6	210	69.4	178
x143	19.4	25700	58.6	51.0	189	63.2	161
x130	14.7	22700	63.4	50.4	169	57.1	145
x119	10.6	20300	70.3	50.4	151	50.6	131
x106	7.48	17400	77.7	49.8	131	44.6	115
x 97	5.86	15800	83.5	49.4	120	41.2	105
x 86	4.10	13600	92.7	48.9	104	36.3	92.8
x 76	2.83	11700	103	48.4	90.7	31.9	81.4
W 18x 71	3.48	4700	59.1	33.7	52.1	25.8	72.7
x 65	2.73	4240	63.4	33.4	47.5	23.8	66.6
x 60	2.17	3850	67.8	33.1	43.5	22.1	61.4
x 55	1.66	3430	73.1	32.9	39.0	19.9	55.9
x 50	1.24	3040	79.6	32.6	34.9	18.0	50.4
W 18x 46	1.22	1710	60.3	26.4	24.2	15.3	45.3
x 40	0.81	1440	67.9	26.1	20.6	13.3	39.2
x 35	0.51	1140	76.6	25.9	16.5	10.7	33.2

TORSION PROPERTIES
W shapes

Designation	Torsional Constant J	Warping Constant C_w	$\sqrt{\dfrac{EC_w}{GJ}}$	Normalized Warping Constant W_{no}	Warping Statical Moment S_w	Statical Moment	
						Q_f	Q_w
	In.^{4}	In.^{6}	In.	In.^{2}	In.^{4}	In.^{3}	In.^{3}
W 16x100	7.73	11900	63.1	41.7	107	39.0	99.0
x 89	5.45	10200	69.8	41.1	93.3	34.4	87.3
x 77	3.57	8590	78.9	40.6	79.3	29.7	75.0
x 67	2.39	7300	88.9	40.1	68.2	25.9	64.9
W 16x 57	2.22	2660	55.8	28.0	35.6	19.0	52.6
x 50	1.52	2270	62.1	27.6	30.8	16.7	46.0
x 45	1.11	1990	68.0	27.4	27.2	15.0	41.1
x 40	0.79	1730	75.2	27.1	23.9	13.4	36.5
x 36	0.54	1460	83.2	26.9	20.2	11.4	32.0
W 16x 31	0.46	739	64.5	21.3	13.0	9.17	27.0
x 26	0.26	565	74.8	21.1	10.0	7.20	22.1
W 14x730	1450	362000	25.4	78.3	1720	319	831
x665	1120	305000	26.5	75.5	1510	287	740
x605	870	258000	27.7	73.0	1320	259	660
x550	670	219000	29.1	70.6	1160	233	588
x500	514	187000	30.7	68.5	1020	209	524
x455	395	160000	32.4	66.5	899	189	468
W 14x426	331	144000	33.6	65.3	827	176	434
x398	273	129000	35.1	64.1	756	163	401
x370	222	116000	36.7	62.9	689	151	368
x342	178	103000	38.6	61.6	623	138	336
x311	136	89100	41.2	60.3	553	125	301
x283	104	77700	44.0	59.1	493	113	271
x257	79.1	67800	47.1	57.9	438	102	243
x233	59.5	59000	50.7	56.9	389	91.7	218
x211	44.6	51500	54.7	55.9	345	82.3	195
x193	34.8	45900	58.4	55.1	312	75.4	177
x176	26.5	40500	63.0	54.4	279	68.0	160
x159	19.8	35600	68.3	53.7	248	61.3	143
x145	15.2	31700	73.6	53.0	224	55.8	130
W 14x132	12.3	25500	73.2	50.2	190	49.9	117
x120	9.37	22700	79.2	49.7	171	45.3	106
x109	7.12	20200	85.8	49.1	154	41.2	95.9
x 99	5.37	18000	93.1	48.7	138	37.2	86.6
x 90	4.06	16000	101	48.3	125	33.7	78.3
W 14x 82	5.08	6710	58.5	34.1	73.8	28.1	69.3
x 74	3.88	5990	63.2	33.7	66.6	25.7	62.8
x 68	3.02	5380	68.0	33.4	60.4	23.5	57.3
x 61	2.20	4710	74.5	33.1	53.3	21.0	51.1

TORSION PROPERTIES
W shapes

Designation	Torsional Constant J	Warping Constant C_w	$\sqrt{\dfrac{EC_w}{GJ}}$	Normalized Warping Constant W_{no}	Warping Statical Moment S_w	Statical Moment	
						Q_f	Q_w
	In.^{4}	In.^{6}	In.	In.^{2}	In.^{4}	In.^{3}	In.^{3}
W 14x 53	1.94	2540	58.1	26.7	35.5	17.3	43.6
x 48	1.46	2240	63.1	26.5	31.6	15.6	39.2
x 43	1.05	1950	69.3	26.2	27.8	13.9	34.8
W 14x 38	0.80	1230	63.2	23.0	20.0	11.5	30.7
x 34	0.57	1070	69.6	22.8	17.5	10.2	27.3
x 30	0.38	887	77.7	22.6	14.7	8.59	23.6
W 14x 26	0.36	405	54.1	16.9	8.94	6.98	20.1
x 22	0.21	314	62.6	16.8	7.02	5.58	16.6
W 12x336	243	57000	24.7	46.4	459	119	301
x305	185	48600	26.1	45.0	403	107	269
x279	143	42000	27.6	44.0	357	96.3	241
x252	108	35800	29.3	42.8	313	86.4	214
x230	83.8	31200	31.1	41.8	279	78.4	193
x210	64.7	27200	33.0	41.0	249	71.1	174
x190	48.8	23600	35.3	40.1	220	64.1	156
x170	35.6	20100	38.3	39.2	192	56.9	137
x152	25.8	17200	41.5	38.4	168	50.4	121
x136	18.5	14700	45.3	37.7	146	44.5	107
x120	12.9	12400	50.0	37.0	126	38.9	93.2
x106	9.13	10700	55.0	36.4	110	34.6	81.9
x 96	6.86	9410	59.6	35.9	98.2	31.3	73.6
x 87	5.10	8270	64.8	35.5	87.2	28.0	66.0
x 79	3.84	7330	70.3	35.2	78.1	25.3	59.5
x 72	2.93	6540	76.0	34.9	70.3	22.9	53.9
x 65	2.18	5780	82.9	34.5	62.7	20.6	48.4
W 12x 58	2.10	3570	66.4	28.9	46.3	18.2	43.2
x 53	1.58	3160	72.0	28.7	41.2	16.3	39.0
W 12x 50	1.78	1880	52.3	23.3	30.2	14.7	36.2
x 45	1.31	1650	57.0	23.1	26.7	13.1	32.4
x 40	0.95	1440	62.5	22.9	23.6	11.8	28.8
W 12x 35	0.74	879	55.4	19.6	16.8	9.86	25.6
x 30	0.46	720	63.9	19.4	13.9	8.30	21.6
x 26	0.30	607	72.4	19.2	11.8	7.15	18.6
W 12x 22	0.29	164	38.1	12.0	5.13	4.87	14.7
x 19	0.18	131	43.4	11.8	4.14	4.01	12.4
x 16	0.10	96.9	49.4	11.7	3.09	3.04	10.0
x 14	0.07	80.4	54.4	11.6	2.59	2.59	8.72

TORSION PROPERTIES
W shapes

Designation	Torsional Constant J	Warping Constant C_w	$\sqrt{\dfrac{EC_w}{GJ}}$	Normalized Warping Constant W_{no}	Warping Statical Moment S_w	Statical Moment	
						Q_f	Q_w
	In.^{4}	In.^{6}	In.	In.^{2}	In.^{4}	In.^{3}	In.^{3}
W 10x112	15.1	6020	32.2	26.3	85.7	30.8	73.7
x100	10.9	5150	35.0	25.8	74.7	27.2	64.9
x 88	7.53	4330	38.6	25.3	64.2	23.8	56.4
x 77	5.11	3630	42.9	24.8	54.9	20.7	48.8
x 68	3.56	3100	47.4	24.4	47.6	18.1	42.6
x 60	2.48	2640	52.6	24.0	41.2	15.9	37.3
x 54	1.82	2320	57.4	23.8	36.6	14.3	33.3
x 49	1.39	2070	62.1	23.6	33.0	13.0	30.2
W 10x 45	1.51	1200	45.3	19.0	23.6	11.5	27.5
x 39	0.98	992	51.3	18.7	19.8	9.77	23.4
x 33	0.58	790	59.2	18.5	16.0	7.98	19.4
W 10x 30	0.62	414	41.5	14.5	10.7	7.09	18.3
x 26	0.40	345	47.1	14.3	9.05	6.08	15.6
x 22	0.24	275	54.5	14.1	7.30	4.95	13.0
W 10x 19	0.23	104	34.0	9.89	3.93	3.76	10.8
x 17	0.16	85.1	37.6	9.80	3.24	3.13	9.33
x 15	0.10	68.3	41.2	9.72	2.62	2.56	8.00
x 12	0.05	50.9	49.1	9.56	1.99	2.00	6.32
W 8x 67	5.06	1440	27.2	16.7	32.3	14.7	35.1
x 58	3.34	1180	30.3	16.3	27.2	12.5	29.9
x 48	1.96	931	35.0	15.8	22.0	10.4	24.5
x 40	1.12	726	40.9	15.5	17.5	8.42	19.9
x 35	0.77	619	45.6	15.3	15.2	7.39	17.3
x 31	0.54	530	50.5	15.1	13.1	6.46	15.2
W 8x 28	0.54	312	38.8	12.4	9.43	5.64	13.6
x 24	0.35	259	44.0	12.2	7.94	4.83	11.6
W 8x 21	0.28	152	37.3	10.4	5.47	4.03	10.2
x 18	0.17	122	42.8	10.3	4.44	3.31	8.52
W 8x 15	0.14	51.8	31.3	7.82	2.47	2.39	6.78
x 13	0.09	40.8	34.8	7.74	1.97	1.93	5.70
x 10	0.04	30.9	43.4	7.57	1.53	1.56	4.43
W 6x 25	0.46	150	29.0	9.01	6.23	3.92	9.46
x 20	0.24	113	34.9	8.78	4.82	3.10	7.45
x 15	0.10	76.5	44.2	8.58	3.34	2.18	5.39
W 6x 16	0.22	38.2	21.0	5.92	2.42	2.28	5.84
x 12	0.09	24.7	26.6	5.75	1.61	1.55	4.15
x 9	0.04	17.7	33.7	5.60	1.19	1.19	3.12
W 5x 19	0.31	50.8	20.4	5.94	3.21	2.44	5.81
x 16	0.19	40.6	23.4	5.81	2.62	2.02	4.82
W 4x 13	0.15	14.0	15.5	3.87	1.36	1.27	3.14

FLEXURAL-TORSIONAL PROPERTIES
Channels

Desigation	Torsional Constant J	Warping Constant C_w	Shear Center Coordinate r_o^*	Flexural Constant H^*
	In.^{4}	In.^{6}	In.	No Units
C 15x50	2.67	492	5.49	.937
x40	1.46	411	5.72	.927
x33.9	1.02	358	5.94	.920
C 12x30	0.87	151	4.55	.919
x25	0.54	130	4.72	.909
x20.7	0.37	112	4.93	.899
C 10x30	1.23	79.3	3.63	.921
x25	0.69	68.4	3.75	.912
x20	0.37	56.9	3.93	.900
x15.3	0.21	45.6	4.19	.883
C 9x20	0.43	39.5	3.46	.899
x15	0.21	31.0	3.69	.882
x13.4	0.17	28.2	3.79	.874
C 8x18.75	0.44	25.1	3.06	.894
x13.75	0.19	19.2	3.27	.874
x11.5	0.13	16.5	3.42	.862
C 7x14.75	0.27	13.1	2.75	.875
x12.25	0.16	11.2	2.87	.862
x 9.8	0.10	9.18	3.02	.846
C 6x13	0.24	7.22	2.37	.858
x10.5	0.13	5.95	2.49	.843
x 8.2	0.08	4.72	2.65	.824
C 5x 9	0.11	2.93	2.10	.814
x 6.7	0.06	2.22	2.26	.790
C 4x 7.25	0.08	1.24	1.75	.768
x 5.4	0.04	0.92	1.89	.741
C 3x 6	0.07	0.46	1.39	.689
x 5	0.04	0.38	1.45	.674
x 4.1	0.03	0.31	1.53	.656

*See LRFD Specification Appendix E3.

FLEXURAL-TORSIONAL PROPERTIES
Channels

[

Designation	Torsional Constant J	Warping Constant C_w	Shear Center Coordinate r_o^*	Flexural Constant H^*
	In.^{4}	In.^{6}	In.	No Units
MC 18x58	2.81	1070	6.56	.944
x51.9	2.03	986	6.70	.939
x45.8	1.45	897	6.88	.933
x42.7	1.23	852	6.97	.930
MC 13x50	2.98	558	5.07	.875
x40	1.57	463	5.33	.860
x35	1.14	413	5.50	.849
x31.8	0.94	380	5.64	.842
MC 12x50	3.24	411	4.77	.859
x45	2.35	374	4.87	.851
x40	1.70	336	5.01	.842
x35	1.25	297	5.18	.832
x31	1.01	268	5.34	.821
x10.6	0.06	11.7	4.27	.983
MC 10x41.1	2.27	270	4.26	.790
x33.6	1.21	224	4.47	.771
x28.5	0.79	194	4.68	.752
x25	0.64	125	4.46	.802
x22	0.51	111	4.63	.790
x 8.4	0.04	7.01	3.68	.972
x 6.5	0.02	2.45	3.43	.990
MC 9x25.4	0.69	104	4.08	.770
23.9	0.60	98.2	4.15	.763
MC 8x22.8	0.57	75.3	3.85	.716
x21.4	0.50	70.9	3.91	.709
x20	0.44	47.9	3.59	.780
x18.7	0.38	45.1	3.65	.773
x 8.5	0.06	8.22	3.24	.910
MC 7x22.7	0.63	58.5	3.53	.662
x19.1	0.41	49.4	3.71	.638
x17.6	0.35	32.5	3.34	.722
MC 6x18	0.38	34.6	3.46	.562
x15.3	0.22	30.1	3.41	.581
x16.3	0.34	22.1	3.11	.643
x15.1	0.29	20.6	3.18	.634
x12	0.15	11.2	2.80	.740

*See LRFD Specification Appendix E3.

FLEXURAL-TORSIONAL PROPERTIES
Single Angles

Designation	Torsional Constant J	Warping Constant C_w	Shear Center Coordinate r_o^*	Flexural Constant H^*
	In.^{4}	In.^{6}	In.	No Units
L 9x4x 5/8	1.06	4.79	4.37	—
9/16	.782	3.53	4.38	—
1/2	.557	2.51	4.39	—
L 8x8x1 1/8	7.13	32.5	4.31	.632
1	5.08	23.4	4.35	.630
7/8	3.46	16.1	4.37	.629
3/4	2.21	10.4	4.41	.627
5/8	1.30	6.16	4.45	.627
9/16	.960	4.55	4.47	.627
1/2	.682	3.23	4.48	.624
L 8x6x1	4.35	16.3	3.89	—
7/8	2.96	11.3	3.93	—
3/4	1.90	7.28	3.96	—
5/8	1.12	4.33	3.99	—
9/16	.822	3.20	4.01	—
1/2	.584	2.28	4.02	—
7/16	.396	1.55	4.04	—
L 8x4x1	3.68	12.9	3.77	—
3/4	1.61	5.75	3.82	—
9/16	.704	2.53	3.86	—
1/2	.501	1.80	3.88	—
L 7x4x 3/4	1.47	3.97	3.33	—
5/8	.873	2.37	3.36	—
1/2	.459	1.25	3.38	—
3/8	.200	.544	3.42	—

*See LRFD Specification Appendix E3.

FLEXURAL-TORSIONAL PROPERTIES
Single Angles

Designation	Torsional Constant J	Warping Constant C_w	Shear Center Coordinate r_o^*	Flexural Constant H^*
	In.^{4}	In.^{6}	In.	No Units
L 6x6 x1	3.68	9.24	3.19	.637
7/8	2.51	6.41	3.22	.632
3/4	1.61	4.17	3.26	.629
5/8	.954	2.50	3.29	.628
9/16	.704	1.85	3.31	.627
1/2	.501	1.32	3.32	.627
7/16	.340	.899	3.34	.627
3/8	.218	.575	3.36	.626
5/16	.129	.338	3.38	.625
L 6x4 x 7/8	2.07	4.04	2.83	—
3/4	1.33	2.64	2.86	—
5/8	.792	1.59	2.89	—
9/16	.585	1.18	2.90	—
1/2	.417	.843	2.92	—
7/16	.284	.575	2.94	—
3/8	.183	.369	2.96	—
5/16	.108	.217	2.97	—
L 6x3½x 1/2	.396	.779	2.88	—
3/8	.174	.341	2.92	—
5/16	.103	.201	2.93	—
L 5x5 x 7/8	2.07	3.53	2.65	.634
3/4	1.33	2.32	2.68	.634
5/8	.792	1.40	2.71	.630
1/2	.417	.744	2.74	.630
7/16	.284	.508	2.77	.629
3/8	.183	.327	2.79	.627
5/16	.108	.193	2.81	.626

*See LRFD Specification Appendix E3.

FLEXURAL-TORSIONAL PROPERTIES
Single Angles

Designation	Torsional Constant J In.^{4}	Warping Constant C_w In.^{6}	Shear Center Coordinate r_o^* In.	Flexural Constant H^* No Units
L 5x3½x¾	1.11	1.52	2.37	—
⅝	.660	.918	2.40	—
½	.348	.491	2.44	—
⁷⁄₁₆	.238	.336	2.45	—
⅜	.153	.217	2.47	—
⁵⁄₁₆	.0905	.128	2.49	—
¼	.0479	.0670	2.50	—
L 5x3 x⅝	.610	.830	2.36	—
½	.322	.444	2.39	—
⁷⁄₁₆	.219	.304	2.41	—
⅜	.141	.196	2.42	—
⁵⁄₁₆	.0832	.116	2.43	—
¼	.0438	.0606	2.45	—
L 4x4 x¾	1.02	1.12	2.11	.639
⅝	.610	.680	2.14	.631
½	.322	.366	2.17	.632
⁷⁄₁₆	.219	.252	2.19	.631
⅜	.141	.162	2.20	.625
⁵⁄₁₆	.0832	.0963	2.22	.623
¼	.0438	.0505	2.23	.627
L 4x3½x⅝	.569	.560	2.01	—
½	.301	.302	2.04	—
⁷⁄₁₆	.206	.208	2.06	—
⅜	.132	.134	2.08	—
⁵⁄₁₆	.0782	.0798	2.09	—
¼	.0412	.0419	2.11	—

*See LRFD Specification Appendix E3.

FLEXURAL-TORSIONAL PROPERTIES
Single Angles

Designation	Torsional Constant J	Warping Constant C_w	Shear Center Coordinate r_o^*	Flexural Constant H^*
	In.^{4}	In.^{6}	In.	No Units
L 4 x3 x⅝	.529	.472	1.91	—
½	.281	.255	1.95	—
⁷⁄₁₆	.192	.176	1.96	—
⅜	.123	.114	1.98	—
⁵⁄₁₆	.0731	.0676	2.00	—
¼	.0386	.0356	2.01	—
L 3½x3½x½	.281	.238	1.89	.631
⁷⁄₁₆	.192	.164	1.91	.629
⅜	.123	.106	1.91	.628
⁵⁄₁₆	.0731	.0634	1.93	.627
¼	.0386	.0334	1.95	.626
L 3½x3 x½	.260	.191	1.76	—
⁷⁄₁₆	.178	.132	1.77	—
⅜	.114	.0858	1.79	—
⁵⁄₁₆	.0680	.0512	1.81	—
¼	.0360	.0270	1.83	—
L 3½x2½x½	.234	.159	1.67	—
⁷⁄₁₆	.160	.110	1.68	—
⅜	.103	.0714	1.70	—
⁵⁄₁₆	.0611	.0426	1.72	—
¼	.0322	.0225	1.73	—
L 3 x3 x½	.234	.144	1.60	.634
⁷⁄₁₆	.160	.100	1.61	.632
⅜	.103	.0652	1.63	.629
⁵⁄₁₆	.0611	.0390	1.65	.628
¼	.0322	.0206	1.66	.627
³⁄₁₆	.0142	.00899	1.68	.626

*See LRFD Specification Appendix E3.

FLEXURAL-TORSIONAL PROPERTIES
Single Angles

Designation	Torsional Constant J In.^{4}	Warping Constant C_w In.^{6}	Shear Center Coordinate r_o^* In.	Flexural Constant H^* No Units
L 3 x2½x½	.213	.112	1.47	—
⁷⁄₁₆	.146	.0777	1.49	—
³⁄₈	.0943	.0507	1.50	—
⁵⁄₁₆	.0560	.0304	1.52	—
¼	.0296	.0161	1.54	—
³⁄₁₆	.0131	.00705	1.55	—
L 3 x2 x½	.192	.0908	1.40	—
⁷⁄₁₆	.132	.0632	1.41	—
³⁄₈	.0855	.0413	1.43	—
⁵⁄₁₆	.0509	.0248	1.45	—
¼	.0270	.0132	1.46	—
³⁄₁₆	.0120	.00576	1.48	—
L 2½x2½x½	.185	.0791	1.31	.639
³⁄₈	.0816	.0362	1.34	.632
⁵⁄₁₆	.0483	.0218	1.36	.630
¼	.0253	.0116	1.37	.628
³⁄₁₆	.0110	.0051	1.39	.627
L 2½x2 x³⁄₈	.0728	.0268	1.22	—
⁵⁄₁₆	.0432	.0162	1.24	—
¼	.0227	.00868	1.25	—
³⁄₁₆	.00990	.00382	1.27	—
L 2 x2 x³⁄₈	.0640	.0174	1.05	.637
⁵⁄₁₆	.0381	.0106	1.07	.633
¼	.0201	.00572	1.09	.630
³⁄₁₆	.0088	.00254	1.10	.628
⅛	.00274	.00079	1.12	.626

*See LRFD Specification Appendix E3.

FLEXURAL-TORSIONAL PROPERTIES
Structural Tees

Designation	Torsional Constant J	Warping Constant C_w	Shear Center Coordinate r_o^*	Flexural Constant H^*
	In.^{4}	In.^{6}	In.	No Units
WT 18 x179.5	54.3	480	7.38	0.797
x164	42.1	363	7.32	0.799
x150	32.0	278	7.30	0.797
x140	26.2	226	7.27	0.796
x130	20.7	181	7.28	0.791
x122.5	17.3	151	7.28	0.788
x115	14.3	125	7.27	0.784
WT 18 x128	26.6	205	7.43	0.703
x116	19.8	151	7.40	0.703
x105	13.9	119	7.49	0.687
x 97	11.1	92.7	7.45	0.687
x 91	9.19	77.6	7.45	0.686
x 85	7.51	63.2	7.44	0.684
x 80	6.17	53.6	7.46	0.678
x 75	5.04	46.0	7.50	0.670
x 67.5	3.48	37.3	7.65	0.644
WT 16.5x177	57.2	468	7.00	0.802
x159	42.1	335	6.94	0.803
x145.5	32.4	256	6.90	0.801
x131.5	24.2	188	6.86	0.802
x120.5	17.9	146	6.91	0.792
x110.5	13.7	113	6.90	0.788
x100.5	10.2	84.9	6.89	0.784
WT 16.5x 84.5	8.83	55.4	6.74	0.714
x 76	6.16	43.0	6.82	0.700
x 70.5	4.84	35.4	6.85	0.691
x 65	3.67	29.3	6.93	0.678
x 59	2.64	23.4	7.02	0.659
WT 15 x117.5	19.9	132	6.25	0.817
x105.5	13.9	96.4	6.27	0.809
x 95.5	10.3	71.2	6.25	0.806
x 86.5	7.61	53.0	6.25	0.802
WT 15 x 74	7.27	37.6	6.10	0.716
x 66	4.85	28.5	6.19	0.698
x 62	3.98	23.9	6.20	0.693
x 58	3.21	20.5	6.24	0.683
x 54	2.49	17.3	6.31	0.669
x 49.5	1.88	14.3	6.38	0.654

*See LRFD Specification Appendix E3.

FLEXURAL-TORSIONAL PROPERTIES
Structural Tees

Designation	Torsional Constant J In.^{4}	Warping Constant C_w In.^{6}	Shear Center Coordinate r_o^* In.	Flexural Constant H^* No Units
WT 13.5x108.5	18.5	105	5.72	0.830
x 97	13.2	74.3	5.66	0.826
x 89	9.74	57.7	5.70	0.815
x 80.5	7.31	42.7	5.67	0.813
x 73	5.44	31.7	5.65	0.810
WT 13.5x 64.5	5.60	24.0	5.48	0.731
x 57	3.65	17.5	5.54	0.716
x 51	2.64	12.6	5.52	0.714
x 47	2.01	10.2	5.57	0.703
x 42	1.40	7.79	5.63	0.685
WT 12 x 88	12.0	55.8	5.09	0.835
x 81	9.22	43.8	5.09	0.831
x 73	6.70	31.9	5.08	0.827
x 65.5	4.74	23.1	5.09	0.818
x 58.5	3.35	16.4	5.08	0.813
x 52	2.35	11.6	5.07	0.809
WT 12 x 51.5	3.54	12.3	4.88	0.733
x 47	2.62	9.57	4.89	0.727
x 42	1.84	6.90	4.89	0.721
x 38	1.34	5.30	4.93	0.709
x 34	.932	4.08	4.99	0.692
WT 12 x 31	.850	3.92	5.13	0.619
x 27.5	.588	2.93	5.18	0.606
WT 10.5x 83	11.9	47.3	4.69	0.861
x 73.5	7.69	32.5	4.64	0.847
x 66	5.62	23.4	4.61	0.845
x 61	4.47	18.4	4.58	0.846
x 55.5	3.40	13.8	4.56	0.846
x 50.5	2.60	10.4	4.54	0.846
WT 10.5x 46.5	3.01	9.33	4.37	0.729
x 41.5	2.16	6.50	4.33	0.732
x 36.5	1.51	4.42	4.31	0.732
x 34	1.22	3.62	4.31	0.727
x 31	.513	2.78	4.31	0.722
WT 10.5x 28.5	.884	2.50	4.36	0.665
x 25	.570	1.89	4.44	0.640
x 22	.383	1.40	4.49	0.623

*See LRFD Specification Appendix E3.

FLEXURAL-TORSIONAL PROPERTIES
Structural Tees

Designation	Torsional Constant J	Warping Constant C_w	Shear Center Coordinate r_o^*	Flexural Constant H^*
	In.^{4}	In.^{6}	In.	No Units
WT 9x71.5	9.69	30.7	4.03	0.874
x65	7.31	22.8	3.99	0.874
x59.5	5.30	17.4	4.03	0.862
x53	3.73	12.1	4.00	0.860
x48.5	2.92	9.29	3.97	0.862
x43	2.04	6.42	3.95	0.860
x38	1.41	4.37	3.92	0.862
WT 9x35.5	1.74	3.96	3.72	0.751
x32.5	1.36	3.01	3.69	0.755
x30	1.08	2.35	3.67	0.756
x27.5	0.829	1.84	3.68	0.749
x25	0.613	1.36	3.66	0.748
WT 9x23	0.609	1.20	3.67	0.694
x20	0.403	.788	3.65	0.692
x17.5	0.252	.598	3.74	0.662
WT 8x50	3.85	10.4	3.62	0.877
x44.5	2.72	7.19	3.60	0.877
x38.5	1.78	4.61	3.56	0.877
x33.5	1.19	3.01	3.53	0.879
WT 8x28.5	1.10	1.99	3.30	0.770
x25	0.76	1.34	3.28	0.770
x22.5	0.655	.974	3.27	0.767
x20	0.396	.673	3.24	0.769
x18	0.271	.516	3.30	0.745
WT 8x15.5	0.229	.366	3.26	0.695
x13	0.130	.243	3.32	0.667

*See LRFD Specification Appendix E3.

T FLEXURAL-TORSIONAL PROPERTIES
Structural Tees

Designation	Torsional Constant J	Warping Constant C_w	Shear Center Coordinate r_o^*	Flexural Constant H^*
	In.^{4}	In.^{6}	In.	No Units
WT 7x365	714	5250	5.47	0.966
x332.5	555	3920	5.36	0.966
x302.5	430	2930	5.25	0.966
x275	331	2180	5.15	0.967
x250	255	1620	5.06	0.967
x227.5	196	1210	4.98	0.967
x213	164	991	4.92	0.968
x199	135	801	4.87	0.968
x185	110	640	4.81	0.968
x171	88.3	502	4.77	0.968
x155.5	67.5	375	4.71	0.968
x141.5	51.8	281	4.66	0.969
x128.5	39.3	209	4.61	0.969
x116.5	29.6	154	4.56	0.970
x105.5	22.2	113	4.52	0.970
x 96.5	17.3	87.2	4.49	0.971
x 88	13.2	65.2	4.46	0.971
x 79.5	9.84	47.9	4.42	0.971
x 72.5	7.56	36.3	4.40	0.971

*See LRFD Specification Appendix E3.

FLEXURAL-TORSIONAL PROPERTIES
Structural Tees

Designation	Torsional Constant J	Warping Constant C_w	Shear Center Coordinate r_o^*	Flexural Constant H^*
	In.^{4}	In.^{6}	In.	No Units
WT 7x66	6.13	26.6	4.21	0.966
x60	4.67	20.0	4.18	0.966
x54.5	3.55	15.0	4.16	0.968
x49.5	2.68	11.1	4.14	0.968
x45	2.03	8.31	4.12	0.968
WT 7x41	2.53	5.63	3.25	0.912
x37	1.94	4.19	3.21	0.917
x34	1.51	3.21	3.19	0.915
x30.5	1.10	2.29	3.18	0.915
WT 7x26.5	0.970	1.46	2.89	0.868
x24	0.726	1.07	2.87	0.866
x21.5	0.524	0.751	2.85	0.866
WT 7x19	0.398	0.554	2.87	0.800
x17	0.284	0.400	2.86	0.793
x15	0.190	0.287	2.90	0.772
WT 7x13	0.179	0.207	2.82	0.713
x11	0.104	0.134	2.86	0.691

*See LRFD Specification Appendix E3.

FLEXURAL-TORSIONAL PROPERTIES
Structural Tees

Designation	Torsional Constant J	Warping Constant C_w	Shear Center Coordinate r_o^*	Flexural Constant H^*
	In.^{4}	In.^{6}	In.	No Units
WT 6x168	120	481	4.07	0.958
x152.5	92.0	356	4.00	0.959
x139.5	70.9	267	3.94	0.957
x126	53.5	195	3.88	0.958
x115	41.6	148	3.84	0.958
x105	32.2	112	3.79	0.959
x 95	24.3	82.1	3.74	0.959
x 85	17.7	58.3	3.69	0.960
x 76	12.8	41.3	3.65	0.960
x 68	9.22	28.9	3.61	0.960
x 60	6.43	19.7	3.58	0.959
x 53	4.55	13.6	3.54	0.961
x 48	3.42	10.1	3.51	0.961
x 43.5	2.54	7.34	3.49	0.960
x 39.5	1.92	5.43	3.46	0.960
x 36	1.46	4.07	3.45	0.961
x 32.5	1.09	2.97	3.43	0.960
WT 6x 29	1.05	2.08	3.01	0.944
x 26.5	0.788	1.53	3.00	0.940
WT 6x 25	0.889	1.23	2.67	0.899
x 22.5	0.656	0.885	2.64	0.898
x 20	0.476	0.620	2.62	0.901
WT 6x 17.5	0.369	0.437	2.56	0.835
x 15	0.228	0.267	2.55	0.830
x 13	0.150	0.174	2.54	0.826
WT 6x 11	0.146	0.137	2.52	0.683
x 9.5	0.0899	0.0934	2.54	0.663
x 8	0.0511	0.0678	2.62	0.624
x 7	0.035	0.0493	2.64	0.610

*See LRFD Specification Appendix E3.

AMERICAN INSTITUTE OF STEEL CONSTRUCTION

FLEXURAL-TORSIONAL PROPERTIES
Structural Tees

Designation	Torsional Constant J	Warping Constant C_w	Shear Center Coordinate r_o^*	Flexural Constant H^*
	In.^{4}	In.^{6}	In.	No Units
WT 5x56	7.50	16.9	3.04	.963
x50	5.41	11.9	3.00	.964
x44	3.75	8.02	2.98	.964
x38.5	2.55	5.31	2.93	.964
x34	1.78	3.62	2.92	.965
x30	1.23	2.46	2.89	.965
x27	0.909	1.78	2.87	.966
x24.5	0.693	1.33	2.85	.966
WT 5x22.5	0.753	0.981	2.44	.940
x19.5	0.487	0.616	2.42	.936
x16.5	0.291	0.356	2.40	.927
WT 5x15	0.310	0.273	2.17	.848
x13	0.201	0.173	2.15	.848
x11	0.119	0.107	2.17	.831
WT 5x 9.5	0.116	0.0796	2.08	.728
x 8.5	0.0776	0.0610	2.12	.702
x 7.5	0.0518	0.0475	2.16	.672
x 6	0.0272	0.0255	2.16	.662

*See LRFD Specification Appendix E3.

FLEXURAL-TORSIONAL PROPERTIES
Structural Tees

Designation	Torsional Constant J In.^{4}	Warping Constant C_w In.^{6}	Shear Center Coordinate r_o^* In.	Flexural Constant H^* No Units
WT 4 x33.5	2.52	3.56	2.41	.962
x29	1.66	2.28	2.39	.961
x24	0.979	1.30	2.34	.966
x20	0.559	0.715	2.31	.961
x17.5	0.385	0.480	2.29	.963
x15.5	0.268	0.327	2.29	.961
WT 4 x14	0.268	0.230	1.97	.935
x12	0.173	0.144	1.96	.936
WT 4 x10.5	0.141	0.0916	1.80	.877
x 9	0.0855	0.0562	1.81	.863
WT 4 x 7.5	0.0679	0.0382	1.72	.762
x 6.5	0.0433	0.0269	1.74	.732
x 5	0.0212	0.0114	1.69	.748
WT 3 x12.5	0.229	0.171	1.76	.952
x10	0.120	0.0858	1.73	.952
x 7.5	0.0504	0.0342	1.71	.937
WT 3 x 8	0.111	0.0426	1.37	.880
x 6	0.0449	0.0178	1.37	.846
x 4.5	0.0202	0.0074	1.34	.852
WT 2.5x 9.5	0.154	0.0775	1.44	.964
x 8	0.0930	0.0453	1.43	.962
WT 2 x 6.5	0.0750	0.0213	1.16	.947

*See LRFD Specification Appendix E3.

FLEXURAL-TORSIONAL PROPERTIES
Structural Tees

Designation	Torsional Constant J In.^{4}	Warping Constant C_w In.^{6}	Shear Center Coordinate r_o^* In.	Flexural Constant H^* No Units
MT 7 x 9	.0599	.0981	3.10	.590
MT 6 x 5.9	.0307	.0337	2.69	.564
MT 5 x 4.5	.0213	.0138	2.21	.584
MT 4 x 3.25	.0146	.00463	1.73	.611
MT 3 x10	.139	.0888	1.61	.955
MT 3 x 2.2	.00993	.00124	1.26	.645
MT 2.5x 9.45	.165	.0732	1.37	.951
MT 2 x 6.5	.0930	.0244	1.08	.957

*See LRFD Specification Appendix E3.

FLEXURAL-TORSIONAL PROPERTIES
Structural Tees

Designation	Torsional Constant J In.^{4}	Warping Constant C_w In.^{6}	Shear Center Coordinate r_o^* In.	Flexural Constant H^* No Units
ST 12 x60.5	6.38	27.5	5.14	.640
12 x53	5.04	15.0	4.88	.685
ST 12 x50	3.76	19.5	5.28	.584
12 x45	3.01	12.1	5.11	.616
12 x40	2.43	6.94	4.88	.657
ST 10 x48	4.15	15.0	4.36	.625
10 x43	3.30	9.17	4.21	.661
ST 10 x37.5	2.28	7.21	4.29	.612
10 x33	1.78	4.02	4.10	.655
ST 9 x35	2.05	7.03	4.02	.583
9 x27.35	1.18	2.26	3.71	.662
ST 7.5x25	1.05	2.02	3.22	.637
7.5x21.45	.767	.995	3.05	.689
ST 6 x25	1.39	1.97	2.60	.663
6 x20.4	.872	.787	2.42	.733
ST 6 x17.5	.538	.556	2.49	.697
6 x15.9	.449	.364	2.39	.731
ST 5 x17.5	.633	.725	2.23	.653
5 x12.7	.300	.173	1.98	.768
ST 4 x11.5	.271	.168	1.73	.707
4 x 9.2	.167	.0642	1.59	.789
ST 3.5x10	.221	.1150	1.55	.703
3.5x 7.65	.120	.0366	1.40	.802
ST 3 x 8.625	.182	.0772	1.36	.706
3 x 6.25	.0838	.0197	1.21	.820
ST 2.5x 7.375	.155	.0513	1.17	.712
2.5x 5	.0568	.0100	1.02	.842
ST 2 x 4.75	.0589	.00995	0.909	.800
2 x 3.85	.0364	.00457	0.841	.872
ST 1.5x 3.75	.0440	.00496	0.736	.832
1.5x 2.85	.0220	.00189	0.673	.913

*See LRFD Specification Appendix E3.

FLEXURAL-TORSIONAL PROPERTIES
Double Angles

| Designation | Long Legs Vertical Back to Back of Angles, In. | | | | | | Short Legs Vertical Back to Back of Angles, In. | | | | | |
| | 0 | | 3/8 | | 3/4 | | 0 | | 3/8 | | 3/4 | |
	r_o^*	H^*	r_o^*	H^*	r_o^*	H^*	r_o^*	H^*	r_o^*	H^*	r_o^*	H^*
L 5x3½x¾	2.50	.697	2.58	.715	2.67	.734	2.61	.943	2.74	.948	2.87	.953
½	2.51	.685	2.59	.703	2.67	.722	2.59	.936	2.71	.941	2.84	.947
⅜	2.52	.682	2.59	.699	2.67	.717	2.58	.932	2.70	.938	2.83	.943
5/16	2.53	.679	2.60	.695	2.68	.713	2.58	.930	2.70	.936	2.82	.942
L 5x3 x½	2.45	.626	2.52	.645	2.59	.665	2.55	.962	2.69	.965	2.82	.969
⅜	2.46	.623	2.52	.641	2.60	.661	2.54	.959	2.67	.963	2.80	.966
5/16	2.47	.621	2.53	.638	2.60	.657	2.54	.957	2.67	.961	2.80	.965
¼	2.48	.618	2.54	.634	2.61	.653	2.53	.956	2.66	.960	2.79	.964
L 4x4 x¾	2.29	.847	2.40	.861	2.52	.873	2.29	.847	2.40	.861	2.52	.873
⅝	2.29	.839	2.40	.853	2.51	.867	2.29	.839	2.40	.853	2.51	.867
½	2.29	.834	2.39	.848	2.50	.862	2.29	.834	2.39	.848	2.50	.862
⅜	2.29	.827	2.39	.841	2.50	.855	2.29	.827	2.39	.841	2.50	.855
5/16	2.29	.824	2.39	.838	2.50	.852	2.29	.824	2.39	.838	2.50	.852
¼	2.29	.823	2.39	.837	2.49	.850	2.29	.823	2.39	.837	2.49	.850
L 4x3½x½	2.15	.783	2.24	.801	2.34	.818	2.17	.881	2.29	.892	2.41	.903
⅜	2.15	.774	2.24	.792	2.34	.809	2.17	.875	2.28	.887	2.40	.898
5/16	2.15	.774	2.24	.791	2.34	.808	2.17	.872	2.28	.884	2.40	.895
¼	2.16	.770	2.24	.787	2.34	.805	2.17	.870	2.28	.882	2.39	.893

*See LRFD Specification Appendix E3.

AMERICAN INSTITUTE OF STEEL CONSTRUCTION

FLEXURAL-TORSIONAL PROPERTIES
Double Angles

Designation	Long Legs Vertical						Short Legs Vertical					
	Back to Back of Angles, In.						Back to Back of Angles, In.					
	0		$\frac{3}{8}$		$\frac{3}{4}$		0		$\frac{3}{8}$		$\frac{3}{4}$	
	$r_o{}^*$	H^*	$r_o{}^*$	H^*	$r_o{}^*$	H^*	$r_o{}^*$	H^*	$r_o{}^*$	H^*	$r_o{}^*$	H^*
L 4 x3 x$\frac{1}{2}$	2.04	.719	2.12	.740	2.22	.762	2.10	.924	2.22	.933	2.35	.940
$\frac{3}{8}$	2.04	.714	2.12	.735	2.21	.757	2.09	.919	2.21	.928	2.34	.935
$\frac{5}{16}$	2.05	.710	2.13	.731	2.22	.752	2.09	.917	2.21	.925	2.33	.933
$\frac{1}{4}$	2.06	.706	2.13	.726	2.22	.747	2.09	.914	2.20	.923	2.33	.931
L 3$\frac{1}{2}$x3$\frac{1}{2}$x$\frac{3}{8}$	2.00	.831	2.10	.847	2.22	.862	2.00	.831	2.10	.847	2.22	.862
$\frac{5}{16}$	2.01	.827	2.10	.843	2.21	.858	2.01	.827	2.10	.843	2.21	.858
$\frac{1}{4}$	2.01	.824	2.10	.839	2.21	.855	2.01	.824	2.10	.839	2.21	.855
L 3$\frac{1}{2}$x3 x$\frac{3}{8}$	1.86	.771	1.95	.791	2.06	.812	1.89	.884	2.00	.897	2.13	.909
$\frac{5}{16}$	1.87	.766	1.96	.787	2.06	.807	1.89	.881	2.00	.894	2.12	.906
$\frac{1}{4}$	1.87	.762	1.96	.782	2.06	.803	1.89	.878	2.00	.891	2.12	.903
L 3$\frac{1}{2}$x2$\frac{1}{2}$x$\frac{3}{8}$	1.76	.696	1.84	.721	1.94	.748	1.82	.932	1.94	.941	2.08	.948
$\frac{5}{16}$	1.77	.691	1.85	.716	1.94	.742	1.81	.930	1.94	.938	2.07	.946
$\frac{1}{4}$	1.77	.691	1.85	.715	1.93	.740	1.81	.927	1.93	.936	2.06	.944
L 3 x3 x$\frac{1}{2}$	1.72	.842	1.83	.860	1.95	.877	1.72	.842	1.83	.860	1.95	.877
$\frac{3}{8}$	1.72	.834	1.82	.852	1.94	.869	1.72	.834	1.82	.852	1.94	.869
$\frac{5}{16}$	1.72	.830	1.82	.848	1.93	.866	1.72	.830	1.82	.848	1.93	.866
$\frac{1}{4}$	1.72	.825	1.82	.844	1.93	.862	1.72	.825	1.82	.844	1.93	.862
$\frac{3}{16}$	1.72	.822	1.82	.841	1.93	.858	1.72	.822	1.82	.841	1.93	.858

*See LRFD Specification Appendix E3.

FLEXURAL-TORSIONAL PROPERTIES
Double Angles

Designation	Long Legs Vertical						Short Legs Vertical					
	Back to Back of Angles, In.						Back to Back of Angles, In.					
	0		⅜		¾		0		⅜		¾	
	r_o^*	H^*	r_o^*	H^*	r_o^*	H^*	r_o^*	H^*	r_o^*	H^*	r_o^*	H^*
L 3 x2½x⅜	1.58	.763	1.67	.789	1.78	.813	1.61	.896	1.73	.910	1.86	.922
¼	1.59	.754	1.67	.779	1.78	.804	1.61	.889	1.72	.903	1.84	.916
3⁄16	1.59	.750	1.67	.775	1.77	.800	1.61	.885	1.72	.899	1.84	.912
L 3 x2 x⅜	1.49	.672	1.57	.704	1.66	.737	1.55	.949	1.68	.956	1.82	.963
5⁄16	1.50	.667	1.57	.698	1.66	.730	1.55	.946	1.68	.954	1.82	.961
¼	1.50	.664	1.57	.694	1.66	.726	1.54	.943	1.67	.951	1.80	.958
3⁄16	1.50	.661	1.57	.690	1.66	.721	1.54	.940	1.66	.949	1.80	.956
L 2½x2½x⅜	1.43	.839	1.54	.861	1.66	.880	1.43	.839	1.54	.861	1.66	.880
5⁄16	1.43	.834	1.54	.856	1.66	.876	1.43	.834	1.54	.856	1.66	.876
¼	1.43	.829	1.53	.851	1.65	.871	1.43	.829	1.53	.851	1.65	.871
3⁄16	1.43	.825	1.53	.847	1.65	.867	1.43	.825	1.53	.847	1.65	.867
L 2½x2 x⅜	1.29	.752	1.39	.785	1.50	.816	1.33	.912	1.45	.927	1.59	.939
5⁄16	1.30	.746	1.39	.779	1.50	.810	1.33	.908	1.45	.923	1.58	.935
¼	1.30	.741	1.39	.773	1.50	.804	1.33	.903	1.45	.919	1.58	.932
3⁄16	1.31	.736	1.39	.767	1.49	.798	1.32	.899	1.44	.915	1.57	.928
L 2 x2 x⅜	1.15	.846	1.26	.873	1.39	.896	1.15	.846	1.26	.873	1.39	.896
5⁄16	1.15	.840	1.26	.867	1.38	.890	1.15	.840	1.26	.867	1.38	.890
¼	1.15	.834	1.25	.861	1.38	.885	1.15	.834	1.25	.861	1.38	.885
3⁄16	1.15	.828	1.25	.855	1.37	.880	1.15	.828	1.25	.855	1.37	.880
⅛	1.15	.822	1.25	.850	1.37	.875	1.15	.822	1.25	.850	1.37	.875

*See LRFD Specification Appendix E3.

SURFACE AREAS AND BOX AREAS
W shapes
Square feet per foot of length

Desig-nation	Case A	Case B	Case C	Case D	Desig-nation	Case A	Case B	Case C	Case D
W 36x359	10.09	11.48	7.63	9.02	W 27x129	6.92	7.76	5.44	6.27
x328	10.03	11.42	7.57	8.95	x114	6.88	7.72	5.39	6.23
x300	9.99	11.40	7.51	8.90	x102	6.85	7.68	5.35	6.18
x280	9.95	11.30	7.47	8.85	x 94	6.82	7.65	5.32	6.15
x260	9.90	11.30	7.42	8.80	x 84	6.78	7.61	5.28	6.11
x245	9.87	11.20	7.39	8.77					
x230	9.84	11.20	7.36	8.73	W 24x176	7.23	8.31	5.28	6.36
					x162	7.22	8.30	5.25	6.33
W 36x256	9.02	10.04	7.26	8.27	x146	7.17	8.24	5.20	6.27
x232	8.96	9.97	7.20	8.21	x131	7.12	8.19	5.15	6.22
x210	8.91	9.93	7.13	8.15	x117	7.08	8.15	5.11	6.18
x194	8.88	9.89	7.09	8.10	x104	7.04	8.11	5.07	6.14
x182	8.85	9.85	7.06	8.07					
x170	8.82	9.82	7.03	8.03	W 24x103	6.18	6.93	4.84	5.59
x160	8.79	9.79	7.00	8.00	x 94	6.16	6.92	4.81	5.56
x150	8.76	9.76	6.97	7.97	x 84	6.12	6.87	4.77	5.52
x135	8.71	9.70	6.92	7.92	x 76	6.09	6.84	4.74	5.49
					x 68	6.06	6.80	4.70	5.45
W 33x354	9.66	11.00	7.27	8.60					
x318	9.58	10.91	7.19	8.52	W 24x 62	5.57	6.16	4.54	5.13
x291	9.52	10.84	7.13	8.46	x 55	5.54	6.13	4.51	5.10
x263	9.46	10.78	7.07	8.39					
x241	9.42	10.70	7.02	8.34	W 21x166	6.65	7.68	4.78	5.82
x221	9.38	10.70	6.97	8.29	x147	6.61	7.66	4.72	5.76
x201	9.33	10.60	6.93	8.24	x132	6.57	7.61	4.68	5.71
					x122	6.54	7.57	4.65	5.68
W 33x169	8.30	9.26	6.60	7.55	x111	6.51	7.54	4.61	5.64
x152	8.27	9.23	6.55	7.51	x101	6.48	7.50	4.58	5.61
x141	8.23	9.19	6.51	7.47					
x130	8.20	9.15	6.47	7.43	W 21x 93	5.54	6.24	4.31	5.01
x118	8.15	9.11	6.43	7.39	x 83	5.50	6.20	4.27	4.96
					x 73	5.47	6.16	4.23	4.92
W 30x235	8.75	10.00	6.47	7.73	x 68	5.45	6.14	4.21	4.90
x211	8.71	9.97	6.42	7.67	x 62	5.42	6.11	4.19	4.87
x191	8.66	9.92	6.37	7.62					
x173	8.62	9.87	6.32	7.57	W 21x 57	5.01	5.56	4.06	4.60
					x 50	4.97	5.51	4.02	4.56
W 30x148	7.53	8.40	5.99	6.86	x 44	4.94	5.48	3.99	4.53
x132	7.49	8.37	5.93	6.81					
x124	7.47	8.34	5.90	6.78	W 18x143	5.87	6.81	4.18	5.12
x116	7.44	8.31	5.88	6.75	x130	5.83	6.76	4.14	5.07
x108	7.41	8.28	5.84	6.72	x119	5.81	6.75	4.10	5.04
x 99	7.37	8.25	5.81	6.68	x106	5.77	6.70	4.06	4.99
					x 97	5.74	6.67	4.03	4.96
W 27x217	8.05	9.22	5.91	7.09	x 86	5.70	6.62	3.99	4.91
x194	7.99	9.15	5.85	7.02	x 76	5.67	6.59	3.95	4.87
x178	7.95	9.12	5.81	6.98					
x161	7.91	9.08	5.77	6.94	W 18x 71	4.85	5.48	3.71	4.35
x146	7.87	9.03	5.73	6.89	x 65	4.82	5.46	3.69	4.32
					x 60	4.80	5.43	3.67	4.30
					x 55	4.78	5.41	3.65	4.27
					x 50	4.76	5.38	3.62	4.25

Case A: Shape perimeter, minus one flange surface.
Case B: Shape perimeter.
Case C: Box perimeter, equal to one flange surface plus twice the depth.
Case D: Box perimeter, equal to two flange surfaces plus twice the depth.

SURFACE AREAS AND BOX AREAS
W shapes
Square feet per foot of length

Desig-nation	Case A	Case B	Case C	Case D	Desig-nation	Case A	Case B	Case C	Case D
W 18x 46	4.41	4.91	3.52	4.02	W 14x132	5.93	7.16	3.67	4.90
x 40	4.38	4.88	3.48	3.99	x120	5.90	7.12	3.64	4.86
x 35	4.34	4.84	3.45	3.95	x109	5.86	7.08	3.60	4.82
					x 99	5.83	7.05	3.57	4.79
W 16x100	5.28	6.15	3.70	4.57	x 90	5.81	7.02	3.55	4.76
x 89	5.24	6.10	3.66	4.52					
x 77	5.19	6.05	3.61	4.47	W 14x 82	4.75	5.59	3.23	4.07
x 67	5.16	6.01	3.57	4.43	x 74	4.72	5.56	3.20	4.04
					x 68	4.69	5.53	3.18	4.01
W 16x 57	4.39	4.98	3.33	3.93	x 61	4.67	5.50	3.15	3.98
x 50	4.36	4.95	3.30	3.89					
x 45	4.33	4.92	3.27	3.86	W 14x 53	4.19	4.86	2.99	3.66
x 40	4.31	4.89	3.25	3.83	x 48	4.16	4.83	2.97	3.64
x 36	4.28	4.87	3.23	3.81	x 43	4.14	4.80	2.94	3.61
W 16x 31	3.92	4.39	3.11	3.57	W 14x 38	3.93	4.50	2.91	3.48
x 26	3.89	4.35	3.07	3.53	x 34	3.91	4.47	2.89	3.45
					x 30	3.89	4.45	2.87	3.43
W 14x730	7.61	9.10	5.23	6.72					
x665	7.46	8.93	5.08	6.55	W 14x 26	3.47	3.89	2.74	3.16
x605	7.32	8.77	4.94	6.39	x 22	3.44	3.86	2.71	3.12
x550	7.19	8.62	4.81	6.24					
x500	7.07	8.49	4.68	6.10	W 12x336	5.77	6.88	3.92	5.03
x455	6.96	8.36	4.57	5.98	x305	5.67	6.77	3.82	4.93
					x279	5.59	6.68	3.74	4.83
W 14x426	6.89	8.28	4.50	5.89	x252	5.50	6.58	3.65	4.74
x398	6.81	8.20	4.43	5.81	x230	5.43	6.51	3.58	4.66
x370	6.74	8.12	4.36	5.73	x210	5.37	6.43	3.52	4.58
x342	6.67	8.03	4.29	5.65					
x311	6.59	7.94	4.21	5.56	W 12x190	5.30	6.36	3.45	4.51
x283	6.52	7.86	4.13	5.48	x170	5.23	6.28	3.39	4.43
x257	6.45	7.78	4.06	5.40	x152	5.17	6.21	3.33	4.37
x233	6.38	7.71	4.00	5.32	x136	5.12	6.15	3.27	4.30
x211	6.32	7.64	3.94	5.25	x120	5.06	6.09	3.21	4.24
x193	6.27	7.58	3.89	5.20	x106	5.02	6.03	3.17	4.19
x176	6.22	7.53	3.84	5.15	x 96	4.98	5.99	3.13	4.15
x159	6.18	7.47	3.79	5.09	x 87	4.95	5.96	3.10	4.11
x145	6.14	7.43	3.76	5.05	x 79	4.92	5.93	3.07	4.08
					x 72	4.89	5.90	3.05	4.05
					x 65	4.87	5.87	3.02	4.02

Case A: Shape perimeter, minus one flange surface.
Case B: Shape perimeter.
Case C: Box perimeter, equal to one flange surface plus twice the depth.
Case D: Box perimeter, equal to two flange surfaces plus twice the depth.

SURFACE AREAS AND BOX AREAS
W shapes
Square feet per foot of length

Desig-nation	Case A	Case B	Case C	Case D	Desig-nation	Case A	Case B	Case C	Case D
W 12x 58	4.39	5.22	2.87	3.70	W 8x67	3.42	4.11	2.19	2.88
x 53	4.37	5.20	2.84	3.68	x58	3.37	4.06	2.14	2.83
					x48	3.32	4.00	2.09	2.77
W 12x 50	3.90	4.58	2.71	3.38	x40	3.28	3.95	2.05	2.72
x 45	3.88	4.55	2.68	3.35	x35	3.25	3.92	2.02	2.69
x 40	3.86	4.52	2.66	3.32	x31	3.23	3.89	2.00	2.67
W 12x 35	3.63	4.18	2.63	3.18	W 8x28	2.87	3.42	1.89	2.43
x 30	3.60	4.14	2.60	3.14	x24	2.85	3.39	1.86	2.40
x 26	3.58	4.12	2.58	3.12					
					W 8x21	2.61	3.05	1.82	2.26
W 12x 22	2.97	3.31	2.39	2.72	x18	2.59	3.03	1.79	2.23
x 19	2.95	3.28	2.36	2.69					
x 16	2.92	3.25	2.33	2.66	W 8x15	2.27	2.61	1.69	2.02
x 14	2.90	3.23	2.32	2.65	x13	2.25	2.58	1.67	2.00
					x10	2.23	2.56	1.64	1.97
W 10x112	4.30	5.17	2.76	3.63					
x100	4.25	5.11	2.71	3.57	W 6x25	2.49	3.00	1.57	2.08
x 88	4.20	5.06	2.66	3.52	x20	2.46	2.96	1.54	2.04
x 77	4.15	5.00	2.62	3.47	x15	2.42	2.92	1.50	2.00
x 68	4.12	4.96	2.58	3.42					
x 60	4.08	4.92	2.54	3.38	W 6x16	1.98	2.31	1.38	1.72
x 54	4.06	4.89	2.52	3.35	x12	1.93	2.26	1.34	1.67
x 49	4.04	4.87	2.50	3.33	x 9	1.90	2.23	1.31	1.64
W 10x 45	3.56	4.23	2.35	3.02	W 5x19	2.04	2.45	1.28	1.70
x 39	3.53	4.19	2.32	2.98	x16	2.01	2.43	1.25	1.67
x 33	3.49	4.16	2.29	2.95					
					W 4x13	1.63	1.96	1.03	1.37
W 10x 30	3.10	3.59	2.23	2.71					
x 26	3.08	3.56	2.20	2.68					
x 22	3.05	3.53	2.17	2.65					
W 10x 19	2.63	2.96	2.04	2.38					
x 17	2.60	2.94	2.02	2.35					
x 15	2.58	2.92	2.00	2.33					
x 12	2.56	2.89	1.98	2.31					

Case A: Shape perimeter, minus one flange surface.
Case B: Shape perimeter.
Case C: Box perimeter, equal to one flange surface plus twice the depth.
Case D: Box perimeter, equal to two flange surfaces plus twice the depth.

STANDARD MILL PRACTICE
General Information

Rolling structural shapes and plates involves such factors as roll wear, subsequent roll dressing, temperature variations, etc., which cause the finished product to vary from published profiles. Such variations are limited by the provisions of the American Society for Testing and Materials Specification A6. Contained in this section is a summary of these provisions, not a reproduction of the complete specification. In its entirety, A6 covers a group of common requirements, which, unless otherwise specified in the purchase order or in an individual specification, shall apply to rolled steel plates, shapes, sheet piling and bars.

In accordance with the Scope clause of ASTM A6, *carbon steel* refers to ASTM Designations A36 and A529; *high-strength low-alloy steel* refers to Designations A242, A441, A572, and A588; *alloy steel* refers to Designation A514.

For further information on mill practices, including permissible variations for rolled tees, zees, and bulb angles in structural and bar sizes, pipe, tubing, sheets and strip, and for other grades of steel, see ASTM A6, A53, A500, A501, A568 and A618, and the AISI Steel Products Manuals and Producers' Catalogs.

The data on spreading rolls to increase areas and weights, and mill cambering of beams, is not a part of A6.

Additional material on mill practice is included in the descriptive material preceding the "Dimensions and Properties" tables for shapes and plates.

Letter symbols representing dimensions on sketches shown herein are in accordance with ASTM A6, AISI and mill catalogs and *not necessarily as defined by the general nomenclature of this manual.*

STANDARD MILL PRACTICE
Methods of increasing areas and weights by spreading rolls

W SHAPES

To vary the area and weight within a given nominal size, the flange width, the flange thickness, and the web thickness are changed as shown in Fig. 1.

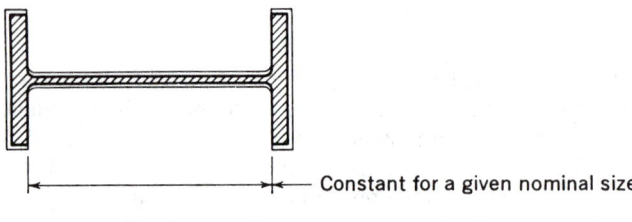

Constant for a given nominal size

Figure 1

S SHAPES AND AMERICAN STANDARD CHANNELS

To vary the area and weight within a given nominal size, the web thickness and the flange width are changed by an equal amount as shown in Figs. 2 and 3.

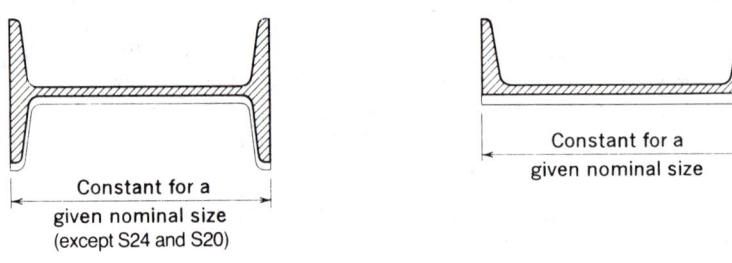

Constant for a given nominal size (except S24 and S20)

Constant for a given nominal size

Figure 2   Figure 3

ANGLES

To vary area and weight for a given leg length, the thickness of each leg is changed. Note that leg length is changed slightly by this method (see Fig. 4).

Figure 4

STANDARD MILL PRACTICE
Cambering of rolled beams

All beams are straightened after rolling to meet permissible variations for sweep and camber listed hereinafter for W shapes and S shapes. The following data refers to the subsequent cold cambering of beams to produce a predetermined dimension.

The maximum lengths that can be cambered depend on the length to which a given section can be rolled, with a maximum of 100 ft. The following table outlines the maximum and minimum induced camber of W shapes and S shapes.

MAXIMUM AND MINIMUM INDUCED CAMBER

Sections Nominal Depth In.	Specified Length of Beam, Ft				
	Over 30 to 42, incl.	Over 42 to 52, incl.	Over 52 to 65, incl.	Over 65 to 85, incl.	Over 85 to 100, incl.
	Max. and Min. Camber Acceptable, In.				
W shapes, 24 and over	1 to 2, incl.	1 to 3, incl.	2 to 4, incl.	3 to 5, incl.	3 to 6, incl.
W shapes, 14 to 21, incl. and S shapes, 12 in. and over	¾ to 2½, incl.	1 to 3, incl.	2 to 4, incl.	2½ to 5, incl.	Inquire

Consult the producer for specific camber and/or lengths outside the above listed available lengths and sections.

Mill camber in beams of less depth than tabulated should not be specified.

A single minimum value for camber, within the ranges shown above for the length ordered, should be specified.

Camber is measured at the mill and will not necessarily be present in the same amount in the section of beam as received due to release of stress induced during the cambering operation. In general, 75% of the specified camber is likely to remain.

Camber will approximate a simple regular curve nearly the full length of the beam, or between any two points specified.

Camber is ordinarily specified by the ordinate at the mid-length of the portion of the beam to be curved. Ordinates at other points should not be specified.

Although mill cambering to achieve reverse or other compound curves is not considered practical, fabricating shop facilities for cambering by heat can accomplish such results as well as form regular curves in excess of the limits tabulated above. Refer to Effect of Heat on Steel, Part 7 of this Manual, for further information.

PERMISSIBLE VARIATIONS FOR CAMBER ORDINATE

Lengths	Plus Variation	Minus Variation
50 ft and less	½ inch	0
Over 50 ft	½ in. plus ⅛ in. for each 10 ft or fraction thereof in excess of 50 ft	0

STANDARD MILL PRACTICE
Positions for measuring camber and sweep

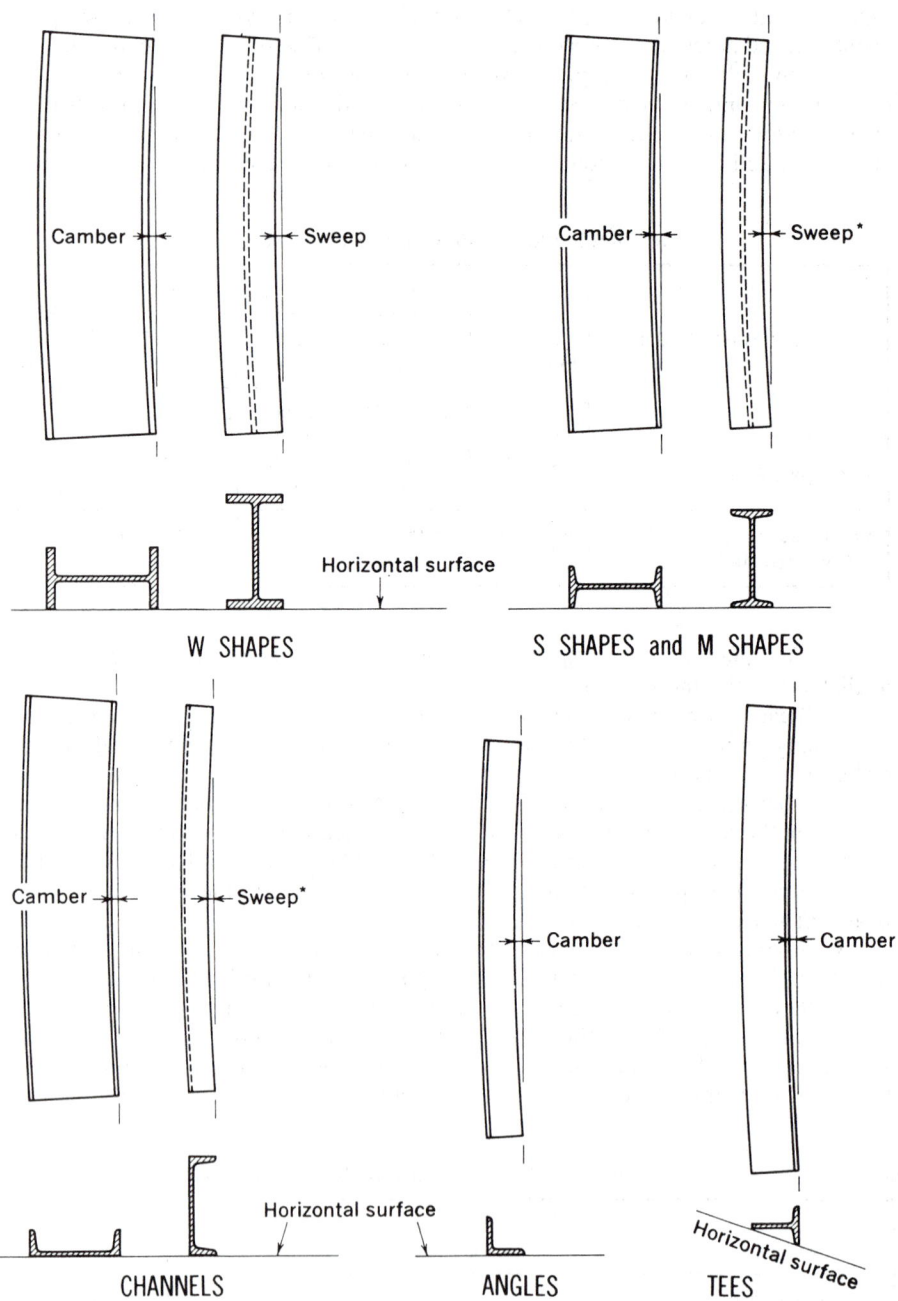

*Due to the extreme variations in flexibility of these shapes, straightness tolerances for sweep are subject to negotiations between manufacturer and purchaser for individual sections involved.

AMERICAN INSTITUTE OF STEEL CONSTRUCTION

STANDARD MILL PRACTICE

W Shapes

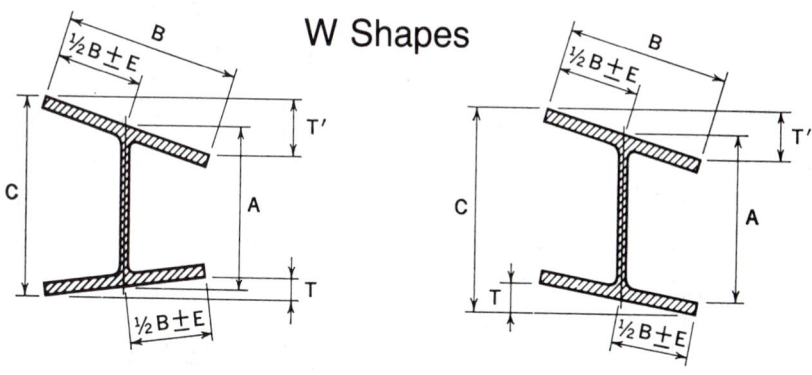

PERMISSIBLE VARIATIONS IN CROSS SECTION

Section Nominal Size, In.	A, Depth, In.		B, Fig. Width, In.		T + T', Flanges, Out of Square, Max, In.	E,[a] Web off Center, Max, In.	C, Max, Depth at any Cross-Section over Theoretical Depth, In.
	Over Theoretical	Under Theoretical	Over Theoretical	Under Theoretical			
To 12, incl.	⅛	⅛	¼	³⁄₁₆	¼	³⁄₁₆	¼
Over 12	⅛	⅛	¼	³⁄₁₆	⁵⁄₁₆	³⁄₁₆	¼

[a]Variation of ⁵⁄₁₆-in. max. for sections over 426 lb./ft.

PERMISSIBLE VARIATIONS IN LENGTH

W Shapes	Variations from Specified Length for Lengths Given, In.			
	30 ft. and Under		Over 30 ft.	
	Over	Under	Over	Under
Beams 24 in. and under in nominal depth	⅜	⅜	⅜ plus ¹⁄₁₆ for each additional 5 ft. or fraction thereof	⅜
Beams over 24 in. nom. depth; all columns	½	½	½ plus ¹⁄₁₆ for each additional 5 ft. or fraction thereof	½

OTHER PERMISSIBLE VARIATIONS

Area and Weight Variation: ±2.5% theoretical or specified amount.

Ends Out-of-Square: 1/64 in. per in. of depth, or of flange width if it is greater than the depth.

Camber and Sweep:

Sizes	Length	Permissible Variation, In.	
		Camber	Sweep
Sizes with flange width equal to or greater than 6 in.	All	$\frac{1}{8}$ in. $\times \dfrac{\text{(total length, ft)}}{10}$	
Sizes with flange width less than 6 in.	All	$\frac{1}{8}$ in. $\times \dfrac{\text{(total length, ft)}}{10}$	$\frac{1}{8}$ in. $\times \dfrac{\text{(total length, ft)}}{5}$
Certain sections with a flange width approx. equal to depth & specified on order as columns[b]	45 ft and under	$\frac{1}{8}$ in. $\times \dfrac{\text{(total length, ft)}}{10}$ with $\frac{3}{8}$ in. max.	
	Over 45 ft	$\frac{3}{8}$ in. $+ \left[\frac{1}{8} \text{ in.} \times \dfrac{\text{(total length, ft} - 45)}{10} \right]$	

[b]Applies only to: W 8 x 31 and heavier, W 12 x 65 and heavier, W 10 x 49 and heavier, W 14 x 90 and heavier. If other sections are specified on the order as columns, the tolerance will be subject to negotiation with the manufacturer.

STANDARD MILL PRACTICE
S shapes, M shapes and channels

PERMISSIBLE VARIATIONS IN CROSS-SECTION

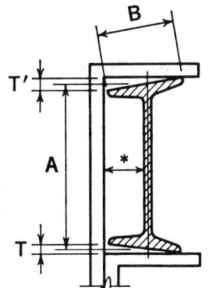 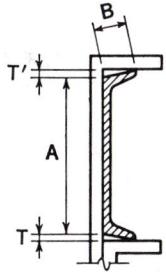

Section	Nominal Size, in.	A, Depth, In.[a]		B, Flange Width, In.		T + T', Out of Square per Inch of B, In.
		Over Theoretical	Under Theoretical	Over Theoretical	Under Theoretical	
S shapes and M shapes	3 to 7, incl.	3/32	1/16	1/8	1/8	1/32
	Over 7 to 14, incl.	1/8	3/32	5/32	5/32	1/32
	Over 14 to 24, incl.	3/16	1/8	3/16	3/16	1/32
Channels	3 to 7, incl	3/32	1/16	1/8	1/8	1/32
	Over 7 to 14, incl.	1/8	3/32	1/8	5/32	1/32
	Over 14	3/16	1/8	1/8	3/16	1/32

[a]A is measured at center line of web for beams; and at back of web for channels.

PERMISSIBLE VARIATIONS IN LENGTH

Section	Variations from Specified Length for Lengths Given, In.									
	To 30 Ft, incl.		Over 30 to 40 Ft, incl.		Over 40 to 50 Ft, incl.		Over 50 to 65 Ft, incl.		Over 65 Ft	
	Over	Under	Over	Under	Over	Under	Over	Under	Over	Under
S shapes, M shapes and Channels	1/2	1/4	3/4	1/4	1	1/4	1 1/8	1/4	1 1/4	1/4

OTHER PERMISSIBLE VARIATIONS

Area and Weight Variation: $\pm 2.5\%$ theoretical or specified amount.

Ends Out-of-Square: S shapes and channels 1/64 in. per in. of depth.

Camber: $\dfrac{1}{8}$ in. $\times \dfrac{\text{total length, ft}}{5}$

AMERICAN INSTITUTE OF STEEL CONSTRUCTION

STANDARD MILL PRACTICE
Tees split from W, M and S shapes
Angles split from channels

PERMISSIBLE VARIATIONS IN DEPTH

Dimension A may be approximately ½ beam or channel depth, or any dimension resulting from off-center splitting, or splitting on two lines as specified on the order.

Depth of Beam from which Tees or Angles are Split	Variations in Depth A Over and Under	
	Tees	Angles
To 6 in., excl.	⅛	⅛
6 to 16, excl.	³⁄₁₆	³⁄₁₆
16 to 20, excl.	¼	¼
20 to 24, excl.	⁵⁄₁₆	. . .
24 and over	⅜	. . .

The above variations for depths of tees or angles include the permissible variations in depth for the beams and channels before splitting.

OTHER PERMISSIBLE VARIATIONS

Other permissible variations in cross-section as well as permissible variations in length, Area and Weight variation, and ends out-of-square will correspond to those of the beam or channel before splitting, except

$$\text{camber} = \text{⅛ in.} \times \frac{\text{total length, ft}}{5}$$

STANDARD MILL PRACTICE
Angles, structural size

PERMISSIBLE VARIATIONS IN CROSS SECTION

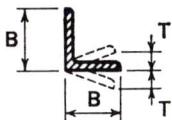

Section	Nominal Size, in.[a]	B — Length of Leg, in.		T. Out of Square Per In. of B, in.
		Over Theoretical	Under Theoretical	
Angles	3 to 4, incl.	⅛	³⁄₃₂	³⁄₁₂₈[b]
	Over 4 to 6, incl.	⅛	⅛	³⁄₁₂₈[b]
	Over 6	³⁄₁₆	⅛	³⁄₁₂₈[b]

[a] For unequal leg angles, longer leg determines classification.
[b] ³⁄₁₂₈ in. per in. = 1½ deg.

PERMISSIBLE VARIATIONS IN LENGTH

Section	Variations from Specified Length for Lengths Given, In.									
	To 30 Ft, incl.		Over 30 to 40 Ft, incl.		Over 40 to 50 Ft, incl.		Over 50 to 65 Ft, incl.		Over 65 Ft	
Angles	Over	Under	Over	Under	Over	Under	Over	Under	Over	Under
	½	¼	¾	¼	1	¼	1⅛	¼	1¼	¼

OTHER PERMISSIBLE VARIATIONS

Area and weight variation: ±2.5% theoretical or specified amount.

Ends out-of-square: ³⁄₁₂₈ in. per in. of leg length, or 1½ deg. Variations based on the longer leg of an unequal angle.

Camber: ⅛ in. × $\dfrac{\text{total length, ft}}{5}$, applied to either leg.

STANDARD MILL PRACTICE
*Angles, bar size

PERMISSIBLE VARIATIONS IN CROSS SECTION

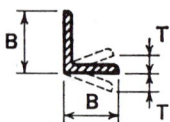

[a]Specified Length of Leg, In.	Variations from Thickness for Thicknesses Given, Over and Under, In.			B Length of Leg, Over and Under, In.	T, Out of Square per Inch of B, In.
	3/16 and Under	Over 3/16 to 3/8, incl.	Over 3/8		
1 and under	0.008	0.010		1/32	3/128 [b]
Over 1 to 2, incl.	0.010	0.010	0.012	3/64	3/128 [b]
Over 2 to 3, excl.	0.012	0.015	0.015	1/16	3/128 [b]

[a]The longer leg of an unequal angle determines the size for permissible variations.
[b]3/128-in. per in. = 1½ degrees.

PERMISSIBLE VARIATIONS IN LENGTH

Section	Variations Over Specified Length for Lengths Given No Variation Under				
	5 to 10 Ft excl.	10 to 20 Ft excl.	20 to 30 Ft excl.	30 to 40 Ft excl.	40 to 65 Ft incl.
All sizes of bar-size angles	5/8	1	1½	2	2½

OTHER PERMISSIBLE VARIATIONS

Camber: ¼ in. in any 5 ft, or ¼ in. $\times \dfrac{\text{total length, ft}}{5}$

Straightness: Because of warpage, permissible variations for straightness do not apply to bars if any subsequent heating operation has been performed.

Ends Out-of-Square: 3/128-in. per in. of leg length or 1½ degrees. Variation based on longer leg of an unequal angle.

---

*A member is "bar size" when its greatest cross-sectional dimension is less than 3 in.

STANDARD MILL PRACTICE
Steel pipe and tubing

DIMENSIONS AND WEIGHT TOLERANCES
Round Tubing and Pipe

ASTM A53	ASTM A501 and ASTM A618

Weight—The weight of the pipe as specified in Table X2 and Table X3 (ASTM Specification A53) shall not vary by more than ± 10 percent.

Note that the weight tolerance of ± 10 percent is determined from the weights of the customary lifts of pipe as produced for shipment by the mill, divided by the number of feet of pipe in the lift. On pipe sizes over 4 in. where individual lengths may be weighed, the weight tolerance is applicable to the individual length.

Diameter—For pipe 2 in. and over in nominal diameter, the outside diameter shall not vary more than ± 1 percent from the standard specified.

Thickness—The minimum wall thickness at any point shall be not more than 12.5 percent under the nominal wall thickness specified.

Outside Dimensions—For round hot formed structural tubing 2 in. and over in nominal size, the outside diameter shall not vary more than ± 1 percent from the standard specified.

Weight (A501 only)—The weight of structural tubing shall not be less than the specified value by more than 3.5 percent.

Mass (A618 only)—The mass of structural tubing shall not be less than the specified value by more than 3.5 percent.

Length—Structural tubing is commonly produced in random mill lengths, in multiple lengths, and in definite cut lengths. When cut lengths are specified for structural tubing, the length tolerances shall be in accordance with the following table:

	22 ft and under		Over 22 to 44 ft, incl.	
	Over	Under	Over	Under
Length tolerance for specified cut lengths, in.	½	¼	¾	¼

Straightness—The permissible variation for straightness of structural tubing shall be ⅛ in. times the number of feet of total length divided by 5.

Square and Rectangular Tubing

ASTM A501, ASTM A500 and ASTM A618

Outside Dimensions—The specified dimensions, measured across the flats at positions at least 2 in. from either end of square or rectangular tubing and including an allowance for convexity or concavity, shall not exceed the plus and minus tolerance shown in the following table:

Largest Outside Dimension, Across Flats, in.	Tolerance[a] Plus and Minus, in.
2½ and under	0.020
Over 2½ to 3½, incl.	0.025
Over 3½ to 5½, incl.	0.030
Over 5½	1 percent

[a]The respective outside dimension tolerances include the allowances for convexity and concavity.

Lengths—Structural tubing is commonly produced in random lengths, in multiple lengths, and in definite cut lengths. When cut lengths are specified for structural tubing, the length tolerances shall be in accordance with the following table:

	22 ft and under		Over 22 to 44 ft, incl.	
	Over	Under	Over	Under
Length tolerance for specified cut lengths, in.	½	¼	¾	¼

Weight (A501 only)—The weight of the structural tubing, as specified in Tables 4 and 5, shall not be less than the specified value by more than 3.5 percent.

Mass (A618 only)—The mass of structural tubing shall not be less than the specified value by more than 3.5 percent.

Straightness—The permissible variation for straightness of structural tubing shall be ⅛ in. times the number of feet of total length divided by 5.

Squareness of Sides—For square or rectangular structural tubing, adjacent sides may deviate from 90 degrees by a tolerance of plus or minus 2 degrees max.

Radius of Corners—For square or rectangular structural tubing, the radius of any outside corner of the section shall not exceed three times the specified wall thickness.

Twist—The tolerances for twist or variation with respect to axial alignment of the section, for square and rectangular structural tubing, shall be as shown in the following table:

Specified Dimension of Longest Side, in.	Maximum Twist per 3 ft of Length, in.
1½ and under	0.050
Over 1½ to 2½, incl.	0.062
Over 2½ to 4 incl.	0.075
Over 4 to 6, incl.	0.087
Over 6 to 8, incl.	0.100
Over 8	0.112

Twist is measured by holding down one end of a square or rectangular tube on a flat surface plate with the bottom side of the tube parallel to the surface plate and noting the height that either corner, at the opposite end of the bottom side of the tube, extends above the surface plate.

Wall Thickness (A500 only)—The tolerance for wall thickness exclusive of the weld area shall be plus and minus 10 percent of the nominal wall thickness specified. The wall thickness is to be measured at the center of the flat.

STANDARD MILL PRACTICE
Rectangular sheared plates and Universal mill plates

PERMISSIBLE VARIATIONS IN WIDTH AND LENGTH FOR SHEARED PLATES
(1½ in. and under in thickness)

PERMISSIBLE VARIATIONS IN LENGTH ONLY FOR UNIVERSAL MILL PLATES
(2½ in. and under in thickness)

Specified Dimensions, In.		Variations over Specified Width and Length for Thickness, In., and Equivalent Weights, Lb. Per Sq. Ft, Given							
		To ⅜, excl.		⅜ to ⅝, excl.		⅝ to 1, excl.		1 to 2, incl.[a]	
Length	Width	To 15.3, excl.		15.3 to 25.5, excl.		25.5 to 40.8, excl.		40.8 to 81.7, incl.	
		Width	Length	Width	Length	Width	Length	Width	Length
To 120, excl.	To 60, excl.	⅜	½	⁷⁄₁₆	⅝	½	¾	⅝	1
	60 to 84, excl.	⁷⁄₁₆	⅝	½	¹¹⁄₁₆	⅝	⅞	¾	1
	84 to 108, excl.	½	¾	⅝	⅞	¾	1	1	1⅛
	108 and over	⅝	⅞	¾	1	⅞	1⅛	1⅛	1¼
120 to 240, excl.	To 60, excl.	⅜	¾	½	⅞	⅝	1	¾	1⅛
	60 to 84, excl.	½	¾	⅝	⅞	¾	1	⅞	1¼
	84 to 108, excl.	⁹⁄₁₆	⅞	¹¹⁄₁₆	¹⁵⁄₁₆	¹³⁄₁₆	1⅛	1	1⅜
	108 and over	⅝	1	¾	1⅛	⅞	1¼	1⅛	1⅜
240 to 360, excl.	To 60, excl.	⅜	1	½	1⅛	⅝	1¼	¾	1½
	60 to 84, excl.	½	1	⅝	1⅛	¾	1¼	⅞	1½
	84 to 108, excl.	⁹⁄₁₆	1	¹¹⁄₁₆	1⅛	⅞	1⅜	1	1½
	108 and over	¹¹⁄₁₆	1⅛	⅞	1¼	1	1⅜	1¼	1¾
360 to 480, excl.	To 60, excl.	⁷⁄₁₆	1⅛	½	1¼	⅝	1⅜	¾	1⅝
	60 to 84, excl.	½	1¼	⅝	1⅜	¾	1½	⅞	1⅝
	84 to 108, excl.	⁹⁄₁₆	1¼	¾	1⅜	⅞	1½	1	1⅞
	108 and over	¾	1⅜	⅞	1½	1	1⅝	1¼	1⅞
480 to 600, excl.	To 60, excl.	⁷⁄₁₆	1¼	½	1½	⅝	1⅝	¾	1⅞
	60 to 84, excl.	½	1⅜	⅝	1½	¾	1⅝	⅞	1⅞
	84 to 108, excl.	⅝	1⅜	¾	1½	⅞	1⅝	1	1⅞
	108 and over	¾	1½	⅞	1⅝	1	1¾	1¼	1⅞
600 to 720, excl.	To 60, excl.	½	1¾	⅝	1⅞	¾	1⅞	⅞	2¼
	60 to 84, excl.	⅝	1¾	¾	1⅞	⅞	1⅞	1	2¼
	84 to 108, excl.	⅝	1¾	¾	1⅞	⅞	1⅞	1⅛	2¼
	108 and over	⅞	1¾	1	2	1⅛	2¼	1¼	2½
720 and over, excl.	To 60, excl.	⁹⁄₁₆	2	¾	2⅛	⅞	2¼	1	2¾
	60 to 84, excl.	¾	2	⅞	2⅛	1	2¼	1⅛	2¾
	84 to 108, excl.	¾	2	⅞	2⅛	1	2¼	1¼	2¾
	108 and over	1	2	1⅛	2⅜	1¼	2½	1⅜	3

[a]Permissible variations in length apply also to Universal Mill plates up to 12 in. width for thicknesses over 2 to 2½ in., incl. except for alloy steels up to 1¾ in. thick.

Notes: Permissible variations under specified width and length, ¼ in.
Table applies to all steels listed in ASTM A6.

STANDARD MILL PRACTICE
Rectangular sheared plates and Universal mill plates

PERMISSIBLE VARIATIONS FROM FLATNESS
(Carbon Steel Only)

Specified Thickness, In.	Variations from Flatness for Specified Widths, In.							
	To 36, excl.	36 to 48, excl.	48 to 60, excl.	60 to 72, excl.	72 to 84, excl.	84 to 96, excl.	96 to 108, excl.	108 to 120, excl.
To ¼, excl.	⁹⁄₁₆	¾	¹⁵⁄₁₆	1¼	1⅜	1½	1⅝	1¾
¼ to ⅜, excl.	½	⅝	¾	¹⁵⁄₁₆	1⅛	1¼	1⅜	1½
⅜ to ½, excl.	½	⁹⁄₁₆	⅝	⅝	¾	⅞	1	1⅛
½ to ¾, excl.	⁷⁄₁₆	½	⁹⁄₁₆	⅝	⅝	¾	1	1
¾ to 1, excl.	⁷⁄₁₆	½	⁹⁄₁₆	⅝	⅝	⅝	¾	⅞
1 to 2, excl.	⅜	½	½	⁹⁄₁₆	⁹⁄₁₆	⅝	⅝	⅝
2 to 4, excl.	⁵⁄₁₆	⅜	⁷⁄₁₆	½	½	½	½	⁹⁄₁₆
4 to 6, excl.	⅜	⁷⁄₁₆	½	½	⁹⁄₁₆	⁹⁄₁₆	⅝	¾
6 to 8, excl.	⁷⁄₁₆	½	½	⅝	¹¹⁄₁₆	¾	⅞	⅞

General Notes:
1. The longer dimension specified is considered the length, and permissible variations in flatness along the length should not exceed the tabular amount for the specified width in plates up to 12 ft. in length.
2. The flatness variations across the width should not exceed the tabular amount for the specified width.
3. When the longer dimension is under 36 in., the permissible variation should not exceed ¼ in. When the longer dimension is from 36 to 72 in., incl., the permissible variation should not exceed 75% of the tabular amount for the specified width, but in no case less than ¼ in.
4. These variations apply to plates which have a specified minimum tensile strength of not more than 60,000 psi or compatible chemistry or hardness. The limits in the table are increased 50% for plates specified to a higher minimum tensile strength or compatible chemistry or hardness.

PERMISSIBLE VARIATIONS IN CAMBER FOR CARBON STEEL SHEARED AND GAS CUT RECTANGULAR PLATES

Maximum permissible camber, in. (all thicknesses) = ⅛ in. × (total length, ft/5)

PERMISSIBLE VARIATIONS IN CAMBER FOR CARBON STEEL UNIVERSAL MILL PLATES, HIGH-STRENGTH AND HIGH-STRENGTH LOW-ALLOY STEEL SHEARED AND GAS CUT RECTANGULAR PLATES, UNIVERSAL MILL PLATES, SPECIAL CUT PLATES

Dimension, In.		Camber for Thicknesses and Widths Given
Thickness	Width	
To 2, incl.	All	⅛ in. × (total length, ft/5)
Over 2 to 15, incl.	To 30, incl.	³⁄₁₆ in. × (total length, ft/5)
Over 2 to 15, incl.	Over 30 to 60, incl.	¼ in. × (total length, ft/5)

AMERICAN INSTITUTE OF STEEL CONSTRUCTION

STANDARD MILL PRACTICE
Rectangular sheared plates and Universal mill plates

PERMISSIBLE VARIATIONS FROM FLATNESS
(High-Strength Low-Alloy and Alloy Steel, Hot Rolled or Thermally Treated)

Specified Thickness, In.	Variations from Flatness for Specified Widths: In.							
	To 36, excl.	36 to 48, excl.	48 to 60, excl.	60 to 72, excl.	72 to 84, excl.	84 to 96, excl.	96 to 108, excl.	108 to 120, excl.
To ¼, excl.	13/16	1⅛	1⅜	1⅞	2	2¼	2⅜	2⅝
¼ to ⅜, excl.	¾	15/16	1⅛	1⅜	1¾	1⅞	2	2¼
⅜ to ½, excl.	¾	⅞	15/16	15/16	1⅛	15/16	1½	1⅝
½ to ¾, excl.	⅝	¾	13/16	⅞	1	1⅛	1¼	1⅜
¾ to 1, excl.	⅝	¾	⅞	⅞	15/16	1	1⅛	15/16
1 to 2, excl.	9/16	⅝	¾	13/16	⅞	15/16	1	1
2 to 4, excl.	½	9/16	11/16	¾	¾	¾	¾	⅞
4 to 6, excl.	9/16	11/16	¾	¾	⅞	⅞	15/16	1⅛
6 to 8, excl.	⅝	¾	¾	15/16	1	1⅛	1¼	15/16

General Notes:
1. The longer dimension specified is considered the length, and variations from a flat surface along the length should not exceed the tabular amount for the specified width in plates up to 12 ft. in length.
2. The flatness variation across the width should not exceed the tabular amount for the specified width.
3. When the longer dimension is under 36 in., the variation should not exceed ⅜ in. When the longer dimension is from 36 to 72 in., incl. the variation should not exceed 75% of the tabular amount for the specified width.

PERMISSIBLE VARIATIONS IN WIDTH FOR UNIVERSAL MILL PLATES
(15 in. and under in thickness)

Specified Width, In.	Variations Over Specified Width for Thickness, in., and Equivalent Weights, lb. per sq. ft, Given					
	To ⅜, excl.	⅜ to ⅝, excl.	⅝ to 1, excl.	1 to 2, incl.	Over 2 to 10, incl.	Over 10 to 15, incl.
	To 15.3, excl.	15.3 to 25.5, excl.	25.5 to 40.8, excl.	40.8 to 81.7, incl.	81.7 to 409.0, incl.	409.0 to 613.0, incl.
Over 8 to 20, excl.	⅛	⅛	3/16	¼	⅜	½
20 to 36, excl.	3/16	¼	5/16	⅜	7/16	9/16
36 and over	5/16	⅜	7/16	½	9/16	⅝

Notes: Permissible variation under specified width, ⅛ in.
Table applies to all steels listed in ASTM A6.

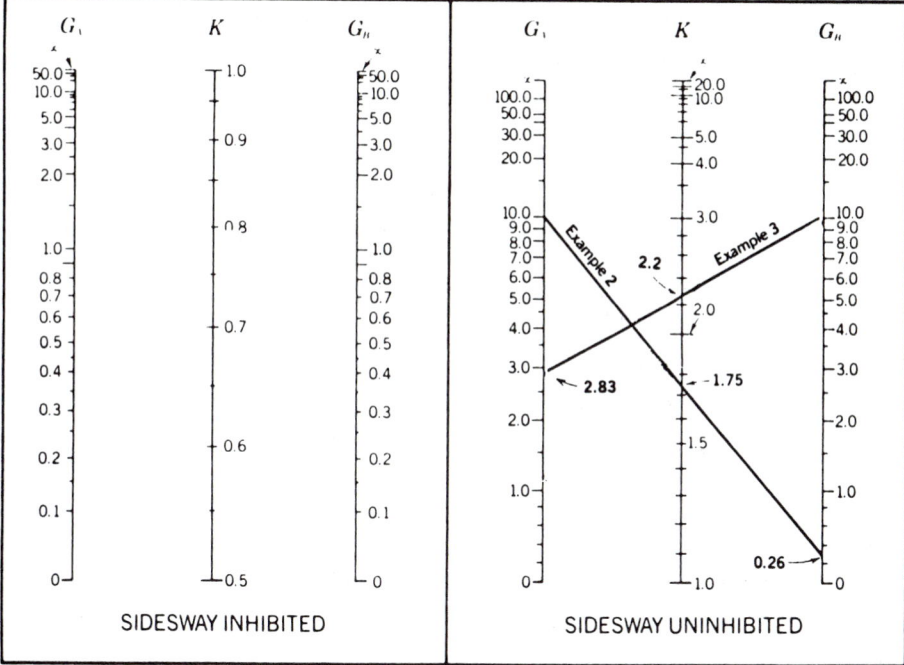

Figure 1. The subscripts A and B refer to the joints at the two ends of the column section being considered. G is defined as

$$G = \frac{\Sigma(I_c/L_c)}{\Sigma(I_g/L_g)}$$

in which Σ indicates a summation of all members rigidly connected to that joint and lying on the plane in which buckling of the column is being considered. I_c is the moment of inertia and L_c the unsupported length of a column section, and I_g is the moment of inertia and L_g the unsupported length of a girder or other restraining member. I_c and I_g are taken about axes perpendicular to the plane of buckling being considered.

For column ends supported by but not rigidly connected to a footing or foundation, G is theoretically infinity, but, unless actually designed as a true friction free pin, may be taken as "10" for practical designs. If the column end is rigidly attached to a properly designed footing, G may be taken as 1.0. Smaller values may be used if justified by analysis.

Approximate effective length relative to X-X axis:
  2.0 × 11 = 22.0 ft

From properties section in Tables, for W12 column:

$r_x/r_y \approx 1.76$

Corresponding effective length relative to the Y-Y axis:

$\frac{22.0}{1.76} \approx 12.5$ ft > 11.0 ft

∴ Effective length for X-X axis is critical.

Enter column load table with an effective length of 12.5 ft:

W12 × 106 column, by interpolation, good for 845 kips > 810 kips **o.k.**

2. Final Selection:

    Try W12 × 106

    Using Fig. 1 (sidesway uninhibited):

        I_x for W12 × 106 column = 933 in.^{4}

        I_x for W30 × 116 girder = 4,930 in.^{4}

        G (at base) = 10 (assume supported but not rigidly connected).

        G (at top) = $\dfrac{933/11}{(4{,}930 \times 2)/30}$ = 0.258, say 0.26.

    Connect points G_A = 10 and G_B = 0.26, read K = 1.75.

    For W12 × 106, r_x/r_y = 1.76.

    Actual effective length relative to Y-Y axis:

        $\dfrac{1.75}{1.76} \times 11.0 = 10.9$ ft < 11.0 ft

    Since effective length for Y-Y axis was critical:

        Use: W12 × 106 column

EXAMPLE 3

Given:

Using the alignment chart, Fig. 1 (sidesway uninhibited) and Table A, design columns for the bent shown, by the inelastic K-factor procedure. Let F_y = 36 ksi. Assume continuous support in the transverse direction.

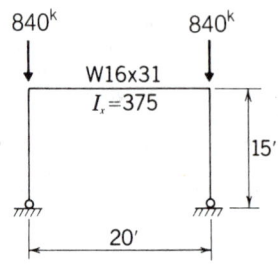

840^k 840^k

W16×31

I_x=375

15'

20'

Solution:

The alignment charts in Fig. 1 are applicable to elastic columns. By multiplying G-values times the stiffness reduction factor E_t/E, the charts may be used for inelastic columns.

Since $E_t/E \approx F_{cr,inelastic}/F_{cr,elastic}$,
the relationship may be written as
$G_{inelastic} = (F_{cr,inelastic}/F_{cr, elastic})\, G_{elastic}$.
By utilizing the calculated stress f_a a direct solution is possible, using the following steps:

1. For known value of factored axial load P_u, select a trial column size.

    Assume W12 × 120

        A = 35.3 in.^{2}, I_x = 1,070 in.^{4}, r_x = 5.51 in.

2. Calculate P_u/A:

        P_u/A = 840/35.3 = 23.8 ksi

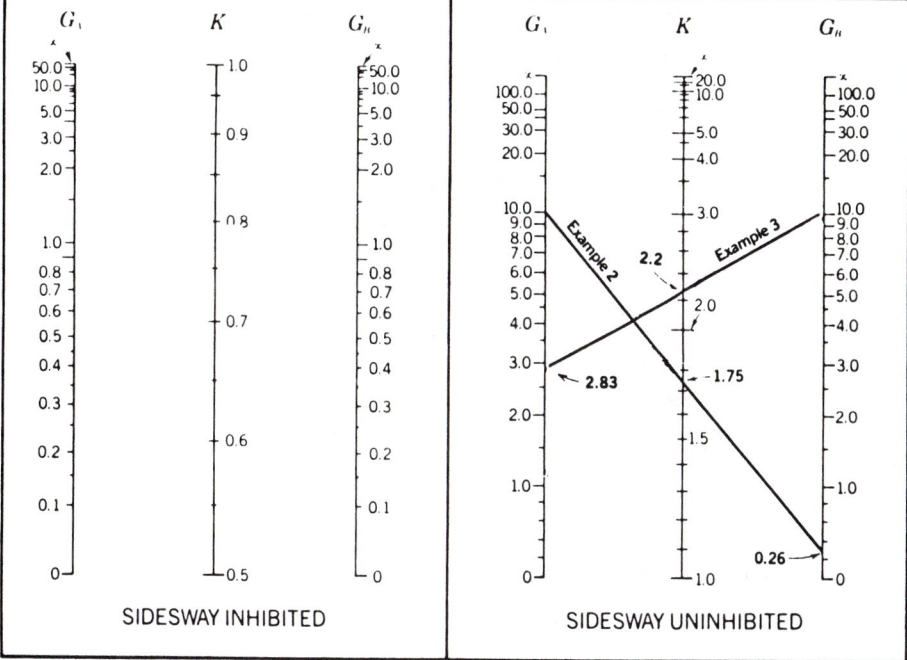

Figure 1.   The subscripts A and B refer to the joints at the two ends of the column section being considered. G is defined as

$$G = \frac{\Sigma(I_c/L_c)}{\Sigma(I_g/L_g)}$$

in which Σ indicates a summation of all members rigidly connected to that joint and lying on the plane in which buckling of the column is being considered. I_c is the moment of inertia and L_c the unsupported length of a column section, and I_g is the moment of inertia and L_g the unsupported length of a girder or other restraining member. I_c and I_g are taken about axes perpendicular to the plane of buckling being considered.

For column ends supported by but not rigidly connected to a footing or foundation, G is theoretically infinity, but, unless actually designed as a true friction free pin, may be taken as "10" for practical designs. If the column end is rigidly attached to a properly designed footing, G may be taken as 1.0. Smaller values may be used if justified by analysis.

Approximate effective length relative to X-X axis:
   $2.0 \times 11 = 22.0$ ft

From properties section in Tables, for W12 column:

$r_x/r_y \approx 1.76$

Corresponding effective length relative to the Y-Y axis:

$\dfrac{22.0}{1.76} \approx 12.5$ ft > 11.0 ft

∴ Effective length for X-X axis is critical.

Enter column load table with an effective length of 12.5 ft:

W12 × 106 column, by interpolation, good for 845 kips $>$ 810 kips   **o.k.**

2. Final Selection:

Try W12 × 106

Using Fig. 1 (sidesway uninhibited):

I_x for W12 × 106 column = 933 in.[4]

I_x for W30 × 116 girder = 4,930 in.[4]

G (at base) = 10 (assume supported but not rigidly connected).

$$G \text{ (at top)} = \frac{933/11}{(4{,}930 \times 2)/30} = 0.258, \text{ say } 0.26.$$

Connect points G_A = 10 and G_B = 0.26, read K = 1.75.

For W12 × 106, r_x/r_y = 1.76.

Actual effective length relative to Y-Y axis:

$$\frac{1.75}{1.76} \times 11.0 = 10.9 \text{ ft} < 11.0 \text{ ft}$$

Since effective length for Y-Y axis was critical:

Use: W12 × 106 column

EXAMPLE 3

Given:

Using the alignment chart, Fig. 1 (sidesway uninhibited) and Table A, design columns for the bent shown, by the inelastic K-factor procedure. Let F_y = 36 ksi. Assume continuous support in the transverse direction.

Solution:

The alignment charts in Fig. 1 are applicable to elastic columns. By multiplying G-values times the stiffness reduction factor E_t/E, the charts may be used for inelastic columns.

Since $E_t/E \approx F_{cr,inelastic}/F_{cr,elastic}$,

the relationship may be written as

$G_{inelastic} = (F_{cr,inelastic}/F_{cr,\ elastic})\ G_{elastic}$.

By utilizing the calculated stress f_a a direct solution is possible, using the following steps:

1. For known value of factored axial load P_u, select a trial column size.

   Assume W12 × 120

   A = 35.3 in.^{2}, I_x = 1,070 in.^{4}, r_x = 5.51 in.

2. Calculate P_u/A:

   P_u/A = 840/35.3 = 23.8 ksi

3. From Table A, determine the Stiffness Reduction Factor SRF; $SRF = 0.745$. For values of P_u/A smaller than those shown in Table A, the column is elastic, and the reduction factor is 1.0.

4. Determine $G_{elastic}$:

$G_{elastic}$ (bottom) $= 10.0$

$$G_{elastic} \text{ (top)} = \frac{1{,}070/15}{375/20} = 3.80$$

5. Calculate $G_{inelastic} = SRF \times G_{elastic}$:

$G_{inelastic}$ (top) $= 0.745\,(3.80) = 2.83$

6. Determine K from Fig. 1 using $G_{inelastic}$:

For $G_{top} = 2.83$ and $G_{bot} = 10$:

Read from Fig. 1, $K = 2.2$

7. Calculate Kl/r:

$Kl/r = 2.2(15)\,(12)/5.51 = 71.9$

8. $P_n = 23.31\,(35.3) = 823$ kips < 840 kips **n.g.**

Try a stronger column

1. Try W12 × 136

$A = 39.9 \text{ in.}^2$, $I_x = 1{,}240 \text{ in.}^4$, $r_x = 5.58 \text{ in.}$

2. $P_u/A = 840/39.9 = 21.1$ ksi

3. From Table A: $SRF = 0.852$

4. $G_{elastic} \text{ (top)} = \dfrac{1{,}240/15}{375/20} = 4.41$

5. $G_{inelastic}$ (top) $= 0.852\,(4.41) = 3.76$

6. $K = 2.4$

7. $Kl/r = 2.4(15)(12)/5.58 = 77.4$

8. $P_n = 22.32\,(39.9) = 891$ kips > 840 kips **o.k.**

Use: W12 × 136

	F_y = 36 ksi
TABLE A	F_y = 50 ksi

Stiffness Reduction Factors

$$F_{cr,inelastic} / F_{cr,elastic}$$

$\dfrac{P_u}{A}$	F_y		$\dfrac{P_u}{A}$	F_y		$\dfrac{P_u}{A}$	F_y	
	36 ksi	50 ksi		36 ksi	50 ksi		36 ksi	50 ksi
50.0	—	—	44.0	—	0.306	38.0	—	0.568
49.9	—	0.005	43.9	—	0.311	37.9	—	0.572
49.8	—	0.011	43.8	—	0.316	37.8	—	0.575
49.7	—	0.016	43.7	—	0.320	37.7	—	0.579
49.6	—	0.022	43.6	—	0.325	37.6	—	0.583
49.5	—	0.027	43.5	—	0.330	37.5	—	0.587
49.4	—	0.032	43.4	—	0.334	37.4	—	0.591
49.3	—	0.038	43.3	—	0.339	37.3	—	0.595
49.2	—	0.043	43.2	—	0.344	37.2	—	0.599
49.1	—	0.049	43.1	—	0.348	37.1	—	0.603
49.0	—	0.054	43.0	—	0.353	37.0	—	0.606
48.9	—	0.059	42.9	—	0.358	36.9	—	0.610
48.8	—	0.065	42.8	—	0.362	36.8	—	0.614
48.7	—	0.070	42.7	—	0.367	36.7	—	0.618
48.6	—	0.075	42.6	—	0.371	36.6	—	0.621
48.5	—	0.080	42.5	—	0.376	36.5	—	0.625
48.4	—	0.086	42.4	—	0.380	36.4	—	0.629
48.3	—	0.091	42.3	—	0.385	36.3	—	0.633
48.2	—	0.096	42.2	—	0.390	36.2	—	0.636
48.1	—	0.101	42.1	—	0.394	36.1	—	0.640
48.0	—	0.107	42.0	—	0.399	36.0	—	0.644
47.9	—	0.112	41.9	—	0.403	35.9	0.008	0.647
47.8	—	0.117	41.8	—	0.408	35.8	0.015	0.651
47.7	—	0.122	41.7	—	0.412	35.7	0.023	0.655
47.6	—	0.127	41.6	—	0.416	35.6	0.030	0.658
47.5	—	0.133	41.5	—	0.421	35.5	0.038	0.662
47.4	—	0.138	41.4	—	0.425	35.4	0.045	0.665
47.3	—	0.143	41.3	—	0.430	35.3	0.052	0.669
47.2	—	0.148	41.2	—	0.434	35.2	0.060	0.672
47.1	—	0.153	41.1	—	0.438	35.1	0.067	0.676
47.0	—	0.158	41.0	—	0.443	35.0	0.075	0.679
46.9	—	0.163	40.9	—	0.447	34.9	0.082	0.683
46.8	—	0.168	40.8	—	0.452	34.8	0.089	0.686
46.7	—	0.174	40.7	—	0.456	34.7	0.096	0.690
46.6	—	0.179	40.6	—	0.460	34.6	0.104	0.693
46.5	—	0.184	40.5	—	0.464	34.5	0.111	0.697
46.4	—	0.189	40.4	—	0.469	34.4	0.118	0.700
46.3	—	0.194	40.3	—	0.473	34.3	0.125	0.704
46.2	—	0.199	40.2	—	0.477	34.2	0.133	0.707
46.1	—	0.204	40.1	—	0.482	34.1	0.140	0.710
46.0	—	0.209	40.0	—	0.486	34.0	0.147	0.714
45.9	—	0.214	39.9	—	0.490	33.9	0.154	0.717
45.8	—	0.219	39.8	—	0.494	33.8	0.161	0.720
45.7	—	0.224	39.7	—	0.498	33.7	0.168	0.724
45.6	—	0.229	39.6	—	0.503	33.6	0.175	0.727
45.5	—	0.234	39.5	—	0.507	33.5	0.182	0.730
45.4	—	0.238	39.4	—	0.511	33.4	0.189	0.733
45.3	—	0.243	39.3	—	0.515	33.3	0.196	0.737
45.2	—	0.248	39.2	—	0.519	33.2	0.203	0.740
45.1	—	0.253	39.1	—	0.523	33.1	0.210	0.743
45.0	—	0.258	39.0	—	0.527	33.0	0.217	0.746
44.9	—	0.263	38.9	—	0.531	32.9	0.224	0.749
44.8	—	0.268	38.8	—	0.536	32.8	0.231	0.753
44.7	—	0.273	38.7	—	0.540	32.7	0.238	0.756
44.6	—	0.277	38.6	—	0.544	32.6	0.244	0.759
44.5	—	0.282	38.5	—	0.548	32.5	0.251	0.762
44.4	—	0.287	38.4	—	0.552	32.4	0.258	0.765
44.3	—	0.292	38.3	—	0.556	32.3	0.265	0.768
44.2	—	0.297	38.2	—	0.560	32.2	0.272	0.771
44.1	—	0.301	38.1	—	0.564	32.1	0.278	0.774

F_y = 36 ksi		
F_y = 50 ksi		

TABLE A (cont'd)
Stiffness Reduction Factors
$$F_{cr,inelastic}/F_{cr,elastic}$$

$\dfrac{P_u}{A}$	F_y		$\dfrac{P_u}{A}$	F_y		$\dfrac{P_u}{A}$	F_y	
	36 ksi	50 ksi		36 ksi	50 ksi		36 ksi	50 ksi
32.0	0.285	0.777	26.0	0.640	0.925	20.0	0.889	0.997
31.9	0.292	0.780	25.9	0.645	0.927	19.9	0.892	0.998
31.8	0.298	0.783	25.8	0.650	0.929	19.8	0.895	0.998
31.7	0.305	0.786	25.7	0.655	0.931	19.7	0.898	0.999
31.6	0.311	0.789	25.6	0.660	0.933	19.6	0.901	0.999
31.5	0.318	0.792	25.5	0.665	0.935	19.5	0.904	0.999
31.4	0.325	0.795	25.4	0.670	0.936	19.4	0.907	1.000
31.3	0.331	0.798	25.3	0.675	0.938	19.3	0.910	—
31.2	0.338	0.801	25.2	0.679	0.940	19.2	0.912	—
31.1	0.344	0.804	25.1	0.684	0.941	19.1	0.915	—
31.0	0.350	0.807	25.0	0.689	0.943	19.0	0.918	—
30.9	0.357	0.809	24.9	0.694	0.945	18.9	0.921	—
30.8	0.363	0.812	24.8	0.699	0.946	18.8	0.923	—
30.7	0.370	0.815	24.7	0.703	0.948	18.7	0.926	—
30.6	0.376	0.818	24.6	0.708	0.950	18.6	0.928	—
30.5	0.382	0.821	24.5	0.713	0.951	18.5	0.931	—
30.4	0.389	0.823	24.4	0.717	0.953	18.4	0.934	—
30.3	0.395	0.826	24.3	0.722	0.954	18.3	0.936	—
30.2	0.401	0.829	24.2	0.727	0.956	18.2	0.938	—
30.1	0.407	0.831	24.1	0.731	0.957	18.1	0.941	—
30.0	0.413	0.834	24.0	0.736	0.959	18.0	0.943	—
29.9	0.420	0.837	23.9	0.740	0.960	17.9	0.945	—
29.8	0.426	0.839	23.8	0.745	0.962	17.8	0.948	—
29.7	0.432	0.842	23.7	0.749	0.963	17.7	0.950	—
29.6	0.438	0.845	23.6	0.753	0.964	17.6	0.952	—
29.5	0.444	0.847	23.5	0.758	0.966	17.5	0.954	—
29.4	0.450	0.850	23.4	0.762	0.967	17.4	0.956	—
29.3	0.456	0.852	23.3	0.766	0.968	17.3	0.958	—
29.2	0.462	0.855	23.2	0.771	0.970	17.2	0.960	—
29.1	0.468	0.857	23.1	0.775	0.971	17.1	0.962	—
29.0	0.474	0.860	23.0	0.779	0.972	17.0	0.964	—
28.9	0.480	0.862	22.9	0.783	0.973	16.9	0.966	—
28.8	0.486	0.865	22.8	0.787	0.974	16.8	0.968	—
28.7	0.492	0.867	22.7	0.791	0.976	16.7	0.970	—
28.6	0.497	0.870	22.6	0.795	0.977	16.6	0.971	—
28.5	0.503	0.872	22.5	0.799	0.978	16.5	0.973	—
28.4	0.509	0.874	22.4	0.803	0.979	16.4	0.975	—
28.3	0.515	0.877	22.3	0.807	0.980	16.3	0.976	—
28.2	0.521	0.879	22.2	0.811	0.981	16.2	0.978	—
28.1	0.526	0.881	22.1	0.815	0.982	16.1	0.979	—
28.0	0.532	0.884	22.0	0.819	0.983	16.0	0.981	—
27.9	0.538	0.886	21.9	0.823	0.984	15.9	0.982	—
27.8	0.543	0.888	21.8	0.827	0.985	15.8	0.984	—
27.7	0.549	0.890	21.7	0.830	0.986	15.7	0.985	—
27.6	0.554	0.893	21.6	0.834	0.987	15.6	0.986	—
27.5	0.560	0.895	21.5	0.838	0.988	15.5	0.987	—
27.4	0.565	0.897	21.4	0.841	0.988	15.4	0.989	—
27.3	0.571	0.899	21.3	0.845	0.989	15.3	0.990	—
27.2	0.576	0.901	21.2	0.849	0.990	15.2	0.991	—
27.1	0.582	0.903	21.1	0.852	0.991	15.1	0.992	—
27.0	0.587	0.906	21.0	0.856	0.992	15.0	0.993	—
26.9	0.593	0.908	20.9	0.859	0.992	14.9	0.994	—
26.8	0.598	0.910	20.8	0.863	0.993	14.8	0.994	—
26.7	0.603	0.912	20.7	0.866	0.994	14.7	0.995	—
26.6	0.608	0.914	20.6	0.869	0.994	14.6	0.996	—
26.5	0.614	0.916	20.5	0.873	0.995	14.5	0.997	—
26.4	0.619	0.918	20.4	0.876	0.995	14.4	0.997	—
26.3	0.624	0.920	20.3	0.879	0.996	14.3	0.998	—
26.2	0.629	0.922	20.2	0.882	0.996	14.2	0.999	—
26.1	0.634	0.923	20.1	0.886	0.997	14.1	0.999	—
						14.0	1.000	—

TABLE B
Values of m^*

$F_y = 36$ ksi
$F_y = 50$ ksi

F_y	36 ksi							50 ksi						
KL (ft)	10	12	14	16	18	20	22 & over	10	12	14	16	18	20	22 & over
1st Approximation														
All shapes	2.9	2.8	2.6	2.6	2.5	2.4	2.3	2.6	2.5	2.4	2.2	2.1	1.9	1.8
Subsequent Approximations														
W, M, S 4	4.3	3.1	2.3	1.9	—	—	—	2.9	2.1	1.7	1.7	—	—	—
W, M, S 5	4.7	3.8	2.9	2.3	1.8	1.7	—	3.6	2.6	1.9	1.7	1.5	1.5	—
W, M, S 6	3.8	3.2	2.8	2.4	2.3	1.9	1.8	3.2	2.7	2.4	2.1	1.9	1.7	1.6
W 8	3.6	3.5	3.4	3.1	2.8	2.4	2.4	3.2	3.0	2.7	2.4	2.1	1.7	1.7
W 10	3.1	3.0	3.0	2.9	2.8	2.5	2.4	2.7	2.7	2.6	2.5	2.3	2.1	1.8
W 12	2.5	2.5	2.4	2.4	2.4	2.4	2.4	2.2	2.2	2.2	2.1	2.1	1.9	1.8
W 14	2.2	2.0	2.0	2.0	2.0	2.0	2.0	1.9	1.8	1.8	1.8	1.8	1.8	1.8

*Values of m are for $C_m = 0.85$. When C_m is other than 0.85, multiply the tabular value of m by $C_m/0.85$.

COMBINED AXIAL AND BENDING LOADING (INTERACTION)

Loads given in the column tables are for concentrically loaded columns. For columns subjected to both axial and bending stress, see Chap. H and Appendix H3 of the LRFD Specification.

The design of a beam-column is a trial and error process in which a trial section is checked for compliance with Formulas H1-1a and H1-1b. A fast method for selecting an economical trial, W, M or S shape section, using an equivalent axial load, is illustrated in the example problem, using Table B and the U values listed in the column properties at the bottom of the column load tables.

The procedure is as follows:

1. With the known value of KL (effective length), select a first approximate value of m from Table B. Let U equal 2.

2. Solve for $P_{ueff} = P_u + M_{ux}m + M_{uy}mU$

where

P_u = actual factored axial load, kips
M_{ux} = factored bending moment about the strong axis, kip-ft
M_{uy} = factored bending moment about the weak axis, kip-ft
m = factor taken from Table B
U = factor taken from column load table

3. From the appropriate column load table, select a tentative section to support P_{ueff}.

4. Based on the section selected in Step 3, select a "subsequent approximate" value of m from Table B and a U value from the column load table.

5. With the values selected in Step 4, solve for P_{ueff}.

6. Repeat Steps 3 and 4 until the values of m and U stabilize.

7. Check section obtained in Step 6 per Formula H1-1a or H1-1b as applies.

EXAMPLE 4

Given: P_u $= 300$ kips
$M_{ntx} = 180$ kip-ft
$M_{ltx} = 0$ (braced frame)
$M_{nty} = 60$ kip-ft
$M_{lty} = 0$ (braced frame)
$Kl_x = Kl_y = 14$ ft
$L_b = 14$ ft
$C_m = 0.85$

Solution:

1. For $KL = 14$ ft, from Table B select a first trial value of $m = 2.6$. Let $U = 2$.

2. $P_{eff} = P_u + mM_{ux} + mUM_{uy} = 300 + 2.6 \times 180 + 2.6 \times 2 \times 60 = 1,080$ kips

3. From column load tables select W14×120 ($\phi_c P_n = 971$) or W12×136 ($\phi_c P_n = 1,050$)

4. Adopt W14 column, so second trial value of m is 2.0. (Note: if a W14 column was required for architectural or other reasons, the selection process could have started with $m = 2.0$). With $m = 2.0$ and $U = 1.35$ (based on a W14×120) from column load table,
$P_{eff} = 300 + 2.0 \times 180 + 2.0 \times 1.35 \times 60 = 822$ kips

5. From column load tables select W14×99 ($\phi_c P_n = 799$)

6. For W14×99, $m = 2.0$, $U = 1.36$. Repeat of steps 3 and 4 not required.

7. Check W14×99

$A = 29.1$ in.^{2}

$r_y = 3.71$ in. $\left(\dfrac{Kl}{r_y}\right) = 45.3$

$r_x = 6.17$ in. $\left(\dfrac{Kl}{r_x}\right) = 27.2$

$\lambda_{cx} = \dfrac{27.2}{\pi}\sqrt{\dfrac{36}{29,000}} = 0.305$

$\lambda_{cy} = \dfrac{45.3}{\pi}\sqrt{\dfrac{36}{29,000}} = 0.508$

$P_{ex} = \dfrac{36 \times 29.1}{(0.305)^2} = 11,261$

$P_{ey} = \dfrac{36 \times 29.1}{(0.508)^2} = 4,059$

$B_x = \dfrac{0.85}{1 - 300/11,261} = 0.873 < 1.0$ use $B_x = 1.0$

$$B_y = \frac{0.85}{1 - 300/4,059} = 0.918 < 1.0 \quad \text{use } B_y = 1.0$$

$$M_{ux} = 1.0 \times 180 = 180 \text{ kip-ft}$$

$$M_{uy} = 1.0 \times 60 = 60 \text{ kip-ft}$$

$$L_p = \frac{300 \times 3.71}{\sqrt{36}} = 185.5 \text{ in.} = 15.5 \text{ ft}$$

Since $L_b = 14.0 < 15.5$, section is compact and

$$\phi_b \, M_{nx} = \phi_b M_{px} = \frac{0.9 \times 36 \times 173}{12} = 467 \text{ kip-ft}$$

$$\phi_b \, M_{ny} = \phi_b M_{py} = \frac{0.9 \times 36 \times 83.6}{12} = 226 \text{ kip-ft}$$

$$\frac{P_u}{P_n} = \frac{300}{799} = 0.38 \geq 0.2 \quad \therefore \text{ Formula H1-1a applies}$$

$$\frac{300}{799} + \frac{8}{9}\left(\frac{180}{467} + \frac{60}{226}\right) = 0.375 + 0.579 = 0.954 < 1.00 \quad \textbf{o.k.}$$

Use: W14 × 99

COLUMN WEB STIFFENERS

Values of P_{wo}, P_{wi}, P_{wb} and P_{fb}, listed in the Properties Section of the column load tables for W, M and S shapes, are useful in determining if a column web requires stiffeners because of forces transmitted into it from the flanges or connecting flange plates of a rigid beam connection to the column flange. When the applied factored beam flange force P_{bf} is equal to or less than the following resisting forces developed within the column section, column web stiffeners are not required.

$$P_{bf} \leq P_{wb} = 3,690\left[\frac{t_w}{h_c}\right] t_w^2 \sqrt{F_{yw}} \qquad \text{from LRFD Specification Formula K1-8*}$$

$$P_{bf} \leq P_{fb} = 5.625 \, t_f^2 \, F_{yf} \qquad \text{from LRFD Specification Formula K1-1}$$

$$P_{bf} \leq P_{wi} t_b + P_{wo} = F_{yw} \, t_w \, (5k + t_b) \qquad \text{from LRFD Specification Formula K1-2}$$

where

P_{wb} = maximum column web resisting force at beam compression flange, kips
P_{fb} = maximum column web resisting force at beam tension flange, kips
P_{wi} = $F_{yw} \, t_w$, kips/in.
P_{wo} = $5 \, F_{yw} \, t_w k$, kips
F_{yw} = yield strength of column web, ksi
h_c = depth of column web clear of fillets, in.
k = distance from outer face of column flange to web toe of column fillet, in.
t_w = thickness of column web, in.
t_b = thickness of beam flange or connection plate delivering concentrated force, in.
t_f = thickness of column flange, in.

---

*For symmetrical shapes $h_c = d_c$

If the factored force of P_{bf} transmitted into the column web exceeds any one of the above three resisting forces, stiffeners are required on the column web. Stiffeners must comply with the provisions of LRFD Specification Sect. K1.

EXAMPLE 5

Given:

The flanges of a W24×84 beam are welded to the flange of a W14×211 column transmitting a factored moment of 900 kip-ft due to factored live and dead loads. Determine if stiffeners are required for the column and, if so, make an appropriate design. Beam and column are F_y = 36 ksi.

Solution:

For W24×84:

d = 24.10 in., t_b = 0.770 in.

$$P_{bf} = \frac{900 \times 12}{24.10} = 448 \text{ kips}$$

From column load table:

P_{wb} = 1,830 kips > 448 kips   **o.k.**   from LRFD Spec. Formula K1-8
P_{fb} = 493 kips > 448 kips   **o.k.**   from LRFD Spec. Formula K1-1
$P_{wo} + t_b P_{wi}$ = 397 + 35(0.770)   from LRFD Spec. Formula K1-2
         = 424 kips < 448 kips   **n.g.**
∴ Web stiffeners required

The following rules are suggested for a stiffener design procedure.

1. The width of each stiffener plus one-half the thickness of the column web shall be not less than one-third the width of the flange or moment connection plate delivering the concentrated force.

2. The thickness of stiffeners shall be not less than $t_b/2$.

3. When the concentrated force delivered occurs on only one column flange, the stiffener length need not exceed one-half the column depth.

4. The weld joining stiffeners to the column web shall be sized to carry the force in the stiffener caused by unbalanced moments on opposite sides of the column.

Stiffener Design:
Beam data: b_f = 9.020 in., t_b = 0.770 in.
Column data: t_w = 0.980 in., t_f = 1.560 in., d = 15.72 in.

Req'd. stiffener area = $\dfrac{448 - 424}{36}$ = 0.67 in.^{2}

Min. width = $\frac{1}{3}b_f - \dfrac{t_w}{2}$ = $\frac{1}{3}$(9.020) - (0.980/2) = 2.517 in.

Min. thickness = $t_b/2$ = 0.770/2 = 0.385 in.

Req'd. thickness = 0.67/2.517 ≑ 0.266 in.   Use ½ in.

Req'd. width = 0.67/0.5 = 1.34 in.   Use 4 in. for practical detailing considerations.

Min. length = $(d/2) - t_f$
$$= (15.72/2) - 1.56 = 6.30 \text{ in.} \quad \text{Use 7 in.}$$

Use: Two PL $4 \times \frac{1}{2} \times 0$ ft-7 in.

Min. weld size = $\frac{5}{16}$ in., based on column web thickness of 0.980 in.
(Table J2.4, LRFD Specification Sect. J2). Use E70 electrodes.

$$\text{Req'd. length (both PL)} = \frac{P}{0.45(70)(0.707)(5/16)}$$

$$= \frac{448 - 424}{6.96} = 3.45 \text{ in.}$$

COLUMNS
W, M and S Shapes

Design strength loads in the tables that follow are tabulated for the effective lengths in feet KL, indicated at the left of each table. They are applicable to axially loaded members with respect to their minor axis in accordance with Sect. E2 of the AISC LRFD Specification. Two strengths are covered, $F_y = 36$ ksi and $F_y = 50$ ksi.

The heavy horizontal lines appearing within the tables indicate $Kl/r = 200$. No values are listed beyond $Kl/r = 200$.

For discussion of effective length, range of l/r, strength about the major axis, combined axial and bending stress and sample problems, see "Columns, General Notes."

Properties and factors are listed at the bottom of the tables for checking strength about the strong axis, combined loading conditions and column stiffener requirements.

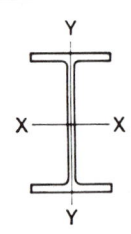

| | | F_y = 36 ksi |
| | | F_y = 50 ksi |

COLUMNS
W shapes
Design axial strength in kips ($\phi = 0.85$)

Designation						W14				
Wt./ft	730		665		605		550		500	
F_y	36	50	36	50	36	50	36	50	36	50
0	6580	9140	6000	8330	5450	7560	4960	6880	4500	6250
11	6310	8620	5740	7850	5210	7110	4740	6460	4290	5850
12	6260	8530	5700	7760	5170	7030	4700	6390	4250	5780
13	6210	8430	5650	7660	5120	6940	4650	6300	4210	5710
14	6150	8320	5590	7560	5070	6850	4600	6220	4170	5620
15	6090	8200	5540	7460	5020	6750	4560	6120	4120	5540
16	6020	8080	5480	7340	4960	6640	4500	6020	4070	5450
17	5960	7960	5410	7220	4900	6530	4450	5920	4020	5350
18	5880	7820	5350	7100	4840	6420	4390	5810	3970	5250
19	5810	7690	5280	6970	4770	6300	4330	5700	3910	5150
20	5730	7550	5200	6840	4700	6170	4270	5590	3850	5040
22	5570	7250	5050	6560	4560	5910	4130	5350	3730	4820
24	5390	6940	4890	6270	4410	5640	3990	5100	3600	4590
26	5210	6610	4720	5970	4250	5360	3840	4840	3460	4350
28	5020	6280	4540	5660	4090	5080	3690	4570	3320	4100
30	4820	5940	4360	5340	3920	4790	3530	4300	3180	3860
32	4620	5600	4170	5030	3740	4490	3370	4030	3030	3610
34	4420	5250	3980	4710	3570	4200	3210	3760	2880	3360
36	4210	4910	3780	4400	3390	3910	3040	3500	2730	3120
38	4000	4580	3590	4090	3210	3630	2880	3240	2580	2880
40	3790	4250	3400	3780	3030	3350	2720	2990	2420	2650
42	3580	3930	3210	3490	2860	3080	2550	2740	2280	2420
44	3380	3620	3020	3200	2680	2820	2390	2500	2130	2210
46	3170	3310	2830	2930	2510	2580	2240	2290	1990	2020
48	2970	3040	2650	2690	2340	2370	2080	2100	1850	1860
50	2780	2800	2470	2480	2180	2180	1940	1940	1710	1710

Effective length in ft KL with respect to least radius of gyration r_y

Properties										
U	1.13	1.25	1.14	1.26	1.15	1.28	1.15	1.28	1.16	1.28
P_{wo} (kips)	3070	4270	2640	3670	2250	3120	1930	2680	1650	2290
P_{wi} (kips/in.)	111	154	102	142	93	130	86	119	79	110
P_{wb} (kips)	53400	66500	44300	52200	33900	39900	26100	30800	20400	24100
P_{fb} (kips)	4880	6780	4140	5750	3500	4870	2960	4100	2480	3440
L_p (ft)	19.5	16.6	19.3	16.3	19.0	16.1	18.7	15.9	18.5	15.7
L_r (ft)	372.1	241.9	341.4	222.0	311.5	202.6	289.1	188.0	263.2	171.2
A (in.^{2})	215		196		178		162		147	
I_x (in.^{4})	14300		12400		10800		9430		8210	
I_y (in.^{4})	4720		4170		3680		3250		2880	
r_y (in.)	4.69		4.62		4.55		4.49		4.43	
Ratio r_x/r_y	1.74		1.73		1.71		1.70		1.69	

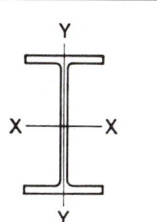

F_y = 36 ksi									
F_y = 50 ksi		COLUMNS W shapes Design axial strength in kips (ϕ = 0.85)							

Designation	W14								
Wt./ft	455		426		398		370		342	
F_y	36	50	36	50	36	50	36	50	36	50
0	4100	5700	3820	5310	3580	4970	3340	4630	3090	4290
11	3910	5330	3640	4960	3410	4640	3170	4320	2940	4000
12	3870	5260	3610	4900	3380	4580	3140	4260	2910	3940
13	3840	5190	3570	4830	3340	4520	3110	4200	2880	3890
14	3790	5110	3530	4760	3300	4450	3070	4140	2840	3830
15	3750	5030	3490	4680	3270	4380	3040	4070	2810	3760
16	3710	4950	3450	4600	3220	4300	3000	4000	2770	3690
17	3660	4860	3400	4520	3180	4220	2960	3920	2740	3620
18	3610	4770	3360	4430	3140	4140	2920	3840	2700	3550
19	3560	4670	3310	4340	3090	4050	2870	3760	2650	3470
20	3500	4570	3260	4250	3040	3960	2820	3680	2610	3400
22	3390	4370	3150	4050	2940	3780	2730	3500	2520	3230
24	3270	4150	3030	3850	2830	3590	2620	3320	2420	3060
26	3140	3930	2910	3640	2720	3390	2520	3140	2320	2890
28	3010	3700	2790	3430	2600	3190	2410	2950	2220	2710
30	2870	3480	2660	3210	2480	2990	2290	2760	2110	2530
32	2740	3250	2530	3000	2360	2780	2180	2560	2010	2360
34	2600	3020	2400	2780	2230	2580	2060	2380	1900	2180
36	2460	2800	2270	2570	2110	2390	1950	2190	1790	2010
38	2320	2580	2140	2370	1990	2190	1830	2010	1680	1840
40	2180	2370	2010	2170	1860	2010	1720	1840	1570	1680
42	2040	2160	1880	1980	1740	1830	1600	1670	1470	1520
44	1910	1970	1760	1800	1620	1660	1490	1520	1370	1390
46	1780	1800	1630	1650	1510	1520	1380	1390	1270	1270
48	1650	1650	1510	1510	1400	1400	1280	1280	1170	1170
50	1520	1520	1400	1400	1290	1290	1180	1180	1080	1080

Effective length in ft KL with respect to least radius of gyration r_y

Properties									
U	1.17	1.29	1.18	1.30	1.18	1.31	1.18	1.31	1.19	1.32
P_{wo} (kips)	1400	1950	1240	1730	1120	1550	987	1370	866	1200
P_{wi} (kips/in.)	73	101	68	94	64	89	60	83	55	77
P_{wb} (kips)	15800	18600	12800	15000	10800	12800	8790	10400	7100	8360
P_{fb} (kips)	2090	2900	1860	2590	1640	2280	1430	1990	1240	1720
L_p (ft)	18.3	15.5	18.1	15.3	18.0	15.2	17.8	15.1	17.7	15.0
L_r (ft)	242.4	157.7	226.4	147.3	213.1	138.7	199.6	129.9	184.8	120.4
A (in.^{2})	134		125		117		109		101	
I_x (in.^{4})	7190		6600		6000		5440		4900	
I_y (in.^{4})	2560		2360		2170		1990		1810	
r_y (in.)	4.38		4.34		4.31		4.27		4.24	
Ratio r_x/r_y	1.67		1.67		1.66		1.66		1.65	

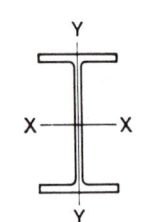

COLUMNS
W shapes
Design axial strength in kips ($\phi = 0.85$)

F_y = 36 ksi

F_y = 50 ksi

Designation					W14					
Wt./ft	311		283		257		233		211	
F_y	36	50	36	50	36	50	36	50	36	50
0	2800	3880	2550	3540	2310	3210	2100	2910	1900	2640
6	2750	3800	2510	3460	2280	3140	2060	2850	1870	2580
7	2740	3770	2500	3440	2260	3120	2050	2820	1860	2550
8	2720	3740	2480	3410	2250	3090	2040	2800	1840	2530
9	2700	3700	2460	3370	2230	3060	2020	2770	1830	2500
10	2680	3660	2440	3330	2210	3020	2000	2730	1810	2470
11	2660	3610	2420	3290	2190	2980	1980	2700	1800	2440
12	2630	3560	2390	3240	2170	2940	1960	2660	1780	2400
13	2600	3510	2370	3200	2150	2890	1940	2620	1760	2370
14	2570	3460	2340	3140	2120	2850	1920	2570	1730	2330
15	2540	3400	2310	3090	2090	2800	1890	2530	1710	2280
16	2510	3330	2280	3030	2060	2740	1870	2480	1690	2240
17	2470	3270	2250	2970	2030	2690	1840	2430	1660	2190
18	2430	3200	2210	2910	2000	2630	1810	2380	1640	2140
19	2390	3130	2180	2840	1970	2570	1780	2320	1610	2090
20	2360	3060	2140	2780	1940	2510	1750	2270	1580	2040
22	2270	2910	2060	2640	1870	2380	1680	2150	1520	1940
24	2180	2750	1980	2500	1790	2250	1620	2030	1460	1830
26	2090	2590	1900	2350	1710	2120	1540	1910	1390	1710
28	2000	2430	1810	2200	1630	1980	1470	1780	1320	1600
30	1900	2270	1720	2050	1550	1840	1400	1660	1260	1490
32	1800	2110	1630	1900	1470	1710	1320	1530	1190	1370
34	1700	1950	1540	1760	1380	1570	1240	1410	1120	1260
36	1600	1790	1450	1620	1300	1440	1170	1290	1050	1160
38	1500	1640	1360	1480	1220	1320	1090	1180	980	1050
40	1410	1490	1270	1340	1140	1190	1020	1070	912	951

Effective length in ft KL with respect to least radius of gyration r_y

Properties										
U	1.20	1.33	1.21	1.34	1.22	1.35	1.23	1.36	1.23	1.37
P_{wo} (kips)	746	1040	639	887	542	753	457	635	397	551
P_{wi} (kips/in.)	51	71	46	65	42	59	39	54	35	49
P_{wb} (kips)	5430	6400	4190	4930	3150	3710	2370	2790	1830	2160
P_{fb} (kips)	1030	1440	868	1200	723	1000	599	832	493	684
L_p (ft)	17.5	14.8	17.4	14.7	17.2	14.6	17.1	14.5	17.0	14.4
L_r (ft)	168.3	109.7	153.9	100.4	140.2	91.6	127.4	83.4	115.8	76.0
A (in.^{2})	91.4		83.3		75.6		68.5		62.0	
I_x (in.^{4})	4330		3840		3400		3010		2660	
I_y (in.^{4})	1610		1440		1290		1150		1030	
r_y (in.)	4.20		4.17		4.13		4.10		4.07	
Ratio r_x/r_y	1.64		1.63		1.62		1.62		1.61	

F_y = 36 ksi
F_y = 50 ksi

COLUMNS
W shapes
Design axial strength in kips ($\phi = 0.85$)

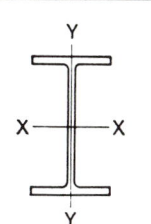

Designation								W14
Wt./ft	193		176		159		145	
F_y	36	50	36	50	36	50	36	50

Effective length in ft KL with respect to least radius of gyration r_y								
0	1740	2410	1580	2200	1430	1980	1310	1810
6	1710	2360	1560	2150	1400	1940	1280	1770
7	1700	2340	1550	2130	1400	1920	1280	1760
8	1680	2320	1540	2110	1390	1900	1270	1740
9	1670	2290	1530	2090	1380	1880	1260	1720
10	1660	2260	1510	2060	1360	1860	1250	1700
11	1640	2230	1500	2030	1350	1830	1230	1670
12	1630	2200	1480	2000	1330	1800	1220	1650
13	1610	2170	1460	1970	1320	1780	1200	1620
14	1590	2130	1450	1940	1300	1740	1190	1590
15	1570	2090	1430	1900	1280	1710	1170	1560
16	1540	2050	1410	1860	1270	1680	1160	1530
17	1520	2000	1380	1820	1250	1640	1140	1500
18	1500	1960	1360	1780	1230	1600	1120	1460
19	1470	1910	1340	1740	1200	1560	1100	1430
20	1440	1870	1310	1700	1180	1520	1080	1390
22	1390	1770	1260	1610	1140	1440	1040	1320
24	1330	1670	1210	1510	1090	1360	992	1240
26	1270	1560	1150	1420	1040	1270	946	1160
28	1210	1460	1100	1320	986	1180	898	1080
30	1150	1350	1040	1220	933	1100	849	998
32	1080	1250	980	1130	880	1010	800	919
34	1020	1150	920	1040	826	928	752	842
36	955	1050	863	946	773	846	703	767
38	892	956	805	859	721	767	655	694
40	830	863	748	775	670	692	608	626

Properties								
U	1.24	1.38	1.25	1.38	1.26	1.39	1.26	1.40
P_{wo} (kips)	340	473	299	415	251	349	214	298
P_{wi} (kips/in.)	32	45	30	42	27	37	24	34
P_{wb} (kips)	1370	1620	1110	1310	803	947	609	718
P_{fb} (kips)	420	583	348	483	287	398	241	334
L_p (ft)	16.9	14.3	16.8	14.2	16.7	14.1	16.6	14.1
L_r (ft)	106.5	70.1	97.6	64.5	88.6	59.0	81.6	54.7
A (in.^{2})	56.8		51.8		46.7		42.7	
I_x (in.^{4})	2400		2140		1900		1710	
I_y (in.^{4})	931		838		748		677	
r_y (in.)	4.05		4.02		4.00		3.98	
Ratio r_x/r_y	1.60		1.60		1.60		1.59	

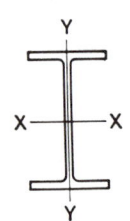

COLUMNS
W shapes
Design axial strength in kips ($\phi = 0.85$)

| $F_y = 36$ ksi |
| $F_y = 50$ ksi |

Designation		W14								
Wt./ft		132		120		109		99		90	
F_y		36	50	36	50	36	50	36	50†	36	50†
Effective length in ft KL with respect to least radius of gyration r_y	0	1190	1650	1080	1500	979	1360	890	1240	810	1130
	6	1160	1600	1060	1460	960	1320	873	1200	795	1100
	7	1160	1590	1050	1450	950	1310	867	1190	789	1080
	8	1150	1570	1040	1430	946	1300	860	1180	783	1070
	9	1140	1550	1030	1410	937	1280	852	1160	775	1060
	10	1120	1530	1020	1390	927	1260	843	1150	767	1040
	11	1110	1510	1010	1370	917	1240	833	1130	758	1030
	12	1100	1480	999	1350	905	1220	823	1110	749	1010
	13	1080	1450	985	1320	893	1200	811	1090	738	989
	14	1070	1420	971	1290	880	1170	799	1060	728	969
	15	1050	1390	956	1270	866	1150	787	1040	716	947
	16	1040	1360	940	1240	852	1120	773	1020	704	925
	17	1020	1330	924	1210	837	1090	759	991	691	902
	18	997	1300	906	1180	821	1060	745	965	678	878
	19	978	1260	888	1140	804	1030	730	938	664	853
	20	958	1220	870	1110	787	1000	714	911	650	828
	22	916	1150	831	1040	752	943	682	854	620	776
	24	872	1070	791	973	715	880	648	796	589	723
	26	826	997	749	902	678	815	614	737	558	670
	28	780	920	706	832	639	751	578	679	525	616
	30	733	844	663	762	600	688	542	621	493	564
	32	686	769	620	694	561	627	507	565	460	512
	34	639	697	577	629	522	567	471	511	428	463
	36	593	627	535	565	483	509	436	458	396	415
	38	547	563	494	507	446	457	402	411	365	372
Properties											
U		1.34	1.48	1.35	1.49	1.35	1.49	1.36	1.50	1.37	1.51
P_{wo} (kips)		196	272	173	240	148	205	125	174	109	151
P_{wi} (kips/in.)		23	32	21	30	19	26	17	24	16	22
P_{wb} (kips)		520	613	399	471	281	331	222	261	165	195
P_{fb} (kips)		215	298	179	249	150	208	123	171	102	142
L_p (ft)		15.7	13.3	15.6	13.2	15.5	13.2	15.5	13.4	15.4	15.0
L_r (ft)		73.6	49.6	67.9	46.2	62.7	43.2	58.2	40.6	54.1	38.4
A (in.^{2})		38.8		35.3		32.0		29.1		26.5	
I_x (in.^{4})		1530		1380		1240		1110		999	
I_y (in.^{4})		548		495		447		402		362	
r_y (in.)		3.76		3.74		3.73		3.71		3.70	
Ratio r_x/r_y		1.67		1.67		1.67		1.66		1.66	

†Flange is noncompact; see discussion preceding column load tables.

F_y = 36 ksi
F_y = 50 ksi

COLUMNS
W shapes
Design axial strength in kips (ϕ = 0.85)

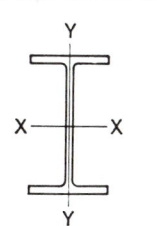

Designation							W14							
Wt./ft	82		74		68		61		53		48		43	
F_y	36	50	36	50	36	50	36	50‡	36	50‡	36	50‡	36‡	50‡
0	737	1020	667	926	612	850	548	761	477	663	431	599	386	536
6	705	963	638	871	585	798	523	714	443	598	400	540	357	482
7	694	942	628	852	576	781	515	698	432	576	390	520	347	463
8	682	918	616	830	565	760	505	680	419	552	378	498	337	443
9	667	892	604	807	553	738	494	660	404	526	365	474	325	422
10	652	863	590	781	540	714	483	638	389	498	351	449	312	399
11	635	833	575	753	526	689	470	615	372	469	336	423	298	375
12	618	801	559	724	511	662	457	591	355	439	320	396	284	350
14	579	732	524	662	479	604	428	539	319	379	287	340	254	301
16	538	661	487	598	444	545	396	486	282	319	253	286	224	252
18	495	588	447	532	408	484	364	431	245	263	220	235	194	206
20	450	517	407	467	371	424	331	377	210	213	188	191	165	167
22	406	447	367	405	334	366	297	326	176	176	157	157	138	138
24	363	381	328	345	297	311	265	276	148	148	132	132	116	116
26	321	325	290	294	262	265	233	236	126	126	113	113	99	99
28	280	280	253	253	229	229	203	203	109	109	97	97	85	85
30	244	244	221	221	199	199	177	177	95	95	85	85	74	74
31	229	229	207	207	187	187	166	166	89	89	79	79	69	69
32	214	214	194	194	175	175	155	155	83	83				
34	190	190	172	172	155	155	138	138						
36	169	169	153	153	138	138	123	123						
38	152	152	138	138	124	124	110	110						

Effective length in ft KL with respect to least radius of gyration r_y

Properties														
U	1.98	2.20	1.99	2.21	2.02	2.23	2.03	2.25	2.55	2.83	2.59	2.87	2.62	2.90
P_{wo} (kips)	149	207	127	176	112	156	97	135	96	133	84	117	72	100
P_{wi} (kips/in.)	18	26	16	23	15	21	14	19	13	19	12	17	11	15
P_{wb} (kips)	257	303	177	209	139	163	102	121	98	116	76	90	55	65
P_{fb} (kips)	148	206	125	173	105	146	84	117	88	123	72	100	57	79
L_p (ft)	10.3	8.8	10.3	8.8	10.3	8.7	10.2	8.7	8.0	6.8	8.0	6.8	7.9	6.7
L_r (ft)	43.0	29.6	40.0	28.0	37.3	26.4	34.7	24.9	28.0	20.1	26.3	19.2	24.7	18.2
A (in.²)	24.1		21.8		20.0		17.9		15.6		14.1		12.6	
I_x (in.⁴)	882		796		723		640		541		485		428	
I_y (in.⁴)	148		134		121		107		57.7		51.4		45.2	
r_y (in.)	2.48		2.48		2.46		2.45		1.92		1.91		1.89	
Ratio r_x/r_y	2.44		2.44		2.44		2.44		3.07		3.06		3.08	

‡Web may be noncompact for combined axial and bending stress; see AISC LRFD Specification Sect. B5.

Note: Heavy line indicates Kl/r of 200.

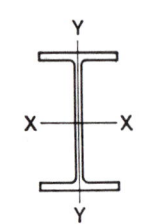

		F_y = 36 ksi
	COLUMNS **W shapes** Design axial strength in kips (ϕ = 0.85)	F_y = 50 ksi

Designation	W12									
Wt./ft	336		305		279		252		230	
F_y	36	50	36	50	36	50	36	50	36	50
0	3020	4200	2740	3810	2510	3480	2270	3150	2070	2880
6	2960	4070	2680	3690	2450	3370	2210	3040	2020	2780
7	2930	4020	2660	3640	2430	3330	2190	3010	2000	2740
8	2900	3970	2630	3590	2400	3280	2170	2960	1980	2710
9	2870	3910	2600	3540	2380	3230	2150	2920	1960	2660
10	2840	3850	2570	3480	2340	3170	2120	2870	1930	2610
11	2800	3780	2530	3420	2310	3110	2090	2810	1900	2560
12	2760	3700	2500	3340	2280	3050	2060	2750	1880	2500
13	2720	3620	2460	3270	2240	2980	2020	2680	1840	2450
14	2670	3540	2410	3190	2200	2910	1980	2620	1810	2380
15	2620	3450	2370	3110	2160	2830	1950	2550	1770	2320
16	2570	3360	2320	3020	2110	2750	1900	2470	1740	2250
17	2520	3260	2270	2940	2070	2670	1860	2400	1700	2180
18	2470	3160	2220	2840	2020	2580	1820	2320	1660	2110
19	2410	3060	2170	2750	1970	2500	1770	2240	1610	2030
20	2350	2960	2120	2660	1920	2410	1730	2160	1570	1960
22	2230	2750	2000	2460	1820	2230	1630	1990	1480	1810
24	2100	2540	1890	2270	1710	2050	1530	1830	1390	1650
26	1980	2320	1770	2070	1600	1870	1430	1660	1300	1500
28	1850	2120	1650	1880	1490	1690	1330	1500	1200	1350
30	1720	1910	1530	1690	1380	1520	1230	1350	1110	1210
32	1590	1720	1410	1520	1270	1350	1130	1200	1020	1070
34	1460	1520	1300	1340	1160	1200	1030	1060	931	951
36	1340	1360	1180	1200	1060	1070	940	945	845	848
38	1220	1220	1080	1080	960	960	848	848	761	761
40	1100	1100	970	970	866	866	765	765	687	687
Properties										
U	1.27	1.41	1.29	1.43	1.29	1.43	1.30	1.44	1.31	1.45
P_{wo} (kips)	1180	1640	1000	1400	878	1220	738	1020	636	883
P_{wi} (kips/in.)	64	89	59	81	55	77	50	70	46	64
P_{wb} (kips)	12700	14900	9740	11500	8230	9700	6160	7250	4810	5670
P_{fb} (kips)	1770	2460	1480	2060	1240	1720	1020	1420	868	1200
L_p (ft)	14.5	12.3	14.3	12.1	14.1	12.0	13.9	11.8	13.8	11.7
L_r (ft)	201.4	131.0	183.0	119.1	168.7	109.8	153.1	99.7	141.1	91.9
A (in.^{2})	98.8		89.6		81.9		74.1		67.7	
I_x (in.^{4})	4060		3550		3110		2720		2420	
I_y (in.^{4})	1190		1050		937		828		742	
r_y (in.)	3.47		3.42		3.38		3.34		3.31	
Ratio r_x/r_y	1.85		1.84		1.82		1.81		1.80	

Effective length in ft KL with respect to least radius of gyration r_y

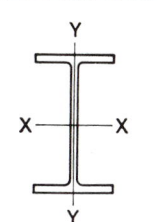

| F_y = 36 ksi |
| F_y = 50 ksi |

COLUMNS
W shapes
Design axial strength in kips (ϕ = 0.85)

Designation							W12						
Wt./ft		210		190		170		152		136		120	
F_y		36	50	36	50	36	50	36	50	36	50	36	50
Effective length in ft KL with respect to least radius of gyration r_y	0	1890	2630	1710	2370	1530	2120	1370	1900	1220	1700	1080	1500
	6	1840	2540	1660	2290	1490	2050	1330	1830	1190	1630	1050	1440
	7	1830	2500	1650	2260	1480	2020	1320	1810	1180	1610	1040	1420
	8	1810	2470	1630	2220	1460	1990	1300	1780	1160	1580	1030	1400
	9	1790	2430	1610	2190	1440	1960	1290	1750	1150	1560	1010	1380
	10	1760	2380	1590	2150	1420	1920	1270	1710	1130	1530	1000	1350
	11	1740	2330	1560	2100	1400	1880	1250	1680	1110	1490	984	1320
	12	1710	2280	1540	2050	1380	1840	1230	1640	1090	1460	966	1280
	13	1680	2230	1510	2000	1350	1790	1210	1600	1070	1420	948	1250
	14	1650	2170	1480	1950	1330	1740	1180	1550	1050	1380	928	1220
	15	1610	2110	1450	1900	1300	1690	1160	1500	1030	1340	908	1180
	16	1580	2040	1420	1840	1270	1640	1130	1460	1000	1290	896	1140
	17	1540	1980	1390	1780	1240	1580	1100	1410	980	1250	864	1100
	18	1500	1910	1350	1720	1210	1530	1070	1360	955	1200	841	1060
	19	1470	1840	1320	1650	1180	1470	1040	1310	928	1160	817	1020
	20	1430	1780	1280	1590	1140	1420	1020	1260	901	1110	793	976
	22	1340	1640	1210	1460	1070	1300	954	1150	846	1020	743	892
	24	1260	1490	1130	1340	1000	1180	891	1050	789	924	692	808
	26	1170	1360	1050	1210	933	1070	827	944	731	832	640	726
	28	1090	1220	973	1090	863	959	763	844	673	742	589	646
	30	1000	1090	895	967	792	852	700	749	617	656	538	569
	32	919	962	819	853	724	750	638	658	561	577	489	500
	34	838	852	745	755	657	664	578	583	508	511	442	443
	36	759	760	674	674	593	593	520	520	456	456	395	395
	38	682	682	605	605	532	532	467	467	409	409	355	355
	40	616	616	546	546	480	480	421	421	369	369	320	320

Properties													
U		1.33	1.47	1.33	1.47	1.35	1.49	1.36	1.51	1.37	1.52	1.38	1.53
P_{wo} (kips)		558	774	465	646	389	540	333	462	276	383	232	322
P_{wi} (kips/in.)		42	59	38	53	35	48	31	44	28	40	26	36
P_{wb} (kips)		3760	4430	2700	3190	2020	2380	1500	1760	1120	1320	815	960
P_{fb} (kips)		731	1020	610	847	493	684	397	551	316	439	247	343
L_p (ft)		13.7	11.6	13.5	11.5	13.4	11.4	13.3	11.3	13.2	11.2	13.0	11.1
L_r (ft)		129.2	84.2	117.3	76.6	105.4	68.9	94.8	62.1	84.6	55.7	75.5	50.0
A (in.²)		61.8		55.8		50.0		44.7		39.9		35.3	
I_x (in.⁴)		2140		1890		1650		1430		1240		1070	
I_y (in.⁴)		664		589		517		454		398		345	
r_y (in.)		3.28		3.25		3.22		3.19		3.16		3.13	
Ratio r_x/r_y		1.80		1.79		1.78		1.77		1.77		1.76	

AMERICAN INSTITUTE OF STEEL CONSTRUCTION

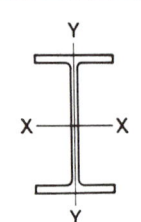

| | **COLUMNS**
W shapes
Design axial strength in kips ($\phi = 0.85$) | | | | | | | | | | | $F_y = 36$ ksi
$F_y = 50$ ksi |

Designation		W12										
Wt./ft		106		96		87		79		72		65
F_y		36	50	36	50	36	50	36	50	36	50	36	50†
	0	955	1330	863	1200	783	1090	710	986	646	897	584	812
	6	928	1280	839	1150	761	1040	689	947	627	861	567	779
	7	919	1260	830	1140	753	1030	682	933	620	848	561	767
	8	908	1240	820	1120	744	1010	674	917	613	834	554	754
	9	896	1210	809	1100	734	994	665	900	604	818	546	739
	10	883	1190	797	1070	723	973	654	881	595	800	538	723
	11	868	1160	784	1050	711	950	643	860	585	781	529	706
	12	853	1130	770	1020	698	926	631	838	574	761	519	687
	13	836	1100	755	995	684	901	619	814	562	740	508	668
	14	819	1070	739	966	669	874	605	790	550	717	497	647
	15	800	1040	722	935	654	846	591	764	537	694	485	626
	16	781	1000	704	904	638	817	576	738	523	670	472	604
	17	761	968	686	871	621	788	561	711	509	645	460	582
	18	741	932	667	838	604	758	545	683	495	620	446	558
	19	719	895	648	805	586	727	529	655	480	594	433	535
	20	698	858	628	771	568	696	512	627	465	569	419	512
	22	653	783	588	703	531	634	479	570	434	517	391	464
	24	608	708	546	635	493	572	444	514	403	465	362	418
	26	562	635	505	569	455	511	409	459	371	415	333	372
	28	516	565	463	505	417	453	375	406	339	367	305	328
	30	472	497	422	443	380	397	341	355	309	321	277	287
	32	428	437	383	390	344	349	308	312	279	282	250	252
	34	386	387	345	345	309	309	277	277	250	250	223	223
	36	345	345	308	308	276	276	247	247	223	223	199	199
	38	310	310	276	276	248	248	221	221	200	200	179	179
	40	279	279	249	249	223	223	200	200	181	181	161	161

Effective length in ft KL with respect to least radius of gyration r_y

Properties												
U	1.39	1.54	1.40	1.55	1.41	1.56	1.42	1.58	1.43	1.58	1.44	1.59
P_{wo} (kips)	185	257	161	223	139	193	122	169	106	148	92	128
P_{wi} (kips/in.)	22	31	20	28	19	26	17	24	15	22	14	20
P_{wb} (kips)	518	611	378	446	311	366	236	278	181	213	135	159
P_{fb} (kips)	198	276	164	228	133	185	109	152	91	126	74	103
L_p (ft)	13.0	11.0	12.9	10.9	12.8	10.9	12.7	10.8	12.7	10.8	12.6	11.8
L_r (ft)	67.2	44.9	61.4	41.3	56.4	38.4	51.8	35.7	48.2	33.6	44.7	31.7
A (in.²)	31.2		28.2		25.6		23.2		21.1		19.1	
I_x (in.⁴)	933		833		740		662		597		533	
I_y (in.⁴)	301		270		241		216		195		174	
r_y (in.)	3.11		3.09		3.07		3.05		3.04		3.02	
Ratio r_x/r_y	1.76		1.76		1.75		1.75		1.75		1.75	

†Flange is noncompact; see discussion preceding column load tables.

AMERICAN INSTITUTE OF STEEL CONSTRUCTION

F_y = 36 ksi
F_y = 50 ksi

COLUMNS
W shapes
Design axial strength in kips ($\phi = 0.85$)

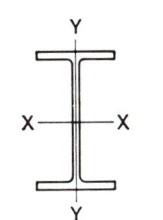

Designation						W12				
Wt./ft	58		53		50		45		40	
F_y	36	50	36	50	36	50	36	50	36	50‡
Effective length in ft KL with respect to least radius of gyration r_y — 0	520	723	477	663	450	625	404	561	361	502
6	498	680	457	623	419	566	376	507	336	453
7	490	666	449	610	408	546	366	489	327	437
8	482	649	441	594	396	524	355	469	317	419
9	472	631	432	577	383	500	343	447	306	399
10	461	611	422	559	369	475	330	424	295	378
11	450	590	411	539	354	448	317	400	282	356
12	437	568	400	518	339	421	302	375	269	334
13	424	545	388	496	322	393	287	350	256	311
14	411	521	375	474	306	365	272	324	242	288
15	397	496	362	451	289	337	257	299	228	266
16	382	471	348	428	271	310	241	274	214	243
18	352	420	320	381	237	257	210	227	187	201
20	321	370	292	334	204	209	180	184	160	163
22	291	322	263	290	173	173	152	152	135	135
24	260	276	235	247	145	145	128	128	113	113
26	231	235	208	210	124	124	109	109	96	96
28	202	202	181	181	107	107	94	94	83	83
30	176	176	158	158	93	93	82	82	72	72
32	155	155	139	139	82	82	72	72	64	64
34	137	137	122	122						
38	110	110	98	98						
41	94	94	85	85						

Properties										
U	1.73	1.92	1.75	1.94	2.18	2.42	2.21	2.45	2.23	2.47
P_{wo} (kips)	89	124	78	108	92	127	75	105	66	92
P_{wi} (kips/in.)	13	18	12	17	13	19	12	17	11	15
P_{wb} (kips)	106	125	94	111	116	136	86	101	59	69
P_{fb} (kips)	83	115	67	93	83	115	67	93	54	75
L_p (ft)	10.5	8.9	10.3	8.8	8.2	6.9	8.1	6.9	8.0	6.8
L_r (ft)	38.4	27.0	35.8	25.6	30.8	21.7	28.5	20.3	26.5	19.3
A (in.^{2})	17.0		15.6		14.7		13.2		11.8	
I_x (in.^{4})	475		425		394		350		310	
I_y (in.^{4})	107		95.8		56.3		50		44.1	
r_y (in.)	2.51		2.48		1.96		1.94		1.93	
Ratio r_x/r_y	2.10		2.11		2.64		2.65		2.66	

‡Web may be noncompact for combined axial and bending stress; see AISC LRFD Specification Sect. B5.
Note: Heavy line indicates Kl/r of 200.

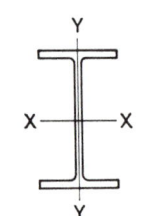

	COLUMNS	F_y = 36 ksi
	W shapes	F_y = 50 ksi

COLUMNS
W shapes
Design axial strength in kips (ϕ = 0.85)

Designation	W10									
Wt./ft	112		100		88		77		68	
F_y	36	50	36	50	36	50	36	50	36	50
0	1010	1400	900	1250	793	1100	692	961	612	850
6	969	1330	865	1180	762	1040	664	908	588	803
7	956	1300	853	1160	751	1020	655	890	579	787
8	941	1270	840	1140	739	999	644	869	569	769
9	924	1240	824	1110	725	973	632	847	558	749
10	906	1210	808	1080	710	945	618	822	547	727
11	886	1170	789	1040	694	916	604	796	534	703
12	865	1130	770	1010	677	884	588	768	520	678
13	842	1090	750	970	659	851	572	738	506	652
14	819	1050	728	931	639	817	555	708	490	625
15	794	1000	706	892	619	782	537	677	475	597
16	768	961	682	851	599	746	519	645	458	569
17	742	915	659	810	577	709	500	612	442	540
18	715	870	634	769	556	672	481	580	424	511
19	688	824	609	727	534	635	461	547	407	482
20	660	778	584	686	511	599	442	515	389	454
22	604	688	534	605	466	527	402	452	354	398
24	548	601	483	527	422	458	363	392	319	344
26	493	518	434	453	378	393	324	335	285	294
28	440	447	386	390	336	339	287	289	252	254
30	389	389	340	340	295	295	252	252	221	221
32	342	342	299	299	259	259	221	221	194	194
34	303	303	265	265	230	230	196	196	172	172
36	270	270	236	236	205	205	175	175	153	153
38	242	242	212	212	184	184	157	157	138	138
40	219	219	191	191	166	166	141	141	124	124

Effective length in ft KL with respect to least radius of gyration r_y

Properties										
U	1.32	1.46	1.33	1.47	1.34	1.48	1.35	1.50	1.36	1.51
P_{wo} (kips)	255	354	214	298	177	246	143	199	116	162
P_{wi} (kips/in.)	27	38	24	34	22	30	19	27	17	24
P_{wb} (kips)	1210	1430	883	1040	623	735	420	495	293	345
P_{fb} (kips)	316	439	254	353	198	276	153	213	120	167
L_p (ft)	11.2	9.5	11.0	9.4	11.0	9.3	10.8	9.2	10.8	9.2
L_r (ft)	86.4	56.5	77.4	50.8	68.4	45.1	60.1	39.9	53.7	36.0
A (in.^{2})	32.9		29.4		25.9		22.6		20.0	
I_x (in.^{4})	716		623		534		455		394	
I_y (in.^{4})	236		207		179		154		134	
r_y (in.)	2.68		2.65		2.63		2.60		2.59	
Ratio r_x/r_y	1.74		1.74		1.73		1.73		1.71	

| F_y = 36 ksi |
| F_y = 50 ksi |

COLUMNS
W shapes
Design axial strength in kips (ϕ = 0.85)

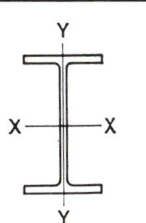

Designation		W10											
Wt./ft		60		54		49		45		39		33	
F_y		36	50	36	50	36	50	36	50	36	50	36	50
Effective length in ft KL with respect to least radius of gyration r_y	0	539	748	483	672	441	612	407	565	352	489	297	413
	6	517	706	464	634	422	577	380	515	328	444	276	373
	7	509	692	457	621	416	565	371	497	320	428	269	360
	8	500	675	449	606	409	551	361	478	311	412	261	345
	9	491	657	440	590	401	536	350	458	301	393	252	329
	10	480	638	431	572	392	520	337	436	290	374	243	312
	11	469	617	420	553	382	502	324	412	278	353	233	294
	12	457	595	409	533	372	484	311	388	266	332	222	276
	13	444	571	398	512	361	465	296	364	254	310	211	257
	14	430	547	385	490	350	444	282	339	241	289	200	239
	15	416	523	373	468	338	424	267	315	228	267	189	220
	16	401	497	360	445	326	403	252	290	215	246	177	202
	17	387	472	346	422	314	382	237	266	201	225	166	184
	18	371	446	332	399	301	361	222	243	188	205	155	167
	19	356	421	318	376	288	340	207	221	175	185	144	150
	20	340	395	304	353	275	319	192	199	162	167	133	135
	22	309	346	276	309	250	278	164	164	138	138	112	112
	24	278	299	248	266	224	239	138	138	116	116	94	94
	26	248	255	221	227	199	204	118	118	99	99	80	80
	28	219	220	195	196	175	176	102	102	85	85	69	69
	30	191	191	170	170	153	153	88	88	74	74	60	60
	32	168	168	150	150	134	134	78	78	65	65	53	53
	33	158	158	141	141	126	126	73	73	61	61		
	34	149	149	133	133	119	119						
	36	133	133	118	118	106	106						
Properties													
U		1.38	1.52	1.38	1.53	1.39	1.54	1.75	1.93	1.77	1.96	1.81	2.00
P_{wo} (kips)		99	138	83	116	73	101	79	109	64	89	55	77
P_{wi} (kips/in.)		15	21	13	19	12	17	13	18	11	16	10	15
P_{wb} (kips)		209	246	143	168	111	131	121	142	88	104	69	81
P_{fb} (kips)		94	130	77	106	64	88	78	108	57	79	38	53
L_p (ft)		10.7	9.1	10.7	9.1	10.6	9.0	8.4	7.1	8.3	7.0	8.1	6.9
L_r (ft)		48.1	32.6	43.9	30.2	40.7	28.3	35.1	24.1	31.2	21.8	27.4	19.7
A (in.^{2})		17.6		15.8		14.4		13.3		11.5		9.71	
I_x (in.^{4})		341		303		272		248		209		170	
I_y (in.^{4})		116		103		93.4		53.4		45		36.6	
r_y (in.)		2.57		2.56		2.54		2.01		1.98		1.94	
Ratio r_x/r_y		1.71		1.71		1.71		2.15		2.16		2.16	

Note: Heavy line indicates Kl/r of 200.

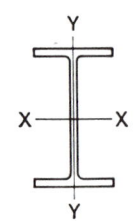

					COLUMNS W shapes Design axial strength in kips ($\phi = 0.85$)						$F_y = 36$ ksi $F_y = 50$ ksi

Designation						W8					
Wt./ft	67		58		48		40		35		31	
F_y	36	50	36	50	36	50	36	50	36	50	36	50
0	603	837	523	727	431	599	358	497	315	438	279	388
6	567	770	492	667	405	549	335	454	295	399	261	354
7	555	746	481	647	396	532	327	439	288	386	255	342
8	541	721	469	624	386	513	319	423	280	372	248	329
9	526	693	455	599	374	492	309	405	272	356	240	315
10	509	662	441	572	362	470	298	386	262	339	232	300
11	492	631	425	544	349	446	287	366	252	321	223	284
12	473	598	409	515	335	422	275	345	242	303	214	268
13	453	564	391	485	321	397	263	324	231	284	204	251
14	433	529	374	455	306	372	251	303	220	265	194	234
15	412	494	355	425	291	347	238	281	208	246	184	217
16	391	460	337	394	276	321	225	260	197	228	174	200
17	370	425	318	365	260	297	212	239	185	209	163	184
18	349	392	300	335	245	272	198	219	174	191	153	168
19	328	359	281	307	229	249	186	200	162	174	143	153
20	307	328	263	279	214	226	173	180	151	157	133	138
22	267	271	228	231	185	187	148	149	129	130	114	114
24	228	228	194	194	157	157	125	125	109	109	96	96
26	194	194	165	165	134	134	107	107	93	93	82	82
28	167	167	143	143	115	115	92	92	80	80	70	70
30	146	146	124	124	100	100	80	80	70	70	61	61
32	128	128	109	109	88	88	70	70	61	61	54	54
33	120	120	103	103	83	83	66	66	58	58	51	51
34	113	113	97	97	78	78	62	62				
35	107	107	91	91								

Effective length in ft KL with respect to least radius of gyration r_y

Properties												
U	1.33	1.48	1.35	1.49	1.37	1.51	1.39	1.54	1.40	1.55	1.41	1.56
P_{wo} (kips)	147	205	120	167	86	119	69	96	56	78	48	67
P_{wi} (kips/in.)	21	29	18	26	14	20	13	18	11	16	10	14
P_{wb} (kips)	648	764	464	547	224	264	163	192	104	123	81	95
P_{fb} (kips)	177	246	133	185	95	132	64	88	50	69	38	53
L_p (ft)	8.8	7.5	8.8	7.4	8.7	7.4	8.5	7.2	8.5	7.2	8.4	7.1
L_r (ft)	64.0	41.9	56.0	36.8	46.7	31.1	39.1	26.4	35.1	24.1	32.0	22.3
A (in.^{2})	19.7		17.1		14.1		11.7		10.3		9.13	
I_x (in.^{4})	272		228		184		146		127		110	
I_y (in.^{4})	88.6		75.1		60.9		49.1		42.6		37.1	
r_y (in.)	2.12		2.10		2.08		2.04		2.03		2.02	
Ratio r_x/r_y	1.75		1.74		1.74		1.73		1.73		1.72	

†Flange is noncompact; see discussion preceding column load tables.
Note: Heavy line indicates Kl/r of 200.

| F_y = 36 ksi | | | | | | | | | |
| **F_y = 50 ksi** | | | | COLUMNS | | | | | |

COLUMNS
W shapes
Design axial strength in kips (ϕ = 0.85)

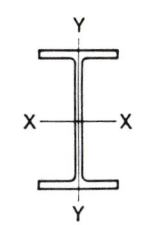

Designation		W8				W6				
Wt./ft		28		24		25		20		15	
F_y		36	50	36	50	36	50	36	50	36[†]	50[†]
Effective length in ft KL with respect to least radius of gyration r_y	0	252	351	217	301	225	312	180	249	136	188
	6	228	303	195	260	200	265	159	211	119	157
	7	219	288	188	247	191	250	152	198	114	147
	8	210	271	180	232	182	233	145	185	108	137
	9	200	253	171	217	172	216	137	171	101	126
	10	189	235	162	200	162	198	128	156	95	114
	11	178	216	152	184	151	180	119	142	88	103
	12	167	197	142	168	140	162	111	127	81	92
	13	155	178	132	151	129	144	102	113	74	81
	14	143	160	122	136	118	128	93	100	67	70
	15	132	142	112	121	107	112	84	87	60	61
	16	121	125	102	106	97	98	76	76	54	54
	17	110	111	93	94	87	87	68	68	48	48
	18	99	99	84	84	78	78	60	60	43	43
	19	89	89	75	75	70	70	54	54	38	38
	20	80	80	68	68	63	63	49	49	35	35
	22	66	66	56	56	52	52	40	40	29	29
	24	56	56	47	47	44	44	34	34	24	24
	25	51	51	44	44	40	40	31	31		
	26	47	47	40	40						
	27	44	44								

Properties										
U		1.74	1.92	1.76	1.95	1.41	1.56	1.44	1.60	1.48	1.64
P_{wo} (kips)		48	67	39	54	47	65	35	49	26	36
P_{wi} (kips/in.)		10	14	9	12	12	16	9	13	8	12
P_{wb} (kips)		81	95	52	61	146	172	78	92	54	64
P_{fb} (kips)		44	61	32	45	42	58	27	37	14	19
L_p (ft)		6.8	5.7	6.7	5.7	6.3	5.4	6.3	5.3	6.6	6.8
L_r (ft)		27.3	18.9	24.4	17.2	31.3	21.0	25.6	17.7	20.7	14.9
A (in.²)		8.25		7.08		7.34		5.87		4.43	
I_x (in.⁴)		98		82.8		53.4		41.4		29.1	
I_y (in.⁴)		21.7		18.3		17.1		13.3		9.32	
r_y (in.)		1.62		1.61		1.52		1.50		1.45	
Ratio r_x/r_y		2.13		2.12		1.78		1.77		1.77	

[†]Flange is noncompact; see discussion preceding column load tables.
Note: Heavy line indicates Kl/r of 200.

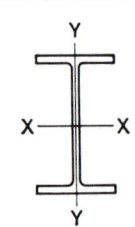

	F_y = 36 ksi
COLUMNS **W shapes** Design axial strength in kips (ϕ = 0.85)	F_y = 50 ksi

Designation		W6					W5				W4	
Wt./ft		16		12		9		19		16		13
F_y		36	50	36	50	36	50	36	50	36	50	36	50
	0	145	201	109	151	82	114	170	235	143	199	117	163
	2	140	193	105	144	79	108	166	229	141	194	114	156
	3	135	182	100	135	75	101	163	222	137	188	109	148
	4	127	168	94	124	71	93	157	212	133	179	104	138
	5	118	152	87	110	65	83	151	201	127	169	97	125
	6	108	134	79	96	59	72	144	187	121	157	89	111
	7	97	116	70	82	52	61	135	172	114	144	81	97
	8	86	98	61	68	45	50	126	156	106	131	72	83
	9	75	81	52	55	39	40	117	140	98	117	63	69
	10	64	66	44	44	33	33	107	124	90	104	55	57
	11	54	54	37	37	27	27	97	108	81	90	47	47
	12	46	46	31	31	23	23	87	93	73	78	39	39
	13	39	39	26	26	19	19	78	80	65	66	34	34
	14	33	33	23	23	17	17	68	69	57	57	29	29
	15	29	29	20	20	14	14	60	60	50	50	25	25
	16	26	26					53	53	44	44	22	22
	17							47	47	39	39		
	18							42	42	35	35		
	19							37	37	31	31		
	20							34	34	28	28		
	21							30	30	25	25		

Left axis labels: Effective length in ft KL with respect to least radius of gyration r_y

Properties												
U	2.17	2.41	2.28	2.52	2.33	2.59	1.33	1.48	1.34	1.49	1.35	1.50
P_{wo} (kips)	35	49	26	36	17	24	39	55	32	45	35	48
P_{wi} (kips/in.)	9	13	8	12	6	9	10	14	9	12	10	14
P_{wb} (kips)	78	92	54	64	22	26	115	136	81	95	164	193
P_{fb} (kips)	33	46	16	22	9	13	37	52	26	36	24	33
L_p (ft)	4.0	3.4	3.8	3.2	3.8	3.2	5.3	4.5	5.3	4.5	4.2	3.5
L_r (ft)	18.3	12.5	14.4	10.2	12.0	8.9	30.3	20.1	26.3	17.6	25.5	16.9

A (in.^{2})	4.74		3.55		2.68		5.54		4.68		3.83	
I_x (in.^{4})	32.1		22.1		16.4		26.2		21.3		11.3	
I_y (in.^{4})	4.43		2.99		2.20		9.13		7.51		3.86	
r_y (in.)	0.966		0.918		0.905		1.28		1.27		1.00	
Ratio r_x/r_y	2.69		2.71		2.73		1.70		1.68		1.72	

Note: Heavy line indicates Kl/r of 200.

F_y = 36 ksi	

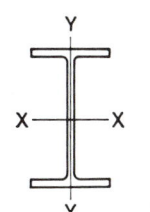

COLUMNS
M shapes
Design axial strength in kips (ϕ = 0.85)

Designation		M6		M5		M4	
Wt./ft		20		18.9		13	
F_y		36	50	36	50	36	50
Effective length in ft *KL* with respect to least radius of gyration r_y	0	180	250	170	236	117	162
	2	177	245	166	229	113	154
	3	174	239	162	221	108	145
	4	169	230	156	209	102	134
	5	164	219	149	196	94	120
	6	157	206	140	180	86	105
	7	149	192	131	164	77	90
	8	141	178	121	147	67	75
	9	132	162	110	129	58	61
	10	122	146	99	112	49	50
	11	113	131	89	96	41	41
	12	103	116	79	81	35	35
	13	94	101	69	69	29	29
	14	84	87	59	59	25	25
	15	76	76	52	52	22	22
	16	67	67	45	45		
	17	59	59	40	40		
	18	53	53	36	36		
	19	47	47	32	32		
	20	43	43				
	21	39	39				
	22	35	35				
	23	32	32				
Properties							
U		1.50	1.66	1.39	1.53	1.38	1.53
P_{wo} (kips)		39	55	50	69	37	52
P_{wi} (kips/in.)		9	13	11	16	9	13
P_{wb} (kips)		75	88	197	233	137	162
P_{fb} (kips)		29	40	35	49	28	39
L_p (ft)		5.8	4.9	5.0	4.2	3.9	3.3
L_r (ft)		26.9	18.2	31.1	20.5	27.8	18.3
A (in.^{2})		5.89		5.55		3.81	
I_x (in.^{4})		39.0		24.1		10.5	
I_y (in.^{4})		11.6		7.86		3.36	
r_y (in.)		1.40		1.19		0.939	
Ratio r_x/r_y		1.84		1.75		1.77	

Note: Heavy line indicates *Kl/r* of 200.

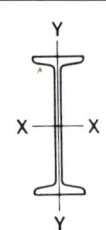

COLUMNS
S shapes
Design axial strength in kips ($\phi = 0.85$)

$F_y = 36$ ksi
$F_y = 50$ ksi

Designation		S6			S5			S4			S3		
Wt./ft		17.25		12.5	14.75		10		9.5		7.7	7.5		5.7
F_y		36	50	36	50	36	50	36	50	36	50	36	50	36	50

Effective length in ft *KL* with respect to least radius of gyration r_y

		S6 36	S6 50	12.5 36	12.5 50	S5 14.75 36	S5 14.75 50	S5 10 36	S5 10 50	S4 9.5 36	S4 9.5 50	S4 7.7 36	S4 7.7 50	S3 7.5 36	S3 7.5 50	S3 5.7 36	S3 5.7 50
	0	155	215	112	156	133	184	90	125	85	119	69	96	68	94	51	71
	2	145	196	106	143	123	165	84	113	78	104	63	85	60	80	46	61
	3	134	175	98	129	111	144	76	99	69	88	57	73	52	66	40	50
	4	119	149	88	111	97	119	67	83	59	70	48	58	43	50	33	38
	5	102	121	77	92	81	93	57	66	48	53	39	44	33	35	25	27
	6	85	94	65	73	65	69	47	50	37	37	31	31	24	24	19	19
	7	69	70	53	55	50	50	37	37	27	27	23	23	18	18	14	14
	8	53	53	42	42	39	39	28	28	21	21	18	18	14	14	11	11
	9	42	42	33	33	30	30	22	22	16	16	13	13				
	10	34	34	27	27	25	25	18	18								
	11	28	28	22	22												

Properties

	S6 36	S6 50	12.5 36	12.5 50	S5 14.75 36	S5 14.75 50	S5 10 36	S5 10 50	S4 9.5 36	S4 9.5 50	S4 7.7 36	S4 7.7 50	S3 7.5 36	S3 7.5 50	S3 5.7 36	S3 5.7 50
U	2.68	2.97	2.88	3.19	2.34	2.59	2.59	2.87	2.17	2.40	2.28	2.52	1.71	1.89	1.86	2.06
P_{wo} (kips)	73	102	37	51	72	100	31	43	44	61	26	36	43	60	21	29
P_{wi} (kips/in.)	17	23	8	12	18	25	8	11	12	16	7	10	13	17	6	9
P_{wb} (kips)	484	570	60	71	720	849	58	69	270	319	56	66	482	568	56	66
P_{fb} (kips)	26	36	26	36	22	30	22	30	17	24	17	24	14	19	14	19
L_p (ft)	2.8	2.4	2.9	2.5	2.6	2.2	2.7	2.3	2.4	2.0	2.4	2.1	2.1	1.8	2.2	1.8
L_r (ft)	19.3	12.8	14.2	9.6	21.9	14.3	13.9	9.3	17.7	11.6	14.0	9.3	21.5	14.0	14.7	9.7

	S6 17.25	S6 12.5	S5 14.75	S5 10	S4 9.5	S4 7.7	S3 7.5	S3 5.7
A (in.^{2})	5.07	3.67	4.34	2.94	2.79	2.26	2.21	1.67
I_x (in.^{4})	26.3	22.1	15.2	12.3	6.79	6.08	2.93	2.52
I_y (in.^{4})	2.31	1.82	1.67	1.22	0.903	0.764	0.586	0.455
r_y (in.)	0.675	0.705	0.620	0.643	0.569	0.581	0.516	0.522
Ratio r_x/r_y	3.38	3.48	3.02	3.19	2.74	2.82	2.23	2.36

Note: Heavy line indicates *Kl/r* of 200.

COLUMNS
Steel Pipe and Structural Tubing

Design strength loads in the tables that follow are tabulated for the effective lengths in feet KL, indicated at the left of each table. They are applicable to axially loaded members with respect to their minor axis in accordance with Sect. E2 of the AISC LRFD Specification. For discussion of effective length, range of l/r, strength about the major axis, combined axial and bending stress and sample problems, see "Columns, General Notes." Properties and factors are listed at the bottom of the tables for checking strength about the strong axis and for checking combined loading conditions.

STEEL PIPE COLUMNS

Design strength loads for unfilled pipe columns are tabulated for $F_y = 36$ ksi. Steel pipe manufactured to ASTM A501 furnishes $F_y = 36$ ksi and ASTM A53, Types E or S, Gr. B furnishes $F_y = 35$ ksi and may be designed at stresses allowed for $F_y = 36$ ksi steel.

The heavy horizontal lines within the table indicate $Kl/r = 200$. No values are listed beyond $Kl/r = 200$.

STRUCTURAL TUBING COLUMNS

Design strength loads for square and rectangular structural tubing columns are tabulated for $F_y = 46$ ksi. Structural tubing is manufactured to $F_y = 46$ ksi under ASTM A500, Gr. B.

All tubes listed in the column load tables satisfy Sect. B5 of the AISC LRFD Specification. The heavy horizontal lines appearing within the tables indicate $Kl/r = 200$. No values are listed beyond $Kl/r = 200$.

COLUMNS
Standard steel pipe
Design axial strength in kips ($\phi = 0.85$)

$F_y = 36$ ksi

Nominal Dia.	12	10	8	6	5	4	3½	3
Wall Thickness	0.375	0.365	0.322	0.280	0.258	0.237	0.226	0.216
Weight per ft	49.56	40.48	28.55	18.97	14.62	10.79	9.11	7.58
F_y					36 ksi			
0	447	364	257	171	132	97	82	68
6	440	357	249	162	122	86	70	56
7	438	354	246	159	118	82	67	52
8	436	351	243	155	115	78	63	48
9	433	348	239	151	111	74	58	43
10	429	344	235	147	106	70	54	39
11	426	340	231	142	102	65	49	35
12	422	336	227	138	97	60	45	30
13	418	331	222	133	92	55	40	26
14	413	326	216	127	86	51	36	23
15	409	321	211	122	81	46	32	20
16	404	315	205	116	76	41	28	17
17	399	309	200	111	71	37	25	15
18	393	303	193	105	66	33	22	14
19	387	297	187	99	61	30	20	12
20	381	291	181	94	56	27	18	
22	369	277	168	83	47	22	15	
24	356	263	155	72	39	19		
25	349	256	149	67	36	17		
26	342	249	142	62	33			
28	328	234	129	53	29			
30	313	219	117	47	25			
31	306	212	111	44	23			
32	298	205	105	41				
34	283	190	93	36				
36	268	176	83	32				
37	260	169	79	31				
38	253	162	75					
40	237	148	67					
Properties								
Area A (in.²)	14.6	11.9	8.40	5.58	4.30	3.17	2.68	2.23
I (in.⁴)	279	161	72.5	28.1	15.2	7.23	4.79	3.02
r (in.)	4.38	3.67	2.94	2.25	1.88	1.51	1.34	1.16

Effective length in ft KL

Note: Heavy line indicates Kl/r of 200.

AMERICAN INSTITUTE OF STEEL CONSTRUCTION

F_y = 36 ksi

COLUMNS
Extra strong steel pipe
Design axial strength in kips (ϕ = 0.85)

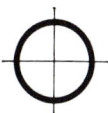

Nominal Dia.	12	10	8	6	5	4	3½	3
Wall Thickness	0.500	0.500	0.500	0.432	0.375	0.337	0.318	0.300
Weight per ft	65.42	54.74	43.39	28.57	20.78	14.98	12.50	10.25
F_y	\multicolumn{8}{c}{36 ksi}							

Effective length in ft KL	12	10	8	6	5	4	3½	3
0	588	493	392	257	187	135	113	92
6	579	483	379	243	172	119	96	75
7	576	479	375	238	168	114	91	69
8	573	475	369	232	162	108	85	64
9	569	470	364	226	156	102	79	58
10	564	465	357	219	149	95	72	52
11	559	460	351	212	143	89	66	46
12	554	453	343	205	135	82	60	40
13	549	447	336	197	128	75	53	34
14	543	440	327	189	121	68	47	30
15	536	433	319	180	113	62	42	26
16	530	425	310	172	105	56	37	23
18	515	409	291	154	91	44	29	18
19	508	400	282	145	83	40	26	16
20	500	391	272	137	76	36	23	
21	492	382	262	128	70	32	21	
22	483	373	252	120	63	30		
24	465	354	231	103	53	25		
26	447	334	211	88	45			
28	428	314	191	76	39			
30	408	294	172	66	34			
32	388	273	154	58				
34	368	253	136	52				
36	348	234	121	46				
38	328	215	109					
40	308	196	98					

Properties

	12	10	8	6	5	4	3½	3
Area A (in.^{2})	19.2	16.1	12.8	8.40	6.11	4.41	3.68	3.02
I (in.^{4})	362	212	106	40.5	20.7	9.61	6.28	3.89
r (in.)	4.33	3.63	2.88	2.19	1.84	1.48	1.31	1.14

Note: Heavy line indicates Kl/r of 200.

	F_y = 36 ksi

COLUMNS
Double-extra strong steel pipe
Design axial strength in kips (ϕ = 0.85)

Nominal Dia.		8	6	5	4	3
Wall Thickness		0.875	0.864	0.750	0.674	0.600
Weight per ft		72.42	53.16	38.55	27.54	18.58
F_y				36 ksi		
	0	652	477	346	248	167
	6	629	448	315	214	131
	7	621	437	305	203	120
	8	612	426	293	191	108
	9	601	413	281	179	96
	10	590	399	268	166	84
	11	578	385	254	152	73
	12	565	369	239	139	62
	13	551	353	224	125	53
	14	536	336	209	112	46
	15	521	319	194	100	40
	16	505	302	179	88	35
	17	489	285	165	78	31
	18	472	268	151	70	
	19	455	251	137	62	
	20	438	234	124	56	
	22	403	201	102	47	
	24	367	170	86		
	26	333	145	73		
	28	299	125	63		
	30	266	109			
	32	235	96			
	34	208	85			
	36	186				
	38	166				
	40	150				
Properties						
A (in.^{2})		21.3	15.6	11.3	8.10	5.47
I (in.^{4})		162	66.3	33.6	15.3	5.99
r (in.)		2.76	2.06	1.72	1.37	1.05

Effective length in ft KL

Note: Heavy line indicates Kl/r of 200.

AMERICAN INSTITUTE OF STEEL CONSTRUCTION

F_y = 46 ksi

COLUMNS
Square structural tubing
Design axial strength in kips (ϕ = 0.85)

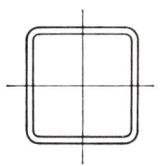

Nominal Size	16 x 16	14 x 14		12 x 12		10 x 10			
Thickness	½	½	⅜	½	⅜	⅝	½	⅜	⁵⁄₁₆
Wt./ft	103.30	89.68	68.31	76.07	58.10	76.33	62.46	47.90	40.35
F_y					46 ksi				
0	1190	1030	786	876	669	876	719	551	465
6	1180	1020	777	862	658	855	703	539	455
7	1170	1020	774	857	655	847	697	534	451
8	1170	1010	770	851	650	839	690	529	447
9	1160	1010	766	845	645	829	682	524	442
10	1160	999	761	838	640	818	674	517	437
11	1150	993	756	830	634	807	664	510	431
12	1150	985	751	821	628	794	655	503	425
13	1140	977	745	812	621	781	644	495	419
14	1130	969	739	803	614	767	633	487	411
15	1120	960	732	792	606	752	621	478	404
16	1120	950	725	781	598	736	608	468	396
17	1110	940	717	770	590	720	595	459	388
18	1100	930	710	758	581	703	582	449	380
19	1090	919	701	746	572	686	568	438	371
20	1080	907	693	733	562	668	553	427	362
21	1070	895	684	719	552	650	539	416	353
22	1060	883	675	706	542	631	524	405	343
23	1040	870	665	692	531	612	508	394	334
24	1030	857	655	677	521	593	493	382	324
25	1020	844	645	663	510	573	477	370	314
26	1010	830	635	648	498	554	462	358	305
27	994	816	624	633	487	534	446	347	295
28	981	802	614	617	476	515	430	335	285
29	967	787	603	602	464	495	414	323	275
30	954	772	592	586	452	476	398	311	265
32	925	742	569	555	428	438	367	287	245
34	896	711	546	523	405	400	337	264	225
36	866	680	522	491	381	364	307	242	206
38	835	648	498	460	357	328	278	220	188
40	803	616	474	429	334	296	251	199	170
Properties									
A (in.²)	30.40	26.40	20.10	22.40	17.10	22.40	18.40	14.10	11.90
I (in.⁴)	1200	791	615	485	380	321	271	214	183
r (in.)	6.29	5.48	5.54	4.66	4.72	3.78	3.84	3.90	3.93

Effective length in ft KL

AMERICAN INSTITUTE OF STEEL CONSTRUCTION

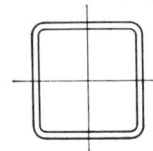

F_y = 46 ksi

COLUMNS
Square structural tubing
Design axial strength in kips (ϕ = 0.85)

Nominal Size	8 x 8					7 x 7				
Thickness	$5/8$	$1/2$	$3/8$	$5/16$	$1/4$	$1/2$	$3/8$	$5/16$	$1/4$	$3/16$
Wt./ft	59.32	48.85	37.60	31.84	25.82	42.05	32.59	27.59	22.42	17.08
F_y	46 ksi									
0	680	563	434	366	297	485	375	317	258	196
6	654	542	418	353	287	461	357	302	246	188
7	644	535	413	349	283	452	351	297	242	185
8	634	526	407	343	279	443	344	291	237	181
9	622	517	400	338	274	432	336	285	232	177
10	609	507	392	331	269	421	327	278	226	173
11	595	496	384	324	264	409	318	270	220	168
12	580	484	375	317	258	396	308	262	214	164
13	564	471	366	309	252	382	298	254	207	159
14	548	458	356	301	245	368	288	245	200	153
15	531	444	345	293	238	353	277	236	193	148
16	513	430	335	284	231	338	265	226	185	142
17	494	415	324	275	224	322	254	217	177	136
18	476	400	312	265	216	307	242	207	170	130
19	456	385	301	256	209	291	230	197	162	124
20	437	369	289	246	201	276	218	187	154	118
21	418	354	277	236	193	260	207	177	146	112
22	398	338	266	226	185	245	195	167	138	107
23	379	322	254	216	177	230	184	158	130	101
24	360	307	242	206	169	215	172	148	122	95
25	341	291	230	197	161	201	161	139	115	89
26	322	276	219	187	153	187	151	130	108	84
27	304	261	207	177	146	173	140	121	101	78
28	286	246	196	168	138	161	130	113	94	73
29	269	232	185	159	131	150	121	105	87	68
30	251	218	174	149	123	140	113	98	81	63
32	221	191	153	132	109	123	100	86	72	56
34	195	169	136	117	97	109	88	76	63	49
36	174	151	121	104	86	97	79	68	57	44
38	156	136	109	93	77	87	71	61	51	40
40	141	122	98	84	70	79	64	55	46	36
Properties										
A (in.^{2})	17.40	14.40	11.10	9.36	7.59	12.40	9.58	8.11	6.59	5.02
I (in.^{4})	153	131	106	90.9	75.1	84.6	68.7	59.5	49.4	38.5
r (in.)	2.96	3.03	3.09	3.12	3.15	2.62	2.68	2.71	2.74	2.77

The left axis label reads: Effective length in ft KL

COLUMNS
Square structural tubing
Design axial strength in kips ($\phi = 0.85$)

$F_y = 46$ ksi

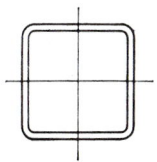

Nominal Size		6 x 6					5 x 5			
Thickness	$\frac{1}{2}$	$\frac{3}{8}$	$\frac{5}{16}$	$\frac{1}{4}$	$\frac{3}{16}$	$\frac{1}{2}$	$\frac{3}{8}$	$\frac{5}{16}$	$\frac{1}{4}$	$\frac{3}{16}$
Wt./ft	35.24	27.48	23.34	19.02	14.53	28.43	22.37	19.08	15.62	11.97
F_y					46 ksi					

Effective length in ft KL

0	407	316	268	219	167	327	257	219	179	138
6	379	295	251	205	157	294	233	199	163	126
7	369	288	245	200	153	282	224	192	158	121
8	358	280	239	195	149	270	215	184	152	117
9	346	271	231	189	145	257	205	176	145	112
10	333	262	223	183	140	242	194	167	138	107
11	320	252	215	176	135	228	183	158	131	101
12	306	241	206	169	130	213	172	148	123	95
13	291	230	197	162	124	197	160	139	115	89
14	276	219	187	154	119	182	149	129	107	84
15	260	207	178	146	113	167	137	119	99	78
16	245	195	168	138	107	152	126	110	92	72
17	229	184	158	131	101	138	115	100	84	66
18	214	172	148	123	95	124	104	91	77	60
19	199	160	139	115	89	111	93	82	69	55
20	184	149	129	107	83	100	84	74	63	50
21	170	138	120	100	78	91	76	67	57	45
22	155	127	111	92	72	83	70	61	52	41
24	131	107	93	78	61	70	59	52	44	34
26	111	91	80	67	52	59	50	44	37	29
28	96	79	69	57	45	51	43	38	32	25
30	84	69	60	50	39	45	37	33	28	22
31	78	64	56	47	37		35	31	26	21
32	73	60	53	44	34				24	19
34	65	53	47	39	30					
36	58	48	41	35	27					
37		45	39	33	26					
38			37	31	24					
39					23					

Properties

A (in.^{2})	10.40	8.08	6.86	5.59	4.27	8.36	6.58	5.61	4.59	3.52
I (in.^{4})	50.5	41.6	36.3	30.3	23.8	27.0	22.8	20.1	16.9	13.4
r (in.)	2.21	2.27	2.30	2.33	2.36	1.80	1.86	1.89	1.92	1.95

Note: Heavy line indicates Kl/r of 200.

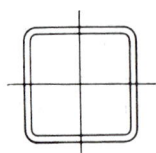

F_y = 46 ksi

COLUMNS
Square structural tubing
Design axial strength in kips (ϕ = 0.85)

Nominal Size		4 x 4				3 x 3		
Thickness	½	⅜	5/16	¼	3/16	5/16	¼	3/16
Wt./ft	21.63	17.27	14.83	12.21	9.42	10.58	8.81	6.87
F_y					46 ksi			
0	248	199	170	140	108	122	101	78
2	244	195	167	138	107	118	98	77
3	238	191	164	135	104	113	94	74
4	230	185	159	131	101	106	89	70
5	219	177	153	126	98	98	83	65
6	208	168	145	120	93	90	76	60
7	195	158	137	114	89	80	68	54
8	180	148	128	107	83	71	61	49
9	166	137	119	100	78	61	53	43
10	151	125	110	92	72	52	45	37
11	136	114	100	84	66	44	38	32
12	121	102	90	76	60	37	32	27
13	107	91	81	68	54	31	27	23
14	93	81	72	61	49	27	24	19
15	81	70	63	54	43	23	21	17
16	71	62	55	47	38	21	18	15
17	63	55	49	42	34	18	16	13
18	56	49	44	37	30		14	12
19	50	44	39	34	27			
20	46	40	35	30	24			
21	41	36	32	28	22			
22	38	33	29	25	20			
23	34	30	27	23	18			
24		27	25	21	17			
25				19	16			

Properties								
A (in.²)	6.36	5.08	4.36	3.59	2.77	3.11	2.59	2.02
I (in.⁴)	12.3	10.7	9.58	8.22	6.59	3.58	3.16	2.60
r (in.)	1.39	1.45	1.48	1.51	1.54	1.07	1.10	1.13

Note: Heavy line indicates Kl/r of 200.

Effective length in ft KL

F_y = 46 ksi

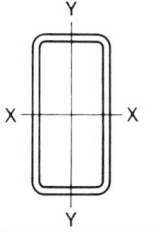

COLUMNS
Rectangular structural tubing
Design axial strength in kips ($\phi = 0.85$)

Nominal Size	16 x 12	16 x 8	14 x 10		12 x 8			12 x 6		
Thickness	1/2	1/2	1/2	3/8	5/8	1/2	3/8	5/8	1/2	3/8
Wt./ft	89.68	76.07	76.07	58.10	76.33	62.46	47.90	67.82	55.66	42.79
F_y					46 ksi					
0	1032	876	876	669	876	719	551	778	641	493
6	1017	848	857	655	845	695	534	731	604	466
7	1012	838	850	650	835	687	527	715	591	456
8	1005	827	843	644	822	677	520	697	577	445
9	998	815	834	638	809	666	512	677	561	434
10	990	801	825	631	794	655	503	655	544	421
11	982	786	815	623	778	642	494	632	525	407
12	973	771	803	615	760	628	484	607	505	393
13	963	754	791	606	742	613	473	581	485	378
14	952	736	779	597	722	598	461	555	464	362
15	941	717	765	587	702	582	449	528	442	346
16	929	698	751	576	681	565	437	500	420	329
17	916	677	737	565	659	547	424	473	398	313
18	903	657	721	554	637	530	410	445	375	296
19	889	635	705	542	614	511	397	418	353	279
20	875	614	689	530	591	493	383	390	331	262
22	845	569	655	505	544	455	355	338	288	230
24	814	525	620	478	497	417	326	288	247	199
26	781	480	584	451	451	380	298	245	211	170
28	746	436	547	424	405	343	270	211	182	146
30	712	393	511	396	362	307	243	184	158	128
32	676	352	474	369	320	273	217	162	139	112
34	640	313	438	341	283	241	192	143	123	99
36	604	279	403	315	252	215	171	128	110	89
38	568	250	369	289	227	193	154	115	99	80
39	550	238	352	276	215	184	146	109	94	75
40	533	226	335	264	205	174	139		89	72

(Left axis label: Effective length in ft KL with respect to least radius of gyration)

Properties

	16 x 12	16 x 8	14 x 10		12 x 8			12 x 6		
A (in.^{2})	26.40	22.40	22.40	17.10	22.40	18.40	14.10	19.90	16.40	12.60
I_x (in.^{4})	962	722	608	476	418	353	279	337	287	228
I_y (in.^{4})	618	244	361	284	221	188	149	112	96.0	77.2
r_x/r_y	1.25	1.72	1.30	1.29	1.37	1.37	1.37	1.74	1.73	1.72
r_y (in.)	4.84	3.30	4.02	4.08	3.14	3.20	3.26	2.37	2.42	2.48

Note: Heavy line indicates Kl/r of 200.

F_y = 46 ksi

COLUMNS
Rectangular structural tubing
Design axial strength in kips (ϕ = 0.85)

Nominal Size	10 x 6				8 x 6				8 x 4			
Thickness	$5/8$	$1/2$	$3/8$	$5/16$	$1/2$	$3/8$	$5/16$	$1/4$	$1/2$	$3/8$	$5/16$	$1/4$
Wt./ft	59.32	48.85	37.69	31.84	42.05	32.58	27.59	22.42	35.24	27.48	23.34	19.02
F_y	46 ksi											
0	680	563	434	366	485	375	317	258	407	316	268	219
6	638	529	409	345	454	352	298	243	351	276	235	192
7	623	517	400	338	444	344	292	238	333	262	224	184
8	606	504	391	330	432	335	284	232	313	248	212	174
9	588	490	380	321	419	325	276	225	292	233	199	164
10	568	474	368	312	404	315	268	218	270	216	185	153
11	547	457	356	302	389	304	258	211	248	200	172	142
12	525	439	343	291	373	292	248	203	226	183	158	131
13	502	421	329	279	357	279	238	195	204	167	144	120
14	478	402	315	267	340	266	227	186	183	151	130	109
15	454	382	300	255	322	253	217	178	162	135	117	98
16	429	362	285	243	305	240	205	169	143	120	104	88
18	380	322	255	218	269	213	183	151	113	95	82	70
20	331	282	225	193	235	187	161	133	91	77	67	56
22	285	244	196	169	201	161	140	116	76	63	55	47
24	241	208	169	146	170	137	119	99	63	53	46	39
25	222	192	155	134	157	126	110	91	58	49	43	36
26	205	177	144	124	145	117	102	85		45	39	33
27	190	164	133	115	134	108	94	78			37	31
28	177	153	124	107	125	101	88	73				
30	154	133	108	93	109	88	76	64				
32	136	117	95	82	96	77	67	56				
34	120	104	84	73	85	68	59	49				
36	107	92	75	65	76	61	53	44				
38	96	83	67	58	68	55	48	40				
39		79	64	55		52	45	38				
40			61	52				36				
Properties												
A (in.^{2})	17.40	14.40	11.10	9.36	12.40	9.58	8.11	6.59	10.40	8.08	6.86	5.59
I_x (in.^{4})	211	181	145	125	103	83.7	72.4	60.1	75.1	61.9	53.9	45.1
I_y (in.^{4})	93.5	80.8	65.4	56.5	65.7	53.5	46.4	38.6	24.6	20.6	18.1	15.3
r_x/r_y	1.50	1.50	1.49	1.49	1.25	1.25	1.25	1.25	1.75	1.73	1.73	1.72
r_y (in.)	2.32	2.37	2.43	2.46	2.31	2.36	2.39	2.42	1.54	1.60	1.62	1.65

Note: Heavy line indicates Kl/r of 200.

Effective length in ft KL with respect to least radius of gyration

F_y = 46 ksi

COLUMNS
Rectangular structural tubing
Design axial strength in kips (ϕ = 0.85)

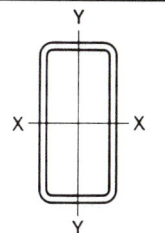

Nominal Size		7 x 5					6 x 4			
Thickness	½	⅜	⁵⁄₁₆	¼	³⁄₁₆	½	⅜	⁵⁄₁₆	¼	³⁄₁₆
Wt./ft	35.24	27.48	23.34	19.02	14.53	28.43	22.37	19.08	15.62	11.97
F_y					46 ksi					
0	407	316	268	219	167	327	257	219	179	138
6	369	288	245	200	154	279	222	190	157	121
7	357	279	238	194	149	263	211	181	149	115
8	342	268	229	187	144	246	198	171	141	109
9	327	257	220	180	138	228	185	160	132	102
10	311	245	210	172	132	210	171	148	123	96
11	294	232	199	164	126	191	157	136	114	89
12	276	219	188	155	119	173	143	125	104	81
13	258	205	177	146	113	155	129	113	95	74
14	240	192	165	137	106	137	116	102	85	67
15	222	178	154	127	99	121	103	91	77	61
16	205	165	143	118	92	106	90	80	68	54
17	187	151	131	109	85	94	80	71	60	48
18	170	138	120	101	79	84	71	63	54	43
19	154	126	110	92	72	75	64	57	48	38
20	139	114	100	84	66	68	58	51	44	35
21	126	103	90	76	60	62	52	46	39	31
22	115	94	82	69	54	56	48	42	36	29
24	97	79	69	58	46	47	40	36	30	24
25	89	73	64	54	42		37	33	28	22
26	82	67	59	50	39			30	26	20
27	76	62	55	46	36					19
29	66	54	47	40	31					
31	58	47	41	35	27					
32		44	39	33	26					
33			37	31	24					

Effective length in ft KL with respect to least radius of gyration

Properties										
A (in.²)	10.40	8.08	6.86	5.59	4.27	8.36	6.58	5.61	4.59	3.52
I_x (in.⁴)	63.5	52.2	45.5	38.0	29.8	35.3	29.7	26.2	22.1	17.4
I_y (in.⁴)	37.2	30.8	26.9	22.6	17.7	18.4	15.6	13.8	11.7	9.32
r_x/r_y	1.31	1.30	1.30	1.30	1.29	1.38	1.38	1.37	1.37	1.37
r_y (in.)	1.90	1.95	1.98	2.01	2.04	1.48	1.54	1.57	1.60	1.63

Note: Heavy line indicates Kl/r of 200.

COLUMNS
Rectangular structural tubing
Design axial strength in kips ($\phi = 0.85$)

F_y = 46 ksi

Nominal Size		6 x 3				5 x 3			
Thickness	3/8	5/16	1/4	3/16	1/2	3/8	5/16	1/4	3/16
Wt./ft	19.82	16.96	13.91	10.70	21.23	17.27	14.83	12.21	9.42
F_y					46 ksi				
0	228	195	160	123	249	199	170	140	108
2	221	189	156	120	240	193	166	137	105
3	214	183	151	116	230	186	160	132	102
4	203	174	144	111	217	176	152	126	97
5	190	164	136	105	201	164	142	118	92
6	175	152	126	98	183	151	132	110	85
7	160	138	116	90	164	137	120	100	78
8	144	125	105	82	145	122	108	91	71
9	127	111	94	74	125	107	95	81	63
10	111	97	83	65	107	93	83	71	56
11	95	84	72	57	89	79	71	61	49
12	81	71	62	50	75	67	60	52	42
13	69	61	53	42	64	57	51	45	36
14	59	52	45	36	55	49	44	38	31
15	52	46	39	32	48	43	39	33	27
16	45	40	35	28	42	38	34	29	23
17	40	36	31	25	37	33	30	26	21
18	36	32	27	22		30	27	23	19
19	32	28	25	20			24	21	17
20			22	18					15

Effective length in ft KL with respect to least radius of gyration

Properties									
A (in.^{2})	5.83	4.98	4.09	3.14	6.36	5.08	4.36	3.59	2.77
I_x (in.^{4})	23.8	21.1	17.9	14.3	16.9	14.7	13.2	11.3	9.06
I_y (in.^{4})	7.78	6.98	6.00	4.83	7.33	6.48	5.85	5.05	4.08
r_x/r_y	1.75	1.74	1.73	1.72	1.52	1.51	1.50	1.50	1.49
r_y (in.)	1.16	1.18	1.21	1.24	1.07	1.13	1.16	1.19	1.21

Note: Heavy line indicates Kl/r of 200.

AMERICAN INSTITUTE OF STEEL CONSTRUCTION

F_y = 46 ksi

COLUMNS
Rectangular structural tubing
Design axial strength in kips ($\phi = 0.85$)

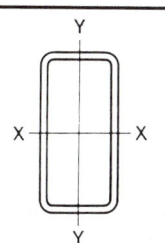

Nominal Size	4 x 3			4 x 2			3 x 2	
Thickness	5/16	1/4	3/16	5/16	1/4	3/16	1/4	3/16
Wt./ft	12.70	10.51	8.15	10.58	8.81	6.87	7.11	5.59
F_y	46 ksi							
0	146	121	93	122	101	79	82	64
2	141	117	91	113	95	74	76	60
3	136	113	88	104	87	69	70	55
4	129	107	84	92	78	62	62	49
5	120	101	79	78	67	54	53	43
6	110	93	73	65	56	46	43	36
7	100	84	66	51	45	37	35	29
8	89	76	60	40	36	30	27	23
9	78	67	53	31	28	24	21	18
10	67	58	47	25	23	19	17	14
11	57	50	40	21	19	16	14	12
12	48	42	34	18	16	13	12	10
13	41	36	29			11		
14	35	31	25					
15	31	27	22					
16	27	24	19					
17	24	21	17					
18	21	19	15					
19		17	14					
Properties								
A (in.^{2})	3.73	3.09	2.39	3.11	2.59	2.02	2.09	1.64
I_x (in.^{4})	7.45	6.45	5.23	5.32	4.69	3.87	2.21	1.86
I_y (in.^{4})	4.71	4.10	3.34	1.71	1.54	1.29	1.15	0.977
r_x/r_y	1.26	1.25	1.25	1.76	1.75	1.73	1.38	1.38
r_y (in.)	1.12	1.15	1.18	0.743	0.770	0.798	0.742	0.771

Effective length in ft KL with respect to least radius of gyration

Note: Heavy line indicates Kl/r of 200.

AMERICAN INSTITUTE OF STEEL CONSTRUCTION

COLUMNS
Double Angles and WT Shapes

Design strengths are tabulated for the effective length KL in feet with respect to both the X-X and Y-Y axes. Design strengths about the X-X axis are in accordance with LRFD Specification Sect. E2. For buckling about the Y-Y axis the shear deformation of the connectors requires the slenderness to be increased in accordance with the formulas for $(Kl/r)_m$ in Sect. E4. Incorporating this slenderness ratio, the design strengths are determined from Sect. E2 or Appendix E3, whichever governs. Discussion under Sect. C2 of the LRFD Specification Commentary points out that for trusses it is usual practice to take $K = 1.0$. No values are listed beyond $KL/r = 200$.

For buckling about the X-X axis, both angles move parallel so that the design strength is not affected by the connectors. The connectors must be spaced so the slenderness a/r_z of the individual angle does not exceed Kl/r_x. For buckling about the Y-Y axis, the design strengths are tabulated for the indicated number n of intermediate connectors spaced at $Kl/(n + 1)$ on the basis of welded or fully tightened bolted connectors. For connectors with snug-tight bolts or different spacings, the design strength must be recalculated using the corresponding modified slenderness and Appendix E3. The number of intermediate connectors given in the table was selected so the design strength about the Y-Y axis is 90% or greater of that for buckling of the two angles acting as a unit. If fewer connectors are used, the capacity must be reduced accordingly.

In designing members fabricated of two angles connected to opposite faces of a gusset plate, Sect. J1 of the LRFD Specification states that eccentricity between the gage lines and gravity axis may be neglected. In the following tables, eccentricity is neglected.

The tabulated loads for double angles referred to in the Y-Y axis assume a $\frac{3}{8}$ in. spacing between angles. These values are conservative when a wider spacing is provided. Example 6 illustrates a method for determining the design strength when a $\frac{3}{4}$-in. gusset plate is used.

Examples 7 and 8 demonstrate how to determine the number of connectors when Kl_x/r_x governs and when the modified $(Kl_y/r_y)_m$ governs.

EXAMPLE 6

Given:

Using $F_y = 36$ ksi steel, determine the design strength with respect to Y-Y axis on a double angle member of $8 \times 8 \times 1$ angles with an effective length equal to 12 ft, and connected to a $\frac{3}{4}$-in. thick gusset plate.

Solution:

$r_y = 3.53$ in. (from Double Angle Column Design Strength Table for
2 L $8 \times 8 \times 1$ with $\frac{3}{8}$-in. plate)

$r_y' = 3.67$ in. (from Part 1, Properties, Two Equal Leg Angles, 2 L $8 \times 8 \times 1$
with $\frac{3}{4}$-in. plate)

$$\frac{r_y}{r_y'} = \frac{3.53}{3.67} = 0.962$$

Equivalent length $= 0.962 \times 12$ ft $= 11.5$ ft

Enter column design strength table for 2 L $8 \times 8 \times 1$ with reference to Y-Y axis for effective lengths between 10 and 15 ft, read 826 and 781 kips, respectively.

$$\text{Equivalent design strength} = 826 - \left[(826 - 781) \times \frac{11.5 - 10}{15 - 10}\right]$$

$$= 813 \text{ kips}$$

EXAMPLE 7

Given:

Using a double angle member of $5 \times 3 \times \frac{1}{2}$ angles (short legs back to back), and 36 ksi steel, with $L_x = 10$ ft and $L_y = 20$ ft and a factored load of 70 kips, determine the number of connectors required. Assume $K = 1.0$.

Solution:

$K_x L_x = 10$ ft, $K_x l_x / r_x = (10 \times 12)/0.829 = 145$
$K_y L_y = 20$ ft, $K_y l_y / r_y = (20 \times 12)/2.5 = 96$

The X-X axis governs. From X-X axis portion of Table
$\phi P_n = 76$ kips > 70 kips **o.k.**
Find number of connectors required based on Sect. E4:

$a/r_z \leq K l_x / r_x$
$L_z = L_x r_z / r_x = (145)(0.648) = 94$ in.

Assume one connector is required; $L_z = (10 \times 12)/2 = 60$ in.
$a/r_i = a/r_z = 60/r_z = 60/0.648 = 92.6$

Check that modified $(K_y l_y / r_y)_m$ does not govern:
Since $a/r_i > 50$, use Formula E4-2.

$(K_y l_y / r_y)_m = \sqrt{(96)^2 + (92.6 - 50)^2}$
$= 105$

$K_x l_x / r_x$ still governs, therefore one connector is required every 5 ft.

EXAMPLE 8

Given:

Using the same steel and shape as Ex. 2, with $L_x = 12$ ft and $L_y = 36$ ft, determine the number of connectors required and the corresponding maximum design strength. Assume $K = 1.0$.

Solution:

$K_x L_x = 12$ ft, $K_x l_x / r_x = (12 \times 12)/0.829 = 174$
$K_y L_y = 36$ ft, $K_y l_y / r_y = (36 \times 12)/2.5 = 173$

$K_x l_x / r_x$ appears to govern, so try 1 connector in the 12-ft length. Check $(K_y l_y / r_y)_m$ with $a/r_i = a/r_z = 72/0.648 = 111$,

$(K_y l_y / r_y)_m = \sqrt{(173)^2 + (111 - 50)^2}$
$= 183$

AMERICAN INSTITUTE OF STEEL CONSTRUCTION

Since $(K_y l_y / r_y)_m$ governs, the Y-Y portion of the table gives a design strength of 50 kips provided 6 connectors are used in the 36-ft length. This gives a spacing of $36/7 = 5.14$ ft. Check if $(K_y L_y / r_y)_m$ governs with

$a/r_z = (5.14 \times 12)/0.648 = 95.2$,

$$(K_y l_y / r_y)_m = \sqrt{(173)^2 + (95.2 - 50)^2}$$
$$= 179$$

$(K_y l_y / r_y)_m$ still governs, so 6 connectors at 5.14 ft $= 5'$-$1\frac{3}{4}$ in. would be appropriate. However, $5'$-$1\frac{3}{4}''$ may not be a practical spacing for this example. The following two alternatives can be considered:

1. Use 8 connectors in the 36-ft length at a spacing of 4 ft. This provides some additional design strength. Conservatively, use a design strength of 50 kips as for 6 connectors.

2. Use 5 connectors in 36-ft at 6-ft spacings. This reduces the design strength, which can be determined as follows:

$$(K_y l_y / r_y)_m = 183 \text{ for 5 connectors}$$
$$= 179 \text{ for 6 connectors}$$

Thus the design strength can be found by increasing the effective length from 36-ft to $(183/179) \times 36 = 36.8$ ft. By interpolation,

$$\phi P_n = 50 - (0.8 \times 5/2) = 48 \text{ kips}$$

Single Angle Struts

Design strengths of single angle struts are not tabulated in this Manual because it is virtually impossible to load such struts concentrically. In theory, concentric loading could be accomplished by milling the ends of an angle and loading it through bearing plates. However, in practice, the actual eccentricity of loading is relatively large; and, its neglect in design may lead to an underdesigned member.

The design of an equal leg single angle strut loaded eccentrically must meet the provisions of Sect. H1 of the AISC LRFD Specification and the design of an unequal leg single angle strut loaded eccentrically must satisfy Sect. H2.

The following example illustrates the design procedure for an equal leg angle loaded eccentrically.

EXAMPLE 9

Given:

An angle 2 × 2 × ¼ is loaded by a gusset plate attached to one leg with an eccentricity of 0.8 in. from the centroid, as shown. Determine the factored compressive load P_u which may be applied. The effective length KL is 40 in.

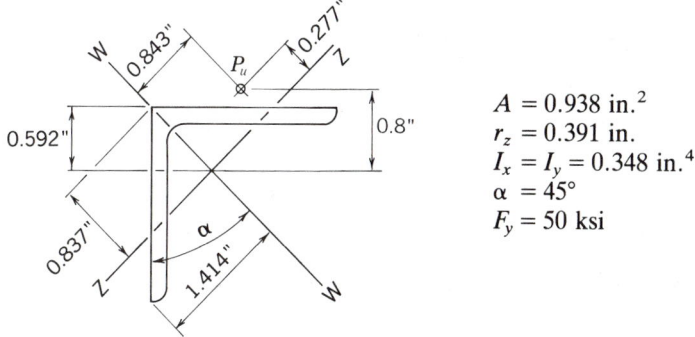

$A = 0.938$ in.^{2}
$r_z = 0.391$ in.
$I_x = I_y = 0.348$ in.^{4}
$\alpha = 45°$
$F_y = 50$ ksi

Solution:

Determine properties for the principal axes Z-Z and W-W using Properties of Geometric Sections, Part 7 of this Manual:

$$K = \pm \frac{abcdt}{4(b+c)} = -\frac{1.75 \times 2 \times 1.75 \times 2 \times 0.25}{4(2+1.75)} = -0.204$$

$$I_w = 0.348 \sin^2(45°) + 0.348 \cos^2(45°) + 0.204 \sin(2 \times 45°) = 0.552 \text{ in.}^4$$

$$r_w = \sqrt{\frac{I_w}{A}} = \sqrt{\frac{0.552}{0.938}} = 0.767 \text{ in.}; \qquad I_z = Ar_z^2 = 0.938(0.391)^2 = 0.143 \text{ in.}^4$$

$$b/t = 2/0.25 = 8 < 76/\sqrt{F_y} = 10.7 \qquad \therefore Q = 1$$

Determine the nominal compressive strength without eccentricity:
For flexural buckling about the weak axis (Z-Z axis):

$$\frac{Kl}{r_z} = \frac{40}{0.391} = 102.3$$

From LRFD Specification, Table 3-50, $\phi F_{cr} = 19.77$ ksi

$$\phi P_n = \phi F_{cr} A = 19.77(0.938) = 18.55 \text{ kips}$$

For flexural-torsional buckling with respect to the axis of symmetry (W-W axis), using Formulas A-E3-6 and A-E3-4:

$\lambda_e = 0.773$, where F_{ey} in Formula A-E3-6 becomes F_{ew}.

$\lambda_e < 1.5, \therefore F_{cr} = [\, e^{-0.419(0.773)^2}]50 = 38.93$ ksi

$\phi P_n = \phi F_{cr} A = 0.85\,(38.93)(0.938) = 31.04$ kips

$\phi P_n = 18.55$ kips governs

Apply Formula H1-1a, assuming $\dfrac{P_u}{\phi P_n} > 0.2$:

$$\frac{P_u}{\phi P_n} + \frac{8}{9}\left(\frac{M_{uz}}{\phi_b M_{nz}} + \frac{M_{uw}}{\phi_b M_{nw}}\right) \le 1.0$$

$$M_{uz} = B_1 M_{nt} = \frac{C_m}{(1 - P_u/P_{ez})} \times 0.277\,P_u \qquad C_m = 1.0$$

$$P_{ez} = \frac{A_g F_y}{\lambda_c^2}; \ \lambda_c^2 = \left(\frac{KL}{r\pi}\right)^2 \frac{F_y}{E} = \left(\frac{40}{0.391\pi}\right)^2 \frac{50}{29{,}000} = 1.828$$

$$P_{ez} = \frac{0.938 \times 50}{1.828} = 25.65 \text{ kips}$$

$$M_{uw} = B_1 M_{nt} = \frac{C_m}{(1 - P_u/P_{ew})} \times 0.843\,P_u; \qquad C_m = 1.0$$

$$P_{ew} = \frac{A_g F_y}{\lambda_c^2}; \ \lambda_c^2 = \left(\frac{40}{0.767\,\pi}\right)^2 \frac{50}{29{,}000} = 0.475$$

$$P_{ew} = \frac{0.938 \times 50}{0.475} = 98.74 \text{ kips}$$

The limiting extreme fiber flexural stress for angles in most practical applications may be assumed, for design purposes, to be F_y.[*]

$M_{nz} = F_y S_z = 50 \times (0.143/0.837) = 8.54$ kips-in.

$M_{nw} = F_y S_w = 50 \times (0.553/1.414) = 19.55$ kips-in.

$$\frac{P_u}{18.55} + \frac{8}{9}\left[\frac{0.277 P_u}{0.90 \times 8.54\left(1 - \dfrac{P_u}{25.65}\right)} + \frac{0.843 P_u}{0.90 \times 19.55\left(1 - \dfrac{P_u}{98.74}\right)}\right] = 1.0$$

$P_u = 6.96$ kips; $\dfrac{P_u}{\phi P_n} = \dfrac{6.96}{18.55} = 0.38 > 0.2$ **o.k.**

---

[*]Leigh, J. M. and M. G. Lay, "The Design of Laterally Unsupported Angles," BHP Technical Bulletin 13(3), November 1969, pp. 24–29.

Leigh, J. M., B. F. Thomas and M. G. Lay, "Safe Load for Laterally Unsupported Angles" (reprint), *AISC Engineering Journal*, 1st Qtr., 1984.

| F_y = 36 ksi |
| F_y = 50 ksi |

COLUMNS
Double angles
Design axial strength in kips ($\phi = 0.85$)
Equal legs
3/8 in. back to back of angles

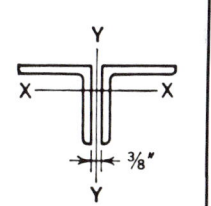

Size	8 x 8												No. of Connectors[a]
Thickness	1⅛		1		⅞		¾		⅝		½		
Wt./ft	113.8		102.0		90.0		77.8		65.4		52.8		
F_y	36	50	36	50	36	50	36	50	36	50	36	50	
X-X AXIS 0	1020	1420	918	1280	811	1130	701	973	586	763	432	550	
10	901	1190	808	1070	715	945	619	819	518	651	387	477	
14	795	1000	715	901	633	798	549	694	461	559	348	417	
18	674	795	607	718	538	638	468	556	394	456	302	348	
22	547	596	495	541	440	481	384	422	324	354	253	278	
26	427	430	388	391	345	349	302	306	257	261	205	212	b
30	323	323	294	294	262	262	230	230	196	196	159	159	
34	251	251	229	229	204	204	179	179	153	153	124	124	
38	201	201	183	183	163	163	143	143	122	122	99	99	
39	191	191	174	174	155	155	136	136	116	116	94	94	
40	182	182	165	165	147	147	129	129	110	110	90	90	
41							123	123	105	105	85	85	
Y-Y AXIS 0	1025	1424	918	1275	811	1126	701	973	586	763	432	550	
10	937	1256	826	1101	713	942	593	772	466	566	315	368	
15	880	1151	781	1018	679	880	570	731	452	544	308	357	
20	794	999	706	885	616	769	522	647	420	496	294	337	2
25	687	817	610	723	534	630	454	532	369	418	267	298	
30	567	626	503	552	439	481	374	407	305	327	228	243	
35	446	452	395	399	345	348	293	296	240	241	184	186	
40	372	372	328	328	286	286	243	243	198	198	154	154	
45	290	290	256	256	223	223	190	190	156	156	122	122	
50	231	231	205	205	179	179	152	152	125	125	99	99	
55	189	189	167	167	146	146	125	125	103	103	81	81	
56	182	182	161	161	140	140	120	120	99	99	78	78	3
57	175	175	155	155	135	135	115	115	95	95	75	75	
58	168	168	149	149	130	130	111	111					
59	162	162											

Effective length in ft KL with respect to indicated axis

Properties of 2 angles—3/8 in. back to back

A (in.²)	33.5	30.0	26.5	22.9	19.2	15.5
r_x (in.)	2.42	2.44	2.45	2.47	2.49	2.50
r_y (in.)	3.55	3.53	3.51	3.49	3.47	3.45

[a]For Y-Y axis, welded or fully tightened bolted connectors only.
[b]For number of connectors, see double angle column discussion.

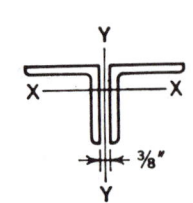

	F_y = 36 ksi
	F_y = 50 ksi

COLUMNS
Double angles
Design axial strength in kips (ϕ = 0.85)
Equal legs
3/8 in. back to back of angles

Size		6 x 6											No. of Connectors[a]
Thickness		1		7/8		3/4		5/8		1/2		3/8		
Wt./ft		74.8		66.2		57.4		48.4		39.2		29.8		
F_y		36	50	36	50	36	50	36	50	36	50	36	50	
X-X AXIS	0	673	935	597	829	517	718	435	604	352	470	243	309	b
	8	579	759	514	675	447	587	376	494	306	389	214	264	
	10	533	675	473	601	412	524	347	442	283	350	200	241	
	12	480	585	427	521	373	457	315	385	257	308	183	216	
	14	425	494	379	441	332	388	280	328	229	265	166	190	
	16	370	407	330	364	290	321	245	272	201	222	147	164	
	18	315	326	282	292	248	259	210	220	173	182	129	138	
	22	218	218	196	196	173	173	147	147	122	122	94	94	
	26	156	156	140	140	124	124	105	105	87	87	68	68	
	30	117	117	105	105	93	93	79	79	65	65	51	51	
	31									61	61	48	48	
Y-Y AXIS	0	673	935	597	829	517	718	435	604	352	470	243	309	
	10	601	798	528	699	451	594	369	481	282	349	175	203	
	12	575	752	506	659	434	562	357	459	275	337	172	200	
	14	546	699	480	613	412	523	340	429	264	321	168	194	
	16	514	642	451	562	387	480	320	394	250	298	163	186	
	18	477	579	418	505	358	431	297	355	233	271	155	174	2
	20	436	512	382	446	327	381	271	313	214	241	145	160	
	22	394	445	345	387	295	330	244	271	193	210	133	143	
	24	352	380	307	329	262	280	217	230	171	180	120	126	
	26	310	348	270	301	230	235	190	194	150	152	107	109	
	28	293	298	255	257	217	219	179	180	141	141	94	94	
	30	257	257	222	222	189	189	156	156	123	123	89	89	
	32	224	224	194	194	165	165	136	136	107	107	78	78	
	34	197	197	170	170	145	145	120	120	95	95	69	69	
	36	174	174	151	151	129	129	106	106	84	84	62	62	
	38	155	155	134	134	115	115	95	95	75	75	55	55	
	40	139	139	120	120	103	103	85	85	67	67	50	50	3
	42	125	125	108	108	93	93	77	77	61	61	45	45	
	43	119	119	103	103	88	88	73	73	58	58	43	43	
	44	113	113	98	98	84	84	69	69	55	55			
	45	108	108	94	94									

Effective length in ft KL with respect to indicated axis

Properties of 2 angles—3/8 in. back to back						
A (in.²)	22.0	19.5	16.9	14.2	11.5	8.72
r_x (in.)	1.80	1.81	1.83	1.84	1.86	1.88
r_y (in.)	2.73	2.70	2.68	2.66	2.64	2.62

[a]For Y-Y axis, welded or fully tightened bolted connectors only.
[b]For number of connectors, see double angle column discussion.

F_y = 36 ksi

F_y = 50 ksi

COLUMNS
Double angles
Design axial strength in kips ($\phi = 0.85$)
Equal legs
⅜ in. back to back of angles

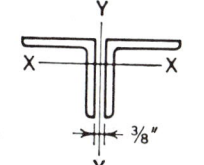

Size		5 x 5									No. of Connectors[a]
Thickness		⅞		¾		½		⅜		5/16	
Wt./ft		54.5		47.2		32.4		24.6		20.6	
F_y		36	50	36	50	36	50	36	50	36	50	
X-X AXIS	0	490	680	425	591	291	404	217	282	169	215	
	6	433	573	377	500	259	344	194	244	152	189	
	8	393	502	344	439	237	304	178	219	141	171	
	10	348	423	305	372	211	259	160	189	127	150	
	12	299	343	263	304	183	213	140	159	113	128	
	14	251	268	222	239	155	169	119	129	97	107	
	16	204	206	181	183	128	130	99	102	82	86	b
	18	162	162	145	145	103	103	80	80	68	68	
	20	132	132	117	117	83	83	65	65	55	55	
	22	109	109	97	97	69	69	54	54	46	46	
	24	91	91	82	82	58	58	45	45	38	38	
	25			75	75	53	53	42	42	35	35	
	26							39	39	33	33	
Y-Y AXIS	0	490	680	425	591	291	404	217	282	169	215	
	6	458	619	393	529	251	330	171	207	124	145	
	8	442	590	381	506	247	321	169	203	122	143	
	10	420	550	363	473	238	306	165	197	120	139	
	12	395	505	341	434	225	283	159	188	117	135	2
	14	366	453	315	389	209	255	149	173	112	127	
	16	332	397	286	340	189	223	137	155	104	117	
	18	296	338	255	290	168	189	122	134	95	103	
	20	259	281	222	240	147	157	107	113	85	89	
	22	223	251	191	213	126	127	92	93	74	75	
	24	206	208	176	177	115	116	84	84	63	68	
	26	176	176	150	150	98	98	72	72	58	58	
	28	150	150	128	128	84	84	62	62	50	50	
	30	129	129	110	110	72	72	53	53	44	44	
	32	112	112	96	96	63	63	47	47	38	38	3
	34	99	99	84	84	56	56	41	41	34	34	
	36	87	87	75	75	49	49	37	37	30	30	
	37	82	82	70	70	46	46	34	34			
	38	83	83	70	70							4

Effective length in ft KL with respect to indicated axis

Properties of 2 angles—⅜ in. back to back					
A (in.²)	16.0	13.9	9.50	7.22	6.05
r_x (in.)	1.49	1.51	1.54	1.56	1.57
r_y (in.)	2.30	2.28	2.24	2.22	2.21

[a]For Y-Y axis, welded or fully tightened bolted connectors only.
[b]For number of connectors, see double angle column discussion.

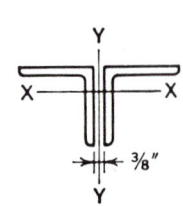

		F_y = 36 ksi
		F_y = 50 ksi

COLUMNS
Double angles
Design axial strength in kips (φ = 0.85)
Equal legs
3/8 in. back to back of angles

Size		4 x 4												No. of Connectors[a]
Thickness		3/4		5/8		1/2		3/8		5/16		1/4		
Wt./ft		37.0		31.4		25.6		19.6		16.4		13.2		
F_y		36	50	36	50	36	50	36	50	36	50	36	50	
X-X AXIS	0	334	463	282	392	230	319	175	243	146	191	108	138	
	4	306	411	259	349	212	285	162	217	135	172	101	126	
	6	275	354	233	301	191	247	146	189	123	151	92	112	
	8	237	288	201	245	166	203	127	156	107	127	81	96	
	10	195	220	167	188	138	157	106	121	90	100	69	78	
	12	154	159	132	137	110	115	85	89	72	76	57	61	b
	14	117	117	100	100	84	84	65	65	56	56	45	46	
	16	89	89	77	77	65	65	50	50	43	43	35	35	
	18	71	71	61	61	51	51	40	40	34	34	28	28	
	19	63	63	54	54	46	46	36	36	30	30	25	25	
	20			49	49	41	41	32	32	27	27	22	22	
Y-Y AXIS	0	334	463	282	392	230	319	175	243	146	191	108	138	
	6	306	411	256	343	204	270	147	190	115	140	78	91	
	8	289	379	242	317	194	252	142	181	112	135	77	89	2
	10	267	340	224	285	179	226	132	165	106	126	74	85	
	12	242	296	202	247	162	196	120	144	97	112	70	79	
	14	212	247	177	205	141	163	105	120	86	95	63	69	
	16	181	198	151	164	120	129	89	95	73	77	55	58	
	18	150	170	125	140	99	110	73	81	60	60	46	47	
	20	135	136	112	112	88	88	65	65	53	53	41	41	
	22	111	111	91	91	72	72	53	53	44	44	34	34	
	24	92	92	76	76	60	60	44	44	36	36	29	29	
	26	77	77	64	64	50	50	37	37	31	31	24	24	3
	28	66	66	54	54	43	43	32	32	26	26	21	21	
	29	65	65	50	50	40	40	30	30	25	25	19	19	
	30	60	60	50	50	37	37	28	28	23	23			
	31	56	56	47	47									4

Effective length in ft KL with respect to indicated axis

Properties of 2 angles—3/8 in. back to back						
A (in.²)	10.9	9.22	7.50	5.72	4.80	3.88
r_x (in.)	1.19	1.20	1.22	1.23	1.24	1.25
r_y (in.)	1.88	1.86	1.83	1.81	1.80	1.79

[a]For Y-Y axis, welded or fully tightened bolted connectors only.
[b]For number of connectors, see double angle column discussion.

F_y = 36 ksi
F_y = 50 ksi

COLUMNS
Double angles
Design axial strength in kips (ϕ = 0.85)
Equal legs
⅜ in. back to back of angles

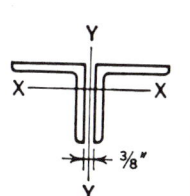

Size		3½ x 3½						No. of Connectors[a]
Thickness		⅜		5/16		¼		
Wt./ft		17.0		14.4		11.6		
F_y		36	50	36	50	36	50	
X-X AXIS	0	152	211	128	175	100	129	
	2	148	204	125	169	97	125	
	4	137	182	115	152	90	114	
	6	120	152	101	127	80	97	
	8	100	117	84	99	67	77	
	10	78	84	67	72	54	58	b
	12	59	59	50	50	41	41	
	14	43	43	37	37	30	30	
	16	33	33	28	28	23	23	
	17	29	29	25	25	21	21	
	18			22	22	18	18	
Y-Y AXIS	0	152	211	128	175	100	129	
	6	130	170	105	133	76	90	
	8	122	155	99	124	73	86	2
	10	110	135	91	109	68	79	
	12	95	111	79	90	60	67	
	14	80	86	66	70	51	54	
	16	64	71	53	58	41	45	
	18	55	55	45	45	36	36	
	20	44	44	36	36	29	29	3
	22	36	36	30	30	24	24	
	24	30	30	25	25	20	20	
	26	25	25	21	21	17	17	

Effective length in ft KL with respect to indicated axis

Properties of 2 angles—⅜ in. back to back			
A (in.²)	4.97	4.18	3.38
r_x (in.)	1.07	1.08	1.09
r_y (in.)	1.61	1.60	1.59

[a]For Y-Y axis, welded or fully tightened bolted connectors only.
[b]For number of connectors, see double angle column discussion.

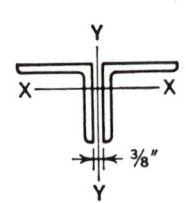

	COLUMNS	**F_y = 36 ksi**
	Double angles Design axial strength in kips (ϕ = 0.85) **Equal legs** $\frac{3}{8}$ in. back to back of angles	**F_y = 50 ksi**

Size		3 x 3									No. of Connectors[a]
Thickness		½		⅜		5/16		¼		3/16	
Wt./ft		18.8		14.4		12.2		9.8		7.42	
F_y		36	50	36	50	36	50	36	50	36	50	
X-X AXIS	0	168	234	129	179	109	151	88	118	61	77	
	2	162	222	125	171	105	144	85	112	59	74	
	4	145	190	112	146	94	124	77	98	54	66	
	5	133	169	103	131	87	111	71	88	50	60	
	6	120	146	93	114	79	97	64	77	46	54	
	7	106	123	83	97	70	82	57	66	41	47	
	8	92	101	72	80	61	68	50	56	37	41	b
	9	79	81	62	64	53	55	43	46	32	34	
	10	66	66	52	52	44	45	37	37	28	28	
	11	54	54	43	43	37	37	31	31	24	24	
	12	46	46	36	36	31	31	26	26	20	20	
	13	39	39	31	31	26	26	22	22	17	17	
	14	34	34	27	27	23	23	19	19	15	15	
	15			23	23	20	20	16	16	13	13	
Y-Y AXIS	0	168	234	129	179	109	151	88	118	61	77	
	2	159	217	118	158	96	126	72	90	45	52	
	4	156	210	116	154	94	124	71	89	44	51	
	6	146	191	110	143	90	116	69	85	43	50	2
	8	131	166	99	124	82	102	64	77	41	47	
	10	113	135	85	101	71	83	56	64	37	42	
	12	93	102	70	76	58	63	46	49	32	34	
	14	79	81	59	60	49	50	35	39	25	28	
	16	61	61	45	45	37	37	30	30	21	21	
	18	47	47	35	35	29	29	23	23	17	17	
	20	38	38	28	28	23	23	19	19	14	14	3
	22	33	33	23	23	19	19	15	15	11	11	
	23	30	30	22	22	18	18	14	14	10	10	
												4

Effective length in ft KL with respect to indicated axis

Properties of 2 angles—⅜ in. back to back					
A (in.²)	5.50	4.22	3.55	2.88	2.18
r_x (in.)	0.898	0.913	0.922	0.930	0.939
r_y (in.)	1.43	1.41	1.40	1.39	1.38

[a]For Y-Y axis, welded or fully tightened bolted connectors only.
[b]For number of connectors, see double angle column discussion.

F_y = 36 ksi		
F_y = 50 ksi		

COLUMNS
Double angles
Design axial strength in kips (ϕ = 0.85)
Equal legs
_{⅜ in. back to back of angles}

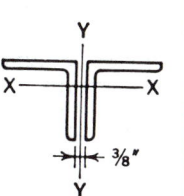

		Size	\multicolumn{8}{c}{2½ x 2½}	No. of Connectors[a]
		Thickness	\multicolumn{2}{c}{⅜}	\multicolumn{2}{c}{5/16}	\multicolumn{2}{c}{¼}	\multicolumn{2}{c}{3/16}	
		Wt./ft	\multicolumn{2}{c}{11.8}	\multicolumn{2}{c}{10.0}	\multicolumn{2}{c}{8.2}	\multicolumn{2}{c}{6.14}	
		F_y	36	50	36	50	36	50	36	50	
Effective length in ft KL with respect to indicated axis	X-X AXIS	0	106	147	90	125	73	101	54	70	
		2	101	137	85	116	69	94	52	66	
		3	94	125	80	106	65	86	48	61	
		4	86	110	73	93	59	76	44	54	
		5	76	93	65	79	53	65	40	47	
		6	66	76	56	65	46	53	35	40	
		7	55	59	47	51	39	42	30	32	b
		8	45	46	39	39	32	33	25	25	
		9	36	36	31	31	26	26	20	20	
		10	29	29	25	25	21	21	16	16	
		11	24	24	21	21	17	17	13	13	
		12	20	20	17	17	14	14	11	11	
	Y-Y AXIS	0	106	147	90	125	73	101	54	70	
		2	99	134	81	109	63	82	42	51	
		3	98	132	81	107	62	81	42	50	
		4	96	128	79	105	61	80	41	49	
		5	92	121	76	100	60	77	41	48	
		6	87	112	72	93	57	73	40	47	2
		7	81	102	68	84	54	66	38	44	
		8	74	90	62	75	49	59	35	40	
		9	67	78	56	65	45	51	32	35	
		10	59	66	49	54	39	43	28	30	
		11	52	59	43	49	34	39	25	26	
		12	48	50	40	41	32	32	23	23	
		13	42	42	35	35	27	27	20	20	
		14	36	36	30	30	23	23	17	17	
		15	31	31	25	25	20	20	15	15	
		16	27	27	22	22	18	18	13	13	3
		17	23	23	19	19	16	16	11	11	
		18	22	22	18	18	14	14	10	10	
		19	20	20	16	16	13	13	9	9	
		20	18	18	15	15					4

\multicolumn{3}{c}{Properties of 2 angles—⅜ in. back to back}								
A (in.^{2})	\multicolumn{2}{c}{3.47}	\multicolumn{2}{c}{2.93}	\multicolumn{2}{c}{2.38}	\multicolumn{2}{c}{1.80}	
r_x (in.)	\multicolumn{2}{c}{0.753}	\multicolumn{2}{c}{0.761}	\multicolumn{2}{c}{0.769}	\multicolumn{2}{c}{0.778}	
r_y (in.)	\multicolumn{2}{c}{1.21}	\multicolumn{2}{c}{1.20}	\multicolumn{2}{c}{1.19}	\multicolumn{2}{c}{1.18}	

[a]For Y-Y axis, welded or fully tightened bolted connectors only.
[b]For number of connectors, see double angle column discussion.

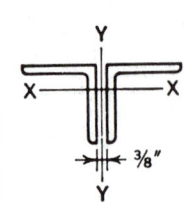

	F_y = 36 ksi
	F_y = 50 ksi

COLUMNS
Double angles
Design axial strength in kips (ϕ = 0.85)
Equal legs
⅜ in. back to back of angles

Size		2 x 2										No. of Connectors[a]
Thickness		⅜		⁵⁄₁₆		¼		³⁄₁₆		⅛		
Wt./ft		9.4		7.84		6.38		4.88		3.30		
F_y		36	50	36	50	36	50	36	50	36	50	
X-X AXIS	0	83	116	70	98	58	80	44	61	27	34	
	2	76	103	65	87	53	71	40	54	25	31	
	3	69	88	58	75	48	62	37	47	23	28	
	4	59	72	50	61	41	51	32	39	20	24	
	5	49	55	42	47	34	39	27	30	17	19	
	6	38	39	33	34	28	29	21	22	14	15	b
	7	29	29	25	25	21	21	16	16	11	11	
	8	22	22	19	19	16	16	13	13	9	9	
	9	18	18	15	15	13	13	10	10	7	7	
	10			12	12	10	10	8	8	6	6	
Y-Y AXIS	0	83	116	70	98	58	80	44	61	27	34	
	2	79	107	66	89	52	69	37	48	19	22	
	3	77	104	64	86	51	68	36	47	19	22	
	4	73	97	61	81	49	64	36	45	19	22	
	5	69	88	58	74	46	59	34	43	18	21	2
	6	63	78	53	65	42	52	31	38	17	20	
	7	56	67	47	56	38	44	28	33	16	18	
	8	49	55	41	46	33	36	24	27	14	16	
	9	41	48	34	40	28	32	20	23	13	13	
	10	38	39	31	32	25	26	18	19	11	12	
	11	32	32	26	26	21	21	15	15	10	10	
	12	26	26	22	22	17	17	13	13	8	8	3
	13	22	22	18	18	15	15	11	11	7	7	
	14	20	20	17	17	13	13	9	9	6	6	
	15	17	17	14	14	11	11	8	8	5	5	
	16	15	15	12	12	10	10	7	7	5	5	4

Effective length in ft KL with respect to indicated axis

Properties of 2 angles—⅜ in. back to back					
A (in.²)	2.72	2.30	1.88	1.43	0.960
r_x (in.)	0.594	0.601	0.609	0.617	0.626
r_y (in.)	1.01	1.00	0.989	0.977	0.965

[a]For Y-Y axis, welded or fully tightened bolted connectors only.
[b]For number of connectors, see double angle column discussion.

F_y = 36 ksi

F_y = 50 ksi

COLUMNS
Double angles
Design axial strength in kips (ϕ = 0.85)
Unequal legs
Long legs $\frac{3}{8}$ in. back to back of angles

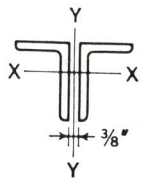

	Size		8 x 6					No. of Connectors[a]		8 x 4					No. of Connectors[a]
	Thickness		1		$\frac{3}{4}$		$\frac{1}{2}$			1		$\frac{3}{4}$		$\frac{1}{2}$		
	Wt./ft		88.4		67.6		46.0			74.8		57.4		39.2		
	F_y	36	50	36	50	36	50		36	50	36	50	36	50	
X-X AXIS	0	796	1100	609	846	376	479	0	673	935	517	718	321	408	
	10	704	932	541	717	339	419	10	597	792	460	611	289	358	
	12	667	865	513	667	323	395	12	567	736	437	569	276	338	
	14	626	792	483	612	306	368	14	533	675	411	523	262	315	
	16	582	715	450	555	287	340	16	496	611	384	474	246	292	
	18	535	637	415	496	267	310	18	457	546	354	425	230	267	
	20	488	560	379	438	247	280	20	417	481	324	376	212	241	
	22	440	485	343	381	226	250	22	378	419	294	328	195	216	
	24	393	415	308	328	205	221	24	338	359	264	282	177	192	
	26	348	353	273	279	184	193	26	300	306	235	241	160	168	
	28	305	305	240	241	165	167	b	28	264	264	207	208	143	146	b
	30	265	265	210	210	146	146	30	230	230	181	181	127	127	
	32	233	233	184	184	128	128	32	202	202	159	159	112	112	
	34	207	207	163	163	113	113	34	179	179	141	141	99	99	
	36	184	184	146	146	101	101	36	160	160	126	126	88	88	
	38	165	165	131	131	91	91	38	143	143	113	113	79	79	
	41	142	142	112	112	78	78	40	129	129	102	102	71	71	
	42			107	107	74	74	42	117	117	92	92	65	65	
								43					62	62	
Y-Y AXIS	0	796	1100	609	846	376	479	0	673	935	517	718	321	408	
	6	730	980	532	700	291	345	6	585	770	431	557	243	286	1
	8	712	948	521	682	287	339	8	540	688	397	497	227	263	
	10	686	900	505	653	281	330	10	477	579	349	416	204	229	
	12	652	838	482	612	273	318	1	12	402	458	292	325	174	188	
	14	607	760	451	557	260	299	14	364	398	261	278	142	163	
	16	555	670	412	492	242	273	16	299	305	212	213	128	129	
	18	497	575	369	422	221	243	18	238	238	167	167	103	103	2
	20	437	546	324	399	197	210	20	190	190	134	134	84	84	
	22	429	469	317	341	192	204	22	155	155	110	110	69	69	
	24	379	394	279	287	171	176	24	139	139	92	92	58	58	
	26	330	333	242	243	151	151	25	128	128	91	91	53	53	
	28	284	284	208	208	131	131	2	26	118	118					
	30	244	244	180	180	114	114									
	32	213	213	157	157	100	100									3
	34	202	202	138	138	88	88									
	36	179	179	132	132	85	85									
	40	144	144	106	106	68	68	3								
	41	137	137	101	101											
	42	130	130													

Effective length in ft KL with respect to indicated axis

Properties of 2 angles—$\frac{3}{8}$ in. back to back

A (in.²)	26.0	19.9	13.5		22.0	16.9	11.5
r_x (in.)	2.49	2.53	2.56		2.52	2.55	2.59
r_y (in.)	2.52	2.48	2.44		1.61	1.55	1.51

[a] For Y-Y axis, welded or fully tightened bolted connectors only.
[b] For number of connectors, see double angle column discussion.

COLUMNS
Double angles
Design axial strength in kips ($\phi = 0.85$)
Unequal legs
Long legs ⅜ in. back to back of angles

F_y = 36 ksi
F_y = 50 ksi

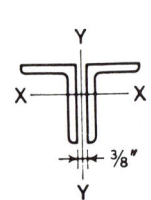

7 x 4

Size		7 x 4						No. of Connectors [a]
Thickness		¾		½		⅜		
Wt./ft		52.4		35.8		27.2		
F_y		36	50	36	50	36	50	
	0	471	655	310	401	205	254	
	8	427	571	283	355	189	230	
	10	404	528	268	332	181	218	
	12	378	481	252	306	171	204	
	14	348	430	233	278	161	188	
	16	318	379	214	248	149	172	
	18	286	327	194	219	137	155	
X-X AXIS	20	255	278	174	190	125	138	
	22	224	232	154	162	113	121	b
	24	194	195	135	137	100	105	
	26	166	166	117	117	89	90	
	28	143	143	100	100	78	78	
	30	125	125	88	88	68	68	
	32	110	110	77	77	59	59	
	34	97	97	68	68	53	53	
	36	87	87	61	61	47	47	
	37	82	82	58	58	44	44	
	0	471	655	310	401	205	254	
	6	405	530	245	295	149	171	1
	8	375	477	229	271	142	161	
	10	333	404	206	236	130	145	
	12	282	321	176	192	115	124	
	14	255	279	158	167	97	110	
Y-Y AXIS	16	210	215	130	131	89	90	
	18	168	168	104	104	73	73	
	20	134	134	84	84	60	60	2
	22	110	110	69	69	50	50	
	24	98	98	58	58	42	42	
	25	91	91	53	53	39	39	
	26	83	83	53	53			3
	27	77	77					

6 x 4

Size		6 x 4								No. of Connectors [a]
Thickness		¾		⅝		½		⅜		
Wt./ft		47.2		40.0		32.4		24.6		
F_y		36	50	36	50	36	50	36	50	
	0	425	591	358	497	291	388	201	256	
	8	371	488	313	412	254	325	179	220	
	10	343	438	290	371	236	294	167	202	
	12	312	385	265	327	215	260	154	182	
	14	279	329	237	281	193	225	140	161	
	16	245	275	209	235	171	191	125	140	
	18	212	225	181	193	148	158	110	119	
X-X AXIS	20	180	182	154	156	126	128	96	100	
	22	150	150	129	129	106	106	82	82	b
	24	126	126	109	109	89	89	69	69	
	26	108	108	93	93	76	76	59	59	
	28	93	93	80	80	65	65	51	51	
	30	81	81	70	70	57	57	44	44	
	31	76	76	65	65	53	53	41	41	
	32							39	39	
	0	425	591	358	497	291	388	201	256	
	6	375	495	309	405	240	300	154	182	1
	8	349	449	288	368	225	276	147	172	
	10	312	384	258	315	202	239	135	154	
	12	267	310	221	254	173	194	118	130	
	14	244	273	201	223	157	170	99	116	
Y-Y AXIS	16	204	213	168	173	130	133	91	93	
	18	166	166	135	135	105	105	75	75	2
	20	132	132	108	108	84	84	61	61	
	22	108	108	88	88	69	69	50	50	
	24	97	97	79	79	62	62	42	42	
	26	82	82	67	67	52	52	38	38	3
	27	76	76	62	62	48	48	35	35	
	28	70	70							

Effective length in ft KL with respect to indicated axis

Properties of 2 angles—⅜ in. back to back

	7 x 4				6 x 4			
A (in.²)	15.4	10.5	7.97		13.9	11.7	9.50	7.22
r_x (in.)	2.22	2.25	2.27		1.88	1.90	1.91	1.93
r_y (in.)	1.62	1.57	1.55		1.69	1.67	1.64	1.62

[a] For Y-Y axis, welded or fully tightened bolted connectors only.
[b] For number of connectors, see double angle column discussion.

| F_y = 36 ksi |
| F_y = 50 ksi |

COLUMNS
Double angles
Design axial strength in kips ($\phi = 0.85$)
Unequal legs
Long legs ⅜ in. back to back of angles

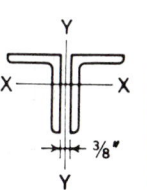

Properties of 2 angles—⅜ in. back to back

6 x 3½

	Size	6 x 3½				No. of Connectors[a]
	Thickness	⅜		5/16		
	Wt./ft	23.4		19.6		
	F_y	36	50	36	50	
X-X AXIS	0	191	243	145	179	0
	8	170	209	130	157	
	10	159	192	123	146	
	12	146	173	114	133	
	14	133	153	105	120	
	16	119	133	95	106	
	18	105	114	85	93	b
	20	91	95	75	79	
	22	78	79	65	67	
	24	66	66	56	56	
	26	56	56	48	48	
	28	49	49	41	41	
	30	42	42	36	36	
	32	37	37	32	32	
Y-Y AXIS	0	191	243	145	179	
	4	150	179	108	124	1
	6	143	169	103	118	
	8	132	152	97	109	
	10	115	127	86	94	
	12	94	109	72	76	
	14	83	85	65	66	
	16	66	66	52	52	2
	18	52	52	42	42	
	20	42	42	34	34	
	22	35	35	28	28	
	23	34	34	28	28	3

Properties of 2 angles—⅜ in. back to back:

	⅜	5/16
A (in.²)	6.84	5.74
r_x (in.)	1.94	1.95
r_y (in.)	1.39	1.38

5 x 3½

	Size	5 x 3½								No. of Connectors[a]
	Thickness	¾		½		⅜		5/16		
	Wt./ft	39.6		27.2		20.8		17.4		
	F_y	36	50	36	50	36	50	36	50	
X-X AXIS	0	355	493	245	340	183	238	143	182	0
	4	337	460	233	318	175	224	137	172	
	6	317	421	219	292	165	207	130	161	
	8	290	372	202	259	152	187	120	146	
	10	259	318	181	223	137	163	109	129	
	12	225	262	158	185	120	138	97	111	
	14	191	209	135	149	103	113	85	93	b
	16	158	161	112	116	87	90	72	76	
	18	127	127	91	91	71	71	60	61	
	20	103	103	74	74	58	58	49	49	
	22	85	85	61	61	48	48	41	41	
	24	72	72	51	51	40	40	34	34	
	25	66	66	47	47	37	37	31	31	
	26			44	44	34	34	29	29	
Y-Y AXIS	0	355	493	245	340	183	238	143	182	
	4	329	444	216	286	151	185	111	132	1
	6	310	409	205	267	145	175	107	127	
	8	284	362	187	234	133	157	100	116	
	10	253	308	160	188	115	130	89	99	
	12	217	249	143	161	94	112	74	87	
	14	179	191	117	122	84	87	67	69	
	16	143	143	93	93	66	66	53	53	2
	18	111	111	72	72	52	52	42	42	
	20	96	96	62	62	42	42	34	34	
	22	79	79	51	51	37	37	30	30	
	24	65	65	43	43	31	31	25	25	3
	25	60	60							

Properties of 2 angles—⅜ in. back to back:

	¾	½	⅜	5/16
A (in.²)	11.6	8.00	6.09	5.12
r_x (in.)	1.55	1.58	1.60	1.61
r_y (in.)	1.53	1.49	1.46	1.45

Effective length in ft KL with respect to indicated axis

[a]For Y-Y axis, welded or fully tightened bolted connectors only.
[b]For number of connectors, see double angle column discussion.

American Institute of Steel Construction

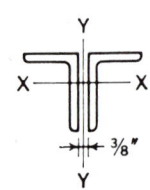

COLUMNS
Double angles
Design axial strength in kips ($\phi = 0.85$)
Unequal legs
Long legs ⅜ in. back to back of angles

F_y = 36 ksi
F_y = 50 ksi

	Size		\multicolumn{8}{c}{5 x 3}		No. of Connectors[a]
	Thickness		½		⅜		5/16		¼		
	Wt./ft		25.6		19.6		16.4		13.2		
	F_y		36	50	36	50	36	50	36	50	
Effective length in ft KL with respect to indicated axis	X-X AXIS	0	230	319	172	223	134	170	95	117	b
		2	227	313	170	220	132	168	95	115	
		4	219	298	164	210	128	161	92	112	
		6	206	274	155	195	122	151	88	105	
		8	189	244	143	176	113	137	82	97	
		10	170	210	129	154	102	121	76	88	
		12	149	175	114	130	91	104	68	78	
		14	127	141	98	107	79	88	61	67	
		16	106	110	82	86	68	71	53	56	
		18	87	87	68	68	56	57	45	46	
		20	70	70	55	55	46	46	38	38	
		22	58	58	45	45	38	38	31	31	
		24	49	49	38	38	32	32	26	26	
		26	42	42	32	32	27	27	22	22	
		27							21	21	
	Y-Y AXIS	0	230	319	172	223	134	170	95	117	
		2	207	277	145	179	107	129	71	81	1
		4	199	262	140	171	104	123	68	78	
		6	183	233	130	156	97	114	65	73	
		8	156	187	112	128	86	97	59	65	
		10	123	152	89	106	70	74	50	53	
		12	106	110	77	78	60	62	44	45	
		14	80	80	57	57	46	46	35	35	2
		16	60	60	44	44	35	35	27	27	
		18	51	51	37	37	28	28	22	22	
		20	41	41	30	30	24	24	19	19	3

\multicolumn{9}{c}{Properties of 2 angles—⅜ in. back to back}
A (in.²)	7.50	5.72	4.80	3.88
r_x (in.)	1.59	1.61	1.61	1.62
r_y (in.)	1.25	1.23	1.22	1.21

[a]For Y-Y axis, welded or fully tightened bolted connectors only.
[b]For number of connectors, see double angle column discussion.

F_y = 36 ksi

F_y = 50 ksi

COLUMNS
Double angles
Design axial strength in kips (ϕ = 0.85)
Unequal legs
Long legs ⅜ in. back to back of angles

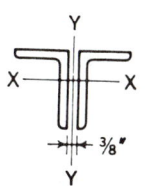

Size		4 x 3½								No. of Connectors[a]
Thickness		½		⅜		5/16		¼		
Wt./ft		23.8		18.2		15.4		12.4		
F_y		36	50	36	50	36	50	36	50	
X-X AXIS	0	214	298	163	227	137	179	101	129	
	2	210	289	160	221	134	174	99	126	
	4	198	266	151	204	127	162	94	118	
	6	179	232	137	178	115	143	87	106	
	8	155	190	120	147	101	120	77	91	
	10	130	148	101	116	85	96	66	75	
	12	104	109	81	86	69	73	55	59	b
	14	80	80	63	63	54	54	44	44	
	16	61	61	48	48	41	41	34	34	
	18	48	48	38	38	33	33	27	27	
	20	39	39	31	31	26	26	22	22	
	21					24	24	20	20	
Y-Y AXIS	0	214	298	163	227	137	179	101	129	
	2	198	267	143	189	113	140	78	92	
	4	195	261	141	184	112	137	77	90	
	6	187	246	136	176	109	132	75	88	
	8	172	220	127	160	103	123	72	83	
	10	155	189	115	139	93	108	67	76	2
	12	134	155	99	114	81	90	60	66	
	14	111	120	82	88	68	71	51	54	
	16	89	90	66	66	54	54	42	42	
	18	76	76	56	56	46	46	33	33	
	20	61	61	45	45	37	37	29	29	
	22	50	50	37	37	30	30	24	24	
	24	41	41	31	31	25	25	20	20	3
	25	38	38	28	28	23	23	18	18	
	26	35	35	26	26					

Effective length in ft KL with respect to indicated axis

Properties of 2 angles—⅜ in. back to back				
A (in.²)	7.00	5.34	4.49	3.63
r_x (in.)	1.23	1.25	1.26	1.27
r_y (in.)	1.58	1.56	1.55	1.54

[a]For Y-Y axis, welded or fully tightened bolted connectors only.
[b]For number of connectors, see double angle column discussion.

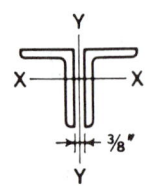

	F_y = 36 ksi
COLUMNS **Double angles** Design axial strength in kips ($\phi = 0.85$) **Unequal legs** Long legs ⅜ in. back to back of angles	**F_y = 50 ksi**

Size		4 x 3							No. of Connectors[a]
Thickness		½		⅜		5/16		¼	
Wt./ft		22.2		17.0		14.4		11.6	
F_y		36	50	36	50	36	50	36	50	
X-X AXIS	0	199	276	152	211	127	166	94	120	
	2	195	269	149	206	125	162	93	117	
	4	184	248	141	190	118	151	88	110	
	6	167	217	128	166	108	133	81	99	
	8	146	179	112	138	94	112	72	85	
	10	122	141	94	109	80	90	62	70	b
	12	99	105	76	81	65	69	51	55	
	14	77	77	60	60	51	51	41	42	
	16	59	59	46	46	39	39	32	32	
	18	46	46	36	36	31	31	25	25	
	20	38	38	29	29	25	25	21	21	
	21			27	27	23	23	19	19	
Y-Y AXIS	0	199	276	152	211	127	166	94	120	
	2	185	250	135	178	107	133	74	88	1
	4	179	239	131	172	105	128	72	86	
	6	166	215	123	157	99	119	69	81	
	8	148	183	110	134	89	104	63	71	
	10	126	146	93	107	76	85	52	56	
	12	102	109	75	79	61	64	46	48	2
	14	78	78	57	57	47	47	36	36	
	16	64	64	43	43	35	35	28	28	
	18	50	50	37	37	30	30	23	23	
	20	40	40	29	29	24	24	19	19	3
	21	36	36	27	27	22	22	17	17	
	22	33	33							

Effective length in ft KL with respect to indicated axis

Properties of 2 angles—⅜ in. back to back				
A (in.²)	6.50	4.97	4.18	3.38
r_x (in.)	1.25	1.26	1.27	1.28
r_y (in.)	1.33	1.31	1.30	1.29

[a]For Y-Y axis, welded or fully tightened bolted connectors only.
[b]For number of connectors, see double angle column discussion.

F_y = 36 ksi
F_y = 50 ksi

COLUMNS
Double angles
Design axial strength in kips (ϕ = 0.85)
Unequal legs
Long legs ⅜ in. back to back of angles

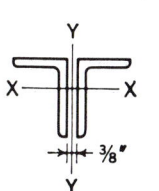

Size		3½ x 3					No. of Connectors[a]		3½ x 2½					No. of Connectors[a]
Thickness		⅜		5/16		¼			⅜		5/16		¼	
Wt./ft		15.8		13.2		10.8			14.4		12.2		9.8	
F_y		36	50	36	50	36	50		36	50	36	50	36	50	
X-X AXIS	0	140	195	118	162	92	119	0	129	179	109	149	85	110	
	2	137	188	115	157	90	116	2	126	173	106	144	83	117	
	4	127	169	107	141	84	106	4	117	156	98	130	77	97	
	6	112	142	94	119	75	91	6	103	131	87	110	69	84	
	8	93	111	79	94	63	73	8	86	103	73	87	59	68	
	10	74	80	63	69	51	55	10	69	75	59	64	47	52	b
	12	56	56	48	48	39	40	12	52	53	45	45	37	37	
	14	41	41	35	35	29	29	14	39	39	33	33	27	27	
	16	32	32	27	27	22	22	16	30	30	25	25	21	21	
	17	28	28	24	24	20	20	18	23	23	20	20	17	17	
	18	25	25	21	21	18	18								
Y-Y AXIS	0	140	195	118	162	92	119	0	129	179	109	149	85	110	
	2	127	169	102	133	75	91	2	117	156	95	123	70	85	
	4	124	164	100	129	73	88	4	112	146	91	116	67	81	
	6	117	151	95	121	70	84	6	100	126	82	101	62	72	2
	8	105	130	86	105	65	75	8	84	99	69	80	53	59	
	10	90	105	74	85	56	63	10	66	70	54	57	42	44	
	12	73	79	60	64	46	49	12	48	48	39	39	30	30	
	14	57	57	47	47	36	36	14	37	37	30	30	24	24	
	16	46	46	38	38	30	30	16	28	28	23	23	18	18	3
	18	36	36	30	30	23	23	18	22	22	18	18	14	14	
	20	29	29	24	24	19	19								
	22	24	24	20	20	15	15								

Effective length in ft KL with respect to indicated axis

Connectors for X-X axis: **2**, then **3**.

Properties of 2 angles—⅜ in. back to back								
A (in.^{2})		4.59		3.87		3.13		
r_x (in.)		1.09		1.10		1.11		
r_y (in.)		1.36		1.35		1.33		
A (in.^{2})							4.22	3.55	2.88
r_x (in.)							1.10	1.11	1.12
r_y (in.)							1.11	1.10	1.09

[a]For Y-Y axis, welded or fully tightened bolted connectors only.
[b]For number of connectors, see double angle column discussion.

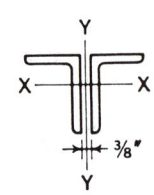

COLUMNS
Double angles
Design axial strength in kips (φ = 0.85)
Unequal legs
Long legs ⅜ in. back to back of angles

F_y = 36 ksi
F_y = 50 ksi

Size	3 x 2½						No. of Connectors[a]		3 x 2								No. of Connectors[a]
Thickness	⅜		¼		3/16				⅜		5/16		¼		3/16		
Wt./ft	13.2		9.0		6.77				11.8		10.0		8.2		6.1		
F_y	36	50	36	50	36	50			36	50	36	50	36	50	36	50	

Effective length in ft KL with respect to indicated axis

X-X AXIS

KL	36	50	36	50	36	50	Conn.	KL	36	50	36	50	36	50	36	50	Conn.
0	118	163	80	107	55	71		0	106	147	90	125	73	97	50	64	
2	113	155	78	103	54	68		2	103	141	87	119	70	94	49	61	
3	109	146	75	97	52	65		3	98	132	83	112	68	88	47	59	
4	102	134	70	90	49	60		4	93	122	78	103	64	81	45	55	
5	94	120	65	81	46	55		5	86	109	73	93	59	74	42	50	
6	86	105	59	71	42	50		6	78	96	66	82	54	65	38	45	
7	76	90	53	62	38	44		7	70	82	59	70	49	57	35	40	
8	67	75	47	52	34	38	b	8	61	69	52	59	43	48	31	35	b
9	58	60	40	43	30	32		9	53	56	45	48	37	40	28	30	
10	49	49	34	35	26	27		10	45	45	39	39	32	32	24	25	
11	40	40	29	29	22	22		11	38	38	32	32	27	27	20	21	
12	34	34	24	24	19	19		12	32	32	27	27	22	22	17	17	
13	29	29	21	21	16	16		13	27	27	23	23	19	19	15	15	
14	25	25	18	18	14	14		14	23	23	20	20	16	16	13	13	
15	22	22	15	15	12	12		15	20	20	17	17	14	14	11	11	
								16							10	10	

Y-Y AXIS

KL	36	50	36	50	36	50	Conn.	KL	36	50	36	50	36	50	36	50	Conn.
0	118	163	80	107	55	71		0	106	147	90	125	73	97	50	64	
2	108	146	68	86	43	51		2	98	132	80	107	62	79	40	47	
3	107	143	67	84	42	50		3	95	126	78	103	61	76	39	46	
4	104	138	66	82	42	49		4	89	116	74	95	58	71	37	44	
5	100	130	64	79	41	48		5	83	104	68	85	54	65	35	41	
6	94	119	61	74	39	46		6	75	91	62	74	49	57	33	37	2
7	87	108	57	67	38	43	2	7	66	76	54	62	43	48	29	32	
8	80	95	52	60	35	39		8	57	61	46	50	36	39	25	27	
9	72	82	47	52	32	35		9	47	48	39	39	30	30	21	21	
10	63	69	41	44	29	30		10	38	38	31	31	24	24	17	17	
11	55	57	35	36	25	26		11	34	34	28	28	22	22	14	14	
12	47	51	30	30	22	22		12	28	28	23	23	18	18	13	13	
13	43	43	27	27	18	18	3	13	24	24	20	20	15	15	11	11	3
14	37	37	24	24	17	17		14	20	20	17	17	13	13	10	10	
15	32	32	20	20	15	15		15	18	18	14	14					
16	28	28	18	18	13	13											
17	24	24	16	16	12	12											
18	22	22	14	14	10	10											
19	19	19															

Properties of 2 angles—⅜ in. back to back

A (in.²)	3.84	2.63	1.99	3.47	2.93	2.38	1.80
r_x (in.)	0.928	0.945	0.954	0.940	0.948	0.957	0.966
r_y (in.)	1.16	1.13	1.12	0.917	0.903	0.891	0.879

[a] For Y-Y axis, welded or fully tightened bolted connectors only.
[b] For number of connectors, see double angle column discussion.

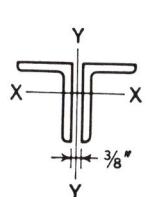

COLUMNS
Double angles
Design axial strength in kips (φ = 0.85)
Unequal legs
Long legs ⅜ in. back to back of angles

Fy = 36 ksi
Fy = 50 ksi

Size			2½ x 2								No. of Connectors[a]
Thickness			⅜		5/16		¼		3/16		
Wt./ft			10.6		9.0		7.2		5.5		
F_y			36	50	36	50	36	50	36	50	
		0	95	131	80	111	65	91	49	63	
		2	90	122	76	104	62	85	46	59	
		3	84	112	72	95	58	78	44	55	
		4	77	99	66	84	53	69	40	49	
		5	69	84	59	72	48	59	36	43	
	X-X AXIS	6	60	69	51	59	42	49	32	36	
		7	50	55	43	47	36	39	27	30	b
		8	42	42	36	37	30	30	23	24	
		9	33	33	29	29	24	24	19	19	
		10	27	27	23	23	19	19	15	15	
		11	22	22	19	19	16	16	12	12	
		12	19	19	16	16	13	13	10	10	
		13					11	11	9	9	
Effective length in ft KL with respect to indicated axis		0	95	131	80	111	65	91	49	63	
		2	89	120	73	99	57	76	39	48	
		3	86	115	72	95	56	73	39	47	
		4	82	107	68	89	54	69	37	45	
		5	76	97	63	80	50	63	35	42	2
		6	69	85	58	71	46	55	33	38	
	Y-Y AXIS	7	61	72	51	60	41	47	29	33	
		8	53	59	44	49	35	38	25	27	
		9	45	47	37	38	29	30	21	22	
		10	40	40	33	33	24	26	17	17	
		11	33	33	27	27	21	21	16	16	
		12	28	28	23	23	18	18	13	13	
		13	23	23	19	19	15	15	11	11	3
		14	20	20	16	16	13	13	9	9	
		15	17	17	14	14	11	11	8	8	
		16	15	15							

Properties of 2 angles—⅜ in. back to back

A (in.²)	3.09	2.62	2.13	1.62
r_x (in.)	0.768	0.776	0.784	0.793
r_y (in.)	0.961	0.948	0.935	0.923

[a]For Y-Y axis, welded or fully tightened bolted connectors only.
[b]For number of connectors, see double angle column discussion.

AMERICAN INSTITUTE OF STEEL CONSTRUCTION

Notes

F_y = 36 ksi

F_y = 50 ksi

COLUMNS
Double angles
Design axial strength in kips ($\phi = 0.85$)
Unequal legs
Short legs 3/8 in. back to back of angles

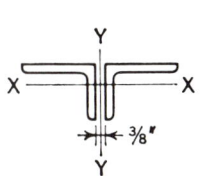

Size	8 x 6						No. of Connectors[a]		8 x 4						No. of Connectors[a]
Thickness	1		3/4		1/2				1		3/4		1/2		
Wt./ft	88.4		67.6		46.0				74.8		57.4		39.2		
F_y	36	50	36	50	36	50			36	50	36	50	36	50	

Effective length in ft KL with respect to indicated axis

X-X AXIS

KL	88.4·36	88.4·50	67.6·36	67.6·50	46.0·36	46.0·50	Conn	KL	74.8·36	74.8·50	57.4·36	57.4·50	39.2·36	39.2·50	Conn
0	796	1100	609	846	376	479		0	673	935	517	718	321	408	
8	676	882	521	680	328	402		4	600	798	463	616	292	361	
12	552	665	428	518	276	322		6	520	654	404	509	259	311	
16	416	449	325	354	217	237		8	426	495	333	390	219	252	
20	288	288	228	228	159	160		10	329	346	260	276	177	192	
24	200	200	159	159	111	111	b	12	240	240	192	192	137	138	b
28	147	147	116	116	82	82		14	176	176	141	141	101	101	
29			109	109	76	76		16	135	135	108	108	78	78	
								17	120	120	96	96	69	69	
								18					61	61	

Y-Y AXIS

KL	88.4·36	88.4·50	67.6·36	67.6·50	46.0·36	46.0·50	Conn	KL	74.8·36	74.8·50	57.4·36	57.4·50	39.2·36	39.2·50	Conn
0	796	1105	609	846	376	479		0	673	935	517	718	321	408	
12	723	968	529	695	288	341		12	622	838	455	601	250	297	
16	687	902	513	667	285	336	2	16	598	793	450	592	249	296	
20	633	805	477	603	277	325		20	559	723	426	549	248	294	4
24	563	684	425	514	258	296		24	509	635	388	482	239	281	
28	484	554	366	417	228	253		28	451	536	343	406	217	249	
32	444	492	334	368	210	228	3	32	388	468	295	354	190	209	
36	373	386	281	289	179	186		36	351	378	266	285	172	185	5
40	306	329	229	229	149	149		40	294	319	222	240	146	159	
44	268	268	201	201	131	131		44	259	259	195	195	129	129	6
48	223	223	167	167	109	109		48	224	224	161	161	107	107	
52	187	187	140	140	92	92	4	52	188	188	141	141	94	94	
56	160	160	120	120	79	79		56	160	160	120	120	80	80	7
60	138	138	103	103	68	68		60	138	138	104	104	69	69	
61	133	133	100	100	66	66		64	125	125	90	90	60	60	
62	135	135	96	96			5	66	117	117	88	88	56	56	8
63	131	131						67	113	113	85	85			
								68	109	109					

Properties of 2 angles—3/8 in. back to back

	8 x 6				8 x 4		
A (in.^{2})	26.0	19.9	13.5		22.0	16.9	11.5
r_x (in.)	1.73	1.76	1.79		1.03	1.05	1.08
r_y (in.)	3.78	3.74	3.69		4.10	4.05	4.00

[a]For Y-Y axis, welded or fully tightened bolted connectors only.
[b]For number of connectors, see double angle column discussion.

COLUMNS
Double angles
Design axial strength in kips (φ = 0.85)
Unequal legs
Short legs ⅜ in. back to back of angles

F_y = 36 ksi
F_y = 50 ksi

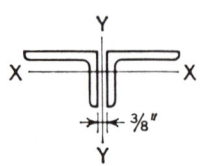

Size		7 x 4					No. of Connectors[a]			6 x 4							No. of Connectors[a]
Thickness		¾		½		⅜				¾		⅝		½		⅜	
Wt./ft		52.4		35.8		27.2				47.2		40.0		32.4		24.6	
F_y		36	50	36	50	36	50			36	50	36	50	36	50	36	50	

X-X AXIS (Effective length in ft KL with respect to indicated axis)

KL	7x4 ¾ 36	7x4 ¾ 50	7x4 ½ 36	7x4 ½ 50	7x4 ⅜ 36	7x4 ⅜ 50	Conn	KL	6x4 ¾ 36	6x4 ¾ 50	6x4 ⅝ 36	6x4 ⅝ 50	6x4 ½ 36	6x4 ½ 50	6x4 ⅜ 36	6x4 ⅜ 50	Conn
0	471	655	310	401	205	254		0	425	591	358	497	291	388	201	256	
4	425	568	282	354	189	230		4	386	516	326	436	265	343	186	231	
6	374	476	250	304	171	203		6	342	437	289	369	236	294	168	203	
8	313	371	212	245	149	171		8	289	345	245	293	201	238	146	170	
10	249	270	171	186	124	137		10	232	255	198	218	164	180	121	135	
12	188	188	132	133	100	104	b	12	178	179	152	154	127	129	97	102	b
14	138	138	98	98	77	77		14	132	132	113	113	95	95	75	75	
16	106	106	75	75	59	59		16	101	101	86	86	73	73	57	57	
18	84	84	59	59	47	47		18	80	80	68	68	57	57	45	45	
								19					52	52	41	41	

Y-Y AXIS

KL	7x4 ¾ 36	7x4 ¾ 50	7x4 ½ 36	7x4 ½ 50	7x4 ⅜ 36	7x4 ⅜ 50	Conn	KL	6x4 ¾ 36	6x4 ¾ 50	6x4 ⅝ 36	6x4 ⅝ 50	6x4 ½ 36	6x4 ½ 50	6x4 ⅜ 36	6x4 ⅜ 50	Conn
0	471	655	310	401	205	254		0	425	591	358	497	291	388	201	256	
8	427	571	253	309	151	175		8	392	528	321	428	248	313	157	187	
12	422	562	252	307	151	174	3	12	372	491	310	407	243	306	156	184	3
16	399	520	249	302	150	173		16	337	428	282	357	226	277	151	178	
20	365	459	233	277	147	169		20	292	350	244	292	196	230	137	157	
24	323	387	203	232	137	155		24	241	268	201	223	162	177	115	126	
28	276	312	182	201	119	139	4	28	204	213	170	177	136	141	92	103	4
32	229	257	151	159	108	115		32	159	159	132	132	106	106	78	78	
36	197	199	131	132	90	98		36	130	130	108	108	82	82	61	61	
40	158	158	105	105	78	78	5	40	104	104	86	86	69	69	51	51	
44	135	135	85	85	64	64		44	85	85	70	70	56	56	42	42	
48	112	112	74	74	56	56		45	81	81	67	67	54	54	40	40	5
52	94	94	63	63	47	47	6	46	77	77	64	64	51	51	38	38	
56	80	80	53	53	40	40		47	73	73	61	61	49	49	36	36	
57	77	77	51	51	39	39		48	70	70	58	58	47	47			
58	74	74						49	67	67							

Properties of 2 angles—⅜ in. back to back

	7x4 ¾	7x4 ½	7x4 ⅜		6x4 ¾	6x4 ⅝	6x4 ½	6x4 ⅜
A (in.²)	15.4	10.5	7.97		13.9	11.7	9.50	7.22
r_x (in.)	1.09	1.11	1.13		1.12	1.13	1.15	1.17
r_y (in.)	3.49	3.44	3.42		2.94	2.92	2.90	2.87

[a]For Y-Y axis, welded or fully tightened bolted connectors only.
[b]For number of connectors, see double angle column discussion.

F_y = 36 ksi

F_y = 50 ksi

COLUMNS
Double angles
Design axial strength in kips ($\phi = 0.85$)
Unequal legs
Short legs 3/8 in. back to back of angles

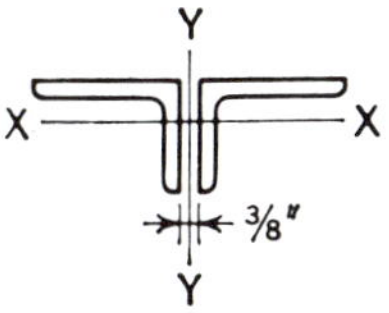

Size			6 x 3½				No. of Connectors[a]
Thickness			3/8		5/16		
Wt./ft			23.4		19.6		
F_y			36	50	36	50	
Effective length in ft *KL* with respect to indicated axis	X-X AXIS	0	191	243	145	179	
		4	170	210	131	158	
		6	148	175	115	135	
		8	121	136	97	109	
		10	94	99	77	82	b
		12	69	69	58	59	
		14	50	50	43	43	
		16	39	39	33	33	
	Y-Y AXIS	0	191	243	145	179	
		8	149	178	106	122	
		12	149	177	106	121	3
		16	145	172	105	120	
		20	131	150	99	112	
		24	108	128	86	94	
		28	94	99	76	80	4
		32	79	80	61	66	
		36	62	62	51	51	5
		40	49	49	41	41	
		44	42	42	35	35	
		48	35	35	29	29	6
		49	34	34	28	28	

Size			5 x 3½								No. of Connectors[a]
Thickness			3/4		1/2		3/8		5/16		
Wt./ft			39.6		27.2		20.8		17.4		
F_y			36	50	36	50	36	50	36	50	
Effective length in ft *KL* with respect to indicated axis	X-X AXIS	0	355	493	245	340	183	238	143	182	
		2	344	472	238	326	178	229	139	176	
		4	313	413	217	288	163	205	129	159	
		6	267	331	187	234	141	170	113	135	
		8	213	243	152	176	116	131	94	107	
		10	160	164	116	121	89	94	74	79	b
		12	114	114	84	84	65	65	56	56	
		14	84	84	62	62	48	48	41	41	
		16	64	64	47	47	37	37	31	31	
		17					32	32	28	28	
	Y-Y AXIS	0	355	493	245	340	183	238	143	182	
		8	326	438	216	285	150	183	110	131	
		10	312	413	210	276	148	181	109	129	
		12	296	383	201	258	145	175	108	127	
		14	277	350	188	235	138	164	105	123	3
		16	255	312	173	210	128	149	99	114	
		18	231	272	156	182	116	131	91	103	
		20	206	231	139	154	103	113	82	90	
		22	180	208	121	128	90	95	73	77	
		24	167	173	112	115	78	85	63	70	
		26	145	146	96	96	72	72	59	59	
		28	124	124	82	82	61	61	51	51	4
		30	106	106	71	71	53	53	44	44	
		32	98	98	61	61	46	46	38	38	
		34	86	86	57	57	43	43	33	33	
		36	76	76	51	51	38	38	31	31	
		38	68	68	45	45	34	34	28	28	
		39	64	64	43	43	32	32	26	26	5
		40	61	61	40	40	30	30			
		41	58	58							

Properties of 2 angles—3/8 in. back to back

	6 x 3½ × 3/8	6 x 3½ × 5/16	5 x 3½ × 3/4	5 x 3½ × 1/2	5 x 3½ × 3/8	5 x 3½ × 5/16
A (in.²)	6.84	5.74	11.6	8.00	6.09	5.12
r_x (in.)	0.988	0.996	0.977	1.01	1.02	1.03
r_y (in.)	2.95	2.94	2.48	2.43	2.41	2.39

[a]For Y-Y axis, welded or fully tightened bolted connectors only.
[b]For number of connectors, see double angle column discussion.

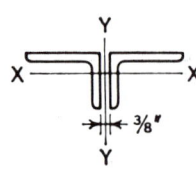

	F_y = 36 ksi
COLUMNS	F_y = 50 ksi

COLUMNS
Double angles
Design axial strength in kips (ϕ = 0.85)
Unequal legs
Short legs 3/8 in. back to back of angles

Size		5 x 3								No. of Connectors[a]
Thickness		½		3/8		5/16		¼		
Wt./ft		25.6		19.6		16.4		13.2		
F_y		36	50	36	50	36	50	36	50	
	0	230	319	172	223	134	170	95	117	
X-X AXIS	2	220	300	165	212	129	162	92	112	
	4	192	249	145	180	115	140	84	99	
	6	154	184	118	137	95	110	71	81	
	8	113	119	88	94	73	79	56	61	b
	10	76	76	61	61	52	52	42	43	
	12	53	53	42	42	36	36	30	30	
	13	45	45	36	36	31	31	25	25	
	14			31	31	26	26	22	22	
	0	230	319	172	223	134	170	95	117	
	8	204	271	141	173	104	123	68	77	
	10	200	263	141	172	103	122	68	77	
	12	191	246	138	168	102	121	67	76	3
	14	178	223	131	157	100	118	67	75	
	16	162	197	121	141	94	109	65	74	
	18	145	169	109	123	86	97	62	69	
	20	128	155	96	113	77	84	57	63	
Y-Y AXIS	22	120	130	90	97	73	78	51	59	
	24	105	115	79	81	64	67	49	52	4
	26	96	97	72	72	56	60	43	45	
	28	82	82	62	62	51	51	40	41	
	30	71	71	53	53	44	44	35	35	
	32	61	61	46	46	38	38	31	31	5
	34	57	57	41	41	34	34	27	27	
	36	50	50	38	38	31	31	24	24	
	38	45	45	34	34	28	28	22	22	
	40	40	40	30	30	25	25	20	20	6
	41	38	38	29	29	24	24	19	19	

(Left margin: Effective length in ft KL with respect to indicated axis)

Properties of 2 angles—3/8 in. back to back

A (in.²)	7.50	5.72	4.80	3.88
r_x (in.)	0.829	0.845	0.853	0.861
r_y (in.)	2.50	2.48	2.47	2.46

[a]For Y-Y axis, welded or fully tightened bolted connectors only.
[b]For number of connectors, see double angle column discussion.

F_y = 36 ksi
F_y = 50 ksi

COLUMNS
Double angles
Design axial strength in kips ($\phi = 0.85$)
Unequal legs
Short legs ⅜ in. back to back of angles

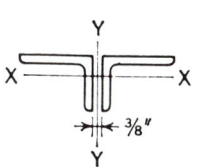

Size		4 x 3½								No. of Connectors[a]
Thickness		½		⅜		5/16		¼		
Wt./ft		23.8		18.2		15.4		12.4		
F_y		36	50	36	50	36	50	36	50	
X-X AXIS	0	214	298	163	227	137	179	101	129	
	2	208	286	159	219	133	172	99	125	
	4	191	255	147	195	123	156	92	114	
	6	166	209	128	162	108	131	81	98	
	8	137	159	106	125	90	103	69	79	
	10	106	112	83	89	71	75	55	60	b
	12	78	78	62	62	53	53	42	43	
	14	57	57	45	45	39	39	31	31	
	16	44	44	35	35	30	30	24	24	
	17	39	39	31	31	26	26	21	21	
Y-Y AXIS	0	214	298	163	227	137	179	101	129	
	4	196	264	141	186	112	137	76	90	
	6	193	257	140	183	111	135	76	89	
	8	184	241	136	175	108	132	75	87	2
	10	171	217	127	160	103	124	73	84	
	12	154	188	115	139	94	110	68	78	
	14	134	156	101	116	83	93	62	69	
	16	114	135	85	92	70	75	53	57	
	18	103	108	77	80	63	65	49	51	
	20	86	86	64	64	53	53	41	41	
	22	70	70	52	52	43	43	34	34	3
	24	58	58	43	43	36	36	28	28	
	26	52	52	39	39	32	32	24	24	
	28	44	44	33	33	27	27	22	22	
	30	38	38	29	29	24	24	19	19	4
	31	36	36	27	27	22	22			

Effective length in ft KL with respect to indicated axis

Properties of 2 angles—⅜ in. back to back				
A (in.²)	7.00	5.34	4.49	3.63
r_x (in.)	1.04	1.06	1.07	1.07
r_y (in.)	1.89	1.87	1.86	1.85

[a]For Y-Y axis, welded or fully tightened bolted connectors only.
[b]For number of connectors, see double angle column discussion.

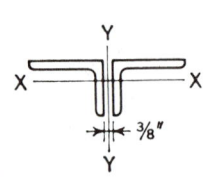

	F_y = 36 ksi
	F_y = 50 ksi

COLUMNS
Double angles
Design axial strength in kips (ϕ = 0.85)
Unequal legs
Short legs ³⁄₈ in. back to back of angles

Size		4 x 3								No. of Connectors[a]
Thickness		½		³⁄₈		⁵⁄₁₆		¼		
Wt./ft		22.2		17.0		14.4		11.6		
F_y		36	50	36	50	36	50	36	50	
X-X AXIS	0	199	276	152	211	127	166	94	120	
	2	191	261	146	200	123	158	91	115	
	4	169	220	130	170	109	136	82	101	
	6	138	166	107	129	90	106	69	81	b
	8	104	112	81	88	69	75	54	59	
	10	72	72	57	57	49	49	40	40	
	12	50	50	40	40	34	34	28	28	
	14	37	37	29	29	25	25	21	21	
Y-Y AXIS	0	199	276	152	211	127	166	94	120	
	4	184	247	133	175	106	130	73	86	
	6	181	243	132	173	105	129	72	85	
	8	173	228	129	168	103	126	72	84	2
	10	162	208	122	155	99	119	70	82	
	12	148	184	112	138	90	105	66	76	
	14	132	157	100	118	78	88	59	66	
	16	115	130	87	97	72	79	55	60	
	18	98	103	74	77	61	64	47	50	3
	20	81	88	61	66	51	54	40	40	
	22	72	72	54	54	45	45	35	35	
	24	60	60	45	45	37	37	29	29	4
	26	50	50	38	38	31	31	25	25	
	28	43	43	32	32	27	27	21	21	
	30	39	39	28	28	23	23	18	18	
	32	34	34	25	25	21	21	17	17	5

Effective length in ft KL with respect to indicated axis

Properties of 2 angles—³⁄₈ in. back to back

A (in.²)	6.50	4.97	4.18	3.38
r_x (in.)	0.864	0.879	0.887	0.896
r_y (in.)	1.96	1.94	1.93	1.92

[a]For Y-Y axis, welded or fully tightened bolted connectors only.
[b]For number of connectors, see double angle column discussion.

F_y = 36 ksi
F_y = 50 ksi

COLUMNS
Double angles
Design axial strength in kips ($\phi = 0.85$)
Unequal legs
Short legs ⅜ in. back to back of angles

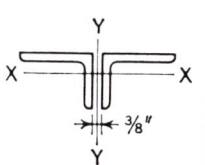

	Size		3½ x 3					No. of Connectors[a]		3½ x 2½					No. of Connectors[a]
	Thickness		⅜		⁵⁄₁₆		¼				⅜		⁵⁄₁₆		¼	
	Wt./ft		15.8		13.2		10.8				14.4		12.2		9.8	
	F_y	KL	36	50	36	50	36	50		KL	36	50	36	50	36	50	
X-X AXIS		0	140	195	118	162	92	119	0	0	129	179	109	149	85	110	0
		2	135	185	114	154	89	114		2	122	165	103	138	81	102	
		4	121	158	102	132	80	100		4	102	129	86	109	68	83	
		6	100	122	85	103	67	79		6	76	86	65	73	52	58	
		8	77	84	65	72	53	58	b	8	50	51	43	43	36	36	b
		10	55	55	47	47	38	39		10	32	32	28	28	23	23	
		12	38	38	33	33	27	27		11	27	27	23	23	19	19	
		14	28	28	24	24	20	20		12			19	19	16	16	
		15			21	21	17	17									
Y-Y AXIS		0	140	195	118	162	92	119		0	129	179	109	149	85	110	
		4	125	166	101	131	73	89		4	116	155	94	121	68	83	
		6	123	161	99	128	73	87		6	114	152	93	120	68	82	
		8	115	149	95	120	71	84	2	8	108	141	89	114	67	80	3
		10	104	129	86	105	66	77		10	99	125	83	102	63	75	
		12	90	105	75	87	58	65		12	88	105	73	87	57	66	
		14	75	89	62	74	49	52		14	75	85	63	70	49	54	
		16	67	69	55	57	43	45		16	62	70	52	58	41	43	
		18	54	54	44	44	35	35	3	18	54	55	45	45	36	36	
		20	43	43	35	35	28	28		20	43	43	36	36	29	29	
		22	37	37	31	31	24	24		22	35	35	29	29	24	24	4
		24	31	31	26	26	20	20	4	24	31	31	26	26	20	20	
		26	26	26	22	22	17	17		26	26	26	22	22	17	17	
		27	24	24	20	20	16	16		28	22	22	19	19	15	15	5
										29	21	21					

Effective length in ft KL with respect to indicated axis

Properties of 2 angles—⅜ in. back to back						
A (in.²)	4.59	3.87	3.13	4.22	3.55	2.88
r_x (in.)	0.897	0.905	0.914	0.719	0.727	0.735
r_y (in.)	1.67	1.66	1.65	1.74	1.73	1.72

[a] For Y-Y axis, welded or fully tightened bolted connectors only.
[b] For number of connectors, see double angle column discussion.

AMERICAN INSTITUTE OF STEEL CONSTRUCTION

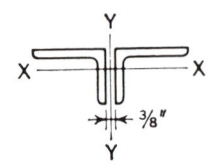

F_y = 36 ksi
F_y = 50 ksi

COLUMNS
Double angles
Design axial strength in kips (ϕ = 0.85)
Unequal legs
Short legs ⅜ in. back to back of angles

Size	3 x 2½						No. of Connectors[a]	3 x 2								No. of Connectors[a]
Thickness	⅜		¼		3/16			⅜		5/16		¼		3/16		
Wt./ft	13.2		9.0		6.77			11.8		10.0		8.2		6.1		
F_y	36	50	36	50	36	50		36	50	36	50	36	50	36	50	

Effective length in ft KL with respect to indicated axis

X-X AXIS

KL	⅜ 36	⅜ 50	¼ 36	¼ 50	3/16 36	3/16 50	Conn	KL	⅜ 36	⅜ 50	5/16 36	5/16 50	¼ 36	¼ 50	3/16 36	3/16 50	Conn
0	118	163	80	107	55	71		0	106	147	90	125	73	97	50	64	
2	111	151	76	100	53	66		2	96	129	82	109	66	86	46	58	
3	104	137	71	91	50	62		3	85	109	72	93	59	74	42	51	
4	94	120	65	81	46	55		4	72	86	61	74	50	59	36	42	b
5	83	100	58	69	41	48		5	58	63	50	55	41	45	30	33	
6	71	81	50	56	36	41		6	44	45	38	39	32	32	24	25	
7	59	63	42	45	31	34	b	7	33	33	28	28	24	24	18	18	
8	48	48	34	35	26	27		8	25	25	22	22	18	18	14	14	
9	38	38	27	27	21	21		9	20	20	17	17	14	14	11	11	
10	31	31	22	22	17	17											
11	25	25	18	18	14	14											
12	21	21	15	15	12	12											

Y-Y AXIS

KL	⅜ 36	⅜ 50	¼ 36	¼ 50	3/16 36	3/16 50	Conn	KL	⅜ 36	⅜ 50	5/16 36	5/16 50	¼ 36	¼ 50	3/16 36	3/16 50	Conn
0	118	163	80	107	55	71		0	106	147	90	125	73	97	50	64	
2	108	146	68	85	42	50		2	98	133	80	107	62	78	39	47	
4	107	143	67	84	42	49		4	98	132	80	106	62	78	39	46	3
6	102	134	66	82	41	49	2	6	94	124	78	103	61	76	39	46	
8	92	117	61	74	40	46		8	86	110	72	92	58	71	38	45	
10	79	94	53	61	36	41		10	76	92	63	76	51	60	35	40	
12	64	78	43	51	30	33		12	63	72	53	59	43	47	30	33	
14	56	58	37	38	27	28	3	14	55	59	45	48	36	39	25	28	4
16	43	43	29	29	21	21		16	43	44	36	36	29	29	21	21	
18	36	36	24	24	18	18		18	36	36	30	30	24	24	17	17	
20	29	29	19	19	14	14	4	20	29	29	24	24	19	19	14	14	
22	23	23	16	16	12	12		22	23	23	19	19	15	15	12	12	5
24	19	19	13	13	10	10		24	19	19	16	16	13	13	10	10	
								25	19	19	15	15	12	12	9	9	6

Properties of 2 angles—⅜ in. back to back

	⅜	¼	3/16		⅜	5/16	¼	3/16
A (in.²)	3.84	2.63	1.99		3.47	2.93	2.38	1.80
r_x (in.)	0.736	0.753	0.761		0.559	0.567	0.574	0.583
r_y (in.)	1.47	1.45	1.44		1.55	1.53	1.52	1.51

[a] For Y-Y axis, welded or fully tightened bolted connectors only.
[b] For number of connectors, see double angle column discussion.

F_y = 36 ksi
F_y = 50 ksi

COLUMNS
Double angles
Design axial strength in kips ($\phi = 0.85$)
Unequal legs
Short legs ⅜ in. back to back of angles

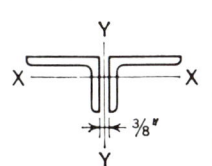

Size		2½ x 2								No. of Connectors[a]
Thickness		⅜		⁵⁄₁₆		¼		³⁄₁₆		
Wt./ft		10.6		9.0		7.2		5.5		
F_y		36	50	36	50	36	50	36	50	
X-X AXIS	0	95	131	80	111	65	91	49	63	
	2	86	116	73	98	60	80	45	57	
	3	77	99	66	84	54	69	40	50	
	4	66	79	56	68	46	56	35	41	
	5	53	60	46	51	38	43	29	32	
	6	42	42	36	37	30	31	23	24	b
	7	31	31	27	27	23	23	18	18	
	8	24	24	21	21	17	17	14	14	
	9	19	19	16	16	14	14	11	11	
	10							9	9	
Y-Y AXIS	0	95	131	80	111	65	91	49	63	
	2	89	121	74	99	57	75	39	47	2
	4	87	116	72	96	56	74	39	47	
	6	80	103	67	86	53	68	37	45	
	8	70	86	57	69	45	55	33	38	
	10	58	66	48	55	36	44	26	31	3
	12	45	47	37	39	30	31	22	23	
	14	36	36	30	30	24	24	18	18	4
	16	27	27	22	22	18	18	13	13	
	18	21	21	17	17	14	14	10	10	
	20	18	18	15	15	11	11	8	8	
	21	16	16	13	13					5

Effective length in ft KL with respect to indicated axis

Properties of 2 angles—⅜ in. back to back				
A (in.²)	3.09	2.62	2.13	1.62
r_x (in.)	0.577	0.584	0.592	0.600
r_y (in.)	1.28	1.26	1.25	1.24

[a]For Y-Y axis, welded or fully tightened bolted connectors only.
[b]For number of connectors, see double angle column discussion.

Notes

| F_y = 36 ksi |
| F_y = 50 ksi |

COLUMNS
Structural tees
cut from W shapes
Design axial strength in kips (ϕ = 0.85)

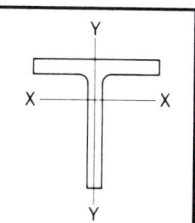

Designation		WT 18									
Wt./ft		150		140		130		122.5		115	
F_y		36	50	36	50	36	50	36	50	36	50
	0	1350	1730	1260	1510	1150	1330	1040	1170	925	1030
	10	1310	1670	1230	1460	1120	1280	1010	1140	903	998
	12	1300	1650	1210	1440	1100	1270	998	1130	893	986
	14	1280	1620	1190	1420	1090	1250	985	1110	882	972
	16	1260	1580	1170	1390	1070	1220	970	1090	869	956
	18	1240	1550	1150	1360	1050	1200	953	1070	854	939
X-X AXIS	20	1210	1510	1130	1330	1030	1170	935	1040	839	919
	22	1180	1460	1100	1290	1010	1140	915	1020	821	899
	24	1150	1420	1080	1250	983	1110	893	993	803	876
	26	1120	1370	1050	1210	957	1070	870	964	783	853
	28	1090	1320	1020	1170	929	1040	846	934	763	828
	30	1060	1260	984	1120	901	1000	821	903	741	802
	32	1020	1210	951	1080	871	964	796	871	719	775
	34	984	1160	917	1030	841	925	769	838	696	748
	36	947	1100	882	986	810	886	742	804	672	720
	38	909	1040	847	939	778	846	714	770	648	691
	40	872	989	812	892	746	806	686	736	624	663
	0	1350	1730	1260	1510	1150	1330	1040	1170	925	1030
	10	1200	1500	1110	1300	995	1120	890	989	787	857
	12	1190	1470	1100	1280	985	1110	881	977	780	848
	14	1170	1440	1080	1260	969	1090	868	962	770	836
	16	1140	1390	1050	1220	948	1060	851	940	756	820
	18	1100	1340	1020	1160	922	1030	829	913	739	799
Y-Y AXIS	20	1060	1280	987	1130	891	989	803	880	717	773
	22	1020	1210	947	1070	856	944	773	843	692	743
	24	976	1140	904	1020	818	896	740	802	664	710
	26	927	1070	860	956	778	846	705	760	635	675
	28	878	998	813	894	736	794	669	715	604	639
	30	827	924	766	832	694	741	632	670	571	601
	32	776	852	718	771	651	689	594	625	539	563
	34	725	781	671	710	608	637	556	580	506	525
	36	675	712	624	650	566	586	518	535	473	487
	38	625	645	578	593	524	536	481	492	440	450
	40	577	583	532	537	483	488	445	450	408	414
Properties											
A (in.²)		44.1		41.2		38.2		36.0		33.8	
r_x (in.)		5.27		5.25		5.26		5.26		5.25	
r_y (in.)		3.83		3.81		3.78		3.75		3.73	

Effective length in ft KL with respect to indicated axis

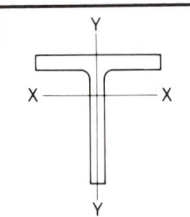

								F_y = 36 ksi

COLUMNS
Structural tees
cut from W shapes
Design axial strength in kips (ϕ = 0.85)

F_y = 50 ksi

Designation								WT 18						
Wt./ft		105		97		91		85		80		75		67.5
F_y		36	50	36	50	36	50	36	50	36	50	36	50	36	50

Effective length in ft KL with respect to indicated axis

X-X AXIS

KL	36	50	36	50	36	50	36	50	36	50	36	50	36	50
0	908	1040	772	851	683	726	587	601	518	521	458	457	385	385
10	887	1010	755	831	670	710	576	590	509	512	451	449	380	380
12	878	1000	748	822	664	704	571	585	505	508	448	446	377	377
14	868	986	740	812	657	696	566	579	500	503	444	442	374	374
16	856	971	731	801	649	687	560	572	495	498	440	438	371	371
18	843	954	720	788	640	677	553	565	489	492	435	433	367	367
20	828	935	709	774	631	667	545	557	483	486	429	428	363	363
22	813	915	696	759	620	655	537	548	476	478	424	422	358	358
24	796	893	683	743	609	642	527	539	468	471	417	416	354	354
26	778	870	668	726	597	629	518	529	460	463	410	409	348	348
28	759	846	653	708	584	614	508	518	452	454	403	402	343	343
30	739	821	637	689	571	599	497	507	443	445	396	395	337	337
32	718	795	621	669	557	584	486	495	433	435	388	387	331	331
36	675	740	586	628	527	551	462	470	413	415	371	370	317	317
40	630	684	549	585	496	516	436	444	392	394	353	352	303	303

Y-Y AXIS

KL	36	50	36	50	36	50	36	50	36	50	36	50	36	50
0	908	1040	772	851	683	726	587	601	518	521	458	457	385	385
10	734	815	622	671	547	573	468	476	407	409	354	353	281	281
12	709	782	603	648	532	556	456	464	398	400	347	346	275	275
14	676	741	578	619	512	534	440	448	385	387	337	336	267	267
16	638	693	548	583	487	507	421	428	370	371	324	324	258	258
18	595	640	514	544	459	476	399	405	352	353	309	309	247	247
20	549	585	478	501	429	442	375	380	332	333	293	292	234	234
22	502	528	440	457	396	407	349	353	310	311	274	274	220	220
24	455	471	401	413	364	371	322	325	287	288	255	255	205	205
26	409	417	363	370	331	336	295	297	264	264	236	235	190	190
28	363	364	325	328	298	301	268	269	241	241	216	216	174	174
30	320	320	289	289	267	267	241	242	218	218	196	196	158	158
32	284	284	256	256	237	237	216	216	196	196	177	177	143	143
34	253	253	228	228	211	211	193	193	175	175	159	159	129	129
36	226	226	205	205	190	190	173	173	158	158	143	143	116	116
39	194	194	176	176	163	163	149	149	136	136	123	123	101	101
41	176	176	159	159	148	148	135	135	123	123	112	112		
42	168	168	152	152	141	141	129	129						
43	161	161												

Properties

	105	97	91	85	80	75	67.5
A (in.²)	30.9	28.5	26.8	25.0	23.5	22.1	19.9
r_x (in.)	5.65	5.62	5.62	5.61	5.61	5.62	5.66
r_y (in.)	2.58	2.56	2.55	2.53	2.50	2.47	2.38

Note: Heavy line indicates Kl/r of 200.

8888888888888888888888888

COLUMNS
Structural tees
cut from W shapes
Design axial strength in kips ($\phi = 0.85$)

$F_y = 36$ ksi
$F_y = 50$ ksi

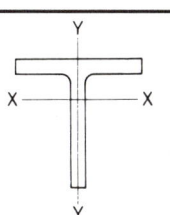

Designation		WT 16.5													
Wt./ft		120.5		110.5		100.5		76		70.5		65		59	
F_y		36	50	36	50	36	50	36	50	36	50	36	50	36	50
X-X AXIS	0	1080	1300	964	1110	810	899	532	548	463	467	402	401	330	330
	10	1050	1260	935	1070	788	872	521	535	453	457	394	393	324	324
	12	1040	1240	923	1050	778	860	516	530	449	453	391	390	321	321
	14	1020	1210	909	1030	767	846	510	524	444	448	387	386	318	318
	16	1000	1190	893	1010	755	831	503	516	439	442	383	382	315	315
	18	980	1160	875	990	740	814	495	508	433	436	378	377	311	311
	20	957	1120	855	965	725	795	487	500	426	429	372	371	307	307
	22	933	1090	834	937	708	774	478	490	419	422	366	365	303	303
	24	907	1050	811	908	690	753	468	480	411	414	360	359	298	298
	26	879	1010	787	878	671	730	458	469	402	405	353	352	293	293
	28	851	975	762	846	652	706	447	457	393	396	345	345	287	287
	30	821	934	736	813	631	681	436	445	384	387	338	337	282	282
	32	790	892	710	779	610	656	424	433	374	377	330	329	276	276
	34	758	849	682	744	588	629	411	420	364	366	321	321	269	269
	36	726	806	654	709	565	603	399	407	354	356	313	312	263	263
	40	661	720	597	639	519	549	373	379	332	334	295	295	249	249
Y-Y AXIS	0	1080	1300	964	1110	810	899	532	548	463	467	402	401	330	330
	10	939	1100	819	919	678	737	429	438	366	368	308	307	243	243
	12	927	1080	810	907	671	729	417	426	357	359	301	300	238	238
	14	910	1060	797	889	661	718	401	409	345	347	291	291	231	231
	16	887	1020	778	866	649	702	382	389	329	331	280	279	223	223
	18	858	985	755	836	632	682	360	367	312	313	266	265	213	213
	20	825	940	727	801	612	658	336	342	292	293	250	250	201	201
	22	789	890	697	762	588	630	311	315	271	272	233	233	188	188
	24	750	837	663	720	562	600	286	289	250	251	216	215	175	175
	26	709	783	628	677	535	567	260	262	228	229	198	198	161	161
	28	667	728	591	632	506	533	235	236	207	207	180	180	147	147
	30	625	672	554	586	476	499	210	210	186	186	162	162	133	133
	34	540	564	480	497	416	429	166	166	148	148	130	130	108	108
	36	498	512	443	454	387	396	149	149	133	133	117	117	97	97
	38	458	462	408	412	357	362	135	135	120	120	106	106	88	88
	39	438	439	390	392	343	346	128	128	114	114	101	101		
	40	419	419	373	373	329	330	122	122	109	109				
	41	399	399	356	356	315	315	116	116						

Effective length in ft KL with respect to indicated axis

Properties

A (in.²)	35.4		32.5		29.5		22.4	
r_x (in.)	4.96		4.96		4.95		5.14	
r_y (in.)	3.63		3.59		3.56		2.47	

20.8	19.2	17.3
5.15	5.18	5.20
2.43	2.39	2.32

Note: Heavy line indicates Kl/r of 200.

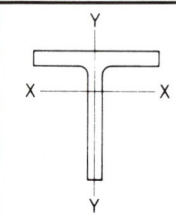

	COLUMNS Structural tees cut from W shapes Design axial strength in kips ($\phi = 0.85$)	$F_y = 36$ ksi
		$F_y = 50$ ksi

Designation		WT 15														
Wt./ft		105.5		95.5		86.5		66		62		58		54		49.5	
F_y		36	50	36	50	36	50	36	50	36	50	36	50	36	50	36	50
X-X AXIS	0	949	1180	844	974	708	791	505	546	447	465	402	412	357	361	304	303
	10	913	1130	812	932	684	761	490	529	434	452	392	401	348	352	297	296
	12	897	1100	799	914	673	748	484	521	429	446	387	396	344	348	294	293
	14	879	1080	783	894	661	732	477	513	423	439	382	391	340	343	290	290
	16	859	1050	765	870	647	715	468	503	416	432	376	384	335	338	286	286
	18	837	1010	746	845	631	696	459	492	408	423	369	377	329	332	282	281
	20	813	976	724	817	615	676	448	480	399	414	361	369	323	326	277	276
	22	787	937	702	787	597	654	437	467	390	404	353	361	316	319	271	271
	24	759	897	677	756	578	630	426	454	380	393	345	352	309	311	266	265
	26	730	854	652	723	557	606	413	439	369	382	336	342	301	304	259	259
	28	701	811	626	690	537	580	400	424	358	370	326	332	293	295	253	252
	30	670	767	599	656	515	554	387	409	347	358	316	322	284	287	246	246
	32	638	722	571	621	493	528	373	393	335	345	306	311	276	278	239	239
	34	607	678	543	586	470	501	358	377	323	332	295	300	266	268	232	231
	36	575	633	515	551	448	474	344	360	311	319	284	289	257	259	224	224
	40	511	547	458	482	402	420	314	326	285	292	262	266	238	240	209	209
Y-Y AXIS	0	949	1180	844	974	708	791	505	546	447	465	402	412	357	361	304	303
	10	828	1000	721	813	595	652	399	423	349	360	308	313	264	266	217	216
	12	816	982	712	801	589	644	383	404	336	346	297	302	255	257	210	209
	14	799	955	698	783	579	632	362	381	319	328	282	287	244	245	201	201
	16	775	920	680	759	566	616	338	353	299	307	265	269	230	231	190	190
	18	747	878	656	729	549	596	311	324	277	283	246	249	214	215	178	178
	20	715	832	629	694	529	571	284	293	254	258	226	228	197	197	164	164
	22	680	781	599	656	506	544	255	261	230	233	205	206	179	179	150	150
	24	643	729	568	616	481	514	228	231	206	208	184	185	161	161	135	135
	26	605	675	534	575	455	483	201	201	182	183	163	163	143	143	121	121
	28	566	622	500	533	428	451	175	175	160	160	143	143	126	126	107	107
	30	527	569	466	491	401	419	154	154	141	141	126	126	111	111	95	95
	32	488	517	432	450	373	386	136	136	125	125	112	112	99	99	84	84
	34	450	467	398	409	345	355	121	121	111	111	100	100	88	88	76	76
	35	431	442	381	390	332	339	115	115	105	105	94	94	84	84	72	72
	36	412	418	365	370	318	324	109	109	100	100	90	90				
	37	394	397	349	351	305	309	103	103	94	94						
	38	376	377	333	334	292	294										
	40	341	341	302	302	266	266										

Effective length in ft KL with respect to indicated axis

Properties								
A (in.²)		31.0	28.1	25.4	19.4	18.2	17.1	15.9	14.5
r_x (in.)		4.43	4.42	4.42	4.66	4.66	4.67	4.69	4.71
r_y (in.)		3.49	3.46	3.43	2.25	2.23	2.19	2.15	2.10

Note: Heavy line indicates Kl/r of 200.

| F_y = 36 ksi |
| F_y = 50 ksi |

COLUMNS
Structural tees
cut from W shapes
Design axial strength in kips (ϕ = 0.85)

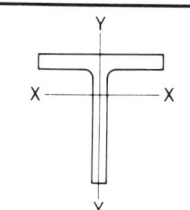

Designation		WT 13.5													
Wt./ft		89		80.5		73		57		51		47		42	
F_y		36	50	36	50	36	50	36	50	36	50	36	50	36	50
X-X AXIS	0	799	1040	725	857	617	698	453	498	358	369	308	311	251	250
	10	761	978	691	810	589	663	436	477	346	356	298	301	244	243
	12	745	951	676	790	577	648	428	468	341	350	294	297	241	240
	14	727	921	660	766	564	631	420	458	334	344	289	292	238	236
	16	706	887	641	741	549	612	410	447	328	337	284	286	234	232
	18	684	850	620	712	532	591	399	434	320	329	278	280	229	228
	20	659	811	598	682	514	568	388	420	312	320	271	273	224	223
	22	633	769	574	650	495	544	375	405	303	310	264	266	219	218
	24	606	726	549	617	474	519	362	390	293	300	256	258	213	212
	26	578	682	523	582	453	492	348	373	283	290	248	250	207	206
	28	549	638	496	548	431	466	334	356	273	279	240	241	201	200
	30	519	593	469	512	409	439	319	339	262	268	231	233	194	193
	32	489	549	442	477	387	412	304	322	251	256	222	223	187	187
	34	459	506	414	443	364	384	289	304	240	244	213	214	180	180
	36	429	463	387	408	342	358	274	286	229	233	204	205	173	173
	38	400	423	361	375	319	331	259	269	217	221	194	195	166	165
	40	371	383	334	343	297	306	243	252	206	209	185	185	159	158
Y-Y AXIS	0	799	1040	725	857	617	698	453	498	358	369	308	311	251	250
	10	695	869	618	709	517	571	358	384	282	288	237	239	186	186
	12	683	848	608	696	510	562	341	364	270	276	229	230	181	180
	14	664	818	593	676	499	549	320	340	256	261	217	219	173	172
	16	640	781	573	649	484	531	296	312	239	243	204	205	163	163
	18	613	737	549	617	466	509	270	282	220	224	189	190	153	152
	20	582	689	522	581	445	483	243	251	201	203	173	174	141	140
	22	549	639	492	543	422	454	217	221	181	182	157	157	128	128
	24	515	587	462	503	397	424	191	192	161	162	140	141	116	115
	26	480	536	430	462	371	393	166	166	142	142	124	125	103	103
	28	445	485	398	422	345	362	144	144	124	124	109	109	91	91
	30	410	436	367	383	319	331	126	126	109	109	96	96	80	80
	32	375	388	335	345	293	301	112	112	96	96	85	85	71	71
	34	341	345	305	308	268	272	99	99	86	86	76	76	64	64
	35	325	326	290	291	256	257	94	94	81	81	72	72		
	36	309	309	276	276	243	244	89	89						
	40	252	252	225	225	199	199								

Effective length in ft KL with respect to indicated axis

Properties							
A (in.²)	26.1	23.7	21.5	16.8	15.0	13.8	12.4
r_x (in.)	3.98	3.96	3.95	4.15	4.14	4.16	4.18
r_y (in.)	3.26	3.24	3.21	2.18	2.15	2.12	2.07

Note: Heavy line indicates Kl/r of 200.

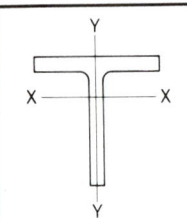

	F_y = 36 ksi
COLUMNS Structural tees cut from W shapes Design axial strength in kips (ϕ = 0.85)	F_y = 50 ksi

Designation		WT 12									
Wt./ft		81		73		65.5		58.5		52	
F_y		36	50	36	50	36	50	36	50	36	50
X-X AXIS	0	731	1020	658	864	591	726	506	580	410	450
	10	687	932	618	797	555	673	477	542	389	424
	12	669	897	602	769	541	651	465	526	379	413
	14	648	858	583	737	524	626	451	507	369	400
	16	624	815	561	702	505	599	435	487	357	386
	18	598	769	538	664	484	569	418	465	344	371
	20	571	720	514	624	462	537	399	442	330	355
	22	542	670	487	583	439	504	380	417	316	338
	24	512	619	460	541	415	471	360	392	301	320
	26	481	568	433	499	390	436	339	366	285	301
	28	450	517	405	457	365	402	318	341	269	283
	30	419	468	377	415	340	369	297	315	252	264
	32	388	421	349	376	315	336	276	289	236	245
	34	357	375	321	337	291	304	255	265	220	227
	36	328	335	295	301	267	273	235	241	204	209
	38	299	300	269	270	244	245	215	217	188	191
	40	271	271	244	244	221	221	196	196	173	175
Y-Y AXIS	0	731	1020	658	864	591	726	506	580	410	450
	10	648	859	572	720	499	591	417	464	330	354
	12	631	827	558	697	489	575	409	454	325	349
	14	608	785	539	665	473	552	398	440	318	340
	16	580	737	515	626	453	524	383	421	308	329
	18	550	684	488	583	429	490	364	398	296	314
	20	518	628	459	538	403	454	344	372	281	297
	22	484	572	428	491	376	417	321	344	264	278
	24	449	516	397	445	348	379	298	316	247	258
	26	414	461	365	399	320	342	274	288	229	237
	28	379	408	334	354	292	305	251	259	211	217
	30	345	358	303	312	264	270	228	232	193	197
	32	312	315	273	275	238	238	205	206	175	177
	34	280	280	244	244	212	212	184	184	158	158
	36	250	250	218	218	190	190	164	164	142	142
	38	225	225	196	196	171	171	148	148	128	128
	40	203	203	178	178	154	154	134	134	116	116

(Effective length in ft KL with respect to indicated axis)

Properties					
A (in.²)	23.9	21.5	19.3	17.2	15.3
r_x (in.)	3.50	3.50	3.52	3.51	3.51
r_y (in.)	3.05	3.01	2.97	2.94	2.91

AMERICAN INSTITUTE OF STEEL CONSTRUCTION

F_y = 36 ksi
F_y = 50 ksi

COLUMNS
Structural tees
cut from W shapes
Design axial strength in kips (ϕ = 0.85)

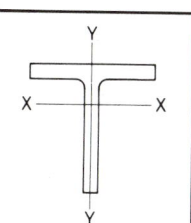

Designation		WT 12											
Wt./ft		47		42		38		34		31		27.5	
F_y		36	50	36	50	36	50	36	50	36	50	36	50
	0	378	419	307	321	254	258	209	208	202	203	155	155
X-X AXIS	10	360	396	293	306	244	247	201	200	194	196	150	150
	12	352	387	287	299	239	243	197	197	191	192	148	148
	14	343	376	280	292	234	237	194	193	187	188	145	145
	16	332	363	273	284	229	231	189	189	183	184	142	142
	18	321	350	265	275	222	225	185	184	178	179	139	139
	20	309	335	256	265	215	218	179	179	173	174	136	136
	22	296	320	246	255	208	210	174	173	168	169	132	132
	24	283	304	236	244	200	202	168	167	162	163	128	128
	26	269	287	225	232	192	194	162	161	156	157	124	124
	28	255	270	215	221	184	185	155	155	149	150	120	120
	30	240	253	204	209	175	176	148	148	143	144	115	115
	32	226	236	192	197	166	167	142	141	136	137	111	111
	34	211	220	181	185	157	158	135	135	130	130	106	106
	36	197	203	170	173	148	149	128	128	123	123	101	101
	38	182	187	159	161	139	140	121	121	116	117	96	96
	40	169	171	148	150	131	131	114	114	109	110	92	92
Y-Y AXIS	0	378	419	307	321	254	258	209	208	202	203	155	155
	10	295	319	237	245	193	195	153	152	128	129	97	97
	12	276	296	224	231	184	185	146	146	115	115	88	88
	14	254	270	208	214	172	173	137	137	100	100	78	78
	16	230	242	190	194	158	159	127	127	85	85	67	67
	18	205	213	171	174	143	144	116	116	70	70	57	57
	20	180	184	152	153	128	129	104	104	58	58	47	47
	22	156	157	133	133	113	113	93	93	49	49	40	40
	23	144	144	123	124	106	106	87	87	45	45		
	24	133	133	114	114	98	98	81	81				
	26	114	114	98	98	85	85	70	70				
	28	99	99	85	85	74	74	61	61				
	30	87	87	75	75	65	65	54	54				
	31	81	81	70	70	61	61	51	51				
	32	76	76	66	66	57	57						
	33	72	72										

Effective length in ft KL with respect to indicated axis

Properties						
A (in.^{2})	13.8		12.4		11.2	
r_x (in.)	3.67		3.67		3.68	
r_y (in.)	1.98		1.95		1.92	

Properties						
A (in.^{2})	10.0		9.11		8.10	
r_x (in.)	3.70		3.79		3.80	
r_y (in.)	1.87		1.38		1.34	

Note: Heavy line indicates Kl/r of 200.

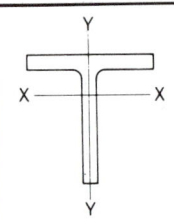

	COLUMNS Structural tees cut from W shapes Design axial strength in kips ($\phi = 0.85$)	**F_y = 36 ksi** **F_y = 50 ksi**

Designation		WT 10.5									
Wt./ft		73.5		66		61		55.5		50.5	
F_y		36	50	36	50	36	50	36	50	36	50
X-X AXIS	0	661	918	594	825	548	757	499	637	452	524
	10	610	821	547	737	505	675	459	573	416	476
	12	589	782	528	701	487	643	443	547	401	456
	14	565	738	506	661	466	606	424	518	384	434
	16	539	691	482	618	444	566	404	486	366	410
	18	510	640	457	573	420	524	382	452	345	384
	20	480	589	429	526	394	481	358	418	324	357
	22	449	536	401	478	368	437	334	382	302	329
	24	417	484	372	431	341	394	310	347	280	301
	26	385	433	343	385	314	351	285	312	258	273
	28	353	384	314	341	288	311	261	278	236	246
	30	322	337	286	299	262	272	237	246	214	220
	32	291	296	259	263	236	239	214	217	193	195
	34	262	263	233	233	212	212	192	192	173	173
	36	234	234	208	208	189	189	171	171	154	154
	38	210	210	186	186	170	170	154	154	139	139
	40	190	190	168	168	153	153	139	139	125	125
Y-Y AXIS	0	661	918	594	825	548	757	499	637	452	524
	10	588	780	521	688	475	622	425	519	377	425
	12	570	747	506	661	463	600	415	504	370	415
	14	546	705	486	625	445	569	401	481	358	400
	16	520	657	463	583	424	532	382	453	343	380
	18	490	606	437	538	400	491	361	421	324	357
	20	459	554	409	491	375	448	338	387	304	331
	22	427	500	380	443	348	405	314	352	283	304
	24	394	448	350	396	321	362	289	317	261	277
	26	361	397	321	351	294	320	264	283	238	250
	28	329	348	292	307	267	280	240	250	217	223
	30	297	304	263	268	241	245	216	219	195	198
	32	267	268	236	237	216	216	193	193	175	175
	34	238	238	210	210	192	192	172	172	155	155
	36	212	212	188	188	172	172	154	154	139	139
	38	191	191	169	169	154	154	138	138	125	125
	40	172	172	152	152	140	140	125	125	113	113

Effective length in ft KL with respect to indicated axis

Properties					
A (in.²)	21.6	19.4	17.9	16.3	14.9
r_x (in.)	3.08	3.06	3.04	3.03	3.01
r_y (in.)	2.95	2.93	2.92	2.90	2.89

AMERICAN INSTITUTE OF STEEL CONSTRUCTION

F_y = 36 ksi
F_y = 50 ksi

COLUMNS
Structural tees
cut from W shapes
Design axial strength in kips (ϕ = 0.85)

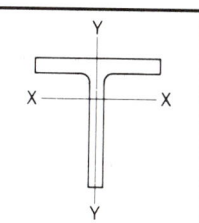

Designation		WT 10.5															
Wt./ft		46.5		41.5		36.5		34		31		28.5		25		22	
F_y		36	50	36	50	36	50	36	50	36	50	36	50	36	50	36	50
X-X AXIS	0	419	562	373	444	297	331	261	283	219	225	203	210	165	167	127	127
	6	409	543	364	430	290	322	255	276	214	221	199	206	162	163	125	125
	8	400	529	356	420	284	316	251	271	211	217	196	203	160	161	123	123
	10	390	511	347	407	278	307	245	264	206	212	192	199	157	158	121	121
	12	378	489	336	392	270	297	239	256	201	207	187	194	153	155	119	119
	14	364	466	323	374	260	286	231	247	195	201	182	188	149	151	116	116
	16	349	439	310	355	250	274	222	237	189	194	176	182	145	146	113	113
	18	332	411	295	335	239	260	213	227	181	186	170	175	140	141	110	110
	20	315	382	279	313	227	246	203	215	174	178	163	167	134	136	106	106
	22	296	353	262	291	215	231	192	203	165	169	155	159	129	130	102	102
	24	277	323	245	269	202	216	181	191	157	160	147	151	123	124	98	98
	26	258	293	228	246	189	200	170	178	148	151	139	143	117	118	94	94
	28	239	264	210	224	176	185	159	165	139	141	131	134	111	111	90	90
	30	220	236	193	203	163	169	148	153	130	132	123	125	104	105	85	85
	32	201	209	176	182	150	154	137	140	121	123	115	117	98	98	81	81
	34	183	185	160	162	137	140	126	128	112	113	107	108	91	92	76	76
	36	165	165	145	145	125	126	115	116	104	104	99	100	85	85	71	71
	38	148	148	130	130	113	113	105	105	95	96	91	92	79	79	67	67
	40	134	134	117	117	102	102	95	95	87	87	83	84	73	73	62	62
Y-Y AXIS	0	419	562	373	444	297	331	261	283	219	225	203	210	165	167	127	127
	6	360	459	314	361	246	269	214	228	176	180	158	163	122	123	90	90
	8	344	431	300	343	237	258	207	220	171	175	148	151	114	115	85	85
	10	320	391	280	316	224	241	196	208	163	167	133	136	104	104	78	78
	12	291	345	256	283	206	220	182	191	152	155	116	118	91	91	70	70
	14	260	296	228	247	186	196	165	172	139	142	98	99	77	77	60	60
	16	228	248	200	211	164	171	147	152	125	127	81	81	64	64	51	51
	18	196	202	171	176	143	146	129	132	110	112	65	65	52	52	42	42
	20	165	165	145	145	122	123	111	112	96	96	54	54	43	43	35	35
	21	151	151	132	132	112	112	102	102	89	89	49	49	39	39	32	32
	22	138	138	120	120	102	102	94	94	82	82	45	45				
	24	116	116	102	102	87	87	80	80	70	70						
	26	99	99	87	87	74	74	68	68	60	60						
	28	86	86	75	75	64	64	59	59	52	52						
	29	80	80	70	70	60	60	55	55	49	49						
	30	75	75	66	66	56	56	52	52								
Properties																	
A (in.²)		13.7		12.2		10.7		10.0		9.13		8.37		7.36		6.49	
r_x (in)		3.25		3.22		3.21		3.20		3.21		3.29		3.30		3.31	
r_y (in.)		1.84		1.83		1.81		1.80		1.77		1.35		1.30		1.26	

Effective length in ft KL with respect to indicated axis

Note: Heavy line indicates Kl/r of 200.

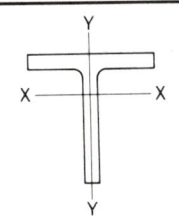

	COLUMNS	F_y = 36 ksi
	Structural tees	F_y = 50 ksi

COLUMNS
Structural tees
cut from W shapes
Design axial strength in kips (ϕ = 0.85)

Designation		WT 9									
Wt./ft		59.5		53		48.5		43		38	
F_y		36	50	36	50	36	50	36	50	36	50
	0	536	744	477	663	438	608	389	507	339	393
	10	479	636	426	567	390	517	346	435	302	343
	12	456	594	406	529	370	482	329	407	287	323
	14	430	548	382	487	349	443	309	376	270	301
	16	402	499	357	443	325	403	288	343	252	278
	18	372	449	331	398	301	361	266	310	232	254
X-X AXIS	20	342	399	304	354	275	319	244	276	213	229
	22	311	350	276	310	250	279	221	243	193	205
	24	280	303	249	268	225	241	198	211	173	181
	26	251	259	222	229	200	205	177	181	154	158
	28	222	224	197	198	177	177	156	156	136	137
	30	195	195	172	172	154	154	136	136	119	119
	34	152	152	134	134	120	120	106	106	93	93
	38	121	121	107	107	96	96	85	85	74	74
	42	99	99	88	88	79	79	69	69	61	61
	43	95	95	84	84						
	0	536	744	477	663	438	608	389	507	339	393
	10	473	626	417	549	378	497	329	408	280	314
	12	453	590	400	519	364	471	318	391	272	304
	14	430	548	380	482	346	439	303	367	260	289
	16	404	503	357	442	325	402	285	339	246	270
	18	377	456	332	400	303	364	265	308	229	249
Y-Y AXIS	20	348	409	306	358	279	325	245	277	211	227
	22	319	362	280	316	255	287	223	246	193	205
	24	289	316	254	276	231	250	202	216	175	182
	26	261	273	228	237	207	215	181	187	157	161
	28	233	236	203	205	185	186	161	162	139	140
	30	206	206	179	179	163	163	142	142	122	122
	34	161	161	140	140	127	127	111	111	96	96
	38	129	129	112	112	102	102	89	89	77	77
	42	106	106	92	92	84	84	73	73	63	63
	43	101	101	88	88	80	80	70	70	60	60
	44	96	96	84	84	76	76				

Effective length in ft KL with respect to indicated axis

Properties											
A (in.^{2})		17.5		15.6		14.3		12.7		11.2	
r_x (in.)		2.60		2.59		2.56		2.55		2.54	
r_y (in.)		2.69		2.66		2.65		2.63		2.61	

Note: Heavy line indicates Kl/r of 200.

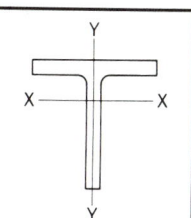

F_y = 36 ksi
F_y = 50 ksi

COLUMNS
Structural tees
cut from W shapes
Design axial strength in kips ($\phi = 0.85$)

Designation		WT 9															
Wt./ft		35.5		32.5		30		27.5		25		23		20		17.5	
F_y		36	50	36	50	36	50	36	50	36	50	36	50	36	50	36	50
X-X AXIS	0	318	426	292	356	261	299	226	253	184	194	172	183	124	124	101	101
	10	288	372	264	314	236	266	206	227	169	177	159	168	116	116	95	95
	12	275	351	252	297	226	253	197	217	163	170	153	161	112	112	92	92
	14	261	327	239	279	214	239	188	206	156	163	147	154	108	108	89	89
	16	246	301	225	259	202	223	178	193	148	154	140	146	104	104	86	86
	18	229	275	210	237	189	206	167	180	140	145	132	138	99	99	82	82
	20	212	248	194	216	175	189	155	166	131	135	124	129	94	94	78	78
	22	195	221	178	194	161	172	143	152	122	126	116	120	89	89	74	74
	24	178	195	162	173	147	155	131	138	113	116	107	111	83	83	70	70
	26	161	171	146	153	133	138	120	124	103	106	99	101	78	78	66	66
	28	144	148	131	134	119	122	108	111	94	96	90	92	72	72	62	62
	30	128	129	116	116	106	107	97	98	85	86	82	83	67	67	57	57
	32	113	113	102	102	94	94	86	86	77	77	74	75	61	61	53	53
	34	100	100	91	91	83	83	76	76	68	68	67	67	56	56	49	49
	36	89	89	81	81	74	74	68	68	61	61	59	59	51	51	45	45
	38	80	80	72	72	66	66	61	61	55	55	53	53	46	46	41	41
	40	72	72	65	65	60	60	55	55	49	49	48	48	41	41	37	37
	42	66	66	59	59	54	54	50	50	45	45	44	44	38	38	34	34
	43	63	63	57	57	52	52	48	48	43	43	42	42	36	36	32	32
	45	57	57	52	52	47	47	44	44	39	39	38	38	33	33	29	29
	46											36	36	31	31	28	28
Y-Y AXIS	0	318	426	292	356	261	299	226	253	184	194	172	183	124	124	101	101
	10	236	285	214	244	191	209	165	178	136	141	110	114	83	83	64	64
	12	211	245	191	213	172	185	149	159	124	128	94	96	72	72	56	56
	14	184	205	167	180	151	160	132	138	111	114	77	78	61	61	48	48
	16	157	166	143	149	130	134	114	118	97	99	62	62	51	51	40	40
	18	132	133	119	120	109	110	96	98	83	84	49	49	41	41	32	32
	20	108	108	98	98	90	90	80	80	70	70	40	40	34	34	26	26
	21	98	98	89	89	82	82	73	73	64	64	37	37	31	31		
	22	90	90	81	81	75	75	67	67	58	58						
	24	76	76	69	69	63	63	56	56	49	49						
	26	65	65	59	59	54	54	48	48	42	42						
	27	60	60	55	55	50	50	45	45	39	39						
	28	56	56	51	51	47	47										

Effective length in ft KL with respect to indicated axis

Properties								
A (in.²)	10.4	9.55	8.82	8.10	7.33	6.77	5.88	5.15
r_x (in.)	2.74	2.72	2.71	2.71	2.70	2.77	2.76	2.79
r_y (in.)	1.70	1.69	1.69	1.67	1.65	1.29	1.27	1.22

Note: Heavy line indicates Kl/r of 200.

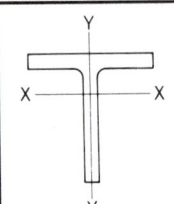

	F_y = 36 ksi
COLUMNS Structural tees cut from W shapes Design axial strength in kips (ϕ = 0.85)	F_y = 50 ksi

Designation			WT 8						
Wt./ft			50		44.5		38.5		33.5	
F_y			36	50	36	50	36	50	36	50
		0	450	625	401	557	346	476	301	361
		10	389	510	346	454	297	386	258	300
		12	365	467	324	415	278	352	241	277
		14	338	420	300	373	257	316	223	251
		16	310	372	275	330	235	279	203	225
		18	280	324	249	287	212	242	183	198
	X-X AXIS	20	251	278	222	246	189	207	163	172
		22	222	234	197	207	166	174	143	148
		24	194	197	172	174	145	146	124	125
		26	167	167	148	148	124	124	106	106
		28	144	144	128	128	107	107	92	92
		30	126	126	111	111	93	93	80	80
		32	111	111	98	98	82	82	70	70
		34	98	98	87	87	73	73	62	62
		36	87	87	77	77	65	65	55	55
		37	83	83	73	73	61	61	52	52
		38	78	78						
		0	450	625	401	557	346	476	301	361
		10	393	518	347	456	295	383	252	292
		12	374	483	331	426	282	359	242	278
		14	351	443	311	391	265	331	229	259
		16	327	401	289	354	247	299	213	238
		18	302	358	266	315	227	267	196	216
	Y-Y AXIS	20	275	316	243	277	207	235	178	193
		22	249	274	219	241	186	203	161	170
		24	223	235	196	206	166	174	143	148
		26	197	201	173	176	147	148	126	128
		28	173	173	152	152	128	128	110	110
		30	151	151	132	132	112	112	96	96
		32	133	133	117	117	99	99	85	85
		34	118	118	103	103	87	87	75	75
		36	105	105	92	92	78	78	67	67
		38	95	95	83	83	70	70	61	61
		41	81	81	71	71	60	60	52	52

Effective length in ft KL with respect to indicated axis

Properties				
A (in.²)	14.7	13.1	11.3	9.84
r_x (in.)	2.28	2.27	2.24	2.22
r_y (in.)	2.51	2.49	2.47	2.46

Note: Heavy line indicates Kl/r of 200.

COLUMNS
Structural tees
cut from W shapes
Design axial strength in kips ($\phi = 0.85$)

F_y = 36 ksi

F_y = 50 ksi

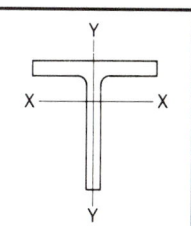

Designation			WT 8												
Wt./ft			28.5		25		22.5		20		18		15.5		13	
F_y			36	50	36	50	36	50	36	50	36	50	36	50	36	50
		0	256	336	223	259	184	205	141	145	122	124	93	93	66	66
		6	245	316	213	245	176	195	136	140	118	120	91	90	65	65
		8	236	301	205	235	170	188	132	136	114	116	88	88	63	63
		10	225	283	196	223	163	179	127	130	110	112	86	85	62	62
		12	212	262	185	208	154	169	121	124	106	107	83	82	60	60
		14	198	240	173	192	145	157	115	117	101	102	79	79	58	58
		16	184	217	160	176	135	145	108	110	95	96	75	75	55	55
		18	168	193	146	159	124	133	100	102	89	90	71	71	53	53
	X-X AXIS	20	152	169	133	141	114	120	92	94	82	83	67	66	50	50
		22	136	147	119	124	103	107	85	86	76	76	62	62	47	47
		24	121	125	105	108	92	95	77	78	69	70	57	57	44	44
		26	106	107	92	93	81	83	69	70	63	63	53	53	41	41
		28	92	92	80	80	72	72	62	62	56	57	48	48	38	38
		30	80	80	70	70	62	62	54	54	50	50	44	44	35	35
		32	70	70	61	61	55	55	48	48	44	44	39	39	32	32
		34	62	62	54	54	49	49	42	42	39	39	35	35	29	29
		36	56	56	49	49	43	43	38	38	35	35	31	31	27	27
		38	50	50	44	44	39	39	34	34	31	31	28	28	24	24
		39	47	47	41	41	37	37	32	32	30	30	27	27	23	23
		40	45	45	39	39					28	28	25	25	22	22
		41													21	21
		0	256	336	223	259	184	205	141	145	122	124	93	93	66	66
		6	218	272	186	209	151	165	116	119	95	97	72	72	49	49
		8	204	248	175	195	143	155	111	114	92	93	67	67	46	46
		10	184	218	158	174	131	141	104	106	86	87	59	59	42	42
		12	162	184	140	150	117	124	94	96	78	79	51	51	36	36
		14	139	151	120	126	102	106	84	85	69	70	42	42	31	31
	Y-Y AXIS	16	117	120	101	103	86	88	73	73	60	61	34	34	25	25
		18	95	95	82	82	71	71	62	62	51	51	27	27	20	20
		19	86	86	74	74	64	64	57	57	47	47	24	24		
		20	78	78	67	67	58	58	51	51	42	42				
		22	64	64	56	56	49	49	43	43	35	35				
		24	54	54	47	47	41	41	36	36	30	30				
		25	50	50	43	43	38	38	33	33	28	28				
		26	46	46	40	40	35	35	31	31						

Effective length in ft KL with respect to indicated axis

Properties							
A (in.^{2})	8.38	7.37	6.63	5.89	5.28	4.56	3.84
r_x (in.)	2.41	2.40	2.39	2.37	2.41	2.45	2.47
r_y (in.)	1.60	1.59	1.57	1.57	1.52	1.17	1.12

Note: Heavy line indicates Kl/r of 200.

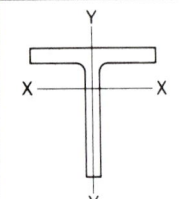

F_y = 36 ksi
F_y = 50 ksi

COLUMNS
Structural tees
cut from W shapes
Design axial strength in kips ($\phi = 0.85$)

Designation			WT 7								
Wt./ft			66		60		54.5		49.5		45	
F_y			36	50	36	50	36	50	36	50	36	50
		0	594	825	542	752	490	680	447	621	404	561
		2	588	813	536	741	484	670	442	611	399	552
		4	570	779	520	710	469	641	428	584	387	528
		6	542	726	493	661	444	594	405	542	366	489
		8	505	658	459	597	412	535	375	487	339	439
		10	461	580	418	525	374	468	340	425	307	383
	X-X AXIS	12	412	497	373	448	332	397	302	360	272	323
		14	361	413	326	371	289	327	262	296	235	265
		16	310	335	279	299	246	261	223	236	200	211
		18	261	266	234	237	205	207	185	186	165	166
		20	215	215	192	192	167	167	151	151	135	135
		22	178	178	158	158	138	138	125	125	111	111
		24	149	149	133	133	116	116	105	105	94	94
		26	127	127	113	113	99	99	89	89	80	80
		27	118	118	105	105	92	92	83	83	74	74
		28	110	110	98	98	85	85				
		0	594	825	542	752	490	680	447	621	404	561
		6	551	743	496	666	441	589	395	522	349	458
		8	550	741	495	664	441	587	394	521	348	456
		10	548	738	494	661	440	586	393	520	348	455
		12	542	727	490	655	438	582	392	517	347	454
		14	531	705	482	639	432	572	389	511	345	450
		16	515	676	468	614	421	552	381	498	340	442
		18	497	644	452	585	407	527	370	477	332	427
	Y-Y AXIS	20	477	609	434	553	391	498	355	451	320	405
		22	456	572	415	519	374	468	340	424	306	381
		24	434	534	395	485	356	437	323	396	291	356
		26	412	496	374	450	337	405	306	367	276	330
		28	389	458	353	414	318	373	288	338	260	304
		30	365	420	331	380	298	342	271	309	244	278
		32	342	383	310	346	279	311	253	281	228	253
		34	318	347	288	313	260	282	235	254	212	228
		36	295	312	267	282	241	253	218	228	196	205
		38	273	280	247	253	222	227	201	205	180	184
		40	251	253	227	228	204	205	184	185	165	166
Properties												
A (in.^{2})			19.4		17.7		16.0		14.6		13.2	
r_x (in.)			1.73		1.71		1.68		1.67		1.66	
r_y (in.)			3.76		3.74		3.73		3.71		3.70	

Effective length in ft KL with respect to indicated axis

Note: Heavy line indicates Kl/r of 200.

F_y = 36 ksi

F_y = 50 ksi

COLUMNS
Structural tees
cut from W shapes
Design axial strength in kips ($\phi = 0.85$)

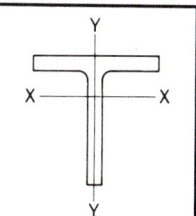

Designation		WT 7												
Wt./ft		41		37		34		30.5		26.5		24		21.5
F_y		36	50	36	50	36	50	36	50	36	50	36	50	36	50
X-X AXIS	0	367	510	334	463	306	425	274	370	239	318	216	265	183	208
	4	354	485	322	440	295	403	264	352	231	303	209	254	177	200
	6	339	456	307	413	281	378	252	330	221	287	200	241	170	191
	8	319	419	288	378	264	346	236	302	208	265	188	224	160	179
	10	294	375	265	337	242	308	217	270	193	239	174	203	149	164
	12	267	327	240	293	219	267	196	235	175	211	158	181	136	148
	14	238	279	213	248	194	226	173	199	157	182	141	157	122	131
	16	208	232	186	205	169	186	151	165	138	153	124	134	107	114
	18	179	188	159	165	144	150	128	133	119	126	107	112	93	97
	20	151	152	133	134	121	121	107	108	101	102	91	92	80	81
	22	126	126	111	111	100	100	89	89	85	85	76	76	67	67
	24	106	106	93	93	84	84	75	75	71	71	64	64	56	56
	26	90	90	79	79	72	72	64	64	61	61	54	54	48	48
	28	78	78	68	68	62	62	55	55	52	52	47	47	41	41
	30	68	68	59	59	54	54	48	48	45	45	41	41	36	36
	31									43	43	38	38	34	34
Y-Y AXIS	0	367	510	334	463	306	425	274	370	239	318	216	265	183	208
	8	332	444	302	403	274	364	242	312	205	258	183	216	154	171
	10	320	422	291	383	265	348	235	301	191	236	172	200	145	160
	12	304	393	276	357	252	324	224	282	175	210	157	179	133	145
	14	286	360	260	327	236	297	211	259	157	182	141	157	121	129
	16	266	325	241	295	219	268	196	235	139	155	125	135	107	113
	18	244	290	222	263	202	238	180	209	121	129	108	114	93	97
	20	222	254	202	231	183	209	163	184	104	105	93	94	80	81
	22	201	220	182	200	165	181	147	160	87	87	78	78	68	68
	24	179	188	163	171	147	154	131	136	73	73	65	65	57	57
	26	158	160	144	146	130	131	115	116	63	63	56	56	49	49
	28	138	138	126	126	113	113	101	101	54	54	48	48	42	42
	30	121	121	110	110	99	99	88	88	47	47	42	42	37	37
	31	113	113	103	103	93	93	82	82	44	44	39	39	34	34
	32	106	106	96	96	87	87	77	77						
	34	94	94	86	86	77	77								
	36	84	84	76	76	69	69								
	40	68	68	62	62	56	56								
	41	65	65	59	59										
Properties															
A (in.^{2})		12.0		10.9		9.99		8.96		7.81		7.07		6.31	
r_x (in.)		1.85		1.82		1.81		1.80		1.88		1.87		1.86	
r_y (in.)		2.48		2.48		2.46		2.45		1.92		1.91		1.89	

Effective length in ft KL with respect to indicated axis

Note: Heavy line indicates Kl/r of 200.

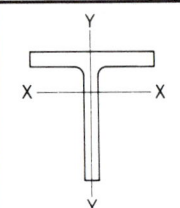

	COLUMNS Structural tees cut from W shapes Design axial strength in kips ($\phi = 0.85$)	F_y = 36 ksi F_y = 50 ksi

Designation			WT 7								
Wt./ft			19		17		15		13		11	
F_y			36	50	36	50	36	50	36	50	36	50
Effective length in ft KL with respect to indicated axis	X-X AXIS	0	159	180	131	142	109	114	87	88	62	62
		2	158	178	130	141	109	114	87	88	62	62
		4	155	174	128	138	107	112	85	86	61	61
		6	150	168	124	134	104	108	83	84	60	60
		8	143	159	118	127	100	104	80	81	58	58
		10	134	148	112	120	95	98	77	78	56	56
		12	125	136	105	111	89	92	73	73	53	53
		14	114	123	96	102	83	85	68	69	51	51
		16	103	110	88	92	76	78	63	64	48	48
		18	92	97	79	82	69	70	58	58	44	44
		20	81	83	70	72	62	63	53	53	41	41
		22	70	71	62	63	55	55	48	48	38	38
		24	60	60	53	54	48	48	42	43	34	34
		26	51	51	46	46	42	42	37	38	31	31
		28	44	44	39	39	36	36	33	33	28	28
		30	38	38	34	34	31	31	28	28	24	24
		32	34	34	30	30	27	27	25	25	22	22
		34	30	30	27	27	24	24	22	22	19	19
		35							21	21	18	18
	Y-Y AXIS	0	159	180	131	142	109	114	87	88	62	62
		6	133	146	107	114	86	89	67	67	46	46
		8	125	137	102	108	82	85	60	60	42	42
		10	113	123	94	99	76	78	51	51	36	36
		12	100	107	84	87	68	70	41	41	30	30
		14	86	90	73	75	60	61	32	32	24	24
		16	72	74	62	63	51	51	25	25	19	19
		17	66	66	56	57	46	47	22	22	17	17
		18	59	59	51	51	42	42	20	20		
		20	48	48	42	42	35	35				
		22	40	40	35	35	29	29				
		24	34	34	29	29	24	24				
		25	31	31	27	27						

Properties					
A (in.^{2})	5.58		5.00	4.42	3.85	3.25
r_x (in.)	2.04		2.04	2.07	2.12	2.14
r_y (in.)	1.55		1.53	1.49	1.08	1.04

Note: Heavy line indicates Kl/r of 200.

F_y = 36 ksi
F_y = 50 ksi

COLUMNS
Structural tees
cut from W shapes
Design axial strength in kips ($\phi = 0.85$)

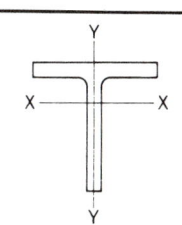

Designation		WT 6									
Wt./ft		29		26.5		25		22.5		20	
F_y		36	50	36	50	36	50	36	50	36	50
X-X AXIS	0	261	362	238	331	225	312	202	280	180	221
	2	257	355	235	325	222	307	200	276	178	218
	4	247	336	226	307	214	292	193	262	172	208
	6	231	306	211	280	202	269	181	241	161	193
	8	210	268	192	246	186	240	167	214	148	174
	10	186	227	171	208	167	207	149	184	132	152
	12	160	184	147	170	147	172	131	153	116	128
	14	135	145	124	134	126	139	111	123	99	106
	16	110	111	102	103	105	109	93	96	82	84
	18	88	88	81	81	86	86	75	75	66	66
	20	71	71	66	66	70	70	61	61	54	54
	22	59	59	54	54	58	58	51	51	44	44
	24	49	49	46	46	48	48	42	42	37	37
	25	45	45	42	42	45	45	39	39	34	34
	26					41	41	36	36	32	32
Y-Y AXIS	0	261	362	238	331	225	312	202	280	180	221
	6	235	314	211	279	203	272	181	240	158	189
	8	233	310	209	276	195	256	174	227	153	182
	10	227	299	205	268	182	233	163	207	144	168
	12	217	280	196	252	167	207	149	184	132	151
	14	204	258	185	232	151	180	135	159	119	133
	16	190	233	172	210	134	153	119	135	105	115
	18	175	208	158	187	117	127	104	112	92	97
	20	160	184	144	164	101	103	89	91	79	80
	22	145	160	130	143	86	86	75	75	66	66
	24	129	137	116	122	72	72	63	63	56	56
	26	115	117	103	104	61	61	54	54	48	48
	28	101	101	90	90	53	53	47	47	41	41
	30	88	88	78	78	46	46	41	41	36	36
	32	77	77	69	69	41	41	36	36	32	32
	34	69	69	61	61						
	36	61	61	54	54						
	38	55	55	49	49						
	41	47	47	42	42						

Effective length in ft KL with respect to indicated axis

Properties					
A (in.^{2})	8.52	7.78	7.34	6.61	5.89
r_x (in.)	1.50	1.51	1.60	1.58	1.57
r_y (in.)	2.51	2.48	1.96	1.94	1.93

Note: Heavy line indicates Kl/r of 200.

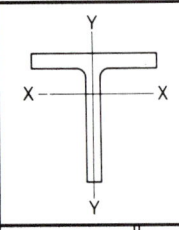

| | | **COLUMNS** Structural tees cut from W shapes Design axial strength in kips ($\phi = 0.85$) | | | | | | | | | | | | F_y = 36 ksi / F_y = 50 ksi |

Designation		WT 6												
Wt./ft		17.5		15		13		11		9.5		8		7	
F_y		36	50	36	50	36	50	36	50	36	50	36	50	36	50
X-X AXIS	0	158	188	120	132	90	92	88	98	68	71	53	54	40	40
	2	157	186	119	131	89	91	88	97	68	70	53	54	40	40
	4	152	179	116	127	87	89	86	95	66	69	52	53	39	39
	6	145	169	111	121	84	86	83	91	64	67	51	51	38	38
	8	135	156	104	113	80	81	78	86	61	63	48	49	37	37
	10	124	140	96	104	74	76	73	80	58	60	46	46	35	35
	12	111	124	87	93	68	69	68	73	54	55	43	43	33	33
	14	98	106	78	82	62	63	61	65	49	50	40	40	31	31
	16	84	89	68	71	55	56	55	58	44	45	36	36	29	29
	18	72	73	59	60	48	49	48	50	40	40	33	33	26	26
	20	59	59	50	50	42	42	42	43	35	35	29	29	24	24
	22	49	49	41	41	36	36	36	36	30	30	26	26	21	21
	24	41	41	35	35	30	30	30	30	26	26	22	22	19	19
	26	35	35	30	30	26	26	26	26	22	22	19	19	17	17
	28	30	30	25	25	22	22	22	22	19	19	16	16	14	14
	29	28	28	24	24	21	21	21	21	18	18	15	15	14	14
	30							19	19	17	17	14	14	13	13
	31							18	18	16	16	13	13	12	12
	32											13	13	11	11
Y-Y AXIS	0	158	188	120	132	90	92	88	98	68	71	53	54	40	40
	2	140	163	104	113	76	78	74	81	55	56	39	39	28	28
	4	138	160	102	111	75	77	69	74	51	53	36	37	26	26
	6	134	154	100	108	74	75	58	62	44	45	32	32	24	24
	8	124	141	94	102	71	72	45	47	35	35	25	25	19	19
	10	112	124	86	92	66	67	32	32	26	26	18	18	14	14
	12	97	105	76	80	59	60	23	23	18	18	13	13	11	11
	13	90	96	71	74	56	57	20	20	16	16				
	14	82	87	65	68	52	53	17	17						
	16	68	69	55	56	45	45								
	18	55	55	45	45	38	38								
	20	45	45	37	37	31	31								
	22	37	37	31	31	26	26								
	24	31	31	26	26	22	22								
	25	29	29	24	24	20	20								
Properties															
A (in.^{2})		5.17		4.40		3.82		3.24		2.79		2.36		2.08	
r_x (in.)		1.76		1.75		1.75		1.90		1.90		1.92		1.92	
r_y (in.)		1.54		1.52		1.51		0.847		0.822		0.773		0.753	

Effective length in ft KL with respect to indicated axis

Note: Heavy line indicates Kl/r of 200.

COLUMNS
Structural tees
cut from W shapes
Design axial strength in kips ($\phi = 0.85$)

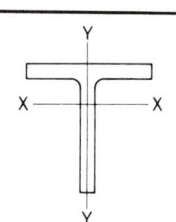

F_y = 36 ksi
F_y = 50 ksi

Designation			WT 5										
Wt./ft		22.5		19.5		16.5		15		13		11	
F_y		36	50	36	50	36	50	36	50	36	50	36	50
X-X AXIS	0	203	282	175	244	148	206	135	188	117	146	99	115
	2	199	274	172	237	146	201	133	184	115	144	98	113
	4	187	253	162	218	137	185	128	173	110	136	94	108
	6	170	220	147	190	125	162	119	157	102	124	87	99
	8	148	182	128	157	109	135	107	136	92	109	79	88
	10	124	142	107	123	92	106	94	114	81	92	69	76
	12	100	105	86	91	75	79	80	91	69	76	59	64
	14	77	77	67	67	58	58	67	70	57	59	49	51
	16	59	59	51	51	45	45	54	54	46	46	40	40
	18	47	47	40	40	35	35	42	42	36	36	32	32
	20	38	38	33	33	29	29	34	34	29	29	26	26
	21					26	26	31	31	27	27	23	23
	22							28	28	24	24	21	21
	24							24	24	20	20	18	18
Y-Y AXIS	0	203	282	175	244	148	206	135	188	117	146	99	115
	2	190	257	161	215	131	174	124	166	104	126	84	95
	4	189	255	160	214	130	172	121	161	102	124	82	93
	6	186	250	158	211	129	170	114	148	97	116	79	88
	8	178	236	153	201	126	164	102	128	87	102	71	79
	10	167	215	143	183	119	151	89	105	76	85	62	67
	12	154	192	132	163	109	135	74	82	63	68	51	54
	14	140	168	119	142	99	117	60	62	51	52	41	42
	16	125	144	106	121	87	99	47	47	40	40	32	32
	18	110	120	93	101	76	82	38	38	32	32	26	26
	20	95	99	80	82	66	67	30	30	26	26	21	21
	22	81	82	68	68	55	55	25	25	21	21	17	17
	24	69	69	57	57	47	47						
	26	59	59	49	49	40	40						
	28	50	50	42	42	34	34						
	30	44	44	37	37	30	30						
	32	39	39	32	32	26	26						
	33	36	36	30	30								

Effective length in ft KL with respect to indicated axis

Properties

	22.5	19.5	16.5	15	13	11
A (in.²)	6.63	5.73	4.85	4.42	3.81	3.24
r_x (in.)	1.24	1.24	1.26	1.45	1.44	1.46
r_y (in.)	2.01	1.98	1.94	1.37	1.36	1.33

Note: Heavy line indicates Kl/r of 200.

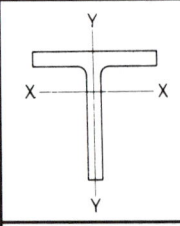

COLUMNS
Structural tees
cut from W shapes
Design axial strength in kips ($\phi = 0.85$)

| F_y = 36 ksi |
| F_y = 50 ksi |

Designation			WT 5						
Wt./ft			9.5		8.5		7.5		6
F_y			36	50	36	50	36	50	36	50
	X-X AXIS	0	86	104	77	90	66	76	43	45
		2	85	103	76	88	65	75	43	44
		4	82	98	73	84	63	72	41	43
		6	77	91	68	79	59	67	39	41
		8	70	81	63	71	54	61	37	38
		10	62	71	56	62	49	54	34	35
		12	54	60	49	53	43	46	30	31
		14	46	49	42	44	37	39	27	27
		16	38	39	34	35	31	31	23	23
		18	30	30	28	28	25	25	19	20
		20	25	25	23	23	20	20	16	16
		22	20	20	19	19	17	17	13	13
		24	17	17	16	16	14	14	11	11
		25	16	16	14	14	13	13	10	10
		26			13	13	12	12	10	10
	Y-Y AXIS	0	86	104	77	90	66	76	43	45
		2	74	87	62	71	50	56	31	32
		4	68	79	58	64	47	51	29	30
		6	57	64	48	52	39	41	25	26
		8	44	46	36	37	29	29	20	20
		10	30	30	25	25	20	20	14	14
		12	21	21	18	18	14	14	10	10
		13	18	18	15	15	12	12	9	9
		14	16	16	13	13				

Effective length in ft KL with respect to indicated axis

Properties					
A (in.²)		2.81	2.50	2.21	1.77
r_x (in.)		1.54	1.56	1.57	1.57
r_y (in.)		0.874	0.844	0.810	0.785

Note: Heavy line indicates Kl/r of 200.

F_y = 36 ksi
F_y = 50 ksi

COLUMNS
Structural tees
cut from W shapes
Design axial strength in kips (ϕ = 0.85)

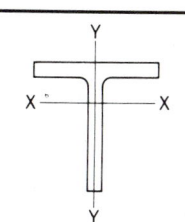

Designation								WT 4						
Wt./ft	14		12		10.5		9		7.5		6.5		5	
F_y	36	50	36	50	36	50	36	50	36	50	36	50	36	50
X-X AXIS														
0	126	175	108	150	94	131	80	112	68	94	59	82	41	46
2	122	168	105	144	92	127	79	108	67	92	58	79	41	45
3	118	160	101	137	89	121	76	104	65	89	56	77	40	44
4	112	148	96	127	86	114	73	98	63	84	54	73	38	42
5	105	135	90	116	81	106	70	91	60	79	52	69	37	40
6	96	121	82	103	76	97	65	83	57	73	49	63	35	38
8	78	90	67	77	64	76	55	67	49	60	43	52	30	33
10	60	62	51	52	51	56	45	50	41	46	36	41	26	27
12	43	43	36	36	39	40	35	35	33	34	29	30	21	21
14	32	32	27	27	29	29	26	26	25	25	22	22	16	16
16	24	24	20	20	22	22	20	20	19	19	17	17	12	12
18					18	18	16	16	15	15	13	13	10	10
19							14	14	14	14	12	12	9	9
20									12	12	11	11	8	8
Y-Y AXIS														
0	126	175	108	150	94	131	80	112	68	94	59	82	41	46
2	117	157	98	130	86	115	70	93	59	78	49	63	32	35
4	116	155	97	129	84	111	69	90	55	70	45	56	30	33
6	112	149	95	125	78	100	65	82	46	54	37	44	26	28
8	104	134	89	114	68	84	57	69	35	37	28	29	20	21
10	94	116	80	99	58	66	48	54	24	24	19	19	14	14
12	83	97	70	83	47	49	38	40	17	17	14	14	10	10
14	71	79	61	67	36	36	29	29	13	13	10	10	8	8
16	60	62	51	53	28	28	23	23						
18	49	49	42	42	22	22	18	18						
20	40	40	34	34	18	18	15	15						
21	36	36	31	31	16	16								
22	33	33	28	28										
24	28	28	24	24										
26	24	24	20	20										
27	22	22												

Effective length in ft KL with respect to indicated axis

Properties							
A (in.^{2})	4.12	3.54	3.08	2.63	2.22	1.92	1.48
r_x (in.)	1.01	0.999	1.12	1.14	1.22	1.23	1.20
r_y (in.)	1.62	1.61	1.26	1.23	0.876	0.843	0.841

Note: Heavy line indicates Kl/r of 200.

Notes

COLUMN BASE PLATES
Design Procedure

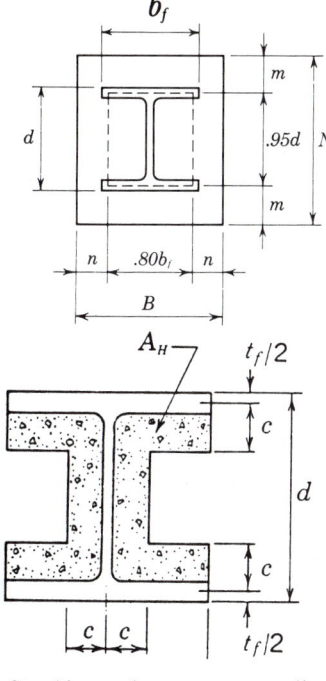

P_u = Total factored load of column, kips
A_1 = Area of base plate, sq. in.
A_2 = Full cross sectional area of concrete support, sq. in.
A_H = Area of H-shaped portion of base plate in light columns, sq. in.
F_y = Specified minimum yield stress of steel, ksi
f_c' = Specified compressive strength of concrete, ksi
t_p = Thickness of base plate, in.
ϕ_c = Resistance factor for concrete = 0.6
ϕ_p = Resistance factor for base plate = 0.9

Steel base plates are generally used under columns for distribution of the column load over a sufficient area of the concrete pier or foundation. The column factored load P_u is assumed to be uniformly distributed over the concrete covered by the base plate.

Unless m and n dimensions are small, the base plate is designed as a cantilever beam, fixed at the edge of a rectangle whose sides are $0.80b_f$ and $0.95d$.

For lightly loaded columns where m and n are small, the load contributary to the area enclosed by the column is assumed to be distributed uniformly over an H-shaped area under the base plate. For this case, the base plate is designed as a cantilever fixed to the web or flange of column. The length c of this cantilever is obtained from the limit state of the bearing strength of concrete within the H-shaped area, where c is measured from the center line of the column web or flange.

Two limit states of strength are considered for base plates. First is the limit state of the bearing strength of concrete supporting the base plate, and second is the limit state of the formation of a plastic hinge in the base plate due to bending.

1. The factored strength of concrete in bearing according to Sect. J9 of the AISC LRFD Specification is $\phi_c P_p$, in kips, where $\phi_c = 0.60$ and P_p is given by:

$$P_p = 0.85f_c'A_1 \tag{1}$$

when the full area of a concrete support is covered by the base plate, and

$$P_p = 0.85f_c'A_1\sqrt{A_2/A_1} \le 1.7f_c'A_1 \tag{2}$$

when less than the full area of a concrete support is covered by the base plate. The general equation of LRFD design, for this case, can be written as:

$$\phi_c P_p \ge P_u \tag{3}$$

The above three equations can be rearranged and written as:

$$\left[\frac{P_u}{0.6(0.85f'_c)}\right]^2 \leq A_1 A_2 \leq 4A_1^2 \tag{4}$$

The first and second terms in Eq. 4 result in:

$$A_1 \geq \frac{1}{A_2}\left[\frac{P_u}{0.6(0.85f'_c)}\right]^2 \tag{5}$$

The first and third terms give minimum base plate area for upper limit of concrete bearing strength:

$$A_1 \geq \frac{P_u}{0.6(1.70f'_c)} \tag{6}$$

Substituting the above equation in Eq. 5, the required pedestal area for this condition is:

$$A_2 \geq \frac{P_u}{0.3f'_c} \tag{7}$$

If conditions permit, the pedestal should be made at least this size for optimum concrete bearing strength.

2. The factored strength of the base plate for the limit state of plastic hinge formation is $\phi_b R_n$, in kips, where $\phi_b = 0.90$ and R_n is the smaller value calculated from the following three equations:

$$R_n = \phi_b \frac{t_p^2}{2m^2} F_y \times N \times B \tag{8}$$

$$R_n = \phi_b \frac{t_p^2}{2n^2} F_y \times N \times B \tag{9}$$

$$R_n = \phi_b \frac{t_p^2}{2c^2} F_y \times N \times B \frac{A_H}{b_f d} \tag{10}$$

Dimensions of base plate are optimized if $m = n$. This condition is approached when $N \approx \sqrt{A_1} + \Delta$, where $\Delta = 0.5(0.95d - 0.80b_f)$ and $B \approx A_1/N$.
For lightly loaded columns where m and n are small, the strength given by Eq. 10 will govern.

Steps in the design of a base plate are:

1. Find A_1 from following three equations and use larger value.

$$A_1 = \frac{1}{A_2}\left[\frac{P_u}{0.6(0.85f'_c)}\right]^2$$

$$A_1 = \frac{P_u}{0.6(1.7f'_c)}$$

$$A_1 = b_f d$$

2. Determine $N \approx \sqrt{A_1} + \Delta > d$ and $B = A_1/N > b_f$.
3. Determine $m = (N - 0.95d)/2$ and $n = (B - 0.80b_f)/2$.

4. Determine load contributary to the area enclosed by the column:

$$P_o = \frac{P_u}{B \times N} b_f d$$

5. Calculate area of H-shaped region,

$$A_H = \frac{P_o}{0.6(0.85 \sqrt{A_2/b_f d} \, f_c')} \geq \frac{P_o}{0.6(1.7f_c')}$$

6. Determine c from:

$$c = \frac{1}{4}\left[d + b_f - t_f - \sqrt{(d + b_f - t_f)^2 - 4(A_H - t_f b_f)} \right]$$

7. Thickness of base plate is the largest t_p calculated from following three equations.

$$t_p = m\sqrt{\frac{2P_u}{0.9F_y BN}}, \quad t_p = n\sqrt{\frac{2P_u}{0.9F_y BN}} \quad \text{or} \quad t_p = c\sqrt{\frac{2P_o}{0.9F_y A_H}}$$

EXAMPLE 10

Column: W12×170   Steel: A36
Dead load = 288 kips   Concrete: $f_c' = 3$ ksi
Live load = 432 kips   Pedestal: 30 in. × 30 in.

Solution:

$P_u = 1.2D + 1.6L = 1.2\,(288) + 1.6\,(432) = 1{,}037$ kips

1. $A_1 = \dfrac{1}{A_2}\left(\dfrac{P_u}{0.6 \times 0.85f_c'}\right)^2 = \dfrac{1}{900}\left(\dfrac{1{,}037}{0.6 \times 0.85 \times 3}\right)^2 = 510.4$ in.^{2} ← governs

   $A_1 = \dfrac{P_u}{0.6 \times 1.70f_c'} = \dfrac{1{,}037}{0.6 \times 1.70 \times 3} = 338.9$ in.^{2}

   $A_1 = d \times b_f = 14.03 \times 12.57 = 176.4$ in.^{2}

2. $\Delta = 0.5\,(0.95d - 0.80b_f) = 0.5\,(0.95 \times 14.03 - 0.80 \times 12.57) = 1.636$ in.

   $N \approx \sqrt{A_1} + \Delta = \sqrt{510.4} + 1.636 = 24.2$ in. $> d$, use $N = 25$ in.

   $B = A_1/N = 510.4/25 = 20.4$ in., use $B = 21$ in.

3. $m = (N - 0.95d)/2 = (25 - 0.95 \times 14.03)/2 = 5.84$ in. ← governs

   $n = (B - 0.80b_f)/2 = (21 - 0.80 \times 12.57)/2 = 5.47$ in.

4. $P_o = \dfrac{P_u}{N \times B} b_f d = \dfrac{1{,}037}{25 \times 21} \times 12.57 \times 14.03 = 348.3$ kips

5. $A_H = \dfrac{P_o}{0.6 \times 0.85 \sqrt{A_2/b_f d} \, f_c'} = \dfrac{348.3}{0.6 \times [0.85 \sqrt{900/(12.57 \times 14.03)} \times 3]}$

   $A_H = 101$ in.$^2 < \dfrac{P_o}{0.6 \times 1.7f_c'} = 114$ in.^{2}

   $A_H = 114$ in.^{2}

6. $c = \dfrac{(d + b_f - t_f) - \sqrt{(d + b_f - t_f)^2 - 4(A_H - t_f b_f)}}{4}$

$c = \dfrac{(14.03 + 12.57 - 1.56) - \sqrt{(25.04)^2 - 4(114 - 1.56 \times 12.57)}}{4} = 2.31$ in.

7. $t_p = m\sqrt{\dfrac{2P_u}{0.9F_y BN}} = 5.84\sqrt{\dfrac{2 \times 1{,}037}{0.9 \times 36 \times 21 \times 25}} = 2.04$ in. ← governs, use 2½

$t_p = c\sqrt{\dfrac{2P_o}{0.9F_y A_H}} = 2.31\sqrt{\dfrac{2 \times 348.3}{0.9 \times 36 \times 114}} = 1.00$ in.

Use: PL $25 \times 2\frac{1}{2} \times 2' - 1''$

EXAMPLE 11

Same as Ex. 7, but Dead load = 240 kips
Live load = 360 kips

Solution:

$P_u = 1.2\,(240) + 1.6\,(360) = 864$ kips

1. $A_1 = \dfrac{1}{900}\left(\dfrac{864}{0.6 \times 0.85 \times 3}\right)^2 = 354.3$ in.^{2} ← governs

$A_1 = \dfrac{P_u}{0.6 \times 1.7f_c'} = \dfrac{864}{0.6 \times 1.7 \times 3} = 282.4$ in.^{2}

$A_1 = d \times b_f = 14.03 \times 12.57 = 176.4$ in.^{2}

2. $N = \sqrt{354.3} + 1.636 = 20.46$ in., use $N = 21$ in.

$B = A_1/N = 354.3/21 = 16.87$ in., use $B = 17$ in.

3. $m = (21 - 0.95 \times 14.03)/2 = 3.84$ ← governs

$n = (17 - 0.80 \times 12.57)/2 = 3.47$

4. $P_o = \dfrac{864}{21 \times 17} \times 12.57 \times 14.03 = 426.8$ kips

5. $A_H = \dfrac{426.8}{0.6 \times 0.85\sqrt{900/(12.57 \times 14.03)} \times 3} = 124$ in.$^2 < \dfrac{426.8}{0.6 \times 1.7f_c'} = 140$ in.^{2}

6. $c = \dfrac{(14.03 + 12.57 - 1.56) - \sqrt{(25.04)^2 - 4(124 - 1.56 \times 12.57)}}{4} = 2.64$ in.

7. $t_p = c\sqrt{\dfrac{2P_o}{0.9F_y A_H}} = 2.64\sqrt{\dfrac{2 \times 426.8}{0.9 \times 36 \times 124}} = 1.22$ in.

$t_p = m\sqrt{\dfrac{2P_u}{0.9F_y BN}} = 3.84\sqrt{\dfrac{2 \times 864}{0.9 \times 36 \times 17 \times 21}} = 1.48$ in. ← governs, use 1½

Use: PL $17 \times 1\frac{1}{2} \times 1' - 9''$

COLUMN BASE PLATES
Finishing

Rolled steel plates are extensively used for column bases. So that they may function properly in transmitting loads to masonry supports, finishing is regulated by specification.

In AISC LRFD Specification, Sect. M2, it is stated:

"Column bases and base plates shall be finished in accordance with the following requirements:

a. Steel bearing plates 2 in. or less in thickness may be used without milling, provided a satisfactory contact bearing is obtained. Steel bearing plates over 2 in. but not over 4 in. in thickness may be straightened by pressing or, if presses are not available, by milling for all bearing surfaces (except as noted in subparagraphs b and c of this section), to obtain a satisfactory contact bearing. Steel bearing plates over 4 in. in thickness shall be milled for all bearing surfaces (except as noted in subparagraphs b and c of this section).

b. Bottom surfaces of bearing plates and column bases which are grouted to insure full bearing contact on foundations need not be milled.

c. Top surfaces of bearing plates need not be milled when full-penetration welds are provided between the column and the bearing plate."

PART 3
Beam and Girder Design

AMERICAN INSTITUTE OF STEEL CONSTRUCTION

Notes

DESIGN STRENGTH OF BEAMS

General

The required strength of rolled I and C sections shall be determined by either plastic or elastic analysis for the appropriate factored load combinations given in Sect. A4 of the LRFD Specification.

Beams shall be proportioned so that no applicable strength limit state is exceeded when subjected to factored load combinations and that no serviceability limit state is exceeded when subjected to service loads. Strength limit states for beams include local buckling and lateral torsional buckling. Serviceability limit states may include, but are not limited to, deflection and vibration.

The flexural design strength for beams must equal or exceed the required strength based on the factored nominal (service) loads. The design strength $\phi_b M_n$ for each applicable limit state shall equal the maximum moment M_u as determined from applicable factored load combinations. Values of $\phi_b M_n$ are tabulated in the pages to follow. These values are based on beam behavior as shown in Fig. 3.1 and explained in the following discussion.

It should be noted that the LRFD Specification expresses values for moments and lengths in kip-in. and in. In other sections of the LRFD Manual, these values are tabulated in kip-ft and ft.

Plastic Analysis

The design flexural strength for plastic analysis is:

$$\phi_b M_n = \phi_b M_p$$

where

$$\phi_b = 0.90$$
$$M_p = Z_x F_y / 12, \text{ kip-ft}$$

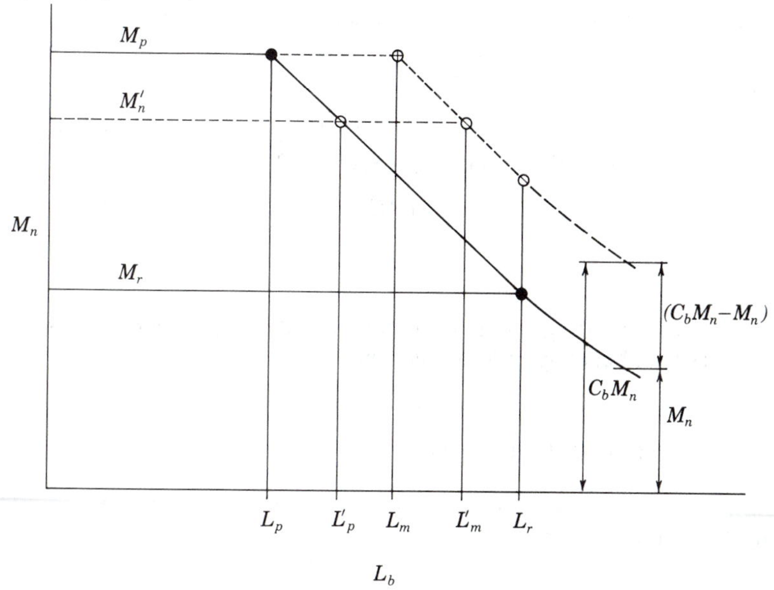

Figure 3.1

The yield strength of material that may be used with plastic analysis is limited to 65 ksi. Plastic analysis is limited to compact I and C shapes as defined in Table B5.1 of the LRFD Specification as:

$\lambda_p = b_f/2t_f \leq 65/\sqrt{F_y}$ for the flanges of I shapes in flexure

$\lambda_p = b_f/t_f \leq 65/\sqrt{F_y}$ for the flanges of C shapes in flexure

and

$\lambda_p = h_c/t_w \leq 640/\sqrt{F_y}$ for beam webs in flexural compression

where

λ_p = limiting slenderness parameter for compact element
b_f = width of flange for I and C shapes, in.
t_f = flange thickness, in.
h_c = twice the distance from the neutral axis to the inside face of the compression flange less the fillet or corner radius, in.
t_w = beam web thickness, in.

In addition, LRFD Specification Sect. F1.1 states: for a section bent about the major axis, the laterally unbraced length of the compression flange at plastic hinge locations associated with the failure mechanism shall not exceed:

$$L_{pd} = \frac{3,600 + 2,200 \ (M_1/M_p)}{F_y} r_y \qquad \text{(F1-1)}$$

where

F_y = specified yield strength of compression flange, ksi
M_1 = smaller moment at end of unbraced length of beam, kip-in.
M_p = plastic moment, kip-in.
r_y = radius of gyration about minor axis, in.
(M_1/M_p) is positive when moment causes reverse curvature

Elastic Analysis

The flexural design strength of rolled I and C shape beams designed using elastic analysis, according to LRFD Specification Sect. F1.2, is:

$$\phi_b M_n$$

where

ϕ_b = 0.90
M_n = nominal resisting moment as determined by the limit state of lateral-torsional buckling or local buckling

Flexural Design Strength for C_b = 1.0

Compact Sections (C_b = 1.0)

$L_b \leq L_p$

The flexural design strength of compact (flange and web local buckling $\lambda \leq \lambda_p$) I or

C rolled shapes (as defined in Sect. B5 of the LRFD Specification) bent about the major or minor axis is:

$$\phi_b M_n = \phi_b M_p = \phi_b Z_x F_y / 12$$

where the distance L_b between points braced against lateral movement of the compression flange or between points braced to prevent twist of the cross section does not exceed the value L_p (see Fig. 3.1).

$$L_p = \frac{300\, r_y}{\sqrt{F_{yf}}} \tag{F1-4}$$

For cases where $C_b > 1.0$, see later discussion.

$L_p < L_b \le L_r$

The flexural design strength of compact I or C rolled shapes bent about the major axis, from Sect. F1.3, is:

$$\phi_b M_n = \phi_b M_p - \phi_b (M_p - M_r)\left(\frac{L_b - L_p}{L_r - L_p}\right) \le \phi_b M_p$$

where the limiting length L_r and the corresponding buckling moment M_r (see Fig. 3.1) are determined as follows:

$$L_r = \frac{r_y X_1}{(F_{yw} - F_r)}\sqrt{1 + \sqrt{1 + X_2 (F_{yw} - F_r)^2}} \tag{F1-6}$$

where

$$X_1 = \frac{\pi}{S_x}\sqrt{\frac{EGJA}{2}} \tag{F1-8}$$

$$X_2 = \frac{4C_w}{I_y}\left(\frac{S_x}{GJ}\right)^2 \tag{F1-9}$$

$\phi_b M_r = \phi_b S_x (F_y - F_r)/12$, kip-ft

S_x = section modulus about major axis, in.[3]
E = modulus of elasticity of steel, 29,000 ksi
G = shear modulus of steel, 11,200 ksi
J = torsional constant, in.[4]
A = cross-sectional area of beam, in.[2]
C_w = warping constant, in.[6]
F_r = compressive residual stress in flange: for rolled shapes F_r = 10 ksi; for welded shapes F_r = 16.5 ksi

Values of J and C_w are tabulated for some shapes in Part 1 of the LRFD Manual. For values not shown, see *Torsional Analysis of Steel Members*.*

---

*AISC, Chicago, Ill., 1983.

Compact and Noncompact Sections (C_b = 1.0)

$L_b > L_r$

According to LRFD Specification Sect. F1.4, the flexural design strength of compact and noncompact I or C rolled shapes bent about the major axis is:

$$\phi_b M_n = \phi_b M_{cr} = \phi_b \left(\frac{\pi}{L_b}\right)\sqrt{EI_y GJ + \left(\frac{\pi E}{L_b}\right)^2 I_y C_w}$$

$$= \phi_b \left(\frac{S_x X_1 \sqrt{2}}{(L_b/r_y)}\right)\sqrt{1 + \frac{X_1^2 X_2}{2(L_b/r_y)^2}} \leq \phi_b M_r$$

Noncompact Sections (C_b = 1.0)

$L_b \leq L_p'$

The flexural design strength $\phi_b M_n'$ (see Fig. 3.1) for noncompact (flange or web local buckling $\lambda_p < \lambda \leq \lambda_r$) I and C rolled shapes bent about the major or minor axis is the smaller value for either local flange buckling or local web buckling as determined by:

$$\phi_b M_n' = \phi_b M_p - \phi_b (M_p - M_r)\left(\frac{\lambda - \lambda_p}{\lambda_r - \lambda_p}\right)$$

For local flange buckling:

$\lambda = b_f/2t_f$ for I-shaped members
$\lambda = b_f/t_f$ for C-shaped members
$\lambda_p = 65/\sqrt{F_y}$
$\lambda_r = 141/\sqrt{F_y - 10}$

For local web buckling:

$\lambda = h_c/t_w$
$\lambda_p = 640/\sqrt{F_y}$
$\lambda_r = 970/\sqrt{F_y}$

where

F_y = specified minimum yield strength, ksi
b_f = flange width, in.
h_c = twice the distance from the neutral axis to the inside face of the compression flange less the fillet or corner radius, in.
t_f = flange thickness, in.
t_w = web thickness, in.

$$L_p' = L_p + (L_r - L_p)\left(\frac{M_p - M_n'}{M_p - M_r}\right)$$

Sections with a width-to-thickness ratio exceeding the specified values for λ_r are slender shapes and must be analyzed using LRFD Specification Appendix B5.3.

$L'_p < L_b \le L_r$

The flexural design strength of noncompact I or C rolled shapes bent about the major axis is determined by:

$$\phi_b M_n = \phi_p M_p - \phi_b (M_p - M_r)\left(\frac{L_b - L_p}{L_r - L_p}\right) \le \phi_b M'_n$$

In the Load Factor Design Selection Table, in the case of noncompact shapes, the values of $\phi_b M'_n$ and L'_p are tabulated as $\phi_b M_p$ and L_p. The formula above may be used with the tabulated values.

Flexural Design Strength for $C_b > 1.0$

C_b is a factor which varies with the moment gradient between bracing points (L_b). For C_b greater than 1.0, the design flexural strength is equal to the tabulated value of the design flexural strength with $C_b = 1.0$ multiplied by the calculated C_b value. The maximum value is $\phi_b M_p$ for compact shapes or $\phi_b M'_n$ for noncompact shapes. The maximum unbraced lengths associated with the maximum flexural design strengths $\phi_b M_p$ and $\phi_b M'_n$ are L_m and L'_m (see Fig. 3.1).

Values for C_b for some typical loading conditions are given in Fig. 3.2.

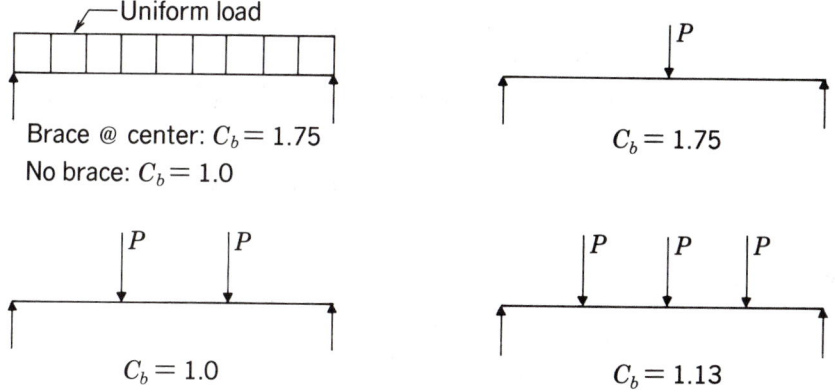

Brace @ center: $C_b = 1.75$
No brace: $C_b = 1.0$

$C_b = 1.75$

$C_b = 1.0$

$C_b = 1.13$

Notes: 1. Loads are equally spaced.
2. Beams are braced at point loads.

Figure 3.2

Compact Sections ($C_b > 1.0$)

$L_b \le L_m$

The flexural design strength for rolled I and C shapes is:

$$\phi_b M_n = \phi_b M_p$$

$L_b > L_m$

The flexural design strength is:

$$\phi_b M_n = C_b\left[\phi_b M_n (\text{for } C_b = 1.0)\right] \le \phi_b M_p$$

AMERICAN INSTITUTE OF STEEL CONSTRUCTION

For $L_m \leq L_r$

$$L_m = L_p + \frac{(C_b M_p - M_p)(L_r - L_p)}{C_b(M_p - M_r)}$$

For $L_m > L_r$

$$L_m = \frac{C_b \pi}{M_p} \sqrt{\frac{EI_y GJ}{2}} \sqrt{1 + \sqrt{1 + \frac{4C_w M_p^2}{I_y C_b^2 G^2 J^2}}}$$

The value of C_b for which L_m or L_m' equals L_r for any rolled shape is:

$$C_b = \frac{F_y Z_x}{(F_y - 10)S_x}$$

Noncompact Sections ($C_b > 1.0$)

$L_b \leq L_m'$

The flexural design strength for rolled I and C shapes is:

$$\phi_b M_n = \phi_b M_n' < \phi_b M_p$$

$L_b > L_m'$

The flexural design strength is:

$$\phi_b M_n = C_b \left[\phi_b M_n \text{ (for } C_b = 1.0) \right] \leq \phi_b M_n'$$

For $L_m' \leq L_r$

$$L_m' = L_p' + \frac{(C_b M_n' - M_n')(L_r - L_p)}{C_b(M_p - M_r)}$$

For $L_m' > L_r$

$$L_m' = \frac{C_b \pi}{M_p} \sqrt{\frac{EI_y GJ}{2}} \sqrt{1 + \sqrt{1 + \frac{4C_w M_p^2}{I_y C_b^2 G^2 J^2}}}$$

LOAD FACTOR DESIGN SELECTION TABLE Z_x
For shapes used as beams

This table facilitates the selection of flexural members designed on the basis of ultimate flexural strength in accordance with Sect. F of the LRFD Specification. It includes only W and M shapes designed as beams. A laterally supported beam can be selected by entering the table with either the required plastic section modulus or with the factored design bending moment, and comparing these with the tabulated values of Z_x or $\phi_b M_p$ respectively.

The table is applicable to adequately braced beams with unbraced lengths not exceeding L_r, i.e. $L_b \leq L_r$. For beams with unbraced lengths much greater than L_p, it may be convenient to use the unbraced beam charts. For most loading conditions, it is convenient to use this selection table. However, for adequately braced, simply supported beams with a uniform load over the entire length, or equivalent symmetrical loading, the tables of Uniform Load Constants can also be used.

In this table, shapes are listed in groups by descending order of plastic section modulus Z_x. Included also for steel of $F_y = 36$ ksi and 50 ksi are values for the maximum flexural design strength $\phi_b M_p$; the limiting buckling moment $\phi_b M_r$; the limiting laterally unbraced compression flange length for full plastic moment capacity and uniform moment ($C_b = 1.0$), L_p; limiting laterally unbraced length for elastic lateral-torsional buckling L_r; and BF, a factor that can be used to calculate the resisting moment $\phi_b M_n$ for beams with unbraced lengths between the limiting bracing lengths L_p and L_r.

For noncompact shapes, as determined by Sect. B5 of the LRFD Specification, the maximum flexural design strength $\phi_b M_{n,max}$ as determined by LRFD Specification Formula A-F1-3 is tabulated as $\phi_b M_p$. The associated maximum unbraced length is tabulated as L_p (see previous discussion under Design Strength of Beams for further explanation).

The symbols used in this table are:

Z_x = plastic section modulus, X-X axis, in.^{3}

$\phi_b M_p$ = maximum design resisting moment, kip-ft

$= \phi_b Z_x F_y / 12$ if shape is compact

$= \phi_b M'_n = \phi_b M_p - \phi_b (M_p - M_r)\left(\dfrac{\lambda - \lambda_p}{\lambda_r - \lambda_p}\right)$ if shape is noncompact

$\phi_b M_r$ = limiting design buckling moment, kip-ft

$= \phi_b S_x (F_y - F_r)/12$

where

F_r = 10 ksi for rolled shapes

= 16.5 ksi for welded shapes

L_p = limiting laterally unbraced length for inelastic LTB, ft
uniform moment case ($C_b = 1$).

L_r = limiting laterally unbraced length for elastic lateral-torsional buckling, ft

BF = a factor that can be used to calculate the design moment capacity for unbraced lengths L_b, between L_p and L_r, kip-ft

$= \dfrac{\phi_b (M_p - M_r)}{L_r - L_p}$

where $\phi_b M_n = C_b \left[\phi_b M_p - BF(L_b - L_p) \right] \leq \phi_b M_p$

Use of the Table

Determine the required plastic section modulus Z_x from the maximum required factored load moment M_u (kip-ft) using the desired steel yield strength.

$$Z_x = \frac{12M_u}{\phi_b F_y}$$

Enter the column headed Z_x and find a value equal to or greater than the plastic section modulus required. Alternately, enter the $\phi_b M_p$ column and find a value of $\phi_b M_p$ equal to or greater than the required factored load moment capacity. The beam opposite these values (Z_x or $\phi_b M_p$) in the shape column, and all beams above it, have sufficient flexural strength based only on these parameters. The first beam appearing in boldface type adjacent to or above the required Z_x or $\phi_b M_p$ is the lightest section that will serve for the steel yield used in the calculations. If the beam must not exceed a certain depth, proceed up the column headed "Shape" until a beam within the required depth is reached.

After a shape has been selected, the following checks should be made. If the lateral bracing of the compressive flange exceeds L_p, but is less than L_r, the design resisting moment may be calculated as follows:

$$\phi_b M_n = C_b \left[\phi_b M_p - BF(L_b - L_p) \right] \leq \phi_b M_p$$

If the bracing length L_b is substantially greater than L_p, i.e. $L_b > L_r$, it is recommended the unbraced beam charts be used. A check should be made of the beam web shear strength by referring to the Uniform Load Constant tables or by use of the formula:

$$\phi_v V_n = \phi_v 0.6\, F_{yw} A_w \text{ (from LRFD Specification Sect. F2)}$$

where

$$\phi_v = 0.90$$

If a deflection limitation exists also, the adequacy of the selected beam should be checked.

Example 1

Given:

Select a beam of $F_y = 36$ ksi steel subjected to a factored uniform bending moment of 184 kip-ft, having its compression flange braced at 6.0 ft intervals. ($C_b = 1.0$)

Solution (Z_x method):

$$Z_x \text{ (req'd)} = \frac{M_u(12)}{\phi_b F_y} = \frac{184\,(12)}{0.9\,(36)} = 68.15 \text{ in.}^3$$

Enter the Load Factor Design Selection Table and find the nearest higher tabulated value of Z_x is 69.6 in., which corresponds to a W14×43. This beam, however, is not in boldface type. Proceed up the shape column and locate the first beam in boldface, W16×40. Note the values tabulated for $\phi_b M_p$ and L_p are 197 kip-ft and 6.5 ft, respectively.

Use: W16 × 40

Alternatively, proceed up the shape column and select a W18 × 40. The tabulated values for $\phi_b M_p$ and L_p are 212 kip-ft and 5.3 ft, respectively. Since the bracing length L_b is larger than L_p and smaller than L_r, the maximum resisting moment may be calculated as follows:

$$\phi_b M_n = C_b \left[\phi_b M_p - BF(L_b - L_p) \right]$$
$$= 1.0 \left[212 - (7.52)(6 - 5.3) \right]$$
$$= 207 \text{ kip-ft}$$

A W18 × 40 is satisfactory.

Alternate Solution (M_p method):

Enter the column of $\phi_b M_p$ values and note the tabulated value nearest and higher than the required factored design moment (M_u) is 188 kip-ft, which corresponds to a W14 × 43. Scanning the $\phi_b M_p$ values for shapes listed higher in the column, a W16 × 40 is found to be the lightest suitable shape with $L_b < L_p$.

Use: W16 × 40

Example 2

Given:

Determine the moment capacity of a W16 × 40 of $F_y = 36$ ksi and $F_y = 50$ ksi steel with the compression flange braced at intervals of 9.0 ft. ($C_b = 1.0$)

Solution:

Enter the Load Factor Design Table and note that for a W16 × 40, $F_y = 36$ ksi:

$\phi_b M_p = 197$ kip-ft
$L_p \quad = 6.5$ ft
$L_r \quad = 19.3$ ft
$BF \quad = 5.53$ kips
$\phi_b M_n = C_b \left[\phi_b M_p - BF(L_b - L_p) \right]$
$\quad = 1.0 \left[197 - 5.53(9 - 6.5) \right]$
$\quad = 197 - 13.6$
$\quad = 183$ kip-ft

Enter the Load Factor Design Selection Table and note that for a W16 × 40, $F_y = 50$ ksi:

$\phi_b M_p = 273$ kip-ft
$L_p \quad = 5.6$ ft
$L_r \quad = 14.7$ ft
$BF \quad = 8.66$ kips
$\phi_b M_n = C_b \left[\phi_b M_p - BF(L_b - L_p) \right]$
$\quad = 1.0 \left[273 - 8.66(9 - 5.6) \right]$
$\quad = 273 - 29.9$
$\quad = 243$ kip-ft

Example 3

Given:

Select a beam of $F_y = 50$ ksi steel subjected to a factored uniform bending moment of 30 kip-ft having its compression flange braced at 4.0 ft intervals and a depth of 8 in. or less. $(C_b = 1.0)$

Solution (Z_x method):

Assume shape is compact and $L_b \leq L_p$.

$$Z_x \text{ (req'd)} = \frac{12M_u}{\phi_b F_y} = \frac{12\,(30)}{0.9\,(50)} = 8.0 \text{ in.}^3$$

Enter the Load Factor Design Selection Table and note that for a W8 × 10, $F_y = 50$ ksi, the shape is noncompact, however, the maximum resisting moment $\phi_b M_n$ listed in the $\phi_b M_p$ column is adequate. Further note:

$\phi_b M_n = 33.0$ kip-ft
$L_p \quad = 3.1$ ft
$L_r \quad = 7.8$ ft
$BF \quad = 2.02$ kips

Since $L_p < L_b \leq L_r$

$$\begin{aligned}
\phi_b M_n &= C_b \left[\phi_b M_n - BF(L_b - L_p) \right] \\
&= 1.0\,[33.0 - 2.02\,(4.0 - 3.1)] \\
&= 33.0 - 1.78 \\
&= 31.2 \text{ kip-ft}
\end{aligned}$$

Use: W8 × 10

Alternate Solution (M_p method):

Enter the Selection Table and note that in the column of $\phi_b M_p$ values for a W8 × 10, $F_y = 50$ ksi, the value of $\phi_b M_p$ is 33.0 kip-ft, which is adequate. Also note, however, $L_p = 3.1$ ft is less than the bracing interval $L_b = 4.0$ ft, and that BF is equal to 2.02 kips. Therefore:

$$\begin{aligned}
\phi_b M_n &= 1.0\,[33.0 - 2.02\,(4 - 3.1)] \\
&= 31.2 \text{ kip-ft}
\end{aligned}$$

Use: W8 × 10

LOAD FACTOR DESIGN SELECTION TABLE
For shapes used as beams
$\phi_b = 0.90$

Z_x

	$F_y = 36$ ksi				Z_x	Shape	$F_y = 50$ ksi				
BF	L_r	L_p	$\phi_b M_r$	$\phi_b M_p$			$\phi_b M_p$	$\phi_b M_r$	L_p	L_r	BF
Kips	Ft	Ft	Kip-ft	Kip-ft	In.^{3}		Kip-ft	Kip-ft	Ft	Ft	Kips
32.5	62.4	16.1	2570	4080	1510	W 36x359	5660	3960	13.7	44.0	56.2
29.4	64.4	15.6	2400	3830	1420	W 33x354	5330	3690	13.2	44.7	51.9
32.2	58.5	16.0	2360	3730	1380	W 36x328	5180	3630	13.6	41.7	54.9
28.9	59.3	15.5	2160	3430	1270	W 33x318	4760	3330	13.1	41.8	49.9
31.6	55.1	16.0	2170	3400	1260	W 36x300	4730	3330	13.5	39.9	52.9
31.0	53.0	15.9	2010	3160	1170	W 36x280	4390	3090	13.5	38.8	51.3
28.2	55.7	15.4	1970	3110	1150	W 33x291	4310	3030	13.0	39.8	48.0
30.3	50.6	15.8	1860	2920	1080	W 36x260	4050	2860	13.4	37.5	49.5
37.0	39.7	11.0	1750	2810	1040	W 36x256	3900	2685	9.4	28.8	62.7
27.7	52.0	15.3	1790	2810	1040	W 33x263	3900	2750	12.9	37.8	46.3
29.6	48.8	15.6	1750	2730	1010	W 36x245	3790	2690	13.3	36.4	47.7
28.7	47.3	15.5	1630	2550	943	W 36x230	3540	2510	13.2	35.6	45.8
27.0	49.2	15.1	1620	2540	939	W 33x241	3520	2490	12.8	36.2	44.2
36.1	37.2	10.9	1580	2530	936	W 36x232	3510	2430	9.3	27.3	59.9
26.0	46.9	15.0	1480	2310	855	W 33x221	3210	2270	12.7	35.0	41.9
24.4	48.6	14.7	1450	2280	845	W 30x235	3170	2240	12.4	33.9	43.5
34.9	35.0	10.8	1400	2250	833	W 36x210	3120	2160	9.1	26.1	56.8
25.0	44.8	14.8	1330	2080	772	W 33x201	2900	2050	12.6	33.8	39.7
34.0	33.5	10.7	1290	2070	767	W 36x194	2880	1990	9.1	25.3	54.6
21.8	47.9	14.5	1290	2020	749	W 30x211	2810	1990	12.3	35.1	36.0
32.7	32.8	10.6	1210	1940	718	W 36x182	2690	1870	9.0	24.9	52.0
18.2	52.0	13.8	1220	1910	708	W 27x217	2660	1870	11.7	36.8	31.3
21.0	45.4	14.4	1170	1820	673	W 30x191	2520	1790	12.2	33.7	33.9
31.5	31.9	10.5	1130	1800	668	W 36x170	2510	1740	8.9	24.4	49.6
28.3	32.6	10.4	1070	1700	629	W 33x169	2360	1650	8.8	24.5	45.4
17.8	48.0	13.7	1080	1700	628	W 27x194	2360	1670	11.6	34.6	30.0
30.7	30.9	10.4	1060	1680	624	W 36x160	2340	1630	8.8	23.7	48.0
20.2	43.2	14.3	1050	1630	605	W 30x173	2270	1620	12.1	32.5	32.0
29.4	30.2	10.3	983	1570	581	W 36x150	2180	1510	8.7	23.4	45.6
17.5	45.2	13.6	979	1530	567	W 27x128	2130	1510	11.5	33.1	28.8
26.8	31.2	10.3	950	1510	559	W 33x152	2100	1460	8.7	23.7	42.3

Z_x LOAD FACTOR DESIGN SELECTION TABLE
For shapes used as beams
$\phi_b = 0.90$

	$F_y = 36$ ksi				Z_x	Shape	$F_y = 50$ ksi				
BF	L_r	L_p	$\phi_b M_r$	$\phi_b M_p$			$\phi_b M_p$	$\phi_b M_r$	L_p	L_r	BF
Kips	Ft	Ft	Kip-ft	Kip-ft	In.^{3}		Kip-ft	Kip-ft	Ft	Ft	Kips
25.7	30.1	10.1	874	1390	514	W 33x141	1930	1340	8.6	23.1	40.2
16.9	42.8	13.5	887	1380	512	W 27x161	1920	1370	11.5	31.7	27.4
14.3	47.8	12.7	878	1380	511	W 24x176	1920	1350	10.7	33.8	24.6
27.5	28.8	9.9	856	1370	509	W 36x135	1910	1320	8.4	22.4	42.2
23.7	30.6	9.5	850	1350	500	W 30x148	1880	1310	8.1	22.8	38.6
14.1	45.2	12.7	807	1260	468	W 24x162	1760	1240	10.8	32.4	23.8
24.5	29.1	10.0	792	1260	467	W 33x130	1750	1220	8.5	22.5	37.9
16.2	40.7	13.4	801	1240	461	W 27x146	1730	1230	11.3	30.6	25.8
22.4	29.0	9.4	741	1180	437	W 30x132	1640	1140	8.0	21.9	35.6
10.8	51.7	12.4	741	1170	432	W 21x166	1620	1140	10.5	35.7	19.1
13.8	42.0	12.5	723	1130	418	W 24x146	1570	1110	10.6	30.6	22.8
23.1	27.8	9.7	700	1120	415	W 33x118	1560	1080	8.2	21.7	35.5
21.6	28.2	9.3	692	1100	408	W 30x124	1530	1070	7.9	21.5	34.1
18.9	30.0	9.2	673	1070	395	W 27x129	1480	1040	7.8	22.3	30.9
21.1	27.1	9.1	642	1020	378	W 30x116	1420	987	7.7	20.8	33.0
10.7	46.4	12.3	642	1010	373	W 21x147	1400	987	10.4	32.8	18.4
13.3	39.3	12.4	642	999	370	W 24x131	1390	987	10.5	29.1	21.5
20.2	26.3	9.0	583	934	346	W 30x108	1300	897	7.6	20.3	31.5
18.0	28.2	9.1	583	926	343	W 27x114	1290	897	7.7	21.3	28.7
10.5	43.1	12.2	575	899	333	W 21x132	1250	885	10.4	30.9	17.7
12.7	37.1	12.3	567	883	327	W 24x117	1230	873	10.4	27.9	20.2
7.82	52.2	11.3	550	869	322	W 18x143	1210	846	9.6	35.5	14.0
19.0	25.5	8.8	525	842	312	W 30x 99	1170	807	7.4	19.8	29.2
10.3	41.0	12.2	532	829	307	W 21x122	1150	819	10.3	29.8	17.1
17.0	26.8	9.0	521	824	305	W 27x102	1140	801	7.6	20.4	26.7
12.0	35.2	12.1	503	780	289	W 24x104	1080	774	10.3	26.8	18.8
7.78	48.0	11.3	499	786	291	W 18x130	1090	768	9.5	33.0	13.8
14.8	27.1	8.3	478	756	280	W 24x103	1050	735	7.0	20.1	24.1
10.1	38.7	12.1	486	753	279	W 21x111	1050	747	10.3	28.5	16.4
16.2	25.9	8.8	474	751	278	W 27x 94	1040	729	7.5	19.9	25.2
7.71	44.1	11.2	450	705	261	W 18x119	979	693	9.5	30.8	13.4
14.3	25.9	8.3	433	686	254	W 24x 94	953	666	7.0	19.4	23.0
9.62	37.1	12.0	443	683	253	W 21x101	949	681	10.2	27.6	15.4
15.0	24.9	8.6	415	659	244	W 27x 84	915	639	7.3	19.3	23.0
7.61	40.4	11.1	398	621	230	W 18x106	863	612	9.4	28.7	13.0
13.6	24.5	8.1	382	605	224	W 24x 84	840	588	6.9	18.6	21.5
11.8	26.6	7.7	374	597	221	W 21x 93	829	576	6.5	19.4	19.6
3.86	67.9	15.6	371	572	212	W 14x120	795	570	13.2	46.2	6.82
7.51	38.1	11.0	367	570	211	W 18x 97	791	564	9.4	27.4	12.6

LOAD FACTOR DESIGN SELECTION TABLE
For shapes used as beams
$\phi_b = 0.90$

Z_x

\multicolumn{5}{c}{$F_y = 36$ ksi}	Z_x	Shape	\multicolumn{5}{c}{$F_y = 50$ ksi}
BF	L_r	L_p	$\phi_b M_r$	$\phi_b M_p$			$\phi_b M_p$	$\phi_b M_r$	L_p	L_r	BF
Kips	Ft	Ft	Kip-ft	Kip-ft	In.^{3}		Kip-ft	Kip-ft	Ft	Ft	Kips
12.7	23.4	8.0	343	540	200	W 24x 76	750	528	6.8	18.0	19.8
6.08	42.1	10.5	341	535	198	W 16x100	743	525	8.9	29.3	10.7
11.3	24.9	7.6	333	529	196	W 21x 83	735	513	6.5	18.5	18.5
3.84	62.7	15.5	337	518	192	W 14x109	720	519	13.2	43.2	6.70
7.28	35.5	11.0	324	502	186	W 18x 86	698	498	9.3	26.1	11.9
2.95	75.5	13.0	318	502	186	W 12x120	698	489	11.1	50.0	5.36
12.1	22.4	7.8	300	478	177	W 24x 68	664	462	6.6	17.4	18.7
6.05	38.6	10.4	302	473	175	W 16x 89	656	465	8.8	27.3	10.3
3.77	58.2	15.5	306	467	173	W 14x 99†	647	471	13.4	40.6	6.46
10.7	23.5	7.5	294	464	172	W 21x 73	645	453	6.4	17.7	17.0
2.95	67.2	13.0	283	443	164	W 12x106	615	435	11.0	44.9	5.32
6.95	33.3	10.9	285	440	163	W 18x 76	611	438	9.2	24.8	11.1
10.4	22.8	7.5	273	432	160	W 21x 68	600	420	6.4	17.3	16.5
3.74	54.1	15.4	279	424	157	W 14x 90†	577	429	15.0	38.4	6.31
13.8	17.2	5.8	255	413	153	W 24x 62	574	393	4.9	13.3	21.4
5.85	34.9	10.3	261	405	150	W 16x 77	563	402	8.7	25.2	9.75
2.91	61.4	12.9	255	397	147	W 12x 96	551	393	10.9	41.3	5.20
8.29	24.4	7.1	248	392	145	W 18x 71	544	381	6.0	17.8	13.8
9.80	21.7	7.4	248	389	144	W 21x 62	540	381	6.3	16.6	15.3
4.15	43.0	10.3	240	375	139	W 14x 82	521	369	8.8	29.6	7.31
12.7	16.6	5.6	222	362	134	W 24x 55	503	342	4.7	12.9	19.6
8.09	23.2	7.0	228	359	133	W 18x 65	499	351	6.0	17.1	13.3
2.90	56.4	12.8	230	356	132	W 12x 87	495	354	10.9	38.4	5.12
5.57	32.3	10.3	228	351	130	W 16x 67	488	351	8.7	23.8	9.02
11.3	17.3	5.6	216	348	129	W 21x 57	484	333	4.8	13.1	18.0
4.10	40.0	10.3	218	340	126	W 14x 74	472	336	8.8	28.0	7.12
7.90	22.4	7.0	211	332	123	W 18x 60	461	324	6.0	16.7	12.8
2.88	51.8	12.7	209	321	119	W 12x 79	446	321	10.8	35.7	5.03
4.05	37.3	10.3	201	311	115	W 14x 68	431	309	8.7	26.4	6.91
7.63	21.4	7.0	192	302	112	W 18x 55	420	295	5.9	16.1	12.2
10.5	16.2	5.4	184	297	110	W 21x 50	413	283	4.6	12.5	16.4
2.86	48.2	12.7	190	292	108	W 12x 72	405	292	10.7	33.6	4.93
6.42	22.8	6.7	180	284	105	W 16x 57	394	277	5.7	16.6	10.7
3.90	34.7	10.2	180	275	102	W 14x 61	383	277	8.7	24.9	6.51
7.29	20.5	6.9	173	273	101	W 18x 50	379	267	5.8	15.6	11.5
2.80	44.7	12.6	171	261	96.8	W 12x 65†	358	264	11.8	31.7	4.72
9.66	15.4	5.3	159	258	95.4	W 21x 44	358	245	4.5	12.0	14.9
6.20	21.2	6.6	158	248	92.0	W 16x 50	345	243	5.6	15.8	10.1
8.12	16.6	5.4	154	245	90.7	W 18x 46	340	236	4.6	12.6	13.0
4.17	28.0	8.0	152	235	87.1	W 14x 53	327	233	6.8	20.1	7.02

†Indicates noncompact shape; $F_y = 50$ ksi

Z_x LOAD FACTOR DESIGN SELECTION TABLE
For shapes used as beams
$\phi_b = 0.90$

$F_y = 36$ ksi					Z_x	Shape	$F_y = 50$ ksi				
BF	L_r	L_p	$\phi_b M_r$	$\phi_b M_p$			$\phi_b M_p$	$\phi_b M_r$	L_p	L_r	BF
Kips	Ft	Ft	Kip-ft	Kip-ft	In.^{3}		Kip-ft	Kip-ft	Ft	Ft	Kips
2.91	38.4	10.5	152	233	86.4	W 12x58	324	234	8.9	27.0	4.96
5.92	20.1	6.5	142	222	82.3	W 16x45	309	218	5.6	15.1	9.43
7.52	15.7	5.3	133	212	78.4	W 18x40	294	205	4.5	12.1	11.7
4.05	26.3	8.0	137	212	78.4	W 14x48	294	211	6.8	19.2	6.70
2.85	35.8	10.3	138	210	77.9	W 12x53	292	212	8.8	25.6	4.77
1.91	48.1	10.7	130	201	74.6	W 10x60	280	200	9.1	32.6	3.38
5.53	19.3	6.5	126	197	72.9	W 16x40	273	194	5.6	14.7	8.67
3.06	30.8	8.2	126	195	72.4	W 12x50	271	194	6.9	21.7	5.25
3.91	24.7	7.9	122	188	69.6	W 14x43	261	188	6.7	18.2	6.32
1.89	43.9	10.7	117	180	66.6	W 10x54	250	180	9.1	30.2	3.30
6.93	14.8	5.1	112	180	66.5	W 18x35	249	173	4.3	11.5	10.7
3.02	28.5	8.1	113	175	64.7	W 12x45	243	174	6.9	20.3	5.07
5.25	18.3	6.3	110	173	64.0	W 16x36	240	170	5.4	14.1	8.08
4.39	20.0	6.5	106	166	61.5	W 14x38	231	164	5.5	14.9	7.07
1.88	40.7	10.6	106	163	60.4	W 10x49	227	164	9.0	28.3	3.25
2.92	26.5	8.0	101	155	57.5	W 12x40	216	156	6.8	19.3	4.82
4.19	19.0	6.4	94.8	147	54.6	W 14x34	205	146	5.4	14.4	6.58
5.70	14.3	4.9	92.0	146	54.0	W 16x31	203	142	4.1	11.0	8.85
3.46	20.6	6.4	88.9	138	51.2	W 12x35	192	137	5.4	15.2	5.67
3.92	17.9	6.2	81.9	128	47.3	W 14x30	177	126	5.3	13.7	6.06
1.93	31.2	8.3	82.1	126	46.8	W 10x39	176	126	7.0	21.8	3.32
5.12	13.3	4.7	74.9	119	44.2	W 16x26	166	115	4.0	10.4	7.88
3.23	19.1	6.3	75.3	116	43.1	W 12x30	162	116	5.4	14.4	5.10
4.44	13.4	4.5	68.8	109	40.2	W 14x26	151	106	3.8	10.3	6.96
1.89	27.4	8.1	68.3	105	38.8	W 10x33	145	105	6.9	19.7	3.15
3.00	18.1	6.3	65.1	100	37.2	W 12x26	140	100	5.3	13.8	4.64
2.44	20.3	5.7	63.2	98.8	36.6	W 10x30	137	97.2	4.8	14.5	4.13
1.23	35.1	8.5	60.8	93.7	34.7	W 8x35	130	93.6	7.2	24.1	2.16
4.06	12.5	4.3	56.5	89.6	33.2	W 14x22	125	87.0	3.7	9.7	6.26
2.34	18.5	5.7	54.4	84.5	31.3	W 10x26	117	83.7	4.8	13.5	3.85
1.20	32.0	8.4	53.6	82.1	30.4	W 8x31	114	82.5	7.1	22.3	2.07
3.88	11.1	3.5	49.5	79.1	29.3	W 12x22	110	76.2	3.0	8.4	6.24
1.27	27.3	6.8	47.4	73.4	27.2	W 8x28	102	72.9	5.7	18.9	2.22
2.18	16.9	5.5	45.2	70.2	26.0	W 10x22	97.5	69.6	4.7	12.7	3.50
4.51	8.8	3.0	41.1	67.2	24.9	W 14x18	93.4	63.3	2.5	6.9	6.93
3.61	10.4	3.4	41.5	66.7	24.7	W 12x19	92.6	63.9	2.9	7.9	5.70
1.24	24.4	6.7	40.8	62.6	23.2	W 8x24	87.0	62.7	5.7	17.2	2.11

LOAD FACTOR DESIGN SELECTION TABLE
For shapes used as beams
$\phi_b = 0.90$

Z_x

BF	L_r	L_p	$\phi_b M_r$	$\phi_b M_p$	Z_x	Shape	$\phi_b M_p$	$\phi_b M_r$	L_p	L_r	BF
Kips	Ft	Ft	Kip-ft	Kip-ft	In.³		Kip-ft	Kip-ft	Ft	Ft	Kips
2.60	12.0	3.6	36.7	58.3	21.6	W 10x19	81.0	56.4	3.1	8.9	4.26
1.46	18.6	5.3	35.5	55.1	20.4	W 8x21	76.5	54.6	4.5	13.3	2.47
3.31	9.6	3.2	33.3	54.3	20.1	W 12x16	75.4	51.3	2.7	7.4	5.12
0.74	31.3	6.3	32.6	51.0	18.9	W 6x25	70.9	50.1	5.4	21.0	1.33
2.46	11.2	3.5	31.6	50.5	18.7	W 10x17	70.1	48.6	3.0	8.4	3.97
2.97	9.2	3.1	29.1	47.0	17.4	W 12x14	65.3	44.7	2.7	7.2	4.56
1.40	16.7	5.1	29.6	45.9	17.0	W 8x18	63.8	45.6	4.4	12.2	2.30
2.34	10.3	3.4	26.9	43.2	16.0	W 10x15	60.0	41.4	2.9	7.9	3.69
0.73	25.6	6.3	26.1	40.2	14.9	W 6x20	55.9	40.2	5.3	17.7	1.27
0.65	26.9	5.8	25.3	39.1	14.5	M 6x20	54.4	39.0	5.0	18.2	1.16
3.51	6.6	2.2	23.4	38.6	14.3	M 12x11.8	53.6	36.0	1.9	5.1	5.40
1.53	12.6	3.7	23.0	36.7	13.6	W 8x15	51.0	35.4	3.1	9.2	2.56
1.98	9.5	3.3	21.3	34.0	12.6	W 10x12†	47.0	32.7	2.9	7.4	3.13
0.82	18.3	4.0	19.9	31.6	11.7	W 6x16	43.9	30.6	3.4	12.5	1.46
0.46	30.3	5.3	19.9	31.3	11.6	W 5x19	43.5	30.6	4.5	20.1	0.83
1.43	11.5	3.5	19.3	30.8	11.4	W 8x13	42.8	29.7	3.0	8.5	2.35
0.42	31.1	5.0	18.8	29.7	11.0	M 5x18.9	41.3	28.9	4.2	20.5	0.76
0.69	20.7	6.7	19.0	28.8	10.8	W 6x15*†	38.6	29.2	6.8	14.9	1.16
0.44	26.3	5.3	16.6	25.9	9.59	W 5x16	36.0	25.5	4.5	17.6	0.80
2.48	5.9	2.0	15.1	24.8	9.19	M 10x 9	34.5	23.3	1.7	4.6	3.84
1.30	10.2	3.5	15.2	23.9	8.87	W 8x10†	33.0	23.4	3.1	7.8	2.03
0.78	14.4	3.8	14.3	22.4	8.30	W 6x12	31.1	21.9	3.3	10.2	1.33
0.30	25.5	4.2	10.6	17.0	6.28	W 4x13	23.6	16.4	3.5	16.9	0.54
0.72	12.0	3.8	10.8	16.8	6.23	W 6x 9	23.4	16.7	3.2	8.9	1.17
0.26	27.8	3.9	10.2	16.3	6.05	M 4x13	22.7	15.7	3.3	18.3	0.47
1.59	5.3	1.8	9.01	14.6	5.42	M 8x 6.5	20.3	13.9	1.5	4.1	2.47
0.94	4.5	1.5	4.68	7.56	2.80	M 6x 4.4	10.5	7.20	1.3	3.5	1.49

*Indicates noncompact shape; $F_y = 36$ ksi.
†Indicates noncompact shape; $F_y = 50$ ksi.

Notes

MOMENT OF INERTIA SELECTION TABLES I_x, I_y
For W and M shapes

These two tables for moment of inertia (I_x and I_y) are provided to facilitate the selection of beams and columns on the basis of their stiffness properties with respect to the X-X axis or Y-Y axis, as applicable, where

I_x = moment of inertia, X-X axis, in.[4]
I_y = moment of inertia, Y-Y axis, in.[4]

In each table the shapes are listed in groups by descending order of moment of inertia for all W and M shapes. The boldface type identifies the shapes that are the lightest in weight in each group.

Enter the column headed I_x (or I_y) and find a value of I_x (or I_y) equal to or greater than the moment of inertia required. The shape opposite this value, and all shapes above it, have sufficient stiffness capacity. Note that the member selected must also be checked for compliance with specification provisions governing its specific application.

I_x MOMENT OF INERTIA SELECTION TABLE
For W and M shapes

Shape	I_x In.^{4}	Shape	I_x In.^{4}	Shape	I_x In.^{4}	Shape	I_x In.^{4}
W 36x359	**24800**	**W 30x116**	**4930**	**W 24x 55**	**1350**	**W 16x26**	**301**
		W 14x342	4900	W 21x 62	1330	W 14x30	291
W 36x328	**22500**	W 27x129	4760	W 18x 76	1330	W 12x35	285
W 33x354	21900	W 24x146	4580	W 16x 89	1300	W 10x49	272
				W 14x109	1240	W 8x67	272
W 36x300	**20300**	**W 30x108**	**4470**	W 12x136	1240	W 10x45	248
		W 14x311	4330	W 21x 57	1170		
W 33x318	**19500**	W 21x166	4280	W 18x 71	1170		
		W 27x114	4090	W 16x 77	1110	**W 14x26**	**245**
W 36x280	**18900**	W 12x336	4060	W 14x 99	1110	W 12x30	238
W 33x291	17700	W 24x131	4020	W 18x 65	1070	W 8x58	228
				W 12x120	1070	W 10x39	209
W 36x260	**17300**			W 14x 90	999		
		W 30x 99	**3990**			**W 12x26**	**204**
		W 14x283	3840				
W 36x256	**16800**	W 21x147	3630	**W 21x 50**	**984**	**W 14x22**	**199**
		W 27x102	3620	W 18x 60	984	W 8x48	184
W 36x245	**16100**	W 12x305	3550	W 16x 67	954	W 10x30	170
W 33x263	15800	W 24x117	3540	W 12x106	933	W 10x33	170
		W 14x257	3400	W 18x 55	890		
W 36x230	**15000**			W 14x 82	882	**W 12x22**	**156**
W 36x232	15000	**W 27x 94**	**3270**				
W 14x730	14300	W 21x132	3220	**W 21x 44**	**843**	**M 14x18**	**148**
W 33x241	14200	W 12x279	3110	W 12x 96	833	W 8x40	146
		W 24x104	3100	W 18x 50	800	W 10x26	144
W 36x210	**13200**	W 14x233	3010	W 14x 74	796	W 12x19	130
W 33x221	12800	W 24x103	3000	W 16x 57	758	W 8x35	127
W 14x665	12400	W 21x122	2960	W 12x 87	740	W 10x22	118
				W 14x 68	723	W 8x31	110
W 36x194	**12100**	**W 27x 84**	**2850**	W 10x112	716		
W 30x235	11700	W 18x143	2750	W 18x 46	712	**W 12x16**	**103**
W 33x201	11500	W 12x252	2720	W 12x 79	662	W 8x28	98.0
		W 24x 94	2700	W 16x 50	659	W 10x19	96.3
W 36x182	**11300**	W 21x111	2670	W 14x 61	640		
W 14x605	10800	W 14x211	2660	W 10x100	623	**W 12x14**	**88.6**
		W 18x130	2460			W 8x24	82.8
W 36x170	**10500**	W 21x101	2420			W 10x17	81.9
W 30x211	10300	W 12x230	2420	**W 18x 40**	**612**	W 8x21	75.3
		W 14x193	2400	W 12x 72	597		
W 36x160	**9750**			W 16x 45	586	**M 12x11.8**	**71.9**
W 14x550	9430	**W 24x 84**	**2370**	W 14x 53	541	W 10x15	68.9
W 33x169	9290	W 18x119	2190	W 10x 88	534	W 8x18	61.9
W 30x191	9170	W 14x176	2140	W 12x 65	533	W 10x12	53.8
		W 12x210	2140			W 6x25	53.4
W 36x150	**9040**			**W 16x 40**	**518**	W 8x15	48.0
W 27x217	8870	**W 24x 76**	**2100**			W 6x20	41.4
W 14x500	8210	W 21x 93	2070			W 8x13	39.6
W 30x173	8200	W 18x106	1910	**W 18x 35**	**510**	M 6x20	39.0
W 33x152	8160	W 14x159	1900	W 14x 48	485		
W 27x194	7820	W 12x190	1890	W 12x 58	475	**M 10x 9**	**38.8**
				W 10x 77	455	W 6x16	32.1
W 36x135	**7800**	**W 24x 68**	**1830**	W 16x 36	448	W 8x10	30.8
W 33x141	7450	W 21x 83	1830	W 14x 43	428	W 6x15	29.1
W 14x455	7190	W 18x 97	1750	W 12x 53	425	W 5x19	26.2
W 27x178	6990	W 14x145	1710	W 12x 50	394	M 5x18.9	24.1
		W 12x170	1650	W 10x 68	394	W 6x12	22.1
W 33x130	**6710**	W 21x 73	1600	W 14x 38	385	W 5x16	21.3
W 30x148	6680						
W 14x426	6600	**W 24x 62**	**1550**	**W 16x 31**	**375**	**M 8x 6.5**	**18.5**
W 27x161	6280	W 18x 86	1530	W 12x 45	350	W 6x 9	16.4
W 14x398	6000	W 14x132	1530	W 10x 60	341	W 4x13	11.3
		W 16x100	1490	W 14x 34	340	M 4x13	10.5
W 33x118	**5900**	W 21x 68	1480	W 12x 40	310		
W 30x132	5770	W 12x152	1430	W 10x 54	303	**M 6x 4.4**	**7.20**
W 24x176	5680	W 14x120	1380				
W 27x146	5630						
W 14x370	5440						
W 30x124	5360						
W 24x162	5170						

MOMENT OF INERTIA SELECTION TABLE
For W and M shapes I_y

Shape	I_y In.^{4}	Shape	I_y In.^{4}	Shape	I_y In.^{4}	Shape	I_y In.^{4}
W 14x730	4720	W 14x109	447	W 10x60	116	W 8x28	21.7
W 14x665	4170	W 27x146	443	W 24x94	109	W 21x44	20.7
W 14x605	3680	W 24x162	443			W 12x30	20.3
		W 21x166	435			W 14x30	19.6
W 14x550	3250	W 36x210	411	W 12x58	107	W 18x40	19.1
W 14x500	2880	W 14x99	402	W 14x61	107		
W 14x455	2560	W 12x136	398	W 27x84	106	W 8x24	18.3
W 14x426	2360	W 24x146	391			W 12x26	17.3
		W 21x147	376	W 10x54	103	W 6x25	17.1
W 14x398	2170	W 36x194	375			W 10x30	16.7
W 14x370	1990	W 14x90	362	W 12x53	95.8	W 18x35	15.3
W 14x342	1810	W 36x182	347	W 24x84	94.4	W 10x26	14.1
		W 12x120	345				
W 14x311	1610	W 24x131	340	W 10x49	93.4	W 6x20	13.3
W 36x359	1570	W 21x132	333	W 21x93	92.9	W 16x31	12.4
W 33x354	1460	W 36x170	320	W 8x67	88.6		
		W 18x143	311	W 24x76	82.5	M 6x20	11.6
W 14x283	1440	W 33x169	310	W 21x83	81.4	W 10x22	11.4
W 36x328	1420	W 21x122	305	W 8x58	75.1	W 8x21	9.77
W 36x300	1300	W 12x106	301	W 21x73	70.6	W 16x26	9.59
		W 24x117	297	W 24x68	70.4		
W 14x257	1290	W 36x160	295	W 21x68	64.7	W 6x15	9.32
W 33x318	1290	W 18x130	278			W 5x19	9.13
W 36x280	1200	W 21x111	274	W 8x48	60.9	W 14x26	8.91
W 12x336	1190	W 33x152	273	W 18x71	60.3	W 8x18	7.97
W 33x291	1160	W 12x96	270	W 14x53	57.7	M 5x18.9	7.86
		W 36x150	270	W 21x62	57.5	W 5x16	7.51
W 14x233	1150	W 24x104	259	W 12x50	56.3	W 14x22	7.00
W 36x260	1090	W 18x119	253	W 18x65	54.8	W 12x22	4.66
W 12x305	1050	W 21x101	248			W 6x16	4.43
		W 33x141	246	W 10x45	53.4	W 10x19	4.29
W 14x211	1030	W 12x87	241	W 14x48	51.4		
W 33x263	1030	W 10x112	236	W 18x60	50.1	W 4x13	3.86
W 36x245	1010	W 30x148	227			W 12x19	3.76
W 36x230	940	W 36x135	225	W 12x45	50.0	W 10x17	3.56
W 12x279	937	W 18x106	220			W 8x15	3.41
W 33x241	932	W 33x130	218	W 8x40	49.1		
		W 12x79	216	W 14x43	45.2	M 4x13	3.36
W 14x193	931	W 10x100	207				
W 30x235	855	W 18x97	201	W 8x35	42.6	W 6x12	2.99
W 33x221	840	W 30x132	196	W 18x50	40.1	W 10x15	2.89
		W 12x72	195	W 16x50	37.2	W 12x16	2.82
W 14x176	838	W 33x118	187			W 8x13	2.73
W 12x252	828	W 16x100	186	W 10x39	45.0	M 14x18	2.64
W 30x211	757	W 27x129	184	W 18x55	44.9	W 12x14	2.36
W 33x201	749	W 30x124	181	W 12x40	44.1		
		W 10x88	179	W 16x57	43.1	W 6x9	2.19
W 14x159	748	W 18x86	175			W 10x12	2.18
W 12x230	742	W 12x65	174	W 8x31	37.1	W 8x10	2.09
W 12x217	704	W 30x116	164	W 10x33	36.6	M 12x11.8	0.980
		W 16x89	163	W 24x62	34.5		
W 14x145	677	W 27x114	159	W 16x45	32.8	M 10x9	0.609
W 30x191	673	W 10x77	154	W 21x57	30.6		
W 12x210	664	W 18x76	152	W 24x55	29.1	M 8x6.5	0.343
W 27x194	618	W 14x82	148	W 16x40	28.9		
W 30x173	598	W 30x108	146	W 14x38	26.7	M 6x4.4	0.165
W 12x190	589	W 27x102	139	W 21x50	24.9		
W 27x178	555	W 16x77	138	W 12x35	24.5		
W 14x132	548	W 10x68	134	W 16x36	24.5		
W 36x256	528	W 14x74	134	W 14x34	23.3		
W 12x170	517	W 30x99	128	W 18x46	22.5		
W 27x161	497	W 27x94	124				
W 14x120	495	W 14x68	121				
W 24x176	479	W 16x67	119				
W 36x232	468	W 24x103	119				
W 12x152	454						

UNIFORM LOAD CONSTANTS
General Notes

The tables of Uniform Load Constants for W, M, S and HP shapes and channels (C, MC) used as simple beams give constants for the direct calculation of the total uniformly distributed design load for laterally supported steel beams. The tables are based on the flexural design strengths specified in Sect. F of the LRFD Specification. Separate tables are presented for $F_y = 36$ ksi and $F_y = 50$ ksi. The tabulated constants take into consideration the weight of the beam, which should be deducted in the calculation to determine the net load that the beam will support.

The tables are also applicable to laterally supported simple beams for concentrated loading conditions. A method to determine the beam load capacity for several cases is shown in this discussion.

It is assumed in all cases the loads are applied normal to the X-X axis shown in the Tables of Properties of Shapes in Part 1 of the LRFD Manual, and that the beam deflects vertically in the plane of bending. If the conditions of loading involve forces outside this plane, design strengths must be determined from the general theory of flexure in accordance with the character of the load and its mode of application.

Lateral Support of Beams

The flexural design strength of a beam is dependent upon lateral support of its compression flange in addition to its section properties. In these tables, the notation L_p is used to denote the maximum unbraced length of the compression flange, in feet, for uniform moment case ($C_b = 1.0$) and for which the design strengths for compact symmetrical shapes are calculated with a flexural design strength of:

$$\phi_b M_n = \phi_b M_p = \phi_b Z_x F_y / 12$$

Noncompact shapes are calculated with a flexural design strength of:

$$\phi_b M'_n = \phi_b M_p - \phi_b (M_p - M_r)\left(\frac{\lambda - \lambda_p}{\lambda_r - \lambda_p}\right)$$

as permitted in LRFD Specification Appendix F1.7. The associated maximum unbraced length for $\phi_b M'_n$ is tabulated as L_p. The notation L_r is the maximum unbraced length of the compression flange for which the flexural design strength for rolled shapes is:

$$\phi_b M_r = \phi_b S_x (F_y - 10)/12$$

These tables are not applicable for beams with unbraced lengths greater than L_r. For such cases, the unbraced beam charts are applicable.

Flexural Stress and Tabulated Uniform Load Constants

For the symmetrical rolled shapes designated W, M, S and HP, the flexural design strengths and resultant load constants are based on the assumption that the compression flanges of the beams are laterally supported at intervals not greater than L_p.

The Uniform Load Constant W_c is obtained from the moment and stress relationship of a simply supported, uniformly loaded beam. The relationship results in the formula:

$$\phi_b W_c = \phi_b (2Z_x F_y / 3), \text{ kip-ft for compact shapes}$$

The following expression may be used for calculating the total uniformly distributed design load W on a simply supported beam or girder:

$$W = \phi_b W_c / L, \text{ kips}$$

For compact shapes, the tabulated constant is based on the yield stress $F_y = 36$ ksi or 50 ksi and the plastic section modulus Z_x (see Appendix F1.7 of the LRFD Specification). For noncompact sections, the tabulated constant is based on the nominal resisting moment as determined by Formula A-F1-3. See LRFD Specification Appendix F1.7.

Shearing Stresses

For relatively short spans, the design strengths for beams and channels may be limited by the shearing capacity of the web instead of the maximum bending capacity of the flanges. This limit is indicated in the tables by the notation L_v. For span lengths less than L_v, the maximum total uniform load the beam can support is twice the design beam shear.

End Bearing

Section K1 of the LRFD Specification requires beams with unstiffened webs be designed so that the maximum end reaction does not exceed:

$$\phi R_n = \phi(2.5k + N)\, F_{yw} t_w$$

where

$\phi = 1.0$
N = length of bearing, in.

nor:

$$\phi_r R_n = \phi_r\, 68 t_w^2 \left[1 + 3\left(\frac{N}{d}\right)\left(\frac{t_w}{t_f}\right)^{1.5} \right] \sqrt{F_y t_f / t_w}$$

where

$\phi_r = 0.75$

Interior Bearing

For interior concentrated loads, the maximum concentrated load may not exceed:

$$\phi R_n = \phi(5k + N)\, F_{yw} t_w$$

where

$\phi = 1.0$

nor:

$$\phi_r R_n = \phi_r\, 135\, t_w^2 \left[1 + 3\left(\frac{N}{d}\right)\left(\frac{t_w}{t_f}\right)^{1.5} \right] \sqrt{F_y t_f / t_w}$$

where

$\phi_r = 0.75$

In all cases, proper lateral support must be provided for the top flanges of beams at the point of load to prevent a decrease in the beam web crippling strength.

Vertical Deflection

For rolled shapes designated W, M, HP, S, C and MC, the maximum vertical deflection may be calculated using the formula:

$$\Delta = ML^2/(C_1 I_x)$$

where

M = service load moment based on uniformly distributed design load, kip-ft
L = span length, ft
I_x = moment of inertia, in.[4]
C_1 = loading constant (see Fig. 3.3)
Δ = deflection, in.

A live-load deflection limit of $L/360$ is assumed.

$$\Delta_{LL} \leq \frac{\text{Span Length}}{360}$$

Deflection can be controlled by limiting the span-depth ratio of a simply supported, uniformly loaded beam as shown in Table 3.1. For large span/depth ratios, vibration may also be a consideration.

TABLE 3.1   RECOMMENDED SPAN/DEPTH RATIOS

Load Ratios		Span/Depth Ratios	
DL/TL	DL/LL	F_y = 36 ksi	F_y = 50 ksi
0.2	0.25	20.0	14.0
0.3	0.43	22.2	16.0
0.4	0.67	25.0	18.0
0.5	1.00	29.0	21.0
0.6	1.50	—	26.0

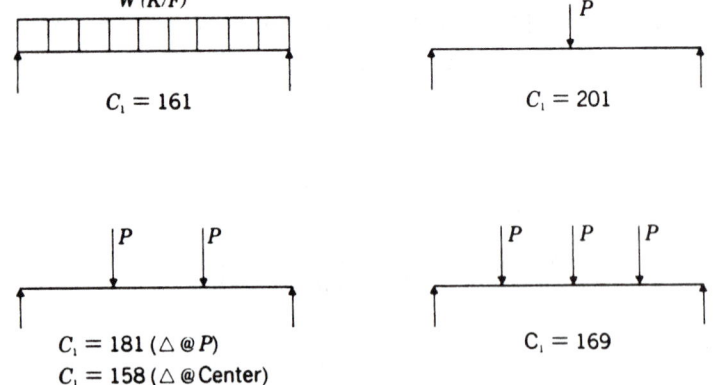

Fig. 3.3   Loading constants

AMERICAN INSTITUTE OF STEEL CONSTRUCTION

UNIFORM LOAD CONSTANTS
Use of Tables

Nomenclature

$\phi_b W_c$ = Uniform load constant, kip-ft
  = $2M_{n, max}/3$
$M_{n, max}$ = $\phi_b M_p$ for compact shapes, kip-in.
  = $\phi_b M_n'$ for noncompact shapes, kip-in.
  $\phi_b = 0.90$
  W = Total uniformly distributed design load, kips
  = $\phi_b W_c/L$
$\phi_v V_n$ = Maximum web shear strength, kips (LRFD Specification Sect. F2).
  For $h/t_w \le 187 \sqrt{k/F_{yw}}$

$$\phi_v V_n = \phi_v(0.6 F_{yw} A_w)$$

  For $187 \sqrt{k/F_{yw}} < h/t_w \le 234 \sqrt{k/F_{yw}}$

$$\phi_v V_n = \phi_v \left(0.6 F_{yw} A_w \frac{187 \sqrt{k/F_{yw}}}{h/t_w} \right)$$

  For $h/t_w > 234 \sqrt{k/F_{yw}}$

$$\phi_v V_n = \phi_v A_w \frac{26,400k}{(h/t_w)^2}$$

k = web plate buckling coefficient
  = 5.0 (for rolled shapes with no stiffeners)
$\phi_v = 0.90$
L_v = Span length below which shear $\phi_v V_n$ in the beam web governs, as compared to greater lengths where flexure constant $\phi_b W_c$ governs, ft
  = $\phi_b W_c/\phi_v(2V_n)$
L_b = Lateral unbraced length; length between points which are either braced against lateral displacement of the compression flange or braced against twist of the cross section, ft
L_p = Limiting laterally unbraced length for full plastic moment capacity with $C_b = 1.0$, ft

$$= \frac{300 r_y}{12 \sqrt{F_y}}, \text{ compact shape}$$

$$= L_p' = L_p + (L_r - L_p) \left(\frac{M_p - M_n'}{M_p - M_r} \right), \text{ noncompact shape}$$

L_r = Limiting laterally unbraced length for elastic lateral-torsional buckling, ft

$$= \frac{r_y X_1}{(F_{yf} - F_r)} \sqrt{1 + \sqrt{1 + X_2(F_{yf} - F_r)^2}} \tag{F1-6}$$

$$X_1 = \frac{\pi}{S_x} \sqrt{\frac{EGJA}{2}} \tag{F1-8}$$

$$X_2 = 4\frac{C_w}{I_y}\left(\frac{S_x}{GJ}\right)^2 \tag{F1-9}$$

J = torsional constant, in.[4]
C_w = warping constant, in.[6]
F_r = compressive residual stress in flange
 = for rolled shapes $F_r = 10$ ksi
 = for welded shapes $F_r = 16.5$ ksi

$\phi_b M_p$ = Plastic moment capacity, kip-ft
 = $\phi_b Z_x F_y / 12$
$\phi_b M_r$ = Limiting buckling moment for elastic lateral torsional buckling, kip-ft
 = $\phi_b S_x (F_y - F_r)/12$
$\phi_b M_n$ = Nominal flexural strength, kip-ft
 For $L_p < L_b \leq L_r$
$$\phi_b M_n = \phi_b M_p - BF(L_b - L_p)$$

BF = Factor used to determine moment capacity for $L_p < L_b \leq L_r$, kips

$$= \frac{(\phi_b M_p - \phi_b M_r)}{L_r - L_p}$$

$\phi_b M_{cr}$ = Limiting moment capacity for doubly symmetric I shapes or singly symmetric channel shapes bent about their major axis with $L_b > L_r$, kip-ft
 = $0.9 \, M_{cr}/12$

$$M_{cr} = C_b \frac{\pi}{L_b} \sqrt{EI_y GJ + \left(\frac{\pi E}{L_b}\right)^2 I_y C_w}$$

$\phi R_1 = \phi 2.5 k F_{yw} t_w$ from first part of Formula K1-3, kips
$\phi R_2 = \phi N F_{yw} t_w$ from second part of Formula K1-3, kip/in.

where

$N = 1.0$
$\phi = 1.0$
k = distance from outer face of flange to web toe of fillet, in.

$$\phi_r R_3 = \phi_r \left[68 t_w^2 \sqrt{\frac{F_y t_f}{t_w}}\right] \text{ from first part of Formula K1-5, kips}$$

$$\phi_r R_4 = \phi_r \left\{68 t_w^2 \left[3\left(\frac{N}{d}\right)\left(\frac{t_w}{t_f}\right)^{1.5}\right]\sqrt{\frac{F_y t_f}{t_w}}\right\} \text{ from second part of Formula K1-5, kip/in.}$$

where

$N = 1.0$

$\phi_r = 0.75$

$R(N = 3\frac{1}{2}\text{in.})$ = Maximum permitted reaction for a bearing length of $3\frac{1}{2}$ in.; the smaller value as determined by LRFD Specification Formulas K1-3 or K1-5.

Z_x = Plastic section modulus, in.[3]

Example 1

Given:

A W16×45 of F_y = 36 ksi steel spans 20 ft. Determine the uniform load capacity, end reaction and total service load deflection. The live load equals the dead load, therefore:

$$\text{factored load} = 1.4 \text{ (total load)}$$

Solution:

Enter the Uniform Load Constants Table for F_y = 36 ksi and note that:

$$\phi_b W_c = 1,780 \text{ kip-ft}$$
$$\phi_v V_n = 108 \text{ kips}$$
$$L_v = 8.2 \text{ ft}$$

1. Total design uniform load = $W = \phi_b W_c / L = 1,780/20 = 89.0$ kips
2. End reaction = $\phi_b W_c / 2L = 1780/2(20) = 44.5$ kips
3. Deflection:
   Service load moment = $WL/8(LF)$
   $$= 89.0(20)/8(1.4)$$
   $$= 158.9 \text{ kip-ft}$$

$$\Delta = \frac{ML^2}{161 I_x} = \frac{158.9(20)^2}{161\,(586)} = 0.67 \text{ in.}$$

where: C_1 = 161 from Fig. 3.3.

Example 2

Given:

A W10×45 beam of F_y = 36 ksi steel spans 6 ft. Determine the load capacity and end reaction.

Solution:

Enter the Uniform Load Constants Table for F_y = 36 ksi and note that:

$$\phi_b W_c = 1190 \text{ kip-ft}$$
$$\phi_v V_n = 68.7 \text{ kips}$$
$$L_v = 8.6 \text{ ft}$$

The beam span is less than L_v; therefore, the total uniform design load W is limited by shear in the web.

$$W = 2(\phi_v V_n) = 2(68.7) = 137.4 \text{ kips}$$

Example 3

Given:

Using F_y = 36 ksi steel, select an 18-in. deep beam to span 30 ft and support two equal concentrated loads at the one- and two-thirds points of the span. The service load intensities are 7.5 kips dead load and 17.0 kips live load. The beam is supported laterally at the points of load application and the ends. Determine the beam size and service load deflection.

Solution:

Refer to the table of Concentrated Load Equivalents and note that:

Equivalent uniform load $= 2.67P$

1. Required uniform load constant:

$$W_uL = 2.67PL = 2.67 \, [(1.6)(17) + (1.2)(7.5)](30)$$
$$= 2.67(36.2)(30)$$
$$= 2,900 \text{ kip-ft}$$

2. Enter the Uniform Load Constants Table for $F_y = 36$ ksi and $W_c \geq 2,900$:

Select W18×71: $\phi_b W_c = 3,130$
$$L_p = 7.1 \text{ ft}$$
$$BF = 8.29 \text{ kips}$$
$$\phi_b M_p = 392 \text{ kip-ft}$$

3. Flexural design strength $\phi_b M_n = C_b[\phi_b M_p - BF(L_b - L_p)]$
$$= 1.0[392 - 8.29(10 - 7.1)]$$
$$= 368 \text{ kip-ft}$$

4. Required flexural strength $= M_u = P_u(L/3)$
$$= 36.2(30/3)$$
$$= 362 \text{ kip-ft} < 368 \text{ kip-ft} \quad \textbf{o.k.}$$

5. Determine service live load deflection:

$$M_{LL} = (P_{LL}/P_u) \, M_u = (17/36.2)362 = 170 \text{ kip-ft}$$

$$\Delta \text{ (at midspan)} = \frac{M_{LL}L^2}{C_1 I_x} = \frac{170 \, (30)^2}{158 \, (1,170)} = 0.83 \text{ in.}$$

$$\Delta \text{ (at point of load)} = \frac{M_{LL}L^2}{C_1 I_x} = \frac{170 \, (30)^2}{181 \, (1,170)} = 0.72 \text{ in.}$$

where

$C_1 = 158$ for deflection at midspan
$C_1 = 181$ for deflection at point of load

Example 4

Given:

A W24×55 of 36 ksi steel spans 20 ft and is braced at 5-ft intervals. Determine the load capacity, end reaction and required end bearing length.

Solution:

1. Enter the Uniform Load Constants Table for $F_y = 36$ ksi and note that:

$$\phi_b W_c = 2{,}890 \text{ kip-ft}$$
$$\phi_v V_n = 181 \text{ kips}$$
$$L_v = 8 \text{ ft}$$
$$\phi R_1 = 46.7 \text{ kips}$$
$$\phi R_2 = 14.2 \text{ kip/in.}$$
$$\phi_r R_3 = 54 \text{ kips}$$
$$\phi_r R_4 = 4.75 \text{ kip/in.}$$

2. Total design uniform load $= W = \phi_b W_c / L = 2{,}890/20 = 144.5$ kips
3. End reaction $= R = W/2 = 144.5/2 = 72.25$ kips
4. Bearing length required is the greater of:

$$L = (R - \phi R_1)/\phi R_2 = (72.25 - 46.7)/14.2 = 1.80 \text{ in.}$$
$$\text{or } L = (R - \phi_r R_3)/\phi_r R_4 = (72.25 - 54)/4.75 = 3.84 \text{ in.}$$

Use 4-in. minimum bearing length.

BEAMS
W shapes
Uniform load constants for beams

$F_y = 36$ ksi

$\phi_b = 0.90 \qquad \phi_v = 0.90 \qquad \phi = 1.00 \qquad \phi_r = 0.75$

For beams laterally unsupported, see page **3-53**

Shape	$\phi_b W_c$	$\phi_v V_n$	L_v	L_p	L_r	$\phi_b M_p$	$\phi_b M_r$	BF	ϕR_1	ϕR_2	ϕR_3	ϕR_4	R(N=3½ in.)	Z_x
	Kip-ft	Kip	Ft	Ft	Ft	Kip-ft	Kip-ft	Kip	Kip	Kip/in.	Kip	Kip/in.	Kip	In.³
W 36x359	32600	814	20.0	16.1	62.5	4080	2570	32.4	315	40.3	514	17.2	456	1510
x328	29800	735	20.3	16.0	58.6	3730	2360	32.1	275	36.7	429	14.2	404	1380
x300	27200	675	20.2	16.0	55.1	3400	2160	31.6	239	34.0	364	12.6	358	1260
x280	25300	628	20.1	15.9	53.0	3160	2010	31.0	214	31.9	319	11.1	326	1170
x260	23300	592	19.7	15.8	50.6	2920	1860	30.3	194	30.2	283	10.4	300	1080
x245	21800	561	19.4	15.6	48.8	2730	1750	29.6	180	28.8	254	9.65	281	1010
x230	20400	530	19.2	15.5	47.3	2550	1630	28.7	162	27.4	228	8.91	258	943
x256	22500	699	16.1	11.0	39.7	2810	1745	37.0	227	34.6	379	12.5	348	1040
x232	20200	628	16.1	10.9	37.2	2530	1580	36.2	196	31.3	311	10.4	305	936
x210	18000	592	15.2	10.8	35.0	2250	1400	34.9	173	29.9	270	10.5	277	833
x194	16600	543	15.3	10.7	33.5	2070	1290	34.0	151	27.5	230	8.94	247	767
x182	15500	512	15.1	10.6	32.8	1940	1210	32.7	139	26.1	205	8.16	230	718
x170	14400	478	15.1	10.5	31.9	1800	1130	31.5	122	24.5	180	7.25	205	668
x160	13500	455	14.8	10.4	30.9	1680	1060	30.7	113	23.4	162	6.86	186	624
x150	12500	436	14.4	10.3	30.2	1570	983	29.4	105	22.5	147	6.65	170	581
x135	11000	415	13.3	9.9	28.8	1370	856	27.5	91.1	21.6	126	7.06	151	509
W 33x354	30700	802	19.1	15.6	64.4	3830	2400	29.4	300	41.8	553	19.3	446	1420
x318	27400	711	19.3	15.5	57.7	3430	2160	28.8	252	37.4	446	15.5	383	1270
x291	24800	650	19.1	15.4	52.0	3110	1970	28.2	221	34.6	379	13.5	342	1150
x263	22500	584	19.2	15.3	52.2	2810	1790	27.6	186	31.3	311	11.2	296	1040
x241	20300	552	18.4	15.1	49.2	2540	1620	27.0	163	29.9	274	11.0	268	939
x221	18500	511	18.1	15.0	46.9	2310	1480	26.0	144	27.9	236	9.88	242	855
x201	16700	468	17.8	14.8	44.8	2080	1330	25.0	125	25.7	198	8.66	215	772
x169	13600	440	15.4	10.4	32.6	1700	1070	28.3	124	24.1	185	6.69	209	629
x152	12100	413	14.6	10.3	31.2	1510	950	26.8	107	22.9	159	6.65	182	559
x141	11100	392	14.2	10.1	30.1	1390	874	25.7	95.3	21.8	141	6.36	163	514
x130	10100	373	13.5	10.0	29.1	1260	792	24.5	88.1	20.9	125	6.33	147	467
x118	8960	351	12.8	9.7	27.8	1120	700	23.1	77.3	19.8	107	6.28	129	415
W 30x235	18300	505	18.1	14.7	48.6	2280	1450	24.4	168	29.9	283	11.2	273	845
x211	16200	466	17.4	14.5	47.9	2020	1290	21.8	148	27.9	239	10.5	246	749
x191	14500	423	17.2	14.4	45.4	1820	1170	21.0	124	25.6	199	9.04	213	673
x173	13100	388	16.9	14.3	43.2	1630	1050	20.2	111	23.6	167	7.96	193	605
x148	10800	388	13.9	9.5	30.6	1350	850	23.7	117	23.4	174	6.97	199	500
x132	9440	362	13.0	9.4	29.0	1180	741	22.4	96.9	22.1	148	7.05	172	437
x124	8810	343	12.8	9.3	28.2	1100	692	21.6	88.8	21.1	132	6.55	155	408
x116	8160	330	12.4	9.1	27.1	1020	642	21.1	82.6	20.3	120	6.49	143	378
x108	7470	316	11.8	9.0	26.3	934	583	20.2	76.6	19.6	107	6.55	130	346
x 99	6740	300	11.2	8.8	25.5	842	525	19.0	67.3	18.7	93.9	6.50	117	312
W 27x217	15300	459	16.7	13.8	51.9	1910	1220	18.2	163	29.9	283	12.3	268	708
x194	13600	410	16.5	13.7	48.0	1700	1080	17.8	139	27.0	230	10.3	234	628
x178	12200	392	15.6	13.6	45.2	1530	979	17.5	122	26.1	206	10.6	214	567
x161	11100	354	15.6	13.5	42.8	1380	887	16.9	108	23.8	171	8.86	191	512
x146	9960	322	15.5	13.4	40.7	1240	801	16.2	91.9	21.8	142	7.62	168	461
x129	8530	328	13.0	9.2	30.0	1070	673	18.9	99.5	22.0	153	6.86	176	395
x114	7410	302	12.3	9.1	28.2	926	583	18.0	83.4	20.5	127	6.70	150	343
x102	6590	271	12.1	9.0	26.8	824	521	17.0	72.4	18.5	103	5.58	123	305
x 94	6000	256	11.7	8.8	25.9	751	474	16.2	63.4	17.6	90.6	5.39	109	278

AMERICAN INSTITUTE OF STEEL CONSTRUCTION

| F_y = 36 ksi | BEAMS W shapes — Uniform load constants for beams $\phi_b = 0.90$ $\phi_v = 0.90$ $\phi = 1.00$ $\phi_r = 0.75$ — For beams laterally unsupported, see page **3-53** | | | | | | | | | | | | |

Shape	$\phi_b W_c$	$\phi_v V_n$	L_v	L_p	L_r	$\phi_b M_p$	$\phi_b M_r$	BF	ϕR_1	ϕR_2	$\phi_r R_3$	$\phi_r R_4$	$R(N=3\frac{1}{2}\text{ in.})$	Z_x
	Kip-ft	Kip	Ft	Ft	Ft	Kip-ft	Kip-ft	Kip	Kip	Kip/in.	Kip	Kip/in.	Kip	In.³
W 27x 84	5270	239	11.0	8.6	24.9	659	415	15.0	56.9	16.6	76.4	5.23	94.7	244
W 24x176	11000	368	15.0	12.7	47.8	1380	878	14.3	143	27.0	230	11.5	238	511
x162	10100	343	14.8	12.7	45.2	1260	807	14.1	127	25.4	200	10.5	216	468
x146	9030	313	14.4	12.5	42.0	1130	723	13.8	110	23.4	167	9.35	192	418
x131	7990	288	13.9	12.4	39.3	999	642	13.3	95.3	21.8	141	8.65	171	370
x117	7060	259	13.6	12.3	37.1	883	567	12.7	80.4	19.8	115	7.41	141	327
x104	6240	234	13.3	12.1	35.2	780	503	12.0	67.5	18.0	93.7	6.36	116	289
x103	6050	262	11.5	8.3	27.1	756	478	14.8	86.6	19.8	124	6.35	146	280
x 94	5490	243	11.3	8.3	25.9	686	433	14.3	75.3	18.5	106	5.89	126	254
x 84	4840	220	11.0	8.1	24.5	605	382	13.6	66.1	16.9	86.5	5.14	104	224
x 76	4320	205	10.6	8.0	23.4	540	343	12.7	56.9	15.8	73.6	4.81	90.5	200
x 68	3820	191	10.0	7.8	22.4	478	300	12.1	51.4	14.9	62.6	4.73	79.1	177
x 62	3300	198	8.3	5.8	17.2	413	255	13.8	53.2	15.5	66.3	5.21	84.5	153
x 55	2890	181	8.0	5.6	16.6	362	222	12.7	46.7	14.2	54.0	4.75	70.6	134
W 21x166	9330	328	14.2	12.4	51.7	1170	741	10.8	143	27.0	232	12.7	238	432
x147	8060	309	13.0	12.3	46.4	1010	642	10.7	122	25.9	200	13.5	212	373
x132	7190	276	13.0	12.2	43.1	899	575	10.5	106	23.4	163	11.2	188	333
x122	6630	253	13.1	12.2	41.0	829	532	10.3	91.1	21.6	139	9.53	167	307
x111	6030	230	13.1	12.1	38.7	753	486	10.1	80.4	19.8	117	8.11	145	279
x101	5460	208	13.2	12.0	37.1	683	443	9.62	70.3	18.0	96.8	6.72	120	253
x 93	4770	244	9.8	7.7	26.6	597	374	11.8	88.1	20.9	130	8.91	161	221
x 83	4230	215	9.9	7.6	24.9	529	333	11.3	72.4	18.5	103	7.01	128	196
x 73	3720	188	9.9	7.5	23.5	464	294	10.7	61.4	16.4	80.8	5.50	100	172
x 68	3460	177	9.8	7.5	22.8	432	273	10.4	55.6	15.5	71.4	5.04	89.1	160
x 62	3110	163	9.5	7.4	21.7	389	248	9.80	49.5	14.4	60.7	4.55	76.6	144
x 57	2790	166	8.4	5.6	17.3	348	216	11.3	50.1	14.6	63.6	4.45	79.2	129
x 50	2380	154	7.7	5.4	16.2	297	184	10.5	44.9	13.7	52.4	4.52	68.2	110
x 44	2060	141	7.3	5.3	15.4	258	159	9.66	37.4	12.6	42.5	4.23	57.3	95.4
W 18x143	6960	277	12.6	11.3	52.1	869	550	7.83	131	26.3	219	13.9	223	322
x130	6290	251	12.5	11.3	48.0	786	499	7.80	113	24.1	184	12.0	197	291
x119	5640	242	11.7	11.2	44.1	705	450	7.71	103	23.6	167	12.8	186	261
x106	4970	215	11.6	11.1	40.4	621	398	7.61	86.3	21.2	134	10.7	161	230
x 97	4560	193	11.8	11.0	38.1	570	367	7.51	75.2	19.3	112	8.69	142	211
x 86	4020	172	11.7	11.0	35.5	502	324	7.28	62.1	17.3	89.3	7.17	114	186
x 76	3520	150	11.7	10.9	33.3	440	285	6.95	52.6	15.3	69.9	5.69	89.8	163
x 71	3130	178	8.8	7.1	24.4	392	248	8.29	66.8	17.8	95.9	7.44	122	145
x 65	2870	161	9.0	7.0	23.2	359	228	8.09	58.2	16.2	80.0	6.08	101	133
x 60	2660	147	9.0	7.0	22.4	332	211	7.90	51.4	14.9	68.2	5.18	86.3	123
x 55	2420	137	8.8	7.0	21.4	302	192	7.63	46.1	14.0	59.2	4.77	75.9	112
x 50	2180	124	8.8	6.9	20.5	273	173	7.29	39.9	12.8	48.9	4.01	62.9	101
x 46	1960	126	7.8	5.4	16.6	245	154	8.12	40.5	13.0	51.4	3.92	65.1	90.7
x 40	1690	110	7.7	5.3	15.7	212	133	7.52	33.7	11.3	39.2	3.05	49.9	78.4
x 35	1440	103	7.0	5.1	14.8	180	112	6.93	30.4	10.8	32.8	3.29	44.3	66.5
W 16x100	4280	193	11.1	10.5	42.1	535	341	6.08	88.8	21.1	136	11.0	163	198
x 89	3780	171	11.1	10.4	38.6	473	302	6.05	73.8	18.9	109	9.06	140	175

BEAMS

$F_y = 36$ ksi

W shapes
Uniform load constants for beams

$\phi_b = 0.90 \qquad \phi_v = 0.90 \qquad \phi = 1.00 \qquad \phi_r = 0.75$

For beams laterally unsupported, see page **3-53**

Shape	$\phi_b W_c$	$\phi_v V_n$	L_v	L_p	L_r	$\phi_b M_p$	$\phi_b M_r$	BF	$\phi_r R_1$	$\phi_r R_2$	$\phi_r R_3$	$\phi_r R_4$	R(N= 3½ in.)	Z_x
	Kip-ft	Kip	Ft	Ft	Ft	Kip-ft	Kip-ft	Kip	Kip	Kip/in.	Kip	Kip/in.	Kip	In.³
W 16x 77	3240	146	11.1	10.3	34.9	405	261	5.85	58.9	16.4	81.9	6.89	106	150
x 67	2810	125	11.2	10.3	32.3	351	228	5.57	48.9	14.2	61.9	5.21	80.2	130
x 57	2270	137	8.3	6.7	22.8	284	180	6.42	53.2	15.5	73.0	6.21	94.7	105
x 50	1990	120	8.3	6.6	21.2	248	158	6.20	44.9	13.7	56.9	4.92	74.1	92.0
x 45	1780	108	8.2	6.5	20.1	222	142	5.92	38.8	12.4	46.6	4.14	61.1	82.3
x 40	1570	94.9	8.3	6.5	19.3	197	126	5.53	32.6	11.0	36.6	3.22	47.9	72.9
x 36	1380	91.0	7.6	6.3	18.3	173	110	5.25	29.9	10.6	32.2	3.46	44.2	64.0
x 31	1170	84.9	6.9	4.9	14.3	146	92.0	5.70	27.8	9.90	29.3	2.73	38.8	54.0
x 26	955	76.3	6.3	4.7	13.3	119	74.9	5.12	23.9	9.00	22.5	2.65	31.7	44.2
W 14x132	5050	184	13.7	15.7	73.6	632	408	3.86	98.0	23.2	161	16.3	179	234
x120	4580	166	13.8	15.6	67.9	572	371	3.86	86.3	21.2	134	13.9	161	212
x109	4150	146	14.2	15.5	62.7	518	337	3.84	73.8	18.9	108	10.8	140	192
x 99	3740	134	14.0	15.5	58.2	467	306	3.77	62.7	17.5	91.3	9.48	124	173
x 90	3390	120	14.1	15.4	54.1	424	279	3.74	54.5	15.8	75.3	7.86	103	157
x 82	3000	142	10.6	10.3	43.0	375	240	4.15	74.6	18.4	103	9.95	138	139
x 74	2720	124	11.0	10.3	40.0	340	218	4.10	63.3	16.2	81.8	7.52	108	126
x 68	2480	113	11.0	10.3	37.3	311	201	4.05	56.0	14.9	69.4	6.49	92.1	115
x 61	2200	101	10.9	10.2	34.7	275	180	3.90	48.5	13.5	56.4	5.40	75.3	102
x 53	1880	100	9.4	8.0	28.0	235	152	4.17	47.9	13.3	55.9	5.06	73.7	87.1
x 48	1690	91.1	9.3	8.0	26.3	212	137	4.05	42.1	12.2	46.8	4.40	62.2	78.4
x 43	1500	81.0	9.3	7.9	24.7	188	122	3.91	36.0	11.0	37.5	3.60	50.1	69.6
x 38	1330	85.0	7.8	6.5	20.0	166	106	4.39	29.6	11.2	37.9	3.77	51.1	61.5
x 34	1180	77.5	7.6	6.4	19.0	147	94.8	4.19	25.6	10.3	31.4	3.34	43.1	54.6
x 30	1020	72.6	7.0	6.2	17.9	128	81.9	3.92	22.8	9.72	26.6	3.39	38.5	47.3
x 26	868	69.0	6.3	4.5	13.4	109	68.8	4.44	21.5	9.18	25.5	2.61	34.7	40.2
x 22	717	61.4	5.8	4.3	12.5	89.6	56.5	4.06	18.1	8.28	19.5	2.43	28.0	33.2
W 12x 87	2850	125	11.4	12.8	56.4	356	230	2.90	69.5	18.5	102	12.4	134	132
x 79	2570	113	11.4	12.7	51.8	321	209	2.88	60.8	16.9	84.5	10.5	120	119
x 72	2330	102	11.4	12.7	48.2	292	190	2.86	53.2	15.5	70.6	8.89	102	108
x 65	2090	91.9	11.4	12.6	44.7	261	171	2.80	46.1	14.0	58.0	7.43	84.0	96.8
x 58	1870	85.3	10.9	10.5	38.4	233	152	2.91	44.6	13.0	52.9	5.49	72.1	86.4
x 53	1680	80.9	10.4	10.3	35.8	210	138	2.85	38.8	12.4	47.0	5.44	66.0	77.9
x 50	1560	87.7	8.9	8.2	30.8	195	126	3.06	45.8	13.3	55.1	5.96	76.0	72.4
x 45	1400	78.5	8.9	8.1	28.5	175	113	3.02	37.7	12.1	45.0	4.98	62.4	64.7
x 40	1240	68.5	9.1	8.0	26.5	155	101	2.92	33.2	10.6	35.2	3.83	48.6	57.5
x 35	1110	72.9	7.6	6.4	20.6	138	88.9	3.46	27.0	10.8	36.3	3.81	49.6	51.2
x 30	931	62.4	7.5	6.3	19.1	116	75.3	3.23	21.9	9.36	26.9	2.97	37.3	43.1
x 26	804	54.6	7.4	6.3	18.1	100	65.1	3.00	18.1	8.28	20.8	2.41	29.2	37.2
x 22	633	62.2	5.1	3.5	11.1	79.1	49.5	3.88	20.5	9.36	26.4	3.08	37.2	29.3
x 19	534	55.6	4.8	3.4	10.4	66.7	41.5	3.61	17.2	8.46	20.6	2.80	30.4	24.7
x 16	434	51.3	4.2	3.2	9.6	54.3	33.3	3.31	14.9	7.92	16.3	3.08	27.0	20.1
x 14	376	46.3	4.1	3.1	9.2	47.0	29.1	2.97	12.4	7.20	13.0	2.74	22.6	17.4
W 10x112	3180	167	9.5	11.2	86.4	397	246	2.01	127	27.2	224	27.8	223	147
x100	2810	147	9.6	11.0	77.4	351	218	2.00	107	24.5	182	23.2	193	130
x 88	2440	127	9.6	11.0	68.4	305	192	1.97	88.5	21.8	143	18.9	165	113
x 77	2110	109	9.7	10.8	60.1	264	168	1.95	71.5	19.1	110	14.8	138	97.6
x 68	1840	95.0	9.7	10.8	53.7	230	148	1.92	58.2	16.9	86.5	11.9	117	85.3

F_y = 36 ksi	BEAMS W shapes													

Uniform load constants for beams

$\phi_b = 0.90 \qquad \phi_v = 0.90 \qquad \phi = 1.00 \qquad \phi_r = 0.75$

For beams laterally unsupported, see page **3-53**

Shape	$\phi_b W_c$	$\phi_v V_n$	L_v	L_p	L_r	$\phi_b M_p$	$\phi_b M_r$	BF	ϕR_1	ϕR_2	$\phi_r R_3$	$\phi_r R_4$	R(N= 3½ in.)	Z_x
	Kip-ft	Kip	Ft	Ft	Ft	Kip-ft	Kip-ft	Kip	Kip	Kip/in.	Kip	Kip/in.	Kip	In.³
W 10x60	1610	83.4	9.7	10.7	48.1	201	130	1.91	49.6	15.1	68.7	9.79	103	74.6
x54	1440	72.6	9.9	10.7	43.9	180	117	1.89	41.6	13.3	54.0	7.49	80.2	66.6
x49	1300	66.0	9.9	10.6	40.7	163	106	1.88	36.3	12.2	45.4	6.46	68.0	60.4
x45	1190	68.7	8.6	8.4	35.1	148	95.7	1.96	39.4	12.6	49.9	6.29	71.9	54.9
x39	1010	60.7	8.3	8.3	31.2	126	82.1	1.93	31.9	11.3	39.4	5.46	58.5	46.8
x33	838	54.9	7.6	8.1	27.4	105	68.3	1.89	27.7	10.4	31.5	5.29	50.0	38.8
x30	791	61.1	6.5	5.7	20.3	98.8	63.2	2.44	25.3	10.8	35.9	4.64	52.2	36.6
x26	676	52.2	6.5	5.7	18.5	84.5	54.4	2.34	20.5	9.36	26.9	3.55	39.3	31.3
x22	562	47.4	5.9	5.5	16.9	70.2	45.2	2.18	16.2	8.64	21.6	3.47	33.7	26.0
x19	467	49.8	4.7	3.6	12.0	58.3	36.7	2.60	18.3	9.00	24.0	3.55	36.5	21.6
x17	404	47.2	4.3	3.5	11.2	50.5	31.6	2.46	16.2	8.64	20.7	3.80	34.0	18.7
x15	346	44.7	3.9	3.4	10.3	43.2	26.9	2.34	14.2	8.28	17.5	4.14	32.0	16.0
x12	272	36.5	3.7	3.3	9.5	34.0	21.3	1.98	10.7	6.84	11.6	3.04	22.2	12.6
W 8x67	1520	99.7	7.6	8.8	64.0	190	118	1.30	73.7	20.5	127	20.2	146	70.2
x58	1290	86.8	7.4	8.8	55.9	161	101	1.27	60.2	18.4	100	17.2	125	59.8
x48	1060	66.1	8.0	8.7	46.7	132	84.4	1.26	42.8	14.4	64.1	10.1	93.2	49.0
x40	860	57.7	7.4	8.5	39.1	107	69.2	1.25	34.4	13.0	49.5	9.27	79.8	39.8
x35	750	48.9	7.7	8.5	35.1	93.7	60.8	1.23	27.9	11.2	37.2	6.80	61.0	34.7
x31	657	44.3	7.4	8.4	32.0	82.1	53.6	1.20	24.0	10.3	30.7	6.11	52.1	30.4
x28	588	44.7	6.6	6.8	27.3	73.4	47.4	1.27	24.0	10.3	31.7	5.67	51.6	27.2
x24	501	37.8	6.6	6.7	24.4	62.6	40.8	1.24	19.3	8.82	23.5	4.26	38.4	23.2
x21	441	40.2	5.5	5.3	18.6	55.1	35.5	1.46	18.3	9.00	24.2	4.33	39.3	20.4
x18	367	36.4	5.0	5.1	16.7	45.9	29.6	1.40	15.5	8.28	19.4	4.16	33.9	17.0
x15	294	38.6	3.8	3.7	12.6	36.7	23.0	1.53	16.5	8.82	20.8	5.28	39.3	13.6
x13	246	35.7	3.5	3.5	11.5	30.8	19.3	1.43	14.2	8.28	17.0	5.48	36.2	11.4
x10	192	26.1	3.7	3.5	10.2	23.9	15.2	1.30	9.56	6.12	9.71	2.79	19.5	8.87
W 6x25	408	39.7	5.1	6.3	31.3	51.0	32.6	0.74	23.4	11.5	37.4	10.4	63.7	18.9
x20	322	31.3	5.1	6.3	25.6	40.2	26.1	0.73	17.5	9.36	24.5	7.13	49.5	14.9
x15**	230	26.8	4.3	6.6	20.7	28.8	19.0	0.69	12.9	8.28	17.2	7.17	41.9	10.8
x16	253	31.7	4.0	4.0	18.3	31.6	19.9	0.82	17.5	9.36	25.8	6.34	48.0	11.7
x12	179	27.0	3.3	3.8	14.4	22.4	14.3	0.78	12.9	8.28	17.9	6.62	41.0	8.30
x 9	135	19.5	3.5	3.8	12.0	16.8	10.8	0.72	8.61	6.12	9.95	3.56	22.4	6.23
W 5x19	251	27.0	4.6	5.3	30.3	31.3	19.9	0.46	19.7	9.72	28.2	8.16	53.8	11.6
x16	207	23.4	4.4	5.3	26.3	25.9	16.6	0.44	16.2	8.64	21.6	7.04	46.2	9.59
W 4x13	136	22.6	3.0	4.2	25.5	17.0	10.6	0.30	17.3	10.1	26.6	14.0	52.6	6.28

**Indicates noncompact shape. For noncompact shapes the maximum resisting moment ($\phi_b M'_n$) and the associated maximum unbraced length are tabulated as $\phi_b M_p$ and L_p.

BEAMS

F_y = 36 ksi

M shapes
Uniform load constants for beams

$\phi_b = 0.90 \qquad \phi_v = 0.90 \qquad \phi = 1.00 \qquad \phi_r = 0.75$

For beams laterally unsupported, see page **3-53**

Shape	$\phi_b W_c$	$\phi_v V_n$	L_v	L_p	L_r	$\phi_b M_p$	$\phi_b M_r$	BF	ϕR_1	ϕR_2	$\phi_r R_3$	$\phi_r R_4$	$R(N=$ 3½ in.)	Z_x
	Kip-ft	Kip	Ft	Ft	Ft	Kip-ft	Kip-ft	Kip	Kip	Kip/in.	Kip	Kip/in.	Kip	In.³
M 14x18	538	58.5	4.6	3.0	8.78	67.2	41.1	4.51	12.1	7.74	15.9	2.41	24.3	24.9
M 12x11.8	309	41.3	3.7	2.2	6.55	38.6	23.4	3.51	8.96	6.37	10.8	1.89	17.4	14.3
M 10x 9	199	30.5	3.3	2.0	5.91	24.8	15.1	2.48	7.95	5.65	8.64	1.72	14.7	9.19
M 8x 6.5	117	21.0	2.8	1.8	5.30	14.6	9.01	1.59	6.08	4.86	6.60	1.49	11.8	5.42
M 6x20	313	29.2	5.4	5.8	26.9	39.1	25.3	0.65	19.7	9.00	23.5	6.31	45.6	14.5
x 4.4	60.5	13.3	2.3	1.5	4.5	7.56	4.68	0.94	4.49	4.10	4.87	1.33	9.51	2.80
M 5x18.9	238	30.7	3.9	5.0	31.1	29.7	18.8	0.42	24.9	11.4	35.1	13.9	64.7	11.0
M 4x13	131	19.8	3.3	3.9	27.8	16.3	10.2	0.26	18.6	9.14	23.9	10.1	50.6	6.05

| $F_y = 36$ ksi | BEAMS — S shapes — Uniform load constants for beams — $\phi_b = 0.90$ $\phi_v = 0.90$ $\phi = 1.00$ $\phi_r = 0.75$ — For beams laterally unsupported, see page 3-53 |

Shape	$\phi_b W_c$ Kip-ft	$\phi_v V_n$ Kip	L_v Ft	L_p Ft	L_r Ft	$\phi_b M_p$ Kip-ft	$\phi_b M_r$ Kip-ft	BF Kip	ϕR_1 Kip	ϕR_2 Kip/in.	ϕR_3 Kip	ϕR_4 Kip/in.	$R(N=3\frac12$ in.) Kip	Z_x In.^{3}
S 24x121	6610	381	8.7	6.4	25.6	826	503	16.8	144	28.8	229	17.6	245	306
x106	6030	295	10.2	6.5	24.2	753	468	16.2	112	22.3	156	8.19	185	279
x100	5180	348	7.5	5.3	20.2	648	388	17.5	117	26.8	184	18.2	211	240
x 90	4800	292	8.2	5.4	19.4	599	365	16.9	98.4	22.5	141	10.7	177	222
x 80	4410	233	9.4	5.6	18.7	551	341	15.9	78.8	18.0	101	5.50	120	204
S 20x 96	4280	316	6.8	5.5	24.3	535	322	11.3	126	28.8	210	25.2	227	198
x 86	3950	260	7.6	5.7	22.9	494	302	11.2	104	23.8	157	14.1	187	183
x 75	3300	247	6.7	4.8	18.8	413	250	11.7	92.9	22.9	138	14.8	173	153
x 66	3020	196	7.7	5.0	17.8	378	232	11.3	73.9	18.2	97.9	7.44	124	140
S 18x 70	2700	249	5.4	4.5	19.3	338	201	9.22	96.0	25.6	152	26.5	186	125
x 54.7	2270	161	7.0	4.8	17.0	284	174	8.93	62.2	16.6	79.6	7.23	105	105
S 15x 50	1670	160	5.19	4.3	17.7	208	126	6.09	68.1	19.8	98.4	16.4	137	77.1
x 42.9	1500	120	6.24	4.5	16.5	187	116	5.90	50.9	14.8	63.6	6.83	87.5	69.3
S 12x 50	1320	160	4.1	4.3	24.3	165	99.1	3.31	88.9	24.7	141	37.6	175	61.2
x 40.8	1150	108	5.3	4.4	20.5	143	88.5	3.42	59.8	16.6	78.0	11.4	118	53.1
x 35	968	99.8	4.9	4.1	16.9	121	74.5	3.62	45.7	15.4	63.2	11.0	99.7	44.8
x 31.8	907	81.6	5.6	4.2	16.1	113	71.0	3.56	37.4	12.6	46.7	6.03	67.8	42.0
S 10x 35	765	115	3.3	3.8	20.9	95.6	57.3	2.23	60.1	21.4	98.2	39.2	135	35.4
x 25.4	613	60.5	5.1	4.0	16.1	76.7	48.2	2.36	31.5	11.2	37.2	5.62	56.9	28.4
S 8x 23	417	68.6	3.0	3.3	17.8	52.1	31.6	1.42	39.7	15.9	58.5	23.1	95.3	19.3
x 18.4	356	42.1	4.2	3.5	15.1	44.5	28.1	1.42	24.4	9.76	28.2	5.36	46.9	16.5
S 7x 20	313	61.2	2.6	3.1	18.3	39.1	23.6	1.02	38.0	16.2	57.8	30.5	94.7	14.5
x 15.3	261	34.3	3.8	3.2	14.4	32.7	20.5	1.08	21.3	9.07	24.2	5.35	43.0	12.1
S 6x 17.25	229	54.2	2.1	2.8	19.3	28.6	17.1	0.70	36.6	16.7	58.1	42.9	95.2	10.6
x 12.5	183	27.1	3.4	2.9	14.2	22.9	14.4	0.75	18.3	8.35	20.5	5.32	39.1	8.47
S 5x 14.75	160	48.0	1.7	2.6	21.9	20.0	11.9	0.42	36.1	17.8	60.7	67.9	98.4	7.42
x 10	122	20.8	2.9	2.7	13.9	15.3	9.59	0.51	15.6	7.70	17.3	5.52	36.6	5.67
S 4x 9.5	87.3	25.3	1.7	2.4	17.7	10.9	6.61	0.28	22.0	11.7	30.8	27.1	63.1	4.04
x 7.7	75.8	15.0	2.5	2.4	14.0	9.48	5.93	0.31	13.0	6.95	14.0	5.63	33.8	3.51
S 3x 7.5	51.0	20.4	1.3	2.1	21.5	6.37	3.80	0.13	21.6	12.6	32.2	50.0	65.6	2.36
x 5.7	42.1	9.9	2.1	2.2	14.7	5.27	3.28	0.16	10.5	6.12	10.9	5.78	31.2	1.95

AMERICAN INSTITUTE OF STEEL CONSTRUCTION

Notes

F_y = 36 ksi													

BEAMS
HP shapes
Uniform load constants for beams

$\phi_b = 0.90 \quad \phi_v = 0.90 \quad \phi = 1.00 \quad \phi_r = 0.75$

For beams laterally unsupported, see page **3-53**

Shape	ϕ_bW_c	ϕ_vV_n	L_v	L_p	L_r	ϕ_bM_p	ϕ_bM_r	BF	ϕR_1	ϕR_2	ϕ_rR_3	ϕ_rR_4	$R(N=3\tfrac{1}{2}\text{ in.})$	Z_x
	Kip-ft	Kip	Ft	Ft	Ft	Kip-ft	Kip-ft	Kip	Kip	Kip/in.	Kip	Kip/in.	Kip	In.³
HP 14x117	4190	222	9.4	15.0	66.0	524	335	3.69	109	29.0	198	41.9	210	194
x102	3650	192	9.5	14.8	59.0	456	293	3.71	87.2	25.4	152	32.6	176	169
x 89**	3080	165	9.3	17.1	53.0	386	255	3.63	72.6	22.1	116	25.1	150	146
x 73**	2370	134	8.9	21.0	46.7	297	209	3.41	54.0	18.2	78.0	17.2	118	118
HP 13x100	3300	196	8.5	13.2	60.1	413	263	3.19	99.0	27.5	179	40.9	195	153
x 87	2830	167	8.5	13.0	53.2	354	228	3.13	82.3	23.9	135	31.3	166	131
x 73**	2340	140	8.4	14.2	46.9	293	193	3.07	63.6	20.3	97.7	23.0	135	110
x 60**	1800	112	8.0	17.8	41.2	225	157	2.94	46.6	16.6	64.7	15.5	105	89.0
HP 12x 84	2590	164	7.9	12.3	54.0	324	207	2.81	84.8	24.7	144	35.1	171	120
x 74	2270	143	8.0	12.2	48.9	284	183	2.74	71.5	21.8	112	27.5	148	105
x 63**	1870	120	7.8	13.6	43.1	234	154	2.71	57.9	18.5	81.2	20.4	123	88.3
x 53**	1510	99.6	7.6	16.4	38.8	188	130	2.59	44.0	15.7	57.9	14.7	98.9	74.0
HP 10x 57	1440	110	6.6	10.2	45.6	180	115	1.84	60.4	20.3	97.7	29.3	132	66.5
x 42**	1020	78.3	6.5	11.7	35.9	127	84.6	1.77	39.7	14.9	53.0	16.1	92.0	48.3
HP 8x 36	726	69.4	5.2	8.1	35.7	90.7	58.1	1.18	37.5	16.0	60.6	22.7	93.6	33.6

**Indicates noncompact shape. For noncompact shapes the maximum resisting moment (ϕ_bM_n') and the associated maximum unbraced length are tabulated as ϕ_bM_p and L_p.

$F_y = 36$ ksi

BEAMS
C shapes
Uniform load constants for beams

$\phi_b = 0.90 \qquad \phi_v = 0.90 \qquad \phi = 1.00 \qquad \phi_r = 0.75$

For beams laterally unsupported, see page **3-53**

Shape	$\phi_b W_c$	$\phi_v V_n$	L_v	L_p	L_r	$\phi_b M_p$	$\phi_b M_r$	BF	ϕR_1	ϕR_2	$\phi_r R_3$	$\phi_r R_4$	$R(N=3\frac{1}{2}\text{ in.})$	Z_x
	Kip-ft	Kip	Ft	Ft	Ft	Kip-ft	Kip-ft	Kip	Kip	Kip/in.	Kip	Kip/in.	Kip	In.^{3}
C 15x50	1470	209	3.5	3.6	19.1	184	105	5.12	92.6	25.8	149	34.6	183	68.3
x40	1240	152	4.1	3.7	15.7	154	90.7	5.30	67.3	18.7	92.5	13.2	133	57.2
x33.9	1090	117	4.7	3.8	14.2	136	81.9	5.20	51.8	14.4	62.4	6.03	83.5	50.4
C 12x30	726	119	3.1	3.2	15.1	90.7	52.6	3.19	51.6	18.4	78.9	20.3	116	33.6
x25	631	90.3	3.5	3.3	13.1	78.8	47.0	3.24	39.2	13.9	52.1	8.85	83.1	29.2
x20.7	549	65.8	4.2	3.3	11.9	68.6	41.9	3.11	28.6	10.2	32.4	3.42	44.4	25.4
C 10x30	575	131	2.2	2.8	19.5	71.8	40.4	1.88	60.6	24.2	112	64.2	145	26.6
x25	497	102	2.4	2.8	15.7	62.1	35.5	2.07	47.3	18.9	77.1	30.6	114	23.0
x20	417	73.7	2.8	2.9	12.7	52.1	30.8	2.17	34.1	13.6	47.1	11.5	81.9	19.3
x15.3	341	46.7	3.7	3.0	10.8	42.7	26.3	2.09	21.6	8.64	23.8	2.91	33.9	15.8
C 9x20	363	78.4	2.3	2.7	14.3	45.4	26.3	1.64	37.8	16.1	59.0	22.2	94.3	16.8
x15	292	49.9	2.9	2.8	11.2	36.5	22.0	1.71	24.0	10.3	29.9	5.72	49.9	13.5
x13.4	270	40.8	3.3	2.8	10.5	33.8	20.7	1.69	19.7	8.39	22.1	3.12	33.1	12.5
C 8x18.75	298	75.7	2.0	2.5	15.6	37.3	21.4	1.20	41.1	17.5	64.9	34.0	102	13.8
x13.75	235	47.1	2.5	2.6	11.5	29.4	17.6	1.33	25.6	10.9	31.9	8.18	60.5	10.9
x11.5	206	34.2	3.0	2.6	10.2	25.8	15.9	1.30	18.6	7.92	19.7	3.13	30.7	9.55
C 7x14.75	209	57.0	1.8	2.4	14.6	26.1	15.2	0.89	33.0	15.1	50.2	26.4	85.8	9.68
x12.25	181	42.7	2.1	2.4	11.9	22.7	13.5	0.96	24.7	11.3	32.6	11.1	64.3	8.40
x 9.8	154	28.6	2.7	2.4	10.0	19.2	11.9	0.97	16.5	7.56	17.8	3.32	29.4	7.12
C 6x13	157	51.0	1.5	2.2	15.9	19.6	11.3	0.60	32.0	15.7	51.8	37.2	87.0	7.26
x10.5	133	36.6	1.8	2.2	12.4	16.6	9.87	0.66	23.0	11.3	31.5	13.8	62.5	6.15
x 8.2	111	23.3	2.4	2.2	10.0	13.9	8.54	0.68	14.6	7.20	16.0	3.57	28.5	5.13
C 5x 9	94.2	31.6	1.5	2.0	13.5	11.8	6.94	0.42	21.9	11.7	32.1	19.7	62.9	4.36
x 6.7	75.8	18.5	2.1	2.1	10.2	9.48	5.85	0.45	12.8	6.84	14.3	3.94	28.1	3.51
C 4x 7.25	60.7	25.0	1.2	1.9	15.0	7.59	4.47	0.24	19.9	11.6	30.3	25.6	60.3	2.81
x 5.4	48.8	14.3	1.7	1.9	10.8	6.10	3.76	0.26	11.4	6.62	13.1	4.83	30.0	2.26
C 3x 6	37.2	20.8	0.9	1.7	20.0	4.64	2.69	0.11	22.0	12.8	34.0	50.6	66.9	1.72
x 5	32.4	15.0	1.1	1.7	15.1	4.05	2.42	0.12	16.0	9.29	21.0	19.2	48.5	1.50
x 4.1	28.1	9.9	1.4	1.7	12.1	3.51	2.14	0.13	10.5	6.12	11.2	5.51	30.5	1.30

AMERICAN INSTITUTE OF STEEL CONSTRUCTION

	F_y = 36 ksi

BEAMS
MC shapes
Uniform load constants for beams
$$\phi_b = 0.90 \qquad \phi_v = 0.90 \qquad \phi = 1.00 \qquad \phi_r = 0.75$$

For beams laterally unsupported, see page **3-53**

Shape	$\phi_b W_c$	$\phi_v V_n$	L_v	L_p	L_r	$\phi_b M_p$	$\phi_b M_r$	BF	ϕR_1	ϕR_2	$\phi_r R_3$	$\phi_r R_4$	$R(N=3\frac{1}{2}\text{ in.})$	Z_x
	Kip-ft	Kip	Ft	Ft	Ft	Kip-ft	Kip-ft	Kip	Kip	Kip/in.	Kip	Kip/in.	Kip	In.^{3}
MC 18x58	2040	245	4.2	4.3	18.7	255	146	7.55	86.6	25:2	142	28.0	175	94.6
x51.9	1870	210	4.5	4.3	17.2	234	136	7.61	74.3	21.6	112	17.6	150	86.5
x45.8	1690	175	4.8	4.4	15.8	212	125	7.55	61.9	18.0	85.5	10.2	121	78.4
x42.7	1610	157	5.1	4.5	15.3	201	120	7.43	55.7	16.2	73.0	7.44	99.1	74.4
MC 13x50	1310	199	3.3	4.4	26.9	163	94.4	3.07	97.4	28.3	167	56.4	197	60.5
x40	1100	142	3.9	4.5	21.1	137	81.9	3.34	69.3	20.2	100	20.3	140	50.9
x35	998	113	4.4	4.6	19.1	125	75.7	3.39	55.3	16.1	71.4	10.3	108	46.2
x31.8	931	94.8	4.9	4.6	17.9	116	71.8	3.35	46.4	13.5	54.9	6.10	76.2	43.1
MC 12x50	1210	195	3.1	4.5	30.8	151	87.6	2.44	98.6	30.1	195	63.6	204	56.1
x45	1120	166	3.4	4.5	26.7	140	81.9	2.60	84.1	25.6	154	39.4	174	51.7
x40	1020	138	3.7	4.6	23.6	128	76.0	2.71	69.7	21.2	116	22.4	144	47.3
x35	924	109	4.2	4.6	21.0	116	70.4	2.76	55.2	16.8	81.7	11.1	114	42.8
x31	849	86.3	4.9	4.7	19.7	106	65.9	2.67	43.7	13.3	57.6	5.54	77.0	39.3
x10.6	251	44.3	2.8	1.5	4.8	31.3	18.0	4.04	11.8	6.84	14.1	1.70	20.0	11.6
MC 10x41.1	840	155	2.7	4.8	34.7	105	61.4	1.46	89.6	28.7	165	80.5	190	38.9
x33.6	721	112	3.2	4.8	26.7	90.2	54.2	1.65	64.7	20.7	101	30.4	137	33.4
x28.5	639	82.6	3.9	4.9	22.5	79.9	49.3	1.74	47.8	15.3	64.3	12.3	101	29.6
x25	557	73.9	3.8	4.2	18.9	69.7	42.9	1.81	42.8	13.7	54.4	8.76	85.0	25.8
x22	510	56.4	4.5	4.2	17.3	63.7	40.0	1.81	32.6	10.4	36.2	3.89	49.9	23.6
x 8.4	170	33.0	2.6	1.5	5.0	21.2	12.5	2.55	10.5	6.12	11.3	1.61	17.0	7.86
x 6.5	124	29.5	2.1	1.0	3.3	15.5	8.62	2.95	5.98	5.47	8.15	1.60	13.7	5.73
MC 9x25.4	501	78.7	3.2	4.2	21.8	62.6	38.2	1.39	48.1	16.2	68.5	16.9	105	23.2
x23.9	480	70.0	3.4	4.2	20.5	59.9	36.9	1.41	42.8	14.4	57.4	11.9	93.2	22.2
MC 8x22.8	406	66.4	3.1	4.3	23.4	50.8	31.2	1.03	45.6	15.4	61.9	17.0	99.4	18.8
x21.4	389	58.3	3.3	4.3	21.9	48.6	30.0	1.05	40.1	13.5	50.9	11.5	87.3	18.0
x20	350	62.2	2.8	3.6	19.3	43.7	26.5	1.10	40.5	14.4	54.7	14.7	90.9	16.2
x18.7	333	54.9	3.0	3.6	18.1	41.6	25.5	1.11	35.7	12.7	45.4	10.1	80.2	15.4
x 8.5	149	27.8	2.7	2.1	7.3	18.7	11.4	1.40	12.1	6.44	12.9	2.12	20.3	6.91
MC 7x22.7	350	68.4	2.6	4.4	28.9	43.7	26.5	0.70	50.9	18.1	77.2	33.4	114	16.2
x19.1	309	47.9	3.2	4.3	23.6	38.6	24.0	0.76	35.6	12.7	45.2	11.4	80.0	14.3
x17.6	274	51.0	2.7	3.7	21.3	34.3	21.1	0.75	35.9	13.5	48.4	14.6	83.1	12.7
MC 6x18	248	44.2	2.8	4.4	27.8	31.0	19.3	0.50	36.2	13.6	49.2	17.5	84.0	11.5
x15.3	212	39.7	2.7	4.4	23.1	26.5	16.5	0.53	26.8	12.2	37.6	15.6	69.6	9.82
x16.3	220	43.7	2.5	3.7	23.9	27.5	16.9	0.53	35.9	13.5	48.4	17.0	83.1	10.2
x15.1	209	36.9	2.8	3.7	22.2	26.2	16.2	0.54	30.2	11.4	37.5	10.2	70.0	9.69
x12	159	36.2	2.2	3.0	16.2	19.9	12.2	0.60	22.7	11.2	32.3	12.2	61.7	7.38

BEAMS
W shapes
Uniform load constants for beams

$\phi_b = 0.90 \qquad \phi_v = 0.90 \qquad \phi = 1.00 \qquad \phi_r = 0.75$

$F_y = 50$ ksi

For beams laterally unsupported, see page **3-53**

Shape	$\phi_b W_c$	$\phi_v V_n$	L_v	L_p	L_r	$\phi_b M_p$	$\phi_b M_r$	BF	ϕR_1	ϕR_2	$\phi_r R_3$	$\phi_r R_4$	$R(N=3\frac{1}{2}$ in.$)$	Z_x
	Kip-ft	Kip	Ft	Ft	Ft	Kip-ft	Kip-ft	Kip	Kip	Kip/in.	Kip	Kip/in.	Kip	In.^{3}
W 36x359	45300	1131	20.0	13.7	44.0	5660	3960	56.2	437	56.0	606	20.2	634	1510
x328	41400	1021	20.3	13.6	41.8	5180	3630	54.7	383	51.0	505	16.7	561	1380
x300	37800	937	20.2	13.5	39.9	4730	3330	52.9	332	47.3	429	14.8	481	1260
x280	35100	873	20.1	13.5	38.8	4390	3090	51.3	297	44.3	376	13.1	422	1170
x260	32400	822	19.7	13.4	37.5	4050	2860	49.4	269	42.0	333	12.3	376	1080
x245	30300	779	19.4	13.3	36.4	3790	2690	47.7	250	40.0	300	11.4	340	1010
x230	28300	737	19.2	13.2	35.6	3540	2510	45.8	226	38.0	268	10.5	305	943
x256	31200	970	16.1	9.37	26.9	3900	2690	54.0	315	48.0	446	14.8	483	1040
x232	28100	872	16.1	9.26	27.3	3510	2430	60.0	272	43.5	367	12.2	409	936
x210	25000	822	15.2	9.12	26.1	3120	2160	56.8	240	41.5	318	12.4	361	833
x194	23000	754	15.3	9.05	25.3	2880	1990	54.6	209	38.3	271	10.5	308	767
x182	21500	711	15.1	9.02	24.9	2690	1870	52.0	193	36.3	242	9.62	275	718
x170	20000	664	15.1	8.94	24.4	2510	1740	49.6	170	34.0	212	8.55	242	668
x160	18700	632	14.8	8.84	23.7	2340	1630	48.0	157	32.5	191	8.09	219	624
x150	17400	605	14.4	8.73	23.4	2180	1510	45.6	146	31.2	173	7.84	200	581
x135	15300	576	13.3	8.41	22.4	1910	1320	42.2	127	30.0	149	8.32	178	509
W 33x354	42600	1113	19.1	13.2	44.9	5330	3690	51.6	417	58.0	651	22.7	620	1420
x318	38100	987	19.3	13.1	41.8	4760	3330	49.9	349	52.0	526	18.3	531	1270
x291	34500	903	19.1	13.0	40.0	4310	3030	47.6	308	48.0	446	15.9	476	1150
x263	31200	811	19.2	12.9	37.9	3900	2750	46.1	258	43.5	367	13.1	411	1040
x241	28200	766	18.4	12.8	36.2	3520	2490	44.2	227	41.5	323	12.9	368	939
x221	25700	710	18.1	12.7	35.0	3210	2270	41.9	200	38.8	278	11.6	319	855
x201	23200	650	17.8	12.6	33.8	2900	2050	39.7	173	35.8	234	10.2	270	772
x169	18900	612	15.4	8.84	24.5	2360	1650	45.5	173	33.5	218	7.89	246	629
x152	16800	574	14.6	8.73	23.7	2100	1460	42.3	149	31.8	187	7.84	215	559
x141	15400	544	14.2	8.59	23.1	1930	1340	40.2	132	30.3	166	7.49	193	514
x130	14000	518	13.5	8.45	22.5	1750	1220	37.9	122	29.0	147	7.46	173	467
x118	12500	488	12.8	8.20	21.7	1560	1080	35.5	107	27.5	127	7.40	152	415
W 30x235	25400	701	18.1	12.4	37.1	3170	2240	37.8	233	41.5	334	13.2	379	845
x211	22500	647	17.4	12.3	35.1	2810	1990	36.0	206	38.8	282	12.4	325	749
x191	20200	588	17.2	12.2	33.7	2520	1790	33.9	172	35.5	235	10.7	272	673
x173	18200	538	16.9	12.1	32.5	2270	1620	32.0	154	32.8	197	9.38	230	605
x148	15000	538	13.9	8.06	22.8	1880	1310	38.5	163	32.5	205	8.21	234	500
x132	13100	503	13.0	7.95	21.9	1640	1140	35.6	135	30.8	174	8.30	203	437
x124	12200	477	12.8	7.88	21.5	1530	1070	34.1	123	29.2	156	7.72	183	408
x116	11300	458	12.4	7.74	20.8	1420	987	33.0	115	28.3	141	7.65	168	378
x108	10400	439	11.8	7.60	20.3	1300	897	31.5	106	27.3	126	7.73	154	346
x 99	9360	416	11.2	7.42	19.8	1170	807	29.2	93.4	26.0	111	7.66	137	312
W 27x217	21200	637	16.7	11.7	36.7	2660	1870	31.3	227	41.5	334	14.5	372	708
x194	18800	569	16.5	11.6	34.5	2360	1670	30.0	193	37.5	271	12.1	314	628
x178	17000	544	15.6	11.5	33.1	2130	1510	28.8	170	36.3	243	12.5	286	567
x161	15400	492	15.6	11.5	31.7	1920	1370	27.4	150	33.0	201	10.4	237	512
x146	13800	447	15.5	11.3	30.6	1730	1230	25.8	128	30.3	168	8.97	199	461
x129	11900	455	13.0	7.81	22.2	1480	1040	30.9	138	30.5	180	8.08	208	395
x114	10300	420	12.3	7.71	21.3	1290	897	28.7	116	28.5	150	7.89	177	343
x102	9150	377	12.1	7.60	20.4	1140	801	26.7	101	25.8	121	6.57	144	305
x 94	8340	356	11.7	7.50	19.9	1040	729	25.2	88.0	24.5	107	6.35	129	278

AMERICAN INSTITUTE OF STEEL CONSTRUCTION

$F_y = 50$ ksi	BEAMS													

BEAMS
W shapes
Uniform load constants for beams
$$\phi_b = 0.90 \qquad \phi_v = 0.90 \qquad \phi = 1.00 \qquad \phi_r = 0.75$$
For beams laterally unsupported, see page **3-53**

Shape	$\phi_b W_c$	$\phi_v V_n$	L_v	L_p	L_r	$\phi_b M_p$	$\phi_b M_r$	BF	ϕR_1	ϕR_2	$\phi_r R_3$	$\phi_r R_4$	$R(N=3\frac{1}{2}\text{ in.})$	Z_x
	Kip-ft	Kip	Ft	Ft	Ft	Kip-ft	Kip-ft	Kip	Kip	Kip/in.	Kip	Kip/in.	Kip	In.^{3}
W 27x 84	7320	332	11.0	7.32	19.3	915	639	23.0	79.1	23.0	90.0	6.16	112	244
W 24x176	15300	511	15.0	10.7	33.8	1920	1350	24.6	199	37.5	271	13.5	318	511
x162	14000	476	14.8	10.8	32.4	1760	1240	23.8	176	35.3	236	12.4	279	468
x146	12500	434	14.4	10.6	30.6	1570	1110	22.8	152	32.5	197	11.0	236	418
x131	11100	400	13.9	10.5	29.1	1390	987	21.5	132	30.3	166	10.2	202	370
x117	9810	360	13.6	10.4	27.9	1230	873	20.2	112	27.5	136	8.73	166	327
x104	8670	325	13.3	10.3	26.8	1080	774	18.8	93.7	25.0	110	7.49	137	289
x103	8400	364	11.5	7.0	20.1	1050	735	24.2	120	27.5	146	7.49	172	280
x 94	7620	338	11.3	7.0	19.4	953	666	23.0	105	25.8	125	6.95	149	254
x 84	6720	306	11.0	6.9	18.6	840	588	21.5	91.8	23.5	102	6.05	123	224
x 76	6000	284	10.6	6.8	18.0	750	528	19.8	79.1	22.0	86.8	5.67	107	200
x 68	5310	266	10.0	6.6	17.4	664	462	18.7	71.3	20.8	73.7	5.57	93.2	177
x 62	4590	276	8.3	4.9	13.3	574	393	21.4	73.9	21.5	78.1	6.14	99.6	153
x 55	4020	251	8.0	4.7	12.9	503	342	19.6	64.8	19.8	63.6	5.60	83.2	134
W 21x166	13000	455	14.2	10.5	35.7	1620	1140	19.0	199	37.5	273	14.9	325	432
x147	11200	429	13.0	10.4	32.8	1400	987	18.4	169	36.0	236	15.9	292	373
x132	9990	383	13.0	10.4	30.9	1250	885	17.7	147	32.5	192	13.1	238	333
x122	9210	351	13.1	10.3	29.8	1150	819	17.1	127	30.0	164	11.2	204	307
x111	8370	319	13.1	10.3	28.5	1050	747	16.4	112	27.5	138	9.56	171	279
x101	7590	288	13.2	10.2	27.6	949	681	15.4	97.7	25.0	114	7.91	142	253
x 93	6630	339	9.8	6.5	19.4	829	576	19.6	122	29.0	154	10.5	190	221
x 83	5880	298	9.9	6.5	18.5	735	513	18.5	101	25.8	122	8.26	151	196
x 73	5160	261	9.9	6.4	17.7	645	453	17.0	85.3	22.8	95.2	6.48	118	172
x 68	4800	245	9.8	6.4	17.3	600	420	16.5	77.3	21.5	84.2	5.94	105	160
x 62	4320	227	9.5	6.3	16.6	540	381	15.3	68.7	20.0	71.5	5.36	90.3	144
x 57	3870	230	8.4	4.8	13.1	484	333	18.0	69.6	20.3	74.9	5.25	93.3	129
x 50	3300	214	7.7	4.6	12.5	413	283	16.4	62.3	19.0	61.8	5.33	80.4	110
x 44	2860	195	7.3	4.5	12.0	358	245	14.9	52.0	17.5	50.1	4.99	67.6	95.4
W 18x143	9660	384	12.6	9.6	35.4	1210	846	14.0	183	36.5	258	16.4	310	322
x130	8730	348	12.5	9.6	33.0	1090	768	13.8	157	33.5	217	14.1	266	291
x119	7830	335	11.7	9.5	30.8	979	693	13.4	143	32.8	197	15.1	250	261
x106	6900	298	11.6	9.4	28.7	863	612	13.0	120	29.5	158	12.6	203	230
x 97	6330	269	11.8	9.4	27.4	791	564	12.6	104	26.8	132	10.2	167	211
x 86	5580	238	11.7	9.3	26.1	698	498	11.9	86.3	24.0	105	8.45	135	186
x 76	4890	209	11.7	9.2	24.8	611	438	11.1	73.0	21.3	82.4	6.71	106	163
x 71	4350	247	8.8	6.0	17.8	544	381	13.8	92.8	24.8	113	8.77	144	145
x 65	3990	223	9.0	6.0	17.1	499	351	13.3	80.9	22.5	94.3	7.16	119	133
x 60	3690	204	9.0	6.0	16.7	461	324	12.8	71.3	20.8	80.4	6.10	102	123
x 55	3360	191	8.8	5.9	16.1	420	295	12.2	64.0	19.5	69.7	5.62	89.4	112
x 50	3030	172	8.8	5.8	15.6	379	267	11.5	55.5	17.8	57.6	4.72	74.1	101
x 46	2720	176	7.8	4.6	12.6	340	236	13.0	56.3	18.0	60.6	4.62	76.8	90.7
x 40	2350	152	7.7	4.5	12.1	294	205	11.7	46.8	15.8	46.2	3.60	58.8	78.4
x 35	2000	143	7.0	4.3	11.5	249	173	10.7	42.2	15.0	38.6	3.88	52.2	66.5
W 16x100	5940	268	11.1	8.9	29.3	743	525	10.7	123	29.2	160	13.0	205	198
x 89	5250	237	11.1	8.8	27.3	656	465	10.3	103	26.2	128	10.7	166	175

BEADS

$F_y = 50$ ksi

BEAMS
W shapes
Uniform load constants for beams

$\phi_b = 0.90 \qquad \phi_v = 0.90 \qquad \phi = 1.00 \qquad \phi_r = 0.75$

For beams laterally unsupported, see page **3-53**

Shape	$\phi_b W_c$	$\phi_v V_n$	L_v	L_p	L_r	$\phi_b M_p$	$\phi_b M_r$	BF	$\phi_r R_1$	$\phi_r R_2$	$\phi_r R_3$	$\phi_r R_4$	$R(N=3\frac{1}{2}$ in.)	Z_x
	Kip-ft	Kip	Ft	Ft	Ft	Kip-ft	Kip-ft	Kip	Kip	Kip/in.	Kip	Kip/in.	Kip	In.³
W 16x 77	4500	203	11.1	8.7	25.2	563	402	9.75	81.8	22.8	96.5	8.12	125	150
x 67	3900	174	11.2	8.7	23.8	488	351	9.02	67.9	19.8	73.0	6.14	94.5	130
x 57	3150	191	8.3	5.7	16.6	394	277	10.7	73.9	21.5	86.0	7.32	112	105
x 50	2760	167	8.3	5.6	15.8	345	243	10.1	62.3	19.0	67.1	5.80	87.3	92.0
x 45	2470	150	8.2	5.6	15.1	309	218	9.43	53.9	17.3	54.9	4.87	72.0	82.3
x 40	2190	132	8.3	5.6	14.7	273	194	8.67	45.3	15.3	43.2	3.80	56.5	72.9
x 36	1920	126	7.6	5.4	14.1	240	170	8.08	41.5	14.7	37.9	4.07	52.1	64.0
x 31	1620	118	6.9	4.1	11.0	203	142	8.85	38.7	13.8	34.5	3.22	45.8	54.0
x 26	1330	106	6.3	4.0	10.4	166	115	7.88	33.2	12.5	26.5	3.12	37.4	44.2
W 14x132	7020	255	13.7	13.3	49.6	878	627	6.89	136	32.3	190	19.2	249	234
x120	6360	231	13.8	13.2	46.2	795	570	6.82	120	29.5	158	16.3	216	212
x109	5760	203	14.2	13.2	43.2	720	519	6.70	103	26.2	127	12.7	172	192
x 99**	5180	185	14.0	13.4	40.6	647	471	6.46	87.1	24.3	108	11.2	147	173
x 90**	4620	167	13.9	15.0	38.4	577	429	6.31	75.6	22.0	88.7	9.26	121	157
x 82	4170	197	10.6	8.8	29.6	521	369	7.31	104	25.5	121	11.7	163	139
x 74	3780	172	11.0	8.8	28.0	472	336	7.12	87.9	22.5	96.5	8.86	127	126
x 68	3450	157	11.0	8.7	26.4	431	309	6.91	77.8	20.8	81.8	7.65	109	115
x 61	3060	141	10.9	8.7	24.9	383	277	6.51	67.4	18.7	66.5	6.37	88.8	102
x 53	2610	139	9.4	6.8	20.1	327	233	7.02	66.5	18.5	65.9	5.96	86.8	87.1
x 48	2350	127	9.3	6.8	19.2	294	211	6.70	58.4	17.0	55.1	5.18	73.3	78.4
x 43	2090	112	9.3	6.7	18.2	261	188	6.32	50.0	15.3	44.2	4.24	59.1	69.6
x 38	1850	118	7.8	5.5	14.9	231	164	7.07	41.2	15.5	44.7	4.44	60.2	61.5
x 34	1640	108	7.6	5.4	14.4	205	146	6.58	35.6	14.3	37.0	3.94	50.8	54.6
x 30	1420	101	7.0	5.3	13.7	177	126	6.06	31.6	13.5	31.4	4.00	45.4	47.3
x 26	1210	95.8	6.3	3.8	10.3	151	106	6.96	29.9	12.8	30.1	3.07	40.8	40.2
x 22	996	85.3	5.8	3.7	9.7	125	87.0	6.26	25.2	11.5	23.0	2.86	33.0	33.2
W 12x 87	3960	174	11.4	10.9	38.4	495	354	5.12	96.6	25.8	120	14.6	171	132
x 79	3570	157	11.4	10.8	35.7	446	321	5.03	84.5	23.5	99.6	12.3	143	119
x 72	3240	142	11.4	10.7	33.6	405	292	4.93	73.9	21.5	83.2	10.5	120	108
x 65**	2860	128	11.2	11.8	31.7	358	264	4.72	64.0	19.5	68.3	8.75	98.9	96.8
x 58	2590	118	10.9	8.9	27.0	324	234	4.96	61.9	18.0	62.3	6.47	85.0	86.4
x 53	2340	112	10.4	8.8	25.6	292	212	4.77	53.9	17.3	55.4	6.41	77.8	77.9
x 50	2170	122	8.9	6.9	21.7	271	194	5.25	63.6	18.5	64.9	7.02	89.5	72.4
x 45	1940	109	8.9	6.9	20.3	243	174	5.07	52.3	16.8	53.0	5.87	73.6	64.7
x 40	1730	95.1	9.1	6.8	19.3	216	156	4.82	46.1	14.7	41.5	4.52	57.3	57.5
x 35	1540	101	7.6	5.4	15.2	192	137	5.67	37.5	15.0	42.7	4.49	58.5	51.2
x 30	1290	86.6	7.5	5.4	14.4	162	116	5.10	30.5	13.0	31.7	3.50	44.0	43.1
x 26	1120	75.9	7.4	5.3	13.8	140	100	4.64	25.2	11.5	24.5	2.83	34.4	37.2
x 22	879	86.4	5.1	3.0	8.4	110	76.2	6.24	28.4	13.0	31.2	3.63	43.9	29.3
x 19	741	77.2	4.8	2.9	7.9	92.6	63.9	5.70	23.9	11.8	24.3	3.30	35.9	24.7
x 16	603	71.2	4.2	2.7	7.4	75.4	51.3	5.12	20.6	11.0	19.2	3.63	31.8	20.1
x 14	522	64.3	4.1	2.7	7.2	65.3	44.7	4.56	17.2	10.0	15.3	3.23	26.6	17.4
W 10x112	4410	232	9.5	9.5	56.5	551	378	3.68	177	37.8	265	32.8	309	147
x100	3900	204	9.6	9.4	50.8	488	336	3.66	149	34.0	214	27.4	268	130
x 88	3390	177	9.6	9.3	45.1	424	295	3.58	123	30.3	169	22.3	229	113
x 77	2930	152	9.7	9.2	39.9	366	258	3.53	99.4	26.5	130	17.5	191	97.6
x 68	2560	132	9.7	9.2	36.0	320	227	3.46	80.8	23.5	102	14.0	151	85.3

**Indicates noncompact shape. For noncompact shapes the maximum resisting moment ($\phi_b M_n'$) and the associated maximum unbraced length are tabulated as $\phi_b M_p$ and L_p.

$F_y = 50$ ksi	BEANS W shapes Uniform load constants for beams

$\phi_b = 0.90 \qquad \phi_v = 0.90 \qquad \phi = 1.00 \qquad \phi_r = 0.75$

For beams laterally unsupported, see page **3-53**

Shape	$\phi_b W_c$	$\phi_v V_n$	L_v	L_p	L_r	$\phi_b M_p$	$\phi_b M_r$	BF	ϕR_1	ϕR_2	$\phi_r R_3$	$\phi_r R_4$	$R(N=3\frac{1}{2}\text{ in.})$	Z_x
	Kip-ft	Kip	Ft	Ft	Ft	Kip-ft	Kip-ft	Kip	Kip	Kip/in.	Kip	Kip/in.	Kip	In.^{3}
W 10x60	2240	116	9.7	9.1	32.6	280	200	3.38	68.9	21.0	80.9	11.5	121	74.6
x54	2000	101	9.9	9.1	30.2	250	180	3.30	57.8	18.5	63.6	8.83	94.6	66.6
x49	1810	91.6	9.9	9.0	28.3	227	164	3.25	50.5	17.0	53.5	7.61	80.1	60.4
x45	1650	95.4	8.6	7.1	24.1	206	147	3.45	54.7	17.5	58.8	7.41	84.7	54.9
x39	1400	84.4	8.3	7.0	21.8	176	126	3.32	44.3	15.8	46.4	6.43	68.9	46.8
x33	1160	76.2	7.6	6.9	19.7	145	105	3.15	38.5	14.5	37.1	6.23	59.0	38.8
x30	1100	84.8	6.5	4.8	14.5	137	97.2	4.13	35.2	15.0	42.3	5.47	61.5	36.6
x26	939	72.5	6.5	4.8	13.5	117	83.7	3.85	28.4	13.0	31.7	4.18	46.4	31.3
x22	780	65.9	5.9	4.7	12.7	97.5	69.6	3.50	22.5	12.0	25.4	4.08	39.7	26.0
x19	648	69.1	4.7	3.1	8.9	81.0	56.4	4.26	25.4	12.5	28.3	4.18	43.0	21.6
x17	561	65.5	4.3	3.0	8.4	70.1	48.6	3.97	22.5	12.0	24.4	4.48	40.0	18.7
x15	480	62.0	3.9	2.9	7.9	60.0	41.4	3.69	19.8	11.5	20.7	4.88	37.7	16.0
x12**	376	50.6	3.7	2.8	7.4	47.0	32.7	3.13	14.8	9.50	13.7	3.58	26.2	12.6
W 8x67	2110	139	7.6	7.5	41.9	263	181	2.38	102	28.5	150	23.8	202	70.2
x58	1790	120	7.4	7.4	36.8	224	156	2.32	83.7	25.5	118	20.2	173	59.8
x48	1470	91.8	8.0	7.4	31.1	184	130	2.27	59.4	20.0	75.5	11.9	117	49.0
x40	1190	80.2	7.4	7.2	26.4	149	107	2.22	47.8	18.0	58.3	10.9	96.5	39.8
x35	1040	68.0	7.7	7.2	24.1	130	93.6	2.16	38.8	15.5	43.8	8.02	71.9	34.7
x31	912	61.6	7.4	7.1	22.3	114	82.5	2.07	33.4	14.3	36.2	7.20	61.4	30.4
x28	816	62.0	6.6	5.7	18.9	102	72.9	2.22	33.4	14.3	37.4	6.68	60.8	27.2
x24	696	52.5	6.6	5.7	17.2	87.0	62.7	2.11	26.8	12.3	27.7	5.02	45.2	23.2
x21	612	55.9	5.5	4.5	13.3	76.5	54.6	2.47	25.4	12.5	28.5	5.10	46.4	20.4
x18	510	50.5	5.0	4.3	12.3	63.8	45.6	2.30	21.6	11.5	22.9	4.90	40.0	17.0
x15	408	53.6	3.8	3.1	9.2	51.0	35.4	2.56	23.0	12.3	24.5	6.23	46.3	13.6
x13	342	49.6	3.5	3.0	8.5	42.8	29.7	2.35	19.8	11.5	20.1	6.46	42.7	11.4
x10**	264	36.2	3.7	3.1	7.8	33.0	23.4	2.03	13.3	8.50	11.4	3.29	22.9	8.87
W 6x25	567	55.1	5.1	5.4	21.0	70.9	50.1	1.33	32.5	16.0	44.0	12.2	86.8	18.9
x20	447	43.5	5.1	5.3	17.7	55.9	40.2	1.27	24.4	13.0	28.9	8.40	58.3	14.9
x15**	309	37.2	4.4	6.8	14.9	38.6	29.2	1.16	18.0	11.5	20.3	8.45	49.9	10.8
x16	351	44.1	4.0	3.4	12.5	43.9	30.6	1.46	24.4	13.0	30.4	7.48	56.6	11.7
x12	249	37.4	3.3	3.2	10.2	31.1	21.9	1.33	18.0	11.5	21.0	7.80	48.3	8.30
x 9	187	27.1	3.5	3.2	8.9	23.4	16.7	1.17	12.0	8.50	11.7	4.19	26.4	6.23
W 5x19	348	37.5	4.6	4.5	20.1	43.5	30.6	0.83	27.4	13.5	33.2	9.62	66.8	11.6
x16	288	32.5	4.4	4.5	17.6	36.0	25.5	0.80	22.5	12.0	25.4	8.29	54.5	9.59
W 4x13	188	31.4	3.0	3.5	16.9	23.6	16.4	0.54	24.1	14.0	31.4	16.5	73.1	6.28

**Indicates noncompact shape. For noncompact shapes the maximum resisting moment ($\phi_b M_n'$) and the associated maximum unbraced length are tabulated as $\phi_b M_p$ and L_p.

BEAMS

$F_y = 50$ ksi

M shapes
Uniform load constants for beams

$$\phi_b = 0.90 \qquad \phi_v = 0.90 \qquad \phi = 1.00 \qquad \phi_r = 0.75$$

For beams laterally unsupported, see page **3-53**

Shape	$\phi_b W_c$	$\phi_v V_n$	L_v	L_p	L_r	$\phi_b M_p$	$\phi_b M_r$	BF	ϕR_1	ϕR_2	$\phi_r R_3$	$\phi_r R_4$	$R(N=3\frac{1}{2}$ in.$)$	Z_x
	Kip-ft	Kip	Ft	Ft	Ft	Kip-ft	Kip-ft	Kip	Kip	Kip/in.	Kip	Kip/in.	Kip	In.³
M 14x18	747	81.3	4.6	2.5	6.9	93.4	63.3	6.93	16.8	10.8	18.7	2.84	28.6	24.9
M 12x11.8	429	57.3	3.7	1.9	5.1	53.6	36.0	5.40	12.4	8.85	12.7	2.22	20.5	14.3
M 10x 9	276	42.4	3.3	1.7	4.6	34.5	23.3	3.84	11.0	7.85	10.2	2.03	17.3	9.19
M 8x 6.5	163	29.2	2.8	1.5	4.1	20.3	13.9	2.47	8.44	6.75	7.78	1.76	13.9	5.42
M 6x20	435	40.5	5.4	4.9	18.2	54.4	39.0	1.16	27.3	12.5	27.8	7.43	53.8	14.5
x 4.4	84.0	18.5	2.3	1.3	3.5	10.5	7.20	1.49	6.23	5.70	5.74	1.56	11.2	2.80
M 5x18.9	330	42.7	3.9	4.2	20.5	41.3	28.9	0.76	34.6	15.8	41.3	16.4	89.9	11.0
M 4x13	182	27.4	3.3	3.3	18.3	22.7	15.7	0.47	25.8	12.7	28.1	11.9	69.9	6.05

| F_y = 50 ksi | | | | | | BEAMS S shapes Uniform load constants for beams | | | | | | | | |

$$\phi_b = 0.90 \qquad \phi_v = 0.90 \qquad \phi = 1.00 \qquad \phi_r = 0.75 \qquad \text{I}$$

For beams laterally unsupported, see page **3-53**

Shape	$\phi_b W_c$	$\phi_v V_n$	L_v	L_p	L_r	$\phi_b M_p$	$\phi_b M_r$	BF	ϕR_1	ϕR_2	$\phi_r R_3$	$\phi_r R_4$	$R(N=$ 3½ in.)	Z_x
	Kip-ft	Kip	Ft	Ft	Ft	Kip-ft	Kip-ft	Kip	Kip	Kip/in.	Kip	Kip/in.	Kip	In.³
S 24x121	9180	529	8.7	5.4	18.1	1150	774	29.3	200	40.0	269	20.7	340	306
x106	8370	410	10.2	5.6	17.4	1050	720	27.6	155	31.0	184	9.66	218	279
x100	7200	483	7.5	4.5	14.6	900	597	29.9	163	37.3	216	21.4	291	240
x 90	6660	405	8.2	4.6	14.2	832	561	28.3	137	31.2	166	12.6	210	222
x 80	6120	324	9.4	4.7	13.9	765	525	26.2	109	25.0	119	6.48	142	204
S 20x 96	5940	438	6.8	4.7	16.9	743	495	20.2	175	40.0	248	29.7	315	198
x 86	5490	362	7.6	4.8	16.2	686	465	19.5	144	33.0	185	16.7	244	183
x 75	4590	343	6.7	4.1	13.5	574	384	20.2	129	31.8	163	17.4	224	153
x 66	4200	273	7.7	4.2	13.0	525	357	19.2	103	25.3	115	8.76	146	140
S 18x 70	3750	346	5.4	3.8	13.6	469	309	16.3	133	35.6	180	31.3	258	125
x 54.7	3150	224	7.0	4.0	12.4	394	268	15.0	86.4	23.0	93.8	8.52	124	105
S 15x 50	2310	223	5.2	3.6	12.5	289	194	10.7	94.5	27.5	116	19.3	184	77.1
x 42.9	2080	166	6.2	3.8	11.9	260	179	10.1	70.6	20.6	74.9	8.05	103	69.3
S 12x 50	1840	223	4.1	3.6	16.3	230	152	6.10	123	34.3	167	44.4	244	61.2
x 40.8	1590	150	5.3	3.7	14.0	199	136	6.14	83.0	23.1	91.9	13.5	139	53.1
x 35	1340	139	4.9	3.5	11.9	168	115	6.35	63.5	21.4	74.5	13.0	120	44.8
x 31.8	1260	113	5.6	3.5	11.4	158	109	6.16	52.0	17.5	55.1	7.11	79.9	42.0
S 10x 35	1060	160	3.3	3.2	14.0	133	88.2	4.12	83.5	29.7	116	46.2	187	35.4
x 25.4	852	84.0	5.1	3.4	11.2	107	74.1	4.12	43.7	15.5	43.8	6.63	67.0	28.4
S 8x 23	579	95.3	3.0	2.8	12.0	72.4	48.6	2.60	55.1	22.1	68.9	27.2	132	19.3
x 18.4	495	58.5	4.2	2.9	10.4	61.9	43.2	2.52	33.9	13.6	33.2	6.32	55.3	16.5
S 7x 20	435	85.0	2.6	2.6	12.2	54.4	36.3	1.89	52.7	22.5	68.2	35.9	131	14.5
x 15.3	363	47.6	3.8	2.7	9.9	45.4	31.5	1.94	29.5	12.6	28.6	6.31	50.6	12.1
S 6x 17.25	318	75.3	2.1	2.4	12.8	39.8	26.3	1.30	50.9	23.3	68.5	50.5	132	10.6
x 12.5	254	37.6	3.4	2.5	9.6	31.8	22.1	1.36	25.4	11.6	24.1	6.27	46.1	8.47
S 5x 14.75	223	66.7	1.7	2.2	14.3	27.8	18.3	0.79	50.2	24.7	71.5	80.0	137	7.42
x 10	170	28.9	2.9	2.3	9.3	21.3	14.8	0.93	21.7	10.7	20.4	6.50	43.2	5.67
S 4x 9.5	121	35.2	1.7	2.0	11.6	15.1	10.2	0.52	30.6	16.3	36.3	32.0	87.6	4.04
x 7.7	105	20.8	2.5	2.1	9.3	13.2	9.12	0.56	18.1	9.65	16.6	6.64	39.8	3.51
S 3x 7.5	70.8	28.3	1.3	1.8	14.0	8.9	5.85	0.25	30.0	17.5	37.9	59.0	91.1	2.36
x 5.7	58.5	13.8	2.1	1.8	9.7	7.3	5.04	0.29	14.6	8.50	12.9	6.81	36.7	1.95

BEAMS
HP shapes
Uniform load constants for beams

$\phi_b = 0.90 \qquad \phi_v = 0.90 \qquad \phi = 1.00 \qquad \phi_r = 0.75$

For beams laterally unsupported, see page **3-53**

$F_y = 50$ ksi

Shape	$\phi_b W_c$	$\phi_v V_n$	L_v	L_p	L_r	$\phi_b M_p$	$\phi_b M_r$	BF	ϕR_1	ϕR_2	$\phi_r R_3$	$\phi_r R_4$	$R(N=3\frac{1}{2}$ in.)	Z_x
	Kip-ft	Kip	Ft	Ft	Ft	Kip-ft	Kip-ft	Kip	Kip	Kip/in.	Kip	Kip/in.	Kip	In.^{3}
HP 14x117**	5810	309	9.4	12.8	45.1	727	516	6.52	151	40.3	234	49.3	292	194
x102**	4930	267	9.3	15.2	41.1	617	450	6.44	121	35.3	179	38.4	245	169
x 89**	4140	230	9.0	17.4	37.6	517	393	6.15	101	30.8	136	29.6	209	146
x 73**	3180	186	8.6	20.5	34.2	397	321	5.56	75.0	25.3	92.0	20.3	163	118
HP 13x100	4590	272	8.5	11.2	40.9	574	405	5.68	137	38.3	211	48.1	271	153
x 87**	3880	233	8.3	12.3	36.9	485	351	5.43	114	33.3	159	36.9	231	131
x 73**	3150	195	8.1	14.7	33.4	393	296	5.18	88.3	28.3	115	27.1	187	110
x 60**	2410	156	7.8	17.5	30.2	302	241	4.80	64.7	23.0	76.3	18.3	140	89.0
HP 12x 84	3600	227	7.9	10.4	36.9	450	318	4.98	118	34.3	169	41.3	238	120
x 74**	3100	198	7.8	11.7	34.0	387	281	4.75	99.3	30.3	133	32.4	205	105
x 63**	2510	166	7.6	13.9	30.7	314	237	4.58	80.5	25.8	95.6	24.0	171	88.3
x 53**	2020	138	7.3	16.1	28.3	252	200	4.23	61.2	21.8	68.2	17.4	129	74.0
HP 10x 57	2000	152	6.6	8.7	31.1	249	176	3.25	83.9	28.3	115	34.6	183	66.5
x 42**	1370	109	6.3	11.9	25.6	171	130	2.99	55.1	20.8	62.5	19.0	128	48.3
HP 8x 36	1010	96.4	5.2	6.9	24.4	126	89.4	2.09	52.1	22.3	71.4	26.7	130	33.6

**Indicates noncompact shape. For noncompact shapes the maximum resisting moment ($\phi_b M_p'$) and the associated maximum unbraced length are tabulated as $\phi_b M_p$ and L_p.

| $F_y = 50$ ksi | **BEAMS** C shapes Uniform load constants for beams $\phi_b = 0.90$ $\phi_v = 0.90$ $\phi = 1.00$ $\phi_r = 0.75$ For beams laterally unsupported, see page **3-53** |

Shape	$\phi_b W_c$	$\phi_v V_n$	L_v	L_p	L_r	$\phi_b M_p$	$\phi_b M_r$	BF	ϕR_1	ϕR_2	$\phi_r R_3$	$\phi_r R_4$	$R(N=3\frac{1}{2}$ in.)	Z_x
	Kip-ft	Kip	Ft	Ft	Ft	Kip-ft	Kip-ft	Kip	Kip	Kip/in.	Kip	Kip/in.	Kip	In.^{3}
C 15x50	2050	290	3.5	3.1	13.0	256	161	9.51	129	35.8	176	40.7	254	68.3
x40	1720	211	4.1	3.1	11.1	215	140	9.46	93.4	26.0	109	15.6	164	57.2
x33.9	1510	162	4.7	3.2	10.2	189	126	9.01	71.9	20.0	73.6	7.10	98.4	50.4
C12x30	1010	165	3.1	2.7	10.4	126	81.0	5.81	71.7	25.5	93.0	23.9	161	33.6
x25	876	125	3.5	2.8	9.3	110	72.3	5.69	54.4	19.4	61.5	10.4	98.0	29.2
x20.7	762	91.4	4.2	2.8	8.6	95.3	64.5	5.29	39.7	14.1	38.2	4.04	52.3	25.4
C10x30	798	182	2.2	2.4	12.9	99.8	62.1	3.57	84.1	33.6	131	75.6	202	26.6
x25	690	142	2.4	2.4	10.6	86.3	54.6	3.85	65.8	26.3	90.8	36.1	158	23.0
x20	579	102	2.8	2.5	8.9	72.4	47.4	3.89	47.4	19.0	55.6	13.5	103	19.3
x15.3	474	64.8	3.7	2.5	7.8	59.3	40.5	3.57	30.0	12.0	28.0	3.43	40.0	15.8
C 9x20	504	109	2.3	2.3	9.7	63.0	40.5	3.03	52.5	22.4	69.5	26.2	131	16.8
x15	405	69.3	2.9	2.3	7.9	50.6	33.9	3.01	33.4	14.3	35.3	6.74	58.8	13.5
x13.4	375	56.6	3.3	2.4	7.5	46.9	31.8	2.94	27.3	11.6	26.1	3.68	39.0	12.5
C 8x18.75	414	105	2.0	2.1	10.4	51.8	33.0	2.26	57.1	24.4	76.5	40.1	142	13.8
x13.75	327	65.4	2.5	2.2	7.9	40.9	27.1	2.40	35.5	15.2	37.6	9.65	71.3	10.9
x11.5	287	47.5	3.0	2.2	7.2	35.8	24.4	2.28	25.8	11.0	23.2	3.69	36.2	9.55
C 7x14.75	290	79.2	1.8	2.0	9.7	36.3	23.3	1.68	45.8	21.0	59.2	31.1	119	9.68
x12.25	252	59.3	2.1	2.0	8.1	31.5	20.8	1.77	34.3	15.7	38.4	13.1	84.1	8.40
x 9.8	214	39.7	2.7	2.1	7.0	26.7	18.2	1.72	23.0	10.5	21.0	3.91	34.7	7.12
C 6x13	218	70.8	1.5	1.9	10.5	27.2	17.4	1.14	44.4	21.9	61.0	43.9	121	7.26
x10.5	185	50.9	1.8	1.9	8.3	23.1	15.2	1.23	31.9	15.7	37.2	16.3	86.8	6.15
x 8.2	154	32.4	2.4	1.9	6.9	19.2	13.1	1.23	20.3	10.0	18.9	4.21	33.6	5.13
C 5x 9	131	43.9	1.5	1.7	8.9	16.4	10.7	0.79	30.5	16.3	37.8	23.2	87.3	4.36
x 6.7	105	25.7	2.1	1.7	6.9	13.2	9.00	0.82	17.8	9.50	16.9	4.64	33.1	3.51
C 4x 7.25	84.3	34.7	1.2	1.6	9.8	10.5	6.87	0.45	27.6	16.1	35.7	30.2	83.8	2.81
x 5.4	67.8	19.9	1.7	1.6	7.2	8.48	5.79	0.48	15.8	9.20	15.5	5.69	35.4	2.26
C 3x 6	51.6	28.8	0.9	1.5	13.0	6.45	4.14	0.20	30.6	17.8	40.0	59.6	92.9	1.72
x 5	45.0	20.9	1.1	1.5	9.9	5.63	3.72	0.23	22.2	12.9	24.7	22.7	67.3	1.50
x 4.1	39.0	13.8	1.4	1.4	8.0	4.88	3.30	0.24	14.6	8.50	13.2	6.49	35.9	1.30

BEAMS
MC shapes
Uniform load constants for beams

$F_y = 50$ ksi

$\phi_b = 0.90 \qquad \phi_v = 0.90 \qquad \phi = 1.00 \qquad \phi_r = 0.75$

For beams laterally unsupported, see page **3-53**

Shape	$\phi_b W_c$	$\phi_v V_n$	L_v	L_p	L_r	$\phi_b M_p$	$\phi_b M_r$	BF	ϕR_1	ϕR_2	$\phi_r R_3$	$\phi_r R_4$	R(N=3½ in.)	Z_x
	Kip-ft	Kip	Ft	Ft	Ft	Kip-ft	Kip-ft	Kip	Kip	Kip/in.	Kip	Kip/in.	Kip	In.³
MC 18x58	2840	340	4.2	3.6	13.1	355	225	13.6	120	35.0	167	33.0	243	94.6
x51.9	2600	292	4.5	3.7	12.3	324	209	13.4	103	30.0	133	20.8	205	86.5
x45.8	2350	243	4.8	3.8	11.6	294	193	12.9	85.9	25.0	101	12.0	143	78.5
x42.7	2230	219	5.1	3.8	11.3	279	185	12.5	77.3	22.5	86.1	8.76	117	74.4
MC 13x50	1820	276	3.3	3.8	18.0	227	145	5.75	135	39.3	197	66.5	273	60.5
x40	1530	197	3.9	3.8	14.5	191	126	6.06	96.3	28.0	118	24.0	194	50.9
x35	1390	157	4.4	3.9	13.3	173	116	6.01	76.8	22.4	84.2	12.2	127	46.2
x31.8	1290	132	4.9	3.9	12.7	162	110	5.84	64.5	18.7	64.7	7.19	89.9	43.1
MC 12x50	1680	271	3.1	3.9	20.3	210	135	4.61	137	41.8	230	75.0	283	56.1
x45	1550	231	3.4	3.9	17.8	194	126	4.88	117	35.6	181	46.5	241	51.7
x40	1420	191	3.7	3.9	15.9	177	117	5.02	96.8	29.5	137	26.5	200	47.3
x35	1280	151	4	3.9	14.3	160	108	5.03	76.6	23.4	96.3	13.1	142	42.8
x31	1180	120	4	4.0	13.7	147	101	4.70	60.7	18.5	67.9	6.52	90.7	39.3
x10.6	348	61.6	2.8	1.2	3.7	43.5	27.7	6.50	16.3	9.50	16.6	2.00	23.6	11.6
MC 10x41.1	1170	215	2.7	4.0	22.8	146	94.5	2.74	124	39.8	194	94.9	264	38.9
x33.6	1000	155	3.2	4.1	17.8	125	83.4	3.05	89.8	28.8	119	35.8	190	33.4
x28.5	888	115	3.9	4.1	15.3	111	75.9	3.15	66.4	21.3	75.8	14.4	126	29.6
x25	774	103	3	3	13.0	96.8	66.0	3.23	59.4	19.0	64.1	10.3	100	25.8
x22	708	78.3	4	3	12.1	88.5	61.5	3.16	45.3	14.5	42.7	4.59	58.8	23.6
x 8.4	236	45.9	2.6	1.3	3.8	29.5	19.2	4.12	14.6	8.50	13.4	1.90	20.0	7.86
x 6.5	172	41.0	2.1	0.9	2.6	21.5	13.3	4.73	8.31	7.60	9.60	1.88	16.2	5.73
MC 9x25.4	696	109	3.2	3.6	14.6	87.0	58.8	2.55	66.8	22.5	80.7	19.9	146	23.2
x23.9	666	97.2	3.4	3.6	13.9	83.3	56.7	2.58	59.4	20.0	67.7	14.0	117	22.2
MC 8x22.8	564	92.2	3.1	3.6	15.6	70.5	48.0	1.89	63.4	21.3	72.9	20.1	138	18.8
x21.4	540	81.0	3.3	3.6	14.7	67.5	46.2	1.93	55.7	18.7	60.0	13.6	108	18.0
x20	486	86.4	2.8	3.1	12.9	60.8	40.8	2.03	56.2	20.0	64.5	17.3	125	16.2
x18.7	462	76.2	3.0	3.1	12.2	57.8	39.3	2.03	49.6	17.6	53.5	11.9	95.1	15.4
x 8.5	207	38.7	2.7	1.8	5.32	25.9	17.5	2.36	16.8	8.95	15.2	2.49	24.0	6.91
MC 7x22.7	486	95.1	2.6	3.7	19.0	60.8	40.8	1.30	70.7	25.2	91.0	39.3	159	16.2
x19.1	429	66.5	3.2	3.7	15.7	53.6	36.9	1.39	49.5	17.6	53.3	13.5	100	14.3
x17.6	381	70.9	2.7	3.1	14.1	47.6	32.4	1.38	49.8	18.7	57.1	17.2	115	12.7
MC 6x18	345	61.4	2.8	3.8	18.3	43.1	29.7	.92	50.3	19.0	58.0	20.7	117	11.5
x15.3	295	55.1	2.7	3.7	15.4	36.8	25.4	.98	37.2	17.0	44.4	18.4	96.7	9.82
x16.3	306	60.8	2.5	3.2	15.7	38.3	26.0	.97	49.8	18.7	57.1	20.0	115	10.2
x15.1	291	51.2	2.8	3.1	14.6	36.3	25.0	.99	42.0	15.8	44.2	12.0	86.1	9.69
x12	221	50.2	2.2	2.6	10.7	27.7	18.7	1.10	31.5	15.5	38.1	14.3	85.7	7.38

AMERICAN INSTITUTE OF STEEL CONSTRUCTION

BEAMS
Design of bearing plates

When a beam is supported by a concrete wall or column, it is essential that the beam reaction be distributed over an area sufficient to keep the average pressure on the concrete within the limits of its design strength. In the absence of code regulations the design bearing load P_p may be selected from LRFD Specification Sect. J9.

The following method of design makes use of the Uniform Load Constants Tables and is recommended for bearing plates on concrete supports.

R = reaction of beam, kips

P_p = design bearing load, kips

A_1 = $B \times N$ = area of steel bearing concentrically on a concrete support, in.²

A_2 = $B_1 \times N_1$ = total cross-sectional area of a concrete support, in.²

$\phi R_1 = \phi 2.5kF_{yw}t_w$ = first part of LRFD Specification Formula K1-3, where $\phi = 1.0$, kips

$\phi R_2 = \phi NF_{yw}t_w$ = second part of LRFD Specification Formula K1-3, where $\phi = 1.0$ and $N = 1.0$, kip/in.

$\phi_r R_3 = \phi_r 68t_w^2 \sqrt{\dfrac{F_{yw}t_f}{t_w}}$ = first part of LRFD Specification Formula K1-5, where $\phi_r = 0.75$, kips

$\phi_r R_4 = \phi_r 68t_w^2 \left[3\left(\dfrac{N}{d}\right)\left(\dfrac{t_w}{t_f}\right)^{1.5} \right] \sqrt{\dfrac{F_{yw}t_f}{t_w}}$ = second part of LRFD Specification Formula K1-5, where $\phi_r = 0.75$ and $N = 1.0$, kip/in.

k = distance from bottom of beam to web toe of fillet, in. (from Dimensions and Properties Tables, LRFD Manual, Part 1)

t = thickness of plate, in.

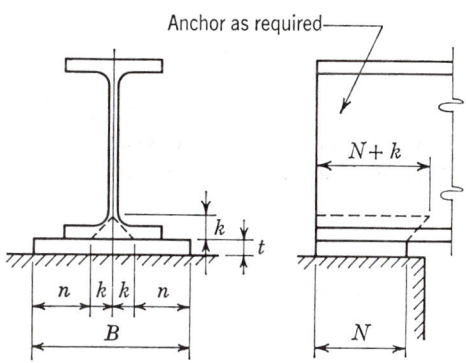

The beam reaction R is assumed to be uniformly distributed to the plate over the area $N \times 2k$ and the bearing plate is assumed to distribute this load uniformly over the concrete support area.

Calculate the minimum bearing length N based on local web yielding (LRFD Specification Formula K1-3) or web crippling (LRFD Specification Formula K1-5). The equation yielding the larger N value controls. By replacing each portion of the above Formulas with a variable, as defined previously, and solving for N, the following equations result:

local web yielding
$$N = \frac{R - \phi R_1}{\phi R_2}, \text{ in.}$$

web crippling
$$N = \frac{R - \phi_r R_3}{\phi_r R_4}, \text{ in.}$$

The values for ϕR_1, ϕR_2, $\phi_r R_3$ and $\phi_r R_4$ are tabulated in the Uniform Load Constants Tables for each shape.

Determine the required bearing plate area, A_1, in.^{2}; by rearranging the formulas given in Sect. J9 of the LRFD Specification and solving for A_1, with $P_p = R$ and $\phi_c = 0.6$.

On full area of concrete support

$$A_1 = \frac{R}{\phi_c 0.85 f_c'}$$

On less than full area of concrete support

$$A_1 = \frac{1}{A_2} \left(\frac{R}{\phi_c 0.85 f_c'} \right)^2$$

Establish N and solve for $B = A_1 / N$. The length of bearing N is usually governed by the available wall thickness or some other structural consideration. Preferably, B and N should be in full inches, and B rounded off so that $B \times N \geq A_1$ (req'd).

Determine $n = (B/2) - k$ and solve for t in the following formula based on cantilever bending of the plate under uniform concrete pressure.

$$t = \sqrt{\frac{2.22R \, n^2}{A_1 \, F_y}}$$

Example 1

Given:

A W18×50 beam, $F_y = 36$ ksi, for which $k = 1\frac{1}{4}$ in., has a factored reaction of 73.5 kips and is to be supported by a 10-in. concrete wall. Using the entire width of the concrete wall for the bearing length N, design a bearing plate for the beam.

Solution:

From the Uniform Load Constants Tables:

$\phi R_1 = 39.9$ kips
$\phi R_2 = 12.8$ kip/in.
$\phi_r R_3 = 48.9$ kips
$\phi_r R_4 = 4.01$ kip/in.

$$N = \frac{R - \phi R_1}{\phi R_2} = \frac{73.5 - 39.9}{12.8} = 2.625 \text{ in.} < 10 \text{ in.}$$

$$N = \frac{R - \phi_r R_3}{\phi_r R_4} = \frac{73.5 - 48.9}{4.01} = 6.13 \text{ in.} < 10 \text{ in.} \quad \textbf{o.k.}$$

$$A_1 \text{ (Req'd)} = \frac{R}{\phi_c 0.85 f_c'} = \frac{73.5}{(0.60 \times 0.85 \times 3)} = 48.0 \text{ in.}^2$$

$B = A_1/N = 48.0/10 = 4.8$ in.; use 8 in. (flange width controls)

$A_1 = B \times N = 8 \times 10 = 80.0 \text{ in.}^2 \geq 48.0 \text{ in.}^2$

$n = 8/2 - 1.25 = 2.75$ in.

$$t = \sqrt{\frac{2.22\ Rn^2}{A_1 F_y}} = \sqrt{\frac{2.22 \times 73.5 \times (2.75)^2}{80.0 \times 36}} = 0.65 \text{ in.; use } \tfrac{3}{4} \text{ in.}$$

Use: Bearing plate $\frac{3}{4} \times 10 \times 0' - 8$

Example 2

Given:

Investigate a $1 \times 6\frac{1}{2} \times 0' - 8$ bearing plate for the beam and support in Example 1. The least distance from the edge of bearing plate to the edge of concrete support, b_1, is 2 in. $f_c' = 3.0$ ksi.

Solution:

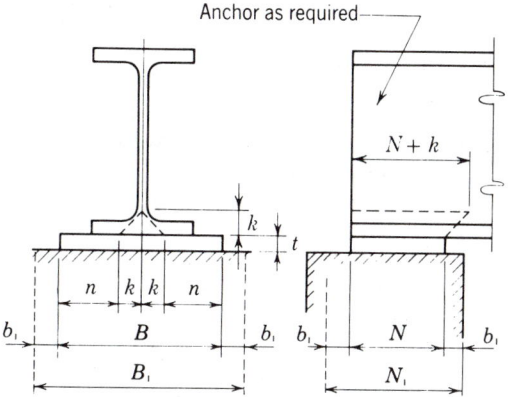

$A_1 = B \times N = 8 \times 6.5 = 52.0 \text{ in.}^2$

Assumed area of concrete support:

$$A_2 = B_1 \times N_1 = [8 + (2 + 2)] \times [6.5 + (2 + 2)] = 126.0 \text{ in.}^2$$

$P_p = \phi_c 0.85 f_c' A_1 \sqrt{A_2/A_1} = 0.60 \times 0.85 \times 3 \times 52 \sqrt{126.0/52.0}$
$\quad\quad\quad = 123.8 \text{ kips} > 73.5 \text{ kips.} \quad \textbf{o.k.}$

and $\sqrt{A_2/A_1} = \sqrt{126/52} = 1.56 \leq 2 \quad \textbf{o.k.}$ (LRFD Specification Sect. J9)

$N = 6.5$ in. > 6.13 in $\quad \textbf{o.k.}$ (see Ex. 1)

$n = B/2 - k = 8/2 - 1.25 = 2.75$ in.

Min. $t = \sqrt{\dfrac{2.22(73.5)(2.75)^2}{(52.0)(36)}} = 0.81$ in. < 1 in $\quad \textbf{o.k.}$

Investigate beam without bearing plate ($t_f = 0.57$ in.):

$$\text{Min. } t = \sqrt{\frac{2.22\ (R)\ (b_f/2 - k)^2}{A_1 F_y}}$$

$$= \sqrt{\frac{2.22\ (73.5)(7.5/2 - 1.25)^2}{(6.5 \times 7.5)(36)}}$$

$$= 0.76 \text{ in.} > 0.57 \text{ in.} = t_f \quad \textbf{n.g.}$$

$\therefore$ Use bearing plate.

DESIGN MOMENTS IN BEAMS
With unbraced length greater than L_p

General Notes

Spacing of lateral bracing at distances greater than L_p creates a problem in which the designer is confronted with a given laterally unbraced length (usually less than the total span) along the compression flange, and a calculated required bending moment. The beam cannot be selected from its plastic section modulus alone, since depth, flange proportions and other properties have an influence on its bending strength.

The following charts show the design moment $\phi_b M_n$ for W and M shapes of $F_y = 36$ ksi and $F_y = 50$ ksi steels, used as beams, with respect to the maximum unbraced length for which this moment is permissible. In bending, ϕ_b of 0.9 is given in Sect. F1.2 of the LRFD Specification. The charts extend over varying unbraced lengths, depending upon the capacity of the beams represented. In general, they extend beyond most unbraced lengths frequently encountered in design practice.

The design moment $\phi_b M_n$, kip-ft, is plotted with respect to the unbraced length with no consideration of the moment due to weight of the beam. Design moments are shown for unbraced lengths in feet, starting at spans less then L_p, of spans between L_p and L_r and of spans beyond L_r.

The unbraced length L_p, in feet, with the limit indicated by a solid symbol, ●, is the maximum unbraced length of the compression flange, with $C_b = 1.0$, for which the design moment is given by $\phi_b M_p$

where

$$L_p = 300 r_y / \sqrt{F_y} \tag{F1-4}$$

$$M_p = Z_x F_y$$

For those noncompact rolled shapes, which meet the requirements of compact sections except that $b_f/2t_f$ exceeds $65/\sqrt{F_y}$, but is less than $141/\sqrt{F_y - F_r}$, the design moment is obtained from Formula A-F1-3 in Appendix F1.7 of the LRFD Specification. This criteria applies to one beam when F_y is equal to 36 ksi and applies to eight beams when F_y is equal to 50 ksi. L_p' for these beams are indicated by a half filled circle ◑.

For the case, $C_b = 1.0$ and noncompact shapes:

$$M_n' = M_p - (M_p - M_r)\left(\frac{\lambda - \lambda_p}{\lambda_r - \lambda_p}\right) \tag{A-F1-3}$$

$$L_p' = L_p + (L_r - L_p)\left(\frac{M_p - M_n'}{M_p - M_r}\right)$$

$$\lambda = b_f/2t_f$$

$$\lambda_p = 65/\sqrt{F_y}$$

$$\lambda_r = 141/\sqrt{F_y - F_r}$$

$$L_r = \frac{r_y X_1}{F_{yf} - F_r}\sqrt{1 + \sqrt{1 + X_2(F_{yf} - F_r)^2}} \tag{F1-6}$$

$$X_1 = \frac{\pi}{S_x}\sqrt{\frac{EGJA}{2}} \tag{F1-8}$$

$$X_2 = \frac{4Cw}{I_y}\left(\frac{S_x}{GJ}\right)^2 \tag{F1-9}$$

$$M_r = (F_{yw} - F_r)S_x \tag{F1-7}$$

$$M_p = Z_x F_y$$

$$F_r = 10 \text{ ksi for rolled shapes}$$

$$= 16.5 \text{ ksi for welded shapes}$$

The unbraced length is the maximum laterally unbraced length of the compression flange corresponding to the design moment. It may be either the total span or any part of the total span between braced points. The curves shown in these charts were computed for beams subjected to loading conditions which produce bending moments within the unbraced length greater than that at both ends of this length. In these cases, C_b is taken as unity in accordance with Sect. F1.3. When a moment gradient exists between points of bracing and the unbraced length is greater than L_p, C_b may be larger than unity. Using this larger value of C_b may provide a more liberal moment capacity for the section chosen. In these cases the design moment can be determined using the provisions of Sect. F1.3 of the LRFD Specification.

$$\phi_b M_n = \phi_b C_b\left[M_p - (M_p - M_r)\left(\frac{L_b - L_p}{L_r - L_p}\right)\right] \le \phi_b M_p$$

The unbraced length L_r, ft, with the limit indicated by an open symbol ○, is the maximum unbraced length of the compression flange beyond which the design moment is governed by Specification Sect. F1.4.

$$\phi_b M_n = \phi_b M_{cr} = \phi_b C_b \frac{\pi}{L_b}\sqrt{EI_y GJ + \left(\frac{\pi E}{L_b}\right)^2 I_y C_w} \le \phi_b C_b M_r$$

In computing the points for the curves, C_b in the above formulas was taken as unity, $E = 29,000$ ksi and $G = 11,200$ ksi. The properties of the beams are taken from the Tables of Dimensions and Properties in Part 1 of this Manual. The beam capacities have been reduced by multiplying the nominal flexural strength M_n by 0.9, the resistance factor ϕ_b for flexure.

To insure that almost all possible span lengths are available, two unbraced length scales are used (0.5 ft and 1.0 ft). Furthermore, the unbraced length scale is extended in some cases to the adjoining facing page. As a result, the length scales extend beyond most unbraced lengths encountered in design practice. The vertical or moment scales are adjusted to provide reasonable accuracy throughout the full range of beam sizes.

Over a limited range of length, a given beam is the lightest available for various combinations of unbraced length and design moment. The charts are designed to assist in selection of the lightest available beam for the given combination.

The solid portion of each curve indicates the most economical section by weight. The dashed portion of each curve indicates ranges in which a lighter weight beam will satisfy the loading conditions. For beams of equal weight, where both would satisfy the loading conditions, the deeper beam, when having a lesser design moment capacity than the shallower beam, is indicated as a dashed curve to assist in making a selection for reduced deflection or a limited depth condition.

In the case of W and M shapes of equal weight and the same nominal depth, the M shape is shown dashed when its design moment capacity is less than the W shape, to indicate that the W shape is usually more readily available.

The curves are plotted without regard to deflection, therefore due care must be exercised in their use. The curves do not extend beyond an arbitrary span/depth limit of 30. In no case is the dashed line or the solid line extended beyond the point where the calculated bending stress is less than 16 ksi for F_y = 36 ksi steel, or 22 ksi for F_y = 50 ksi steel.

The following examples illustrate the use of the charts for selection of proper sized beams with an unbraced length greater than L_p.

Example 1

Given:

Using F_y = 36 ksi steel, determine the size of a "simple" framed girder with a span of 35 ft, which supports two equal concentrated load points only. The factored loads produce a maximum calculated moment of 330 kip-ft in the center 15-ft section between the loads.

Solution:

For this loading condition, C_b = 1.0.

Center section of 15 ft is longest unbraced length.

With total span equal to 35 ft and M_u = 330 kip-ft, assume approximate weight of beam at 70 lbs/ft (equal to 0.07 kips/ft).

$$\text{Total } M_u = 330 + \left[\frac{0.07 \times (35)^2}{8} \times 1.2 \right] = 343 \text{ kip-ft}$$

Entering chart, with unbraced length equal to 15 ft on the bottom scale (abscissa), proceed upward to meet the horizontal line corresponding to a design moment equal to 343 kip-ft on the left-hand scale (ordinate). Any beam listed above and to the right of the point so located satisfies the design moment requirement. In this case, the lightest section satisfying this criteria is a W21 × 68, for which the total design moment with an unbraced length of 15 ft is 354 kip-ft.

Use: W21 × 68

Note: If depth is limited, a W14 × 82 could be selected, provided deflection conditions are not critical.

Example 2

Given:

A "fixed end" girder with a span of 60 ft supports a concentrated load at the center. The compression flange is laterally supported at the concentrated load point and at the inflection points. The factored load produces a maximum calculated moment of 330 kip-ft at the load point and at the supports. Determine the size of the beam using F_y = 36 ksi steel.

Solution:

For this loading condition, C_b = 1.75 with an unbraced length of 15 ft. With the total span equal to 60 ft and M_u = 330 kip-ft, assume approximate weight of beam at 60 lbs/ft (0.06 kips/ft).

$$\text{Total } M_u = 330 + \left[\frac{0.06 \times (60)^2}{24} \times 1.2\right] = 341 \text{ kip-ft at the centerline and 352 kip-ft at the supports.}$$

Compute $M_{effective}$ by dividing the required design moment by C_b

$$M_{effective} = 352/1.75 = 201 \text{ kip-ft}$$

Enter charts with unbraced length equal to 15 ft and proceed upward to 201 kip-ft. Any beam listed above and to the right of the point satisfies the design moment.

The lightest section satisfying the criteria of a design moment of 201 kip-ft at an unbraced length of 15 ft and $\phi_b M_p$ greater than 352 kip-ft is a W21 × 62. The design moment for a W21 × 62 with an unbraced length of 15 ft is 314 kip-ft and $\phi_b M_p$ is 389 kip-ft.

A 21-in. nominal depth beam spanning 60 ft should be checked for deflection since the span/depth ratio exceeds 30. Any deeper beam at an unbraced length of 15 ft with a design moment capacity greater than the W21 × 62 satisfying a deflection criterion would be satisfactory.

BEAM DESIGN MOMENTS ($\phi = 0.9$, $C_b = 1$, $F_y = 36$ ksi)

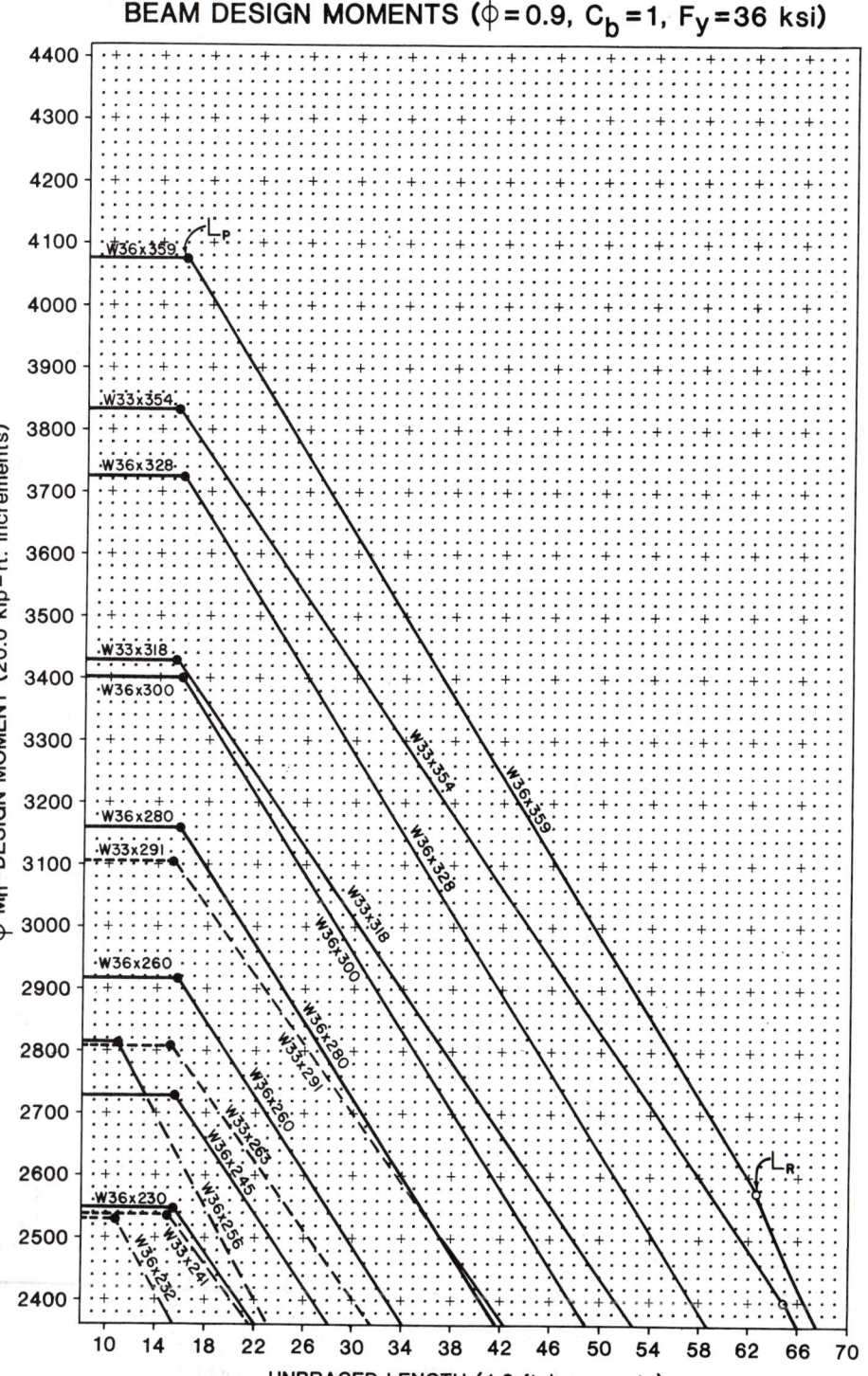

BEAM DESIGN MOMENTS (ϕ=0.9, C_b=1, F_y=36 ksi)

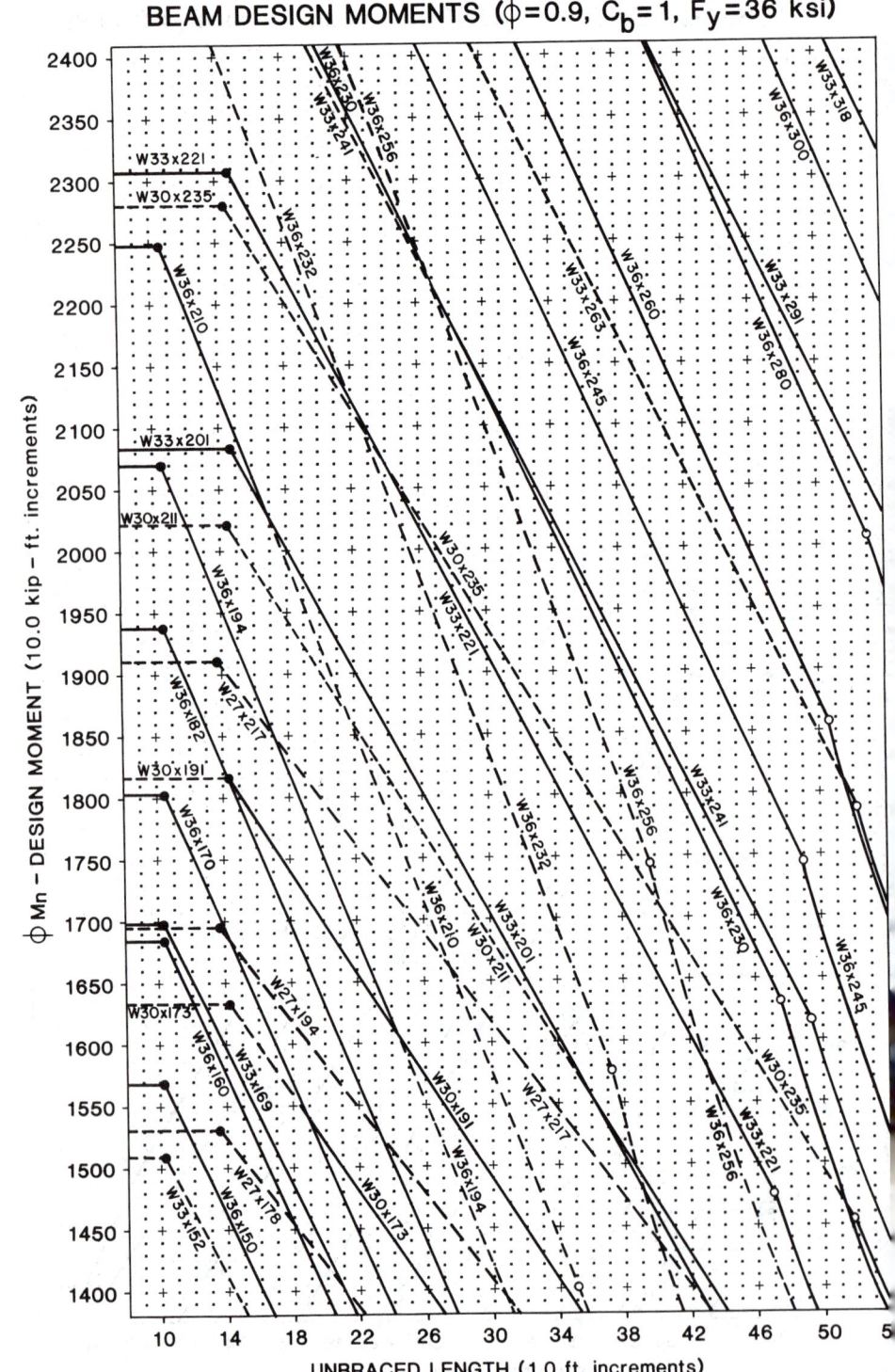

ϕM_n – DESIGN MOMENT (10.0 kip – ft. increments)

UNBRACED LENGTH (1.0 ft. increments)

BEAM DESIGN MOMENTS ($\phi = 0.9$, $C_b = 1$, $F_y = 36$ ksi)

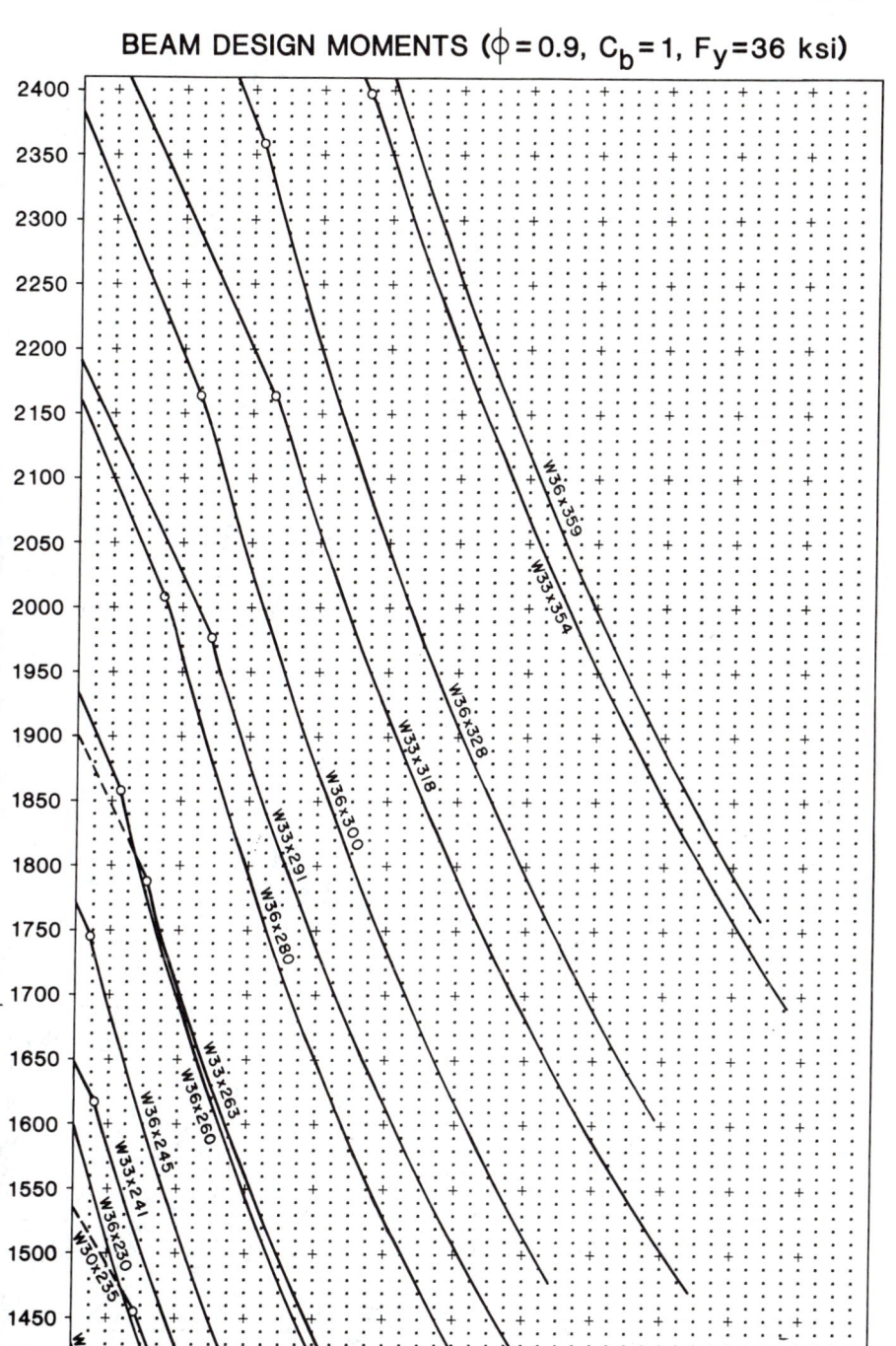

UNBRACED LENGTH (1.0 ft. increments)

BEAM DESIGN MOMENTS ($\phi = 0.9$, $C_b = 1$, $F_y = 36$ ksi)

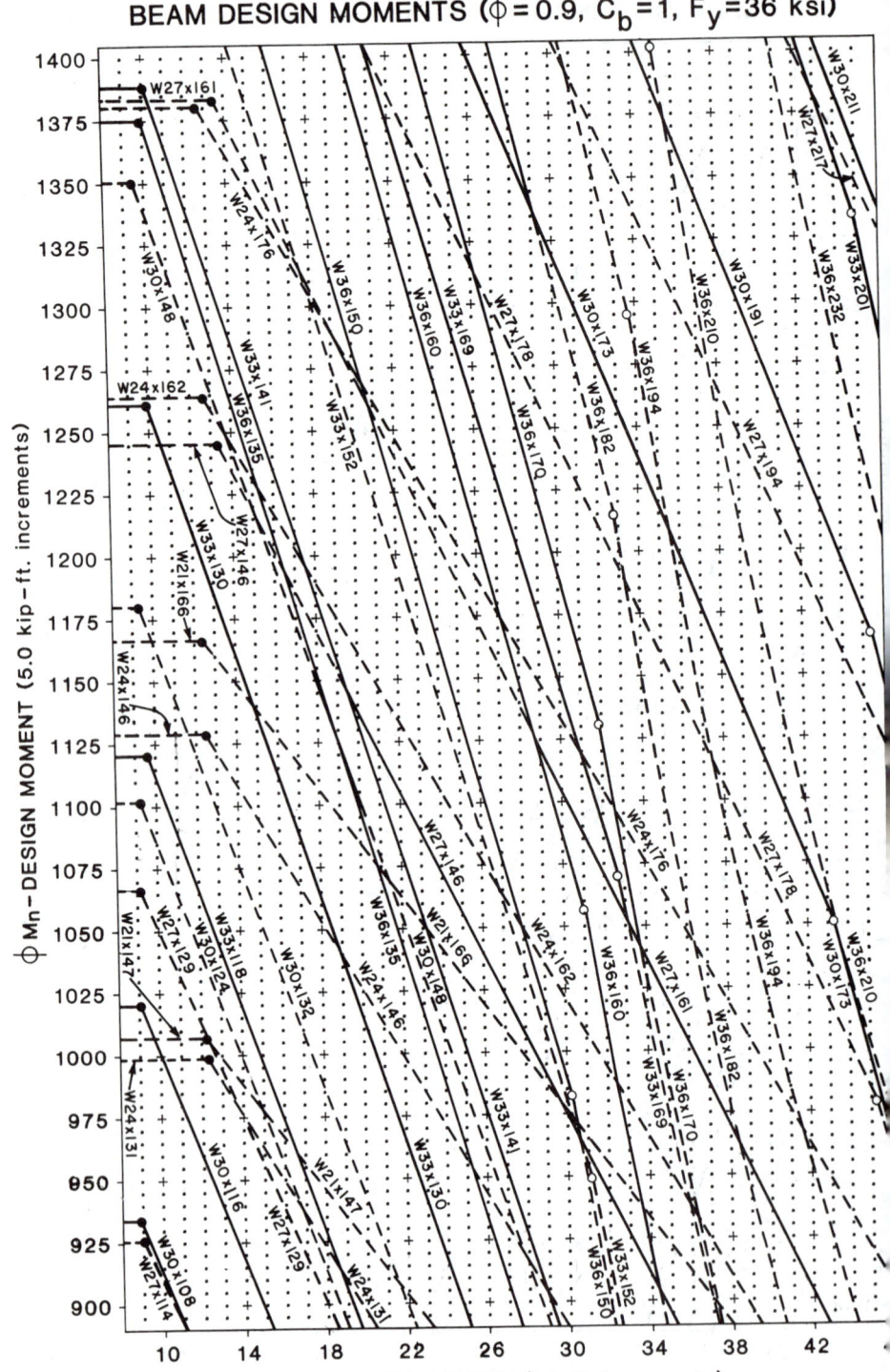

ϕM_n—DESIGN MOMENT (5.0 kip-ft. increments)

UNBRACED LENGTH (1.0 ft. increments)

BEAM DESIGN MOMENTS ($\phi = 0.9$, $C_b = 1$, $F_y = 36$ ksi)

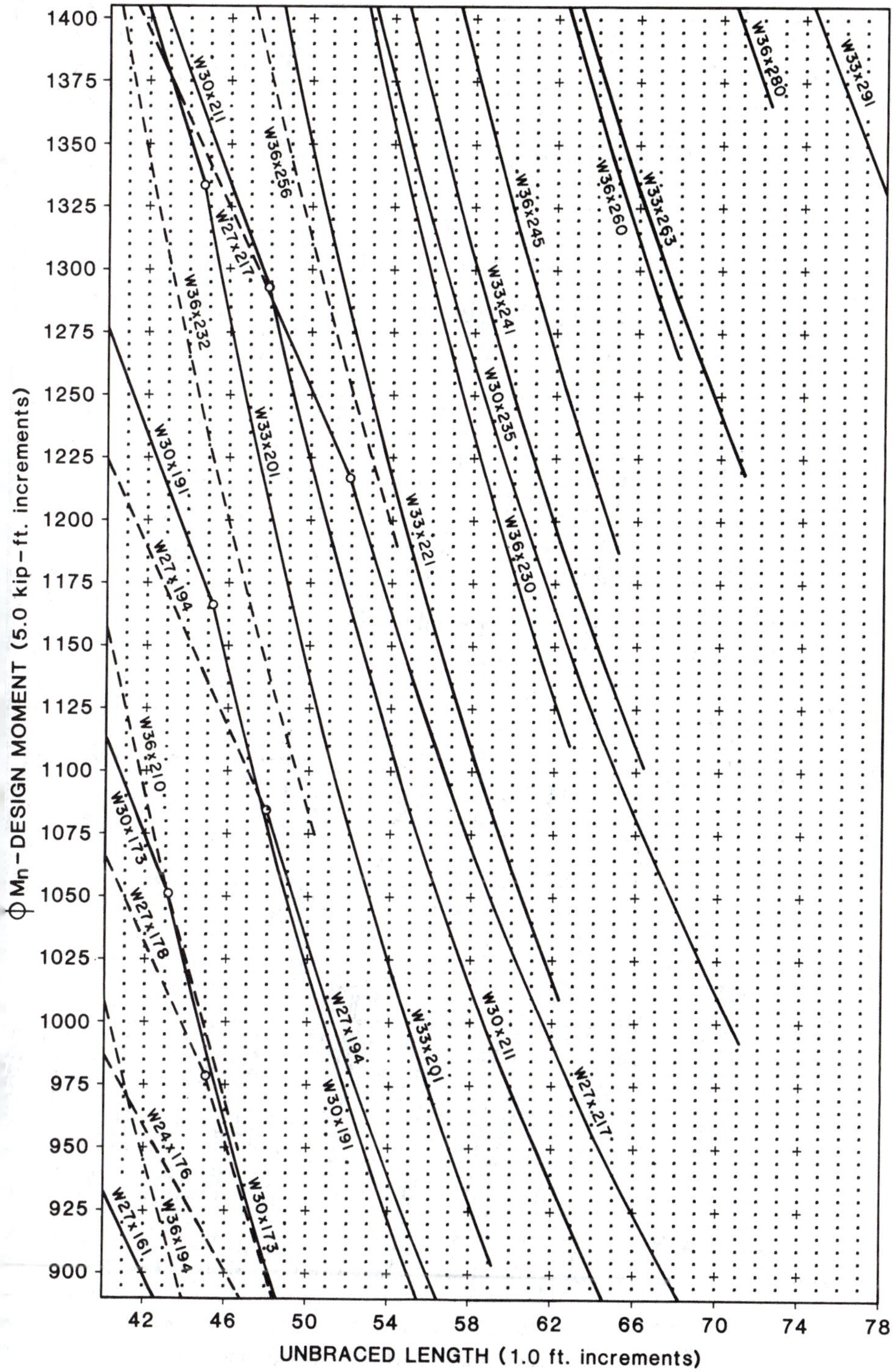
φMn - DESIGN MOMENT (5.0 kip - ft. increments)
UNBRACED LENGTH (1.0 ft. increments)
1400
1375
1350
1325
1300
1275
1250
1225
1200
1175
1150
1125
1100
1075
1050
1025
1000
975
950
925
900
42
46
50
54
58
62
66
70
74
78
W36x280
W33x291
W30x211
W36x256
W27x217
W36x245
W36x260
W33x263
W36x232
W33x241
W30x235
W33x201
W30x191
W33x221
W36x230
W27x194
W36x210
W30x173
W27x178
W27x194
W33x201
W30x211
W27x217
W30x191
W24x176
W36x194
W30x173
W27x161

BEAM DESIGN MOMENTS ($\phi = 0.9$, $C_b = 1$, $F_y = 36$ ksi)

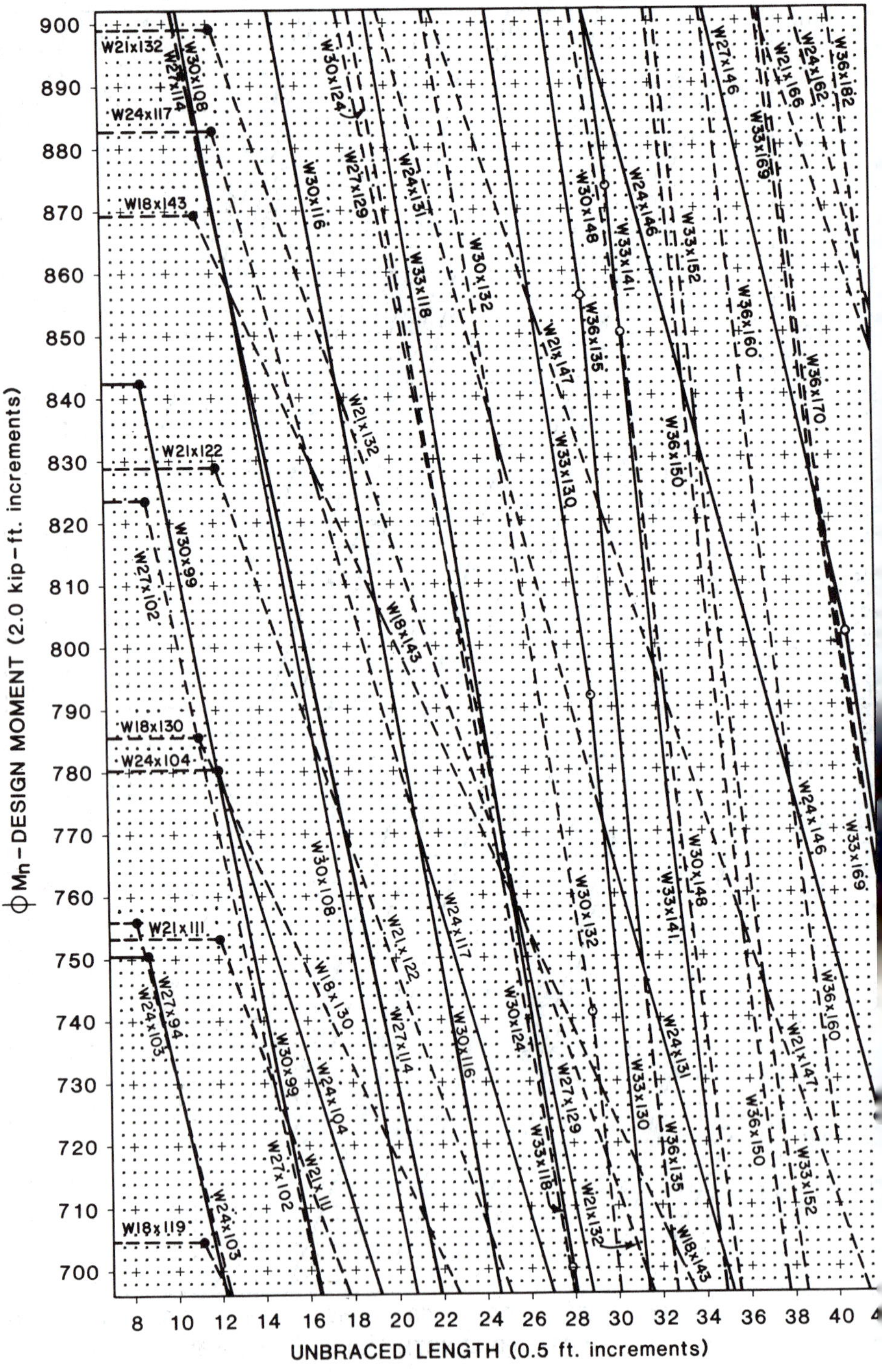

ϕM_n – DESIGN MOMENT (2.0 kip–ft. increments) vs **UNBRACED LENGTH (0.5 ft. increments)**

BEAM DESIGN MOMENTS (ϕ = 0.9, C_b = 1, F_y = 36 ksi)

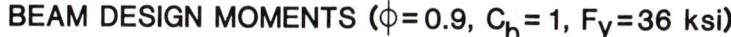

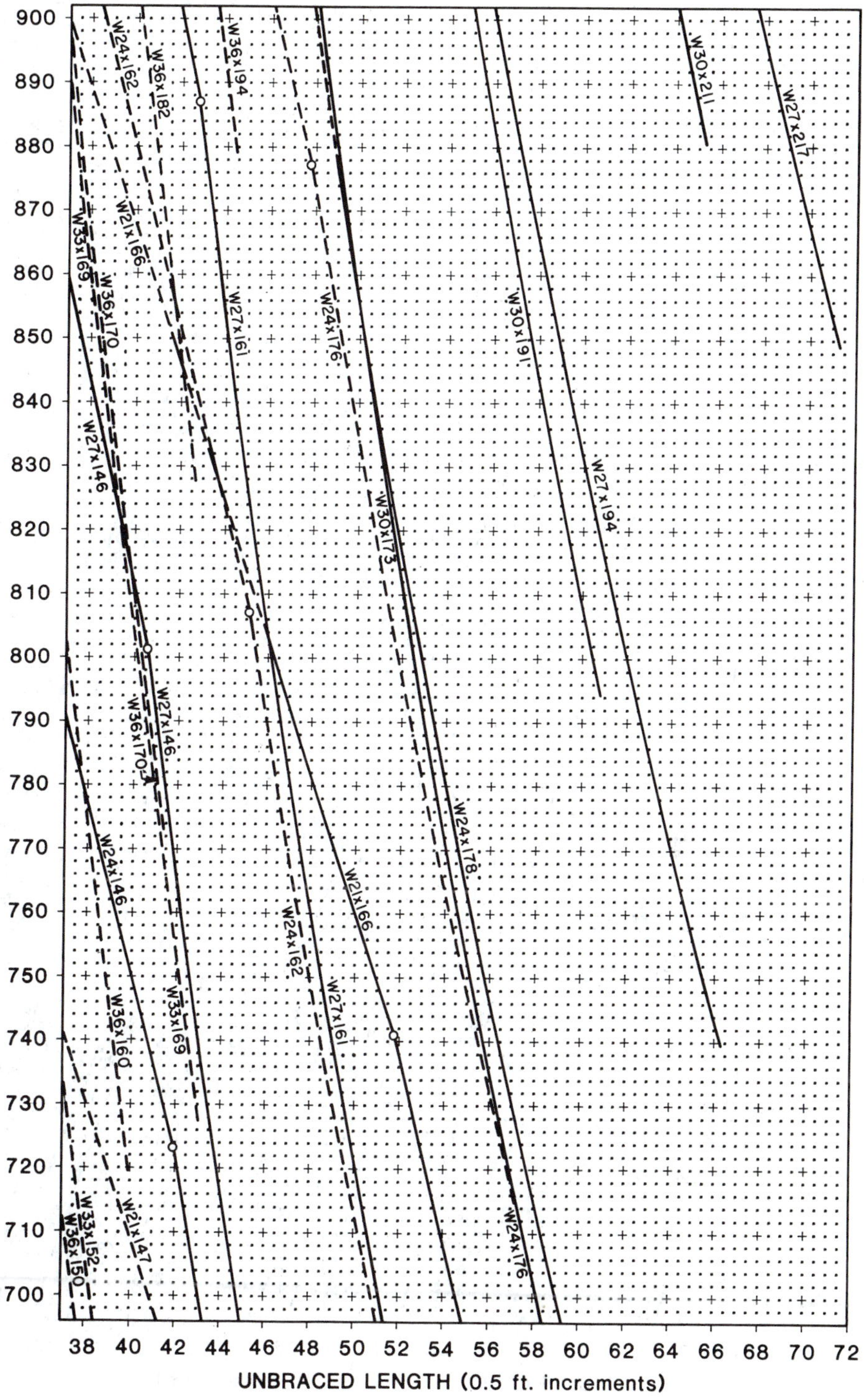
900
890
880
870
860
850
840
830
820
810
800
790
780
770
760
750
740
730
720
710
700
38 40 42 44 46 48 50 52 54 56 58 60 62 64 66 68 70 72
UNBRACED LENGTH (0.5 ft. increments)
W24x162
W36x182
W36x194
W30x211
W27x217
W21x166
W33x169
W36x170
W27x161
W24x176
W30x191
W27x146
W27x194
W30x173
W27x146
W36x170
W24x178
W24x146
W21x166
W24x162
W36x160
W33x169
W27x161
W33x152
W36x150
W21x147
W24x176

BEAM DESIGN MOMENTS (ϕ=0.9, C_b=1, F_y=36 ksi)

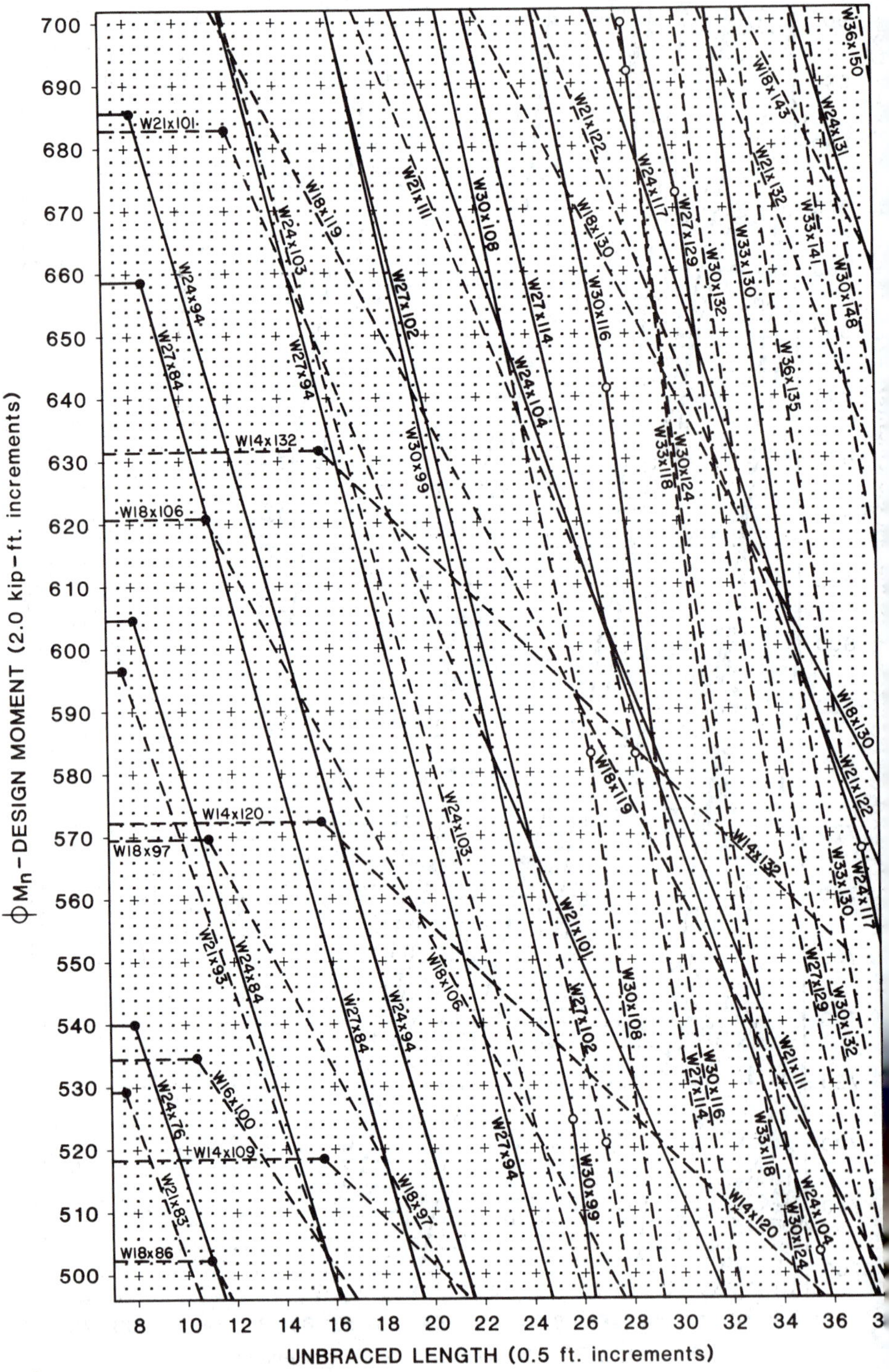

Y-axis: ϕM_n-DESIGN MOMENT (2.0 kip-ft. increments)

X-axis: UNBRACED LENGTH (0.5 ft. increments)

BEAM DESIGN MOMENTS ($\phi = 0.9$, $C_b = 1$, $F_y = 36$ ksi)

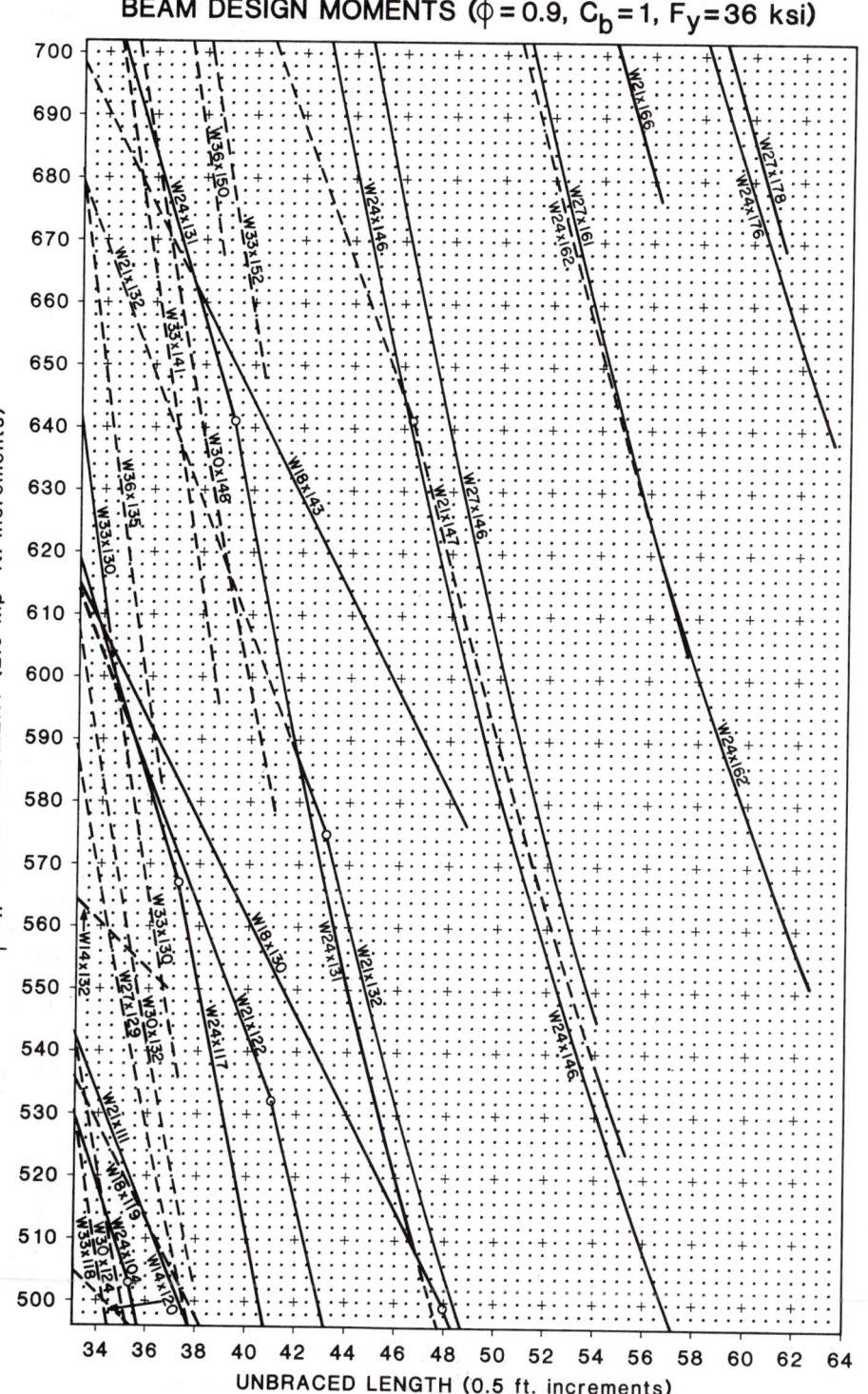

ϕM_n – DESIGN MOMENT (2.0 kip-ft. increments)

UNBRACED LENGTH (0.5 ft. increments)

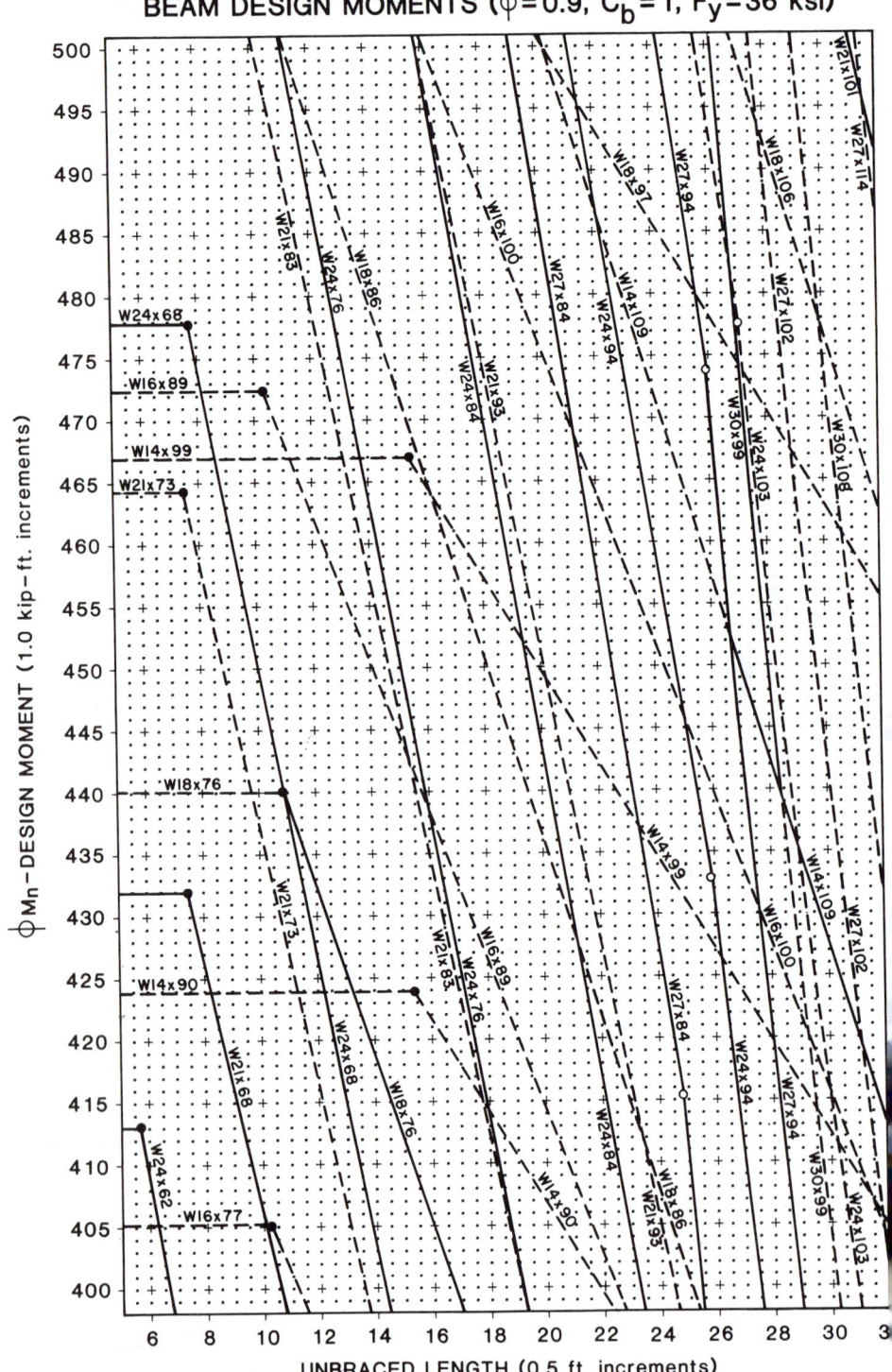

BEAM DESIGN MOMENTS ($\phi = 0.9$, $C_b = 1$, $F_y = 36$ ksi)

BEAM DESIGN MOMENTS ($\phi = 0.9$, $C_b = 1$, $F_y = 36$ ksi)

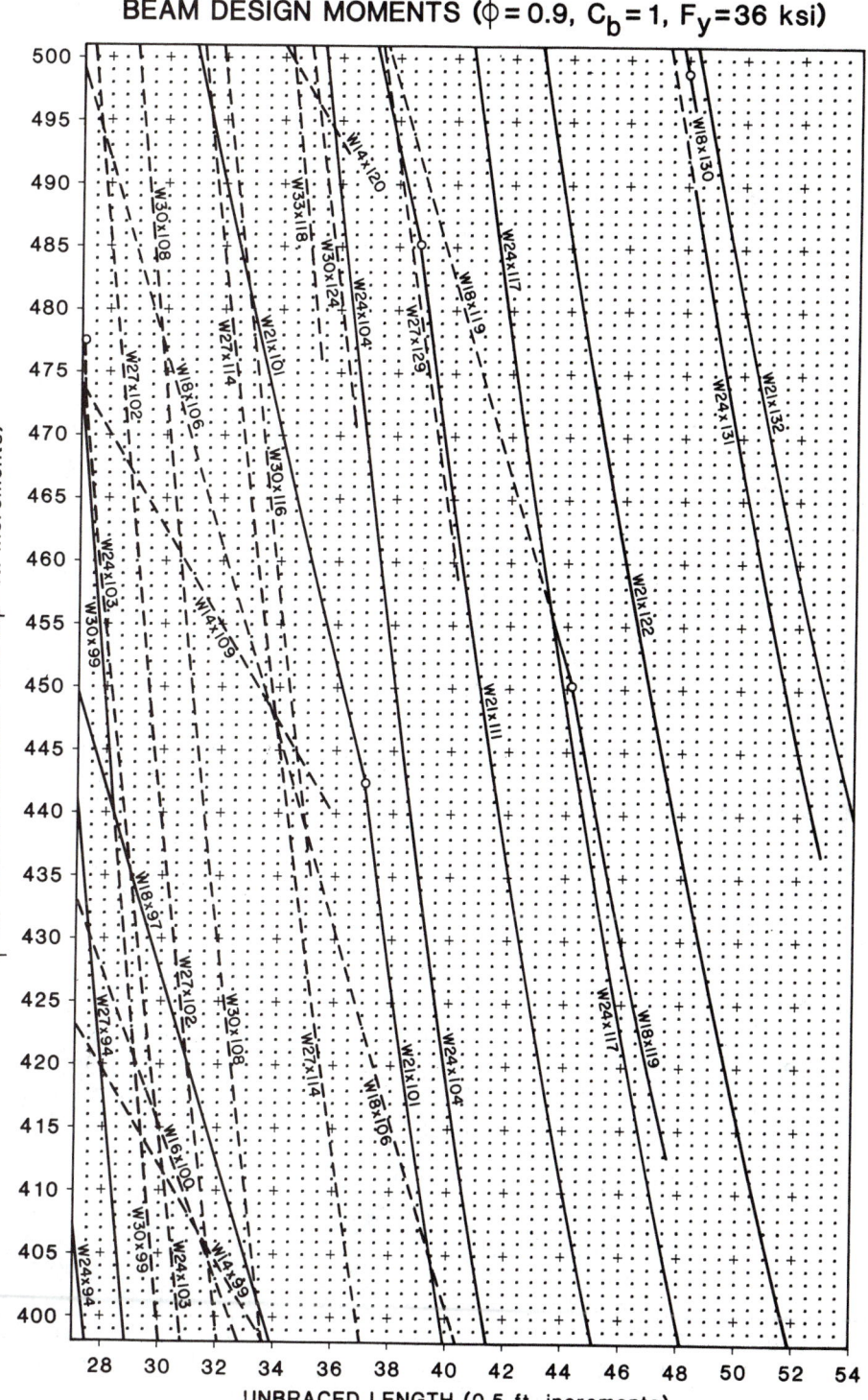

ϕM_n – DESIGN MOMENT (1.0 kip-ft. increments)

UNBRACED LENGTH (0.5 ft. increments)

BEAM DESIGN MOMENTS ($\phi = 0.9$, $C_b = 1$, $F_y = 36$ ksi)

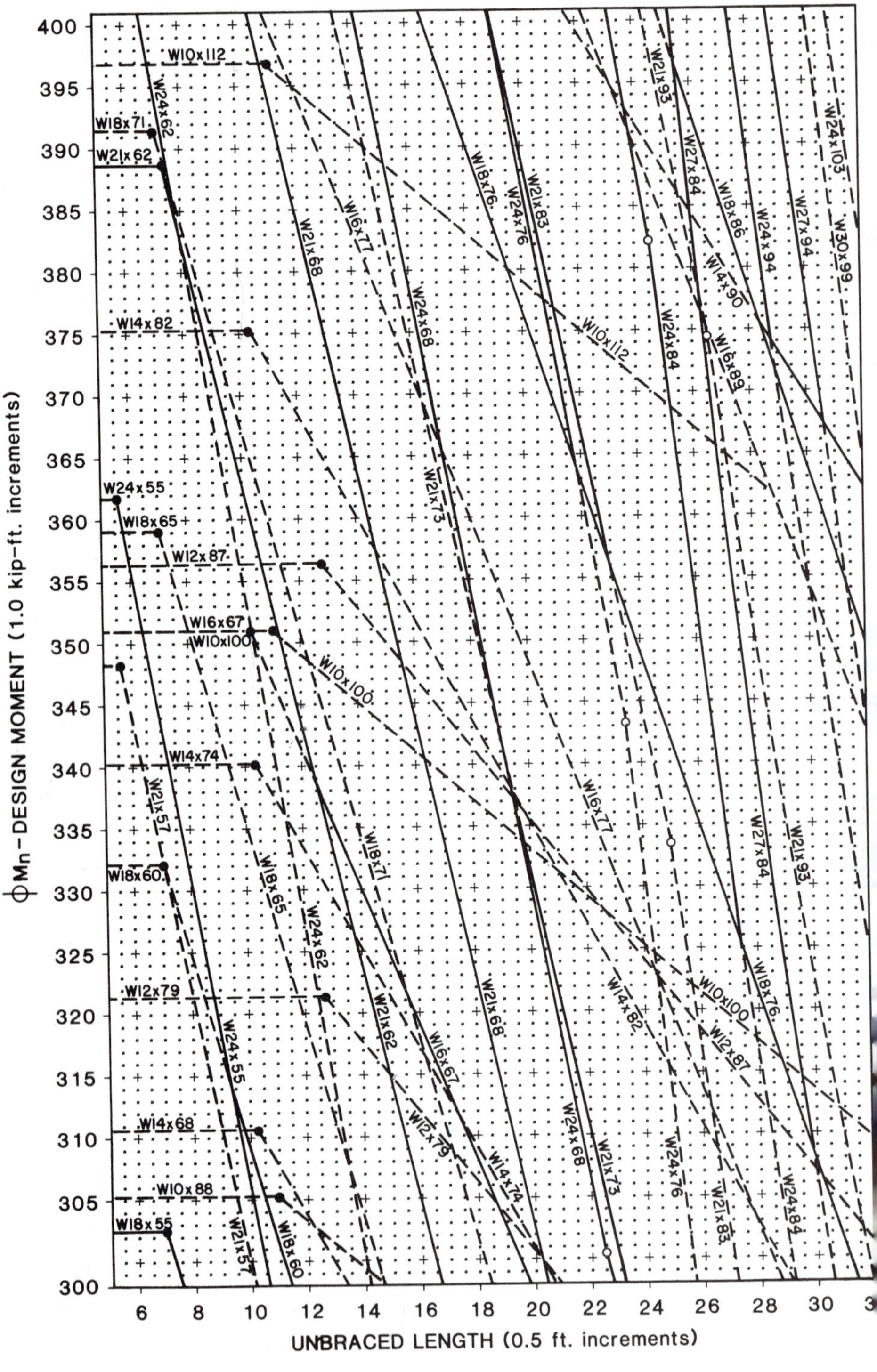

Chart plotting ϕM_n – DESIGN MOMENT (1.0 kip-ft. increments) on the vertical axis (300 to 400) versus UNBRACED LENGTH (0.5 ft. increments) on the horizontal axis (6 to 32).

BEAM DESIGN MOMENTS (ϕ=0.9, C_b=1, F_y=36 ksi)

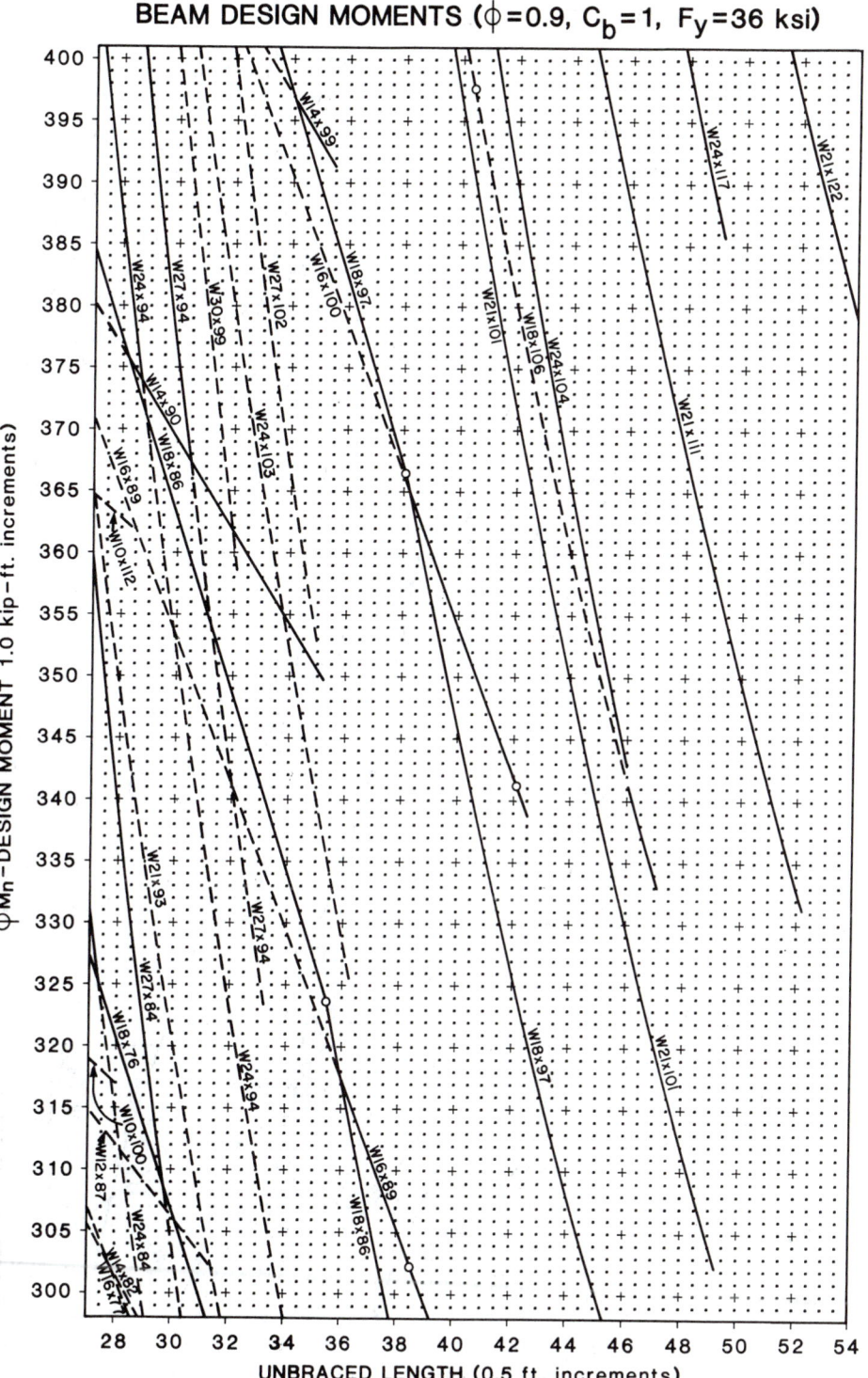

BEAM DESIGN MOMENTS ($\phi = 0.9$, $C_b = 1$, $F_y = 36$ ksi)

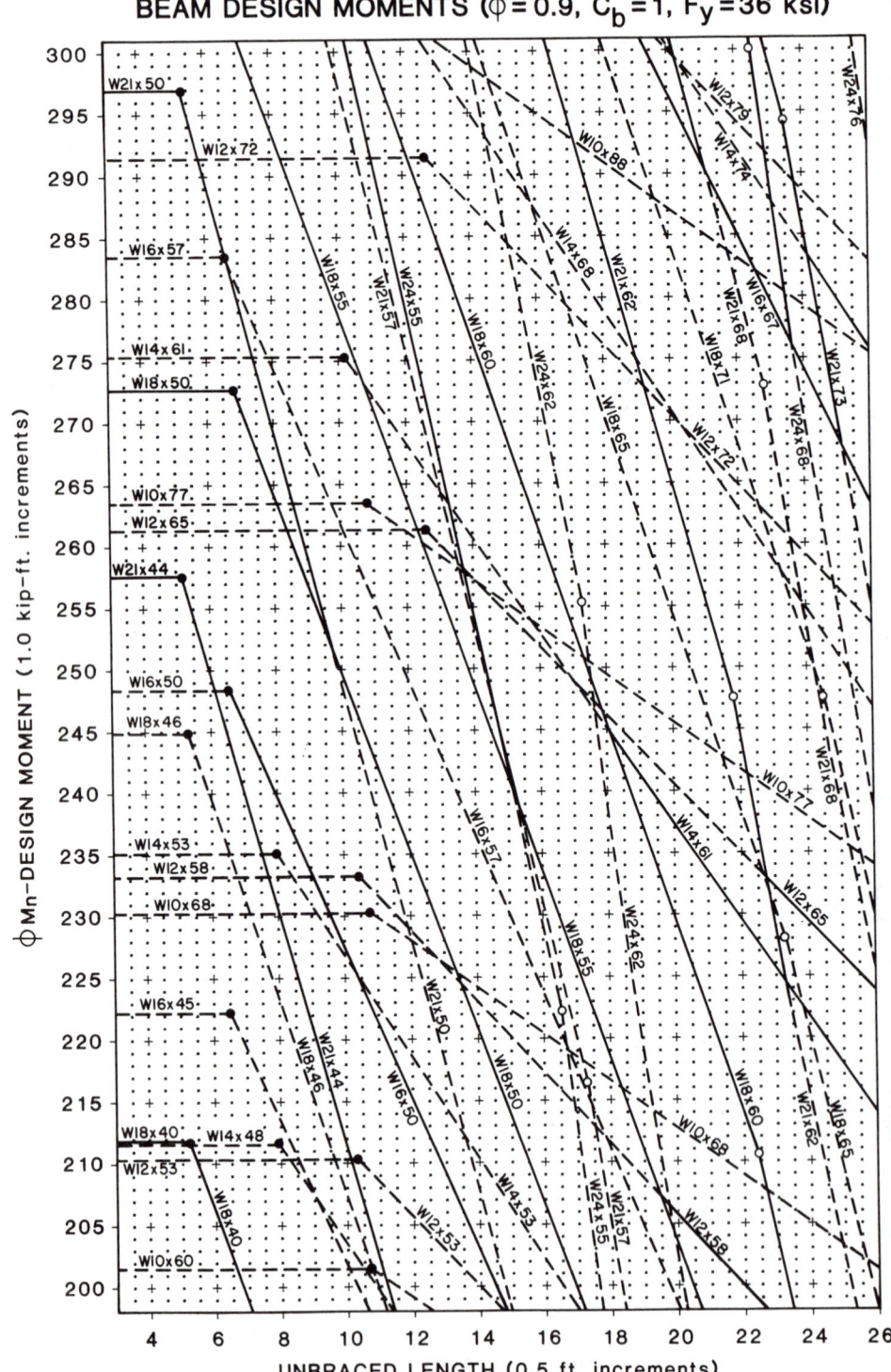

ϕM_n—DESIGN MOMENT (1.0 kip-ft. increments)

UNBRACED LENGTH (0.5 ft. increments)

AMERICAN INSTITUTE OF STEEL CONSTRUCTION

BEAM DESIGN MOMENTS ($\phi = 0.9$, $C_b = 1$, $F_y = 36$ ksi)

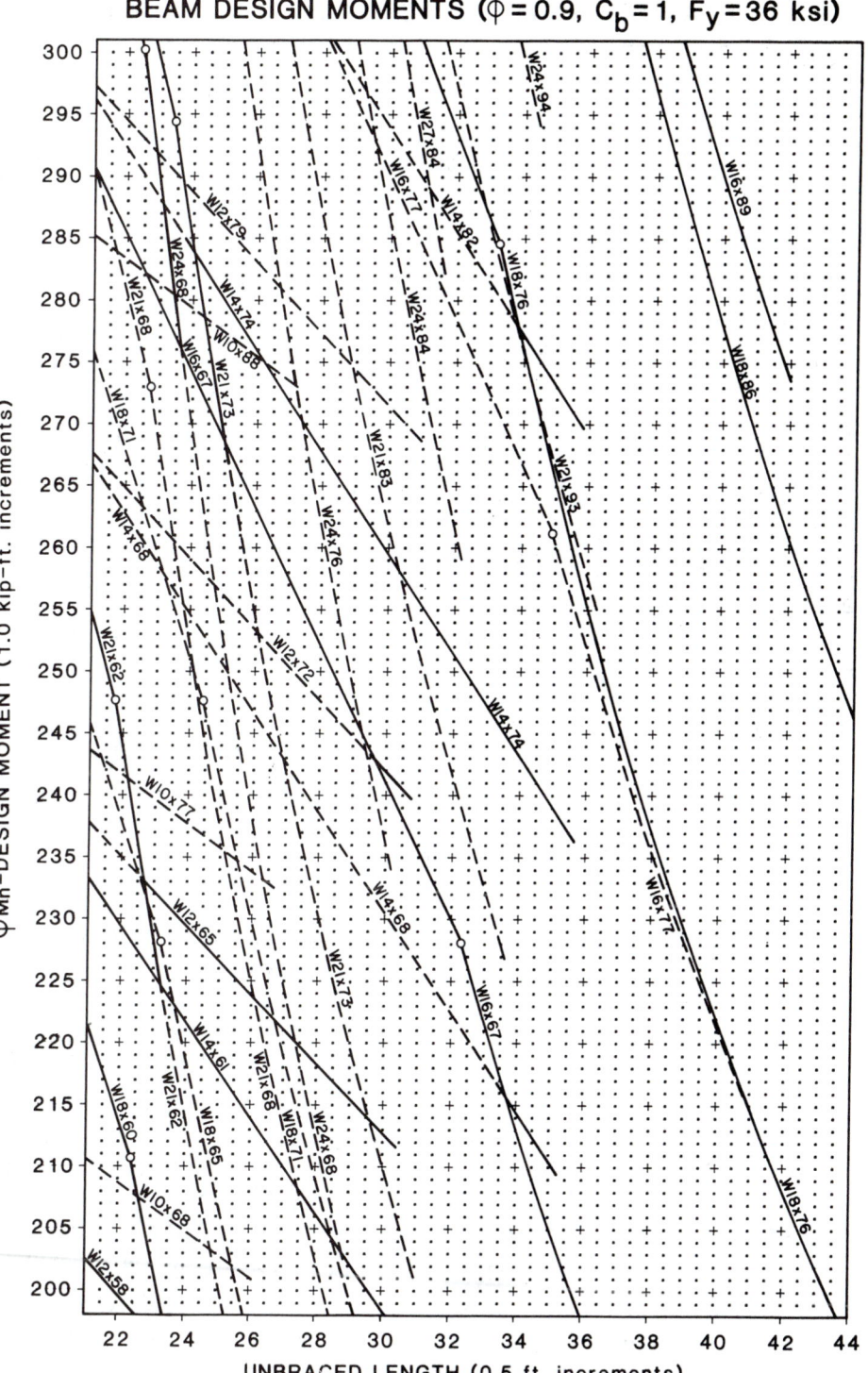

ϕM_n–DESIGN MOMENT (1.0 kip-ft. increments)

UNBRACED LENGTH (0.5 ft. increments)

BEAM DESIGN MOMENTS (ϕ = 0.9, C_b = 1, F_y = 36 ksi)

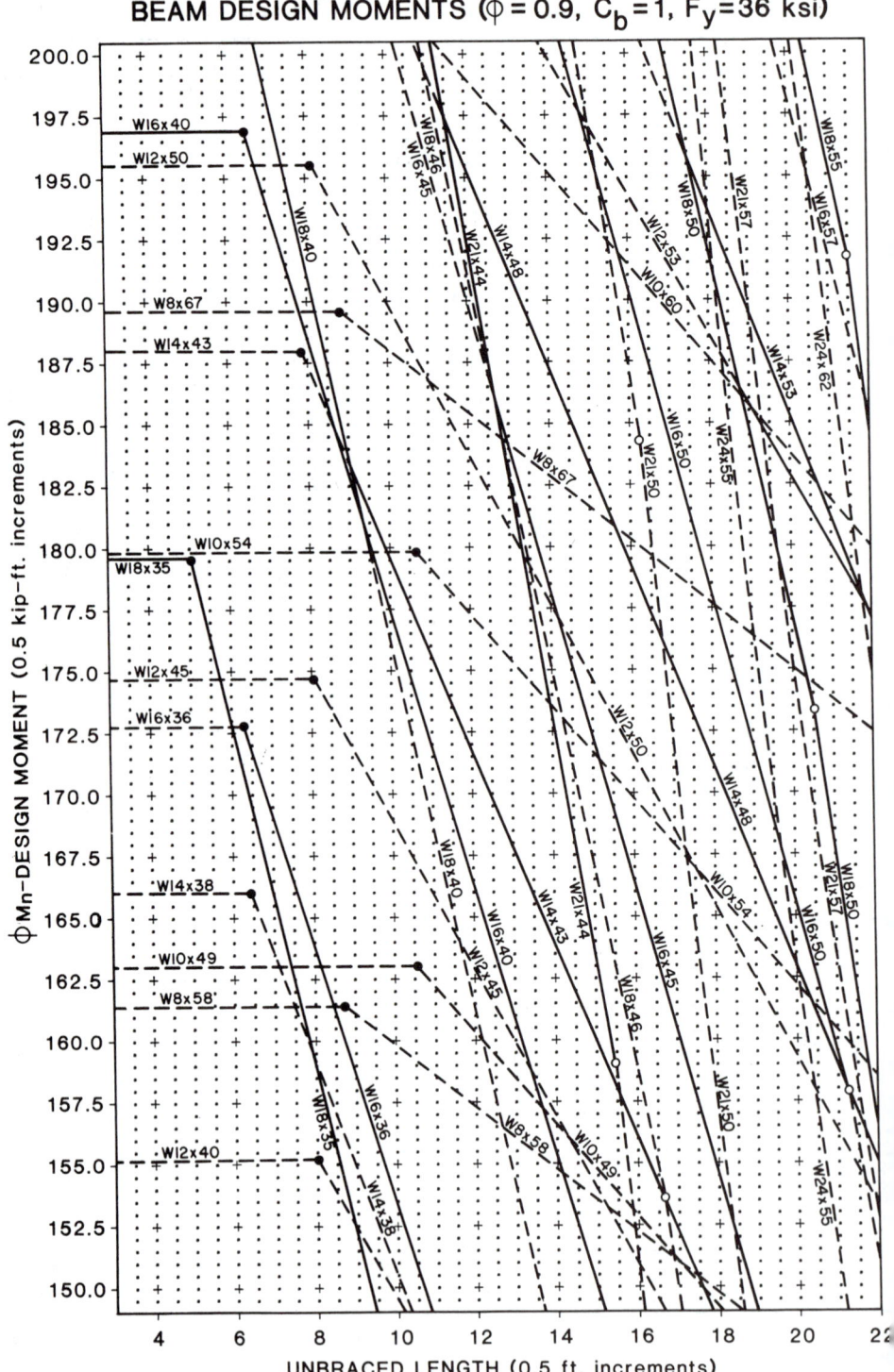

BEAM DESIGN MOMENTS ($\phi = 0.9$, $C_b = 1$, $F_y = 36$ ksi)

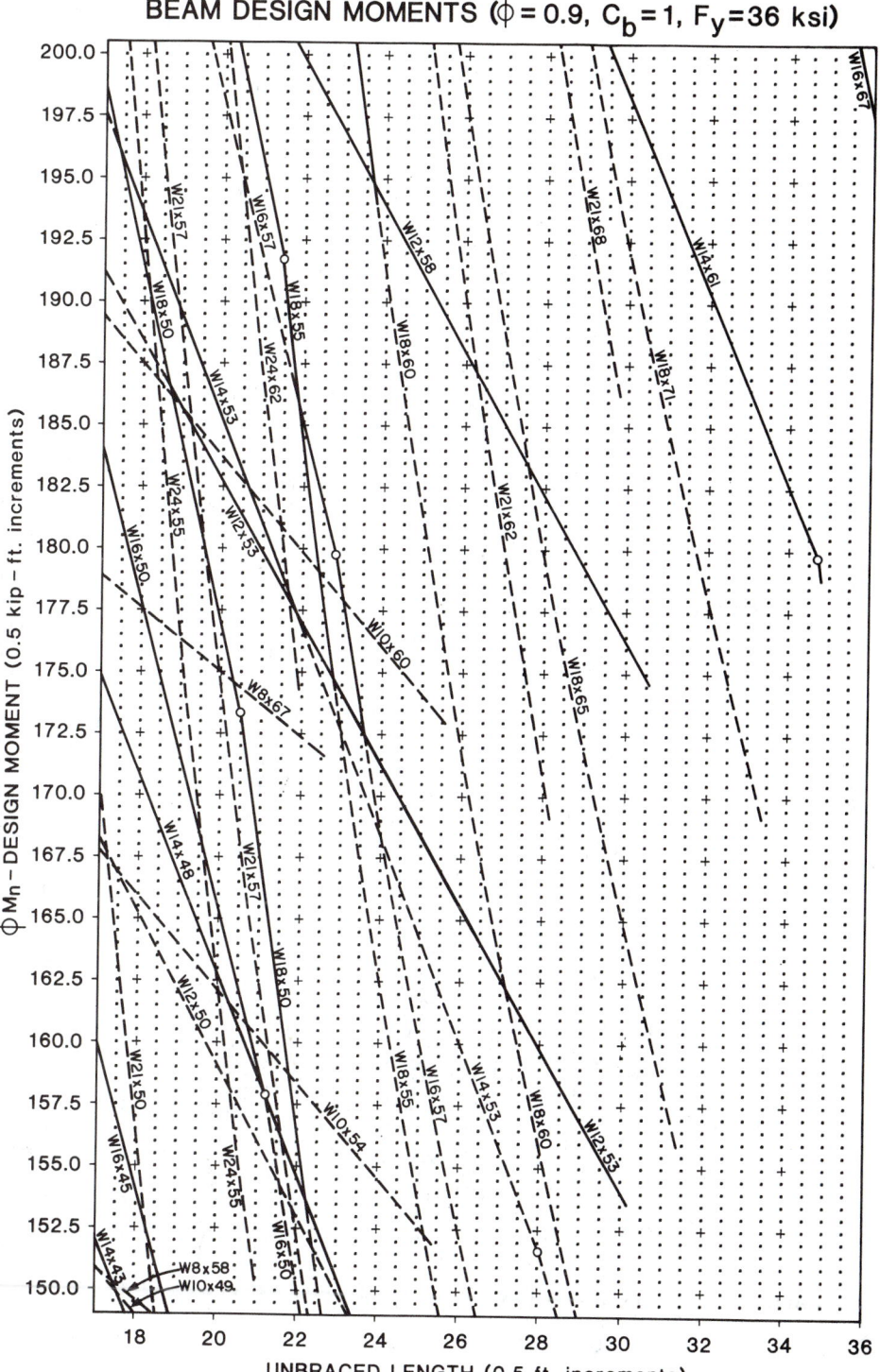

ϕM_n – DESIGN MOMENT (0.5 kip – ft. increments)

UNBRACED LENGTH (0.5 ft. increments)

BEAM DESIGN MOMENTS ($\phi = 0.9$, $C_b = 1$, $F_y = 36$ ksi)

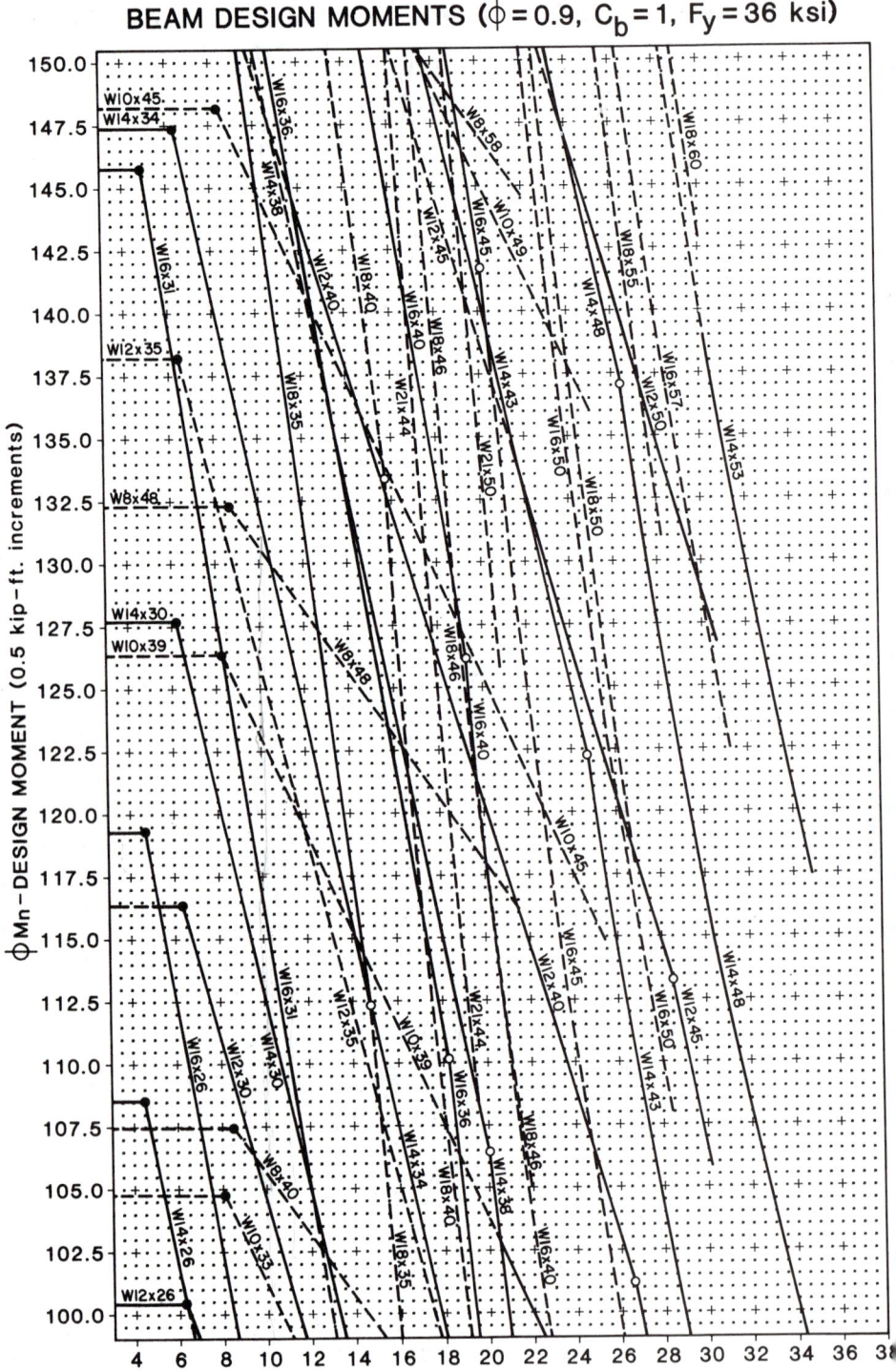

ϕM_n – DESIGN MOMENT (0.5 kip-ft. increments)

UNBRACED LENGTH (0.5 ft. increments)

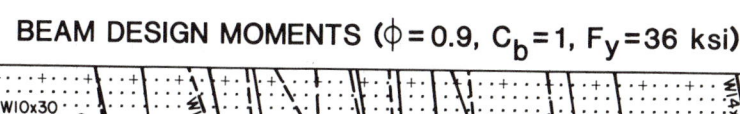

BEAM DESIGN MOMENTS ($\phi = 0.9$, $C_b = 1$, $F_y = 36$ ksi)

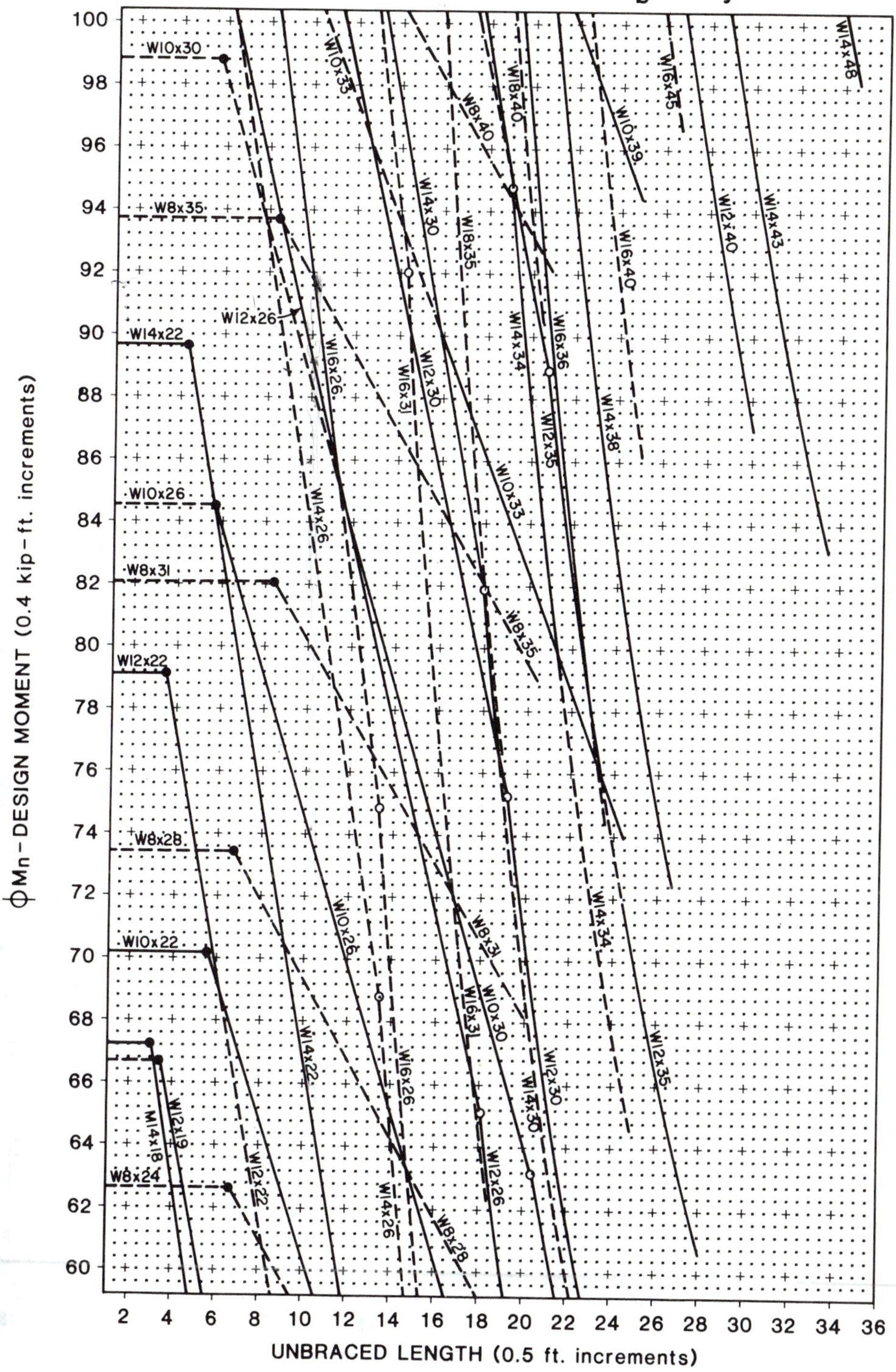

ϕM_n – DESIGN MOMENT (0.4 kip-ft. increments)

UNBRACED LENGTH (0.5 ft. increments)

AMERICAN INSTITUTE OF STEEL CONSTRUCTION

BEAM DESIGN MOMENTS ($\phi = 0.9$, $C_b = 1$, $F_y = 36$ ksi)

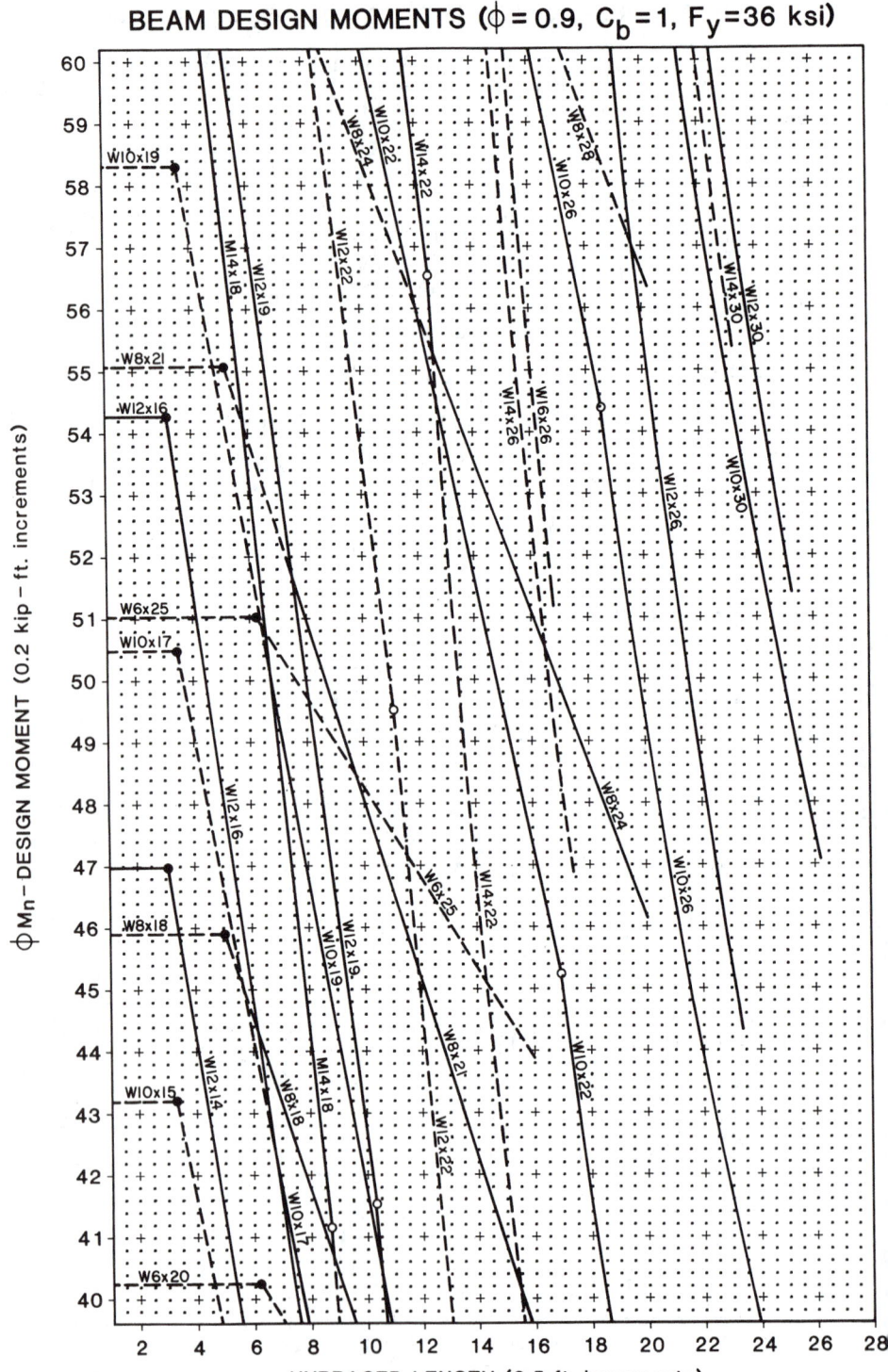

ϕM_n – DESIGN MOMENT (0.2 kip – ft. increments)

UNBRACED LENGTH (0.5 ft. increments)

BEAM DESIGN MOMENTS ($\phi = 0.9$, $C_b = 1$, $F_y = 36$ ksi)

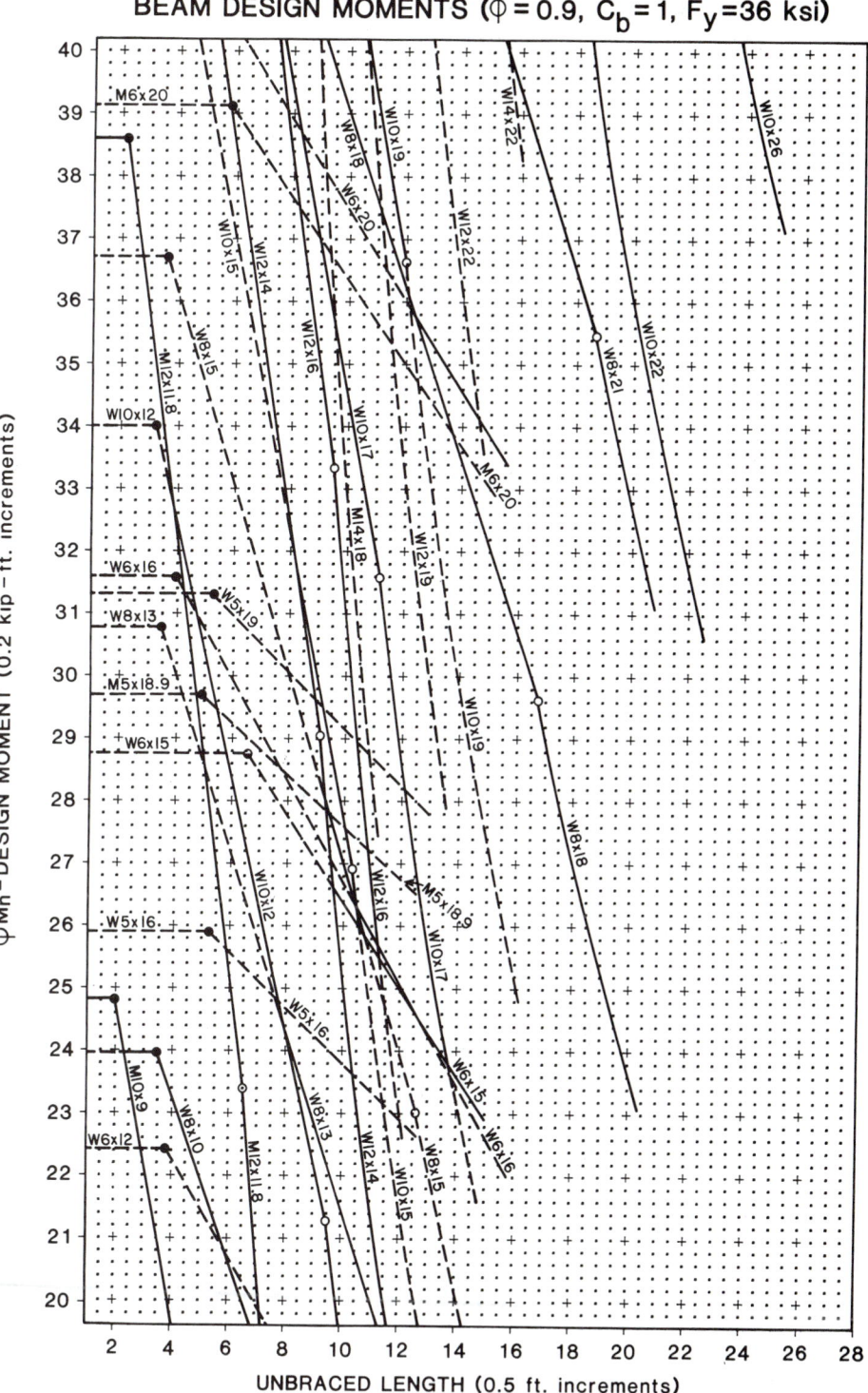

BEAM DESIGN MOMENTS ($\phi = 0.9$, $C_b = 1$, $F_y = 36$ ksi)

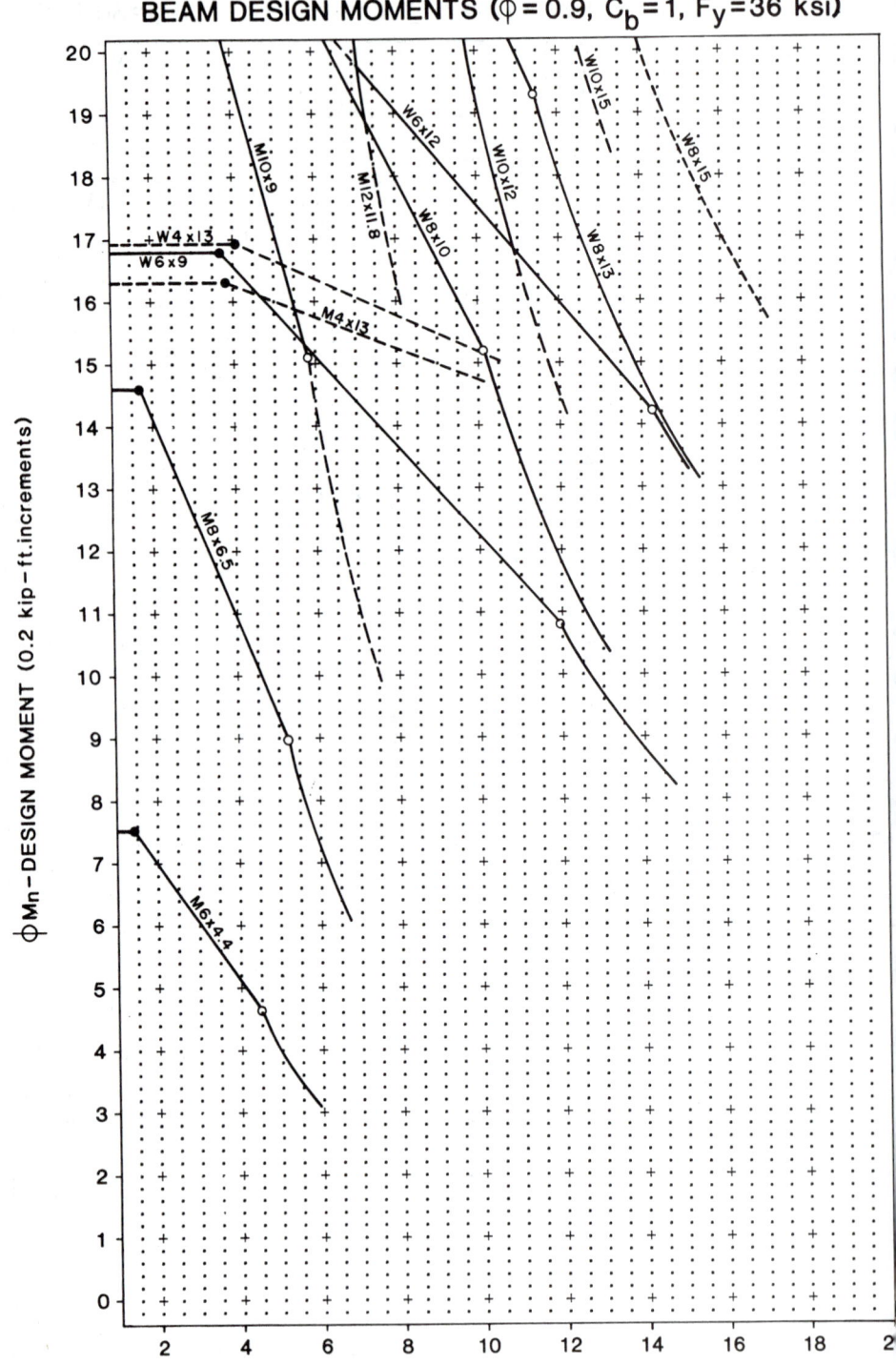

ϕM_n–DESIGN MOMENT (0.2 kip–ft.increments)

UNBRACED LENGTH (0.5 ft. increments)

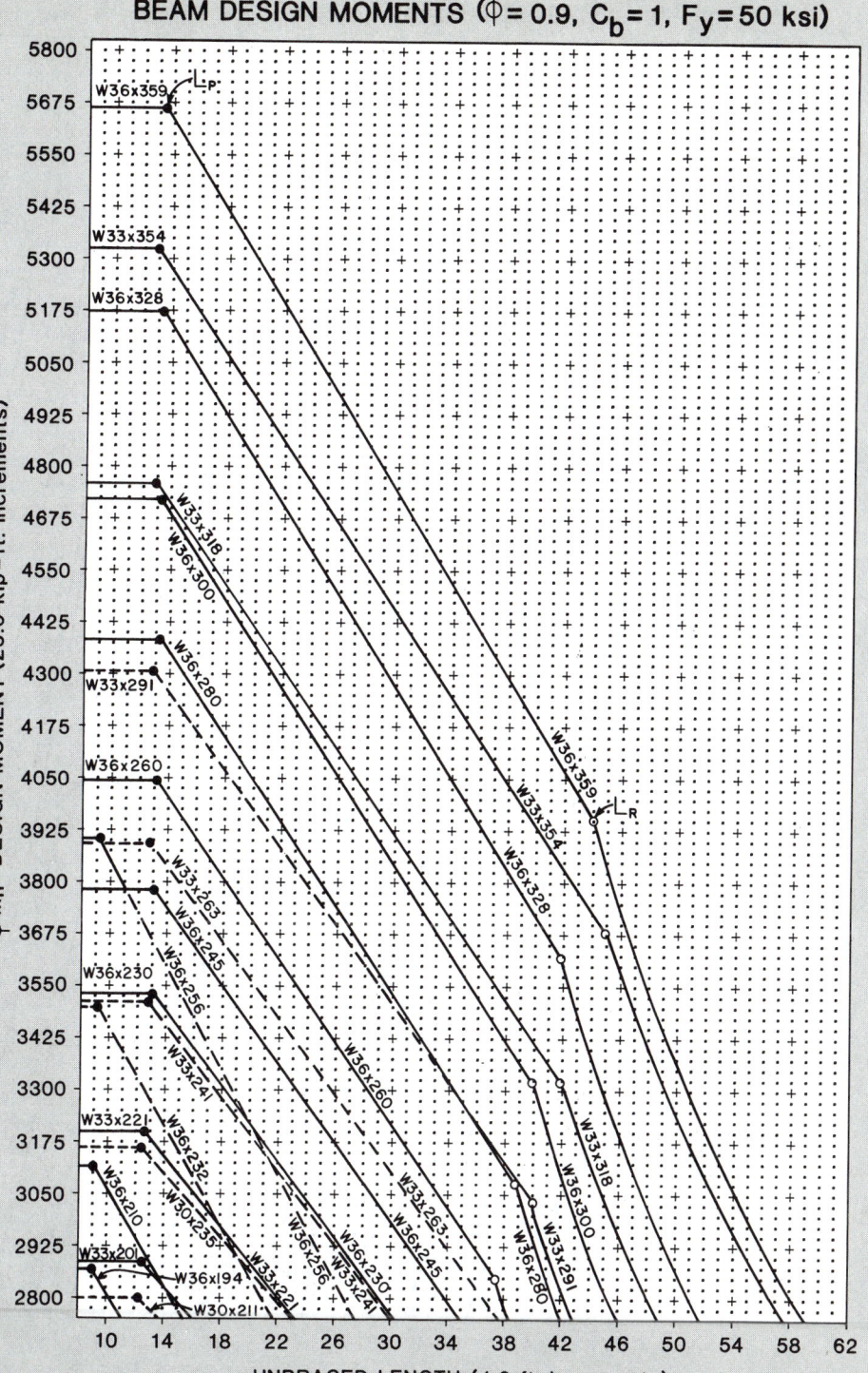

BEAM DESIGN MOMENTS (ϕ = 0.9, C_b = 1, F_y = 50 ksi)

ϕM_n - DESIGN MOMENT (25.0 kip-ft. increments)

UNBRACED LENGTH (1.0 ft. increments)

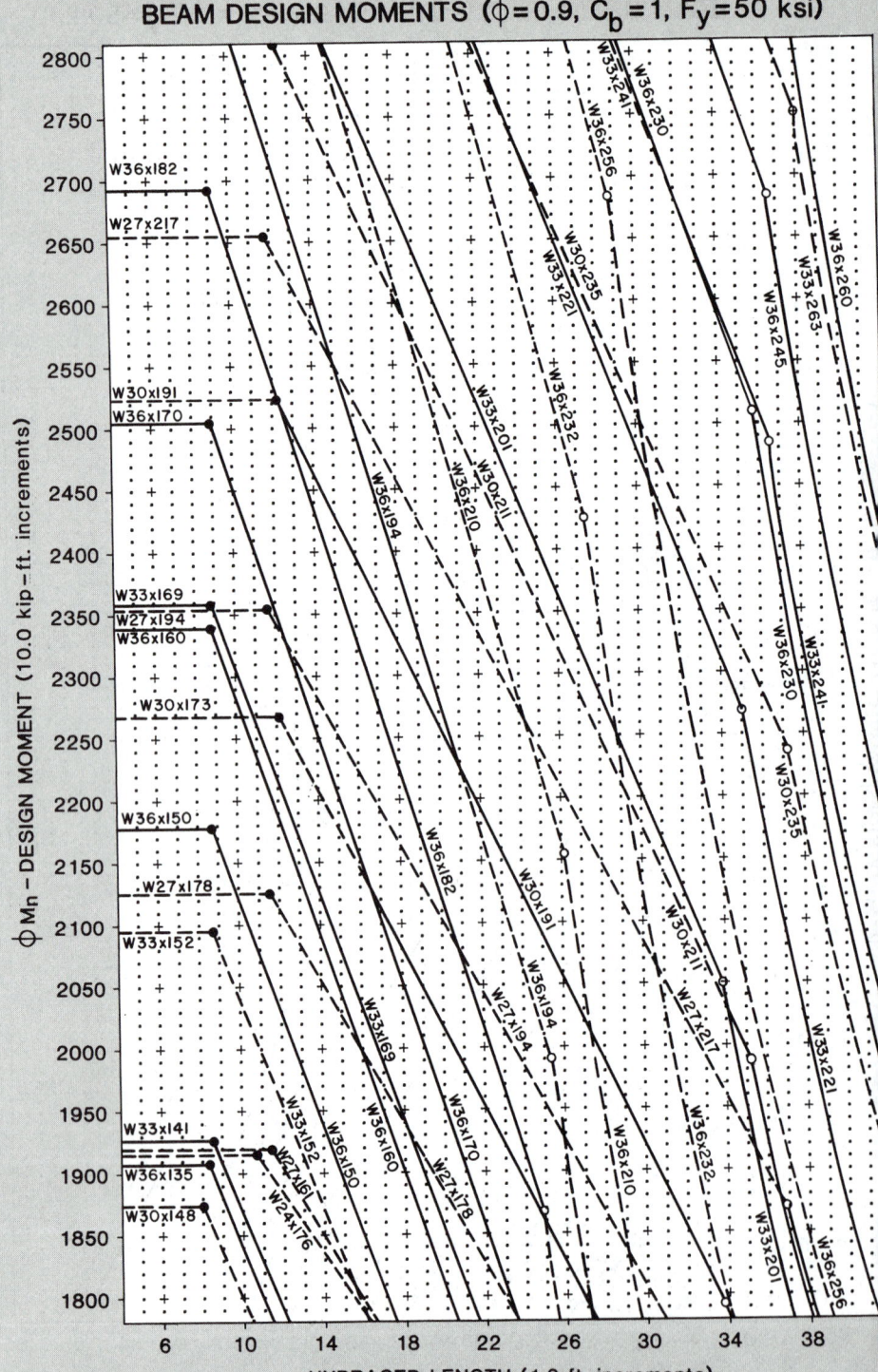

BEAM DESIGN MOMENTS (ϕ = 0.9, C_b = 1, F_y = 50 ksi)

ϕM_n – DESIGN MOMENT (10.0 kip–ft. increments)

UNBRACED LENGTH (1.0 ft. increments)

AMERICAN INSTITUTE OF STEEL CONSTRUCTION

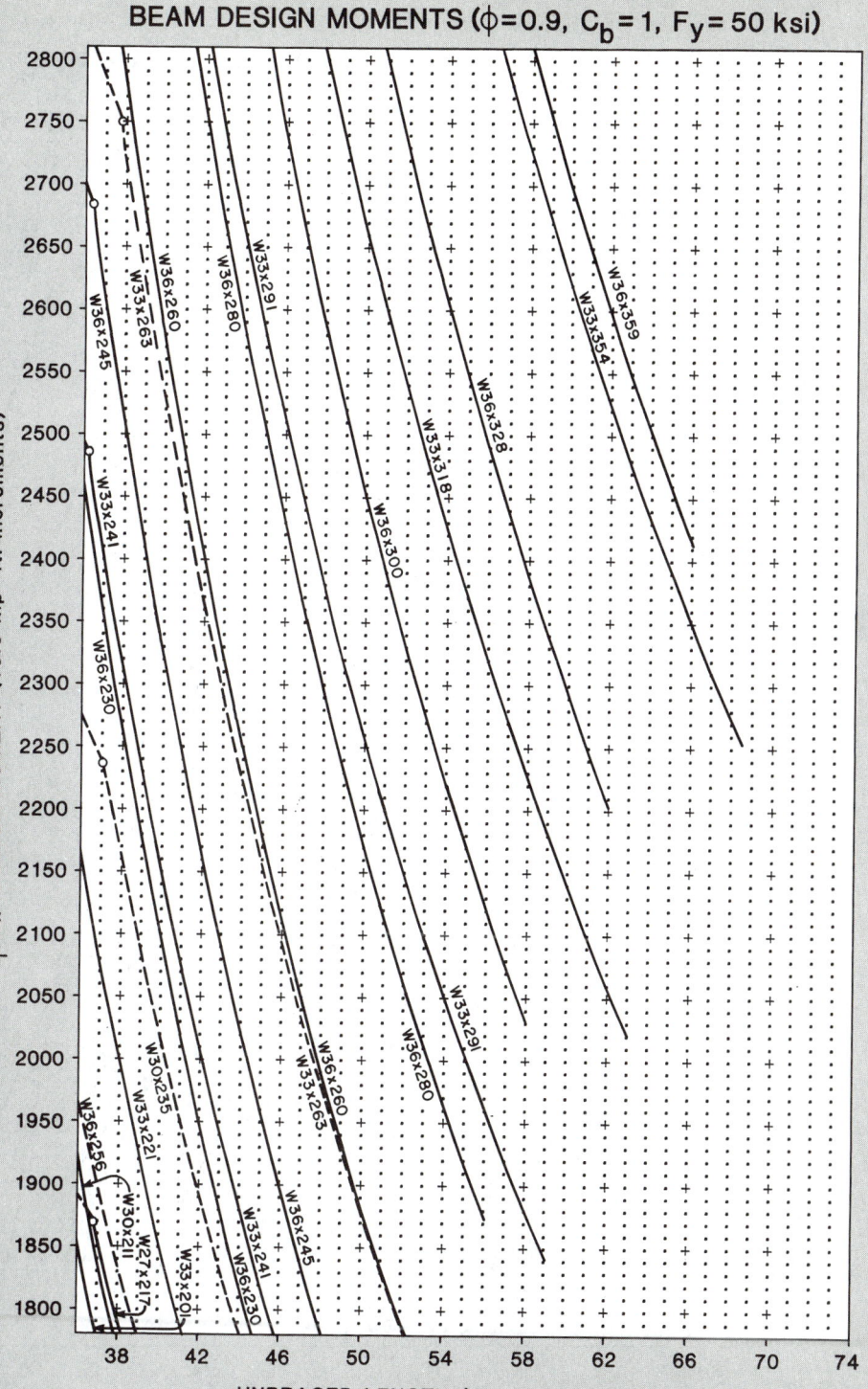

AMERICAN INSTITUTE OF STEEL CONSTRUCTION

BEAM DESIGN MOMENTS ($\phi = 0.9$, $C_b = 1$, $F_y = 50$ ksi)

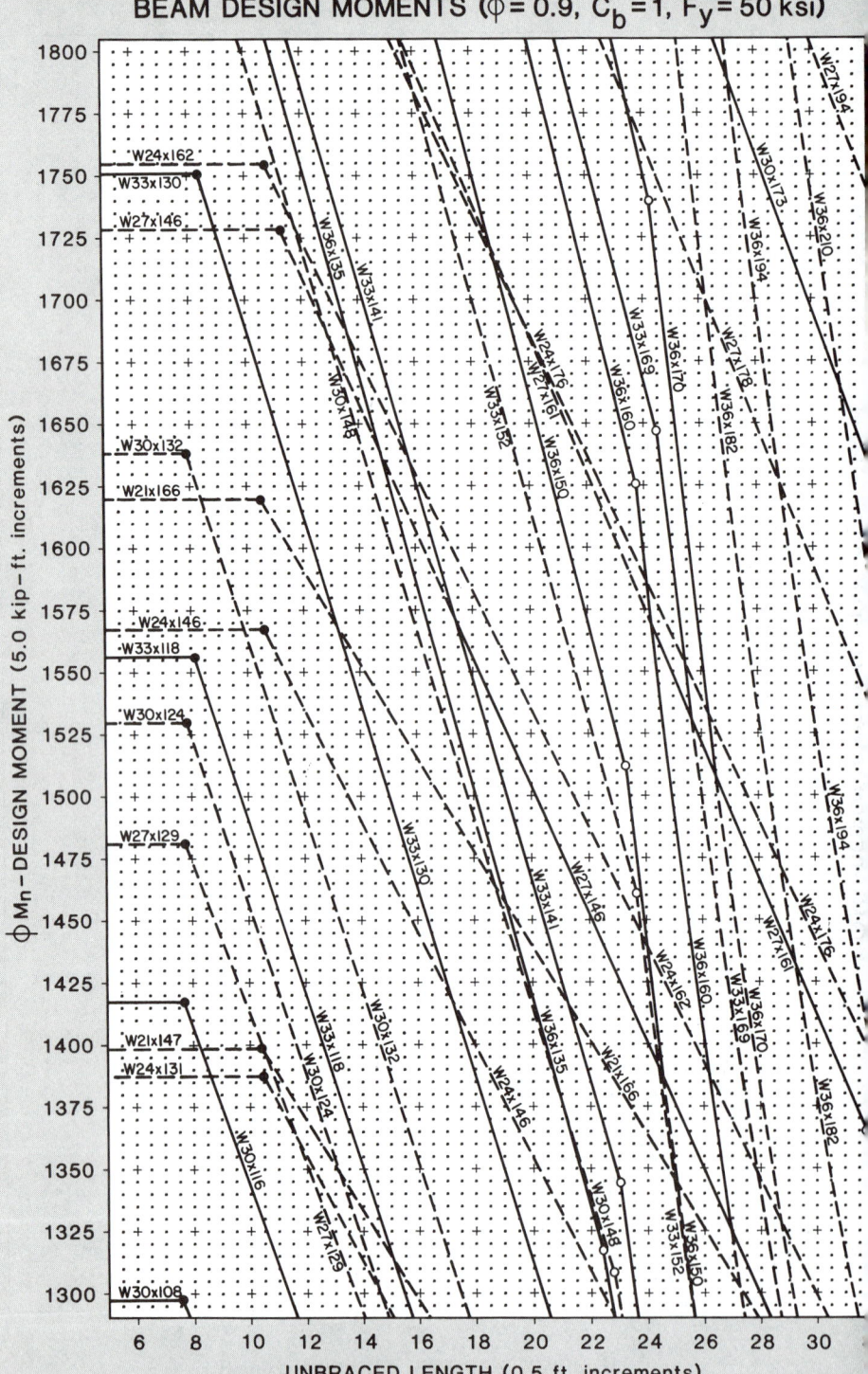

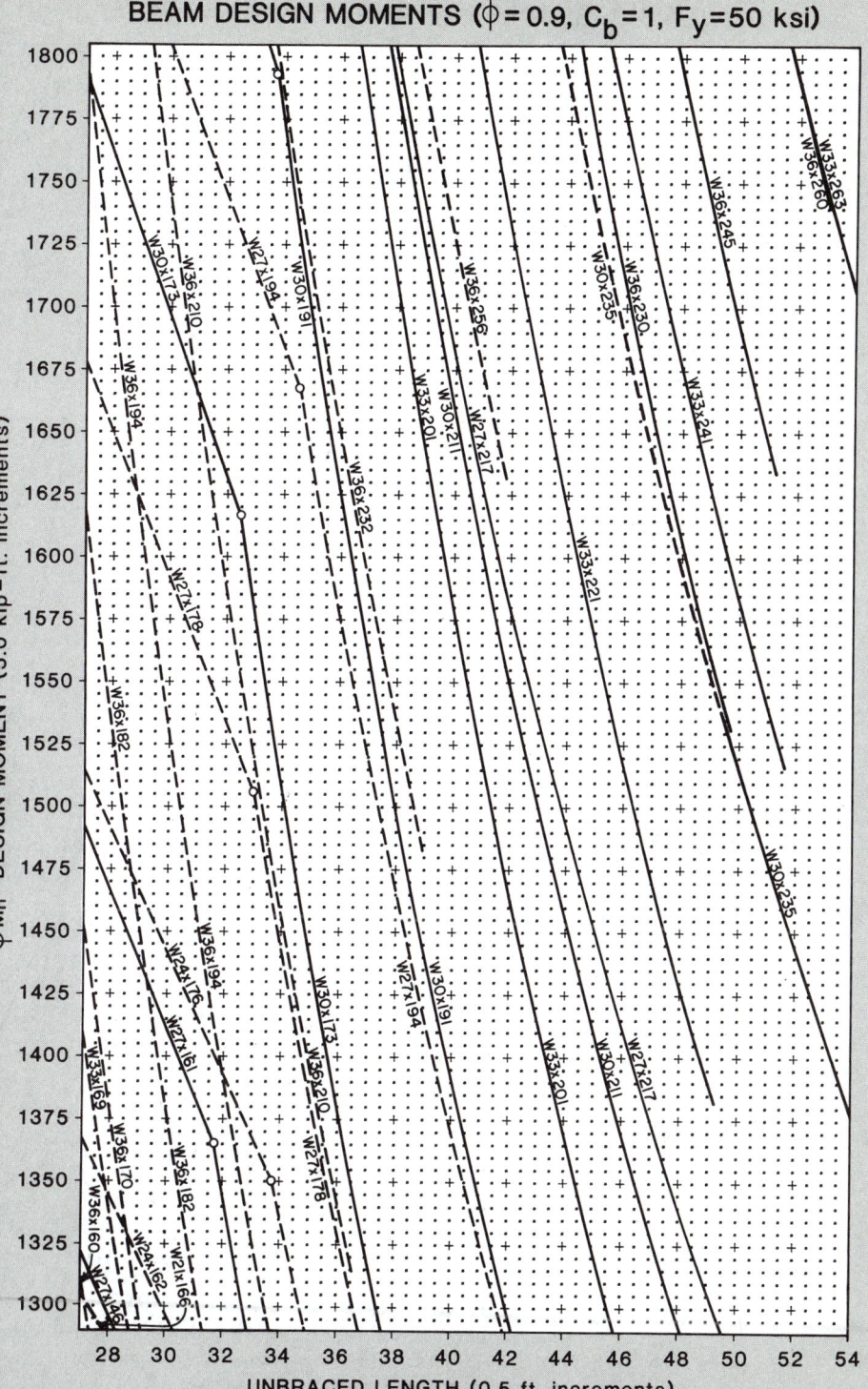

BEAM DESIGN MOMENTS ($\phi = 0.9$, $C_b = 1$, $F_y = 50$ ksi)

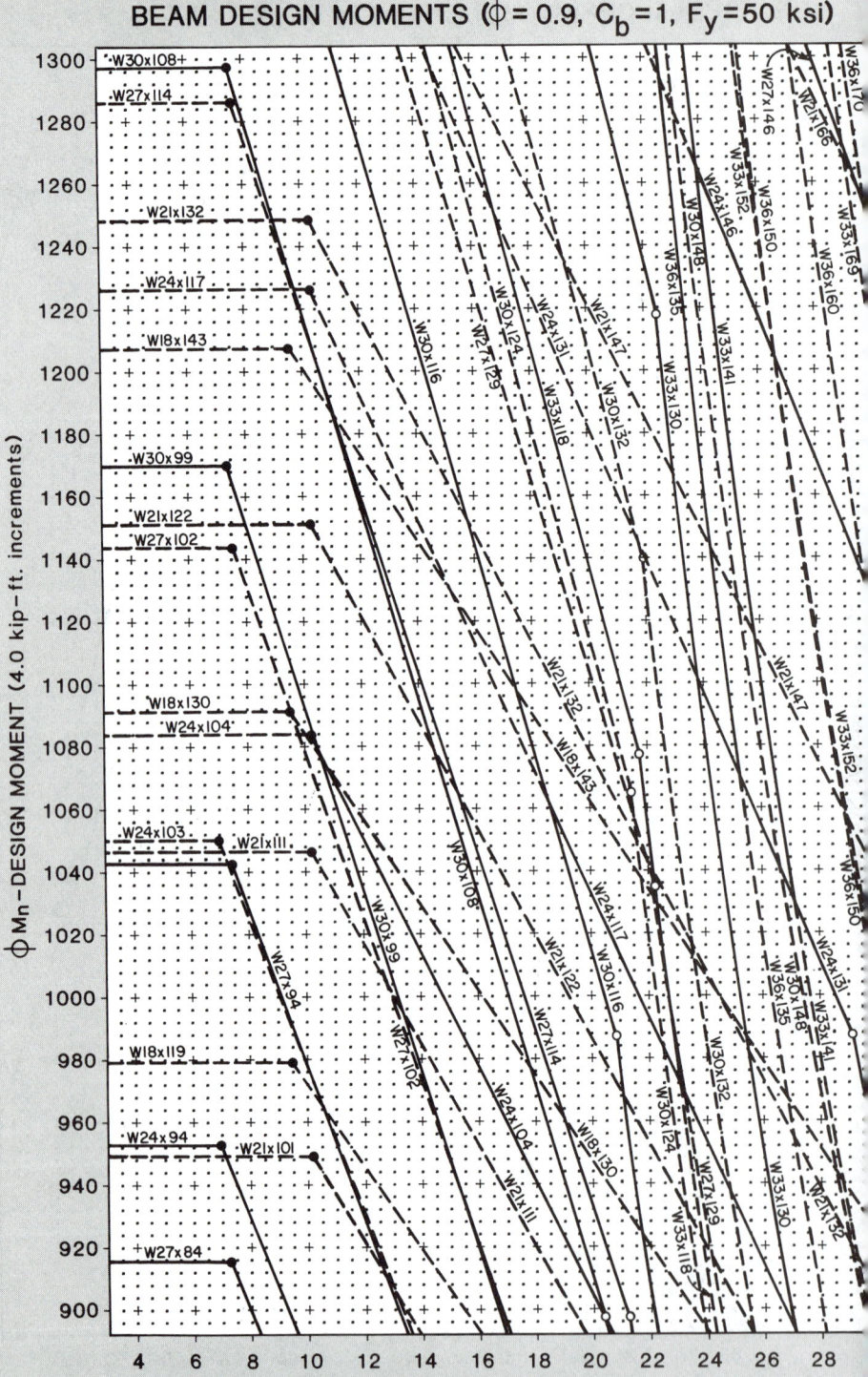

ϕM_n -DESIGN MOMENT (4.0 kip-ft. increments)

UNBRACED LENGTH (0.5 ft. increments)

BEAM DESIGN MOMENTS ($\phi = 0.9$, $C_b = 1$, $F_y = 50$ ksi)

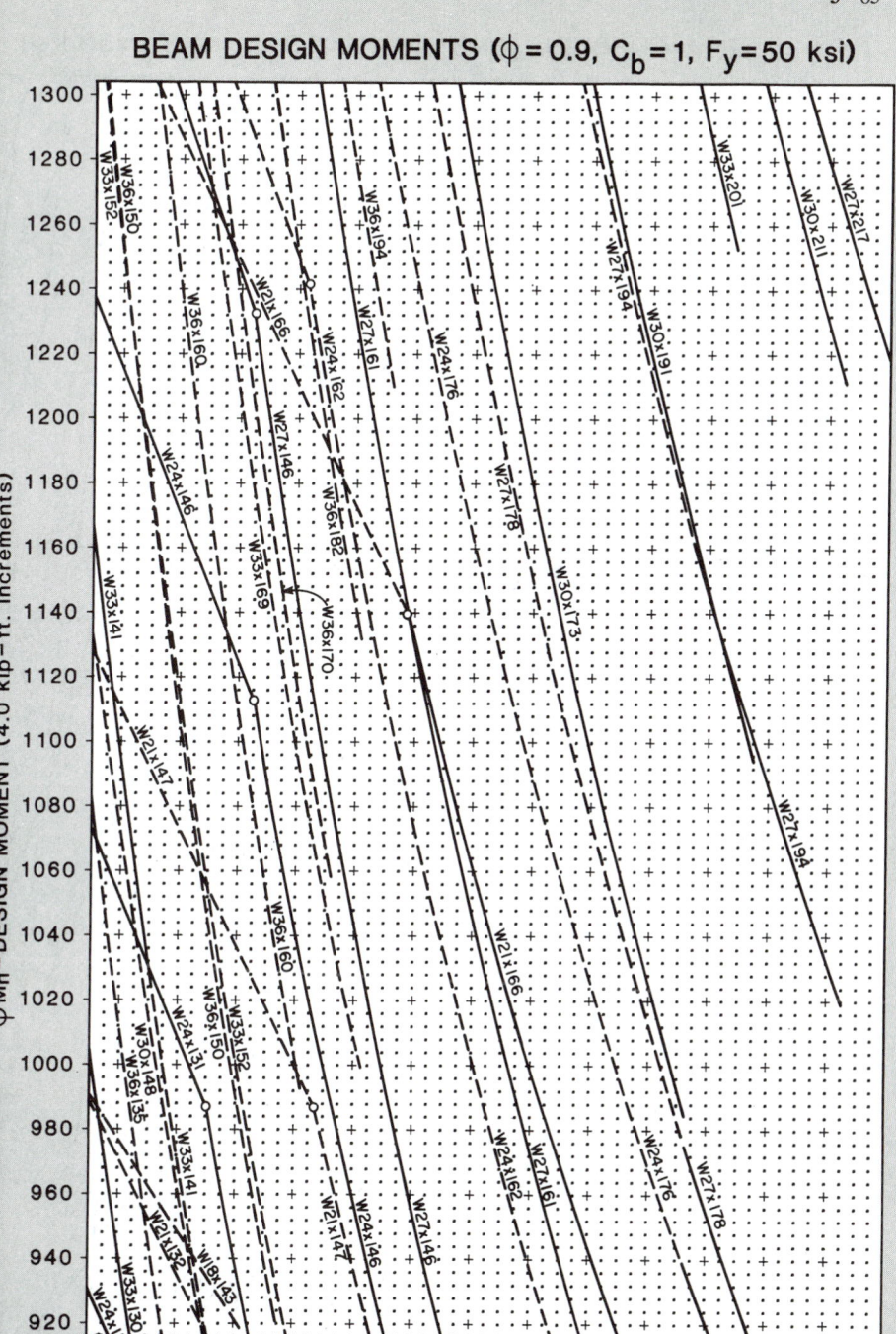

ϕM_n–DESIGN MOMENT (4.0 kip–ft. increments)

UNBRACED LENGTH (0.5 ft. increments)

BEAM DESIGN MOMENTS (ϕ = 0.9, C_b = 1, F_y = 50 ksi)

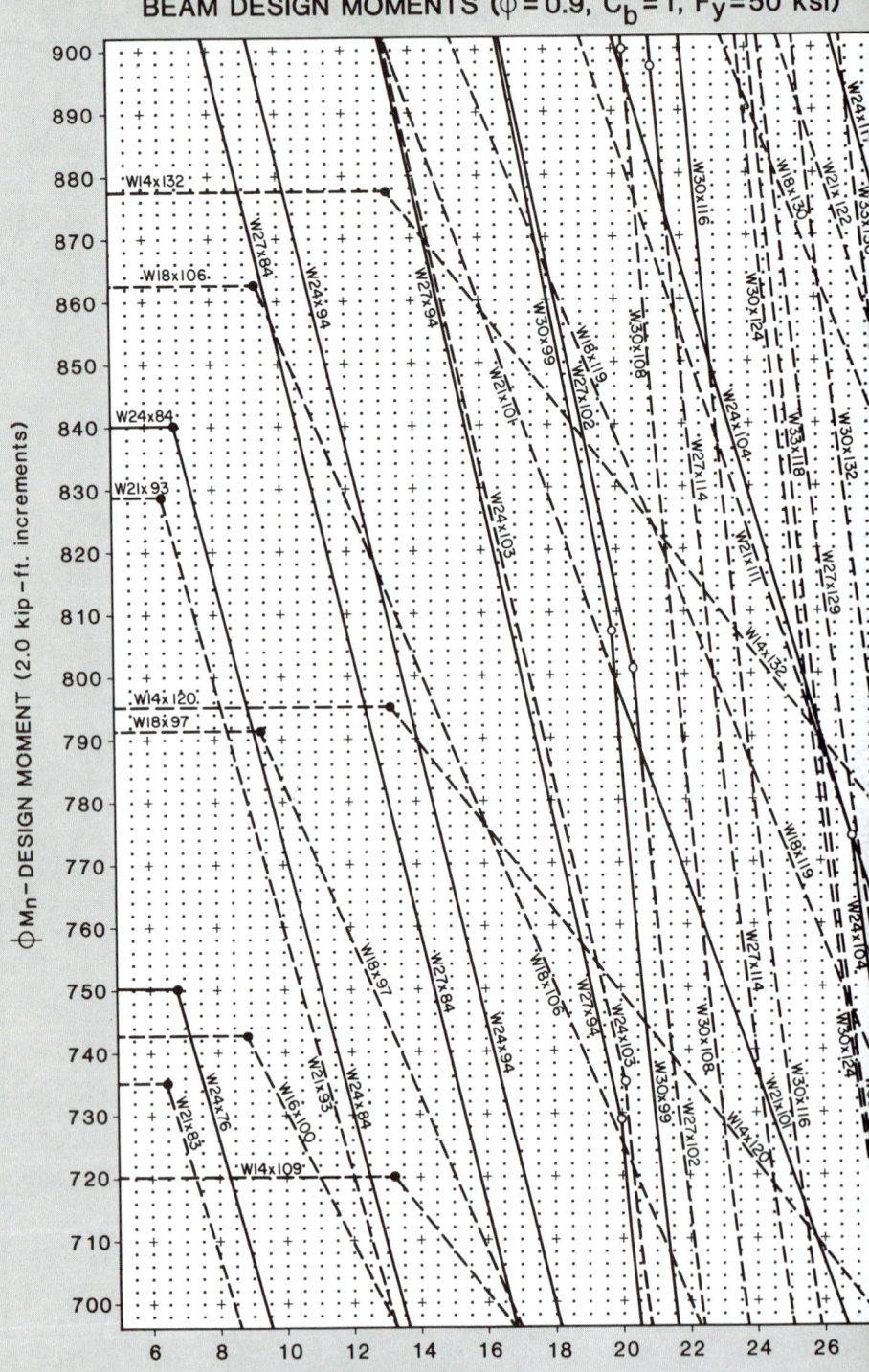

ϕM_n - DESIGN MOMENT (2.0 kip-ft. increments)

UNBRACED LENGTH (0.5 ft. increments)

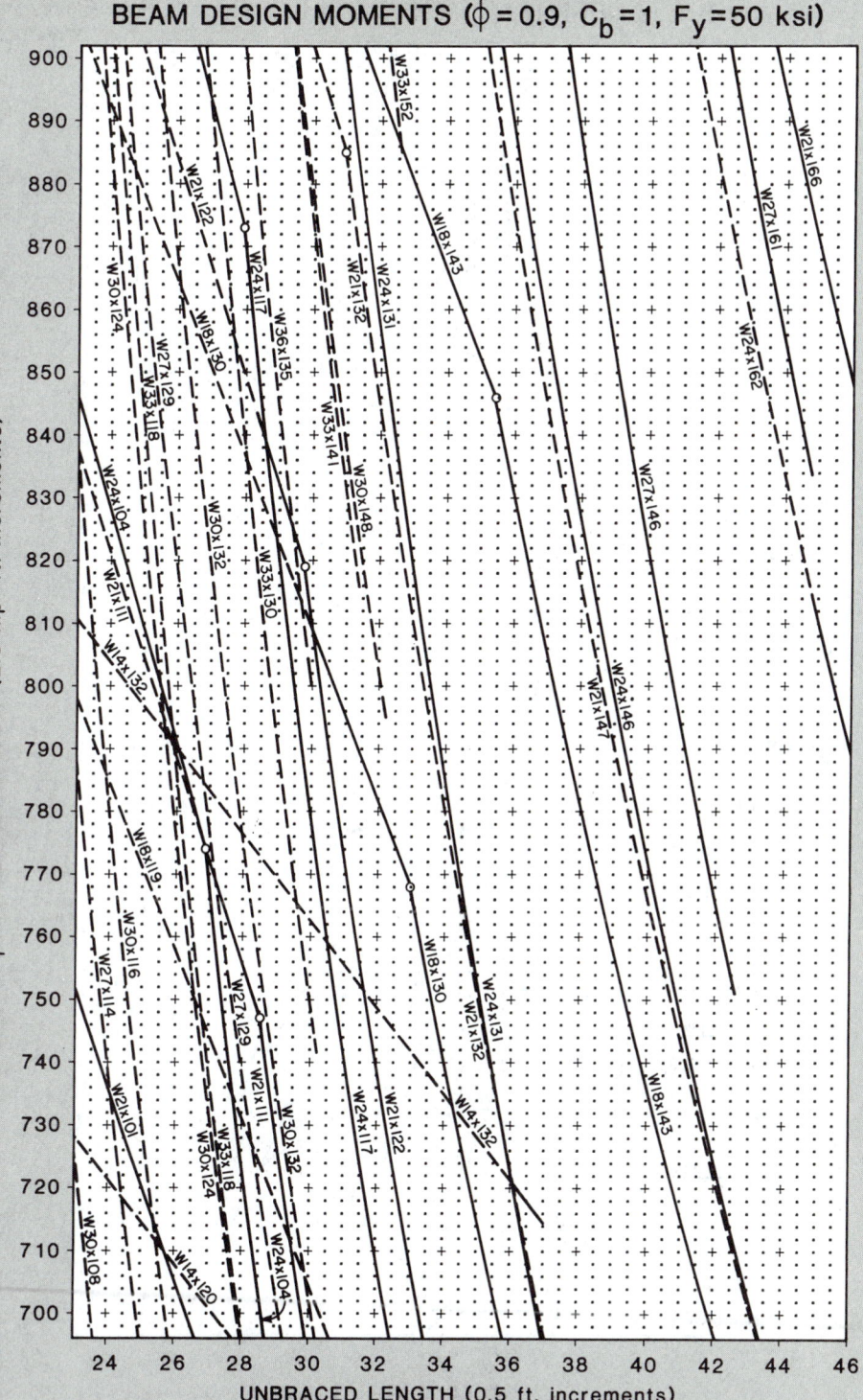

BEAM DESIGN MOMENTS ($\phi = 0.9$, $C_b = 1$, $F_y = 50$ ksi)

BEAM DESIGN MOMENTS ($\phi=0.9$, $C_b=1$, $F_y=50$ ksi)

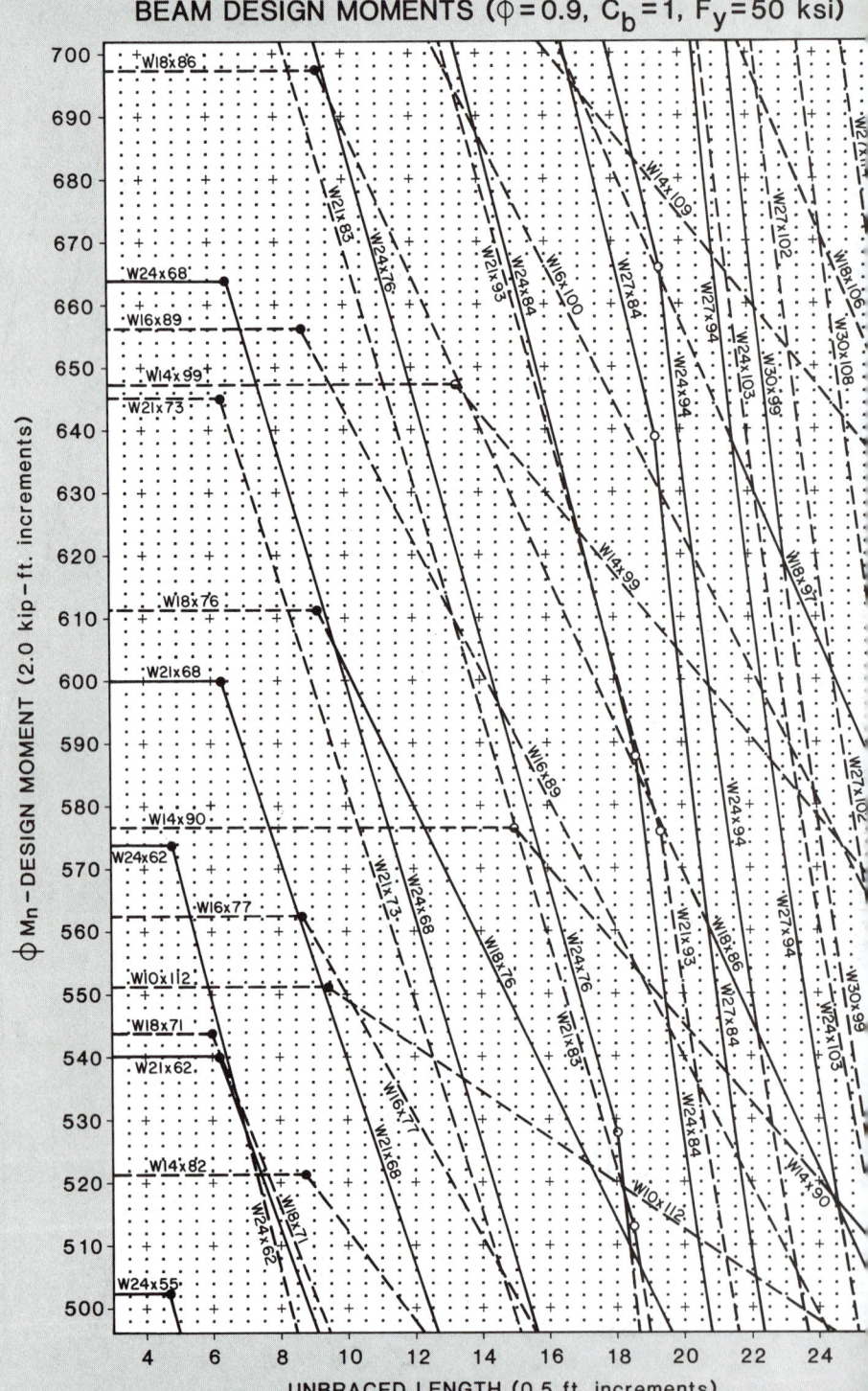

ϕM_n – DESIGN MOMENT (2.0 kip – ft. increments)

UNBRACED LENGTH (0.5 ft. increments)

BEAM DESIGN MOMENTS ($\phi = 0.9$, $C_b = 1$, $F_y = 50$ ksi)

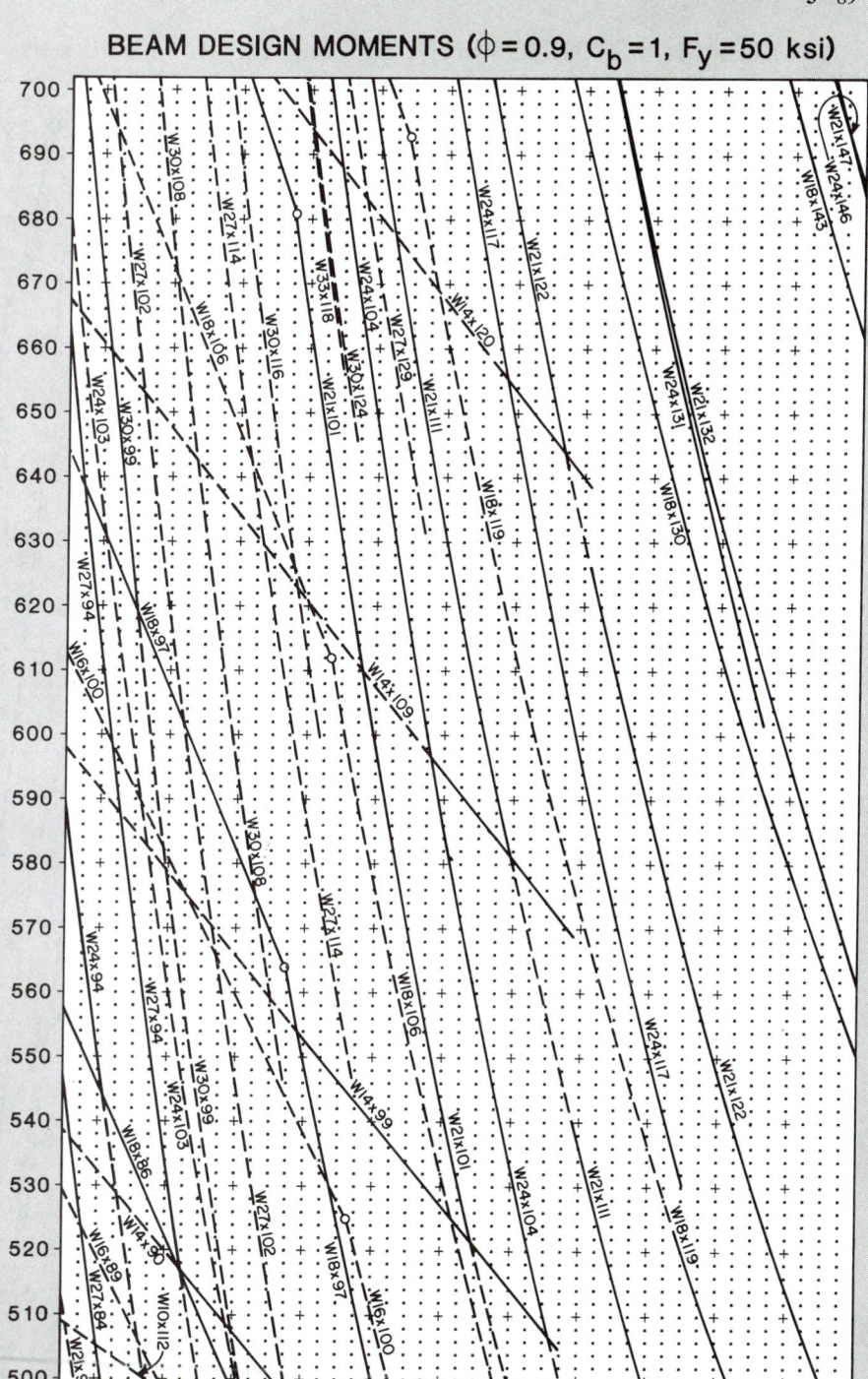

UNBRACED LENGTH (0.5 ft. increments)

BEAM DESIGN MOMENTS ($\phi = 0.9$, $C_b = 1$, $F_y = 50$ ksi)

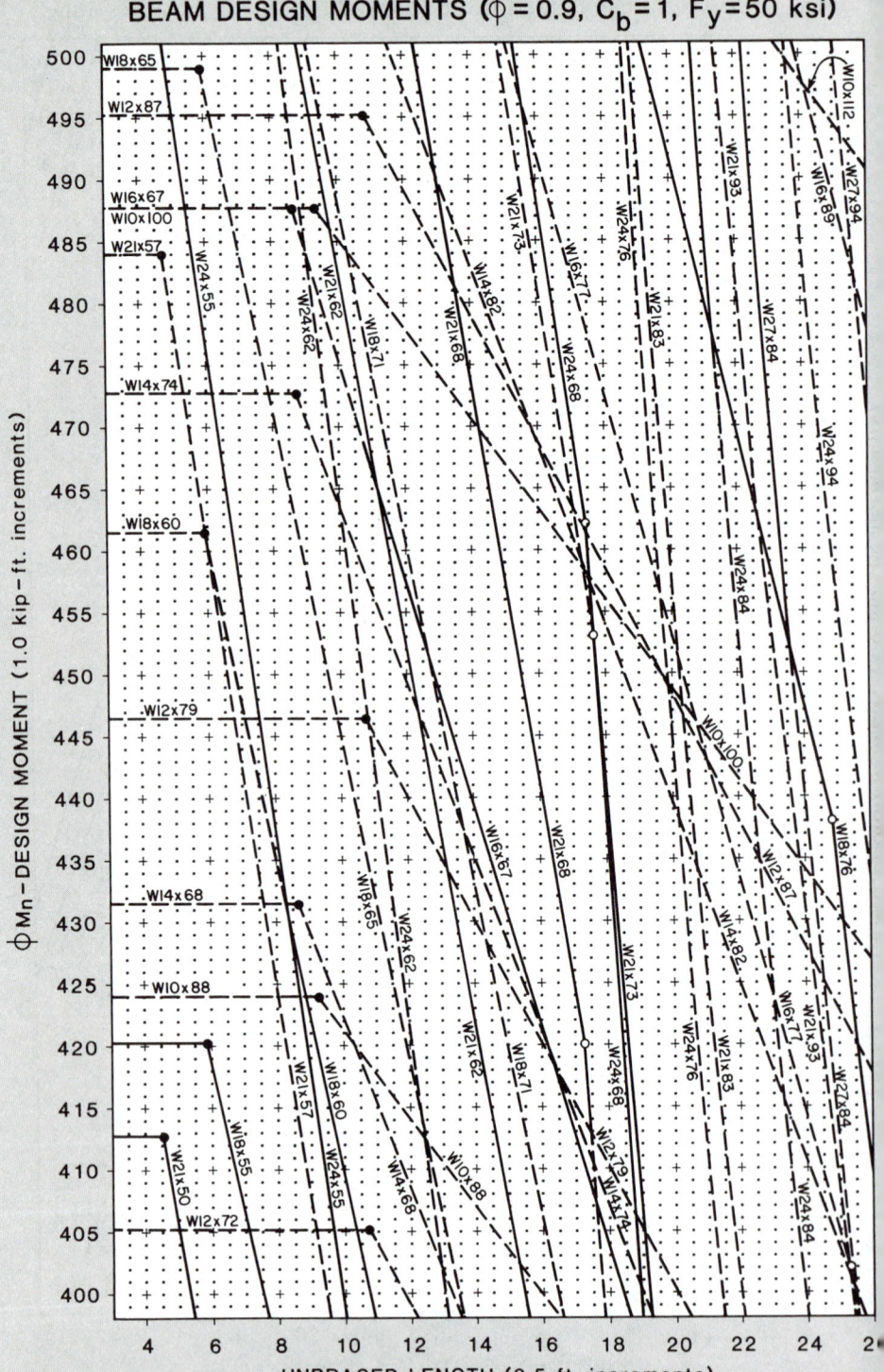

ϕM_n – DESIGN MOMENT (1.0 kip–ft. increments)

UNBRACED LENGTH (0.5 ft. increments)

BEAM DESIGN MOMENTS ($\phi = 0.9$, $C_b = 1$, $F_y = 50$ ksi)

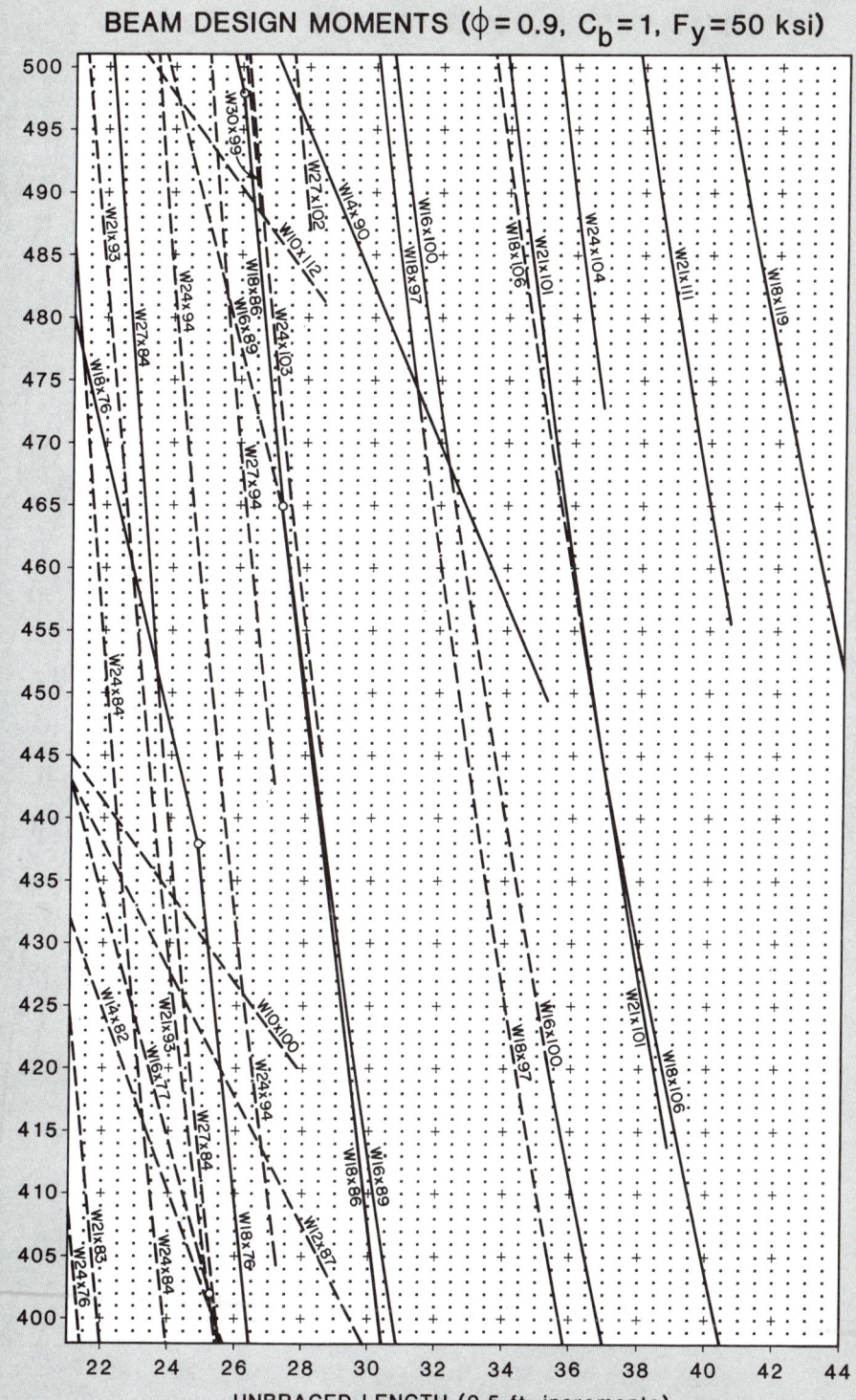

UNBRACED LENGTH (0.5 ft. increments)

BEAM DESIGN MOMENTS ($\phi = 0.9$, $C_b = 1$, $F_y = 50$ ksi)

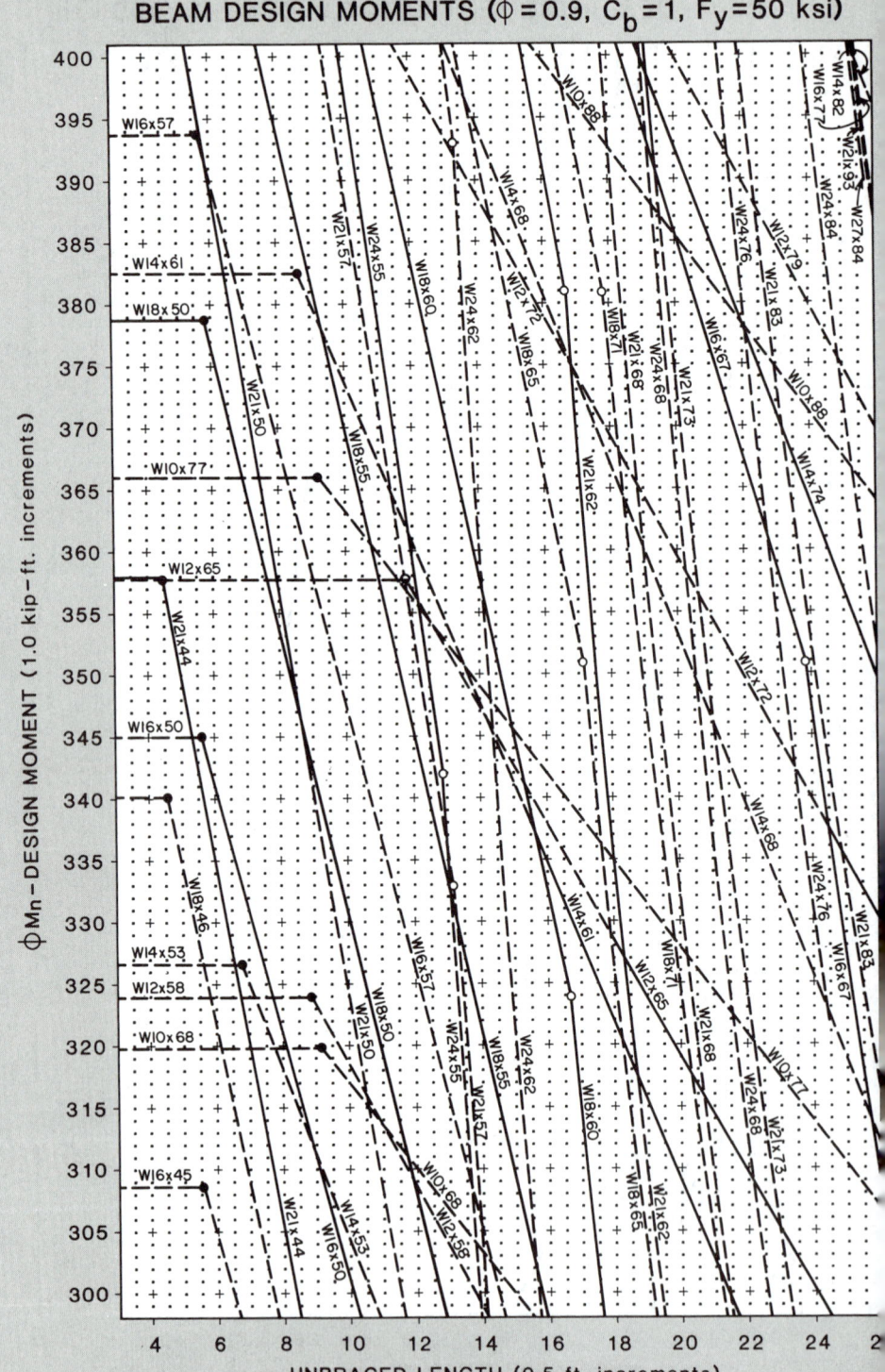

BEAM DESIGN MOMENTS ($\phi = 0.9$, $C_b = 1$, $F_y = 50$ ksi)

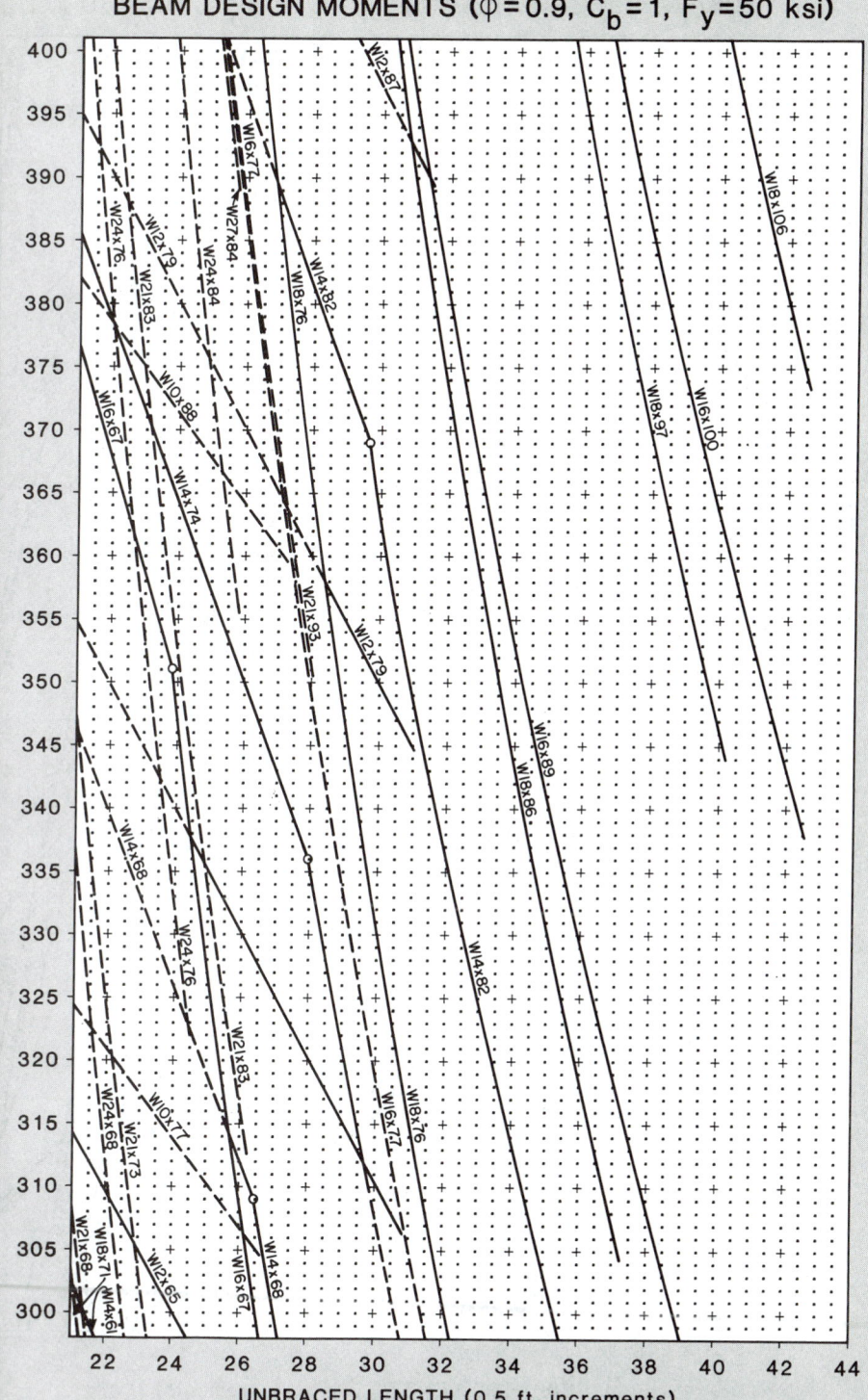

UNBRACED LENGTH (0.5 ft. increments)

BEAM DESIGN MOMENTS (ϕ = 0.9, C_b = 1, F_y = 50 ksi)

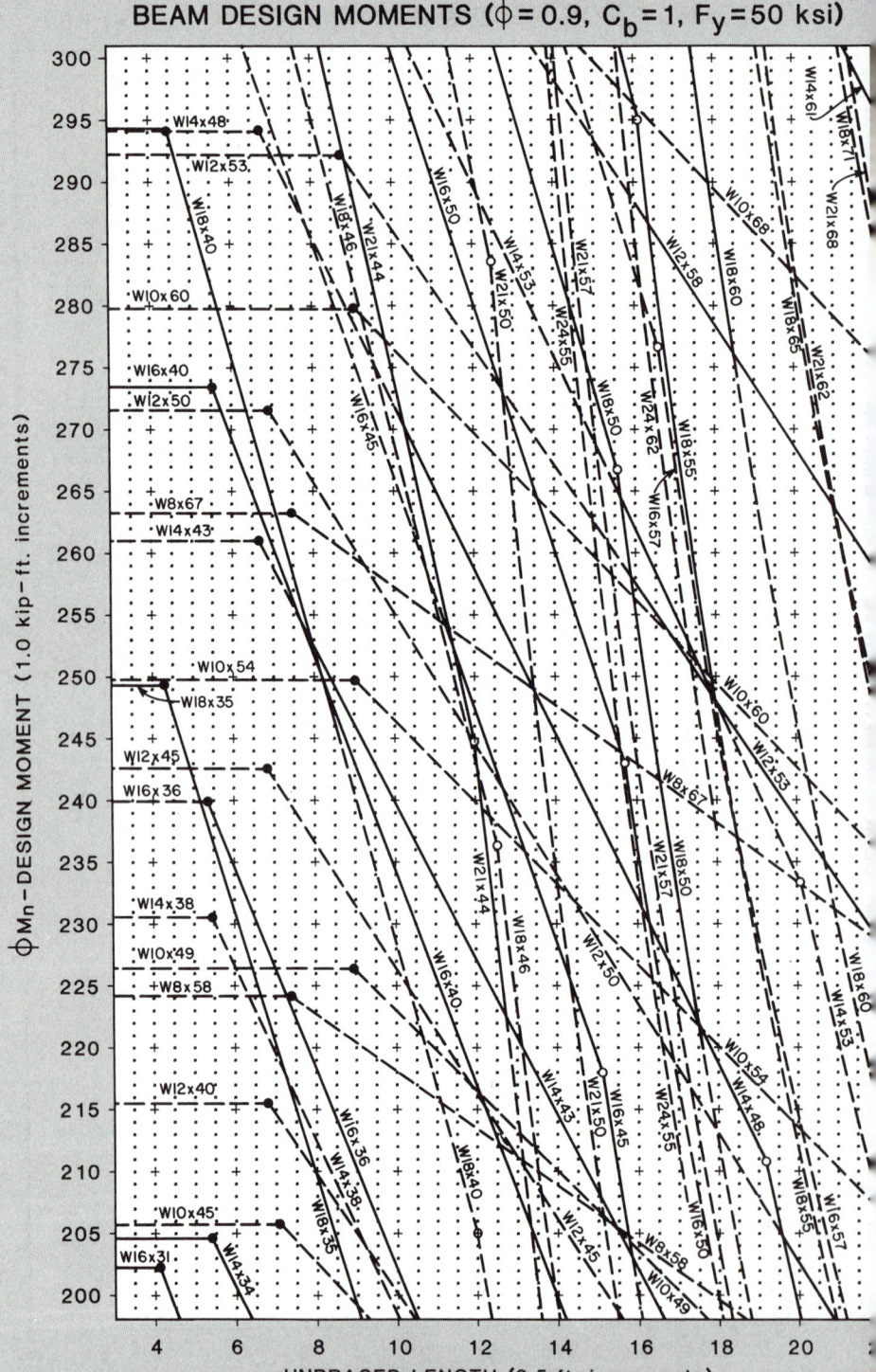

BEAM DESIGN MOMENTS ($\phi = 0.9$, $C_b = 1$, $F_y = 50$ ksi)

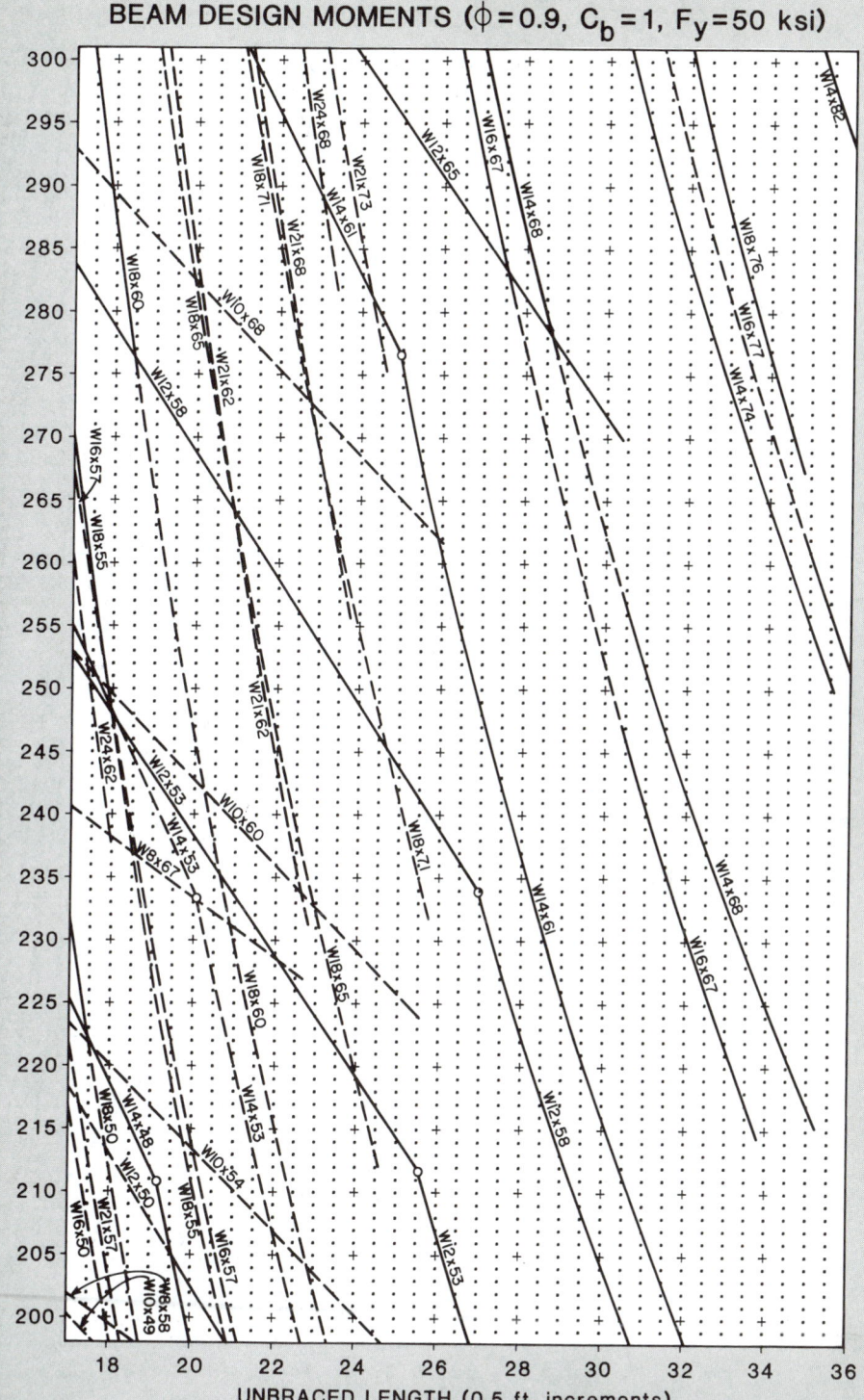

UNBRACED LENGTH (0.5 ft. increments)

AMERICAN INSTITUTE OF STEEL CONSTRUCTION

BEAM DESIGN MOMENTS (ϕ = 0.9, C_b = 1, F_y = 50 ksi)

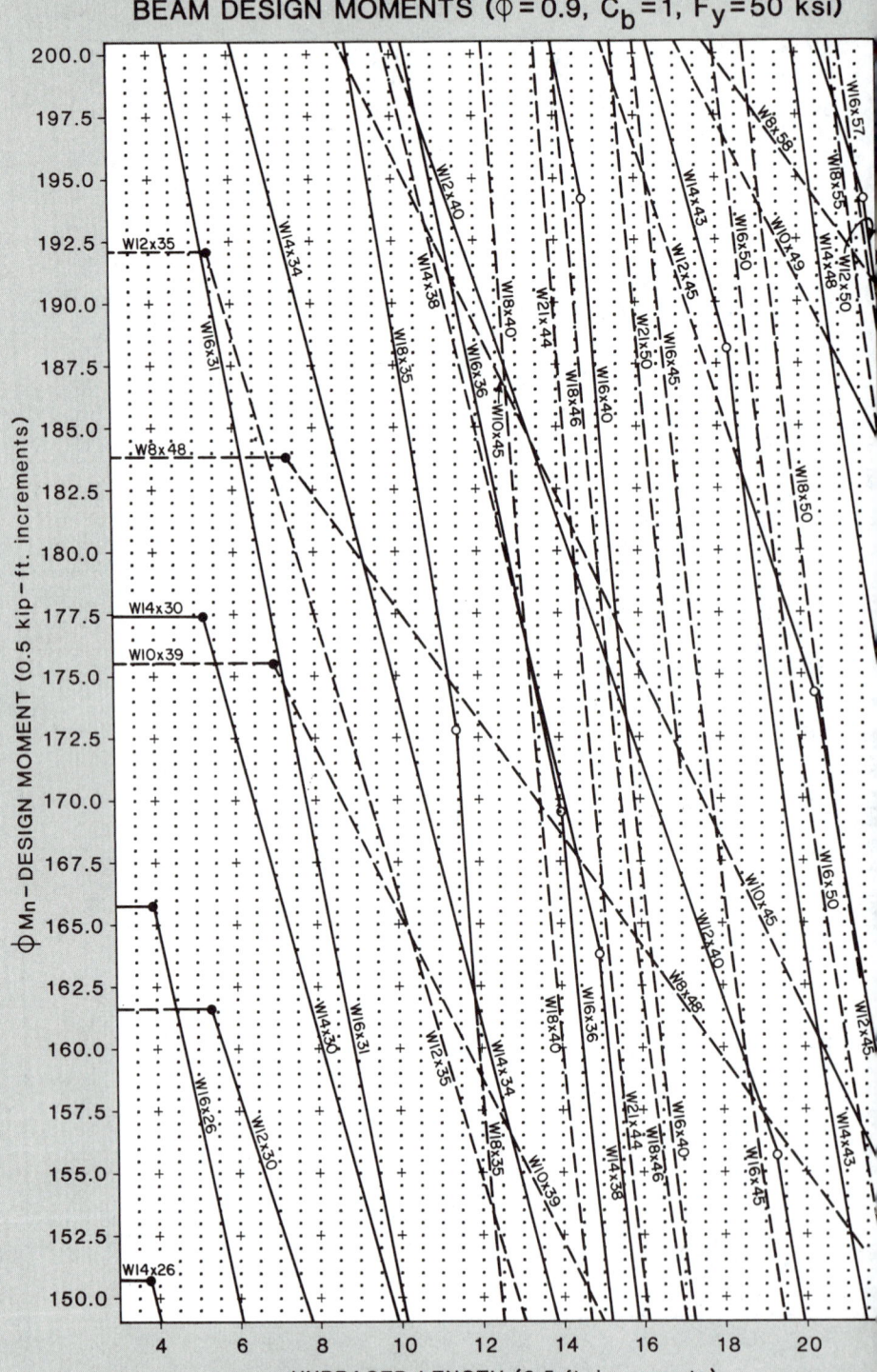

ϕM_n – DESIGN MOMENT (0.5 kip – ft. increments)

UNBRACED LENGTH (0.5 ft. increments)

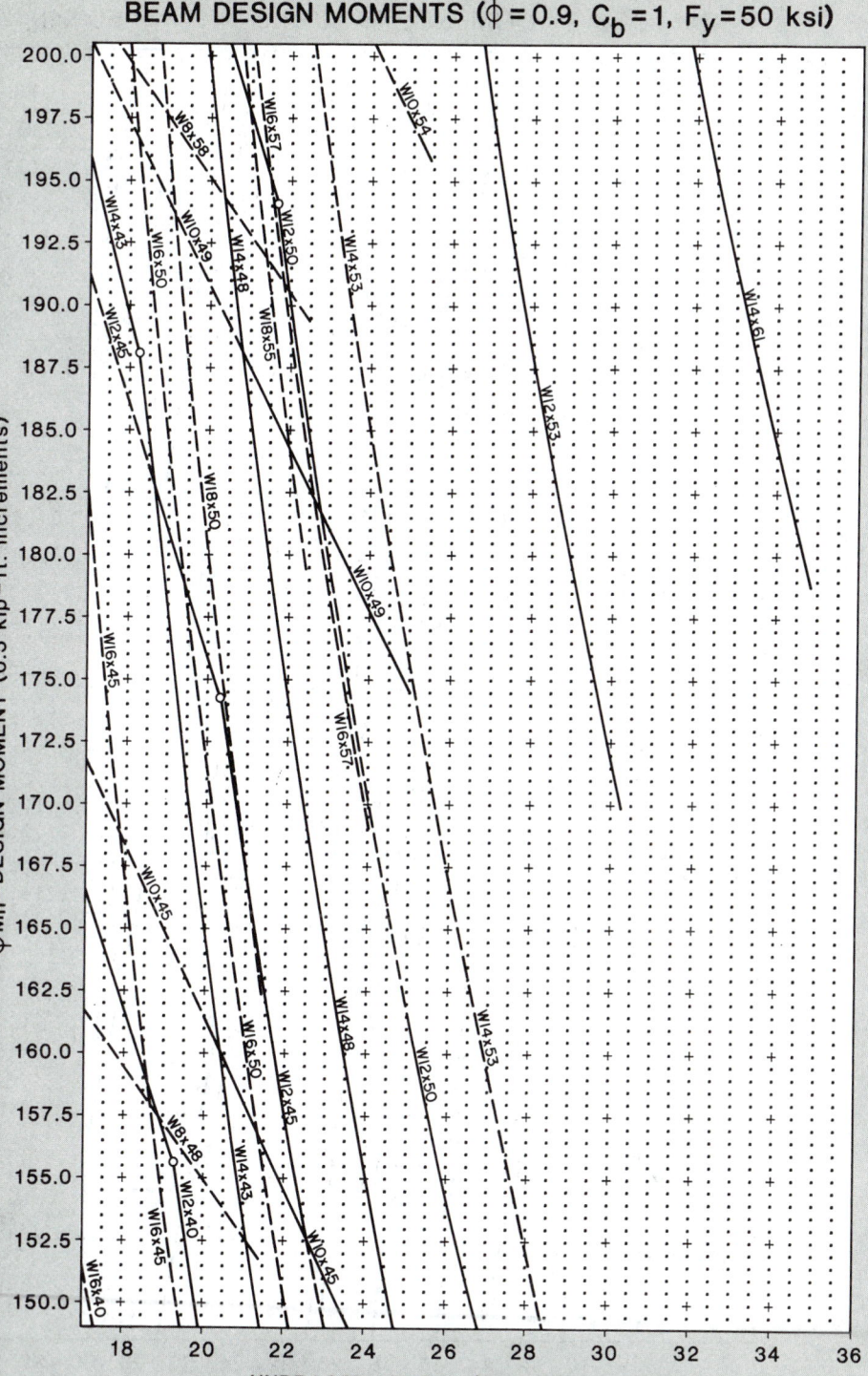

BEAM DESIGN MOMENTS ($\phi = 0.9$, $C_b = 1$, $F_y = 50$ ksi)

AMERICAN INSTITUTE OF STEEL CONSTRUCTION

BEAM DESIGN MOMENTS ($\phi = 0.9$, $C_b = 1$, $F_y = 50$ ksi)

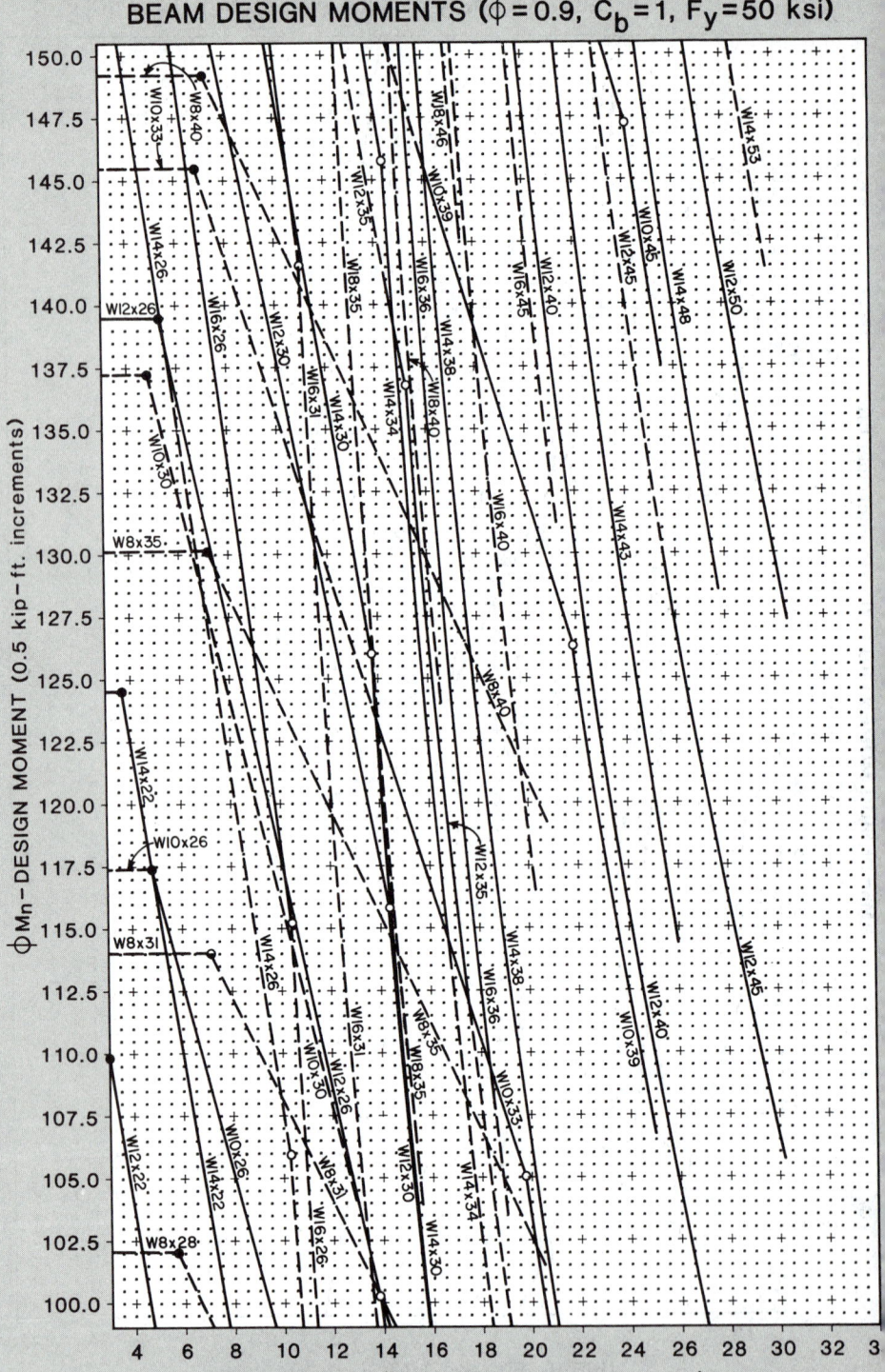

ϕM_n-DESIGN MOMENT (0.5 kip–ft. increments)

UNBRACED LENGTH (0.5 ft. increments)

BEAM DESIGN MOMENTS (ϕ=0.9, C$_b$=1, F$_y$=50 ksi)

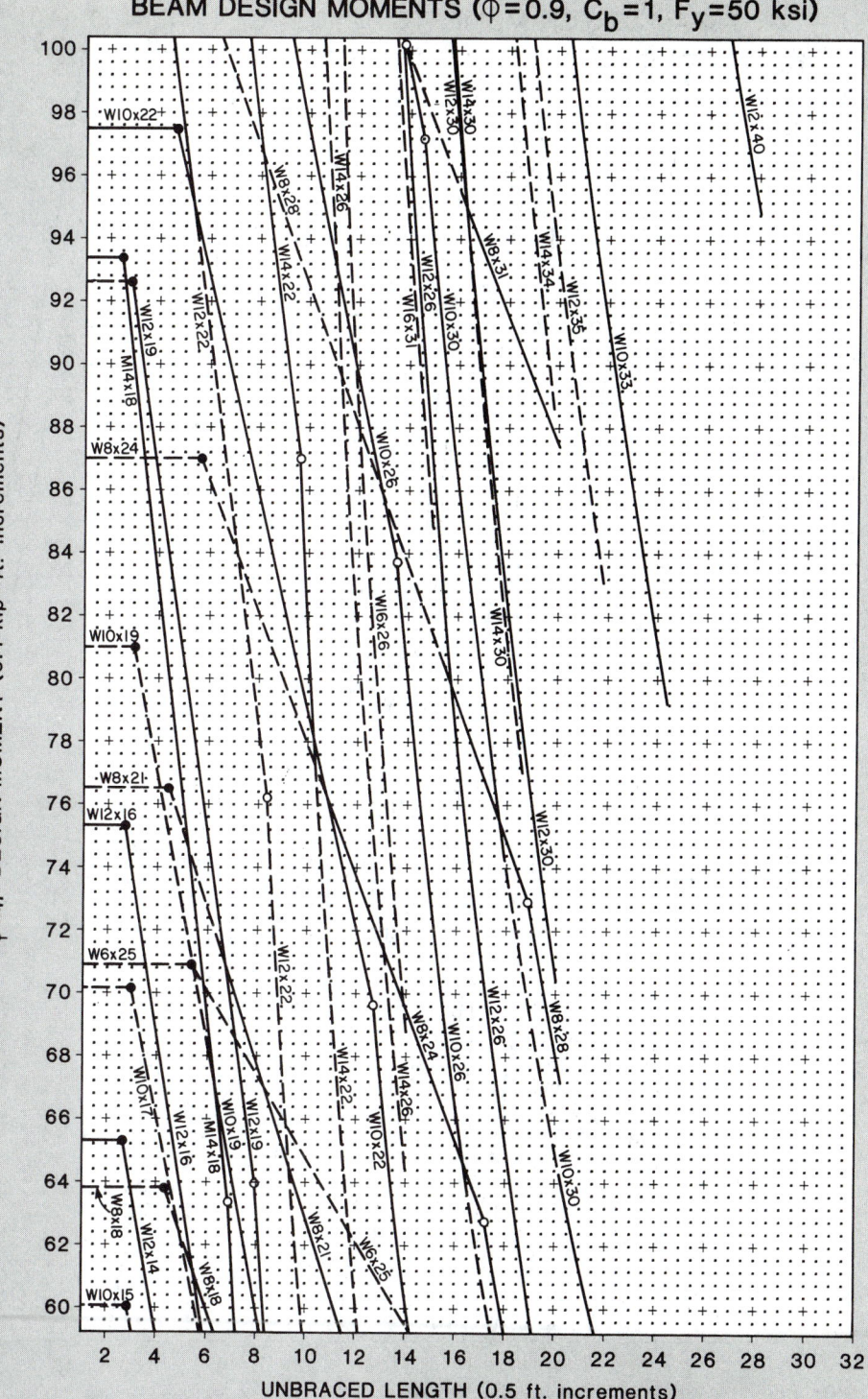

ϕM$_n$–DESIGN MOMENT (0.4 kip–ft. increments)

UNBRACED LENGTH (0.5 ft. increments)

AMERICAN INSTITUTE OF STEEL CONSTRUCTION

BEAM DESIGN MOMENTS ($\phi = 0.9$, $C_b = 1$, $F_y = 50$ ksi)

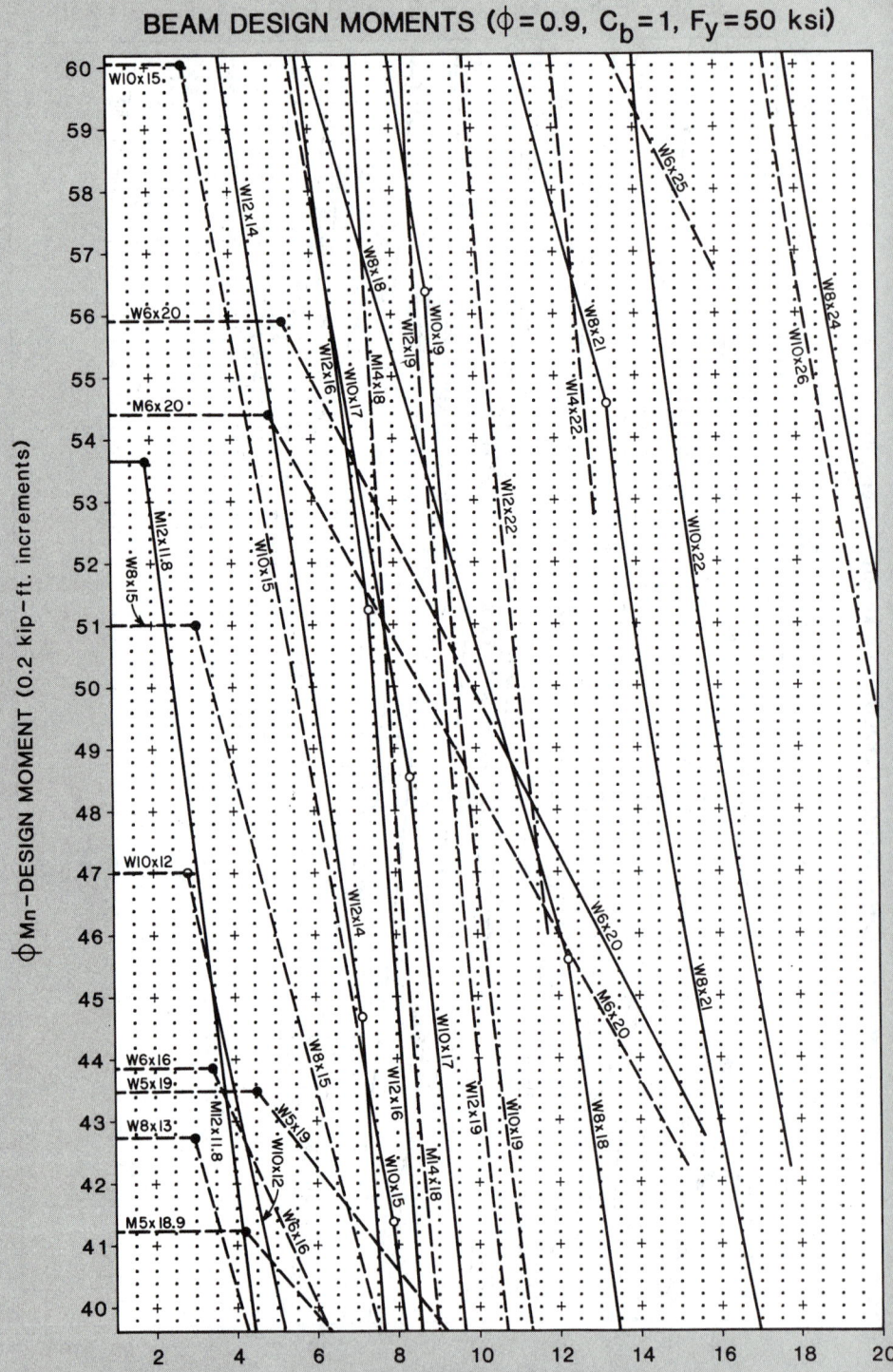

ϕM_n – DESIGN MOMENT (0.2 kip–ft. increments)

UNBRACED LENGTH (0.5 ft. increments)

BEAM DESIGN MOMENTS (ϕ=0.9, C_b=1, F_y=50 ksi)

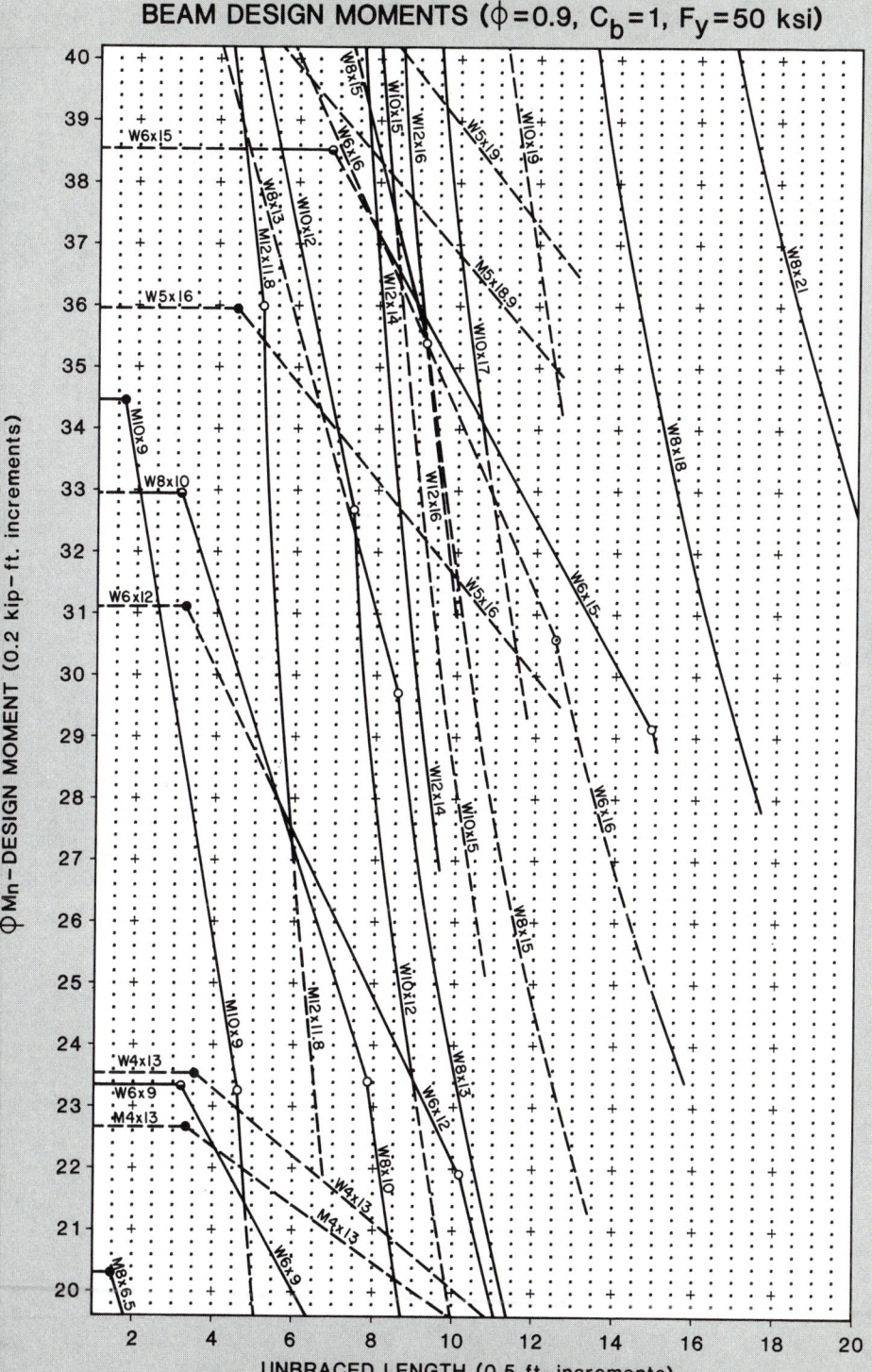

ϕM_n–DESIGN MOMENT (0.2 kip–ft. increments)

UNBRACED LENGTH (0.5 ft. increments)

BEAM DESIGN MOMENTS ($\phi = 0.9$, $C_b = 1$, $F_y = 50$ ksi)

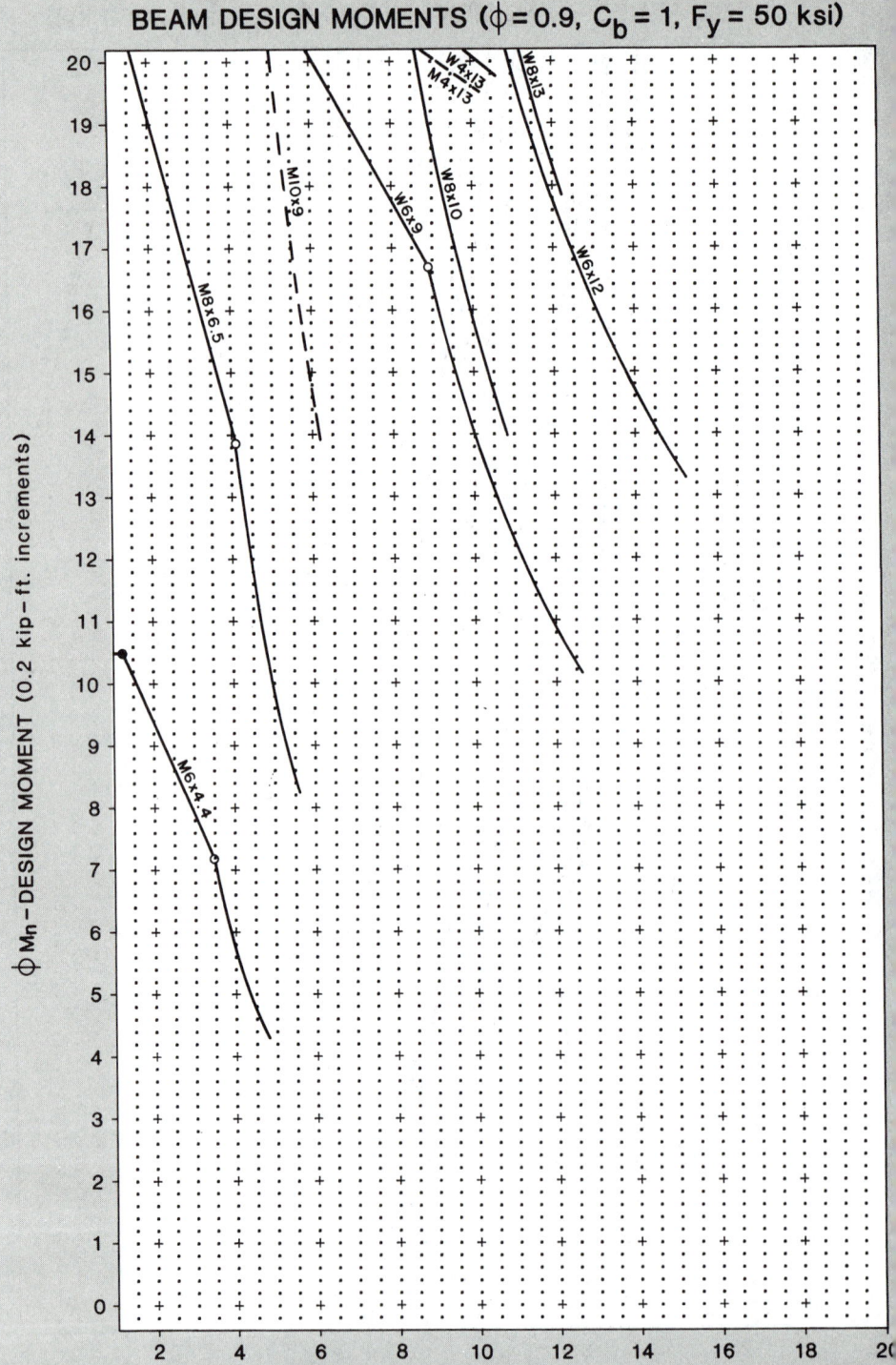

ϕM_n - DESIGN MOMENT (0.2 kip - ft. increments)

UNBRACED LENGTH (0.5 ft. increments)

PLATE GIRDERS
Design

GENERAL NOTES

The distinction between a beam and a plate girder, according to the LRFD Specification, must be made before the design of a plate girder can be undertaken. A beam can be a rolled or a welded shape, but it does not have stiffeners and its web width-thickness ratio h_c/t_w must be less than or equal to $970/\sqrt{F_{yf}}$. Plate girders may have stiffeners or h_c/t_w is greater than $970/\sqrt{F_{yf}}$, or both.

The items that must be considered in a plate girder design include: flexural strength, bearing under concentrated loads, shear strength, and flexure-shear interaction (for tension field action only). From these checks, the adequacy of the section and the need for stiffeners can be determined. This section contains design examples and flow charts to explain the treatment of these items, as well as stiffener design, in the LRFD Specification.

FLEXURAL AND SHEAR STRENGTH
General

In plate girder design, the flexural strength of the trial section must be determined to insure that an adequate section modulus is provided. Although there are preliminary steps, the flexural capacity, using elastic design, is ultimately determined from LRFD Specification Sects. F1.2 through F1.5 if the section is compact. For sections with more slender webs, either LRFD Specification Appendix F1.7 or Appendix G2 is used, depending on the section's classification as a beam or a plate girder.*

The shear strength calculation is required to ascertain if there is a need for intermediate stiffeners. The applicable formulas are found in LRFD Specification Sect. F2, or Appendix G3 if the tension field action design method is implemented. Note, however, that Appendix G cannot be used if h/t_w exceeds the limits given in Appendix G1.

Flowcharts

The flowcharts included here for determining flexural and shear strength consist of diamond-shaped decision points that eventually lead to the appropriate LRFD Specification formula. Each formula in the flowcharts is identified to its right according to the formula labels given in the LRFD Specification. Also, all references to Sections and Appendices come from the LRFD Specification. In the flexural strength flowchart, the compactness checks may be disregarded since it is possible to go directly from the elastic design box to the $h/t_w \le 970/\sqrt{F_{yf}}$ decision point. This is feasible because LRFD Specification Appendices F1.7 and G2 include their own compactness determination.

Also note the "Go to Ⓑ " statement, used twice in the shear strength flowchart, means to jump to the subroutine labeled Ⓑ at the end of the flowchart, and calculate C_v as indicated. After finding C_v, return to the point in the flowchart immediately following the "Go to Ⓑ " statement.

*Zahn, C. J. Plate Girder Design Using LRFD AISC *Engineering Journal* *Vol. 24, No. 1, 1st Quarter 1987.*

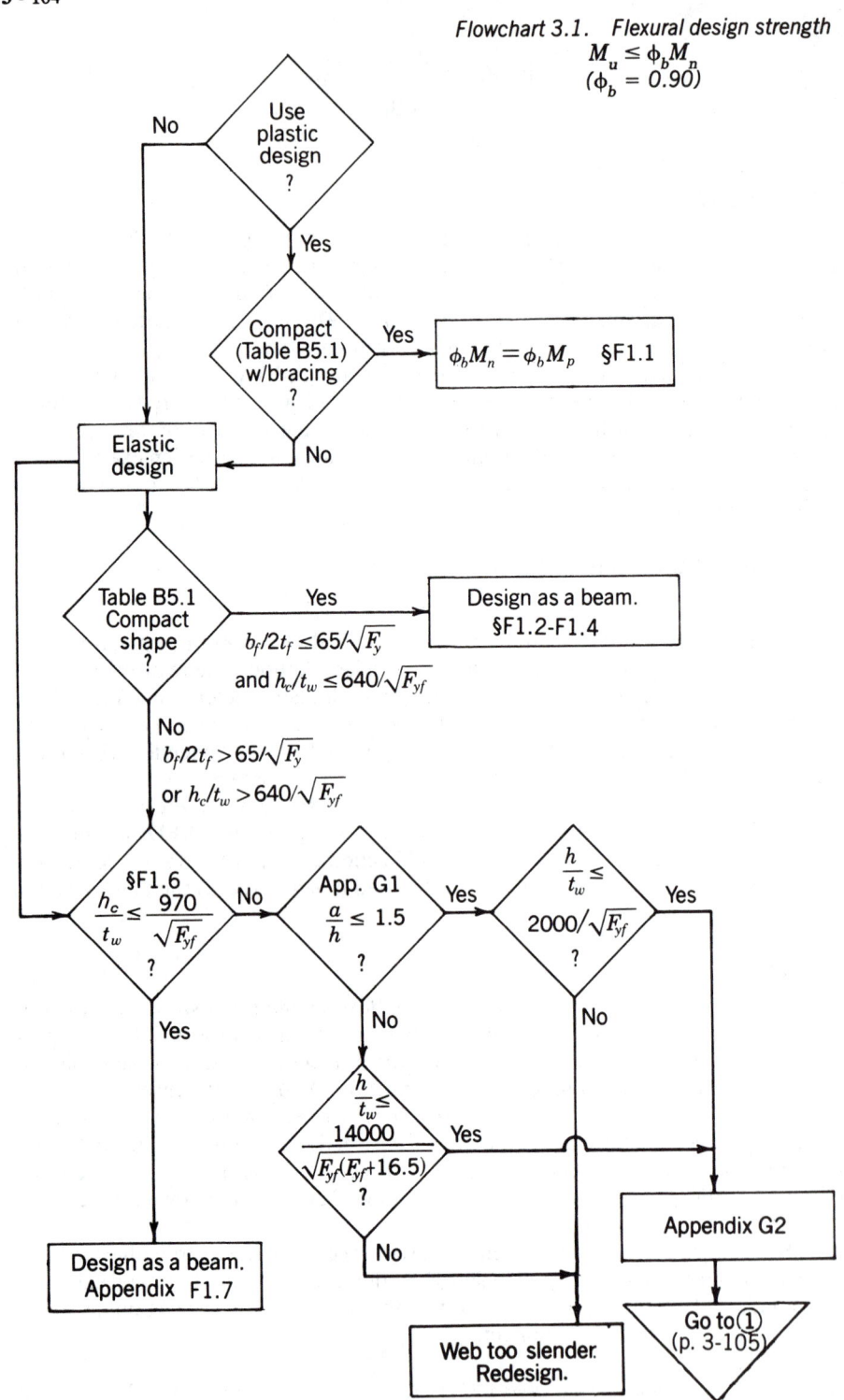

Flowchart 3.1. Flexural design strength
$$M_u \le \phi_b M_n$$
$$(\phi_b = 0.90)$$

(continued)

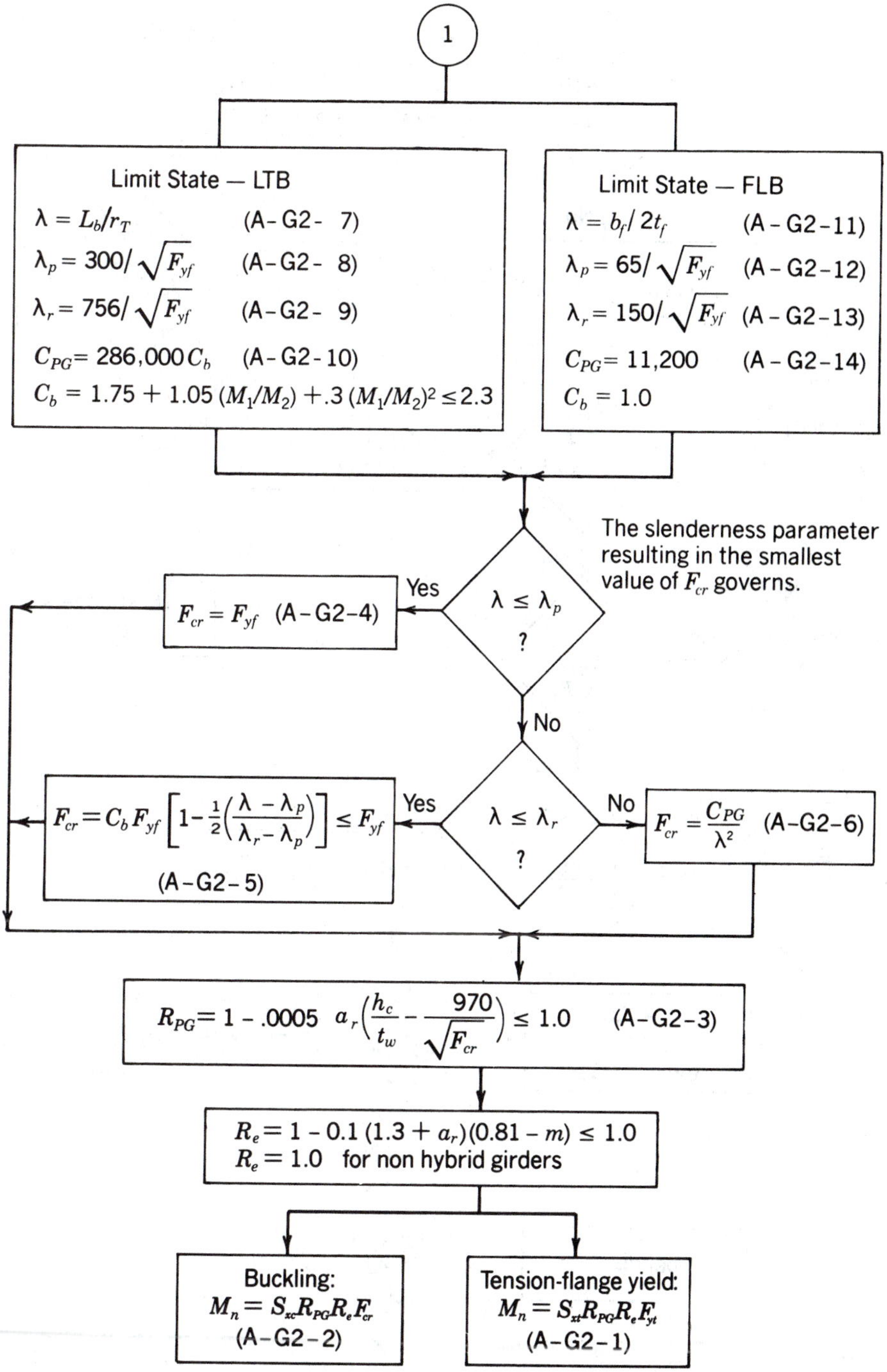

1

Limit State — LTB

$$\lambda = L_b/r_T \qquad \text{(A-G2- 7)}$$

$$\lambda_p = 300/\sqrt{F_{yf}} \qquad \text{(A-G2- 8)}$$

$$\lambda_r = 756/\sqrt{F_{yf}} \qquad \text{(A-G2- 9)}$$

$$C_{PG} = 286,000\,C_b \qquad \text{(A-G2-10)}$$

$$C_b = 1.75 + 1.05\,(M_1/M_2) + .3\,(M_1/M_2)^2 \le 2.3$$

Limit State — FLB

$$\lambda = b_f/2t_f \qquad \text{(A-G2-11)}$$

$$\lambda_p = 65/\sqrt{F_{yf}} \qquad \text{(A-G2-12)}$$

$$\lambda_r = 150/\sqrt{F_{yf}} \qquad \text{(A-G2-13)}$$

$$C_{PG} = 11,200 \qquad \text{(A-G2-14)}$$

$$C_b = 1.0$$

The slenderness parameter resulting in the smallest value of F_{cr} governs.

$\lambda \le \lambda_p$?

Yes → $F_{cr} = F_{yf}$ (A-G2-4)

No ↓

$\lambda \le \lambda_r$?

Yes → $F_{cr} = C_b F_{yf}\left[1 - \frac{1}{2}\left(\dfrac{\lambda - \lambda_p}{\lambda_r - \lambda_p}\right)\right] \le F_{yf}$ (A-G2-5)

No → $F_{cr} = \dfrac{C_{PG}}{\lambda^2}$ (A-G2-6)

$$R_{PG} = 1 - .0005\ a_r\left(\frac{h_c}{t_w} - \frac{970}{\sqrt{F_{cr}}}\right) \le 1.0 \qquad \text{(A-G2-3)}$$

$$R_e = 1 - 0.1\,(1.3 + a_r)(0.81 - m) \le 1.0$$
$$R_e = 1.0 \quad \text{for non hybrid girders}$$

Buckling:
$$M_n = S_{xc} R_{PG} R_e F_{cr}$$
(A-G2-2)

Tension-flange yield:
$$M_n = S_{xt} R_{PG} R_e F_{yt}$$
(A-G2-1)

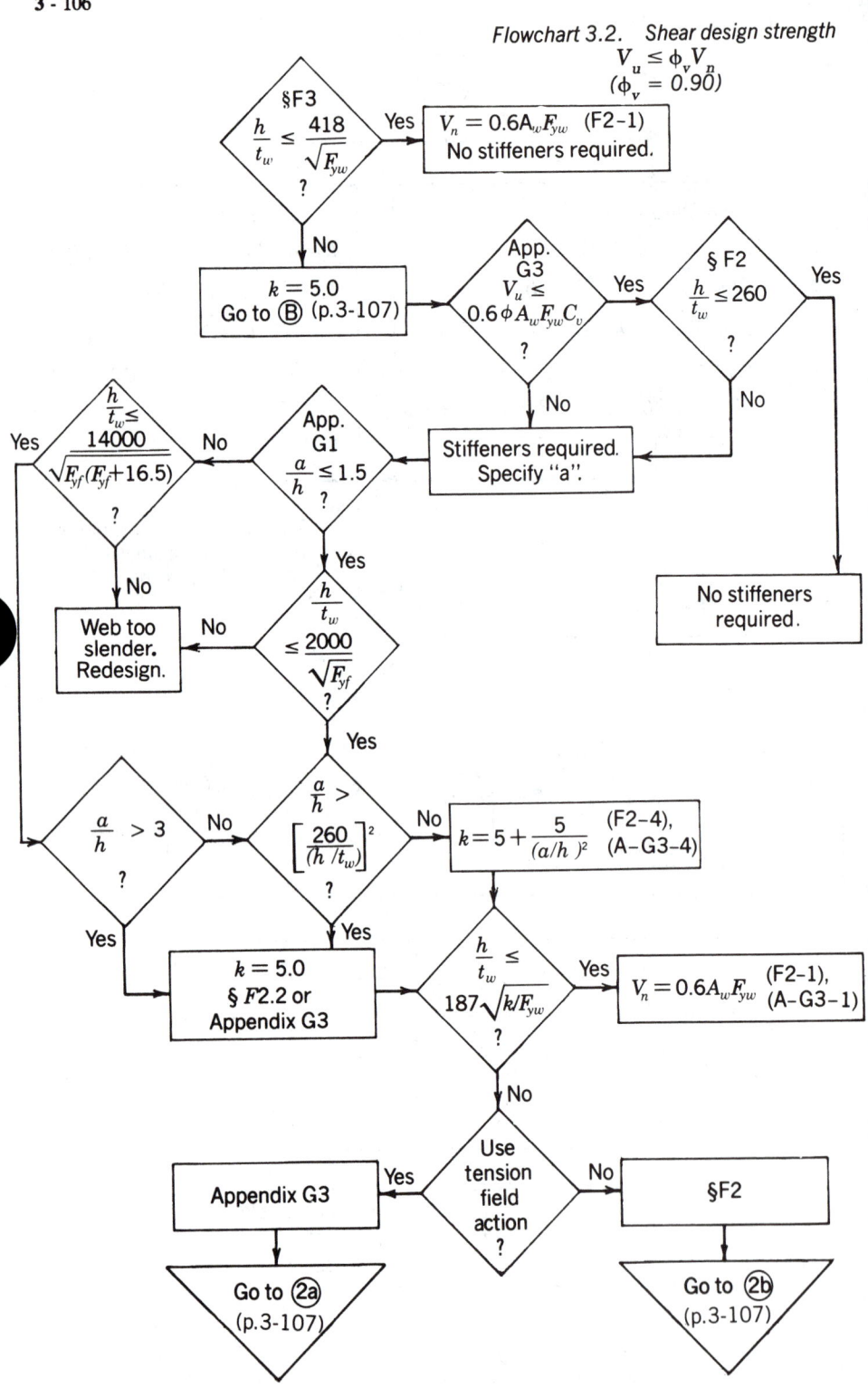

Flowchart 3.2.   Shear design strength
$$V_u \leq \phi_v V_n$$
$$(\phi_v = 0.90)$$

(continued)

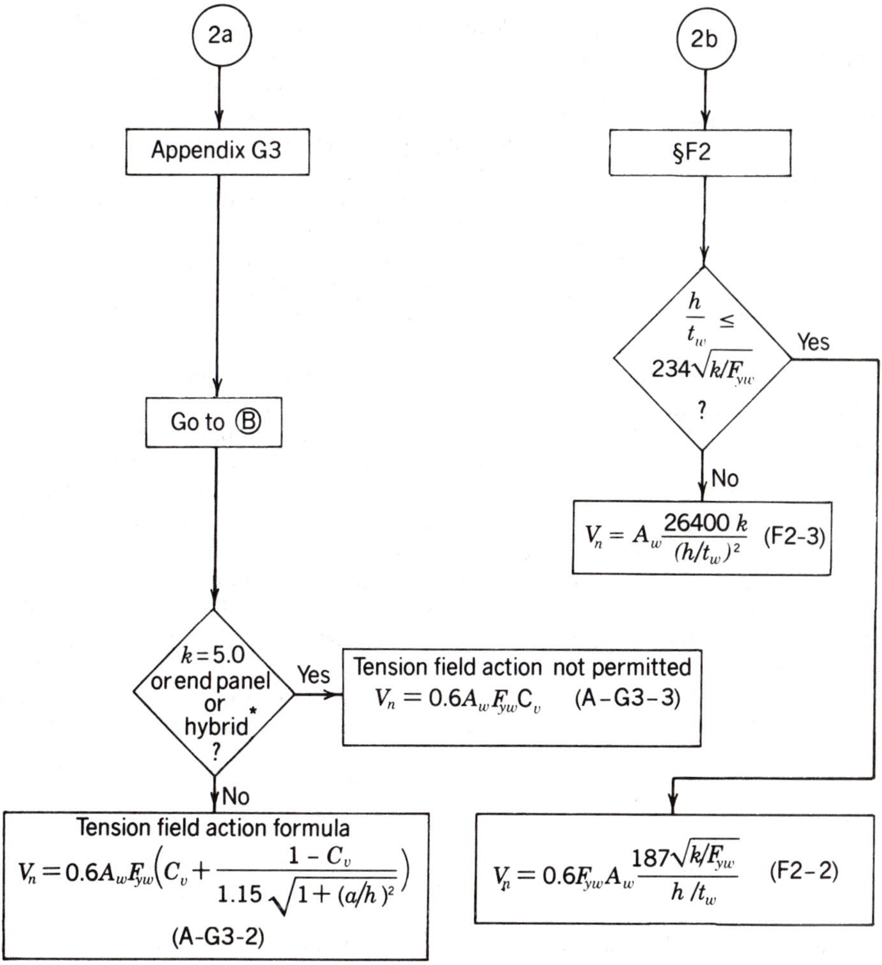

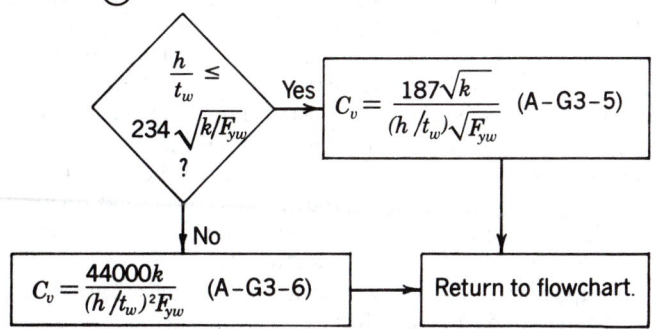

```
    (2a)                                    (2b)
     │                                        │
     ▼                                        ▼
┌──────────────┐                        ┌──────────────┐
│ Appendix G3  │                        │     §F2      │
└──────────────┘                        └──────────────┘
     │                                        │
     │                                        ▼
     │                                    ◇ h/t_w ≤ 234√(k/F_yw) ? ──Yes──┐
     ▼                                        │                           │
┌──────────────┐                             No                          │
│  Go to Ⓑ    │                              ▼                           │
└──────────────┘                    ┌─────────────────────────────┐     │
     │                              │ V_n = A_w (26400 k)/(h/t_w)² │     │
     │                              │                      (F2-3)  │     │
     │                              └─────────────────────────────┘     │
     ▼                                                                    │
◇ k=5.0 or end panel or hybrid* ? ──Yes──┐                               │
     │                                    ▼                               │
     No                      ┌──────────────────────────────────┐       │
     │                       │ Tension field action not permitted │      │
     ▼                       │ V_n = 0.6 A_w F_yw C_v   (A-G3-3)  │      │
┌───────────────────────┐   └──────────────────────────────────┘       │
│ Tension field action  │                                                │
│ formula               │                   ┌────────────────────────────▼──┐
└───────────────────────┘                   │                               │
```

$$V_n = 0.6 A_w F_{yw}\left(C_v + \frac{1-C_v}{1.15\sqrt{1+(a/h)^2}}\right) \quad \text{(A-G3-2)}$$

$$V_n = 0.6 F_{yw} A_w \frac{187\sqrt{k/F_{yw}}}{h/t_w} \quad \text{(F2-2)}$$

$◇ \frac{h}{t_w} \le 234\sqrt{k/F_{yw}} ?$

* Tension field action is not permissible for web-tapered girders

Ⓑ –Subroutine to determine the value of C_v (Appendix G3)

$◇ \dfrac{h}{t_w} \le 234\sqrt{k/F_{yw}} ?$ ──Yes──▶ $C_v = \dfrac{187\sqrt{k}}{(h/t_w)\sqrt{F_{yw}}} \quad \text{(A-G3-5)}$

No

$$C_v = \frac{44000k}{(h/t_w)^2 F_{yw}} \quad \text{(A-G3-6)}$$

──▶ Return to flowchart.

AMERICAN INSTITUTE OF STEEL CONSTRUCTION

TABLE OF DIMENSIONS AND PROPERTIES OF BUILT-UP WIDE FLANGE SECTIONS

This table serves as a guide for selecting welded built-up wide-flange sections of economical proportions. It provides dimensions and properties for a wide range of sections with nominal depths from 45 to 92 in. No preference is intended for the tabulated flange plate dimensions, as compared to other flange plates having the same area. Substitution of wider but thinner flange plates, without a change in flange area, will result in a slight reduction in section modulus.

In analyzing overall economy, weight savings must be balanced against higher fabrication costs incurred in splicing the flanges. In some cases, it may prove economical to reduce the size of flange plates at one or more points near the girder ends, where the bending moment is substantially less. Economy through reduction of flange plate sizes is most likely to be realized with long girders, where flanges must be spliced in any case.

Only one thickness of web plate is given for each depth of girder. When the design is primarily dominated by shear in the web, rather than moment capacity, overall economy may dictate selection of a thicker web plate. The resultant increase in section modulus can be obtained by multiplying the value S', given in the table, by the number of sixteenths of an inch increase in web thickness, and adding the value obtained to the section modulus value S for the girder profile shown in the table. The increased plastic section modulus can be calculated in the same way.

Overall economy may often be obtained by using a web plate of such thickness that intermediate stiffeners are not required. This is not always the case, however. The girder sections listed in the table will provide a "balanced" design with respect to bending moment and web shear without excessive use of intermediate stiffeners.

The maximum design end shear strength permissible without transverse stiffeners is given in the table column labeled $\phi_v V_n$. These values came from the formula,

$$\phi_v V_n = 0.6\ \phi_v A_w F_{yw} C_v \qquad \text{Appendix G3}$$

where

$$C_v = \frac{44{,}000k}{(h/t_w)^2 F_{yw}} \text{ with } k = 5.0$$

It is evident from this formula that a thicker web plate increases the design shear strength.

DESIGN EXAMPLES

Design of a plate girder by the moment of inertia method should begin with a preliminary design or selection of a trial section. The initial choice may require one or more adjustments before a final cross section is obtained that satisfies all the provisions of the LRFD Specification with maximum economy. In the following design examples, all applicable provisions of the LRFD Specification are listed at the right of each page.

In addition, references to Tables 10 and 11 in the LRFD Specification are listed in the right margin of each page. These tables may be used in place of the formulas for $\phi_v V_n$. Values for $\phi_v V_n / A_w$ are given in ksi for plate girders. Tables 10-36 and 10-50 do not include the tension field action equation and, therefore, are based on LRFD Specification Sect. F2. For design with tension field action, Tables 11-36 and 11-50, based on Appendix G3, are applicable. Table 11 also includes the required gross area of pairs of stiffeners, as a percent of ($h \times t_w$), from LRFD Specification Formula A-G4-2.

Example 1 illustrates a recommended procedure for designing a welded plate girder of constant depth. The selection of a suitable trial cross section is obtained by the flange area method, and then checked by the moment of inertia method.

Example 2 shows a recommended procedure for designing a welded hybrid girder of constant depth.

Example 3 illustrates use of the Table of Dimensions and Properties of Built-up Wide-Flange Sections, to obtain an efficient trial profile. The 52-in. depth specified for this example demonstrates how tabular data may be used for girder depths intermediate to those listed. Another design requirement in this example is the omission of intermediate web stiffeners.

Example 1

Design a welded plate girder to support a factored uniform load of 5 kips per ft and two concentrated factored loads of 110 kips located 17 ft from each end. The compression flange of the girder will be laterally supported only at points of concentrated load.

Given:

Maximum bending moment: 3,310 kip-ft
Maximum vertical shear: 230 kips
Span: 48 ft
Maximum depth: 72 in.
Steel: $F_y = 36$ ksi

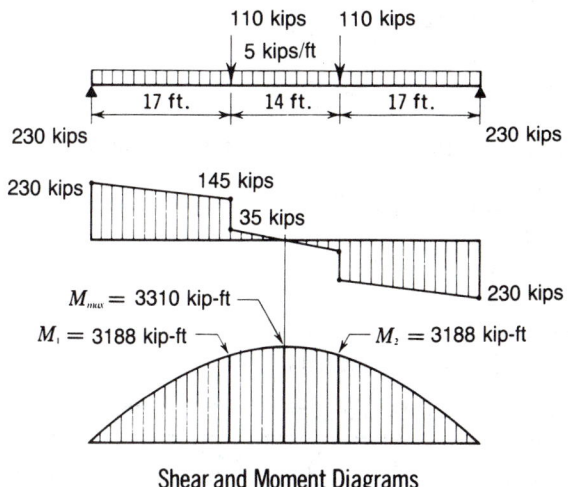

Shear and Moment Diagrams

Solution:

A. Preliminary web design:

1. Assume web depth, $h = h_c = 70$ in.
   For noncompact web,
   $$640/\sqrt{F_{yf}} < h_c/t_w \leq 970/\sqrt{F_{yf}} = 162$$
   Corresponding thickness of web = $70/162 = 0.43$ in.

2. Assuming $a/h > 1.5$, minimum thickness of web = $70/322 =$ 0.22 in.

   Choose thinnest web.

   Try web plate $5/16 \times 70$:

   $$A_w = 21.9 \text{ in.}$$
   $$h/t_w = h_c/t_w = 70/0.313 = 224$$

   Since $0.31 < 0.43$ in. as calculated above, expect R_{PG} to be less than 1.0.

B. Preliminary flange design:

1. Required flange area:

   An approximate formula for the area of one flange is:

   $$A_f \approx \frac{M_u}{F_y h} = \frac{3,310(12)}{36(70)} = 15.8 \text{ in.}^2$$

   Try $7/8 \times 18$ plate: $A_f = 15.8 \text{ in.}^2$

2. Check for compactness for no reduction in critical stress:

   $$\frac{b_f}{2t_f} = \frac{18}{2(.875)} = 10.3 \leq 65/\sqrt{36} = 10.8 \quad \textbf{o.k.}$$

C. Trial girder section:

   Web $5/16 \times 70$; 2 flange plates $7/8 \times 18$

1. Find section modulus by moment of inertia method:

Section	A In.^{2}	y In.	$\Sigma A_y{}^2$ In.^{4}	I_o In.^{4}	I_{gr} In.^{4}
1 web $5/16 \times 70$	21.9			8932	8932
1 flange $7/8 \times 18$	15.8	35.44	39684	2	39686
1 flange $7/8 \times 18$	15.8				
Moment of inertia					48618

   Section modulus furnished: $48,618/35.875 = 1,355 \text{ in.}^3$

2. Check flexural strength using elastic design:

   Since $h_c/t_w > 970/\sqrt{F_{yf}}$, Appendix G2 applies.

Moment of inertia of flange plus ⅙ web about Y-Y axis:

$I_{oy} = \frac{7}{8} \times (18)^3/12 = 425$ in.^{4}

$A_f + \frac{1}{6} A_w = 15.8 + \frac{1}{6}(21.9) = 19.45$ in.^{2}

$r_T = \sqrt{425/19.45} = 4.67$ in.

a. Check limitations of Appendix G:

　Assume $a/h \leq 1.5$

$$(h/t_w)\ max = \frac{2,000}{\sqrt{F_{yf}}} = 333 > 224\quad \textbf{o.k.}$$

(A-G1-1)

b. Check strength of 14-ft panel: $M_u = 3,310$ kip-ft

The moment within a significant portion of the unbraced
14-ft segment is $\geq$ larger of the segment end moments

$\therefore C_b = 1.0$

Sect. F1.4

For the limit state of lateral-torsional buckling:

Appendix G2

$$\lambda = \frac{L_b}{r_T} = \frac{14 \times 12}{4.67} = 36.0$$

(A-G2-7)

$$\lambda_p = 300/\sqrt{F_{yf}} = 50.0$$

(A-G2-8)

$$\lambda_r = 756/\sqrt{F_{yf}} = 126.0$$

(A-G2-9)

Since $\lambda \leq \lambda_p$, $F_{cr} = F_{yf} = 36.0$ ksi

(A-G2-4)

For the limit state of flange local buckling:

$$\lambda = b_f/2t_f = 18/(2 \times 0.875) = 10.3$$

(A-G2-11)

$$\lambda_p = 65/\sqrt{F_{yf}} = 10.8$$

(A-G2-12),

$$\lambda_r = 150/\sqrt{F_{yf}} = 25.0$$

(A-G2-13)

Since $\lambda \leq \lambda_p$, $F_{cr} = F_{yf} = 36.0$ ksi

(A-G2-4)

Design flexural strength:

$$R_{PG} = 1 - 0.0005 \left(\frac{21.9}{15.8}\right)\left(\frac{70}{0.313} - \frac{970}{\sqrt{36}}\right) = 0.957$$

(A-G2-3)

With $F_{cr} = 36.0$ ksi use Formula A-G2-1 or A-G2-2 as
applicable:

$$M_n = 1,355\ (1/12)(0.957)(1.0)(36) = 3,890\ \text{kip-ft}$$

(A-G2-1) or
(A-G2-2)

$$\therefore \phi M_n = 0.90(3,890) = 3,501\ \text{kip-ft} > 3,310\quad \textbf{o.k.}$$

c. Check strength of 17-ft panels:

$M_u = 3,188$ kip-ft

$$C_b = 1.75 + 1.05\frac{M_1}{M_2} + 0.3\left(\frac{M_1}{M_2}\right)^2 \leq 2.3$$

Appendix G2

where $M_1 = 0$; then $\dfrac{M_1}{M_2} = 0$　$\therefore C_b = 1.75$

For the limit state of lateral-torsional buckling:

$$\lambda = \frac{L_b}{r_T} = \frac{17(12)}{4.67} = 43.7$$ (A-G2-7)

$$\lambda_p = 300/\sqrt{F_{yf}} = 50.0;$$ (A-G2-8)

$$\lambda_r = 756/\sqrt{F_{yf}} = 126.0$$ (A-G2-9)

Since $\lambda \le \lambda_p$, $F_{cr} = 36.0$ ksi (A-G2-4)

For the limit state of flange local buckling:

$F_{cr} = F_{yf} = 36$ ksi (as for the 14-ft panel) (A-G2-4)

$R_{PG} = 0.957$ (as for the 14-ft panel) (A-G2-3)

Again, with $F_{cr} = 36$ ksi use Formula A-G2-1 or A-G2-2 as applicable:

$M_n = 3{,}890$ kip-ft (see Step C.2.a.)

$\phi_b M_n = 0.90 \times 3{,}890 = 3{,}501$ kip-ft $> 3{,}188$ **o.k.** (A-G2-1) or
(A-G2-2)

Use: Web: One plate $\frac{5}{16} \times 70$
Flanges: Two plates $\frac{7}{8} \times 18$

D. Stiffener requirements:

1. Bearing stiffeners:

a. Check bearing at end reactions: Sect. K1

Assume point bearing ($N = 0$) and $\frac{5}{16}$-in. web-to-flange welds.

Check local web yielding:

$R_n = (5k + N) F_{yw} t_w$; $k = \frac{7}{8} + \frac{5}{16} = 1.188$ in. (K1-2)

$\phi R_n = 1.0 \ [5(1.188) + 0](36)(\frac{5}{16})$

$= 66.8$ kips < 230 **n.g.**

$\therefore$ Provide bearing stiffeners at unframed girder ends.

(*Note:* If local web yielding criteria is satisfied, criteria set forth in Sects. K1.4 and K1.5 should also be checked.) Sect. K1.8

b. Bearing stiffeners are also required at concentrated load points since $66.8 < 110$ **n.g.**

2. Intermediate stiffeners:

a. Check shear strength in unstiffened end panel: Appendix G4

$h/t_w = 224 > 418/\sqrt{F_{yw}} = 70$

$a/h = (17)(12/70) = 2.9$

$V_u/A_w = 230/21.9 = 10.5$ ksi

Tension field action is not permitted for end panels, or when $a/h > 3.0$ or $[260/(h/t_w)]^2$. Here, $2.9 > (260/224)^2 = 1.35$. In either of these cases, Formulas A-G3-3 and F2-3 are both applicable, as they are equivalent formulas.

Using F2-3, with $k = 5.0$

$$\frac{\phi_v V_n}{A_w} = \frac{0.9(26400)(5)}{(224)^2} = 2.4 < 10.5 \text{ ksi}$$

$\therefore$ Provide intermediate stiffeners.

b. End panel stiffener spacing

Let $\dfrac{\phi_v V_n}{A_w} = 10.5$ ksi and solve for a/h.

Result: $a/h = 0.58$

$a \leq (0.58)(70) = 40.5$ in.

Use: 36 in.

c. Check for additional stiffeners:

Shear at first intermediate stiffener:

$V_u = 230 - [5(36/12)] = 215$ kips

$$\frac{V_u}{A_w} = \frac{215}{21.9} = 9.8 \text{ ksi}$$

Distance between first intermediate stiffener and concentrated load:

$a = (17)(12) - 36 = 168$ in.

$a/h = 168/70 = 2.4$

Use $k = 5$, since $a/h > 1.35$

and again $\dfrac{\phi_v V_n}{A_w} = 2.4 < 9.8$ ksi

$\therefore$ Provide intermediate stiffeners spaced at $168/2 = 84$ in.

$a/h = 84/70 = 1.2$

Maximum a/h for tension field action:

$$\left[\frac{260}{(h/t_w)}\right]^2 = \left(\frac{260}{224}\right)^2 = 1.35 > 1.2$$

Design for tension field action:

For $a/h = 1.2$ and $h/t_w = 224$:

$$k = 5 + \frac{5}{(1.2)^2} = 8.5$$

$h/t_w = 224 > 187\sqrt{8.5}/\sqrt{36} = 91$

Appendix G3

$$C_v = \frac{44,000(8.5)}{(224)^2(36)} = 0.21$$

(A-G3-6)

$$\frac{\phi_v V_n}{A_w} = (0.9)(0.6)(36)\left[0.21 + \frac{1-0.21}{1.15\sqrt{1+(1.2)^2}}\right]$$

(A-G3-2) or
Table 11-36

$$= 12.6 > 9.8 \text{ ksi} \quad \textbf{o.k.}$$

d. Check center 14 ft panel:

$h/t_w = 224;\ a/h = (14)(12)/70 = 2.4 > 1.35$

$k = 5.0;\ C_v = 0.12$

$$\frac{\phi_v V_n}{A_w} = 2.4 \text{ ksi}$$

(F2-3) or
(A-G3-3) or
Table 10-36

$$\frac{V_u}{A_w} = \frac{35}{21.9} = 1.6 \text{ ksi} < 2.4 \text{ ksi} \quad \textbf{o.k.}$$

3. Flexure − shear interaction:

Check V_u/M_u at intermediate stiffener and concentrated load
locations in tension field panel:

Appendix G 5

Location	V_u	M_u	V_n	M_n	$\dfrac{0.6V_n}{M_n}$	$\dfrac{V_u}{M_u}$	$\dfrac{V_n}{0.75M_n}$
3 ft.	230	668	256	3890	0.04	0.34	0.09
10 ft.	215	2050	307	3890	0.05	0.105	0.105
17 ft.	180	3188	307	3890	0.05	0.06	0.105

Since $\dfrac{0.6V_n}{M_n} \le \dfrac{V_u}{M_u} \le \dfrac{V_n}{0.75M_n}$ at 10 ft

and 17 ft, interaction must be checked.

At 10 ft:

$$\frac{M_u}{M_n} + 0.625\frac{V_u}{V_n} = \frac{2,050}{3,890} + 0.625\left(\frac{215}{307}\right)$$

(A-G5-1)

$$= 0.96 < 1.375\phi = 1.24 \quad \textbf{o.k.}$$

At 17 ft:

$$\frac{3,188}{3,890} + 0.625\left(\frac{180}{307}\right) = 1.19 < 1.24 \quad \textbf{o.k.}$$

Summary: Space stiffeners as shown:

| 3'-0 | 2@7'-0 | 14'-0 | 2@7'-0 | 3'-0 |

E. Stiffener design:

1. For intermediate stiffeners:

   a. Area required (single plate stiffener):

   For a single plate stiffener, or when $\dfrac{V_u}{\phi V_n} < 1$, use Formula A-G4-2 instead of Table 11.

$$A_{st} = \frac{F_{yw}}{F_{yst}}\left[0.15Dh_ct_w(1 - C_v)\frac{V_u}{\phi V_n} - 18t_w^2\right] \geq 0.0 \qquad \text{(A-G4-2)}$$

   where h_c = 70 in.
   t_w = 0.3125 in.
   D = 2.4
   C_v = 0.21
   V_u = 215 kips
   ϕV_n = 276 kips

$$A_{st} = 0.15(2.4)(70)(0.3125)(1 - 0.21)\left(\frac{215}{276}\right)$$
$$- 18(0.3125)^2 = 3.09 \text{ in.}^2$$

   Try one bar ½ × 7

$$A_{st} = 3.5 \text{ in.}^2 > 3.09 \quad \textbf{o.k.}$$

   b. Check width-thickness ratio:

   $7/0.5 = 14 < 95/\sqrt{F_y} = 15.8$ **o.k.**   Table B5.1

   c. Check moment of inertia:

   $I_{req'd} = at_w^3 j$ Appendix G4

   $j = \dfrac{2.5}{(1.2)^2} - 2 = -0.3 < 0.5; j = 0.5$ (A-G4-1) or Table 11-36

   $I_{req'd} = 84\,(\tfrac{5}{16})^3(0.5) = 1.28 \text{ in.}^4$
   $I_{furn} = \tfrac{1}{3}\,(0.5)\,(7)^3 = 57.2 \text{ in.}^4$

   57.2 in.^{4} > 1.28 in.^{4} **o.k.**

3 - 116

d. Minimum length required:

It is suggested that intermediate stiffeners be stopped short of the tension flange and the weld by which they are attached to the web not closer than 4 times nor more than 6 times the web thickness from the near toe of the web-to-flange weld.*

Sect. F3

$70 - \frac{5}{16} - (4)(\frac{5}{16}) = 68.4$ in.

$70 - \frac{5}{16} - (6)(\frac{5}{16}) = 67.8$ in.

Use for intermediate stiffeners:
One plate ½ × 7 × 5'-8", fillet-welded to the compression flange and web.

2. For bearing stiffeners:

At end of girder, design for end reaction.

Try two $\frac{9}{16} \times 8$ in. bars.

a. Check width-thickness ratio (local buckling check):

$8/0.5625 = 14.2 < 95/\sqrt{F_y} = 15.8$ **o.k.**

Table B5.1

b. Check compressive strength:

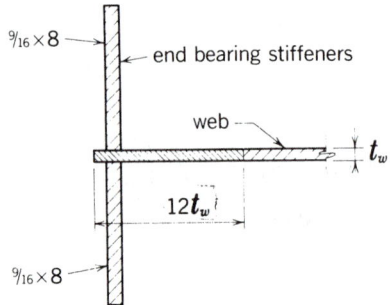

$\frac{9}{16} \times 8$ — end bearing stiffeners

web

t_w

$12t_w$

$\frac{9}{16} \times 8$

Effective Area

$I = (\frac{9}{16})\frac{(16.31)^3}{12} = 203$ in.^{4}

$A_{eff} = (2)(8)(\frac{9}{16}) + [(12)(\frac{5}{16})^2] = 10.17$ in.^{2}

$r = \sqrt{\frac{203}{10.17}} = 4.47$ in.

---

*When single stiffeners are used, they shall be attached to the compression flange, if it consists of a rectangular plate, to resist any uplift tendency due to torsion in the plate. When lateral bracing is attached to a stiffener, or a pair of stiffeners, these, in turn, shall be connected to the compression flange to transmit 1 percent of the total flange stress, unless the flange is composed only of angles.

$KL = 0.75\ h = (0.75)\ 70 = 52.5$ in. Sect. K1.8

$\dfrac{Kl}{r} = \dfrac{52.5}{4.47} = 11.7$

$\lambda_c = 0.13$

Design stress: $\phi F_{cr} = 30.38$ ksi (E2-2) or
Table 3-36

Design strength:

$\phi P_n = \phi F_{cr} A_g = (30.38)10.17 = 309$ kips

309 kips > 230 kips **o.k.**

c. Check bearing criterion Sect. J8.1

Design strength:

$\phi R_n = (0.75)2.0\ F_y A_{pb}$

$A_{pb} = 2(16 - 0.5)(\%_{16}) = 17.4$

(The 0.5 accounts for cutout for welds.)

$\phi R_n = 940 > 230$ **o.k.**

Use for bearing stiffeners: Two plates $\%_{16} \times 8 \times 5'\text{-}9\%''$ with close bearing on flange receiving reaction or concentrated loads.

Use same size stiffeners for bearing under concentrated loads.[*]

Example 2

Design a hybrid girder to support a factored uniform load of 3 kips per ft and three concentrated factored loads of 300 kips located at the quarter points. The girder depth must be limited to 5 ft. The compression flange will be laterally supported throughout its length.

Given:

Maximum bending moment: 14,400 kip-ft
Maximum vertical shear: 570 kips
Span: 80 ft
Maximum depth: 60 in.
Steel: Flanges: $F_y = 50$ ksi
   Web: $F_y = 36$ ksi

---

[*]In this example, bearing stiffeners were designed for end bearing; however, $25t_w$ may be used in determining effective area of web for bearing stiffeners under concentrated loads at interior panels (Sect. K1.8).

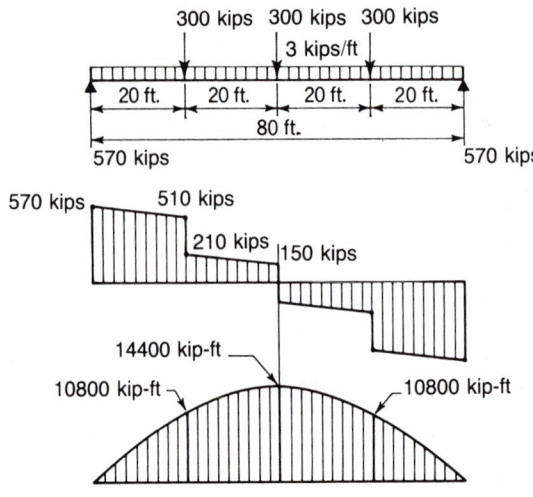

Shear and Moment Diagrams

Solution:

	LRFD Specification Reference

A. Preliminary web design:

Assume web depth, $h = h_c = 54$ in.

For $a/h > 1.5$ minimum thickness of web: $54/243 = 0.22$ in.   (A-G1-2)

For $R_{PG} = 1.0$, $h_c/t_w \leq 970/\sqrt{F_{yf}} = 137$ (A-G2-3)

Corresponding web thickness $= 54/137 = 0.39$.

Minimum t_w required for maximum $\dfrac{\phi_v V_n}{A_w}$ of 19.4 ksi: (F2-1) or Table 10-36

$$t_w = \frac{V_u}{19.4h} = \frac{570}{19.4 \times 54} = 0.54 \text{ in.}$$

Try web plate $\frac{9}{16} \times 54$; $A_w = 30.38$ in.^{2}

$$\frac{V_u}{A_w} = 570/30.38 = 18.8 \text{ ksi} < 19.4 \text{ ksi} \quad \textbf{o.k.}$$

$h/t_w = 54/0.563 = 96$

B. Preliminary flange design:

1. An approximate formula for the area of one flange is:

$$A_f \approx \frac{M_u}{F_{yf} h} = \frac{14,400(12)}{(50)(54)} = 64.0 \text{ in.}^2$$

Try $2\frac{5}{8} \times 24$ plate: $A_f = 63.0$ in.^{2}

2. Check adequacy against local buckling:

$b_f/2t_f = 24/(2)(2.625) = 4.6 < 65/\sqrt{F_y} = 9.2$ **o.k.**   Table B5.1

Flange is compact.

C. Trial girder section:

Web $\frac{9}{16} \times 54$; 2 flange plates $2\frac{5}{8} \times 24$

1. Determine plastic section moduli:

$$Z_f = 2\left[(2.625)(24)\left(\frac{54}{2} + \frac{2.625}{2}\right)\right] = 3,567 \text{ in.}^3$$

$$Z_w = 2\left[(\frac{9}{16})\left(\frac{54}{2}\right)\left(\frac{54}{4}\right)\right] = 410 \text{ in.}^3$$

2. Check flexural strength:

Compression flange is supported laterally for its full length and the section is compact.

$M_n = M_p = F_{yf} Z_f + F_{yw} Z_w$

$M_n = [(50)(3,567) + (36)(410)] \, 1/12 = 16,093 \text{ kip-ft}$

$\phi_b M_n = (0.90)16,093 = 14,483 \text{ kip-ft} > 14,400 \text{ kip-ft}$ **o.k.**

Appendix
F1.7

Use: Web: One plate $\frac{9}{16} \times 54$ ($F_y = 36$ ksi)
Flanges: Two plates $2\frac{5}{8} \times 24$ ($F_y = 50$ ksi)

D. Stiffener requirements:

1. Bearing stiffeners:

a. Check bearing at end reactions:

Sect. K1

Assume point bearing ($N = 0$) and $\frac{5}{16}$ in. web-to-flange welds.

Check local web yielding:

$R_n = (5k + N)F_{yw}t_w$; $k = 2\frac{5}{8} + \frac{5}{16} = 2\frac{15}{16}$ in.

(K1-2)

$\phi R_n = 1.0[(5)(2\frac{15}{16}) + 0](36)(\frac{9}{16}) = 297$ kips

297 kips < 570 kips **n.g.**

Note: If local web yielding criteria is satisfied, applicable criteria set forth in Sects. K1.4 and K1.5 should also be checked.

b. Bearing stiffeners at points of concentrated loads are also required.

2. Intermediate stiffeners:

The LRFD Specification does not permit design of hybrid girders on the basis of tension field action. Therefore, determine the need for intermediate stiffeners by use of Formulas F2-1, F2-2, F2-3, Table 10-36 or by the limit $418/\sqrt{F_{yw}}$ on h/t_w.

Appendix G3

Sect. F3

a. Check shear strength without intermediate stiffeners:

h/t_w = 96, a/h exceeds 3.0

Sect. F2.2

$\therefore k = 5.0$

$$\frac{V_u}{A_w} = \frac{570}{30.38} = 18.8 \text{ ksi}$$

Since $h/t_w = 96 > 234\sqrt{5/36} = 87$, use Formula F2-3:

$$\frac{\phi_v V_n}{A_w} = 13.0 \text{ ksi} < 18.8 \text{ ksi}$$

(F2-3) or
Table 10-36

$\therefore$ Intermediate stiffeners required.

b. End panel stiffener spacing

$$\frac{\phi_v V_n}{A_w} = 18.8 \text{ ksi}$$

(F2-2) or
Table 10-36

$\therefore a/h = 1.1$

Max. $a_1 = 1.1 (54) = 59.4$ in. (use 57 in.)

c. Check for additional stiffeners:

Shear at 57 in. from centerline bearing:

$$V_u = 570 - \left[3\left(\frac{57}{12}\right)\right] = 556 \text{ kips}$$

$$\frac{V_u}{A_w} = \frac{556}{30.38} = 18.3 \text{ ksi}$$

$$\frac{\phi_v V_n}{A_w} = 18.3 \text{ ksi}$$

(F2-2) or
Table 10-36

$\therefore a/h = 1.2$

Max. $a_2 = 1.2(54) = 64.8$ in. (use 61 in.)

Shear at 57 + 61 = 118 in. from centerline bearing:

$$V_u = 570 - \left[(3)\left(\frac{118}{12}\right)\right] = 541 \text{ kips}$$

$$\frac{V_u}{A_w} = \frac{541}{30.38} = 17.8 \text{ ksi}$$

$$\frac{\phi_v V_n}{A_w} = 17.8 \text{ ksi}$$

(F2-2) or
Table 10-36

$\therefore a/h = 1.3$

Max. $a_3 = 1.3(54) = 70.2$ in. (use 61 in.)

Shear at $57 + 61 + 61 = 179$ in. from centerline bearing:

$$V_u = 570 - \left[3\left(\frac{179}{12}\right)\right] = 525 \text{ kips}$$

$$\frac{V_u}{A_w} = \frac{525}{30.38} = 17.3 \text{ ksi}$$

(F2-2) or
Table 10-36

$$\frac{\phi_v V_n}{A_w} = 17.3 \text{ ksi}$$

$$\therefore a/h = 1.4$$

Max. $a_4 = 1.4 (54) = 75.6$

$[240 - (57 + 61 + 61)] = 61$ in. (use 61 in.)

d. Check need for stiffeners between concentrated loads:

$h/t_w = 96$, a/h is over 3, $k = 5$

$$\frac{V_u}{A_w} = \frac{210}{30.38} = 6.9 \text{ ksi}$$

$$\frac{\phi_v V_n}{A_w} = 13.0 \text{ ksi} > 6.9 \text{ ksi} \quad \textbf{o.k.}$$

(F2-3) or
Table 10-36

$\therefore$ Intermediate stiffeners not required between the concentrated loads.

SUMMARY

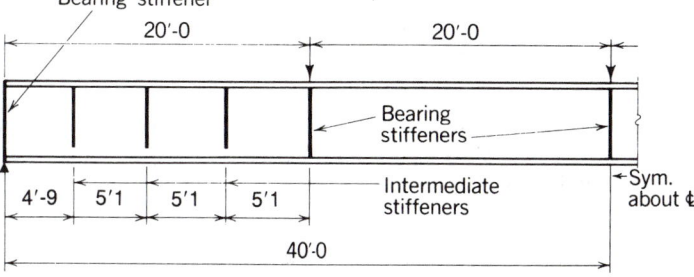

E. Stiffener design:

1. Bearing stiffeners:

   See Step E.2, Ex. 1, for design procedure.

   Use for bearing stiffeners: Two plates ¾ × 11 × 4'-5¾" with close bearing on flange receiving reaction or concentrated loads.

2. Intermediate stiffeners:

Assume $\frac{5}{16} \times 4$ in., F_y = 36 ksi, one side only.

a. Check width-thickness ratio:

$4/0.313 = 12.8 < 95/\sqrt{F_y} = 15.8$ **o.k.**

Table B5.1

b. Check moment of inertia:

$I_{req'd} = 61 \, (\frac{5}{16})^3(0.5) = 5.43$ in.[4]

Appendix G4

$I_{furn} = \frac{1}{3}(0.313)(4.28)^3 = 8.18$ in.[4]

8.18 in.[4] > 5.43 in.[4] **o.k.**

c. Length required (see Step E.1.d, Ex. 1):

Sect. F3

$54 - \frac{5}{16} - (6)(\frac{5}{16}) = 50\frac{5}{16}$

$54 - \frac{5}{16} - (4)(\frac{5}{16}) = 51\frac{7}{16}$ (use 51 in.)

Use for intermediate stiffeners: One plate $\frac{5}{16} \times 4 \times 4'\text{-}3''$, fillet-welded to the compression flange and web, one side of web only.

Example 3

Design the section of a nominal 52-in. deep welded girder with no intermediate stiffeners to support a factored uniform load of 3.8 kips per linear foot on an 85-ft span. The girder will be framed between columns and its compression flange will be laterally supported for its entire length.

Given:

Maximum bending moment: 3,432 kip-ft
Maximum vertical shear: 162 kips
Span: 85 ft
Nominal depth: 52 in.
Steel: F_y = 36 ksi

Solution:

For compact web and flange, $M_n = F_y Z$

Design for a compact web and flange.

Design for plastic moment, $F_y Z$:

Required plastic section modulus:

$$Z_{req'd} = \frac{M_u}{\phi F_y} = \frac{3,432 \times 12}{0.90 \times 36} = 1,271 \text{ in.}^3$$

Enter Table of Built-up Wide-Flange Sections, Dimensions and Properties:

For girder having $\frac{3}{8} \times 48$ web with $1\frac{1}{4} \times 16$ flange plates:

$Z = 1,200$ in.$^3 < 1,271$ in.^{3}

For girder having ⅜ × 52 web with 1¼ × 18 flange plates:

$Z = 1,450$ in.$^3 > 1,271$ in.^{3}

LRFD
Specification
Reference

A. Determine web required:

For compact web, $h_c/t_w \leq 640/\sqrt{F_{yf}} = 107$

Table B5.1

Assume $h = h_c = 50$ in.

Minimum $t_w = 50/107 = 0.46$ in.

Try: web = ½ × 50; $A_w = 25$ in.^{2}

$h/t_w = 50/0.50 = 100$

Since $h/t_w = 100 > 418/\sqrt{36} = 70$, transverse stiffeners may be required. If V_u is less than or equal to $\phi_v V_n$ calculated from Formulas F2-3 or A-G3-3, however, no stiffeners are necessary.

Appendix G4
or
Sect. F3

Note: Formulas F2-3 and A-G3-3 are equivalent.

From Table 10-36 under column headed "Over 3,"

(F2-3) or
(A-G3-3)

$\dfrac{\phi_v V_n}{A_w}$ without intermediate stiffeners = 11.9 ksi

$\dfrac{V_u}{A_w} = \dfrac{162}{25} = 6.5$ ksi < 11.9 ksi **o.k.**

∴ No intermediate stiffeners are necessary.

B. Determine flange required.

$A_f \approx \dfrac{M_u}{F_y h} = \dfrac{(3,432)(12)}{36 \times 50} = 22.9$ in.^{2}

Try 1⅛ × 18 plate: $A_f = 20.3$ in.^{2}

$b_f/2t_f = 18/(2)(1.125) = 8.0 < 65/\sqrt{F_y} = 10.8$ **o.k.**

Table B5.1

Flange is compact.

C. Check plastic section modulus:

$Z_{furn} = 2\left[(18)(1.125)\left(\dfrac{50}{2}\right)\left(\dfrac{1.125}{2}\right) + \left(\dfrac{1}{2}\right)\left(\dfrac{50}{2}\right)\dfrac{50}{4}\right] = 1,348$ in.^{3}

$1,348$ in.$^3 > 1,271$ in.^{3} **o.k.**

Use: Web: One plate ½ × 50
*Flanges: Two plates 1⅛ × 18

Note: Because this girder will be framed between columns, the usual end bearing stiffeners are not required.

Sect. K1.8

---

*Because this girder is longer than 60 ft, some economy may be gained by decreasing the flange size in areas of smaller moment, i.e. near ends of girder.

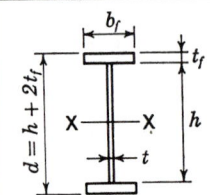

BUILT-UP WIDE-FLANGE SECTIONS
Dimensions and properties

Nominal Size	Wt. per Ft	Area	Depth d	Flange Width b_f	Flange Thick t_f	Web Depth h	Web Thick t	Axis X-X I	S	$^aS'$	Z	$^bZ'$	$^c\phi_v V_n$
In.	Lb.	In.^{2}	In.	In.	In.	In.	In.	In.^{4}	In.^{3}	In.^{3}	In.^{3}	In.^{3}	Kips
61 x 20	429	126	65.00	20	2½	60	7/16	106000	3250	34.6	3520	56.3	165.8
h/t = 137	361	106	64.00	20	2	60	7/16	84800	2650	35.2	2870	56.3	165.8
	327	96.2	63.50	20	1¾	60	7/16	74600	2350	35.4	2560	56.3	165.8
	293	86.2	63.00	20	1½	60	7/16	64600	2090	35.7	2240	56.3	165.8
	259	76.2	62.50	20	1¼	60	7/16	54800	1750	36.0	1930	56.3	165.8
	225	66.2	62.00	20	1	60	7/16	45100	1450	36.3	1610	56.3	165.8
	208	61.2	61.75	20	⅞	60	7/16	40300	1310	36.4	1460	56.3	165.8
	191	56.2	61.50	20	¾	60	7/16	35600	1160	36.6	1310	56.3	165.8
57 x 18	389	115	61.00	18	2½	56	7/16	83500	2740	30.0	2980	49.0	177.6
h/t = 128	328	96.5	60.00	18	2	56	7/16	67000	2230	30.5	2430	49.0	177.6
	298	87.5	59.50	18	1¾	56	7/16	58900	1980	30.7	2160	49.0	177.6
	267	78.5	59.00	18	1½	56	7/16	51000	1730	31.0	1900	49.0	177.6
	236	69.5	58.50	18	1¼	56	7/16	43300	1480	31.3	1630	49.0	177.6
	206	60.5	58.00	18	1	56	7/16	35600	1230	31.5	1370	49.0	177.6
	190	56.0	57.75	18	⅞	56	7/16	31900	1100	31.7	1240	49.0	177.6
	175	51.5	57.50	18	¾	56	7/16	28100	979	31.8	1110	49.0	177.6
	160	47.0	57.25	18	⅝	56	7/16	24400	854	32.0	980	49.0	177.6
53 x 18	342	100	56.50	18	2¼	52	3/8	64000	2270	25.9	2450	42.3	120.5
h/t = 138	311	91.5	56.00	18	2	52	3/8	56900	2030	26.2	2200	42.3	120.5
	280	82.5	55.50	18	1¾	52	3/8	49900	1800	26.4	1950	42.3	120.5
	250	73.5	55.00	18	1½	52	3/8	43000	1566	26.6	1700	42.3	120.5
	219	64.5	54.50	18	1¼	52	3/8	36300	1330	26.9	1450	42.3	120.5
	189	55.5	54.00	18	1	52	3/8	29700	1100	27.1	1210	42.3	120.5
	173	51.0	53.75	18	⅞	52	3/8	26400	983	27.2	1090	42.3	120.5
	158	46.5	53.50	18	¾	52	3/8	23200	866	27.4	966	42.3	120.5
	143	42.0	53.25	18	⅝	52	3/8	20000	750	27.5	846	42.3	120.5
49 x 16	306	90.0	52.50	16	2¼	48	3/8	48900	1860	21.9	2030	36.0	130.5
h/t = 128	279	82.0	52.00	16	2	48	3/8	43500	1670	22.2	1820	36.0	130.5
	252	74.0	51.50	16	1¾	48	3/8	38100	1480	22.4	1610	36.0	130.5
	224	66.0	51.00	16	1½	48	3/8	32900	1290	22.6	1400	36.0	130.5
	197	58.0	50.50	16	1¼	48	3/8	27700	1100	22.8	1200	36.0	130.5
	170	50.0	50.00	16	1	48	3/8	22700	907	23.0	1000	36.0	130.5
	156	46.0	49.75	16	⅞	48	3/8	20200	811	23.2	900	36.0	130.5
	143	42.0	49.50	16	¾	48	3/8	17700	716	23.3	801	36.0	130.5
	129	38.0	49.25	16	⅝	48	3/8	15300	620	23.4	702	36.0	130.5
45 x 16	237	69.8	47.50	16	1¾	44	5/16	31500	1330	18.7	1430	30.3	82.4
h/t = 141	210	61.8	47.00	16	1½	44	5/16	27100	1150	18.9	1240	30.3	82.4
	183	53.8	46.50	16	1¼	44	5/16	22700	976	19.1	1060	30.3	82.4

BUILT-UP WIDE-FLANGE SECTIONS
Dimensions and properties

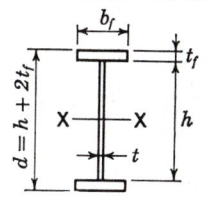

Nominal Size	Wt. per Ft	Area	Depth d	Flange Width b_f	Flange Thick t_f	Web Depth h	Web Thick t	Axis X-X I	S	$^aS'$	Z	$^bZ'$	$^c\phi_v V_n$
In.	Lb.	In.²	In.	In.	In.	In.	In.	In.⁴	In.³	In.³	In.³	In.³	Kips
92 x 30	823	242	96.00	30	3	90	11/16	431000	8980	79.1	9760	127	428.9
h/t = 131	721	212	95.00	30	2½	90	11/16	363000	7640	79.9	8330	127	428.9
	619	182	94.00	30	2	90	11/16	296000	6290	80.8	6910	127	428.9
	568	167	93.50	30	1¾	90	11/16	263000	5620	81.2	6210	127	428.9
	517	152	93.00	30	1½	90	11/16	230000	4950	81.7	5510	127	428.9
	466	137	92.50	30	1¼	90	11/16	198000	4280	82.1	4810	127	428.9
	415	122	92.00	30	1	90	11/16	166000	3610	82.5	4120	127	428.9
86 x 28	750	220	90.00	28	3	84	5/8	349000	7750	68.6	8410	110	345.3
h/t = 134	654	192	89.00	28	2½	84	5/8	293000	6580	69.4	7160	110	345.3
	559	164	88.00	28	2	84	5/8	238000	5410	70.2	5920	110	345.3
	512	150	87.50	28	1¾	84	5/8	211000	4820	70.6	5300	110	345.3
	464	136	87.00	28	1½	84	5/8	184000	4240	71.0	4690	110	345.3
	416	122	86.50	28	1¼	84	5/8	158000	3650	71.4	4090	110	345.3
	369	108	86.00	28	1	84	5/8	132000	3070	71.8	3480	110	345.3
80 x 26	696	205	84.00	26	3	78	5/8	281000	6680	58.8	7270	95.1	371.8
h/t = 125	608	179	83.00	26	2½	78	5/8	235000	5670	59.6	6180	95.1	371.8
	519	153	82.00	26	2	78	5/8	191000	4660	60.3	5110	95.1	371.8
	475	140	81.50	26	1¾	78	5/8	169000	4160	60.7	4580	95.1	371.8
	431	127	81.00	26	1½	78	5/8	148000	3650	61.0	4050	95.1	371.8
	387	114	80.50	26	1¼	78	5/8	127000	3150	61.4	3530	95.1	371.8
	343	101	80.00	26	1	78	5/8	106000	2690	61.8	3000	95.1	371.8
	320	94.2	79.75	26	7/8	78	5/8	95500	2390	62.0	2750	95.1	371.8
74 x 24	627	184	78.00	24	3	72	9/16	220000	5640	49.8	6130	81.0	293.7
h/t = 128	546	160	77.00	24	2½	72	9/16	184000	4780	50.5	5200	81.0	293.7
	464	136	76.00	24	2	72	9/16	149000	3920	51.2	4280	81.0	293.7
	423	124	75.50	.24	1¾	72	9/16	132000	3490	51.5	3830	81.0	293.7
	382	112	75.00	24	1½	72	9/16	115000	3060	51.8	3380	81.0	293.7
	342	100	74.50	24	1¼	72	9/16	98000	2630	52.2	2930	81.0	293.7
	301	88.5	74.00	24	1	72	9/16	81400	2200	52.5	2480	81.0	293.7
	280	82.5	73.75	24	7/8	72	9/16	73300	1990	52.7	2260	81.0	293.7
68 x 22	561	165	72.00	22	3	66	½	169000	4700	41.6	5100	68.1	225.0
h/t = 132	486	143	71.00	22	2½	66	½	141000	3970	42.2	4310	68.1	225.0
	411	121	70.00	22	2	66	½	114000	3250	42.8	3540	68.1	225.0
	374	110	69.50	22	1¾	66	½	100000	2890	43.1	3150	68.1	225.0
	337	99.0	69.00	22	1½	66	½	87200	2530	43.4	2770	68.1	225.0
	299	88.0	68.50	22	1¼	66	½	74200	2170	43.7	2390	68.1	225.0
	262	77.0	68.00	22	1	66	½	61400	1800	44.0	2020	68.1	225.0
	243	71.5	67.75	22	7/8	66	½	55000	1620	44.2	1830	68.1	225.0
	224	66.0	67.50	22	¾	66	½	48700	1440	44.4	1650	68.1	225.0

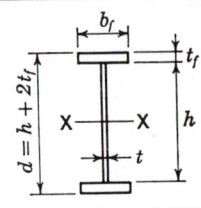

BUILT-UP WIDE-FLANGE SECTIONS
Dimensions and properties

Nominal Size	Wt. per Ft	Area	Depth d	Flange Width b_f	Flange Thick t_f	Web Depth h	Web Thick t	Axis X-X I	S	$^aS'$	Z	$^bZ'$	$^c\phi_v V_n$
In.	Lb.	In.^{2}	In.	In.	In.	In.	In.	In.^{4}	In.^{3}	In.^{3}	In.^{3}	In.^{3}	Kips
	156	45.8	46.00	16	1	44	5/16	18400	801	19.3	871	30.3	82.4
	142	41.8	45.75	16	7/8	44	5/16	16300	713	19.4	780	30.3	82.4
	128	37.8	45.50	16	3/4	44	5/16	14200	626	19.5	688	30.3	82.4
	115	33.8	45.25	16	5/8	44	5/16	12200	538	19.6	598	30.3	82.4

$^aS'$ = Additional section modulus corresponding to 1/16" increase in web thickness.
$^bZ'$ = Additional plastic section modulus corresponding to 1/16" increase in web thickness.
$^c\phi_v V_n$ = Maximum design end shear strength permissible without transverse stiffeners for tabulated web plate (LRFD Specification Sect. F3 and Appendix Sect. G4). $\phi_v = 0.90$.

Notes:
The width-thickness ratios for girders in this table are noncompact shapes in accordance with LRFD Specification Sect. B5 for $F_y = 36$ ksi steel. For steels of higher yield strengths, check flanges for compliance with this section.
This table does not consider local effects on the web due to concentrated loads .
See LRFD Specification Sect. K1 and Appendix Sect. G2.
See LRFD Specification Appendix G4 for design of stiffeners.
Welds not included in tabulated weight per foot.

BEAM DIAGRAMS AND FORMULAS

Nomenclature

E = modulus of elasticity of steel at 29,000 ksi

I = moment of inertia of beam (in.^{4}).

L = total length of beam between reaction points (ft).

M_{max} = maximum moment (kip-in.).

M_1 = maximum moment in left section of beam (kip-in.).

M_2 = maximum moment in right section of beam (kip-in.).

M_3 = maximum positive moment in beam with combined end moment conditions (kip-in.).

M_x = moment at distance x from end of beam (kip-in.).

P = concentrated load (kips).

P_1 = concentrated load nearest left reaction (kips).

P_2 = concentrated load nearest right reaction, and of different magnitude than P_1 (kips).

R = end beam reaction for any condition of symmetrical loading (kips).

R_1 = left end beam reaction (kips).

R_2 = right end or intermediate beam reaction (kips).

R_3 = right end beam reaction (kips).

V = maximum vertical shear for any condition of symmetrical loading (kips).

V_1 = maximum vertical shear in left section of beam (kips).

V_2 = vertical shear at right reaction point, or to left of intermediate reaction point of beam (kips).

V_3 = vertical shear at right reaction point, or to right of intermediate reaction point of beam (kips).

V_x = vertical shear at distance x from end of beam (kips).

W = total load on beam (kips).

a = measured distance along beam (in.).

b = measured distance along beam which may be greater or less than "a" (in.).

l = total length of beam between reaction points (in.).

w = uniformly distributed load per unit of length (kips per in.).

w_1 = uniformly distributed load per unit of length nearest left reaction (kips per in.).

w_2 = uniformly distributed load per unit of length nearest right reaction, and of different magnitude than w_1 (kips per in.).

x = any distance measured along beam from left reaction (in.).

x_1 = any distance measured along overhang section of beam from nearest reaction point (in.).

Δ_{max} = maximum deflection (in.).

Δ_a = deflection at point of load (in.).

Δ_x = deflection at any point x distance from left reaction (in.).

Δ_{x_1} = deflection of overhang section of beam at any distance from nearest reaction point (in.).

BEAM DIAGRAMS AND FORMULAS
Frequently used formulas

The formulas given below are frequently required in structural designing. They are included herein for the convenience of those engineers who have infrequent use for such formulas and hence may find reference necessary. Variation from the standard nomenclature on page 3-127 is noted.

BEAMS

Flexural stress at extreme fiber:
$$f = Mc/I = M/S$$

Flexural stress at any fiber:
$$f = My/I \qquad y = \text{distance from neutral axis to fiber.}$$

Average vertical shear (for maximum see below):
$$v = V/A = V/dt \text{ (for beams and girders)}$$

Horizontal shearing stress at any section A-A:
$$v = VQ/Ib \qquad Q = \text{statical moment about the neutral axis of the entire sec-}$$
tion of that portion of the cross-section lying outside of section A-A,
$$b = \text{width at section A-A}$$

(Intensity of vertical shear is equal to that of horizontal shear acting normal to it at the same point and both are usually a maximum at mid-height of beam.)

Shear and deflection at any point:
$$EI\frac{d^2y}{dx^2} = M \qquad x \text{ and } y \text{ are abscissa and ordinate respectively of a point on the}$$
neutral axis, referred to axes of rectangular coordinates through a selected point of support.

(First integration gives slopes; second integration gives deflections. Constants of integration must be determined.)

CONTINUOUS BEAMS (the theorem of three moments)

Uniform load:
$$M_a\frac{l_1}{I_1} + 2M_b\left(\frac{l_1}{I_1} + \frac{l_2}{I_2}\right) + M_c\frac{l_2}{I_2} = -\frac{1}{4}\left(\frac{w_1 l_1^3}{I_1} + \frac{w_2 l_2^3}{I_2}\right)$$

Concentrated loads:
$$M_a\frac{l_1}{I_1} + 2M_b\left(\frac{l_1}{I_1} + \frac{l_2}{I_2}\right) + M_c\frac{l_2}{I_2} = -\frac{P_1 a_1 b_1}{I_1}\left(1 + \frac{a_1}{l_1}\right) - \frac{P_2 a_2 b_2}{I_2}\left(1 + \frac{b_2}{l_2}\right)$$

Considering any two consecutive spans in any continuous structure:

M_a, M_b, M_c = moments at left, center and right supports respectively, of any pair of adjacent spans.

l_1 and l_2 = length of left and right spans, respectively, of the pair.
I_1 and I_2 = moment of inertia of left and right spans, respectively.
w_1 and w_2 = load per unit of length on left and right spans, respectively.
P_1 and P_2 = concentrated loads on left and right spans, respectively.
a_1 and a_2 = distance of concentrated loads from left support, in left and right spans, respectively.
b_1 and b_2 = distance of concentrated loads from right support in left and right spans, respectively.

The above equations are for beams with moment of inertia constant in each span but differing in different spans, continuous over three or more supports. By writing such an equation for each successive pair of spans and introducing the known values (usually zero) of end moments, all other moments can be found.

BEAM DIAGRAMS AND FORMULAS
Table of Concentrated Load Equivalents

n	Loading	Coeff.	Simple Beam	Beam Fixed One End Supported at Other	Beam Fixed Both Ends
∞		a	0.1250	0.0703	0.0417
		b	—	0.1250	0.0833
		c	0.5000	0.3750	—
		d	—	0.6250	0.5000
		e	0.0130	0.0054	0.0026
		f	1.0000	1.0000	0.6667
		g	1.0000	0.4151	0.3000
2		a	0.2500	0.1563	0.1250
		b	—	0.1875	0.1250
		c	0.5000	0.3125	—
		d	—	0.6875	0.5000
		e	0.0208	0.0093	0.0052
		f	2.0000	1.5000	1.0000
		g	0.8000	0.4770	0.4000
3		a	0.3333	0.2222	0.1111
		b	—	0.3333	0.2222
		c	1.0000	0.6667	—
		d	—	1.3333	1.0000
		e	0.0355	0.0152	0.0077
		f	2.6667	2.6667	1.7778
		g	1.0222	0.4381	0.3333
4		a	0.5000	0.2656	0.1875
		b	—	0.4688	0.3125
		c	1.5000	1.0313	—
		d	—	1.9688	1.5000
		e	0.0495	0.0209	0.0104
		f	4.0000	3.7500	2.5000
		g	0.9500	0.4281	0.3200
5		a	0.6000	0.3600	0.2000
		b	—	0.6000	0.4000
		c	2.0000	1.4000	—
		d	—	2.6000	2.0000
		e	0.0630	0.0265	0.0130
		f	4.8000	4.8000	3.2000
		g	1.0080	0.4238	0.3120

Maximum positive moment (kip-ft.):
  a × P × L

Maximum negative moment (kip-ft.):
  b × P × L

Pinned end reaction (kips): c × P

Fixed end reaction (kips): d × P

Maximum deflection (in.): e × Pl^3/EI

Equivalent simple span uniform load (kips):
  f × P

Deflection coeff. for equivalent simple
  span uniform load: g

Number of equal load spaces: n

Span of beam (ft.): L

Span of beam (in.): l

AMERICAN INSTITUTE OF STEEL CONSTRUCTION

BEAM DIAGRAMS AND FORMULAS
For various static loading conditions
For meaning of symbols, see page 3-127

---

1. SIMPLE BEAM—UNIFORMLY DISTRIBUTED LOAD

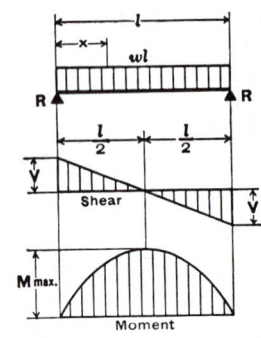

Total Equiv. Uniform Load . . . $= wl$

$R = V$ . . . . $= \dfrac{wl}{2}$

V_x . . . . $= w\left(\dfrac{l}{2}-x\right)$

M max. $\left(\text{at center}\right)$ . . . . $= \dfrac{wl^2}{8}$

M_x . . . . $= \dfrac{wx}{2}(l-x)$

Δmax. $\left(\text{at center}\right)$ . . . . $= \dfrac{5\,wl^4}{384\,EI}$

Δ_x . . . . $= \dfrac{wx}{24EI}(l^3-2lx^2+x^3)$

---

2. SIMPLE BEAM—LOAD INCREASING UNIFORMLY TO ONE END

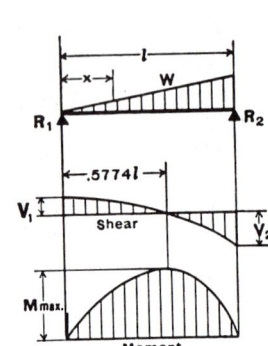

Total Equiv. Uniform Load . . . $= \dfrac{16W}{9\sqrt{3}} = 1.0264W$

$R_1 = V_1$ . . . . $= \dfrac{W}{3}$

$R_2 = V_2$ max. . . . . $= \dfrac{2W}{3}$

V_x . . . . $= \dfrac{W}{3} - \dfrac{Wx^2}{l^2}$

M max. $\left(\text{at }x=\dfrac{l}{\sqrt{3}}=.5774l\right)$ . . $= \dfrac{2Wl}{9\sqrt{3}} = .1283\,Wl$

M_x . . . . $= \dfrac{Wx}{3l^2}(l^2-x^2)$

Δmax. $\left(\text{at }x=l\sqrt{1-\sqrt{\dfrac{8}{15}}}=.5193l\right)$ $= .01304\dfrac{Wl^3}{EI}$

Δ_x . . . . $= \dfrac{Wx}{180EI\,l^2}(3x^4-10l^2x^2+7l^4)$

---

3. SIMPLE BEAM—LOAD INCREASING UNIFORMLY TO CENTER

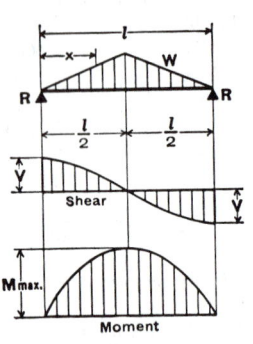

Total Equiv. Uniform Load . . . $= \dfrac{4W}{3}$

$R = V$ . . . . $= \dfrac{W}{2}$

V_x $\left(\text{when }x<\dfrac{l}{2}\right)$ . . . . $= \dfrac{W}{2l^2}(l^2-4x^2)$

M max. $\left(\text{at center}\right)$ . . . . $= \dfrac{Wl}{6}$

M_x $\left(\text{when }x<\dfrac{l}{2}\right)$ . . . . $= Wx\left(\dfrac{1}{2}-\dfrac{2x^2}{3l^2}\right)$

Δmax. $\left(\text{at center}\right)$ . . . . $= \dfrac{Wl^3}{60EI}$

Δ_x $\left(\text{when }x<\dfrac{l}{2}\right)$ . . . . $= \dfrac{Wx}{480\,EI\,l^2}(5l^2-4x^2)^2$

BEAM DIAGRAMS AND FORMULAS
For various static loading conditions
For meaning of symbols, see page 3-127

---

4. SIMPLE BEAM—UNIFORM LOAD PARTIALLY DISTRIBUTED

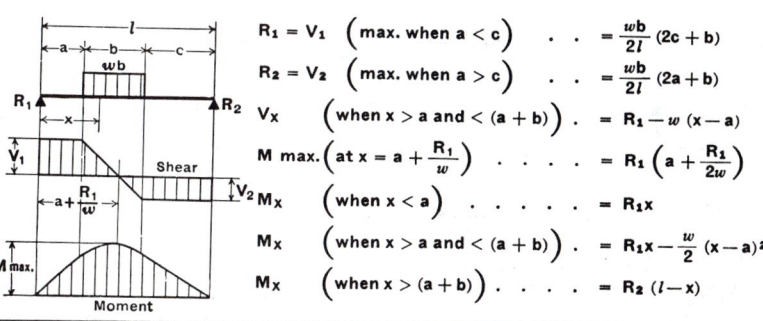

$R_1 = V_1 \left(\text{max. when a < c} \right) \quad \cdots \quad = \dfrac{wb}{2l}(2c + b)$

$R_2 = V_2 \left(\text{max. when a > c} \right) \quad \cdots \quad = \dfrac{wb}{2l}(2a + b)$

$V_x \quad \left(\text{when } x > a \text{ and } < (a + b) \right) . \quad = R_1 - w(x - a)$

$M \text{ max.} \left(\text{at } x = a + \dfrac{R_1}{w} \right) \quad \cdots \quad = R_1 \left(a + \dfrac{R_1}{2w} \right)$

$M_x \quad \left(\text{when } x < a \right) \quad \cdots \quad \cdots \quad = R_1 x$

$M_x \quad \left(\text{when } x > a \text{ and } < (a + b) \right) . \quad = R_1 x - \dfrac{w}{2}(x - a)^2$

$M_x \quad \left(\text{when } x > (a + b) \right) . \quad \cdots \quad = R_2 (l - x)$

---

5. SIMPLE BEAM—UNIFORM LOAD PARTIALLY DISTRIBUTED AT ONE END

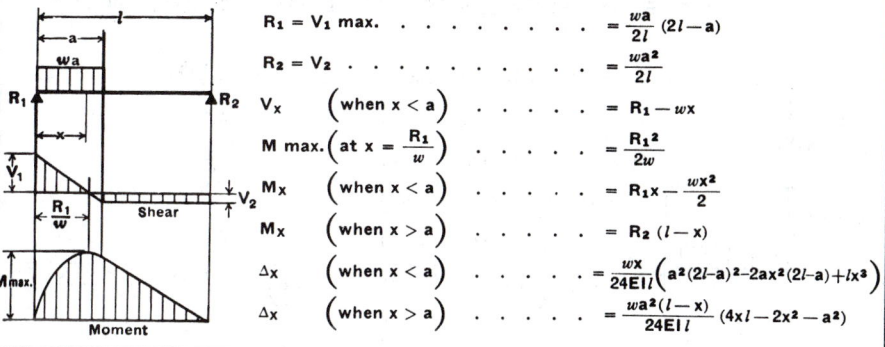

$R_1 = V_1 \text{ max.} \quad \cdots \quad \cdots \quad \cdots \quad = \dfrac{wa}{2l}(2l - a)$

$R_2 = V_2 \quad \cdots \quad \cdots \quad \cdots \quad \cdots \quad = \dfrac{wa^2}{2l}$

$V_x \quad \left(\text{when } x < a \right) \quad \cdots \quad \cdots \quad = R_1 - wx$

$M \text{ max.} \left(\text{at } x = \dfrac{R_1}{w} \right) \quad \cdots \quad \cdots \quad = \dfrac{R_1^2}{2w}$

$M_x \quad \left(\text{when } x < a \right) \quad \cdots \quad \cdots \quad = R_1 x - \dfrac{wx^2}{2}$

$M_x \quad \left(\text{when } x > a \right) \quad \cdots \quad \cdots \quad = R_2 (l - x)$

$\Delta_x \quad \left(\text{when } x < a \right) \quad \cdots \quad = \dfrac{wx}{24EIl}\left(a^2(2l - a)^2 - 2ax^2(2l - a) + lx^3 \right)$

$\Delta_x \quad \left(\text{when } x > a \right) \quad \cdots \quad = \dfrac{wa^2(l - x)}{24EIl}(4xl - 2x^2 - a^2)$

---

6. SIMPLE BEAM—UNIFORM LOAD PARTIALLY DISTRIBUTED AT EACH END

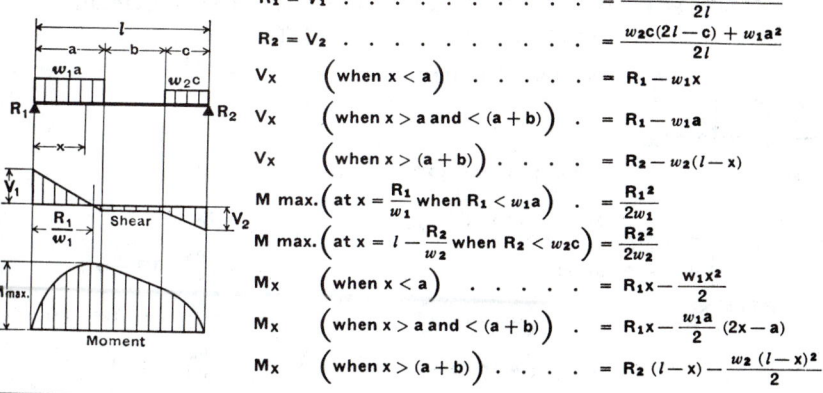

$R_1 = V_1 \quad \cdots \quad \cdots \quad \cdots \quad = \dfrac{w_1 a(2l - a) + w_2 c^2}{2l}$

$R_2 = V_2 \quad \cdots \quad \cdots \quad \cdots \quad = \dfrac{w_2 c(2l - c) + w_1 a^2}{2l}$

$V_x \quad \left(\text{when } x < a \right) \quad \cdots \quad \cdots \quad = R_1 - w_1 x$

$V_x \quad \left(\text{when } x > a \text{ and } < (a + b) \right) . \quad = R_1 - w_1 a$

$V_x \quad \left(\text{when } x > (a + b) \right) \quad \cdots \quad = R_2 - w_2 (l - x)$

$M \text{ max.} \left(\text{at } x = \dfrac{R_1}{w_1} \text{ when } R_1 < w_1 a \right) \quad = \dfrac{R_1^2}{2w_1}$

$M \text{ max.} \left(\text{at } x = l - \dfrac{R_2}{w_2} \text{ when } R_2 < w_2 c \right) = \dfrac{R_2^2}{2w_2}$

$M_x \quad \left(\text{when } x < a \right) \quad \cdots \quad \cdots \quad = R_1 x - \dfrac{w_1 x^2}{2}$

$M_x \quad \left(\text{when } x > a \text{ and } < (a + b) \right) . \quad = R_1 x - \dfrac{w_1 a}{2}(2x - a)$

$M_x \quad \left(\text{when } x > (a + b) \right) \quad \cdots \quad = R_2 (l - x) - \dfrac{w_2 (l - x)^2}{2}$

BEAM DIAGRAMS AND FORMULAS
For various static loading conditions
For meaning of symbols, see page 3-127

---

7. SIMPLE BEAM—CONCENTRATED LOAD AT CENTER

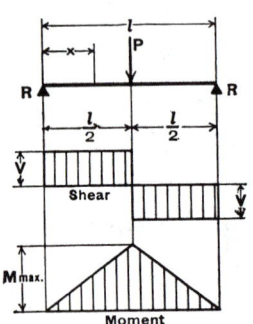

Total Equiv. Uniform Load . . . . . $= 2P$

$R = V$ . . . . $= \dfrac{P}{2}$

M max. $\left(\text{at point of load}\right)$ . . . . . . $= \dfrac{Pl}{4}$

M_x $\left(\text{when } x < \dfrac{l}{2}\right)$ . . . . . $= \dfrac{Px}{2}$

Δmax. $\left(\text{at point of load}\right)$ . . . . . $= \dfrac{Pl^3}{48EI}$

Δ_x $\left(\text{when } x < \dfrac{l}{2}\right)$ . . . . . $= \dfrac{Px}{48EI}(3l^2 - 4x^2)$

---

8. SIMPLE BEAM—CONCENTRATED LOAD AT ANY POINT

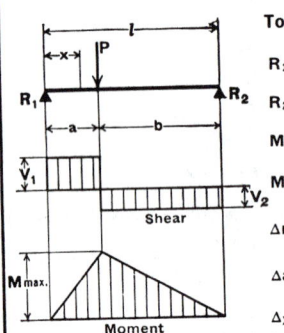

Total Equiv. Uniform Load . . . . . $= \dfrac{8\,Pab}{l^2}$

$R_1 = V_1 \left(\text{max. when } a < b\right)$ . . . . $= \dfrac{Pb}{l}$

$R_2 = V_2 \left(\text{max. when } a > b\right)$ . . . . $= \dfrac{Pa}{l}$

M max. $\left(\text{at point of load}\right)$ . . . . $= \dfrac{Pab}{l}$

M_x $\left(\text{when } x < a\right)$ . . . . $= \dfrac{Pbx}{l}$

Δmax. $\left(\text{at } x = \sqrt{\dfrac{a(a+2b)}{3}} \text{ when } a > b\right)$ $= \dfrac{Pab(a+2b)\ \sqrt{3a(a+2b)}}{27\ EI\ l}$

Δa $\left(\text{at point of load}\right)$ . . . . $= \dfrac{Pa^2b^2}{3EI\ l}$

Δ_x $\left(\text{when } x < a\right)$ . . . . $= \dfrac{Pbx}{6EI\ l}(l^2 - b^2 - x^2)$

---

9. SIMPLE BEAM—TWO EQUAL CONCENTRATED LOADS SYMMETRICALLY PLACED

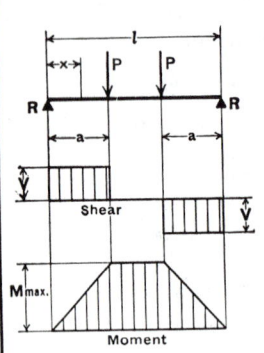

Total Equiv. Uniform Load . . . . . $= \dfrac{8\,Pa}{l}$

$R = V$ . . . . $= P$

M max. $\left(\text{between loads}\right)$ . . . . . $= Pa$

M_x $\left(\text{when } x < a\right)$ . . . . . $= Px$

Δmax. $\left(\text{at center}\right)$ . . . . . . $= \dfrac{Pa}{24EI}(3l^2 - 4a^2)$

Δ_x $\left(\text{when } x < a\right)$ . . . . . $= \dfrac{Px}{6EI}(3la - 3a^2 - x^2)$

Δ_x $\left(\text{when } x > a \text{ and } < (l-a)\right)$ . . $= \dfrac{Pa}{6EI}(3lx - 3x^2 - a^2)$

BEAM DIAGRAMS AND FORMULAS
For various static loading conditions
For meaning of symbols, see page 3-127

---

10. SIMPLE BEAM—TWO EQUAL CONCENTRATED LOADS UNSYMMETRICALLY PLACED

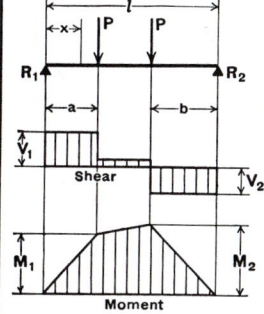

$$R_1 = V_1 \left(\text{max. when } a < b\right) \quad \cdots \quad = \frac{P}{l}(l - a + b)$$

$$R_2 = V_2 \left(\text{max. when } a > b\right) \quad \cdots \quad = \frac{P}{l}(l - b + a)$$

$$V_x \quad \left(\text{when } x > a \text{ and} < (l-b)\right) \cdots \quad = \frac{P}{l}(b - a)$$

$$M_1 \quad \left(\text{max. when } a > b\right) \quad \cdots \quad = R_1 a$$

$$M_2 \quad \left(\text{max. when } a < b\right) \quad \cdots \quad = R_2 b$$

$$M_x \quad \left(\text{when } x < a\right) \quad \cdots \quad = R_1 x$$

$$M_x \quad \left(\text{when } x > a \text{ and} < (l-b)\right) \cdots \quad = R_1 x - P(x - a)$$

---

11. SIMPLE BEAM—TWO UNEQUAL CONCENTRATED LOADS UNSYMMETRICALLY PLACED

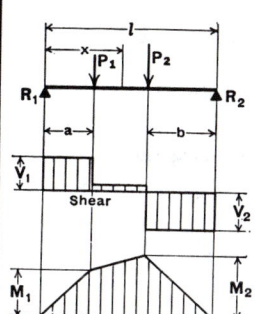

$$R_1 = V_1 \quad \cdots \quad = \frac{P_1(l - a) + P_2 b}{l}$$

$$R_2 = V_2 \quad \cdots \quad = \frac{P_1 a + P_2(l - b)}{l}$$

$$V_x \quad \left(\text{when } x > a \text{ and} < (l-b)\right) \cdots \quad = R_1 - P_1$$

$$M_1 \quad \left(\text{max. when } R_1 < P_1\right) \quad \cdots \quad = R_1 a$$

$$M_2 \quad \left(\text{max. when } R_2 < P_2\right) \quad \cdots \quad = R_2 b$$

$$M_x \quad \left(\text{when } x < a\right) \quad \cdots \quad = R_1 x$$

$$M_x \quad \left(\text{when } x > a \text{ and} < (l-b)\right) \cdots \quad = R_1 x - P_1(x - a)$$

---

12. BEAM FIXED AT ONE END, SUPPORTED AT OTHER—UNIFORMLY DISTRIBUTED LOAD

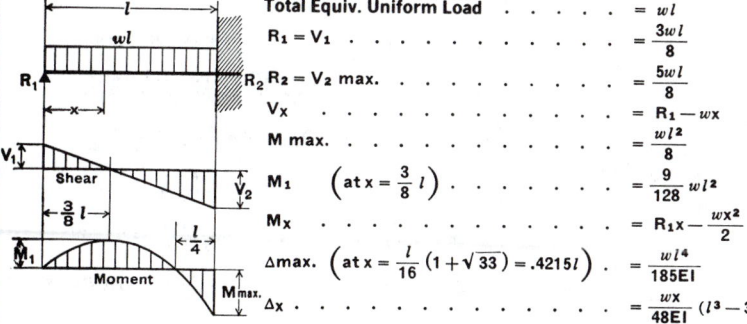

$$\text{Total Equiv. Uniform Load} \quad \cdots \quad = wl$$

$$R_1 = V_1 \quad \cdots \quad = \frac{3wl}{8}$$

$$R_2 = V_2 \text{ max.} \quad \cdots \quad = \frac{5wl}{8}$$

$$V_x \quad \cdots \quad = R_1 - wx$$

$$M \text{ max.} \quad \cdots \quad = \frac{wl^2}{8}$$

$$M_1 \quad \left(\text{at } x = \frac{3}{8}l\right) \quad \cdots \quad = \frac{9}{128}wl^2$$

$$M_x \quad \cdots \quad = R_1 x - \frac{wx^2}{2}$$

$$\Delta \text{max.} \quad \left(\text{at } x = \frac{l}{16}\left(1 + \sqrt{33}\right) = .4215l\right) \quad = \frac{wl^4}{185EI}$$

$$\Delta_x \quad \cdots \quad = \frac{wx}{48EI}(l^3 - 3lx^2 + 2x^3)$$

BEAM DIAGRAMS AND FORMULAS
For various static loading conditions
For meaning of symbols, see page 3-127

---

13. BEAM FIXED AT ONE END, SUPPORTED AT OTHER— CONCENTRATED LOAD AT CENTER

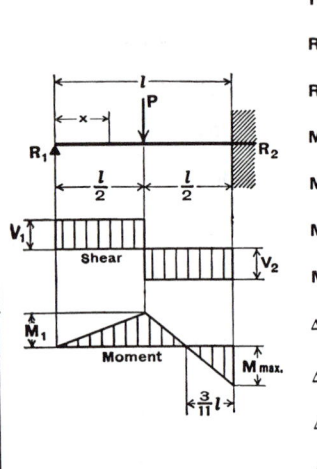

Total Equiv. Uniform Load $\dots\dots = \dfrac{3P}{2}$

$R_1 = V_1 \dots\dots\dots\dots\dots = \dfrac{5P}{16}$

$R_2 = V_2$ max. $\dots\dots\dots\dots = \dfrac{11P}{16}$

M max. $\left(\text{at fixed end}\right) \dots\dots = \dfrac{3Pl}{16}$

$M_1 \left(\text{at point of load}\right) \dots\dots = \dfrac{5Pl}{32}$

$M_x \left(\text{when } x < \dfrac{l}{2}\right) \dots\dots = \dfrac{5Px}{16}$

$M_x \left(\text{when } x > \dfrac{l}{2}\right) \dots\dots = P\left(\dfrac{l}{2} - \dfrac{11x}{16}\right)$

Δmax. $\left(\text{at } x = l\sqrt{\dfrac{1}{5}} = .4472l\right) \dots = \dfrac{Pl^3}{48EI\sqrt{5}} = .009317\dfrac{Pl^3}{EI}$

$\Delta_x \left(\text{at point of load}\right) \dots\dots = \dfrac{7Pl^3}{768EI}$

$\Delta_x \left(\text{when } x < \dfrac{l}{2}\right) \dots\dots = \dfrac{Px}{96EI}(3l^2 - 5x^2)$

$\Delta_x \left(\text{when } x > \dfrac{l}{2}\right) \dots\dots = \dfrac{P}{96EI}(x-l)^2(11x-2l)$

---

14. BEAM FIXED AT ONE END, SUPPORTED AT OTHER— CONCENTRATED LOAD AT ANY POINT

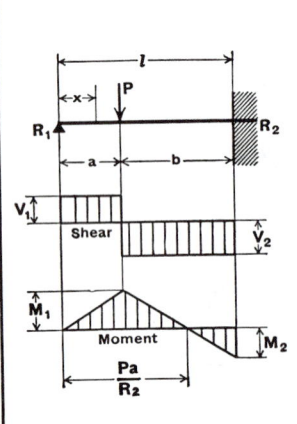

$R_1 = V_1 \dots\dots\dots\dots\dots = \dfrac{Pb^2}{2l^3}(a+2l)$

$R_2 = V_2 \dots\dots\dots\dots\dots = \dfrac{Pa}{2l^3}(3l^2 - a^2)$

$M_1 \left(\text{at point of load}\right) \dots\dots = R_1a$

$M_2 \left(\text{at fixed end}\right) \dots\dots = \dfrac{Pab}{2l^2}(a+l)$

$M_x \left(\text{when } x < a\right) \dots\dots = R_1x$

$M_x \left(\text{when } x > a\right) \dots\dots = R_1x - P(x-a)$

Δmax. $\left(\text{when } a < .414l \text{ at } x = l\dfrac{l^2+a^2}{3l^2-a^2}\right) = \dfrac{Pa}{3EI}\dfrac{(l^2-a^2)^3}{(3l^2-a^2)^2}$

Δmax. $\left(\text{when } a > .414l \text{ at } x = l\sqrt{\dfrac{a}{2l+a}}\right) = \dfrac{Pab^2}{6EI}\sqrt{\dfrac{a}{2l+a}}$

$\Delta_a \left(\text{at point of load}\right) \dots\dots = \dfrac{Pa^2b^3}{12EIl^3}(3l+a)$

$\Delta_x \left(\text{when } x < a\right) \dots\dots = \dfrac{Pb^2x}{12EIl^3}(3al^2 - 2lx^2 - ax^2)$

$\Delta_x \left(\text{when } x > a\right) \dots\dots = \dfrac{Pa}{12EIl^3}(l-x)^2(3l^2x - a^2x - 2a^2l)$

BEAM DIAGRAMS AND FORMULAS
For various static loading conditions
For meaning of symbols, see page 3-127

---

15. BEAM FIXED AT BOTH ENDS—UNIFORMLY DISTRIBUTED LOADS

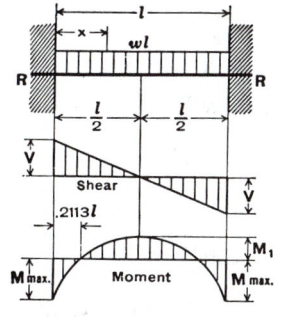

Total Equiv. Uniform Load . . . . $= \dfrac{2wl}{3}$

$R = V$ . . . . $= \dfrac{wl}{2}$

V_x . . . . $= w\left(\dfrac{l}{2} - x\right)$

M max. $\Big($ at ends $\Big)$ . . . . . $= \dfrac{wl^2}{12}$

M_1 $\Big($ at center $\Big)$ . . . . . $= \dfrac{wl^2}{24}$

M_x . . . . $= \dfrac{w}{12}(6lx - l^2 - 6x^2)$

Δmax. $\Big($ at center $\Big)$ . . . . . $= \dfrac{wl^4}{384EI}$

Δ_x . . . . $= \dfrac{wx^2}{24EI}(l - x)^2$

---

16. BEAM FIXED AT BOTH ENDS—CONCENTRATED LOAD AT CENTER

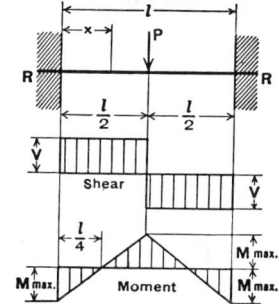

Total Equiv. Uniform Load . . . . $= P$

$R = V$ . . . . $= \dfrac{P}{2}$

M max. $\Big($ at center and ends $\Big)$ . . . $= \dfrac{Pl}{8}$

M_x $\Big($ when $x < \dfrac{l}{2}\Big)$ . . . . . $= \dfrac{P}{8}(4x - l)$

Δmax. $\Big($ at center $\Big)$ . . . . . $= \dfrac{Pl^3}{192EI}$

Δ_x $\Big($ when $x < \dfrac{l}{2}\Big)$ . . . . . $= \dfrac{Px^2}{48EI}(3l - 4x)$

---

17. BEAM FIXED AT BOTH ENDS—CONCENTRATED LOAD AT ANY POINT

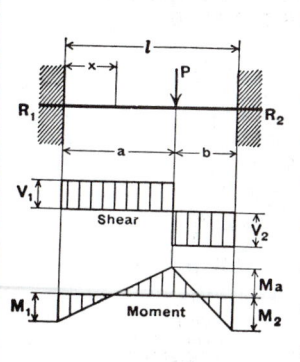

$R_1 = V_1 \Big($ max. when $a < b\Big)$ . . . $= \dfrac{Pb^2}{l^3}(3a + b)$

$R_2 = V_2 \Big($ max. when $a > b\Big)$ . . . $= \dfrac{Pa^2}{l^3}(a + 3b)$

M_1 $\Big($ max. when $a < b\Big)$ . . . $= \dfrac{Pab^2}{l^2}$

M_2 $\Big($ max. when $a > b\Big)$ . . . $= \dfrac{Pa^2b}{l^2}$

M_a $\Big($ at point of load $\Big)$ . . . $= \dfrac{2Pa^2b^2}{l^3}$

M_x $\Big($ when $x < a\Big)$ . . . . . $= R_1x - \dfrac{Pab^2}{l^2}$

Δmax. $\Big($ when $a > b$ at $x = \dfrac{2al}{3a + b}\Big)$. $= \dfrac{2Pa^3b^2}{3EI(3a + b)^2}$

Δ_a $\Big($ at point of load $\Big)$ . . . $= \dfrac{Pa^3b^3}{3EIl^3}$

Δ_x $\Big($ when $x < a\Big)$ . . . . . $= \dfrac{Pb^2x^2}{6EIl^3}(3al - 3ax - bx)$

---

3 - 136

BEAM DIAGRAMS AND FORMULAS
For various static loading conditions
For meaning of symbols, see page 3-127

---

18. CANTILEVER BEAM—LOAD INCREASING UNIFORMLY TO FIXED END

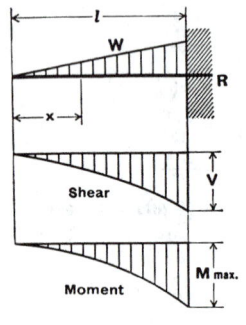

Total Equiv. Uniform Load . . . . $= \frac{8}{3}W$

$R = V$ . . . . $= W$

V_x . . . . $= W\frac{x^2}{l^2}$

M max. $\left(\text{at fixed end}\right)$ . . . . . $= \frac{Wl}{3}$

M_x . . . . $= \frac{Wx^3}{3l^2}$

Δmax. $\left(\text{at free end}\right)$ . . . . $= \frac{Wl^3}{15EI}$

Δ_x . . . . $= \frac{W}{60EIl^2}(x^5 - 5l^4x + 4l^5)$

---

19. CANTILEVER BEAM—UNIFORMLY DISTRIBUTED LOAD

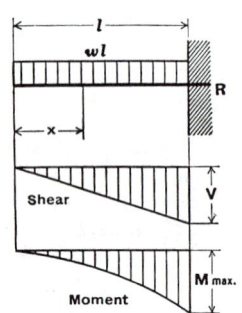

Total Equiv. Uniform Load . . . . $= 4wl$

$R = V$ . . . . $= wl$

V_x . . . . $= wx$

M max. $\left(\text{at fixed end}\right)$ . . . . . $= \frac{wl^2}{2}$

M_x . . . . $= \frac{wx^2}{2}$

Δmax. $\left(\text{at free end}\right)$ . . . . $= \frac{wl^4}{8EI}$

Δ_x . . . . $= \frac{w}{24EI}(x^4 - 4l^3x + 3l^4)$

---

20. BEAM FIXED AT ONE END, FREE TO DEFLECT VERTICALLY BUT NOT ROTATE AT OTHER—UNIFORMLY DISTRIBUTED LOAD

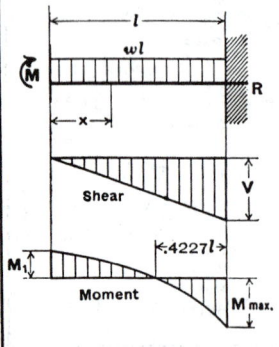

Total Equiv. Uniform Load . . . . $= \frac{8}{3}wl$

$R = V$ . . . . $= wl$

V_x . . . . $= wx$

M max. $\left(\text{at fixed end}\right)$ . . . . . $= \frac{wl^2}{3}$

M_1 $\left(\text{at deflected end}\right)$ . . . . $= \frac{wl^2}{6}$

M_x . . . . $= \frac{w}{6}(l^2 - 3x^2)$

Δmax. $\left(\text{at deflected end}\right)$ . . . . $= \frac{wl^4}{24EI}$

Δ_x . . . . $= \frac{w(l^2 - x^2)^2}{24EI}$

BEAM DIAGRAMS AND FORMULAS
For various static loading conditions
For meaning of symbols, see page 3-127

---

21. CANTILEVER BEAM—CONCENTRATED LOAD AT ANY POINT

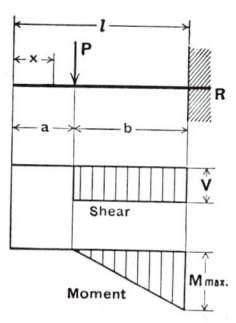

Total Equiv. Uniform Load . . . . $= \dfrac{8Pb}{l}$

$R = V$ . . . . $= P$

M max. $\left(\text{at fixed end}\right)$ . . . . . $= Pb$

M_x $\left(\text{when } x > a\right)$ . . . . . $= P(x - a)$

Δmax. $\left(\text{at free end}\right)$ . . . . . $= \dfrac{Pb^2}{6EI}(3l - b)$

Δa $\left(\text{at point of load}\right)$ . . . . $= \dfrac{Pb^3}{3EI}$

Δ_x $\left(\text{when } x < a\right)$ . . . . . $= \dfrac{Pb^2}{6EI}(3l - 3x - b)$

Δ_x $\left(\text{when } x > a\right)$ . . . . . $= \dfrac{P(l-x)^2}{6EI}(3b - l + x)$

---

22. CANTILEVER BEAM—CONCENTRATED LOAD AT FREE END

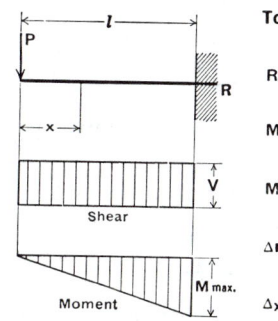

Total Equiv. Uniform Load . . . . $= 8P$

$R = V$ . . . . $= P$

M max. $\left(\text{at fixed end}\right)$ . . . . . $= Pl$

M_x . . . . $= Px$

Δmax. $\left(\text{at free end}\right)$ . . . . $= \dfrac{Pl^3}{3EI}$

Δ_x . . . . $= \dfrac{P}{6EI}(2l^3 - 3l^2x + x^3)$

---

23. BEAM FIXED AT ONE END, FREE TO DEFLECT VERTICALLY BUT NOT ROTATE AT OTHER—CONCENTRATED LOAD AT DEFLECTED END

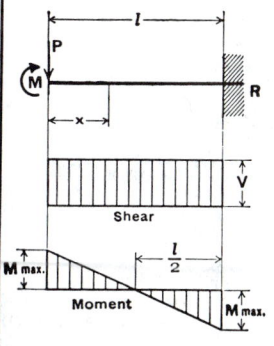

Total Equiv. Uniform Load . . . . $= 4P$

$R = V$ . . . . $= P$

M max. $\left(\text{at both ends}\right)$ . . . . $= \dfrac{Pl}{2}$

M_x . . . . $= P\left(\dfrac{l}{2} - x\right)$

Δmax. $\left(\text{at deflected end}\right)$ . . . . $= \dfrac{Pl^3}{12EI}$

Δ_x . . . . $= \dfrac{P(l-x)^2}{12EI}(l + 2x)$

BEAM DIAGRAMS AND FORMULAS
For various static loading conditions
For meaning of symbols, see page 3-127

---

24. BEAM OVERHANGING ONE SUPPORT—UNIFORMLY DISTRIBUTED LOAD

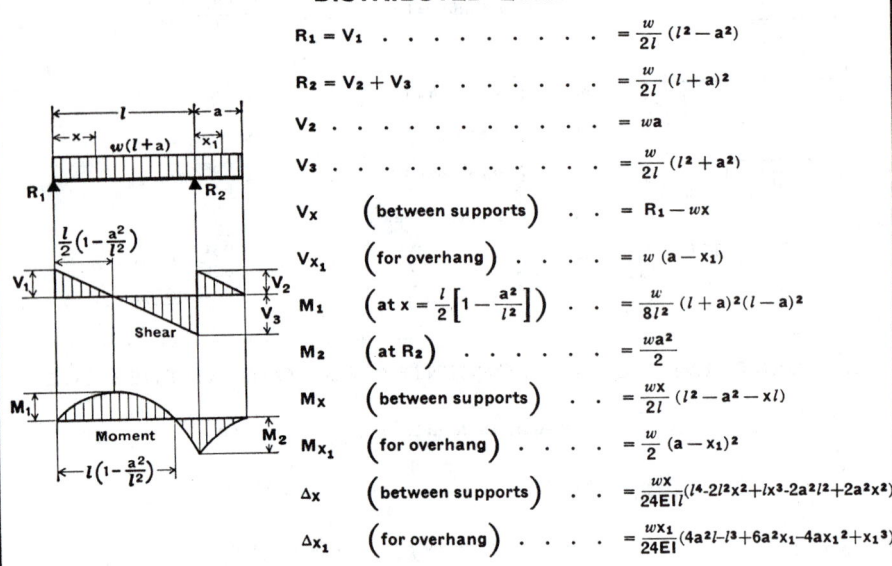

$$R_1 = V_1 \quad \cdots \cdots \cdots \cdots = \frac{w}{2l}(l^2 - a^2)$$

$$R_2 = V_2 + V_3 \quad \cdots \cdots \cdots = \frac{w}{2l}(l + a)^2$$

$$V_2 \quad \cdots \cdots \cdots \cdots \cdots = wa$$

$$V_3 \quad \cdots \cdots \cdots \cdots \cdots = \frac{w}{2l}(l^2 + a^2)$$

$$V_x \quad \left(\text{between supports}\right) \quad \cdots = R_1 - wx$$

$$V_{x_1} \quad \left(\text{for overhang}\right) \quad \cdots = w(a - x_1)$$

$$M_1 \quad \left(\text{at } x = \frac{l}{2}\left[1 - \frac{a^2}{l^2}\right]\right) \quad \cdots = \frac{w}{8l^2}(l + a)^2(l - a)^2$$

$$M_2 \quad \left(\text{at } R_2\right) \quad \cdots \cdots \cdots = \frac{wa^2}{2}$$

$$M_x \quad \left(\text{between supports}\right) \quad \cdots = \frac{wx}{2l}(l^2 - a^2 - xl)$$

$$M_{x_1} \quad \left(\text{for overhang}\right) \quad \cdots = \frac{w}{2}(a - x_1)^2$$

$$\Delta_x \quad \left(\text{between supports}\right) \quad = \frac{wx}{24EIl}(l^4 - 2l^2x^2 + lx^3 - 2a^2l^2 + 2a^2x^2)$$

$$\Delta_{x_1} \quad \left(\text{for overhang}\right) \quad \cdots = \frac{wx_1}{24EI}(4a^2l - l^3 + 6a^2x_1 - 4ax_1^2 + x_1^3)$$

---

25. BEAM OVERHANGING ONE SUPPORT—UNIFORMLY DISTRIBUTED LOAD ON OVERHANG

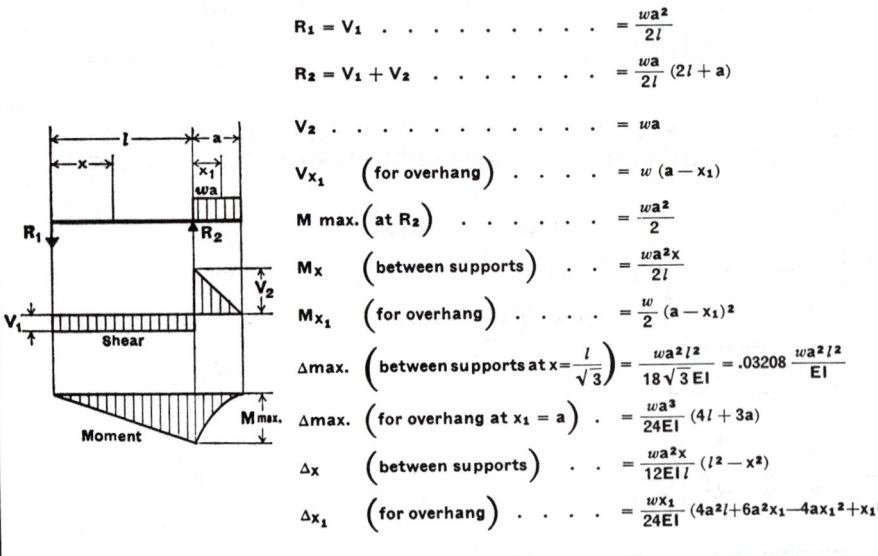

$$R_1 = V_1 \quad \cdots \cdots \cdots \cdots = \frac{wa^2}{2l}$$

$$R_2 = V_1 + V_2 \quad \cdots \cdots \cdots = \frac{wa}{2l}(2l + a)$$

$$V_2 \quad \cdots \cdots \cdots \cdots \cdots = wa$$

$$V_{x_1} \quad \left(\text{for overhang}\right) \quad \cdots = w(a - x_1)$$

$$M \text{ max.} \left(\text{at } R_2\right) \quad \cdots \cdots \cdots = \frac{wa^2}{2}$$

$$M_x \quad \left(\text{between supports}\right) \quad \cdots = \frac{wa^2x}{2l}$$

$$M_{x_1} \quad \left(\text{for overhang}\right) \quad \cdots = \frac{w}{2}(a - x_1)^2$$

$$\Delta \text{max.} \left(\text{between supports at } x = \frac{l}{\sqrt{3}}\right) = \frac{wa^2l^2}{18\sqrt{3}EI} = .03208\frac{wa^2l^2}{EI}$$

$$\Delta \text{max.} \left(\text{for overhang at } x_1 = a\right) \quad = \frac{wa^3}{24EI}(4l + 3a)$$

$$\Delta_x \quad \left(\text{between supports}\right) \quad \cdots = \frac{wa^2x}{12EIl}(l^2 - x^2)$$

$$\Delta_{x_1} \quad \left(\text{for overhang}\right) \quad \cdots = \frac{wx_1}{24EI}(4a^2l + 6a^2x_1 - 4ax_1^2 + x_1^3)$$

BEAM DIAGRAMS AND FORMULAS
For various static loading conditions
For meaning of symbols, see page 3-127

---

26. BEAM OVERHANGING ONE SUPPORT—CONCENTRATED LOAD AT END OF OVERHANG

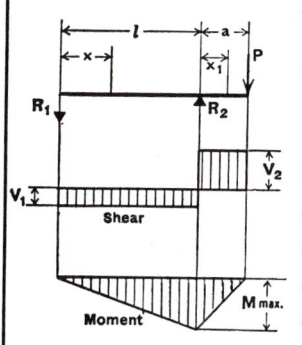

$$R_1 = V_1 \ldots \ldots \ldots \ldots = \frac{Pa}{l}$$

$$R_2 = V_1 + V_2 \ldots \ldots \ldots = \frac{P}{l}(l+a)$$

$$V_2 \ldots \ldots \ldots \ldots \ldots = P$$

$$M \text{ max.} \left(\text{at } R_2 \right) \ldots \ldots \ldots = Pa$$

$$M_x \left(\text{between supports} \right) \ldots = \frac{Pax}{l}$$

$$M_{x_1} \left(\text{for overhang} \right) \ldots \ldots = P(a-x_1)$$

$$\Delta\text{max.} \left(\text{between supports at } x = \frac{l}{\sqrt{3}} \right) = \frac{Pal^2}{9\sqrt{3}\,EI} = .06415\frac{Pal^2}{EI}$$

$$\Delta\text{max.} \left(\text{for overhang at } x_1 = a \right) = \frac{Pa^2}{3EI}(l+a)$$

$$\Delta_x \left(\text{between supports} \right) \ldots = \frac{Pax}{6EIl}(l^2-x^2)$$

$$\Delta_{x_1} \left(\text{for overhang} \right) \ldots \ldots = \frac{Px_1}{6EI}(2al+3ax_1-x_1^2)$$

---

27. BEAM OVERHANGING ONE SUPPORT—UNIFORMLY DISTRIBUTED LOAD BETWEEN SUPPORTS

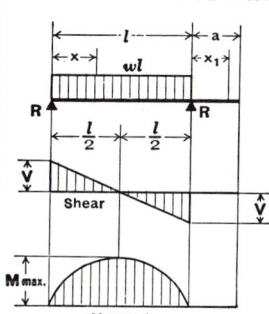

$$\text{Total Equiv. Uniform Load} \ldots = wl$$

$$R = V \ldots \ldots \ldots \ldots \ldots = \frac{wl}{2}$$

$$V_x \ldots \ldots \ldots \ldots \ldots \ldots = w\left(\frac{l}{2}-x\right)$$

$$M \text{ max.} \left(\text{at center} \right) \ldots \ldots = \frac{wl^2}{8}$$

$$M_x \ldots \ldots \ldots \ldots \ldots \ldots = \frac{wx}{2}(l-x)$$

$$\Delta\text{max.} \left(\text{at center} \right) \ldots \ldots = \frac{5wl^4}{384EI}$$

$$\Delta_x \ldots \ldots \ldots \ldots \ldots \ldots = \frac{wx}{24EI}(l^3-2lx^2+x^3)$$

$$\Delta_{x_1} \ldots \ldots \ldots \ldots \ldots \ldots = \frac{wl^3 x_1}{24EI}$$

---

28. BEAM OVERHANGING ONE SUPPORT—CONCENTRATED LOAD AT ANY POINT BETWEEN SUPPORTS

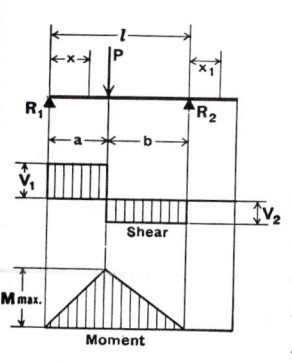

$$\text{Total Equiv. Uniform Load} \ldots = \frac{8Pab}{l^2}$$

$$R_1 = V_1 \left(\text{max. when } a < b \right) \ldots = \frac{Pb}{l}$$

$$R_2 = V_2 \left(\text{max. when } a > b \right) \ldots = \frac{Pa}{l}$$

$$M \text{ max.} \left(\text{at point of load} \right) \ldots = \frac{Pab}{l}$$

$$M_x \left(\text{when } x < a \right) \ldots \ldots = \frac{Pbx}{l}$$

$$\Delta\text{max.} \left(\text{at } x = \sqrt{\frac{a(a+2b)}{3}} \text{ when } a > b \right) = \frac{Pab(a+2b)\sqrt{3a(a+2b)}}{27EIl}$$

$$\Delta a \left(\text{at point of load} \right) \ldots = \frac{Pa^2 b^2}{3EIl}$$

$$\Delta_x \left(\text{when } x < a \right) \ldots \ldots = \frac{Pbx}{6EIl}(l^2-b^2-x^2)$$

$$\Delta_x \left(\text{when } x > a \right) \ldots \ldots = \frac{Pa(l-x)}{6EIl}(2lx-x^2-a^2)$$

$$\Delta_{x_1} \ldots \ldots \ldots \ldots \ldots \ldots = \frac{Pabx_1}{6EIl}(l+a)$$

BEAM DIAGRAMS AND FORMULAS
For various static loading conditions
For meaning of symbols, see page 3-127

---

29. CONTINUOUS BEAM—TWO EQUAL SPANS—UNIFORM LOAD ON ONE SPAN

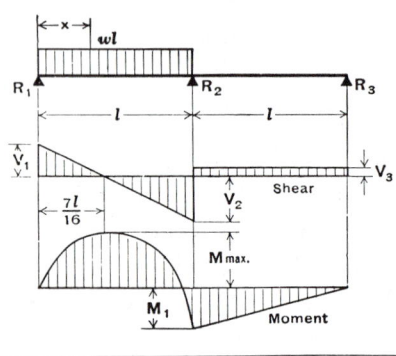

Total Equiv. Uniform Load $= \frac{49}{64} wl$

$R_1 = V_1 \ldots = \frac{7}{16} wl$

$R_2 = V_2 + V_3 \ldots = \frac{5}{8} wl$

$R_3 = V_3 \ldots = -\frac{1}{16} wl$

$V_2 \ldots = \frac{9}{16} wl$

M max. $\left(\text{at } x = \frac{7}{16} l\right) = \frac{49}{512} wl^2$

$M_1 \left(\text{at support } R_2\right) = \frac{1}{16} wl^2$

$M_x \left(\text{when } x < l\right) = \frac{wx}{16}(7l - 8x)$

Δ Max. (0.472 l from R_1) $= 0.0092\, wl^4/EI$

---

30. CONTINUOUS BEAM—TWO EQUAL SPANS—CONCENTRATED LOAD AT CENTER OF ONE SPAN

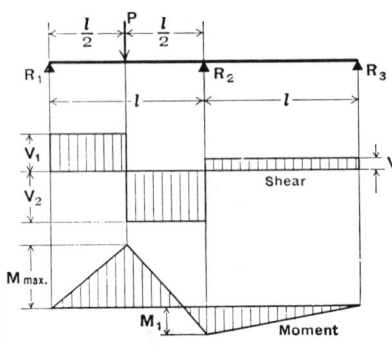

Total Equiv. Uniform Load $= \frac{13}{8} P$

$R_1 = V_1 \ldots = \frac{13}{32} P$

$R_2 = V_2 + V_3 \ldots = \frac{11}{16} P$

$R_3 = V_3 \ldots = -\frac{3}{32} P$

$V_2 \ldots = \frac{19}{32} P$

M max. $\left(\text{at point of load}\right) = \frac{13}{64} Pl$

$M_1 \left(\text{at support } R_2\right) = \frac{3}{32} Pl$

Δ Max. (0.480 l from R_1) $= 0.015\, Pl^3/EI$

---

31. CONTINUOUS BEAM—TWO EQUAL SPANS—CONCENTRATED LOAD AT ANY POINT

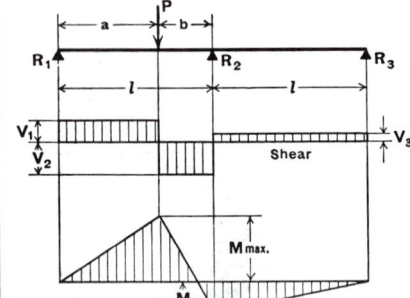

$R_1 = V_1 \ldots = \frac{Pb}{4l^3}\left(4l^2 - a(l+a)\right)$

$R_2 = V_2 + V_3 \ldots = \frac{Pa}{2l^3}\left(2l^2 + b(l+a)\right)$

$R_3 = V_3 \ldots = -\frac{Pab}{4l^3}(l+a)$

$V_2 \ldots = \frac{Pa}{4l^3}\left(4l^2 + b(l+a)\right)$

M max. $\left(\text{at point of load}\right) = \frac{Pab}{4l^3}\left(4l^2 - a(l+a)\right)$

$M_1 \left(\text{at support } R_2\right) = \frac{Pab}{4l^2}(l+a)$

BEAM DIAGRAMS AND FORMULAS
For various static loading conditions
For meaning of symbols, see page 3-127

---

32. BEAM—UNIFORMLY DISTRIBUTED LOAD AND VARIABLE END MOMENTS

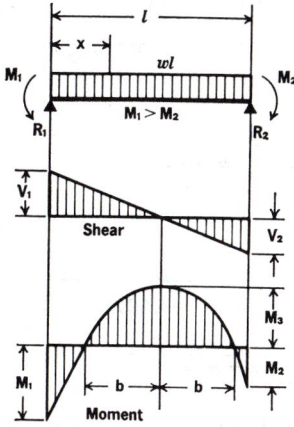

$$R_1 = V_1 = \frac{wl}{2} + \frac{M_1 - M_2}{l}$$

$$R_2 = V_2 = \frac{wl}{2} - \frac{M_1 - M_2}{l}$$

$$V_x = w\left(\frac{l}{2} - x\right) + \frac{M_1 - M_2}{l}$$

$$M_3 \left(\text{at } x = \frac{l}{2} + \frac{M_1 - M_2}{wl}\right)$$

$$= \frac{wl^2}{8} - \frac{M_1 + M_2}{2} + \frac{(M_1 - M_2)^2}{2wl^2}$$

$$M_x = \frac{wx}{2}(l - x) + \left(\frac{M_1 - M_2}{l}\right) x - M_1$$

$$b\binom{\text{To locate}}{\text{inflection points}} = \sqrt{\frac{l^2}{4} - \left(\frac{M_1 + M_2}{w}\right) + \left(\frac{M_1 - M_2}{wl}\right)^2}$$

$$\Delta_x = \frac{wx}{24EI}\left[x^3 - \left(2l + \frac{4M_1}{wl} - \frac{4M_2}{wl}\right)x^2 + \frac{12M_1}{w}x + l^3 - \frac{8M_1 l}{w} - \frac{4M_2 l}{w}\right]$$

---

33. BEAM—CONCENTRATED LOAD AT CENTER AND VARIABLE END MOMENTS

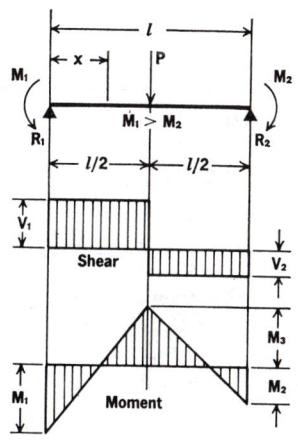

$$R_1 = V_1 = \frac{P}{2} + \frac{M_1 - M_2}{l}$$

$$R_2 = V_2 = \frac{P}{2} - \frac{M_1 - M_2}{l}$$

$$M_3 \text{ (At center)} = \frac{Pl}{4} - \frac{M_1 + M_2}{2}$$

$$M_x \left(\text{When } x < \frac{l}{2}\right) = \left(\frac{P}{2} + \frac{M_1 - M_2}{l}\right) x - M_1$$

$$M_x \left(\text{When } x > \frac{l}{2}\right) = \frac{P}{2}(l - x) + \frac{(M_1 - M_2)x}{l} - M_1$$

$$\Delta_x \left(\text{When } x < \frac{l}{2}\right) = \frac{Px}{48EI}\left(3l^2 - 4x^2 - \frac{8(l - x)}{Pl}[M_1(2l - x) + M_2(l + x)]\right)$$

BEAM DIAGRAMS AND FORMULAS
For various static loading conditions
For meaning of symbols, see page 3-127

34.   CONTINUOUS BEAM—THREE EQUAL SPANS—ONE END SPAN UNLOADED

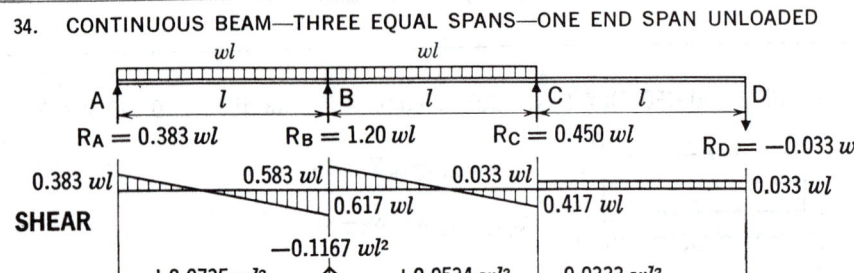

$R_A = 0.383\ wl$ $R_B = 1.20\ wl$ $R_C = 0.450\ wl$ $R_D = -0.033\ wl$

SHEAR

0.383 wl   0.583 wl   0.033 wl   0.033 wl
0.617 wl   0.417 wl

$-0.1167\ wl^2$

$+0.0735\ wl^2$ $+0.0534\ wl^2$ $-0.0333\ wl^2$

MOMENT

0.383 l   0.583 l

△ **Max. (0.430 l from A) = 0.0059** wl^4/EI

35.   CONTINUOUS BEAM—THREE EQUAL SPANS—END SPANS LOADED

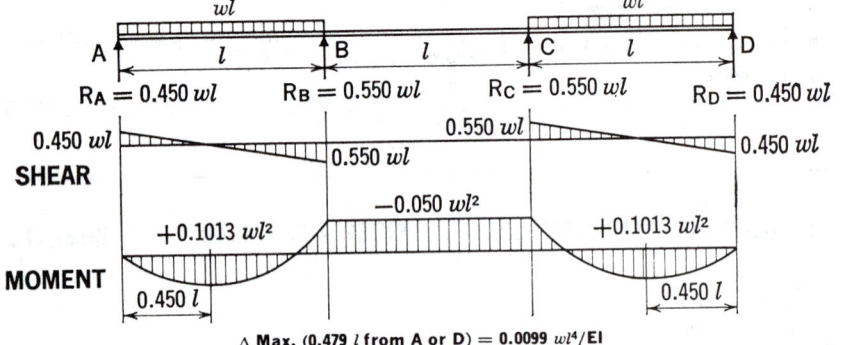

$R_A = 0.450\ wl$ $R_B = 0.550\ wl$ $R_C = 0.550\ wl$ $R_D = 0.450\ wl$

0.450 wl   0.550 wl   0.450 wl

SHEAR

0.550 wl

$-0.050\ wl^2$

$+0.1013\ wl^2$ $+0.1013\ wl^2$

MOMENT

0.450 l   0.450 l

△ **Max. (0.479 l from A or D) = 0.0099** wl^4/EI

36.   CONTINUOUS BEAM—THREE EQUAL SPANS—ALL SPANS LOADED

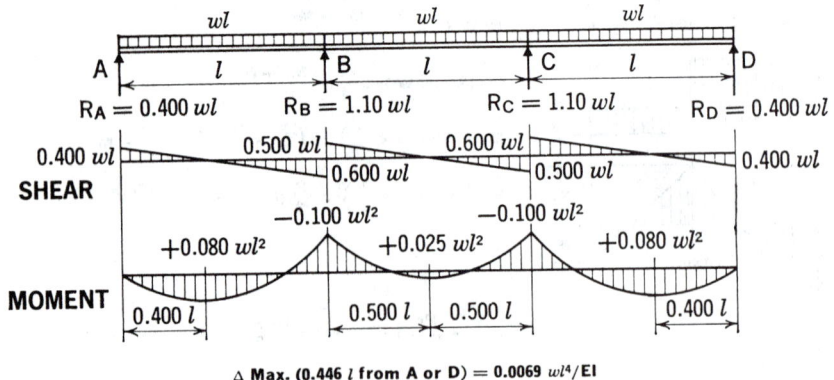

$R_A = 0.400\ wl$ $R_B = 1.10\ wl$ $R_C = 1.10\ wl$ $R_D = 0.400\ wl$

0.400 wl   0.500 wl   0.600 wl   0.400 wl
0.600 wl   0.500 wl

SHEAR

$-0.100\ wl^2$ $-0.100\ wl^2$

$+0.080\ wl^2$ $+0.025\ wl^2$ $+0.080\ wl^2$

MOMENT

0.400 l   0.500 l   0.500 l   0.400 l

△ **Max. (0.446 l from A or D) = 0.0069** wl^4/EI

BEAM DIAGRAMS AND FORMULAS
For various static loading conditions

For meaning of symbols, see page 3-127

37. CONTINUOUS BEAM—FOUR EQUAL SPANS—THIRD SPAN UNLOADED

Δ Max. (0.475 l from E) = 0.0094 wl^4/EI

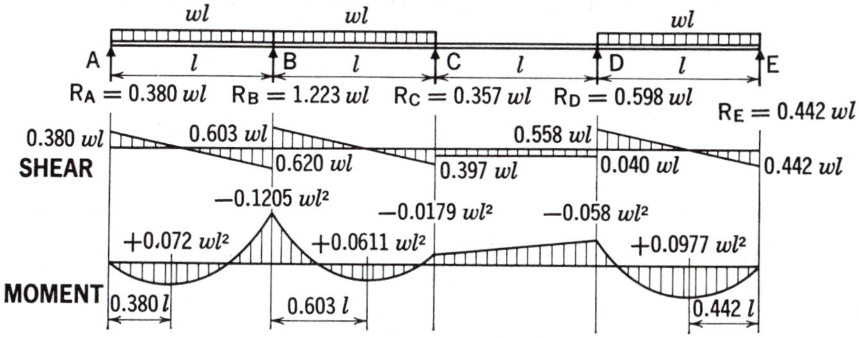

38. CONTINUOUS BEAM—FOUR EQUAL SPANS—LOAD FIRST AND THIRD SPANS

Δ Max. (0.477 l from A) = 0.0097 wl^4/EI

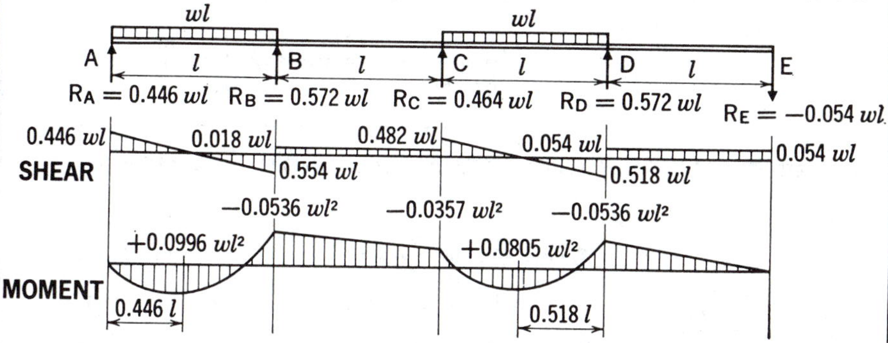

39. CONTINUOUS BEAM—FOUR EQUAL SPANS—ALL SPANS LOADED

Δ Max. (0.440 l from A and E) = 0.0065 wl^4/EI

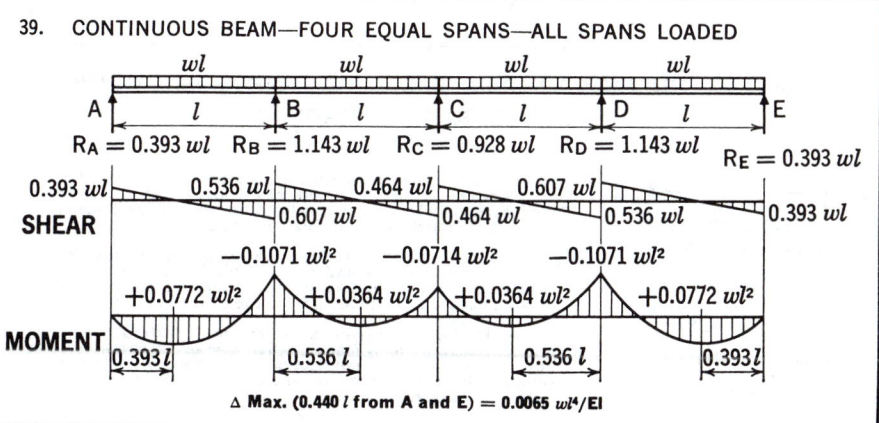

BEAM DIAGRAMS AND FORMULAS
For various concentrated moving loads

The values given in these formulas do not include impact which varies according to the requirements of each case. For meaning of symbols, see page 3-127

---

40. SIMPLE BEAM—ONE CONCENTRATED MOVING LOAD

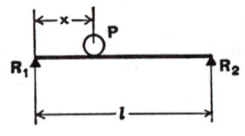

$$R_1 \text{ max.} = V_1 \text{ max.} \left(\text{at } x = 0 \right) \quad \ldots \quad = P$$

$$M \text{ max.} \left(\text{at point of load, when } x = \frac{l}{2} \right) \quad . \quad = \frac{Pl}{4}$$

---

41. SIMPLE BEAM—TWO EQUAL CONCENTRATED MOVING LOADS

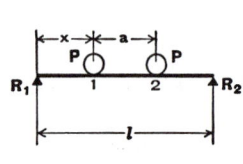

$$R_1 \text{ max.} = V_1 \text{ max.} \left(\text{at } x = 0 \right) \quad \ldots \quad = P \left(2 - \frac{a}{l} \right)$$

$$M \text{ max.} \begin{cases} \begin{bmatrix} \text{when } a < (2 - \sqrt{2})\, l = .586 l \\ \text{under load 1 at } x = \frac{1}{2} \left(l - \frac{a}{2} \right) \end{bmatrix} = \frac{P}{2l} \left(l - \frac{a}{2} \right)^2 \\[2em] \begin{bmatrix} \text{when } a > (2 - \sqrt{2})\, l = .586 l \\ \text{with one load at center of span} \\ \text{(case 40)} \end{bmatrix} \frac{Pl}{4} \end{cases}$$

---

42. SIMPLE BEAM—TWO UNEQUAL CONCENTRATED MOVING LOADS

$$P_1 > P_2$$

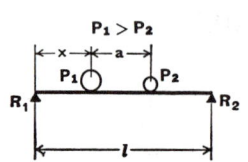

$$R_1 \text{ max.} = V_1 \text{ max.} \left(\text{at } x = 0 \right) \quad \ldots \quad = P_1 + P_2 \frac{l - a}{l}$$

$$M \text{ max.} \begin{cases} \begin{bmatrix} \text{under } P_1, \text{ at } x = \frac{1}{2} \left(l - \frac{P_2 a}{P_1 + P_2} \right) \end{bmatrix} = \left(P_1 + P_2 \right) \frac{x^2}{l} \\[2em] \begin{bmatrix} \text{M max. may occur with larger} \\ \text{load at center of span and other} \\ \text{load off span (case 40)} \end{bmatrix} = \frac{P_1 l}{4} \end{cases}$$

GENERAL RULES FOR SIMPLE BEAMS CARRYING MOVING CONCENTRATED LOADS

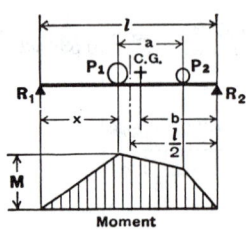

Moment

The maximum shear due to moving concentrated loads occurs at one support when one of the loads is at that support. With several moving loads, the location that will produce maximum shear must be determined by trial.

The maximum bending moment produced by moving concentrated loads occurs under one of the loads when that load is as far from one support as the center of gravity of all the moving loads on the beam is from the other support.

In the accompanying diagram, the maximum bending moment occurs under load P_1 when $x = b$. It should also be noted that this condition occurs when the center line of the span is midway between the center of gravity of loads and the nearest concentrated load.

BEAM DIAGRAMS AND FORMULAS
Design properties of cantilevered beams
Equal loads, equally spaced

No. Spans	System
2	
3	
4	
5	
≥6 (even)	
≥7 (odd)	

n	∞	2	3	4	5
Typical Span Loading	P	P/2 P P/2	P/2 P P P/2	P/2 P P P P/2	P/2 P P P P P/2

		∞	2	3	4	5
Moments	M_1	0.086 × PL	0.167 × PL	0.250 × PL	0.333 × PL	0.429 × PL
	M_2	0.096 × PL	0.188 × PL	0.278 × PL	0.375 × PL	0.480 × PL
	M_3	0.063 × PL	0.125 × PL	0.167 × PL	0.250 × PL	0.300 × PL
	M_4	0.039 × PL	0.083 × PL	0.083 × PL	0.167 × PL	0.171 × PL
	M_5	0.051 × PL	0.104 × PL	0.139 × PL	0.208 × PL	0.249 × PL
Reactions	A	0.414 × P	0.833 × P	1.250 × P	1.667 × P	2.071 × P
	B	1.172 × P	2.333 × P	3.500 × P	4.667 × P	5.857 × P
	C	0.438 × P	0.875 × P	1.333 × P	1.750 × P	2.200 × P
	D	1.063 × P	2.125 × P	3.167 × P	4.250 × P	5.300 × P
	E	1.086 × P	2.167 × P	3.250 × P	4.333 × P	5.429 × P
	F	1.109 × P	2.208 × P	3.333 × P	4.417 × P	5.557 × P
	G	0.977 × P	1.958 × P	2.917 × P	3.917 × P	4.871 × P
	H	1.000 × P	2.000 × P	3.000 × P	4.000 × P	5.000 × P
Cantilever Dimensions	a	0.172 × L	0.250 × L	0.200 × L	0.182 × L	0.176 × L
	b	0.125 × L	0.200 × L	0.143 × L	0.143 × L	0.130 × L
	c	0.220 × L	0.333 × L	0.250 × L	0.222 × L	0.229 × L
	d	0.204 × L	0.308 × L	0.231 × L	0.211 × L	0.203 × L
	e	0.157 × L	0.273 × L	0.182 × L	0.176 × L	0.160 × L
	f	0.147 × L	0.250 × L	0.167 × L	0.167 × L	0.150 × L

AMERICAN INSTITUTE OF STEEL CONSTRUCTION

PART 4
Composite Design

Notes

COMPOSITE BEAMS

GENERAL NOTES

The Composite Beam Tables can be used for the design and analysis of simple composite steel beams. Values for the design flexural strength ϕM_n for rolled I-shaped beams with yield strengths of 36 ksi and 50 ksi are tabulated, as well as lower bound moments of inertia. The values tabulated are independent of the concrete flange properties. The strength evaluation of the concrete flange portion of the composite section is left to the design engineer. The preparation of these tables is based upon the fact that the location of the plastic neutral axis (PNA) is uniquely determined by the horizontal shear force ΣQ_n at the interface between the steel section and the concrete slab. With the knowledge of the location of the PNA and the distance to the centroid of the concrete flange force ΣQ_n, the design flexural strengths for the rolled section ϕM_n can be computed.

DESIGN FLEXURAL STRENGTH (POSITIVE)

The design flexural strength of simple steel beams with composite concrete flanges is computed from the equilibrium of internal forces using the plastic stress distribution as shown in Fig. 4.1:

$$\phi M_n = \phi T_{Tot} y = \phi C_{Tot} y$$

where

ϕ = 0.85
T_{Tot} = sum of tensile forces = F_y (tensile force beam area)
C_{Tot} = sum of compressive forces = concrete flange force + F_y (compressive force beam area)
y = moment arm between centroid of tensile force and the resultant compressive force.

The Commentary Sect. I3 of the LRFD Specification contains a summary of the formulas required for analysis and design of composite steel beam and concrete slab

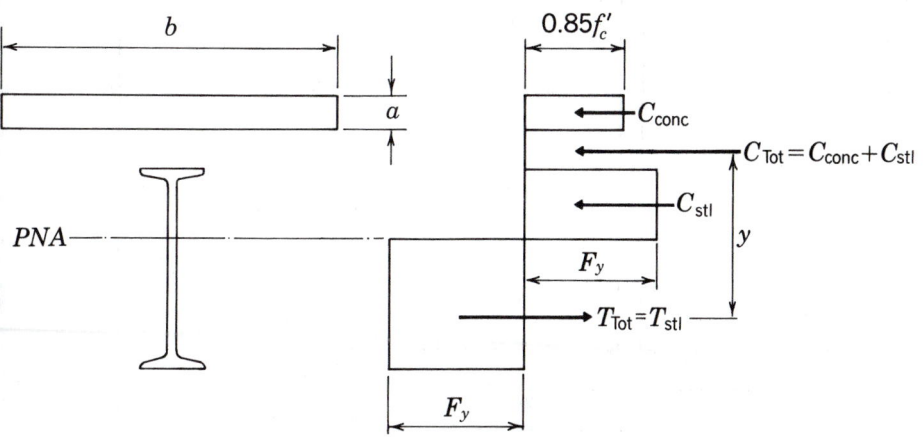

Fig. 4.1. Plastic stress distribution

AMERICAN INSTITUTE OF STEEL CONSTRUCTION

sections. The model of the steel beam (Fig. 4.2) used in the calculation of design strengths differs slightly from that used in the formulas contained in the Commentary. A summary of the model properties follows:

A_s = cross-sectional area of steel section, in.^{2}
A_f = flange area = $b_f \times t_f$, in.^{2}
A_w = web area = $(d - 2k)t_w$, in.^{2}
K_{dep} = $k - t_f$, in.
K_{area} = $(A_s - 2A_f - A_w)/2$, in.^{2}

Limitations for tabulated values include the following:

$(d - 2k)/t_w \leq 640/\sqrt{F_y}$

and

ΣQ_n (min.) = $0.25 A_s F_y$

The limitation of ΣQ_n (min.) is not required by the specification, but is deemed to be a practical minimum value. Design strength moment values are tabulated for plastic neutral axis (PNA) locations at top and intermediate quarter points of the steel beam top flange. In addition, PNA locations are computed at the point where ΣQ_n equals $0.25 A_s F_y$, and the point where ΣQ_n is the average of the minimum value ($0.25 A_s F_y$) and the value of ΣQ_n when the PNA is at the bottom of the top flange (see Fig. 4.3). To use the tables, select a valid value of ΣQ_n, determine the appropriate value of $Y2$ and read the design flexural strength moment ϕM_n directly. Values for $Y1$ are also

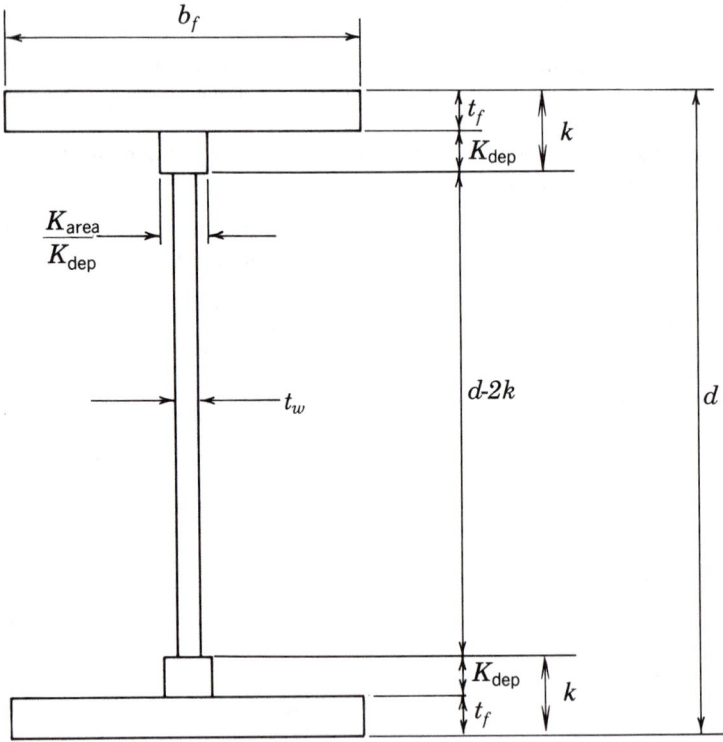

Fig. 4.2. *Composite beam model*

AMERICAN INSTITUTE OF STEEL CONSTRUCTION

tabulated for convenience. The parameters $Y1$ and $Y2$ are defined as follows:

> $Y1$ = distance from PNA to beam top flange
> $Y2$ = distance from concrete flange force to beam top flange

Valid values for ΣQ_n are the smaller of the following three expressions (LRFD Specification Sect. I5):

> $0.85 f'_c A_c$
> $A_s F_y$
> $n Q_n$

where

> f'_c = specified compressive strength of concrete, ksi
> A_c = area of concrete slab within effective width, in.^{2}
> A_s = area of steel cross section, in.^{2}
> F_y = specified yield stress, ksi
> n = number of shear connectors between point of maximum positive moment and the point of zero moment to either side
> Q_n = shear capacity of single shear connector, kips

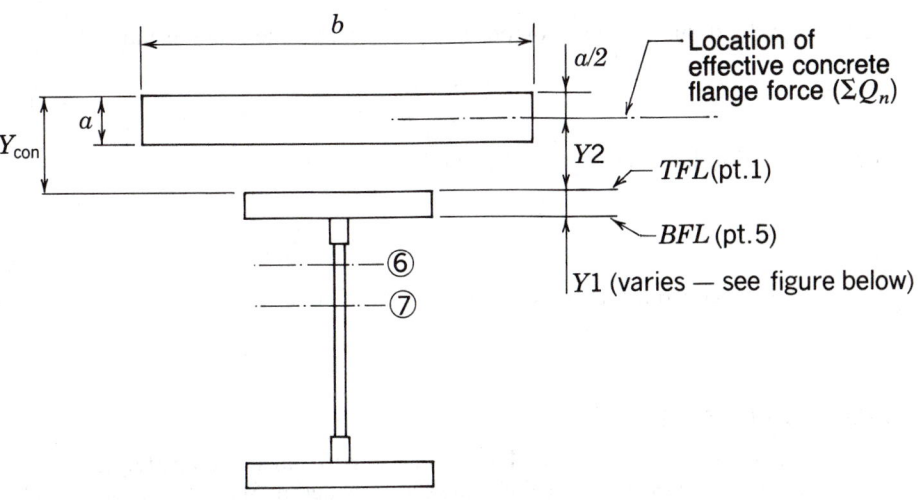

$Y1$ = Distance from top of steel flange to any of the seven tabulated PNA locations.

$$\Sigma Q_n (@ \text{ point } ⑥) = \frac{\Sigma Q_n(@ \text{pt. 5}) + \Sigma Q_n(@ \text{ pt.7})}{2}$$

$$\Sigma Q_n (@ \text{ point } ⑦) = .25 A_s F_y$$

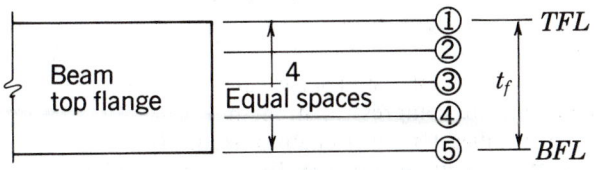

PNA FLANGE LOCATIONS

Fig. 4.3. Composite beam table parameters

CONCRETE FLANGE

According to LRFD Specification Sect. I3.1 the effective width of the concrete slab on each side of the beam centerline shall not exceed:

    a. one-eighth of the beam span, center to center, of supports;
    b. one-half the distance to the centerline of the adjacent beams; or
    c. the distance to the edge of the slab.

The maximum concrete flange force is equal to $0.85 f_c' A_c$ where A_c is based on the actual slab thickness, t_c. However, often the maximum concrete flange force exceeds the maximum capacity of the specified steel beam. In that case, the effective concrete flange force is determined from a value of ΣQ_n, which will be the smaller of $A_s F_y$ or nQ_n. That is, the effective concrete flange force is:

$$\Sigma Q_n = C_{conc} = 0.85 f_c' (b) (a)$$

where

    C_{conc} = effective concrete flange force, kips
    b = effective concrete flange width, in.
    a = effective concrete flange thickness, in.

The basis of the design of most composite beams will be the relationship:

$$a = \frac{\Sigma Q_n}{0.85 f_c' b} \qquad \text{(C-I3-4)}$$

From this relationship, the value of $Y2$ can be computed as

$$Y2 = Y_{con} - a/2.$$

where

    Y_{con} = distance from top of steel beam to top of concrete, in.

SHEAR CONNECTORS

Shear connectors shall be headed steel studs not less than four stud diameters in length after installation, or hot-rolled steel channels as specified in Sect. A3 of the LRFD Specification. Shear connectors shall be embedded in concrete slabs made with ASTM C33 aggregate or with rotary kiln produced aggregates conforming to ASTM C330, with concrete unit weight not less than 90 pcf.

    The nominal strength of one stud shear connector embedded in a solid concrete slab is:

$$Q_n = 0.5 A_{sc} \sqrt{f_c' E_c} \leq A_{sc} F_u \qquad \text{(I5-1)}$$

where

    A_{sc} = cross-sectional area of a stud shear connector, in.^{2}
    f_c' = specified concrete compressive strength, ksi
    F_u = minimum specified tensile strength of stud, ksi
    E_c = modulus of elasticity of concrete, ksi
        = $(w^{1.5}) \sqrt{f_c'}$
    w = unit weight of concrete, lbs./cu. ft

The nominal shear strengths of ¾-in. headed studs embedded in concrete slabs are listed in Table 4.1

TABLE 4.1
Nominal Stud Shear Strength, Q_n (kips) for ¾-in. Headed Studs

f_c'	w	Q_n
Ksi	Lbs/cu ft	Kips
3.0	115	17.7
3.0	145	21.0
3.5	115	19.8
3.5	145	23.6
4.0	115	21.9
4.0	145	26.1

Note, the effective shear capacity of studs used in conjunction with composite or non-composite metal forms may be affected by the shape of the deck and spacing of the studs. See LRFD Specification Sect. I3.5 and I5.6.

STRENGTH DURING CONSTRUCTION

When temporary shores are not used during construction, the steel section must have sufficient strength to support the applied loads prior to the concrete attaining 75% of the specified concrete strength f_c' (LRFD Specification Sect. I3.4). The effect of deflection on unshored steel beams during construction should be considered.

LATERAL SUPPORT

Adequate lateral support for the compression flange of the steel section will be provided by the concrete slab after hardening. During construction, however, lateral support must be provided, or the design strength must be reduced in accordance with Sect. F1 of the LRFD Specification. Steel deck with adequate attachment to the compression flange, or properly constructed concrete forms will usually provide the necessary lateral support. For construction using fully encased beams, particular attention should be given to lateral support during construction.

DESIGN SHEAR STRENGTH

The design shear strength of composite beams is determined by the strength of the steel web, in accordance with the requirements of Sect. F2 of the LRFD Specification.

LOWER BOUND MOMENT OF INERTIA

A table of lower bound moments of inertia of composite sections is included to assist in the evaluation of deflection with regard to serviceability. If calculated deflections using the lower bound moment of inertia are acceptable, a complete elastic analysis of the composite section may be avoided.

The lower bound moment of inertia is based on the area of the beam and an equivalent concrete area of $\Sigma Q_n/F_y$. The analysis includes only the horizontal shear force transferred by the shear connectors supplied; and thus, neglects the contribution of the concrete flange not considered in the plastic distribution of forces (see Fig. 4.4). The value for the lower bound moment of inertia can be calculated as follows:

$$I_{LB} = I_x + A_s \left(Y_{ENA} - \frac{d}{2}\right)^2 + \left(\frac{\Sigma Q_n}{F_y}\right)(d + Y2 - Y_{ENA})^2$$

where

Y_{ENA} = distance from bottom of beam to elastic neutral axis (ENA).

$$= \frac{\left[\dfrac{A_s d}{2} + \left(\dfrac{\Sigma Q_n}{F_y}\right)(d + Y2)\right]}{\left[A_s + \left(\dfrac{\Sigma Q_n}{F_y}\right)\right]}$$

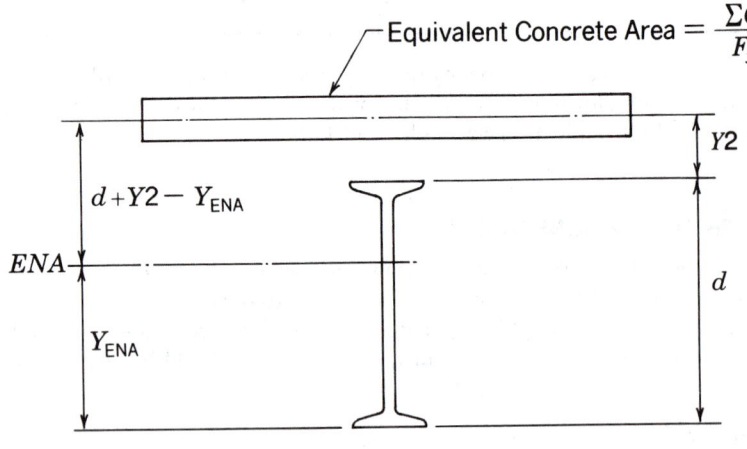

Fig. 4.4. Moment of inertia

COMPOSITE BEAM REACTIONS

Design reactions for symmetrically loaded composite beams may be computed using the Composite Beam Tables. Two situations will be considered. First, an upper bound value for a beam reaction may be computed neglecting the composite concrete flange properties other than concrete strength. Second, a more refined value for a beam reaction can be computed if the properties of the composite concrete flange are determined initially.

When the properties of the composite concrete flange have not been computed, a conservative value for the maximum horizontal shear between the composite concrete slab and the steel section (ΣQ_n) may be taken as the smaller of $A_s F_y$ or nQ_n. Here, n is the number of headed studs between the reaction point and point of maximum moment. The value of Q_n may be taken from Table 4.1 or determined from LRFD Specification Sect. I5. A value for ϕM_n of the composite section may be obtained from the Composite Beam Tables using the sum of horizontal shear ΣQ_n as described above. In this case, $Y2$ is defined as the distance from the top of the steel beam to the top of the concrete slab. The design reaction may be determined from the value of ϕM_n as discussed in the following paragraph.

When the properties of the concrete flange have been computed (effective width and depth), a slightly different method is used to find ϕM_n. The stud efficiency can be determined in accordance with Sect. I5 of the LRFD Specification, or Table 4.1 can be used for ¾-in. dia. stud shear connectors. The value for the sum of the horizontal shear force ΣQ_n can be taken as the smaller of nQ_n, $A_s F_y$ or $0.85f_c'A_c$, where f_c' is the concrete cylinder strength (ksi), and A_c is the maximum permitted concrete flange area (LRFD Specification Sect. I5.2). The distance $Y2$ is the distance from the top of the steel beam to the top of the concrete slab less $[\Sigma Q_n/(0.85f_c'b)]/2$. Using these values for ΣQ_n and $Y2$, the value for ϕM_n can be selected from the Composite Beam Tables.

The design beam reaction for a symmetrically loaded composite beam may be computed from known values of ϕM_n and the span length as

$$R = C_c \phi M_n / L$$

where

R = design beam reaction, kips
C_c = coefficient from Fig. 4.5
ϕM_n = composite beam flexural design strength, kip-ft
L = span length, ft.

PRELIMINARY SECTION SELECTION

When using the Composite Beam Tables, the approximate beam weight required for several different beam depths may be calculated as follows:

$$\text{Beam weight} = \left[\frac{M_u(12)}{(d/2 + Y_{con} - a/2)\phi F_y}\right] 3.4, \text{lbs}$$

where

M_u = required flexural strength, kip-ft
d = beam depth, in.
Y_{con} = distance from top of steel beam to top of concrete slab, in.
a = effective concrete slab thickness, in.
F_y = steel yield stress, ksi
ϕ = 0.85
3.4 = ratio of the weight of a beam to its area, lb/in.^{2}

For convenience in the preliminary selection phase the nominal section depth may be used. A value for $a/2$ must also be selected. For relatively light sections and loads, this value can be assumed to be 1 in. With the PNA at the top of the steel beam, i.e. $\Sigma Q_n = A_s F_y$, the flexural design strength is:

$$\phi M_n = \phi A_s F_y (d/2 + Y_{con} - a/2)/12$$

where

ϕM_n = flexural design strength, kip-ft
A_s = steel beam cross-sectional area, in.^{2}

DESIGN EXAMPLES

Example 1:

Given:

Determine the beam, with F_y = 36 ksi, required to support a service live load of

1.0 kips/ft and a service dead load of 0.65 kips/ft. The beam span and spacing is 30 ft and 10 ft. The concrete slab is 3¼-in. l.w. concrete (f'_c = 3.5 ksi, 115 pcf) supported by a 3-in. deep composite metal deck with an average rib width of 6 in. and a rib depth of 3 in. The ribs are oriented perpendicular to the beam centerline. Shored construction is specified. Also determine the number of ¾-in. dia. headed studs required and the service live load deflection.

Solution:

A. Load tabulation:

	Service load (kips/ft)	(L.F.)	Factored load (kips/ft)
LL	1.0	(1.6)	1.6
DL	0.65	(1.2)	0.78
Total	1.65	(1.44)	2.38

B. Flexural design strength:

Beam moments

$$M_u = 2.38 \ (30)^2/8 = 267.8 \text{ kip-ft}$$
$$M_{LL} = 1.0 \ (30)^2/8 = 112.5 \text{ kip-ft}$$

The maximum effective flange width b is $2L/8 = 2(30)/8 = 7.5$ ft $= 90$ in.

C. Select section and determine properties:

At this point, go directly to the Composite Beam Tables and select a section or compute a preliminary trial section size using the formula:

$$\text{WT.} = \left[\frac{M_u(12)}{(d/2 + Y_{con} - a/2) \, \phi F_y} \right] 3.4$$

where

$$Y_{con} = 3 + 3.25 = 6.25 \text{ in.}$$
$$a/2 = 1 \text{ in. (estimate)}$$
$$\phi = 0.85$$
$$F_y = 36 \text{ ksi}$$

d	$\dfrac{M_u(12)(3.4)}{\phi F_y}$	$d/2$	$(Y_{con} - a/2)$	WT.
16	357.1	8	5.25	26.95
18	357.1	9	5.25	25.06

From the results above, a W16 × 31 would be the most likely candidate, however, since the weight required for a 16-in. section is less than one pound more than the W16 × 26, a quick check for the adequacy of the W16 × 26 is worthwhile. Let $\Sigma Q_n = A_s F_y = 7.68 \ (36) = 276$ kips.

$$a_{req'd} = \frac{\Sigma Q_n}{0.85 f'_c b} = \frac{276}{0.85 \ (3.5) \ (90)} = 1.03 \text{ in.}$$

$$Y2 = 6.25 - 1.03/2 = 5.74 \text{ in.}$$

By interpolation from the Composite Beam Tables for a W16 × 26 and a value of $Y2$ equal to 5.74 in.,

$$\phi M_n = 261 + (0.24/0.50) \, (271 - 261)$$
$$= 266 \text{ kip-ft} < 267.8 \text{ kip-ft} \quad \textbf{n.g.}$$

From the Composite Beam Tables, try W16 × 31, $F_y = 36$ ksi, $Y2 = 5.5$ in., and $\Sigma Q_n = 241$ kips, $Y1 = 0.22$ in. and $\phi M_n = 278$ kip-ft.

The required effective flange thickness is

$$a = \frac{\Sigma Q_n}{0.85 f_c' b} = \frac{241}{0.85(3.5)(7.5)(12)} = 0.90 \text{ in.}$$

$$Y2 = Y_{con} - (a/2) = 3 + 3.25 - (0.90/2) = 5.8 \text{ in.}$$

The selected section is adequate for $Y2 = 5.5$ in.

D. Compute number of studs required:

The stud reduction is calculated to be:

$$\text{Reduction factor} = \frac{0.85}{\sqrt{N_r}} \, (w_r/h_r) \, (H_s/h_r - 1.0) \le 1.0 \qquad \text{(I3-1)}$$

$$= \frac{0.85}{\sqrt{1}} \, (6/3) \, (5/3 - 1.0) = 1.13$$

where

N_r = number of stud connectors in one rib; not to exceed three in computations, although more than three may be installed

w_r = average width of rib, in.

h_r = nominal rib height, in.

H_s = length of stud connector after welding, in.; not to exceed the value $(h_r + 3)$ in computations, although actual length may be greater. Also, must not be less than four stud diameters.

The value for $H_s = 5.0$ was selected to insure the stud capacity reduction factor is 1.0.

The number of studs required is:

with $Q_n = 19.8$ kips (Table 4.1)
$$2 \, (\Sigma Q_n)/Q_n = 2(241)/19.8 = 24.3$$

Use: 26-¾ in. dia. × 5-in. headed studs.

E. Check deflection:

For the selected section, a W16 × 31, $F_y = 36$ ksi, $Y2 = 5.5$ in. and $Y1 = 0.22$ in.; from the Elastic Moment of Inertia Tables one can find the lower bound moment of inertia is 1,070 in.^{4} Thus, the service live load deflection can be calculated as follows (see LRFD Manual Part 3):

$$\Delta_{LL} = \frac{M_{LL}L^2}{161 \, I_{LB}} = \frac{112.5 \, (30)^2}{161 \, (1,070)} = 0.59 \text{ in.} \approx \frac{L}{610} < \frac{L}{360} \quad \textbf{o.k.}$$

F. Shear check:

$$V_u = 2.38(15) = 37.5 \text{ kips}$$
$$\phi V_n = \phi 0.6 F_{yw} A_w$$
$$= (0.9)(0.6)(36)(15.88)(.25)$$
$$= 77.2 \text{ kips}$$

Use: W16 × 31, $F_y = 36$ ksi

Example 2:

Given:

Determine the beam, with $F_y = 50$ ksi, required to support a service live load of 250 psf and a service dead load of 90 psf. The beam span is 40 ft and the beam spacing is 10 ft. Assume 3 in. metal deck is used with a 4.5 in. concrete slab of 4 ksi normal weight concrete (145 pcf). The stud reduction factor is 1.0. Unshored construction is specified. Determine the beam size and service dead and live load deflections. Also select a non-composite section (no shear connectors).

Solution:

A. Load tabulation:

	Service Load (kips/ft)	(L.F.)	Factored Load (kips/ft)
LL 10(0.25) =	2.5	(1.6)	4.0
DL 10(0.09) =	0.9	(1.2)	1.08
Total	3.4	(1.49)	5.08

B. Beam moments:

$$M_u = 5.08 \,(40)^2/8 = 1,016 \text{ kip-ft}$$
$$M_{LL} = 2.5 \,\,(40)^2/8 = 500 \text{ kip-ft}$$
$$M_{DL} = 0.9 \,\,(40)^2/8 = 180 \text{ kip-ft}$$

C. Select section and determine properties:

Assume $a = 2$ in., therefore, we will take $Y2 = 7.5 - 2/2 = 6.5$ in. From the Composite Beam Tables, for $F_y = 50$ ksi and $Y2 = 6.5$ in., a W21 × 62, W24 × 55 or W24 × 62 are possible sizes.

Try W24 × 55: $\quad F_y = 50$ ksi
$$Y2 = 6.5 \text{ in.}$$
$$Q_n = 810 \text{ kips}$$
$$\phi M_n = 1,050 \text{ kip-ft}$$

Compute $Y2$ for $\Sigma Q_n = 810$ kips:

$$b = 2\,(40)(12)/8 = 120 \text{ in.}$$

$$a = \frac{\Sigma Q_n}{.85 f_c' b} = \frac{810}{.85(4)(120)} = 1.99 \text{ in.}$$

$$Y2 = 7.5 - 1.99/2 = 6.5 \text{ in.}$$

D. Compute the number of studs required:

$$Q_n = 26.1 \text{ kips (Table 4.1)}$$
$$\text{No. studs} = (2)\Sigma Q_n / Q_n = 2(810)/26.1 = 62.1$$

A W24 × 55, $F_y = 50$ ksi with 64-¾ in. dia. headed studs satisfies the requirements for flexural design strength.

E. Construction phase strength check:

Conservatively, the concrete weight will be considered to be live load during the construction phase. Use 75 psf for the wet concrete live load. In addition, a construction live load of 20 psf will be assumed.

Load Tabulation:

		Service Load (kips/ft)	L.F.	Factored Load (kips/ft)
LL	10(0.02)	0.20	(1.6)	0.32
Concrete	10(0.075)	0.75	(1.6)	1.20
DL	10(0.09 − 0.075)	0.15	(1.2)	0.18
Total		1.10	(1.55)	1.70

$$M_u = 1.70 \, (40)^2/8 = 340 \text{ kip-ft}$$

From the Composite Beam Tables for a W24 × 55 with $F_y = 50$ ksi, and assuming adequate lateral support is provided by the attachment of the steel deck to the compression flange,

$$\phi M_n = \phi M_p = 503 \text{ kip-ft} > 340 \text{ kip-ft}$$

F. Service load deflections:

Assume that the wet concrete load moment is equal to the service dead load moment. With $I_x = 1,350$ in.[4] for a W24 × 55,

$$\Delta_{DL} = \frac{180 \, (40)^2}{161 \, (1,350)} = 1.33 \text{ in.}$$

For the W24 × 55 with $Y2 = 6.5$ in., the lower bound moment of inertia can be found in the Lower Bound Elastic Moment of Inertia Tables; $I_{LB} = 4,060$ in.[4]

$$\Delta_{LL} = \frac{500 \, (40)^2}{161 \, (4,060)} = 1.22 \text{ in.}$$

$$= \frac{L}{393} < \frac{L}{360} \quad \textbf{o.k.}$$

Specify a beam camber of 1¼ in. to overcome the dead load deflection.

G. Check shear:

$$V_u = 5.08 \, (40)/2 = 102 \text{ kips}$$

$$\phi V = \phi \, (0.6) \, F_{yw} A_w$$
$$= (0.9)(0.6)(50)(23.57)(0.395)$$
$$= 251 \text{ kips} > 102 \text{ kips} \quad \textbf{o.k.}$$

H. Final section selection:

    Use: W24 × 55, F_y = 50 ksi, camber 1¼ in.,
          64-¾ in. dia. studs (32 each side of midspan)

I. Non-composite section:

Considering the given problem without shear connectors (i.e. non-composite), a steel section can be selected from the ϕM_p values tabulated under each section in either the Composite Beam Tables or the Load Factor Design Selection Tables.

For M_u = 1,016 kip-ft, select a W27 × 94, F_y = 50 ksi, with a ϕM_p flexural design strength equal to 1,040 kip-ft.

$$\Delta_{DL} = \frac{180\,(40)^2}{161\,(3{,}270)} = 0.55 \text{ in.}$$

$$\Delta_{LL} = \frac{500\,(40)^2}{161\,(3{,}270)} = 1.52 \text{ in.} > L/360$$

For $\Delta = L/360 = 1.33$ in.,

$$I_{req'd} = \frac{500\,(40)^2}{161\,(1.33)} = 3{,}736 \text{ in.}^4$$

Use: W30 × 99, F_y = 50 ksi, $\phi\,M_n$ = 1,170 kip-ft

Example 3:

Given:

A W21 × 44, F_y = 50 ksi, steel girder spans 30 ft and supports intermediate beams at the third points. A total of sixty-eight ¾-in. diameter headed studs are applied to the beam as follows: thirty-three between each support and the beams at the one-third points, and two between the intermediate beams. The slab consists of 3¼-in. l.w. concrete (115 pcf) with a specified design strength of 3.5 ksi over a 3-in. deep composite metal deck with an average rib width of 6 in. The ribs are oriented parallel to the beam centerline. Determine the design beam reactions.

Solution:

For studs in a single row the spacing between the support and first intermediate beam would be 10(12)/33 = 3.64 in. which is greater than the specified minimum of four stud diameters (LRFD Specification Sect. I5.6). Since w_r/h_r = 6/3 = 2 is greater than 1.5, the stud reduction factor is not necessary (LRFD Specification Sect. I3.5c). Therefore, the stud shear capacity is:

    nQ_n = 33(19.8) = 653 kips

The above value exceeds the tabulated maximum value of ΣQ_n = 650 kips from the Composite Beam Tables. For ΣQ_n = 650 kips, the required effective concrete flange thickness can be calculated to be

$$a = \frac{650}{0.85(3.5)(7.5)(12)} = 2.43 \text{ in.}$$

$Y2 = 3 + 3.25 - 2.43/2 = 5.04 \text{ in.}$

Use $Y2 = 5.04$ in.

Beam reaction:

For a W21 × 44, $F_y = 50$ ksi, $Y2 = 5.0$ in.

$\phi M_n = 706$ kip-ft

$\begin{aligned} R &= C_c \phi M_n / L \\ &= 3(706)/30 \\ &= 70.6 \text{ kips} \end{aligned}$

where

R = design reaction, kips

C_c = coefficient from Fig. 4.5

ϕM_n = flexural design strength of beam, kip-ft

L = span length, ft

Note: The beam weight was neglected in this example.

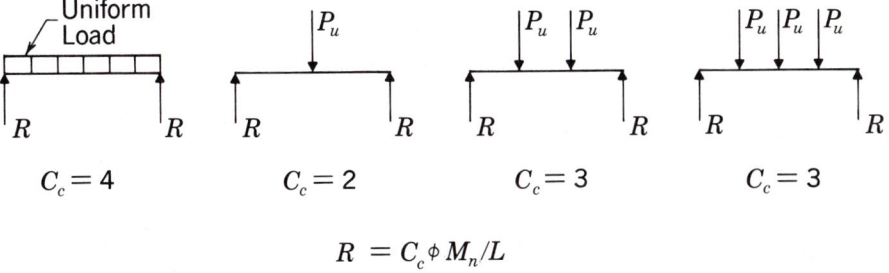

$$R = C_c \phi M_n / L$$

Fig. 4.5. Beam reaction coefficients

COMPOSITE DESIGN
COMPOSITE BEAM SELECTION TABLE
W Shapes

$F_y = 36$ ksi

$\phi = 0.85$ $\phi_b = 0.90$

Shape	$\phi_b M_p$	PNA[c]	Y1[a]	ΣQ_n	ϕM_n (kip-ft) Y2[b] (in.)										
	Kip-ft		In.	Kips	2	2.5	3	3.5	4	4.5	5	5.5	6	6.5	7
W 36x300	3400	TFL	0.00	3180	4590	4700	4810	4920	5040	5150	5260	5370	5490	5600	5710
		2	0.42	2680	4510	4600	4700	4790	4890	4980	5080	5170	5270	5360	5460
		3	0.84	2170	4410	4490	4570	4640	4720	4800	4880	4950	5030	5110	5180
		4	1.26	1670	4310	4360	4420	4480	4540	4600	4660	4720	4780	4840	4900
		BFL	1.68	1160	4180	4220	4260	4310	4350	4390	4430	4470	4510	4550	4590
		6	3.97	979	4120	4150	4190	4220	4260	4290	4330	4360	4400	4430	4470
		7	6.69	795	4020	4050	4080	4110	4140	4160	4190	4220	4250	4280	4300
W 36x280	3160	TFL	0.00	2970	4260	4360	4470	4570	4680	4780	4890	4990	5100	5200	5310
		2	0.39	2500	4180	4270	4360	4450	4540	4630	4710	4800	4890	4980	5070
		3	0.79	2030	4100	4170	4240	4310	4390	4460	4530	4600	4670	4740	4820
		4	1.18	1560	4000	4050	4110	4160	4220	4280	4330	4390	4440	4500	4550
		BFL	1.57	1090	3890	3930	3960	4000	4040	4080	4120	4160	4200	4230	4270
		6	3.88	916	3830	3860	3890	3930	3960	3990	4020	4060	4090	4120	4150
		7	6.62	742	3740	3770	3790	3820	3850	3870	3900	3920	3950	3980	4000
W 36x260	2920	TFL	0.00	2750	3930	4020	4120	4220	4320	4410	4510	4610	4710	4800	4900
		2	0.36	2330	3860	3940	4030	4110	4190	4270	4350	4440	4520	4600	4680
		3	0.72	1900	3780	3850	3920	3980	4050	4120	4190	4250	4320	4390	4450
		4	1.08	1470	3700	3750	3800	3850	3900	3960	4010	4060	4110	4160	4210
		BFL	1.44	1040	3600	3630	3670	3710	3740	3780	3820	3850	3890	3930	3960
		6	3.86	863	3540	3570	3600	3630	3660	3690	3720	3750	3780	3820	3850
		7	6.75	689	3450	3470	3500	3520	3550	3570	3600	3620	3640	3670	3690
W 36x245	2730	TFL	0.00	2600	3680	3780	3870	3960	4050	4140	4240	4330	4420	4510	4600
		2	0.34	2190	3620	3700	3780	3860	3930	4010	4090	4170	4240	4320	4400
		3	0.68	1790	3550	3620	3680	3740	3810	3870	3930	4000	4060	4120	4190
		4	1.01	1390	3470	3520	3570	3620	3670	3720	3770	3820	3870	3910	3960
		BFL	1.35	991	3380	3420	3450	3490	3520	3560	3590	3630	3660	3700	3730
		6	3.81	820	3330	3360	3380	3410	3440	3470	3500	3530	3560	3590	3620
		7	6.77	649	3240	3260	3280	3310	3330	3350	3380	3400	3420	3440	3470
W 36x230	2550	TFL	0.00	2430	3440	3530	3610	3700	3780	3870	3960	4040	4130	4210	4300
		2	0.32	2060	3380	3450	3530	3600	3670	3750	3820	3890	3970	4040	4110
		3	0.63	1690	3320	3380	3440	3500	3560	3620	3670	3730	3790	3850	3910
		4	0.95	1310	3240	3290	3340	3380	3430	3480	3520	3570	3610	3660	3710
		BFL	1.26	939	3160	3190	3230	3260	3290	3330	3360	3390	3430	3460	3490
		6	3.81	774	3110	3140	3160	3190	3220	3250	3270	3300	3330	3360	3380
		7	6.83	608	3020	3040	3070	3090	3110	3130	3150	3170	3200	3220	3240

[a]Y1 = distance from top of the steel beam to plastic neutral axis.
[b]Y2 = distance from top of the steel beam to concrete flange force.
[c]See Fig. 4.3 for PNA locations.

COMPOSITE DESIGN
COMPOSITE BEAM SELECTION TABLE
W Shapes

F_y = 36 ksi

$\phi = 0.85$ $\phi_b = 0.90$

Shape	$\phi_b M_p$ (Kip-ft)	PNA[c]	$Y1$[a] (In.)	ΣQ_n (Kips)	ϕM_n (kip-ft) $Y2$[b] (in.) 2	2.5	3	3.5	4	4.5	5	5.5	6	6.5	7
W 36x210	2250	TFL	0.00	2220	3210	3280	3360	3440	3520	3600	3680	3760	3840	3920	3990
		2	0.34	1930	3160	3230	3300	3370	3430	3500	3570	3640	3710	3770	3840
		3	0.68	1630	3110	3170	3220	3280	3340	3400	3450	3510	3570	3630	3680
		4	1.02	1330	3050	3090	3140	3190	3240	3280	3330	3380	3420	3470	3520
		BFL	1.36	1030	2980	3020	3050	3090	3130	3160	3200	3240	3270	3310	3350
		6	5.06	794	2890	2920	2950	2980	3010	3030	3060	3090	3120	3150	3170
		7	9.04	556	2740	2760	2780	2800	2820	2840	2860	2880	2900	2920	2940
W 36x194	2070	TFL	0.00	2050	2940	3020	3090	3160	3230	3310	3380	3450	3520	3600	3670
		2	0.32	1780	2900	2960	3030	3090	3150	3220	3280	3340	3400	3470	3530
		3	0.63	1500	2850	2910	2960	3010	3070	3120	3170	3220	3280	3330	3380
		4	0.95	1230	2800	2840	2890	2930	2970	3020	3060	3100	3150	3190	3230
		BFL	1.26	953	2740	2770	2810	2840	2870	2910	2940	2970	3010	3040	3080
		6	4.94	733	2660	2690	2710	2740	2760	2790	2820	2840	2870	2890	2920
		7	8.93	513	2520	2540	2560	2580	2590	2610	2630	2650	2670	2680	2700
W 36x182	1940	TFL	0.00	1930	2760	2820	2890	2960	3030	3100	3170	3230	3300	3370	3440
		2	0.29	1670	2720	2780	2840	2890	2950	3010	3070	3130	3190	3250	3310
		3	0.59	1420	2670	2720	2770	2820	2870	2920	2970	3020	3070	3120	3170
		4	0.89	1160	2620	2660	2710	2750	2790	2830	2870	2910	2950	2990	3030
		BFL	1.18	904	2570	2600	2630	2660	2700	2730	2760	2790	2820	2860	2890
		6	4.89	693	2490	2520	2540	2570	2590	2620	2640	2670	2690	2720	2740
		7	8.92	482	2360	2380	2400	2410	2430	2450	2460	2480	2500	2520	2530
W 36x170	1800	TFL	0.00	1800	2560	2620	2690	2750	2820	2880	2940	3010	3070	3130	3200
		2	0.28	1560	2520	2580	2640	2690	2750	2800	2860	2910	2970	3020	3080
		3	0.55	1320	2480	2530	2580	2620	2670	2720	2770	2810	2860	2910	2950
		4	0.83	1090	2440	2480	2520	2550	2590	2630	2670	2710	2750	2780	2820
		BFL	1.10	847	2390	2420	2450	2480	2510	2540	2570	2600	2630	2660	2690
		6	4.84	649	2320	2340	2370	2390	2410	2440	2460	2480	2500	2530	2550
		7	8.89	450	2200	2210	2230	2240	2260	2280	2290	2310	2320	2340	2360
W 36x160	1680	TFL	0.00	1690	2400	2460	2520	2580	2640	2700	2760	2820	2880	2940	3000
		2	0.26	1470	2360	2420	2470	2520	2570	2630	2680	2730	2780	2830	2890
		3	0.51	1250	2330	2370	2420	2460	2500	2550	2590	2640	2680	2730	2770
		4	0.77	1030	2290	2320	2360	2400	2430	2470	2510	2540	2580	2610	2650
		BFL	1.02	811	2240	2270	2300	2330	2360	2380	2410	2440	2470	2500	2530
		6	4.82	617	2170	2200	2220	2240	2260	2280	2310	2330	2350	2370	2390
		7	8.97	423	2050	2070	2080	2100	2110	2130	2140	2160	2170	2190	2200

[a] $Y1$ = distance from top of the steel beam to plastic neutral axis.
[b] $Y2$ = distance from top of the steel beam to concrete flange force.
[c] See Fig. 4.3 for PNA locations.

F_y = 36 ksi

COMPOSITE DESIGN
COMPOSITE BEAM SELECTION TABLE
W Shapes

$\phi = 0.85$ $\phi_b = 0.90$

Shape	$\phi_b M_p$ Kip-ft	PNA[c]	$Y1$[a] In.	ΣQ_n Kips	ϕM_n (kip-ft) — $Y2$[b] (in.) 2	2.5	3	3.5	4	4.5	5	5.5	6	6.5	7
W 36x150	1570	TFL	0.00	1590	2250	2300	2360	2410	2470	2530	2580	2640	2700	2750	2810
		2	0.24	1390	2220	2260	2310	2360	2410	2460	2510	2560	2610	2660	2710
		3	0.47	1190	2180	2220	2270	2310	2350	2390	2430	2480	2520	2560	2600
		4	0.71	983	2140	2180	2210	2250	2280	2320	2350	2390	2420	2460	2490
		BFL	0.94	781	2100	2130	2160	2190	2210	2240	2270	2300	2330	2350	2380
		6	4.83	589	2040	2060	2080	2100	2120	2140	2160	2190	2210	2230	2250
		7	9.08	398	1920	1930	1950	1960	1970	1990	2000	2020	2030	2040	2060
W 36x135	1370	TFL	0.00	1430	2000	2050	2100	2150	2200	2260	2310	2360	2410	2460	2510
		2	0.20	1260	1980	2020	2070	2110	2160	2200	2240	2290	2330	2380	2420
		3	0.40	1090	1950	1990	2030	2060	2100	2140	2180	2220	2260	2300	2330
		4	0.59	919	1920	1950	1980	2020	2050	2080	2110	2150	2180	2210	2240
		BFL	0.79	749	1890	1910	1940	1970	1990	2020	2050	2070	2100	2130	2150
		6	4.96	553	1820	1840	1860	1880	1900	1920	1940	1960	1980	2000	2020
		7	9.50	357	1690	1710	1720	1730	1740	1760	1770	1780	1790	1810	1820
W 33x221	2310	TFL	0.00	2340	3140	3230	3310	3390	3470	3560	3640	3720	3810	3890	3970
		2	0.32	1980	3090	3160	3230	3300	3370	3440	3510	3580	3650	3720	3790
		3	0.64	1610	3020	3080	3140	3200	3250	3310	3370	3420	3480	3540	3600
		4	0.96	1250	2950	3000	3040	3090	3130	3170	3220	3260	3310	3350	3400
		BFL	1.28	889	2870	2900	2940	2970	3000	3030	3060	3090	3120	3160	3190
		6	3.76	737	2820	2850	2880	2900	2930	2960	2980	3010	3030	3060	3090
		7	6.48	585	2750	2770	2790	2810	2830	2850	2870	2890	2910	2930	2960
W 33x201	2080	TFL	0.00	2130	2840	2910	2990	3070	3140	3220	3290	3370	3440	3520	3590
		2	0.29	1800	2790	2850	2920	2980	3050	3110	3170	3240	3300	3360	3430
		3	0.58	1480	2730	2790	2840	2890	2940	2990	3050	3100	3150	3200	3260
		4	0.86	1150	2670	2710	2750	2790	2830	2870	2920	2960	3000	3040	3080
		BFL	1.15	824	2600	2630	2660	2690	2720	2750	2780	2810	2830	2860	2890
		6	3.67	678	2560	2580	2600	2630	2650	2680	2700	2720	2750	2770	2800
		7	6.51	532	2480	2500	2520	2540	2560	2580	2600	2620	2630	2650	2670
W 33x141	1390	TFL	0.00	1500	1980	2030	2080	2140	2190	2240	2300	2350	2400	2460	2510
		2	0.24	1300	1950	1990	2040	2090	2130	2180	2220	2270	2320	2360	2410
		3	0.48	1100	1920	1950	1990	2030	2070	2110	2150	2190	2230	2270	2300
		4	0.72	900	1880	1910	1940	1970	2010	2040	2070	2100	2130	2170	2200
		BFL	0.96	700	1840	1860	1890	1910	1940	1960	1990	2010	2040	2060	2090
		6	4.31	537	1790	1810	1820	1840	1860	1880	1900	1920	1940	1960	1980
		7	8.05	374	1690	1710	1720	1730	1740	1760	1770	1780	1800	1810	1820

[a]$Y1$ = distance from top of the steel beam to plastic neutral axis.
[b]$Y2$ = distance from top of the steel beam to concrete flange force.
[c]See Fig. 4.3 for PNA locations.

| F_y = 36 ksi | | COMPOSITE DESIGN COMPOSITE BEAM SELECTION TABLE W Shapes $\phi = 0.85$ $\phi_b = 0.90$ | | | | | | | | | | | | | |

Shape	$\phi_b M_p$	PNA[c]	$Y1^a$	ΣQ_n	ϕM_n (kip-ft)											
					$Y2^b$ (in.)											
	Kip-ft		In.	Kips	2	2.5	3	3.5	4	4.5	5	5.5	6	6.5	7
W 33x130	1260	TFL	0.00	1380	1810	1860	1910	1960	2010	2060	2100	2150	2200	2250	2300
		2	0.21	1200	1780	1830	1870	1910	1950	2000	2040	2080	2130	2170	2210
		3	0.43	1020	1760	1790	1830	1860	1900	1940	1970	2010	2050	2080	2120
		4	0.64	847	1720	1750	1780	1810	1840	1870	1900	1930	1960	1990	2020
		BFL	0.86	670	1690	1710	1740	1760	1780	1810	1830	1860	1880	1900	1930
		6	4.39	507	1640	1660	1670	1690	1710	1730	1750	1760	1780	1800	1820
		7	8.29	345	1540	1550	1570	1580	1590	1600	1610	1630	1640	1650	1660
W 33x118	1120	TFL	0.00	1250	1630	1680	1720	1760	1810	1850	1900	1940	1980	2030	2070
		2	0.19	1100	1610	1650	1690	1720	1760	1800	1840	1880	1920	1960	2000
		3	0.37	943	1580	1620	1650	1680	1720	1750	1780	1820	1850	1880	1920
		4	0.56	790	1560	1580	1610	1640	1670	1700	1720	1750	1780	1810	1840
		BFL	0.74	638	1530	1550	1570	1600	1620	1640	1660	1690	1710	1730	1750
		6	4.44	475	1480	1490	1510	1530	1540	1560	1580	1590	1610	1630	1650
		7	8.54	312	1380	1390	1400	1410	1420	1430	1450	1460	1470	1480	1490
W 30x116	1020	TFL	0.00	1230	1480	1530	1570	1610	1660	1700	1740	1790	1830	1880	1920
		2	0.21	1070	1460	1500	1530	1570	1610	1650	1690	1720	1760	1800	1840
		3	0.43	910	1430	1460	1500	1530	1560	1590	1630	1660	1690	1720	1750
		4	0.64	749	1400	1430	1460	1480	1510	1540	1560	1590	1620	1640	1670
		BFL	0.85	589	1370	1390	1410	1440	1460	1480	1500	1520	1540	1560	1580
		6	3.98	448	1330	1350	1360	1380	1390	1410	1430	1440	1460	1470	1490
		7	7.44	308	1250	1260	1270	1290	1300	1310	1320	1330	1340	1350	1360
W 30x108	934	TFL	0.00	1140	1370	1410	1450	1490	1530	1570	1610	1650	1690	1730	1770
		2	0.19	998	1350	1380	1420	1450	1490	1520	1560	1590	1630	1660	1700
		3	0.38	855	1320	1350	1380	1410	1440	1470	1500	1530	1570	1600	1630
		4	0.57	711	1300	1320	1350	1370	1400	1420	1450	1470	1500	1520	1550
		BFL	0.76	568	1270	1290	1310	1330	1350	1370	1390	1410	1430	1450	1470
		6	4.04	427	1230	1240	1260	1270	1290	1300	1320	1330	1350	1360	1380
		7	7.64	285	1150	1160	1170	1180	1190	1200	1210	1220	1230	1240	1250
W 30x 99	842	TFL	0.00	1050	1250	1290	1320	1360	1400	1430	1470	1510	1550	1580	1620
		2	0.17	922	1230	1260	1300	1330	1360	1390	1430	1460	1490	1520	1560
		3	0.34	796	1210	1240	1270	1290	1320	1350	1380	1410	1440	1460	1490
		4	0.50	670	1190	1210	1240	1260	1280	1310	1330	1350	1380	1400	1430
		BFL	0.67	543	1170	1180	1200	1220	1240	1260	1280	1300	1320	1340	1360
		6	4.07	403	1120	1140	1150	1170	1180	1190	1210	1220	1240	1250	1270
		7	7.83	262	1040	1050	1060	1070	1080	1090	1100	1110	1120	1130	1140

[a] $Y1$ = distance from top of the steel beam to plastic neutral axis.
[b] $Y2$ = distance from top of the steel beam to concrete flange force.
[c] See Fig. 4.3 for PNA locations.

				ϕM_n (kip-ft)										

F_y = 36 ksi

COMPOSITE DESIGN
COMPOSITE BEAM SELECTION TABLE
W Shapes
$$\phi = 0.85 \qquad \phi_b = 0.90$$

Shape	$\phi_b M_p$	PNA[c]	$Y1$[a]	ΣQ_n	\multicolumn 2	2.5	3	3.5	4	4.5	5	5.5	6	6.5	7
	Kip-ft		In.	Kips	Y2[b] (in.)										
W 27x102	824	TFL	0.00	1080	1190	1230	1270	1300	1340	1380	1420	1460	1500	1530	1570
		2	0.21	930	1170	1200	1230	1270	1300	1330	1360	1400	1430	1460	1500
		3	0.42	781	1140	1170	1200	1230	1250	1280	1310	1340	1360	1390	1420
		4	0.62	631	1120	1140	1160	1180	1210	1230	1250	1270	1290	1320	1340
		BFL	0.83	482	1090	1100	1120	1140	1160	1170	1190	1210	1220	1240	1260
		6	3.41	376	1060	1070	1080	1100	1110	1120	1140	1150	1160	1180	1190
		7	6.26	270	1010	1010	1020	1030	1040	1050	1060	1070	1080	1090	1100
W 27x 94	751	TFL	0.00	997	1090	1130	1160	1200	1230	1270	1300	1340	1370	1410	1450
		2	0.19	863	1070	1100	1130	1160	1190	1230	1260	1290	1320	1350	1380
		3	0.37	729	1050	1080	1100	1130	1150	1180	1210	1230	1260	1280	1310
		4	0.56	595	1030	1050	1070	1090	1110	1130	1150	1170	1200	1220	1240
		BFL	0.75	461	1000	1020	1030	1050	1070	1080	1100	1120	1130	1150	1170
		6	3.39	355	972	985	997	1010	1020	1040	1050	1060	1070	1090	1100
		7	6.39	249	921	929	938	947	956	965	974	982	991	1000	1010
W 27x 84	659	TFL	0.00	893	971	1000	1030	1070	1100	1130	1160	1190	1220	1260	1290
		2	0.16	778	954	982	1010	1040	1060	1090	1120	1150	1170	1200	1230
		3	0.32	663	936	959	983	1010	1030	1050	1080	1100	1120	1150	1170
		4	0.48	549	916	936	955	975	994	1010	1030	1050	1070	1090	1110
		BFL	0.64	434	896	911	926	942	957	972	988	1000	1020	1030	1050
		6	3.44	329	866	878	890	901	913	925	936	948	960	971	983
		7	6.62	223	814	822	830	838	846	854	861	869	877	885	893
W 24x 76	540	TFL	0.00	806	797	826	855	883	912	940	969	997	1030	1050	1080
		2	0.17	696	781	806	830	855	880	904	929	954	978	1000	1030
		3	0.34	586	764	784	805	826	847	867	888	909	930	950	971
		4	0.51	476	745	762	778	795	812	829	846	863	880	896	913
		BFL	0.68	366	724	737	750	763	776	789	802	815	828	841	854
		6	3.00	284	703	713	723	733	743	753	763	773	783	793	803
		7	5.60	202	666	673	680	687	694	702	709	716	723	730	737
W 24x 68	478	TFL	0.00	724	711	736	762	788	813	839	864	890	916	941	967
		2	0.15	629	697	719	741	764	786	808	830	853	875	897	920
		3	0.29	535	682	701	720	739	758	777	796	815	833	852	871
		4	0.44	440	666	682	697	713	729	744	760	775	791	807	822
		BFL	0.59	346	649	662	674	686	698	711	723	735	747	760	772
		6	3.05	263	628	637	646	656	665	674	684	693	702	712	721
		7	5.81	181	590	596	603	609	616	622	628	635	641	648	654

[a]$Y1$ = distance from top of the steel beam to plastic neutral axis.
[b]$Y2$ = distance from top of the steel beam to concrete flange force.
[c]See Fig. 4.3 for PNA locations.

AMERICAN INSTITUTE OF STEEL CONSTRUCTION

F_y = 36 ksi

COMPOSITE DESIGN
COMPOSITE BEAM SELECTION TABLE
W Shapes
$$\phi = 0.85 \qquad \phi_b = 0.90$$

Shape	$\phi_b M_p$	PNA[c]	$Y1$[a]	ΣQ_n	ϕM_n (kip-ft) $Y2$[b] (in.)										
	Kip-ft		In.	Kips	2	2.5	3	3.5	4	4.5	5	5.5	6	6.5	7
W 24x62	413	TFL	0.00	655	644	667	690	713	737	760	783	806	829	853	876
		2	0.15	580	633	653	674	694	715	736	756	777	797	818	838
		3	0.29	506	621	639	657	675	693	711	728	746	764	782	800
		4	0.44	431	608	624	639	654	669	685	700	715	731	746	761
		BFL	0.59	356	595	608	620	633	646	658	671	683	696	709	721
		6	3.47	260	568	577	587	596	605	614	623	633	642	651	660
		7	6.58	164	520	526	532	538	543	549	555	561	567	572	578
W 24x55	362	TFL	0.00	583	569	590	611	631	652	673	693	714	735	755	776
		2	0.13	520	560	579	597	615	634	652	671	689	707	726	744
		3	0.25	456	550	566	583	599	615	631	647	663	679	696	712
		4	0.38	392	540	554	568	582	595	609	623	637	651	665	679
		BFL	0.51	328	529	540	552	564	575	587	599	610	622	634	645
		6	3.45	237	504	512	520	529	537	546	554	562	571	579	588
		7	6.66	146	458	463	468	474	479	484	489	494	499	505	510
W 21x62	389	TFL	0.00	659	583	606	630	653	676	700	723	746	770	793	816
		2	0.15	568	570	590	610	630	650	670	690	710	730	751	771
		3	0.31	476	555	572	589	606	623	640	656	673	690	707	724
		4	0.46	385	540	553	567	581	594	608	622	635	649	663	676
		BFL	0.62	294	523	534	544	555	565	575	586	596	607	617	628
		6	2.53	229	507	516	524	532	540	548	556	564	572	581	589
		7	4.78	165	482	487	493	499	505	511	517	522	528	534	540
W 21x57	348	TFL	0.00	601	534	555	576	597	619	640	661	683	704	725	747
		2	0.16	525	522	541	559	578	597	615	634	652	671	689	708
		3	0.33	448	510	526	542	558	574	589	605	621	637	653	669
		4	0.49	371	497	510	523	536	550	563	576	589	602	615	628
		BFL	0.65	294	483	493	504	514	525	535	546	556	566	577	587
		6	2.90	222	464	472	480	488	496	504	511	519	527	535	543
		7	5.38	150	433	438	443	449	454	459	465	470	475	481	486
W 21x50	297	TFL	0.00	529	465	484	503	522	540	559	578	597	615	634	653
		2	0.13	466	456	473	489	506	522	539	555	572	588	605	621
		3	0.27	403	446	461	475	489	504	518	532	546	561	575	589
		4	0.40	341	436	448	460	472	484	496	508	520	532	545	557
		BFL	0.54	278	425	435	445	454	464	474	484	494	504	513	523
		6	2.92	205	406	413	421	428	435	442	450	457	464	472	479
		7	5.58	132	374	379	383	388	393	397	402	407	411	416	421

[a]$Y1$ = distance from top of the steel beam to plastic neutral axis.
[b]$Y2$ = distance from top of the steel beam to concrete flange force.
[c]See Fig. 4.3 for PNA locations.

AMERICAN INSTITUTE OF STEEL CONSTRUCTION

				F_y = 36 ksi

COMPOSITE DESIGN
COMPOSITE BEAM SELECTION TABLE
W Shapes

$\phi = 0.85$ \qquad\qquad $\phi_b = 0.90$

Shape	$\phi_b M_p$	PNA[c]	$Y1^a$	ΣQ_n	ϕM_n (kip-ft)											
					$Y2^b$ (in.)											
	Kip-ft		In.	Kips	2	2.5	3	3.5	4	4.5	5	5.5	6	6.5	7
W 21x44	258	TFL	0.00	468	409	425	442	458	475	492	508	525	541	558	574
		2	0.11	415	401	416	430	445	460	475	489	504	519	533	548
		3	0.23	363	393	406	419	432	444	457	470	483	496	509	521
		4	0.34	310	384	395	406	417	428	439	450	461	472	483	494
		BFL	0.45	257	376	385	394	403	412	421	430	439	448	458	467
		6	2.90	187	358	364	371	378	384	391	398	404	411	417	424
		7	5.69	127	326	331	335	339	343	347	351	355	360	364	368
W 18x60	332	TFL	0.00	634	499	522	544	566	589	611	634	656	679	701	723
		2	0.17	539	485	504	523	542	561	581	600	619	638	657	676
		3	0.35	445	470	486	501	517	533	549	564	580	596	612	627
		4	0.52	350	454	466	478	491	503	516	528	540	553	565	578
		BFL	0.70	256	436	445	454	463	472	481	491	500	509	518	527
		6	2.19	207	424	432	439	446	454	461	468	476	483	490	498
		7	3.82	158	407	413	418	424	430	435	441	447	452	458	463
W 18x55	302	TFL	0.00	583	457	477	498	519	539	560	581	601	622	643	663
		2	0.16	498	444	462	479	497	515	532	550	568	585	603	620
		3	0.32	412	431	445	460	474	489	504	518	533	547	562	577
		4	0.47	327	416	428	439	451	462	474	486	497	509	520	532
		BFL	0.63	242	401	409	418	426	435	443	452	461	469	478	486
		6	2.16	194	389	396	403	410	417	424	430	437	444	451	458
		7	3.86	146	372	377	383	388	393	398	403	408	414	419	424
W 18x50	273	TFL	0.00	529	412	431	450	468	487	506	525	543	562	581	600
		2	0.14	452	401	417	433	449	465	481	497	513	529	545	561
		3	0.29	375	389	402	415	429	442	455	469	482	495	508	522
		4	0.43	299	376	387	397	408	418	429	439	450	461	471	482
		BFL	0.57	222	362	370	378	386	394	402	409	417	425	433	441
		6	2.07	177	352	358	365	371	377	383	390	396	402	408	415
		7	3.82	132	336	341	346	350	355	360	365	369	374	379	383
W 18x46	245	TFL	0.00	486	380	397	414	431	449	466	483	500	517	535	552
		2	0.15	420	370	385	400	415	430	444	459	474	489	504	519
		3	0.30	354	360	372	385	397	410	422	435	447	460	472	485
		4	0.45	288	348	359	369	379	389	399	410	420	430	440	450
		BFL	0.61	222	337	345	352	360	368	376	384	392	400	407	415
		6	2.40	172	324	330	336	343	349	355	361	367	373	379	385
		7	4.34	122	305	310	314	318	322	327	331	335	340	344	348

[a]$Y1$ = distance from top of the steel beam to plastic neutral axis.
[b]$Y2$ = distance from top of the steel beam to concrete flange force.
[c]See Fig. 4.3 for PNA locations.

AMERICAN INSTITUTE OF STEEL CONSTRUCTION

					\multicolumn{13}{c}{}

F_y = 36 ksi

COMPOSITE DESIGN
COMPOSITE BEAM SELECTION TABLE
W Shapes
$\phi = 0.85$ $\phi_b = 0.90$

Shape	$\phi_b M_p$	PNA[c]	$Y1$[a]	ΣQ_n	\multicolumn{13}{c}{ϕM_n (kip-ft) — $Y2$[b] (in.)}
	Kip-ft		In.	Kips	2	2.5	3	3.5	4	4.5	5	5.5	6	6.5	7
W 18x40	212	TFL	0.00	425	329	345	360	375	390	405	420	435	450	465	480
		2	0.13	368	321	334	347	360	373	386	399	412	425	438	451
		3	0.26	311	312	323	334	345	356	367	378	389	400	411	423
		4	0.39	254	303	312	321	330	339	348	357	366	375	384	393
		BFL	0.53	197	293	300	307	314	321	328	335	342	349	356	363
		6	2.26	152	282	288	293	298	304	309	315	320	325	331	336
		7	4.27	106	265	269	273	277	280	284	288	292	295	299	303
W 18x35	180	TFL	0.00	371	285	298	311	324	338	351	364	377	390	403	416
		2	0.11	325	278	290	301	313	324	336	347	359	370	382	393
		3	0.21	279	271	281	291	301	311	321	331	340	350	360	370
		4	0.32	233	264	272	280	289	297	305	313	322	330	338	346
		BFL	0.43	187	256	263	269	276	283	289	296	303	309	316	323
		6	2.37	140	245	250	255	260	265	270	275	280	285	290	295
		7	4.56	92.7	227	230	233	237	240	243	246	250	253	256	260
W 16x36	173	TFL	0.00	382	268	282	295	309	322	336	349	363	377	390	404
		2	0.11	328	261	272	284	295	307	319	330	342	353	365	377
		3	0.22	273	252	262	272	281	291	301	310	320	330	339	349
		4	0.32	219	244	251	259	267	275	282	290	298	306	314	321
		BFL	0.43	165	234	240	246	252	258	264	270	275	281	287	293
		6	1.79	130	227	232	236	241	245	250	255	259	264	268	273
		7	3.44	95.4	215	219	222	226	229	232	236	239	243	246	249
W 16x31	146	TFL	0.00	328	231	243	254	266	278	289	301	313	324	336	347
		2	0.11	285	225	235	245	255	265	275	285	295	305	315	326
		3	0.22	241	218	227	235	244	252	261	269	278	286	295	303
		4	0.33	197	211	218	225	232	239	246	253	260	267	274	281
		BFL	0.44	153	204	209	214	220	225	231	236	242	247	253	258
		6	2.00	118	196	200	204	208	212	217	221	225	229	233	237
		7	3.79	82.1	183	186	189	192	195	198	201	204	207	209	212
W 16x26	119	TFL	0.00	276	193	203	212	222	232	242	252	261	271	281	291
		2	0.09	242	188	196	205	214	222	231	239	248	257	265	274
		3	0.17	208	183	190	197	205	212	220	227	234	242	249	256
		4	0.26	174	177	184	190	196	202	208	214	220	227	233	239
		BFL	0.35	140	172	177	182	187	192	197	202	206	211	216	221
		6	2.04	104	164	168	171	175	179	182	186	190	194	197	201
		7	4.01	69.1	151	154	156	159	161	164	166	169	171	173	176

[a]$Y1$ = distance from top of the steel beam to plastic neutral axis.
[b]$Y2$ = distance from top of the steel beam to concrete flange force.
[c]See Fig. 4.3 for PNA locations.

F_y = 36 ksi

COMPOSITE DESIGN
COMPOSITE BEAM SELECTION TABLE
W Shapes
$\phi = 0.85$ $\phi_b = 0.90$

Shape	$\phi_b M_p$	PNA[c]	$Y1$[a]	ΣQ_n	ϕM_n (kip-ft) $Y2$[b] (in.)										
	Kip-ft		In.	Kips	2	2.5	3	3.5	4	4.5	5	5.5	6	6.5	7
W 14x38	166	TFL	0.00	403	258	273	287	301	316	330	344	358	373	387	401
		2	0.13	340	249	261	273	285	298	310	322	334	346	358	370
		3	0.26	278	240	249	259	269	279	289	299	308	318	328	338
		4	0.39	215	229	237	244	252	260	267	275	283	290	298	305
		BFL	0.52	152	218	224	229	234	240	245	251	256	261	267	272
		6	1.38	126	213	218	222	226	231	235	240	244	249	253	258
		7	2.53	101	206	209	213	217	220	224	227	231	234	238	242
W 14x34	147	TFL	0.00	360	229	242	255	267	280	293	306	318	331	344	357
		2	0.11	305	221	232	243	254	264	275	286	297	308	318	329
		3	0.23	250	213	222	230	239	248	257	266	275	283	292	301
		4	0.34	194	204	211	218	224	231	238	245	252	259	266	273
		BFL	0.46	139	194	199	204	209	214	219	224	229	234	239	244
		6	1.41	115	189	193	197	202	206	210	214	218	222	226	230
		7	2.60	90.0	182	186	189	192	195	198	202	205	208	211	214
W 14x30	128	TFL	0.00	319	201	213	224	235	246	258	269	280	292	303	314
		2	0.10	272	195	204	214	223	233	243	252	262	272	281	291
		3	0.19	225	187	195	203	211	219	227	235	243	251	259	267
		4	0.29	179	180	186	193	199	205	212	218	224	231	237	243
		BFL	0.39	132	172	177	182	186	191	196	200	205	210	214	219
		6	1.48	106	167	171	174	178	182	186	189	193	197	201	204
		7	2.82	79.7	159	162	165	168	171	173	176	179	182	185	187
W 14x26	109	TFL	0.00	277	176	185	195	205	215	225	234	244	254	264	274
		2	0.11	239	170	179	187	195	204	212	221	229	238	246	255
		3	0.21	201	164	171	179	186	193	200	207	214	221	228	235
		4	0.32	163	158	164	170	175	181	187	193	199	204	210	216
		BFL	0.42	125	152	156	161	165	170	174	178	183	187	192	196
		6	1.67	97.0	146	149	153	156	160	163	167	170	173	177	180
		7	3.19	69.2	137	140	142	145	147	149	152	154	157	159	162
W 14x22	89.6	TFL	0.00	234	147	155	163	172	180	188	196	205	213	221	230
		2	0.08	203	142	150	157	164	171	178	186	193	200	207	215
		3	0.17	173	138	144	150	156	162	169	175	181	187	193	199
		4	0.25	143	133	138	143	148	153	159	164	169	174	179	184
		BFL	0.34	113	128	132	136	140	144	148	152	156	160	164	168
		6	1.69	85.7	123	126	129	132	135	138	141	144	147	150	153
		7	3.34	58.4	114	116	118	120	122	124	126	128	130	132	135

[a]$Y1$ = distance from top of the steel beam to plastic neutral axis.
[b]$Y2$ = distance from top of the steel beam to concrete flange force.
[c]See Fig. 4.3 for PNA locations.

AMERICAN INSTITUTE OF STEEL CONSTRUCTION

F_y = 36 ksi														

COMPOSITE DESIGN
COMPOSITE BEAM SELECTION TABLE
W Shapes
$\phi = 0.85$ $\phi_b = 0.90$

Shape	$\phi_b M_p$	PNA[c]	$Y1$[a]	ΣQ_n	ϕM_n (kip-ft) $Y2$[b] (in.)										
	Kip-ft		In.	Kips	2	2.5	3	3.5	4	4.5	5	5.5	6	6.5	7
W 12x30	116	TFL	0.00	316	183	194	206	217	228	239	250	262	273	284	295
		2	0.11	265	176	185	194	204	213	223	232	241	251	260	269
		3	0.22	213	168	175	183	190	198	205	213	221	228	236	243
		4	0.33	162	159	165	171	177	182	188	194	199	205	211	217
		BFL	0.44	110	151	155	158	162	166	170	174	178	182	186	190
		6	1.12	94.5	148	151	154	158	161	164	168	171	174	178	181
		7	1.94	79.1	144	147	149	152	155	158	161	163	166	169	172
W 12x26	100	TFL	0.00	275	158	168	178	187	197	207	217	226	236	246	256
		2	0.10	231	152	160	168	176	184	193	201	209	217	225	234
		3	0.19	187	145	152	158	165	171	178	185	191	198	205	211
		4	0.29	142	138	143	148	153	158	163	168	173	178	183	188
		BFL	0.38	97.8	131	134	138	141	145	148	151	155	158	162	165
		6	1.08	83.3	128	131	134	137	140	143	146	149	151	154	157
		7	1.95	68.9	124	127	129	132	134	136	139	141	144	146	149
W 12x22	79.1	TFL	0.00	233	135	143	151	160	168	176	184	193	201	209	217
		2	0.11	202	130	137	145	152	159	166	173	180	188	195	202
		3	0.21	172	126	132	138	144	150	156	162	168	174	180	186
		4	0.32	141	121	126	131	136	141	146	151	156	160	165	170
		BFL	0.43	110	115	119	123	127	131	135	139	143	147	150	154
		6	1.66	84.1	110	113	116	119	122	125	128	131	134	137	140
		7	3.04	58.3	102	104	106	108	110	112	114	116	119	121	123
W 12x19	66.7	TFL	0.00	201	115	122	129	136	143	150	157	164	172	179	186
		2	0.09	175	111	117	124	130	136	142	148	155	161	167	173
		3	0.18	150	107	113	118	123	129	134	139	145	150	155	160
		4	0.26	125	103	108	112	117	121	125	130	134	139	143	148
		BFL	0.35	99.6	99.2	103	106	110	113	117	120	124	127	131	134
		6	1.66	74.9	94.0	96.7	99.3	102	105	107	110	113	115	118	121
		7	3.12	50.1	86.3	88.1	89.9	91.7	93.5	95.2	97.0	98.8	101	102	104
W 12x16	54.3	TFL	0.00	170	96.0	102	108	114	120	126	132	138	144	150	156
		2	0.07	151	93.3	98.6	104	109	115	120	125	131	136	141	147
		3	0.13	131	90.5	95.1	99.8	104	109	114	118	123	128	132	137
		4	0.20	112	87.5	91.5	95.5	99.5	103	107	111	115	119	123	127
		BFL	0.26	93.4	84.5	87.8	91.1	94.5	97.8	101	104	108	111	114	118
		6	1.71	67.9	79.2	81.6	84.0	86.4	88.8	91.2	93.6	96.1	98.5	101	103
		7	3.32	42.4	71.1	72.6	74.1	75.6	77.1	78.6	80.1	81.6	83.1	84.6	86.1

[a]$Y1$ = distance from top of the steel beam to plastic neutral axis.
[b]$Y2$ = distance from top of the steel beam to concrete flange force.
[c]See Fig. 4.3 for PNA locations.

AMERICAN INSTITUTE OF STEEL CONSTRUCTION

$F_y = 36$ ksi

COMPOSITE DESIGN
COMPOSITE BEAM SELECTION TABLE
W Shapes

$$\phi = 0.85 \qquad\qquad \phi_b = 0.90$$

Shape	$\phi_b M_p$ Kip-ft	PNA[c]	Y1[a] In.	ΣQ_n Kips	ϕM_n (kip-ft) Y2[b] (in.) 2	2.5	3	3.5	4	4.5	5	5.5	6	6.5	7
W 12x14	47.0	TFL	0.00	150	84.4	89.7	95.0	100	106	111	116	122	127	132	137
		2	0.06	134	82.1	86.8	91.5	96.3	101	106	110	115	120	125	129
		3	0.11	118	79.7	83.9	88.0	92.2	96.4	101	105	109	113	117	121
		4	0.17	102	77.3	80.9	84.5	88.1	91.6	95.2	98.8	102	106	110	113
		BFL	0.23	85.4	74.8	77.8	80.8	83.8	86.9	89.9	92.9	95.9	99.0	102	105
		6	1.69	61.4	69.8	72.0	74.2	76.4	78.5	80.7	82.9	85.1	87.2	89.4	91.6
		7	3.36	37.4	62.2	63.5	64.8	66.1	67.5	68.8	70.1	71.4	72.8	74.1	75.4
W 10x26	84.5	TFL	0.00	274	139	149	158	168	178	188	197	207	217	226	236
		2	0.11	228	132	140	149	157	165	173	181	189	197	205	213
		3	0.22	183	125	132	138	145	151	158	164	171	177	184	190
		4	0.33	137	118	123	128	133	137	142	147	152	157	162	166
		BFL	0.44	91.2	110	114	117	120	123	126	130	133	136	139	143
		6	0.90	79.8	108	111	114	117	119	122	125	128	131	134	136
		7	1.51	68.5	106	108	110	113	115	118	120	123	125	127	130
W 10x22	70.2	TFL	0.00	234	117	126	134	142	150	159	167	175	183	192	200
		2	0.09	196	112	119	126	133	140	147	154	161	167	174	181
		3	0.18	159	106	112	117	123	129	134	140	146	151	157	163
		4	0.27	122	100	105	109	113	118	122	126	131	135	139	144
		BFL	0.36	84.6	94.2	97.2	100	103	106	109	112	115	118	121	124
		6	0.95	71.5	91.8	94.3	96.9	99.4	102	104	107	110	112	115	117
		7	1.70	58.4	88.7	90.8	92.9	94.9	97.0	99.1	101	103	105	107	109
W 10x19	58.3	TFL	0.00	202	102	109	116	124	131	138	145	152	159	167	174
		2	0.10	174	97.9	104	110	116	123	129	135	141	147	153	159
		3	0.20	145	93.5	98.7	104	109	114	119	124	130	135	140	145
		4	0.30	117	89.0	93.1	97.2	101	106	110	114	118	122	126	130
		BFL	0.40	88.0	84.2	87.4	90.5	93.6	96.7	99.8	103	106	109	112	115
		6	1.27	69.3	80.5	83.0	85.4	87.9	90.4	92.8	95.3	97.7	100	103	105
		7	2.31	50.6	75.5	77.3	79.1	80.9	82.7	84.5	86.3	88.1	89.8	91.6	93.4
W 10x17	50.5	TFL	0.00	180	89.8	96.1	102	109	115	122	128	134	141	147	153
		2	0.08	156	86.3	91.8	97.4	103	108	114	119	125	130	136	142
		3	0.17	132	82.7	87.4	92.1	96.8	101	106	111	115	120	125	129
		4	0.25	108	79.0	82.9	86.7	90.5	94.3	98.2	102	106	110	114	117
		BFL	0.33	84.4	75.2	78.1	81.1	84.1	87.1	90.1	93.1	96.1	99.1	102	105
		6	1.31	64.6	71.3	73.6	75.8	78.1	80.4	82.7	85.0	87.3	89.6	91.9	94.2
		7	2.46	44.9	65.8	67.4	69.0	70.6	72.2	73.8	75.4	77.0	78.6	80.2	81.7

[a]$Y1$ = distance from top of the steel beam to plastic neutral axis.
[b]$Y2$ = distance from top of the steel beam to concrete flange force.
[c]See Fig. 4.3 for PNA locations.

$F_y = 36$ ksi

COMPOSITE DESIGN
COMPOSITE BEAM SELECTION TABLE
W Shapes
$\phi = 0.85$ $\qquad\qquad$ $\phi_b = 0.90$

Shape	$\phi_b M_p$	PNA[c]	Y1[a]	ΣQ_n	ϕM_n (kip-ft)										
					Y2[b] (in.)										
	Kip-ft		In.	Kips	2	2.5	3	3.5	4	4.5	5	5.5	6	6.5	7
W 10x15	43.2	TFL	0.00	159	78.7	84.3	89.9	95.5	101	107	112	118	124	129	135
		2	0.07	139	75.9	80.8	85.7	90.7	95.6	101	105	110	115	120	125
		3	0.14	120	73.0	77.2	81.5	85.7	90.0	94.2	98.4	103	107	111	115
		4	0.20	100	70.0	73.5	77.1	80.7	84.2	87.8	91.3	94.9	98.4	102	106
		BFL	0.27	81.0	66.9	69.8	72.6	75.5	78.4	81.2	84.1	87.0	89.9	92.7	95.6
		6	1.35	60.3	62.9	65.0	67.1	69.3	71.4	73.5	75.7	77.8	80.0	82.1	84.2
		7	2.60	39.7	57.0	58.4	59.9	61.3	62.7	64.1	65.5	66.9	68.3	69.7	71.1
W 10x12	34.0	TFL	0.00	127	62.6	67.1	71.6	76.1	80.7	85.2	89.7	94.2	98.7	103	108
		2	0.05	112	60.5	64.4	68.4	72.4	76.4	80.4	84.4	88.3	92.3	96.3	100
		3	0.11	97.5	58.2	61.7	65.2	68.6	72.1	75.5	79.0	82.4	85.9	89.3	92.8
		4	0.16	82.5	56.0	58.9	61.8	64.8	67.7	70.6	73.5	76.5	79.4	82.3	85.2
		BFL	0.21	67.6	53.7	56.1	58.5	60.9	63.2	65.6	68.0	70.4	72.8	75.2	77.6
		6	1.30	49.7	50.3	52.0	53.8	55.5	57.3	59.1	60.8	62.6	64.3	66.1	67.9
		7	2.61	31.9	45.3	46.4	47.5	48.6	49.8	50.9	52.0	53.2	54.3	55.4	56.5
W 8x28	73.4	TFL	0.00	297	127	137	148	158	169	179	190	200	211	222	232
		2	0.12	242	119	127	136	145	153	162	170	179	188	196	205
		3	0.23	188	110	117	124	130	137	144	150	157	164	170	177
		4	0.35	133	102	106	111	116	120	125	130	135	139	144	149
		BFL	0.47	78.2	92.3	95.0	97.8	101	103	106	109	112	114	117	120
		6	0.53	76.2	91.9	94.6	97.3	100	103	105	108	111	114	116	119
		7	0.59	74.3	91.5	94.2	96.8	99.4	102	105	107	110	113	115	118
W 8x24	62.6	TFL	0.00	255	108	117	126	135	144	153	162	171	180	189	198
		2	0.10	208	101	108	116	123	130	138	145	152	160	167	175
		3	0.20	161	93.8	99.5	105	111	117	122	128	134	139	145	151
		4	0.30	115	86.3	90.4	94.4	98.5	103	107	111	115	119	123	127
		BFL	0.40	67.8	78.5	80.9	83.3	85.7	88.2	90.6	93.0	95.4	97.8	100	103
		6	0.47	65.8	78.2	80.5	82.8	85.2	87.5	89.8	92.2	94.5	96.8	99.2	101
		7	0.55	63.7	77.8	80.1	82.3	84.6	86.9	89.1	91.4	93.6	95.9	98.1	100
W 8x21	55.1	TFL	0.00	222	96.4	104	112	120	128	136	144	151	159	167	175
		2	0.10	184	90.9	97.4	104	110	117	123	130	137	143	150	156
		3	0.20	146	85.2	90.3	95.5	101	106	111	116	121	126	132	137
		4	0.30	108	79.1	82.9	86.8	90.6	94.4	98.2	102	106	110	114	117
		BFL	0.40	70.0	72.8	75.3	77.8	80.2	82.7	85.2	87.7	90.1	92.6	95.1	97.6
		6	0.70	62.7	71.5	73.7	75.9	78.1	80.4	82.6	84.8	87.0	89.3	91.5	93.7
		7	1.06	55.4	70.0	72.0	73.9	75.9	77.9	79.8	81.8	83.8	85.7	87.7	89.6

[a] Y1 = distance from top of the steel beam to plastic neutral axis.
[b] Y2 = distance from top of the steel beam to concrete flange force.
[c] See Fig. 4.3 for PNA locations.

COMPOSITE DESIGN
COMPOSITE BEAM SELECTION TABLE
W Shapes

$F_y = 36$ ksi

$\phi = 0.85$ $\phi_b = 0.90$

Shape	$\phi_b M_p$	PNA[c]	Y1[a]	ΣQ_n	ϕM_n (kip-ft) / Y2[b] (in.)										
	Kip-ft		In.	Kips	2	2.5	3	3.5	4	4.5	5	5.5	6	6.5	7
W 8x18	45.9	TFL	0.00	189	81.4	88.1	94.8	102	108	115	122	128	135	142	148
		2	0.08	158	76.9	82.5	88.1	93.7	99.3	105	111	116	122	127	133
		3	0.17	127	72.2	76.7	81.2	85.7	90.2	94.7	99.2	104	108	113	117
		4	0.25	95.8	67.3	70.7	74.1	77.5	80.9	84.3	87.7	91.1	94.5	97.9	101
		BFL	0.33	64.6	62.3	64.6	66.9	69.2	71.4	73.7	76.0	78.3	80.6	82.9	85.2
		6	0.71	56.0	60.7	62.7	64.7	66.7	68.7	70.7	72.6	74.6	76.6	78.6	80.6
		7	1.21	47.3	58.9	60.6	62.3	64.0	65.6	67.3	69.0	70.7	72.4	74.0	75.7
W 8x15	36.7	TFL	0.00	160	68.6	74.2	79.9	85.5	91.2	96.9	103	108	114	120	125
		2	0.08	137	65.3	70.1	75.0	79.8	84.7	89.5	94.4	99.2	104	109	114
		3	0.16	114	61.9	65.9	69.9	74.0	78.0	82.1	86.1	90.2	94.2	98.3	102
		4	0.24	91.5	58.3	61.6	64.8	68.0	71.3	74.5	77.8	81.0	84.2	87.5	90.7
		BFL	0.32	68.8	54.6	57.1	59.5	61.9	64.4	66.8	69.3	71.7	74.1	76.6	79.0
		6	0.97	54.4	52.0	53.9	55.8	57.7	59.7	61.6	63.5	65.4	67.4	69.3	71.2
		7	1.79	40.0	48.5	49.9	51.3	52.8	54.2	55.6	57.0	58.4	59.8	61.2	62.7
W 8x13	30.8	TFL	0.00	138	58.7	63.6	68.5	73.4	78.3	83.2	88.1	93.0	97.9	103	108
		2	0.06	120	56.1	60.3	64.6	68.8	73.0	77.3	81.5	85.8	90.0	94.3	98.5
		3	0.13	102	53.3	56.9	60.5	64.1	67.7	71.3	74.9	78.5	82.1	85.7	89.3
		4	0.19	83.2	50.5	53.5	56.4	59.4	62.3	65.3	68.2	71.1	74.1	77.0	80.0
		BFL	0.26	64.8	47.6	49.9	52.2	54.5	56.8	59.1	61.4	63.7	66.0	68.3	70.6
		6	1.00	49.7	44.9	46.6	48.4	50.1	51.9	53.7	55.4	57.2	58.9	60.7	62.5
		7	1.91	34.6	41.2	42.4	43.6	44.8	46.1	47.3	48.5	49.7	51.0	52.2	53.4
W 8x10	23.9	TFL	0.00	107	44.9	48.6	52.4	56.2	60.0	63.7	67.5	71.3	75.1	78.8	82.6
		2	0.05	92.0	42.8	46.0	49.3	52.6	55.8	59.1	62.3	65.6	68.9	72.1	75.4
		3	0.10	77.5	40.6	43.4	46.1	48.9	51.6	54.4	57.1	59.9	62.6	65.3	68.1
		4	0.15	62.9	38.5	40.7	42.9	45.1	47.4	49.6	51.8	54.1	56.3	58.5	60.7
		BFL	0.21	48.4	36.2	37.9	39.6	41.4	43.1	44.8	46.5	48.2	49.9	51.6	53.4
		6	0.88	37.5	34.3	35.6	36.9	38.3	39.6	40.9	42.2	43.6	44.9	46.2	47.6
		7	1.77	26.6	31.7	32.7	33.6	34.5	35.5	36.4	37.4	38.3	39.3	40.2	41.1

[a]Y1 = distance from top of the steel beam to plastic neutral axis.
[b]Y2 = distance from top of the steel beam to concrete flange force.
[c]See Fig. 4.3 for PNA locations.

F_y = 50 ksi

COMPOSITE DESIGN
COMPOSITE BEAM SELECTION TABLE
W Shapes
$\phi = 0.85$ $\phi_b = 0.90$

Shape	$\phi_b M_p$	PNA[c]	Y1[a]	ΣQ_n	ϕM_n (kip-ft) Y2[b] (in.) 2	2.5	3	3.5	4	4.5	5	5.5	6	6.5	7
	Kip-ft		In.	Kips											
W 36x300	4730	TFL	0.00	4420	6370	6530	6680	6840	7000	7150	7310	7460	7620	7780	7930
		2	0.42	3720	6260	6390	6520	6660	6790	6920	7050	7180	7310	7450	7580
		3	0.84	3020	6130	6240	6340	6450	6560	6660	6770	6880	6990	7090	7200
		4	1.26	2320	5980	6060	6140	6230	6310	6390	6470	6550	6640	6720	6800
		BFL	1.68	1620	5810	5860	5920	5980	6040	6090	6150	6210	6270	6320	6380
		6	3.97	1360	5720	5770	5820	5870	5910	5960	6010	6060	6110	6150	6200
		7	6.69	1100	5590	5630	5670	5710	5740	5780	5820	5860	5900	5940	5980
W 36x280	4390	TFL	0.00	4120	5910	6060	6200	6350	6500	6640	6790	6930	7080	7230	7370
		2	0.39	3470	5810	5930	6060	6180	6300	6430	6550	6670	6790	6920	7040
		3	0.79	2820	5690	5790	5890	5990	6090	6190	6290	6390	6490	6590	6690
		4	1.18	2170	5550	5630	5710	5780	5860	5940	6010	6090	6170	6240	6320
		BFL	1.57	1510	5400	5450	5510	5560	5610	5670	5720	5770	5830	5880	5930
		6	3.88	1270	5320	5360	5410	5450	5500	5540	5590	5630	5680	5720	5770
		7	6.62	1030	5190	5230	5270	5300	5340	5380	5410	5450	5490	5520	5560
W 36x260	4050	TFL	0.00	3830	5450	5590	5720	5860	6000	6130	6270	6400	6540	6670	6810
		2	0.36	3230	5360	5480	5590	5710	5820	5930	6050	6160	6280	6390	6510
		3	0.72	2630	5250	5350	5440	5530	5630	5720	5810	5910	6000	6090	6190
		4	1.08	2040	5130	5200	5280	5350	5420	5490	5570	5640	5710	5780	5850
		BFL	1.44	1440	4990	5050	5100	5150	5200	5250	5300	5350	5400	5450	5510
		6	3.86	1200	4920	4960	5000	5040	5090	5130	5170	5210	5260	5300	5340
		7	6.75	956	4790	4830	4860	4890	4930	4960	4990	5030	5060	5100	5130
W 36x245	3790	TFL	0.00	3610	5120	5240	5370	5500	5630	5760	5880	6010	6140	6270	6390
		2	0.34	3050	5030	5140	5250	5360	5460	5570	5680	5790	5900	6000	6110
		3	0.68	2490	4930	5020	5110	5200	5290	5370	5460	5550	5640	5730	5810
		4	1.01	1930	4820	4890	4960	5030	5090	5160	5230	5300	5370	5440	5510
		BFL	1.35	1380	4690	4740	4790	4840	4890	4940	4990	5040	5080	5130	5180
		6	3.81	1140	4620	4660	4700	4740	4780	4820	4860	4900	4940	4980	5020
		7	6.77	901	4500	4530	4560	4590	4620	4660	4690	4720	4750	4780	4820
W 36x230	3540	TFL	0.00	3380	4780	4900	5020	5140	5260	5370	5490	5610	5730	5850	5970
		2	0.32	2860	4700	4800	4900	5000	5100	5200	5310	5410	5510	5610	5710
		3	0.63	2340	4610	4690	4770	4860	4940	5020	5100	5190	5270	5350	5440
		4	0.95	1820	4500	4570	4630	4700	4760	4830	4890	4960	5020	5090	5150
		BFL	1.26	1300	4390	4440	4480	4530	4570	4620	4670	4710	4760	4810	4850
		6	3.81	1070	4320	4350	4390	4430	4470	4510	4540	4580	4620	4660	4690
		7	6.83	845	4200	4230	4260	4290	4320	4350	4380	4410	4440	4470	4500

[a] Y1 = distance from top of the steel beam to plastic neutral axis.
[b] Y2 = distance from top of the steel beam to concrete flange force.
[c] See Fig. 4.3 for PNA locations.

$F_y = 50$ ksi

COMPOSITE DESIGN
COMPOSITE BEAM SELECTION TABLE
W Shapes

$\phi = 0.85$ $\phi_b = 0.90$

Shape	$\phi_b M_p$	PNA[c]	$Y1$[a]	ΣQ_n	\multicolumn{11}{c}{ϕM_n (kip-ft) — $Y2$[b] (in.)}										
	Kip-ft		In.	Kips	2	2.5	3	3.5	4	4.5	5	5.5	6	6.5	7
W 36x210	3120	TFL	0.00	3090	4450	4560	4670	4780	4890	5000	5110	5220	5330	5440	5550
		2	0.34	2680	4390	4480	4580	4670	4770	4860	4960	5050	5150	5240	5340
		3	0.68	2260	4320	4400	4480	4560	4640	4720	4800	4880	4960	5040	5120
		4	1.02	1850	4230	4300	4360	4430	4490	4560	4620	4690	4760	4820	4890
		BFL	1.36	1430	4140	4190	4240	4290	4340	4390	4440	4490	4540	4600	4650
		6	5.06	1100	4020	4060	4100	4130	4170	4210	4250	4290	4330	4370	4410
		7	9.04	773	3810	3830	3860	3890	3920	3940	3970	4000	4030	4050	4080
W 36x194	2880	TFL	0.00	2850	4090	4190	4290	4390	4490	4590	4690	4790	4890	5000	5100
		2	0.32	2470	4030	4120	4200	4290	4380	4470	4550	4640	4730	4820	4900
		3	0.63	2090	3960	4040	4110	4180	4260	4330	4410	4480	4550	4630	4700
		4	0.95	1710	3890	3950	4010	4070	4130	4190	4250	4310	4370	4430	4490
		BFL	1.26	1320	3800	3850	3900	3940	3990	4040	4080	4130	4180	4220	4270
		6	4.94	1020	3690	3730	3770	3800	3840	3880	3910	3950	3980	4020	4060
		7	8.93	713	3500	3530	3550	3580	3600	3630	3650	3680	3700	3730	3750
W 36x182	2690	TFL	0.00	2680	3830	3920	4020	4110	4210	4300	4400	4490	4590	4680	4780
		2	0.29	2320	3770	3860	3940	4020	4100	4190	4270	4350	4430	4510	4600
		3	0.59	1970	3710	3780	3850	3920	3990	4060	4130	4200	4270	4340	4410
		4	0.89	1610	3640	3700	3760	3810	3870	3930	3990	4040	4100	4160	4210
		BFL	1.18	1260	3570	3610	3660	3700	3740	3790	3830	3880	3920	3970	4010
		6	4.89	963	3460	3500	3530	3570	3600	3640	3670	3700	3740	3770	3810
		7	8.92	670	3280	3300	3330	3350	3370	3400	3420	3450	3470	3490	3520
W 36x170	2500	TFL	0.00	2500	3560	3650	3730	3820	3910	4000	4090	4180	4270	4350	4440
		2	0.28	2170	3510	3580	3660	3740	3810	3890	3970	4040	4120	4200	4270
		3	0.55	1840	3450	3520	3580	3650	3710	3780	3840	3910	3970	4040	4100
		4	0.83	1510	3390	3440	3490	3550	3600	3650	3710	3760	3810	3870	3920
		BFL	1.10	1180	3320	3360	3400	3440	3480	3530	3570	3610	3650	3690	3730
		6	4.84	901	3220	3250	3290	3320	3350	3380	3410	3450	3480	3510	3540
		7	8.89	625	3050	3070	3090	3120	3140	3160	3180	3200	3230	3250	3270
W 36x160	2340	TFL	0.00	2350	3330	3410	3500	3580	3660	3750	3830	3910	4000	4080	4160
		2	0.26	2040	3280	3360	3430	3500	3570	3650	3720	3790	3860	3940	4010
		3	0.51	1740	3230	3290	3360	3420	3480	3540	3600	3660	3720	3790	3850
		4	0.77	1430	3180	3230	3280	3330	3380	3430	3480	3530	3580	3630	3680
		BFL	1.02	1130	3110	3150	3190	3230	3270	3310	3350	3390	3430	3470	3510
		6	4.82	857	3020	3050	3080	3110	3140	3170	3200	3230	3260	3290	3320
		7	8.97	588	2850	2870	2890	2910	2930	2960	2980	3000	3020	3040	3060

[a] $Y1$ = distance from top of the steel beam to plastic neutral axis.
[b] $Y2$ = distance from top of the steel beam to concrete flange force.
[c] See Fig. 4.3 for PNA locations.

$F_y = 50$ ksi

COMPOSITE DESIGN
COMPOSITE BEAM SELECTION TABLE
W Shapes

$\phi = 0.85$ $\phi_b = 0.90$

Shape	$\phi_b M_p$	PNA[c]	$Y1$[a]	ΣQ_n	ϕM_n (kip-ft) $Y2$[b] (in.)										
	Kip-ft		In.	Kips	2	2.5	3	3.5	4	4.5	5	5.5	6	6.5	7
W 36x150	2180	TFL	0.00	2210	3120	3200	3280	3350	3430	3510	3590	3670	3750	3820	3900
		2	0.24	1930	3080	3150	3210	3280	3350	3420	3490	3560	3620	3690	3760
		3	0.47	1650	3030	3090	3150	3210	3260	3320	3380	3440	3500	3560	3610
		4	0.71	1370	2980	3030	3080	3120	3170	3220	3270	3320	3370	3410	3460
		BFL	0.94	1080	2920	2960	3000	3040	3080	3110	3150	3190	3230	3270	3310
		6	4.83	818	2830	2860	2890	2920	2950	2980	3010	3040	3060	3090	3120
		7	9.08	553	2660	2680	2700	2720	2740	2760	2780	2800	2820	2840	2860
W 36x135	1910	TFL	0.00	1990	2780	2850	2920	2990	3060	3130	3200	3270	3340	3410	3480
		2	0.20	1750	2750	2810	2870	2930	2990	3060	3120	3180	3240	3300	3360
		3	0.40	1510	2710	2760	2810	2870	2920	2970	3030	3080	3140	3190	3240
		4	0.59	1280	2670	2710	2760	2800	2850	2890	2940	2980	3030	3070	3120
		BFL	0.79	1040	2620	2660	2690	2730	2770	2800	2840	2880	2920	2950	2990
		6	4.97	769	2530	2560	2580	2610	2640	2660	2690	2720	2750	2770	2800
		7	9.50	496	2350	2370	2390	2400	2420	2440	2460	2470	2490	2510	2530
W 33x221	3210	TFL	0.00	3250	4370	4480	4600	4710	4830	4940	5060	5170	5290	5400	5520
		2	0.32	2750	4290	4390	4480	4580	4680	4780	4870	4970	5070	5160	5260
		3	0.64	2240	4200	4280	4360	4440	4520	4600	4680	4760	4840	4920	4990
		4	0.96	1740	4100	4160	4220	4290	4350	4410	4470	4530	4590	4650	4720
		BFL	1.28	1230	3990	4030	4080	4120	4160	4210	4250	4300	4340	4380	4430
		6	3.76	1020	3920	3960	3990	4030	4070	4100	4140	4170	4120	4250	4280
		7	6.48	813	3820	3850	3870	3900	3930	3960	3990	4020	4050	4080	4100
W 33x201	2900	TFL	0.00	2960	3940	4050	4150	4260	4360	4470	4570	4680	4780	4890	4990
		2	0.29	2500	3870	3960	4050	4140	4230	4320	4410	4500	4580	4670	4760
		3	0.58	2050	3800	3870	3940	4010	4090	4160	4230	4300	4380	4450	4520
		4	0.86	1600	3710	3770	3820	3880	3940	3990	4050	4110	4160	4220	4280
		BFL	1.15	1140	3610	3650	3690	3730	3780	3820	3860	3900	3940	3980	4020
		6	3.67	942	3550	3580	3620	3650	3680	3720	3750	3780	3820	3850	3880
		7	6.51	739	3450	3480	3500	3530	3550	3580	3610	3630	3660	3680	3710
W 33x141	1930	TFL	0.00	2080	2750	2820	2900	2970	3040	3120	3190	3260	3340	3410	3480
		2	0.24	1800	2710	2770	2830	2900	2960	3030	3090	3150	3220	3280	3340
		3	0.48	1530	2660	2710	2770	2820	2880	2930	2980	3040	3090	3150	3200
		4	0.72	1250	2610	2650	2700	2740	2790	2830	2870	2920	2960	3010	3050
		BFL	0.96	973	2550	2590	2620	2660	2690	2730	2760	2790	2830	2860	2900
		6	4.31	746	2480	2510	2530	2560	2590	2610	2640	2670	2690	2720	2750
		7	8.05	520	2350	2370	2390	2410	2420	2440	2460	2480	2500	2520	2530

[a] $Y1$ = distance from top of the steel beam to plastic neutral axis.
[b] $Y2$ = distance from top of the steel beam to concrete flange force.
[c] See Fig. 4.3 for PNA locations.

$$F_y = 50 \text{ ksi}$$

COMPOSITE DESIGN
COMPOSITE BEAM SELECTION TABLE
W Shapes

$$\phi = 0.85 \qquad\qquad \phi_b = 0.90$$

Shape	$\phi_b M_p$	PNA[c]	$Y1^a$	ΣQ_n	ϕM_n (kip-ft) $Y2^b$ (in.)										
					2	2.5	3	3.5	4	4.5	5	5.5	6	6.5	7
	Kip-ft		In.	Kips											
W 33x130	1750	TFL	0.00	1920	2520	2580	2650	2720	2790	2850	2920	2990	3060	3130	3190
		2	0.21	1670	2480	2540	2600	2660	2720	2770	2830	2890	2950	3010	3070
		3	0.43	1420	2440	2490	2540	2590	2640	2690	2740	2790	2840	2890	2940
		4	0.64	1180	2390	2440	2480	2520	2560	2600	2640	2690	2730	2770	2810
		BFL	0.86	931	2350	2380	2410	2450	2480	2510	2540	2580	2610	2640	2680
		6	4.39	705	2770	2300	2320	2350	2370	2400	2420	2450	2470	2500	2520
		7	8.29	479	2140	2160	2170	2190	2210	2230	2240	2260	2280	2290	2310
W 33x118	1560	TFL	0.00	1740	2260	2330	2390	2450	2510	2570	2630	2700	2760	2820	2880
		2	0.19	1520	2230	2290	2340	2400	2450	2500	2560	2610	2660	2720	2770
		3	0.37	1310	2200	2250	2290	2340	2380	2430	2480	2520	2570	2620	2660
		4	0.56	1100	2160	2200	2240	2280	2320	2360	2400	2430	2470	2510	2550
		BFL	0.74	885	2120	2150	2190	2220	2250	2280	2310	2340	2370	2400	2440
		6	4.44	660	2050	2070	2100	2120	2140	2170	2190	2210	2240	2260	2280
		7	8.54	434	1920	1930	1950	1960	1980	1990	2010	2020	2040	2050	2070
W 30x116	1420	TFL	0.00	1710	2060	2120	2180	2240	2300	2360	2420	2480	2540	2600	2670
		2	0.21	1490	2030	2080	2130	2180	2240	2290	2340	2400	2450	2500	2550
		3	0.43	1260	1990	2030	2080	2120	2170	2210	2260	2300	2350	2390	2440
		4	0.64	1040	1950	1990	2020	2060	2100	2130	2170	2210	2240	2280	2320
		BFL	0.85	818	1910	1940	1960	1990	2020	2050	2080	2110	2140	2170	2200
		6	3.98	623	1850	1870	1890	1910	1940	1960	1980	2000	2020	2050	2070
		7	7.44	428	1740	1760	1770	1790	1800	1820	1830	1850	1860	1880	1890
W 30x108	1300	TFL	0.00	1590	1900	1960	2010	2070	2120	2180	2240	2290	2350	2400	2460
		2	0.19	1390	1870	1920	1970	2020	2070	2110	2160	2210	2260	2310	2360
		3	0.38	1190	1840	1880	1920	1960	2010	2050	2090	2130	2170	2220	2260
		4	0.57	988	1800	1840	1870	1910	1940	1980	2010	2050	2080	2120	2150
		BFL	0.76	789	1760	1790	1820	1850	1880	1900	1930	1960	1990	2020	2040
		6	4.04	593	1710	1730	1750	1770	1790	1810	1830	1850	1870	1890	1920
		7	7.64	396	1600	1610	1620	1640	1650	1670	1680	1690	1710	1720	1740
W 30x 99	1170	TFL	0.00	1460	1730	1790	1840	1890	1940	1990	2040	2090	2150	2200	2250
		2	0.17	1280	1710	1750	1800	1840	1890	1930	1980	2030	2070	2120	2160
		3	0.34	1100	1680	1720	1760	1800	1840	1880	1920	1950	1990	2030	2070
		4	0.50	930	1650	1680	1720	1750	1780	1810	1850	1880	1910	1950	1980
		BFL	0.67	755	1620	1640	1670	1700	1730	1750	1780	1810	1830	1860	1890
		6	4.07	559	1560	1580	1600	1620	1640	1660	1680	1700	1720	1740	1760
		7	7.83	364	1450	1460	1480	1490	1500	1510	1530	1540	1550	1570	1580

[a] $Y1$ = distance from top of the steel beam to plastic neutral axis.
[b] $Y2$ = distance from top of the steel beam to concrete flange force.
[c] See Fig. 4.3 for PNA locations.

$F_y = 50$ ksi

COMPOSITE DESIGN
COMPOSITE BEAM SELECTION TABLE
W Shapes

$\phi = 0.85$ $\phi_b = 0.90$

Shape	$\phi_b M_p$ Kip-ft	PNA[c]	$Y1^a$ In.	ΣQ_n Kips	ϕM_n (kip-ft) $Y2^b$ (in.) 2	2.5	3	3.5	4	4.5	5	5.5	6	6.5	7
W 27x102	1140	TFL	0.00	1500	1650	1700	1760	1810	1860	1920	1970	2020	2080	2130	2180
		2	0.21	1290	1620	1670	1710	1760	1800	1850	1900	1940	1990	2030	2080
		3	0.42	1080	1590	1630	1660	1700	1740	1780	1820	1860	1890	1930	1970
		4	0.62	877	1550	1580	1610	1640	1670	1700	1740	1770	1800	1830	1860
		BFL	0.83	669	1510	1530	1560	1580	1600	1630	1650	1680	1700	1720	1750
		6	3.41	522	1470	1490	1500	1520	1540	1560	1580	1600	1620	1630	1650
		7	6.26	375	1400	1410	1420	1440	1450	1460	1480	1490	1500	1520	1530
W 27x 94	1040	TFL	0.00	1390	1520	1570	1610	1660	1710	1760	1810	1860	1910	1960	2010
		2	0.19	1200	1490	1530	1570	1620	1660	1700	1740	1790	1830	1870	1910
		3	0.37	1010	1460	1490	1530	1570	1600	1640	1670	1710	1750	1780	1820
		4	0.56	827	1430	1460	1490	1510	1540	1570	1600	1630	1660	1690	1720
		BFL	0.75	641	1390	1410	1440	1460	1480	1510	1530	1550	1570	1600	1620
		6	3.39	493	1350	1370	1390	1400	1420	1440	1460	1470	1490	1510	1530
		7	6.39	346	1280	1290	1300	1320	1330	1340	1350	1360	1380	1390	1400
W 27x 84	915	TFL	0.00	1240	1350	1390	1440	1480	1520	1570	1610	1660	1700	1740	1790
		2	0.16	1080	1330	1360	1400	1440	1480	1520	1550	1590	1630	1670	1710
		3	0.32	921	1300	1330	1370	1400	1430	1460	1500	1530	1560	1590	1630
		4	0.48	762	1270	1300	1330	1350	1380	1410	1430	1460	1490	1520	1540
		BFL	0.64	603	1240	1270	1290	1310	1330	1350	1370	1390	1410	1440	1460
		6	3.44	456	1200	1220	1240	1250	1270	1280	1300	1320	1330	1350	1360
		7	6.62	310	1130	1140	1150	1160	1170	1190	1200	1210	1220	1230	1240
W 24x 76	750	TFL	0.00	1120	1110	1150	1190	1230	1270	1310	1350	1390	1420	1460	1500
		2	0.17	967	1080	1120	1150	1190	1220	1260	1290	1320	1360	1390	1430
		3	0.34	814	1060	1090	1120	1150	1180	1200	1230	1260	1290	1320	1350
		4	0.51	662	1030	1060	1080	1100	1130	1150	1170	1200	1220	1250	1270
		BFL	0.68	509	1010	1020	1040	1060	1080	1100	1110	1130	1150	1170	1190
		6	3.00	394	976	990	1000	1020	1030	1050	1060	1070	1090	1100	1120
		7	5.60	280	925	935	945	955	964	974	984	994	1000	1010	1020
W 24x 68	664	TFL	0.00	1010	987	1020	1060	1090	1130	1160	1200	1240	1270	1310	1340
		2	0.15	874	968	999	1030	1060	1090	1120	1150	1180	1220	1250	1280
		3	0.29	743	947	973	1000	1030	1050	1080	1100	1130	1160	1180	1210
		4	0.44	612	925	947	969	990	1010	1030	1060	1080	1100	1120	1140
		BFL	0.59	481	902	919	936	953	970	987	1000	1020	1040	1060	1070
		6	3.05	366	872	885	898	910	923	936	949	962	975	988	1000
		7	5.81	251	819	828	837	846	855	864	873	882	891	899	908

[a]$Y1$ = distance from top of the steel beam to plastic neutral axis.
[b]$Y2$ = distance from top of the steel beam to concrete flange force.
[c]See Fig. 4.3 for PNA locations.

				ϕM_n (kip-ft)										

F_y = 50 ksi

COMPOSITE DESIGN
COMPOSITE BEAM SELECTION TABLE
W Shapes

$\phi = 0.85$ $\phi_b = 0.90$

Shape	$\phi_b M_p$	PNA[c]	Y1[a]	ΣQ_n											
					2	2.5	3	3.5	4	4.5	5	5.5	6	6.5	7
	Kip-ft		In.	Kips											
W 24x62	574	TFL	0.00	910	894	926	958	991	1020	1060	1090	1120	1150	1180	1220
		2	0.15	806	879	907	936	964	993	1020	1050	1080	1110	1140	1160
		3	0.29	702	862	887	912	937	962	987	1010	1040	1060	1090	1110
		4	0.44	598	845	866	887	909	930	951	972	993	1010	1040	1060
		BFL	0.59	495	827	844	862	879	897	914	932	949	967	984	1000
		6	3.47	361	789	802	815	827	840	853	866	879	891	904	917
		7	6.58	228	723	731	739	747	755	763	771	779	787	795	803
W 24x55	503	TFL	0.00	810	791	820	848	877	906	934	963	992	1020	1050	1080
		2	0.13	722	778	804	829	855	880	906	931	957	982	1010	1030
		3	0.25	633	764	787	809	832	854	876	899	921	944	966	989
		4	0.38	545	750	769	788	808	827	846	866	885	904	923	943
		BFL	0.51	456	734	751	767	783	799	815	831	848	864	880	896
		6	3.45	329	700	711	723	735	746	758	770	781	793	805	816
		7	6.66	203	636	643	651	658	665	672	679	686	694	701	708
W 21x62	540	TFL	0.00	915	810	842	875	907	939	972	1000	1040	1070	1100	1130
		2	0.15	788	791	819	847	875	903	931	959	987	1010	1040	1070
		3	0.31	662	771	795	818	841	865	888	912	935	959	982	1010
		4	0.46	535	750	769	788	807	826	845	863	882	901	920	939
		BFL	0.62	408	727	741	756	770	785	799	814	828	843	857	872
		6	2.53	318	705	716	727	739	750	761	772	784	795	806	818
		7	4.78	229	669	677	685	693	701	709	717	726	734	742	750
W 21x57	484	TFL	0.00	835	741	771	800	830	859	889	919	948	978	1010	1040
		2	0.16	728	725	751	777	803	829	854	880	906	932	958	983
		3	0.33	622	708	730	753	775	797	819	841	863	885	907	929
		4	0.49	515	690	709	727	745	763	782	800	818	836	855	873
		BFL	0.65	409	671	685	700	714	729	743	758	772	787	801	816
		6	2.90	309	645	656	667	677	688	699	710	721	732	743	754
		7	5.38	209	601	608	616	623	631	638	645	653	660	668	675
W 21x50	413	TFL	0.00	735	646	672	698	724	750	777	803	829	855	881	907
		2	0.13	648	634	657	679	702	725	748	771	794	817	840	863
		3	0.27	560	620	640	660	679	699	719	739	759	779	799	818
		4	0.40	473	606	622	639	656	673	689	706	723	740	756	773
		BFL	0.54	386	590	604	618	631	645	659	672	686	700	713	727
		6	2.92	285	564	574	584	594	604	615	625	635	645	655	665
		7	5.58	184	519	526	532	539	545	552	559	565	572	578	585

[a]Y1 = distance from top of the steel beam to plastic neutral axis.
[b]Y2 = distance from top of the steel beam to concrete flange force.
[c]See Fig. 4.3 for PNA locations.

AMERICAN INSTITUTE OF STEEL CONSTRUCTION

F_y = 50 ksi

COMPOSITE DESIGN
COMPOSITE BEAM SELECTION TABLE
W Shapes

$$\phi = 0.85 \qquad\qquad \phi_b = 0.90$$

Shape	$\phi_b M_p$	PNA[c]	$Y1$[a]	ΣQ_n	ϕM_n (kip-ft) $Y2$[b] (in.)										
	Kip-ft		In.	Kips	2	2.5	3	3.5	4	4.5	5	5.5	6	6.5	7
W 21x44	358	TFL	0.00	650	568	591	614	637	660	683	706	729	752	775	798
		2	0.11	577	557	577	598	618	639	659	680	700	720	741	761
		3	0.23	504	546	564	581	599	617	635	653	671	689	706	724
		4	0.34	431	534	549	564	580	595	610	626	641	656	671	687
		BFL	0.45	358	522	534	547	560	572	585	598	610	623	636	648
		6	2.90	260	497	506	515	525	534	543	552	561	571	580	589
		7	5.69	163	453	459	465	471	476	482	488	494	499	505	511
W 18x60	461	TFL	0.00	880	693	724	755	787	818	849	880	911	942	974	1000
		2	0.17	749	674	700	727	753	780	806	833	859	886	912	939
		3	0.35	617	653	675	696	718	740	762	784	806	828	850	871
		4	0.52	486	630	647	665	682	699	716	733	751	768	785	802
		BFL	0.70	355	606	618	631	644	656	669	681	694	706	719	732
		6	2.19	287	590	600	610	620	630	640	651	661	671	681	691
		7	3.82	220	566	573	581	589	597	605	612	620	628	636	644
W 18x55	420	TFL	0.00	810	634	663	692	720	749	778	806	835	864	892	921
		2	0.16	691	617	641	666	690	715	739	764	788	813	837	862
		3	0.32	573	598	618	639	659	679	699	720	740	760	781	801
		4	0.47	454	578	594	610	626	642	658	674	691	707	723	739
		BFL	0.63	336	556	568	580	592	604	616	628	640	652	663	675
		6	2.16	269	541	550	560	569	579	588	598	607	617	626	636
		7	3.86	203	517	524	531	539	546	553	560	567	574	582	589
W 18x50	379	TFL	0.00	735	572	598	624	651	677	703	729	755	781	807	833
		2	0.14	628	557	479	601	624	646	668	690	712	735	757	779
		3	0.29	521	540	558	577	595	614	632	651	669	688	706	725
		4	0.43	415	522	537	552	566	581	596	610	625	640	654	669
		BFL	0.57	308	503	514	525	536	547	558	569	580	590	601	612
		6	2.07	246	489	498	506	515	524	532	541	550	559	567	576
		7	3.82	184	467	474	480	487	493	500	506	513	519	526	532
W 18x46	340	TFL	0.00	675	527	551	575	599	623	647	671	695	719	743	766
		2	0.15	583	514	535	555	576	597	617	638	659	679	700	720
		3	0.30	492	499	517	534	552	569	587	604	621	639	656	674
		4	0.45	400	484	498	512	527	541	555	569	583	597	612	626
		BFL	0.61	308	468	478	489	500	511	522	533	544	555	566	577
		6	2.40	239	450	459	467	476	484	493	501	510	518	526	535
		7	4.34	169	424	430	436	442	448	454	460	466	472	478	484

[a]$Y1$ = distance from top of the steel beam to plastic neutral axis.
[b]$Y2$ = distance from top of the steel beam to concrete flange force.
[c]See Fig. 4.3 for PNA locations.

$$F_y = 50 \text{ ksi}$$

COMPOSITE DESIGN
COMPOSITE BEAM SELECTION TABLE
W Shapes

$$\phi = 0.85 \qquad\qquad \phi_b = 0.90$$

Shape	$\phi_b M_p$	PNA[c]	$Y1^a$	ΣQ_n	ϕM_n (kip-ft)										
					Y2[b] (in.)										
	Kip-ft		In.	Kips	2	2.5	3	3.5	4	4.5	5	5.5	6	6.5	7
W 18x40	294	TFL	0.00	590	458	479	499	520	541	562	583	604	625	646	667
		2	0.13	511	446	464	482	500	518	537	555	573	591	609	627
		3	0.26	432	434	449	464	480	495	510	526	541	556	572	587
		4	0.39	353	421	433	446	458	471	483	496	508	521	533	546
		BFL	0.53	274	407	417	426	436	446	456	465	475	485	494	504
		6	2.26	211	392	400	407	415	422	429	437	444	452	459	467
		7	4.27	148	369	374	379	384	389	395	400	405	410	416	421
W 18x35	249	TFL	0.00	515	396	414	432	451	469	487	505	523	542	560	578
		2	0.11	451	387	403	418	434	450	466	482	498	514	530	546
		3	0.21	388	377	391	404	418	432	445	459	473	487	500	514
		4	0.32	324	367	378	389	401	412	424	435	447	458	470	481
		BFL	0.43	260	356	365	374	383	393	402	411	420	430	439	448
		6	2.37	194	340	347	354	361	368	375	382	389	395	402	409
		7	4.56	129	315	320	324	329	333	338	342	347	351	356	361
W 16x36	240	TFL	0.00	530	373	392	410	429	448	467	485	504	523	542	560
		2	0.11	455	362	378	394	410	426	442	459	475	491	507	523
		3	0.22	380	350	364	377	391	404	418	431	445	458	471	485
		4	0.32	305	338	349	360	371	381	392	403	414	425	435	446
		BFL	0.43	230	326	334	342	350	358	366	374	383	391	399	407
		6	1.79	181	315	322	328	334	341	347	354	360	366	373	379
		7	3.44	133	299	304	309	313	318	323	327	332	337	342	346
W 16x31	203	TFL	0.00	456	321	337	353	370	386	402	418	434	450	466	483
		2	0.11	395	312	326	340	354	368	382	396	410	424	438	452
		3	0.22	334	303	315	327	338	350	362	374	386	398	410	421
		4	0.33	274	293	303	312	322	332	342	351	361	371	380	390
		BFL	0.44	213	283	290	298	305	313	321	328	336	343	351	358
		6	2.00	163	272	278	283	289	295	301	307	312	318	324	330
		7	3.79	114	255	259	263	267	271	275	279	283	287	291	295
W 16x26	166	TFL	0.00	384	268	281	295	309	322	336	349	363	377	390	404
		2	0.09	337	261	273	285	297	309	321	332	344	356	368	380
		3	0.17	289	254	264	274	284	295	305	315	325	336	346	356
		4	0.26	242	246	255	263	272	281	289	298	306	315	323	332
		BFL	0.35	194	239	245	252	259	266	273	280	287	294	301	307
		6	2.04	145	228	233	238	243	248	253	259	264	269	274	279
		7	4.01	96.0	210	214	217	220	224	227	231	234	237	241	244

[a] $Y1$ = distance from top of the steel beam to plastic neutral axis.
[b] $Y2$ = distance from top of the steel beam to concrete flange force.
[c] See Fig. 4.3 for PNA locations.

F_y = 50 ksi

COMPOSITE DESIGN
COMPOSITE BEAM SELECTION TABLE
W Shapes

$\phi = 0.85$ $\phi_b = 0.90$

Shape	$\phi_b M_p$ Kip-ft	PNA[c]	Y1[a] In.	ΣQ_n Kips	ϕM_n (kip-ft) Y2[b] (in.) 2	2.5	3	3.5	4	4.5	5	5.5	6	6.5	7
W 14x38	231	TFL	0.00	560	359	379	399	418	438	458	478	498	518	537	557
		2	0.13	473	346	363	380	396	413	430	447	463	480	497	514
		3	0.26	386	333	346	360	374	387	401	415	428	442	456	469
		4	0.39	299	318	329	340	350	361	371	382	392	403	414	424
		BFL	0.52	211	303	311	318	326	333	341	348	356	363	371	378
		6	1.38	176	296	302	308	315	321	327	333	339	346	352	358
		7	2.53	140	286	291	296	301	306	311	316	321	326	331	335
W 14x34	205	TFL	0.00	500	318	336	354	372	389	407	425	442	460	478	495
		2	0.11	423	307	322	337	352	367	382	397	412	427	442	457
		3	0.23	347	295	308	320	332	345	357	369	381	394	406	418
		4	0.34	270	283	293	302	312	321	331	340	350	359	369	379
		BFL	0.46	193	270	277	284	290	297	304	311	318	325	332	338
		6	1.41	159	263	269	274	280	286	291	297	302	308	314	319
		7	2.60	125	253	258	262	267	271	275	280	284	289	293	298
W 14x30	177	TFL	0.00	443	280	295	311	327	342	358	374	389	405	421	436
		2	0.10	378	270	284	297	310	324	337	350	364	377	391	404
		3	0.19	313	260	271	283	294	305	316	327	338	249	360	371
		4	0.29	248	250	259	268	276	285	294	303	312	320	329	338
		BFL	0.39	183	239	246	252	259	265	272	278	285	291	298	304
		6	1.48	147	232	237	242	248	253	258	263	268	274	279	284
		7	2.82	111	221	225	229	233	237	241	245	249	253	256	260
W 14x26	151	TFL	0.00	385	244	258	271	285	298	312	326	339	353	366	380
		2	0.11	332	236	248	260	271	283	295	307	318	330	342	354
		3	0.21	279	228	238	248	258	268	278	287	297	307	317	327
		4	0.32	226	220	228	236	244	252	260	268	276	284	292	300
		BFL	0.42	173	211	217	223	229	235	242	248	254	260	266	272
		6	1.67	135	203	207	212	217	222	227	231	236	241	246	250
		7	3.19	96.1	191	194	197	201	204	208	211	214	218	221	225
W 14x22	125	TFL	0.00	325	204	215	227	238	250	261	273	284	296	307	319
		2	0.08	283	198	208	218	228	238	248	258	268	278	288	298
		3	0.17	241	192	200	209	217	226	234	243	251	260	268	277
		4	0.25	199	185	192	199	206	213	220	227	234	241	248	255
		BFL	0.34	157	178	184	189	195	200	206	212	217	223	228	234
		6	1.69	119	170	174	179	183	187	191	196	200	204	208	212
		7	3.34	81.1	158	161	164	167	170	172	175	178	181	184	187

[a] Y1 = distance from top of the steel beam to plastic neutral axis.
[b] Y2 = distance from top of the steel beam to concrete flange force.
[c] See Fig. 4.3 for PNA locations.

AMERICAN INSTITUTE OF STEEL CONSTRUCTION

$F_y = 50$ ksi

COMPOSITE DESIGN
COMPOSITE BEAM SELECTION TABLE
W Shapes

$\phi = 0.85$ $\phi_b = 0.90$

Shape	$\phi_b M_p$	PNA[c]	Y1[a]	ΣQ_n	ϕM_n (kip-ft) Y2[b] (in.)										
	Kip-ft		In.	Kips	2	2.5	3	3.5	4	4.5	5	5.5	6	6.5	7
W 12x30	162	TFL	0.00	440	254	270	285	301	317	332	348	363	379	394	410
		2	0.11	368	244	257	270	283	296	309	322	335	348	361	374
		3	0.22	296	233	243	254	264	275	285	296	306	317	327	338
		4	0.33	224	221	229	237	245	253	261	269	277	285	293	301
		BFL	0.44	153	209	215	220	225	231	236	242	247	252	258	263
		6	1.12	131	205	210	214	219	224	228	233	238	242	247	252
		7	1.94	110	200	204	207	211	215	219	223	227	231	235	239
W 12x26	140	TFL	0.00	383	220	233	247	260	274	287	301	315	328	342	355
		2	0.10	321	211	222	234	245	256	268	279	290	302	313	324
		3	0.19	259	201	211	220	229	238	247	257	266	275	284	293
		4	0.29	198	192	199	206	213	220	227	234	241	248	255	262
		BFL	0.38	136	181	186	191	196	201	206	210	215	220	225	230
		6	1.08	116	178	182	186	190	194	198	202	206	210	215	219
		7	1.95	95.6	173	176	179	183	186	190	193	196	200	203	206
W 12x22	110	TFL	0.00	324	187	199	210	222	233	245	256	267	279	290	302
		2	0.11	281	181	191	201	211	221	231	241	251	261	271	281
		3	0.21	238	174	183	191	200	208	217	225	233	242	250	259
		4	0.32	196	168	174	181	188	195	202	209	216	223	230	237
		BFL	0.43	153	160	166	171	177	182	187	193	198	204	209	214
		6	1.66	117	153	157	161	165	169	173	178	182	186	190	194
		7	3.04	81.0	142	145	147	150	153	156	159	162	165	167	170
W 12x19	92.6	TFL	0.00	279	159	169	179	189	199	209	219	228	238	248	258
		2	0.09	243	154	163	172	180	189	197	206	215	223	232	241
		3	0.18	208	149	156	164	171	179	186	193	201	208	215	223
		4	0.26	173	144	150	156	162	168	174	180	187	193	199	205
		BFL	0.35	138	138	143	148	152	157	162	167	172	177	182	187
		6	1.66	104	131	134	138	142	145	149	153	156	160	164	167
		7	3.12	69.6	120	122	125	127	130	132	135	137	140	142	145
W 12x16	75.4	TFL	0.00	236	133	142	150	158	167	175	183	192	200	208	217
		2	0.07	209	130	137	144	152	159	167	174	181	189	196	204
		3	0.13	183	126	132	139	145	152	158	164	171	177	184	190
		4	0.20	156	122	127	133	138	144	149	155	160	166	171	177
		BFL	0.26	130	117	122	127	131	136	140	145	150	154	159	163
		6	1.71	94.3	110	113	117	120	123	127	130	133	137	140	143
		7	3.32	58.9	98.7	101	103	105	107	109	111	113	115	117	120

[a] Y1 = distance from top of the steel beam to plastic neutral axis.
[b] Y2 = distance from top of the steel beam to concrete flange force.
[c] See Fig. 4.3 for PNA locations.

AMERICAN INSTITUTE OF STEEL CONSTRUCTION

F_y = 50 ksi

COMPOSITE DESIGN
COMPOSITE BEAM SELECTION TABLE
W Shapes

$$\phi = 0.85 \qquad \phi_b = 0.90$$

Shape	$\phi_b M_p$	PNA[c]	Y1[a]	ΣQ_n	ϕM_n (kip-ft) Y2[b] (in.)										
	Kip-ft		In.	Kips	2	2.5	3	3.5	4	4.5	5	5.5	6	6.5	7
W 12x14	65.2	TFL	0.00	208	117	125	132	139	147	154	161	169	176	184	191
		2	0.06	186	114	121	127	134	140	147	153	160	167	173	180
		3	0.11	163	111	116	122	128	134	140	145	151	157	163	169
		4	0.17	141	107	112	117	122	127	132	137	142	147	152	157
		BFL	0.23	119	104	108	112	116	121	125	129	133	137	142	146
		6	1.69	85.3	97.0	100	103	106	109	112	115	118	121	124	127
		7	3.36	52.0	86.3	88.2	90.0	91.8	93.7	95.5	97.4	99.2	101	103	105
W 10x26	117	TFL	0.00	381	193	207	220	234	247	260	274	287	301	314	328
		2	0.11	317	184	195	206	218	229	240	251	262	274	285	296
		3	0.22	254	174	183	192	201	210	219	228	237	246	255	264
		4	0.33	190	164	171	177	184	191	198	204	211	218	225	231
		BFL	0.44	127	153	158	162	167	171	176	180	185	189	194	198
		6	0.90	111	150	154	158	162	166	170	174	178	182	186	189
		7	1.51	95.1	147	150	153	157	160	163	167	170	174	177	180
W 10x22	97.5	TFL	0.00	325	163	174	186	197	209	220	232	243	255	266	278
		2	0.09	273	155	165	175	184	194	204	213	223	233	242	252
		3	0.18	221	148	155	163	171	179	187	194	202	210	218	226
		4	0.27	169	139	145	151	157	163	169	175	181	187	193	199
		BFL	0.36	117	131	135	139	143	148	152	156	160	164	168	173
		6	0.95	99.3	127	131	135	138	142	145	149	152	156	159	163
		7	1.70	81.1	123	126	129	132	135	138	140	143	146	149	152
W 10x19	81.0	TFL	0.00	281	142	152	162	172	182	191	201	211	221	231	241
		2	0.10	241	136	145	153	162	170	179	187	196	204	213	221
		3	0.20	202	130	137	144	151	158	166	173	180	187	194	201
		4	0.30	162	124	129	135	141	147	152	158	164	169	175	181
		BFL	0.40	122	117	121	126	130	134	139	143	147	152	156	160
		6	1.27	96.2	112	115	119	122	125	129	132	136	139	143	146
		7	2.31	70.3	105	107	110	112	115	117	120	122	125	127	130
W 10x17	70.1	TFL	0.00	249	125	134	142	151	160	169	178	187	195	204	213
		2	0.08	216	120	128	135	143	151	158	166	174	181	189	197
		3	0.17	183	115	121	128	134	141	147	154	160	167	173	180
		4	0.25	150	110	115	120	126	131	136	142	147	152	158	163
		BFL	0.33	117	104	109	113	117	121	125	129	133	138	142	146
		6	1.31	89.8	99.0	102	105	109	112	115	118	121	124	128	131
		7	2.46	62.4	91.4	93.7	95.9	98.1	100	102	105	107	109	111	114

[a] Y1 = distance from top of the steel beam to plastic neutral axis.
[b] Y2 = distance from top of the steel beam to concrete flange force.
[c] See Fig. 4.3 for PNA locations.

					F_y = 50 ksi

COMPOSITE DESIGN
COMPOSITE BEAM SELECTION TABLE
W Shapes
$$\phi = 0.85 \qquad \phi_b = 0.90$$

Shape	$\phi_b M_p$	PNA[c]	Y1[a]	ΣQ_n	ϕM_n (kip-ft) Y2[b] (in.)									
	Kip-ft		In.	Kips	2	2.5	3	3.5	4	4.5	5	5.5	6	6.5	7
W 10x15	60.0	TFL	0.00	221	109	117	125	133	140	148	156	164	172	180	187
		2	0.07	194	105	112	119	126	133	140	146	153	160	167	174
		3	0.14	167	101	107	113	119	125	131	137	143	149	154	160
		4	0.20	139	97.2	102	107	112	117	122	127	132	137	142	147
		BFL	0.27	112	92.9	96.9	101	105	109	113	117	121	125	129	133
		6	1.35	83.8	87.3	90.3	93.2	96.2	99.2	102	105	108	111	114	117
		7	2.60	55.1	79.2	81.2	83.1	85.1	87.0	89.0	90.9	92.9	94.8	96.8	98.7
W 10x12	47.3	TFL	0.00	177	86.9	93.2	99.5	106	112	118	125	131	137	143	150
		2	0.05	156	84.0	89.5	95.0	101	106	112	117	123	128	134	139
		3	0.11	135	80.9	85.7	90.5	95.3	100	105	110	114	119	124	129
		4	0.16	115	77.8	81.8	85.9	89.9	94.0	98.1	102	106	110	114	118
		BFL	0.21	93.8	74.5	77.9	81.2	84.5	87.8	91.2	94.5	97.8	101	104	108
		6	1.30	69.0	69.8	72.3	74.7	77.1	79.6	82.0	84.5	86.9	89.4	91.8	94.3
		7	2.61	44.3	62.9	64.4	66.0	67.6	69.1	70.7	72.3	73.8	75.4	77.0	78.5
W 8x28	102	TFL	0.00	413	176	191	205	220	235	249	264	278	293	308	322
		2	0.12	337	165	177	189	201	213	225	237	249	260	272	284
		3	0.23	261	153	163	172	181	190	200	209	218	227	236	246
		4	0.35	185	141	148	154	161	167	174	180	187	193	200	206
		BFL	0.47	109	128	132	136	140	144	147	151	155	159	163	167
		6	0.53	106	128	131	135	139	143	146	150	154	158	161	165
		7	0.59	103	127	131	134	138	142	145	149	153	156	160	164
W 8x24	87.0	TFL	0.00	354	150	162	175	187	200	212	225	237	250	262	275
		2	0.10	289	140	150	161	171	181	191	202	212	222	232	243
		3	0.20	224	130	138	146	154	162	170	178	186	194	202	210
		4	0.30	159	120	126	131	137	142	148	154	159	165	171	176
		BFL	0.40	94.2	109	112	116	119	122	126	129	132	136	139	142
		6	0.47	91.3	109	112	115	118	122	125	128	131	134	138	141
		7	0.55	88.5	108	111	114	117	121	124	127	130	133	136	139
W 8x21	76.5	TFL	0.00	308	134	145	156	167	178	188	199	210	221	232	243
		2	0.10	255	126	135	144	153	162	172	181	190	199	208	217
		3	0.20	203	118	125	133	140	147	154	161	169	176	183	190
		4	0.30	150	110	115	120	126	131	136	142	147	152	158	163
		BFL	0.40	97.2	101	105	108	111	115	118	122	125	129	132	136
		6	0.70	87.1	99.3	102	105	109	112	115	118	121	124	127	130
		7	1.06	77.0	97.2	100	103	105	108	111	114	116	119	122	125

[a]Y1 = distance from top of the steel beam to plastic neutral axis.
[b]Y2 = distance from top of the steel beam to concrete flange force.
[c]See Fig. 4.3 for PNA locations.

AMERICAN INSTITUTE OF STEEL CONSTRUCTION

F_y = 50 ksi

COMPOSITE DESIGN
COMPOSITE BEAM SELECTION TABLE
W Shapes

$\phi = 0.85$ $\phi_b = 0.90$

Shape	$\phi_b M_p$	PNA[c]	$Y1$[a]	ΣQ_n	ϕM_n (kip-ft) $Y2$[b] (in.)										
	Kip-ft		In.	Kips	2	2.5	3	3.5	4	4.5	5	5.5	6	6.5	7
W 8x18	63.7	TFL	0.00	263	113	122	132	141	150	160	169	178	188	197	206
		2	0.08	220	107	115	122	130	138	146	154	161	169	177	185
		3	0.17	176	100	107	113	119	125	132	138	144	150	157	163
		4	0.25	133	93.5	98.2	103	108	112	117	122	127	131	136	141
		BFL	0.33	89.8	86.5	89.7	92.9	96.0	99.2	102	106	109	112	115	118
		6	0.71	77.8	84.4	87.1	89.9	92.6	95.4	98.1	101	104	106	109	112
		7	1.21	65.8	81.9	84.2	86.5	88.8	91.2	93.5	95.8	98.2	100	103	105
W 8x15	51.0	TFL	0.00	222	95.2	103	111	119	127	135	142	150	158	166	174
		2	0.08	190	90.6	97.4	104	111	118	124	131	138	145	151	158
		3	0.16	159	85.9	91.5	97.1	103	108	114	120	125	131	137	142
		4	0.24	127	81.0	85.5	90.0	94.5	99.0	103	108	113	117	122	126
		BFL	0.32	95.5	75.9	79.3	82.7	86.0	89.4	92.8	96.2	99.6	103	106	110
		6	0.97	75.5	72.2	74.8	77.5	80.2	82.9	85.5	88.2	90.9	93.6	96.2	98.9
		7	1.79	55.5	67.4	69.3	71.3	73.3	75.2	77.2	79.2	81.1	83.1	85.1	87.0
W 8x13	42.7	TFL	0.00	192	81.5	88.3	95.1	102	109	116	122	129	136	143	150
		2	0.06	167	77.9	83.8	89.7	95.6	101	107	113	119	125	131	137
		3	0.13	141	74.1	79.1	84.1	89.1	94.1	99.0	104	109	114	119	124
		4	0.19	116	70.2	74.3	78.4	82.4	86.5	90.6	94.7	98.8	103	107	111
		BFL	0.26	90.0	66.2	69.3	72.5	75.7	78.9	82.1	85.3	88.5	91.7	94.8	98.0
		6	0.99	69.0	62.3	64.7	67.2	69.6	72.1	74.5	77.0	79.4	81.8	84.3	86.7
		7	1.91	48.0	57.2	58.9	60.6	62.3	64.0	65.7	67.4	69.1	70.8	72.5	74.2
W 8x10	33.3	TFL	0.00	148	62.3	67.6	72.8	78.0	83.3	88.5	93.8	99.0	104	109	115
		2	0.05	128	59.4	64.0	68.5	73.0	77.5	82.1	86.6	91.1	95.6	100	105
		3	0.10	108	56.5	60.3	64.1	67.9	71.7	75.5	79.3	83.1	86.9	90.8	94.6
		4	0.15	87.4	53.4	56.5	59.6	62.7	65.8	68.9	72.0	75.1	78.2	81.3	84.4
		BFL	0.21	67.2	50.3	52.7	55.1	57.4	59.8	62.2	64.6	67.0	69.3	71.7	74.1
		6	0.88	52.1	47.6	49.5	51.3	53.1	55.0	56.8	58.7	60.5	62.4	64.2	66.1
		7	1.77	37.0	44.0	45.4	46.7	48.0	49.3	50.6	51.9	53.2	54.5	55.8	57.2

[a] $Y1$ = distance from top of the steel beam to plastic neutral axis.
[b] $Y2$ = distance from top of the steel beam to concrete flange force.
[c] See Fig. 4.3 for PNA locations.

LOWER BOUND
ELASTIC MOMENT OF INERTIA
FOR PLASTIC COMPOSITE SECTIONS

I_{LB}

Shape	PNA[c]	$Y1$[a]	I_{LB} (in.^{4})										
			$Y2$[b] (in.)										
		In.	2	2.5	3	3.5	4	4.5	5	5.5	6	6.5	7
W 36x300	TFL	0.00	38600	39500	40500	41400	42400	43400	44400	45500	46500	47600	48700
	2	0.42	37000	37900	38700	39600	40500	41400	42300	43300	44300	45300	46300
	3	0.84	35200	35900	36700	37400	38200	39000	39900	40700	41600	42500	43400
	4	1.26	32900	33500	34200	34800	35500	36200	36900	37600	38300	39100	39900
	BFL	1.68	30100	30600	31100	31600	32100	32700	33200	33800	34400	34900	35500
	6	3.97	28900	29400	29800	30200	30700	31200	31700	32200	32700	33200	33700
	7	6.69	27600	28000	28400	28700	29100	29500	29900	30400	30800	31200	31700
W 36x280	TFL	0.00	35800	36700	37500	38400	39300	40200	41200	42200	43100	44200	45200
	2	0.39	34400	35100	35900	36700	37600	38400	39300	40200	41100	42000	42900
	3	0.79	32600	33300	34000	34700	35500	36200	37000	37800	38600	39400	40300
	4	1.18	30600	31100	31700	32300	33000	33600	34300	34900	35600	36300	37000
	BFL	1.57	28000	28400	28900	29400	29900	30400	30900	31400	31900	32500	33000
	6	3.88	26900	27300	27700	28100	28500	29000	29400	29900	30300	30800	31300
	7	6.62	25700	26000	26300	26700	27100	27400	27800	28200	28600	29000	29400
W 36x260	TFL	0.00	32800	33600	34400	35200	36000	36900	37800	38700	39600	40500	41500
	2	0.36	31500	32200	32900	33700	34500	35200	36000	36900	37700	38500	39400
	3	0.72	29900	30600	31200	31900	32600	33300	34000	34700	35500	36200	37000
	4	1.08	28100	28600	29200	29700	30300	30900	31500	32100	32800	33400	34100
	BFL	1.44	25800	26200	26700	27100	27600	28000	28500	29000	29500	30000	30500
	6	3.86	24700	25100	25500	25800	26200	26600	27100	27500	27900	28400	28800
	7	6.75	23500	23800	24100	24500	24800	25100	25500	25800	26200	26600	27000
W 36x245	TFL	0.00	30600	31300	32100	32800	33600	34400	35200	36100	36900	37800	38700
	2	0.34	29400	30000	30700	31400	32100	32900	33600	34400	35200	36000	36800
	3	0.68	27900	28500	29100	29800	30400	31100	31700	32400	33100	33800	34600
	4	1.01	26200	26700	27200	27800	28300	28900	29500	30000	30600	31300	31900
	BFL	1.35	24100	24500	24900	25300	25800	26200	26700	27100	27600	28100	28600
	6	3.81	23100	23400	23800	24100	24500	24900	25300	25700	26100	26500	27000
	7	6.77	21900	22200	22500	22800	23100	23400	23800	24100	24400	24800	25100
W 36x230	TFL	0.00	28500	29100	29800	30600	31300	32000	32800	33600	34400	35200	36000
	2	0.32	27300	28000	28600	29300	29900	30600	31300	32000	32800	33500	34300
	3	0.63	26000	26600	27100	27700	28300	28900	29600	30200	30900	31500	32200
	4	0.95	24400	24900	25400	25900	26400	26900	27500	28000	28600	29200	29700
	BFL	1.26	22500	22900	23300	23700	24100	24500	24900	25400	25800	26300	26700
	6	3.81	21500	21800	22200	22500	22900	23200	23600	24000	24400	24800	25200
	7	6.83	20400	20700	20900	21200	21500	21800	22100	22400	22800	23100	23400

[a] $Y1$ = distance from top of the steel beam to plastic neutral axis.
[b] $Y2$ = distance from top of the steel beam to concrete flange force.
[c] See Fig. 4.3 for PNA locations.

LOWER BOUND
ELASTIC MOMENT OF INERTIA
FOR PLASTIC COMPOSITE SECTIONS

I_{LB}

Shape	PNA[c]	Y1[a]	I_{LB} (in.[4])										
			Y2[b] (in.)										
		In.	2	2.5	3	3.5	4	4.5	5	5.5	6	6.5	7
W 36x210	TFL	0.00	26000	26600	27300	27900	28600	29300	30000	30800	31500	32300	33000
	2	0.34	25100	25700	26300	26900	27500	28200	28800	29500	30200	30900	31600
	3	0.68	24000	24500	25100	25700	26200	26800	27400	28100	28700	29300	30000
	4	1.02	22800	23200	23700	24200	24700	25300	25800	26300	26900	27500	28100
	BFL	1.36	21300	21700	22100	22500	23000	23400	23900	24300	24800	25300	25800
	6	5.06	19900	20300	20600	21000	21300	21700	22100	22400	22800	23200	23600
	7	9.04	18300	18600	18800	19100	19400	19700	19900	20200	20500	20800	21100
W 36x194	TFL	0.00	23800	24400	25000	25600	26200	26800	27500	28200	28900	29600	30300
	2	0.32	22900	23500	24000	24600	25200	25800	26400	27000	27700	28300	29000
	3	0.63	22000	22500	23000	23500	24000	24600	25100	25700	26300	26900	27500
	4	0.95	20800	21300	21700	22200	22700	23100	23600	24100	24600	25200	25700
	BFL	1.26	19500	19900	20300	20600	21000	21500	21900	22300	22700	23200	23600
	6	4.94	18200	18600	18900	19200	19500	19900	20200	20600	20900	21300	21700
	7	8.93	16800	17000	17200	17500	17700	18000	18300	18500	18800	19100	19400
W 36x182	TFL	0.00	22200	22700	23300	23900	24500	25100	25700	26300	26900	27600	28300
	2	0.29	21400	21900	22500	23000	23500	24100	24700	25200	25800	26400	27100
	3	0.59	20500	21000	21500	22000	22400	23000	23500	24000	24600	25100	25700
	4	0.89	19500	19900	20300	20700	21200	21600	22100	22600	23100	23500	24000
	BFL	1.18	18300	18600	19000	19300	19700	20100	20500	20900	21300	21700	22100
	6	4.89	17100	17300	17600	17900	18300	18600	18900	19200	19600	19900	20300
	7	8.92	15700	15900	16100	16300	16600	16800	17100	17300	17600	17800	18100
W 36x170	TFL	0.00	20600	21100	21600	22100	22700	23300	23800	24400	25000	25600	26200
	2	0.28	19900	20300	20800	21300	21800	22300	22900	23400	24000	24500	25100
	3	0.55	19000	19500	19900	20400	20800	21300	21800	22300	22800	23300	23800
	4	0.83	18100	18500	18900	19300	19700	20100	20500	21000	21400	21900	22300
	BFL	1.10	17000	17300	17600	18000	18300	18700	19000	19400	19800	20200	20600
	6	4.84	15800	16100	16400	16700	17000	17300	17600	17900	18200	18500	18800
	7	8.89	14500	14700	14900	15200	15400	15600	15800	16100	16300	16500	16800
W 36x160	TFL	0.00	19200	19600	20100	20600	21100	21700	22200	22700	23300	23900	24400
	2	0.26	18500	18900	19400	19900	20300	20800	21300	21800	22300	22900	23400
	3	0.51	17700	18200	18600	19000	19400	19900	20300	20800	21300	21700	22200
	4	0.77	16900	17200	17600	18000	18400	18800	19200	19600	20000	20400	20900
	BFL	1.02	15800	16200	16500	16800	17100	17500	17800	18200	18500	18900	19300
	6	4.82	14800	15000	15300	15600	15800	16100	16400	16700	17000	17300	17600
	7	8.97	13500	13700	13900	14100	14300	14500	14700	14900	15200	15400	15600

[a] Y1 = distance from top of the steel beam to plastic neutral axis.
[b] Y2 = distance from top of the steel beam to concrete flange force.
[c] See Fig. 4.3 for PNA locations.

LOWER BOUND ELASTIC MOMENT OF INERTIA FOR PLASTIC COMPOSITE SECTIONS

I_{LB}

Shape	PNA[c]	$Y1$[a]	I_{LB} (in.[4])										
			$Y2$[b] (in.)										
		In.	2	2.5	3	3.5	4	4.5	5	5.5	6	6.5	7
W 36x150	TFL	0.00	17800	18300	18700	19200	19700	20200	20700	21200	21700	22200	22800
	2	0.24	17200	17600	18100	18500	18900	19400	19900	20300	20800	21300	21800
	3	0.47	16500	16900	17300	17700	18100	18500	19000	19400	19800	20300	20800
	4	0.71	15700	16100	16400	16800	17200	17500	17900	18300	18700	19100	19500
	BFL	0.94	14800	15100	15400	15700	16000	16400	16700	17000	17400	17700	18100
	6	4.83	13800	14000	14300	14500	14800	15000	15300	15600	15900	16200	16500
	7	9.08	12500	12700	12900	13100	13300	13500	13700	13900	14100	14300	14500
W 36x135	TFL	0.00	15600	16000	16400	16800	17200	17600	18100	18600	19000	19500	20000
	2	0.20	15100	15400	15800	16200	16600	17000	17400	17900	18300	18800	19200
	3	0.40	14500	14900	15200	15600	15900	16300	16700	17100	17500	17900	18300
	4	0.59	13900	14200	14500	14800	15200	15500	15900	16200	16600	17000	17300
	BFL	0.79	13100	13400	13700	14000	14300	14600	14900	15200	15500	15800	16200
	6	4.96	12100	12400	12600	12800	13100	13300	13500	13800	14100	14300	14600
	7	9.50	10900	11100	11200	11400	11600	11700	11900	12100	12300	12500	12700
W 33x221	TFL	0.00	24500	25100	25800	26400	27100	27800	28500	29200	29900	30700	31500
	2	0.32	23500	24100	24700	25300	25900	26500	27200	27800	28500	29200	29900
	3	0.64	22300	22900	23400	23900	24500	25000	25600	26200	26800	27400	28000
	4	0.96	20900	21400	21800	22300	22800	23200	23700	24200	24700	25300	25800
	BFL	1.28	19200	19600	19900	20300	20700	21000	21400	21800	22200	22700	23100
	6	3.76	18400	18700	19000	19300	19600	20000	20300	20700	21000	21400	21700
	7	6.48	17500	17700	18000	18200	18500	18800	19100	19400	19700	20000	20300
W 33x201	TFL	0.00	22000	22600	23100	23700	24300	25000	25600	26200	26900	27600	28300
	2	0.29	21100	21600	22200	22700	23300	23800	24400	25000	25600	26300	26900
	3	0.58	20100	20600	21000	21500	22000	22500	23000	23600	24100	24700	25300
	4	0.86	18900	19300	19700	20100	20500	20900	21400	21800	22300	22800	23300
	BFL	1.15	17400	17700	18000	18300	18700	19000	19400	19700	20100	20500	20900
	6	3.67	16600	16800	17100	17400	17700	18000	18300	18600	18900	19300	19600
	7	6.51	15700	15900	16200	16400	16600	16900	17100	17400	17700	17900	18200
W 33x141	TFL	0.00	14700	15100	15500	15900	16300	16800	17200	17700	18100	18600	19100
	2	0.24	14200	14500	14900	15300	15700	16100	16500	16900	17400	17800	18300
	3	0.48	13600	13900	14200	14600	15000	15300	15700	16100	16500	16900	17300
	4	0.72	12900	13200	13500	13800	14100	14400	14800	15100	15500	15800	16200
	BFL	0.96	12100	12300	12600	12800	13100	13400	13700	14000	14200	14600	14900
	6	4.31	11300	11500	11700	11900	12100	12400	12600	12800	13100	13300	13600
	7	8.05	10300	10500	10700	10800	11000	11200	11300	11500	11700	11900	12100

[a]$Y1$ = distance from top of the steel beam to plastic neutral axis.
[b]$Y2$ = distance from top of the steel beam to concrete flange force.
[c]See Fig. 4.3 for PNA locations.

LOWER BOUND
ELASTIC MOMENT OF INERTIA
FOR PLASTIC COMPOSITE SECTIONS

LB

Shape	PNA[c]	$Y1$[a]	I_{LB} (in.[4])										
			$Y2$[b] (in.)										
		In.	2	2.5	3	3.5	4	4.5	5	5.5	6	6.5	7
W 33x130	TFL	0.00	13300	13700	14000	14400	14800	15200	15600	16000	16400	16900	17300
	2	0.21	12800	13200	13500	13900	14200	14600	15000	15400	15800	16200	16600
	3	0.43	12300	12600	12900	13300	13600	13900	14300	14600	15000	15400	15800
	4	0.64	11700	12000	12300	12600	12900	13200	13500	13800	14100	14500	14800
	BFL	0.86	11000	11300	11500	11700	12000	12300	12500	12800	13100	13400	13700
	6	4.39	10300	10400	10600	10900	11100	11300	11500	11700	11900	12200	12400
	7	8.29	9340	9490	9640	9790	9940	10100	10300	10400	10600	10800	11000
W 33x118	TFL	0.00	11800	12100	12500	12800	13100	13500	13900	14200	14600	15000	15400
	2	0.19	11400	11700	12000	12300	12700	13000	13300	13700	14100	14400	14800
	3	0.37	11000	11300	11500	11800	12100	12400	12800	13100	13400	13700	14100
	4	0.56	10500	10700	11000	11200	11500	11800	12100	12400	12700	13000	13300
	BFL	0.74	9880	10100	10300	10600	10800	11000	11300	11500	11800	12100	12300
	6	4.44	9150	9330	9510	9700	9890	10100	10300	10500	10700	10900	11100
	7	8.54	8260	8390	8520	8660	8800	8940	9090	9240	9390	9550	9710
W 30x116	TFL	0.00	9870	10200	10500	10800	11100	11400	11800	12100	12500	12800	13200
	2	0.21	9530	9800	10100	10400	10700	11000	11300	11600	11900	12300	12600
	3	0.43	9130	9380	9640	9910	10200	10500	10700	11000	11300	11700	12000
	4	0.64	8670	8900	9130	9360	9600	9850	10100	10400	10600	10900	11200
	BFL	0.85	8130	8320	8520	8720	8930	9140	9360	9580	9810	10000	10300
	6	3.98	7570	7730	7890	8060	8230	8400	8580	8770	8960	9150	9350
	7	7.44	6910	7030	7150	7270	7400	7530	7670	7810	7950	8090	8240
W 30x108	TFL	0.00	9000	9280	9560	9840	10100	10400	10800	11100	11400	11700	12100
	2	0.19	8700	8960	9220	9480	9760	10000	10300	10600	10900	11300	11600
	3	0.38	8350	8590	8830	9070	9330	9590	9850	10100	10400	10700	11000
	4	0.57	7950	8160	8380	8600	8820	9060	9300	9540	9790	10100	10300
	BFL	0.76	7480	7660	7850	8040	8240	8440	8650	8860	9080	9300	9530
	6	4.04	6940	7090	7240	7400	7560	7720	7890	8070	8240	8430	8610
	7	7.64	6280	6390	6500	6620	6740	6860	6980	7110	7240	7380	7510
W 30x 99	TFL	0.00	8110	8360	8610	8880	9150	9420	9710	10000	10300	10600	10900
	2	0.17	7850	8080	8320	8560	8820	9080	9340	9620	9900	10200	10500
	3	0.34	7550	7760	7980	8210	8440	8680	8930	9180	9440	9700	9970
	4	0.50	7200	7400	7600	7800	8010	8230	8450	8680	8910	9150	9390
	BFL	0.67	6800	6970	7150	7330	7510	7700	7900	8100	8300	8510	8720
	6	4.07	6280	6420	6560	6700	6850	7010	7170	7330	7490	7660	7840
	7	7.83	5640	5740	5840	5940	6050	6160	6280	6390	6510	6640	6760

[a] $Y1$ = distance from top of the steel beam to plastic neutral axis.
[b] $Y2$ = distance from top of the steel beam to concrete flange force.
[c] See Fig. 4.3 for PNA locations.

I_{LB} LOWER BOUND ELASTIC MOMENT OF INERTIA FOR PLASTIC COMPOSITE SECTIONS

Shape	PNA[c]	Y1[a]	I_{LB} (in.^{4}) Y2[b] (in.)										
		In.	2	2.5	3	3.5	4	4.5	5	5.5	6	6.5	7
W 27x102	TFL	0.00	7240	7480	7730	7980	8240	8500	8780	9060	9350	9650	9950
	2	0.21	6970	7190	7420	7650	7890	8140	8390	8660	8920	9200	9480
	3	0.42	6660	6860	7070	7280	7490	7720	7950	8190	8430	8680	8930
	4	0.62	6290	6470	6650	6830	7030	7220	7430	7630	7850	8070	8290
	BFL	0.83	5860	6000	6150	6310	6470	6630	6800	6980	7150	7340	7520
	6	3.41	5490	5610	5740	5870	6000	6140	6280	6430	6580	6730	6890
	7	6.26	5070	5160	5260	5360	5470	5570	5680	5800	5910	6030	6150
W 27x 94	TFL	0.00	6580	6800	7020	7250	7490	7740	7990	8250	8510	8790	9070
	2	0.19	6340	6540	6750	6970	7190	7420	7650	7890	8140	8390	8650
	3	0.37	6070	6250	6440	6640	6840	7040	7260	7480	7700	7930	8170
	4	0.56	5740	5910	6080	6250	6430	6610	6800	6990	7190	7400	7600
	BFL	0.75	5360	5500	5640	5790	5940	6100	6260	6420	6590	6760	6940
	6	3.39	5010	5120	5240	5360	5490	5620	5750	5890	6030	6170	6320
	7	6.39	4590	4680	4770	4860	4960	5060	5160	5260	5370	5480	5590
W 27x 84	TFL	0.00	5770	5970	6170	6370	6580	6800	7030	7260	7500	7740	7990
	2	0.16	5570	5750	5940	6130	6330	6530	6740	6960	7180	7400	7630
	3	0.32	5340	5510	5680	5850	6030	6220	6410	6610	6810	7020	7230
	4	0.48	5080	5220	5370	5530	5690	5860	6030	6210	6390	6570	6760
	BFL	0.64	4760	4890	5020	5150	5290	5440	5580	5730	5890	6050	6210
	6	3.44	4420	4530	4630	4750	4860	4980	5100	5220	5350	5480	5610
	7	6.62	4020	4100	4180	4260	4340	4430	4520	4610	4710	4810	4910
W 24x 76	TFL	0.00	4280	4440	4610	4780	4950	5130	5320	5510	5710	5920	6130
	2	0.17	4120	4270	4420	4580	4740	4910	5090	5260	5450	5640	5830
	3	0.34	3940	4070	4210	4350	4500	4650	4810	4970	5140	5310	5490
	4	0.51	3720	3840	3960	4090	4220	4350	4490	4640	4780	4930	5090
	BFL	0.68	3460	3560	3670	3770	3880	4000	4110	4230	4360	4480	4610
	6	3.00	3240	3320	3410	3490	3590	3680	3780	3880	3980	4090	4200
	7	5.60	2970	3040	3100	3170	3240	3310	3390	3470	3550	3630	3710
W 24x 68	TFL	0.00	3760	3900	4050	4200	4360	4520	4690	4860	5040	5220	5410
	2	0.15	3630	3760	3900	4040	4180	4330	4490	4650	4810	4980	5160
	3	0.29	3470	3590	3720	3850	3980	4120	4260	4410	4560	4710	4870
	4	0.44	3290	3400	3510	3630	3740	3870	3990	4120	4260	4390	4540
	BFL	0.59	3080	3170	3270	3370	3470	3570	3680	3790	3910	4020	4140
	6	3.05	2860	2940	3020	3100	3180	3270	3360	3450	3540	3640	3740
	7	5.81	2600	2660	2720	2780	2840	2910	2970	3040	3110	3190	3260

[a] $Y1$ = distance from top of the steel beam to plastic neutral axis.
[b] $Y2$ = distance from top of the steel beam to concrete flange force.
[c] See Fig. 4.3 for PNA locations.

LOWER BOUND
ELASTIC MOMENT OF INERTIA
FOR PLASTIC COMPOSITE SECTIONS

I LB

Shape	PNA[c]	Y1[a]	I_{LB} (in.^{4})										
			Y2[b] (in.)										
		In.	2	2.5	3	3.5	4	4.5	5	5.5	6	6.5	7
W 24x62	TFL	0.00	3300	3430	3560	3700	3840	3990	4140	4300	4460	4620	4790
	2	0.15	3190	3320	3440	3570	3700	3840	3980	4130	4280	4440	4590
	3	0.29	3080	3190	3300	3420	3550	3670	3810	3940	4080	4230	4370
	4	0.44	2940	3040	3150	3260	3370	3480	3600	3730	3860	3990	4120
	BFL	0.59	2780	2870	2970	3060	3160	3270	3370	3480	3600	3710	3830
	6	3.47	2540	2620	2690	2770	2850	2940	3020	3110	3200	3290	3390
	7	6.58	2250	2300	2350	2410	2470	2530	2590	2650	2710	2780	2850
W 24x55	TFL	0.00	2890	3000	3120	3240	3370	3500	3630	3770	3910	4060	4210
	2	0.13	2800	2910	3020	3130	3250	3370	3500	3630	3760	3900	4040
	3	0.25	2700	2800	2900	3010	3120	3230	3350	3470	3600	3730	3860
	4	0.38	2590	2680	2770	2870	2970	3080	3190	3300	3410	3530	3650
	BFL	0.51	2460	2540	2630	2710	2800	2900	2990	3090	3200	3300	3410
	6	3.45	2240	2310	2370	2440	2520	2590	2670	2750	2830	2920	3000
	7	6.66	1970	2010	2060	2110	2160	2210	2260	2320	2370	2430	2490
W 21x62	TFL	0.00	2760	2880	3000	3120	3250	3390	3530	3670	3820	3970	4130
	2	0.15	2650	2760	2870	2990	3110	3230	3360	3500	3630	3780	3920
	3	0.31	2530	2630	2730	2830	2940	3060	3170	3290	3420	3550	3680
	4	0.46	2380	2470	2560	2650	2750	2850	2950	3060	3170	3280	3400
	BFL	0.62	2210	2280	2360	2440	2520	2600	2690	2770	2870	2960	3060
	6	2.53	2070	2130	2190	2260	2320	2390	2460	2540	2620	2690	2780
	7	4.78	1900	1950	2000	2050	2100	2150	2210	2270	2330	2390	2450
W 21x57	TFL	0.00	2480	2590	2700	2810	2930	3060	3180	3320	3450	3590	3740
	2	0.16	2390	2490	2590	2700	2810	2930	3050	3170	3300	3430	3560
	3	0.33	2290	2380	2480	2570	2680	2780	2890	3000	3120	3240	3360
	4	0.49	2170	2250	2340	2420	2520	2610	2710	2810	2910	3020	3130
	BFL	0.65	2030	2100	2170	2250	2330	2410	2490	2580	2670	2760	2860
	6	2.90	1880	1940	2000	2060	2120	2190	2260	2330	2400	2480	2560
	7	5.38	1690	1740	1780	1830	1880	1920	1980	2030	2080	2140	2200
W 21x50	TFL	0.00	2120	2210	2310	2410	2510	2620	2730	2850	2960	3090	3210
	2	0.13	2050	2130	2220	2320	2410	2520	2620	2730	2840	2950	3070
	3	0.27	1960	2040	2130	2220	2310	2400	2500	2590	2700	2800	2910
	4	0.40	1870	1940	2020	2100	2180	2260	2350	2440	2530	2630	2730
	BFL	0.54	1760	1830	1890	1960	2040	2110	2190	2270	2350	2430	2520
	6	2.92	1620	1670	1720	1780	1840	1900	1960	2020	2090	2160	2230
	7	5.58	1440	1470	1510	1550	1590	1640	1680	1730	1780	1830	1880

[a]Y1 = distance from top of the steel beam to plastic neutral axis.
[b]Y2 = distance from top of the steel beam to concrete flange force.
[c]See Fig. 4.3 for PNA locations.

LOWER BOUND ELASTIC MOMENT OF INERTIA FOR PLASTIC COMPOSITE SECTIONS

Shape	PNA[c]	Y1[a] In.	I_{LB} (in.^{4}) Y2[b] (in.) 2	2.5	3	3.5	4	4.5	5	5.5	6	6.5	7
W 21x44	TFL	0.00	1830	1910	2000	2090	2180	2270	2370	2470	2580	2680	2800
	2	0.11	1770	1850	1930	2010	2100	2190	2280	2370	2470	2570	2680
	3	0.23	1710	1780	1850	1930	2010	2090	2180	2270	2360	2450	2550
	4	0.34	1630	1700	1760	1830	1910	1980	2060	2140	2220	2310	2400
	BFL	0.45	1540	1600	1660	1730	1790	1860	1930	2000	2070	2150	2230
	6	2.90	1410	1450	1500	1550	1610	1660	1720	1770	1830	1900	1960
	7	5.69	1240	1270	1300	1340	1380	1410	1450	1490	1540	1580	1620
W 18x60	TFL	0.00	2070	2170	2280	2390	2500	2620	2740	2860	3000	3130	3270
	2	0.17	1980	2080	2170	2270	2380	2480	2600	2710	2830	2960	3090
	3	0.35	1880	1960	2050	2140	2230	2330	2430	2540	2640	2750	2870
	4	0.52	1760	1830	1900	1980	2060	2150	2230	2320	2420	2510	2610
	BFL	0.70	1610	1670	1730	1790	1850	1920	1990	2070	2140	2220	2300
	6	2.19	1520	1570	1620	1670	1730	1790	1850	1910	1970	2040	2110
	7	3.82	1420	1460	1500	1540	1590	1640	1690	1740	1790	1840	1900
W 18x55	TFL	0.00	1880	1970	2070	2170	2270	2380	2490	2610	2730	2850	2980
	2	0.16	1800	1890	1970	2070	2160	2260	2360	2470	2580	2700	2810
	3	0.32	1710	1790	1870	1950	2030	2120	2220	2310	2410	2510	2620
	4	0.47	1600	1670	1740	1810	1880	1960	2040	2120	2210	2300	2390
	BFL	0.63	1470	1520	1580	1640	1700	1760	1830	1900	1970	2040	2110
	6	2.16	1380	1430	1480	1530	1580	1630	1690	1750	1810	1870	1930
	7	3.86	1290	1320	1360	1400	1440	1490	1530	1580	1620	1670	1730
W 18x50	TFL	0.00	1690	1770	1860	1950	2040	2140	2240	2340	2450	2560	2680
	2	0.14	1620	1700	1770	1860	1940	2030	2130	2220	2320	2430	2530
	3	0.29	1540	1610	1680	1750	1830	1910	1990	2080	2170	2260	2360
	4	0.43	1440	1500	1560	1630	1700	1770	1840	1910	1990	2070	2160
	BFL	0.57	1320	1370	1420	1480	1530	1590	1650	1710	1780	1840	1910
	6	2.07	1250	1290	1330	1380	1420	1470	1520	1570	1630	1680	1740
	7	3.82	1160	1190	1220	1260	1300	1340	1380	1420	1460	1510	1550
W 18x46	TFL	0.00	1530	1610	1690	1770	1860	1950	2040	2140	2240	2340	2450
	2	0.15	1470	1540	1620	1690	1770	1860	1940	2030	2130	2220	2320
	3	0.30	1400	1470	1540	1610	1680	1750	1830	1910	2000	2080	2170
	4	0.45	1320	1380	1440	1500	1560	1630	1700	1770	1850	1920	2000
	BFL	0.61	1230	1270	1320	1380	1430	1490	1550	1610	1670	1730	1800
	6	2.40	1140	1180	1220	1270	1310	1360	1410	1460	1510	1560	1620
	7	4.34	1040	1070	1100	1140	1170	1210	1240	1280	1320	1360	1410

[a]Y1 = distance from top of the steel beam to plastic neutral axis.
[b]Y2 = distance from top of the steel beam to concrete flange force.
[c]See Fig. 4.3 for PNA locations.

LOWER BOUND ELASTIC MOMENT OF INERTIA FOR PLASTIC COMPOSITE SECTIONS

Shape	PNA[c]	$Y1^a$	I_{LB} (in.[4])										
			$Y2^b$ (in.)										
		In.	2	2.5	3	3.5	4	4.5	5	5.5	6	6.5	7
W 18x40	TFL	0.00	1320	1390	1450	1530	1600	1680	1760	1840	1930	2020	2110
	2	0.13	1270	1330	1390	1460	1530	1600	1680	1760	1840	1920	2010
	3	0.26	1210	1270	1320	1390	1450	1510	1580	1650	1730	1800	1880
	4	0.39	1140	1190	1240	1300	1350	1410	1470	1530	1600	1670	1740
	BFL	0.53	1060	1100	1150	1190	1240	1290	1340	1390	1450	1510	1560
	6	2.26	985	1020	1060	1090	1130	1170	1220	1260	1310	1350	1400
	7	4.27	895	921	949	978	1010	1040	1070	1100	1140	1180	1210
W 18x35	TFL	0.00	1120	1170	1230	1300	1360	1430	1500	1570	1650	1720	1800
	2	0.11	1080	1130	1190	1240	1300	1370	1430	1500	1570	1640	1720
	3	0.21	1030	1080	1130	1180	1240	1300	1360	1420	1490	1550	1620
	4	0.32	978	1020	1070	1120	1170	1220	1270	1330	1390	1450	1510
	BFL	0.43	917	955	995	1040	1080	1130	1170	1220	1270	1320	1380
	6	2.37	842	874	906	940	976	1010	1050	1090	1130	1170	1220
	7	4.56	753	775	799	824	850	877	905	934	964	995	1030
W 16x36	TFL	0.00	971	1020	1080	1140	1200	1270	1330	1400	1480	1550	1630
	2	0.11	931	981	1030	1090	1140	1200	1270	1330	1400	1470	1540
	3	0.22	884	929	977	1030	1080	1130	1190	1250	1310	1370	1430
	4	0.32	830	869	910	954	999	1050	1090	1150	1200	1250	1310
	BFL	0.43	764	797	831	867	904	943	984	1030	1070	1120	1160
	6	1.79	714	742	770	801	832	865	899	935	972	1010	1050
	7	3.44	657	679	701	725	750	776	802	830	859	889	921
W 16x31	TFL	0.00	826	872	921	972	1030	1080	1140	1200	1260	1330	1390
	2	0.11	793	837	882	929	979	1030	1080	1140	1200	1260	1320
	3	0.22	756	796	837	880	925	972	1020	1070	1120	1180	1240
	4	0.33	713	748	784	823	863	904	948	993	1040	1090	1140
	BFL	0.44	662	691	722	755	789	824	861	899	939	980	1020
	6	2.00	613	637	663	690	718	747	778	810	843	877	912
	7	3.79	555	574	593	614	635	657	680	704	729	755	782
W 16x26	TFL	0.00	673	712	753	795	840	886	935	985	1040	1090	1150
	2	0.09	649	685	723	763	804	848	893	940	989	1040	1090
	3	0.17	621	654	689	726	764	804	845	888	933	980	1030
	4	0.26	589	618	650	683	717	753	790	829	870	911	955
	BFL	0.35	551	577	604	633	663	694	727	760	796	832	870
	6	2.04	505	526	549	572	597	622	649	676	705	734	765
	7	4.01	450	465	482	499	517	535	554	575	595	617	639

[a] $Y1$ = distance from top of the steel beam to plastic neutral axis.
[b] $Y2$ = distance from top of the steel beam to concrete flange force.
[c] See Fig. 4.3 for PNA locations.

AMERICAN INSTITUTE OF STEEL CONSTRUCTION

LOWER BOUND ELASTIC MOMENT OF INERTIA FOR PLASTIC COMPOSITE SECTIONS

I$_{LB}$

Shape	PNA[c]	Y1[a]	I_{LB} (in.^{4})										
			Y2[b] (in.)										
		In.	2	2.5	3	3.5	4	4.5	5	5.5	6	6.5	7
W 14x38	TFL	0.00	844	896	951	1010	1070	1130	1200	1270	1340	1410	1490
	2	0.13	805	853	903	956	1010	1070	1130	1190	1260	1330	1400
	3	0.26	759	802	846	893	943	994	1050	1100	1160	1220	1290
	4	0.39	704	740	778	818	861	905	950	998	1050	1100	1150
	BFL	0.52	636	665	695	727	760	794	831	868	908	948	991
	6	1.38	604	629	655	683	712	742	773	806	840	876	913
	7	2.53	568	589	611	634	659	684	710	738	766	796	827
W 14x34	TFL	0.00	744	790	839	890	944	1000	1060	1120	1180	1250	1320
	2	0.11	711	753	798	844	894	945	999	1060	1110	1170	1240
	3	0.23	671	709	749	790	834	880	929	979	1030	1080	1140
	4	0.34	623	656	690	726	763	803	844	887	931	978	1030
	BFL	0.46	565	591	618	647	677	708	741	775	810	847	885
	6	1.41	535	557	581	606	631	659	687	716	747	779	812
	7	2.60	502	520	540	560	582	604	628	652	677	704	731
W 14x30	TFL	0.00	643	684	726	771	819	868	920	974	1030	1090	1150
	2	0.10	615	653	692	734	777	823	870	920	971	1020	1080
	3	0.19	583	616	652	689	728	769	812	857	903	951	1000
	4	0.29	544	573	604	636	670	706	743	782	822	864	907
	BFL	0.39	497	521	546	573	600	629	659	691	724	758	793
	6	1.48	467	487	508	531	554	579	605	631	659	688	719
	7	2.82	432	448	465	483	502	522	542	564	586	610	634
W 14x26	TFL	0.00	553	589	626	665	706	750	795	841	890	941	994
	2	0.11	531	563	598	634	672	712	754	798	843	890	939
	3	0.21	504	534	565	598	633	669	707	747	788	830	875
	4	0.32	473	500	527	556	587	619	652	687	723	761	800
	BFL	0.42	437	459	482	506	532	559	587	616	646	678	711
	6	1.67	405	423	443	463	485	507	530	555	580	606	634
	7	3.19	368	382	397	413	430	447	465	484	503	523	545
W 14x22	TFL	0.00	454	484	515	548	582	619	656	696	736	779	823
	2	0.08	437	464	493	524	556	590	625	661	699	739	780
	3	0.17	416	442	468	496	526	556	588	622	657	693	731
	4	0.25	393	416	439	464	490	518	546	576	607	640	673
	BFL	0.34	365	385	405	427	449	473	497	523	550	577	606
	6	1.69	336	352	369	386	405	424	444	466	488	510	534
	7	3.34	301	313	325	339	352	367	382	398	414	431	449

[a] Y1 = distance from top of the steel beam to plastic neutral axis.
[b] Y2 = distance from top of the steel beam to concrete flange force.
[c] See Fig. 4.3 for PNA locations.

LOWER BOUND
ELASTIC MOMENT OF INERTIA
FOR PLASTIC COMPOSITE SECTIONS

I_{LB}

Shape	PNA[c]	$Y1$[a]	I_{LB} (in.^{4})										
			$Y2$[b] (in.)										
		In.	2	2.5	3	3.5	4	4.5	5	5.5	6	6.5	7
W 12x30	TFL	0.00	531	568	608	649	693	738	786	837	889	944	1000
	2	0.11	505	539	575	612	652	694	738	783	831	881	933
	3	0.22	474	504	536	569	604	641	679	720	762	806	852
	4	0.33	436	461	488	516	545	576	609	643	678	715	753
	BFL	0.44	389	408	429	450	472	496	521	547	574	602	631
	6	1.12	373	390	408	427	447	468	490	513	537	562	589
	7	1.94	355	370	386	402	420	438	457	477	498	520	543
W 12x36	TFL	0.00	456	488	521	557	595	635	676	720	765	812	861
	2	0.10	434	463	494	526	561	597	635	674	716	759	804
	3	0.19	407	433	460	489	520	552	585	621	657	695	735
	4	0.29	375	397	420	445	470	497	526	555	586	618	652
	BFL	0.38	336	353	370	389	409	430	452	474	498	523	549
	6	1.08	321	336	351	368	386	404	423	444	465	487	509
	7	1.95	305	317	331	345	360	376	393	410	428	447	467
W 12x22	TFL	0.00	371	399	428	458	490	524	559	596	635	675	717
	2	0.11	356	382	408	437	466	498	531	565	601	638	677
	3	0.21	339	362	386	412	439	468	498	529	562	596	631
	4	0.32	318	339	360	383	408	433	459	487	516	547	578
	BFL	0.43	294	312	330	350	370	392	414	438	463	488	515
	6	1.66	270	285	300	316	333	351	370	389	410	431	453
	7	3.04	242	253	265	277	290	303	317	332	347	364	380
W 12x19	TFL	0.00	312	335	360	386	413	442	472	503	536	571	606
	2	0.09	300	321	344	368	394	421	449	478	509	541	574
	3	0.18	286	306	327	349	372	397	423	450	478	507	538
	4	0.26	270	287	306	326	347	369	392	417	442	468	496
	BFL	0.35	251	266	282	300	318	337	357	378	400	423	446
	6	1.66	229	241	255	269	284	299	316	333	351	370	389
	7	3.12	203	212	222	232	243	255	267	279	293	306	321
W 12x16	TFL	0.00	254	273	294	315	338	362	388	414	442	471	501
	2	0.07	245	263	282	303	324	347	371	396	422	449	477
	3	0.13	234	251	269	288	309	330	352	375	399	424	450
	4	0.20	223	239	255	272	291	310	330	351	373	396	420
	BFL	0.26	210	224	238	254	270	287	305	324	344	364	386
	6	1.71	189	200	212	224	238	251	266	281	297	313	330
	7	3.32	163	171	179	188	197	207	217	227	239	250	262

[a] $Y1$ = distance from top of the steel beam to plastic neutral axis.
[b] $Y2$ = distance from top of the steel beam to concrete flange force.
[c] See Fig. 4.3 for PNA locations.

LOWER BOUND
ELASTIC MOMENT OF INERTIA
FOR PLASTIC COMPOSITE SECTIONS

I_{LB} (in.^{4})

Shape	PNA[c]	Y1[a]	Y2[b] (in.)										
		In.	2	2.5	3	3.5	4	4.5	5	5.5	6	6.5	7
W 12x14	TFL	0.00	220	237	255	275	295	316	338	362	386	411	438
	2	0.06	213	229	246	264	283	303	324	346	369	393	418
	3	0.11	204	219	235	252	270	289	308	329	350	372	396
	4	0.17	195	209	223	239	255	272	290	309	329	349	371
	BFL	0.23	184	197	210	224	238	254	270	287	305	323	342
	6	1.69	165	175	186	197	209	221	234	247	262	276	292
	7	3.36	141	148	155	163	171	180	188	198	208	218	228
W 10x26	TFL	0.00	339	368	398	430	464	499	537	577	618	662	707
	2	0.11	322	347	375	404	435	467	501	537	575	615	656
	3	0.22	300	323	347	372	400	428	458	490	523	558	594
	4	0.33	274	293	313	334	357	381	406	432	460	489	519
	BFL	0.44	242	256	271	287	304	321	340	360	381	403	425
	6	0.90	232	245	258	273	288	304	321	339	358	378	398
	7	1.51	222	233	245	258	272	286	301	317	334	351	369
W 10x22	TFL	0.00	281	305	330	357	386	416	448	482	517	554	592
	2	0.09	267	289	312	336	363	390	419	450	482	516	551
	3	0.18	250	269	290	312	335	360	385	413	441	471	502
	4	0.27	230	246	263	282	302	322	344	367	391	417	443
	BFL	0.36	205	217	231	245	260	277	293	311	330	350	370
	6	0.95	194	205	217	230	244	258	273	288	305	322	340
	7	1.70	183	193	203	214	225	237	250	263	277	292	308
W 10x19	TFL	0.00	239	259	282	305	330	356	384	413	444	476	509
	2	0.10	228	247	267	289	312	337	362	389	417	447	478
	3	0.20	215	233	251	271	292	314	337	361	387	413	441
	4	0.30	200	216	232	249	267	286	307	328	350	374	398
	BFL	0.40	183	195	209	223	238	254	271	288	307	326	347
	6	1.27	169	180	191	203	216	229	243	258	274	290	307
	7	2.31	153	162	170	180	190	200	211	223	235	248	261
W 10x17	TFL	0.00	206	224	244	265	286	310	334	360	387	415	444
	2	0.08	197	214	232	252	272	294	316	340	365	391	419
	3	0.17	187	203	219	237	255	275	296	317	340	364	389
	4	0.25	175	189	204	219	236	253	272	291	311	332	354
	BFL	0.33	161	173	185	199	213	227	243	260	277	295	314
	6	1.31	148	157	168	179	190	202	215	229	243	258	274
	7	2.46	132	139	147	155	164	173	183	193	204	215	227

[a]Y1 = distance from top of the steel beam to plastic neutral axis.
[b]Y2 = distance from top of the steel beam to concrete flange force.
[c]See Fig. 4.3 for PNA locations.

LOWER BOUND
ELASTIC MOMENT OF INERTIA
FOR PLASTIC COMPOSITE SECTIONS

I_{LB}

Shape	PNA[c]	$Y1$[a]	I_{LB} (in.⁴)										
			$Y2$[b] (in.)										
		In.	2	2.5	3	3.5	4	4.5	5	5.5	6	6.5	7
W 10x15	TFL	0.00	177	193	210	228	247	268	289	312	335	360	386
	2	0.07	170	185	201	218	236	255	275	296	318	341	365
	3	0.14	162	175	190	206	222	240	258	278	298	320	342
	4	0.20	153	165	178	192	207	223	240	257	275	295	315
	BFL	0.27	142	153	164	176	189	203	218	233	249	266	283
	6	1.35	128	137	147	157	167	178	190	203	216	229	244
	7	2.60	112	118	125	133	140	148	157	166	176	185	196
W 10x12	TFL	0.00	139	152	165	180	195	211	229	247	265	285	306
	2	0.05	134	146	158	172	186	202	218	235	252	271	290
	3	0.11	128	139	150	163	176	190	205	221	237	254	272
	4	0.16	121	131	141	153	165	178	191	205	220	236	252
	BFL	0.21	113	122	131	141	152	163	175	187	200	214	229
	6	1.30	102	109	116	124	133	142	152	162	173	184	195
	7	2.61	87.9	92.9	98.4	104	110	117	124	131	138	146	155
W 8x28	TFL	0.00	248	274	302	332	364	398	434	473	513	555	600
	2	0.12	233	256	281	308	337	368	400	435	471	509	549
	3	0.23	214	234	256	279	304	330	358	388	419	452	487
	4	0.35	191	207	224	243	262	284	306	330	355	381	408
	BFL	0.47	161	171	183	196	209	223	238	254	271	289	307
	6	0.53	159	170	181	194	207	221	235	251	268	285	303
	7	0.59	158	168	180	192	204	218	233	248	264	281	299
W 8x24	TFL	0.00	209	231	255	280	307	336	367	400	434	470	508
	2	0.10	196	216	237	260	285	311	339	368	399	431	465
	3	0.20	180	198	216	236	257	279	303	329	355	383	413
	4	0.30	161	175	189	205	222	240	259	280	301	323	347
	BFL	0.40	136	145	155	166	177	189	202	216	231	246	262
	6	0.47	134	143	153	164	175	187	200	213	227	242	257
	7	0.55	133	142	151	162	173	184	197	210	223	238	253
W 8x21	TFL	0.00	191	211	232	255	279	305	333	362	392	424	458
	2	0.10	181	198	218	238	260	284	309	335	362	391	422
	3	0.20	167	183	200	218	237	258	279	302	327	352	379
	4	0.30	151	164	178	193	209	226	244	263	283	304	326
	BFL	0.40	131	140	151	162	173	186	199	213	227	243	259
	6	0.70	126	135	145	155	165	177	189	201	215	229	244
	7	1.06	122	130	138	147	157	167	178	190	202	215	228

[a] $Y1$ = distance from top of the steel beam to plastic neutral axis.
[b] $Y2$ = distance from top of the steel beam to concrete flange force.
[c] See Fig. 4.3 for PNA locations.

LOWER BOUND
ELASTIC MOMENT OF INERTIA
FOR PLASTIC COMPOSITE SECTIONS

I$_{LB}$

Shape	PNA[c]	Y1[a] In.	I_{LB} (in.⁴) Y2[b] (in.)										
			2	2.5	3	3.5	4	4.5	5	5.5	6	6.5	7
W 8x18	TFL	0.00	159	175	193	213	233	255	278	303	329	356	384
	2	0.08	150	165	182	199	218	238	259	281	305	329	355
	3	0.17	140	153	167	183	199	217	236	255	276	298	321
	4	0.25	127	138	150	163	177	192	207	224	241	259	278
	BFL	0.33	111	120	129	139	149	160	172	184	198	211	226
	6	0.71	106	114	122	131	140	150	161	172	184	196	209
	7	1.21	101	107	114	122	130	139	148	158	169	179	191
W 8x15	TFL	0.00	129	143	158	175	192	210	230	251	272	295	319
	2	0.08	123	136	150	165	181	198	216	235	255	276	299
	3	0.16	116	128	140	154	168	183	200	217	235	254	274
	4	0.24	107	117	128	140	153	166	181	196	211	228	246
	BFL	0.32	97.0	105	114	124	135	146	158	170	183	197	211
	6	0.97	89.3	96.4	104	112	121	130	140	151	162	174	186
	7	1.79	80.6	86.2	92.2	98.7	106	113	121	129	138	147	157
W 8x13	TFL	0.00	109	121	134	147	162	178	195	213	231	251	272
	2	0.06	104	115	127	140	154	168	184	200	218	236	255
	3	0.13	98.0	108	119	131	144	157	171	186	202	219	236
	4	0.19	91.4	100	110	121	132	144	156	170	184	198	214
	BFL	0.26	83.6	91.3	99.6	108	118	128	139	150	162	175	188
	6	1.00	76.1	82.4	89.3	96.6	104	113	122	131	141	151	162
	7	1.91	67.2	72.0	77.2	82.7	88.7	95.0	102	109	116	124	132
W 8x10	TFL	0.00	83.1	92.3	102	113	124	136	149	163	177	192	208
	2	0.05	79.3	87.8	97.0	107	117	129	141	153	166	180	195
	3	0.10	74.8	82.6	90.9	99.9	109	120	131	142	154	167	180
	4	0.15	69.6	76.5	83.8	91.7	100	109	119	129	140	151	162
	BFL	0.21	63.5	69.2	75.4	82.0	89.2	96.7	105	113	122	132	142
	6	0.88	58.0	62.8	68.0	73.5	79.5	85.8	92.5	99.6	107	115	123
	7	1.77	51.7	55.4	59.4	63.6	68.2	73.0	78.2	83.6	89.4	95.4	102

[a]Y1 = distance from top of the steel beam to plastic neutral axis.
[b]Y2 = distance from top of the steel beam to concrete flange force.
[c]See Fig. 4.3 for PNA locations.

AMERICAN INSTITUTE OF STEEL CONSTRUCTION

COMPOSITE COLUMNS
General Notes

COMPOSITE COLUMN LOAD TABLES

Load tables for composite columns are presented for a range of W shapes, pipe and structural tubing. Tabular loads are computed in accordance with Section I2.2 of the LRFD Specification for axially loaded members having effective unsupported lengths indicated at the left of each table. The effective length KL is the actual unbraced length, in feet, multiplied by factor K, which depends on the rotational restraint at the ends of the unbraced length and the means available to resist lateral movements.

Load tables are provided for W shapes encased in reinforced normal weight concrete of square or rectangular cross section, and for steel pipe and structural tubing filled with normal weight concrete. The following W shapes are included:

Nominal depth, in.	Weight, lb. per ft
14	426–43
12	336–50
10	112–39
8	67–35

Two values of yield stress, F_y equal to 36 and 50 ksi, and three values of concrete cylinder strength, f_c' equal to 3.5, 5 and 8 ksi, are included for W shapes with all reinforcing steel of Grade 60. The tables for steel pipe columns include nominal pipe diameters of 4 to 12 in., yield stress of $F_y = 36$ ksi and concrete cylinder strength f_c' equal to 3.5 and 5 ksi. The tables for tubular columns include tubes of nominal side dimensions 4 to 16 in., yield stress F_y equal to 46 ksi and concrete cylinder strengths f_c' equal to 3.5 and 5 ksi.

All axial design strengths are tabulated in kips. Strength values are omitted when Kl/r exceeds 200. Resistance factor $\phi = 0.85$ was used in computing the axial design strengths of all composite columns. The K-factor may be selected using as a guide Table C-C2.1 in the Commentary of the LRFD Specification. Interpolation between the idealized cases is a matter of engineering judgment. More precise values for K may be obtained, if desired, from the alignment chart in Fig. C-C2.2 in the LRFD Commentary once sections have been selected for several framing members.

In all tables, the design strengths are given for effective lengths with respect to the minor axis. When the minor axis is braced at closer intervals than the major axis, the capacity of the column must be investigated with reference to both the major (X-X) and minor (Y-Y) axes. The ratio r_{mx}/r_{my} included in the tables provides a convenient method for investigating the strength of a column with respect to its major axis.

To obtain an effective length with respect to the minor axis equivalent in load carrying capacity to the actual effective length about the major axis, divide the major axis effective length by the r_{mx}/r_{my} ratio. Compare this length with the actual effective length about the minor axis. The longer of the two lengths will control the design and the design strength may be taken from the table opposite the longer of the two effective lengths.

Properties useful to the designer are listed at the bottom of the Column Load Tables. They are helpful in considering buckling about the major axis as discussed above and in the design of members under combined axial compression and bending as discussed subsequently. Both of these cases are illustrated with design examples. The properties have the following units:

modified radius of gyration r_m, in.
modified nominal flexural strength M_n, kip-ft
Euler buckling term $P_e(KL)^2$, kip-ft^2

Subscripts x and y refer to the major and minor axes. Resistance factor $\phi_b = 0.9$ was used in computing the flexural design strength $\phi_b M_n$.

Additional notes relating specifically to the W shape, steel pipe and structural tube column tables precede each of these groups of tables.

Example 1

Given:

Using tables on the pages that follow, design the smallest composite column with a W shape of $F_y = 36$ ksi to support a factored concentric load of 1,000 kips. The effective length with respect to its minor axis is 16 ft and that with respect to its major axis is 31 ft.

Solution:

Use composite column tables for W shapes with $f'_c = 8$ ksi since the strongest concrete requires the smallest size column. An inspection of the tables reveals that the tabulated values of r_{mx}/r_{my} do not exceed 1.22 and in most cases are equal to 1.0 (square columns).

Assuming that r_{mx}/r_{my} is equal to 1.0, enter the tables with equivalent effective column length of $KL = 31$ ft. Select an 18 in. $\times$ 18 in. column with a W10 $\times$ 54 and $r_{mx}/r_{my} = 1.0$. By interpolation, the column has an axial design strength ϕP_n of 1,033 kips.

Use: 18 in. $\times$ 18 in. column with W10 $\times$ 54 of $F_y = 36$ ksi, $f'_c = 8$ ksi, 4-#8 Gr. 60 longitudinal bars and # 3 Gr. 60 ties spaced 12 in. o.c.

Note: A W14 $\times$ 145 section is required for an unencased steel column (see Ex. 1 in Part 2—Columns). The outside dimensions of a W14 $\times$ 145 are 14.78 in. $\times$ 15.5 in. without fireproofing.

Example 2

Given:

Redesign the column from Ex. 1 using (a) rectangular and (b) square structural tubing, both filled with structural concrete.

Solution:

(a) Enter the Composite Column Tables for rectangular structural tubing filled with 5 ksi concrete at an effective column length of $KL = 16$ ft

Select 14 $\times$ 10 tubing with ½-in. thick walls; $\phi P_n = 1,090$ kips.

$r_{mx}/r_{my} = 1.30$

Equivalent effective length for *X-X* axis:

$$31/1.30 = 23.8 \text{ ft}$$

Since 23.8 ft > 16 ft, *X-X* axis controls.

Re-enter the table at an effective length of $KL = 23.8$ ft. It is apparent that 14 x 10 tube will not satisfy the axial load of 1,000 kips. Select 16 × 12 steel tube with ½ in. walls:

$$r_{mx}/r_{my} = 1.25$$

$$KL = 31/1.25 = 24.8 \text{ ft}$$

By interpolation, the column provides an axial design strength of 1,206 kips. The same tubing filled with 3.5 ksi concrete is good for 1,094 kips > 1,000 kips.

Use: 16-in. × 12-in. × ½-in. tubing filled with 3.5-ksi concrete.

(b) Enter the composite column tables for square structural tubing filled with 5-ksi concrete at effective column length $KL = 31$ ft.

Select 14 × 14 tubing with ½-in. thick walls good for 1,135 kips. The same tubing filled with 3.5 ksi concrete is good for 1,035 kips > 1,000 kips.

Use: 14-in. × 14-in. × ½-in. tubing filled with 3.5-ksi concrete.

Note: The weight of both the 16 × 12 and 14 × 14 steel tubing with ½-in. thick walls is 89.68 lb. per ft.

COMBINED AXIAL COMPRESSION AND BENDING (INTERACTION)

Loads given in the Composite Column Tables are for concentrically loaded columns. For columns subjected to combined compression and flexure, the nominal flexural design strength $\phi_b M_n$ determined from Formula C-I4-1 of the LRFD Commentary and the elastic buckling load P_e times the square of the effective column length are given at the bottom of the tables. With these two quantities and the loads tabulated for concentric loading, the column may be designed by successive approximations based on LRFD Specification Sect. I4. The procedure is illustrated in Ex. 3 for a column subjected to a large bending moment combined with a moderate axial load and in Ex. 4 for a column subjected to a large axial load combined with a moderate bending moment.

Example 3

Given:

Design a composite encased W shape column to resist a factored axial load of 200 kips and a factored moment about the *X-X* axis of 240 kip-ft. The unsupported length of the column is 12 ft, $F_y = 50$ ksi, $f'_c = 3.5$ ksi and $C_m = 1.0$. The loads were obtained by first order elastic analysis and there is no lateral translation of column ends.

Solution:

Since the moment is large in relation to the axial load, assume that $P_u/\phi P_n = \frac{1}{3}$ and $B_1 = 1$. From Formula H1-1a

$$\phi_b M_{nx} = (8/9) \times 240 \times (3/2) = 320 \text{ kip-ft}$$

From the Composite Column Tables, find a column with $\phi_b M_{nx}$ close to 320 kip-ft.

Try W10 × 45 with 18-in. × 18-in. encasement: $\phi_b M_{nx}$ = 351 kip-ft, ϕP_n = 1,130 kips and P_{ex} = 117 × 10⁴/12² = 8,125 kips.

$$\therefore P_u/\phi P_n = 200/1,130 = 0.1770 < 0.2$$

From Formula H1-2 and H1-3, with M_{lt} = 0 since there is no lateral translation,

$$M_{ux} = \frac{240}{1 - \dfrac{200}{8,125}} = 246.1 \text{ kip-ft}$$

and from Formula H1-1b

$$\phi_b M_{nx} = \frac{246.1}{1 - \dfrac{1}{2}\left(\dfrac{200}{1,130}\right)} = 270.0 \text{ kip-ft}$$

Since 270 kip-ft is much less than the 351 kip-ft provided, select a smaller size. Try W8 × 48 with 16-in. × 16-in. encasement: $\phi_b M_{nx}$ = 291 kip-in., ϕP_n = 1,010 kips and P_{ex} = 89.3 × 10⁴/12² = 6,201 kips.

$$M_{ux} = \frac{240}{1 - \dfrac{200}{6,201}} = 248.0 \text{ kip-ft}$$

From Formula H1-1b,

$$\frac{200}{2 \times 1,010} + \frac{248}{291} = 0.951 < 1.0 \quad \textbf{o.k.}$$

Use: 16-in. × 16-in. column with W8 × 48 of F_y = 50 ksi, f_c' = 3.5 ksi, 4-#7 Gr. 60 longitudinal bars and #3 Gr. 60 ties spaced at 10 in.

Example 4:

Given:

Design a composite encased W shape column to resist a factored axial load of 1,100 kips and factored moment of 200 kip-ft. Use 50-ksi structural steel and 5-ksi concrete. The unsupported column length is 11 ft and C_m = 0.85. Assume that sidesway is prevented.

Solution:

Since the axial load is large in relation to the moment, assume that

$$\frac{8}{9} \frac{M_u}{\phi_b M_{nx}} = 0.5$$

From Formula H1-1a

$$\phi P_n = 1,100/0.5 = 2,200 \text{ kips}$$

From the Composite Column Tables, find a column with ϕP_n close to 2,200 kips at $KL = 11$ ft

Try W12 × 106 with 20-in. × 20-in. encasement: $\phi P_n = 2,270$ kips, $\phi_b M_{nx} = 899$ kip-ft and $P_{ex} = 294 \times 10^4/11^2 = 24,300$ kips.

From Formula H1-3

$$B_1 = \frac{0.85}{1 - \dfrac{1,100}{24,300}} = 0.890 < 1.0$$

$\therefore B_1 = 1.0$. From Formula H1-1a

$$\frac{P_u}{\phi P_n} = 1 - \frac{8}{9} \times \frac{200}{899} = 0.802$$

and

$$\phi P_n = 1100/0.802 = 1,372 \text{ kips}$$

Try W10 × 45 with 18-in. × 18-in. encasement: $\phi P_n = 1,360$ kips, $\phi_b M_{nx} = 356$ kip-ft and $P_{ex} = 125 \times 10^4/11^2 = 10,330$ kips.

From Formula H1-3

$$B_1 = \frac{0.85}{1 - \dfrac{1,100}{10,330}} = 0.951 < 1.0$$

$\therefore B_1 = 1.0$ and

$$\frac{P_u}{\phi P_n} = 1 - \frac{8}{9} \times \frac{200}{356} = 0.501$$

and

$$\phi P_n = 1,100/0.501 = 2,196 \text{ kips}$$

Since the convergence is slow, estimate a new ϕP_n as

$$\phi P_n = \frac{2,196 + 1,372}{2} = 1,784 \text{ kips}$$

Try W10 × 77 with 18-in. × 18-in. encasement: $\phi P_n = 1,720$ kips, $\phi_b M_{nx} = 555$ kip-ft and $P_{ex} = 178 \times 10^4/11^2 = 14,710$ kips.

$$B_1 = \frac{0.85}{1 - \dfrac{1,100}{14,710}} = 0.919 < 1.0$$

and

$$\frac{1,100}{1,720} + \frac{8}{9} \times \frac{200}{555} = 0.960 < 1.0 \quad \textbf{o.k.}$$

Use: 18-in. × 18-in. column with W10 × 77 of $F_y = 50$ ksi, $f'_c = 5$ ksi, 4-# 8 Gr. 60 longitudinal bars and #3 Gr. 60 ties spaced at 12 in.

AMERICAN INSTITUTE OF STEEL CONSTRUCTION

Notes

COMPOSITE COLUMNS
W Shapes Encased in Concrete

Concentric design strengths in the tables that follow are tabulated for the effective length KL in feet, listed at the left of each table. They are applicable to axially loaded members with respect to their minor axis in accordance with Sect. I2.2 of the LRFD Specification. Two steel yield stresses, $F_y = 36$ ksi and $F_y = 50$ ksi, and three concrete strengths, $f_c' = 3.5$ ksi, $f_c' = 5$ ksi and $f_c' = 8$ ksi, are covered. The tables apply to normal weight concrete. All reinforcing steel is Gr. 60; however, $F_{yr} = 55$ ksi is used in the calculation of ϕP_n in accordance with LRFD Specification Sect. I2.1.

Each W shape is embedded in concrete of a square or rectangular cross section reinforced with four longitudinal reinforcing bars placed in the four corners and with lateral ties spaced as required in Sect. I2.1. For the design of additional confinement reinforcement, see LRFD Specification Sect. I2.1b. The size of the concrete section was selected so as to provide at least the minimum required cover over the reinforcing bars in the column with the largest shape of the group of five listed on any one page.

For discussion of the effective length, range of Kl/r, strength about the major axis, combined axial and bending strength, and for sample problems, see "Composite Columns, General Notes."

The properties listed at the bottom of each table are for use in checking strength about the strong axis and in design for combined axial load and bending.

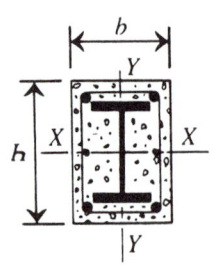

F_y = 36 ksi

F_y = 50 ksi

COMPOSITE COLUMNS
W Shapes
f'_c = 3.5 ksi
All reinforcing steel is Grade 60
Axial design strength in kips

Size $b \times h$			24 in. × 26 in.							
Reinf. bars			4-#11 bars							
Ties			#4 bars spaced 16 in. c. to c.							

	Designation		**W 14**							
Steel Shape	Wt./ft		426		398		370		342		311	
	F_y	36	50	36	50	36	50	36	50	36	50
Effective length in ft KL with respect to least radius of gyration r_{my}	0	4910	6400	4680	6070	4450	5740	4220	5420	3940	5030
	6	4880	6340	4650	6020	4420	5700	4190	5370	3910	4990
	7	4870	6330	4640	6000	4410	5680	4180	5360	3910	4970
	8	4850	6300	4630	5980	4400	5660	4170	5340	3890	4950
	9	4840	6280	4610	5960	4380	5640	4160	5320	3880	4930
	10	4820	6250	4600	5930	4370	5620	4140	5300	3870	4910
	11	4810	6220	4580	5910	4350	5590	4130	5270	3850	4890
	12	4790	6190	4560	5880	4340	5560	4110	5240	3840	4860
	13	4770	6160	4540	5840	4320	5530	4090	5210	3820	4830
	14	4750	6120	4520	5810	4300	5490	4070	5180	3800	4800
	15	4720	6080	4500	5770	4270	5460	4050	5150	3780	4770
	16	4700	6040	4470	5730	4250	5420	4030	5110	3760	4740
	17	4670	5990	4450	5690	4230	5380	4000	5070	3740	4700
	18	4640	5950	4420	5640	4200	5340	3980	5030	3710	4660
	19	4610	5900	4390	5590	4170	5290	3950	4990	3690	4620
	20	4580	5850	4360	5540	4140	5240	3930	4940	3660	4580
	22	4520	5740	4300	5440	4080	5140	3870	4850	3610	4490
	24	4440	5620	4230	5330	4020	5040	3800	4750	3550	4400
	26	4370	5490	4160	5210	3950	4920	3740	4640	3480	4300
	28	4290	5360	4080	5080	3870	4800	3660	4520	3420	4190
	30	4200	5220	4000	4950	3790	4680	3590	4410	3340	4080
	32	4110	5080	3910	4810	3710	4550	3510	4280	3270	3960
	34	4020	4930	3820	4670	3630	4410	3430	4150	3190	3840
	36	3930	4780	3730	4530	3540	4270	3340	4020	3110	3720
	38	3830	4620	3640	4380	3450	4130	3260	3890	3030	3590
	40	3720	4460	3540	4230	3350	3990	3170	3750	2940	3460

Properties										
$\phi_b M_{nx}$ (kip-ft)	2970	3850	2780	3610	2600	3380	2420	3150	2210	2890
$\phi_b M_{ny}$ (kip-ft)	1750	2170	1660	2070	1560	1960	1460	1840	1360	1720
$P_{ex}(K_x L_x)^2/10^4$ (kip-ft^2)	1650	1650	1550	1550	1460	1460	1360	1360	1250	1250
$P_{ey}(K_y L_y)^2/10^4$ (kip-ft^2)	1400	1400	1320	1320	1240	1240	1160	1160	1060	1060
r_{my} (in.)	7.20		7.20		7.20		7.20		7.20	
r_{mx}/r_{my} (in./in.)	1.08		1.08		1.08		1.08		1.08	

| F_y = 36 ksi |
| F_y = 50 ksi |

COMPOSITE COLUMNS
W Shapes
f'_c = 3.5 ksi
All reinforcing steel is Grade 60
Axial design strength in kips

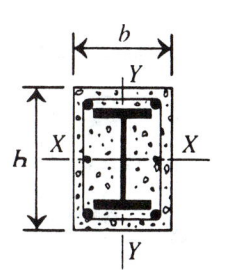

Size $b \times h$		24 in. × 24 in.									
Reinf. bars		4-#11 bars									
Ties		#4 bars spaced 16 in. c. to c.									
Designation		**W 14**									
Wt./ft		283		257		233		211		193	
F_y		36	50	36	50	36	50	36	50	36	50
	0	3620	4610	3400	4300	3200	4010	3010	3750	2860	3530
	6	3600	4570	3380	4260	3170	3980	2990	3710	2840	3500
	7	3590	4560	3370	4250	3170	3960	2980	3700	2830	3490
	8	3580	4540	3360	4230	3160	3950	2970	3690	2820	3480
	9	3570	4530	3350	4220	3150	3930	2960	3670	2810	3460
	10	3560	4510	3340	4200	3130	3910	2950	3660	2800	3450
	11	3540	4480	3320	4180	3120	3900	2940	3640	2790	3430
	12	3530	4460	3310	4150	3110	3870	2920	3620	2780	3410
	13	3510	4430	3290	4130	3090	3850	2910	3590	2760	3390
	14	3490	4410	3280	4100	3080	3830	2890	3570	2750	3370
	15	3470	4380	3260	4080	3060	2800	2880	3550	2730	3340
	16	3460	4340	3240	4050	3040	3770	2860	3520	2720	3320
	17	3430	4310	4220	4010	3020	3740	2840	3490	2700	3290
	18	3410	4280	3200	3980	3000	3710	2820	3460	2680	3260
	19	3390	4240	3180	3950	2980	3680	2800	3430	2660	3230
	20	3360	4200	3150	3910	2960	3640	2780	3400	2640	3200
	22	3310	4120	3100	3830	2910	3570	2740	3330	2590	3140
	24	3260	4030	3050	3750	2860	3490	2690	3250	2550	3060
	26	3200	3940	2990	3660	2810	3410	2630	3180	2500	2990
	28	3140	3840	2930	3570	2750	3320	2580	3090	2440	2910
	30	3070	3740	2870	3470	2690	3230	2520	3010	2390	2830
	32	3000	3630	2810	3370	2630	3140	2460	2920	2330	2740
	34	2930	3520	2740	3270	2560	3040	2400	2820	2270	2660
	36	2850	3410	2670	3160	2490	2940	2330	2730	2210	2560
	38	2780	3290	2590	3050	2420	2830	2270	2630	2140	2470
	40	2700	3170	2520	2940	2350	2730	2200	2530	2080	2380

Effective length in ft KL with respect to least radius of gyration r_{my}

Properties											
$\phi_b M_{nx}$ (kip-ft)		1970	2580	1810	2360	1650	2160	1510	1980	1400	1830
$\phi_b M_{ny}$ (kip-ft)		1250	1570	1150	1460	1070	1360	991	1260	924	1170
$P_{ex}(K_x L_x)^2/10^4$ (kip-ft²)		971	971	894	894	822	822	757	757	704	704
$P_{ey}(K_y L_y)^2/10^4$ (kip-ft²)		971	971	894	894	822	822	757	757	704	704
r_{my} (in.)		7.20		7.20		7.20		7.20		7.20	
r_{mx}/r_{my} (in./in.)		1.00		1.00		1.00		1.00		1.00	

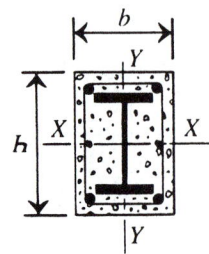

	F_y = 36 ksi
	F_y = 50 ksi

COMPOSITE COLUMNS
W Shapes
f'_c = 3.5 ksi
All reinforcing steel is Grade 60
Axial design strength in kips

Size $b \times h$		24 in. × 24 in.								
Reinf. bars		4-#10 bars								
Ties		#3 bars spaced 16 in. c. to c.								
Designation		**W 14**								
Wt./ft		176		159		145		132		120	
F_y		36	50	36	50	36	50	36	50	36	50
	0	2680	3290	2530	3090	2420	2920	2300	2770	2200	2620
	6	2660	3270	2510	3060	2400	2900	2290	2740	2180	2600
	7	2650	3250	2510	3050	2390	2890	2280	2730	2180	2590
	8	2640	3240	2500	3040	2380	2880	2270	2720	2170	2580
	9	2630	3230	2490	3020	2370	2860	2260	2710	2160	2570
	10	2620	3210	2480	3010	2370	2850	2250	2690	2150	2550
	11	2610	3200	2470	2990	2350	2830	2240	2680	2140	2540
	12	2600	3180	2460	2980	2340	2820	2230	2660	2130	2520
	13	2590	3160	2440	2960	2330	2800	2220	2650	2120	2510
	14	2570	3140	2430	2940	2320	2780	2210	2630	2110	2490
	15	2560	3120	2420	2920	2300	2760	2190	2610	2090	2470
	16	2540	3090	2400	2890	2290	2740	2180	2590	2080	2450
	17	2530	3070	2380	2870	2270	2720	2160	2560	2070	2430
	18	2510	3040	2370	2840	2260	2690	2150	2540	2050	2410
	19	2490	3010	2350	2820	2240	2670	2130	2520	2030	2380
	20	2470	2980	2330	2790	2220	2640	2110	2490	2010	2360
	22	2430	2920	2290	2730	2180	2580	2070	2440	1980	2310
	24	2380	2860	2250	2670	2140	2520	2030	2380	1940	2250
	26	2340	2790	2200	2600	2090	2460	1990	2320	1890	2190
	28	2290	2710	2150	2530	2050	2390	1940	2250	1850	2130
	30	2230	2630	2100	2460	2000	2320	1890	2190	1800	2070
	32	2180	2550	2050	2380	1940	2250	1840	2120	1750	2000
	34	2120	2470	1990	2310	1890	2170	1790	2050	1700	1930
	36	2060	2390	1940	2230	1840	2100	1740	1970	1650	1860
	38	2000	2300	1880	2140	1780	2020	1680	1900	1600	1790
	40	1940	2210	1820	2060	1720	1940	1630	1820	1540	1720

Effective length in ft KL with respect to least radius of gyration r_{my}

Properties										
$\phi_b M_{nx}$ (kip-ft)	1260	1660	1150	1520	1060	1400	983	1290	908	1190
$\phi_b M_{ny}$ (kip-ft)	837	1070	771	987	719	921	656	835	611	777
$P_{ex}(K_x L_x)^2/10^4$ (kip-ft^2)	654	654	603	603	563	563	523	523	488	488
$P_{ey}(K_y L_y)^2/10^4$ (kip-ft^2)	654	654	603	603	563	563	523	523	488	488
r_{my} (in.)	7.20		7.20		7.20		7.20		7.20	
r_{mx}/r_{my} (in./in.)	1.00		1.00		1.00		1.00		1.00	

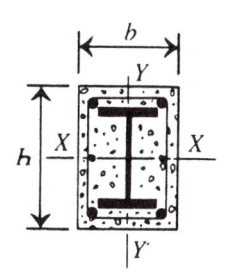

		F_y = 36 ksi
		F_y = 50 ksi

COMPOSITE COLUMNS
W Shapes
f'_c = 3.5 ksi
All reinforcing steel is Grade 60
Axial design strength in kips

Size $b \times h$		22 in. × 22 in.								
Reinf. bars		4-#10 bars								
Ties		#3 bars spaced 14 in. c. to c.								
Designation		**W 14**								
Wt./ft		109		99		90		82		74	
F_y		36	50	36	50	36	50	36	50	36	50
	0	1940	2320	1860	2210	1780	2100	1720	2000	1650	1910
	6	1920	2300	1840	2180	1770	2080	1700	1980	1630	1890
	7	1920	2290	1840	2170	1760	2070	1690	1970	1630	1880
	8	1910	2280	1830	2160	1750	2060	1680	1960	1620	1870
	9	1900	2270	1820	2150	1750	2050	1680	1950	1610	1860
	10	1890	2250	1810	2140	1740	2030	1670	1940	1600	1850
	11	1880	2240	1800	2120	1730	2020	1660	1920	1590	1830
	12	1870	2220	1790	2110	1710	2000	1650	1910	1580	1720
	13	1860	2200	1780	2090	1700	1990	1640	1890	1570	1800
	14	1850	2190	1760	2070	1690	1970	1620	1880	1560	1790
	15	1830	2170	1750	2050	1680	1950	1610	1860	1540	1770
	16	1820	2140	1740	2030	1660	1930	1600	1840	1530	1750
	17	1800	2120	1720	2010	1650	1910	1580	1820	1520	1730
	18	1780	2100	1700	1990	1630	1890	1560	1800	1500	1710
	19	1770	2070	1690	1970	1610	1870	1550	1780	1480	1690
	20	1750	2050	1670	1940	1600	1840	1530	1750	1470	1670
	22	1710	2000	1630	1890	1560	1790	1490	1710	1430	1620
	24	1670	1940	1590	1830	1520	1740	1460	1650	1390	1570
	26	1630	1880	1550	1780	1480	1680	1420	1600	1350	1520
	28	1580	1820	1510	1720	1440	1630	1370	1540	1310	1460
	30	1530	1750	1450	1650	1390	1570	1330	1490	1270	1410
	32	1480	1680	1410	1590	1340	1500	1280	1430	1220	1350
	34	1430	1620	1360	1520	1300	1440	1230	1360	1180	1290
	36	1380	1550	1310	1460	1250	1380	1190	1300	1130	1230
	38	1330	1480	1260	1390	1200	1310	1140	1240	1080	1170
	40	1280	1410	1210	1320	1150	1250	1090	1180	1030	1110
Properties										
$\phi_b M_{nx}$ (kip-ft)		801	1050	740	966	684	892	654	846	603	779
$\phi_b M_{ny}$ (kip-ft)		533	677	498	631	465	587	400	493	373	459
$P_{ex}(K_x L_x)^2/10^4$ (kip-ft²)		364	364	340	340	318	318	297	297	278	278
$P_{ey}(K_y L_y)^2/10^4$ (kip-ft²)		364	364	340	340	318	318	297	297	278	278
r_{my} (in.)		6.60		6.60		6.60		6.60		6.60	
r_{mx}/r_{my} (in./in.)		1.00		1.00		1.00		1.00		1.00	

Left margin: Steel Shape; Effective length in ft KL with respect to least radius of gyration r_{my}

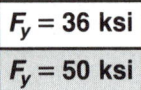

| $F_y = 36$ ksi |
| $F_y = 50$ ksi |

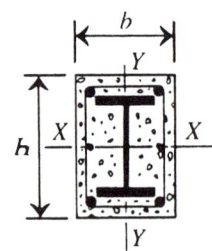

COMPOSITE COLUMNS
W Shapes
$f_c' = 3.5$ ksi
All reinforcing steel is Grade 60
Axial design strength in kips

Size $b \times h$		18 in. × 22 in.								
Reinf. bars		4-#9 bars								
Ties		#3 bars spaced 12 in. c. to c.								
	Designation	**W 14**								
Steel Shape	Wt./ft	68		61		53		48		43	
	F_y	36	50	36	50	36	50	36	50	36	50
Effective length in ft *KL* with respect to least radius of gyration r_{my}	0	1410	1640	1350	1560	1280	1470	1240	1400	1190	1340
	6	1390	1620	1330	1530	1260	1440	1220	1380	1170	1320
	7	1380	1610	1320	1520	1250	1430	1210	1370	1170	1310
	8	1370	1590	1310	1510	1240	1420	1200	1360	1160	1300
	9	1360	1580	1300	1500	1230	1410	1190	1350	1150	1290
	10	1350	1570	1290	1480	1220	1390	1180	1330	1140	1270
	11	1340	1550	1280	1470	1210	1380	1170	1320	1130	1260
	12	1320	1530	1260	1450	1200	1360	1160	1300	1110	1240
	13	1310	1510	1250	1430	1190	1340	1140	1280	1100	1230
	14	1300	1490	1240	1410	1170	1320	1130	1270	1090	1210
	15	1280	1470	1220	1390	1160	1300	1110	1250	1070	1190
	16	1260	1450	1200	1370	1140	1280	1100	1230	1060	1170
	17	1250	1430	1190	1350	1120	1260	1080	1210	1040	1150
	18	1230	1400	1170	1320	1110	1240	1060	1180	1020	1130
	19	1210	1380	1150	1300	1090	1220	1050	1160	1000	1110
	20	1190	1350	1130	1270	1070	1190	1030	1140	985	1080
	22	1150	1300	1090	1220	1030	1140	988	1090	946	1030
	24	1100	1240	1050	1170	987	1090	946	1040	905	984
	26	1060	1180	1000	1110	943	1030	903	983	863	932
	28	1010	1120	957	1050	898	977	859	928	819	880
	30	963	1060	910	991	852	920	814	873	775	826
	32	914	993	862	931	806	863	768	818	730	773
	34	864	930	814	871	759	806	722	763	685	719
	36	815	868	766	812	712	749	677	708	641	667
	38	765	807	718	753	666	694	632	655	597	616
	40	717	747	671	696	621	640	588	603	554	566
Properties											
$\phi_b M_{nx}$ (kip-ft)		535	693	490	633	448	575	416	533	383	488
$\phi_b M_{ny}$ (kip-ft)		279	348	259	323	229	281	216	264	201	246
$P_{ex}(K_x L_x)^2/10^4$ (kip-ft^2)		246	246	228	228	208	208	196	196	183	183
$P_{ey}(K_y L_y)^2/10^4$ (kip-ft^2)		164	164	153	153	140	140	131	131	123	123
r_{my} (in.)		5.40		5.40		5.40		5.40		5.40	
r_{mx}/r_{my} (in./in.)		1.22		1.22		1.22		1.22		1.22	

| F_y = 36 ksi | | | | | | | | | |
| F_y = 50 ksi | | | | | | | | | |

COMPOSITE COLUMNS
W Shapes
f_c' = 3.5 ksi
All reinforcing steel is Grade 60
Axial design strength in kips

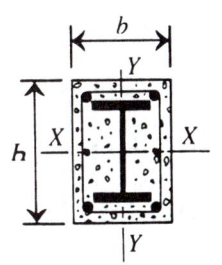

Size $b \times h$		22 in. × 24 in.								
Reinf. bars		4-#10 bars								
Ties		#3 bars spaced 14 in. c. to c.								
	Designation	**W 12**								
Steel Shape	Wt./ft	336		305		279		252		230	
	F_y	36	50	36	50	36	50	36	50	36	50
Effective length in ft KL with respect to least radius of gyration r_{my}	0	3950	5120	3680	4750	3460	4430	3230	4120	3050	3860
	6	3920	5070	3650	4700	3430	4390	3210	4080	3030	3820
	7	3910	5060	3640	4680	3420	4370	3200	4060	3020	3800
	8	3890	5030	3630	4670	3410	4360	3190	4040	3010	3790
	9	3880	5010	3620	4640	3400	4340	3180	4020	3000	3770
	10	3870	4990	3600	4620	3390	4310	3160	4000	2980	3750
	11	3850	4960	3590	4590	3370	4290	3150	3980	2970	3730
	12	3830	4930	3570	4570	3350	4260	3130	3950	2950	3700
	13	3810	4890	3550	4530	3340	4230	3120	3930	2940	3680
	14	3790	4860	3530	4500	3320	4200	3100	3900	2920	3650
	15	3770	4820	3510	4470	3300	4170	3080	3870	2900	3620
	16	3740	4780	3490	4430	3270	4130	3060	3830	2880	3590
	17	3720	4740	3460	4390	3250	4100	3040	3800	2860	3550
	18	3690	4690	3440	4350	3230	4060	3010	3760	2840	3520
	19	3660	4650	3410	4300	3200	4020	2990	3720	2810	3480
	20	3630	4600	3380	4260	3180	3970	2960	3680	2790	3450
	22	3570	4500	3320	4160	3120	3880	2910	3600	2740	3360
	24	3500	4390	3260	4060	3060	3780	2850	3510	2680	3280
	26	3430	4270	3190	3950	2990	3680	2790	3410	2620	3190
	28	3350	4150	3120	3840	2920	3570	2730	3310	2560	3090
	30	3270	4020	3040	3720	2850	3460	2660	3200	2500	2990
	32	3190	3890	2970	3590	2780	3350	2590	3100	2430	2890
	34	3110	3750	2880	3470	2700	3230	2510	2980	2360	2780
	36	3020	3610	2800	3340	2620	3110	2440	2870	2290	2680
	38	2930	3470	2710	3210	2540	2980	2360	2750	2210	2570
	40	2830	3330	2630	3070	2450	2860	2280	2640	2140	2460
Properties											
$\phi_b M_{nx}$ (kip-ft)		2110	2730	1920	2490	1770	2290	1610	2090	1480	1930
$\phi_b M_{ny}$ (kip-ft)		1180	1450	1090	1350	1020	1270	942	1180	879	1100
$P_{ex}(K_x L_x)^2/10^4$ (kip-ft^2)		1120	1120	1020	1020	946	946	868	868	803	803
$P_{ey}(K_y L_y)^2/10^4$ (kip-ft^2)		938	938	860	860	795	795	729	729	675	675
r_{my} (in.)		6.60		6.60		6.60		6.60		6.60	
r_{mx}/r_{my} (in./in.)		1.09		1.09		1.09		1.09		1.09	

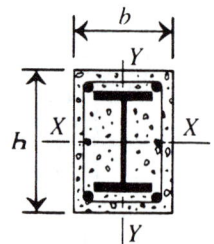

	$F_y = 36$ ksi
COMPOSITE COLUMNS W Shapes $f'_c = 3.5$ ksi All reinforcing steel is Grade 60 Axial design strength in kips	$F_y = 50$ ksi

Size $b \times h$		20 in. × 22 in.								
Reinf. bars		4-#10 bars								
Ties		#3 bars spaced 13 in. c. to c.								
Steel Shape Designation		**W 12**								
Wt./ft		210		190		170		152		136
F_y		36	50	36	50	36	50	36	50	36	50

Effective length in ft KL with respect to least radius of gyration r_{my}	0	2720	3460	2550	3210	2380	2980	2230	2760	2090	2570
	6	2700	3420	2530	3170	2360	2940	2210	2730	2070	2530
	7	2690	3400	2520	3160	2350	2930	2200	2710	2060	2520
	8	2680	3380	2510	3140	2340	2910	2190	2700	2050	2510
	9	2670	3360	2490	3130	2330	2900	2180	2680	2040	2490
	10	2650	3340	2480	3110	2320	2880	2170	2670	2030	2480
	11	2640	3320	2470	3080	2300	2850	2150	2650	2020	2460
	12	2620	3290	2450	3060	2290	2830	2140	2620	2010	2440
	13	2600	3270	2440	3030	2270	2810	2120	2600	1990	2410
	14	2580	3240	2420	3000	2260	2780	2110	2580	1970	2390
	15	2560	3200	2400	2970	2240	2750	2090	2550	1960	2370
	16	2540	3170	2380	2940	2220	2720	2070	2520	1940	2340
	17	2520	3140	2360	2910	2200	2690	2050	2490	1920	2310
	18	2500	3100	2330	2880	2180	2660	2030	2460	1900	2280
	19	2470	3060	2310	2840	2150	2630	2010	2430	1880	2250
	20	2450	3020	2290	2800	2130	2590	1990	2400	1860	2220
	22	2390	2930	2230	2720	2080	2510	1940	2330	1810	2150
	24	2330	2840	2180	2640	2030	2430	1890	2250	1760	2080
	26	2270	2750	2120	2550	1970	2350	1840	2170	1710	2010
	28	2210	2650	2060	2450	1910	2260	1780	2090	1660	1930
	30	2140	2550	1990	2360	1850	2170	1720	2010	1600	1850
	32	2070	2440	1930	2260	1790	2080	1660	1920	1550	1770
	34	2000	2340	1860	2160	1720	1990	1600	1830	1490	1690
	36	1930	2230	1790	2060	1660	1890	1540	1740	1430	1600
	38	1850	2120	1720	1960	1590	1800	1470	1650	1370	1520
	40	1780	2010	1650	1850	1520	1700	1410	1560	1300	1440

Properties										
$\phi_b M_{nx}$ (kip-ft)	1310	1690	1190	1550	1080	1410	983	1280	892	1160
$\phi_b M_{ny}$ (kip-ft)	759	946	704	883	646	813	594	748	547	689
$P_{ex}(K_x L_x)^2/10^4$ (kip-ft^2)	608	608	557	557	508	508	463	463	423	423
$P_{ey}(K_y L_y)^2/10^4$ (kip-ft^2)	502	502	460	460	420	420	383	383	349	349
r_{my} (in.)	6.00		6.00		6.00		6.00		6.00	
r_{mx}/r_{my} (in./in.)	1.10		1.10		1.10		1.10		1.10	

| F_y = 36 ksi |
| F_y = 50 ksi |

COMPOSITE COLUMNS
W Shapes
f'_c = 3.5 ksi
All reinforcing steel is Grade 60
Axial design strength in kips

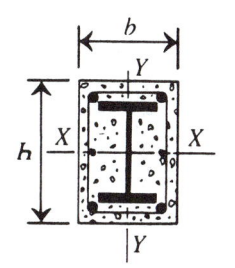

Size $b \times h$		20 in. × 20 in.								
Reinf. bars		4-#9 bars								
Ties		#3 bars spaced 13 in. c. to c.								
Steel Shape	Designation	**W 12**								
	Wt./ft	120		106		96		87		79	
	F_y	36	50	36	50	36	50	36	50	36	50
Effective length in ft KL with respect to least radius of gyration r_{my}	0	1850	2270	1740	2110	1650	1990	1580	1880	1510	1780
	6	1840	2250	1720	2080	1630	1960	1560	1850	1490	1760
	7	1830	2240	1710	2070	1630	1950	1550	1850	1480	1750
	8	1820	2220	1700	2060	1620	1940	1540	1830	1470	1740
	9	1810	2210	1690	2050	1610	1930	1540	1820	1470	1730
	10	1800	2190	1680	2030	1600	1910	1530	1810	1460	1710
	11	1790	2180	1670	2020	1590	1900	1520	1800	1450	1700
	12	1780	2160	1660	2000	1580	1880	1500	1780	1440	1690
	13	1760	2140	1650	1980	1570	1860	1490	1760	1420	1670
	14	1750	2120	1640	1960	1550	1850	1480	1740	1410	1650
	15	1740	2100	1620	1940	1540	1830	1470	1730	1400	1630
	16	1720	2070	1610	1920	1520	1800	1450	1710	1380	1610
	17	1700	2050	1590	1900	1510	1780	1440	1680	1370	1590
	18	1690	2020	1570	1870	1490	1760	1420	1660	1350	1570
	19	1670	2000	1560	1850	1470	1730	1400	1640	1340	1550
	20	1650	1970	1540	1820	1460	1710	1390	1610	1320	1530
	22	1610	1910	1500	1760	1420	1660	1350	1560	1280	1480
	24	1570	1850	1460	1700	1380	1600	1310	1510	1250	1420
	26	1520	1780	1410	1640	1340	1540	1270	1450	1210	1370
	28	1470	1710	1370	1580	1290	1480	1230	1390	1160	1310
	30	1420	1640	1320	1510	1250	1420	1180	1330	1120	1260
	32	1370	1570	1270	1450	1200	1350	1140	1270	1080	1200
	34	1320	1500	1220	1380	1150	1290	1090	1210	1030	1140
	36	1270	1430	1170	1310	1100	1220	1040	1150	984	1080
	38	1210	1350	1120	1240	1050	1160	993	1080	937	1020
	40	1160	1280	1070	1170	1000	1090	945	1020	891	957
Properties										
$\phi_b M_{nx}$ (kip-ft)		746	977	669	878	613	803	566	739	521	680
$\phi_b M_{ny}$ (kip-ft)		474	599	429	545	398	505	372	471	347	438
$P_{ex}(K_x L_x)^2/10^4$ (kip-ft^2)		311	311	282	282	261	261	243	243	226	226
$P_{ey}(K_y L_y)^2/10^4$ (kip-ft^2)		311	311	282	282	261	261	243	243	226	226
r_{my} (in.)		6.00		6.00		6.00		6.00		6.00	
r_{mx}/r_{my} (in./in.)		1.00		1.00		1.00		1.00		1.00	

	F_y = 36 ksi
	F_y = 50 ksi

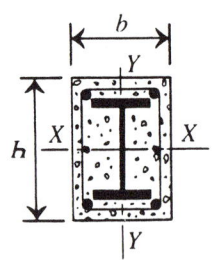

COMPOSITE COLUMNS
W Shapes
f'_c = 3.5 ksi
All reinforcing steel is Grade 60
Axial design strength in kips

Size $b \times h$		20 in. × 20 in.								
Reinf. bars		4-#9 bars								
Ties		#3 bars spaced 13 in. c. to c.								
Steel Shape	Designation	**W 12**								
	Wt./ft	72		65		58		53		50	
	F_y	36	50	36	50	36	50	36	50	36	50

Effective length in ft KL with respect to least radius of gyration r_{my}		36	50	36	50	36	50	36	50	36	50
	0	1450	1700	1390	1620	1330	1530	1290	1470	1260	1440
	6	1430	1670	1370	1590	1310	1510	1270	1450	1240	1410
	7	1420	1660	1360	1580	1300	1500	1260	1440	1240	1410
	8	1410	1650	1360	1570	1300	1490	1260	1430	1230	1400
	9	1410	1640	1350	1560	1290	1480	1250	1420	1220	1390
	10	1400	1630	1340	1550	1280	1470	1240	1410	1220	1380
	11	1390	1620	1330	1540	1270	1460	1230	1400	1210	1360
	12	1380	1600	1320	1520	1260	1440	1220	1390	1200	1350
	13	1370	1590	1310	1510	1250	1430	1210	1370	1180	1340
	14	1350	1570	1300	1490	1240	1410	1200	1360	1170	1320
	15	1340	1550	1280	1480	1220	1390	1190	1340	1160	1310
	16	1330	1530	1270	1460	1210	1380	1170	1320	1150	1290
	17	1310	1510	1260	1440	1200	1360	1160	1300	1130	1270
	18	1300	1490	1240	1420	1180	1340	1140	1290	1120	1250
	19	1280	1470	1220	1400	1170	1320	1130	1270	1100	1230
	20	1260	1450	1210	1370	1150	1300	1110	1250	1090	1210
	22	1230	1400	1170	1330	1120	1250	1080	1200	1050	1170
	24	1190	1350	1140	1280	1080	1210	1040	1160	1020	1120
	26	1150	1300	1100	1230	1040	1160	1000	1110	980	1080
	28	1110	1240	1060	1180	1000	1110	965	1060	941	1030
	30	1070	1190	1020	1120	962	1060	925	1010	902	980
	32	1020	1130	973	1070	920	1000	884	958	861	930
	34	979	1070	930	1010	878	949	842	907	819	879
	36	934	1020	886	958	835	896	800	855	778	828
	38	889	958	842	902	792	843	758	803	736	778
	40	843	901	797	847	749	790	716	752	694	728
Properties											
$\phi_b M_{nx}$ (kip-ft)		483	629	445	577	410	530	384	494	374	480
$\phi_b M_{ny}$ (kip-ft)		325	409	302	380	264	328	252	311	237	289
$P_{ex}(K_x L_x)^2/10^4$ (kip-ft²)		211	211	197	197	183	183	173	173	167	167
$P_{ey}(K_y L_y)^2/10^4$ (kip-ft²)		211	211	197	197	183	183	173	173	167	167
r_{my} (in.)		6.00		6.00		6.00		6.00		6.00	
r_{mx}/r_{my} (in./in.)		1.00		1.00		1.00		1.00		1.00	

AMERICAN INSTITUTE OF STEEL CONSTRUCTION

F_y = 36 ksi

F_y = 50 ksi

COMPOSITE COLUMNS
W Shapes
f_c' = 3.5 ksi
All reinforcing steel is Grade 60
Axial design strength in kips

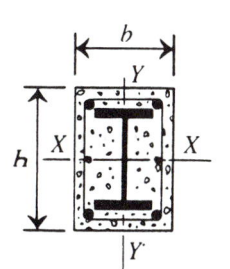

Size $b \times h$		18 in. × 18 in.								
Reinf. bars		4-#8 bars								
Ties		#3 bars spaced 12 in. c. to c.								
	Designation	**W 10**								
	Wt./ft	112		100		88		77		68	
	F_y	36	50	36	50	36	50	36	50	36	50
	0	1620	2020	1520	1870	1420	1730	1330	1600	1250	1490
	6	1600	1980	1500	1840	1400	1700	1310	1570	1230	1470
	7	1600	1970	1500	1830	1400	1690	1300	1560	1230	1460
	8	1590	1960	1490	1820	1390	1680	1300	1550	1220	1450
	9	1580	1950	1480	1810	1380	1670	1290	1540	1210	1430
	10	1570	1930	1470	1790	1370	1650	1280	1520	1200	1420
	11	1560	1910	1460	1780	1360	1640	1270	1510	1190	1410
	12	1540	1890	1450	1760	1350	1620	1260	1490	1180	1390
	13	1530	1870	1430	1740	1340	1600	1240	1480	1170	1380
	14	1520	1850	1420	1720	1320	1580	1230	1460	1160	1360
	15	1500	1830	1400	1700	1310	1560	1220	1440	1150	1340
	16	1480	1800	1390	1670	1290	1540	1200	1420	1130	1320
	17	1470	1780	1370	1650	1280	1520	1190	1400	1120	1300
	18	1450	1750	1360	1620	1260	1490	1170	1370	1100	1280
	19	1430	1720	1340	1600	1240	1470	1160	1350	1080	1260
	20	1410	1690	1320	1570	1230	1440	1140	1330	1070	1230
	22	1370	1630	1280	1510	1190	1390	1100	1280	1030	1190
	24	1330	1570	1240	1450	1150	1330	1060	1220	996	1130
	26	1280	1500	1190	1390	1110	1270	1020	1170	957	1080
	28	1230	1430	1150	1320	1060	1210	981	1110	917	1030
	30	1180	1360	1100	1260	1020	1150	939	1050	876	973
	32	1130	1290	1050	1190	972	1090	895	994	833	918
	34	1080	1220	1000	1120	926	1030	850	935	791	862
	36	1030	1150	955	1060	879	963	806	876	748	807
	38	979	1080	905	988	832	901	761	818	705	752
	40	927	1000	856	922	785	839	717	761	663	699
	Properties										
$\phi_b M_{nx}$ (kip-ft)		576	756	522	685	467	612	415	543	373	488
$\phi_b M_{ny}$ (kip-ft)		366	464	336	426	305	387	276	350	251	318
$P_{ex}(K_x L_x)^2/10^4$ (kip-ft²)		228	228	208	208	189	189	170	170	155	155
$P_{ey}(K_y L_y)^2/10^4$ (kip-ft²)		228	228	208	208	189	189	170	170	155	155
r_{my} (in.)		5.40		5.40		5.40		5.40		5.40	
r_{mx}/r_{my} (in./in.)		1.00		1.00		1.00		1.00		1.00	

Left side label: Steel Shape

Vertical label: Effective length in ft KL with respect to least radius of gyration r_{my}

AMERICAN INSTITUTE OF STEEL CONSTRUCTION

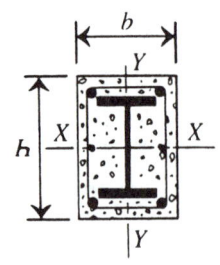

	F_y = 36 ksi
	F_y = 50 ksi

COMPOSITE COLUMNS
W Shapes
f'_c = 3.5 ksi
All reinforcing steel is Grade 60
Axial design strength in kips

Size $b \times h$		18 in. × 18 in.								
Reinf. bars		4-#8 bars								
Ties		#3 bars spaced 12 in. c. to c.								
	Designation	**W 10**								
	Wt./ft	60		54		49		45		39	
	F_y	36	50	36	50	36	50	36	50	36	50
Effective length in ft KL with respect to least radius of gyration r_{my}	0	1180	1390	1130	1320	1090	1260	1060	1220	1010	1140
	6	1170	1370	1110	1300	1070	1240	1040	1200	991	1120
	7	1160	1360	1110	1290	1070	1230	1040	1190	985	1120
	8	1150	1350	1100	1280	1060	1220	1030	1180	978	1110
	9	1140	1340	1090	1270	1050	1210	1020	1170	970	1100
	10	1140	1330	1080	1260	1040	1200	1010	1160	962	1090
	11	1130	1310	1070	1240	1040	1190	1000	1140	952	1070
	12	1120	1300	1060	1230	1020	1170	994	1130	942	1060
	13	1100	1280	1050	1210	1010	1160	983	1120	932	1050
	14	1090	1270	1040	1200	1000	1140	971	1100	920	1030
	15	1080	1250	1030	1180	989	1130	958	1090	908	1020
	16	1060	1230	1010	1160	976	1110	945	1070	895	1000
	17	1050	1210	1000	1140	962	1090	931	1050	881	983
	18	1040	1190	986	1120	947	1070	917	1030	867	965
	19	1020	1170	971	1100	932	1050	902	1010	852	946
	20	1000	1150	955	1080	917	1030	886	992	837	927
	22	969	1100	921	1040	884	989	854	950	805	887
	24	933	1050	886	992	849	944	820	907	771	845
	26	896	1000	849	944	813	898	784	861	736	801
	28	857	952	811	895	775	850	747	815	700	757
	30	817	900	772	845	737	802	709	768	664	712
	32	776	848	733	795	698	753	671	721	626	667
	34	735	795	693	745	659	705	633	674	589	622
	36	694	743	653	695	620	657	595	627	552	578
	38	653	692	613	646	582	610	557	581	516	535
	40	612	641	574	598	543	564	519	537	480	492
Properties										
$\phi_b M_{nx}$ (kip-ft)		337	439	307	399	286	370	272	351	245	314
$\phi_b M_{ny}$ (kip-ft)		230	290	212	267	199	249	179	222	165	203
$P_{ex}(K_x L_x)^2/10^4$ (kip-ft^2)		142	142	131	131	123	123	117	117	107	107
$P_{ey}(K_y L_y)^2/10^4$ (kip-ft^2)		142	142	131	131	123	123	117	117	107	107
r_{my} (in.)		5.40		5.40		5.40		5.40		5.40	
r_{mx}/r_{my} (in./in.)		1.00		1.00		1.00		1.00		1.00	

| F_y = 36 ksi |
| F_y = 50 ksi |

COMPOSITE COLUMNS
W Shapes
f'_c = 3.5 ksi
All reinforcing steel is Grade 60
Axial design strength in kips

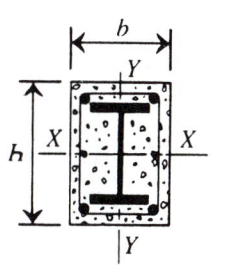

Size $b \times h$		16 in. × 16 in.									
Reinf. bars		4-#7 bars									
Ties		#3 bars spaced 10 in. c. to c.									
Steel Shape Designation		**W 8**									
Wt./ft		67		58		48		40		35	
F_y		36	50	36	50	36	50	36	50	36	50
Effective length in ft KL with respect to least radius of gyration r_{my}	0	1100	1330	1020	1230	938	1110	868	1010	828	951
	6	1080	1310	1010	1200	920	1080	851	985	811	929
	7	1070	1300	1000	1190	914	1070	845	977	805	921
	8	1070	1290	992	1180	907	1060	839	968	799	912
	9	1060	1270	984	1170	899	1050	831	957	791	902
	10	1050	1260	975	1160	890	1040	822	946	783	891
	11	1040	1240	965	1140	881	1030	813	933	773	879
	12	1030	1230	954	1130	870	1010	803	920	763	866
	13	1010	1210	942	1110	859	997	792	905	753	852
	14	1000	1190	930	1090	847	981	780	890	741	837
	15	988	1170	917	1080	834	963	768	874	729	822
	16	973	1150	903	1060	821	945	755	857	717	805
	17	958	1130	888	1040	807	927	742	839	703	788
	18	943	1110	873	1010	793	907	728	821	690	770
	19	926	1080	857	992	778	887	713	802	675	752
	20	909	1060	841	970	762	866	698	782	661	733
	22	874	1010	807	923	730	823	667	742	630	695
	24	836	958	772	874	696	778	634	700	598	654
	26	798	904	735	824	661	732	601	657	565	613
	28	758	850	697	773	625	685	566	614	532	572
	30	717	795	658	722	588	638	532	570	498	530
	32	677	740	619	671	552	592	497	527	464	489
	34	636	686	580	621	515	546	462	485	431	449
	36	595	633	542	572	479	501	428	444	398	410
	38	554	581	504	524	444	458	395	404	366	373
	40	515	532	466	478	410	416	363	366	335	337
Properties											
$\phi_b M_{nx}$ (kip-ft)		298	390	264	345	223	291	194	251	174	224
$\phi_b M_{ny}$ (kip-ft)		197	250	178	225	153	193	136	171	124	155
$P_{ex}(K_x L_x)^2/10^4$ (kip-ft²)		114	114	103	103	89.3	89.3	78.5	78.5	72.3	72.3
$P_{ey}(K_y L_y)^2/10^4$ (kip-ft²)		114	114	103	103	89.3	89.3	78.5	78.5	72.3	72.3
r_{my} (in.)		4.80		4.80		4.80		4.80		4.80	
r_{mx}/r_{my} (in./in.)		1.00		1.00		1.00		1.00		1.00	

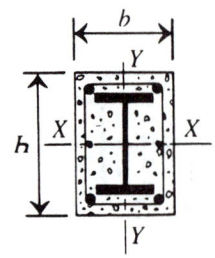

| $F_y = 36$ ksi |
| $F_y = 50$ ksi |

COMPOSITE COLUMNS
W Shapes
$f'_c = 5$ ksi
All reinforcing steel is Grade 60
Axial design strength in kips

Size $b \times h$		24 in. × 26 in.								
Reinf. bars		4-#11 bars								
Ties		#4 bars spaced 16 in. c. to c.								
	Designation	**W 14**								
	Wt./ft	426		398		370		342		311	
	F_y	36	50	36	50	36	50	36	50	36	50
	0	5290	6770	5060	6450	4840	6130	4610	5810	4340	5430
	6	5250	6720	5030	6400	4800	6080	4580	5760	4310	5380
	7	5240	6700	5020	6380	4790	6060	4570	5750	4300	5370
	8	5220	6670	5000	6360	4780	6040	4560	5730	4290	5350
	9	5210	6650	4990	6330	4760	6020	4540	5700	4270	5320
	10	5190	6620	4970	6300	4750	5990	4520	5680	4260	5300
	11	5170	6580	4950	6270	4730	5960	4510	5650	4240	5270
	12	5150	6550	4930	6240	4710	5930	4490	5620	4220	5240
	13	5120	6510	4910	6200	4690	5890	4470	5580	4200	5210
	14	5100	6470	4880	6160	4660	5850	4440	5550	4180	5180
	15	5070	6430	4850	6120	4640	5810	4420	5510	4160	5140
	16	5040	6380	4830	6070	4610	5770	4390	5470	4130	5100
	17	5010	6330	4800	6030	4580	5720	4360	5420	4100	5060
	18	4980	6280	4770	5980	4550	5680	4330	5380	4070	5010
	19	4950	6220	4730	5930	4520	5630	4300	5330	4050	4970
	20	4910	6170	4700	5870	4490	5570	4270	5280	4010	4920
	22	4840	6050	4630	5760	4410	5460	4200	5170	3950	4820
	24	4760	5920	4550	5630	4340	5340	4130	5060	3880	4710
	26	4670	5780	4460	5500	4260	5220	4050	4940	3800	4600
	28	4580	5640	4380	5360	4170	5090	3970	4810	3720	4480
	30	4480	5490	4280	5220	4080	4950	3880	4680	3640	4350
	32	4380	5330	4190	5060	3990	4800	3790	4540	3550	4220
	34	4280	5170	4080	4910	3890	4650	3690	4400	3460	4090
	36	4170	5000	3980	4750	3790	4500	3600	4250	3370	3950
	38	4060	4830	3870	4590	3680	4340	3500	4100	3270	3810
	40	3940	4660	3760	4420	3580	4180	3390	3950	3170	3660

Effective length in ft KL with respect to least radius of gyration r_{my}

Properties										
$\phi_b M_{nx}$ (kip-ft)	3090	4070	2890	3810	2690	3550	2490	3300	2280	3010
$\phi_b M_{ny}$ (kip-ft)	1850	2380	1750	2250	1640	2120	1540	1980	1420	1830
$P_{ex}(K_x L_x)^2/10^4$ (kip-ft^2)	1670	1670	1580	1580	1480	1480	1390	1390	1280	1280
$P_{ey}(K_y L_y)^2/10^4$ (kip-ft^2)	1420	1420	1340	1340	1260	1260	1180	1180	1090	1090
r_{my} (in.)	7.20		7.20		7.20		7.20		7.20	
r_{mx}/r_{my} (in./in.)	1.08		1.08		1.08		1.08		1.08	

AMERICAN INSTITUTE OF STEEL CONSTRUCTION

F_y = 36 ksi
F_y = 50 ksi

COMPOSITE COLUMNS
W Shapes
f'_c = 5 ksi
All reinforcing steel is Grade 60
Axial design strength in kips

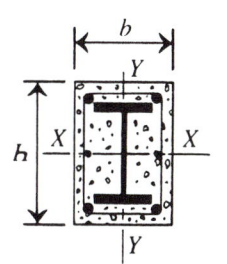

Size $b \times h$		24 in. × 24 in.									
Reinf. bars		4-#11 bars									
Ties		#4 bars spaced 16 in. c. to c.									
Steel Shape	Designation	**W 14**									
	Wt./ft	283		257		233		211		193	
	F_y	36	50	36	50	36	50	36	50	36	50
Effective length in ft KL with respect to least radius of gyration r_{my}	0	3990	4980	3780	4680	3580	4390	3400	4130	3250	3930
	6	3970	4940	3750	4640	3550	4350	3370	4100	3220	3890
	7	3960	4920	3740	4620	3540	4340	3360	4080	3220	3880
	8	3940	4910	3730	4600	3530	4320	3350	4070	3200	3860
	9	3930	4890	3720	4580	3520	4300	3340	4050	3190	3840
	10	3920	4860	3700	4560	3500	4280	3320	4030	3180	3820
	11	3900	4840	3690	4540	3490	4260	3310	4010	3160	3800
	12	3880	4810	3670	4510	3470	4230	3290	3980	3150	3780
	13	3860	4780	3650	4480	3450	4210	3280	3960	3130	3750
	14	3840	4750	3630	4450	3440	4180	3260	3930	3110	3730
	15	3820	4720	3610	4420	3410	4150	3240	3900	3090	3700
	16	3800	4680	3590	4390	3390	4110	3210	3870	3070	3670
	17	3770	4640	3560	4350	3370	4080	3190	3830	3050	3640
	18	3750	4600	3540	4310	3340	4040	3170	3800	3030	3600
	19	3720	4560	3510	4270	3320	4010	3140	3760	3000	3570
	20	3690	4520	3480	4230	3290	3970	3120	3720	2980	3530
	22	3630	4420	3420	4140	3230	3880	3060	3640	2920	3450
	24	3560	4320	3360	4050	3170	3790	3000	3560	2860	3370
	26	3490	4220	3290	3950	3110	3690	2940	3460	2800	3280
	28	3420	4110	3220	3840	3040	3590	2870	3370	2730	3190
	30	3340	3990	3150	3730	2970	3490	2800	3270	2670	3090
	32	3260	3870	3070	3610	2890	3380	2730	3160	2590	2990
	34	3180	3750	2990	3500	2810	3270	2650	3060	2520	2890
	36	3090	3620	2900	3380	2730	3150	2570	2950	2440	2780
	38	3000	3490	2820	3250	2650	3030	2490	2830	2360	2670
	40	2910	3360	2730	3130	2560	2920	2410	2720	2290	2570

Properties											
$\phi_b M_{nx}$ (kip-ft)		2020	2680	1850	2450	1690	2230	1540	2040	1420	1880
$\phi_b M_{ny}$ (kip-ft)		1300	1680	1200	1550	1110	1430	1020	1320	949	1220
$P_{ex}(K_x L_x)^2/10^4$ (kip-ft^2)		993	993	916	916	845	845	780	780	728	728
$P_{ey}(K_y L_y)^2/10^4$ (kip-ft^2)		993	993	916	916	845	845	780	780	728	728
r_{my} (in.)		7.20		7.20		7.20		7.20		7.20	
r_{mx}/r_{my} (in./in.)		1.00		1.00		1.00		1.00		1.00	

	F_y = 36 ksi
	F_y = 50 ksi

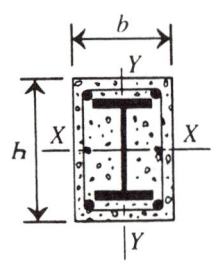

COMPOSITE COLUMNS
W Shapes
f'_c = 5 ksi
All reinforcing steel is Grade 60
Axial design strength in kips

Size $b \times h$		24 in. $\times$ 24 in.								
Reinf. bars		4-#10 bars								
Ties		#3 bars spaced 16 in. c. to c.								
Steel Shape	Designation	W 14								
	Wt./ft	176		159		145		132		120	
	F_y	36	50	36	50	36	50	36	50	36	50

Effective length in ft KL with respect to least radius of gyration r_{my}		176 (36)	176 (50)	159 (36)	159 (50)	145 (36)	145 (50)	132 (36)	132 (50)	120 (36)	120 (50)
	0	3080	3690	2930	3490	2820	3330	2710	3170	2610	3030
	6	3050	3660	2910	3450	2800	3290	2690	3140	2590	3000
	7	3040	3640	2900	3440	2790	3280	2680	3130	2580	2990
	8	3030	3630	2890	3430	2780	3270	2670	3110	2570	2980
	9	3020	3610	2880	3410	2770	3250	2660	3100	2560	2960
	10	3010	3590	2870	3390	2750	3240	2650	3080	2550	2950
	11	2990	3570	2850	3370	2740	3220	2630	3060	2530	2930
	12	2980	3550	2840	3350	2730	3200	2620	3040	2520	2910
	13	2960	3530	2820	3330	2710	3170	2600	3020	2500	2890
	14	2940	3500	2800	3310	2690	3150	2580	3000	2490	2860
	15	2920	3480	2780	3280	2670	3120	2570	2970	2470	2840
	16	2900	3450	2760	3250	2650	3100	2550	2950	2450	2810
	17	2880	3420	2740	3220	2630	3070	2530	2920	2430	2790
	18	2860	3380	2720	3190	2610	3040	2500	2890	2410	2760
	19	2840	3350	2700	3160	2590	3010	2480	2860	2390	2730
	20	2810	3320	2670	3130	2560	2980	2460	2830	2360	2700
	22	2760	3240	2620	3050	2510	2910	2410	2760	2310	2630
	24	2700	3160	2570	2980	2460	2830	2350	2690	2260	2560
	26	2640	3080	2510	2900	2400	2750	2300	2620	2200	2490
	28	2580	2990	2450	2810	2340	2670	2240	2540	2150	2410
	30	2510	2900	2380	2720	2280	2590	2180	2450	2080	2330
	32	2450	2800	2320	2630	2210	2500	2110	2370	2020	2250
	34	2370	2710	2250	2540	2140	2410	2040	2280	1950	2160
	36	2300	2610	2170	2440	2070	2320	1980	2190	1890	2080
	38	2230	2510	2100	2350	2000	2220	1910	2100	1820	1990
	40	2150	2400	2030	2250	1930	2130	1830	2010	1750	1900

Properties										
$\phi_b M_{nx}$ (kip-ft)	1280	1700	1170	1550	1080	1430	996	1320	920	1210
$\phi_b M_{ny}$ (kip-ft)	860	1110	789	1020	734	950	670	861	623	799
$P_{ex}(K_x L_x)^2/10^4$ (kip-ft^2)	678	678	627	627	587	587	547	547	512	512
$P_{ey}(K_y L_y)^2/10^4$ (kip-ft^2)	678	678	627	627	587	587	547	547	512	512
r_{my} (in.)	7.20		7.20		7.20		7.20		7.20	
r_{mx}/r_{my} (in./in.)	1.00		1.00		1.00		1.00		1.00	

| F_y = 36 ksi |
| F_y = 50 ksi |

COMPOSITE COLUMNS
W Shapes
f'_c = 5 ksi
All reinforcing steel is Grade 60
Axial design strength in kips

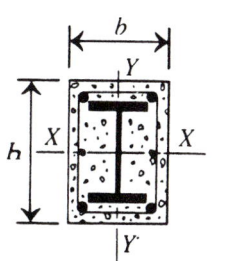

Size $b \times h$		\multicolumn{10}{c}{22 in. × 22 in.}
Reinf. bars		\multicolumn{10}{c}{4-#10 bars}
Ties		\multicolumn{10}{c}{#3 bars spaced 14 in. c. to c.}
Steel Shape	Designation	\multicolumn{10}{c}{**W 14**}
	Wt./ft	\multicolumn{2}{c}{109}	\multicolumn{2}{c}{99}	\multicolumn{2}{c}{90}	\multicolumn{2}{c}{82}	\multicolumn{2}{c}{74}
	F_y	36	50	36	50	36	50	36	50	36	50

Effective length in ft KL with respect to least radius of gyration r_{my}

KL	109 (36)	109 (50)	99 (36)	99 (50)	90 (36)	90 (50)	82 (36)	82 (50)	74 (36)	74 (50)
0	2290	2670	2200	2550	2130	2450	2060	2350	2000	2260
6	2260	2630	2180	2520	2110	2410	2040	2320	1980	2230
7	2250	2620	2170	2510	2100	2400	2030	2310	1970	2220
8	2240	2610	2160	2490	2090	2390	2020	2300	1960	2200
9	2230	2590	2150	2480	2080	2380	2010	2280	1950	2190
10	2220	2580	2140	2460	2070	2360	2000	2270	1930	2170
11	2200	2560	2120	2440	2050	2340	1980	2250	1920	2160
12	2190	2540	2110	2420	2040	2320	1970	2230	1910	2140
13	2170	2520	2090	2400	2020	2300	1950	2210	1890	2120
14	2160	2490	2080	2380	2000	2280	1940	2190	1870	2100
15	2140	2470	2060	2360	1990	2260	1920	2160	1850	2070
16	2120	2440	2040	2330	1970	2230	1900	2140	1840	2050
17	2100	2410	2020	2300	1950	2200	1880	2110	1820	2030
18	2080	2380	2000	2270	1920	2180	1860	2090	1790	2000
19	2050	2350	1970	2250	1900	2150	1840	2060	1770	1970
20	2030	2320	1950	2210	1880	2120	1810	2030	1750	1940
22	1980	2260	1900	2150	1830	2050	1760	1970	1700	1880
24	1930	2190	1850	2080	1780	1990	1710	1900	1650	1820
26	1870	2110	1790	2010	1720	1920	1660	1830	1600	1750
28	1810	2030	1740	1930	1670	1840	1600	1760	1540	1680
30	1750	1950	1680	1860	1610	1770	1540	1690	1480	1610
32	1690	1870	1610	1780	1550	1690	1480	1610	1420	1540
34	1620	1790	1550	1700	1480	1610	1420	1540	1360	1460
36	1560	1710	1490	1620	1420	1530	1360	1460	1300	1390
38	1490	1620	1420	1530	1350	1450	1290	1380	1230	1310
40	1420	1540	1350	1450	1290	1380	1230	1300	1170	1240

Properties										
$\phi_b M_{nx}$ (kip-ft)	811	1070	748	982	691	905	663	864	611	793
$\phi_b M_{ny}$ (kip-ft)	543	695	507	647	472	600	409	510	380	473
$P_{ex}(K_x L_x)^2/10^4$ (kip-ft^2)	381	381	357	357	335	335	315	315	295	295
$P_{ey}(K_y L_y)^2/10^4$ (kip-ft^2)	381	381	357	357	335	335	315	315	295	295
r_{my} (in.)	\multicolumn{2}{c}{6.60}	\multicolumn{2}{c}{6.60}	\multicolumn{2}{c}{6.60}	\multicolumn{2}{c}{6.60}	\multicolumn{2}{c}{6.60}
r_{mx}/r_{my} (in./in.)	\multicolumn{2}{c}{1.00}	\multicolumn{2}{c}{1.00}	\multicolumn{2}{c}{1.00}	\multicolumn{2}{c}{1.00}	\multicolumn{2}{c}{1.00}

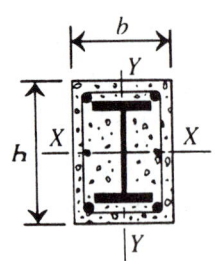

	$F_y = 36$ ksi
	$F_y = 50$ ksi

COMPOSITE COLUMNS
W Shapes
$f'_c = 5$ ksi
All reinforcing steel is Grade 60
Axial design strength in kips

Size $b \times h$		18 in. × 22 in.								
Reinf. bars		4-#9 bars								
Ties		#3 bars spaced 12 in. c. to c.								

	Designation	W 14								
	Wt./ft	68		61		53		48		43	
	F_y	36	50	36	50	36	50	36	50	36	50
Effective length in ft KL with respect to least radius of gyration r_{my}	0	1690	1930	1630	1850	1570	1750	1530	1690	1480	1630
	6	1660	1890	1600	1810	1540	1720	1500	1660	1450	1600
	7	1650	1880	1590	1800	1530	1700	1490	1650	1440	1590
	8	1640	1860	1580	1780	1520	1690	1470	1630	1430	1570
	9	1630	1850	1570	1760	1500	1670	1460	1610	1420	1560
	10	1610	1830	1550	1740	1490	1650	1450	1600	1400	1540
	11	1600	1810	1540	1720	1470	1630	1430	1580	1390	1520
	12	1580	1780	1520	1700	1460	1610	1410	1560	1370	1500
	13	1560	1760	1500	1680	1440	1590	1390	1530	1350	1470
	14	1540	1730	1480	1650	1420	1570	1370	1510	1330	1450
	15	1520	1710	1460	1630	1400	1540	1350	1480	1310	1420
	16	1500	1680	1440	1600	1370	1510	1330	1460	1290	1400
	17	1470	1650	1410	1570	1350	1480	1310	1430	1260	1370
	18	1450	1620	1390	1540	1330	1450	1280	1400	1240	1340
	19	1420	1580	1360	1510	1300	1420	1260	1370	1220	1310
	20	1400	1550	1340	1470	1270	1390	1230	1340	1190	1280
	22	1340	1480	1280	1410	1220	1320	1180	1270	1140	1220
	24	1280	1410	1230	1340	1160	1260	1120	1200	1080	1150
	26	1220	1330	1170	1260	1100	1190	1060	1130	1020	1080
	28	1160	1260	1110	1190	1040	1110	1000	1060	962	1010
	30	1100	1180	1040	1110	984	1040	943	993	902	944
	32	1040	1100	982	1040	922	969	883	922	842	876
	34	972	1030	920	965	861	897	823	853	783	808
	36	909	950	858	892	801	827	763	785	725	742
	38	847	876	797	821	742	759	705	719	668	678
	40	786	805	738	752	684	694	649	656	612	616

Properties										
$\phi_b M_{nx}$ (kip-ft)		543	708	496	644	454	586	421	542	387	496
$\phi_b M_{ny}$ (kip-ft)		285	360	264	332	234	290	220	272	205	252
$P_{ex}(K_x L_x)^2/10^4$ (kip-ft²)		260	260	242	242	223	223	210	210	198	198
$P_{ey}(K_y L_y)^2/10^4$ (kip-ft²)		174	174	162	162	149	149	141	141	132	132
r_{my} (in.)		5.40		5.40		5.40		5.40		5.40	
r_{mx}/r_{my} (in./in.)		1.22		1.22		1.22		1.22		1.22	

F_y = 36 ksi
F_y = 50 ksi

COMPOSITE COLUMNS
W Shapes
f'_c = 5 ksi
All reinforcing steel is Grade 60
Axial design strength in kips

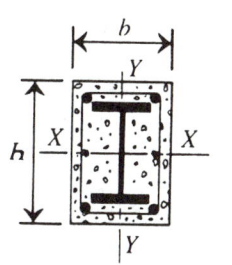

Size $b \times h$		22 in. × 24 in.									
Reinf. bars		4-#10 bars									
Ties		#3 bars spaced 14 in. c. to c.									
	Designation	**W 12**									
Steel Shape	Wt./ft	336		305		279		252		230	
	F_y	36	50	36	50	36	50	36	50	36	50
Effective length in ft KL with respect to least radius of gyration r_{my}	0	4270	5450	4010	5080	3800	4770	3580	4460	3400	4200
	6	4240	5390	3980	5030	3770	4720	3550	4410	3370	4160
	7	4230	5370	3970	5010	3750	4700	3540	4400	3360	4140
	8	4210	5350	3960	4990	3740	4680	3520	4380	3350	4120
	9	4200	5320	3940	4960	3730	4660	3510	4350	3330	4100
	10	4180	5300	3920	4940	3710	4640	3490	4330	3320	4080
	11	4160	5260	3900	4910	3690	4610	3480	4300	3300	4050
	12	4140	5230	3880	4870	3670	4580	3460	4270	3280	4030
	13	4110	5190	3860	4840	3650	4540	3440	4240	3260	4000
	14	4090	5150	3840	4800	3630	4510	3420	4210	3240	3960
	15	4060	5110	3810	4760	3610	4470	3390	4170	3220	3930
	16	4040	5070	3790	4720	3580	4430	3370	4140	3190	3890
	17	4010	5020	3760	4680	3550	4390	3340	4100	3170	3860
	18	2980	4970	3730	4630	3520	4340	3310	4050	3140	3820
	19	3940	4920	3700	4580	3490	4300	3290	4010	3110	3770
	20	3910	4870	3670	4530	3460	4250	3260	3960	3090	3730
	22	3840	4750	3600	4420	3400	4150	3190	3870	3020	3640
	24	3760	4630	3520	4310	3320	4040	3120	3760	2960	3540
	26	3680	4500	3450	4190	3250	3920	3050	3650	2890	3430
	28	3590	4370	3360	4060	3170	3800	2970	3540	2810	3320
	30	3500	4230	3280	3930	3090	3680	2890	3420	2730	3210
	32	3410	4080	3190	3790	3000	3550	2810	3300	2650	3090
	34	3310	3930	3090	3650	2910	3410	2720	3170	2570	2970
	36	3210	3780	3000	3510	2820	3280	2630	3040	2480	2850
	38	3110	3630	2900	3360	2720	3140	2540	2910	2400	2730
	40	3000	3470	2800	3220	2630	3000	2450	2780	2310	2600

Properties

$\phi_b M_{nx}$ (kip-ft)	2190	2890	1990	2630	1830	2410	1660	2190	1530	2010
$\phi_b M_{ny}$ (kip-ft)	1260	1600	1150	1480	1080	1380	990	1270	919	1180
$P_{ex}(K_xL_x)^2/10^4$ (kip-ft^2)	1140	1140	1040	1040	966	966	888	888	824	824
$P_{ey}(K_yL_y)^2/10^4$ (kip-ft^2)	954	954	877	877	812	812	746	746	692	692
r_{my} (in.)	6.60		6.60		6.60		6.60		6.60	
r_{mx}/r_{my} (in./in.)	1.09		1.09		1.09		1.09		1.09	

AMERICAN INSTITUTE OF STEEL CONSTRUCTION

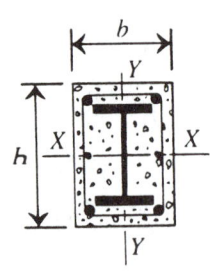

	COMPOSITE COLUMNS **W Shapes** $f'_c = 5$ ksi All reinforcing steel is Grade 60 Axial design strength in kips		$F_y = 36$ ksi
			$F_y = 50$ ksi

Size $b \times h$	20 in. × 22 in.									
Reinf. bars	4-#10 bars									
Ties	#3 bars spaced 13 in. c. to c.									
Designation	**W 12**									
Wt./ft	210		190		170		152		136	
F_y	36	50	36	50	36	50	36	50	36	50

Steel Shape

Effective length in ft KL with respect to least radius of gyration r_{my}

KL	36	50	36	50	36	50	36	50	36	50
0	3010	3740	2840	3500	2680	3270	2530	3060	2390	2870
6	2980	3700	2810	3460	2650	3230	2500	3020	2370	2830
7	2970	3680	2800	3440	2640	3210	2490	3000	2360	2810
8	2950	3660	2790	3420	2630	3200	2480	2990	2340	2800
9	2940	3640	2770	3400	2610	3180	2470	2970	2330	2780
10	2920	3610	2760	3380	2600	3150	2450	2950	2320	2760
11	2910	3590	2740	3350	2580	3130	2430	2920	2300	2740
12	2890	3560	2720	3320	2560	3100	2420	2900	2280	2710
13	2870	3520	2700	3290	2540	3070	2400	2870	2270	2690
14	2840	3490	2680	3260	2520	3040	2380	2840	2250	2660
15	2820	3450	2660	3230	2500	3010	2360	2810	2220	2630
16	2790	3420	2630	3190	2480	2970	2330	2780	2200	2600
17	2770	3380	2610	3150	2450	2940	2310	2740	2180	2560
18	2740	3330	2580	3110	2430	2900	2280	2700	2150	2530
19	2710	3290	2550	3070	2400	2860	2260	2670	2130	2490
20	2680	3240	2520	3030	2370	2820	2230	2630	2100	2450
22	2620	3150	2460	2940	2310	2730	2170	2540	2040	2370
24	2550	3040	2390	2840	2250	2640	2110	2460	1980	2290
26	2480	2940	2320	2740	2180	2540	2040	2360	1920	2200
28	2400	2830	2250	2630	2110	2440	1970	2270	1850	2110
30	2320	2710	2180	2520	2030	2340	1900	2170	1780	2020
32	2240	2590	2100	2410	1960	2230	1830	2070	1710	1920
34	2160	2470	2020	2300	1880	2130	1760	1970	1640	1830
36	2070	2350	1930	2180	1800	2020	1680	1870	1570	1730
38	1980	2230	1850	2070	1720	1910	1600	1760	1490	1630
40	1900	2110	1770	1950	1640	1800	1530	1660	1420	1530

Properties										
$\phi_b M_{nx}$ (kip-ft)	1350	1770	1230	1610	1110	1460	1010	1320	910	1190
$\phi_b M_{ny}$ (kip-ft)	796	1020	734	941	671	860	614	787	563	721
$P_{ex}(K_x L_x)^2/10^4$ (kip-ft^2)	622	622	572	572	523	523	478	478	438	438
$P_{ey}(K_y L_y)^2/10^4$ (kip-ft^2)	514	514	472	472	432	432	395	395	362	362
r_{my} (in.)	6.00		6.00		6.00		6.00		6.00	
r_{mx}/r_{my} (in./in.)	1.10		1.10		1.10		1.10		1.10	

| F_y = 36 ksi |
| F_y = 50 ksi |

COMPOSITE COLUMNS
W Shapes
f'_c = 5 ksi
All reinforcing steel is Grade 60
Axial design strength in kips

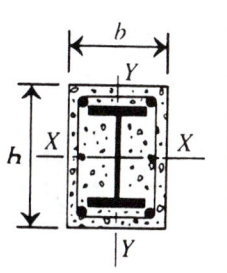

Size $b \times h$		20 in. × 20 in.									
Reinf. bars		4-#9 bars									
Ties		#3 bars spaced 13 in. c. to c.									
Steel Shape — Designation		**W 12**									
	Wt./ft	120		106		96		87		79	
	F_y	36	50	36	50	36	50	36	50	36	50
Effective length in ft KL with respect to least radius of gyration r_{my}	0	2130	2550	2020	2390	1930	2270	1860	2160	1790	2070
	6	2110	2520	1990	2350	1910	2230	1830	2130	1770	2040
	7	2100	2500	1980	2340	1900	2220	1830	2120	1760	2020
	8	2090	2490	1970	2330	1890	2210	1820	2110	1750	2010
	9	2080	2470	1960	2310	1880	2190	1810	2090	1740	2000
	10	2060	2450	1950	2290	1870	2180	1790	2070	1730	1980
	11	2050	2430	1930	2270	1850	2160	1780	2060	1710	1960
	12	2030	2410	1920	2250	1840	2140	1760	2040	1700	1940
	13	2020	2390	1900	2230	1820	2120	1750	2020	1680	1920
	14	2000	2360	1890	2210	1800	2090	1730	1990	1670	1900
	15	1980	2340	1870	2180	1790	2070	1710	1970	1650	1880
	16	1960	2310	1850	2150	1770	2040	1700	1940	1630	1850
	17	1940	2280	1830	2130	1750	2010	1680	1920	1610	1830
	18	1920	2250	1810	2100	1730	1990	1650	1890	1590	1800
	19	1890	2220	1780	2070	1700	1960	1630	1860	1570	1770
	20	1870	2180	1760	2030	1680	1920	1610	1830	1540	1740
	22	1820	2110	1710	1970	1630	1860	1560	1770	1500	1680
	24	1770	2040	1660	1890	1580	1790	1510	1700	1450	1620
	26	1710	1960	1600	1820	1530	1720	1460	1630	1390	1550
	28	1650	1880	1550	1740	1470	1640	1400	1560	1340	1480
	30	1590	1800	1490	1660	1410	1570	1340	1480	1280	1410
	32	1530	1710	1430	1580	1350	1490	1290	1410	1220	1330
	34	1460	1620	1360	1500	1290	1410	1230	1330	1170	1260
	36	1400	1540	1300	1420	1230	1330	1170	1260	1110	1190
	38	1330	1450	1240	1340	1170	1250	1110	1180	1050	1110
	40	1260	1370	1170	1260	1100	1180	1040	1110	989	1040

Properties											
$\phi_b M_{nx}$ (kip-ft)		760	1000	680	899	622	820	573	754	528	693
$\phi_b M_{ny}$ (kip-ft)		489	628	440	566	407	522	380	486	353	450
$P_{ex}(K_x L_x)^2/10^4$ (kip-ft^2)		322	322	294	294	273	273	255	255	238	238
$P_{ey}(K_y L_y)^2/10^4$ (kip-ft^2)		322	322	294	294	273	273	255	255	238	238
r_{my} (in.)		6.00		6.00		6.00		6.00		6.00	
r_{mx}/r_{my} (in./in.)		1.00		1.00		1.00		1.00		1.00	

AMERICAN INSTITUTE OF STEEL CONSTRUCTION

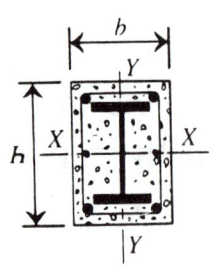

	F_y = 36 ksi
	F_y = 50 ksi

COMPOSITE COLUMNS
W Shapes
f'_c = 5 ksi
All reinforcing steel is Grade 60
Axial design strength in kips

Size $b \times h$		20 in. × 20 in.									
Reinf. bars		4-#9 bars									
Ties		#3 bars spaced 13 in. c. to c.									
	Designation	**W 12**									
	Wt./ft	72		65		58		53		50	
	F_y	36	50	36	50	36	50	36	50	36	50
Effective length in ft KL with respect to least radius of gyration r_{my}	0	1730	1980	1680	1900	1620	1820	1580	1760	1550	1730
	6	1710	1950	1650	1870	1590	1790	1550	1730	1530	1700
	7	1700	1940	1640	1860	1590	1780	1550	1720	1520	1690
	8	1690	1930	1630	1850	1580	1770	1540	1710	1510	1680
	9	1680	1910	1620	1840	1560	1750	1530	1700	1500	1660
	10	1670	1900	1610	1820	1550	1740	1510	1680	1490	1650
	11	1650	1880	1600	1800	1540	1720	1500	1670	1470	1630
	12	1640	1860	1580	1780	1520	1700	1490	1650	1460	1610
	13	1620	1840	1570	1760	1510	1680	1470	1630	1440	1590
	14	1610	1820	1550	1740	1490	1660	1450	1610	1430	1570
	15	1590	1800	1530	1720	1480	1640	1440	1590	1410	1550
	16	1570	1770	1520	1700	1460	1620	1420	1560	1390	1530
	17	1550	1750	1500	1670	1440	1590	1400	1540	1370	1510
	18	1530	1720	1480	1650	1420	1570	1380	1510	1350	1480
	19	1510	1690	1450	1620	1400	1540	1360	1490	1330	1450
	20	1490	1660	1430	1590	1370	1510	1330	1460	1310	1430
	22	1440	1600	1380	1530	1330	1460	1290	1400	1260	1370
	24	1390	1540	1340	1470	1280	1400	1240	1350	1210	1310
	26	1340	1480	1280	1410	1230	1330	1190	1280	1160	1250
	28	1280	1410	1230	1340	1170	1270	1130	1220	1110	1190
	30	1230	1340	1180	1270	1120	1200	1080	1160	1060	1130
	32	1170	1270	1120	1200	1060	1140	1030	1090	1000	1060
	34	1110	1200	1060	1130	1010	1070	970	1020	946	996
	36	1060	1120	1000	1060	951	1000	915	959	891	931
	38	997	1050	948	996	895	935	860	895	836	868
	40	939	984	891	929	840	871	805	831	782	806
Properties											
$\phi_b M_{nx}$ (kip-ft)		489	640	449	586	414	537	387	501	378	488
$\phi_b M_{ny}$ (kip-ft)		330	419	307	388	268	335	255	318	241	297
$P_{ex}(K_x L_x)^2/10^4$ (kip-ft^2)		223	223	209	209	195	195	185	185	179	179
$P_{ey}(K_y L_y)^2/10^4$ (kip-ft^2)		223	223	209	209	195	195	185	185	179	179
r_{my} (in.)		6.00		6.00		6.00		6.00		6.00	
r_{mx}/r_{my} (in./in.)		1.00		1.00		1.00		1.00		1.00	

| **Fy = 36 ksi** |
| **Fy = 50 ksi** |

COMPOSITE COLUMNS
W Shapes
$f'_c = 5$ ksi
All reinforcing steel is Grade 60
Axial design strength in kips

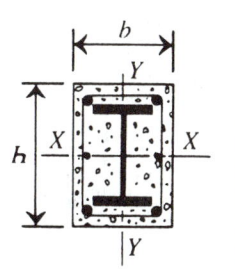

Size $b \times h$		18 in. × 18 in.									
Reinf. bars		4-#8 bars									
Ties		#3 bars spaced 12 in. c. to c.									
	Designation	**W 10**									
	Wt./ft	112		100		88		77		68	
	F_y	36	50	36	50	36	50	36	50	36	50
	0	1840	2240	1750	2100	1650	1960	1560	1820	1480	1720
	6	1820	2200	1720	2060	1620	1920	1530	1790	1460	1690
	7	1810	2190	1710	2050	1610	1910	1520	1780	1450	1680
	8	1800	2170	1700	2030	1600	1900	1510	1770	1440	1660
	9	1790	2150	1690	2020	1590	1880	1500	1750	1430	1650
	10	1770	2130	1680	2000	1580	1860	1490	1730	1420	1630
	11	1760	2110	1660	1980	1570	1840	1480	1720	1400	1620
	12	1740	2090	1650	1960	1550	1820	1460	1700	1390	1600
	13	1730	2070	1630	1930	1540	1800	1450	1670	1370	1580
	14	1710	2040	1610	1910	1520	1780	1430	1650	1360	1550
	15	1690	2010	1600	1880	1500	1750	1410	1630	1340	1530
	16	1670	1980	1580	1850	1480	1720	1390	1600	1320	1510
	17	1650	1950	1560	1830	1460	1700	1370	1580	1300	1480
	18	1630	1920	1530	1790	1440	1670	1350	1550	1280	1450
	19	1600	1890	1510	1760	1420	1640	1330	1520	1260	1430
	20	1580	1850	1490	1730	1400	1610	1310	1490	1240	1400
	22	1530	1780	1440	1660	1350	1540	1260	1430	1190	1340
	24	1480	1710	1390	1590	1300	1470	1210	1360	1140	1270
	26	1420	1630	1330	1520	1250	1400	1160	1300	1090	1210
	28	1360	1550	1280	1440	1190	1330	1110	1230	1040	1140
	30	1300	1470	1220	1360	1140	1260	1050	1160	990	1080
	32	1240	1380	1160	1280	1080	1180	1000	1090	936	1010
	34	1180	1300	1100	1210	1020	1110	944	1020	883	942
	36	1120	1220	1040	1130	963	1030	889	946	829	876
	38	1060	1140	982	1050	906	962	834	878	776	811
	40	995	1060	923	975	850	891	779	811	723	748

Effective length in ft KL with respect to least radius of gyration r_{my}

Properties											
$\phi_b M_{nx}$ (kip-ft)		589	780	532	704	475	627	421	555	378	497
$\phi_b M_{ny}$ (kip-ft)		379	488	346	445	313	402	282	361	256	327
$P_{ex}(K_x L_x)^2/10^4$ (kip-ft²)		236	236	216	216	196	196	178	178	163	163
$P_{ey}(K_y L_y)^2/10^4$ (kip-ft²)		236	236	216	216	196	196	178	178	163	163
r_{my} (in.)		5.40		5.40		5.40		5.40		5.40	
r_{mx}/r_{my} (in./in.)		1.00		1.00		1.00		1.00		1.00	

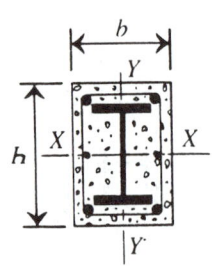

	F_y = 36 ksi
	F_y = 50 ksi

COMPOSITE COLUMNS
W Shapes
f'_c = 5 ksi
All reinforcing steel is Grade 60
Axial design strength in kips

Size $b \times h$		18 in. $\times$ 18 in.								
Reinf. bars		4-#8 bars								
Ties		#3 bars spaced 12 in. c. to c.								
	Designation	**W 10**								
	Wt./ft	60		54		49		45		39	
	F_y	36	50	36	50	36	50	36	50	36	50
	0	1420	1620	1360	1550	1330	1500	1290	1450	1240	1380
	6	1390	1590	1340	1520	1300	1470	1270	1420	1220	1350
	7	1380	1580	1330	1510	1290	1460	1260	1410	1210	1340
	8	1370	1570	1320	1500	1280	1440	1250	1400	1200	1330
	9	1360	1560	1310	1490	1270	1430	1240	1390	1190	1320
	10	1350	1540	1300	1470	1260	1420	1230	1370	1180	1300
	11	1340	1520	1290	1450	1250	1400	1220	1360	1170	1290
	12	1320	1500	1270	1430	1230	1380	1200	1340	1150	1270
	13	1310	1480	1260	1410	1220	1360	1190	1320	1140	1250
	14	1290	1460	1240	1390	1200	1340	1170	1300	1120	1230
	15	1270	1440	1220	1370	1190	1320	1150	1280	1100	1210
	16	1260	1420	1210	1350	1170	1300	1140	1250	1090	1190
	17	1240	1390	1190	1320	1150	1270	1120	1230	1070	1160
	18	1220	1370	1170	1300	1130	1250	1100	1210	1050	1140
	19	1200	1340	1150	1270	1110	1220	1080	1180	1030	1120
	20	1170	1310	1120	1250	1090	1200	1060	1160	1000	1090
	22	1130	1250	1080	1190	1040	1140	1010	1100	961	1040
	24	1080	1190	1030	1130	995	1080	965	1040	915	982
	26	1030	1130	984	1070	947	1020	917	987	867	925
	28	981	1070	934	1010	897	963	868	927	819	868
	30	929	1000	883	947	847	903	818	868	770	810
	32	877	938	832	884	796	842	768	808	720	753
	34	825	874	781	822	746	782	718	749	671	696
	36	772	811	729	761	695	723	668	692	623	641
	38	721	749	679	702	646	665	620	636	576	587
	40	670	689	630	644	598	609	572	581	530	535

Effective length in ft KL with respect to least radius of gyration r_{my}

Properties										
$\phi_b M_{nx}$ (kip-ft)		341	446	310	405	288	375	275	356	247	318
$\phi_b M_{ny}$ (kip-ft)		234	297	215	272	201	254	182	227	167	207
$P_{ex}(K_x L_x)^2/10^4$ (kip-ft²)		149	149	139	139	131	131	125	125	115	115
$P_{ey}(K_y L_y)^2/10^4$ (kip-ft²)		149	149	139	139	131	131	125	125	115	115
r_{my} (in.)		5.40		5.40		5.40		5.40		5.40	
r_{mx}/r_{my} (in./in.)		1.00		1.00		1.00		1.00		1.00	

| F_y = 36 ksi |
| F_y = 50 ksi |

COMPOSITE COLUMNS
W Shapes
f'_c = 5 ksi
All reinforcing steel is Grade 60
Axial design strength in kips

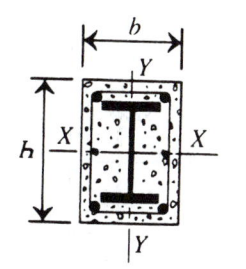

Size $b \times h$						16 in. × 16 in.				
Reinf. bars						4-#7 bars				
Ties						#3 bars spaced 10 in. c. to c.				
Steel Shape / Designation						**W 8**				
Wt./ft		67		58		48		40		35
F_y		36	50	36	50	36	50	36	50	36	50
Effective length in ft KL with respect to least radius of gyration r_{my}	0	1280	1510	1200	1410	1120	1290	1050	1190	1010	1140
	6	1250	1480	1180	1380	1100	1260	1030	1160	991	1110
	7	1250	1470	1170	1360	1090	1250	1020	1150	982	1100
	8	1240	1450	1160	1350	1080	1230	1010	1140	973	1090
	9	1220	1440	1150	1340	1070	1220	1000	1130	962	1070
	10	1210	1420	1140	1320	1060	1200	990	1110	951	1060
	11	1200	1400	1130	1300	1040	1190	977	1100	938	1040
	12	1180	1380	1110	1280	1030	1170	963	1080	924	1020
	13	1170	1360	1100	1260	1010	1150	948	1060	909	1010
	14	1150	1340	1080	1240	999	1130	932	1040	893	986
	15	1130	1310	1060	1220	982	1110	916	1020	877	965
	16	1120	1290	1050	1190	964	1080	898	996	859	944
	17	1100	1260	1030	1170	946	1060	880	973	841	922
	18	1080	1230	1010	1140	927	1040	861	949	822	898
	19	1060	1210	987	1120	907	1010	841	925	803	875
	20	1030	1180	966	1090	886	984	821	900	783	850
	22	989	1120	922	1030	844	929	779	848	741	800
	24	942	1050	876	971	799	873	736	795	698	748
	26	894	990	829	910	754	816	692	741	655	696
	28	844	926	781	849	707	759	647	686	610	643
	30	794	861	733	787	661	702	602	632	566	591
	32	743	796	684	727	614	646	557	579	522	540
	34	693	733	636	667	569	591	513	528	480	491
	36	644	672	589	610	524	537	470	478	438	444
	38	595	612	542	554	480	486	429	431	398	398
	40	548	555	498	501	438	438	389	389	360	360

Properties											
$\phi_b M_{nx}$ (kip-ft)		303	400	268	353	226	296	196	255	175	227
$\phi_b M_{ny}$ (kip-ft)		202	259	182	233	155	198	138	175	125	157
$P_{ex}(K_x L_x)^2/10^4$ (kip-ft^2)		119	119	107	107	94.1	94.1	83.4	83.4	77.2	77.2
$P_{ey}(K_y L_y)^2/10^4$ (kip-ft^2)		119	119	107	107	94.1	94.1	83.4	83.4	77.2	77.2
r_{my} (in.)		4.80		4.80		4.80		4.80		4.80	
r_{mx}/r_{my} (in./in.)		1.00		1.00		1.00		1.00		1.00	

AMERICAN INSTITUTE OF STEEL CONSTRUCTION

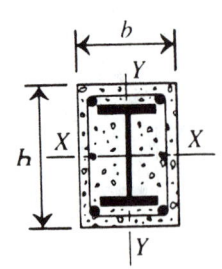

F_y = 36 ksi
F_y = 50 ksi

COMPOSITE COLUMNS
W Shapes
f'_c = 8 ksi
All reinforcing steel is Grade 60
Axial design strength in kips

Size $b \times h$		24 in. × 26 in.								
Reinf. bars		4-#11 bars								
Ties		#4 bars spaced 16 in. c. to c.								
	Designation	W 14								
	Wt./ft	426		398		370		342		311	
	F_y	36	50	36	50	36	50	36	50	36	50
Effective length in ft KL with respect to least radius of gyration r_{my}	0	6040	7530	5830	7220	5620	6910	5400	6610	5150	6240
	6	6000	7460	5780	7150	5570	6850	5360	6540	5110	6180
	7	5980	7430	5770	7130	5560	6820	5350	6520	5090	6150
	8	5960	7410	5750	7100	5540	6800	5330	6490	5070	6130
	9	5940	7370	5730	7070	5520	6770	5310	6460	5060	6100
	10	5920	7340	5710	7040	5500	6730	5290	6430	5030	6070
	11	5890	7300	5680	7000	5470	6700	5260	6400	5010	6030
	12	5870	7260	5660	6960	5450	6660	5240	6360	4980	6000
	13	5840	7210	5630	6910	5420	6610	5210	6310	4960	5960
	14	5800	7160	5590	6860	5390	6570	5180	6270	4930	5910
	15	5770	7110	5560	6810	5350	6520	5140	6220	4890	5860
	16	5730	7050	5530	6760	5320	6460	5110	6170	4860	5820
	17	5690	6990	5490	6700	5280	6410	5070	6120	4820	5760
	18	5650	6930	5450	6640	5240	6350	5030	6060	4790	5710
	19	5610	6870	5410	6580	5200	6290	4990	6000	4750	5650
	20	5570	6800	5360	6510	5160	6230	4950	5940	4700	5590
	22	5470	6660	5270	6370	5070	6090	4860	5810	4620	5460
	24	5370	6500	5170	6220	4970	5940	4770	5670	4520	5330
	26	5260	6340	5060	6070	4860	5790	4660	5520	4420	5190
	28	5150	6170	4950	5900	4750	5630	4550	5360	4310	5040
	30	5030	5990	4830	5730	4640	5460	4440	5200	4200	4880
	32	4900	5800	4710	5550	4520	5290	4320	5030	4090	4720
	34	4770	5610	4580	5360	4390	5110	4200	4850	3970	4550
	36	4640	5420	4450	5170	4260	4920	4070	4680	3840	4380
	38	4500	5220	4320	4980	4130	4740	3940	4500	3720	4210
	40	4360	5010	4180	4780	4000	4550	3810	4310	3590	4030
Properties											
$\phi_b M_{nx}$ (kip-ft)		3190	4270	2970	3980	2770	3700	2560	3430	2330	3120
$\phi_b M_{ny}$ (kip-ft)		1940	2550	1830	2410	1710	2260	1600	2100	1470	1930
$P_{ex}(K_xL_x)^2/10^4$ (kip-ft²)		1710	1710	1620	1620	1530	1530	1430	1430	1320	1320
$P_{ey}(K_yL_y)^2/10^4$ (kip-ft²)		1460	1460	1380	1380	1300	1300	1220	1220	1130	1130
r_{my} (in.)		7.20		7.20		7.20		7.20		7.20	
r_{mx}/r_{my} (in./in.)		1.08		1.08		1.08		1.08		1.08	

F_y = 36 ksi

F_y = 50 ksi

COMPOSITE COLUMNS
W Shapes
f'_c = 8 ksi
All reinforcing steel is Grade 60
Axial design strength in kips

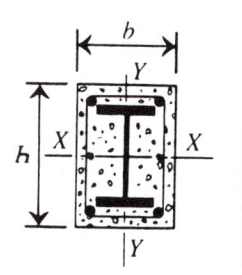

Size $b \times h$		24 in. × 24 in.									
Reinf. bars		4-#11 bars									
Ties		#4 bars spaced 16 in. c. to c.									
Designation		**W 14**									
Wt./ft		283		257		233		211		193	
F_y		36	50	36	50	36	50	36	50	36	50
Effective length in ft KL with respect to least radius of gyration r_{my}	0	4740	5730	4530	5430	4350	5160	4170	4910	4040	4710
	6	4700	5670	4500	5380	4310	5110	4140	4860	4000	4660
	7	4690	5650	4480	5360	4290	5090	4120	4840	3980	4640
	8	4670	5630	4470	5340	4280	5070	4110	4820	3970	4620
	9	4650	5600	4450	5310	4260	5040	4090	4790	3950	4600
	10	4630	5570	4430	5280	4240	5010	4070	4770	3930	4570
	11	4610	5540	4410	5250	4220	4980	4050	4740	3910	4540
	12	4590	5510	4380	5220	4200	4950	4020	4710	3890	4510
	13	4560	5470	4360	5180	4170	4920	4000	4670	3860	4480
	14	4530	5430	4330	5140	4140	4880	3970	4630	3830	4440
	15	4500	5390	4300	5100	4110	4840	3940	4590	3810	4400
	16	4470	5340	4270	5060	4080	4790	3910	4550	3780	4360
	17	4440	5290	4240	5010	4050	4750	3880	4510	3740	4320
	18	4400	5240	4200	4960	4020	4700	3850	4460	3710	4270
	19	4370	5190	4170	4910	3980	4650	3810	4410	3670	4220
	20	4330	5130	4130	4860	3940	4600	3770	4360	3640	4170
	22	4250	5020	4050	4740	3860	4490	3690	4250	3560	4070
	24	4160	4890	3960	4620	3780	4370	3610	4140	3470	3960
	26	4060	4760	3870	4490	3690	4250	3520	4020	3380	3840
	28	3970	4620	3770	4360	3590	4120	3430	3890	3290	3710
	30	3860	4480	3670	4220	3490	3980	3330	3760	3190	3590
	32	3760	4330	3570	4070	3390	3840	3220	3630	3090	3450
	34	3640	4170	3460	3930	3280	3700	3120	3490	2990	3320
	36	3530	4010	3340	3770	3170	3550	3010	3340	2880	3180
	38	3410	3850	3230	3620	3060	3400	2900	3200	2770	3040
	40	3300	3690	3110	3460	2950	3250	2790	3060	2660	2900
Properties											
$\phi_b M_{nx}$ (kip-ft)		2070	2770	1890	2530	1720	2300	1570	2090	1440	1920
$\phi_b M_{ny}$ (kip-ft)		1350	1770	1240	1620	1140	1490	1050	1370	972	1270
$P_{ex}(K_x L_x)^2/10^4$ (kip-ft²)		1030	1030	952	952	882	882	817	817	765	765
$P_{ey}(K_y L_y)^2/10^4$ (kip-ft²)		1030	1030	952	952	882	882	817	817	765	765
r_{my} (in.)		7.20		7.20		7.20		7.20		7.20	
r_{mx}/r_{my} (in./in.)		1.00		1.00		1.00		1.00		1.00	

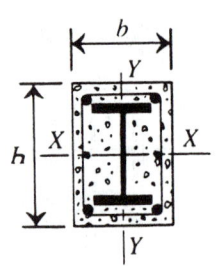

		F_y = 36 ksi
		F_y = 50 ksi

COMPOSITE COLUMNS
W Shapes
f_c' = 8 ksi
All reinforcing steel is Grade 60
Axial design strength in kips

Size $b \times h$		\multicolumn{10}{c}{24 in. × 24 in.}

Size $b \times h$	24 in. × 24 in.
Reinf. bars	4-#10 bars
Ties	#3 bars spaced 16 in. c. to c.

| | Designation | \multicolumn{10}{c}{W 14} |

Steel Shape		W 14									
	Wt./ft	\multicolumn{2}{c}{176}	\multicolumn{2}{c}{159}	\multicolumn{2}{c}{145}	\multicolumn{2}{c}{132}	\multicolumn{2}{c}{120}
	F_y	36	50	36	50	36	50	36	50	36	50
	0	3870	4490	3730	4290	3630	4140	3520	3990	3430	3850
	6	3830	4440	3700	4240	3590	4090	3490	3940	3390	3800
	7	3820	4420	3680	4220	3580	4070	3470	3920	3380	3790
	8	3800	4400	3670	4200	3560	4050	3460	3900	3370	3770
	9	3790	4370	3650	4180	3540	4030	3440	3880	3350	3750
	10	3770	4350	3630	4160	3530	4000	3420	3850	3330	3720
	11	3750	4320	3610	4130	3500	3980	3400	3830	3310	3690
	12	3720	4290	3590	4100	3480	3950	3380	3800	3280	3670
	13	3700	4260	3560	4070	3460	3910	3350	3770	3260	3630
	14	3670	4220	3540	4030	3430	3880	3330	3730	3230	3600
	15	3640	4180	3510	3990	3400	3840	3300	3700	3200	3560
	16	3610	4140	3480	3950	3370	3800	3270	3660	3170	3530
	17	3580	4100	3450	3910	3340	3760	3240	3620	3140	3490
	18	3550	4060	3410	3870	3310	3720	3200	3580	3110	3450
	19	3510	4010	3380	3820	3270	3680	3170	3530	3070	3400
	20	3480	3960	3340	3780	3240	3630	3130	3490	3040	3360
	22	3400	3860	3270	3680	3160	3530	3050	3390	2960	3260
	24	3320	3750	3180	3570	3080	3430	2970	3290	2880	3160
	26	3230	3640	3100	3460	2990	3320	2880	3180	2790	3050
	28	3140	3520	3010	3340	2900	3200	2790	3070	2700	2940
	30	3040	3400	2910	3220	2800	3080	2700	2950	2600	2830
	32	2950	3270	2810	3100	2710	2960	2600	2830	2510	2710
	34	2840	3140	2710	2970	2610	2840	2500	2710	2410	2590
	36	2740	3010	2610	2840	2500	2710	2400	2580	2310	2470
	38	2630	2870	2500	2710	2400	2580	2300	2460	2200	2340
	40	2530	2740	2400	2580	2300	2450	2190	2330	2100	2220

Properties

	176		159		145		132		120	
$\phi_b M_{nx}$ (kip-ft)	1300	1740	1190	1580	1090	1450	1010	1340	930	1230
$\phi_b M_{ny}$ (kip-ft)	879	1150	805	1050	748	975	681	884	633	818
$P_{ex}(K_x L_x)^2/10^4$ (kip-ft^2)	716	716	665	665	625	625	587	587	552	552
$P_{ey}(K_y L_y)^2/10^4$ (kip-ft^2)	716	716	665	665	625	625	587	587	552	552
r_{my} (in.)	\multicolumn{2}{c}{7.20}	\multicolumn{2}{c}{7.20}	\multicolumn{2}{c}{7.20}	\multicolumn{2}{c}{7.20}	\multicolumn{2}{c}{7.20}
r_{mx}/r_{my} (in./in.)	\multicolumn{2}{c}{1.00}	\multicolumn{2}{c}{1.00}	\multicolumn{2}{c}{1.00}	\multicolumn{2}{c}{1.00}	\multicolumn{2}{c}{1.00}

Effective length in ft KL with respect to least radius of gyration r_{my}

F_y = 36 ksi
F_y = 50 ksi

COMPOSITE COLUMNS
W Shapes
f'_c = 8 ksi
All reinforcing steel is Grade 60
Axial design strength in kips

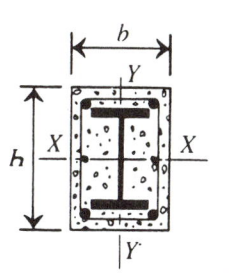

Size $b \times h$			22 in. × 22 in.								
Reinf. bars			4-#10 bars								
Ties			#3 bars spaced 14 in. c. to c.								
Steel Shape	Designation		W 14								
	Wt./ft		109		99		90		82		74
	F_y		36	50	36	50	36	50	36	50	36	50
Effective length in ft KL with respect to least radius of gyration r_{my}		0	2970	3350	2890	3240	2820	3140	2760	3050	2700	2960
		6	2930	3300	2850	3190	2780	3090	2720	3000	2660	2910
		7	2920	3280	2840	3170	2770	3070	2710	2980	2640	2890
		8	2900	3260	2820	3150	2750	3050	2690	2960	2630	2870
		9	2880	3240	2810	3130	2740	3030	2670	2940	2610	2850
		10	2860	3220	2790	3110	2720	3010	2650	2920	2590	2830
		11	2840	3190	2770	3080	2700	2980	2630	2890	2570	2800
		12	2820	3160	2740	3050	2670	2950	2610	2860	2540	2770
		13	2790	3130	2720	3020	2650	2920	2580	2830	2520	2740
		14	2770	3090	2690	2990	2620	2890	2550	2800	2490	2710
		15	2740	3060	2660	2950	2590	2850	2520	2760	2460	2670
		16	2710	3020	2630	2910	2560	2810	2490	2720	2430	2640
		17	2680	2980	2600	2870	2530	2770	2460	2680	2400	2600
		18	2640	2940	2560	2830	2490	2730	2430	2640	2360	2560
		19	2610	2900	2530	2790	2460	2690	2390	2600	2330	2510
		20	2570	2850	2490	2740	2420	2650	2350	2560	2290	2470
		22	2500	2750	2420	2650	2340	2550	2280	2460	2210	2380
		24	2420	2650	2340	2550	2260	2450	2200	2370	2130	2280
		26	2330	2550	2250	2450	2180	2350	2110	2270	2040	2180
		28	2240	2440	2160	2340	2090	2250	2020	2160	1950	2080
		30	2150	2330	2070	2230	2000	2140	1930	2050	1860	1970
		32	2060	2220	1980	2120	1910	2030	1840	1950	1770	1860
		34	1960	2100	1880	2000	1810	1920	1740	1840	1680	1760
		36	1870	1980	1790	1890	1720	1810	1650	1730	1580	1650
		38	1770	1870	1690	1780	1620	1700	1560	1620	1490	1540
		40	1670	1760	1600	1670	1530	1590	1460	1510	1400	1440
Properties											
$\phi_b M_{nx}$ (kip-ft)			819	1080	755	996	697	917	671	879	617	805
$\phi_b M_{ny}$ (kip-ft)			551	712	514	661	478	612	417	526	386	485
$P_{ex}(K_x L_x)^2/10^4$ (kip-ft^2)			409	409	385	385	363	363	343	343	324	324
$P_{ey}(K_y L_y)^2/10^4$ (kip-ft^2)			409	409	385	385	363	363	343	343	324	324
r_{my} (in.)			6.60		6.60		6.60		6.60		6.60	
r_{mx}/r_{my} (in./in.)			1.00		1.00		1.00		1.00		1.00	

	F_y = 36 ksi
	F_y = 50 ksi

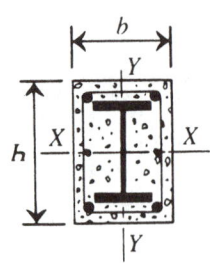

COMPOSITE COLUMNS
W Shapes
f'_c = 8 ksi
All reinforcing steel is Grade 60
Axial design strength in kips

Size $b \times h$		18 in. $\times$ 22 in.								
Reinf. bars		4-#9 bars								
Ties		#3 bars spaced 12 in. c. to c.								
	Designation	W 14								
	Wt./ft	68		61		53		48		43	
	F_y	36	50	36	50	36	50	36	50	36	50
Effective length in ft KL with respect to least radius of gyration r_{my}	0	2260	2500	2200	2420	2140	2330	2100	2270	2060	2210
	6	2210	2440	2160	2360	2100	2270	2050	2210	2010	2160
	7	2200	2420	2140	2340	2080	2250	2040	2190	2000	2140
	8	2180	2400	2120	2320	2060	2230	2020	2170	1980	2110
	9	2160	2370	2100	2290	2040	2200	1990	2140	1950	2090
	10	2130	2340	2070	2260	2010	2170	1970	2110	1930	2060
	11	2110	2310	2050	2230	1980	2140	1940	2080	1900	2030
	12	2080	2280	2020	2200	1950	2110	1910	2050	1870	1990
	13	2050	2240	1990	2160	1920	2070	1880	2010	1840	1950
	14	2010	2200	1960	2120	1890	2030	1850	1970	1800	1920
	15	1980	2160	1920	2080	1860	1990	1810	1930	1770	1880
	16	1940	2120	1890	2040	1820	1950	1780	1890	1730	1830
	17	1910	2070	1850	1990	1780	1900	1740	1850	1690	1790
	18	1870	2020	1810	1950	1740	1860	1700	1800	1650	1740
	19	1830	1980	1770	1900	1700	1810	1660	1750	1610	1700
	20	1790	1930	1730	1850	1660	1760	1610	1710	1570	1650
	22	1700	1820	1640	1750	1570	1660	1530	1610	1480	1550
	24	1610	1720	1550	1640	1480	1560	1440	1500	1390	1450
	26	1520	1610	1460	1540	1390	1450	1340	1400	1300	1340
	28	1430	1500	1360	1430	1300	1350	1250	1300	1200	1240
	30	1330	1390	1270	1320	1200	1240	1160	1190	1110	1140
	32	1240	1280	1180	1220	1110	1140	1070	1090	1020	1040
	34	1140	1180	1090	1110	1020	1040	977	992	932	944
	36	1050	1080	998	1010	933	944	890	898	846	851
	38	966	977	911	917	849	850	807	807	764	764
	40	882	882	827	827	767	767	728	728	689	689
Properties											
$\phi_b M_{nx}$ (kip-ft)		549	720	501	654	459	596	426	551	390	502
$\phi_b M_{ny}$ (kip-ft)		290	370	268	341	238	298	223	279	207	258
$P_{ex}(K_x L_x)^2/10^4$ (kip-ft^2)		283	283	265	265	246	246	234	234	221	221
$P_{ey}(K_y L_y)^2/10^4$ (kip-ft^2)		189	189	178	178	165	165	156	156	148	148
r_{my} (in.)		5.40		5.40		5.40		5.40		5.40	
r_{mx}/r_{my} (in./in.)		1.22		1.22		1.22		1.22		1.22	

| F_y = 36 ksi |
| F_y = 50 ksi |

COMPOSITE COLUMNS
W Shapes
f'_c = 8 ksi
All reinforcing steel is Grade 60
Axial design strength in kips

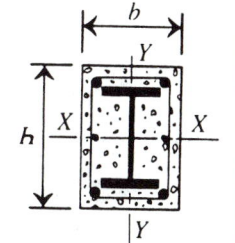

Size $b \times h$		22 in. × 24 in.									
Reinf. bars		4-#10 bars									
Ties		#3 bars spaced 14 in. c. to c.									
Steel Shape	Designation	**W 12**									
	Wt./ft	336		305		279		252		230	
	F_y	36	50	36	50	36	50	36	50	36	50
Effective length in ft KL with respect to least radius of gyration r_{my}	0	4920	6100	4680	5740	4470	5450	4260	5150	4100	4900
	6	4880	6030	4630	5680	4430	5380	4220	5090	4050	4840
	7	4860	6000	4620	5650	4410	5360	4210	5060	4040	4820
	8	4840	5980	4600	5630	4400	5340	4190	5040	4020	4800
	9	4820	5950	4580	5600	4380	5310	4170	5010	4000	4770
	10	4800	5910	4560	5570	4360	5270	4150	4980	3980	4740
	11	4770	5870	4530	5530	4330	5240	4130	4950	3960	4710
	12	4750	5830	4510	5490	4310	5200	4100	4910	3930	4670
	13	4720	5790	4480	5450	4280	5160	4070	4870	3910	4630
	14	4690	5740	4450	5400	4250	5120	4040	4830	3880	4590
	15	4650	5690	4420	5350	4220	5070	4010	4780	3840	4540
	16	4620	5640	4380	5300	4180	5020	3980	4730	3810	4500
	17	4580	5580	4340	5250	4140	4970	3940	4680	3780	4450
	18	4540	5520	4310	5190	4110	4910	3910	4630	3740	4400
	19	4500	5460	4260	5130	4070	4850	3870	4570	3700	4340
	20	4460	5390	4220	5070	4030	4790	3830	4510	3660	4290
	22	4360	5260	4130	4930	3940	4670	3740	4390	3580	4170
	24	4270	5110	4040	4790	3840	4530	3650	4260	3480	4040
	26	4160	4960	3940	4650	3740	4390	3550	4120	3390	3910
	28	4050	4790	3830	4490	3640	4240	3450	3980	3290	3770
	30	3940	4630	3720	4330	3530	4080	3340	3830	3180	3620
	32	3820	4450	3600	4170	3420	3920	3230	3680	3070	3480
	34	3700	4280	3480	4000	3300	3760	3120	3520	2960	3330
	36	3570	4100	3360	3830	3180	3600	3000	3370	2850	3170
	38	3440	3920	3240	3650	3060	3430	2880	3210	2730	3020
	40	3310	3730	3110	3480	2940	3260	2760	3050	2620	2870
Properties											
$\phi_b M_{nx}$ (kip-ft)		2270	3030	2050	2750	1880	2510	1710	2280	1570	2090
$\phi_b M_{ny}$ (kip-ft)		1320	1730	1210	1580	1130	1470	1030	1350	954	1250
$P_{ex}(K_xL_x)^2/10^4$ (kip-ft²)		1170	1170	1080	1080	999	999	921	921	857	857
$P_{ey}(K_yL_y)^2/10^4$ (kip-ft²)		980	980	904	904	839	839	774	774	720	720
r_{my} (in.)		6.60		6.60		6.60		6.60		6.60	
r_{mx}/r_{my} (in./in.)		1.09		1.09		1.09		1.09		1.09	

AMERICAN INSTITUTE OF STEEL CONSTRUCTION

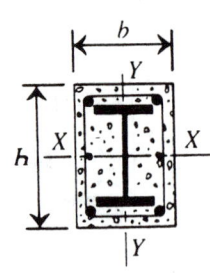

		F_y = 36 ksi
		F_y = 50 ksi

COMPOSITE COLUMNS
W Shapes
f'_c = 8 ksi
All reinforcing steel is Grade 60
Axial design strength in kips

Size $b \times h$			20 in. × 22 in.								
Reinf. bars			4-#10 bars								
Ties			#3 bars spaced 13 in. c. to c.								
	Designation		**W 12**								
	Wt./ft		210		190		170		152		136	
	F_y		36	50	36	50	36	50	36	50	36	50
	0		3580	4320	3420	4080	3270	3860	3130	3660	3000	3470
	6		3540	4250	3380	4020	3230	3800	3080	3600	2960	3420
	7		3520	4230	3360	4000	3210	3780	3070	3580	2940	3400
	8		3500	4210	3350	3980	3190	3760	3050	3560	2930	3380
	9		3480	4180	3330	3950	3170	3730	3030	3530	2910	3350
	10		3460	4150	3310	3920	3150	3700	3010	3500	2880	3320
	11		3440	4110	3280	3890	3130	3670	2990	3470	2860	3290
	12		3410	4070	3260	3850	3100	3630	2960	3440	2840	3260
	13		3380	4030	3230	3810	3080	3600	2940	3400	2810	3220
	14		3350	3990	3200	3770	3050	3560	2910	3360	2780	3180
	15		3320	3940	3170	3730	3010	3510	2880	3320	2750	3140
	16		3290	3900	3130	3680	2980	3470	2840	3270	2720	3100
	17		3250	3850	3100	3630	2950	3420	2810	3230	2680	3050
	18		3220	3790	3060	3580	2910	3370	2770	3180	2650	3000
	19		3180	3740	3020	3520	2870	3320	2730	3130	2610	2950
	20		3140	3680	2980	3470	2830	3260	2690	3070	2570	2900
	22		3050	3560	2900	3350	2750	3150	2610	2960	2490	2800
	24		2960	3430	2810	3230	2660	3030	2520	2850	2400	2680
	26		2860	3300	2710	3100	2570	2900	2430	2730	2310	2570
	28		2760	3160	2610	2960	2470	2770	2340	2600	2210	2440
	30		2660	3010	2510	2830	2370	2640	2240	2470	2120	2320
	32		2550	2870	2410	2690	2270	2510	2140	2340	2020	2190
	34		2440	2720	2300	2540	2160	2370	2030	2210	1920	2070
	36		2330	2570	2190	2400	2060	2240	1930	2080	1820	1940
	38		2220	2430	2080	2260	1950	2100	1830	1950	1710	1820
	40		2110	2280	1980	2120	1850	1970	1730	1830	1610	1690

(Left column label: Effective length in ft KL with respect to least radius of gyration r_{my})

Properties											
$\phi_b M_{nx}$ (kip-ft)		1380	1840	1250	1670	1130	1510	1020	1360	926	1230
$\phi_b M_{ny}$ (kip-ft)		828	1080	760	991	692	901	631	821	578	749
$P_{ex}(K_x L_x)^2/10^4$ (kip-ft²)		645	645	595	595	546	546	502	502	462	462
$P_{ey}(K_y L_y)^2/10^4$ (kip-ft²)		533	533	492	492	452	452	415	415	382	382
r_{my} (in.)		6.00		6.00		6.00		6.00		6.00	
r_{mx}/r_{my} (in./in.)		1.10		1.10		1.10		1.10		1.10	

F_y = 36 ksi

F_y = 50 ksi

COMPOSITE COLUMNS
W Shapes
f'_c = 8 ksi
All reinforcing steel is Grade 60
Axial design strength in kips

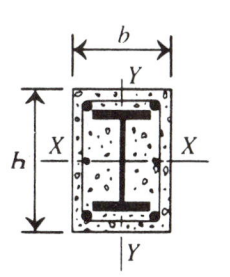

Size $b \times h$		20 in. × 20 in.									
Reinf. bars		4-#9 bars									
Ties		#3 bars spaced 13 in. c. to c.									
Steel Shape	Designation	**W 12**									
	Wt./ft	120		106		96		87		79	
	F_y	36	50	36	50	36	50	36	50	36	50
Effective length in ft KL with respect to least radius of gyration r_{my}	0	2680	3100	2570	2950	2490	2830	2430	2730	2360	2640
	6	2650	3050	2540	2900	2460	2780	2390	2680	2320	2590
	7	2630	3040	2520	2880	2440	2760	2370	2670	2310	2570
	8	2620	3010	2510	2860	2430	2740	2360	2650	2290	2550
	9	2600	2990	2490	2840	2410	2720	2340	2620	2280	2530
	10	2580	2970	2470	2810	2390	2700	2320	2600	2260	2510
	11	2560	2940	2450	2780	2370	2670	2300	2570	2240	2480
	12	2540	2910	2430	2750	2350	2640	2280	2540	2210	2450
	13	2510	2880	2400	2720	2320	2610	2250	2510	2190	2420
	14	2490	2840	2380	2690	2300	2580	2230	2480	2160	2390
	15	2460	2800	2350	2650	2270	2540	2200	2440	2130	2350
	16	2430	2770	2320	2610	2240	2500	2170	2410	2100	2320
	17	2400	2730	2290	2570	2210	2460	2140	2370	2070	2280
	18	2370	2680	2260	2530	2180	2420	2110	2330	2040	2240
	19	2330	2640	2220	2490	2140	2380	2070	2290	2010	2200
	20	2300	2590	2190	2450	2110	2340	2040	2240	1970	2150
	22	2220	2500	2110	2350	2030	2240	1960	2150	1900	2060
	24	2140	2400	2040	2250	1960	2150	1890	2060	1820	1970
	26	2060	2290	1960	2150	1880	2050	1800	1960	1740	1870
	28	1980	2180	1870	2050	1790	1940	1720	1860	1660	1770
	30	1890	2070	1790	1940	1710	1840	1640	1750	1570	1670
	32	1800	1960	1700	1830	1620	1730	1550	1650	1480	1570
	34	1710	1850	1610	1720	1530	1630	1460	1550	1400	1470
	36	1620	1730	1520	1610	1440	1520	1380	1440	1310	1370
	38	1530	1620	1430	1500	1360	1420	1290	1340	1230	1270
	40	1440	1510	1340	1400	1270	1320	1200	1240	1140	1170
Properties											
$\phi_b M_{nx}$ (kip-ft)		773	1030	690	917	629	835	580	767	534	704
$\phi_b M_{ny}$ (kip-ft)		501	653	450	584	415	537	387	499	359	461
$P_{ex}(K_x L_x)^2/10^4$ (kip-ft^2)		340	340	312	312	291	291	273	273	257	257
$P_{ey}(K_y L_y)^2/10^4$ (kip-ft^2)		340	340	312	312	291	291	273	273	257	257
r_{my} (in.)		6.00		6.00		6.00		6.00		6.00	
r_{mx}/r_{my} (in./in.)		1.00		1.00		1.00		1.00		1.00	

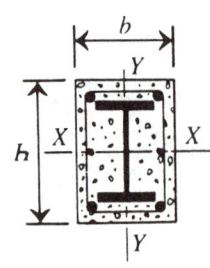

| F_y = 36 ksi |
| F_y = 50 ksi |

COMPOSITE COLUMNS
W Shapes
f'_c = 8 ksi
All reinforcing steel is Grade 60
Axial design strength in kips

Size $b \times h$		20 in. × 20 in.								
Reinf. bars		4-#9 bars								
Ties		#3 bars spaced 13 in. c. to c.								
	Designation	**W 12**								
Steel Shape	Wt./ft	72		65		58		53		50	
	F_y	36	50	36	50	36	50	36	50	36	50
Effective length in ft KL with respect to least radius of gyration r_{my}	0	2310	2560	2250	2480	2200	2400	2160	2350	2140	2310
	6	2270	2510	2210	2430	2160	2350	2120	2300	2100	2260
	7	2250	2490	2200	2420	2140	2340	2110	2280	2080	2250
	8	2240	2470	2180	2400	2130	2320	2090	2260	2060	2230
	9	2220	2450	2170	2380	2110	2290	2070	2240	2050	2210
	10	2200	2430	2150	2350	2090	2270	2050	2220	2030	2180
	11	2180	2400	2120	2320	2070	2240	2030	2190	2000	2160
	12	2160	2370	2100	2300	2040	2220	2000	2160	1980	2130
	13	2130	2340	2080	2270	2020	2190	1980	2130	1950	2100
	14	2100	2310	2050	2230	1990	2150	1950	2100	1930	2060
	15	2080	2270	2020	2200	1960	2120	1920	2070	1900	2030
	16	2050	2240	1990	2160	1930	2080	1890	2030	1860	1990
	17	2010	2200	1960	2130	1900	2050	1860	1990	1830	1960
	18	1980	2160	1920	2090	1870	2010	1820	1950	1800	1920
	19	1950	2120	1890	2040	1830	1970	1790	1910	1760	1880
	20	1910	2080	1860	2000	1790	1920	1750	1870	1730	1840
	22	1840	1990	1780	1910	1720	1840	1680	1780	1650	1750
	24	1760	1900	1700	1820	1640	1750	1600	1690	1570	1660
	26	1680	1800	1620	1730	1560	1650	1520	1600	1490	1570
	28	1600	1700	1540	1630	1480	1560	1440	1510	1410	1470
	30	1510	1600	1450	1530	1390	1460	1350	1410	1320	1380
	32	1430	1500	1370	1430	1310	1360	1270	1310	1240	1280
	34	1340	1400	1280	1340	1220	1270	1180	1220	1160	1190
	36	1260	1300	1200	1240	1140	1170	1100	1130	1070	1100
	38	1170	1210	1120	1150	1060	1080	1020	1040	992	1010
	40	1090	1110	1040	1050	978	991	938	949	913	921

Properties										
$\phi_b M_{nx}$ (kip-ft)	494	649	453	593	417	544	390	507	382	495
$\phi_b M_{ny}$ (kip-ft)	335	428	311	396	271	342	258	324	244	304
$P_{ex}(K_x L_x)^2/10^4$ (kip-ft^2)	242	242	229	229	214	214	204	204	198	198
$P_{ey}(K_y L_y)^2/10^4$ (kip-ft^2)	242	242	229	229	214	214	204	204	198	198
r_{my} (in.)	6.00		6.00		6.00		6.00		6.00	
r_{mx}/r_{my} (in./in.)	1.00		1.00		1.00		1.00		1.00	

F_y = 36 ksi
F_y = 50 ksi

COMPOSITE COLUMNS
W Shapes
f'_c = 8 ksi
All reinforcing steel is Grade 60
Axial design strength in kips

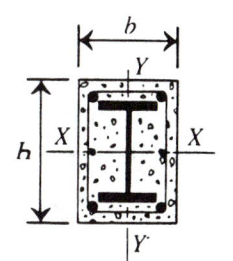

Size $b \times h$										
	Size $b \times h$				18 in. × 18 in.					
	Reinf. bars				4-#8 bars					
	Ties				#3 bars spaced 12 in. c. to c.					
Steel Shape	Designation				W 10					
	Wt./ft	112		100		88		77		68	
	F_y	36	50	36	50	36	50	36	50	36	50

Effective length in ft KL with respect to least radius of gyration r_{my}		112 (36)	112 (50)	100 (36)	100 (50)	88 (36)	88 (50)	77 (36)	77 (50)	68 (36)	68 (50)
	0	2280	2680	2190	2540	2100	2410	2010	2280	1940	2180
	6	2250	2630	2160	2490	2060	2360	1970	2230	1910	2130
	7	2230	2610	2140	2470	2050	2340	1960	2220	1890	2120
	8	2220	2590	2130	2450	2030	2320	1950	2200	1880	2100
	9	2200	2560	2110	2430	2020	2300	1930	2170	1860	2080
	10	2180	2540	2090	2410	2000	2270	1910	2150	1840	2050
	11	2160	2510	2070	2380	1980	2250	1890	2120	1820	2020
	12	2140	2480	2050	2350	1950	2220	1870	2090	1800	2000
	13	2120	2450	2020	2320	1930	2190	1840	2060	1770	1970
	14	2090	2410	2000	2280	1900	2150	1820	2030	1750	1930
	15	2060	2370	1970	2250	1880	2120	1790	2000	1720	1900
	16	2030	2340	1940	2210	1850	2080	1760	1960	1690	1860
	17	2000	2290	1910	2170	1820	2040	1730	1920	1660	1830
	18	1970	2250	1880	2130	1790	2000	1700	1880	1630	1790
	19	1940	2210	1850	2080	1750	1960	1670	1840	1600	1750
	20	1900	2160	1810	2040	1720	1920	1630	1800	1560	1710
	22	1830	2070	1740	1950	1650	1830	1560	1710	1490	1620
	24	1760	1970	1670	1850	1580	1730	1490	1620	1420	1530
	26	1680	1870	1590	1750	1500	1640	1410	1530	1340	1440
	28	1600	1760	1510	1650	1420	1540	1340	1430	1270	1350
	30	1520	1660	1430	1550	1340	1440	1260	1340	1190	1260
	32	1430	1550	1350	1450	1260	1340	1180	1240	1110	1160
	34	1350	1450	1270	1350	1180	1250	1100	1150	1030	1070
	36	1270	1340	1190	1250	1100	1150	1020	1060	957	985
	38	1180	1240	1110	1150	1020	1060	946	970	883	900
	40	1100	1140	1030	1050	948	968	872	885	811	817

Properties											
$\phi_b M_{nx}$ (kip-ft)		599	800	541	721	482	640	426	565	382	505
$\phi_b M_{ny}$ (kip-ft)		389	509	354	462	320	416	287	371	260	335
$P_{ex}(K_x L_x)^2/10^4$ (kip-ft²)		248	248	228	228	208	208	190	190	175	175
$P_{ey}(K_y L_y)^2/10^4$ (kip-ft²)		248	248	228	228	208	208	190	190	175	175
r_{my} (in.)		5.40		5.40		5.40		5.40		5.40	
r_{mx}/r_{my} (in./in.)		1.00		1.00		1.00		1.00		1.00	

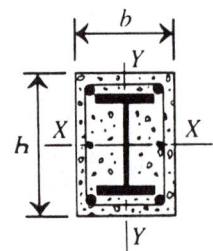

	F_y = 36 ksi
	F_y = 50 ksi

COMPOSITE COLUMNS
W Shapes
f'_c = 8 ksi
All reinforcing steel is Grade 60
Axial design strength in kips

Size $b \times h$		18 in. × 18 in.								
Reinf. bars		4-#8 bars								
Ties		#3 bars spaced 12 in. c. to c.								
	Designation	W 10								
	Wt./ft	60		54		49		45		39	
	F_y	36	50	36	50	36	50	36	50	36	50
	0	1880	2090	1830	2020	1790	1970	1770	1920	1720	1850
	6	1840	2040	1790	1970	1760	1920	1730	1880	1680	1810
	7	1830	2020	1780	1960	1740	1900	1710	1860	1660	1790
	8	1810	2010	1760	1940	1730	1880	1700	1840	1650	1770
	9	1790	1980	1750	1910	1710	1860	1680	1820	1630	1750
	10	1770	1960	1730	1890	1690	1840	1660	1800	1610	1730
	11	1750	1930	1700	1870	1670	1810	1640	1770	1590	1700
	12	1730	1910	1680	1840	1640	1780	1610	1740	1560	1670
	13	1710	1880	1660	1810	1620	1750	1590	1710	1540	1640
	14	1680	1840	1630	1780	1590	1720	1560	1680	1510	1610
	15	1650	1810	1600	1740	1560	1690	1530	1650	1480	1580
	16	1620	1770	1570	1710	1530	1650	1500	1610	1450	1540
	17	1590	1740	1540	1670	1500	1620	1470	1580	1420	1510
	18	1560	1700	1510	1630	1470	1580	1440	1540	1390	1470
	19	1530	1660	1480	1590	1440	1540	1410	1500	1350	1430
	20	1490	1620	1440	1550	1400	1500	1370	1460	1320	1390
	22	1420	1540	1370	1470	1330	1420	1300	1380	1250	1310
	24	1350	1450	1300	1380	1260	1330	1230	1290	1170	1230
	26	1280	1360	1230	1300	1180	1250	1150	1210	1100	1140
	28	1200	1270	1150	1210	1110	1160	1080	1120	1020	1060
	30	1120	1180	1070	1120	1030	1070	1000	1040	946	974
	32	1050	1090	996	1030	956	987	925	951	871	891
	34	970	1000	921	946	882	903	851	868	798	811
	36	895	916	847	863	809	822	779	788	727	733
	38	822	834	776	784	739	743	709	711	659	659
	40	752	754	707	707	671	671	642	642	595	595
Properties										
$\phi_b M_{nx}$ (kip-ft)		344	452	313	410	290	379	277	361	249	321
$\phi_b M_{ny}$ (kip-ft)		237	304	217	277	204	258	184	231	169	210
$P_{ex}(K_x L_x)^2/10^4$ (kip-ft^2)		162	162	152	152	144	144	138	138	128	128
$P_{ey}(K_y L_y)^2/10^4$ (kip-ft^2)		162	162	152	152	144	144	138	138	128	128
r_{my} (in.)		5.40		5.40		5.40		5.40		5.40	
r_{mx}/r_{my} (in./in.)		1.00		1.00		1.00		1.00		1.00	

Steel Shape

Effective length in ft KL with respect to least radius of gyration r_{my}

F_y = 36 ksi
F_y = 50 ksi

COMPOSITE COLUMNS
W Shapes
f'_c = 8 ksi
All reinforcing steel is Grade 60
Axial design strength in kips

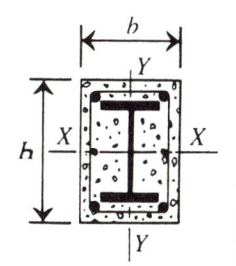

Size $b \times h$		16 in. × 16 in.								
Reinf. bars		4-#7 bars								
Ties		#3 bars spaced 10 in. c. to c.								
Steel Shape	Designation	**W 8**								
	Wt./ft	67		58		48		40		35	
	F_y	36	50	36	50	36	50	36	50	36	50
Effective length in ft KL with respect to least radius of gyration r_{my}	0	1640	1870	1570	1770	1490	1650	1420	1560	1390	1510
	6	1600	1820	1530	1720	1450	1610	1380	1520	1350	1460
	7	1590	1800	1520	1710	1440	1590	1370	1500	1330	1450
	8	1570	1790	1500	1690	1420	1570	1360	1480	1320	1430
	9	1550	1760	1480	1660	1400	1550	1340	1460	1300	1410
	10	1530	1740	1470	1640	1380	1530	1320	1440	1280	1380
	11	1510	1710	1440	1620	1360	1500	1300	1410	1260	1360
	12	1490	1680	1420	1590	1340	1470	1270	1380	1240	1330
	13	1470	1650	1400	1560	1320	1450	1250	1360	1210	1300
	14	1440	1620	1370	1530	1290	1410	1220	1320	1180	1270
	15	1420	1590	1350	1490	1260	1380	1200	1290	1160	1240
	16	1390	1550	1320	1460	1240	1350	1170	1260	1130	1210
	17	1360	1520	1290	1420	1210	1310	1140	1220	1100	1170
	18	1330	1480	1260	1380	1180	1280	1110	1190	1070	1140
	19	1300	1440	1230	1350	1150	1240	1080	1150	1040	1100
	20	1270	1400	1200	1310	1120	1200	1050	1120	1010	1060
	22	1200	1310	1130	1230	1050	1120	981	1040	940	989
	24	1130	1230	1060	1140	983	1040	914	961	873	912
	26	1060	1140	996	1060	915	963	847	883	806	836
	28	993	1060	926	977	846	884	779	807	739	761
	30	922	972	857	895	778	805	713	731	673	687
	32	852	888	789	815	712	729	648	659	609	617
	34	784	806	722	737	648	656	586	589	548	549
	36	717	729	657	663	586	586	525	525	490	490
	38	653	654	595	595	526	526	471	471	439	439
	40	590	590	537	537	475	475	425	425	397	397
Properties										
$\phi_b M_{nx}$ (kip-ft)		307	408	272	359	228	300	197	258	177	230
$\phi_b M_{ny}$ (kip-ft)		206	268	186	240	157	202	140	178	127	160
$P_{ex}(K_x L_x)^2/10^4$ (kip-ft²)		127	127	115	115	102	102	91.3	91.3	85.1	85.1
$P_{ey}(K_y L_y)^2/10^4$ (kip-ft²)		127	127	115	115	102	102	91.3	91.3	85.1	85.1
r_{my} (in.)		4.80		4.80		4.80		4.80		4.80	
r_{mx}/r_{my} (in./in.)		1.00		1.00		1.00		1.00		1.00	

4 - 98

COMPOSITE COLUMNS
Concrete Filled Steel Pipe
and Structural Tubing

Concentric design strengths in the tables that follow are tabulated for the effective lengths KL in feet, shown at the left of each table. They are applicable to axially loaded members with respect to their minor axis in accordance with Section I2.2 of the LRFD Specification. The tables apply to normal-weight concrete.

For discussion of the effective length, range of Kl/r, strength about the major axis, combined axial and bending strength and for sample problems, see "Composite Columns, General Notes."

The properties listed at the bottom of each table are for use in checking strength about the strong axis and in design for combined axial compression and bending.

The heavy horizontal lines within the tables indicate $Kl/r = 200$. No values are listed beyond these lines.

STEEL PIPE FILLED WITH CONCRETE

Design strengths for filled pipe are tabulated for $F_y = 36$ ksi and f'_c equal to 3.5 and 5 ksi. Steel pipe is manufactured to $F_y = 36$ ksi under ASTM A501 and to $F_y = 35$ ksi under ASTM A53 Types E or S, Grade B. Both are designed for 36 ksi yield stress.

STRUCTURAL TUBING FILLED WITH CONCRETE

Design strengths for square and rectangular structural tubing filled with concrete are tabulated for $F_y = 46$ ksi and f'_c equal to 3.5 and 5 ksi. Structural tubing is manufactured to $F_y = 46$ ksi under ASTM A500, Grade B.

COMPOSITE COLUMNS
Steel pipe
$f'_c = 3.5$ ksi
Axial design strength in kips

$F_y = 36$ ksi

	Nominal Diameter (in.)		12		10		8		
Steel Pipe	Wall thickness (in.)		0.500	0.375	0.500	0.365	0.875	0.500	0.322
	Wt./ft		65.42	49.56	54.74	40.48	72.42	43.39	28.55
	F_y					36 ksi			
Effective length in feet KL with respect to radius of gyration		0	862	733	681	564	746	507	384
		6	847	720	665	550	718	489	370
		7	842	716	660	545	708	482	365
		8	836	711	653	540	697	475	359
		9	829	705	646	534	684	467	353
		10	822	698	638	527	671	458	346
		11	814	691	629	520	656	448	338
		12	805	684	620	512	640	438	330
		13	796	675	610	503	623	427	322
		14	786	667	599	494	606	415	313
		15	775	658	587	485	588	403	304
		16	764	648	576	475	569	390	294
		17	752	638	563	464	549	377	284
		18	739	627	550	454	529	364	274
		19	727	616	537	442	509	351	264
		20	713	604	523	431	488	337	254
		22	686	580	495	407	447	309	232
		24	656	555	466	383	405	281	211
		26	626	529	436	358	365	254	191
		28	595	502	406	333	325	228	170
		30	563	475	376	308	288	202	151
		32	531	448	347	284	253	178	133
		34	499	420	318	260	224	158	118
		36	467	393	290	236	200	141	105
		38	436	366	263	214	179	126	94.3
		40	405	339	237	193	162	114	85.1
	Properties								
	r_m (in.)		4.33	4.38	3.63	3.67	2.76	2.88	2.94
	$\phi_b M_n$ (kip-ft)		203	155	142	106	143	89.2	60.0
	$P_e(KL)^2/10^4$ (kip-ft^2)		89.8	75.1	51.0	41.4	34.8	24.5	18.3

Note: Heavy line indicates Kl/r of 200.

$F_y = 36$ ksi

COMPOSITE COLUMNS
Steel pipe
$f'_c = 3.5$ ksi
Axial design strength in kips

Steel Pipe	Nominal Diameter (in.)	6			5			4		
	Wall thickness (in.)	0.864	0.432	0.280	0.750	0.375	0.258	0.674	0.337	0.237
	Wt./ft	53.16	28.57	18.97	38.55	20.78	14.62	27.54	14.98	10.79
	F_y	36 ksi								
Effective length in feet KL with respect to radius of gyration	0	525	323	244	379	233	182	268	164	129
	6	491	303	229	344	213	167	230	143	113
	7	479	297	224	332	206	161	218	136	107
	8	466	289	218	319	199	156	205	129	102
	9	451	281	212	305	191	149	191	121	95.3
	10	436	271	205	290	182	142	176	112	88.7
	11	419	262	197	274	173	135	162	104	82.0
	12	401	252	190	258	163	128	147	95.1	75.2
	13	383	241	182	241	154	120	132	86.5	68.5
	14	364	230	173	225	144	112	118	78.1	61.9
	15	345	219	165	208	134	105	105	70.0	55.5
	16	326	207	156	191	124	97.0	92.3	62.2	49.4
	17	306	196	147	175	114	89.4	81.7	55.1	43.7
	18	287	184	139	160	105	82.0	72.9	49.1	39.0
	19	268	173	130	145	95.7	74.9	65.4	44.1	35.0
	20	249	161	121	131	86.8	67.9	59.1	39.8	31.6
	22	213	139	105	108	71.7	56.1	48.8	32.9	26.1
	24	180	119	89.3	90.7	60.3	47.2		27.6	21.9
	26	153	101	76.1	77.3	51.4	40.2			
	28	132	87.2	65.6	66.6	44.3	34.7			
	30	115	75.9	57.1		38.6	30.2			
	32	101	66.7	50.2						
	34	89.5	59.1	44.5						
	36		52.7	39.7						

Properties									
r_m (in.)	2.06	2.19	2.25	1.72	1.84	1.88	1.37	1.48	1.51
$\phi_b M_n$ (kip-ft)	78.0	44.8	30.5	47.3	27.3	19.6	26.9	15.8	11.6
$P_e(KL)^2/10^4$ (kip-ft^2)	13.9	9.13	6.92	6.99	4.66	3.65	3.15	2.15	1.70

Note: Heavy line indicates Kl/r of 200.

F_y = 36 ksi

COMPOSITE COLUMNS
Steel pipe
f'_c = 5 ksi
Axial design strength in kips

Steel Pipe	Nominal Diameter (in.)		12		10		8		
	Wall thickness (in.)		0.500	0.375	0.500	0.365	0.875	0.500	0.322
	Wt./ft		65.42	49.56	54.74	40.48	72.42	43.39	28.55
	F_y					36 ksi			
		0	979	855	762	649	786	557	438
Effective length in feet KL with respect to radius of gyration		6	961	839	743	632	755	535	420
		7	955	833	736	626	745	528	414
		8	947	827	728	619	732	519	407
		9	939	819	720	611	719	510	399
		10	930	811	710	603	704	499	391
		11	920	802	699	594	688	488	382
		12	909	792	688	584	671	476	372
		13	897	781	676	573	653	463	361
		14	885	770	663	562	633	449	351
		15	872	758	649	550	614	435	339
		16	858	746	635	537	593	421	327
		17	844	733	620	524	572	406	315
		18	828	719	605	511	550	391	303
		19	813	705	589	497	528	375	291
		20	797	691	573	483	506	360	278
		22	763	660	540	454	461	328	253
		24	727	628	505	424	417	297	228
		26	691	596	471	394	374	266	203
		28	653	562	436	364	332	236	180
		30	615	528	401	334	292	208	158
		32	577	495	367	305	256	183	139
		34	539	461	334	276	227	162	123
		36	502	428	302	249	203	145	109
		38	465	395	272	224	182	130	98.2
		40	429	363	245	202	164	117	88.7
Properties									
r_m (in.)			4.33	4.38	3.63	3.67	2.76	2.88	2.94
$\phi_b M_n$ (kip-ft)			203	155	142	106	143	89.2	60.0
$P_e (KL)^2/10^4$ (kip-ft^2)			93.3	78.9	52.7	43.2	35.3	25.2	19.1

Note: Heavy line indicates Kl/r of 200.

		F_y = 36 ksi

COMPOSITE COLUMNS
Steel pipe
$f_c' = 5$ ksi
Axial design strength in kips

	Nominal Diameter (in.)		6			5			4	
	Wall thickness (in.)		0.864	0.432	0.280	0.750	0.375	0.258	0.674	0.337	0.237
	Wt./ft		53.16	28.57	18.97	38.55	20.78	14.62	27.54	14.98	10.79
	F_y						36 ksi				
		0	545	351	275	393	253	204	276	176	143
		6	509	329	257	356	230	185	237	153	124
		7	496	321	251	343	222	179	224	145	117
		8	482	312	244	330	214	172	210	137	110
		9	467	302	236	315	204	164	196	128	103
		10	450	292	228	299	195	156	180	118	95.5
		11	432	281	219	282	184	148	165	109	87.7
		12	414	269	209	265	173	139	149	99.3	80.0
		13	394	257	200	247	162	130	134	89.9	72.3
		14	374	245	190	230	151	121	120	80.7	64.8
		15	354	232	180	212	140	112	106	71.9	57.7
		16	334	219	169	195	129	103	93.0	63.5	50.8
		17	313	206	159	178	119	94.3	82.4	56.2	45.0
		18	293	193	149	162	108	85.9	73.5	50.1	40.2
		19	273	180	139	146	98.4	77.8	66.0	45.0	36.1
		20	253	168	129	132	88.8	70.2	59.5	40.6	32.5
		22	216	144	110	109	73.4	58.0	49.2	33.6	26.9
		24	181	121	92.6	91.5	61.7	48.8		28.2	22.6
		26	155	104	78.9	78.0	52.6	41.6			
		28	133	89.3	68.0	67.3	45.3	35.8			
		30	116	77.8	59.2		39.5	31.2			
		32	102	68.3	52.1						
		34	90.4	60.5	46.1						
		36		54.0	41.1						

Effective length in feet KL with respect to radius of gyration

		Properties									
	r_m (in.)		2.06	2.19	2.25	1.72	1.84	1.88	1.37	1.48	1.51
	$\phi_b M_n$ (kip-ft)		78.0	44.8	30.5	47.3	27.3	19.6	26.9	15.8	11.6
	$P_e (KL)^2/10^4$ (kip-ft^2)		14.0	9.35	7.18	7.06	4.77	3.78	3.18	2.19	1.75

Note: Heavy line indicates Kl/r of 200.

F_y = 46 ksi

COMPOSITE COLUMNS
Square structural tubing
f'_c = 3.5 ksi
Axial design strength in kips

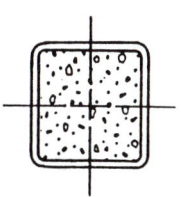

Steel Tube Nominal Size	16×16	14×14		12×12		10×10			
Thickness (in.)	½	½	⅜	½	⅜	⅝	½	⅜	⁵⁄₁₆
Wt./ft	103.30	89.68	68.31	76.07	58.10	76.33	62.46	47.90	40.35
F_y					46 ksi				
0	1760	1460	1230	1180	987	1070	922	766	686
6	1740	1440	1210	1160	970	1040	899	747	669
7	1730	1430	1210	1150	964	1030	891	741	663
8	1730	1420	1200	1140	957	1020	881	733	656
9	1720	1420	1190	1130	950	1010	871	724	648
10	1710	1410	1190	1120	941	996	859	715	640
11	1700	1400	1180	1110	931	981	847	705	630
12	1690	1380	1170	1100	921	965	833	693	620
13	1680	1370	1160	1090	910	948	819	681	610
14	1660	1360	1150	1070	898	930	803	669	598
15	1650	1350	1130	1060	886	910	787	655	586
16	1640	1330	1120	1040	873	890	770	641	574
17	1620	1320	1110	1030	859	870	752	627	561
18	1610	1300	1100	1010	844	848	734	612	547
19	1590	1280	1080	990	830	826	715	596	534
20	1570	1260	1070	972	814	803	696	580	519
22	1540	1230	1030	933	782	756	656	547	490
24	1500	1190	1000	892	748	708	615	513	459
26	1460	1150	966	850	713	659	573	479	429
28	1420	1100	930	807	676	610	531	444	398
30	1370	1060	892	763	640	561	489	410	367
32	1330	1010	854	718	602	514	448	376	337
34	1280	967	815	674	565	467	409	343	307
36	1230	920	775	629	528	423	370	311	279
38	1190	873	736	586	492	380	333	280	251
40	1140	826	696	543	456	343	301	253	227

Effective length in feet KL with respect to radius of gyration

Properties									
r_m (in.)	6.29	5.48	5.54	4.66	4.72	3.78	3.84	3.90	3.93
$\phi_b M_n$ (kip-ft)	604	456	352	329	255	268	223	174	148
$P_e(KL)^2/10^4$ (kip-ft²)	319	202	170	120	101	73.6	64.6	54.3	48.6

Note: Heavy line indicates Kl/r of 200.

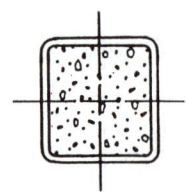

		F_y = 46 ksi

COMPOSITE COLUMNS
Square structural tubing
f'_c = 3.5 ksi
Axial design strength in kips

Steel Tube	Nominal Size	8x8					7x7				
	Thickness (in.)	5/8	1/2	3/8	5/16	1/4	1/2	3/8	5/16	1/4	3/16
	Wt./ft	59.32	48.85	37.60	31.84	25.82	42.05	32.59	27.59	22.42	17.08
	F_y	46 ksi									

Effective length in feet KL with respect to radius of gyration											
	0	796	685	566	503	439	574	473	420	364	307
	6	763	658	544	484	422	544	449	399	346	292
	7	752	648	536	477	416	533	441	391	340	286
	8	739	637	527	469	409	522	431	383	333	280
	9	724	625	518	461	402	509	421	374	325	274
	10	709	612	507	451	394	494	409	364	316	266
	11	692	598	495	441	385	479	397	353	307	259
	12	673	583	483	430	375	463	384	342	297	250
	13	654	567	470	419	365	446	371	330	287	242
	14	634	550	456	407	355	429	356	317	276	232
	15	613	532	442	394	344	411	342	304	265	223
	16	592	514	427	381	332	392	327	291	253	213
	17	569	495	412	367	321	373	312	278	242	204
	18	547	476	397	354	309	354	296	264	230	194
	19	524	457	381	340	297	335	281	251	218	184
	20	501	437	365	326	284	317	266	237	207	174
	22	454	398	333	297	259	279	235	210	183	154
	24	408	359	301	269	235	244	206	184	161	135
	26	364	321	270	241	211	210	178	160	140	118
	28	321	285	240	214	187	181	154	138	120	101
	30	281	250	211	189	165	158	134	120	105	88.3
	32	247	219	185	166	145	139	118	105	92.1	77.6
	34	218	194	164	147	128	123	104	93.4	81.6	68.7
	36	195	173	146	131	115	109	92.9	83.3	72.8	61.3
	38	175	156	131	118	103	98.2	83.4	74.8	65.3	55.0
	40	158	140	119	106	92.8	88.6	75.3	67.5	59.0	49.6

Properties											
r_m (in.)		2.96	3.03	3.09	3.12	3.15	2.62	2.68	2.71	2.74	2.77
$\phi_b M_n$ (kip-ft)		163	137	108	92.2	75.5	102	81.0	69.4	57.1	44.0
$P_e (KL)^2/10^4$ (kip-ft²)		33.8	30.1	25.4	22.8	19.9	19.0	16.2	14.5	12.7	10.7

Note: Heavy line indicates Kl/r of 200.

F_y = 46 ksi

COMPOSITE COLUMNS
Square structural tubing
$f'_c = 3.5$ ksi
Axial design strength in kips

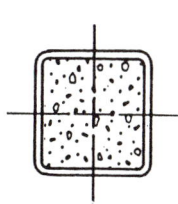

Steel Tube	Nominal Size	6x6					5x5				
	Thickness (in.)	½	⅜	5/16	¼	3/16	½	⅜	5/16	¼	3/16
	Wt./ft	35.24	27.48	23.34	19.02	14.53	28.43	22.37	19.08	15.62	11.97
	F_y	46 ksi									
Effective length in feet KL with respect to radius of gyration	0	468	385	341	295	247	367	302	267	231	192
	6	434	359	318	275	230	328	272	241	208	173
	7	422	349	310	268	225	315	262	232	201	167
	8	410	339	301	260	218	301	251	223	193	160
	9	395	328	291	252	211	285	239	212	184	153
	10	380	316	280	243	204	269	226	201	174	145
	11	364	303	269	233	195	252	212	189	164	137
	12	347	289	257	223	187	235	198	177	154	128
	13	330	275	245	213	178	217	184	165	143	120
	14	312	261	232	202	169	200	170	152	133	111
	15	293	246	219	191	160	183	156	140	122	102
	16	275	231	206	180	151	166	143	128	112	94.0
	17	257	217	194	168	141	150	130	117	102	85.7
	18	239	202	181	157	132	134	117	106	92.6	77.8
	19	221	188	168	146	123	121	105	95.0	83.3	70.1
	20	204	174	156	136	114	109	94.7	85.7	75.2	63.2
	22	171	147	132	115	97.0	90.0	78.3	70.8	62.2	52.3
	24	144	123	111	97.1	81.6	75.6	65.8	59.5	52.2	43.9
	26	123	105	94.6	82.7	69.5	64.4	56.1	50.7	44.5	37.4
	28	106	90.7	81.5	71.3	59.9	55.5	48.3	43.7	38.4	32.3
	30	92.2	79.0	71.0	62.1	52.2	48.4	42.1	38.1	33.4	28.1
	32	81.0	69.4	62.4	54.6	45.9				29.4	24.7
	34	71.8	61.5	55.3	48.4	40.6					
	36	64.0	54.8	49.3	43.1	36.3					
	38			44.3	38.7	32.5					
Properties											
r_m (in.)		2.21	2.27	2.30	2.33	2.36	1.80	1.86	1.89	1.92	1.95
$\phi_b M_n$ (kip-ft)		72.0	57.8	49.8	41.2	31.9	47.3	38.6	33.5	27.9	21.7
$P_e (KL)^2 / 10^4$ (kip-ft²)		11.1	9.54	8.58	7.50	6.30	5.84	5.08	4.60	4.04	3.39

Note: Heavy line indicates Kl/r of 200.

4 - 106

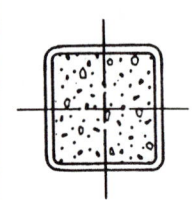

		$F_y = 46$ ksi

COMPOSITE COLUMNS
Square structural tubing
$f'_c = 3.5$ ksi
Axial design strength in kips

Steel Tube	Nominal Size	4×4				
	Thickness (in.)	½	⅜	5/16	¼	3/16
	Wt./ft	21.63	17.27	14.83	12.21	9.42
	F_y	46 ksi				

Effective length in feet KL with respect to radius of gyration		½	⅜	5/16	¼	3/16
	0	271	225	199	171	141
	6	225	189	169	146	121
	7	211	178	159	137	114
	8	195	166	148	129	107
	9	179	153	137	119	99.2
	10	162	140	125	109	91.2
	11	145	126	114	99.5	83.2
	12	129	113	102	89.8	75.2
	13	114	100	91.3	80.3	67.4
	14	98.8	88.3	80.6	71.1	59.9
	15	86.1	77.0	70.5	62.4	52.7
	16	75.7	67.7	61.9	54.8	46.3
	17	67.0	60.0	54.9	48.6	41.0
	18	59.8	53.5	48.9	43.3	36.6
	19	53.7	48.0	43.9	38.9	32.8
	20	48.4	43.3	39.6	35.1	29.6
	22	40.0	35.8	32.8	29.0	24.5
	24		30.1	27.5	24.4	20.6

Properties						
r_m (in.)		1.39	1.45	1.48	1.51	1.54
$\phi_b M_n$ (kip-ft)		27.7	23.2	20.4	17.1	13.5
$P_e (KL)^2/10^4$ (kip-ft²)		2.60	2.33	2.13	1.88	1.59

Note: Heavy line indicates Kl/r of 200.

AMERICAN INSTITUTE OF STEEL CONSTRUCTION

F_y = 46 ksi

COMPOSITE COLUMNS
Square structural tubing
$f'_c = 5$ ksi
Axial design strength in kips

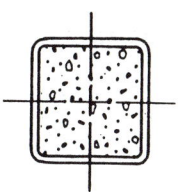

	Nominal Size	16x16	14x14		12x12		10x10			
Steel Tube	Thickness (in.)	½	½	⅜	½	⅜	⅝	½	⅜	5⁄16
	Wt./ft	103.30	89.68	68.31	76.07	58.10	76.33	62.46	47.90	40.35
	F_y					46 ksi				
Effective length in feet KL with respect to radius of gyration	0	2000	1640	1420	1310	1120	1150	1010	859	781
	6	1980	1620	1400	1290	1100	1120	983	836	760
	7	1970	1610	1390	1280	1100	1110	973	828	753
	8	1960	1600	1380	1270	1090	1100	962	819	744
	9	1950	1590	1370	1260	1080	1080	950	808	735
	10	1940	1580	1360	1240	1070	1070	937	797	724
	11	1930	1570	1350	1230	1060	1050	922	785	713
	12	1920	1550	1340	1220	1040	1030	907	771	700
	13	1900	1540	1330	1200	1030	1010	890	757	687
	14	1890	1520	1310	1180	1010	994	872	742	673
	15	1870	1500	1300	1170	999	973	853	726	659
	16	1850	1490	1280	1150	983	950	834	709	644
	17	1840	1470	1270	1130	966	927	814	692	628
	18	1820	1450	1250	1110	949	903	793	674	611
	19	1800	1430	1230	1090	931	878	771	656	595
	20	1780	1410	1210	1060	912	853	749	637	577
	22	1730	1360	1180	1020	873	801	703	598	542
	24	1690	1320	1130	972	831	747	657	558	506
	26	1640	1270	1090	923	789	693	610	518	469
	28	1590	1220	1050	873	746	639	562	478	432
	30	1530	1160	1000	822	702	585	516	438	396
	32	1480	1110	953	771	657	533	470	399	361
	34	1420	1050	905	720	614	483	426	362	326
	36	1360	999	857	669	570	434	383	325	293
	38	1310	944	809	620	528	390	344	292	263
	40	1250	889	762	572	486	352	311	264	238
						Properties				
r_m (in.)		6.29	5.48	5.54	4.66	4.72	3.78	3.84	3.90	3.93
$\phi_b M_n$ (kip-ft)		604	456	352	329	255	268	223	174	148
$P_e(KL)^2/10^4$ (kip-ft²)		334	211	180	125	106	75.5	66.7	56.6	51.0

Note: Heavy line indicates Kl/r of 200.

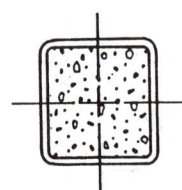

F_y = 46 ksi

COMPOSITE COLUMNS
Square structural tubing
f'_c = 5 ksi
Axial design strength in kips

	Nominal Size	8×8					7×7				
Steel Tube	Thickness (in.)	5/8	1/2	3/8	5/16	1/4	1/2	3/8	5/16	1/4	3/16
	Wt./ft	59.32	48.85	37.60	31.84	25.82	42.05	32.59	27.59	22.42	17.08
	F_y	46 ksi									
	0	845	738	622	562	500	612	515	464	410	355
	6	809	707	597	539	479	579	488	439	388	335
	7	797	696	588	531	472	568	478	430	380	329
	8	782	684	578	522	463	555	467	420	372	321
	9	766	671	566	511	454	540	455	410	362	313
	10	749	656	554	500	444	524	442	398	352	304
	11	730	640	541	488	433	508	428	386	341	294
	12	710	623	526	475	422	490	413	372	329	284
	13	689	605	511	461	409	471	398	358	317	273
	14	667	586	495	447	397	452	382	344	304	262
	15	644	566	479	432	383	432	365	329	291	250
	16	621	546	462	417	370	412	349	314	277	238
	17	597	525	444	401	355	391	331	299	263	226
	18	572	504	427	385	341	370	314	283	250	214
	19	547	482	409	369	327	350	297	268	236	202
	20	522	461	391	352	312	329	280	252	222	190
	22	472	417	354	319	282	289	246	222	196	167
	24	422	375	318	287	253	251	214	193	170	145
	26	374	333	283	255	225	215	184	165	146	124
	28	329	293	250	225	198	185	158	143	126	107
	30	286	256	218	196	173	161	138	124	109	93.1
	32	252	225	192	173	152	142	121	109	96.2	81.9
	34	223	199	170	153	135	125	107	96.8	85.2	72.5
	36	199	178	151	136	120	112	95.7	86.3	76.0	64.7
	38	178	160	136	122	108	100	85.9	77.5	68.2	58.0
	40	161	144	123	110	97.3	90.6	77.5	69.9	61.5	52.4

Effective length in feet KL with respect to radius of gyration

Properties											
r_m (in.)		2.96	3.03	3.09	3.12	3.15	2.62	2.68	2.71	2.74	2.77
$\phi_b M_n$ (kip-ft)		163	137	108	92.2	75.5	102	81.0	69.4	57.1	44.0
$P_e (KL)^2/10^4$ (kip-ft²)		34.5	30.9	26.3	23.7	20.9	19.5	16.6	15.0	13.2	11.2

Note: Heavy line indicates Kl/r of 200.

F_y = 46 ksi

COMPOSITE COLUMNS
Square structural tubing
f'_c = 5 ksi
Axial design strength in kips

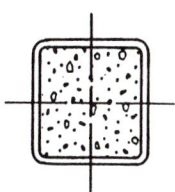

Steel Tube	Nominal Size	6x6					5x5				
	Thickness (in.)	½	⅜	⁵⁄₁₆	¼	³⁄₁₆	½	⅜	⁵⁄₁₆	¼	³⁄₁₆
	Wt./ft	35.24	27.48	23.34	19.02	14.53	28.43	22.37	19.08	15.62	11.97
	F_y	46 ksi									
Effective length in feet KL with respect to radius of gyration	0	494	415	372	328	281	384	322	288	252	215
	6	458	385	345	304	261	342	288	259	227	193
	7	445	374	336	296	254	328	277	249	218	185
	8	431	363	326	287	246	313	265	238	208	177
	9	415	350	315	277	237	296	251	226	198	169
	10	399	337	302	266	228	279	237	213	187	159
	11	381	322	290	255	218	261	222	200	176	149
	12	363	307	276	243	208	242	207	187	164	140
	13	344	291	262	231	197	224	192	173	152	129
	14	324	275	248	218	186	205	177	160	141	119
	15	305	259	233	205	175	187	162	146	129	109
	16	285	243	219	193	164	170	147	133	118	99.8
	17	266	227	204	180	153	153	133	121	106	90.4
	18	246	211	190	167	143	136	119	109	95.9	81.3
	19	228	195	176	155	132	122	107	97.4	86.0	73.0
	20	209	180	162	143	122	111	96.8	87.9	77.7	65.9
	22	175	151	136	120	102	91.4	80.0	72.7	64.2	54.5
	24	147	127	114	101	85.6	76.8	67.2	61.1	53.9	45.8
	26	125	108	97.5	85.9	72.9	65.4	57.3	52.0	45.9	39.0
	28	108	93.0	84.1	74.0	62.9	56.4	49.4	44.9	39.6	33.6
	30	93.9	81.0	73.2	64.5	54.8	49.1	43.0	39.1	34.5	29.3
	32	82.6	71.2	64.4	56.7	48.1				30.3	25.7
	34	73.1	63.1	57.0	50.2	42.6					
	36	65.2	56.3	50.9	44.8	38.0					
	38			45.6	40.2	34.1					

Properties											
r_m (in.)		2.21	2.27	2.30	2.33	2.36	1.80	1.86	1.89	1.92	1.95
$\phi_b M_n$ (kip-ft)		72.0	57.8	49.8	41.2	31.9	47.3	38.6	33.5	27.9	21.7
$P_e (KL)^2/10^4$ (kip-ft²)		11.3	9.78	8.84	7.79	6.61	5.93	5.19	4.72	4.17	3.54

Note: Heavy line indicates Kl/r of 200.

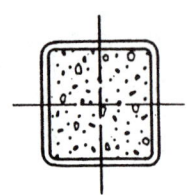

	$F_y = 46$ ksi

COMPOSITE COLUMNS
Square structural tubing
$f'_c = 5$ ksi
Axial design strength in kips

	Nominal Size		4x4				
Steel Tube	Thickness (in.)		½	⅜	⁵⁄₁₆	¼	³⁄₁₆
	Wt./ft		21.63	17.27	14.83	12.21	9.42
	F_y		46 ksi				
		0	280	236	211	184	156
		6	232	198	178	156	132
		7	217	186	167	146	124
		8	200	172	155	136	115
		9	183	158	143	126	107
Effective length in feet KL with respect to radius of gyration		10	166	144	131	115	97.6
		11	148	130	118	104	88.5
		12	131	116	106	93.7	79.5
		13	115	103	93.9	83.3	70.8
		14	102	89.9	82.5	73.3	62.4
		15	87.1	78.3	71.9	64.0	54.5
		16	76.5	68.8	63.2	56.3	47.9
		17	67.8	61.0	56.0	49.8	42.4
		18	60.5	54.4	49.9	44.5	37.9
		19	54.3	48.8	44.8	39.9	34.0
		20	49.0	44.1	40.5	36.0	30.7
		22	40.5	36.4	33.4	29.8	25.3
		24		30.6	28.1	25.0	21.3

Properties					
r_m (in.)	1.39	1.45	1.48	1.51	1.54
$\phi_b M_n$ (kip-ft)	27.7	23.2	20.4	17.1	13.5
$P_e (KL)^2/10^4$ (kip-ft²)	2.63	2.36	2.17	1.93	1.65

Note: Heavy line indicates Kl/r of 200.

F_y = 46 ksi

COMPOSITE COLUMNS
Rectangular structural tubing
$f_c' = 3.5$ ksi
Axial design strength in kips

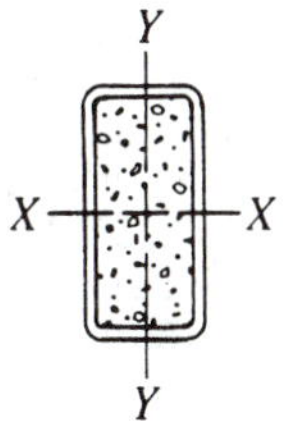

Steel Tube	Nominal Size	16x12	16x8	14x10		12x8			12x6		
	Thickness (in.)	½	½	½	⅜	⅝	½	⅜	⅝	½	⅜
	Wt./ft	89.68	76.07	76.07	58.10	76.33	62.46	47.90	67.82	55.66	42.79
	F_y	46 ksi									

Effective length in feet KL with respect to radius of gyration

KL	16x12	16x8	14x10 (½)	14x10 (⅜)	12x8 (⅝)	12x8 (½)	12x8 (⅜)	12x6 (⅝)	12x6 (½)	12x6 (⅜)
0	1450	1140	1170	977	1060	912	756	908	778	641
6	1420	1100	1140	955	1020	879	730	850	731	603
7	1420	1090	1130	947	1010	868	720	831	714	590
8	1410	1070	1120	938	992	855	710	809	696	575
9	1400	1050	1110	928	974	840	698	784	675	558
10	1380	1040	1100	916	955	824	685	758	653	541
11	1370	1010	1080	904	935	807	670	729	630	522
12	1360	993	1070	891	913	788	655	700	605	502
13	1340	969	1050	876	889	768	639	669	579	481
14	1320	945	1030	861	865	748	622	637	553	459
15	1310	919	1010	845	839	726	604	604	525	437
16	1290	892	990	829	813	704	586	572	498	415
17	1270	864	969	811	785	680	567	538	470	392
18	1250	836	948	793	757	657	547	505	442	370
19	1230	807	925	774	729	633	528	473	414	347
20	1210	777	902	755	700	608	507	441	387	325
22	1160	717	854	715	642	558	467	379	334	282
24	1110	657	804	674	583	509	426	321	284	241
26	1070	597	754	632	526	460	385	273	242	205
28	1010	539	703	589	470	412	346	235	209	177
30	962	482	652	547	417	366	308	205	182	154
32	910	428	601	505	367	323	272	180	160	135
34	857	379	552	464	325	286	241	160	142	120
36	804	338	504	424	290	255	215	142	126	107
38	752	303	458	385	260	229	193	128	113	96.0
40	701	274	414	348	235	207	174		102	86.6

Properties

	16x12	16x8	14x10 (½)	14x10 (⅜)	12x8 (⅝)	12x8 (½)	12x8 (⅜)	12x6 (⅝)	12x6 (½)	12x6 (⅜)
r_{my} (in.)	4.84	3.30	4.02	4.08	3.14	3.20	3.26	2.37	2.42	2.48
r_{mx}/r_{my}	1.25	1.72	1.30	1.29	1.37	1.37	1.37	1.74	1.73	1.72
$\phi_b M_{nx}$ (kip-ft)	497	390	363	281	300	250	195	251	210	165
$\phi_b M_{ny}$ (kip-ft)	409	241	288	224	226	189	147	153	129	101
$P_{ex}(K_x L_x)^2/10^4$ (kip-ft²)	245	174	149	125	95.3	83.4	69.9	74.8	65.7	55.0
$P_{ey}(K_y L_y)^2/10^4$ (kip-ft²)	157	58.8	88.8	74.7	50.4	44.4	37.4	24.8	22.0	18.6

Note: Heavy line indicates Kl/r of 200.

AMERICAN INSTITUTE OF STEEL CONSTRUCTION

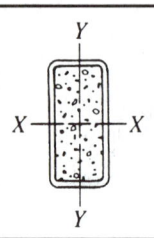

	F_y = 46 ksi

COMPOSITE COLUMNS
Rectangular structural tubing
$f'_c = 3.5$ ksi
Axial design strength in kips

	Nominal Size	10x6				8x6			
	Thickness (in.)	5/8	1/2	3/8	5/16	1/2	3/8	5/16	1/4
	Wt./ft	59.32	48.85	37.69	31.84	42.05	32.58	27.59	22.42
	F_y				46 ksi				
	0	786	675	556	493	571	470	417	362
	6	734	632	521	463	533	440	390	339
	7	717	617	510	453	520	430	381	331
	8	697	601	496	441	505	418	371	322
	9	675	582	482	428	489	405	360	312
	10	651	563	466	414	472	391	347	302
	11	626	542	449	399	453	376	334	291
	12	599	520	431	384	434	360	321	279
	13	572	496	413	367	413	344	306	266
	14	544	473	393	350	393	327	291	254
	15	515	449	374	333	371	310	276	241
	16	486	424	354	316	350	293	261	227
	17	456	399	334	298	329	276	246	214
	18	427	375	314	280	307	258	231	201
	19	399	350	294	263	286	241	216	188
	20	370	327	275	246	266	224	201	175
	22	316	280	237	212	226	192	172	150
	24	267	237	201	181	190	162	146	127
	26	227	202	172	154	162	138	124	108
	28	196	174	148	133	140	119	107	93.5
	30	171	152	129	116	122	104	93.2	81.5
	32	150	133	113	102	107	91.2	81.9	71.6
	34	133	118	100	90.0	94.9	80.8	72.5	63.4
	36	118	105	89.5	80.3	84.7	72.1	64.7	56.6
	38	106	94.6	80.3	72.0	76.0	64.7	58.1	50.8
	40			72.5	65.0				45.8

(Effective length in feet KL with respect to radius of gyration)

Properties									
r_{my} (in.)		2.32	2.37	2.43	2.46	2.31	2.36	2.39	2.42
r_{mx}/r_{my}		1.50	1.50	1.49	1.49	1.25	1.25	1.25	1.25
$\phi_b M_{nx}$ (kip-ft)		187	157	124	106	111	88.3	75.6	62.2
$\phi_b M_{ny}$ (kip-ft)		130	110	86.9	74.3	91.0	72.4	62.1	51.1
$P_{ex}(K_x L_x)^2/10^4$ (kip-ft^2)		46.4	41.0	34.5	30.8	23.1	19.6	17.6	15.3
$P_{ey}(K_y L_y)^2/10^4$ (kip-ft^2)		20.6	18.3	15.6	14.0	14.7	12.5	11.2	9.84

Note: Heavy line indicates Kl/r of 200.

F_y = 46 ksi

COMPOSITE COLUMNS
Rectangular structural tubing
f'_c = 3.5 ksi
Axial design strength in kips

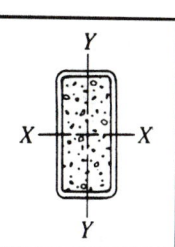

Steel Tube	Nominal Size		8x4				7x5				
	Thickness (in.)		½	⅜	⁵⁄₁₆	¼	½	⅜	⁵⁄₁₆	¼	³⁄₁₆
	Wt./ft		35.24	27.48	23.34	19.02	35.24	27.48	23.34	19.02	14.53
	F_y						46 ksi				

Effective length in feet KL with respect to radius of gyration											
	0	457	375	331	285	465	383	338	292	244
	6	393	325	288	248	420	347	308	266	223
	7	372	308	273	236	405	335	297	257	215
	8	349	290	258	223	389	322	286	248	207
	9	325	271	241	209	371	308	273	237	198
	10	300	252	224	194	351	293	260	225	189
	11	274	231	206	179	331	277	246	214	179
	12	249	211	189	164	310	260	232	201	168
	13	224	191	171	149	289	243	217	188	158
	14	200	171	154	135	268	226	202	176	147
	15	177	153	138	121	247	209	187	163	137
	16	156	135	122	107	227	192	172	150	126
	17	138	119	108	94.8	207	176	158	138	116
	18	123	107	96.3	84.5	187	160	144	126	106
	19	110	95.6	86.4	75.9	169	145	131	114	96.3
	20	99.5	86.3	78.0	68.5	152	131	118	103	87.0
	22	82.3	71.3	64.4	56.6	126	108	97.5	85.4	71.9
	24	69.1	59.9	54.2	47.5	106	91.0	82.0	71.8	60.4
	26		51.1	46.1	40.5	90.1	77.5	69.8	61.2	51.5
	28					77.7	66.8	60.2	52.7	44.4
	30					67.7	58.2	52.5	45.9	38.7
	32						51.2	46.1	40.4	34.0

Properties										
r_{my} (in.)		1.54	1.60	1.62	1.65	1.90	1.95	1.98	2.01	2.04
r_{mx}/r_{my}		1.75	1.73	1.73	1.72	1.31	1.30	1.30	1.30	1.29
$\phi_b M_{nx}$ (kip-ft)		85.3	68.6	59.1	48.8	79.5	63.8	55.0	45.4	35.1
$\phi_b M_{ny}$ (kip-ft)		51.8	41.9	36.3	30.1	62.8	50.5	43.6	36.0	27.9
$P_{ex}(K_x L_x)^2/10^4$ (kip-ft²)		16.3	13.9	12.5	10.9	13.9	11.9	10.7	9.33	7.83
$P_{ey}(K_y L_y)^2/10^4$ (kip-ft²)		5.34	4.63	4.18	3.67	8.17	7.03	6.33	5.55	4.67

Note: Heavy line indicates Kl/r of 200.

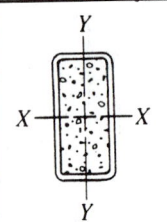

	F_y = 46 ksi

COMPOSITE COLUMNS
Rectangular structural tubing
f'_c = 3.5 ksi
Axial design strength in kips

Steel Tube	Nominal Size		6x4				
	Thickness (in.)		½	⅜	⁵⁄₁₆	¼	³⁄₁₆
	Wt./ft		28.43	22.37	19.05	15.62	11.97
	F_y		46 ksi				

Effective length in feet KL with respect to radius of gyration						
	0	365	300	265	228	189
	6	309	257	228	197	164
	7	292	243	216	187	155
	8	272	228	203	176	146
	9	252	212	189	164	137
	10	231	196	175	152	127
	11	210	179	160	140	117
	12	189	162	146	127	106
	13	168	146	131	115	96.3
	14	149	130	118	103	86.5
	15	130	115	104	91.6	77.1
	16	115	101	91.7	80.7	68.0
	17	101	89.6	81.3	71.5	60.3
	18	90.5	79.9	72.5	63.8	53.7
	19	81.2	71.7	65.1	57.3	48.2
	20	73.3	64.7	58.7	51.7	43.5
	22	61.1	53.5	48.5	42.7	36.0
	24	51.3	44.9	40.8	35.9	30.2
	26			34.7	30.6	25.8

Properties					
r_{my} (in.)	1.48	1.54	1.57	1.60	1.63
r_{mx}/r_{my}	1.39	1.38	1.37	1.37	1.37
$\phi_b M_{nx}$ (kip-ft)	53.1	43.3	37.5	31.2	24.3
$\phi_b M_{ny}$ (kip-ft)	39.7	32.6	28.3	23.6	18.4
$P_{ex}(K_x L_x)^2/10^4$ (kip-ft^2)	7.59	6.59	5.95	5.21	4.37
$P_{ey}(K_y L_y)^2/10^4$ (kip-ft^2)	3.92	3.47	3.15	2.77	2.34

Note: Heavy line indicates Kl/r of 200.

F_y = 46 ksi

COMPOSITE COLUMNS
Rectangular structural tubing
f'_c = 5 ksi
Axial design strength in kips

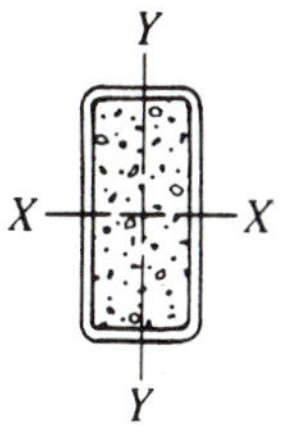

Steel Tube	Nominal Size	16x12	16x8	14x10		12x8			12x6		
	Thickness (in.)	½	½	½	⅜	⅝	½	⅜	⅝	½	⅜
	Wt./ft	89.68	76.07	76.07	58.10	76.33	62.46	47.90	67.82	55.66	42.79
	F_y					46 ksi					

Effective length in feet KL with respect to radius of gyration											
0	1630	1250	1300	1110	1140	995	845	963	837	705
6	1600	1210	1260	1080	1090	957	813	900	784	660
7	1590	1190	1250	1070	1080	944	801	878	766	645
8	1580	1170	1240	1060	1060	929	789	854	745	628
9	1560	1150	1230	1050	1040	912	774	827	722	609
10	1550	1130	1210	1040	1020	894	759	798	697	588
11	1530	1110	1190	1020	998	874	742	767	671	566
12	1520	1080	1170	1000	973	853	724	734	643	543
13	1500	1060	1150	986	947	830	705	701	614	519
14	1480	1030	1130	968	920	806	685	666	585	495
15	1460	997	1110	948	891	782	664	631	555	469
16	1430	966	1090	928	862	756	642	595	524	444
17	1410	934	1060	907	831	730	620	559	493	418
18	1390	901	1040	885	800	703	597	524	463	392
19	1360	868	1010	862	769	676	574	489	432	367
20	1340	834	982	839	737	648	550	454	402	342
22	1280	766	926	791	673	592	503	388	345	294
24	1230	698	869	742	609	537	456	327	292	249
26	1170	630	811	692	546	482	410	279	248	212
28	1110	565	752	642	486	429	365	240	214	183
30	1050	502	694	592	428	379	322	209	187	159
32	986	442	637	542	376	333	283	184	164	140
34	925	392	581	495	333	295	251	163	145	124
36	864	350	527	448	297	263	224	145	130	111
38	803	314	475	404	267	236	201	130	116	99.3
40	744	283	429	365	241	213	181		105	89.6

Properties											
r_{my} (in.)	4.84	3.30	4.02	4.08	3.14	3.20	3.26	2.37	2.42	2.48
r_{mx}/r_{my}	1.25	1.72	1.30	1.29	1.37	1.37	1.37	1.74	1.73	1.72
$\phi_b M_{nx}$ (kip-ft)	497	390	363	281	300	250	195	251	210	165
$\phi_b M_{ny}$ (kip-ft)	409	241	288	224	226	189	147	153	129	101
$P_{ex}(K_x L_x)^2/10^4$ (kip-ft^2)	256	180	155	131	97.6	86.0	72.7	76.3	67.3	56.8
$P_{ey}(K_y L_y)^2/10^4$ (kip-ft^2)	164	60.8	92.1	78.2	51.7	45.7	38.9	25.3	22.5	19.2

Note: Heavy line indicates Kl/r of 200.

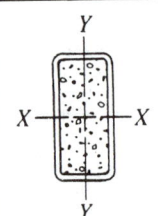

COMPOSITE COLUMNS
Rectangular structural tubing
$f'_c = 5$ ksi
Axial design strength in kips

$F_y = 46$ ksi

	Nominal Size		10x6				8x6			
Steel Tube	Thickness (in.)		$^5/_8$	$^1/_2$	$^3/_8$	$^5/_{16}$	$^1/_2$	$^3/_8$	$^5/_{16}$	$^1/_4$
	Wt./ft		59.32	48.85	37.69	31.84	42.05	32.58	27.59	22.42
	F_y					46 ksi				
Effective length in feet KL with respect to radius of gyration		0	831	723	608	548	609	511	460	406
		6	774	675	569	512	567	477	429	379
		7	755	659	555	500	552	465	418	369
		8	733	640	540	486	536	451	406	359
		9	709	620	523	471	518	437	393	347
		10	683	598	504	455	499	421	379	334
		11	656	575	485	437	478	404	363	321
		12	627	550	465	419	457	386	347	307
		13	597	525	444	400	435	368	331	292
		14	567	498	422	380	412	349	314	277
		15	535	472	399	360	389	329	297	262
		16	504	445	377	340	365	310	279	247
		17	473	418	354	320	342	291	262	231
		18	441	391	332	299	319	272	245	216
		19	411	364	310	279	296	253	228	201
		20	381	338	288	260	274	234	211	186
		22	323	288	246	222	232	199	179	158
		24	272	243	208	187	195	167	151	133
		26	231	207	177	160	166	142	128	113
		28	200	179	153	138	143	123	111	97.6
		30	174	156	133	120	125	107	96.5	85.0
		32	153	137	117	105	109	93.9	84.8	74.7
		34	135	121	104	93.4	97.0	83.2	75.1	66.2
		36	121	108	92.3	83.3	86.5	74.2	67.0	59.0
		38	108	96.9	82.9	74.8	77.6	66.6	60.1	53.0
		40			74.8	67.5				47.8
Properties										
r_{my} (in.)			2.32	2.37	2.43	2.46	2.31	2.36	2.39	2.42
r_{mx}/r_{my}			1.50	1.50	1.49	1.49	1.25	1.25	1.25	1.25
$\phi_b M_{nx}$ (kip-ft)			187	157	124	106	111	88.3	75.6	62.2
$\phi_b M_{ny}$ (kip-ft)			130	110	86.9	74.3	91.0	72.4	62.1	51.1
$P_{ex}(K_x L_x)^2/10^4$ (kip-ft^2)			47.3	42.0	35.6	32.0	23.7	20.2	18.2	16.0
$P_{ey}(K_y L_y)^2/10^4$ (kip-ft^2)			21.0	18.8	16.1	14.5	15.0	12.9	11.6	10.3

Note: Heavy line indicates Kl/r of 200.

F_y = 46 ksi

COMPOSITE COLUMNS
Rectangular structural tubing
$f'_c = 5$ ksi
Axial design strength in kips

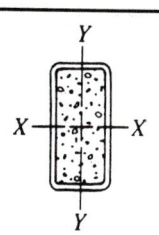

Steel Tube	Nominal Size		8×4				7×5			
	Thickness (in.)	½	⅜	5⁄16	¼	½	⅜	5⁄16	¼	3⁄16
	Wt./ft	35.24	27.48	23.34	19.02	35.24	27.48	23.34	19.02	14.53
	F_y					46 ksi				
	0	480	400	358	313	491	411	369	324	277
	6	410	345	309	271	442	372	333	293	251
	7	388	326	293	257	426	358	322	283	242
	8	363	307	275	241	407	343	308	271	232
	9	337	286	257	225	388	327	294	259	221
	10	310	264	237	209	367	310	279	245	210
	11	283	242	218	192	345	293	263	232	198
	12	256	220	198	174	323	274	247	217	185
	13	230	198	179	158	300	256	230	203	173
	14	204	177	160	141	278	237	214	188	160
	15	180	157	142	126	255	218	197	174	148
	16	158	138	125	111	233	200	181	159	136
	17	140	122	111	98.0	212	182	165	145	124
	18	125	109	98.9	87.4	191	165	150	132	112
	19	112	97.8	88.8	78.4	172	149	135	119	101
	20	101	88.2	80.1	70.8	155	134	122	107	91.1
	22	83.6	72.9	66.2	58.5	128	111	100	88.6	75.3
	24	70.2	61.3	55.6	49.2	108	93.2	84.4	74.5	63.3
	26		52.2	47.4	41.9	91.7	79.5	71.9	63.4	53.9
	28					79.1	68.5	62.0	54.7	46.5
	30					68.9	59.7	54.0	47.7	40.5
	32						52.5	47.5	41.9	35.6

(Effective length in feet KL with respect to radius of gyration)

Properties										
r_{my} (in.)		1.54	1.60	1.62	1.65	1.90	1.95	1.98	2.01	2.04
r_{mx}/r_{my}		1.75	1.73	1.73	1.72	1.31	1.30	1.30	1.30	1.29
$\phi_b M_{nx}$ (kip-ft)		85.3	68.6	59.1	48.8	79.5	63.8	55.0	45.4	35.1
$\phi_b M_{ny}$ (kip-ft)		51.8	41.9	36.3	30.1	62.8	50.5	43.6	36.0	27.9
$P_{ex}(K_x L_x)^2/10^4$ (kip-ft²)		16.6	14.2	12.8	11.2	14.2	12.2	11.0	9.68	8.20
$P_{ey}(K_y L_y)^2/10^4$ (kip-ft²)		5.43	4.73	4.30	3.80	8.32	7.21	6.52	5.75	4.89

Note: Heavy line indicates Kl/r of 200.

	F_y = 46 ksi

COMPOSITE COLUMNS
Rectangular structural tubing
f'_c = 5 ksi
Axial design strength in kips

Steel Tube	Nominal Size		6x4				
	Thickness (in.)		½	⅜	5⁄16	¼	3⁄16
	Wt./ft		28.43	22.37	19.05	15.62	11.97
	F_y		46 ksi				

Effective length in feet KL with respect to radius of gyration						
0	381	318	284	249	211
6	322	271	243	213	181
7	303	256	230	202	171
8	282	240	215	189	161
9	260	222	200	176	149
10	238	204	184	162	138
11	216	186	168	148	126
12	193	168	152	134	114
13	172	151	137	120	102
14	151	134	121	107	91.2
15	132	117	107	94.7	80.5
16	116	103	94.0	83.2	70.8
17	103	91.4	83.3	73.7	62.7
18	91.8	81.5	74.3	65.8	55.9
19	82.4	73.2	66.7	59.0	50.2
20	74.4	66.0	60.2	53.3	45.3
22	61.5	54.6	49.7	44.0	37.4
24	51.7	45.9	41.8	37.0	31.5
26			35.6	31.5	26.8

Properties						
r_{my} (in.)		1.48	1.54	1.57	1.60	1.63
r_{mx}/r_{my}		1.39	1.38	1.37	1.37	1.37
$\phi_b M_{nx}$ (kip-ft)		53.1	43.3	37.5	31.2	24.3
$\phi_b M_{ny}$ (kip-ft)		39.7	32.6	28.3	23.6	18.4
$P_{ex}(K_x L_x)^2/10^4$ (kip-ft^2)		7.70	6.73	6.10	5.38	4.55
$P_{ey}(K_y L_y)^2/10^4$ (kip-ft^2)		3.97	3.54	3.23	2.86	2.43

Note: Heavy line indicates Kl/r of 200.

PART 5
Connections

AMERICAN INSTITUTE OF STEEL CONSTRUCTION

BOLTS, THREADED PARTS AND RIVETS
Tension
Design loads in kips

TABLE I-A. BOLTS AND RIVETS
Tension on gross (nominal) area

ASTM Designation	ϕF_t Ksi	\<th colspan="8">Nominal Diameter d, In.</th>
		5/8	3/4	7/8	1	1 1/8	1 1/4	1 3/8	1 1/2
		\<td colspan="8">Area (Based on Nominal Diameter), In.²</td>
		0.3068	0.4418	0.6013	0.7854	0.9940	1.227	1.485	1.767
A307 bolts	33.8	10.4	14.9	20.3	26.5	33.5	41.4	50.1	59.6
A325 bolts	67.5	20.7	29.8	40.6	53.0	67.1	82.8	100.2	119.3
A490 bolts	84.4	25.9	37.3	50.7	66.3	83.9	103.5	125.3	149.1
A502-1 rivets	33.8	10.4	14.9	20.3	26.5	33.5	41.4	50.1	59.6
A502-2, 3 rivets	45.0	13.8	19.9	27.1	35.3	44.7	55.2	66.8	79.5

The above table lists ASTM specified materials that are generally for use as structural fasteners.
For dynamic and fatigue loading, only A325 or A490 high-strength bolts should be specified. See LRFD Specification, Appendix K4, Sect. K4.3.
For designing combined shear and tension loads, see LRFD Specification Sect. J3.4 or J3.5.

TABLE I-B. THREADED FASTENERS
Tension on gross (nominal) area

ASTM Designation	F_y Ksi	F_u Ksi	ϕF_t Ksi	\<th colspan="8">Nominal Diameter d, In.</th>
				5/8	3/4	7/8	1	1 1/8	1 1/4	1 3/8	1 1/2
				\<td colspan="8">Area (Based on Nominal Diameter), In.²</td>
				0.3068	0.4418	0.6013	0.7854	0.9940	1.227	1.485	1.767
A36	36	58	32.6	10.0	14.4	19.6	25.6	32.4	40.0	48.4	57.6
A572, Gr. 50	50	65	36.6	11.2	16.2	22.0	28.7	36.4	44.9	54.4	64.7
A588	50	70	39.4	12.1	17.4	23.7	30.9	39.2	48.3	58.5	69.6
A449 $d < 1$	92	120	67.5	20.7	29.8	40.6	53.0	—	—	—	—
$1 < d < 1\frac{1}{2}$	81	105	59.1	—	—	—	—	58.7	72.5	87.8	104.4

The above table lists ASTM specified materials available in round bar stock that are generally intended for use in threaded applications such as tie rods, cross bracing and similar uses.
The design tensile capacity of the threaded portion of an upset rod shall be larger than the body area times F_y.
F_u = specified minimum tensile strength of the fastener material.
ϕF_t = 0.75x0.75F_u = design tensile stress in threaded fastener.

BOLTS, AND THREADED PARTS
ASTM Specifications

TABLE I-C. MATERIAL FOR ANCHOR BOLTS AND TIE RODS

| | ASTM Specification | Strength, Ksi | | | Maximum Diameter In. | Type of Material [b] | Headed or Unheaded |
		Proof Load	Yield (Min.)	Tensile (Min.)			
Bolts and Studs	A307	—	—	60	4	C	H
	A325[a]	85	92	120	½ to 1, incl.	C, QT	H
		74	81	105	1⅛ to 1½ incl.		
	A354 Gr. BD	120	130	150	¼ to 2½ incl.	A, QT	H, U
		105	115	140	over 2½ to 4 incl.		
	A354 Gr. BC	105	109	125	¼ to 2½ incl.	A, QT	H, U
		95	99	115	over 2½ to 4 incl.		
	A449	85	92	120	¼ to 1 incl.	C, QT	H, U
		74	81	105	1⅛ to 1½ incl.		
		55	58	90	1¾ to 3 incl.		
	A490	120	—	150	½ to 1½ incl.	A, QT	H
	A687	—	105	150[c]	⅝ to 3 incl.	A, QT, NT	U
Threaded Round Stock	A36	—	36	58	8	C	U
	A572 Gr. 50	—	50	65	2	HSLA	U
	A572 Gr. 42	—	42	60	6	HSLA	U
	A588	—	50	70	To 4 incl.	HSLA, ACR	U
		—	46	67	over 4 to 5 incl.		
		—	42	63	over 5 to 8 incl.		

[a]Available with weathering (atmospheric corrosion resistance) characteristics comparable to ASTM A242 and A588 steel.
[b]C = carbon
QT = quenched and tempered
A = alloy
NT = notch tough (Charpy V-notch 15 ft-lb. @ −20°F)
HSLA = high-strength low alloy
ACR = atmospheric corrosion-resistant
[c]Maximum (ultimate tensile strength)
Notes:
ASTM specified material for anchor bolts, tie rods and similar applications can be obtained from either specifications for threaded bolts and studs normally used as connectors or for structural material available in round stock that may then be threaded. The material supplier should be consulted for availability of size and length.
Suitable nuts by grade may be obtained from ASTM Specification A563.
Anchor bolt material that is quenched and tempered should not be welded or heated to facilitate erection.
Threaded rod with properties meeting A325, A490 or A449 Specifications may be obtained by the use of an appropriate steel (such as AISI C1040 or C4140), quenched and tempered after fabrication.

Shear
Design load in kips

ASTM Designation		Connection Type[a]	Hole Type[b]	ϕF_v Ksi	Loading[c]	Nominal Diameter d, In.							
						5/8	3/4	7/8	1	1 1/8	1 1/4	1 3/8	1 1/2
						Area (Based on Nominal Diameter), In.²							
						0.3068	0.4418	0.6013	0.7854	0.9940	1.227	1.485	1.767
Bolts	A307	—	STD NSL	16.2	S D	4.97 9.94	7.16 14.3	9.74 19.5	12.7 25.4	16.1 32.2	19.9 39.8	24.1 48.1	28.6 57.3
	A325	SC[d] Class A	STD	17.0	S D	5.22 10.4	7.51 15.0	10.2 20.4	13.4 26.7	16.9 33.8	20.9 41.7	25.2 50.5	30.0 60.1
			OVS SSL	15.0	S D	4.60 9.20	6.63 13.3	9.02 18.0	11.8 23.6	14.9 29.8	18.4 36.8	22.3 44.6	26.5 53.0
			LSL	12.0	S D	3.68 7.36	5.30 10.6	7.22 14.4	9.42 18.8	11.9 23.9	14.7 29.4	17.8 35.6	21.2 42.4
		N	STD NSL	35.1	S D	10.8 21.5	15.5 31.0	21.1 42.2	27.6 55.1	34.9 69.8	43.1 86.1	52.1 104.2	62.0 124.1
		X	STD NSL	46.8	S D	14.4 28.7	20.7 41.4	28.1 56.3	36.8 73.5	46.5 93.0	57.4 114.9	69.5 139.0	82.7 165.4
	A490	SC[d] Class A	STD	21.0	S D	6.44 12.9	9.28 18.6	12.6 25.3	16.5 33.0	20.9 41.7	25.8 51.5	31.2 62.4	37.1 74.2
			OVS SSL	18.0	S D	5.52 11.0	7.95 15.9	10.8 21.6	14.1 28.3	17.9 35.8	22.1 44.2	26.7 53.5	31.8 63.6
			LSL	15.0	S D	4.60 9.20	6.63 13.3	9.02 18.0	11.8 23.6	14.9 29.8	18.4 36.8	22.3 44.6	26.5 53.0
		N	STD NSL	43.9	S D	13.5 26.9	19.4 38.8	26.4 52.8	34.5 68.9	43.6 87.2	53.8 107.7	65.1 130.3	77.5 155.1
		X	STD NSL	58.5	S D	17.9 35.9	25.8 51.7	35.2 70.4	45.9 91.9	58.2 116.3	71.8 143.6	86.9 173.7	103.4 206.8

[a]SC = Slip-critical connection.
   N = Bearing-type connection with threads **included** in shear plane.
   X = Bearing-type connection with threads **excluded** from shear plane.
[b]STD = Standard round holes (d + 1/16″)   OVS = Oversize round holes
   LSL = Long-slotted holes   SSL = Short-slotted holes
   NSL = Long- or short-slotted hole normal to load direction
        (required in bearing-type connection).
[c]S = Single shear   D = Double shear
[d]Slip-critical connections are a serviceability criterion to prevent slip at service (working) loads.
   Tabulated values are service load values and should be compared to the working loads, *not* to
   the factored loads.
When bearing-type connections used to splice tension members have a fastener pattern whose
length, measured parallel to the line of force, exceeds 50 in., tabulated values shall be reduced
by 20%. See Commentary Sect. J3.3.

Shear
Design load in kips

TABLE I-D. SHEAR

ASTM Designation		Con-nection Type[a]	Hole Type[b]	ϕF_v Ksi	Load-ing[c]	Nominal Diameter d, In.							
						5/8	3/4	7/8	1	1 1/8	1 1/4	1 3/8	1 1/2
						Area (Based on Nominal Diameter), In.²							
						0.3068	0.4418	0.6013	0.7854	0.9940	1.227	1.485	1.767
Rivets	A502-1	—	STD	23.4	S	7.2	10.3	14.1	18.4	23.3	28.7	34.7	41.4
					D	14.4	20.7	28.1	36.8	46.5	57.4	69.5	82.7
	A502-2 A502-3	—	STD	31.2	S	9.6	13.8	18.8	24.5	31.0	38.3	46.3	55.1
					D	19.1	27.6	37.5	49.0	62.0	76.6	92.7	110.3
Threaded Parts	A36 ($F_u = 58$ ksi)	N	STD	17.0	S	5.2	7.5	10.2	13.3	16.9	20.8	25.2	30.0
					D	10.4	15.0	20.4	26.6	33.7	41.6	50.4	60.0
		X	STD	22.6	S	6.9	10.0	13.6	17.8	22.5	27.8	33.6	40.0
					D	13.9	20.0	27.2	35.5	45.0	55.5	67.2	79.9
	A572, Gr. 50 ($F_u = 65$ ksi)	N	STD	19.0	S	5.8	8.4	11.4	14.9	18.9	23.3	28.2	33.6
					D	11.7	16.8	22.9	29.9	37.8	46.7	56.5	67.2
		X	STD	25.4	S	7.8	11.2	15.2	19.9	25.2	31.1	37.6	44.8
					D	15.6	22.4	30.5	39.8	50.4	62.2	75.3	89.6
	A588 ($F_u = 70$ ksi)	N	STD	20.5	S	6.3	9.0	12.3	16.1	20.4	25.1	30.4	36.2
					D	12.6	18.1	24.6	32.2	40.7	50.3	60.8	72.4
		X	STD	27.3	S	8.4	12.1	16.4	21.4	27.1	33.5	40.5	48.2
					D	16.8	24.1	32.8	42.9	54.3	67.0	81.1	96.5

For threaded parts of materials not listed, use $\phi F_v = 0.65 \times 0.45 F_u$ when threads are included in a shear plane, and $\phi F_v = 0.65 \times 0.60 F_u$ when threads are excluded from a shear plane.
[a]N = Bearing-type connection with threads **included** in shear plane.
X = Bearing-type connection with threads **excluded** from shear plane.
[b]STD = Standard round holes (d + 1/16″).
[c]S = Single shear. D = Double shear.

BOLTS AND THREADED PARTS
Bearing
Design loads in kips

TABLE I-E. BEARING
Slip-critical and Bearing-type Connections

Material Thickness	F_u = 58 Ksi Bolt dia.			F_u = 65 Ksi Bolt dia.			F_u = 70 Ksi Bolt dia.			F_u = 100 Ksi Bolt dia.		
	¾	⅞	1	¾	⅞	1	¾	⅞	1	¾	⅞	1
⅛	9.8	11.4	13.0	11.0	12.8	14.6	11.8	13.8	15.2	16.9	19.7	22.5
³⁄₁₆	14.7	17.1	19.6	16.5	19.2	21.9	17.7	20.7	23.6	25.3	29.5	33.8
¼	19.6	22.8	26.1	21.9	25.6	29.2	23.6	27.6	31.5	33.8	39.4	45.0
⁵⁄₁₆	24.5	28.5	32.6	27.4	32.0	36.6	29.5	34.5	39.4	42.2	49.2	56.2
⅜	29.4	34.3	39.1	32.9	38.4	43.9	35.4	41.3	47.2	50.6	59.1	67.5
⁷⁄₁₆	34.3	40.0	45.7	38.4	44.8	51.2	41.3	48.2	55.1		68.9	78.8
½	39.1	45.7	52.2	43.9	51.2	58.5	47.2	55.1	63.0			90.0
⁹⁄₁₆	44.0	51.4	58.7	49.4	57.6	65.8		62.0	70.9			
⅞	48.9	57.1	65.2		64.0	73.1		68.9	78.8			
¹¹⁄₁₆		62.8	71.8		70.4	80.4			86.6			
¾		68.5	78.3			87.8						
¹³⁄₁₆			84.8									
⅞			91.4									
¹⁵⁄₁₆												
1	78.3	91.3	104.4	87.8	102.4	117.0	94.5	110.2	126.0	135.0	157.5	180.0

Notes:

This table is applicable to all mechanical fasteners in both slip-critical and bearing-type connections utilizing standard holes. Standard holes shall have a diameter nominally ¹⁄₁₆ in. larger than the nominal bolt diameter (d + ¹⁄₁₆ in.).

Tabulated bearing values are based on $\phi F_p = \phi \times 2.4 \times F_u$, where $\phi = 0.75$.

F_u = specified minimum tensile strength of the connected part.

In connections transmitting axial force whose length between extreme fasteners measured parallel to the line of force exceeds 50 in., tabulated values shall be reduced 20%.

Connections using high-strength bolts in slotted holes with the load applied in a direction other than approximately normal (between 80 and 100 degrees) to the axis of the hole and connections with bolts in oversize holes shall be designed for resistance against slip at working load in accordance with LRFD Specification Sect. J3.

Tabulated values apply when the distance l parallel to the line of force from the center of the bolt to the edge of the connected part is not less than 1½ d and the distance from the center of a bolt to the center of an adjacent bolt is not less than 3d. See LRFD Commentary Sect. J3.9.

Under certain conditions, values greater than the tabulated values may be justified under LRFD Specification Sect. J.

Values are limited to the double-shear bearing capacity of A490-X bolts.

AMERICAN INSTITUTE OF STEEL CONSTRUCTION

Notes

BOLTS AND RIVETS
Bearing
Design loads in kips

TABLE I-F. EDGE DISTANCE

	Edge Distance[b] l_v In.	Design Loads, Kips[a] (for one fastener, 1″ thick material)			
		$F_u = 58$	$F_u = 65$	$F_u = 70$	$F_u = 100$
	1	43.5	48.8	52.5	75.0
spacing = 3 n = no. of bolts	1⅛	49.0	54.9	59.0	84.4
	1¼	54.4	61.0	65.6	93.5
	<1½	65.3	73.1	78.8	113

COPED	Bolt Dia.	1½ d In.	Values when edge distance is 1½ d or greater[c]			
	1	1½	104	117	126	180
	⅞	1⁵⁄₁₆	91.3	102	110	158
	¾	1⅛	78.3	87.8	94.5	135

[a]Total design load = (tabular value) × t × n, kips
where

   t = thickness of critical connected part, in.

   n = number of fasteners in connection to beam web

[b]$l_v \geq P/0.75 F_u t$ (LRFD Spec. Sect. J3.10) distance from center of hole to free edge of connected part in direction of force, in.
where

   F_u = specified minimum tensile strength of web material, ksi

   P = force transmitted by one fastener to the critical connected part, kips

[c]$P = 1.8 F_u d$ (LRFD Spec. Sect. J3.6), kips/1 in. of material.

BOLTS AND RIVETS
Bearing

TABLE I-G.1 COEFFICIENTS FOR WEB TEAR-OUT (BLOCK SHEAR)
Based on standard holes and 3″ fastener spacing
Shear yield, Tension fracture
$F_y = 36$ ksi

Coefficient C_1

l_v, In.	l_h, In.												
	1	1⅛	1¼	1⅜	1½	1⅝	1¾	1⅞	2	2¼	2½	2¾	3
1¼	63.8	69.2	74.6	80.1	85.5	90.9	96.4	102	107	118	129	140	151
1⅜	65.8	71.2	76.7	82.1	87.5	93.0	98.4	104	109	120	131	142	153
1½	67.8	73.2	78.7	84.1	89.6	95.0	100	106	111	122	133	144	155
1⅝	69.8	75.3	80.7	86.1	91.6	97.0	102	108	113	124	135	146	157
1¾	71.9	77.3	82.7	88.2	93.6	99.0	104	110	115	126	137	148	159
1⅞	73.9	79.3	84.8	90.2	95.6	101	107	112	117	128	139	150	161
2	75.9	81.3	86.8	92.2	97.7	103	109	114	119	130	141	152	163
2¼	80.0	85.4	90.8	96.3	102	107	113	118	123	134	145	156	167
2½	84.0	89.4	94.9	100	106	111	117	122	128	138	149	160	171
2¾	88.1	93.5	98.9	104	110	115	121	126	132	142	153	164	175
3	92.1	97.5	103	108	114	119	125	130	136	146	157	168	179

Coefficient C_2

n	Bolt Dia., In.		
	¾	⅞	1
2	37.5	35.8	34.1
3	86.1	84.4	82.7
4	135	133	131
5	183	182	180
6	232	230	228
7	281	279	277
8	329	327	326
9	378	376	374
10	426	425	423

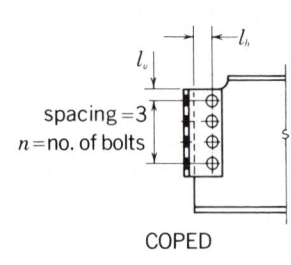

spacing = 3
n = no. of bolts

COPED

Notes:

R_{BS} = Resistance to block shear, kips
$= \phi [0.6 F_y A_{vg} + F_u A_{nt}]$ (from LRFD Commentary Sect. J4)
$= (0.45 F_y l_v + 0.75 F_u l_h) t + \{0.45 F_y [(n-1)s + d_h/2] - 0.75 F_u d_h/2\} t$
$= (C_1 + C_2) t$

where
A_{nt} = net tension area, in.²
A_{vg} = gross shear area, in.²
F_u = specified min. tensile strength, ksi
F_y = specified yield stress, ksi
d_h = diameter of hole (diameter of fastener + 1/16″), in.
l_h = distance from center of hole to beam end, in.
l_v = distance from center of hole to edge of web, in.
n = number of fasteners
s = fastener spacing, in.
t = thickness of web, in.
ϕ = resistance factor, 0.75

The governing value of R_{BS} is the *greater* of the two determined from this table and Table I-G.2.
Tabular values are based on the following:
LRFD Specification Sects. J3.10 and J4
LRFD Commentary Sects. J3.10 and J4

BOLTS AND RIVETS
Bearing

TABLE I-G.2 COEFFICIENTS FOR WEB TEAR-OUT
(BLOCK SHEAR)
Tension yield, Shear fracture
$F_y = 36$ ksi

Coefficient C_1

l_v, In.	l_h, In.												
	1	1⅛	1¼	1⅜	1½	1⅝	1¾	1⅞	2	2¼	2½	2¾	3
1¼	59.6	63.0	66.4	69.8	73.1	76.5	79.9	83.3	86.6	93.4	100	107	114
1⅜	62.9	66.3	69.6	73.0	76.4	79.8	83.1	86.5	89.9	96.6	103	110	117
1½	66.2	69.5	72.9	76.3	79.7	83.0	86.4	89.8	93.2	99.9	107	113	120
1⅝	69.4	72.8	76.2	79.5	82.9	86.3	89.7	93.0	96.4	103	110	117	123
1¾	72.7	76.1	79.4	82.8	86.2	89.6	92.9	96.3	99.7	106	113	120	127
1⅞	75.9	79.3	82.7	86.1	89.4	92.8	96.2	99.6	103	110	116	123	130
2	79.2	82.6	86.0	89.3	92.7	96.1	99.5	103	106	113	120	126	133
2¼	85.7	89.1	92.5	95.9	99.2	103	106	109	113	119	126	133	140
2½	92.3	95.6	99.0	102	106	109	113	116	119	126	133	140	146
2¾	98.8	102	106	109	112	116	119	122	126	133	139	146	153
3	105	109	112	115	119	122	126	129	132	139	146	153	159

Coefficient C_2

n	Bolt Dia., In.		
	¾	⅞	1
2	46.5	41.6	36.7
3	104	95.4	87.3
4	161	149	138
5	218	203	188
6	275	257	239
7	332	311	290
8	389	365	340
9	446	418	391
10	503	472	441

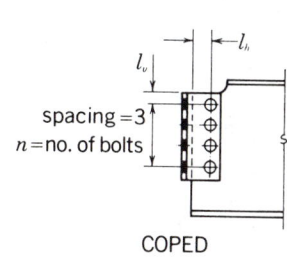

spacing = 3
n = no. of bolts

COPED

Notes:

R_{BS} = Resistance to block shear, kips

$\quad = \phi [0.6 F_u A_{ns} + F_y A_{tg}]$ (from AISC LRFD Commentary Sect. J4)

$\quad = (0.45 F_u l_v + 0.75 F_y l_h) t + \{0.45 F_u [(n - 1)(s + d_h)] - d_h/2\} t$

$\quad = (C_1 + C_2) t$

where

$\quad A_{tg}$ = gross tension area, in.^{2}

$\quad A_{ns}$ = net shear area, in.^{2}

$\quad F_u$ = specified min. tensile strength, ksi

$\quad F_y$ = specified yield stress, ksi

$\quad d_h$ = diameter of hole (diameter of fastener + ¹⁄₁₆″), in.

$\quad l_h$ = distance from center of hole to beam end, in.

$\quad l_v$ = distance from center of hole to edge of web, in.

$\quad n$ = number of fasteners

$\quad s$ = fastener spacing, in.

$\quad t$ = thickness of web, in.

$\quad \phi$ = resistance factor, 0.75

The governing value of R_{BS} is the *greater* of the two determined from this table and Table I-G.1. Tabular values are based on the following:

$\quad$ LRFD Specification Sects. J3.10 and J4

$\quad$ LRFD Commentary Sects. J3.10 and J4

BOLTS AND RIVETS
Bearing

TABLE I-G.1 COEFFICIENTS FOR WEB TEAR-OUT
(BLOCK SHEAR)
Based on standard holes and 3″ fastener spacing
Shear yield, Tension fracture
$F_y = 50$ ksi

Coefficient C_1

l_v, In.	l_h, In.												
	1	1⅛	1¼	1⅜	1½	1⅝	1¾	1⅞	2	2¼	2½	2¾	3
1¼	76.9	83.0	89.1	95.2	101	107	113	120	126	138	150	162	174
1⅜	79.7	85.8	91.9	98.0	104	110	116	122	128	141	153	165	177
1½	82.5	88.6	94.7	101	107	113	119	125	131	143	156	168	180
1⅝	85.3	91.4	97.5	104	110	116	122	128	134	146	158	171	183
1¾	88.1	94.2	100	106	113	119	125	131	137	149	161	173	186
1⅞	90.9	97.0	103	109	115	121	128	134	140	152	164	176	188
2	93.8	99.8	106	112	118	124	130	136	143	155	167	179	191
2¼	99.4	105	112	118	124	130	136	142	148	160	173	185	197
2½	105	111	117	123	129	135	142	148	154	166	178	190	203
2¾	111	117	123	129	135	141	147	153	159	172	184	196	208
3	116	122	128	135	141	147	153	159	165	177	189	202	214

Coefficient C_2

n	Bolt Dia., In.		
	¾	⅞	1
2	56.8	55.2	53.6
3	124	123	121
4	192	190	189
5	259	258	256
6	327	325	324
7	394	393	391
8	462	460	459
9	529	528	526
10	597	595	594

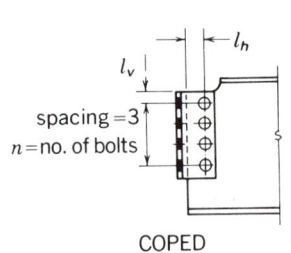

spacing = 3
n = no. of bolts

COPED

Notes:

R_{BS} = resistance to block shear, kips

$\quad = \phi [0.6 F_y A_{vg} + F_u A_{nt}]$ (from LRFD Commentary Sect. J4)

$\quad = (0.45 F_y l_v + 0.75 F_u l_h) t + \{0.45 F_y [(n-1)s + d_h/2] - 0.75 F_u d_h/2\} t$

$\quad = (C_1 + C_2) t$

where

A_{nt} = net tension area, in.^{2}
A_{vg} = gross shear area, in.^{2}
F_u = specified min. tensile strength, ksi
F_y = specified yield stress, ksi
d_h = diameter of hole (diameter of fastener + $\frac{1}{16}$″), in.
l_h = distance from center of hole to beam end, in.
l_v = distance from center of hole to edge of web, in.
n = number of fasteners
s = fastener spacing, in.
t = thickness of web, in.
ϕ = resistance factor, 0.75

The governing value of R_{BS} is the *greater* of the two determined from this table and Table I-G.2.
Tabular values are based on the following:
$\quad$ LRFD Specification Sects. J3.10 and J4
$\quad$ LRFD Commentary Sects. J3.10 and J4

BOLTS AND RIVETS
Bearing

TABLE I-G.2 COEFFICIENTS FOR WEB TEAR-OUT (BLOCK SHEAR)
Tension yield, Shear fracture
$F_y = 50$ ksi

Coefficient C_1

l_v, In.	l_h, In.												
	1	1⅛	1¼	1⅜	1½	1⅝	1¾	1⅞	2	2¼	2½	2¾	3
1¼	74.1	78.8	83.4	88.1	92.8	97.5	102	107	112	121	130	140	149
1⅜	77.7	82.4	87.1	91.8	96.5	101	106	111	115	125	134	143	153
1½	81.4	86.1	90.8	95.4	100	105	110	114	119	128	138	147	156
1⅝	85.0	89.7	94.4	99.1	104	108	113	118	123	132	141	151	160
1¾	88.7	93.4	98.1	103	107	112	117	122	126	136	145	154	164
1⅞	92.3	97.0	102	106	111	116	120	125	130	139	149	158	167
2	96.0	101	105	110	115	119	124	129	134	143	152	162	171
2¼	103	108	113	117	122	127	131	136	141	150	160	169	178
2½	111	115	120	125	129	134	139	143	148	158	167	176	186
2¾	118	123	127	132	137	141	146	151	155	165	174	184	193
3	125	130	135	139	144	149	153	158	163	172	182	191	200

Coefficient C_2

n	Bolt Dia., In.		
	¾	⅞	1
2	52.1	46.6	41.1
3	116	107	97.8
4	180	167	154
5	244	228	211
6	308	288	268
7	372	348	324
8	436	409	381
9	500	469	438
10	564	529	495

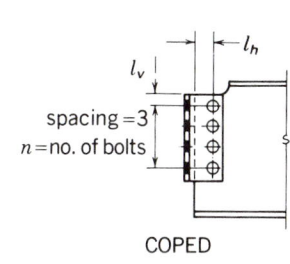

spacing = 3
n = no. of bolts

COPED

Notes:

R_{BS} = resistance to block shear, kips

$= \phi [0.6 F_u A_{ns} + F_y A_{tg}]$ (from LRFD Commentary Sect. J4)

$= (0.45 F_u l_v + 0.75 F_y l_h) t + \{0.45 F_u [(n-1)(s - d_h) - d_h/2] t$

$= (C_1 + C_2) t$

where

A_{tg} = gross tension area, in.^{2}
A_{ns} = net shear area, in.^{2}
F_u = specified min. tensile strength, ksi
F_y = specified yield stress, ksi
d_h = diameter of hole (diameter of fastener + $\frac{1}{16}$″), in.
l_h = distance from center of hole to beam end, in.
l_v = distance from center of hole to edge of web, in.
n = number of fasteners
s = fastener spacing, in.
t = thickness of web, in.
ϕ = resistance factor, 0.75

The governing value of R_{BS} is the *greater* of the two determined from this table and Table I-G.1.
Tabular values are based on the following:
    LRFD Specification Sects. J3.10 and J4
    LRFD Commentary Sects. J3.10 and J4

Notes

FRAMED BEAM CONNECTIONS
Bolted

TABLE II

BEAM REACTIONS

For economical connections, the beam reactions should be shown on the contract drawings. If these reactions are not shown, connections *must* be selected to support one-half the total uniform load capacity shown in the tables of Uniform Load Constants, Part 3 of this Manual, for the given beam, span and grade of steel specified. The effects of any concentrated loads must be taken into account.

Beam reactions *must* be shown on contract drawings for composite construction and continuous framing.

TYPE OF CONNECTION

Tables are developed for design reactions for simply supported beams. No eccentricity or moment resistance is considered in determining the tabulated values. The inherent rigidity of the connections is a factor the designer should be aware of and consider where critical. The arbitrary thickness limitation of ⅝″ for framing angles in the tables was selected to assure flexibility.

In applying the provisions of LRFD Specification Sect. J4, the calculation of area effective in resisting failure due to shear (and tension, when present) is accomplished by deducting the area of the standard holes along the minimum net failure surface. A standard hole is defined as the nominal fastener diameter plus 1/16 inch. Attention is called to the difference between this net area provision and that described in LRFD Specification Sect. B2, which applies to the minimum net *tensile* failure surfaces.

BOLTS

Bolts approved for use in steel building structures are listed in LRFD Specification Sect. A3.3, *Bolts*. Application should comply with Sect. J3, *Bolts, Threaded Parts and Rivets* and Sect. J1.9, *Limitations on Bolted and Welded Connections*.

The type of application of a high-strength bolt is indicated as follows:

A325-SC and A490-SC: Slip-critical connection.

A325-N and A490-N: Bearing-type connection with threads included in the shear planes.

A325-X and A490-X: Bearing-type connection with threads excluded from the shear planes.

Note that SC-N-X is not part of the ASTM Specification designation, but is descriptive of the design assumptions.

TABLE II-A

This table is for bolts in bearing-type connections having standard or slotted holes and for bolts in slip-critical connections having standard holes and Class A, clean scale surface condition.

Design loads are based on shear capacity of the type and number of bolts in the connection. Footnotes indicate where net shear in the angles govern over bolt shear (see Table II-C).

TABLE II-B

This table is for bolts in slip-critical connections having oversize, short slotted, or long slotted bolts and Class A, clean mill scale condition.

Design loads in the table are based on slip capacity of the type and number of bolts in the connection and type of hole.

For slip-shear values of other classes of surface conditions in slip-critical connections, see Specification for Structural Joints Using ASTM A325 or A490 Bolts, Load and Resistance Factor Design.

TABLE II-C

This table is for checking the shear capacity on the net section of the connection angles.

Design reactions are based on $0.45F_u$ (26.1 ksi for A36 material) on the net section vertically through the bolt holes, per LRFD Specification Sect. J4, using:

$$R_v = 2t[(L \text{ or } L') - n(d + \frac{1}{16})] \, 0.45F_u, \text{ kips}$$

WEB TEAR-OUT (BLOCK SHEAR)

When high fastener values are used on relatively thin material, failure can occur by a combination of:

Net shear along a vertical plane through the fastener holes plus tension along a horizontal plane on the area effective in resisting tension failure,

or

Gross shear along the vertical plane and net tension along the horizontal plane. Research has demonstrated that both models are conservative so the larger of the two may be used in design. The failure mechanism is referred to as web tear-out or block shear.

Tables I-G.1 and I-G.2 list coefficients to be used for determining block shear capacity for several common beam framing conditions of l_v and l_h and three different bolt diameters in standard size holes at 3-in. spacing.

The block shear or web tear-out design reaction is determined by adding coefficient C_1, for the given l_v and l_h, to coefficient C_2 for number and size of bolts, then multiplying this sum by the given value of F_u and the web thickness. This is done for both the tension yield, shear fracture and shear yield, tension fracture cases. The larger of these two cases governs. For conditions that differ from those tabulated, the general equations shown in Tables I-G.1 and I-G.2 can be utilized.

For oversize and slotted holes, refer to LRFD Specification Sects. J3 and J4.

SELECTION OF CONNECTION

The LRFD Specification requires that several conditions be checked to determine the design capacity for a framed beam connection. These conditions are: bolt shear, bolt bearing on connecting material, beam web tear-out (block shear), shear on the net area of the connection angles or connection plate, and local bending stresses.

In Tables II-A and II-B, bolt shear or slip-critical design loads have been tabulated. Note that certain values are limited by net shear on the angles. The footnotes identify the values that are so limited and refer to Table II-C.

When the beam flange is not coped and standard size holes at 3″ spacing are used, the design load for the beam web in kips is obtained from Table I-E.

When the beam flange is coped and standard size holes at 3″ spacing are used, the design load in kips is checked by using the coefficients in Table I-G.

For bolts in slip-critical connections having oversize or slotted holes and class A, clean mill scale surface condition, the slip load is given in Table II-B. If the beam is coped, block shear must be checked using the coefficients in Table I-G and appropriate adjustments must be made to reflect the length of slot as it affects l_v or l_h.

For bearing-type connections using slotted holes, it is required that the long direction of the slot be perpendicular to the direction of the load. For an uncoped beam with slots in the web, the bolt shear design values are determined from Table II-A and the bolt bearing design value may be determined from Table I-F. For coped beams, beam web tear-out must also be checked using Table I-G. Intermediate values in Tables I-F and I-G can be obtained by proportioning. Use Table II-C to check the net shear capacity.

For slip-critical connections using slotted holes or oversized holes in the connection angles, slip-critical values are determined from Table II-B. When slots are used in the beam web and the long direction of the slot is perpendicular to the direction of the load, bolt bearing values and beam web tear-out may be determined as given above for the bearing-type connection. For slots with the long direction of the slot parallel to the direction of the load and for oversize holes, see the LRFD Specification.

Connection angle lengths L (L' for staggered holes) and number of bolts at 3″ spacing with 1¼″ end distance are tabulated in Tables II-A and II-B.

DETAILS

1. In Tables II-A and II-B, connection angle lengths vary from 5½″ to 29½″ (7″ to 31″ for staggered hole arrangement) in multiples of 3″. The length of the angles for a connection must be compatible with the beam T-dimension for uncoped beams. It is recommended that the minimum length of connection angle be at least one-half the T-dimension, to provide stability during erection.

2. Vertical fastener spacing is arbitrarily chosen as 3″ for these tables. This may be varied within the parameters established by the LRFD Specification.

3. End distance on angles is set at 1¼″, as permitted by LRFD Specification Table J3.7. When using oversize or slotted holes, the edge or end distance must be adjusted over that for standard holes as required.

4. Gages for supporting members should be selected to meet the requirements of LRFD Specification Sect. J3.

5. Clearance for assembly is essential in all cases. Framing angles with staggered holes are permitted as alternates to provide clearance and to permit smaller gages on legs of connection angles.

COMBINATION OF WELDED AND BOLTED FRAMED BEAM CONNECTIONS

See Framed Beam Connections—Welded, Table III, for appropriate Case I or Case II combination welded and bolted connections with tabulated reaction values.

OTHER FRAMED CONNECTIONS

These tables are not intended to preclude the use of other adequately designed connections.

EXAMPLE 1:

Given:

Beam: W18 × 50, t_w = 0.355″
   ASTM A36 (F_y = 36 ksi and F_u = 58 ksi)
Top flange coped 2″ deep
Factored beam reaction = 65 kips
Bolts: ¾″ dia. A325-N in ¹³⁄₁₆″ dia. holes

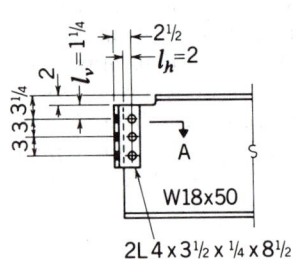

2L 4 x 3½ x ¼ x 8½

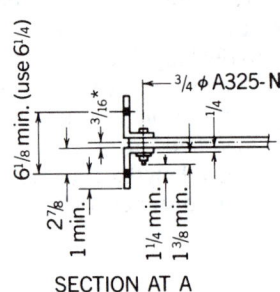

SECTION AT A

Solution:

From Table II-A, select a connection with 3 rows (n = 3) of ¾″ diam. A325-N bolts and shear value of 93.0 kips using ⁵⁄₁₆″ thick angles. This exceeds the 65-kip reaction.

The net shear in the connection angles may be checked using Table II-C. However, this is only critical for those values in Tables II-A and II-B which are footnoted. The value of 93.0 in Table II-A is not footnoted; therefore, net shear in the ⁵⁄₁₆″ angles is not critical. Since bolt shear is good for ⁵⁄₁₆″ angles, it is also good for ¼″ angles. From Table II-C, with d = ¾ and n = 3, the design shear in a pair of ¼″ angles is 79.1 kips. This exceeds the 65-kip reaction.

LRFD Specification Sect. J3.9 limits minimum bolt spacing to 2⅔ diameters or ¾ × 2⅔ = 2″. The preferred pitch of 3″ will be used.

Using the normal 1¼-in. edge distance to the cope (l_v = 1¼″) and a 3″ fastener spacing of ¾″ bolts, enter Table I-F at F_u = 58 ksi and read coefficient 78.3 kips per in. of thickness for all 3 bolts. The design bearing strength is then (78.3 × 3) × 0.355 = 83.4 kips > 65 kips and 1¼ in. is adequate for beam web and angles.

LRFD Specification Table J3.7 edge requirements are also met.

Since this beam is coped, beam web tear-out (block shear) must be checked. Enter Table I-G.1 with F_y = 36 ksi, l_v = 1¼″ and l_h = 2 in.

*This dimension (see sketch, Section at A) is determined to be one-half the decimal web thickness rounded to the next higher ¹⁄₁₆ in. Example: 0.355/2 = 0.1775; use ³⁄₁₆″. This will produce a spacing of holes slightly closer than detailed to permit spreading, rather than closing, at time of erection to supporting member.

Read coefficient $C_1 = 107$. Coefficient C_2 for ¾" bolts and $n = 3$ is 86.1.

$R_{BS} = (107 + 86.1) \times 0.355 = 68.6$ kips

Using Table I-G.2, $C_1 = 86.6$ and $C_2 = 104$.

$R_{BS} = (86.6 + 104) \times 0.355 = 67.7$ kips

The larger value, 68.6 kips, governs. 68.6 kips > 65 kips

The connection as sketched is adequate. Attention must also be given to the following:

    a. Insertion and tightening clearances; see Table of Assembling Clearances (LRFD Manual Part 5).

    b. Need for reinforcement at the coped section in bending.

    c. Adequacy of this connection with respect to the supporting member.

    d. If a smaller gage on the outstanding leg of the connection angles is required, the bolts must be staggered with those in the beam web. An even stagger will permit the use of a 5" gage using angles 3½ × 3½ × ¼ × 0'10". The stagger portion of the Table of Stagger for Tightening (LRFD Manual Part 5) is helpful in verifying clearances.

EXAMPLE 2:

Given:

    Beam: W36×230, $t_w = 0.760"$
           A36 ($F_y = 36$ ksi and $F_u = 58$ ksi)
    Factored beam reaction = 510 kips
    Connection angles: A36 ($F_y = 36$ ksi and $F_u = 58$ ksi)
    Bolts: ⅞" dia. A490-X in ¹⁵⁄₁₆" dia. holes

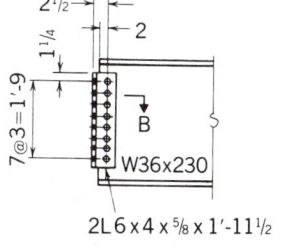

2L 6 × 4 × ⅝ × 1'-11½

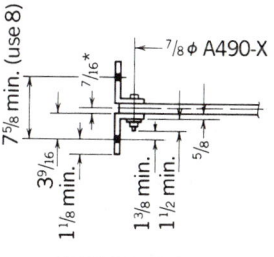

SECTION AT B

Solution:

From Table II-A, select a connection with 8 rows ($n = 8$) of ⅞" dia. A490-X bolts and load value of 563 kips, which exceeds the 510-kip reaction. Note that this value is footnoted (c) and is therefore limited by shear on the net section of the ⅝" thick angles at length $L = 23½"$. From Table II-C, the net shear capacity with ⅞" dia. bolts and $n = 8$ on angles ⅝" thick and 23½" long is 522 kips > 510 kips   **o.k.**

---

*This dimension (see sketch, Section at B) is one-half the decimal web thickness rounded to the next higher ¹⁄₁₆", as in Ex. 1.

Bearing capacity may be checked using Table I-F.

$8 \times 91.3 \times 0.760 = 555$ kips > 510 kips **o.k..**

Check capacity of A36 angles, using Table I-F.

$1 \times 54.4 \times \frac{5}{8} = 34.0$ (top bolt)

$7 \times 91.3 \times \frac{5}{8} = \underline{399.4}$ (remaining bolts)

$\phantom{7 \times 91.3 \times \frac{5}{8} =} 433.4 > 510/2$ kips **o.k.**

Since this beam is not coped, beam web tear-out does not require checking.

The connection as sketched is adequate. Attention must also be given to:

a. Insertion and driving clearance; see Table of Assembling Clearances (LRFD Manual Part 5).

b. Adequacy of this connection with respect to the supporting member.

c. If a smaller gage is required on the outstanding legs of the connection angles, the bolts must be staggered with those in the beam web. An even stagger will permit the gage to be reduced to 7″ and still maintain driving clearances while maintaining the 3″ pitch. See Table of Stagger for Tightening (LRFD Manual Part 5).

EXAMPLE 3:

Given:

Beam: W24 × 76, $t_w = 0.440''$
    A572 Gr. 50 ($F_y = 50$ ksi and $F_u = 65$ ksi)
Connection angles: A36 ($F_y = 36$ ksi and $F_u = 58$ ksi)
Factored beam reaction = 234 kips
Bolts: 1″ dia. A325-N
Holes: $1\frac{1}{16} \times 1\frac{5}{16}$ (short slots), long axis perpendicular to transmitted force

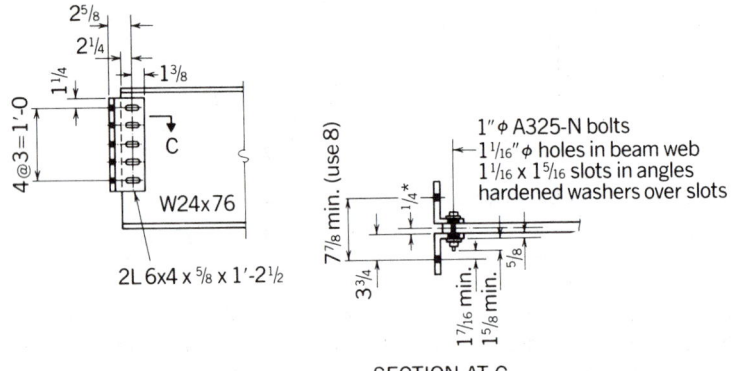

SECTION AT C

---

*This dimension (see sketch, Section at C) is one-half the decimal web thickness rounded to the next higher ¹⁄₁₆″, as in Ex. 1.

Solution:

From Table II-A, select a connection with 5 rows ($n = 5$) of 1 in. dia. A325-N bolts and shear value of 276 kips using ⅝" thick angles. This exceeds the 234-kip reaction.

Bearing capacity may be checked using Table I-F.

$5 \times 117 \times 0.44 = 257$ kips > 234 kips **o.k.**

Check capacity of the A36 connection angles, using Table I-F.

$1 \times 54.4 \times ⅝ = 34.0$ (top bolt)

$$4 \times 104 \times ⅝ = \underline{260.0} \text{ (remaining bolts)}$$
$$294.0 > 234/2 \text{ kips} \quad \textbf{o.k.}$$

Since this beam is not coped, beam web tear-out (block shear) does not require checking.

The net shear capacity of the angles need not be checked, using Table II-C because the 276 kips is not footnoted.

Note that short slots are permitted in any and all plies of slip-critical or bearing-type connections using high strength bolts. To simplify the fabrication, the slots are put in the angles, rather than the beam web. Hardened washers are required to be placed over these slots, because they are in an outer ply.

The connection as sketched is adequate. Attention must also be given to the following:

    a. Insertion and driving clearances; see Table of Assembling Clearances (LRFD Manual Part 5).

    b. Adequacy of this connection with respect to the supporting member.

    c. An even stagger of web bolts with those in the outstanding legs will permit use of a 7" gage and still maintain driving clearances while maintaining the 3" pitch. See the Table of Stagger for Tightening (LRFD Manual Part 5).

EXAMPLE 4:

Given:

The same conditions as Ex. 3, except that the beam is coped 1¾" deep at the top flange. Find the design reaction on the modified beam.

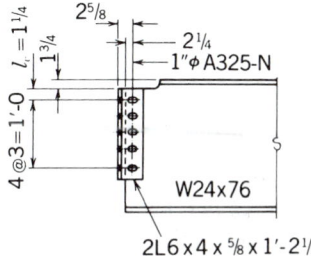

2L 6 x 4 x ⅝ x 1'-2½

Solution:

The design reaction of the coped beam will be the least value determined by the bolt shear, bolt bearing, beam web tear-out (block shear) or shear capacity of the connection angles.

1. Bolt shear: From Table II-A, R = 276 kips.

2. Bolt bearing on beam web:

    From Table I-F for F_u = 65 ksi, l_v = 1¼", d = 1" and 3" pitch:

    Design load, top bolt = 61.0 kips/in. thickness

    Design load, remaining bolts = 117 kips/in. thickness

    R = (61.0 + 4 × 117) × 0.440 = 233 kips

3. Bolt bearing on connection angles:

    From Table I-F, for F_u = 58 ksi, l_v = 1¼", d = 1" and 3" pitch:

    Design load, top bolt = 54.4 kips/in. thickness

    Design load, remaining bolts = 104 kips/in. thickness

    R = (54.4 + 4 × 104) × 2(⅝) = 588 kips

4. Web tear-out:

    From Table I-G.1, for F_y = 36 ksi, l_v = 1¼", and l_h = 2¼":

    C_1 = 118 and C_2 = 180, R_{BS} = (118 + 180) × 0.440 = 131 kips

    From Table I-G.2, for F_y = 36 ksi, l_v = 1¼", and l_h = 2¼":

    C_1 = 93.4 and C_2 = 188, R_{BS} = (93.4 + 188) × 0.440 = 124 kips

    The larger value, 131 kips, governs. R = 131 kips.

5. Shear capacity of angles:

    From Table II-C, for 1" bolts, angle thickness = ⅝", L = 14½" (n = 5):

    Shear capacity = 300 kips

The design reaction is limited to 131 kips due to web tear-out and the conditions assumed would require a redesign of this connection since 131 kips < 234 kips.

EXAMPLE 5:

Given:

Beam: W36 × 300, t_w = 0.945 in.
   A36 (F_y = 36 ksi and F_u = 58 ksi)
Top flange coped 3" deep
Factored beam reaction = 458 kips
Bolts: 1⅛" dia. A490-X in 1³⁄₁₆" dia. holes
Surface condition: Class A, clean mill scale

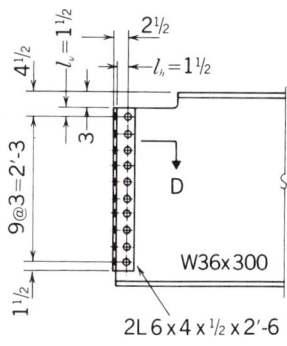
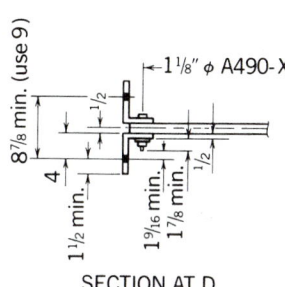

SECTION AT D

Solution:

Since values are not tabulated in this Manual for 1⅛″ dia. fasteners, they must be derived from the LRFD Specification. The following is offered as a guide and outline.

1. Bolt shear:

    Area of 1⅛″ dia. bolt = 0.994 in.^{2}

    From Table I-D, ϕF_v = 58.5 ksi

    Single shear capacity = 58.2 kips/bolt

    No. bolts req'd = 458/(58.2 × 2) = 3.9 ≈ 4

    Since this beam is coped, try 10 bolts because web tear-out is likely to be critical.

2. Bolt spacing:

    Minimum spacing = 2⅔d = 3 in.

    Use 3-in. spacing, which is standard.

3. Edge distance:

    Use 1½″ (minimum for gas cut edge, LRFD Specification Table J3.7)

4. Bolt bearing on beam web:

    Bearing capacity may be checked using LRFD Specification Sect. J3.6.

    Top bolt: (0.75)(58)(1.5)(0.945) = 61.7

    Remaining bolts: 1.8(58)(1.125)(0.945)(9) = 998.9

    1060.6 > 458 kips **o.k.**

5. Since this beam is coped, failure by shear and tension through the fastener holes must be checked.

    From the expression shown in Table I-G.1:

    R_{BS} = [0.45 $F_y(l_v + (n-1)s + d_h/2) + 0.75 F_u(l_h - d_h/2)$] t

    = [0.45 × 36(1½ + (10 − 1)3 + ¹⁹⁄₃₂) + 0.75 × 58(1½ − ¹⁹⁄₃₂)] 0.945

    = 483 kips

From the expression shown in Table I-G.2:

$$R_{BS} = [0.45 \, F_u(l_v + (n - 1)(s - d_h) - d_h/2) + 0.75F_y l_h] \, t$$
$$= [0.45 \times 58(1\frac{1}{2} + (10 - 1)(3 - 1\frac{3}{16}) - \frac{19}{32}) + 0.75 \times 36 \times 1\frac{1}{2}] \, 0.945$$
$$= 463 \text{ kips}$$

The larger value, 483 kips, governs. 483 kips > 458 kips **o.k.**

6. Connection angles:

Try 2 L $6 \times 4 \times \frac{1}{2} \times 2'$-6″, A36.

Design bearing:

Top bolt: $\qquad 0.75(58)(1.5)(\frac{1}{2}) = 32.6$

Remaining bolts: $1.8(58)(1.125)(\frac{1}{2})(9) = \dfrac{528.5}{561.1} > \dfrac{458}{2}$ kips **o.k.**

Shear on plane through fasteners:

$$\phi F_n A_{ns} = \phi(0.60 F_u) \, A_{ns}$$
$$A_{ns} = 2t[L - n(d + \frac{1}{16})]$$
$$= (2 \times \frac{1}{2}) [30 - (10 \times 1\frac{3}{16})] = 18.1 \text{ in.}^2$$
$$\phi F_n A_{ns} = 0.75(0.60 \times 58) \times 18.1 = 472 \text{ kips} > 458 \text{ kips} \quad \textbf{o.k.}$$

7. The connection as sketched is adequate. Attention must also be given to the following:

   a. Insertion and driving clearances; see Table of Assembling Clearances (LRFD Manual Part 5).

   b. Need for reinforcement at the coped section in bending.

   c. Adequacy of this connection with respect to the supporting member.

   d. If a smaller gage is required on the outstanding legs, a new design is required, since the maximum number of fasteners has been utilized in the beam web. Consideration can be given to larger diameter bolts and to the use of a second gage line in the web. The effect of web tear-out (block shear) on a connection with two gage lines has not been determined at this time.

EXAMPLE 6

Given:

Beam: W36 × 230; t_w = 0.760 in., F_y = 36 ksi
Connections Angles: F_y = 36 ksi, F_u = 58 ksi
Bolts: ASTM A490

Reaction: Dead load gravity reaction: 80 kips
Live load gravity reaction: 120 kips
resulting from vibrating machinery
Wind load: 40 kips axial load

Solution:

Bolts are checked as slip-critical at service loads because of the vibrating live load. Ultimate shear and tension capacity of the bolts are checked as bearing-type-N against factored loads.

Serviceability check:

Try 8 − ⅞ dia. A490-SC bolts, standard holes.
From Table II-A, slip resistance is 202 kips.
Force due to Service Load (SL) is calculated for two cases:

Case 1 (includes wind):
$$SL = 0.75 \sqrt{(D + L)^2 + W^2}$$
$$SL = 0.75 \sqrt{(80 + 120)^2 + (40)^2}$$
$$SL = 153 \text{ kips} < 202 \text{ kips} \quad \textbf{o.k.}$$

Case 2 (no wind):
$$SL = 1.0 \, (D + L)$$
$$SL = 1.0 \, (80 + 120)$$
$$SL = 200 \text{ kips} < 202 \text{ kips} \quad \textbf{o.k.}$$

Strength check:

Try 8 − ⅞ dia. A490-N bolts, standard holes.
From Table II-A, capacity is 422 kips.
Force due to Factored Load (FL) is calculated using appropriate load factors:

Case 1 (includes wind):
$$FL = \sqrt{(1.2D + 0.5L)^2 + (1.3W)^2}$$
$$FL = \sqrt{(1.2 \times 80 + 0.5 \times 120)^2 + (1.3 \times 40)^2}$$
$$FL = 164 \text{ kips} < 422 \text{ kips}$$

Case 2 (no wind):
$$FL = 1.2D + 1.6L$$
$$FL = (1.2 \times 80) + (1.6 \times 120)$$
$$FL = 288 \text{ kips} < 422 \text{ kips} \quad \textbf{o.k.}$$

Check bearing strength on web assuming no cope. From Table I-E,

$$288/8 = 36 \text{ kips} < 91.3 \times 0.760 = 69.4 \text{ kips} \quad \textbf{o.k.}$$

Design framing angles:

From Table II-C, $8 - \frac{7}{8}$ bolts, $\frac{3}{8}$ in. thick angles, design strength = 313 kips > 288 kips   **o.k.**

EXAMPLE 7

Given:

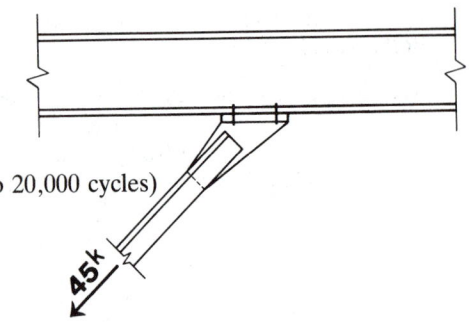

4 – A325 bolts
Dead load: 5 kips
Live load: 40 kips (vibrating up to 20,000 cycles)
Determine bolt size

Solution:

Bolts are checked as slip-critical at service loads because of the vibrating live load. Ultimate shear and tension of the bolts are checked as bearing-type-N against factored loads.

Serviceability check:

Try $4 - \frac{7}{8}$ dia. A325-SC bolts, standard holes.
From Table I-A: $\phi T = 40.6$ kips
From Table I-D: $\phi V = 21.1$ kips (A325-N)
$\qquad \phi V(SC) = 10.2$ kips
Pre-tension = 39.0 kips

Force due to Service Load (SL):
SL (tens.) = $(45 \times \sin 45°)/4 = 7.95$ kips per bolt
SL (shear) = $(45 \times \cos 45°)/4 = 7.95$ kips per bolt

Calculate reduced design slip resistance $\phi V'$ (SC) due to tension in bolt. Prying action is a strength consideration and is not checked at service loads.

$\phi V'(SC) = \phi V(SC) \times (1 - T/T_b)$
$\phi V'(SC) = 10.2 \times (1 - 7.95/39.0) = 8.12$ kips > 7.95 kips   **o.k.**

Strength check:

Force due to Factored Load (FL):
FL = $1.2D + 1.6L$
FL = $1.2 \times 5 + 1.6 \times 40 = 70$ kips
Factored Tension per bolt:
$\quad 70 \times \cos 45°/4 = 12.4$ kips
Factored Shear per bolt:
$\quad 70 \times \sin 45°/4 = 12.4$ kips

$\phi F'_t = 85 - 1.8 \times f_v \leq 68$
$\phi F'_t = 85 - (1.8 \times 12.4)/0.6013 = 47.9$ ksi
$f_t \quad = 12.4/0.6013 = 20.6$ ksi < 47.9 ksi   **o.k.**
$f_v \quad = 12.4/0.6013 = 20.6$ ksi < 35.1 ksi   **o.k.**

Margin for additional bolt tension due to prying action:
$(47.9 - 20.6)/20.6 = 133\%$ Assumed **o.k.** See Design Example, Hanger type Connection LRFD Manual, Part 5.

FRAMED BEAM CONNECTIONS
Bolted
TABLE II Design loads in kips

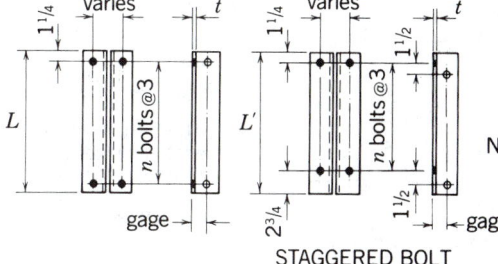

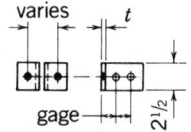

Note: For $L = 2\frac{1}{2}$ use one half
the tabular load value
shown for $L = 5\frac{1}{2}$, for the
same bolt type, diameter,
and thickness.

STAGGERED BOLT
ALTERNATE

TABLE II-A Bolt Shear[a,b]

For A307 bolts in standard or slotted holes and for A325 and A490 bolts in **slip-critical** connections with standard holes and Class A, clean mill scale surface condition.

Bolt Type				A307			A325-SC[d]			A490-SC[d]		
ϕF_v, Ksi				16.2			17			21		
Bolt dia., d In.				$\frac{3}{4}$	$\frac{7}{8}$	1	$\frac{3}{4}$	$\frac{7}{8}$	1	$\frac{3}{4}$	$\frac{7}{8}$	1
Angle Thickness t, In.				$\frac{1}{4}$	$\frac{1}{4}$	$\frac{5}{16}$	$\frac{1}{4}$	$\frac{5}{16}$	$\frac{1}{2}$	$\frac{5}{16}$	$\frac{1}{2}$	$\frac{5}{8}$
L In.	L' In.	n										
$29\frac{1}{2}$	31	10		143	195	254	150	204	267	186	253	330
$26\frac{1}{2}$	28	9		129	175	229	135	184	240	167	227	297
$23\frac{1}{2}$	25	8		115	156	204	120	164	214	148	202	264
$20\frac{1}{2}$	22	7		100	136	178	105	143	187	130	177	231
$17\frac{1}{2}$	19	6		85.9	117	153	90.1	123	160	111	152	198
$14\frac{1}{2}$	16	5		71.6	97.4	127	75.1	102	134	92.8	126	165
$11\frac{1}{2}$	13	4		57.3	77.9	102	60.1	81.8	107	74.2	101	132
$8\frac{1}{2}$	10	3		42.9	58.4	76.3	45.1	61.3	80.1	55.7	75.8	99.0
$5\frac{1}{2}$	7	2		28.6	39.0	50.9	30.0	40.9[c]	53.4	37.1	50.5	66.0

Note:
For slip-critical connections with oversize or slotted holes, see Table II-B.

Notes:
[a]Tabulated load values are based on double shear of bolts. See LRFD Specification Sect. J3.
[b]The enclosed beam web should be checked for design bearing load.
[c]Capacity is governed by net shear on angles for length L. For reduced values see Table II-C.
[d]Slip-critical connections are a serviceability criterion to prevent slip at service (working) loads. Tabulated values are service load values and should be compared to the working loads, not factored loads. Capacity of the connection should be checked as a bearing-type connection at factored loads. If capacity is governed by net shear on angles, see Table II-C.

FRAMED BEAM CONNECTIONS
Bolted
TABLE II Design loads in kips

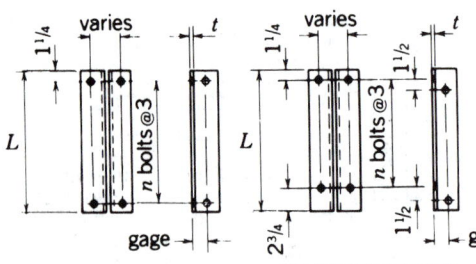

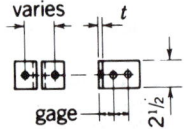

Note: For $L = 2\frac{1}{2}$ use one half
the tabular load value
shown for $L = 5\frac{1}{2}$, for the
same bolt type, diameter,
and thickness.

STAGGERED BOLT
ALTERNATE

TABLE II-A Bolt Shear[a,b]
For bolts in **bearing-type** connections with standard or slotted holes

Bolt Type		A325-N			A490-N			A325-X			A490-X		
ϕF_v, Ksi		35.1			43.9			46.8			58.5		
Bolt dia. d, In.		$\frac{3}{4}$	$\frac{7}{8}$	1	$\frac{3}{4}$	$\frac{7}{8}$	1	$\frac{3}{4}$	$\frac{7}{8}$	1	$\frac{3}{4}$	$\frac{7}{8}$	1
Angle Thickness t, In.		$\frac{5}{16}$	$\frac{3}{8}$	$\frac{5}{8}$	$\frac{3}{8}$	$\frac{1}{2}$	$\frac{5}{8}$	$\frac{3}{8}$	$\frac{5}{8}$	$\frac{5}{8}$	$\frac{1}{2}$	$\frac{5}{8}$	$\frac{5}{8}$
L In.	L' In. n												
$29\frac{1}{2}$	31 10	310	422[c]	551	388	528[c]	690[d]	414	563	735[d]	517	704[c]	919[d]
$26\frac{1}{2}$	28 9	279	380[c]	496	349	475[c]	621[d]	372	507	662[d]	465	633[c]	827[d]
$23\frac{1}{2}$	25 8	248	338[c]	441	310	422[c]	552[d]	331	450	588[d]	414	563[c]	735[d]
$20\frac{1}{2}$	22 7	217	295[c]	386	272	370[c]	483[d]	289	394	515[d]	362	492[c]	643[d]
$17\frac{1}{2}$	19 6	186	253[c]	331	233	317[c]	414[d]	248[c]	338	441[d]	310	422[c]	551[d]
$14\frac{1}{2}$	16 5	155	211[c]	276	194	264[c]	345[d]	207[c]	281	368[d]	258	352[c]	459[d]
$11\frac{1}{2}$	13 4	124	169[c]	221	155	211[c]	276[c]	165[c]	225	294[d]	207	281[c]	368[d]
$8\frac{1}{2}$	10 3	93.0	127[c]	165	116	158[c]	207[c]	124[c]	169	221[c]	155	211[c]	276[d]
$5\frac{1}{2}$	7 2	62.0	84.4[c]	110	77.6[c]	106[c]	138[c]	82.7[c]	113	147[c]	103[c]	141[c]	184[d]

Notes:
[a]Tabulated load values are based on double shear of bolts. See LRFD Specification Sect. J3.
[b]The enclosed beam web should be checked for design bearing load.
[c]Capacity is governed by net shear on angles for lengths L. For reduced value see Table II-C.
[d]Capacity is governed by net shear on angles for lengths L and L'. For reduced value see Table II-C.

FRAMED BEAM CONNECTIONS
Bolted
TABLE II Design loads in kips

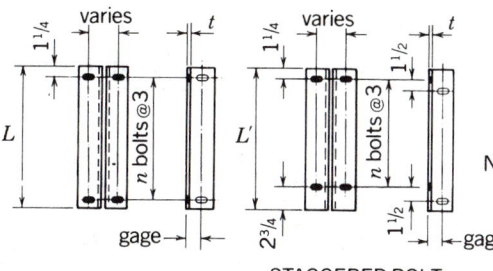

Note: For $L = 2\frac{1}{2}$ use one half
the tabular load value
shown for $L = 5\frac{1}{2}$, for the
same bolt type, diameter,
and thickness.

STAGGERED BOLT
ALTERNATE

TABLE II-B Bolt Shear[a,b]
For bolts in **slip-critical** connections with oversize, short-slotted, or long-slotted holes
and Class A, clean mill scale surface condition.

Hole Type			Long-slotted Holes						Oversize and Short-slotted Holes					
Bolt Type			A325-SC[d]			A490-SC[d]			A325-SC[d]			A490-SC[d]		
F_v			12			15			15			18		
Bolt dia. d, In.			$\frac{3}{4}$	$\frac{7}{8}$	1	$\frac{3}{4}$	$\frac{7}{8}$	1	$\frac{3}{4}$	$\frac{7}{8}$	1	$\frac{3}{4}$	$\frac{7}{8}$	1
Angle Thickness t, In.			$\frac{1}{4}$	$\frac{1}{4}$	$\frac{5}{16}$	$\frac{1}{4}$	$\frac{5}{16}$	$\frac{1}{2}$	$\frac{1}{4}$	$\frac{5}{16}$	$\frac{1}{2}$	$\frac{5}{16}$	$\frac{3}{8}$	$\frac{5}{8}$
L In.	L' In.	n												
$29\frac{1}{2}$	31	10	106	144	188	133	180	236	133	180	236	159	216	283
$26\frac{1}{2}$	28	9	95.4	129	170	119	162	212	119	162	212	143	195	254
$23\frac{1}{2}$	25	8	84.8	115	151	106	144	188	106	144	188	127	173	226
$20\frac{1}{2}$	22	7	74.2	101	132	92.8	126	165	92.8	126	165	111	152	198
$17\frac{1}{2}$	19	6	63.6	86.6	113	79.5	108	141	79.5	108	141	95.4	130	170
$14\frac{1}{2}$	16	5	53.0	72.2	94.2	66.3	90.0	118	66.3	90.0	118	79.5	108	141
$11\frac{1}{2}$	13	4	42.4	57.7	75.4	53.0	72.2	94.2	53.0	72.2	94.2	63.6	86.6	113
$8\frac{1}{2}$	10	3	31.8	43.3	56.5	39.8	54.1	70.7	39.8	54.1	70.7	47.7	64.9	84.8
$5\frac{1}{2}$	7	2	21.2	28.9	37.7[c]	26.5	36.1	47.1	26.5	36.1	47.1	31.8	43.3	56.5

Notes:
[a]Tabulated load values are based on double shear of bolts. See LRFD Specification Sect. J3.
[b]The enclosed beam web should be checked for design bearing load.
[c]Capacity is governed by net shear on angles for length L. For reduced value see Table II-C.
[d]Slip-critical connections are a serviceability criterion to prevent slip at service (working) loads.
Tabulated values are service load values and should be compared to the working loads, not
factored loads. Capacity of the connection should be checked as a bearing-type connection
with factored loads. If capacity is governed by net shear on angles, see Table II-C.

FRAMED BEAM CONNECTIONS
Bolted
TABLE II Design loads in kips

TABLE II-C													
Design Shear in Connection Angles for A36 Material													

Bolt Dia. d, In.		¾				⅞					1				
Angle Thickness t, In.		¼	⁵⁄₁₆	⅜	½	¼	⁵⁄₁₆	⅜	½	⅝	¼	⁵⁄₁₆	⅜	½	⅝
L In.	n														
29½	10	279	349	418	558	263	328	394	525	657	246	308	369	493	616
26½	9	250	313	376	501	236	295	354	471	589	221	276	332	442	553
23½	8	222	277	333	444	209	261	313	418	522	196	245	294	392	489
20½	7	193	242	290	387	182	227	273	364	455	170	213	256	341	426
17½	6	165	206	247	330	155	194	232	310	387	145	181	218	290	363
14½	5	136	170	204	272	128	160	192	256	320	120	150	180	240	300
11½	4	108	135	161	215	101	126	152	202	253	94.6	118	142	189	237
8½	3	79.1	98.9	119	158	74.2	92.8	111	148	186	69.3	86.7	104	139	173
5½	2	50.6	63.2	75.9	101	47.3	59.1	71.0	94.6	118	44.0	55.1	66.1	88.1	110
L' In.	n														
31	10	299	373	448	597	282	353	423	564	706	266	332	399	532	665
28	9	270	337	405	540	255	319	383	511	638	241	301	361	481	602
25	8	241	302	362	483	228	285	343	457	571	215	269	323	431	538
22	7	213	266	319	426	201	252	302	403	504	190	238	285	380	475
19	6	184	230	276	369	175	218	262	349	436	165	206	247	330	412
16	5	156	195	234	312	148	185	221	295	369	139	174	209	279	349
13	4	127	159	191	254	121	151	181	241	302	114	143	171	228	285
10	3	98.7	123	148	197	93.8	117	141	188	234	88.9	111	133	178	222
7	2	70.1	87.7	105	140	66.9	83.6	100	134	167	63.6	79.5	95.4	127	159

Notes:
Table based on design shear of $\phi(0.6) F_u$ (26.1 ksi for A36 angles) on the net section of two angles. $\phi = 0.75$
Net section based on diameter of fastener + ¹⁄₁₆ in.

FRAMED BEAM CONNECTIONS
Welded—E70XX electrodes for combination with Table II connections

TABLE III

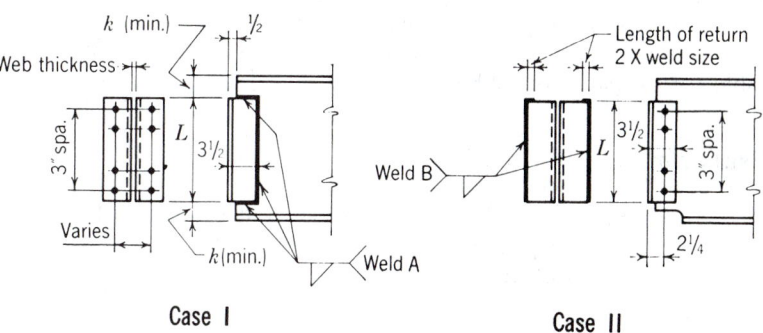

Case I

Case II

Table III is arranged to permit substitution of welds for bolts in the connections shown in Table II which fall within the weld capacities. Welds A replace fasteners in the beam web legs (Case I). Welds B replace fasteners in the outstanding legs (Case II).

To accommodate usual gages, angle leg widths will generally be $4 \times 3\frac{1}{2}$, with the 4″ leg outstanding. Width of web legs in Case I may be reduced optionally from $3\frac{1}{2}$″ to 3″. Width of outstanding legs in Case II may be reduced optionally from 4″ to 3″ for values of $L = 5\frac{1}{2}$″ through 1′-5½″.

Angle thickness is equal to weld size plus $\frac{1}{16}$″, or thickness of angle from applicable Table II, whichever is greater.

Angle length L must be as tabulated in Table III.

Holes for erection bolts may be placed as required in legs that are to be field welded (optional).

When bolts are used, investigate bearing capacity of supporting member.

Although it is permissible to use Welds A and B in combination to obtain all-welded connections, it is recommended that such connections be chosen from Table IV. This table will usually provide greater economy and allow increased flexibility in selection of angle lengths and connection capacities.

Design capacity for Weld A utilizes instantaneous center solutions based upon the same criteria developed for Tables XVIII to XXV Eccentric Loads on Weld Groups. However, capacity for Weld B is computed using traditional vector analysis techniques.

EXAMPLES: CASE I

(a) *Given:*

Beam: W36 × 150; $t_w = 0.625$ in.
                      $F_y = 36$ ksi
Factored reaction: 300 kips
Bolts: ⅞-in. dia., ASTM A325-N
Welds: E70XX

Solution:

Enter Table III under Weld A and note that a value that satisfies the reaction is 380 kips. This requires 5⁄16″ welds and 23½″ long angles. Use ⅜″ thick angles to meet the weld requirement stipulated in LRFD Specification Sect. J2.2. The 0.625″ web thickness is less than the minimum required 0.69″, so the reduction in capacity is $(0.625/0.69) \times 380$ kips = 344 kips.

Note in Table II-A that the angle length provides for 8 rows of ⅞″ dia. ASTM A325-N bolts with a capacity of 338 kips. The ⅜″ required angle thickness is the same as the ⅜″ angle thickness required due to Weld A. Footnote C refers to Table II-C; capacity is 313 kips.

Detail Data: Two L4 × 3½ × ⅜ × 1′-11½″
F_y = 36 ksi
Sixteen ⅞″ dia. ASTM A325-N bolts (threads included in shear plane)
5⁄16″ fillet weld, E70XX

(b) *Given:*

Beam: W16 × 26; t_w = 0.25
F_y = 36 ksi
Reaction: 70 kips
Bolts: ⅞″ dia. ASTM A307
Welds: E70XX

Solution:

See Table II-A and note 4 rows of bolts with 11½″ long angles are compatible with a 16″ deep section. Capacity of the ⅞″ dia. ASTM A307 bolts with ¼″ thick angles is 77.9 kips.

Note in Table III that 119 kips capacity is designated for 3⁄16″ Weld A and 11½″ long angles. The 0.25″ web thickness is less than the minimum 0.41″ listed. The reduced capacity is $(0.25/0.41) \times 119$ kips = 72.6 kips. The ¼″ angle thickness required for bolts is satisfactory for the 3⁄16″ weld.

Detail Data: Two L4 × 3½ × ¼ × 0′-11½″
F_y = 36 ksi
Eight ⅞″ dia. ASTM A307 bolts
3⁄16″ fillet weld, E70XX

EXAMPLES: CASE II

(c) *Given:*

Beam: W36 × 150; t_w = 0.625″
F_y = 36 ksi
Reaction: 230 kips
Bolts: ⅞″ dia. ASTM A490-X
Welds: E70XX

Solution:

Enter Table III under Weld B and note that the value most nearly satisfying the reaction is 234 kips. This requires $5/16''$ Weld B and $20\frac{1}{2}''$ long $3/8''$ thick angles. Table II-A shows a bolt capacity for a 7 fastener connection of 492 kips, which is more than the 230 kips required. Therefore a $20\frac{1}{2}''$ long angle is selected from Table III. Table II-C lists a capacity of 273 kips for a $20\frac{1}{2}''$ long $3/8''$ thick angle.

Check bearing capacity of web with Table I-F, assuming beam is not coped and pitch is 3 in.

$$7 \times 91.3 \times 0.625 = 399 \text{ kips} \quad \textbf{o.k.}$$

Detail Data: Two L4 × 3½ × ⅜ × 1'-8½"
F_y = 36 ksi
Seven ⅞" dia. ASTM A490-X bolts
$5/16''$ fillet weld, E70XX

(d) *Given:*

Beam: W16 × 31; t_w = 0.275"
F_y = 50 ksi
Reaction: 60 kips
Bolts: ¾" dia. ASTM A325-N (threads included in shear plane)
Welds: E70XX

Solution:

Enter Table III under Weld B and note that the value most nearly satisfying the reaction is 60.5 kips. This requires $5/16''$ Weld B and 8½" long, ⅜" thick angles.

Enter Table II-A for 3 rows of fasteners and note that the angle length is compatible with beam size. Capacity of three ¾" dia. ASTM A325-N bolts is 93.0 kips.

Check bearing on beam web = 60/3 = 20 kips per bolt in double shear. From Table I-E, the bearing design load is 87.8 kips for 1" thickness at 3" spacing; 87.8 × 0.275 = 24.1 kips per bolt permissible. From Table I-F, single shear bearing capacity of a ⅜", 36 ksi angle with a 1¼" edge distance is 88.1 kips.

Detail Data: Two L4 × 3½ × ⅜ × 0'-8½"
F_y = 36 ksi
Three ¾" dia. ASTM A325-N bolts
$5/16''$ fillet weld, E70XX

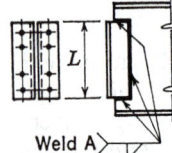

FRAMED BEAM CONNECTIONS
Welded—E70XX electrodes for combination with Table II connections
TABLE III Design loads in kips

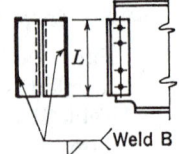

Weld A Weld B

Weld A		Weld B		Angle Length	Minimum Web[a] Thickness for Welds A		Maximum Number of Fasteners in One Vertical Row (Table II)
Capacity, Kips	Size,[b] In.	Capacity,[c] Kips	Size, In.	L In.	$F_y = 36$ Ksi $\phi F_v = 20.2$ Ksi	$F_y = 50$ Ksi $\phi F_v = 28.0$	
468	5⁄16	444	3⁄8	29½	.69	.50	
375	¼	371	5⁄16	29½	.55	.40	10
281	3⁄16	296	¼	29½	.41	.30	
424	5⁄16	392	3⁄8	26½	.69	.50	
339	¼	326	5⁄16	26½	.55	.40	9
254	3⁄16	260	¼	26½	.41	.30	
380	5⁄16	335	3⁄8	23½	.69	.50	
304	¼	279	5⁄16	23½	.55	.40	8
228	3⁄16	224	¼	23½	.41	.30	
336	5⁄16	281	3⁄8	20½	.69	.50	
269	¼	234	5⁄16	20½	.55	.40	7
201	3⁄16	188	¼	20½	.41	.30	
290	5⁄16	228	3⁄8	17½	.69	.50	
232	¼	189	5⁄16	17½	.55	.40	6
174	3⁄16	152	¼	17½	.41	.30	
245	5⁄16	173	3⁄8	14½	.69	.50	
196	¼	144	5⁄16	14½	.55	.40	5
147	3⁄16	115	¼	14½	.41	.30	
198	5⁄16	120	3⁄8	11½	.69	.50	
158	¼	100	5⁄16	11½	.55	.40	4
119	3⁄16	80.1	¼	11½	.41	.30	
150	5⁄16	72.3	3⁄8	8½	.69	.50	
120	¼	60.5	5⁄16	8½	.55	.40	3
90.0	3⁄16	48.3	¼	8½	.41	.30	
101	5⁄16	32.9	3⁄8	5½	.69	.50	
80.6	¼	27.5	5⁄16	5½	.55	.40	2
60.5	3⁄16	21.9	¼	5½	.41	.30	

[a]When the beam web thickness is less than the minimum, multiply the connection capacity furnished by Weld A by the ratio of the actual web thickness to the tabulated minimum thickness. Thus, if 5⁄16″ Weld A, with a connection capacity of 245 kips and a 14½″ long angle, is considered for a beam of web thickness of 0.375″ with $F_y = 36$ ksi, the connection capacity must be multiplied by 0.375/0.69, giving 133 kips.

[b]Should the thickness of material to which connection angles are welded exceed the limits set by LRFD Specification J2.2b for weld sizes specified, increase the weld size as required, but not to exceed the angle thickness.

[c]When welds are used on outstanding legs, connection capacity may be limited by the shear capacity of the supporting member as stipulated by LRFD Specification Sect. J5.2. See Examples d and e for Table IV.

Note 1: Connection Angles: Two L4x3½xthicknessx L; $F_y = 36$ ksi. See discussion preceding examples for Table III limiting values of thickness and optional width of legs.

Note 2: Capacities shown in this table apply only when the material welded is $F_y = 36$ ksi or $F_y = 50$ ksi steel.

FRAMED BEAM CONNECTIONS
Welded—E70XX electrodes

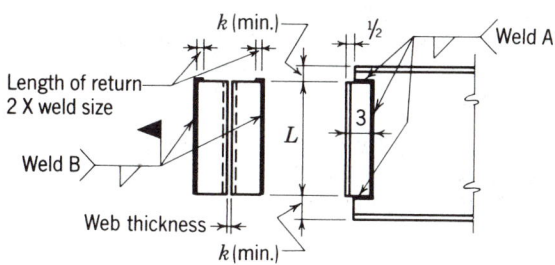

TABLE IV

Table IV lists capacities and details for angle connections welded to both the beam web and the supporting member.

Holes for erection bolts may be placed as required in legs that are to be field-welded (optional).

Design capacity for Weld A utilizes instantaneous center solutions based on the same criteria developed for Tables XVIII to XXV, Eccentric Loads on Weld Groups. However, Weld B capacity is computed using traditional vector analysis technique.

EXAMPLES

(a) *Given:*

Beam:	W36 × 150; t_w = 0.625″; T = 32⅛″
	F_y = 36 ksi
Weld:	E70XX
Factored reaction:	300 kips

Solution:

Enter Table IV and select a Weld A capacity of 328 kips (weld size = ¼″). Weld B has a capacity of 317 kips and is satisfactory. The angle length (26″) is less than T for the W36 × 150 and is satisfactory. The beam web thickness (0.625″) exceeds the minimum web thickness (0.55″), so no reduction in Weld A capacity is required.

Detail Data: Two L4 × 3 × ⅜ × 2′-2″; F_y = 36 ksi
                     Weld A = ¼″, E70XX
                     Weld B = ⁵⁄₁₆″, E70XX

(b) *Given:*

Same data as Ex. (a) except the factored reaction is 220 kips.

Solution:

Enter Table IV and select a Weld A capacity of 257 kips (weld size = ¼"). Weld B has a capacity of 228 kips and is satisfactory. The angle length (20") is less than T and is satisfactory. The beam web thickness (0.625") exceeds the minimum web thickness (0.55"), so no reduction in Weld A capacity is required.

Unless framing details require this short angle length, longer angles with less deposited weld metal may be desirable. The 26" long angles with Weld A capacity of 246 kips (weld size = ³⁄₁₆") and Weld B capacity of 254 kips are also satisfactory and may be selected.

Detail Data: Two L4 × 3 × ⁵⁄₁₆ × 2'-2"; F_y = 36 ksi
              Weld A = ³⁄₁₆", E70XX
              Weld B = ¼", E70XX

(c) *Given:*

Beam: W16 × 26; t_w = 0.25"; T = 13⅝"
           F_y = 50 ksi
Weld: E70XX
Reaction: 50 kips

Solution:

Enter Table IV and select a Weld B capacity of 53.3 kips (weld size = ¼"). Angle length (8") is less than T and is satisfactory. Weld A has a capacity of 83 kips using ³⁄₁₆" weld size. However, the beam web thickness (0.25") is less than the minimum web thickness (0.30"), so Weld A capacity is reduced to .25/.30 times 83, or 69.2 kips.

Detail Data: Two L3 × 3 × ⁵⁄₁₆ × 0'-8"; F_y = 36 ksi
              Weld A = ³⁄₁₆", E70XX
              Weld B = ¼", E70XX

WELDS TO SUPPORTING MEMBERS

Selection of connections tabulated herein is based on and limited by the requirement that Weld B will be applied in accordance with LRFD Specification Sect. J2.2, which stipulates minimum welds for various material thicknesses.

With respect to Weld B, it should be noted that supporting members with limited shear capacity, or which support opposed connections, may be subject to a reduction in connection capacity. See LRFD Specification Sect. J5.2.

For Weld A, LRFD Specification Formula J5-3 is used to determine the minimum web thickness. For E70 electrodes weld design strength per ¹⁄₁₆ in. leg dimension is $0.75 \times 0.6 \times 70 \times 0.707 \times ¹⁄₁₆ = 1.3919$ kips/in. Material design strength is $0.8 \times 0.7 \times F_y \times t = 0.56 F_y \times t$. The minimum web thickness t is

$$t = \frac{D \times 1.3919 \times 2}{0.56 F_y}$$

where

D = weld size in sixteenths of an inch.

EXAMPLES

(d) *Given:*

Weld B = ⁵⁄₁₆″ fillet weld, E70XX, fully loaded on one side of ¼″ thick supporting member web of $F_y = 36$ ksi steel.

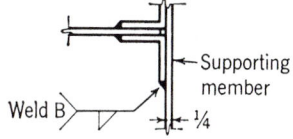

Weld B

Solution:

Shear value of one ⁵⁄₁₆″ fillet weld = 0.3125″ × 0.707 × 31.5 ksi = 6.96 kips/lin. in. Shear value of ¼″ thick web = 0.25″ × 20.2 ksi = 5.05 kips/lin. in.

Because of this deficiency in web shear capacity, the total capacity selected from the Weld B column for ⁵⁄₁₆″ weld must be reduced by the ratio 5.05/6.96.

(e) *Given:*

Two floor beams with end reactions of 23.0 kips each are to be supported by a beam of $F_y = 36$ ksi steel with a ⁵⁄₁₆″ thick web.

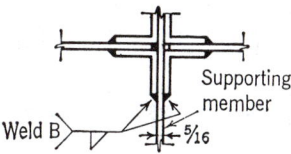

Weld B

Solution:

¼″ Weld B with 5″ long angles has a capacity of 23.6 kips when fully loaded. Maximum shear developed in the two ¼″ fillet Welds B on opposite sides of the supporting beam web = 2 × 0.25 × 0.707 × 31.5 × 23/23.6 = 10.85 kips/lin. in. Shear capacity of ⁵⁄₁₆″ web = 0.3125 × 20.2 = 6.3 kips/lin. in. A longer connection is required to reduce the web shear. Required Weld B capacity is 23.6 kips × 10.85/6.3 = 40.6 kips. Two 7″ long angles with ¼″ Weld B have a tabulated capacity of 42.5 kips and are adequate.

FRAMED BEAM CONNECTIONS
Welded—E70XX electrodes
TABLE IV

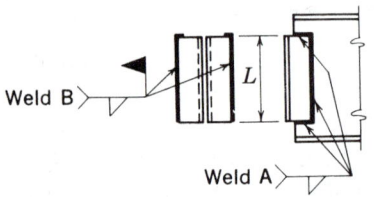

Weld B

L

Weld A

Weld A		Weld B		Angle Length	Angle Size	Minimum Web[a] Thickness for Weld A	
Capacity Kips	Size[b] In.	Capacity[c] Kips	Size[b] In.	L, In.	F_y = 36 Ksi	F_y = 36 Ksi ϕF_v = 20.2 Ksi	F_y = 50 Ksi ϕF_v = 28 Ksi
496	5/16	489	3/8	32	4x3x1/2	.69	.50
397	1/4	407	5/16	32	4x3x3/8	.55	.40
297	3/16	326	1/4	32	4x3x5/16	.41	.30
467	5/16	453	3/8	30	4x3x1/2	.69	.50
374	1/4	377	5/16	30	4x3x3/8	.55	.40
280	3/16	302	1/4	30	4x3x5/16	.41	.30
439	5/16	417	3/8	28	4x3x1/2	.69	.50
351	1/4	347	5/16	28	4x3x3/8	.55	.40
263	3/16	278	1/4	28	4x3x5/16	.41	.30
410	5/16	381	3/8	26	4x3x1/2	.69	.50
328	1/4	317	5/16	26	4x3x3/8	.55	.40
246	3/16	254	1/4	26	4x3x5/16	.41	.30
380	5/16	345	3/8	24	4x3x1/2	.69	.50
304	1/4	287	5/16	24	4x3x3/8	.55	.40
228	3/16	230	1/4	24	4x3x5/16	.41	.30
351	5/16	309	3/8	22	4x3x1/2	.69	.50
281	1/4	257	5/16	22	4x3x3/8	.55	.40
211	3/16	206	1/4	22	4x3x5/16	.41	.30
322	5/16	272	3/8	20	4x3x1/2	.69	.50
257	1/4	228	5/16	20	4x3x3/8	.55	.40
193	3/16	182	1/4	20	4x3x5/16	.41	.30
292	5/16	236	3/8	18	4x3x1/2	.69	.50
234	1/4	197	5/16	18	4x3x3/8	.55	.40
175	3/16	158	1/4	18	4x3x5/16	.41	.30
262	5/16	222	3/8	16	3x3x1/2	.69	.50
209	1/4	185	5/16	16	3x3x3/8	.55	.40
157	3/16	148	1/4	16	3x3x5/16	.41	.30
232	5/16	186	3/8	14	3x3x1/2	.69	.50
185	1/4	155	5/16	14	3x3x3/8	.55	.40
139	3/16	124	1/4	14	3x3x5/16	.41	.30
201	5/16	149	3/8	12	3x3x1/2	.69	.50
161	1/4	125	5/16	12	3x3x3/8	.55	.40
121	3/16	99.8	1/4	12	3x3x5/16	.41	.30

For footnotes, see next page.

FRAMED BEAM CONNECTIONS
Welded—E70XX electrodes
TABLE IV

Weld A		Weld B		Angle Length	Angle Size	Minimum Web[a] Thickness for Weld A	
Capacity Kips	Size[b] In.	Capacity[c] Kips	Size[b] In.	L, In.	F_y = 36 Ksi	F_y = 36 Ksi ϕF_v = 20.2 Ksi	F_y = 50 Ksi ϕF_v = 28 Ksi
170	5/16	114	3/8	10	3x3x1/2	.69	.50
136	1/4	95.0	5/16	10	3x3x3/8	.55	.40
102	3/16	75.8	1/4	10	3x3x5/16	.44	.30
154	5/16	96.5	3/8	9	3x3x1/2	.69	.50
123	1/4	80.6	5/16	9	3x3x3/8	.55	.40
92	3/16	64.4	1/4	9	3x3x5/16	.44	.30
138	5/16	79.8	3/8	8	3x3x1/2	.69	.50
110	1/4	66.6	5/16	8	3x3x3/8	.55	.40
83	3/16	53.3	1/4	8	3x3x5/16	.44	.30
122	5/16	63.8	3/8	7	3x3x1/2	.69	.50
97	1/4	53.3	5/16	7	3x3x3/8	.55	.40
73	3/16	42.5	1/4	7	3x3x5/16	.44	.30
105	5/16	48.9	3/8	6	3x3x1/2	.69	.50
84	1/4	40.7	5/16	6	3x3x3/8	.55	.40
63	3/16	32.6	1/4	6	3x3x5/16	.44	.30
89	5/16	35.1	3/8	5	3x3x1/2	.69	.50
71	1/4	29.3	5/16	5	3x3x3/8	.55	.40
53	3/16	23.6	1/4	5	3x3x5/16	.44	.30
73	5/16	23.3	3/8	4	3x3x1/2	.69	.50
58	1/4	19.4	5/16	4	3x3x3/8	.55	.40
44	3/16	15.6	1/4	4	3x3x5/16	.44	.30

[a]When the beam web thickness is less than the minimum, multiply the connection capacity furnished by Welds A by the ratio of the actual thickness to the tabulated minimum thickness. Thus, if 5/16″ Weld A, with a connection capacity of 138 kips and an 8″ long angle, is considered for a beam of web thickness 0.305″ and F_y = 36 ksi, the connection capacity must be multiplied by 0.305/0.69, giving 61.0 kips.

[b]Should the thickness of material to which connection angles are welded exceed the limits set by LRFD Specification Sect. J2.2b for weld sizes specified, increase the weld size as required, but not to exceed the angle thickness.

[c]For welds on outstanding legs, connection capacity may be limited by the shear capacity of the supporting members, as stipulated by LRFD Specification Sect. J5.2. See Examples d and e for Table IV.

Note: Capacities shown in this table apply only when connection angles are F_y = 36 ksi steel and the material to which they are welded is either F_y = 36 ksi or F_y = 50 ksi steel.

SEATED BEAM CONNECTIONS
Bolted

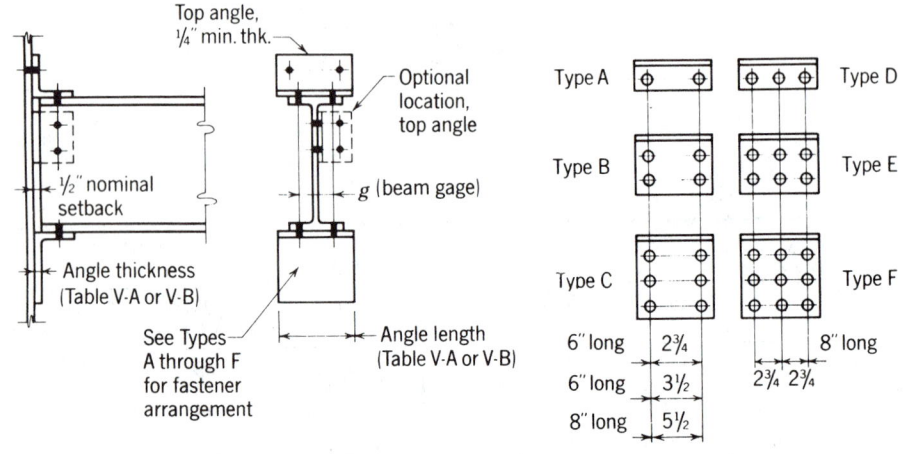

Seat Angle Types

TABLE V

Seated connections should be used only when the beam is supported by a top angle placed as shown above, or in the optional location as indicated.

Nominal beam setback is ½". Design loads in Tables V-A and V-B are based on ¾" setback, which provides for possible mill underrun in beam length.

ASTM A307 bolts may be used in seated connections, provided the stipulations of LRFD Specification Sect. J1.9 are observed.

Design loads in Table V-A are based on F_y = 36 ksi steel in both beam and seat angle. These values will be conservative when used with beams of F_y greater than 36 ksi. For beams with F_y equal to or greater than 50 ksi, use Table V-B.

Design loads in Table V-B are based on F_y = 36 ksi steel in the seat angle and F_y = 50 ksi steel in the beams. For beams with F_y greater than 50 ksi, these values will be conservative.

Vertical spacing of fasteners and gages in seat angles may be arranged to suit conditions, provided they conform to LRFD Specification Sects. J3.9 and J3.10 with regard to minimum spacing and minimum edge distances. Where thick angles are used, driving clearances may require an increase in the outstanding leg gage.

In the event the thin web of a supporting member limits its bearing capacity, it will be necessary to reduce values listed in Table V-C.

For the most economical connection, the reaction values of the beams should be shown on the contract drawings. If the reactions are not shown, the connections must be selected to support the beam end reaction calculated from the tables of Uniform Load Constants (Part 3) for the given shape, span, and steel specified for the beam in question. The effect of concentrated loads near an end connection **must** also be considered.

EXAMPLES

(a) *Given:*

Beam:	W16 × 50 (⅜″ web)
	F_y = 36 ksi material
Factored reaction:	55 kips
Bolts:	⅞″ dia. A325-N
Column gage:	5½″ in column web

Solution:

Enter Table V-A under 8″ angle length, for a ⅜″ beam web; select a ¾″ angle thickness (capacity = 55.6 kips). Enter Table V-C opposite ⅞″ dia. A325-N; note that a Type D connection (capacity = 63.3 kips) is required. From Table V-D, with a Type D connection, a 4x4 angle is available in ¾″ thickness.

Detail Data: Seat: One L4 × 4 × ¾ × 0′-8″ with three ⅞″ dia. A325-N bolts.

Top angle or side support: To be chosen to suit conditions.

(b) *Given:*

Same as Ex. (a), except connect to a column flange with column gage = 5½″.

Solution:

As in Ex. (a), a ¾″ angle thickness is adequate. Enter Table V-C opposite ⅞″ dia. A325-N; note that a Type B connection (capacity = 84.4 kips) is required. From Table V-D, with a Type B connection, a 6 × 4 angle is available in ¾″ thickness.

Detail Data: Seat: One L6 × 4 × ¾ × 0′-8″ (4″ OSL) with four ⅞″ dia. A325-N bolts.

Top angle or side support: To be chosen to suit conditions.

SEATED BEAM CONNECTIONS
Bolted
TABLE V Design loads in Kips

TABLE V-A Outstanding Leg Capacity, kips (based on OSL = 4.0 In.)

		Angle Length, In.	6 In.					8 In.				
		Angle Thickness, In.	3/8	1/2	5/8	3/4	1	3/8	1/2	5/8	3/4	1
F_y (ksi)	36	3/16	13.6	18.5	22.6	26.8	29.8	15.7	20.3	25.0	29.6	29.8
		1/4	15.7	23.3	30.0	34.9	43.3	18.1	26.5	32.6	38.1	43.3
		5/16	17.5	26.4	35.4	44.5	57.3	20.3	30.0	39.8	49.3	60.7
	Beam Web Thickness, In.	3/8	19.2	29.2	39.5	49.9	70.3	22.2	33.1	44.3	55.6	75.0
		7/16	20.8	31.9	43.3	55.0	78.4	24.0	36.1	48.6	61.2	86.5
		1/2	22.2	34.4	47.1	59.9	85.8	25.6	38.9	52.6	66.5	94.4
		9/16	23.5	36.8	50.6	64.6	93.0	27.2	41.6	56.5	71.6	102

TABLE V-B Outstanding Leg Capacity, Kips (based on OSL = 4.0 In.)

		Angle Length, In.	6 In.					8 In.				
		Angle Thickness, In.	3/8	1/2	5/8	3/4	1	3/8	1/2	5/8	3/4	1
F_y (ksi)	50	3/16	16.0	23.2	28.3	33.4	41.5	18.5	25.3	31.0	36.7	41.5
		1/4	18.5	28.0	37.7	44.3	56.6	21.3	31.8	41.2	47.9	60.2
		5/16	20.7	31.7	43.1	54.7	73.7	23.9	36.0	48.3	60.9	78.8
	Beam Web Thickness, In.	3/8	22.6	35.2	48.3	61.5	88.2	26.1	39.9	53.9	68.2	96.9
		7/16	24.5	38.5	53.1	68.0	98.1	28.2	43.5	59.3	75.2	108
		1/2	26.1	41.7	57.8	74.3	108	30.2	47.0	64.3	82.0	118
		9/16	27.7	44.7	62.4	80.4	117	32.0	50.3	69.2	88.5	128

TABLE V-C Fastener Capacity, Kips

Fastener[a] Designation	Fastener Dia., In.	A	B	C	D	E	F
A307	3/4	14.3	28.6	42.9	21.5	42.9	64.4
	7/8	19.5	39.0	58.4	29.2	58.4	87.7
	1	25.4	50.9	76.3	38.2	76.3	115
A325-N	3/4	31.0	62.0	93.0	46.5	93.0	140
	7/8	42.2	84.4	127.0	63.3	127	—
	1	55.1	110	165	82.7	165	—
A325-X	3/4	41.4	82.7	124	62.0	124	186
	7/8	56.3	113	—	84.4	—	—
	1	73.5	147	—	110	—	—
A490-N	3/4	38.8	77.6	116	58.2	116	175
	7/8	52.8	106	158	79.2	158	—
	1	69.0	138	—	103	—	—
A490-X	3/4	51.7	103	155	77.5	155	233
	7/8	70.4	141	—	106	—	—
	1	91.9	184	—	138	—	—

TABLE V-D Available Seat Angle and Thickness Range

Type	Angle Size, In.	t, In.
A, D	4x3	3/8 – 1/2
	4x3½	3/8 – 1/2
	4x4	3/8 – 3/4
B, E	6x4	3/8 – 3/4
	7x4	3/8 – 3/4
	8x4	1/2 – 1
C, F	8x4[b]	1/2 – 1

[b]Suitable for use with 3/4" and 7/8" fasteners only.

[a]A325-N and A490-N: Bearing-type connections with threads included in shear plane.
A325-X and A490-X: Bearing-type connections with threads excluded from shear plane.

SEATED BEAM CONNECTIONS
Welded—E70XX electrodes

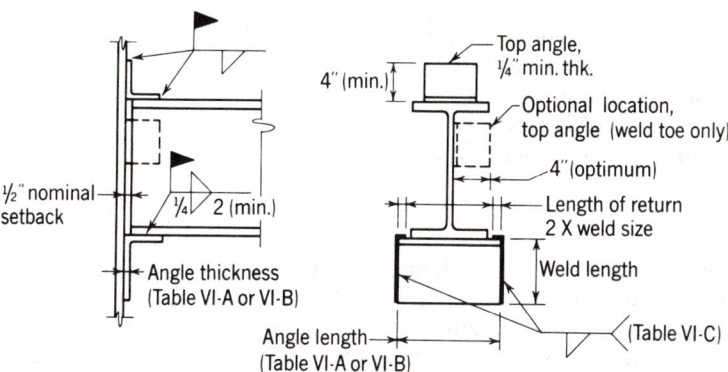

TABLE VI

Seated connections should be used only when the beam is supported by a top angle placed as shown above, or in the optional location as indicated.

Design loads in Table VI are based on the use of E70XX electrodes. The table may be used for other electrodes, provided the tabular values are adjusted for the electrodes used (e.g., for E60XX electrodes, multiply tabular values by 60/70, or 0.86, etc.) and the welds and base metal meet the required strength level provisions of LRFD Specification Sect. J2.

Welds attaching beams to seat or top angles may be replaced by bolts. ASTM A307 bolts may be used in seated connections, provided the stipulations of LRFD Specification Sect. J1.7 are observed.

In addition to the welds shown, temporary erection bolts may be used to attach beams to seats (optional).

Nominal beam setback is ½". Design loads in Tables VI-A and VI-B are based on ¾" setback, which provides for possible mill underrun in beam length.

Design loads in Table VI-A are based on F_y = 36 ksi material in both beam and seat angle. These values will be conservative when used for beams with F_y greater than 36 ksi.

Design loads in Table VI-B are based on F_y = 36 ksi material in the seat angle with beam material of F_y = 50 ksi. These values will be conservative when used with beams of F_y greater than 50 ksi.

Design weld capacities in Table VI-C are computed using traditional vector analysis.

Should combinations of material thickness and weld size selected from Tables VI-A or VI-B and VI-C exceed the limits set by LRFD Specification Sect. J2.2, increase the weld size or material thickness as required.

No reduction of the tabulated weld capacity is required when unstiffened seats line up on opposite sides of the supporting web.

For the most economical connection, the reaction values of the beams should be shown on the contract drawings. If the reactions are not shown, the connections must be selected to support the beam end reaction calculated from the tables of Uniform Load Constants for the given shape, span, and steel specified for the beam in question. The effect of concentrated loads near an end connection must also be considered.

5 - 44

EXAMPLE

Given:

Beam: W21 × 62 (⅜″ web).
           Attach beam flange to seat with bolts.
           F_y = 36 ksi material
Reaction: 50 kips
Welds: E70XX electrodes
Column: Column web will permit use of 8″ long seat angle.

Solution:

Enter Table VI-A opposite ⅜″ web thickness; under 8″ angle length, read 55.6 kips. Note that a ¾″ angle thickness is required. Enter Table VI-C and note that satisfactory weld capacities appear under 5 through 8 inch leg angles, all of which are shown to be available in ¾″ thickness. In this case the 5 × 3½ and 6 × 4 angles are ruled out because of the rather heavy welds required. (⁷⁄₁₆″ and ⅜″ respectively).

Angles 8 × 4 (capacity = 53.4 kips, ¼″ weld) and 7 × 4 (capacity = 53.4 kips, ⁵⁄₁₆″ weld) are equally suitable. Angle 7 × 4 is chosen because the material savings will usually offset the cost differential between welds of ¹⁄₁₆″ thickness differential, provided that each weld can be made with the same number of passes (⁵⁄₁₆″ welds and smaller are single pass welds).

Detail Data: One L7 × 4 × ¾ × 0′-8″ with ⁵⁄₁₆″ welds (E70XX).
               Top or side angle, as required.

Had it been required to weld the beam to the seat, the ¾″ seat angle thickness would dictate a ¼″ weld (see LRFD Specification Sect. J2.2b), which is compatible with the ⅝″ beam flange thickness (see LRFD Specification Sect. J2.2b). Block beam flange to permit welding to the 8″ seat angle or use a longer seat angle if space permits.

544

33

SEATED BEAM CONNECTIONS
Welded—E70XX electrodes
TABLE VI Design loads in kips

TABLE VI-A Outstanding Leg Capacity, Kips (based on OSL = 3½ or 4 in.)

		Angle Length, In.	6 In.					8 In.				
		Angle Thickness, In.	3/8	1/2	5/8	3/4	1	3/8	1/2	5/8	3/4	1
F_y (ksi)	36	Beam Web Thickness In. — 3/16	13.6	18.5	22.6	26.8	29.8	15.7	20.3	25.0	29.6	29.8
		1/4	15.7	23.3	30.0	34.9	43.3	18.1	26.5	32.6	38.1	43.3
		5/16	17.5	26.4	35.4	44.5	57.3	20.3	30.0	39.8	49.3	60.7
		3/8	19.2	29.2	39.5	49.9	70.3	22.2	33.1	44.3	55.6	75.0
		7/16	20.8	31.9	43.3	55.0	78.4	24.0	36.1	48.6	61.2	86.5
		1/2	22.2	34.4	47.1	59.9	85.8	25.6	38.9	52.6	66.5	94.4
		9/16	23.5	36.8	50.6	64.6	93.0	27.2	41.6	56.5	71.6	102

Note: Values above heavy lines apply only for 4-in. outstanding legs.

TABLE VI-B Outstanding Leg Capacity, Kips (based on OSL = 3½ or 4 in.)

		Angle Length, In.	6 In.					8 In.				
		Angle Thickness, In.	3/8	1/2	5/8	3/4	1	3/8	1/2	5/8	3/4	1
F_y (ksi)	50	Beam Web Thickness, In. — 3/16	16.0	23.2	28.3	33.4	41.5	18.5	25.3	31.0	36.7	41.5
		1/4	18.5	28.0	37.7	44.3	56.6	21.3	31.8	41.2	47.9	60.2
		5/16	20.7	31.7	43.1	54.7	73.7	23.9	36.0	48.3	60.9	78.8
		3/8	22.6	35.2	48.3	61.5	88.2	26.1	39.9	53.9	68.2	96.9
		7/16	24.5	38.5	53.1	68.0	98.1	28.2	43.5	59.3	75.2	108
		1/2	26.1	41.7	57.8	74.3	108	30.2	47.0	64.3	82.0	118
		9/16	27.7	44.7	62.4	80.4	117	32.0	50.3	69.2	88.5	128

Note: Values above heavy lines apply only for 4-in. outstanding legs.

TABLE VI-C Weld Capacity, Kips

Weld Size, In.	E70XX Electrodes				
	Seat Angle Size (long leg vertical)				
	4x3½	5x3½	6x4	7x4	8x4
1/4	17.3	25.8	32.7	42.8	53.4
5/16	21.5	32.3	41.0	53.4	66.8
3/8	25.8	38.7	49.1	64.1	80.1
7/16	30.2	45.2	57.3	74.7	93.5
1/2	—	51.6	65.4	83.4	107
5/8	—	64.5	81.8	107	134
11/16	—	71.0	90.0	117	—
3/4	—	—	—	—	—
Range of Available Seat Angle Thicknesses					
Minimum	3/8	3/8	3/8	3/8	1/2
Maximum	1/2	3/4	3/4	3/4	1

STIFFENED SEATED BEAM CONNECTIONS
Bolted

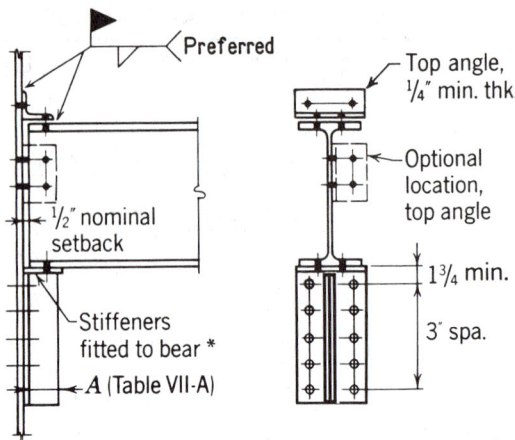

TABLE VII

Seated connections should be used only when the beam is supported by a top angle placed as shown above, or in the optional location as indicated.

Design capacities in Table VII-A are based on factored nominal bearing using steel of $F_y = 36$ ksi or $F_y = 50$ ksi in the stiffener angles. Capacities of fastener groups in Table VII-B are based on single shear. Connection capacity is based on the lesser of these two values in conjunction with the local web yielding value of the supported beam.

Effective length of stiffener bearing is assumed ½″ less than length of outstanding leg.

Maximum gage in legs of stiffeners connected to columns is 2½″.

ASTM A307 bolts may be used in seated connections, provided the stipulations of LRFD Specification Sect. J1.9 are observed.

Vertical spacing of fasteners in stiffener angles may be arranged to suit conditions, provided they conform to Sects. J3.9 and J3.10 with respect to minimum spacing and minimum edge distances.

Beam seats, traditionally provided in column webs for simple beams, permit advantages:

1. Beams require only routine punching
2. Ample erection clearance is provided
3. Erection is fast, safe and simple
4. Bay length of structure is easy to keep.

To permit selection of the most economical beam connection, the beam reactions should be shown on the contract drawings. If they are not shown, the connections must be selected to support the beam end reaction calculated from the tables of Uniform Load Constants for the given shape, span, and steel of the beam in question. The effect of concentrated loads near an end connection **must** also be considered.

For loads in excess of tabulated capacities, or for thin webs of supporting members, it is necessary to design special seated connections.

---

*A structural tee may be used instead of a pair of angles.

EXAMPLE

Given:

Design a stiffened seated beam connection of F_y = 36 ksi steel to support a W30 × 99, also F_y = 36 ksi, with an end reaction of 126 kips. Use ⅞" dia. ASTM A325-N bolts to attach the seat to a column web with a 5½" gage and thickness of ½". Assume a top angle is required.

Solution:

1. From the F_y = 36 ksi Uniform Load Constant Tables, under W30 × 99, note that

    ϕR_1 = 67.3, ϕR_2 = 18.7, ϕR_3 = 93.9, ϕR_4 = 6.50
    L   = (126 − 67.3)/18.7 = 3.14"
    L   = (126 − 93.9)/6.50 = 4.9"

    From Table VII-A, under F_y = 36 ksi, it will be seen that a 4.9" length of bearing requires that 5" OSL stiffener angles be used. In the outstanding leg column under 5 in., note that the 126 kip reaction requires stiffener angles of ⁵⁄₁₆" thickness. Use a seat angle of ⅜" thickness extending beyond the stiffener angle; this requires a 6" leg outstanding.

2. In Table VII-B for a ⅞" dia. A325-N fastener, 3 rows of bolts with a capacity of 127 kips will be required for a 126 kip reaction.

3. Bearing on web = 126/6 = 21 kips/bolt < 45.7 (Table I-E, F_y = 58 ksi, Fastener Spacing = 3")

Detail Data:

Steps 1 and 2 indicate the use of a Type A connection with 3 rows of ⅞" dia. A325-N bolts. Assuming it is possible to employ the suggested spacing of fasteners, detail material will be as follows:

Steel: F_y = 36 ksi
2 Stiffeners: L5 × 5 × ⁵⁄₁₆ × 0'-8⅝"
1 Seat plate: PL⅜ × 6 × 0'-10"
1 Top angle: L5 × 5 × ⅜ × 0'-8"

STIFFENED SEATED BEAM CONNECTIONS
Bolted
TABLE VII

TABLE VII-A Stiffener Angle Capacity, Kips
(For Type A)

Stiffener Material		F_y = 36 Ksi (R_n = 54.0 × A_{pb})[a]			F_y = 50 Ksi (R_n = 75.0 × A_{pb})[a]		
Stiffener Outstanding Leg A, In.		3½	4	5	3½	4	5
Max. Length Beam Bearing, In.		3½	4	5	3½	4	5
Thickness of Stiffener Outstanding Legs, In.	5/16	101	118	152	141	164	211
	3/8	122	142	182	169	197	253
	½	162	189	243	225	263	338
	5/8	203	236	304	281	328	422
	¾	243	284	365	338	394	506

Use 3/8" thick seat plate wide enough to extend beyond outstanding legs of stiffener.
[a]See LRFD Specification Sect. J8.1.

TABLE VII-B Fastener Capacity, Kips
(For Type A)

Fastener[a] Specification	Fastener Diameter, In.	Number of Fasteners in One Vertical Row				
		3	4	5	6	7
A307	¾	42.9	57.3	71.6	85.9	100
	7/8	58.4	77.9	97.4	117	136
	1	76.3	102	127	153	178
A325-N	¾	93.0	124	155	186	217
	7/8	127	169	211	253	295
	1	165	221	276	330	386
A325-X	¾	124	165	207	248	289
	7/8	169	225	281	338	394
	1	221	294	368	441	515
A490-N	¾	116	155	194	233	272
	7/8	158	211	264	317	370
	1	207	276	345	414	483
A490-X	¾	155	207	258	310	362
	7/8	211	281	352	422	492
	1	276	368	459	551	643

[a]A325-N and A490-N: Bearing-type connections with threads included in shear plane.
A325-X and A490-X: Bearing-type connections with threads excluded from shear plane.

STIFFENED SEATED BEAM CONNECTIONS
Welded—E70XX electrodes

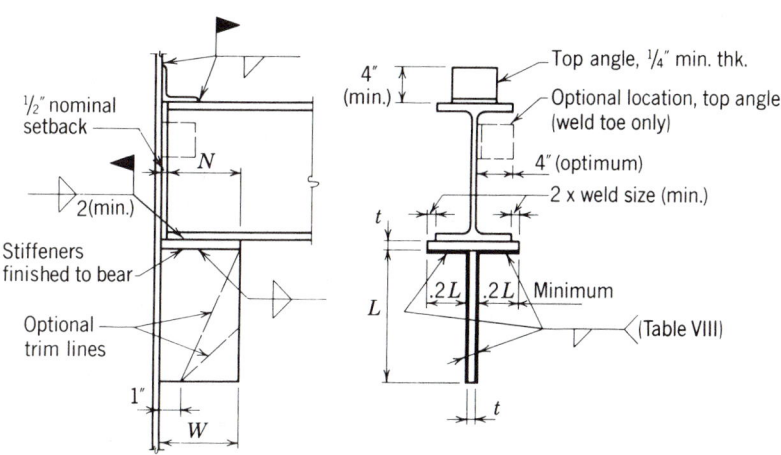

TABLE VIII

Seated connections should be used only when the beam is supported by a top angle placed as shown above, or in the optional location as indicated.

Design loads in Table VIII are based on the use of E70XX electrodes. The table may be used for other electrodes, provided that the tabular values are adjusted for the electrodes used (e.g., for E60XX electrodes, multiply tabular values by 60/70 or 0.86, etc.) and the welds and base metal meet the provisions of LRFD Specification Sect. J2.

Design weld capacities in Table VIII are computed using traditional vector analysis.

Based on F_y = 36 ksi bracket material, minimum stiffener plate thickness, t, for supported beams with unstiffened webs should not be less than the supported beam web thickness for F_y = 36 ksi beams, and not less than 1.4 times the beam web thickness for beams with F_y = 50 ksi. Based on bracket material of F_y = 50 ksi or greater, the minimum stiffener plate thickness t for supported beams with unstiffened webs should be the beam web thickness multiplied by the ratio of F_y of the beam to F_y of the bracket [e.g., F_y (beam) = 65 ksi; F_y (bracket) = 50 ksi; $t = t_w$ (beam) × 65/50, minimum]. The minimum stiffener plate thickness, t, should be at least two times the required E70XX weld size when F_y of the bracket is 36 ksi, and should be at least 1.5 times the required E70XX weld size when F_y of the bracket is 50 ksi.

Thickness t of the horizontal seat plate, or tee flange, should not be less than 3/8".

If seat and stiffener are separate plates, finish stiffener to bear against seat. Welds connecting the two plates should have a strength equal to, or greater than, the horizontal welds to the support under the seat plate.

Welds attaching beam to seat may be replaced by bolts.

ASTM A307 bolts may be used in seated connections, provided the stipulations of LRFD Specification Sect. J1.9 are observed.

For stiffener seats in line on opposite sides of a column web of F_y = 36 ksi material, select E70XX weld size no greater than 0.50 of column web thickness. For column web of F_y = 50 ksi, select E70XX weld size no greater than 0.67 of column web thickness.

<u>Notes</u>

Should combinations of material thickness and weld size selected from Table VIII exceed the limits set by LRFD Specification Sect. J2, increase the weld size or material thickness as required.

In addition to the welds shown, temporary erection bolts may be used to attach beams to seats (optional).

To permit selection of the most economical connection, the reaction values should be given on the contract drawings. If the reaction values are not given, the connections must be selected to support the beam end reaction calculated from the Tables of Uniform Load Constants (Part 3) for the given shape, span and steel specification of the beam in question. The effect of concentrated loads near an end connection **must** also be considered.

EXAMPLES

(a) *Given:*

Beam: W30 × 116 (flange = 10.495" × 0.85"; web = 0.565")
                ASTM A36 steel (F_y = 36 ksi)
Welds: E70XX
Reaction: 150 kips

Design a two-plate welded stiffener seat.

Solution:

From the F_y = 36 ksi Uniform Load Constants tables:

ϕR_1 = 82.6, ϕR_2 = 20.3
ϕR_3 = 120, ϕR_4 = 6.49
(150 − 82.6)/20.3 = 3.32
(150 − 120)/6.49 = 4.62

Stiffener width: W = 4.62 + 0.5 (setback) = 5.12"

Use: W = 6"

Enter Table VIII with W = 6" and a reaction of 150 kips; select a $5/16$" weld with L = 15", which has a capacity of 154 kips. From this, the minimum length of weld between seat plate and support is 2 × 0.2L = 6". This also establishes the minimum weld between the seat plate and the stiffener as 6" total, or 3" on each side of stiffener.

Stiffener plate thickness t to develop welds is 2 × $5/16$ = $5/8$", or 0.625". This is greater than the beam web thickness of 0.565"; thus, the stiffener plate thickness need not be increased.

Use: $5/8$" plate for the stiffener and $3/8$" plate for seat.

Welds attaching the beam flange to the seat must be $5/16$" to conform to LRFD Specification Sect. J2.2, due to the 0.85" flange thickness of the W30 × 116 beam.

Seat plate width to permit field welding of beam to seat = flange width + (4 × weld size) = 10.5 + 4($5/16$) = 11.75"

Use: $3/8$ × 6 × 12" plate

This is adequate for the required minimum weld length.

Detail Data:

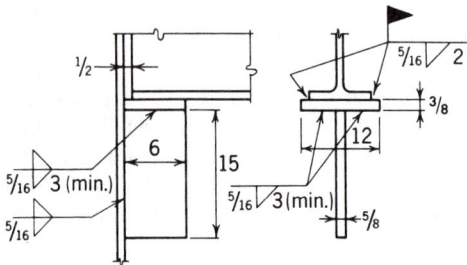

Use: L4 × 4 × ⅜ × 0'-4" top angle (F_y = 36 ksi) with 5/16" welds along toes of angle only.

(b) *Given:*

Beam:	W21 × 68 (flange = 8.27" × 0.685"; web = 0.43")
	ASTM A572, Gr. 50 steel (F_y = 50 ksi)
Welds:	E70XX electrodes
Reaction:	125 kips

Design a two-plate welded stiffener seat using ASTM A36 steel.

Solution:

From the F_y = 50 ksi Uniform Load Constants Tables:

$$\phi R_1 = 77.3, \ \phi R_2 = 21.5$$
$$\phi R_3 = 84.2, \ \phi R_4 = 5.94$$
$$L \quad = (125 - 77.3)/21.5 = 2.22$$
$$L \quad = (125 - 84.2)/5.94 = 6.9$$

Stiffener width:

$$W = 6.9 + 0.5 \ (\text{setback}) = 7.4''$$

Use: W = 8"

Enter Table VIII with W = 8" and a reaction of 125 kips; satisfying these requirements are a 5/16" weld, L = 15" (127 kips), or a ⅜" weld, L = 14" (135 kips), or an even larger weld size. Generally, the 5/16 in. weld is the better selection, as this can be made in one pass using manual welding. Select 5/16" weld. From this, the minimum length of 5/16" weld between seat plate and support is 2 × 0.2L = 6".

Use: 6" weld length. This also establishes the minimum weld between the seat plate and the stiffener as 6" total, or 3" on each side.

Stiffener plate thickness t to develop welds is 2 × 5/16 = ⅝", or 0.625". The minimum thickness t for a bracket of F_y = 36 ksi with a beam of F_y = 50 ksi is 1.4 times the beam web thickness = 1.4 × 0.43 = 0.602".

Use: ⅝" plate for the stiffener and ⅜" plate for the seat.

Welds attaching the beam flange to the seat can be ¼" for a flange of 0.685", as per LRFD Specification Table J2.5.

Seat plate width, to permit field welding of beam to seat = flange width + (4 × weld size) = 8.27 + (4 × ¼) = 9.27".

Use: ⅜ × 8 × 10" plate.

This is adequate for the required minimum weld length.

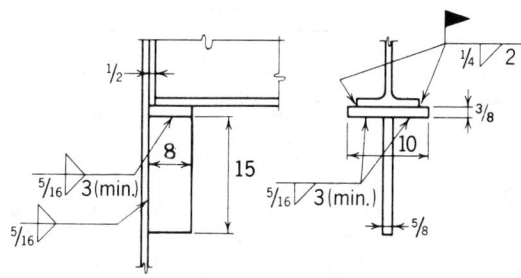

Detail Data:

Use: L4 × 4 × ⅜ × 0'4" top angle (F_y = 36 ksi) with ⁵⁄₁₆" welds along toes of angle only.

STIFFENED SEATED BEAM CONNECTIONS
Welded—E70XX electrodes
TABLE VIII Ultimate loads in kips

L, In.	Width of Seat W, In.											
	4				5				6			
	Weld Size, In.				Weld Size, In.				Weld Size, In.			
	1/4	5/16	3/8	7/16	5/16	3/8	7/16	1/2	5/16	3/8	7/16	1/2
6	34.0	42.5	51.1	59.6	35.2	42.2	49.3	56.3	29.9	35.9	41.9	47.8
7	44.9	56.1	67.3	78.6	46.9	56.2	65.6	75.0	40.1	48.1	56.1	64.1
8	56.7	70.8	85.0	99.2	59.8	71.7	83.7	95.6	51.4	61.7	72.0	82.2
9	69.2	86.5	104	121	73.7	88.5	103	118	63.8	76.6	89.3	102
10	82.3	103	123	144	88.5	106	124	142	77.2	92.6	108	123
11	95.8	120	144	168	104	125	146	167	91.3	110	128	146
12	110	137	165	192	120	144	168	192	106	127	149	170
13	124	155	186	217	137	164	192	219	122	146	170	195
14	138	173	207	242	154	185	216	246	138	165	193	220
15	152	191	229	267	171	206	240	274	154	185	216	247
16	167	209	250	292	189	227	265	302	171	205	240	274
17	181	227	272	318	207	248	290	331	188	226	264	301
18	196	245	294	343	225	270	315	360	206	247	288	329
19	211	263	316	369	243	291	340	388	223	268	313	357
20	225	281	338	394	261	313	365	417	241	289	337	386
21	240	300	359	419	279	335	391	446	259	311	362	414
22	254	318	381	445	297	357	416	476	277	332	388	443
23	269	336	403	470	315	378	442	505	295	354	413	472
24	283	354	425	495	334	400	467	534	313	376	438	501
25	297	372	446	520	352	422	492	563	331	397	464	530
26	312	390	468	546	370	444	518	592	349	419	489	559
27	326	408	489	571	388	466	543	621	368	441	515	588

Note: Loads shown are for E70XX electrodes. For E60XX electrodes, multiply tabular loads by 0.86, or enter table with 1.17 times the given reaction. For E80XX electrodes, multiply tabular loads by 1.14, or enter table with 0.875 times the given reaction.

STIFFENED SEATED BEAM CONNECTIONS
Welded—E70XX electrodes

TABLE VIII Ultimate loads in kips

L, In.	Width of Seat W, In.											
	7				8				9			
	Weld Size, In.				Weld Size, In.				Weld Size, In.			
	5/16	3/8	7/16	1/2	5/16	3/8	1/2	5/8	5/16	3/8	1/2	5/8
11	81.0	97.2	113	130	72.5	87.1	116	145	65.6	78.7	105	131
12	94.7	114	133	151	85.1	102	136	170	77.1	92.5	123	154
13	109	131	153	174	98.3	118	157	197	89.3	107	143	179
14	124	149	174	198	112	135	180	224	102	123	164	204
15	139	167	195	223	127	152	203	253	116	139	185	232
16	155	186	217	249	142	170	227	283	130	156	208	260
17	172	206	240	275	157	189	251	314	144	173	231	289
18	188	226	264	301	173	208	277	346	159	191	255	319
19	205	246	287	329	189	227	303	378	175	210	280	350
20	223	267	312	356	206	247	329	411	191	229	305	381
21	240	288	336	384	222	267	356	445	207	248	331	413
22	258	309	361	412	240	287	383	479	223	268	357	446
23	275	330	385	440	257	308	411	514	240	288	384	480
24	293	352	410	469	274	329	439	548	257	308	411	513
25	311	373	435	498	292	350	467	584	274	329	438	548
26	329	395	461	526	309	371	495	619	291	349	466	582
27	347	417	486	555	327	393	524	655	308	370	494	617
28	365	438	511	584	345	414	552	690	326	391	522	652
29	383	460	537	613	363	436	581	726	344	412	550	687
30	402	482	562	643	381	457	610	762	362	434	578	723
31	420	504	588	672	399	479	639	799	379	455	607	759
32	438	526	613	701	417	501	668	835	397	477	636	795

Note: Loads shown are for E70XX electrodes. For E60XX electrodes, multiply tabular loads by 0.86, or enter table with 1.17 times the given reaction. For E80XX electrodes, multiply tabular loads by 1.14, or enter table with 0.875 times the given reaction.

Notes

END PLATE SHEAR CONNECTIONS

TABLE IX

This type of connection consists of a plate, less than the beam depth in length, perpendicular to the longitudinal axis of the beam, welded to the beam web with fillet welds each side of the beam web. The end plate connection compares favorably to the double-angle connection and, for like thicknesses, gage lines and length of connection will furnish end rotation capacity and strength of connection closely approximating that of the double-angle framing connection, within the range listed in the table.

Fabrication of this type of connection requires close control in cutting the beam to length; adequate consideration must be given to squaring the beam ends such that both end plates are parallel and the effect of beam camber does not result in out-of-square end plates, which make erection and field fit-up difficult. Shims may be required on runs of beams to compensate for mill and shop tolerances.

For adequate end rotation capacity, it is suggested that end plates be designed for a plate thickness range of ¼" to ⅜", inclusive. To develop full capacity of the fasteners and welds, the end plate and web thicknesses must equal or exceed the values listed in the table. If the material thickness supplied by either the plate or the web is less than required, the fastener or weld capacity must be reduced by the ratio of thickness supplied to thickness required.

The gage g should be 3½" to 5½" for average plate thicknesses, with an edge distance of 1¼". Lesser values of edge distance should be avoided. Plates ¼" thick, of $F_y = 36$ ksi steel, and a gage of 3" should provide adequate end rotation capacity in the connection. All end plate material thicknesses listed in the table are for $F_y = 36$ ksi. Use of higher values of F_y should be based on engineering investigation that confirms that adequate end rotation capacity is available.

Weld values listed are for two fillet welds, and are based on the use of E70XX electrodes. These weld values have been reduced by considering the effective weld length equal to the plate length minus twice the weld size. Welds should not be returned across the web at the top or bottom of the end plates.

EXAMPLE 1

Given:

Select an end plate connection for a W14 × 30 beam.

$F_y = 36$ ksi and factored end reaction = 36 kips.

Solution:

Beam web thickness: 0.270"

Gage: 3½"

From Table IX, for beam depth limits 12" through 18", select a plate length of 8½" with three ¾" dia. A307 bolts per vertical row, with a listed capacity of 42.9 kips and a required minimum plate thickness of $t = 0.14$".

Select a ¼" plate: 0.25" > 0.14" **o.k.**

From Table IX:

Weld capacity: 8½" of 3/16" fillet = 67.8 kips

Minimum web thickness = 0.41"

$$\frac{0.270}{0.41} \times 67.8 = 44.6 \text{ kips} > 36 \text{ kips} \quad \textbf{o.k.}$$

Use: End plate 6" wide × 8½" long × ¼" thick, with six ¾" dia. A307 bolts on 3½"
gage. Weld the plate to the beam web with 3/16" fillet welds on each side of the
web.

EXAMPLE 2

Given:

Select an end plate connection for two W12 × 58 beams framing into both sides of a
W30 × 191 girder. Beam reaction = 50 kips for each of the W12 × 58 beams and F_y =
50 ksi for both beams and girder.

Solution:

Beam web thickness: 0.360"

Usual gage: 5½"

Girder web thickness: 0.710"

From Table IX for beam depth limits 8" through 12", select a plate length of 5½" with
two ¾" dia. A325-N bolts per vertical row, with a listed capacity of 62.0 kips and
minimum t = 0.31".

Use a 5/16" end plate thickness t = 0.3125".

From Table IX:

Weld capacity: 5½" of ¼ in. fillet weld = 55.7 kips

Minimum web t = 0.40"

$$\frac{0.360}{0.40} \times 55.7 = 50.1 \text{ kips} > 50 \text{ kips} \quad \textbf{o.k.}$$

Girder web thickness must be checked separately, using fastener spacing bearing
criteria:

Total load on girder web in double shear = 2 × 50 = 100 kips

Using Table I-E, based on ¾" dia. bolts, 3" pitch, and F_u = 65 ksi:

$$t = \frac{100}{4 \times 87.8} = 0.285" < 0.701"$$

∴ Girder web is adequate

Use: End plate 8" wide × 5½" long × ¼" thick, with four ¾" dia., A325-N bolts on 5½"
gage. Weld the plate to the beam web with ¼" fillet welds on each side of the web.

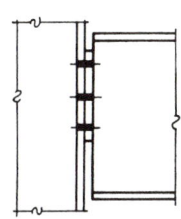

END PLATE SHEAR CONNECTIONS

Welded—E70XX electrodes

TABLE IX

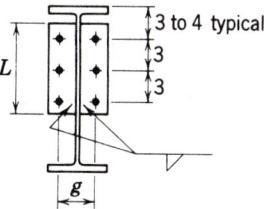

3 to 4 typical

Bolts per Vertical Line	Bolt Designation	¾" Dia.		⅞" Dia.		Plate Length L, In.	Beam Depth Limits, In.
		Total Capacity, Kips	Min Plate Thickness t, In.	Total Capacity, Kips	Min Plate Thickness t, In.		
1	A307	14.3	0.13	19.5	0.18	3	5-8
	A325-N	31.0	0.27	42.2	0.39		
	A325-X	41.4	0.36	—	—		
2	A307	28.6	0.14	39.0	0.21	5½	8–12
	A325-N	62.0	0.31	84.4	0.45		
	A325-X	82.7	0.41	—	—		
3	A307	42.9	0.14	58.4	0.19	8½	12–18
	A325-N	93.0	0.30	—	—		
	A325-X	124	0.40	—	—		
4	A307	57.3	0.13	77.9	0.19	11½	15–24
	A325-N	124	0.29	169	0.41		
	A325-X	165	0.38	—	—		
5	A307	71.6	0.13	97.4	0.19	14½	18–30
	A325-N	155	0.29	211	0.41		
	A325-X	206	0.38	—	—		
6	A307	85.9	0.13	117	0.19	17½	21–36
	A325-N	186	0.28	253	0.42		
	A325-X	248	0.38	—	—		

WELD CAPACITY, Kips

Weld Size	Thickness, In.		Plate Length L, In.					
	$F_y = 36$	$F_y = 50$	3	5½	8½	11½	14½	17½
3/16	.41	.30	21.9	42.8	67.8	92.9	118	143
¼	.55	.40	27.9	55.7	89.1	123	156	189
5/16	.69	.50	33.0	67.8	110	152	194	234
⅜	.82	.60	37.7	79.4	129	180	230	281

ECCENTRIC LOADS ON FASTENER GROUPS

TABLES X–XVII

ULTIMATE STRENGTH METHOD*

When fastener groups are loaded in shear by an external load that does not act through the center of gravity of the group, the load is eccentric and will tend to cause a relative rotation and translation of the connected material. This condition is equivalent to that of pure rotation about a single point. This point is called the *instantaneous center of rotation* (at distance r_o from center of gravity of the group). Both r_o and l are measured perpendicular to the applied force line of action.

The individual resistance force of each fastener can then be assumed to act on a line perpendicular to a ray passing through the instantaneous center and that fastener's location (see Fig. 1).

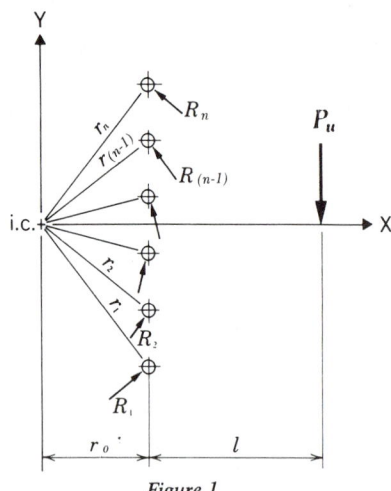

Figure 1

The ultimate shear strength of fastener groups can be obtained from the load-deformation relationship of a single fastener, which is expressed as:

$$R = R_{ult}(1 - e^{-10\Delta})^{0.55}$$

where

R = shear force in a single fastener at any given deformation (see Fig. 2)

R_{ult} = ultimate shear load of a single fastener

Δ = total deformation of a fastener, including shearing, bending, and bearing deformation, plus local bearing deformation of the plate

e = base of natural logarithm = 2.718 . . .

By applying a maximum deformation Δ_{max} to the fastener (or fasteners) most remote from the instantaneous center, the maximum shear force for that fastener can be computed. For other fasteners, deformations are computed to vary linearly from the instantaneous center, and shear forces can be obtained from the above relationship.

---

*Crawford, S. F., and Kulak, G. L., *Eccentrically Loaded Bolted Connections*, Journal of the Structural Div., ASCE, Vol. 97, ST3, March 1971, pp. 765–783.

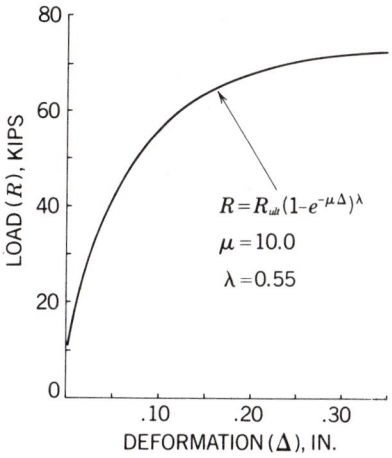

$$R = R_{ult}(1-e^{-\mu\Delta})^{\lambda}$$
$$\mu = 10.0$$
$$\lambda = 0.55$$

Figure 2

The total resistance force of all fasteners then combine to resist the eccentric external load and if the correct location of the instantaneous center has been selected, the three equations of statics will be satisfied.

Although the development of fastener load-deformation relationships as expressed above is based on connections which may experience slip under load, both load tests and analytical studies* indicate that extending these procedures to slip-resistant connections is conservative.

Tables X to XVII have been prepared based upon the solution of the instantaneous center problem for each fastener pattern and each eccentric condition. Each fastener configuration has also been analyzed for inclined loads at 0°, 45° and 75° increments. A convergence criterion of 1% was employed for the tabulated iterative solutions. Straight line interpolation between these angles may be significantly unconservative. *Therefore, unless a direct analysis is performed, use only the values for the next lower angle for design.* The load-deformation relationship data is based upon the following values obtained experimentally for ¾″ dia. ASTM A325 bolts:

$$R_{ult} = 74 \text{ kips}$$

$$\Delta_{max} = 0.34 \text{ in.}$$

The non-dimensional coefficient C is obtained by dividing the ultimate load P by R_{ult}. The values derived can be safely used with any fastener diameter and are conservative when used with ASTM A490 bolts. By use of these tables, margins of safety are provided equivalent to those bolts used in joints less than 50 in. long, subject to shear produced by concentric load only in both slip-critical or bearing-type connections.

For any fastener group listed in these tables, the coefficient C times the design value of one fastener equals the total factored load P_u located at an eccentric distance from the centroid of the fastener group ($P_u = C \times \phi r_v$). Thus, by dividing P_u by the design fastener value ϕr_v the minimum coefficient is obtained, and a fastener group can be selected for which the coefficient is of that magnitude or greater.

---

*Kulak, G. L., *Eccentrically Loaded Slip-Resistant Connections*, Engineering Journal, AISC, Vol. 12, No. 2, 1975.

EXAMPLE 1

Given:

Find the maximum load that can be supported by the bracket shown in Fig. 3. Column and brackets are F_y = 36 ksi. Use ⅞" dia. A325-N bolts, and assume the column flange is at least adequately thick so bolt shear will govern.

$n = 6$, $b = 3$ in., $D = 5½$ in., $l = \chi_o = 16$ in.

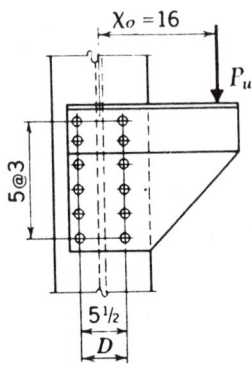

Solution:

From Table XII, with $n = 6$, $b = 3$ in. and $\chi_o = 16$ in.:

$C = 3.55$

Using $\phi r_v = 21.1$ kips (from Table I-D)

$P_u = 3.55 \times 21.1 = 74.9$ kips

Figure 3

ECCENTRIC LOADS ON FASTENER GROUPS
TABLE X Coefficients C

ANGLE = 0°

Required minimum $C = \dfrac{P_u}{(\phi r_v)}$

$P_u = C \times \phi r_v$
n = total number of fasteners in the vertical row
P_u = factored load acting with lever arm l, in.
ϕr_v = design load on one fastener by Specification
X_o = horizontal distance from bolt centroid to P_u
ℓ = load eccentricity (not tabulated, can be calculated by geometry)
C = coefficients tabulated below

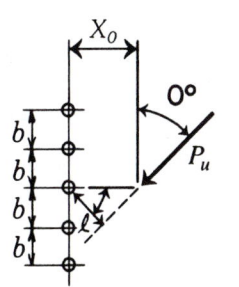

X_o, In.	n										
	2	3	4	5	6	7	8	9	10	11	12
b = 3 In.											
2	1.18	2.23	3.32	4.40	5.45	6.48	7.51	8.52	9.53	10.54	11.54
3	.88	1.77	2.83	3.91	4.99	6.07	7.13	8.17	9.21	10.24	11.26
4	.69	1.40	2.38	3.42	4.49	5.58	6.66	7.73	8.79	9.85	10.89
5	.56	1.15	2.01	2.97	4.01	5.08	6.16	7.24	8.32	9.39	10.46
6	.48	1.00	1.73	2.59	3.57	4.60	5.66	6.74	7.82	8.90	9.98
7	.41	.83	1.51	2.28	3.18	4.16	5.19	6.24	7.32	8.40	9.48
8	.36	.73	1.34	2.04	2.86	3.76	4.75	5.77	6.83	7.89	8.97
9	.32	.65	1.21	1.83	2.59	3.43	4.35	5.32	6.36	7.41	8.48
10	.29	.59	1.09	1.66	2.36	3.14	4.00	4.93	5.91	6.95	8.00
12	.24	.49	.92	1.40	2.00	2.68	3.44	4.27	5.16	6.10	7.08
14	.21	.42	.79	1.21	1.74	2.33	3.01	3.75	4.56	5.42	6.32
16	.18	.37	.70	1.06	1.53	2.06	2.67	3.33	4.06	4.85	5.69
18	.16	.33	.62	.95	1.37	1.84	2.39	3.00	3.66	4.38	5.15
20	.15	.30	.56	.85	1.24	1.67	2.17	2.72	3.33	3.99	4.70
24	.12	.25	.47	.71	1.03	1.40	1.82	2.29	2.81	3.37	3.99
28	.11	.21	.40	.61	.89	1.20	1.57	1.97	2.42	2.92	3.45
32	.09	.19	.35	.54	.78	1.05	1.38	1.73	2.13	2.57	3.04
36	.08	.17	.31	.48	.69	.94	1.22	1.54	1.90	2.29	2.72
b = 6 In.											
2	1.63	2.72	3.75	4.77	5.77	6.77	7.76	8.75	9.74	10.73	11.72
3	1.39	2.48	3.56	4.60	5.63	6.65	7.65	8.66	9.66	10.65	11.64
4	1.18	2.23	3.32	4.40	5.45	6.48	7.51	8.52	9.53	10.54	11.54
5	1.01	1.99	3.07	4.16	5.23	6.29	7.33	8.36	9.39	10.40	11.41
6	.88	1.77	2.83	3.91	4.99	6.07	7.13	8.17	9.21	10.24	11.26
7	.77	1.57	2.59	3.66	4.75	5.83	6.90	7.96	9.01	10.05	11.09
8	.69	1.40	2.38	3.42	4.49	5.58	6.66	7.73	8.79	9.85	10.89
9	.62	1.26	2.18	3.19	4.25	5.33	6.41	7.49	8.56	9.63	10.69
10	.56	1.15	2.01	2.97	4.01	5.08	6.16	7.24	8.32	9.39	10.46
12	.48	1.00	1.73	2.59	3.57	4.60	5.66	6.74	7.82	8.90	9.98
14	.41	.83	1.51	2.28	3.18	4.16	5.19	6.24	7.32	8.40	9.48
16	.36	.73	1.34	2.04	2.86	3.76	4.75	5.77	6.83	7.89	8.97
18	.32	.65	1.21	1.83	2.59	3.43	4.35	5.32	6.36	7.41	8.48
20	.29	.59	1.09	1.66	2.36	3.14	4.00	4.93	5.91	6.95	8.00
24	.24	.49	.92	1.40	2.00	2.68	3.44	4.27	5.16	6.10	7.08
28	.21	.42	.79	1.21	1.74	2.33	3.01	3.75	4.56	5.42	6.32
32	.18	.37	.70	1.06	1.53	2.06	2.67	3.33	4.06	4.85	5.69
36	.16	.33	.62	.95	1.37	1.84	2.39	3.00	3.66	4.38	5.15

ECCENTRIC LOADS ON FASTENER GROUPS
TABLE X Coefficients *C*

ANGLE = 45°

Required minimum $C = \dfrac{P_u}{(\phi r_v)}$

$P_u = C \times \phi r_v$

n = total number of fasteners in the vertical row

P_u = factored load acting with lever arm l, in.

ϕr_v = design load on one fastener by Specification

X_o = horizontal distance from bolt centroid to P_u

ℓ = load eccentricity (not tabulated, can be calculated by geometry)

C = coefficients tabulated below

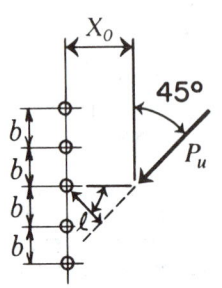

X_o, In.	n 2	3	4	5	6	7	8	9	10	11	12
b = 3 In.											
2	1.17	2.25	3.32	4.40	5.43	6.46	7.47	8.48	9.48	10.48	11.48
3	.92	1.89	2.89	3.93	4.99	6.07	7.11	8.14	9.17	10.19	11.20
4	.75	1.63	2.54	3.52	4.55	5.60	6.66	7.74	8.79	9.82	10.85
5	.64	1.42	2.25	3.17	4.15	5.17	6.21	7.26	8.31	9.41	10.46
6	.55	1.25	2.01	2.88	3.81	4.78	5.79	6.82	7.86	8.92	9.97
7	.49	1.11	1.81	2.63	3.51	4.44	5.41	6.41	7.43	8.47	9.52
8	.44	1.01	1.64	2.41	3.25	4.14	5.07	6.04	7.04	8.05	9.08
9	.40	.90	1.49	2.22	3.02	3.87	4.77	5.71	6.67	7.66	8.67
10	.36	.82	1.37	2.06	2.82	3.63	4.50	5.40	6.34	7.30	8.29
12	.31	.69	1.17	1.79	2.47	3.22	4.02	4.87	5.75	6.66	7.60
14	.27	.59	1.03	1.58	2.20	2.88	3.62	4.41	5.24	6.11	7.00
16	.24	.52	.91	1.41	1.97	2.60	3.29	4.03	4.81	5.63	6.48
18	.22	.47	.82	1.28	1.78	2.36	3.01	3.70	4.43	5.21	6.02
20	.20	.42	.74	1.16	1.63	2.17	2.77	3.41	4.10	4.84	5.61
24	.17	.35	.63	.99	1.38	1.85	2.38	2.95	3.56	4.22	4.92
28	.15	.30	.54	.85	1.20	1.62	2.08	2.59	3.14	3.74	4.37
32	.13	.26	.48	.75	1.06	1.43	1.84	2.30	2.81	3.34	3.92
36	.11	.23	.43	.67	.94	1.28	1.65	2.07	2.53	3.02	3.55
b = 6 In.											
2	1.62	2.71	3.74	4.74	5.74	6.74	7.73	8.72	9.70	10.69	11.67
3	1.35	2.48	3.56	4.59	5.61	6.61	7.61	8.61	9.61	10.60	11.59
4	1.17	2.25	3.32	4.40	5.43	6.46	7.47	8.48	9.48	10.48	11.48
5	1.03	2.06	3.10	4.16	5.24	6.27	7.30	8.32	9.34	10.34	11.35
6	.92	1.89	2.89	3.93	4.99	6.07	7.11	8.14	9.17	10.19	11.20
7	.83	1.75	2.70	3.72	4.77	5.82	6.91	7.95	8.98	10.01	11.03
8	.75	1.63	2.54	3.52	4.55	5.60	6.66	7.74	8.79	9.82	10.85
9	.69	1.52	2.39	3.34	4.34	5.38	6.43	7.49	8.58	9.62	10.66
10	.64	1.42	2.25	3.17	4.15	5.17	6.21	7.26	8.31	9.41	10.46
12	.55	1.25	2.01	2.88	3.81	4.78	5.79	6.82	7.86	8.92	9.97
14	.49	1.11	1.81	2.63	3.51	4.44	5.41	6.41	7.43	8.47	9.52
16	.44	1.01	1.64	2.41	3.25	4.14	5.07	6.04	7.04	8.05	9.08
18	.40	.90	1.49	2.22	3.02	3.87	4.77	5.71	6.67	7.66	8.67
20	.36	.82	1.37	2.06	2.82	3.63	4.50	5.40	6.34	7.30	8.29
24	.31	.69	1.17	1.79	2.47	3.22	4.02	4.87	5.75	6.66	7.60
28	.27	.59	1.03	1.58	2.20	2.88	3.62	4.41	5.24	6.11	7.00
32	.24	.52	.91	1.41	1.97	2.60	3.29	4.03	4.81	5.63	6.48
36	.22	.47	.82	1.28	1.78	2.36	3.01	3.70	4.43	5.21	6.02

ECCENTRIC LOADS ON FASTENER GROUPS
TABLE X Coefficients *C*

ANGLE = 75°

Required minimum $C = \dfrac{P_u}{(\phi r_v)}$

$P_u = C \times \phi r_v$
n = total number of fasteners in the vertical row
P_u = factored load acting with lever arm l, in.
ϕr_v = design load on one fastener by Specification
X_o = horizontal distance from bolt centroid to P_u
ℓ = load eccentricity (not tabulated, can be calculated by geometry)
C = coefficients tabulated below

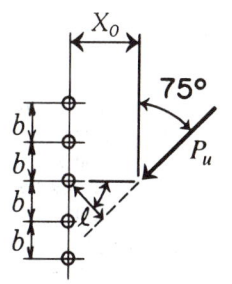

X_o, In.	n=2	3	4	5	6	7	8	9	10	11	12
b = 3 In.											
2	1.67	2.76	3.78	4.78	5.77	6.76	7.75	8.73	9.72	10.70	11.69
3	1.34	2.53	3.63	4.65	5.66	6.66	7.66	8.65	9.64	10.63	11.62
4	1.20	2.18	3.44	4.49	5.52	6.54	7.55	8.55	9.54	10.54	11.53
5	1.12	2.04	2.99	4.31	5.36	6.39	7.41	8.42	9.43	10.43	11.43
6	1.04	1.94	2.86	3.79	5.17	6.22	7.26	8.28	9.30	10.31	11.31
7	.90	1.82	2.73	3.65	4.58	6.03	7.08	8.12	9.15	10.17	11.18
8	.85	1.72	2.62	3.53	4.44	5.68	6.89	7.94	8.98	10.01	11.03
9	.78	1.63	2.52	3.40	4.31	5.23	6.53	7.75	8.80	9.84	10.88
10	.73	1.56	2.43	3.29	4.18	5.10	6.02	7.38	8.61	9.66	10.70
12	.65	1.42	2.24	3.10	3.96	4.84	5.75	6.66	7.59	9.07	10.33
14	.58	1.30	2.07	2.90	3.75	4.62	5.49	6.41	7.32	8.24	9.17
16	.53	1.20	1.92	2.71	3.55	4.41	5.28	6.15	7.06	7.97	8.88
18	.48	1.11	1.78	2.55	3.36	4.20	5.06	5.93	6.80	7.71	8.62
20	.44	1.04	1.66	2.40	3.18	4.00	4.85	5.71	6.59	7.45	8.36
24	.39	.90	1.46	2.13	2.86	3.65	4.47	5.30	6.16	7.02	7.90
28	.34	.79	1.29	1.91	2.60	3.35	4.13	4.94	5.77	6.61	7.47
32	.30	.70	1.17	1.73	2.38	3.09	3.83	4.61	5.41	6.23	7.07
36	.28	.62	1.06	1.59	2.20	2.86	3.56	4.30	5.07	5.87	6.69
b = 6 In.											
2	1.88	2.89	3.88	4.86	5.85	6.83	7.81	8.79	9.78	10.76	11.74
3	1.80	2.83	3.83	4.82	5.81	6.80	7.78	8.77	9.75	10.73	11.72
4	1.67	2.76	3.78	4.78	5.77	6.76	7.75	8.73	9.72	10.70	11.69
5	1.41	2.67	3.71	4.72	5.72	6.71	7.71	8.70	9.68	10.67	11.65
6	1.34	2.53	3.63	4.65	5.66	6.66	7.66	8.65	9.64	10.63	11.62
7	1.27	2.28	3.54	4.58	5.60	6.60	7.60	8.60	9.60	10.59	11.58
8	1.20	2.18	3.44	4.49	5.52	6.54	7.55	8.55	9.54	10.54	11.53
9	1.14	2.11	3.25	4.40	5.44	6.47	7.48	8.49	9.49	10.49	11.48
10	1.12	2.04	2.99	4.31	5.36	6.39	7.41	8.42	9.43	10.43	11.43
12	1.04	1.94	2.86	3.79	5.17	6.22	7.26	8.28	9.30	10.31	11.31
14	.90	1.82	2.73	3.65	4.58	6.03	7.08	8.12	9.15	10.17	11.18
16	.85	1.72	2.62	3.53	4.44	5.68	6.89	7.94	8.98	10.01	11.03
18	.78	1.63	2.52	3.40	4.31	5.23	6.53	7.75	8.80	9.84	10.88
20	.73	1.56	2.43	3.29	4.18	5.10	6.02	7.38	8.61	9.66	10.70
24	.65	1.42	2.24	3.10	3.96	4.84	5.75	6.66	7.59	9.07	10.33
28	.58	1.30	2.07	2.90	3.75	4.62	5.49	6.41	7.32	8.24	9.17
32	.53	1.20	1.92	2.71	3.55	4.41	5.28	6.15	7.06	7.97	8.88
36	.48	1.11	1.78	2.55	3.36	4.20	5.06	5.93	6.80	7.71	8.62

ECCENTRIC LOADS ON FASTENER GROUPS
TABLE XI Coefficients C

ANGLE $= 0°$

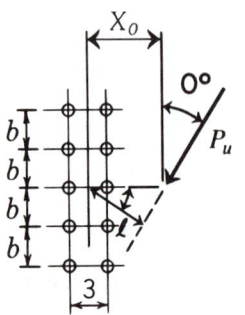

Required minimum $C = \dfrac{P_u}{(\phi r_v)}$

$P_u = C \times \phi r_v$

n = total number of fasteners in the vertical row

P_u = factored load acting with lever arm l, in.

ϕr_v = design load on one fastener by Specification

X_o = horizontal distance from bolt centroid to P_u

ℓ = load eccentricity (not tabulated, can be calculated by geometry)

C = coefficients tabulated below

	X_o, In.	\multicolumn{12}{c	}{n}
		1	2	3	4	5	6	7	8	9	10	11	12
$b = 3$ In.	2	.84	2.54	4.50	6.62	8.74	10.84	12.92	14.97	17.00	19.03	21.04	23.04
	3	.65	2.03	3.68	5.69	7.80	9.94	12.07	14.19	16.29	18.36	20.42	22.47
	4	.54	1.67	3.06	4.87	6.87	8.97	11.11	13.26	15.40	17.52	19.64	21.73
	5	.45	1.42	2.59	4.21	6.03	8.04	10.14	12.28	14.43	16.58	18.72	20.85
	6	.39	1.22	2.25	3.69	5.33	7.20	9.21	11.30	13.44	15.59	17.74	19.90
	7	.35	1.08	1.98	3.27	4.75	6.48	8.37	10.38	12.47	14.59	16.74	18.90
	8	.31	.96	1.78	2.93	4.27	5.87	7.63	9.55	11.56	13.63	15.75	17.90
	9	.28	.86	1.61	2.65	3.87	5.34	6.97	8.76	10.72	12.73	14.80	16.92
	10	.26	.79	1.46	2.42	3.53	4.90	6.42	8.10	9.92	11.89	13.90	15.97
	12	.22	.67	1.24	2.06	3.01	4.19	5.51	7.01	8.64	10.39	12.24	14.24
	14	.19	.58	1.08	1.78	2.62	3.66	4.82	6.15	7.61	9.20	10.90	12.69
	16	.17	.51	.95	1.57	2.32	3.24	4.27	5.47	6.79	8.23	9.78	11.43
	18	.15	.45	.85	1.41	2.07	2.90	3.83	4.92	6.11	7.43	8.85	10.37
	20	.14	.41	.77	1.27	1.88	2.63	3.48	4.46	5.55	6.76	8.07	9.48
	24	.12	.34	.65	1.07	1.58	2.21	2.93	3.76	4.69	5.73	6.84	8.06
	28	.10	.29	.56	.92	1.36	1.90	2.53	3.25	4.05	4.96	5.93	7.00
	32	.09	.26	.49	.80	1.19	1.67	2.22	2.86	3.56	4.36	5.23	6.18
	36	.08	.23	.43	.72	1.06	1.49	1.98	2.55	3.18	3.89	4.67	5.52
$b = 6$ In.	2	.84	3.25	5.40	7.48	9.51	11.52	13.52	15.51	17.49	19.47	21.45	23.43
	3	.65	2.79	4.94	7.08	9.18	11.24	13.27	15.29	17.29	19.29	21.28	23.27
	4	.54	2.41	4.45	6.62	8.76	10.87	12.94	14.99	17.02	19.05	21.06	23.06
	5	.45	2.10	3.98	6.13	8.29	10.43	12.54	14.63	16.69	18.74	20.78	22.80
	6	.39	1.85	3.56	5.64	7.80	9.96	12.10	14.22	16.31	18.39	20.45	22.49
	7	.35	1.64	3.19	5.19	7.30	9.46	11.62	13.76	15.88	17.99	20.07	22.14
	8	.31	1.47	2.87	4.77	6.82	8.96	11.12	13.28	15.43	17.55	19.66	21.76
	9	.28	1.34	2.61	4.40	6.37	8.47	10.62	12.79	14.94	17.09	19.22	21.34
	10	.26	1.22	2.39	4.07	5.95	8.00	10.13	12.28	14.45	16.60	18.75	20.88
	12	.22	1.04	2.04	3.52	5.21	7.14	9.18	11.29	13.44	15.60	17.77	19.92
	14	.19	.90	1.78	3.10	4.61	6.37	8.32	10.36	12.46	14.60	16.76	18.92
	16	.17	.80	1.57	2.75	4.12	5.74	7.53	9.50	11.53	13.63	15.76	17.91
	18	.15	.71	1.41	2.48	3.72	5.21	6.88	8.70	10.68	12.71	14.80	16.92
	20	.14	.64	1.28	2.25	3.38	4.77	6.31	8.02	9.86	11.86	13.89	15.97
	24	.12	.54	1.07	1.90	2.86	4.06	5.40	6.91	8.56	10.33	12.20	14.15
	28	.10	.46	.93	1.64	2.47	3.52	4.70	6.05	7.52	9.13	10.84	12.65
	32	.09	.41	.81	1.44	2.18	3.11	4.16	5.36	6.69	8.15	9.71	11.38
	36	.08	.36	.73	1.29	1.44	2.78	3.72	4.81	6.02	7.34	8.78	10.31

ECCENTRIC LOADS ON FASTENER GROUPS
TABLE XI Coefficients C

ANGLE = 45°

Required minimum $C = \dfrac{P_u}{(\phi r_v)}$

$P_u = C \times \phi r_v$
n = total number of fasteners in the vertical row
P_u = factored load acting with lever arm l, in.
ϕr_v = design load on one fastener by Specification
X_o = horizontal distance from bolt centroid to P_u
ℓ = load eccentricity (not tabulated, can be calculated by geometry)
C = coefficients tabulated below

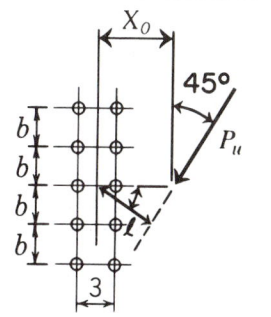

X_o, In.	n 1	2	3	4	5	6	7	8	9	10	11	12
$b = 3$ In.												
2	1.17	2.79	4.74	6.81	8.86	10.90	12.93	14.95	16.96	18.96	20.95	22.95
3	.92	2.32	4.06	5.96	7.99	10.12	12.19	14.25	16.30	18.34	20.37	22.39
4	.75	1.99	3.57	5.31	7.20	9.20	11.27	13.43	15.51	17.59	19.65	21.71
5	.64	1.74	3.18	4.78	6.54	8.44	10.43	12.48	14.56	16.75	18.84	20.92
6	.55	1.54	2.85	4.33	5.98	7.78	9.68	11.66	13.70	15.77	17.86	20.08
7	.49	1.39	2.57	3.93	5.49	7.20	9.02	10.93	12.90	14.92	16.99	19.07
8	.44	1.25	2.34	3.60	5.06	6.70	8.44	10.27	12.17	14.14	16.15	18.20
9	.40	1.14	2.13	3.30	4.69	6.25	7.91	9.68	11.52	13.42	15.38	17.39
10	.36	1.05	1.96	3.05	4.37	5.85	7.44	9.14	10.92	12.77	14.67	16.63
12	.31	.90	1.68	2.65	3.83	5.17	6.63	8.20	9.86	11.60	13.41	15.27
14	.27	.79	1.47	2.33	3.40	4.62	5.95	7.40	8.97	10.61	12.32	14.09
16	.24	.70	1.31	2.08	3.06	4.16	5.38	6.74	8.21	9.75	11.37	13.05
18	.22	.62	1.17	1.89	2.77	3.78	4.91	6.18	7.55	9.00	10.53	12.13
20	.20	.57	1.07	1.72	2.53	3.45	4.51	5.70	6.98	8.34	9.80	11.32
24	.17	.48	.90	1.46	2.15	2.94	3.87	4.91	6.04	7.26	8.57	9.95
28	.15	.41	.77	1.26	1.86	2.57	3.39	4.31	5.31	6.41	7.60	8.85
32	.13	.36	.68	1.11	1.64	2.27	3.01	3.83	4.73	5.73	6.81	7.95
36	.11	.32	.61	1.00	1.47	2.04	2.70	3.44	4.26	5.18	6.16	7.21
$b = 6$ In.												
2	1.17	3.35	5.44	7.48	9.49	11.48	13.47	15.45	17.43	19.40	21.37	23.34
3	.92	2.84	5.04	7.13	9.18	11.21	13.22	15.22	17.21	19.21	21.19	23.17
4	.75	2.51	4.57	6.71	8.80	10.87	12.91	14.94	16.95	18.96	20.95	22.95
5	.64	2.25	4.18	6.23	8.38	10.47	12.55	14.60	16.64	18.66	20.68	22.69
6	.55	2.03	3.86	5.82	7.89	10.05	12.14	14.22	16.28	18.33	20.36	22.39
7	.49	1.85	3.59	5.46	7.47	9.55	11.72	13.81	15.89	17.96	20.02	22.06
8	.44	1.70	3.35	5.14	7.08	9.12	11.21	13.31	15.48	17.57	19.64	21.70
9	.40	1.57	3.13	4.85	6.72	8.71	10.77	12.86	14.97	17.15	19.24	21.31
10	.36	1.46	2.94	4.58	6.40	8.33	10.35	12.42	14.52	16.63	18.82	20.91
12	.31	1.28	2.60	4.11	5.81	7.65	9.59	11.59	13.64	15.73	17.83	19.94
14	.27	1.13	2.33	3.71	5.31	7.06	8.91	10.84	12.83	14.87	16.94	19.04
16	.24	1.02	2.10	3.37	4.88	6.55	8.32	10.17	12.10	14.08	16.11	18.17
18	.22	.92	1.90	3.08	4.50	6.09	7.78	9.57	11.43	13.35	15.33	17.34
20	.20	.84	1.74	2.83	4.18	5.68	7.31	9.02	10.82	12.69	14.61	16.58
24	.17	.72	1.47	2.43	3.65	5.00	6.48	8.08	9.76	11.52	13.34	15.21
28	.15	.63	1.28	2.13	3.23	4.45	5.80	7.28	8.86	10.51	12.24	14.02
32	.13	.56	1.13	1.90	2.89	4.00	5.24	6.62	8.09	9.65	11.28	12.97
36	.11	.50	1.01	1.71	2.61	3.62	4.77	6.05	7.43	8.89	10.44	12.05

ECCENTRIC LOADS ON FASTENER GROUPS
TABLE XI Coefficients C

ANGLE = 75°

Required minimum $C = \dfrac{P_u}{(\phi r_v)}$

$P_u = C \times \phi r_v$

n = total number of fasteners in the vertical row

P_u = factored load acting with lever arm l, in.

ϕr_v = design load on one fastener by Specification

X_o = horizontal distance from bolt centroid to P_u

ℓ = load eccentricity (not tabulated, can be calculated by geometry)

C = coefficients tabulated below

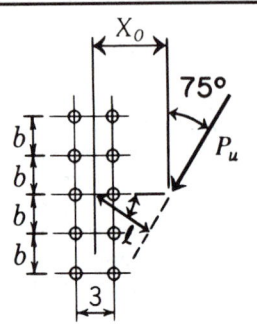

X_o, In.	n											
	1	2	3	4	5	6	7	8	9	10	11	12
b = 3 In.												
2	1.84	3.70	5.64	7.61	9.58	11.56	13.53	15.50	17.47	19.44	21.41	23.38
3	1.71	3.48	5.40	7.38	9.37	11.36	13.35	15.33	17.31	19.29	21.27	23.24
4	1.58	3.24	5.11	7.09	9.10	11.11	13.12	15.12	17.11	19.10	21.09	23.07
5	1.45	3.01	4.69	6.76	8.79	10.82	12.85	14.87	16.88	18.88	20.88	22.87
6	1.32	2.80	4.42	6.16	7.95	10.48	12.53	14.57	16.60	18.62	20.64	22.64
7	1.20	2.62	4.20	5.89	7.64	9.44	12.19	14.25	16.30	18.34	20.36	22.38
8	1.11	2.46	3.99	5.65	7.37	9.17	10.98	13.89	15.96	18.02	20.06	22.10
9	1.02	2.31	3.81	5.43	7.13	8.89	10.68	13.25	15.60	17.67	19.74	21.79
10	.95	2.18	3.63	5.23	6.92	8.66	10.42	12.23	14.93	17.31	19.39	21.46
12	.82	1.95	3.33	4.86	6.50	8.20	9.93	11.71	13.51	15.35	18.28	20.73
14	.73	1.77	3.07	4.53	6.11	7.78	9.50	11.22	13.01	14.79	16.61	18.49
16	.65	1.61	2.83	4.23	5.75	7.37	9.06	10.78	12.50	14.30	16.09	17.90
18	.59	1.48	2.63	3.96	5.42	6.99	8.64	10.34	12.07	13.78	15.58	17.38
20	.54	1.37	2.45	3.72	5.12	6.63	8.24	9.91	11.62	13.36	15.07	16.87
24	.45	1.18	2.15	3.30	4.60	6.02	7.54	9.15	10.81	12.50	14.21	15.95
28	.39	1.04	1.91	2.95	4.16	5.49	6.93	8.45	10.04	11.71	13.39	15.09
32	.35	.93	1.72	2.67	3.79	5.04	6.41	7.86	9.37	10.98	12.61	14.28
36	.31	.83	1.55	2.43	3.47	4.66	5.95	7.32	8.78	10.30	11.90	13.53
b = 6 In.												
2	1.84	3.80	5.78	7.76	9.72	11.69	13.66	15.62	17.59	19.55	21.52	23.48
3	1.71	3.68	5.68	7.67	9.65	11.63	13.60	15.57	17.53	19.50	21.47	23.43
4	1.58	3.52	5.56	7.57	9.56	11.54	13.52	15.50	17.47	19.44	21.41	23.37
5	1.45	3.19	5.40	7.44	9.45	11.44	13.43	15.41	17.39	19.37	21.34	23.31
6	1.32	3.02	5.23	7.29	9.32	11.33	13.33	15.32	17.31	19.28	21.26	23.24
7	1.20	2.88	4.71	7.13	9.18	11.20	13.21	15.21	17.20	19.19	21.18	23.15
8	1.11	2.75	4.55	6.94	9.02	11.06	13.09	15.10	17.10	19.09	21.08	23.07
9	1.02	2.63	4.40	6.61	8.84	10.91	12.95	14.97	16.98	18.98	20.97	22.96
10	.95	2.52	4.28	6.09	8.66	10.74	12.80	14.83	16.85	18.86	20.86	22.86
12	.82	2.32	4.04	5.82	7.66	10.37	12.46	14.52	16.57	18.60	20.62	22.63
14	.73	2.15	3.83	5.57	7.38	9.24	12.09	14.18	16.25	18.30	20.34	22.36
16	.65	2.00	3.63	5.36	7.15	8.95	11.42	13.81	15.90	17.97	20.03	22.07
18	.59	1.87	3.45	5.17	6.89	8.69	10.52	13.11	15.52	17.62	19.70	21.76
20	.54	1.75	3.28	4.97	6.67	8.44	10.25	12.08	14.80	17.24	19.34	21.42
24	.45	1.55	2.99	4.57	6.28	7.99	9.75	11.56	13.37	15.22	18.18	20.67
28	.39	1.40	2.74	4.24	5.87	7.58	9.30	11.05	12.86	14.66	16.52	18.37
32	.35	1.27	2.52	3.95	5.50	7.16	8.88	10.61	12.35	14.16	15.97	17.79
36	.31	1.16	2.34	3.68	5.17	6.78	8.46	10.17	11.91	13.64	15.46	17.27

ECCENTRIC LOADS ON FASTENER GROUPS
TABLE XII Coefficients C

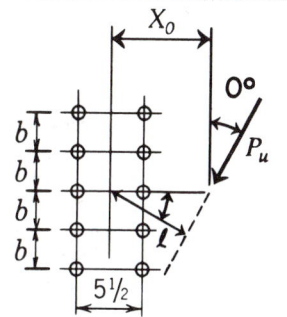

ANGLE = 0°

Required minimum $C = \dfrac{P_u}{(\phi r_v)}$

$P_u = C \times \phi r_v$

n = total number of fasteners in the vertical row

P_u = factored load acting with lever arm l, in.

ϕr_v = design load on one fastener by Specification

X_o = horizontal distance from bolt centroid to P_u

ℓ = load eccentricity (not tabulated, can be calculated by geometry)

C = coefficients tabulated below

X_o, In.	n 1	2	3	4	5	6	7	8	9	10	11	12
$b = 3$ In.												
2	1.14	2.75	4.63	6.70	8.77	10.84	12.90	14.94	16.98	19.00	21.01	23.02
3	.94	2.32	3.94	5.84	7.88	9.97	12.07	14.17	16.26	18.33	20.39	22.44
4	.80	1.99	3.39	5.10	7.01	9.05	11.15	13.27	15.38	17.50	19.60	21.69
5	.70	1.74	2.97	4.51	6.25	8.18	10.22	12.31	14.44	16.57	18.70	20.82
6	.62	1.54	2.64	4.03	5.61	7.40	9.34	11.38	13.48	15.60	17.74	19.88
7	.55	1.38	2.36	3.63	5.07	6.73	8.55	10.51	12.54	14.64	16.76	18.90
8	.50	1.25	2.14	3.30	4.61	6.15	7.86	9.71	11.67	13.71	15.80	17.92
9	.46	1.14	1.96	3.01	4.22	5.66	7.24	8.99	10.86	12.83	14.87	16.96
10	.42	1.04	1.80	2.78	3.89	5.23	6.71	8.35	10.13	12.02	14.00	16.03
12	.37	.90	1.55	2.39	3.36	4.53	5.82	7.29	8.88	10.61	12.43	14.35
14	.32	.79	1.37	2.10	2.96	3.99	5.13	6.44	7.87	9.44	11.09	12.85
16	.29	.70	1.22	1.87	2.64	3.55	4.58	5.76	7.05	8.48	9.99	11.62
18	.26	.63	1.10	1.68	2.38	3.20	4.14	5.21	6.38	7.68	9.08	10.58
20	.24	.58	1.00	1.53	2.16	2.91	3.77	4.75	5.82	7.02	8.31	9.69
24	.20	.49	.84	1.29	1.83	2.46	3.19	4.03	4.94	5.97	7.07	8.28
28	.18	.42	.73	1.12	1.58	2.13	2.77	3.49	4.29	5.19	6.15	7.21
32	.16	.38	.64	.98	1.39	1.88	2.44	3.08	3.79	4.58	5.44	6.38
36	.14	.33	.58	.88	1.24	1.68	2.18	2.75	3.39	4.10	4.87	5.72
$b = 6$ In.												
2	1.14	3.28	5.40	7.46	9.50	11.51	13.51	15.50	17.48	19.46	21.44	23.42
3	.94	2.86	4.95	7.07	9.16	11.22	13.25	15.27	17.28	19.28	21.27	23.26
4	.80	2.52	4.49	6.62	8.74	10.84	12.92	14.97	17.01	19.03	21.04	23.05
5	.70	2.24	4.05	6.14	8.28	10.41	12.52	14.61	16.67	18.72	20.76	22.78
6	.62	2.00	3.66	5.68	7.80	9.94	12.08	14.19	16.29	18.37	20.43	22.47
7	.55	1.81	3.32	5.24	7.32	9.46	11.60	13.74	15.86	17.97	20.05	22.12
8	.50	1.64	3.02	4.85	6.86	8.97	11.11	13.26	15.40	17.53	19.64	21.73
9	.46	1.50	2.77	4.50	6.42	8.49	10.62	12.77	14.92	17.06	19.19	21.30
10	.42	1.38	2.56	4.18	6.02	8.03	10.13	12.28	14.43	16.58	18.73	20.86
12	.37	1.19	2.21	3.66	5.31	7.19	9.20	11.30	13.43	15.59	17.75	19.90
14	.32	1.05	1.95	3.24	4.72	6.46	8.36	10.38	12.46	14.59	16.75	18.91
16	.29	.93	1.74	2.90	4.24	5.83	7.59	9.54	11.55	13.63	15.75	17.90
18	.26	.84	1.57	2.62	3.84	5.31	6.95	8.75	10.71	12.72	14.80	16.92
20	.24	.76	1.43	2.39	3.50	4.87	6.39	8.09	9.91	11.88	13.90	15.97
24	.20	.64	1.21	2.02	2.98	4.16	5.49	6.99	8.62	10.37	12.23	14.23
28	.18	.55	1.05	1.76	2.59	3.63	4.80	6.13	7.59	9.18	10.88	12.68
32	.16	.49	.93	1.55	2.29	3.21	4.25	5.45	6.77	8.21	9.76	11.42
36	.14	.44	.83	1.38	2.05	2.88	3.81	4.90	6.09	7.41	8.83	10.36

ECCENTRIC LOADS ON FASTENER GROUPS
TABLE XII Coefficients C

ANGLE = 45°

Required minimum $C = \dfrac{P_u}{(\phi r_v)}$

$P_u = C \times \phi r_v$

n = total number of fasteners in the vertical row

P_u = factored load acting with lever arm l, in.

ϕr_v = design load on one fastener by Specification

X_o = horizontal distance from bolt centroid to P_u

ℓ = load eccentricity (not tabulated, can be calculated by geometry)

C = coefficients tabulated below

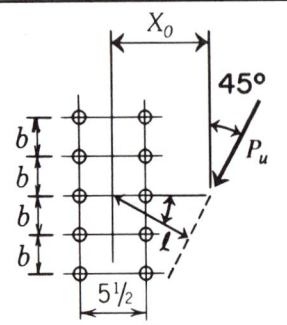

X_o, In.	n 1	2	3	4	5	6	7	8	9	10	11	12
b = 3 In.												
2	1.55	3.26	5.12	7.05	9.02	11.01	13.00	15.00	16.99	18.98	20.97	22.96
3	1.30	2.76	4.46	6.36	8.31	10.31	12.33	14.35	16.38	18.40	20.41	22.42
4	1.11	2.44	3.97	5.67	7.52	9.46	11.55	13.59	15.63	17.68	19.72	21.76
5	.98	2.17	3.60	5.19	6.90	8.73	10.67	12.68	14.82	16.88	18.94	21.00
6	.87	1.95	3.28	4.77	6.38	8.11	9.95	11.89	13.89	15.93	17.99	20.18
7	.78	1.78	3.01	4.40	5.91	7.56	9.32	11.18	13.11	15.10	17.13	19.20
8	.71	1.63	2.77	4.07	5.50	7.07	8.76	10.54	12.40	14.33	16.32	18.34
9	.65	1.50	2.57	3.78	5.13	6.64	8.26	9.97	11.77	13.64	15.56	17.54
10	.60	1.40	2.39	3.52	4.81	6.25	7.81	9.46	11.19	13.00	14.87	16.80
12	.52	1.22	2.09	3.09	4.27	5.58	7.01	8.54	10.16	11.87	13.64	15.47
14	.46	1.08	1.85	2.75	3.82	5.03	6.33	7.76	9.28	10.89	12.57	14.31
16	.41	1.01	1.66	2.48	3.46	4.56	5.77	7.10	8.53	10.05	11.63	13.29
18	.37	.87	1.50	2.25	3.15	4.16	5.29	6.53	7.88	9.30	10.81	12.38
20	.34	.80	1.37	2.06	2.89	3.83	4.88	6.05	7.30	8.65	10.07	11.58
24	.29	.68	1.16	1.76	2.47	3.29	4.21	5.24	6.35	7.56	8.85	10.22
28	.25	.59	1.01	1.53	2.16	2.88	3.70	4.61	5.61	6.70	7.87	9.11
32	.22	.52	.90	1.36	1.91	2.55	3.29	4.11	5.02	6.01	7.07	8.20
36	.20	.47	.80	1.22	1.71	2.29	2.96	3.71	4.53	5.44	6.41	7.45
b = 6 In.												
2	1.55	3.50	5.50	7.50	9.50	11.49	13.47	15.45	17.43	19.40	21.37	23.34
3	1.30	3.07	5.14	7.18	9.21	11.22	13.23	15.22	17.22	19.21	21.19	23.17
4	1.11	2.74	4.70	6.79	8.85	10.89	12.92	14.94	16.95	18.96	20.95	22.95
5	.98	2.49	4.33	6.33	8.45	10.51	12.57	14.61	16.65	18.67	20.68	22.69
6	.87	2.28	4.02	5.93	7.97	10.10	12.18	14.24	16.30	18.34	20.37	22.39
7	.78	2.10	3.76	5.59	7.55	9.61	11.76	13.84	15.91	17.97	20.02	22.06
8	.71	1.94	3.53	5.28	7.17	9.19	11.26	13.42	15.51	17.58	19.65	21.70
9	.65	1.81	3.32	5.00	6.83	8.79	10.82	12.90	15.08	17.17	19.25	21.32
10	.60	1.69	3.14	4.74	6.51	8.42	10.41	12.47	14.55	16.75	18.83	20.92
12	.52	1.50	2.81	4.29	5.95	7.75	9.66	11.65	13.69	15.76	17.86	20.07
14	.46	1.34	2.53	3.89	5.45	7.17	9.00	10.91	12.88	14.91	16.98	19.06
16	.41	1.21	2.29	3.55	5.02	6.66	8.41	10.25	12.16	14.13	16.14	18.20
18	.37	1.11	2.09	3.26	4.65	6.22	7.89	9.65	11.50	13.41	15.37	17.38
20	.34	1.01	1.92	3.01	4.33	5.82	7.41	9.11	10.90	12.75	14.66	16.62
24	.29	.87	1.65	2.61	3.79	5.14	6.60	8.17	9.84	11.59	13.39	15.25
28	.25	.76	1.44	2.30	3.37	4.58	5.92	7.38	8.95	10.59	12.30	14.07
32	.22	.67	1.27	2.05	3.03	4.13	5.35	6.72	8.18	9.73	11.35	13.03
36	.20	.60	1.14	1.85	2.74	3.74	4.88	6.16	7.52	8.98	10.51	12.12

ECCENTRIC LOADS ON FASTENER GROUPS
TABLE XII Coefficients C

ANGLE = 75°

Required minimum $C = \dfrac{P_u}{(\phi r_v)}$

$P_u = C \times \phi r_v$

n = total number of fasteners in the vertical row

P_u = factored load acting with lever arm l, in.

ϕr_v = design load on one fastener by Specification

X_o = horizontal distance from bolt centroid to P_u

ℓ = load eccentricity (not tabulated, can be calculated by geometry)

C = coefficients tabulated below

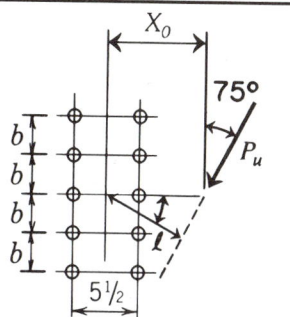

	X_o, In.	n											
		1	2	3	4	5	6	7	8	9	10	11	12
b = 3 In.	2	1.92	3.83	5.75	7.69	9.64	11.60	13.56	15.52	17.49	19.45	21.42	23.38
	3	1.87	3.74	5.62	7.54	9.48	11.44	13.41	15.37	17.34	19.31	21.28	23.25
	4	1.82	3.62	5.46	7.35	9.28	11.24	13.21	15.19	17.16	19.14	21.12	23.10
	5	1.75	3.49	5.27	7.13	9.05	11.01	12.98	14.96	16.95	18.94	20.92	22.90
	6	1.68	3.35	5.07	6.89	8.79	10.74	12.72	14.71	16.71	18.70	20.70	22.69
	7	1.60	3.21	4.87	6.64	8.51	10.45	12.43	14.43	16.43	18.44	20.45	22.45
	8	1.53	3.07	4.67	6.38	8.21	10.14	12.11	14.12	16.13	18.15	20.17	22.18
	9	1.46	2.93	4.48	6.09	7.77	9.51	11.78	13.79	15.81	17.84	19.87	21.89
	10	1.38	2.81	4.30	5.87	7.51	9.20	10.94	12.74	15.46	17.50	19.54	21.58
	12	1.25	2.58	3.98	5.50	7.08	8.72	10.41	12.14	13.97	15.79	18.84	20.90
	14	1.14	2.36	3.71	5.15	6.70	8.30	9.96	11.67	13.40	15.16	16.95	18.77
	16	1.04	2.19	3.46	4.85	6.34	7.91	9.54	11.21	12.92	14.67	16.42	18.23
	18	.96	2.03	3.24	4.57	6.01	7.54	9.13	10.78	12.46	14.19	15.93	17.70
	20	.89	1.89	3.04	4.32	5.71	7.19	8.76	10.37	12.04	13.76	15.45	17.20
	24	.77	1.66	2.70	3.88	5.17	6.57	8.05	9.61	11.21	12.87	14.56	16.29
	28	.68	1.47	2.42	3.51	4.71	6.03	7.44	8.93	10.47	12.08	13.73	15.41
	32	.60	1.32	2.19	3.19	4.32	5.56	6.90	8.32	9.81	11.36	12.96	14.60
	36	.55	1.20	1.99	2.92	3.98	5.15	6.43	7.78	9.21	10.70	12.24	13.84
b = 6 In.	2	1.92	3.85	5.80	7.77	9.73	11.70	13.66	15.63	17.59	19.55	21.52	23.48
	3	1.87	3.77	5.72	7.69	9.66	11.63	13.60	15.57	17.54	19.50	21.47	23.43
	4	1.82	3.67	5.62	7.60	9.58	11.56	13.53	15.50	17.47	19.44	21.41	23.38
	5	1.75	3.56	5.51	7.49	9.48	11.46	13.44	15.42	17.40	19.37	21.34	23.31
	6	1.68	3.44	5.37	7.36	9.36	11.36	13.34	15.33	17.31	19.29	21.26	23.24
	7	1.60	3.31	5.22	7.22	9.23	11.24	13.23	15.23	17.21	19.20	21.18	23.16
	8	1.53	3.18	5.06	7.06	9.09	11.10	13.11	15.11	17.11	19.10	21.09	23.07
	9	1.46	3.05	4.75	6.90	8.93	10.96	12.98	14.99	16.99	18.99	20.98	22.97
	10	1.38	2.94	4.59	6.37	8.76	10.80	12.84	14.86	16.87	18.88	20.87	22.87
	12	1.25	2.72	4.35	6.09	7.89	10.46	12.52	14.56	16.60	18.62	20.63	22.64
	14	1.14	2.54	4.13	5.83	7.58	9.39	12.17	14.23	16.29	18.33	20.36	22.38
	16	1.04	2.38	3.92	5.59	7.32	9.12	10.94	13.87	15.95	18.01	20.06	22.09
	18	.96	2.24	3.74	5.38	7.08	8.85	10.64	13.22	15.58	17.66	19.73	21.78
	20	.89	2.11	3.57	5.18	6.87	8.62	10.38	12.20	14.90	17.29	19.38	21.45
	24	.77	1.89	3.27	4.81	6.45	8.16	9.89	11.68	13.48	15.32	18.26	20.71
	28	.68	1.70	3.00	4.47	6.06	7.74	9.46	11.18	12.98	14.77	16.62	18.46
	32	.60	1.55	2.77	4.17	5.70	7.33	9.02	10.75	12.47	14.27	16.07	17.87
	36	.55	1.42	2.57	3.91	5.37	6.94	8.60	10.30	12.04	13.75	15.56	17.35

ECCENTRIC LOADS ON FASTENER GROUPS
TABLE XIII Coefficients C

Required minimum $C = \dfrac{P_u}{(\phi r_v)}$

ANGLE = 0°

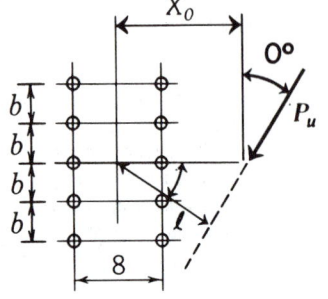

$P_u = C \times \phi r_v$
n = total number of fasteners in the vertical row
P_u = factored load acting with lever arm l, in.
ϕr_v = design load on one fastener by Specification
X_o = horizontal distance from bolt centroid to P_u
ℓ = load eccentricity (not tabulated, can be calculated by geometry)
C = coefficients tabulated below

X_o, In.	n											
	1	2	3	4	5	6	7	8	9	10	11	12
b = 3 In.												
2	1.31	2.91	4.74	6.85	8.85	10.88	12.91	14.94	16.97	18.99	21.00	23.00
3	1.16	2.54	4.15	5.99	8.02	10.06	12.12	14.19	16.26	18.32	20.37	22.42
4	.98	2.24	3.66	5.33	7.20	9.18	11.23	13.32	15.41	17.50	19.59	21.67
5	.92	2.00	3.27	4.80	6.50	8.37	10.35	12.40	14.49	16.60	18.71	20.81
6	.79	1.80	2.96	4.35	5.91	7.65	9.53	11.51	13.57	15.66	17.77	19.89
7	.71	1.63	2.70	3.97	5.40	7.02	8.79	10.69	12.68	14.73	16.82	18.93
8	.65	1.50	2.46	3.65	4.97	6.48	8.13	9.93	11.84	13.83	15.89	17.98
9	.60	1.38	2.27	3.37	4.59	6.01	7.56	9.26	11.08	13.00	14.99	17.05
10	.56	1.28	2.11	3.13	4.27	5.59	7.05	8.65	10.38	12.22	14.15	16.15
12	.49	1.11	1.84	2.73	3.73	4.90	6.19	7.63	9.18	10.86	12.65	14.52
14	.44	.99	1.64	2.42	3.32	4.36	5.51	6.80	8.20	9.73	11.37	13.11
16	.39	.89	1.47	2.17	2.98	3.91	4.95	6.13	7.40	8.80	10.30	11.90
18	.36	.81	1.33	1.97	2.70	3.55	4.50	5.57	6.73	8.01	9.39	10.87
20	.33	.74	1.22	1.80	2.47	3.25	4.12	5.10	6.17	7.35	8.62	10.00
24	.28	.63	1.04	1.54	2.11	2.77	3.51	4.35	5.28	6.30	7.39	8.59
28	.25	.55	.91	1.34	1.83	2.41	3.06	3.79	4.60	5.50	6.46	7.51
32	.22	.49	.81	1.18	1.62	2.13	2.71	3.36	4.08	4.87	5.73	6.67
36	.20	.44	.73	1.06	1.46	1.91	2.43	3.01	3.66	4.37	5.15	5.99
b = 6 In.												
2	1.31	3.37	5.42	7.46	9.49	11.50	13.50	15.49	17.47	19.46	21.43	23.41
3	1.16	2.94	4.99	7.08	9.15	11.21	13.24	15.26	17.27	19.27	21.26	23.25
4	.98	2.63	4.55	6.64	8.74	10.83	12.90	14.95	16.99	19.01	21.03	23.03
5	.92	2.37	4.15	6.18	8.29	10.41	12.51	14.59	16.66	18.70	20.74	22.77
6	.79	2.15	3.78	5.74	7.82	9.95	12.07	14.18	16.27	18.35	20.41	22.45
7	.71	1.97	3.47	5.33	7.36	9.47	11.60	13.73	15.84	17.94	20.03	22.10
8	.65	1.81	3.19	4.96	6.92	8.99	11.12	13.26	15.39	17.51	19.61	21.71
9	.60	1.67	2.95	4.63	6.50	8.53	10.64	12.77	14.91	17.05	19.17	21.28
10	.56	1.55	2.75	4.33	6.11	8.09	10.16	12.28	14.43	16.57	18.71	20.84
12	.49	1.35	2.41	3.82	5.43	7.27	9.25	11.33	13.44	15.59	17.74	19.89
14	.44	1.20	2.14	3.41	4.87	6.57	8.44	10.42	12.49	14.60	16.74	18.90
16	.39	1.08	1.92	3.07	4.40	5.97	7.71	9.60	11.59	13.65	15.76	17.90
18	.36	1.01	1.75	2.79	4.00	5.46	7.08	8.86	10.77	12.76	14.82	16.93
20	.33	.89	1.60	2.56	3.67	5.02	6.53	8.21	10.02	11.93	13.93	15.99
24	.28	.76	1.37	2.19	3.14	4.32	5.63	7.11	8.72	10.45	12.29	14.27
28	.25	.66	1.19	1.90	2.75	3.78	4.93	6.26	7.70	9.27	10.96	12.74
32	.22	.59	1.06	1.69	2.44	3.35	4.38	5.58	6.88	8.31	9.85	11.49
36	.20	.52	.95	1.51	2.19	3.01	3.94	5.02	6.21	7.52	8.93	10.44

ECCENTRIC LOADS ON FASTENER GROUPS
TABLE XIII Coefficients *C*

ANGLE = 45°

Required minimum $C = \dfrac{P_u}{(\phi r_v)}$

$P_u = C \times \phi r_v$

n = total number of fasteners in the vertical row

P_u = factored load acting with lever arm l, in.

ϕr_v = design load on one fastener by Specification

X_o = horizontal distance from bolt centroid to P_u

ℓ = load eccentricity (not tabulated, can be calculated by geometry)

C = coefficients tabulated below

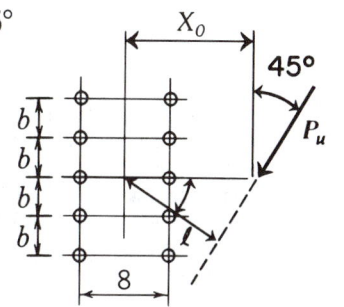

X_o, In.	n 1	2	3	4	5	6	7	8	9	10	11	12
b = 3 In.												
2	1.75	3.54	5.37	7.26	9.19	11.14	13.10	15.07	17.05	19.03	21.01	22.99
3	1.52	3.16	4.91	6.73	8.61	10.55	12.51	14.50	16.49	18.48	20.48	22.48
4	1.35	2.79	4.35	6.08	7.97	9.87	11.82	13.80	15.81	17.82	19.83	21.85
5	1.21	2.53	3.98	5.57	7.28	9.09	11.00	13.05	15.05	17.07	19.10	21.13
6	1.09	2.30	3.69	5.17	6.78	8.49	10.30	12.20	14.16	16.28	18.30	20.34
7	1.00	2.12	3.41	4.83	6.35	7.97	9.69	11.51	13.40	15.36	17.36	19.39
8	.92	1.96	3.17	4.51	5.96	7.51	9.16	10.90	12.72	14.62	16.57	18.57
9	.85	1.82	2.96	4.23	5.60	7.08	8.68	10.35	12.11	13.94	15.83	17.78
10	.79	1.70	2.78	3.97	5.28	6.70	8.23	9.86	11.55	13.32	15.16	17.05
12	.69	1.50	2.46	3.54	4.73	6.04	7.46	8.97	10.56	12.23	13.96	15.76
14	.61	1.35	2.21	3.18	4.27	5.49	6.80	8.20	9.70	11.28	12.92	14.63
16	.55	1.22	2.00	2.89	3.89	5.01	6.23	7.54	8.95	10.45	12.01	13.63
18	.51	1.11	1.83	2.64	3.57	4.61	5.74	6.98	8.30	9.71	11.19	12.74
20	.46	1.02	1.68	2.43	3.29	4.25	5.32	6.48	7.73	9.06	10.46	11.94
24	.40	.88	1.44	2.09	2.84	3.68	4.62	5.66	6.77	7.96	9.24	10.59
28	.35	.77	1.26	1.83	2.49	3.24	4.08	5.00	6.01	7.09	8.25	9.48
32	.31	.68	1.12	1.63	2.22	2.89	3.65	4.48	5.39	6.38	7.44	8.56
36	.28	.62	1.01	1.47	2.00	2.61	3.29	4.05	4.88	5.79	6.76	7.80
b = 6 In.												
2	1.75	3.62	5.57	7.54	9.52	11.50	13.48	15.45	17.43	19.40	21.37	23.34
3	1.52	3.34	5.27	7.25	9.25	11.25	13.24	15.24	17.22	19.21	21.19	23.17
4	1.35	2.98	4.92	6.90	8.92	10.94	12.95	14.96	16.97	18.97	20.96	22.95
5	1.21	2.72	4.52	6.52	8.54	10.58	12.61	14.64	16.67	18.68	20.69	22.69
6	1.09	2.52	4.21	6.09	8.09	10.18	12.23	14.28	16.32	18.36	20.38	22.40
7	1.00	2.34	3.96	5.75	7.68	9.71	11.83	13.89	15.95	18.00	20.04	22.08
8	.92	2.20	3.73	5.45	7.31	9.30	11.34	13.48	15.55	17.62	19.67	21.72
9	.85	2.04	3.53	5.19	6.98	8.90	10.92	12.98	15.14	17.21	19.28	21.34
10	.79	1.92	3.35	4.94	6.68	8.54	10.51	12.55	14.62	16.79	18.87	20.95
12	.69	1.72	3.03	4.50	6.13	7.90	9.78	11.74	13.76	15.82	17.91	20.11
14	.61	1.55	2.75	4.12	5.65	7.33	9.13	11.01	12.97	14.98	17.03	19.11
16	.55	1.41	2.51	3.78	5.23	6.83	8.55	10.36	12.25	14.20	16.21	18.25
18	.51	1.29	2.31	3.49	4.85	6.39	8.04	9.78	11.60	13.49	15.44	17.44
20	.46	1.19	2.13	3.24	4.53	6.00	7.57	9.25	11.01	12.84	14.74	16.68
24	.40	1.03	1.84	2.82	4.00	5.32	6.77	8.32	9.97	11.69	13.49	15.33
28	.35	.90	1.62	2.50	3.56	4.77	6.09	7.53	9.08	10.71	12.40	14.16
32	.31	.81	1.44	2.24	3.21	4.31	5.52	6.87	8.23	9.85	11.46	13.13
36	.28	.72	1.30	2.02	2.91	3.92	5.04	6.30	7.66	9.10	10.63	12.22

ECCENTRIC LOADS ON FASTENER GROUPS
TABLE XIII Coefficients C

ANGLE = 75°

Required minimum $C = \dfrac{P_u}{(\phi r_v)}$

$P_u = C \times \phi r_v$

n = total number of fasteners in the vertical row

P_u = factored load acting with lever arm l, in.

ϕr_v = design load on one fastener by Specification

X_o = horizontal distance from bolt centroid to P_u

ℓ = load eccentricity (not tabulated, can be calculated by geometry)

C = coefficients tabulated below

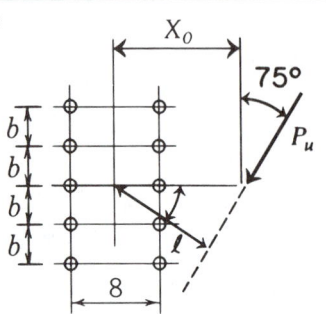

X_o, In.	n											
	1	2	3	4	5	6	7	8	9	10	11	12
2	1.94	3.87	5.81	7.74	9.69	11.64	13.59	15.55	17.51	19.47	21.43	23.40
3	1.92	3.82	5.73	7.65	9.58	11.52	13.47	15.42	17.38	19.35	21.31	23.28
4	1.89	3.76	5.63	7.52	9.44	11.37	13.31	15.27	17.23	19.20	21.16	23.13
5	1.85	3.68	5.52	7.38	9.27	11.19	13.13	15.08	17.05	19.01	20.98	22.96
6	1.81	3.60	5.40	7.22	9.09	10.99	12.92	14.87	16.84	18.81	20.78	22.76
7	1.76	3.51	5.26	7.05	8.88	10.77	12.69	14.64	16.60	18.58	20.56	22.54
8	1.71	3.41	5.12	6.86	8.67	10.53	12.44	14.38	16.35	18.32	20.31	22.30
9	1.66	3.31	4.97	6.68	8.44	10.28	12.17	14.11	16.07	18.05	20.04	22.03
10	1.61	3.21	4.83	6.49	8.22	10.02	11.89	13.82	15.78	17.76	19.75	21.75
12	1.51	3.02	4.55	6.12	7.77	9.42	11.21	13.21	15.15	17.12	19.12	21.13
14	1.41	2.83	4.28	5.78	7.35	8.94	10.57	12.31	14.01	15.75	17.51	20.45
16	1.32	2.66	4.04	5.47	6.98	8.53	10.15	11.78	13.51	15.20	16.92	18.77
18	1.23	2.49	3.81	5.19	6.65	8.17	9.73	11.36	13.00	14.71	16.41	18.13
20	1.15	2.34	3.61	4.94	6.34	7.82	9.37	10.95	12.59	14.24	15.94	17.66
24	1.02	2.09	3.25	4.48	5.80	7.20	8.67	10.21	11.80	13.40	15.06	16.72
28	.91	1.88	2.94	4.09	5.33	6.66	8.05	9.52	11.05	12.63	14.23	15.88
32	.82	1.71	2.68	3.76	4.92	6.17	7.51	8.91	10.38	11.90	13.47	15.07
36	.75	1.56	2.46	3.47	4.56	5.75	7.02	8.36	9.77	11.24	12.76	14.33
2	1.94	3.88	5.82	7.78	9.74	11.70	13.66	15.63	17.59	19.56	21.52	23.48
3	1.92	3.83	5.76	7.72	9.60	11.64	13.61	15.57	17.54	19.50	21.47	23.43
4	1.89	3.77	5.69	7.64	9.61	11.57	13.54	15.51	17.48	19.44	21.41	23.38
5	1.85	3.70	5.60	7.55	9.52	11.49	13.46	15.43	17.41	19.38	21.35	23.31
6	1.81	3.62	5.50	7.45	9.41	11.39	13.37	15.35	17.32	19.30	21.27	23.24
7	1.76	3.53	5.39	7.33	9.30	11.28	13.27	15.25	17.23	19.21	21.19	23.17
8	1.71	3.44	5.27	7.20	9.17	11.16	13.15	15.14	17.13	19.11	21.10	23.08
9	1.66	3.35	5.15	7.06	9.04	11.03	13.03	15.03	17.02	19.01	21.00	22.98
10	1.61	3.25	5.02	6.92	8.89	10.89	12.90	14.90	16.90	18.90	20.89	22.88
12	1.51	3.07	4.76	6.61	8.57	10.58	12.60	14.62	16.64	18.65	20.66	22.66
14	1.41	2.89	4.48	6.16	7.89	10.24	12.27	14.31	16.34	18.37	20.39	22.41
16	1.32	2.73	4.28	5.90	7.61	9.40	11.92	13.97	16.02	18.06	20.10	22.13
18	1.23	2.59	4.08	5.68	7.37	9.12	10.87	13.61	15.67	17.73	19.78	21.82
20	1.15	2.46	3.91	5.47	7.13	8.83	10.59	12.40	15.30	17.37	19.44	21.50
24	1.02	2.22	3.59	5.10	6.71	8.38	10.10	11.86	13.66	15.49	18.70	20.78
28	.91	2.02	3.32	4.77	6.33	7.96	9.66	11.38	13.14	14.92	16.72	18.61
32	.82	1.85	3.08	4.47	5.97	7.57	9.23	10.94	12.65	14.43	16.21	18.01
36	.75	1.70	2.87	4.19	5.64	7.19	8.81	10.50	12.22	13.92	15.71	17.49

Note: First block b = 3 In. Second block b = 6 In.

ECCENTRIC LOADS ON FASTENER GROUPS
TABLE XIV Coefficients C

ANGLE = 0°

Required minimum $C = \dfrac{P_u}{(\phi r_v)}$

$P_u = C \times \phi r_v$

n = total number of fasteners in the vertical row

P_u = factored load acting with lever arm l, in.

ϕr_v = design load on one fastener by Specification

X_o = horizontal distance from bolt centroid to P_u

ℓ = load eccentricity (not tabulated, can be calculated by geometry)

C = coefficients tabulated below

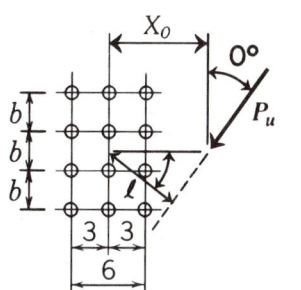

X_o, In.	n											
	1	2	3	4	5	6	7	8	9	10	11	12
2	1.72	4.08	6.89	9.96	13.09	16.21	19.31	22.39	25.44	28.48	31.50	34.51
3	1.42	3.40	5.79	8.66	11.73	14.89	18.06	21.21	24.35	27.47	30.56	33.63
4	1.21	2.90	4.97	7.54	10.41	13.49	16.65	19.84	23.03	26.20	29.36	32.51
5	1.06	2.51	4.35	6.64	9.26	12.16	15.23	18.40	21.59	24.80	28.00	31.19
6	.92	2.21	3.85	5.91	8.28	10.98	13.91	16.98	20.14	23.34	26.55	29.77
7	.81	1.96	3.45	5.31	7.46	9.96	12.71	15.66	18.73	21.88	25.07	28.29
8	.72	1.77	3.11	4.80	6.78	9.09	11.65	14.45	17.40	20.47	23.62	26.81
9	.64	1.60	2.83	4.38	6.20	8.34	10.73	13.37	16.19	19.15	22.22	25.36
10	.58	1.46	2.59	4.02	5.71	7.70	9.92	12.40	15.08	17.93	20.90	23.97
12	.49	1.24	2.21	3.44	4.91	6.65	8.59	10.80	13.20	15.79	18.55	21.43
14	.42	1.08	1.92	3.01	4.30	5.83	7.57	9.53	11.68	14.01	16.51	19.17
16	.37	.95	1.70	2.66	3.82	5.19	6.75	8.51	10.45	12.58	14.87	17.32
18	.33	.85	1.52	2.39	3.43	4.67	6.08	7.68	9.44	11.40	13.50	15.75
20	.29	.77	1.37	2.16	3.11	4.24	5.53	6.99	8.61	10.40	12.34	14.43
24	.25	.65	1.15	1.82	2.62	3.57	4.67	5.92	7.30	8.84	10.50	12.31
28	.21	.56	.99	1.57	2.26	3.08	4.03	5.12	6.33	7.67	9.13	10.71
32	.18	.49	.87	1.38	1.98	2.71	3.55	4.51	5.58	6.77	8.06	9.47
36	.16	.43	.77	1.23	1.77	2.42	3.17	4.03	4.98	6.05	7.21	8.48
2	1.72	4.88	8.07	11.18	14.23	17.26	20.26	23.24	26.22	29.19	32.16	35.12
3	1.42	4.24	7.39	10.59	13.73	16.81	19.87	22.90	25.91	28.91	31.90	34.89
4	1.21	3.72	6.69	9.90	13.10	16.25	19.37	22.44	25.50	28.54	31.56	34.57
5	1.06	3.29	6.02	9.18	12.40	15.60	18.76	21.90	25.00	28.07	31.13	34.17
6	.92	2.93	5.43	8.47	11.67	14.89	18.10	21.27	24.42	27.54	30.63	33.70
7	.81	2.63	4.91	7.81	10.94	14.16	17.38	20.59	23.78	26.94	30.06	33.17
8	.72	2.38	4.47	7.22	10.24	13.42	16.65	19.87	23.09	26.28	29.45	32.58
9	.64	2.17	4.09	6.68	9.58	12.70	15.90	19.13	22.37	25.58	28.77	31.95
10	.58	2.00	3.78	6.21	8.97	12.01	15.17	18.39	21.62	24.86	28.07	31.27
12	.49	1.71	3.27	5.42	7.91	10.74	13.77	16.92	20.13	23.36	26.60	29.83
14	.42	1.49	2.87	4.78	7.03	9.64	12.50	15.53	18.67	21.87	25.10	28.34
16	.37	1.32	2.55	4.28	6.29	8.69	11.34	14.26	17.29	20.42	23.61	26.83
18	.33	1.19	2.30	3.86	5.70	7.91	10.38	13.08	16.03	19.06	22.17	25.35
20	.29	1.08	2.09	3.51	5.20	7.25	9.54	12.09	14.82	17.79	20.82	23.93
24	.25	.91	1.76	2.97	4.42	6.19	8.18	10.44	12.89	15.52	18.31	21.32
28	.21	.78	1.52	2.57	3.84	5.39	7.14	9.15	11.35	13.73	16.29	18.99
32	.18	.69	1.33	2.27	3.39	4.76	6.33	8.13	10.11	12.28	14.61	17.10
36	.16	.61	1.19	2.03	3.03	4.27	5.68	7.31	9.10	11.08	13.22	15.51

($b = 3$ In. for upper block; $b = 6$ In. for lower block)

ECCENTRIC LOADS ON FASTENER GROUPS
TABLE XIV Coefficients C

ANGLE = 45°

Required minimum $C = \dfrac{P_u}{(\phi r_v)}$

$P_u = C \times \phi r_v$

n = total number of fasteners in the vertical row

P_u = factored load acting with lever arm l, in.

ϕr_v = design load on one fastener by Specification

X_o = horizontal distance from bolt centroid to P_u

ℓ = load eccentricity (not tabulated, can be calculated by geometry)

C = coefficients tabulated below

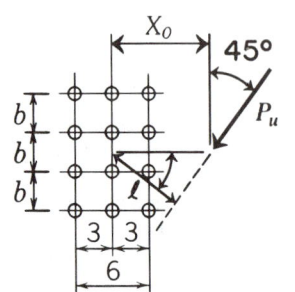

X_o, In.	\multicolumn{12}{c}{n}											
	1	2	3	4	5	6	7	8	9	10	11	12
$b = 3$ In.												
2	2.25	4.74	7.54	10.46	13.44	16.45	19.45	22.46	25.46	28.45	31.44	34.42
3	1.89	4.06	6.53	9.30	12.34	15.36	18.41	21.46	24.51	27.55	30.58	33.60
4	1.63	3.57	5.84	8.37	11.12	14.06	17.21	20.29	23.37	26.45	29.53	32.59
5	1.42	3.18	5.27	7.63	10.20	12.95	15.88	18.90	21.99	25.23	28.34	31.44
6	1.25	2.85	4.78	6.99	9.41	12.01	14.79	17.71	20.72	23.79	26.90	30.20
7	1.11	2.57	4.36	6.42	8.70	11.18	13.84	16.63	19.54	22.54	25.61	28.71
8	1.01	2.34	4.00	5.93	8.08	10.45	12.99	15.68	18.48	21.39	24.38	27.43
9	.90	2.13	3.68	5.49	7.53	9.80	12.24	14.82	17.52	20.34	23.24	26.22
10	.82	1.96	3.40	5.10	7.05	9.21	11.55	14.04	16.65	19.38	22.20	25.10
12	.69	1.68	2.95	4.46	6.22	8.20	10.35	12.66	15.10	17.67	20.34	23.09
14	.59	1.47	2.60	3.95	5.55	7.36	9.34	11.49	13.77	16.20	18.73	21.35
16	.52	1.31	2.31	3.54	5.00	6.66	8.49	10.49	12.65	14.93	17.32	19.81
18	.47	1.17	2.08	3.20	4.54	6.07	7.77	9.64	11.67	13.82	16.08	18.45
20	.42	1.07	1.89	2.92	4.16	5.57	7.16	8.91	10.81	12.84	14.98	17.24
24	.35	.90	1.60	2.48	3.54	4.77	6.16	7.71	9.39	11.20	13.15	15.20
28	.30	.77	1.38	2.15	3.08	4.16	5.40	6.78	8.28	9.92	11.68	13.55
32	.26	.68	1.22	1.90	2.73	3.69	4.80	6.03	7.40	8.88	10.48	12.19
36	.23	.61	1.08	1.70	2.44	3.31	4.31	5.43	6.67	8.03	9.49	11.06
$b = 6$ In.												
2	2.25	5.18	8.22	11.23	14.23	17.22	20.20	23.17	26.13	29.09	32.05	35.01
3	1.89	4.52	7.66	10.73	13.79	16.82	19.83	22.83	25.82	28.80	31.78	34.75
4	1.63	4.05	6.98	10.14	13.24	16.31	19.37	22.40	25.42	28.43	31.42	34.41
5	1.42	3.68	6.42	9.44	12.63	15.74	18.83	21.90	24.95	27.99	31.01	34.02
6	1.25	3.36	5.96	8.84	11.90	15.12	18.24	21.34	24.43	27.49	30.54	33.58
7	1.11	3.09	5.57	8.31	11.28	14.37	17.61	20.74	23.85	26.94	30.02	33.08
8	1.01	2.86	5.22	7.85	10.71	13.73	16.85	20.10	23.23	26.35	29.45	32.54
9	.90	2.65	4.90	7.43	10.19	13.13	16.20	19.32	22.60	25.73	28.86	31.96
10	.82	2.47	4.61	7.04	9.71	12.57	15.58	18.67	21.80	25.09	28.23	31.36
12	.69	2.16	4.12	6.35	8.85	11.58	14.45	17.43	20.50	23.61	26.76	30.09
14	.59	1.92	3.70	5.76	8.11	10.70	13.45	16.32	19.29	22.34	25.44	28.57
16	.52	1.73	3.34	5.25	7.47	9.94	12.57	15.33	18.20	21.15	24.18	27.27
18	.47	1.57	3.05	4.82	6.91	9.26	11.78	14.43	17.21	20.07	23.02	26.04
20	.42	1.43	2.79	4.44	6.43	8.66	11.07	13.62	16.30	19.09	21.95	24.90
24	.35	1.22	2.39	3.84	5.62	7.64	9.84	12.21	14.72	17.34	20.05	22.85
28	.30	1.06	2.08	3.37	4.98	6.81	8.82	11.02	13.37	15.84	18.41	21.07
32	.26	.94	1.84	3.01	4.47	6.13	7.98	10.03	12.23	14.55	16.98	19.51
36	.23	.85	1.65	2.71	4.05	5.56	7.27	9.18	11.24	13.42	15.73	18.14

ECCENTRIC LOADS ON FASTENER GROUPS
TABLE XIV Coefficients C

ANGLE = 75°

Required minimum $C = \dfrac{P_u}{(\phi r_v)}$

$P_u = C \times \phi r_v$

n = total number of fasteners in the vertical row

P_u = factored load acting with lever arm l, in.

ϕr_v = design load on one fastener by Specification

X_o = horizontal distance from bolt centroid to P_u

ℓ = load eccentricity (not tabulated, can be calculated by geometry)

C = coefficients tabulated below

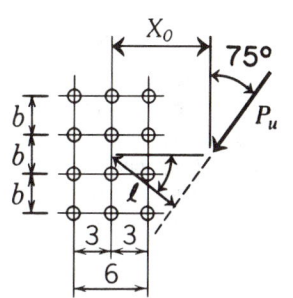

X_o, In.	\multicolumn{12}{c}{n}											
	1	2	3	4	5	6	7	8	9	10	11	12
$b = 3$ In.												
2	2.86	5.71	8.59	11.50	14.43	17.38	20.32	23.27	26.22	29.17	32.12	35.07
3	2.77	5.54	8.36	11.25	14.17	17.12	20.08	23.04	26.00	28.96	31.92	34.87
4	2.66	5.32	8.07	10.93	13.85	16.81	19.77	22.75	25.72	28.69	31.67	34.63
5	2.54	5.09	7.75	10.56	13.47	16.43	19.41	22.40	25.39	28.38	31.36	34.34
6	2.41	4.84	7.41	10.16	13.04	16.00	19.00	22.00	25.01	28.01	31.01	34.01
7	2.27	4.60	7.07	9.74	12.59	15.54	18.54	21.56	24.58	27.61	30.63	33.64
8	2.14	4.37	6.75	9.24	11.75	14.43	18.04	21.07	24.12	27.16	30.20	33.23
9	2.01	4.15	6.45	8.86	11.39	14.01	16.61	20.56	23.62	26.68	29.74	32.79
10	1.89	3.94	6.17	8.53	11.00	13.56	16.18	18.88	23.08	26.16	29.24	32.31
12	1.68	3.57	5.67	7.96	10.36	12.85	15.41	18.07	20.75	23.48	28.16	31.26
14	1.49	3.25	5.25	7.44	9.78	12.21	14.72	17.32	19.93	22.58	25.27	28.00
16	1.34	2.97	4.87	6.98	9.24	11.61	14.08	16.61	19.21	21.83	24.52	27.19
18	1.21	2.73	4.54	6.56	8.74	11.06	13.46	15.96	18.50	21.11	23.74	26.42
20	1.10	2.52	4.24	6.18	8.28	10.53	12.90	15.34	17.86	20.45	23.00	25.67
24	.93	2.19	3.75	5.52	7.48	9.59	11.84	14.19	16.62	19.12	21.68	24.20
28	.80	1.93	3.35	4.97	6.79	8.78	10.92	13.16	15.51	17.93	20.42	22.96
32	.71	1.72	3.02	4.51	6.20	8.08	10.11	12.26	14.51	16.84	19.26	21.74
36	.63	1.55	2.74	4.12	5.70	7.48	9.40	11.45	13.61	15.86	18.19	20.60
$b = 6$ In.												
2	2.86	5.75	8.69	11.64	14.59	17.54	20.49	23.44	26.38	29.33	32.28	35.22
3	2.77	5.62	8.57	11.53	14.49	17.44	20.40	23.35	26.30	29.25	32.20	35.15
4	2.66	5.46	8.41	11.38	14.36	17.33	20.29	23.25	26.20	29.16	32.11	35.06
5	2.54	5.27	8.22	11.21	14.20	17.18	20.16	23.13	26.09	29.05	32.01	34.96
6	2.41	5.06	8.00	11.01	14.02	17.02	20.01	22.99	25.96	28.93	31.89	34.85
7	2.27	4.84	7.76	10.79	13.82	16.84	19.84	22.83	25.81	28.80	31.77	34.73
8	2.14	4.59	7.20	10.54	13.60	16.63	19.65	22.66	25.66	28.64	31.62	34.60
9	2.01	4.39	6.98	10.28	13.35	16.41	19.45	22.47	25.48	28.48	31.47	34.45
10	1.89	4.21	6.75	9.43	13.09	16.17	19.23	22.27	25.29	28.30	31.30	34.30
12	1.68	3.91	6.39	9.02	11.74	15.65	18.75	21.82	24.88	27.91	30.94	33.95
14	1.49	3.63	6.06	8.63	11.29	14.07	18.21	21.32	24.41	27.48	30.53	33.56
16	1.34	3.39	5.75	8.28	10.91	13.60	16.34	20.78	23.90	26.99	30.07	33.13
18	1.21	3.17	5.47	7.96	10.55	13.21	15.94	19.78	23.34	26.47	29.57	32.66
20	1.10	2.98	5.22	7.66	10.22	12.83	15.53	18.24	22.30	25.91	29.04	32.15
24	.93	2.65	4.76	7.10	9.59	12.17	14.78	17.46	20.17	22.93	27.35	31.05
28	.80	2.38	4.37	6.60	8.99	11.52	14.11	16.72	19.41	22.10	24.88	27.65
32	.71	2.16	4.03	6.15	8.45	10.91	13.46	16.06	18.65	21.35	24.06	26.77
36	.63	1.97	3.74	5.75	7.96	10.33	12.83	15.39	18.00	20.58	23.29	25.99

ECCENTRIC LOADS ON FASTENER GROUPS
TABLE XV Coefficients *C*

ANGLE = 0°

Required minimum $C = \dfrac{P_u}{(\phi r_v)}$

$P_u = C \times \phi r_v$

n = total number of fasteners in the vertical row

P_u = factored load acting with lever arm l, in.

ϕr_v = design load on one fastener by Specification

X_0 = horizontal distance from bolt centroid to P_u

ℓ = load eccentricity (not tabulated, can be calculated by geometry)

C = coefficients tabulated below

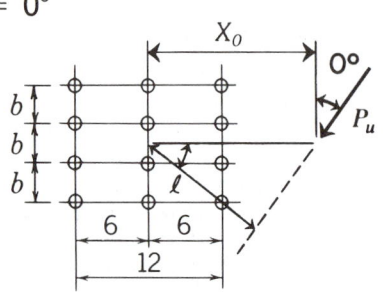

X_0, In.	n											
	1	2	3	4	5	6	7	8	9	10	11	12
b = 3 In.												
2	2.29	4.62	7.24	10.41	13.36	16.35	19.37	22.39	25.41	28.44	31.45	34.45
3	1.92	4.06	6.43	9.11	12.16	15.15	18.20	21.27	24.35	27.42	30.49	33.55
4	1.72	3.65	5.80	8.24	10.92	13.89	16.90	19.98	23.09	26.20	29.32	32.43
5	1.55	3.31	5.27	7.51	9.99	12.72	15.63	18.65	21.74	24.86	28.00	31.15
6	1.42	3.02	4.82	6.88	9.16	11.71	14.46	17.37	20.39	23.48	26.62	29.77
7	1.31	2.78	4.44	6.34	8.46	10.83	13.41	16.18	19.10	22.13	25.23	28.37
8	1.21	2.56	4.10	5.88	7.85	10.07	12.48	15.11	17.91	20.84	23.87	26.96
9	1.13	2.38	3.81	5.47	7.32	9.39	11.66	14.15	16.81	19.63	22.57	25.60
10	1.06	2.22	3.56	5.10	6.84	8.79	10.93	13.29	15.82	18.52	21.35	24.30
12	.92	1.95	3.12	4.48	6.04	7.78	9.70	11.81	14.09	16.56	19.18	21.94
14	.81	1.72	2.78	4.00	5.39	6.95	8.69	10.61	12.68	14.93	17.33	19.89
16	.72	1.54	2.49	3.59	4.85	6.27	7.85	9.60	11.50	13.56	15.77	18.13
18	.64	1.39	2.25	3.25	4.40	5.70	7.15	8.75	10.50	12.41	14.44	16.63
20	.58	1.26	2.05	2.97	4.02	5.22	6.56	8.03	9.65	11.42	13.31	15.34
24	.49	1.06	1.74	2.52	3.43	4.45	5.60	6.88	8.29	9.82	11.47	13.24
28	.42	.92	1.50	2.19	2.98	3.87	4.88	6.00	7.24	8.59	10.05	11.62
32	.37	.81	1.32	1.93	2.63	3.42	4.32	5.32	6.42	7.62	8.93	10.33
36	.33	.72	1.18	1.72	2.35	3.06	3.87	4.77	5.76	6.84	8.02	9.29
b = 6 In.												
2	2.29	5.15	8.15	11.18	14.21	17.22	20.22	23.21	26.19	29.16	32.13	35.09
3	1.92	4.48	7.53	10.61	13.70	16.77	19.82	22.85	25.86	28.87	31.86	34.84
4	1.72	4.08	6.89	9.96	13.09	16.21	19.31	22.39	25.44	28.48	31.50	34.51
5	1.55	3.71	6.31	9.29	12.42	15.57	18.71	21.83	24.93	28.01	31.06	34.11
6	1.42	3.40	5.79	8.66	11.73	14.89	18.06	21.21	24.35	27.47	30.56	33.63
7	1.31	3.14	5.35	8.07	11.05	14.18	17.36	20.54	23.71	26.86	29.99	33.09
8	1.21	2.90	4.97	7.54	10.41	13.49	16.65	19.84	23.03	26.20	29.36	32.51
9	1.13	2.69	4.64	7.07	9.81	12.81	15.93	19.12	22.32	25.51	28.70	31.86
10	1.06	2.51	4.35	6.64	9.26	12.16	15.23	18.40	21.59	24.80	28.00	31.19
12	.92	2.21	3.85	5.91	8.28	10.98	13.91	16.98	20.14	23.34	26.55	29.77
14	.81	1.96	3.45	5.31	7.46	9.96	12.71	15.66	18.73	21.88	25.07	28.29
16	.72	1.77	3.11	4.80	6.78	9.09	11.65	14.45	17.40	20.47	23.62	26.81
18	.64	1.60	2.83	4.38	6.20	8.34	10.73	13.37	16.19	19.15	22.22	25.36
20	.58	1.46	2.59	4.02	5.71	7.70	9.92	12.40	15.08	17.93	20.90	23.97
24	.49	1.24	2.21	3.44	4.91	6.65	8.59	10.80	13.20	15.79	18.55	21.43
28	.42	1.08	1.92	3.01	4.30	5.83	7.57	9.53	11.68	14.01	16.51	19.17
32	.37	.95	1.70	2.66	3.82	5.19	6.75	8.51	10.45	12.58	14.87	17.32
36	.33	.85	1.52	2.39	3.43	4.67	6.08	7.68	9.44	11.40	13.50	15.75

ECCENTRIC LOADS ON FASTENER GROUPS
TABLE XV Coefficients C

ANGLE = 45°

Required minimum $C = \dfrac{P_u}{(\phi r_v)}$

$P_u = C \times \phi r_v$

n = total number of fasteners in the vertical row
P_u = factored load acting with lever arm l, in.
ϕr_v = design load on one fastener by Specification
X_o = horizontal distance from bolt centroid to P_u
ℓ = load eccentricity (not tabulated, can be calculated by geometry)
C = coefficients tabulated below

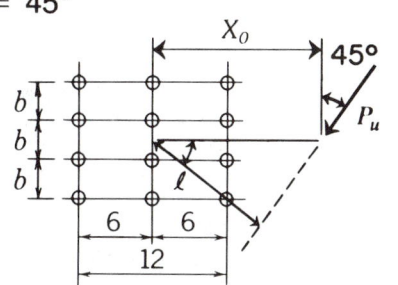

X_o, In.	n 1	2	3	4	5	6	7	8	9	10	11	12
b = 3 In.												
2	2.71	5.44	8.22	11.04	13.90	16.80	19.73	22.67	25.62	28.57	31.54	34.50
3	2.48	5.04	7.66	10.37	13.16	16.03	18.94	21.88	24.83	27.80	30.78	33.76
4	2.25	4.57	6.98	9.61	12.31	15.11	17.99	20.92	23.88	26.86	29.86	32.86
5	2.06	4.18	6.42	8.80	11.34	14.15	16.96	19.86	22.81	25.79	28.80	31.82
6	1.89	3.86	5.96	8.20	10.59	13.13	15.82	18.63	21.69	24.66	27.66	30.68
7	1.75	3.59	5.57	7.70	9.99	12.43	14.94	17.62	20.43	23.32	26.29	29.50
8	1.63	3.35	5.22	7.25	9.43	11.75	14.21	16.74	19.43	22.23	25.12	28.09
9	1.52	3.13	4.90	6.83	8.91	11.14	13.49	15.95	18.54	21.24	24.04	26.93
10	1.42	2.94	4.61	6.45	8.44	10.58	12.84	15.24	17.74	20.35	23.06	25.86
12	1.25	2.60	4.12	5.78	7.61	9.58	11.70	13.94	16.30	18.76	21.32	23.97
14	1.11	2.33	3.70	5.22	6.90	8.73	10.71	12.82	15.03	17.36	19.80	22.32
16	1.01	2.10	3.34	4.74	6.30	8.00	9.85	11.83	13.93	16.13	18.45	20.86
18	.90	1.90	3.05	4.34	5.78	7.37	9.10	10.96	12.95	15.04	17.24	19.54
20	.82	1.74	2.79	3.99	5.33	6.82	8.44	10.20	12.08	14.07	16.16	18.36
24	.69	1.47	2.39	3.43	4.60	5.91	7.35	8.92	10.61	12.41	14.32	16.32
28	.59	1.28	2.08	2.99	4.04	5.20	6.49	7.90	9.43	11.07	12.81	14.65
32	.52	1.13	1.84	2.66	3.59	4.64	5.80	7.08	8.47	9.96	11.56	13.26
36	.47	1.01	1.65	2.38	3.23	4.18	5.24	6.40	7.67	9.05	10.52	12.08
b = 6 In.												
2	2.71	5.51	8.40	11.34	14.30	17.26	20.22	23.18	26.14	29.10	32.05	35.01
3	2.48	5.17	8.01	10.94	13.91	16.89	19.88	22.86	25.84	28.81	31.78	34.75
4	2.25	4.74	7.54	10.46	13.44	16.45	19.45	22.46	25.46	28.45	31.44	34.42
5	2.06	4.36	6.99	9.93	12.91	15.93	18.96	21.99	25.01	28.03	31.04	34.04
6	1.89	4.06	6.53	9.30	12.34	15.36	18.41	21.46	24.51	27.55	30.58	33.60
7	1.75	3.80	6.15	8.80	11.67	14.77	17.82	20.89	23.96	27.02	30.07	33.12
8	1.63	3.57	5.84	8.37	11.12	14.06	17.21	20.29	23.37	26.45	29.53	32.59
9	1.52	3.36	5.54	7.99	10.64	13.49	16.48	19.54	22.76	25.85	28.95	32.03
10	1.42	3.18	5.27	7.63	10.20	12.95	15.88	18.90	21.99	25.23	28.34	31.44
12	1.25	2.85	4.78	6.99	9.41	12.01	14.79	17.71	20.72	23.79	26.90	30.20
14	1.11	2.57	4.36	6.42	8.70	11.18	13.84	16.63	19.54	22.54	25.61	28.71
16	1.01	2.34	4.00	5.93	8.08	10.45	12.99	15.68	18.48	21.39	24.38	27.43
18	.90	2.13	3.68	5.49	7.53	9.80	12.24	14.82	17.52	20.34	23.24	26.22
20	.82	1.96	3.40	5.10	7.05	9.21	11.55	14.04	16.65	19.38	22.20	25.10
24	.69	1.68	2.95	4.46	6.22	8.20	10.35	12.66	15.10	17.67	20.34	23.09
28	.59	1.47	2.60	3.95	5.55	7.36	9.34	11.49	13.77	16.20	18.73	21.35
32	.52	1.31	2.31	3.54	5.00	6.66	8.49	10.49	12.65	14.93	17.32	19.81
36	.47	1.17	2.08	3.20	4.54	6.07	7.77	9.64	11.67	13.82	16.08	18.45

ECCENTRIC LOADS ON FASTENER GROUPS

TABLE XV Coefficients C

ANGLE = 75°

Required minimum $C = \frac{P_u}{(\phi r_v)}$

$P_u = C \times \phi r_v$

n = total number of fasteners in the vertical row

P_u = factored load acting with lever arm l, in.

ϕr_v = design load on one fastener by Specification

X_o = horizontal distance from bolt centroid to P_u

ℓ = load eccentricity (not tabulated, can be calculated by geometry)

C = coefficients tabulated below

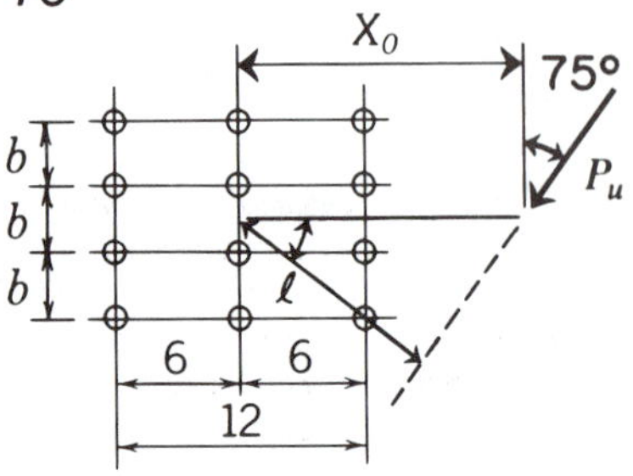

	X_O, In.	n = 1	2	3	4	5	6	7	8	9	10	11	12
b = 3 In.	2	2.92	5.83	8.74	11.65	14.56	17.49	20.41	23.35	26.28	29.22	32.16	35.10
	3	2.90	5.78	8.65	11.54	14.43	17.34	20.26	23.18	26.11	29.05	31.99	34.94
	4	2.86	5.71	8.55	11.40	14.27	17.16	20.06	22.98	25.91	28.85	31.79	34.73
	5	2.82	5.62	8.42	11.23	14.07	16.93	19.83	22.74	25.66	28.60	31.54	34.50
	6	2.77	5.53	8.28	11.04	13.85	16.69	19.56	22.46	25.38	28.32	31.27	34.22
	7	2.72	5.42	8.12	10.84	13.60	16.41	19.27	22.16	25.07	28.01	30.96	33.91
	8	2.66	5.30	7.95	10.62	13.34	16.12	18.95	21.82	24.73	27.67	30.62	33.57
	9	2.60	5.18	7.77	10.39	13.06	15.80	18.61	21.47	24.37	27.30	30.25	33.21
	10	2.54	5.06	7.58	10.15	12.77	15.47	18.25	21.09	23.97	26.90	29.85	32.82
	12	2.41	4.80	7.21	9.66	12.18	14.79	17.50	20.28	23.14	26.04	28.99	31.96
	14	2.27	4.54	6.83	9.18	11.60	14.12	16.74	19.45	22.25	25.13	28.06	31.02
	16	2.14	4.29	6.48	8.72	11.05	13.47	15.87	18.44	20.93	23.46	26.21	28.81
	18	2.01	4.05	6.13	8.30	10.54	12.88	15.22	17.70	20.14	22.73	25.25	27.96
	20	1.89	3.82	5.82	7.89	10.07	12.33	14.69	17.03	19.52	21.96	24.54	27.06
	24	1.68	3.41	5.24	7.18	9.22	11.36	13.59	15.91	18.25	20.68	23.18	25.65
	28	1.49	3.06	4.75	6.56	8.49	10.52	12.65	14.86	17.16	19.53	21.90	24.37
	32	1.34	2.77	4.33	6.02	7.84	9.77	11.80	13.93	16.14	18.42	20.78	23.15
	36	1.21	2.52	3.97	5.56	7.27	9.10	11.05	13.09	15.21	17.42	19.70	22.04
b = 6 In.	2	2.92	5.83	8.75	11.68	14.62	17.56	20.50	23.45	26.39	29.33	32.28	35.23
	3	2.90	5.78	8.68	11.60	14.53	17.48	20.42	23.37	26.31	29.26	32.21	35.15
	4	2.86	5.71	8.59	11.50	14.43	17.38	20.32	23.27	26.22	29.17	32.12	35.07
	5	2.82	5.63	8.48	11.38	14.31	17.26	20.21	23.16	26.12	29.07	32.02	34.98
	6	2.77	5.54	8.36	11.25	14.17	17.12	20.08	23.04	26.00	28.96	31.92	34.87
	7	2.72	5.43	8.22	11.09	14.02	16.97	19.93	22.90	25.87	28.83	31.80	34.76
	8	2.66	5.32	8.07	10.93	13.85	16.81	19.77	22.75	25.72	28.69	31.67	34.63
	9	2.60	5.21	7.92	10.75	13.67	16.62	19.60	22.58	25.56	28.54	31.51	34.50
	10	2.54	5.09	7.75	10.56	13.47	16.43	19.41	22.40	25.39	28.38	31.36	34.34
	12	2.41	4.84	7.41	10.16	13.04	16.00	19.00	22.00	25.01	28.01	31.01	34.01
	14	2.27	4.60	7.07	9.74	12.59	15.54	18.54	21.56	24.58	27.61	30.63	33.64
	16	2.14	4.37	6.75	9.24	11.75	14.43	18.04	21.07	24.12	27.16	30.20	33.23
	18	2.01	4.15	6.45	8.86	11.39	14.01	16.61	20.56	23.62	26.68	29.74	32.79
	20	1.89	3.94	6.17	8.53	11.00	13.56	16.18	18.88	23.08	26.16	29.24	32.31
	24	1.68	3.57	5.67	7.96	10.36	12.85	15.41	18.07	20.75	23.48	28.16	31.26
	28	1.49	3.25	5.25	7.44	9.78	12.21	14.72	17.32	19.93	22.58	25.27	28.00
	32	1.34	2.97	4.87	6.98	9.24	11.61	14.08	16.61	19.21	21.83	24.52	27.19
	36	1.21	2.73	4.54	6.56	8.74	11.06	13.46	15.96	18.50	21.11	23.74	26.42

ECCENTRIC LOADS ON FASTENER GROUPS
TABLE XVI Coefficients *C*

ANGLE = 0°

Required minimum $C = \dfrac{P_u}{(\phi r_v)}$

$P_u = C \times \phi r_v$
n = total number of fasteners in the vertical row
P_u = factored load acting with lever arm l, in.
ϕr_v = design load on one fastener by Specification
X_o = horizontal distance from bolt centroid to P_u
ℓ = load eccentricity (not tabulated, can be calculated by geometry)
C = coefficients tabulated below

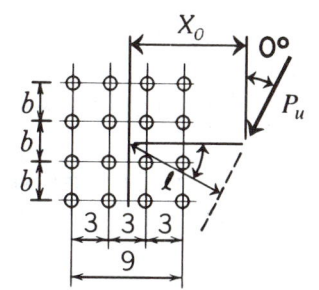

X_o, In.	n											
	1	2	3	4	5	6	7	8	9	10	11	12
b = 3 In.												
2	2.60	5.71	9.37	13.42	17.50	21.61	25.71	29.79	33.86	37.91	41.94	45.95
3	2.23	4.92	8.05	11.77	15.75	19.88	24.05	28.23	32.39	36.53	40.66	44.76
4	1.94	4.30	7.09	10.39	14.09	18.09	22.23	26.43	30.65	34.86	39.06	43.24
5	1.69	3.79	6.31	9.29	12.66	16.42	20.42	24.56	28.77	33.01	37.26	41.49
6	1.49	3.37	5.65	8.37	11.45	14.94	18.73	22.74	26.88	31.10	35.35	39.61
7	1.32	3.03	5.10	7.59	10.43	13.67	17.22	21.05	25.06	29.20	33.42	37.67
8	1.19	2.74	4.64	6.93	9.56	12.57	15.88	19.51	23.36	27.38	31.52	35.73
9	1.07	2.50	4.24	6.36	8.81	11.61	14.70	18.13	21.80	25.68	29.71	33.84
10	1.00	2.29	3.90	5.86	8.15	10.77	13.67	16.90	20.39	24.10	28.00	32.03
12	.83	1.96	3.34	5.06	7.06	9.37	11.96	14.83	17.96	21.35	24.95	28.72
14	.73	1.72	2.92	4.44	6.22	8.27	10.59	13.17	15.99	19.07	22.37	25.87
16	.65	1.52	2.59	3.95	5.54	7.39	9.48	11.82	14.39	17.19	20.20	23.43
18	.58	1.37	2.33	3.56	4.99	6.67	8.57	10.70	13.05	15.62	18.38	21.37
20	.53	1.24	2.12	3.23	4.53	6.07	7.81	9.77	11.93	14.30	16.85	19.61
24	.45	1.05	1.79	2.73	3.83	5.14	6.62	8.30	10.15	12.19	14.41	16.80
28	.39	.90	1.55	2.36	3.31	4.45	5.73	7.20	8.82	10.61	12.55	14.66
32	.34	.79	1.36	2.07	2.92	3.92	5.05	6.35	7.79	9.38	11.11	12.98
36	.31	.71	1.21	1.85	2.60	3.50	4.51	5.68	6.96	8.39	9.95	11.64
b = 6 In.												
2	2.60	6.59	10.75	14.88	18.95	22.97	26.98	30.96	34.94	38.90	42.86	46.81
3	2.23	5.77	9.87	14.09	18.26	22.38	26.45	30.49	34.51	38.51	42.50	46.48
4	1.94	5.12	8.96	13.18	17.42	21.62	25.77	29.88	33.95	38.00	42.03	46.05
5	1.69	4.58	8.13	12.24	16.49	20.75	24.96	29.14	33.28	37.38	41.46	45.51
6	1.49	4.13	7.38	11.34	15.54	19.81	24.07	28.30	32.50	36.66	40.78	44.88
7	1.32	3.74	6.74	10.50	14.59	18.84	23.13	27.40	31.64	35.85	40.03	44.17
8	1.19	3.42	6.20	9.74	13.69	17.88	22.15	26.45	30.72	34.97	39.19	43.38
9	1.07	3.14	5.73	9.06	12.84	16.94	21.18	25.47	29.76	34.04	38.30	42.53
10	1.00	2.89	5.32	8.45	12.06	16.04	20.21	24.48	28.78	33.08	37.36	41.62
12	.83	2.50	4.63	7.43	10.68	14.39	18.38	22.54	26.80	31.10	35.41	39.71
14	.73	2.20	4.09	6.60	9.53	12.97	16.72	20.72	24.88	29.12	33.41	37.73
16	.65	1.95	3.65	5.93	8.59	11.76	15.26	19.06	23.07	27.21	31.44	35.73
18	.58	1.76	3.29	5.37	7.81	10.73	13.99	17.58	21.41	25.41	29.55	33.77
20	.53	1.60	2.99	4.90	7.15	9.85	12.86	16.21	19.83	23.75	27.76	31.88
24	.45	1.35	2.53	4.16	6.10	8.44	11.07	14.05	17.28	20.77	24.46	28.44
28	.39	1.17	2.19	3.61	5.31	7.37	9.69	12.34	15.25	18.41	21.79	25.37
32	.34	1.03	1.93	3.19	4.69	6.53	8.61	10.99	13.60	16.48	19.57	22.87
36	.31	.92	1.72	2.85	4.20	5.85	7.73	9.89	12.26	14.89	17.72	20.76

ECCENTRIC LOADS ON FASTENER GROUPS
TABLE XVI Coefficients C

ANGLE = 45°

Required minimum $C = \dfrac{P_u}{(\phi r_v)}$

$P_u = C \times \phi r_v$

n = total number of fasteners in the vertical row

P_u = factored load acting with lever arm l, in.

ϕr_v = design load on one fastener by Specification

X_o = horizontal distance from bolt centroid to P_u

ℓ = load eccentricity (not tabulated, can be calculated by geometry)

C = coefficients tabulated below

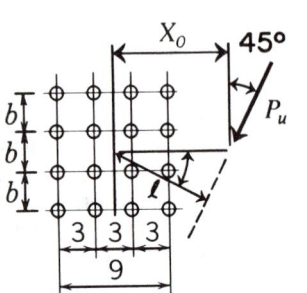

X_o, In.	n											
	1	2	3	4	5	6	7	8	9	10	11	12
b = 3 In.
2	3.32	6.81	10.46	14.26	18.15	22.09	26.05	30.02	33.99	37.97	41.94	45.92
3	2.89	5.96	9.30	13.02	16.85	20.78	24.76	28.77	32.80	36.82	40.84	44.85
4	2.54	5.31	8.37	11.69	15.32	19.31	23.27	27.30	31.35	35.41	39.48	43.54
5	2.25	4.78	7.63	10.77	14.12	17.74	21.57	25.53	29.77	33.85	37.94	42.04
6	2.01	4.33	6.99	9.94	13.15	16.54	20.16	23.98	27.94	31.99	36.31	40.43
7	1.81	3.93	6.42	9.20	12.24	15.51	18.94	22.59	26.41	30.35	34.39	38.49
8	1.64	3.60	5.93	8.55	11.44	14.54	17.85	21.36	25.03	28.85	32.78	36.81
9	1.49	3.30	5.49	7.96	10.71	13.69	16.89	20.26	23.79	27.48	31.30	35.22
10	1.37	3.05	5.10	7.44	10.05	12.92	15.99	19.25	22.66	26.23	29.93	33.75
12	1.17	2.65	4.46	6.55	8.93	11.56	14.41	17.45	20.65	24.01	27.51	31.13
14	1.03	2.33	3.95	5.83	8.00	10.43	13.07	15.90	18.92	22.10	25.41	28.85
16	.91	2.08	3.54	5.24	7.24	9.47	11.92	14.58	17.42	20.42	23.57	26.84
18	.82	1.89	3.20	4.75	6.60	8.66	10.94	13.43	16.10	18.94	21.93	25.04
20	.74	1.72	2.92	4.35	6.05	7.97	10.09	12.43	14.95	17.63	20.47	23.45
24	.63	1.46	2.48	3.71	5.18	6.85	8.72	10.80	13.04	15.44	18.01	20.73
28	.54	1.26	2.15	3.23	4.52	5.99	7.66	9.51	11.52	13.71	16.03	18.51
32	.48	1.11	1.90	2.86	4.01	5.32	6.82	8.49	10.31	12.30	14.42	16.68
36	.43	1.00	1.70	2.56	3.59	4.78	6.14	7.65	9.31	11.13	13.08	15.16
b = 6 In.
2	3.32	7.10	11.04	15.02	19.00	22.97	26.93	30.89	34.84	38.79	42.73	46.67
3	2.89	6.44	10.37	14.39	18.42	22.44	26.45	30.44	34.42	38.40	42.36	46.32
4	2.54	5.75	9.61	13.65	17.72	21.79	25.84	29.88	33.90	37.90	41.89	45.88
5	2.25	5.27	8.80	12.75	16.94	21.05	25.14	29.22	33.28	37.32	41.35	45.36
6	2.01	4.85	8.20	11.97	16.00	20.24	24.37	28.49	32.58	36.66	40.72	44.77
7	1.81	4.49	7.70	11.29	15.18	19.27	23.55	27.69	31.82	35.94	40.03	44.11
8	1.64	4.16	7.25	10.69	14.44	18.43	22.55	26.86	31.01	35.16	39.29	43.39
9	1.49	3.87	6.83	10.15	13.76	17.64	21.69	25.83	30.18	34.34	38.49	42.63
10	1.37	3.62	6.45	9.65	13.15	16.91	20.87	24.97	29.12	33.49	37.66	41.82
12	1.17	3.19	5.78	8.74	12.04	15.60	19.39	23.33	27.40	31.54	35.72	40.14
14	1.03	2.84	5.22	7.97	11.07	14.46	18.08	21.87	25.81	29.85	33.96	38.13
16	.91	2.56	4.74	7.30	10.23	13.46	16.92	20.57	24.36	28.28	32.30	36.40
18	.82	2.33	4.34	6.72	9.48	12.57	15.88	19.39	23.06	26.86	30.77	34.77
20	.74	2.13	3.99	6.21	8.83	11.77	14.95	18.32	21.86	25.55	29.35	33.26
24	.63	1.82	3.43	5.38	7.75	10.41	13.32	16.45	19.77	23.24	26.84	30.55
28	.54	1.59	2.99	4.74	6.88	9.30	11.98	14.87	17.98	21.26	24.66	28.19
32	.48	1.41	2.66	4.23	6.18	8.38	10.85	13.55	16.47	19.54	22.77	26.13
36	.43	1.27	2.38	3.82	5.60	7.62	9.90	12.43	15.15	18.05	21.10	24.30

ECCENTRIC LOADS ON FASTENER GROUPS
TABLE XVI Coefficients C

ANGLE = 75°

Required minimum $C = \dfrac{P_u}{(\phi r_v)}$

$P_u = C \times \phi r_v$
n = total number of fasteners in the vertical row
P_u = factored load acting with lever arm l, in.
ϕr_v = design load on one fastener by Specification
X_o = horizontal distance from bolt centroid to P_u
ℓ = load eccentricity (not tabulated, can be calculated by geometry)
C = coefficients tabulated below

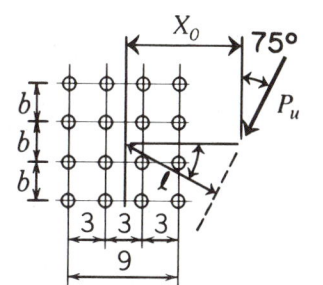

X_o, In.	n 1	2	3	4	5	6	7	8	9	10	11	12
2	3.86	7.70	11.55	15.42	19.32	23.22	27.14	31.06	34.99	38.92	42.84	46.77
3	3.79	7.56	11.35	15.18	19.04	22.95	26.86	30.79	34.72	38.65	42.58	46.52
4	3.70	7.38	11.09	14.86	18.70	22.59	26.51	30.44	34.37	38.33	42.27	46.21
5	3.60	7.17	10.79	14.50	18.30	22.18	26.09	30.03	33.97	37.93	41.89	45.84
6	3.48	6.94	10.46	14.10	17.85	21.70	25.61	29.56	33.52	37.49	41.46	45.43
7	3.35	6.70	10.11	13.67	17.37	21.18	25.08	29.03	33.00	36.99	40.98	44.96
8	3.22	6.45	9.76	13.22	16.85	20.63	24.51	28.46	32.44	36.44	40.44	44.45
9	3.09	6.19	9.40	12.77	16.33	20.05	23.91	27.85	31.83	35.85	39.87	43.89
10	2.95	5.95	9.06	12.33	15.66	19.09	23.28	27.20	31.19	35.21	39.25	43.29
12	2.69	5.48	8.40	11.51	14.69	17.98	21.36	24.95	28.49	33.85	37.91	41.98
14	2.45	5.05	7.81	10.77	13.85	17.11	20.41	23.79	27.33	30.83	34.38	40.55
16	2.24	4.65	7.29	10.12	13.14	16.27	19.55	22.87	26.33	29.77	33.26	36.93
18	2.06	4.31	6.82	9.54	12.47	15.55	18.72	22.02	25.36	28.81	32.26	35.83
20	1.90	4.02	6.40	9.03	11.86	14.85	17.98	21.19	24.51	27.86	31.31	34.77
24	1.64	3.52	5.69	8.13	10.77	13.59	16.57	19.67	22.86	26.15	29.52	32.88
28	1.43	3.12	5.11	7.36	9.83	12.49	15.32	18.29	21.38	24.56	27.83	31.17
32	1.27	2.80	4.63	6.71	9.02	11.53	14.22	17.07	20.04	23.13	26.31	29.55
36	1.14	2.54	4.22	6.15	8.31	10.69	13.25	15.98	18.84	21.81	24.89	28.05

$b = 3$ In.

X_o, In.	1	2	3	4	5	6	7	8	9	10	11	12
2	3.86	7.72	11.62	15.54	19.47	23.39	27.32	31.25	35.18	39.11	43.04	46.96
3	3.79	7.60	11.48	15.40	19.34	23.27	27.21	31.14	35.07	39.00	42.93	46.87
4	3.70	7.44	11.31	15.23	19.18	23.12	27.06	31.01	34.94	38.88	42.82	46.75
5	3.60	7.26	11.10	15.03	18.98	22.94	26.90	30.85	34.80	38.74	42.68	46.62
6	3.48	7.05	10.86	14.79	18.76	22.73	26.70	30.67	34.63	38.58	42.53	46.48
7	3.35	6.83	10.60	14.53	18.51	22.50	26.49	30.47	34.44	38.41	42.37	46.32
8	3.22	6.61	10.32	14.24	18.24	22.25	26.25	30.25	34.23	38.21	42.19	46.15
9	3.09	6.38	10.03	13.94	17.94	21.97	25.99	30.01	34.00	37.99	41.97	45.95
10	2.95	6.15	9.54	13.61	17.63	21.67	25.72	29.75	33.76	37.76	41.76	45.74
12	2.69	5.73	8.98	12.48	16.94	21.02	25.10	29.17	33.22	37.26	41.28	45.29
14	2.45	5.34	8.54	11.94	15.43	20.30	24.42	28.52	32.61	36.68	40.74	44.77
16	2.24	5.00	8.10	11.43	14.89	18.48	22.09	27.82	31.95	36.06	40.14	44.21
18	2.06	4.70	7.73	10.99	14.41	17.93	21.48	26.58	31.23	35.37	39.49	43.59
20	1.90	4.42	7.37	10.57	13.94	17.41	20.95	24.56	29.91	34.64	38.80	42.93
24	1.64	3.95	6.75	9.83	13.10	16.48	19.97	23.52	27.11	30.77	36.61	41.48
28	1.43	3.57	6.21	9.16	12.32	15.64	19.07	22.53	26.09	29.66	33.29	37.04
32	1.27	3.25	5.75	8.56	11.60	14.83	18.19	21.64	25.09	28.66	32.25	35.86
36	1.14	2.98	5.34	8.02	10.95	14.08	17.36	20.75	24.21	27.65	31.23	34.82

$b = 6$ In.

ECCENTRIC LOADS ON FASTENER GROUPS
TABLE XVII Coefficients C

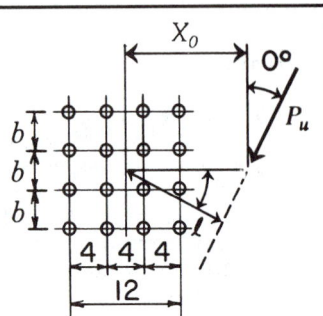

$$\text{ANGLE} = 0°$$

Required minimum $C = \dfrac{P_u}{(\phi r_v)}$

$P_u = C \times \phi r_v$
n = total number of fasteners in the vertical row
P_u = factored load acting with lever arm l, in.
ϕr_v = design load on one fastener by Specification
X_o = horizontal distance from bolt centroid to P_u
ℓ = load eccentricity (not tabulated, can be
         calculated by geometry)
C = coefficients tabulated below

X_o, In.	n 1	2	3	4	5	6	7	8	9	10	11	12
b = 3 In.												
2	2.82	5.99	9.58	13.73	17.69	21.71	25.75	29.80	33.84	37.88	41.90	45.91
3	2.50	5.31	8.46	12.05	16.04	20.06	24.15	28.27	32.40	36.51	40.61	44.71
4	2.23	4.74	7.57	10.81	14.46	18.35	22.39	26.52	30.69	34.86	39.03	43.19
5	2.01	4.27	6.86	9.82	13.12	16.77	20.66	24.72	28.86	33.05	37.26	41.46
6	1.81	3.87	6.24	8.96	11.99	15.39	19.07	22.98	27.04	31.19	35.39	39.61
7	1.64	3.52	5.71	8.22	11.05	14.21	17.66	21.38	25.30	29.36	33.51	37.71
8	1.49	3.22	5.24	7.58	10.22	13.17	16.40	19.92	23.68	27.61	31.67	35.83
9	1.36	2.96	4.83	7.01	9.48	12.25	15.29	18.62	22.19	25.98	29.92	33.99
10	1.25	2.74	4.48	6.51	8.83	11.44	14.31	17.45	20.84	24.47	28.28	32.24
12	1.07	2.37	3.89	5.68	7.74	10.06	12.64	15.47	18.52	21.83	25.34	29.04
14	.95	2.08	3.42	5.03	6.86	8.95	11.28	13.84	16.62	19.63	22.85	26.28
16	.83	1.86	3.05	4.50	6.15	8.04	10.15	12.49	15.04	17.80	20.75	23.92
18	.75	1.67	2.75	4.06	5.56	7.29	9.22	11.36	13.70	16.25	18.98	21.90
20	.68	1.52	2.50	3.70	5.07	6.66	8.43	10.40	12.57	14.92	17.46	20.18
24	.58	1.29	2.12	3.14	4.30	5.66	7.18	8.88	10.75	12.79	15.00	17.38
28	.50	1.12	1.85	2.72	3.74	4.92	6.24	7.73	9.37	11.17	13.12	15.22
32	.45	1.00	1.63	2.40	3.30	4.34	5.51	6.84	8.29	9.90	11.64	13.51
36	.40	.88	1.46	2.15	2.95	3.89	4.94	6.13	7.43	8.88	10.44	12.14
b = 6 In.												
2	2.82	6.77	10.81	14.88	18.93	22.95	26.95	30.94	34.91	38.88	42.84	46.79
3	2.50	5.92	9.96	14.10	18.24	22.35	26.42	30.46	34.48	38.48	42.48	46.46
4	2.23	5.34	9.10	13.22	17.41	21.59	25.73	29.84	33.91	37.97	42.00	46.01
5	2.01	4.84	8.31	12.32	16.51	20.73	24.93	29.10	33.23	37.34	41.41	45.47
6	1.81	4.42	7.61	11.45	15.58	19.81	24.05	28.26	32.45	36.61	40.73	44.83
7	1.64	4.05	7.02	10.66	14.67	18.86	23.11	27.36	31.59	35.79	39.98	44.12
8	1.49	3.73	6.51	9.94	13.80	17.92	22.15	26.42	30.68	34.92	39.13	43.33
9	1.36	3.45	6.06	9.30	12.98	17.01	21.19	25.45	29.73	34.00	38.25	42.47
10	1.25	3.21	5.66	8.73	12.23	16.13	20.25	24.48	28.76	33.04	37.32	41.57
12	1.07	2.80	4.98	7.74	10.91	14.54	18.47	22.58	26.80	31.08	35.37	39.67
14	.95	2.48	4.43	6.92	9.81	13.17	16.86	20.80	24.91	29.12	33.39	37.69
16	.83	2.22	3.98	6.25	8.90	12.00	15.43	19.18	23.14	27.24	31.45	35.71
18	.75	2.01	3.60	5.68	8.13	10.99	14.19	17.72	21.50	25.47	29.57	33.77
20	.68	1.83	3.29	5.21	7.47	10.13	13.11	16.43	20.02	23.83	27.81	31.91
24	.58	1.55	2.79	4.45	6.40	8.72	11.33	14.28	17.46	20.91	24.65	28.50
28	.50	1.35	2.42	3.87	5.59	7.64	9.96	12.59	15.45	18.58	21.93	25.48
32	.45	1.19	2.14	3.43	4.95	6.79	8.87	11.23	13.83	16.67	19.73	23.00
36	.40	1.06	1.92	3.07	4.44	6.10	7.98	10.12	12.48	15.09	17.90	20.92

ECCENTRIC LOADS ON FASTENER GROUPS
TABLE XVII Coefficients C

ANGLE = 45°

Required minimum $C = \dfrac{P_u}{(\phi r_v)}$

$P_u = C \times \phi r_v$
n = total number of fasteners in the vertical row
P_u = factored load acting with lever arm l, in.
ϕr_v = design load on one fastener by Specification
X_o = horizontal distance from bolt centroid to P_u
ℓ = load eccentricity (not tabulated, can be calculated by geometry)
C = coefficients tabulated below

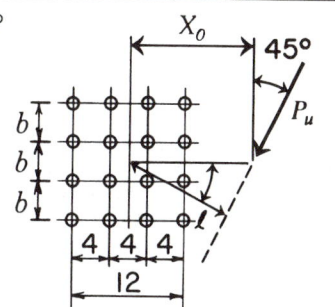

X_o, In.						n						
	1	2	3	4	5	6	7	8	9	10	11	12
2	3.56	7.15	10.83	14.59	18.42	22.31	26.22	30.15	34.10	38.05	42.01	45.96
3	3.21	6.50	10.00	13.61	17.35	21.19	25.09	29.04	33.00	36.98	40.96	44.95
4	2.89	5.88	9.05	12.42	16.13	19.89	23.76	27.69	31.67	35.67	39.69	43.71
5	2.62	5.37	8.31	11.45	14.81	18.40	22.34	26.23	30.20	34.20	38.24	42.28
6	2.39	4.93	7.70	10.69	13.88	17.22	20.80	24.56	28.45	32.65	36.68	40.74
7	2.19	4.55	7.15	9.98	13.03	16.27	19.63	23.21	26.96	30.85	34.82	39.13
8	2.01	4.21	6.65	9.34	12.25	15.35	18.63	22.02	25.63	29.38	33.25	37.22
9	1.86	3.90	6.21	8.76	11.54	14.51	17.66	20.96	24.43	28.05	31.80	35.67
10	1.72	3.64	5.82	8.24	10.89	13.75	16.79	20.00	23.35	26.84	30.48	34.24
12	1.49	3.18	5.14	7.33	9.76	12.40	15.24	18.24	21.40	24.70	28.13	31.69
14	1.32	2.82	4.59	6.58	8.81	11.26	13.90	16.72	19.70	22.83	26.09	29.47
16	1.17	2.53	4.14	5.95	8.01	10.28	12.75	15.39	18.21	21.17	24.28	27.50
18	1.06	2.29	3.76	5.42	7.33	9.44	11.74	14.23	16.89	19.71	22.66	25.74
20	1.00	2.10	3.44	4.98	6.75	8.72	10.87	13.22	15.73	18.40	21.21	24.16
24	.82	1.79	2.94	4.27	5.81	7.53	9.44	11.53	13.78	16.19	18.74	21.43
28	.71	1.56	2.56	3.73	5.09	6.61	8.32	10.20	12.22	14.41	16.74	19.20
32	.63	1.38	2.27	3.31	4.52	5.89	7.43	9.12	10.96	12.96	15.09	17.35
36	.56	1.24	2.03	2.97	4.07	5.30	6.70	8.24	9.92	11.75	13.71	15.80

$b = 3$ In.

X_o	1	2	3	4	5	6	7	8	9	10	11	12
2	3.56	7.28	11.15	15.08	19.04	22.99	26.95	30.90	34.84	38.79	42.73	46.67
3	3.21	6.77	10.59	14.52	18.50	22.49	26.48	30.46	34.43	38.40	42.37	46.32
4	2.89	6.16	9.92	13.85	17.85	21.88	25.90	29.91	33.92	37.91	41.90	45.88
5	2.62	5.67	9.16	13.11	17.11	21.16	25.22	29.28	33.32	37.35	41.36	45.37
6	2.39	5.27	8.55	12.26	16.33	20.39	24.48	28.56	32.64	36.70	40.75	44.78
7	2.19	4.91	8.07	11.59	15.43	19.59	23.68	27.79	31.89	35.99	40.07	44.13
8	2.01	4.59	7.64	11.01	14.70	18.63	22.71	26.97	31.10	35.22	39.33	43.43
9	1.86	4.30	7.23	10.51	14.05	17.86	21.87	25.97	30.28	34.42	38.55	42.67
10	1.72	4.04	6.85	10.02	13.45	17.15	21.06	25.12	29.24	33.58	37.73	41.87
12	1.49	3.59	6.19	9.14	12.38	15.88	19.61	23.51	27.54	31.65	35.81	40.21
14	1.32	3.22	5.62	8.38	11.44	14.77	18.33	22.07	25.97	29.98	34.07	38.22
16	1.17	2.91	5.14	7.71	10.61	13.78	17.19	20.79	24.54	28.43	32.43	36.50
18	1.06	2.66	4.72	7.12	9.87	12.91	16.18	19.64	23.26	27.02	30.91	34.89
20	1.00	2.44	4.36	6.61	9.22	12.12	15.26	18.58	22.09	25.73	29.51	33.39
24	.82	2.10	3.77	5.76	8.12	10.76	13.65	16.74	20.01	23.45	27.02	30.70
28	.71	1.84	3.31	5.09	7.23	9.65	12.30	15.18	18.24	21.49	24.87	28.37
32	.63	1.63	2.94	4.55	6.51	8.72	11.16	13.85	16.73	19.79	22.98	26.32
36	.56	1.47	2.65	4.11	5.91	7.94	10.21	12.71	15.42	18.30	21.33	24.50

$b = 6$ In.

ECCENTRIC LOADS ON FASTENER GROUPS
TABLE XVII Coefficients C

ANGLE = 75°

Required minimum $C = \dfrac{P_u}{(\phi r_v)}$

$P_u = C \times \phi r_v$
n = total number of fasteners in the vertical row
P_u = factored load acting with lever arm l, in.
ϕr_v = design load on one fastener by Specification
X_o = horizontal distance from bolt centroid to P_u
ℓ = load eccentricity (not tabulated, can be calculated by geometry)
C = coefficients tabulated below

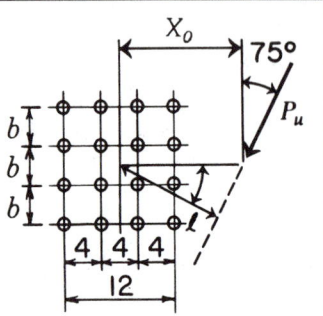

X_o, In.	n=1	2	3	4	5	6	7	8	9	10	11	12
b = 3 In.												
2	3.89	7.76	11.63	15.51	19.39	23.29	27.20	31.11	35.03	38.95	42.87	46.80
3	3.85	7.67	11.50	15.34	19.19	23.07	26.97	30.87	34.79	38.71	42.63	46.56
4	3.79	7.56	11.33	15.12	18.94	22.80	26.68	30.59	34.50	38.42	42.35	46.28
5	3.73	7.43	11.13	14.86	18.65	22.47	26.34	30.23	34.15	38.07	42.00	45.95
6	3.65	7.27	10.91	14.58	18.31	22.11	25.95	29.84	33.75	37.68	41.62	45.56
7	3.57	7.11	10.66	14.26	17.94	21.70	25.53	29.40	33.31	37.24	41.18	45.14
8	3.48	6.93	10.40	13.93	17.55	21.26	25.06	28.92	32.82	36.75	40.71	44.67
9	3.38	6.75	10.13	13.58	17.14	20.80	24.57	28.40	32.30	36.23	40.19	44.16
10	3.29	6.56	9.86	13.23	16.72	20.32	24.05	27.86	31.74	35.67	39.63	43.61
12	3.09	6.17	9.30	12.52	15.87	19.35	22.97	26.71	30.56	34.47	38.43	42.42
14	2.89	5.79	8.77	11.84	15.04	18.25	21.68	25.03	28.44	32.16	37.13	41.12
16	2.69	5.43	8.26	11.20	14.27	17.39	20.68	23.95	27.43	30.81	34.25	37.96
18	2.51	5.08	7.78	10.61	13.57	16.67	19.79	23.10	26.37	29.84	33.21	36.65
20	2.34	4.77	7.35	10.06	12.92	15.93	19.00	22.22	25.53	28.82	32.27	35.66
24	2.06	4.23	6.58	9.10	11.80	14.66	17.63	20.72	23.86	27.12	30.46	33.77
28	1.83	3.78	5.95	8.30	10.85	13.54	16.38	19.33	22.40	25.58	28.76	32.06
32	1.64	3.42	5.41	7.62	10.00	12.55	15.25	18.09	21.04	24.10	27.25	30.43
36	1.48	3.11	4.96	7.02	9.26	11.68	14.26	16.98	19.82	22.77	25.82	28.96
b = 6 In.												
2	3.89	7.76	11.65	15.56	19.48	23.40	27.33	31.26	35.18	39.11	43.04	46.97
3	3.85	7.68	11.55	15.45	19.37	23.29	27.22	31.15	35.08	39.01	42.94	46.87
4	3.79	7.57	11.41	15.31	19.22	23.15	27.09	31.02	34.96	38.89	42.82	46.76
5	3.73	7.45	11.25	15.13	19.05	22.99	26.93	30.87	34.81	38.75	42.69	46.63
6	3.65	7.30	11.07	14.94	18.86	22.80	26.75	30.71	34.65	38.60	42.55	46.49
7	3.57	7.15	10.87	14.72	18.64	22.59	26.55	30.51	34.47	38.43	42.38	46.33
8	3.48	6.98	10.65	14.48	18.40	22.36	26.33	30.30	34.27	38.24	42.21	46.16
9	3.38	6.81	10.42	14.22	18.14	22.11	26.09	30.07	34.05	38.03	42.00	45.97
10	3.29	6.63	10.18	13.95	17.86	21.84	25.83	29.83	33.82	37.81	41.79	45.77
12	3.09	6.27	9.69	13.38	17.26	21.24	25.26	29.28	33.30	37.32	41.33	45.32
14	2.89	5.92	9.14	12.54	16.11	20.59	24.62	28.67	32.73	36.77	40.80	44.83
16	2.69	5.59	8.70	11.99	15.43	19.01	23.94	28.01	32.09	36.16	40.22	44.27
18	2.51	5.29	8.34	11.56	14.94	18.44	21.93	27.30	31.40	35.51	39.60	43.67
20	2.34	5.00	7.96	11.11	14.43	17.86	21.37	24.98	30.68	34.80	38.92	43.03
24	2.06	4.50	7.29	10.35	13.57	16.92	20.36	23.91	27.49	31.13	37.46	41.62
28	1.83	4.09	6.73	9.67	12.80	16.07	19.44	22.91	26.42	29.97	33.56	37.20
32	1.64	3.73	6.25	9.06	12.09	15.27	18.58	21.96	25.44	28.96	32.55	36.13
36	1.48	3.43	5.82	8.51	11.43	14.53	17.76	21.11	24.56	27.98	31.51	35.10

ECCENTRIC LOADS ON WELD GROUPS

TABLES XVIII–XXV

ULTIMATE STRENGTH METHOD*

When weld groups are loaded in shear by an external load that does not act through the center of gravity of the group, the load is eccentric and will tend to cause a relative rotation and translation between the parts connected by the weld. The point about which rotation tends to take place is called the *instantaneous center of rotation*. Its location is dependent upon the eccentricity, geometry of the weld group, and deformation of the weld at different angles of the resultant elemental force relative to the weld axis.

The individual resistance force of each unit weld element can then be assumed to act on a line perpendicular to a ray passing through the instantaneous center and that element's location (see Fig. 1).

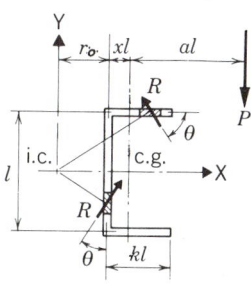

Figure 1

The ultimate shear strength of weld groups can be obtained from the load deformation relationship of a single unit weld element. This relationship was originally given by Butler, Pal and Kulak for E60 electrodes.* New curves for E70 electrodes were obtained by Kulak and Timler.** The latter curves can be expressed in non-dimensional form as:

$$R = R_{ult}(1 - e^{(-k_1\Delta/\Delta_0)})^{k_2}$$

This weld strength is limited to $0.6\ F_{EXX}$ according to LRFD Specification Sect. J2, Table J2.3,

where

R	= shear force in a single element at any given deformation
R_{ult}	= ultimate shear load of a single element
k_1, k_2	= regression coefficients
Δ	= deformation of weld element
Δ_0	= maximum deformation for θ equal to $0°$ ($= 0.11$ in.)
e	= base of natural logarithm $= 2.718 \ldots$

Unlike the load-deformation relationship for bolts, strength and deformation performance in welds are dependent on the angle θ that the resultant elemental force makes with the axis of the weld element (see Fig. 1).

---

*Butler, Pal and Kulak, *Eccentrically Loaded Weld Connections*, Journal of the Structural Div. ASCE, Vol. 98, No. ST5, May 1972, pp. 989–1005.
**Kulak and Timler, "Tests on Eccentrically Loaded Fillet Welds," Dept. of Civil Eng., University of Alberta, Edmonton, December 1984.

The critical weld element is usually (but not always) the weld element furthest from the instantaneous center. The critical deformation can be calculated as:

$$\Delta_{max} = \Delta_0\left(\frac{\theta}{5} + 1\right)^{-0.47}, \text{ where } \theta \text{ is expressed in degrees.}$$

The deformation of other weld elements varies linearly with distance from instantaneous center:

$$\Delta = \frac{r}{r_{max}} \Delta_{max}$$

The values of R_{ult}, k_1 and k_2 depend on the value of the angle θ and can be obtained from the following relations:

$$R_{ult} = \frac{(10 + \theta)}{10 + 0.5820} (0.791 F_{EXX} t)$$

$$k_1 = 8.274 e^{0.0114\theta}$$
$$k_2 = 0.4 e^{0.0146\theta}$$
$$F_{EXX} = \text{weld electrode strength}$$
$$t = \text{weld throat dimension}$$

Since the maximum weld strength according to the LRFD Specification is limited to $0.6 F_{EXX}$, all the curves are conservatively lowered to this level as shown in Fig. 2. Load vs. deformation curves for values of $\theta = 0°$, $10°$, $30°$ and $90°$ and E70 electrodes are shown in Fig. 2 as dashed lines along with the solid maximum deformation and strength limits for design. The ductility of the weld group is governed by Δ_{max} of the element that first reaches its limit. This idealization resembles the elastic-plastic assumptions of plastic design.

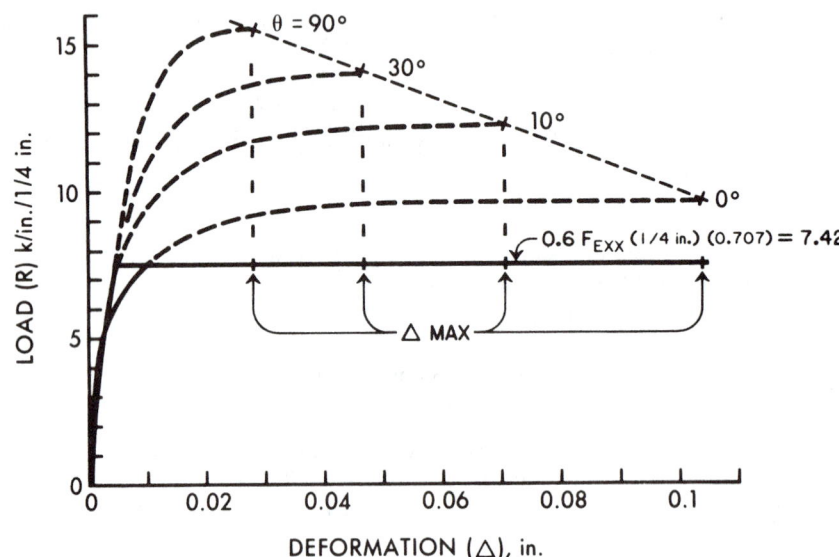

Figure 2

The total resistance of all the weld elements combine to resist the eccentric ultimate load, and if the correct location of the instantaneous center has been selected, the three equations of statics will be satisfied.

TABLES XVIII–XXV

The nominal weld strength has been directly limited to the LRFD Specification limit of $0.6 F_{EXX}$ in the instantaneous center analysis, and the ultimate weld group strength is further reduced to design level by the required resistance factor ϕ of 0.75.

Each weld configuration has also been analyzed for inclined loads at 0°, 45° and 75°. A convergence criterion of 1% was employed for the tabulated iterative solutions. Straight line interpolation between these angles may be significantly unconservative. *Therefore, unless a direct analysis is performed, use only the values tabulated for the next lower angle.*

The LRFD Specification indicates that the fusion face of weld metal and base material is not critical in determining weld strength. (See Commentary Sect. J2, Fig. C-J2.6) The tables are, therefore, valid for weld metal with a strength level equal to or matching the base material.

To obtain the capacity of a weld group carrying an eccentric load:

$$P_u = CC_1 Dl$$

where

P_u = factored load, kips
C = tabular value (includes $\phi = 0.75$)
C_1 = coefficient for electrode used (see table below)
$\quad = 1.0$ for E70XX electrodes
l = length of vertical weld, in.
D = number of sixteenths of an inch, weld size

Electrode	E60	E70	E80	E90	E100	E110
F_{EXX} (ksi)	60	70	80	90	100	110
C_1	0.857	1.0	1.03	1.16	1.21	1.34

Tables XIX through XXVI are based on welds made with E70XX electrodes and matching base metal (AWS Table 4.1.1). They also recognize that for equal leg fillet welds the area of the fusion surface is always larger than the leg dimension times the weld length: therefore, the values are based upon the strength through the throat of the weld, $(0.75 \times 0.6 \times F_{EXX} \times 0.707 \times \frac{1}{16})$. When electrodes other than E70XX are used with matching or stronger base metals, multiply tabulated weld C coefficients by C_1 values tabulated in table above. Note that C_1 for E80 and E90 includes an additional reduction factor of 0.9, and that for E100 and E110 a factor of 0.85 to account for the uncertainty of extrapolation to higher strength electrodes.

ALTERNATE METHOD

In addition to the ultimate strength method previously described, the elastic method may be used to design/analyze eccentrically loaded fastener groups not conforming to the LRFD Manual tables. By assuming each weld element as a line coincident with the edge of a fillet weld, each unit element would be assumed to support:

1. An equal share of the vertical component of the load.
2. An equal share of the horizontal component of the load.
3. A proportional share (dependent on the element's distance from the centroid of the group) of the eccentric moment portion of the load.

The vectorial resolution of these stresses at the element most remote from the group's centroid is designated the critical element. This elastic method, although providing a simplified and conservative approach, does not render a consistent factor of safety and in some cases results in excessively conservative designs of connections.

EXAMPLE 1

Given:

Weld group shown in Fig. 1, with $l = 10$ in., $kl = 5$ in., eccentric load angle $= 0°$ (vertical) and $xl + al = 10$ in. Find the maximum factored load P_u for a ⅜-in. weld using E70XX electrodes by using Table XXII.

Solution:

$$k = \frac{kl}{l} = \frac{5}{10} = 0.5; \, xl + al = 10 \text{ in.}$$

Enter Table XXII (angle 0°): For $k = 0.5$, $x = 0.125$

$xl = 0.125 \times 10 = 1.25$ in.
$al = 10 - xl = 8.75$ in.; $a = 0.875$

Interpolating between $a = 0.8$ and $a = 0.9$ for $k = 0.5$:

$C = 1.135$
$D = 6$ (⅜-in. weld)
$C_1 = 1.0$ for E70XX electrodes
$P_u = C_1 C D l = 1.0 \times 1.135 \times 6 \times 10 = 68.1$ kips

EXAMPLE 2

Given:

Same weld group and question as in Ex. 1, except that the eccentric load angle $= 75°$. Find maximum P_u and its components.

Solution:

$k = 0.5$
$a = 0.875$

Enter Table XXII (angle $= 75°$):

Interpolating between $a = 0.8$ and 0.9 for $k = 0.5$:

$C \quad = 2.17$
$P_u \quad = 1.0 \times 2.17 \times 6 \times 10 = 130.2$ kips
$P_u \cos 75° = 33.7$ kips (vertical component)
$P_u \sin 75° = 125.8$ kips (horizontal component)

ECCENTRIC LOADS ON WELD GROUPS
TABLE XVIII   Coefficients C

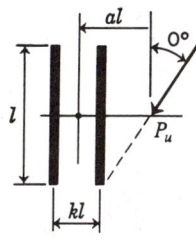

$$\text{ANGLE} = 0°$$

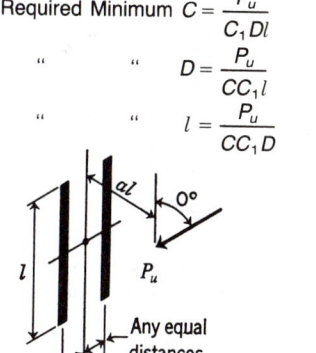

Required Minimum $C = \dfrac{P_u}{C_1 D l}$

"   " $D = \dfrac{P_u}{C C_1 l}$

"   " $l = \dfrac{P_u}{C C_1 D}$

P_u = factored eccentric load in kips
l = length of each weld, in.
D = number of sixteenths of an inch in fillet weld size
C = coefficients tabulated below (includes $\phi = 0.75$)
C_1 = coefficient for electrode used (see Table on page 5-89)
   = 1.0 for E70XX electrodes
$P_u = C C_1 D l$

Any equal distances

SPECIAL CASE*
(Load not in plane of weld group)
Use C-values given in column headed $k = 0$

a	k															
	0	0.1	0.2	0.3	0.4	0.5	0.6	0.7	0.8	0.9	1.0	1.2	1.4	1.6	1.8	2.0
.05	2.743	2.743	2.744	2.745	2.747	2.751	2.754	2.757	2.761	2.763	2.766	2.770	2.773	2.775	2.777	2.778
.10	2.627	2.628	2.631	2.636	2.645	2.656	2.668	2.681	2.692	2.703	2.713	2.728	2.740	2.748	2.755	2.760
.15	2.459	2.461	2.468	2.480	2.497	2.517	2.540	2.564	2.587	2.608	2.628	2.661	2.685	2.704	2.718	2.729
.20	2.264	2.268	2.281	2.300	2.325	2.354	2.387	2.420	2.455	2.487	2.518	2.571	2.612	2.644	2.668	2.687
.25	2.064	2.071	2.089	2.117	2.150	2.186	2.226	2.267	2.309	2.351	2.391	2.464	2.523	2.570	2.606	2.635
.30	1.874	1.883	1.907	1.942	1.982	2.026	2.070	2.116	2.162	2.209	2.213	2.344	2.421	2.483	2.533	2.572
.40	1.549	1.561	1.593	1.638	1.691	1.746	1.799	1.850	1.900	1.941	1.987	2.080	2.137	2.157	2.202	2.422
.50	1.299	1.312	1.348	1.400	1.460	1.522	1.581	1.638	1.691	1.741	1.786	1.871	1.956	2.005	2.049	2.111
.60	1.107	1.123	1.160	1.216	1.279	1.345	1.408	1.468	1.525	1.577	1.625	1.713	1.790	1.865	1.933	1.969
.70	.962	.975	1.015	1.072	1.138	1.204	1.268	1.329	1.387	1.442	1.493	1.583	1.663	1.733	1.797	1.858
.80	.849	.862	.902	.960	1.025	1.090	1.154	1.215	1.272	1.326	1.377	1.475	1.554	1.626	1.691	1.748
.90	.758	.771	.811	.868	.932	.996	1.060	1.120	1.177	1.229	1.279	1.373	1.464	1.533	1.601	1.658
1.00	.684	.698	.737	.791	.853	.916	.979	1.039	1.095	1.148	1.197	1.287	1.371	1.458	1.517	1.580
1.20	.572	.586	.622	.672	.729	.789	.847	.905	.961	1.014	1.062	1.149	1.227	1.301	1.369	1.454
1.40	.491	.504	.538	.583	.635	.690	.745	.800	.854	.905	.954	1.038	1.114	1.185	1.250	1.312
1.60	.430	.443	.474	.514	.561	.612	.665	.715	.767	.817	.864	.947	1.023	1.090	1.153	1.211
1.80	.383	.395	.423	.460	.503	.550	.598	646	.695	.742	.788	.870	.945	1.012	1.071	1.127
2.00	.345	.356	.382	.416	.455	.498	.544	.590	.634	.680	.724	.804	.878	.944	1.003	1.057
2.20	.314	.324	.348	.379	.415	.455	.498	.541	.583	.626	.668	.749	.819	.885	.944	.997
2.40	.287	.298	.319	.348	.382	.419	.459	.499	.541	.580	.620	.698	.767	.832	.891	.943
2.60	.265	.275	.295	.322	.353	.388	.425	.463	.502	.540	.578	.653	.720	.784	.843	.896
2.80	.246	.255	.274	.299	.329	.361	.396	.432	.469	.506	.541	.613	.681	.741	.799	.852
3.00	.230	.239	.256	.280	.307	.338	.371	.405	.439	.475	.508	.577	.643	.702	.759	.812

*Valid only when the connection material between the welds is solid and does not bend in the plane of the welds.

ECCENTRIC LOADS ON WELD GROUPS
TABLE XVIII   Coefficients C

ANGLE = 45°

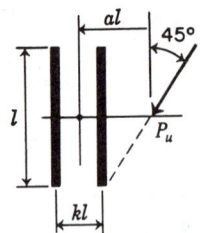

Required Minimum $C = \dfrac{P_u}{C_1 Dl}$

" " $D = \dfrac{P_u}{CC_1 l}$

" " $l = \dfrac{P_u}{CC_1 D}$

P_u = factored eccentric load in kips
l = length of each weld, in.
D = number of sixteenths of an inch
  in fillet weld size
C = coefficients tabulated below
  (includes $\phi = 0.75$)
C_1 = coefficient for electrode used
  (see Table on page 5-89)
  = 1.0 for E70XX electrodes

$P_u = CC_1 Dl$

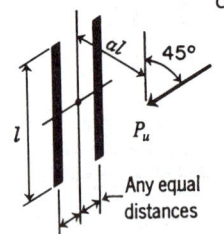

Any equal distances

SPECIAL CASE*
(Load not in plane of weld group)
Use C-values given
in column headed $k = 0$

a	k															
	0	0.1	0.2	0.3	0.4	0.5	0.6	0.7	0.8	0.9	1.0	1.2	1.4	1.6	1.8	2.0
.05	2.753	2.754	2.757	2.760	2.763	2.766	2.769	2.772	2.773	2.775	2.776	2.778	2.780	2.781	2.781	2.782
.10	2.666	2.669	2.677	2.689	2.702	2.714	2.725	2.734	2.742	2.748	2.753	2.761	2.766	2.770	2.772	2.775
.15	2.536	2.541	2.558	2.581	2.607	2.632	2.655	2.675	2.691	2.704	2.715	2.732	2.744	2.752	2.758	2.763
.20	2.382	2.390	2.413	2.447	2.486	2.526	2.563	2.595	2.623	2.646	2.664	2.693	2.713	2.728	2.738	2.746
.25	2.217	2.233	2.260	2.300	2.350	2.403	2.455	2.501	2.540	2.574	2.602	2.645	2.675	2.697	2.713	2.725
.30	2.064	2.075	2.103	2.156	2.211	2.272	2.336	2.395	2.447	2.492	2.529	2.588	2.630	2.660	2.683	2.700
.40	1.790	1.804	1.840	1.891	1.956	2.022	2.093	2.159	2.242	2.307	2.364	2.456	2.523	2.572	2.610	2.639
.50	1.567	1.581	1.620	1.680	1.749	1.819	1.889	1.959	2.025	2.099	2.175	2.307	2.399	2.469	2.523	2.564
.60	1.382	1.397	1.442	1.508	1.580	1.655	1.727	1.798	1.866	1.929	1.997	2.140	2.257	2.355	2.425	2.480
.70	1.231	1.247	1.294	1.364	1.441	1.518	1.593	1.663	1.730	1.794	1.856	1.986	2.120	2.227	2.321	2.389
.80	1.104	1.122	1.173	1.243	1.322	1.400	1.475	1.547	1.614	1.679	1.740	1.858	1.985	2.106	2.205	2.294
.90	1.001	1.019	1.071	1.141	1.219	1.296	1.372	1.445	1.514	1.578	1.639	1.753	1.867	1.989	2.097	2.187
1.00	.914	.932	.983	1.052	1.129	1.206	1.283	1.356	1.424	1.489	1.550	1.662	1.770	1.881	1.994	2.091
1.20	.776	.795	.844	.909	.981	1.057	1.130	1.202	1.272	1.337	1.398	1.510	1.612	1.705	1.802	1.901
1.40	.673	.691	.736	.797	.864	.935	1.005	1.076	1.144	1.209	1.270	1.384	1.485	1.572	1.661	1.749
1.60	.593	.611	.653	.707	.770	.837	.906	.971	1.037	1.101	1.162	1.274	1.375	1.462	1.546	1.628
1.80	.530	.548	.586	.636	.693	.756	.821	.883	.947	1.008	1.068	1.179	1.274	1.365	1.448	1.526
2.00	.479	.496	.531	.577	.630	.689	.749	.811	.869	.929	.986	1.095	1.190	1.280	1.363	1.439
2.20	.437	.452	.485	.528	.578	.632	.689	.747	.803	.859	.915	1.021	1.119	1.204	1.285	1.361
2.40	.402	.416	.446	.486	.533	.583	.637	.692	.747	.799	.853	.956	1.051	1.135	1.216	1.291
2.60	.374	.385	.413	.451	.494	.541	.592	.644	.696	.746	.798	.897	.991	1.073	1.153	1.227
2.80	.345	.358	.385	.420	.460	.505	.552	.601	.651	.702	.749	.844	.936	1.021	1.095	1.168
3.00	.323	.334	.360	.393	.431	.473	.518	.564	.612	.660	.708	.797	.886	.969	1.043	1.115

*Valid only when the connection material between the welds is solid and does not bend in
the plane of the welds.

ECCENTRIC LOADS ON WELD GROUPS
TABLE XVIII   Coefficients C

ANGLE = 75°

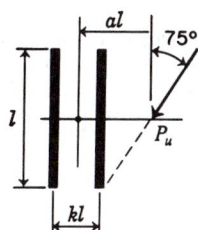

Required Minimum $C = \dfrac{P_u}{C_1 Dl}$

"   " $D = \dfrac{P_u}{CC_1 l}$

"   " $l = \dfrac{P_u}{CC_1 D}$

P_u = factored eccentric load in kips
l = length of each weld, in.
D = number of sixteenths of an inch in fillet weld size
C = coefficients tabulated below (includes $\phi = 0.75$)
C_1 = coefficient for electrode used (see Table on page 5-89)
 = 1.0 for E70XX electrodes

$P_u = CC_1 Dl$

SPECIAL CASE*
(Load not in plane of weld group)
Use C-values given in column headed $k = 0$

a	k																
	0	0.1	0.2	0.3	0.4	0.5	0.6	0.7	0.8	0.9	1.0	1.2	1.4	1.6	1.8	2.0
.05	2.779	2.779	2.780	2.780	2.781	2.782	2.782	2.783	2.783	2.783	2.783	2.784	2.784	2.784	2.784	2.784
.10	2.763	2.764	2.766	2.769	2.772	2.774	2.776	2.778	2.779	2.780	2.781	2.782	2.782	2.783	2.783	2.783
.15	2.736	2.738	2.743	2.750	2.756	2.762	2.766	2.770	2.772	2.775	2.776	2.778	2.780	2.781	2.781	2.782
.20	2.699	2.703	2.712	2.724	2.735	2.745	2.753	2.759	2.764	2.767	2.770	2.774	2.776	2.778	2.779	2.780
.25	2.653	2.658	2.673	2.691	2.708	2.723	2.735	2.745	2.752	2.758	2.762	2.768	2.772	2.775	2.777	2.778
.30	2.598	2.606	2.626	2.651	2.676	2.698	2.715	2.728	2.738	2.746	2.752	2.761	2.767	2.771	2.773	2.775
.40	2.464	2.478	2.513	2.556	2.598	2.635	2.664	2.686	2.704	2.718	2.729	2.744	2.754	2.760	2.765	2.769
.50	2.203	2.235	2.380	2.444	2.506	2.559	2.602	2.635	2.662	2.682	2.699	2.722	2.737	2.747	2.755	2.760
.60	2.105	2.133	2.215	2.320	2.404	2.474	2.531	2.577	2.613	2.641	2.663	2.696	2.717	2.732	2.742	2.750
.70	1.979	2.009	2.084	2.184	2.296	2.384	2.455	2.512	2.558	2.594	2.623	2.666	2.694	2.713	2.727	2.738
.80	1.873	1.903	1.981	2.083	2.191	2.293	2.376	2.444	2.499	2.544	2.580	2.632	2.668	2.693	2.711	2.724
.90	1.779	1.814	1.892	1.989	2.095	2.203	2.296	2.374	2.438	2.490	2.533	2.596	2.640	2.670	2.692	2.708
1.00	1.697	1.731	1.811	1.906	2.011	2.118	2.218	2.304	2.375	2.435	2.484	2.558	2.609	2.645	2.672	2.692
1.20	1.546	1.582	1.665	1.760	1.864	1.969	2.072	2.166	2.249	2.321	2.381	2.475	2.542	2.591	2.627	2.654
1.40	1.419	1.456	1.535	1.631	1.735	1.839	1.942	2.039	2.128	2.208	2.277	2.388	2.470	2.531	2.576	2.611
1.60	1.306	1.343	1.420	1.516	1.619	1.723	1.825	1.922	2.015	2.099	2.174	2.299	2.394	2.467	2.522	2.565
1.80	1.207	1.244	1.320	1.412	1.512	1.616	1.718	1.815	1.910	1.997	2.076	2.211	2.317	2.400	2.464	2.515
2.00	1.121	1.156	1.230	1.319	1.418	1.519	1.620	1.719	1.813	1.901	1.983	2.125	2.240	2.332	2.405	2.463
2.20	1.044	1.079	1.147	1.236	1.330	1.431	1.531	1.629	1.723	1.811	1.895	2.042	2.165	2.264	2.344	2.409
2.40	.976	1.010	1.078	1.161	1.254	1.351	1.449	1.546	1.639	1.728	1.812	1.963	2.091	2.196	2.283	2.354
2.60	.916	.949	1.014	1.093	1.183	1.277	1.374	1.469	1.562	1.650	1.734	1.887	2.019	2.130	2.222	2.299
2.80	.863	.893	.956	1.033	1.119	1.211	1.305	1.399	1.490	1.578	1.662	1.816	1.950	2.065	2.162	2.244
3.00	.815	.843	.903	.977	1.060	1.150	1.242	1.333	1.423	1.510	1.594	1.748	1.884	2.002	2.103	2.189

*Valid only when the connection material between the welds is solid and does not bend in the plane of the welds.

AMERICAN INSTITUTE OF STEEL CONSTRUCTION

ECCENTRIC LOADS ON WELD GROUPS
TABLE XIX   Coefficients C

ANGLE = 0°

P_u = factored eccentric load in kips
l = length of each weld, in.
D = number of sixteenths of an inch in fillet weld size
C = coefficients tabulated below (includes $\phi = 0.75$)
C_1 = coefficient for electrode used (see Table on page 5-89)
  = 1.0 for E70XX electrodes

$$P_u = CC_1Dl$$

Required Minimum $C = \dfrac{P_u}{C_1Dl}$

"   " $D = \dfrac{P_u}{CC_1l}$

"   " $l = \dfrac{P_u}{CC_1D}$

a	\multicolumn{16}{c}{k}
	0	0.1	0.2	0.3	0.4	0.5	0.6	0.7	0.8	0.9	1.0	1.2	1.4	1.6	1.8	2.0
.05	2.702	2.705	2.715	2.726	2.738	2.747	2.755	2.761	2.765	2.769	2.771	2.775	2.777	2.779	2.780	2.781
.10	2.469	2.484	2.521	2.566	2.608	2.643	2.672	2.694	2.710	2.723	2.733	2.748	2.757	2.763	2.767	2.770
.15	2.077	2.124	2.204	2.341	2.423	2.493	2.548	2.592	2.626	2.653	2.674	2.704	2.724	2.737	2.747	2.754
.20	1.854	1.905	2.005	2.119	2.228	2.324	2.404	2.470	2.522	2.564	2.598	2.647	2.680	2.702	2.719	2.730
.25	1.691	1.741	1.839	1.947	2.054	2.162	2.257	2.339	2.407	2.464	2.510	2.579	2.626	2.660	2.684	2.702
.30	1.549	1.599	1.692	1.798	1.908	2.015	2.119	2.211	2.290	2.358	2.415	2.503	2.566	2.610	2.643	2.668
.40	1.318	1.364	1.450	1.547	1.657	1.767	1.874	1.974	2.066	2.149	2.222	2.341	2.430	2.497	2.548	2.587
.50	1.139	1.180	1.259	1.350	1.454	1.561	1.667	1.769	1.866	1.955	2.037	2.176	2.286	2.373	2.441	2.495
.60	.995	1.034	1.103	1.192	1.287	1.390	1.492	1.593	1.690	1.782	1.867	2.018	2.143	2.245	2.327	2.394
.70	.882	.916	.982	1.060	1.151	1.247	1.345	1.443	1.538	1.629	1.716	1.872	2.006	2.119	2.213	2.291
.80	.790	.820	.880	.954	1.038	1.128	1.221	1.314	1.406	1.495	1.581	1.739	1.878	1.998	2.100	2.187
.90	.713	.742	.796	.865	.943	1.027	1.115	1.203	1.292	1.378	1.462	1.619	1.760	1.884	1.992	2.086
1.00	.650	.677	.727	.789	.862	.941	1.024	1.108	1.192	1.275	1.357	1.511	1.652	1.778	1.890	1.988
1.20	.551	.573	.617	.671	.735	.804	.878	.953	1.029	1.105	1.181	1.326	1.463	1.589	1.703	1.807
1.40	.477	.496	.533	.583	.639	.701	.766	.834	.903	.972	1.041	1.177	1.307	1.429	1.542	1.646
1.60	.420	.438	.471	.514	.564	.620	.679	.740	.802	.866	.929	1.055	1.177	1.293	1.403	1.506
1.80	.376	.391	.421	.460	.505	.555	.608	.664	.721	.779	.838	.954	1.068	1.178	1.284	1.383
2.00	.339	.353	.381	.415	.457	.502	.551	.602	.654	.708	.762	.870	.977	1.081	1.181	1.277
2.20	.309	.322	.347	.379	.417	.458	.503	.550	.599	.648	.698	.799	.899	.997	1.092	1.183
2.40	.284	.296	.319	.348	.383	.422	.463	.506	.551	.597	.644	.738	.831	.924	1.014	1.102
2.60	.263	.274	.295	.322	.354	.390	.429	.469	.511	.554	.597	.685	.773	.860	.946	1.030
2.80	.244	.254	.274	.299	.329	.363	.399	.437	.476	.516	.557	.639	.722	.805	.886	.966
3.00	.228	.237	.256	.280	.308	.339	.373	.408	.445	.483	.521	.599	.677	.756	.833	.909

ECCENTRIC LOADS ON WELD GROUPS
TABLE XIX   Coefficients C

ANGLE = 45°

P_u = factored eccentric load in kips
l = length of each weld, in.
D = number of sixteenths of an inch in fillet weld size
C = coefficients tabulated below (includes ϕ = 0.75)
C_1 = coefficient for electrode used (see Table on page 5-89)
= 1.0 for E70XX electrodes

$$P_u = CC_1Dl$$

Required Minimum $C = \dfrac{P_u}{C_1Dl}$

"　　　"　$D = \dfrac{P_u}{CC_1l}$

"　　　"　$l = \dfrac{P_u}{CC_1D}$

a	\multicolumn{16}{c}{k}
	0	0.1	0.2	0.3	0.4	0.5	0.6	0.7	0.8	0.9	1.0	1.2	1.4	1.6	1.8	2.0
.05	2.753	2.754	2.757	2.760	2.763	2.766	2.769	2.772	2.773	2.775	2.776	2.778	2.780	2.781	2.781	2.782
.10	2.666	2.669	2.677	2.689	2.702	2.714	2.725	2.734	2.742	2.748	2.753	2.761	2.766	2.770	2.772	2.775
.15	2.536	2.541	2.558	2.581	2.607	2.632	2.655	2.675	2.691	2.704	2.715	2.732	2.744	2.752	2.758	2.763
.20	2.382	2.390	2.413	2.447	2.486	2.526	2.563	2.595	2.623	2.646	2.664	2.693	2.713	2.728	2.738	2.746
.25	2.217	2.233	2.260	2.300	2.350	2.403	2.455	2.501	2.540	2.574	2.602	2.645	2.675	2.697	2.713	2.725
.30	2.064	2.075	2.103	2.156	2.211	2.272	2.336	2.395	2.447	2.492	2.529	2.588	2.630	2.660	2.683	2.700
.40	1.790	1.804	1.840	1.891	1.956	2.022	2.093	2.159	2.242	2.307	2.364	2.456	2.523	2.572	2.610	2.639
.50	1.567	1.581	1.620	1.680	1.749	1.819	1.889	1.959	2.025	2.099	2.175	2.307	2.399	2.469	2.523	2.564
.60	1.382	1.397	1.442	1.508	1.580	1.655	1.727	1.798	1.866	1.929	1.997	2.140	2.257	2.355	2.425	2.480
.70	1.231	1.247	1.294	1.364	1.441	1.518	1.593	1.663	1.730	1.794	1.856	1.986	2.120	2.227	2.321	2.389
.80	1.104	1.122	1.173	1.243	1.322	1.400	1.475	1.547	1.614	1.679	1.740	1.858	1.985	2.106	2.205	2.294
.90	1.001	1.019	1.071	1.141	1.219	1.296	1.372	1.445	1.514	1.578	1.639	1.753	1.867	1.989	2.097	2.187
1.00	.914	.932	.983	1.052	1.129	1.206	1.283	1.356	1.424	1.489	1.550	1.662	1.770	1.881	1.994	2.091
1.20	.776	.795	.844	.909	.981	1.057	1.130	1.202	1.272	1.337	1.398	1.510	1.612	1.705	1.802	1.901
1.40	.673	.691	.736	.797	.864	.935	1.005	1.076	1.144	1.209	1.270	1.384	1.485	1.572	1.661	1.749
1.60	.593	.611	.653	.707	.770	.837	.906	.971	1.037	1.101	1.162	1.274	1.375	1.462	1.546	1.628
1.80	.530	.548	.586	.636	.693	.756	.821	.883	.947	1.008	1.068	1.179	1.274	1.365	1.448	1.526
2.00	.479	.496	.531	.577	.630	.689	.749	.811	.869	.929	.986	1.095	1.190	1.280	1.363	1.439
2.20	.437	.452	.485	.528	.578	.632	.689	.747	.803	.859	.915	1.021	1.119	1.204	1.285	1.361
2.40	.402	.416	.446	.486	.533	.583	.637	.692	.747	.799	.853	.956	1.051	1.135	1.216	1.291
2.60	.374	.385	.413	.451	.494	.541	.592	.644	.696	.746	.798	.897	.991	1.073	1.153	1.227
2.80	.345	.358	.385	.420	.460	.505	.552	.601	.651	.702	.749	.844	.936	1.021	1.095	1.168
3.00	.323	.334	.360	.393	.431	.473	.518	.564	.612	.660	.708	.797	.886	.969	1.043	1.115

ECCENTRIC LOADS ON WELD GROUPS
TABLE XIX   Coefficients C

ANGLE = 75°

P_u = factored eccentric load in kips
l = length of each weld, in.
D = number of sixteenths of an inch in fillet weld size
C = coefficients tabulated below (includes $\phi = 0.75$)
C_1 = coefficient for electrode used (see Table on page 5-89)
= 1.0 for E70XX electrodes

$$P_u = CC_1Dl$$

Required Minimum $C = \dfrac{P_u}{C_1Dl}$

"   " $D = \dfrac{P_u}{CC_1l}$

"   " $l = \dfrac{P_u}{CC_1D}$

a	k															
	0	0.1	0.2	0.3	0.4	0.5	0.6	0.7	0.8	0.9	1.0	1.2	1.4	1.6	1.8	2.0
.05	2.781	2.781	2.781	2.781	2.782	2.782	2.782	2.782	2.783	2.783	2.783	2.783	2.783	2.784	2.784	2.784
.10	2.772	2.772	2.773	2.773	2.774	2.775	2.776	2.777	2.778	2.779	2.779	2.780	2.781	2.782	2.782	2.783
.15	2.758	2.758	2.758	2.760	2.761	2.764	2.766	2.768	2.770	2.772	2.773	2.776	2.778	2.779	2.780	2.781
.20	2.737	2.738	2.739	2.741	2.744	2.748	2.752	2.756	2.759	2.762	2.765	2.769	2.772	2.775	2.776	2.778
.25	2.712	2.712	2.714	2.717	2.722	2.728	2.734	2.740	2.746	2.750	2.755	2.761	2.766	2.769	2.772	2.774
.30	2.682	2.682	2.685	2.690	2.696	2.704	2.713	2.721	2.729	2.736	2.742	2.751	2.758	2.763	2.767	2.770
.40	2.608	2.610	2.614	2.622	2.633	2.646	2.660	2.674	2.687	2.699	2.709	2.726	2.738	2.747	2.753	2.759
.50	2.522	2.524	2.531	2.542	2.558	2.576	2.597	2.617	2.636	2.653	2.669	2.694	2.712	2.726	2.736	2.744
.60	2.425	2.429	2.438	2.453	2.473	2.497	2.524	2.550	2.576	2.600	2.621	2.655	2.681	2.701	2.716	2.727
.70	2.323	2.328	2.340	2.359	2.384	2.413	2.445	2.477	2.509	2.539	2.566	2.611	2.646	2.672	2.692	2.707
.80	2.219	2.224	2.240	2.263	2.292	2.325	2.362	2.400	2.437	2.473	2.506	2.562	2.606	2.639	2.664	2.684
.90	2.115	2.121	2.139	2.167	2.200	2.237	2.277	2.319	2.362	2.403	2.441	2.508	2.561	2.602	2.634	2.658
1.00	2.013	2.020	2.041	2.072	2.109	2.150	2.194	2.239	2.285	2.330	2.374	2.451	2.513	2.562	2.600	2.630
1.20	1.821	1.830	1.856	1.894	1.939	1.985	2.034	2.083	2.133	2.173	2.200	2.328	2.408	2.473	2.525	2.565
1.40	1.648	1.659	1.688	1.734	1.785	1.837	1.889	1.941	1.986	2.035	2.084	2.155	2.192	2.374	2.440	2.493
1.60	1.498	1.510	1.543	1.591	1.647	1.705	1.761	1.815	1.868	1.913	1.961	2.058	2.121	2.149	2.199	2.412
1.80	1.367	1.380	1.415	1.466	1.525	1.587	1.646	1.704	1.757	1.809	1.854	1.945	2.032	2.096	2.114	2.159
2.00	1.250	1.267	1.303	1.357	1.419	1.483	1.546	1.605	1.660	1.712	1.761	1.849	1.936	2.020	2.041	2.086
2.20	1.152	1.165	1.206	1.261	1.326	1.392	1.456	1.517	1.574	1.627	1.676	1.765	1.848	1.924	1.978	2.022
2.40	1.067	1.080	1.121	1.178	1.243	1.311	1.376	1.437	1.496	1.551	1.601	1.691	1.772	1.849	1.921	1.962
2.60	.992	1.005	1.046	1.104	1.171	1.239	1.304	1.366	1.424	1.480	1.533	1.624	1.705	1.778	1.846	1.920
2.80	.926	.940	.981	1.040	1.107	1.174	1.239	1.302	1.359	1.415	1.468	1.563	1.644	1.717	1.784	1.848
3.00	.868	.882	.923	.982	1.049	1.116	1.181	1.243	1.301	1.356	1.408	1.507	1.589	1.662	1.728	1.790

ECCENTRIC LOADS ON WELD GROUPS
TABLE XX   Coefficients C

ANGLE = 0°

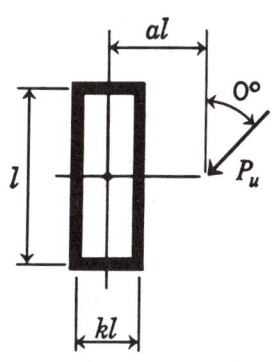

P_u = factored eccentric load in kips
l = length of each weld, in.
D = number of sixteenths of an inch in fillet weld size
C = coefficients tabulated below (includes ϕ = 0.75)
C_1 = coefficient for electrode used (see Table on page 5-89)
= 1.0 for E70XX electrodes

$$P_u = CC_1Dl$$

Required Minimum $C = \dfrac{P_u}{C_1Dl}$

"   " $D = \dfrac{P_u}{CC_1l}$

"   " $l = \dfrac{P_u}{CC_1D}$

Note: When load P_u is perpendicular to longer side l use Table XXI.

a	k															
	0	0.1	0.2	0.3	0.4	0.5	0.6	0.7	0.8	0.9	1.0	1.2	1.4	1.6	1.8	2.0
.05	2.743	3.024	3.304	3.583	3.862	4.141	4.421	4.700	4.979	5.258	5.537	6.095	6.654	7.212	7.770	8.328
.10	2.627	2.916	3.200	3.481	3.762	4.042	4.323	4.603	4.884	5.165	5.446	6.008	6.571	7.133	7.695	8.257
.15	2.459	2.755	3.044	3.328	3.610	3.891	4.173	4.455	4.737	5.020	5.303	5.870	6.437	7.005	7.572	8.139
.20	2.264	2.565	2.856	3.142	3.424	3.706	3.987	4.268	4.551	4.834	5.118	5.688	6.261	6.834	7.407	7.980
.25	2.064	2.364	2.656	2.941	3.223	3.502	3.781	4.060	4.341	4.622	4.905	5.475	6.050	6.627	7.205	7.784
.30	1.874	2.169	2.457	2.740	3.019	3.295	3.570	3.845	4.121	4.399	4.679	5.244	5.816	6.394	6.975	7.558
.40	1.549	1.821	2.095	2.366	2.635	2.901	3.167	3.431	3.697	3.963	4.231	4.774	5.327	5.891	6.464	7.045
.50	1.299	1.544	1.797	2.052	2.307	2.561	2.815	3.068	3.322	3.576	3.831	4.347	4.871	5.406	5.929	6.481
.60	1.107	1.329	1.560	1.797	2.037	2.277	2.518	2.760	3.002	3.245	3.490	3.983	4.481	4.987	5.502	6.011
.70	.962	1.161	1.371	1.591	1.816	2.043	2.271	2.502	2.734	2.967	3.202	3.677	4.159	4.646	5.140	5.639
.80	.849	1.025	1.220	1.425	1.635	1.850	2.067	2.287	2.510	2.735	2.961	3.421	3.887	4.361	4.840	5.322
.90	.758	.918	1.098	1.288	1.487	1.690	1.897	2.107	2.321	2.533	2.751	3.194	3.646	4.110	4.576	5.049
1.00	.684	.831	.998	1.176	1.363	1.555	1.752	1.953	2.157	2.359	2.568	2.995	3.432	3.878	4.339	4.799
1.20	.572	.698	.843	1.001	1.166	1.338	1.515	1.697	1.883	2.068	2.260	2.655	3.061	3.478	3.906	4.350
1.40	.491	.601	.730	.870	1.017	1.171	1.331	1.495	1.665	1.838	2.010	2.374	2.751	3.141	3.542	3.953
1.60	.430	.528	.643	.768	.901	1.040	1.184	1.334	1.488	1.647	1.810	2.141	2.491	2.854	3.229	3.616
1.80	.383	.471	.574	.687	.807	.934	1.066	1.202	1.344	1.489	1.639	1.950	2.276	2.616	2.960	3.324
2.00	.345	.425	.519	.622	.731	.847	.968	1.093	1.223	1.358	1.496	1.784	2.088	2.404	2.734	3.076
2.20	.314	.387	.473	.567	.668	.774	.886	1.002	1.122	1.246	1.375	1.643	1.925	2.222	2.531	2.853
2.40	.287	.355	.434	.521	.614	.713	.816	.924	1.035	1.151	1.271	1.521	1.785	2.063	2.354	2.658
2.60	.265	.328	.401	.482	.569	.660	.756	.857	.961	1.069	1.181	1.415	1.663	1.924	2.198	2.485
2.80	.246	.305	.373	.449	.529	.615	.705	.799	.896	.997	1.102	1.322	1.555	1.801	2.060	2.331
3.00	.230	.285	.349	.419	.495	.575	.659	.748	.839	.935	1.033	1.240	1.460	1.693	1.938	2.195

ECCENTRIC LOADS ON WELD GROUPS
TABLE XX   Coefficients C

ANGLE = 45°

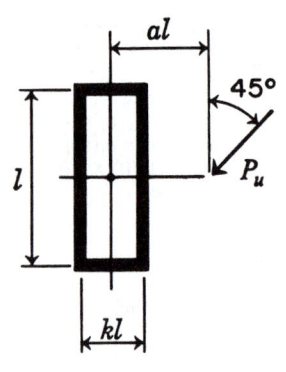

P_u = factored eccentric load in kips
l = Length of each weld, in.
D = number of sixteenths of an inch in fillet weld size
C = coefficients tabulated below (includes ϕ = 0.75)
C_1 = coefficient for electrode used (see Table on page 5-89)
  = 1.0 for E70XX electrodes
$$P_u = CC_1Dl$$

Required Minimum $C = \dfrac{P_u}{C_1 Dl}$

" " $D = \dfrac{P_u}{CC_1 l}$

" " $l = \dfrac{P_u}{CC_1 D}$

Note: When load P_u is perpendicular to longer side l use Table XXI.

a	\multicolumn{17}{c}{k}
	0	0.1	0.2	0.3	0.4	0.5	0.6	0.7	0.8	0.9	1.0	1.2	1.4	1.6	1.8	2.0
.05	2.753	3.034	3.315	3.595	3.876	4.155	4.435	4.715	4.994	5.273	5.553	6.111	6.669	7.227	7.784	8.342
.10	2.666	2.954	3.240	3.526	3.811	4.094	4.378	4.661	4.943	5.225	5.506	6.069	6.630	7.191	7.751	8.311
.15	2.536	2.831	3.125	3.417	3.708	3.997	4.286	4.573	4.860	5.146	5.431	6.000	6.567	7.132	7.697	8.260
.20	2.382	2.682	2.981	3.278	3.575	3.870	4.164	4.457	4.749	5.040	5.329	5.906	6.480	7.051	7.621	8.189
.25	2.217	2.523	2.822	3.121	3.421	3.720	4.020	4.318	4.615	4.910	5.205	5.790	6.372	6.951	7.527	8.100
.30	2.064	2.359	2.661	2.956	3.256	3.557	3.859	4.161	4.462	4.762	5.061	5.656	6.246	6.832	7.415	7.995
.40	1.790	2.070	2.352	2.635	2.921	3.222	3.520	3.821	4.125	4.431	4.736	5.345	5.950	6.551	7.147	7.740
.50	1.567	1.829	2.098	2.368	2.639	2.913	3.191	3.474	3.766	4.083	4.390	5.003	5.617	6.228	6.836	7.441
.60	1.382	1.629	1.884	2.143	2.405	2.670	2.936	3.206	3.481	3.761	4.050	4.656	5.274	5.888	6.502	7.114
.70	1.231	1.459	1.703	1.952	2.208	2.462	2.720	2.982	3.246	3.514	3.785	4.342	4.922	5.549	6.164	6.777
.80	1.104	1.319	1.550	1.792	2.036	2.284	2.534	2.786	3.042	3.300	3.561	4.092	4.646	5.217	5.815	6.443
.90	1.001	1.201	1.420	1.651	1.887	2.126	2.368	2.612	2.860	3.110	3.363	3.874	4.401	4.954	5.520	6.110
1.00	.914	1.100	1.309	1.530	1.756	1.986	2.220	2.455	2.695	2.936	3.181	3.676	4.186	4.712	5.264	5.825
1.20	.776	.941	1.129	1.329	1.537	1.749	1.966	2.182	2.409	2.630	2.863	3.330	3.807	4.301	4.810	5.341
1.40	.673	.820	.990	1.172	1.362	1.558	1.757	1.962	2.165	2.375	2.590	3.029	3.485	3.950	4.430	4.924
1.60	.593	.726	.880	1.046	1.219	1.400	1.585	1.774	1.963	2.160	2.361	2.774	3.207	3.646	4.099	4.567
1.80	.530	.651	.791	.943	1.102	1.267	1.440	1.616	1.797	1.977	2.165	2.553	2.955	3.378	3.808	4.252
2.00	.479	.589	.717	.857	1.004	1.158	1.317	1.481	1.650	1.823	2.000	2.360	2.741	3.142	3.551	3.973
2.20	.437	.538	.656	.785	.922	1.064	1.213	1.366	1.524	1.686	1.853	2.197	2.557	2.932	3.321	3.723
2.40	.402	.495	.605	.724	.851	.984	1.123	1.266	1.414	1.567	1.724	2.049	2.389	2.744	3.115	3.499
2.60	.374	.459	.560	.672	.790	.915	1.044	1.179	1.319	1.462	1.610	1.918	2.241	2.579	2.931	3.297
2.80	.345	.427	.522	.626	.737	.854	.976	1.103	1.234	1.370	1.510	1.801	2.107	2.429	2.765	3.115
3.00	.323	.399	.488	.586	.690	.801	.916	1.036	1.160	1.288	1.420	1.697	1.988	2.294	2.615	2.950

ECCENTRIC LOADS ON WELD GROUPS
TABLE XX   Coefficients C

ANGLE = 75°

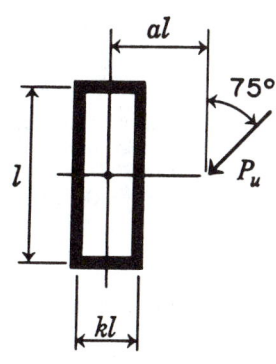

P_u = factored eccentric load in kips
l = length of each weld, in.
D = number of sixteenths of an inch in fillet weld size
C = coefficients tabulated below (includes ϕ = 0.75)
C_1 = coefficient for electrode used (see Table on page 5-89)
= 1.0 for E70XX electrodes
$$P_u = CC_1 Dl$$

Required Minimum $C = \dfrac{P_u}{C_1 Dl}$

"   " $D = \dfrac{P_u}{CC_1 l}$

"   " $l = \dfrac{P_u}{CC_1 D}$

Note: When load P_u is perpendicular to longer side l use Table XXI.

a	k															
	0	0.1	0.2	0.3	0.4	0.5	0.6	0.7	0.8	0.9	1.0	1.2	1.4	1.6	1.8	2.0
.05	2.779	3.058	3.337	3.616	3.894	4.173	4.452	4.730	5.009	5.288	5.566	6.123	6.680	7.237	7.794	8.351
.10	2.763	3.043	3.324	3.604	3.884	4.164	4.443	4.723	5.002	5.281	5.560	6.118	6.676	7.233	7.790	8.348
.15	2.736	3.019	3.302	3.585	3.866	4.148	4.429	4.709	4.990	5.270	5.550	6.109	6.667	7.226	7.784	8.342
.20	2.699	2.986	3.272	3.558	3.842	4.126	4.409	4.691	4.973	5.254	5.535	6.096	6.656	7.216	7.775	8.333
.25	2.653	2.944	3.235	3.524	3.811	4.098	4.383	4.668	4.951	5.234	5.517	6.080	6.642	7.203	7.763	8.323
.30	2.598	2.894	3.189	3.483	3.775	4.064	4.353	4.639	4.925	5.210	5.494	6.060	6.624	7.187	7.749	8.309
.40	2.464	2.771	3.078	3.382	3.683	3.981	4.276	4.569	4.860	5.150	5.438	6.011	6.580	7.147	7.713	8.276
.50	2.203	2.625	2.944	3.260	3.572	3.879	4.182	4.482	4.780	5.075	5.368	5.949	6.525	7.097	7.667	8.235
.60	2.105	2.384	2.688	3.122	3.444	3.761	4.074	4.381	4.685	4.986	5.285	5.875	6.458	7.037	7.612	8.184
.70	1.979	2.274	2.590	2.874	3.306	3.632	3.953	4.269	4.580	4.887	5.191	5.791	6.383	6.968	7.548	8.125
.80	1.873	2.156	2.458	2.792	3.107	3.496	3.825	4.147	4.465	4.778	5.088	5.698	6.298	6.890	7.477	8.059
.90	1.779	2.057	2.353	2.656	2.995	3.341	3.692	4.021	4.344	4.663	4.978	5.597	6.206	6.805	7.398	7.986
1.00	1.697	1.969	2.260	2.552	2.864	3.204	3.559	3.891	4.220	4.543	4.863	5.491	6.107	6.714	7.313	7.906
1.20	1.546	1.811	2.094	2.380	2.667	2.969	3.293	3.639	3.970	4.299	4.624	5.266	5.895	6.515	7.126	7.731
1.40	1.419	1.673	1.949	2.229	2.505	2.786	3.087	3.406	3.733	4.059	4.386	5.033	5.672	6.302	6.923	7.537
1.60	1.306	1.552	1.820	2.091	2.362	2.631	2.909	3.206	3.518	3.838	4.159	4.803	5.445	6.080	6.709	7.331
1.80	1.207	1.442	1.700	1.965	2.229	2.491	2.757	3.034	3.328	3.634	3.948	4.582	5.220	5.857	6.489	7.117
2.00	1.121	1.346	1.595	1.851	2.108	2.364	2.620	2.884	3.161	3.452	3.753	4.373	5.002	5.637	6.269	6.899
2.20	1.044	1.260	1.498	1.746	1.996	2.246	2.496	2.749	3.012	3.289	3.577	4.179	4.796	5.422	6.052	6.681
2.40	.976	1.181	1.412	1.651	1.892	2.135	2.376	2.626	2.878	3.141	3.416	3.997	4.601	5.217	5.839	6.466
2.60	.916	1.112	1.333	1.563	1.797	2.034	2.267	2.513	2.756	3.007	3.269	3.828	4.417	5.020	5.634	6.255
2.80	.863	1.050	1.261	1.483	1.710	1.935	2.166	2.401	2.638	2.885	3.136	3.673	4.243	4.833	5.438	6.050
3.00	.815	.993	1.195	1.410	1.629	1.848	2.072	2.300	2.532	2.768	3.014	3.529	4.080	4.656	5.248	5.853

ECCENTRIC LOADS ON WELD GROUPS
TABLE XXI   Coefficients C

ANGLE = 0°

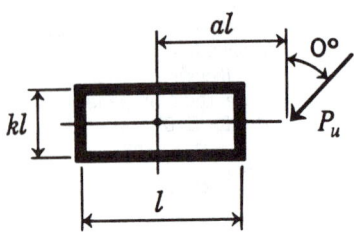

P_u = factored eccentric load in kips
l = length of each weld, in.
D = number of sixteenths of an inch in fillet weld size
C = coefficients tabulated below (includes $\phi = 0.75$)
C_1 = coefficient for electrode used (see Table on page 5-89)
 = 1.0 for E70XX electrodes

$$P_u = CC_1 Dl$$

Required Minimum $C = \dfrac{P_u}{C_1 Dl}$

" " $D = \dfrac{P_u}{CC_1 l}$

" " $l = \dfrac{P_u}{CC_1 D}$

Note: When load P_u is perpendicular to longer side l use Table XXI.

a	\multicolumn{17}{c}{k}
	0	0.1	0.2	0.3	0.4	0.5	0.6	0.7	0.8	0.9	1.0	1.2	1.4	1.6	1.8	2.0
.05	2.702	2.988	3.275	3.560	3.845	4.128	4.411	4.693	4.975	5.256	5.537	6.098	6.658	7.217	7.776	8.335
.10	2.469	2.778	3.086	3.391	3.692	3.990	4.285	4.578	4.869	5.158	5.446	6.019	6.588	7.155	7.720	8.283
.15	2.077	2.357	2.686	3.135	3.461	3.779	4.092	4.400	4.704	5.005	5.303	5.892	6.475	7.053	7.628	8.199
.20	1.854	2.150	2.459	2.811	3.131	3.523	3.853	4.177	4.495	4.809	5.118	5.727	6.326	6.918	7.504	8.085
.25	1.691	1.974	2.274	2.566	2.893	3.241	3.597	3.932	4.261	4.586	4.905	5.533	6.149	6.755	7.353	7.944
.30	1.549	1.824	2.119	2.403	2.692	3.005	3.352	3.688	4.022	4.352	4.679	5.321	5.951	6.570	7.180	7.783
.40	1.318	1.572	1.847	2.119	2.388	2.661	2.951	3.260	3.582	3.906	4.231	4.881	5.525	6.161	6.790	7.411
.50	1.139	1.370	1.626	1.885	2.143	2.399	2.659	2.931	3.219	3.521	3.831	4.462	5.099	5.737	6.373	7.004
.60	.995	1.207	1.441	1.685	1.930	2.175	2.422	2.670	2.930	3.204	3.490	4.088	4.703	5.328	5.958	6.589
.70	.882	1.075	1.293	1.517	1.748	1.976	2.210	2.447	2.688	2.941	3.202	3.760	4.346	4.949	5.563	6.185
.80	.790	.966	1.165	1.376	1.592	1.808	2.030	2.254	2.483	2.717	2.961	3.476	4.028	4.605	5.198	5.803
.90	.713	.875	1.061	1.256	1.458	1.662	1.872	2.085	2.302	2.524	2.751	3.232	3.748	4.296	4.864	5.448
1.00	.650	.799	.971	1.153	1.343	1.538	1.733	1.936	2.142	2.353	2.568	3.021	3.502	4.018	4.561	5.123
1.20	.551	.680	.828	.987	1.155	1.329	1.508	1.691	1.873	2.064	2.260	2.665	3.096	3.552	4.041	4.555
1.40	.477	.590	.721	.861	1.010	1.166	1.327	1.492	1.662	1.837	2.010	2.380	2.774	3.183	3.620	4.085
1.60	.420	.520	.637	.762	.896	1.036	1.182	1.332	1.487	1.646	1.810	2.149	2.507	2.883	3.279	3.699
1.80	.376	.465	.569	.683	.804	.931	1.064	1.201	1.343	1.489	1.639	1.952	2.282	2.630	2.995	3.380
2.00	.339	.421	.515	.619	.729	.845	.966	1.093	1.223	1.357	1.496	1.786	2.091	2.414	2.754	3.110
2.20	.309	.384	.470	.565	.666	.773	.885	1.001	1.122	1.246	1.375	1.643	1.928	2.228	2.545	2.878
2.40	.284	.353	.432	.520	.613	.712	.815	.923	1.035	1.151	1.271	1.521	1.787	2.068	2.364	2.676
2.60	.263	.326	.400	.481	.568	.660	.756	.856	.961	1.069	1.181	1.415	1.664	1.928	2.206	2.499
2.80	.244	.303	.372	.447	.528	.614	.704	.798	.896	.997	1.102	1.322	1.556	1.804	2.066	2.343
3.00	.228	.283	.348	.418	.494	.575	.659	.747	.839	.934	1.033	1.240	1.461	1.695	1.942	2.204

ECCENTRIC LOADS ON WELD GROUPS
TABLE XXI   Coefficients C

ANGLE = 45°

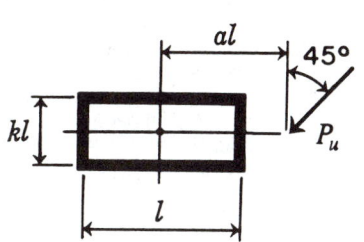

P_u = factored eccentric load in kips
l = length of each weld, in.
D = number of sixteenths of an inch
     in fillet weld size
C = coefficients tabulated below
     (includes ϕ = 0.75)
C_1 = coefficient for electrode used
     (see Table on page 5-89)
     = 1.0 for E70XX electrodes

$$P_u = CC_1Dl$$

Required Minimum $C = \dfrac{P_u}{C_1Dl}$

"   " $D = \dfrac{P_u}{CC_1l}$

"   " $l = \dfrac{P_u}{CC_1D}$

Note: When load P_u is perpendicular to
longer side l use Table XXI.

a	\multicolumn{16}{c}{k}
	0	0.1	0.2	0.3	0.4	0.5	0.6	0.7	0.8	0.9	1.0	1.2	1.4	1.6	1.8	2.0
.05	2.753	3.034	3.315	3.595	3.876	4.155	4.435	4.715	4.994	5.273	5.553	6.111	6.669	7.227	7.784	8.342
.10	2.666	2.954	3.240	3.526	3.811	4.094	4.378	4.661	4.943	5.225	5.506	6.069	6.630	7.191	7.751	8.311
.15	2.536	2.831	3.125	3.417	3.708	3.997	4.286	4.573	4.860	5.146	5.431	6.000	6.567	7.132	7.697	8.260
.20	2.382	2.682	2.981	3.278	3.575	3.870	4.164	4.457	4.749	5.040	5.329	5.906	6.480	7.051	7.621	8.189
.25	2.217	2.523	2.822	3.121	3.421	3.720	4.020	4.318	4.615	4.910	5.205	5.790	6.372	6.951	7.527	8.100
.30	2.064	2.359	2.661	2.956	3.256	3.557	3.859	4.161	4.462	4.762	5.061	5.656	6.246	6.832	7.415	7.995
.40	1.790	2.070	2.352	2.635	2.921	3.222	3.520	3.821	4.125	4.431	4.736	5.345	5.950	6.551	7.147	7.740
.50	1.567	1.829	2.098	2.368	2.639	2.913	3.191	3.474	3.766	4.083	4.390	5.003	5.617	6.228	6.836	7.441
.60	1.382	1.629	1.884	2.143	2.405	2.670	2.936	3.206	3.481	3.761	4.050	4.656	5.274	5.888	6.502	7.114
.70	1.231	1.459	1.703	1.952	2.208	2.462	2.720	2.982	3.246	3.514	3.785	4.342	4.922	5.549	6.164	6.777
.80	1.104	1.319	1.550	1.792	2.036	2.284	2.534	2.786	3.042	3.300	3.561	4.092	4.646	5.217	5.815	6.443
.90	1.001	1.201	1.420	1.651	1.887	2.126	2.368	2.612	2.860	3.110	3.363	3.874	4.401	4.954	5.520	6.110
1.00	.914	1.100	1.309	1.530	1.756	1.986	2.220	2.455	2.695	2.936	3.181	3.676	4.186	4.712	5.264	5.825
1.20	.776	.941	1.129	1.329	1.537	1.749	1.966	2.182	2.409	2.630	2.863	3.330	3.807	4.301	4.810	5.341
1.40	.673	.820	.990	1.172	1.362	1.558	1.757	1.962	2.165	2.375	2.590	3.029	3.485	3.950	4.430	4.924
1.60	.593	.726	.880	1.046	1.219	1.400	1.585	1.774	1.963	2.160	2.361	2.774	3.207	3.646	4.099	4.567
1.80	.530	.651	.791	.943	1.102	1.267	1.440	1.616	1.797	1.977	2.165	2.553	2.955	3.378	3.808	4.252
2.00	.479	.589	.717	.857	1.004	1.158	1.317	1.481	1.650	1.823	2.000	2.360	2.741	3.142	3.551	3.973
2.20	.437	.538	.656	.785	.922	1.064	1.213	1.366	1.524	1.686	1.853	2.197	2.557	2.932	3.321	3.723
2.40	.402	.495	.605	.724	.851	.984	1.123	1.266	1.414	1.567	1.724	2.049	2.389	2.744	3.115	3.499
2.60	.374	.459	.560	.672	.790	.915	1.044	1.179	1.319	1.462	1.610	1.918	2.241	2.579	2.931	3.297
2.80	.345	.427	.522	.626	.737	.854	.976	1.103	1.234	1.370	1.510	1.801	2.107	2.429	2.765	3.115
3.00	.323	.399	.488	.586	.690	.801	.916	1.036	1.160	1.288	1.420	1.697	1.988	2.294	2.615	2.950

ECCENTRIC LOADS ON WELD GROUPS
TABLE XXI   Coefficients C

ANGLE = 75°

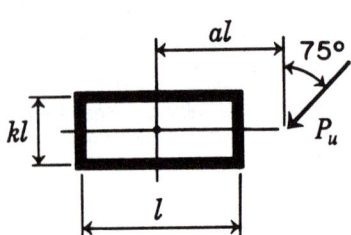

P_u = factored eccentric load in kips
l = length of each weld, in.
D = number of sixteenths of an inch in fillet weld size
C = coefficients tabulated below (includes ϕ = 0.75)
C_1 = coefficient for electrode used (see Table on page 5-89) = 1.0 for E70XX electrodes

$$P_u = CC_1 Dl$$

Required Minimum $C = \dfrac{P_u}{C_1 Dl}$

"   " $D = \dfrac{P_u}{CC_1 l}$

"   " $l = \dfrac{P_u}{CC_1 D}$

Note: When load P_u is perpendicular to longer side l use Table XXI.

a	k															
	0	0.1	0.2	0.3	0.4	0.5	0.6	0.7	0.8	0.9	1.0	1.2	1.4	1.6	1.8	2.0
.05	2.781	3.060	3.338	3.617	3.895	4.174	4.452	4.731	5.009	5.288	5.566	6.123	6.680	7.237	7.794	8.351
.10	2.772	3.052	3.331	3.609	3.888	4.167	4.445	4.724	5.003	5.281	5.560	6.117	6.675	7.232	7.789	8.346
.15	2.758	3.038	3.317	3.597	3.876	4.155	4.434	4.713	4.992	5.271	5.550	6.107	6.665	7.223	7.781	8.338
.20	2.737	3.019	3.300	3.579	3.859	4.138	4.418	4.697	4.976	5.256	5.535	6.094	6.652	7.211	7.769	8.327
.25	2.712	2.996	3.277	3.557	3.837	4.117	4.397	4.677	4.957	5.237	5.517	6.076	6.636	7.195	7.754	8.313
.30	2.682	2.967	3.250	3.531	3.811	4.092	4.373	4.653	4.933	5.214	5.494	6.055	6.615	7.176	7.736	8.296
.40	2.608	2.898	3.183	3.466	3.748	4.030	4.311	4.593	4.874	5.156	5.438	6.001	6.564	7.127	7.690	8.252
.50	2.522	2.815	3.103	3.387	3.671	3.953	4.236	4.519	4.801	5.084	5.368	5.934	6.500	7.066	7.631	8.196
.60	2.425	2.723	3.012	3.298	3.582	3.865	4.149	4.432	4.716	5.000	5.285	5.854	6.423	6.992	7.561	8.129
.70	2.323	2.623	2.914	3.201	3.485	3.769	4.052	4.336	4.620	4.905	5.191	5.763	6.335	6.908	7.479	8.051
.80	2.219	2.520	2.812	3.098	3.382	3.665	3.949	4.232	4.517	4.802	5.088	5.662	6.237	6.812	7.388	7.962
.90	2.115	2.415	2.707	2.993	3.276	3.558	3.840	4.123	4.407	4.692	4.978	5.553	6.130	6.708	7.286	7.865
1.00	2.013	2.311	2.602	2.887	3.169	3.450	3.730	4.011	4.293	4.577	4.863	5.437	6.015	6.596	7.177	7.758
1.20	1.821	2.111	2.397	2.679	2.957	3.233	3.509	3.785	4.063	4.342	4.624	5.194	5.770	6.352	6.937	7.523
1.40	1.648	1.928	2.207	2.482	2.755	3.025	3.295	3.565	3.837	4.111	4.386	4.944	5.514	6.092	6.676	7.265
1.60	1.498	1.764	2.033	2.301	2.567	2.831	3.096	3.360	3.623	3.891	4.159	4.703	5.259	5.827	6.406	6.992
1.80	1.367	1.619	1.876	2.137	2.396	2.654	2.911	3.170	3.428	3.687	3.948	4.477	5.017	5.570	6.109	6.682
2.00	1.250	1.492	1.738	1.989	2.241	2.493	2.743	2.995	3.247	3.500	3.753	4.267	4.790	5.326	5.851	6.408
2.20	1.152	1.380	1.616	1.857	2.101	2.346	2.590	2.835	3.081	3.328	3.577	4.079	4.587	5.105	5.620	6.155
2.40	1.067	1.279	1.507	1.739	1.975	2.213	2.451	2.691	2.931	3.173	3.416	3.908	4.406	4.911	5.424	5.937
2.60	.992	1.192	1.410	1.633	1.862	2.093	2.325	2.559	2.794	3.031	3.269	3.752	4.242	4.738	5.238	5.748
2.80	.926	1.116	1.324	1.539	1.760	1.984	2.210	2.439	2.669	2.902	3.136	3.611	4.091	4.579	5.073	5.573
3.00	.868	1.048	1.247	1.454	1.668	1.886	2.107	2.330	2.556	2.784	3.014	3.480	3.954	4.434	4.920	5.413

ECCENTRIC LOADS ON WELD GROUPS
TABLE XXII   Coefficients C

ANGLE = 0°

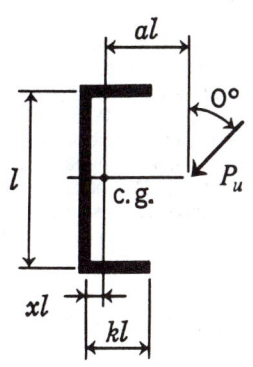

P_u = factored eccentric load in kips
l = length of each weld, in.
D = number of sixteenths of an inch in fillet weld size
C = coefficients tabulated below (includes $\phi = 0.75$)
C_1 = coefficient for electrode used (see Table on page 5-89)
= 1.0 for E70XX electrodes
xl = distance from vertical weld to center of gravity of weld group

$$P_u = CC_1Dl$$

Required Minimum $C = \dfrac{P_u}{C_1Dl}$

" " $D = \dfrac{P_u}{CC_1l}$

" " $l = \dfrac{P_u}{CC_1D}$

a	\multicolumn{16}{c}{k}															
	0	0.1	0.2	0.3	0.4	0.5	0.6	0.7	0.8	0.9	1.0	1.2	1.4	1.6	1.8	2.0
.05	1.371	1.652	1.930	2.208	2.486	2.764	3.041	3.319	3.597	3.875	4.153	4.709	5.266	5.823	6.380	6.938
.10	1.314	1.599	1.877	2.153	2.429	2.704	2.979	3.255	3.531	3.808	4.084	4.640	5.196	5.754	6.312	6.871
.15	1.229	1.518	1.795	2.068	2.339	2.611	2.882	3.154	3.427	3.701	3.976	4.529	5.084	5.642	6.202	6.762
.20	1.132	1.421	1.694	1.962	2.228	2.494	2.760	3.027	3.295	3.565	3.836	4.383	4.936	5.493	6.053	6.616
.25	1.032	1.318	1.585	1.845	2.104	2.362	2.621	2.881	3.143	3.408	3.674	4.213	4.760	5.314	5.873	6.436
.30	.937	1.215	1.473	1.725	1.975	2.225	2.475	2.728	2.983	3.240	3.500	4.027	4.566	5.113	5.669	6.231
.40	.774	1.028	1.265	1.497	1.726	1.957	2.190	2.425	2.663	2.904	3.148	3.647	4.159	4.685	5.223	5.773
.50	.649	.876	1.090	1.300	1.509	1.720	1.935	2.152	2.373	2.598	2.826	3.293	3.774	4.270	4.760	5.278
.60	.553	.756	.948	1.137	1.328	1.521	1.718	1.920	2.125	2.334	2.547	2.985	3.437	3.903	4.371	4.858
.70	.481	.662	.834	1.006	1.179	1.357	1.538	1.724	1.915	2.110	2.310	2.723	3.152	3.596	4.053	4.521
.80	.424	.585	.742	.899	1.057	1.221	1.388	1.560	1.738	1.922	2.110	2.503	2.913	3.339	3.780	4.232
.90	.379	.525	.668	.811	.957	1.107	1.262	1.423	1.590	1.763	1.941	2.314	2.709	3.121	3.546	3.983
1.00	.342	.475	.606	.737	.872	1.012	1.156	1.307	1.464	1.627	1.796	2.154	2.533	2.930	3.342	3.768
1.20	.286	.398	.510	.623	.740	.861	.989	1.122	1.263	1.410	1.563	1.890	2.238	2.605	2.988	3.386
1.40	.246	.343	.440	.539	.642	.749	.863	.983	1.110	1.243	1.384	1.682	2.003	2.342	2.697	3.068
1.60	.215	.301	.386	.474	.566	.663	.765	.874	.990	1.112	1.240	1.514	1.808	2.121	2.450	2.795
1.80	.191	.268	.344	.423	.506	.594	.687	.787	.893	1.005	1.123	1.374	1.645	1.935	2.241	2.562
2.00	.172	.241	.311	.382	.458	.538	.624	.715	.813	.916	1.025	1.257	1.508	1.776	2.061	2.361
2.20	.157	.219	.283	.349	.418	.492	.571	.656	.746	.841	.942	1.157	1.390	1.640	1.906	2.186
2.40	.144	.201	.260	.320	.384	.453	.526	.605	.689	.778	.871	1.072	1.289	1.522	1.771	2.039
2.60	.133	.187	.240	.296	.356	.420	.488	.562	.640	.723	.810	.997	1.201	1.419	1.657	1.904
2.80	.123	.173	.223	.275	.331	.391	.455	.524	.597	.675	.757	.932	1.123	1.332	1.552	1.786
3.00	.115	.161	.208	.257	.310	.366	.426	.491	.560	.633	.710	.875	1.057	1.252	1.459	1.680
x	.000	.008	.029	.056	.089	.125	.164	.204	.246	.289	.333	.424	.516	.610	.704	.800

ECCENTRIC LOADS ON WELD GROUPS
TABLE XXII   Coefficients C

ANGLE $= 45°$

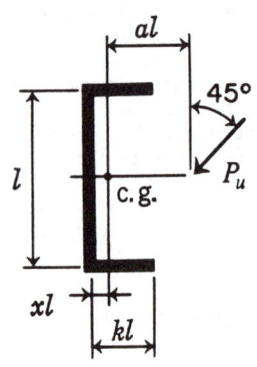

P_u = factored eccentric load in kips
l = length of each weld, in.
D = number of sixteenths of an inch in fillet weld size
C = coefficients tabulated below (includes $\phi = 0.75$)
C_1 = coefficient for electrode used (see Table on page 5-89)
   = 1.0 for E70XX electrodes
xl = distance from vertical weld to center of gravity of weld group

$$P_u = CC_1Dl$$

Required Minimum $C = \dfrac{P_u}{C_1Dl}$

"   " $D = \dfrac{P_u}{CC_1l}$

"   " $l = \dfrac{P_u}{CC_1D}$

a	\multicolumn{17}{c}{k}
	0	0.1	0.2	0.3	0.4	0.5	0.6	0.7	0.8	0.9	1.0	1.2	1.4	1.6	1.8	2.0
.05	1.377	1.657	1.935	2.214	2.492	2.771	3.049	3.328	3.606	3.885	4.163	4.721	5.278	5.835	6.393	6.950
.10	1.333	1.617	1.896	2.175	2.453	2.731	3.010	3.288	3.567	3.847	4.126	4.684	5.243	5.802	6.361	6.920
.15	1.268	1.556	1.835	2.113	2.391	2.668	2.947	3.225	3.505	3.784	4.064	4.625	5.186	5.747	6.309	6.870
.20	1.191	1.480	1.758	2.034	2.310	2.586	2.863	3.141	3.421	3.701	3.981	4.544	5.108	5.672	6.237	6.801
.25	1.109	1.398	1.672	1.944	2.216	2.489	2.764	3.041	3.319	3.599	3.880	4.445	5.011	5.579	6.147	6.715
.30	1.032	1.311	1.583	1.848	2.114	2.383	2.654	2.928	3.205	3.483	3.764	4.329	4.898	5.469	6.041	6.613
.40	.895	1.158	1.406	1.652	1.901	2.154	2.421	2.684	2.953	3.225	3.502	4.064	4.635	5.210	5.789	6.368
.50	.783	1.027	1.256	1.483	1.713	1.947	2.185	2.433	2.685	2.946	3.226	3.775	4.341	4.916	5.498	6.082
.60	.691	.917	1.130	1.340	1.553	1.770	1.992	2.220	2.454	2.695	2.942	3.463	4.017	4.608	5.187	5.773
.70	.616	.825	1.022	1.219	1.416	1.620	1.827	2.042	2.262	2.489	2.722	3.208	3.723	4.272	4.847	5.457
.80	.552	.747	.931	1.113	1.300	1.490	1.687	1.890	2.102	2.317	2.540	3.006	3.496	4.003	4.547	5.114
.90	.501	.681	.852	1.025	1.199	1.379	1.565	1.760	1.960	2.167	2.381	2.830	3.299	3.786	4.295	4.834
1.00	.457	.624	.785	.946	1.111	1.281	1.460	1.644	1.835	2.034	2.240	2.672	3.124	3.594	4.083	4.594
1.20	.388	.534	.676	.819	.966	1.121	1.282	1.450	1.626	1.810	2.000	2.400	2.822	3.261	3.716	4.189
1.40	.336	.465	.591	.721	.854	.993	1.140	1.294	1.455	1.625	1.801	2.174	2.570	2.980	3.408	3.853
1.60	.297	.411	.525	.642	.763	.890	1.024	1.166	1.315	1.472	1.637	1.983	2.350	2.741	3.141	3.559
1.80	.265	.368	.472	.578	.688	.805	.930	1.060	1.199	1.345	1.498	1.820	2.164	2.525	2.911	3.305
2.00	.240	.334	.428	.525	.626	.735	.849	.971	1.100	1.236	1.379	1.679	2.001	2.343	2.701	3.081
2.20	.218	.305	.391	.481	.575	.675	.782	.896	1.016	1.143	1.275	1.558	1.860	2.182	2.521	2.876
2.40	.201	.280	.360	.443	.531	.624	.724	.830	.943	1.062	1.186	1.452	1.736	2.040	2.360	2.703
2.60	.187	.259	.334	.411	.493	.581	.674	.774	.880	.991	1.109	1.357	1.627	1.913	2.217	2.544
2.80	.173	.241	.311	.383	.460	.542	.630	.724	.824	.929	1.039	1.275	1.529	1.800	2.090	2.400
3.00	.161	.226	.291	.359	.431	.508	.592	.680	.774	.873	.978	1.201	1.442	1.700	1.975	2.269
x	.000	.008	.029	.056	.089	.125	.164	.204	.246	.289	.333	.424	.516	.610	.704	.800

ECCENTRIC LOADS ON WELD GROUPS
TABLE XXII   Coefficients C

ANGLE = 75°

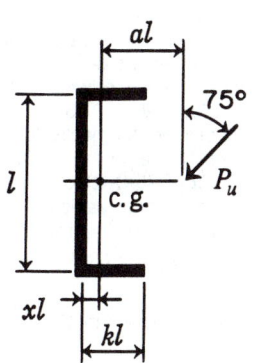

P_u = factored eccentric load in kips
l = length of each weld, in.
D = number of sixteenths of an inch in fillet weld size
C = coefficients tabulated below (includes $\phi = 0.75$)
C_1 = coefficient for electrode used (see Table on page 5-89)
  = 1.0 for E70XX electrodes
xl = distance from vertical weld to center of gravity of weld group

$$P_u = CC_1Dl$$

Required Minimum $C = \dfrac{P_u}{C_1Dl}$

"   " $D = \dfrac{P_u}{CC_1l}$

"   " $l = \dfrac{P_u}{CC_1D}$

a	\								k								
	0	0.1	0.2	0.3	0.4	0.5	0.6	0.7	0.8	0.9	1.0	1.2	1.4	1.6	1.8	2.0
.05	1.389	1.668	1.947	2.225	2.504	2.782	3.060	3.339	3.617	3.896	4.174	4.731	5.288	5.845	6.402	6.959
.10	1.381	1.661	1.940	2.218	2.497	2.775	3.054	3.333	3.611	3.890	4.169	4.726	5.284	5.841	6.398	6.956
.15	1.368	1.649	1.928	2.207	2.486	2.765	3.044	3.323	3.602	3.881	4.160	4.718	5.276	5.834	6.392	6.949
.20	1.350	1.632	1.912	2.191	2.470	2.749	3.029	3.308	3.588	3.867	4.147	4.706	5.265	5.824	6.382	6.941
.25	1.326	1.611	1.892	2.171	2.451	2.730	3.010	3.290	3.570	3.851	4.131	4.691	5.251	5.811	6.371	6.930
.30	1.299	1.586	1.868	2.147	2.427	2.707	2.987	3.268	3.549	3.830	4.111	4.673	5.234	5.796	6.356	6.916
.40	1.232	1.525	1.808	2.088	2.369	2.650	2.931	3.214	3.496	3.779	4.061	4.627	5.192	5.757	6.320	6.883
.50	1.102	1.451	1.735	2.016	2.297	2.579	2.862	3.146	3.430	3.715	4.000	4.570	5.139	5.707	6.274	6.840
.60	1.052	1.299	1.559	1.933	2.215	2.498	2.782	3.067	3.354	3.640	3.927	4.502	5.076	5.648	6.219	6.789
.70	.989	1.256	1.483	1.740	2.125	2.408	2.693	2.979	3.267	3.556	3.846	4.425	5.003	5.580	6.156	6.729
.80	.936	1.190	1.443	1.684	1.945	2.216	2.597	2.885	3.174	3.465	3.756	4.340	4.923	5.505	6.085	6.663
.90	.890	1.134	1.372	1.623	1.902	2.161	2.427	2.786	3.076	3.368	3.661	4.248	4.836	5.423	6.007	6.589
1.00	.848	1.085	1.312	1.550	1.803	2.083	2.345	2.640	2.974	3.267	3.561	4.152	4.744	5.334	5.923	6.510
1.20	.773	1.000	1.214	1.433	1.666	1.912	2.180	2.465	2.755	3.048	3.356	3.950	4.457	5.144	5.741	6.335
1.40	.709	.925	1.130	1.338	1.559	1.789	2.029	2.287	2.563	2.850	3.141	3.744	4.342	4.943	5.545	6.145
1.60	.653	.859	1.055	1.254	1.465	1.684	1.913	2.151	2.404	2.673	2.954	3.543	4.137	4.738	5.341	5.945
1.80	.604	.801	.988	1.178	1.379	1.590	1.809	2.036	2.274	2.527	2.789	3.358	3.940	4.534	5.136	5.741
2.00	.561	.748	.927	1.111	1.303	1.504	1.714	1.933	2.159	2.396	2.648	3.186	3.753	4.336	4.934	5.536
2.20	.522	.701	.873	1.049	1.233	1.426	1.629	1.838	2.056	2.282	2.520	3.030	3.579	4.151	4.738	5.334
2.40	.488	.659	.824	.993	1.170	1.355	1.550	1.753	1.962	2.179	2.405	2.891	3.418	3.975	4.550	5.138
2.60	.458	.621	.779	.942	1.111	1.287	1.476	1.672	1.877	2.085	2.301	2.766	3.270	3.810	4.372	4.950
2.80	.431	.586	.739	.894	1.056	1.228	1.410	1.599	1.795	1.995	2.207	2.651	3.136	3.656	4.203	4.767
3.00	.407	.555	.702	.851	1.007	1.172	1.347	1.530	1.720	1.915	2.118	2.546	3.012	3.513	4.043	4.597
x	.000	.008	.029	.056	.089	.125	.164	.204	.246	.289	.333	.424	.516	.610	.704	.800

ECCENTRIC LOADS ON WELD GROUPS
TABLE XXIII   Coefficients C

ANGLE $= 0°$

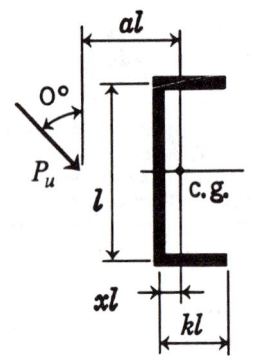

P_u = factored eccentric load in kips
l = length of each weld, in.
D = number of sixteenths of an inch in fillet weld size
C = coefficients tabulated below (includes $\phi = 0.75$)
C_1 = coefficient for electrode used (see Table on page 5-89)
    = 1.0 for E70XX electrodes
xl = distance from vertical weld to center of gravity of weld group

$$P_u = CC_1Dl$$

Required Minimum $C = \dfrac{P_u}{C_1Dl}$

    "   " $D = \dfrac{P_u}{CC_1l}$

    "   " $l = \dfrac{P_u}{CC_1D}$

a	k															
	0	0.1	0.2	0.3	0.4	0.5	0.6	0.7	0.8	0.9	1.0	1.2	1.4	1.6	1.8	2.0
.05	1.371	1.652	1.931	2.209	2.486	2.764	3.042	3.319	3.597	3.875	4.153	4.710	5.266	5.823	6.381	6.938
.10	1.314	1.600	1.879	2.156	2.432	2.708	2.983	3.259	3.534	3.811	4.087	4.642	5.198	5.755	6.313	6.872
.15	1.229	1.521	1.801	2.077	2.350	2.623	2.894	3.166	3.439	3.712	3.986	4.536	5.090	5.646	6.205	6.765
.20	1.132	1.426	1.706	1.980	2.250	2.518	2.785	3.053	3.320	3.589	3.858	4.402	4.950	5.504	6.062	6.622
.25	1.032	1.325	1.602	1.873	2.139	2.403	2.665	2.927	3.189	3.452	3.716	4.249	4.790	5.337	5.891	6.450
.30	.937	1.223	1.496	1.762	2.024	2.283	2.540	2.797	3.053	3.311	3.569	4.090	4.619	5.157	5.703	6.257
.40	.774	1.038	1.295	1.547	1.797	2.045	2.292	2.538	2.784	3.031	3.278	3.776	4.280	4.793	5.315	5.848
.50	.649	.886	1.121	1.356	1.591	1.826	2.061	2.296	2.532	2.768	3.006	3.484	3.968	4.459	4.957	5.465
.60	.553	.765	.978	1.194	1.413	1.633	1.855	2.078	2.303	2.529	2.757	3.216	3.683	4.156	4.636	5.120
.70	.481	.670	.862	1.060	1.261	1.467	1.675	1.886	2.099	2.315	2.532	2.973	3.423	3.879	4.344	4.816
.80	.424	.592	.768	.948	1.134	1.325	1.520	1.718	1.920	2.125	2.332	2.754	3.186	3.627	4.076	4.534
.90	.379	.531	.690	.855	1.027	1.204	1.387	1.573	1.763	1.957	2.154	2.557	2.972	3.397	3.831	4.275
1.00	.342	.480	.626	.778	.936	1.101	1.271	1.447	1.626	1.809	1.996	2.381	2.778	3.187	3.607	4.036
1.20	.286	.403	.526	.656	.793	.936	1.085	1.240	1.400	1.564	1.732	2.081	2.445	2.823	3.213	3.615
1.40	.246	.346	.452	.565	.685	.811	.943	1.081	1.224	1.371	1.523	1.840	2.173	2.521	2.883	3.258
1.60	.215	.303	.396	.496	.602	.714	.832	.956	1.084	1.217	1.355	1.644	1.949	2.270	2.606	2.955
1.80	.191	.270	.353	.442	.536	.637	.744	.855	.972	1.093	1.218	1.483	1.763	2.060	2.372	2.697
2.00	.172	.243	.317	.398	.483	.575	.671	.773	.879	.990	1.105	1.348	1.607	1.882	2.172	2.476
2.20	.157	.221	.289	.362	.440	.523	.611	.705	.802	.904	1.010	1.235	1.475	1.731	2.001	2.286
2.40	.144	.202	.264	.332	.403	.480	.561	.647	.737	.832	.930	1.138	1.362	1.600	1.853	2.120
2.60	.133	.189	.244	.306	.372	.443	.518	.598	.682	.770	.861	1.055	1.264	1.487	1.724	1.975
2.80	.123	.174	.227	.284	.346	.411	.482	.556	.634	.716	.801	.983	1.179	1.388	1.612	1.848
3.00	.115	.162	.211	.265	.322	.384	.450	.519	.592	.669	.749	.920	1.104	1.301	1.512	1.735
x	.000	.008	.029	.056	.089	.125	.164	.204	.246	.289	.333	.424	.516	.610	.704	.800

ECCENTRIC LOADS ON WELD GROUPS
TABLE XXIII   Coefficients C

ANGLE $= 45°$

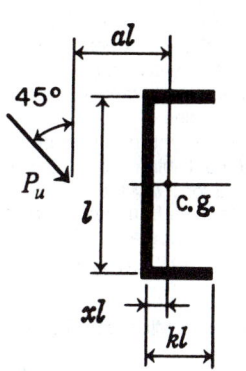

P_u = factored eccentric load in kips
l = length of each weld, in.
D = number of sixteenths of an inch in fillet weld size
C = coefficients tabulated below (includes $\phi = 0.75$)
C_1 = coefficient for electrode used (see Table on page 5-89)
$\quad$ = 1.0 for E70XX electrodes
xl = distance from vertical weld to center of gravity of weld group

$$P_u = CC_1Dl$$

Required Minimum $C = \dfrac{P_u}{C_1Dl}$

$\quad$ " $\quad$ " $\quad D = \dfrac{P_u}{CC_1l}$

$\quad$ " $\quad$ " $\quad l = \dfrac{P_u}{CC_1D}$

a	k															
	0	0.1	0.2	0.3	0.4	0.5	0.6	0.7	0.8	0.9	1.0	1.2	1.4	1.6	1.8	2.0
.05	1.377	1.657	1.935	2.214	2.492	2.771	3.049	3.328	3.606	3.885	4.163	4.721	5.278	5.835	6.393	6.950
.10	1.333	1.617	1.896	2.174	2.453	2.731	3.009	3.288	3.567	3.846	4.126	4.684	5.243	5.802	6.361	6.920
.15	1.268	1.555	1.834	2.112	2.389	2.667	2.946	3.225	3.504	3.784	4.064	4.625	5.186	5.748	6.309	6.871
.20	1.191	1.480	1.757	2.032	2.307	2.584	2.861	3.140	3.420	3.700	3.981	4.544	5.109	5.673	6.238	6.802
.25	1.109	1.398	1.670	1.940	2.211	2.484	2.760	3.038	3.317	3.598	3.880	4.445	5.013	5.581	6.149	6.717
.30	1.032	1.312	1.576	1.843	2.108	2.376	2.647	2.923	3.201	3.481	3.763	4.331	4.901	5.472	6.045	6.616
.40	.895	1.162	1.411	1.652	1.890	2.136	2.392	2.658	2.943	3.220	3.500	4.067	4.641	5.218	5.796	6.376
.50	.783	1.034	1.271	1.501	1.726	1.952	2.187	2.431	2.686	2.955	3.226	3.786	4.355	4.932	5.514	6.097
.60	.691	.924	1.151	1.371	1.590	1.808	2.029	2.258	2.496	2.744	2.998	3.526	4.078	4.642	5.219	5.800
.70	.616	.833	1.046	1.257	1.468	1.679	1.893	2.111	2.335	2.567	2.807	3.307	3.829	4.375	4.934	5.509
.80	.552	.754	.955	1.156	1.358	1.562	1.769	1.979	2.193	2.413	2.640	3.115	3.615	4.134	4.674	5.232
.90	.501	.688	.876	1.067	1.260	1.456	1.655	1.857	2.065	2.276	2.493	2.944	3.422	3.920	4.439	4.977
1.00	.457	.631	.806	.987	1.172	1.360	1.552	1.747	1.946	2.150	2.358	2.791	3.249	3.727	4.225	4.743
1.20	.388	.539	.694	.855	1.022	1.194	1.371	1.552	1.737	1.926	2.120	2.523	2.945	3.389	3.850	4.333
1.40	.336	.469	.607	.751	.902	1.059	1.221	1.388	1.560	1.736	1.917	2.291	2.687	3.104	3.533	3.982
1.60	.297	.415	.538	.667	.805	.948	1.097	1.251	1.410	1.574	1.742	2.093	2.462	2.856	3.259	3.680
1.80	.265	.371	.483	.600	.724	.856	.993	1.135	1.283	1.436	1.593	1.921	2.267	2.638	3.018	3.416
2.00	.240	.336	.437	.544	.658	.779	.905	1.040	1.178	1.320	1.467	1.775	2.101	2.445	2.806	3.181
2.20	.218	.306	.399	.498	.602	.714	.833	.956	1.084	1.217	1.355	1.644	1.951	2.274	2.616	2.974
2.40	.201	.282	.367	.458	.555	.658	.769	.883	1.003	1.128	1.257	1.528	1.818	2.124	2.447	2.787
2.60	.187	.261	.339	.424	.514	.610	.713	.821	.933	1.049	1.171	1.427	1.700	1.990	2.296	2.619
2.80	.173	.242	.316	.394	.479	.570	.665	.766	.871	.981	1.095	1.337	1.595	1.870	2.161	2.467
3.00	.161	.227	.295	.369	.447	.533	.623	.717	.817	.920	1.028	1.257	1.502	1.763	2.040	2.332
x	.000	.008	.029	.056	.089	.125	.164	.204	.246	.289	.333	.424	.516	.610	.704	.800

ECCENTRIC LOADS ON WELD GROUPS
TABLE XXIII   Coefficients C

ANGLE = 75°

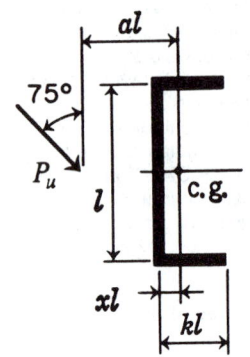

P_u = factored eccentric load in kips
l = length of each weld, in.
D = number of sixteenths of an inch in fillet weld size
C = coefficients tabulated below (includes ϕ = 0.75)
C_1 = coefficient for electrode used (see Table on page 5-89)
= 1.0 for E70XX electrodes
xl = distance from vertical weld to center of gravity of weld group

$$P_u = CC_1Dl$$

Required Minimum $C = \dfrac{P_u}{C_1Dl}$

" " $D = \dfrac{P_u}{CC_1l}$

" " $l = \dfrac{P_u}{CC_1D}$

a	k 0	0.1	0.2	0.3	0.4	0.5	0.6	0.7	0.8	0.9	1.0	1.2	1.4	1.6	1.8	2.0
.05	1.389	1.668	1.947	2.225	2.504	2.782	3.060	3.339	3.617	3.896	4.174	4.731	5.288	5.845	6.402	6.959
.10	1.381	1.661	1.940	2.218	2.497	2.775	3.054	3.333	3.611	3.890	4.169	4.726	5.284	5.841	6.398	6.956
.15	1.368	1.649	1.928	2.207	2.486	2.765	3.043	3.322	3.601	3.881	4.160	4.718	5.276	5.834	6.392	6.949
.20	1.350	1.632	1.912	2.191	2.470	2.749	3.029	3.308	3.588	3.867	4.147	4.706	5.265	5.824	6.382	6.941
.25	1.326	1.611	1.892	2.171	2.450	2.730	3.010	3.290	3.570	3.850	4.131	4.691	5.251	5.811	6.371	6.930
.30	1.299	1.586	1.867	2.147	2.426	2.706	2.987	3.268	3.548	3.829	4.111	4.673	5.234	5.796	6.356	6.917
.40	1.232	1.524	1.806	2.086	2.367	2.648	2.930	3.212	3.495	3.778	4.061	4.627	5.192	5.757	6.320	6.883
.50	1.102	1.449	1.731	2.012	2.293	2.576	2.859	3.144	3.428	3.714	3.999	4.569	5.139	5.707	6.274	6.841
.60	1.052	1.291	1.538	1.926	2.208	2.492	2.777	3.063	3.350	3.638	3.926	4.501	5.075	5.648	6.219	6.789
.70	.989	1.247	1.481	1.735	2.013	2.399	2.686	2.974	3.263	3.553	3.843	4.424	5.003	5.581	6.156	6.730
.80	.936	1.187	1.423	1.668	1.960	2.227	2.588	2.878	3.169	3.461	3.754	4.339	4.923	5.506	6.086	6.664
.90	.890	1.133	1.362	1.606	1.883	2.148	2.444	2.778	3.070	3.364	3.658	4.248	4.837	5.424	6.009	6.591
1.00	.848	1.086	1.312	1.542	1.801	2.082	2.371	2.638	2.970	3.264	3.560	4.152	4.745	5.337	5.926	6.512
1.20	.773	1.003	1.221	1.441	1.672	1.921	2.190	2.473	2.760	3.065	3.360	3.954	4.552	5.150	5.746	6.340
1.40	.709	.928	1.140	1.352	1.570	1.802	2.048	2.311	2.586	2.879	3.167	3.756	4.353	4.953	5.554	6.153
1.60	.653	.863	1.067	1.273	1.481	1.698	1.931	2.178	2.438	2.711	2.991	3.565	4.156	4.754	5.355	5.958
1.80	.604	.804	1.000	1.199	1.401	1.609	1.827	2.059	2.305	2.562	2.830	3.389	3.966	4.558	5.156	5.759
2.00	.561	.751	.939	1.131	1.328	1.526	1.735	1.955	2.187	2.432	2.687	3.225	3.788	4.369	4.960	5.559
2.20	.522	.704	.884	1.069	1.259	1.454	1.652	1.861	2.082	2.314	2.558	3.073	3.621	4.187	4.770	5.363
2.40	.488	.661	.834	1.012	1.195	1.383	1.577	1.777	1.987	2.208	2.441	2.935	3.464	4.018	4.587	5.172
2.60	.458	.623	.789	.959	1.137	1.319	1.507	1.700	1.902	2.112	2.334	2.809	3.318	3.856	4.414	4.987
2.80	.431	.589	.747	.911	1.083	1.258	1.441	1.629	1.823	2.025	2.237	2.692	3.183	3.704	4.249	4.811
3.00	.407	.557	.709	.867	1.032	1.203	1.379	1.562	1.750	1.945	2.148	2.584	3.058	3.562	4.093	4.643
x	.000	.008	.029	.056	.089	.125	.164	.204	.246	.289	.333	.424	.516	.610	.704	.800

ECCENTRIC LOADS ON WELD GROUPS
TABLE XXIV   Coefficients C

ANGLE = 0°

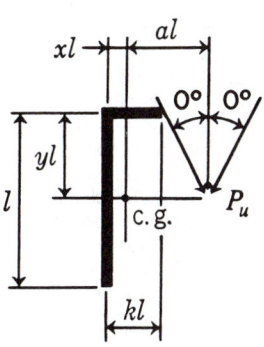

P_u = factored eccentric load in kips
l = length of each weld, in.
D = number of sixteenths of an inch in fillet weld size
C^* = coefficients tabulated below (includes ϕ = 0.75)
C_1 = coefficient for electrode used (see Table on page 5-89)
= 1.0 for E70XX electrodes
xl = distance from vertical weld to center of gravity of weld group
yl = distance from horizontal weld to center of gravity of weld group

$$P_u = CC_1 Dl$$

Required Minimum $C = \dfrac{P_u}{C_1 Dl}$

"   " $D = \dfrac{P_u}{CC_1 l}$

"   " $l = \dfrac{P_u}{CC_1 D}$

a	k															
	0	0.1	0.2	0.3	0.4	0.5	0.6	0.7	0.8	0.9	1.0	1.2	1.4	1.6	1.8	2.0
.05	1.371	1.512	1.651	1.789	1.927	2.065	2.204	2.342	2.481	2.620	2.760	3.039	3.318	3.598	3.878	4.158
.10	1.314	1.456	1.594	1.730	1.866	2.001	2.137	2.274	2.411	2.550	2.689	2.970	3.253	3.537	3.821	4.105
.15	1.229	1.373	1.509	1.642	1.774	1.906	2.038	2.171	2.307	2.444	2.582	2.864	3.151	3.439	3.729	4.019
.20	1.132	1.276	1.408	1.537	1.663	1.790	1.918	2.047	2.179	2.313	2.450	2.731	3.019	3.312	3.608	3.904
.25	1.032	1.173	1.301	1.424	1.545	1.666	1.788	1.913	2.041	2.171	2.305	2.582	2.869	3.164	3.463	3.765
.30	.937	1.074	1.196	1.313	1.427	1.542	1.659	1.779	1.901	2.027	2.157	2.427	2.710	3.002	3.303	3.608
.40	.774	.898	1.008	1.111	1.213	1.316	1.422	1.531	1.643	1.757	1.871	2.119	2.379	2.654	2.941	3.188
.50	.649	.759	.856	.948	1.038	1.130	1.225	1.324	1.426	1.531	1.637	1.864	2.104	2.356	2.620	2.885
.60	.553	.652	.738	.819	.900	.982	1.068	1.158	1.251	1.348	1.449	1.664	1.888	2.125	2.370	2.625
.70	.481	.567	.643	.717	.790	.864	.942	1.024	1.110	1.202	1.297	1.500	1.716	1.942	2.177	2.420
.80	.424	.501	.570	.636	.702	.769	.841	.917	.998	1.083	1.173	1.366	1.573	1.792	2.019	2.254
.90	.379	.448	.511	.569	.630	.692	.758	.830	.906	.986	1.071	1.255	1.452	1.664	1.883	2.110
1.00	.342	.405	.462	.516	.571	.629	.691	.758	.829	.905	.986	1.160	1.351	1.553	1.764	1.982
1.20	.286	.339	.387	.434	.481	.531	.586	.644	.709	.777	.850	1.010	1.182	1.366	1.560	1.762
1.40	.246	.292	.333	.374	.415	.460	.508	.561	.619	.681	.747	.891	1.049	1.219	1.398	1.586
1.60	.215	.256	.293	.329	.365	.405	.449	.497	.549	.605	.665	.798	.942	1.098	1.263	1.436
1.80	.191	.227	.261	.293	.326	.362	.402	.445	.493	.545	.600	.721	.853	.996	1.150	1.312
2.00	.172	.205	.235	.264	.294	.327	.364	.404	.448	.495	.545	.656	.778	.912	1.054	1.204
2.20	.157	.188	.214	.241	.268	.299	.332	.369	.410	.453	.500	.602	.716	.840	.972	1.112
2.40	.144	.171	.196	.221	.246	.275	.305	.340	.377	.418	.461	.557	.662	.777	.900	1.033
2.60	.133	.158	.181	.204	.228	.254	.283	.315	.350	.388	.428	.517	.615	.723	.839	.963
2.80	.123	.146	.168	.190	.212	.237	.263	.293	.326	.362	.399	.483	.574	.676	.785	.902
3.00	.115	.137	.157	.177	.198	.221	.246	.275	.306	.339	.374	.453	.538	.635	.737	.848
x	.000	.005	.017	.035	.057	.083	.113	.144	.178	.213	.250	.327	.408	.492	.579	.667
y	.500	.455	.417	.385	.357	.333	.313	.294	.278	.263	.250	.227	.208	.192	.179	.167

*Tabulated values are the minimum of either − or + load angles.

ECCENTRIC LOADS ON WELD GROUPS
TABLE XXIV Coefficients C

ANGLE = 45°

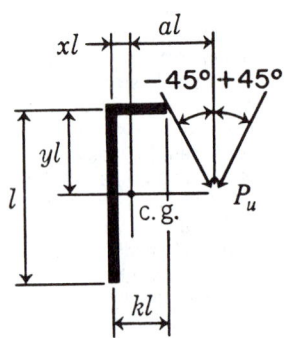

P_u = factored eccentric load in kips
l = length of each weld, in.
D = number of sixteenths of an inch in fillet weld size
C^* = coefficients tabulated below (includes ϕ = 0.75)
C_1 = coefficient for electrode used (see Table on page 5-89)
  = 1.0 for E70XX electrodes
xl = distance from vertical weld to center of gravity of weld group
yl = distance from horizontal weld to center of gravity of weld group

$$P_u = CC_1Dl$$

Required Minimum $C = \dfrac{P_u}{C_1 Dl}$

"   " $D = \dfrac{P_u}{CC_1 l}$

"   " $l = \dfrac{P_u}{CC_1 D}$

a	\multicolumn{17}{c}{k}
	0	0.1	0.2	0.3	0.4	0.5	0.6	0.7	0.8	0.9	1.0	1.2	1.4	1.6	1.8	2.0
.05	1.377	1.516	1.655	1.794	1.933	2.071	2.211	2.350	2.490	2.629	2.769	3.049	3.329	3.609	3.888	4.168
.10	1.333	1.474	1.612	1.749	1.886	2.024	2.162	2.302	2.442	2.538	2.725	3.009	3.293	3.577	3.860	4.142
.15	1.268	1.410	1.547	1.681	1.815	1.950	2.087	2.226	2.367	2.510	2.654	2.944	3.235	3.525	3.813	4.100
.20	1.191	1.334	1.468	1.598	1.728	1.859	1.993	2.130	2.271	2.415	2.561	2.857	3.156	3.454	3.749	4.043
.25	1.109	1.253	1.384	1.509	1.634	1.761	1.890	2.023	2.162	2.304	2.451	2.752	3.059	3.365	3.669	3.970
.30	1.032	1.169	1.297	1.418	1.542	1.663	1.787	1.915	2.049	2.188	2.332	2.634	2.947	3.262	3.574	3.884
.40	.895	1.022	1.137	1.254	1.369	1.480	1.594	1.712	1.834	1.960	2.092	2.372	2.644	2.929	3.350	3.677
.50	.783	.899	1.002	1.105	1.210	1.321	1.436	1.546	1.663	1.779	1.901	2.159	2.437	2.739	3.068	3.307
.60	.691	.798	.892	.982	1.075	1.174	1.280	1.394	1.513	1.625	1.742	1.989	2.250	2.526	2.821	3.130
.70	.616	.714	.800	.882	.966	1.055	1.153	1.258	1.372	1.493	1.605	1.842	2.096	2.364	2.638	2.930
.80	.552	.644	.723	.798	.875	.959	1.049	1.147	1.255	1.369	1.485	1.714	1.959	2.218	2.484	2.764
.90	.501	.584	.658	.728	.800	.877	.962	1.054	1.156	1.264	1.378	1.601	1.838	2.089	2.351	2.623
1.00	.457	.535	.603	.669	.736	.808	.887	.976	1.071	1.173	1.281	1.501	1.732	1.976	2.232	2.496
1.20	.388	.456	.516	.573	.633	.697	.769	.847	.933	1.024	1.120	1.326	1.549	1.781	2.024	2.277
1.40	.336	.397	.450	.501	.554	.613	.677	.748	.825	.907	.994	1.181	1.384	1.607	1.850	2.090
1.60	.297	.350	.398	.444	.493	.546	.604	.669	.739	.813	.894	1.063	1.250	1.454	1.678	1.917
1.80	.265	.314	.357	.399	.443	.492	.545	.604	.667	.737	.811	.966	1.138	1.327	1.533	1.754
2.00	.240	.283	.323	.362	.402	.447	.496	.550	.609	.673	.740	.884	1.045	1.219	1.409	1.616
2.20	.218	.259	.295	.331	.368	.410	.455	.505	.560	.619	.681	.815	.964	1.127	1.306	1.498
2.40	.201	.238	.272	.305	.340	.378	.420	.466	.518	.573	.630	.756	.895	1.048	1.216	1.395
2.60	.187	.220	.252	.283	.315	.351	.390	.434	.481	.533	.587	.704	.834	.979	1.136	1.305
2.80	.173	.205	.234	.263	.294	.327	.364	.405	.450	.497	.549	.659	.781	.918	1.065	1.225
3.00	.161	.192	.219	.246	.275	.307	.341	.380	.422	.467	.515	.620	.734	.863	1.002	1.154
x	.000	.005	.017	.035	.057	.083	.113	.144	.178	.213	.250	.327	.408	.492	.579	.667
y	.500	.455	.417	.385	.357	.333	.313	.294	.278	.263	.250	.227	.208	.192	.179	.167

*Tabulated values are the minimum of either − or + load angles.

ECCENTRIC LOADS ON WELD GROUPS
TABLE XXIV   Coefficients C

ANGLE = 75°

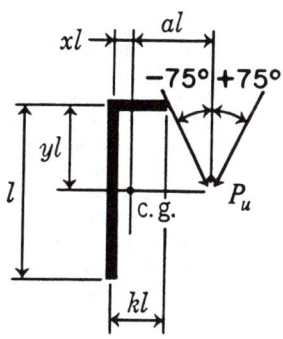

P_u = factored eccentric load in kips
l = length of each weld, in.
D = number of sixteenths of an inch in fillet weld size
C^* = coefficients tabulated below (includes ϕ = 0.75)
C_1 = coefficient for electrode used (see Table on page 5-89)
= 1.0 for E70XX electrodes
xl = distance from vertical weld to center of gravity of weld group
yl = distance from horizontal weld to center of gravity of weld group

$$P_u = CC_1Dl$$

Required Minimum $C = \dfrac{P_u}{C_1Dl}$

"   " $D = \dfrac{P_u}{CC_1l}$

"   " $l = \dfrac{P_u}{CC_1D}$

a	\multicolumn{16}{c}{k}
	0	0.1	0.2	0.3	0.4	0.5	0.6	0.7	0.8	0.9	1.0	1.2	1.4	1.6	1.8	2.0
.05	1.389	1.529	1.668	1.807	1.946	2.086	2.225	2.364	2.504	2.643	2.782	3.061	3.340	3.618	3.897	4.175
.10	1.381	1.521	1.660	1.799	1.939	2.078	2.218	2.357	2.497	2.637	2.777	3.056	3.336	3.615	3.894	4.173
.15	1.368	1.508	1.648	1.787	1.926	2.066	2.206	2.346	2.486	2.627	2.768	3.048	3.329	3.609	3.889	4.168
.20	1.350	1.491	1.630	1.769	1.908	2.048	2.189	2.330	2.471	2.613	2.755	3.038	3.320	3.601	3.882	4.162
.25	1.326	1.469	1.608	1.746	1.886	2.026	2.167	2.309	2.452	2.595	2.738	3.023	3.308	3.591	3.873	4.155
.30	1.299	1.442	1.581	1.719	1.859	1.999	2.141	2.285	2.429	2.573	2.718	3.006	3.293	3.579	3.862	4.145
.40	1.232	1.377	1.516	1.653	1.792	1.933	2.077	2.223	2.371	2.519	2.667	2.963	3.256	3.547	3.835	4.121
.50	1.102	1.225	1.437	1.572	1.710	1.851	1.997	2.146	2.298	2.451	2.604	2.909	3.210	3.507	3.800	4.090
.60	1.052	1.168	1.291	1.434	1.575	1.690	1.905	2.057	2.213	2.371	2.530	2.845	3.155	3.459	3.759	4.054
.70	.989	1.112	1.251	1.370	1.506	1.641	1.759	1.908	2.045	2.281	2.446	2.773	3.093	3.405	3.711	4.011
.80	.936	1.054	1.167	1.307	1.435	1.562	1.696	1.840	1.995	2.133	2.283	2.693	3.023	3.344	3.657	3.963
.90	.890	1.003	1.108	1.225	1.370	1.500	1.630	1.766	1.910	2.066	2.236	2.608	2.948	3.278	3.598	3.911
1.00	.848	.957	1.057	1.166	1.293	1.444	1.572	1.705	1.843	1.988	2.146	2.501	2.870	3.208	3.536	3.854
1.20	.773	.878	.970	1.068	1.179	1.307	1.457	1.592	1.725	1.863	2.010	2.331	2.691	3.060	3.401	3.731
1.40	.709	.809	.895	.987	1.089	1.202	1.331	1.479	1.618	1.753	1.895	2.199	2.536	2.897	3.259	3.599
1.60	.653	.749	.831	.915	1.011	1.117	1.232	1.364	1.512	1.653	1.791	2.085	2.406	2.752	3.108	3.464
1.80	.604	.694	.773	.853	.943	1.042	1.149	1.268	1.403	1.552	1.695	1.982	2.293	2.625	2.972	3.330
2.00	.561	.648	.723	.798	.883	.975	1.077	1.188	1.311	1.448	1.598	1.887	2.189	2.510	2.846	3.191
2.20	.522	.605	.677	.749	.829	.917	1.013	1.116	1.230	1.359	1.497	1.799	2.092	2.404	2.730	3.067
2.40	.488	.567	.636	.704	.781	.864	.954	1.053	1.160	1.278	1.411	1.700	2.001	2.305	2.622	2.951
2.60	.458	.534	.599	.665	.737	.816	.903	.996	1.097	1.208	1.331	1.606	1.910	2.212	2.521	2.842
2.80	.431	.503	.567	.630	.698	.773	.856	.945	1.041	1.145	1.261	1.521	1.813	2.123	2.426	2.739
3.00	.407	.476	.537	.596	.663	.734	.814	.897	.990	1.088	1.198	1.446	1.723	2.021	2.336	2.643
x	.000	.005	.017	.035	.057	.083	.113	.144	.178	.213	.250	.327	.408	.492	.579	.667
y	.500	.455	.417	.385	.357	.333	.313	.294	.278	.263	.250	.227	.208	.192	.179	.167

*Tabulated values are the minimum of either − or + load angles.

ECCENTRIC LOADS ON WELD GROUPS
TABLE XXV Coefficients *C*

ANGLE = 0°

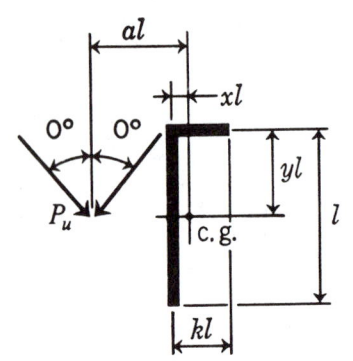

P_u = factored eccentric load in kips
l = length of each weld, in.
D = number of sixteenths of an inch in fillet weld size
C^* = coefficients tabulated below (includes ϕ = 0.75)
C_1 = coefficient for electrode used (see Table on page 5-89) = 1.0 for E70XX electrodes
xl = distance from vertical weld to center of gravity of weld group
yl = distance from horizontal weld to center of gravity of weld group

$$P_u = CC_1Dl$$

Required Minimum $C = \dfrac{P_u}{C_1Dl}$

"   " $D = \dfrac{P_u}{CC_1l}$

"   " $l = \dfrac{P_u}{CC_1D}$

a	\multicolumn{16}{c}{k}
	0	0.1	0.2	0.3	0.4	0.5	0.6	0.7	0.8	0.9	1.0	1.2	1.4	1.6	1.8	2.0
.05	1.371	1.512	1.651	1.789	1.927	2.066	2.204	2.342	2.481	2.620	2.759	3.038	3.318	3.598	3.878	4.158
.10	1.314	1.457	1.595	1.732	1.867	2.002	2.137	2.272	2.409	2.547	2.686	2.967	3.251	3.535	3.819	4.104
.15	1.229	1.375	1.513	1.647	1.778	1.907	2.037	2.168	2.300	2.435	2.573	2.855	3.143	3.433	3.724	4.015
.20	1.132	1.278	1.415	1.546	1.672	1.796	1.919	2.043	2.169	2.299	2.432	2.711	3.000	3.297	3.595	3.894
.25	1.032	1.177	1.311	1.438	1.560	1.679	1.797	1.915	2.034	2.157	2.275	2.539	2.835	3.134	3.439	3.745
.30	.937	1.078	1.209	1.332	1.450	1.565	1.679	1.792	1.901	2.018	2.137	2.383	2.647	2.930	3.186	3.450
.40	.774	.903	1.024	1.138	1.249	1.358	1.466	1.570	1.680	1.791	1.905	2.144	2.390	2.647	2.910	3.186
.50	.649	.764	.873	.977	1.080	1.183	1.286	1.387	1.493	1.601	1.712	1.944	2.185	2.434	2.690	2.950
.60	.553	.656	.753	.848	.942	1.038	1.136	1.232	1.334	1.439	1.546	1.771	2.005	2.247	2.497	2.751
.70	.481	.571	.659	.744	.831	.919	1.011	1.102	1.199	1.299	1.402	1.619	1.846	2.082	2.324	2.572
.80	.424	.505	.584	.660	.740	.822	.907	.993	1.085	1.180	1.277	1.486	1.705	1.932	2.168	2.410
.90	.379	.452	.522	.591	.666	.741	.820	.901	.987	1.076	1.171	1.367	1.579	1.800	2.027	2.262
1.00	.342	.408	.472	.535	.604	.674	.747	.823	.904	.989	1.077	1.266	1.467	1.679	1.901	2.128
1.20	.286	.341	.395	.449	.508	.568	.632	.700	.770	.846	.925	1.095	1.279	1.473	1.680	1.892
1.40	.246	.293	.339	.386	.436	.490	.546	.607	.669	.736	.808	.961	1.128	1.308	1.497	1.697
1.60	.215	.257	.297	.338	.382	.429	.479	.533	.590	.651	.714	.854	1.007	1.171	1.347	1.532
1.80	.191	.228	.264	.301	.340	.382	.427	.475	.527	.582	.640	.767	.906	1.059	1.220	1.394
2.00	.172	.206	.238	.271	.306	.344	.385	.429	.476	.526	.579	.695	.824	.964	1.115	1.275
2.20	.157	.189	.216	.246	.278	.313	.350	.390	.433	.480	.528	.636	.754	.884	1.024	1.174
2.40	.144	.171	.198	.226	.255	.287	.321	.358	.398	.440	.486	.585	.695	.816	.946	1.087
2.60	.133	.158	.183	.208	.235	.265	.296	.331	.368	.407	.449	.541	.644	.757	.879	1.011
2.80	.123	.147	.170	.193	.218	.246	.275	.307	.341	.378	.418	.504	.600	.706	.820	.944
3.00	.115	.137	.158	.180	.204	.229	.257	.287	.319	.353	.390	.471	.561	.661	.769	.885
x	.000	.005	.017	.035	.057	.083	.113	.144	.178	.213	.250	.327	.408	.492	.579	.667
y	.500	.455	.417	.385	.357	.333	.313	.294	.278	.263	.250	.227	.208	.192	.179	.167

*Tabulated values are the minimum of either − or + load angles.

ECCENTRIC LOADS ON WELD GROUPS
TABLE XXV   Coefficients *C*

ANGLE = 45°

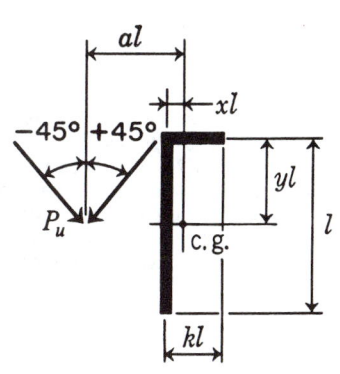

P_u = factored eccentric load in kips
l = length of each weld, in.
D = number of sixteenths of an inch in fillet weld size
C^* = coefficients tabulated below (includes $\phi = 0.75$)
C_1 = coefficient for electrode used (see Table on page 5-89) = 1.0 for E70XX electrodes
xl = distance from vertical weld to center of gravity of weld group
yl = distance from horizontal weld to center of gravity of weld group

$$P_u = CC_1 Dl$$

Required Minimum $C = \dfrac{P_u}{C_1 Dl}$

"   " $D = \dfrac{P_u}{CC_1 l}$

"   " $l = \dfrac{P_u}{CC_1 D}$

a	*k*															
---	0	0.1	0.2	0.3	0.4	0.5	0.6	0.7	0.8	0.9	1.0	1.2	1.4	1.6	1.8	2.0
.05	1.377	1.516	1.655	1.794	1.932	2.071	2.210	2.350	2.489	2.629	2.769	3.049	3.329	3.609	3.888	4.168
.10	1.333	1.474	1.611	1.747	1.884	2.021	2.160	2.300	2.441	2.583	2.725	3.010	3.294	3.578	3.861	4.143
.15	1.268	1.409	1.542	1.674	1.807	1.942	2.080	2.221	2.364	2.509	2.654	2.946	3.237	3.527	3.815	4.102
.20	1.191	1.330	1.457	1.581	1.707	1.838	1.975	2.117	2.263	2.411	2.561	2.862	3.162	3.459	3.754	4.047
.25	1.109	1.244	1.360	1.468	1.582	1.654	1.794	1.996	2.145	2.297	2.451	2.761	3.071	3.377	3.680	3.979
.30	1.032	1.163	1.268	1.356	1.449	1.565	1.707	1.854	1.985	2.174	2.332	2.650	2.968	3.283	3.594	3.900
.40	.895	1.020	1.118	1.195	1.279	1.382	1.503	1.636	1.782	1.935	2.092	2.413	2.751	3.077	3.399	3.718
.50	.783	.899	.996	1.077	1.156	1.248	1.355	1.479	1.612	1.752	1.901	2.213	2.532	2.869	3.196	3.521
.60	.691	.800	.894	.976	1.054	1.139	1.236	1.346	1.470	1.602	1.742	2.043	2.350	2.669	2.992	3.326
.70	.616	.717	.804	.886	.965	1.045	1.135	1.236	1.349	1.473	1.605	1.889	2.187	2.498	2.813	3.131
.80	.552	.647	.731	.810	.887	.965	1.049	1.142	1.246	1.361	1.485	1.754	2.043	2.341	2.649	2.960
.90	.501	.587	.667	.742	.817	.893	.973	1.060	1.157	1.263	1.380	1.634	1.911	2.203	2.498	2.804
1.00	.457	.538	.612	.683	.755	.829	.907	.989	1.079	1.178	1.278	1.529	1.792	2.074	2.362	2.657
1.20	.388	.459	.524	.589	.654	.721	.793	.868	.949	1.037	1.133	1.349	1.589	1.847	2.121	2.397
1.40	.336	.398	.457	.514	.573	.636	.702	.772	.845	.924	1.010	1.203	1.422	1.660	1.913	2.180
1.60	.297	.352	.404	.456	.510	.567	.627	.691	.760	.832	.910	1.085	1.284	1.503	1.739	1.987
1.80	.265	.315	.362	.409	.458	.511	.566	.626	.688	.756	.827	.987	1.169	1.371	1.589	1.822
2.00	.240	.285	.327	.370	.416	.463	.515	.570	.629	.691	.757	.904	1.072	1.259	1.462	1.678
2.20	.218	.260	.299	.338	.380	.424	.473	.523	.578	.636	.698	.834	.989	1.162	1.352	1.555
2.40	.201	.239	.275	.311	.349	.391	.436	.483	.534	.589	.646	.773	.917	1.078	1.256	1.446
2.60	.187	.221	.254	.288	.324	.363	.404	.449	.496	.548	.601	.720	.855	1.006	1.172	1.351
2.80	.173	.205	.236	.268	.301	.338	.376	.419	.463	.511	.562	.674	.801	.942	1.098	1.267
3.00	.161	.192	.221	.250	.282	.316	.352	.392	.434	.480	.528	.633	.752	.886	1.032	1.192
x	.000	.005	.017	.035	.057	.083	.113	.144	.178	.213	.250	.327	.408	.492	.579	.667
y	.500	.455	.417	.385	.357	.333	.313	.294	.278	.263	.250	.227	.208	.192	.179	.167

*Tabulated values are the minimum of either − or + load angles.

ECCENTRIC LOADS ON WELD GROUPS

TABLE XXV   Coefficients C

ANGLE = 75°

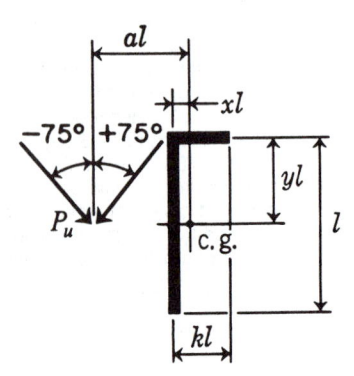

P_u = factored eccentric load in kips
l = length of each weld, in.
D = number of sixteenths of an inch in fillet weld size
C^* = coefficients tabulated below (includes ϕ = 0.75)
C_1 = coefficient for electrode used (see Table on page 5-89)
   = 1.0 for E70XX electrodes
xl = distance from vertical weld to center of gravity of weld group
yl = distance from horizontal weld to center of gravity of weld group

$$P_u = CC_1Dl$$

Required Minimum $C = \dfrac{P_u}{C_1Dl}$

"   " $D = \dfrac{P_u}{CC_1l}$

"   " $l = \dfrac{P_u}{CC_1D}$

a	k															
	0	0.1	0.2	0.3	0.4	0.5	0.6	0.7	0.8	0.9	1.0	1.2	1.4	1.6	1.8	2.0
.05	1.389	1.529	1.668	1.807	1.946	2.086	2.225	2.364	2.504	2.643	2.782	3.061	3.340	3.618	3.897	4.175
.10	1.381	1.521	1.660	1.799	1.939	2.078	2.218	2.357	2.497	2.637	2.777	3.056	3.336	3.615	3.894	4.173
.15	1.368	1.508	1.648	1.787	1.926	2.066	2.206	2.346	2.487	2.627	2.768	3.049	3.329	3.609	3.889	4.169
.20	1.350	1.491	1.630	1.769	1.908	2.049	2.189	2.331	2.472	2.614	2.755	3.038	3.320	3.602	3.882	4.163
.25	1.326	1.468	1.608	1.746	1.886	2.027	2.169	2.311	2.454	2.597	2.740	3.025	3.309	3.592	3.874	4.155
.30	1.299	1.442	1.581	1.720	1.859	2.001	2.144	2.287	2.432	2.576	2.721	3.008	3.295	3.580	3.863	4.146
.40	1.232	1.376	1.515	1.653	1.794	1.937	2.082	2.230	2.378	2.526	2.674	2.968	3.260	3.550	3.837	4.123
.50	1.102	1.238	1.406	1.573	1.715	1.860	2.009	2.160	2.312	2.465	2.617	2.919	3.218	3.513	3.804	4.093
.60	1.052	1.154	1.277	1.467	1.536	1.774	1.926	2.082	2.238	2.395	2.552	2.863	3.168	3.469	3.765	4.059
.70	.989	1.102	1.212	1.318	1.474	1.621	1.791	1.998	2.159	2.320	2.481	2.800	3.112	3.419	3.721	4.019
.80	.936	1.049	1.144	1.261	1.409	1.576	1.716	1.885	2.076	2.241	2.406	2.732	3.052	3.365	3.672	3.974
.90	.890	1.000	1.094	1.199	1.328	1.481	1.650	1.817	1.993	2.161	2.329	2.661	2.987	3.306	3.619	3.926
1.00	.848	.956	1.050	1.147	1.266	1.409	1.569	1.734	1.903	2.081	2.251	2.588	2.920	3.244	3.562	3.874
1.20	.773	.879	.968	1.061	1.163	1.286	1.427	1.585	1.751	1.919	2.098	2.441	2.780	3.114	3.441	3.761
1.40	.709	.810	.898	.986	1.079	1.188	1.315	1.458	1.615	1.780	1.948	2.296	2.639	2.979	3.312	3.639
1.60	.653	.750	.834	.918	1.008	1.106	1.221	1.353	1.497	1.654	1.822	2.158	2.501	2.842	3.180	3.513
1.80	.604	.696	.778	.859	.943	1.035	1.141	1.261	1.396	1.542	1.698	2.031	2.367	2.709	3.048	3.384
2.00	.561	.649	.727	.804	.886	.973	1.070	1.181	1.307	1.445	1.594	1.911	2.242	2.579	2.918	3.255
2.20	.522	.606	.682	.756	.835	.918	1.008	1.112	1.228	1.357	1.497	1.802	2.125	2.455	2.792	3.128
2.40	.488	.568	.640	.712	.788	.867	.954	1.049	1.157	1.279	1.411	1.702	2.015	2.339	2.671	3.005
2.60	.458	.535	.603	.672	.745	.822	.904	.993	1.096	1.209	1.334	1.611	1.913	2.230	2.555	2.885
2.80	.431	.504	.570	.636	.706	.781	.858	.944	1.040	1.146	1.265	1.528	1.818	2.127	2.445	2.771
3.00	.407	.477	.540	.603	.670	.742	.817	.899	.989	1.090	1.202	1.452	1.732	2.031	2.342	2.661
x	.000	.005	.017	.035	.057	.083	.113	.144	.178	.213	.250	.327	.408	.492	.579	.667
y	.500	.455	.417	.385	.357	.333	.313	.294	.278	.263	.250	.227	.208	.192	.179	.167

*Tabulated values are the minimum of either − or + load angles.

ONE-SIDED CONNECTIONS

In designing a one-sided connection, it is customary to consider vertical shear or bearing in all fasteners and the effect of eccentricity in the outstanding leg fasteners. Shown below is a table of coefficients for one-sided framed beam connections and an example of its use. The shear capacity of the framing angle should be checked against the net area (LRFD Specification Sect. J4), using hole diameter = nominal bolt diameter + $\frac{1}{16}$".

n	Coefficient C	
	Case I	Case II
1	—	0.54
2	0.95	1.67
3	1.87	3.06
4	2.94	4.86
5	4.02	6.84
6	5.10	8.93
7	6.17	11.1
8	7.22	13.2
9	8.26	15.4
10	9.29	17.5

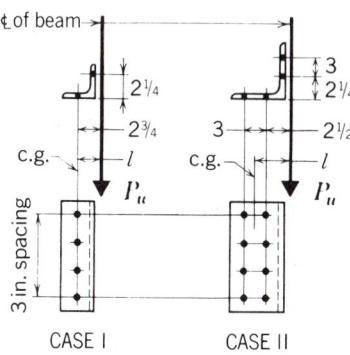

$P_u = C\phi r_v$ or $C = P_u/\phi r_v$
n = total number of fasteners in one vertical row
C = coefficient
P_u = factored load, kips
ϕr_v = design shear or bearing value for one fastener, kips
l = actual arm between centerline of beam and center of gravity of fasteners, in.

Do not exceed gages shown for web leg. For outstanding leg gages, other than those shown, coefficients may be interpolated from Tables X and XI.

Select angle thickness to provide sufficient net shear capacity, or limit connection capacity to permissible shear value of angle used. Use minimum angle thickness of $\frac{3}{8}$" for $\frac{3}{4}$" dia. and $\frac{7}{8}$" dia. fasteners, and $\frac{1}{2}$" for 1" dia. fasteners. It will be permissible to design a connection using combinations of leg widths as well as fastener specification and diameters.

EXAMPLE 1

Given:

Select a one-sided connection for a W18 × 50, F_y = 36 ksi, with a factored end reaction of 100 kips. Use $\frac{7}{8}$" dia. ASTM A325-N bolts in the beam web leg and outstanding leg.

Solution:

1. Outstanding leg:

    Single shear value ϕr_v of $\frac{7}{8}$" dia.
    A325-N bolt from shear design load
    Table I-D = 21.1 kips

    $C = 100/21.1 = 4.74$

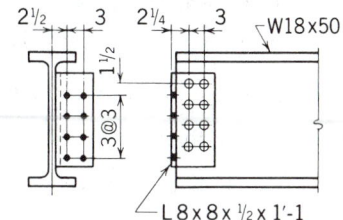

The next larger value of C in the coefficient table above requires six A325 bolts in Case I or eight A325 bolts in Case II. Since the beam depth will not allow Case I to be used, use 8 bolts, as shown. Bolts must be staggered to assure driving clearance for bolts in the outstanding leg.

2. Web leg:

Single shear value ϕr_v of $\frac{7}{8}''$ dia. A325-N bolt from Table I-D = 21.1 kips.

$t_w = 0.355''$

Bearing value from Table I-F = 91.3 kips/1″ thickness for one fastener.

Reaction $R = 91.3 \times 0.355 = 32.4$ kips

Single shear at 21.1 kips governs, requiring $100/21.1 = 4.74$ bolts

Use: 8 bolts arranged in pattern dictated by outstanding leg.

3. Angle Size: Try $8 \times 8 \times \frac{1}{2} \times 1'\text{-}1''$ angle.

Design load (net shear) = $[13 - 4(\frac{7}{8} + \frac{1}{16})] \; \frac{1}{2} \times 0.45 \times 58$

$= 121$ kips > 100 kips   **o.k.**

EXAMPLE 2

Given:

Same as Ex. 1, except weld the web leg of the connection and factored reaction = 75 kips.

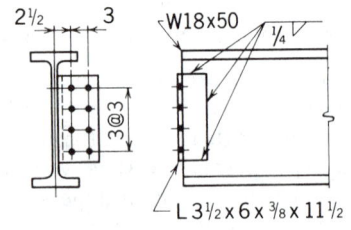

Solution:

1. Outstanding leg: Same as Ex. 1.

2. Web leg: The required weld may be determined from Framed Beam Connections, Welded, E70XX electrodes, Table III.

Since the capacities shown in Table III are for two angles, it will be convenient to double the given reaction and select weld sizes directly from the tables, and angle lengths directly or by interpolation.

Since $R = 75$ kips, the tabular capacity needed is 150 kips.

From Table III, which covers $F_y = 36$ ksi connections and E70XX electrodes, the capacity of Weld A at $\frac{1}{4}''$ on an $11\frac{1}{2}''$ angle is 158 kips.

Check beam web thickness: $0.355'' > 0.55/2 = 0.275''$ **o.k.**

Use: E70XX $\frac{1}{4}''$ weld (as shown).

3. Angle size: Use 6″ for outstanding leg and $3\frac{1}{2}''$ for web leg.

Assume a $\frac{3}{8}''$ thick angle and check for net shear capacity:

$[11.5 - 4(\frac{7}{8} + \frac{1}{16})] \times \frac{3}{8}'' \times 0.45 \times 58$ ksi $= 75.9$ kips > 75 kips   **o.k.**

Use: $6 \times 3\frac{1}{2} \times \frac{3}{8} \times 11\frac{1}{2}''$ angle.

ECCENTRIC CONNECTIONS

BEAM-TO-COLUMN CONNECTIONS

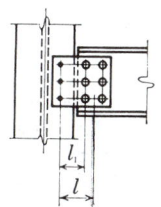

To reduce the moment in the column full eccentricity in the fasteners connecting plate to beam should be figured. Lever arm l should be used. A coefficient for this fastener group, for ordinary cases, can be found in Tables X–XVII.

Field fasteners connecting plate to column should be used (least number) if beam can be erected and if there are no interfering details in the web of the column. The plate should figure for a moment with lever arm l_1.

SYMMETRICAL BEAM-TO-COLUMN CONNECTIONS

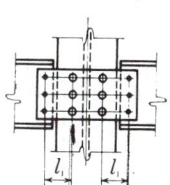

A single plate across the column may be used. If the reactions of the two beams are equal, there is no eccentricity to figure on either beams or columns. The case of live load on one beam only must, however, be considered. Where for this or other reason the beam reactions are unequal, figure the fasteners in the column for the sum of the reactions and the difference of the moments, taken to the center of the connection. See Table XI.

Plate should figure for greater moment with lever arm l_1. See table of Net Section Moduli of Bracket Plates on page 5-118.

ZEE CONNECTIONS

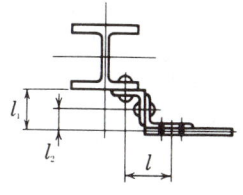

In general use for light loads only. Eccentricity in fasteners connecting connection angle to the beam should be figured, using the lever arm l. Eccentricity in fasteners connecting connection angle to column, with a lever arm of l_1, should be figured. The thickness of the angle should be ample to resist the bending moment.

Eccentricity in fasteners connecting the two connection angles should be figured if the lever arm l_2 is 2½ in. or more. The connection should be designed so field work is at a minimum. There can be many variations of this type of connection depending on the length of l_1. The eccentricity should be considered in all cases.

TRUSS CONNECTIONS

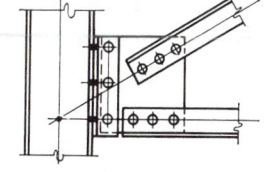

The intersection of the working lines should be located to hold the effect of moment on connection and column to a minimum. It is often advantageous to detail so the working lines intersect at the corner of the gusset plate. In this case there is no moment on the gusset plate and the resulting moment on the beam may not be a controlling factor.

BRACKET PLATES
Net section moduli

Diameter of holes assumed ⅛ in. larger than nominal diameter of fastener

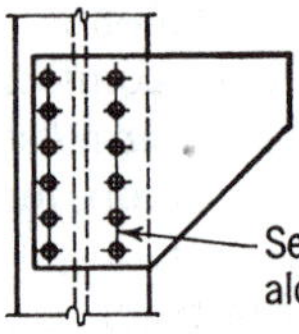

— Section moduli taken along this line

Fasteners spaced 3 in. vertically

No. of Fasteners in One Vertical Line	Depth of Plate in In.	¾ In. Fasteners					⅞ In. Fasteners					1 In. Fasteners				
		Thickness of Plate, In.					Thickness of Plate, In.					Thickness of Plate, In.				
		¼	⅜	½	⅝	¾	⅜	½	⅝	¾	⅞	½	⅝	¾	⅞	1
2	6	1.2	1.8	2.3	2.9	3.5	1.7	2.3	2.9	3.4	4.0	2.2	2.7	3.2	3.8	4.3
3	9	2.5	3.8	5.0	6.3	7.5	3.6	4.8	5.9	7.1	8.3	4.5	5.6	6.8	7.9	9.0
4	12	4.4	6.3	8.7	11	13	6.2	8.2	10	12	14	7.8	9.7	12	14	16
5	15	6.8	10	14	17	20	10	13	16	19	22	12	15	18	21	24
6	18	9.6	15	19	24	29	14	18	23	27	32	17	21	26	30	34
7	21	13	20	26	33	39	19	25	31	37	43	23	29	35	41	47
8	24	17	26	34	43	51	24	32	40	48	56	30	38	45	53	61
9	27	22	32	43	54	65	31	41	51	61	71	38	48	57	67	77
10	30	27	40	53	67	80	38	50	63	75	88	47	59	71	83	94
12	36	38	58	77	96	115	54	72	90	108	126	68	85	102	119	136
14	42	52	78	104	130	157	74	98	123	147	172	92	115	138	161	184
16	48	68	102	136	170	204	96	128	160	192	224	120	150	180	211	241
18	54	86	129	172	215	259	122	162	203	243	284	152	190	228	266	304
20	60	106	160	213	266	319	150	200	250	300	350	188	235	282	329	376
22	66	129	193	257	322	386	182	242	303	363	424	227	284	341	398	454
24	72	153	230	306	383	459	216	288	360	432	504	270	338	406	473	541
26	78	180	270	359	449	539	254	338	423	507	592	317	397	476	555	634
28	84	208	313	417	521	625	294	392	490	588	686	368	460	552	644	736
30	90	240	359	478	598	718	338	450	563	675	788	422	528	633	739	845
32	96	272	408	544	680	816	384	512	640	768	896	480	600	721	841	961
34	102	308	461	614	768	922	434	578	723	867	1012	542	678	813	949	1085
36	108	344	517	689	861	1033	486	648	810	972	1134	608	760	912	1064	1216

Interpolate for intermediate thickness of plates.
General equation for net section modulus of bracket plates:

$$S_{net} = \frac{t_p d^2}{6} - \frac{b^2 n(n^2 - 1)[t_p \times (\text{Bolt Dia.} + 0.125)]}{6d}$$

where

t_p = plate thickness, in.
d = plate depth, in.
n = number of fasteners in one vertical row
b = fastener spacing vertically, inches.

When calculating stress from factored loads, the limiting stress may be assumed to be 0.9 F_y. In the compression zone, plate buckling should be checked.

HANGER TYPE CONNECTIONS
Fasteners loaded in tension and shear

In the design of hanger type connections, prying action must be considered. The table below is useful for making a preliminary selection of a trial fitting, using $F_y = 36$ ksi. The fitting must then be checked for bending stresses and tension in the bolts, using a final design procedure.

PRELIMINARY SELECTION TABLE

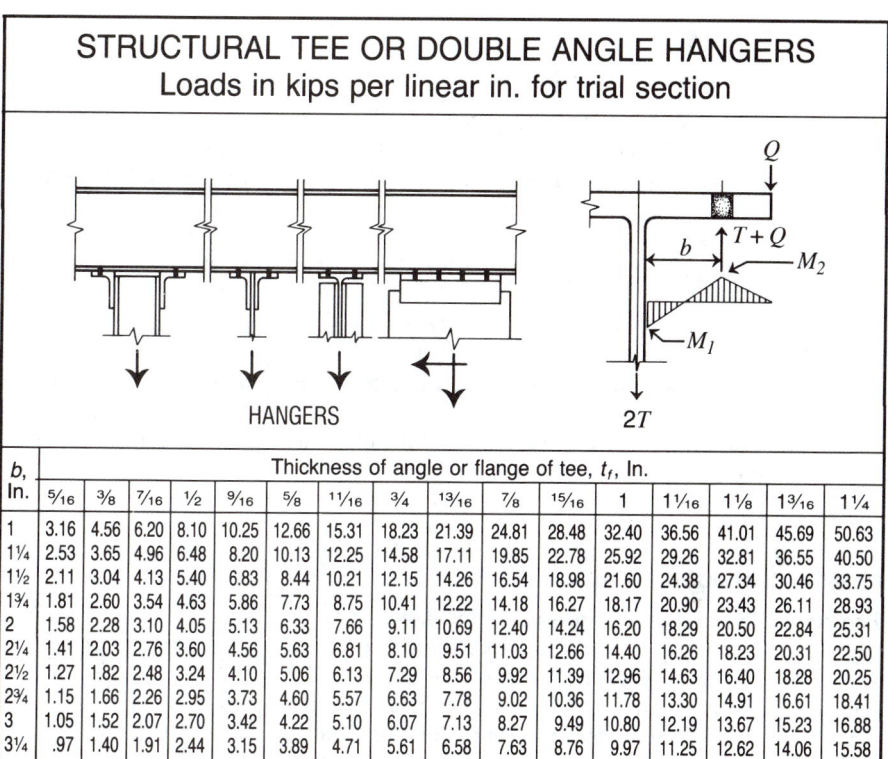

STRUCTURAL TEE OR DOUBLE ANGLE HANGERS
Loads in kips per linear in. for trial section

b, In.	Thickness of angle or flange of tee, t_f, In.															
	$\frac{5}{16}$	$\frac{3}{8}$	$\frac{7}{16}$	$\frac{1}{2}$	$\frac{9}{16}$	$\frac{5}{8}$	$\frac{11}{16}$	$\frac{3}{4}$	$\frac{13}{16}$	$\frac{7}{8}$	$\frac{15}{16}$	1	$1\frac{1}{16}$	$1\frac{1}{8}$	$1\frac{3}{16}$	$1\frac{1}{4}$
1	3.16	4.56	6.20	8.10	10.25	12.66	15.31	18.23	21.39	24.81	28.48	32.40	36.56	41.01	45.69	50.63
$1\frac{1}{4}$	2.53	3.65	4.96	6.48	8.20	10.13	12.25	14.58	17.11	19.85	22.78	25.92	29.26	32.81	36.55	40.50
$1\frac{1}{2}$	2.11	3.04	4.13	5.40	6.83	8.44	10.21	12.15	14.26	16.54	18.98	21.60	24.38	27.34	30.46	33.75
$1\frac{3}{4}$	1.81	2.60	3.54	4.63	5.86	7.73	8.75	10.41	12.22	14.18	16.27	18.17	20.90	23.43	26.11	28.93
2	1.58	2.28	3.10	4.05	5.13	6.33	7.66	9.11	10.69	12.40	14.24	16.20	18.29	20.50	22.84	25.31
$2\frac{1}{4}$	1.41	2.03	2.76	3.60	4.56	5.63	6.81	8.10	9.51	11.03	12.66	14.40	16.26	18.23	20.31	22.50
$2\frac{1}{2}$	1.27	1.82	2.48	3.24	4.10	5.06	6.13	7.29	8.56	9.92	11.39	12.96	14.63	16.40	18.28	20.25
$2\frac{3}{4}$	1.15	1.66	2.26	2.95	3.73	4.60	5.57	6.63	7.78	9.02	10.36	11.78	13.30	14.91	16.61	18.41
3	1.05	1.52	2.07	2.70	3.42	4.22	5.10	6.07	7.13	8.27	9.49	10.80	12.19	13.67	15.23	16.88
$3\frac{1}{4}$	.97	1.40	1.91	2.44	3.15	3.89	4.71	5.61	6.58	7.63	8.76	9.97	11.25	12.62	14.06	15.58

To select a preliminary size of tee or double angle, the above table assumes equal critical moments at the fastener line and at the face of the tee stem or angle leg.

$$M' = T' \times b/2 = \phi M'_p = 0.9 \times \frac{1}{4} \times 1 \times t^2 \times 36$$

Hanger capacity $= 2T' = 32.40 \; t^2/b$

where

$2T'$ = factored load on two angles or a structural tee, in kips per linear inch, using maximum bending strength

t = thickness of angle or tee flange, in.

b = distance from bolt centerline to the face of angle leg or tee stem, in.

AMERICAN INSTITUTE OF STEEL CONSTRUCTION

DESIGN AND ANALYSIS METHODS

Nomenclature

T = factored applied tension per bolt (exclusive of initial tightening), kips
Q = factored prying force per bolt at design load, kips
  = $B_c - T$
B = design tensile strength of bolt, kips
B_c = factored load per bolt including prying action, kips
  = $T + Q$
t_c = flange or angle thickness required to develop B in bolts with no prying action, in.

$$= \sqrt{\frac{4.44Bb'}{pF_y}}$$

F_y = yield strength of the flange material, ksi
p = length of flange, parallel to stem or leg, tributary to each bolt, in.
a = distance from bolt centerline to edge of tee flange or angle leg but not more than $1.25b$, in.
d = bolt dia., in.
d' = width of bolt hole parallel to tee stem or angle leg, in.
b' = $b - d/2$, in.
a' = $a + d/2$, in.
ρ = b'/a'
α = $(0 \leq \alpha \leq 1.0)$. Ratio of moment at bolt line to moment at stem line
  = $M_2/\delta M_1$
δ = ratio of net area (at bolt line) and the gross area (at the face of the stem or angle leg)
  = $1 - d'/p$

DESIGN CONSIDERATIONS

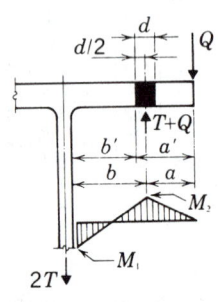

The actual distribution of stress in the tee flange or angle leg and the extent of prying action are extremely complex. Significant deformation of the fitting under design load is seldom tolerable. Flange stiffness, rather than bending strength, is the key to satisfactory performance. Therefore, dimension b should be made as small as the wrench clearance will permit. Since dimension b is only slightly larger than the thickness of the fitting, the classical moment diagram (right) does not truly represent all the restraining forces at the bolt line, and overestimates the actual prying force. In addition, local deformation of the fitting ("quilting") under the pretension force of high-strength bolts also accounts for a less critical prying force than indicated by earlier investigations.

Good correlation between estimated connection strength and observed test results has been obtained using the procedures outlined below[1]. Since these procedures are

1. *Fisher, J. W. and J. H. A. Struik*   Guide to Design Criteria for Bolted and Riveted Joints   *John Wiley and Sons, New York, 1974. pp. 273–277.*

formulated in terms of factored loads, they are not applicable to situations where service loads must be used (i.e., fatigue, deflection and drift limitations). For these situations use the working stress procedure in the 8th Edition Manual or in Refs. 2 or 3.

SOLUTION PROCEDURES

Method 1 (Design)

Given: p, F_y, B, T

Find: t

Step 1. Determine the number and size of bolts required such that $T \leq B$

Step 2. Using preliminary selection table to estimate required flange thickness, choose a trial section with flange thickness t, and calculate b, a ($\leq 1.25b$), b', a', ρ

Step 3. Calculate $\beta = \dfrac{1}{\rho}\left(\dfrac{B}{T} - 1\right)$

if $\beta \geq 1$ set $\alpha = 1$

if $\beta < 1$ set $\alpha = $ lesser of

$\dfrac{1}{\delta}\left(\dfrac{\beta}{1 - \beta}\right)$ and 1.0.

Step 4. Calculate

$$t_{req'd} = \sqrt{\frac{4.44 T b'}{p F_y (1 + \delta\alpha)}}$$

if $t_{req'd} \leq t$, design is satisfactory

if $t_{req'd} > t$, choose a heavier section

with $t \geq t_{req'd}$, or change geometry (b and p), and repeat Steps 3 and 4

Step 5. If the factored prying force Q is required, it can be calculated as follows:

$$\alpha = \frac{1}{\delta}\left[\frac{T/B}{(t/t_c)^2} - 1\right]$$

If $\alpha < 0$ set $\alpha = 0$

$$Q = B\delta\alpha\rho\left(\frac{t}{t_c}\right)^2$$

Step 6. In applications where the prying force Q must be reduced to an insignificant amount, skip Steps 3, 4 and 5, set $\alpha = 0$, and calculate

$$t_{req'd} = \sqrt{\frac{4.44\ T b'}{p\ F_y}}$$

2. *Astaneh, A.* Procedure for Design Analysis of Hanger-Type Connections *AISC Engineering Journal*, Vol. 22, No. 2, 2nd Quarter, 1985. pp.63–66.
3. *Thornton, W. A.* Prying Action—A General Treatment *AISC Engineering Journal*, Vol. 22, No. 2, 2nd Quarter, 1985. pp. 67–75.

Method 2 (Analysis)

Given: t, a, b, p, B, T

Find: $T_{max.}$

Step 1. Check $T \leq B$. If true, proceed; if not, use more or stronger bolts

Step 2. Then, calculate

$$\alpha = \frac{1}{\rho(1 + \delta)} \left[\left(\frac{t_c}{t} \right)^2 - 1 \right]$$

if $\alpha > 1$,

$$T_{max.} = B \left(\frac{t}{t_c} \right)^2 (1 + \delta)$$

if $0 \leq \alpha \leq 1$

$$T_{max.} = B \left(\frac{t}{t_c} \right)^2 (1 + \delta\alpha)$$

if $\alpha < 0$,

$$T_{max.} = B$$

Step 3. If $T_{max.} \geq T$ design is satisfactory

If $T_{max.} < T$, choose section with thicker flange, or change geometry (b, p), and repeat Step 2

Step 4. If the prying force is required, use formulas of Step 5 of Method 1.

EXAMPLE 1

Select a WT section hanger using A36 steel and ¾" dia. A325-N bolts to support a factored load of 80 kips suspended from the bottom flange of a W36 × 160. Bolts to be located on a 4" beam gage and the fitting length is 9" max.

Solution: (Method 1)

1. $B = 0.75 \times 90 \times 0.4418 = 29.8$ kips (from Table IA)

    No. of bolts required $= \dfrac{80}{29.8} = 2.68$; try 4 bolts

    $T = \dfrac{80}{4} = 20$ kips < 29.8 kips

2. From preliminary selection table with

    $\dfrac{2 \times 20}{4.5} = 8.89$ kips/in.

    and $b = 2 - 0.25 = 1.75$ in., choose preliminary thickness $t_{pre} = $ ¹¹⁄₁₆" to ¾".
    Tentatively select a WT9 × 30 with $t = 0.695$"

    $b = \dfrac{4 - 0.415}{2} = 1.793$ in. $> 1¼$ in. wrench clearance

$$a = \frac{7.555 - 4}{2} = 1.778$$

check $1.25 \, b = 1.25 \times 1.793 = 2.241$

since $2.241 > 1.778$, use $a = 1.778$ in.

$b' = 1.793 - 0.375 = 1.418$

$a' = 1.778 + 0.375 = 2.153$

$p = 4.5$

$d' = {}^{13}\!/_{16} = 0.8125$

$\delta = 1 - 0.8125/4.5 = 0.819$

$\rho = 1.418/2.153 = 0.659$

3. Calculate $\beta = \dfrac{1}{0.659}\left(\dfrac{29.8}{20} - 1\right) = 0.744$

$$\alpha = \frac{1}{0.819}\left(\frac{0.744}{1 - 0.744}\right) = 3.549$$

$$\therefore \alpha = 1$$

4. Calculate $t_{req'd} = \sqrt{\dfrac{4.44 \times 20 \times 1.418}{4.5 \times 36 \times 1.819}} = 0.654$ in.

Since 0.654 in. < 0.695 in., WT9 $\times$ 30 **o.k.**

5. If the prying force is required;

$$t_c = \sqrt{\frac{4.44 \times 29.8 \times 1.418}{4.5 \times 36}} = 1.076 \text{ in.}$$

$$\alpha = \frac{1}{0.819}\left[\frac{20/29.8}{(0.695/1.076)^2} - 1\right] = 0.743$$

$$Q = 29.8 \times 0.819 \times 0.743 \times 0.659 \times \left(\frac{0.695}{1.076}\right)^2 = 4.99 \text{ kips}$$

EXAMPLE 2

Redesign the hanger fitting of Ex. 1 for a loading condition where virtual elimination of the prying force is required.

The solution is given by Step 6 of Method 1. Note that because $\alpha = 0$, the solution does not depend on a. Assuming a WT hanger stem thickness of ½ in.,

$b = 2 - 0.25 = 1.75$, $b' = 1.75 - 0.375 = 1.3750$

$$t_{req'd} = \sqrt{\frac{4.44 \times 20 \times 1.375}{4.5 \times 36}} = 0.868 \text{ in.}$$

Choose a WT9 × 48.5

$t = 0.870'' > 0.868''$ **o.k.**

$t_w = 0.535'' > 0.5''$ **o.k.**

Use: WT9 × 48.5

EXAMPLE 3

Select a double angle connection using A36 steel and ¾-in. dia. A325-N bolts to carry a factored load of 100 kips at a bevel of 6 to 12. Fasteners to be located on a 5½-in. column gage.

Solution: (Method 2)

$$\text{Shear} = \frac{1}{\sqrt{5}} \times 100 = 44.7 \text{ kips}$$

$$\text{Tension} = \frac{2}{\sqrt{5}} \times 100 = 89.4 \text{ kips}$$

Assume ½-in. gusset plate

1. Try 6 bolts.

   $$\text{Shear per bolt } V = \frac{44.7}{6} = 7.45 \text{ kips}$$

   $$\text{Tension per bolt } T = \frac{89.4}{6} = 14.9 \text{ kips}$$

   From interaction (Table J3.3)

   $B = (85 \times 0.4418) - (1.8 \times 7.45) = 24.1 \text{ kips} < 30.0 \text{ kips}$

   Since 14.9 kips < 24.1 kips, bolts **o.k.** for tension.

   Bolt design shear strength $= 0.65 \times 54 \times 0.4418 = 15.5$ kips

   Since 7.45 kips < 15.5 kips, bolts **o.k.** for shear.

2. Assume $p = 4.5$ in.

   Enter preliminary selection table with

   $$\frac{2T}{p} = \frac{2 \times 14.9}{4.5} = 6.62 \text{ kips/in., and}$$

   b equal to approximately 2 in.

   Select $t_{pre} = \frac{5}{8}$ to $\frac{11}{16}$

   Try two angles $4 \times 4 \times \frac{5}{8} \times 1'\text{-}1\frac{1}{2}''$

   $b = 1.875$ in.

   $a = 1.5$ in. $(< 1.25b)$

$b' = 1.875 - 0.375 = 1.50$ in.

$a' = 1.5 + 0.375 = 1.875$ in.

$\rho = 1.50/1.875 = 0.800$

$\delta = 0.819$

$t_c = \sqrt{\dfrac{4.44 \times 24.1 \times 1.50}{4.5 \times 36}} = 0.995$ in.

Calculate $\alpha = \dfrac{1}{.800 \times 1.819}\left[\left(\dfrac{.995}{.625}\right)^2 - 1\right] = 1.055$

3. $T_{max.} = 24.1 \times \left(\dfrac{0.625}{0.995}\right)^2 (1 + 0.819) = 17.3$ kips

Since 17.3 kips > 14.9 kips, angles $4 \times 4 \times \frac{5}{8}$ are **o.k.**

TENSION MEMBERS
Net areas

	Two angles—net area											
Angle Designation	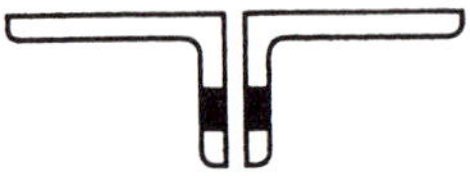 2 Holes out				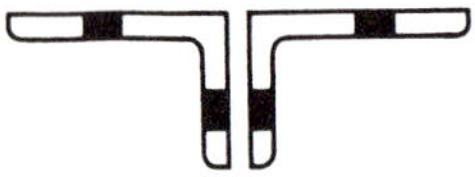 4 Holes out				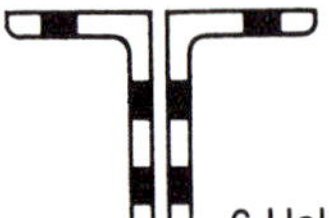 6 Holes out			
	Fastener Dia., In.				Fastener Dia., In.				Fastener Dia., In.			
	$3/4$	$7/8$	1	$1 1/8$	$3/4$	$7/8$	1	$1 1/8$	$3/4$	$7/8$	1	$1 1/8$
L 8x8x1⅛	31.5	31.2	30.9	30.7	29.5	29.0	28.4	27.8	27.6	26.7	25.9	25.0
1	28.3	28.0	27.8	27.5	26.5	26.0	25.5	25.0	24.8	24.0	23.3	22.5
⅞	24.9	24.7	24.5	24.3	23.4	23.0	22.5	22.1	21.9	21.2	20.6	19.9
¾	21.6	21.4	21.2	21.0	20.3	19.9	19.5	19.1	18.9	18.4	17.8	17.3
⅝	18.1	18.0	17.8	17.7	17.0	16.7	16.4	16.1	15.9	15.5	15.0	14.5
½	14.6	14.5	14.4	14.3	13.8	13.5	13.3	13.0	12.9	12.5	12.1	11.8
L 8x6x1	24.3	24.0	23.8	23.5	22.5	22.0	21.5	21.0	20.8	20.0	19.3	18.5
¾	18.6	18.4	18.2	18.0	17.3	16.9	16.5	16.1	15.9	15.4	14.8	14.3
½	12.6	12.5	12.4	12.3	11.8	11.5	11.3	11.0	10.9	10.5	10.1	9.75
L 8x4x1	20.3	20.0	19.8	19.5	18.5	18.0	17.5	17.0	16.8	16.0	15.3	14.5
¾	15.6	15.4	15.2	15.0	14.3	13.9	13.5	13.1	12.9	12.4	11.8	11.3
½	10.6	10.5	10.4	10.3	9.75	9.50	9.25	9.00	8.87	8.50	8.13	7.75
L 7x4x ¾	14.1	13.9	13.7	13.5	12.8	12.4	12.0	11.6	11.4	10.9	—	—
½	9.62	9.50	9.37	9.25	8.75	8.50	8.25	8.00	7.87	7.50	—	—
⅜	7.32	7.23	7.14	7.04	6.67	6.48	6.27	6.08	6.01	5.73	—	—
L 6x6x1	20.3	20.0	19.8	19.5	18.5	18.0	17.5	17.0	16.8	16.0	—	—
⅞	17.9	17.7	17.5	17.3	16.4	16.0	15.5	15.1	14.9	14.2	—	—
¾	15.6	15.4	15.2	15.0	14.3	13.9	13.5	13.1	12.9	12.4	—	—
⅝	13.1	13.0	12.8	12.7	12.0	11.7	11.4	11.1	10.9	10.5	—	—
½	10.6	10.5	10.4	10.3	9.75	9.50	9.25	9.00	8.87	8.50	—	—
⅜	8.06	7.97	7.88	7.78	7.41	7.22	7.03	6.84	6.75	6.47	—	—
L 6x4x ¾	12.6	12.4	12.2	12.0	11.3	10.9	10.5	10.1	9.94	—	—	—
⅝	10.6	10.5	10.3	10.2	9.53	9.22	8.91	8.60	8.44	—	—	—
½	8.62	8.50	8.37	8.25	7.75	7.50	7.25	7.00	6.88	—	—	—
⅜	6.56	6.47	6.38	6.28	5.91	5.72	5.53	5.34	5.25	—	—	—

Net areas are computed in accordance with LRFD Specification, Sect. B2.

TENSION MEMBERS
Net areas

Angle Designation	Two angles—net area											
	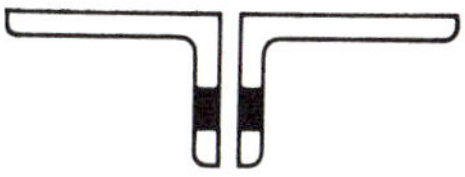 2 Holes out				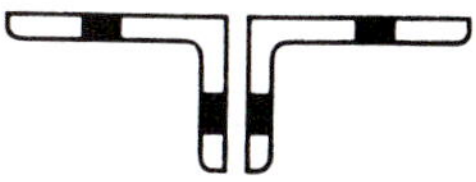 4 Holes out				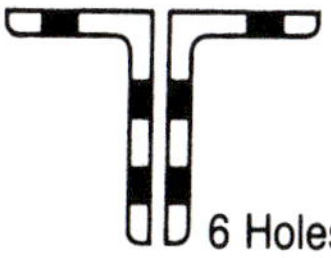 6 Holes out			
	Fastener Dia., In.				Fastener Dia., In.				Fastener Dia., In.			
	$3/4$	$7/8$	1	$1\,1/8$	$3/4$	$7/8$	1	$1\,1/8$	$3/4$	$7/8$	1	$1\,1/8$
L 6x3½x⅜	6.18	6.09	6.00	—	5.53	5.34	—	—	4.87	—	—	—
⁵/₁₆	5.19	5.11	5.04	—	4.65	4.49	—	—	4.10	—	—	—
L 5x5 x⅞	14.4	14.2	14.0	13.8	12.9	12.5	12.0	11.6	—	—	—	—
¾	12.6	12.4	12.2	12.0	11.3	10.9	10.5	10.1	—	—	—	—
½	8.62	8.50	8.37	8.25	7.75	7.50	7.25	7.00	—	—	—	—
⅜	6.56	6.47	6.38	6.28	5.91	5.72	5.53	5.34	—	—	—	—
L 5x3½x¾	10.3	10.1	9.93	—	8.99	8.62	—	—	—	—	—	—
½	7.12	7.00	6.87	—	6.25	6.00	—	—	—	—	—	—
⅜	5.44	5.35	5.25	—	4.79	4.60	—	—	—	—	—	—
⁵/₁₆	4.57	4.49	4.42	—	4.03	3.87	—	—	—	—	—	—
L 5x3 x½	6.62	6.50	—	—	5.75	5.50	—	—	—	—	—	—
⅜	5.06	4.97	—	—	4.41	4.22	—	—	—	—	—	—
⁵/₁₆	4.25	4.17	—	—	3.71	3.55	—	—	—	—	—	—
L 4x4 x¾	9.57	9.38	9.19	9.00	8.26	7.88	—	—	—	—	—	—
⅝	8.13	7.97	7.81	7.66	7.03	6.72	—	—	—	—	—	—
½	6.62	6.50	6.37	6.25	5.75	5.50	—	—	—	—	—	—
⅜	5.06	4.97	4.88	4.78	4.41	4.22	—	—	—	—	—	—
⁵/₁₆	4.25	4.17	4.10	4.02	3.71	3.55	—	—	—	—	—	—
L 4x3½x½	6.12	6.00	5.87	—	5.25	5.00	—	—	—	—	—	—
⅜	4.68	4.59	4.50	—	4.03	3.84	—	—	—	—	—	—
⁵/₁₆	3.95	3.87	3.80	—	3.41	3.25	—	—	—	—	—	—
L 4x3 x½	5.62	5.50	—	—	4.75	4.50	—	—	—	—	—	—
⅜	4.30	4.21	—	—	3.65	3.46	—	—	—	—	—	—
⁵/₁₆	3.63	3.55	—	—	3.09	2.93	—	—	—	—	—	—
¼	2.94	2.88	—	—	2.50	2.38	—	—	—	—	—	—

Net areas are computed in accordance with LRFD Specification, Sect. B2.

AMERICAN INSTITUTE OF STEEL CONSTRUCTION

REDUCTION OF AREA FOR HOLES
Area in sq. in. = diameter of hole × thickness of material

Thickness In.	Dia. of Hole, In.							
	³⁄₄	¹³⁄₁₆	⁷⁄₈	¹⁵⁄₁₆	1	1¹⁄₁₆	1¹⁄₈	1³⁄₁₆
³⁄₁₆	.141	.152	.164	.176	.188	.199	.211	.223
¹⁄₄	.188	.203	.219	.234	.250	.266	.281	.297
⁵⁄₁₆	.234	.254	.273	.293	.313	.332	.352	.371
³⁄₈	.281	.305	.328	.352	.375	.398	.422	.445
⁷⁄₁₆	.328	.355	.383	.410	.438	.465	.492	.520
¹⁄₂	.375	.406	.438	.469	.500	.531	.563	.594
⁹⁄₁₆	.422	.457	.492	.527	.563	.598	.633	.668
⁵⁄₈	.469	.508	.547	.586	.625	.664	.703	.742
¹¹⁄₁₆	.516	.559	.602	.645	.688	.730	.773	.816
³⁄₄	.563	.609	.656	.703	.750	.797	.844	.891
¹³⁄₁₆	.609	.660	.711	.762	.813	.863	.914	.965
⁷⁄₈	.656	.711	.766	.820	.875	.930	.984	1.04
¹⁵⁄₁₆	.703	.762	.820	.879	.938	.996	1.05	1.11
1	.750	.813	.875	.938	1.00	1.06	1.13	1.19
¹⁄₁₆	.797	.863	.930	.996	1.06	1.13	1.20	1.26
¹⁄₈	.844	.914	.984	1.05	1.13	1.20	1.27	1.34
³⁄₁₆	.891	.965	1.04	1.11	1.19	1.26	1.34	1.41
¹⁄₄	.938	1.02	1.09	1.17	1.25	1.33	1.41	1.48
⁵⁄₁₆	.984	1.07	1.15	1.23	1.31	1.39	1.48	1.56
³⁄₈	1.03	1.12	1.20	1.29	1.38	1.46	1.55	1.63
⁷⁄₁₆	1.08	1.17	1.26	1.35	1.44	1.53	1.62	1.71
¹⁄₂	1.13	1.22	1.31	1.41	1.50	1.59	1.69	1.78
⁹⁄₁₆	1.17	1.27	1.37	1.46	1.56	1.66	1.76	1.86
⁵⁄₈	1.22	1.32	1.42	1.52	1.63	1.73	1.83	1.93
¹¹⁄₁₆	1.27	1.37	1.48	1.58	1.69	1.79	1.90	2.00
³⁄₄	1.31	1.42	1.53	1.64	1.75	1.86	1.97	2.08
¹³⁄₁₆	—	—	1.59	1.70	1.81	1.93	2.04	2.15
⁷⁄₈	—	—	1.64	1.76	1.88	1.99	2.11	2.23
¹⁵⁄₁₆	—	—	1.70	1.82	1.94	2.06	2.18	2.30
2	—	—	1.75	1.88	2.00	2.13	2.25	2.38
¹⁄₁₆	—	—	1.80	1.93	2.06	2.19	2.32	2.45
¹⁄₈	—	—	1.86	1.99	2.13	2.26	2.39	2.52
³⁄₁₆	—	—	1.91	2.05	2.19	2.32	2.46	2.60
¹⁄₄	—	—	1.97	2.11	2.25	2.39	2.53	2.67
⁵⁄₁₆	—	—	2.02	2.17	2.31	2.46	2.60	2.75
³⁄₈	—	—	2.08	2.23	2.38	2.52	2.67	2.82
⁷⁄₁₆	—	—	2.13	2.29	2.44	2.59	2.74	2.89
¹⁄₂	—	—	2.19	2.34	2.50	2.66	2.81	2.97
⁵⁄₈	—	—	2.30	2.46	2.63	2.79	2.95	3.12
³⁄₄	—	—	2.41	2.58	2.75	2.92	3.09	3.27
⁷⁄₈	—	—	2.52	2.70	2.88	3.05	3.23	3.41
3	—	—	2.63	2.81	3.00	3.19	3.38	3.56

NET SECTION OF TENSION MEMBERS

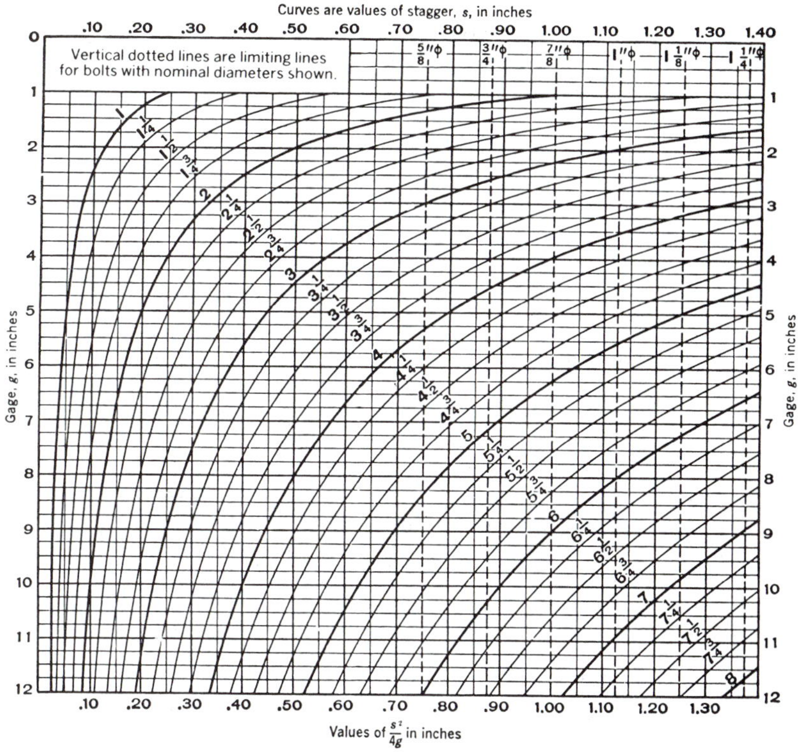

Curves are values of stagger, s, in inches

Vertical dotted lines are limiting lines for bolts with nominal diameters shown.

Gage, g, in inches

Values of $\frac{s^2}{4g}$ in inches

The above chart will simplify the application of the rule for net width, Sect. B2 of the LRFD Specification. Entering the chart at left or right with the gage g and proceeding horizontally to intersection with the curve for the pitch s, then vertically to top or bottom, the value of $s^2/4g$ may be read directly.

Step 1 of the example below illustrates the application of the rule and the use of the chart. Step 2 illustrates the application of the 85% of gross area limitation applicable to connection fittings.

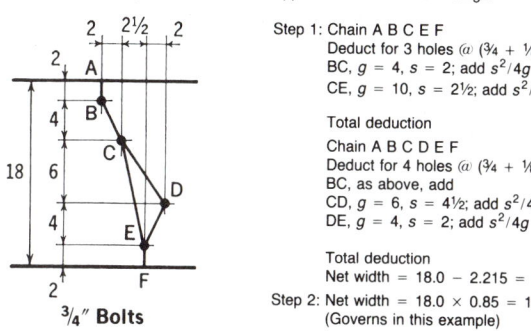

Step 1: Chain A B C E F

Deduct for 3 holes @ (¾ + ⅛)	= −2.625
BC, $g = 4$, $s = 2$; add $s^2/4g$	= +0.25
CE, $g = 10$, $s = 2½$; add $s^2/4g$	= +0.16
Total deduction	= −2.215″

Chain A B C D E F	
Deduct for 4 holes @ (¾ + ⅛)	= −3.50
BC, as above, add	= +0.25
CD, $g = 6$, $s = 4½$; add $s^2/4g$	= +0.85
DE, $g = 4$, $s = 2$; add $s^2/4g$	= +0.25
Total deduction	= −2.15″

Net width = 18.0 − 2.215 = 15.785″.

Step 2: Net width = 18.0 × 0.85 = 15.3″
(Governs in this example).

In comparing the path CDE with the path CE, it is seen that if the sum of the two values of $s^2/4g$ for CD and DE exceed the single value of $s^2/4g$ for CE, by more than the deduction for one hole, then the path CDE is not critical as compared with CE.

Evidently if the value of $s^2/4g$ for one leg CD of the path CDE is greater than the deduction for one hole, the path CDE cannot be critical as compared with CE. The vertical dotted lines in the chart serve to indicate, for the respective bolt diameters noted at the top thereof, that any value of $s^2/4g$ to the right of such line is derived from a non-critical chain which need not be further considered.

MOMENT CONNECTIONS
Welded

Many Type PR (partially restrained) connections must be designed to develop specified resisting moments. The following method is recommended for design of such a connection subjected to gravity loading.

M = factored connection moment, kip-ft

R = factored end reaction of beam, kips

ϕr_v = design shear or bearing value for one fastener, kips

ϕF_{vp} = design shear in plate, ksi

D = number of $\frac{1}{16}$ in. in weld size

$1.392D$ = value of E70XX weld, kips per linear in. of fillet weld per $\frac{1}{16}$ in. leg

T = horizontal force top and bottom of beam, kips

A_p = area of plate, top or bottom, in.^{2}

A_{st} = area of stiffeners, in.^{2}

A_{bc} = planar area of web at beam-to-column connection, in.^{2}

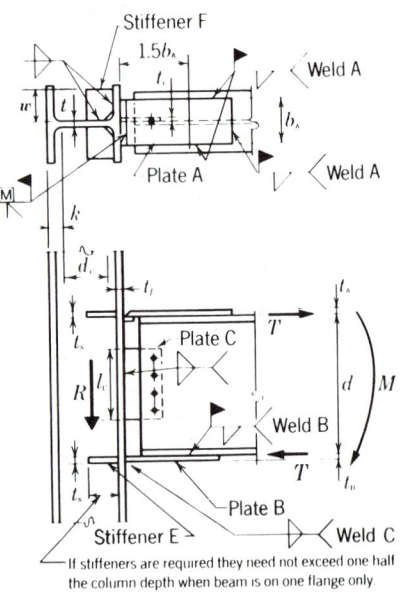

The moment is assumed to be resisted by plates A and B welded to the top and bottom of beam and to the column. The shear is assumed to be transferred to the column by the vertical web plate C, using fasteners in the beam web and shop welds to the column.

An unwelded length of 1.5 times the top flange width b_A is assumed in this example, to permit the elongation of the plate that is necessary to obtain the desired partially restrained action.

DESIGN PROCEDURE

A. Determine horizontal force $T = (M \times 12)/d$

B. Design top plate A; determine length and size of weld A and C.

$$A_p = T/\phi F_t$$

Length of weld = $T/1.392D$ (for E70XX Electrodes)

Select bottom plate B and determine length and size of weld B.
Area of plate B should be $\geq$ area of plate A.

C. Design the web connection:

1. No. of fasteners required for shear $= R/\phi r_v$ (Table I-D).

2. Check bearing on beam web (Table I-F or I-E).

3. Design shear plate C:

   a. Check net section in shear

   b. Check bearing on plate C (Table I-E).

   c. Determine weld size.

   $$\text{Min. } D = \frac{R}{2 \times 1.392l} \text{ (for full length welds both sides)}$$

   If intermittent or less than full length welds are used, min. plate thickness and weld size must be adjusted to satisfy LRFD Specification Sects. J2.4 and J2.2.

   d. Check minimum plate thickness for weld used (Table III).

D. Investigate column web shear (LRFD Commentary K1.7)

   Column web reinforcement required if

   $$t_w < \frac{20.1(M_1 + M_2) + 1.6V_u d_b}{F_{yc} A_{bc}}$$

   When there are connections on each side of the column, M_1 and M_2 refer to the moments on the left and right hand side of the column, respectively.

E. Check column for web stiffeners

   Column web stiffeners are required (for nomenclature, see Part 2, Columns, General Notes on Column Web Stiffeners):

   At both flanges if $P_{bf} > t_b P_{wi} + P_{wo}$

   At compression flange if $P_{bf} > P_{wb}$

   At tension flange if $P_{bf} > P_{fb}$

   Required stiffeners must comply with the provisions of Sect. K1.

   Selected stiffeners must comply with width-thickness ratio limitation.

F. Determine weld requirements of stiffeners to column web and flange.

EXAMPLE

Given:

Design a partially restrained connection of a W18 × 50 beam framed to the flange of a W14 × 109 column. The factored end moment of 225 kip-ft and factored end reaction of 75 kips are from live and dead load only. Factored column web shear is 19 kips. Material is A36, bolts are ¾″ dia. A325-N, weld is E70XX.

From Properties tables, Part 1:

Beam (W18 × 50): d = 17.99 in., b_f = 7½ in.
t_w = 0.355 in., t_f = 0.570 in.

Column (W14 × 109): d = 14.32 in., b_f = 14⅝ in.
t_w = 0.525 in., t_f = 0.860 in.

Solution:

A. Horizontal force at beam flange:

$T = (M \times 12)/d = (225 \times 12)/18 = 150$ kips

B. $A_p = T/\phi F_t = 150/(0.9 \times 36) = 4.63$ in.^{2}

Top flange plate A:

Select 6 in. wide plate:

$t_A = 4.63/6 = 0.77$ in. (use ¾ × 6 in. flat)

Bottom flange plate B:

Select 9 in. wide plate:

$t_B = 4.63/9 = 0.51$ in. (use ½ × 9 in. flat)

Design Welds A and B:

Select 5⁄16 in. fillet weld, E70XX.

Length of weld = $150/(5 \times 1.392) = 22$ in. minimum.

Weld A: Use 6 in. across and 8 in. along each side.

Weld B: Use 11 in. along each side.

Design Weld C:

Weld C: $150/(2 \times 9 \times 1.392) = 6$; use ⅜ -in. fillet.

C. Design web connection:

1. Connect beam web to column shear plate with ¾"dia. A325-N bolts.

No. bolts required = $R/\phi r_v = 75/15.5 = 4.8$; use 5 bolts

2. Check bearing on beam web:

For F_u = 58 ksi and t_w = 0.355 in. (using Table I-F):

Design load = $78.3 \times 0.355 \times 5 = 139$ kips > 75 kips

3. Design shear plate C:

a. Try a shear plate with 3-in. pitch, 1¼-in. end distance, l = 12 in.:
l_{net} = 12 − 4 (¾ + ¹⁄16) = 8.75 in.

$\phi F_{vp} = 0.75 \times 0.6 \times 58 = 26.1$ ksi (Formula J4-1)

t_c = $75/(26.1 \times 8.75) = 0.33$ in.

Try a ⅜-in. plate.

b. Check bearing:

From Table I-F, for $F_u = 58$ ksi, $l_v = 1\frac{1}{4}$ in.:

Design load: $78.3 \times \frac{3}{8} = 29.4$ kips/bolt at 3-in. spacing

5 bolts @ 29.4 = 147 kips

c. Determine required weld to column flange:

$$D_{min} = \frac{75}{2 \times 1.392 \times 12} = 2.24$$

Since the column flange thickness is over $\frac{3}{4}$ in., use $\frac{5}{16}$-in. fillet weld (LRFD Specification Table J2.5).

Use: Shear connection plate $\frac{3}{8} \times 5 \times 12$ in. welded to the column flange with $\frac{5}{16}$ in. fillet welds full length each side.

D. Investigate column web shear, $V_u = 19$ kips:

A_{bc} = planar area of column web

$= (17.99 + \frac{3}{4} + \frac{1}{2}) \times 14.32 = 276$ in.^{2}

$$t \geq \frac{(20.1 \times 225) - (1.6 \times 19 \times 17.99)}{36 \times 276} = 0.40 \text{ in.} < 0.525 \text{ in.}$$

∴ column web need not be reinforced

E. Determine need for column web stiffeners.

Horizontal force at stiffeners:

$$\frac{M \times 12}{d + \frac{1}{2}(t_A + t_B)} = \frac{225 \times 12}{17.99 + \frac{1}{2}(\frac{3}{4} + \frac{1}{2})}$$

$$= 145 \text{ kips}$$

From Column Load Tables for W14 × 109,

$P_{wo} = 148$ kips (yielding)

$P_{wi} = 19$ kips/in. (yielding)

$P_{wb} = 281$ kips (buckling)

$P_{fb} = 150$ kips (tension)

$P_{wo} + t_f P_{wi} = 148 + (0.57 \times 19) = 159$ kips > 145 kips; stiffeners not req'd.

$P_{wb} = 281$ kips > 145 kips; stiffeners not req'd.

$P_{fb} = 150$ kips > 145 kips; stiffeners not req'd.

MOMENT CONNECTIONS
Shop welded—field bolted

Many framing systems are designed as Type FR (fully restrained) and the connections must be designed to develop the frame moments. The following example illustrates the design of a moment connection that may be used in fully restrained construction. For nomenclature, see Moment Connections, Welded.

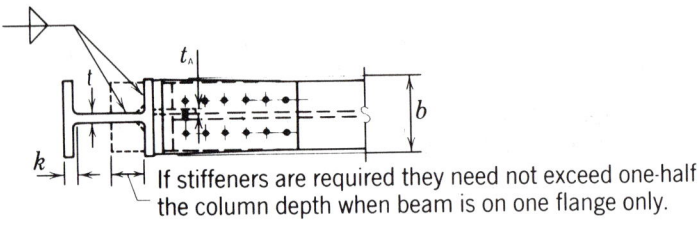

If stiffeners are required they need not exceed one-half the column depth when beam is on one flange only.

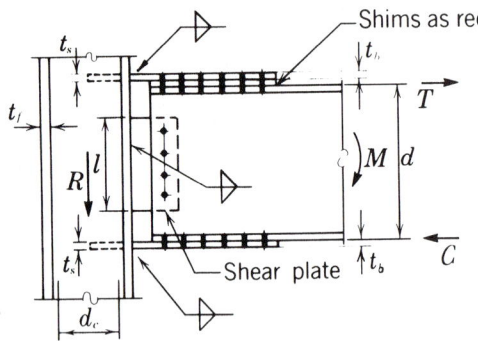

Shims as required

Shear plate

The moment is assumed to be resisted by flange plates shop welded to the column and field connected to the beam flanges. The shear is assumed to be transferred to the column by a vertical plate shop welded to the column and field connected to the beam web.

DESIGN PROCEDURE

A. Determine beam flange area reduction for fastener holes (LRFD Specification Sect. B1).

B. Determine horizontal force at beam flange:

$$T = (M \times 12)/d$$

C. Design flange plates:

Gross section: $f_t = T/A_{gross} \leq 0.9F_y$
Net section: $f_t = T/A_{net} \leq 0.75F_u$

D. Determine the number of fasteners required to develop the horizontal force in the flanges:

No. of fasteners $= T/\phi r_v$

E.   Design the web connection:

No. of fasteners required for shear = $R/\phi r_v$ (Table I-D).

Check bearing on beam web (Table I-E).

Check shear on plate.

Check bearing on plate.

Determine weld requirements (LRFD Specification Table J2.3).

F.   Check column web shear:

Column web reinforcement may be required by LRFD Commentary K1.7.

G.   Check column for web stiffeners:

Column web stiffeners are required (for nomenclature, see Part 2, Columns, General Notes on Column Web Stiffeners):

At both flanges if $P_{bf} > t_b P_{wi} + P_{wo}$

At compression flange if $P_{bf} > P_{wb}$

At tension flange if $P_{bf} > P_{fb}$

where

P_{bf} = factored beam flange or connection plate force in a restrained connection, kips

Selected stiffeners must comply with width-thickness ratio limitation.

H.   Determine weld requirements of stiffeners to column web and flange.

Note: Oversize holes in the moment flange plates assist in the field assembly of this type connection by compensating for the rolling, fabrication and erection tolerances.

EXAMPLE

Given:

Design a moment connection for a W18 × 50 beam framed to a W14 × 99 column. The factored design moment is 218 kip-ft and the factored end reaction is 43.5 kips. Factored column web shear is 18 kips. All material is ASTM A36 steel. Use A325 bolts and E70XX electrodes. Use standard holes.

From Properties and Dimensions Tables, Part 1:

Beam (W18 × 50): d = 17.99 in., b_f = 7.495 in.
t_w = 0.355 in., t_f = 0.570 in.

Column (W14 × 99): d = 14.16 in., b_f = 14.565 in.
t_w = 0.485 in., t_f = 0.780 in.
k = 1^{7}⁄$_{16}$ in., T = 11¼ in.

Solution:

$M = 218$ kip-ft

$Z_{req} = (218 \times 12)/(0.9 \times 36) = 80.7$ in.^{3}

Assume 2 rows of ⅞-in. dia. A325 bolts in a bearing-type connection.

A. Beam flange area reduction:

A_f (gross) $= 7.495 \times 0.570 = 4.27$ in.^{2}

A_f (net) $= 4.27 - 2(0.875 + 0.125)(0.57) = 3.13$ in.^{2}

$$\frac{4.27 - 3.13}{4.27} \times 100\% = 26.7\%$$
$$\underline{\quad\quad\quad -15.0\%}$$
$$11.7\% \text{ excess}$$

Z (net) $\approx 101 - (2 \times 0.117 \times 4.27 \times 9)$
$= 92$ in.$^3 > 80.7$ in.^{3} (beam is **o.k.**)

B. Horizontal force at beam flange:

$T = (M \times 12)/d = (218 \times 12)/17.99 = 145$ kips

C. Design flange plates:

Try plate ¾ × 6¾
Gross section

$$f_t = \frac{145}{¾ \times 6¾} = 28.6 < 0.9 \times 36 = 32.4 \text{ ksi} \quad \textbf{o.k.}$$

Net section
Net area $= ¾[6¾ - 2(⅞ + ⅛)]$
$= 3.56 < 0.85 \times 5.06 = 4.30$ in.^{2}

$$f_t = \frac{145}{3.56} = 40.7 < 0.75 \times 58 = 43.5 \text{ ksi} \quad \textbf{o.k.} \quad \text{(Formula D1-2)}$$

D. Flange connection:

Assume ⅞"-dia. A325-N bolts in a bearing-type connection in standard holes,

No. bolts required:

$T/\phi r_v = 145/21.1 = 6.9$ (Table I-D)

Use: 8—⅞ in. dia. A325-N bolts.

E. Web connection:

Assume ⅞" dia. A325 bolts in a bearing-type connection with threads included in the shear plane.

No. of bolts required for shear:

$R/\phi r_v = 43.5/21.1 = 2.1$ (Table I-D)

Try 4 bolts.

Check bearing on beam web:

Assume 1½ in. end distance and 3 in. pitch.

From Table I-E, for $F_u = 58$ ksi, $l_v = 1½$ in. and $t = 0.390$ in.:

Design load = 91.3 (for $t = 1$ in.) × 0.390 × 4 = 142 kips > 43.5 kips

Use: 4 ⅞"-dia. A325-N bolts in a bearing-type connection.

Design shear plate:

Try a shear plate with 3-in. pitch, 1½-in. end distance, $l = 12$ in.

$l_{net} = 12 - 4(⅞ + 1/16) = 8.25$ in.

$F_{vp} = 0.45 × 58 = 26.1$ ksi

$t_p = 43.5/(26.1 × 8.25) = 0.20$ in.

Try a 5/16-in. plate

Check bearing:

From Table I-E, for $F_u = 58$ ksi, $l_v = 1½$ in. and $t = 5/16$ in.:

$4 × 91.3 × 0.3125 = 114$ kips **o.k.**

Determine required weld to column flange:

$$D_{min} = \frac{43.5}{2 × 1.392 × 12} = 1.30$$

Since the column flange thickness is over ¾ in., use 5/16 in. fillet weld (LRFD Specification Table J2.5).

Use: Shear connector plate 5/16 × 5 × 12 in. welded to the column flange with 5/16 in. fillet welds full length each side. Designate on the drawings that the weld is to be built out to obtain 5/16-in. throat thickness.

F. Investigate column web shear; $V_u = 18$ kips:

$A_{bc} = [18.11 + (2 × ⅞)] × 14.16 = 281$ in.^{2}

$$t \geq \frac{20.1(M_1 + M_2) - 1.6V_u d_b}{F_{yc}A_{bc}} = \frac{20.1 × 218 - 1.6 × 18 × 17.99}{36 × 281}$$

$$= 0.382 \text{ in.} < 0.485 \text{ in.} \quad \textbf{o.k.}$$

Column web need not be reinforced.

G. Determine need for column web stiffeners.

Horizontal force at stiffeners:

$$\frac{M × 12}{d + t_b} = \frac{218 × 12}{17.99 + 0.75} = 140 \text{ kips}$$

From Column Load Tables for W14 × 99,

$P_{wo} = 125$ kips (yielding)

$P_{wi} = 17$ kips/in. (yielding)

$P_{wb} = 222$ kips (buckling)

$P_{fb} = 123$ kips (tension)

$P_{wo} + t_f P_{wi} = 125 + (0.57 \times 17) = 135$ kips < 140 kips; stiffeners req'd.

$P_{wb} = 222$ kips > 140 kips; stiffeners not req'd.

$P_{fb} = 123$ kips < 140 kips; stiffeners req'd.

Stiffeners are required at both the compression flange and the tension flange. For stiffener design guidelines, see Ex. 5, LRFD Manual, Part 2.

Min. stiffener width = $7.495/3 - 0.485/2$
= 2.26 in. Use 2½ in.

Min. stiffener thickness = $0.57/2 = 0.29$ in.
Use 5/16 in.

Min. stiffener length = $14.16/2 - .78 = 6.3$ in.
Use 6½ in.

Stiffener area req'd = $\dfrac{140 - 123}{36.0} = 0.5$ in.^{2}

Stiffener area supplied = $2½ \times 5/16 = .78$ in.^{2} **o.k.**

H. Stiffener weld requirements

Use fillet welds, E70 electrodes

From Table J2.5,

Min. weld size to column web: 3/16 in.

Min. weld size to column flange: 5/16 in.

Weld length calculation:

Force on compression flange: $140 - 135 = 5$ kips

Force on tension flange: $140 - 123 = 17$ kips

Tension flange governs.

Load in stiffener = $17/2 = 8.5$ kips each side.

Weld length (web) = $\dfrac{8.5}{2 \times 3 \times 1.392} = 1.0$ in.

Make weld full length of stiffener:

$6½ - 3/4$ (clip) = 5¾ in. both sides.

Weld length (flange) = $\dfrac{8.5}{2 \times 5 \times 1.392} = 0.61$ in.

Make weld full width of stiffener:

2½ − ¾ (clip) = 1¾ in. both sides.

Check shear stress in stiffener base metal:

$$f_v = \frac{8.5}{\text{⁵⁄₁₆} \times 5\text{¾}}$$

= 4.73 ksi < 0.56 F_y = 20.2 ksi   **o.k.**   (Formula J5-3)

Since the design example as stated does not have reversible end moments, it would be permissible to furnish the two compression flange stiffeners to bear instead of welding to the column flange. See LRFD Sect. J8.1.

8.5 kips < 0.75 × 2.0 × 36 × ⁵⁄₁₆ × 2½ = 42 kips   **o.k.**

MOMENT CONNECTIONS
Field welded—field bolted

Many framing systems are designed as Type FR (fully restrained) and the connections must be designed to develop the frame moments. The following example illustrates the design of a moment connection that may be used in fully restrained construction. For nomenclature, see Moment Connections, Welded.

Full plastic moment of beam can be developed by welding the flanges. The shear is assumed to be transferred to the column by a vertical plate shop welded to the column and field connected to the beam web.

DESIGN PROCEDURE

A. Check beam strength.

B. Check beam web connection.

C. Investigate column high shear.

D. Determine need for stiffeners.

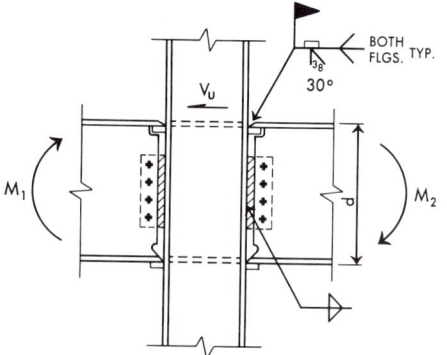

EXAMPLE

Given:

Design a moment connection for a W24 × 55 beam framed to a W14 × 74 column. The factored design moments are $M_1 = M_2 = 275$ kip-ft, and the factored end reaction is 50 kips. Factored column web shear is 27.5 kips. All material is ASTM A36 steel. Use ⅞ dia. A325-N bolts and E70 electrodes.

From Properties and Dimensions Tables, Part 1:

Beam (W24 × 55): $d = 23.57$ in., $b_f = 7.005$ in.
$t_w = 0.395$ in., $t_f = 0.505$ in.

Column (W14 × 99): $d = 14.16$ in., $b_f = 14.57$ in.
$t_w = 0.485$ in., $t_f = 0.78$ in.
$k = 1\frac{7}{16}$ in., $T = 11\frac{1}{4}$ in.

Research performed at the University of California and Lehigh University has demonstrated that the full plastic moment capacity of the girder can be developed by welding only the flanges. Therefore, a full penetration weld will be used for connecting the girder flanges to the column. A plate will be used to transfer the shear.

A. Check beam strength:

$Z_{req'd} = 275 \times 12/0.9 \times 36 = 102$ in.^{3} **o.k.**

B. Design web connection:

$50/21.1 = 2.4$, try 4 bolts

Check bearing on beam web

From Table I-E

$4 \times 91.3 \times 0.395 = 144$ kips > 50 kips **o.k.**

Design shear plate

Try 5/16 plate, 3 in. pitch, l = 12 in.

A_{net} = .3125 × [12 − 4(7/8 + 1/16)] = 2.58 in.^{2}

0.45 × 58 × 2.58 = 67.3 kips > 50 kips   **o.k.**

Check bearing on shear plate

From Table I-E

4 × 28.5 = 114 kips > 50 kips   **o.k.**

Weld to column flange

D_{min} = 50/(2 × 1.392 × 12) = 1.5 in.

Use 5/16 fillet. See Table J2.5.

C. Investigate column web high shear, V_u = 27.5 kips

M_1 = M_2 = 153 kip-ft, wind (net = 153 + 153).

M_1 = M_2 = 122 kip-ft, gravity (net = 0)

A_{bc} = 23.57 × 14.16 = 333.8 in.^{2}

$$\frac{20.1(153 + 153) − (1.6 × 27.5 × 23.57)}{36 × 333.8} = 0.426 \text{ in.} < 0.485 \text{ in.} \quad \textbf{o.k.}$$

D. Determine column web stiffener requirements

Horizontal force at stiffeners:

$$\frac{M × 12}{d − t_f} = \frac{275 × 12}{23.57 − 0.505} = 143 \text{ kips}$$

From Column Load Tables for W14 × 99,

P_{wo} = 125 kips (yielding)

P_{wi} = 17 kips/in. (yielding)

P_{wb} = 222 kips (buckling)

P_{fb} = 123 kips (tension)

P_{wo} + $t_f P_{wi}$ = 125 + (0.505 × 17) = 134 kips < 143 kips; stiffeners req'd.

P_{wb} = 222 kips > 143 kips; stiffeners not req'd.

P_{fb} = 123 kips < 143 kips; stiffeners req'd.

Stiffeners are required at both the compression flange and the tension flange. For stiffener design guidelines, see Ex. 5, LRFD Manual, Part 2.

Stiffener width = 7.005/3 − 0.485/2 = 2.09 in. Use 2½ in.

Stiffener thickness = 0.505/2 = 0.25 in. Use ¼ in.

Stiffener length = (14.16/2) − 0.78 = 6.3 in. Use 6½ in.

Stiffener area req'd. = (143 − 123)/36 = 0.56 in.^{2}

Stiffener area supplied = 2½ × ¼ = 0.625 in.^{2} **o.k.**

E. Stiffener weld requirements

Use fillet welds, E70 electrodes

From Table J2.5,

Min. weld size to column web: $\frac{3}{16}$ in.

Min. weld size to column flange: $\frac{5}{16}$ in.

Weld length calculation:

Force on compression flange: $143 - 134 = 9$ kips

Force on tension flange: $143 - 123 = 20$ kips

Tension flange governs.

Load in stiffener $= 20/2 = 10$ kips

Weld length (web) $= 10/(2 \times 3 \times 1.392) = 1.2$ in.

Make weld full length of stiffener:

$6\frac{1}{2} - \frac{3}{4}$ (clip) $= 5\frac{3}{4}$ in., both sides.

Weld length (flange) $= 10/(2 \times 5 \times 1.392) = 0.72$ in.

$2\frac{1}{2} - \frac{3}{4}$ (clip) $= 1\frac{3}{4}$ in., both sides.

Check shear stress in stiffener base metal:

$$f_v = \frac{10}{\frac{1}{4} \times 5\frac{3}{4}}$$

$= 6.96$ ksi $< 0.56 \times 36 = 20.2$ ksi   **o.k.** (Formula J5-4)

It may be permissible to furnish the two compression flange stiffeners to bear instead of welding to the column flange. See LRFD Sect. J8.1.

10 kips $< 0.75 \times 2.0 \times 36 \times \frac{1}{4} \times 2\frac{1}{2} = 33.8$ kips   **o.k.**

MOMENT CONNECTIONS
End plate

DESIGN PROCEDURE (STATIC LOADING ONLY)

When the high-strength bolts in an end plate moment connection are located a distance P_f above and approximately the same distance below the beam tension flange, the force applied to each bolt by the beam end moment can be considered equal. Neglecting minor moment resistance produced by any intermediate bolts needed to participate in resisting the beam reaction, the applied tensile force in the flange may be computed as:

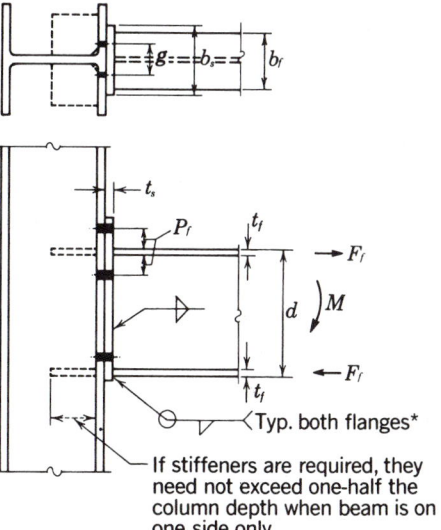

$$F_f = \frac{M}{(d - t_f)}$$

where

t_f = tension flange thickness, in.

d = depth of beam, in.

M = factored beam end moment, kip-in.

F_f = nominal flange force, kips

If stiffeners are required, they need not exceed one-half the column depth when beam is on one side only.

*For fillet welds size greater than ½ in, full-penetration welds with reinforcement should be considered.

If the size of a beam designed for a given loading is substituted for a more readily available stronger one, or for one of greater stiffness, the end connection need only be designed for the actual end moment M, unless otherwise specified.

The welds connecting the end plate to the beam flanges should be either complete-penetration groove type welds or fillet welds completely surrounding the beam flange. The size of fillet welds at a tension flange should be large enough to develop the force F_f, resulting from factored live plus dead moments, while the web welds should be adequate to resist the beam reaction. The welds in the web will usually be fillet welds designed to match the web thickness.

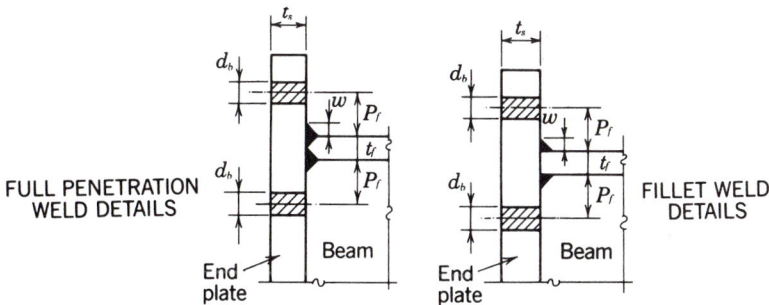

FULL PENETRATION WELD DETAILS

FILLET WELD DETAILS

The required nominal area per bolt is computed as:

$$A_b = F_f / 2n \phi F_t$$

where

> n = number of bolts on one transverse line
>
> ϕF_t = design tensile stress per bolt, ksi

The *effective* span P_e, used to compute the bending moment in the end plate, may be taken as:

$$P_e = P_f - (d_b/4) - 0.707w$$

where

> P_f = distance from center line of bolt to nearer surface of the tension flange, $d_b + \frac{1}{2}''$ is generally enough to provide wrench clearance, in.
>
> w = fillet weld size or reinforcement of groove weld, in.
>
> d_b = nominal bolt dia., in.

The end plate is designed to resist the moment M_e, using $F_b = 0.9F_y$ (LRFD Specification Sect. F1):

$$M_e = \alpha_m F_f P_e / 4$$

where

> $\alpha_m = C_a C_b (A_f / A_w)^{1/3} (P_e / d_b)^{1/4}$
>
> C_a = constant. See Table A.
>
> $C_b = \sqrt{(b_f / b_s)}$
>
> A_f = area of tension flange
>
> A_w = web area, clear of flanges. See Table B for A_f / A_w values for W shapes.

Required plate width:

$$b_s \leq 1.15 b_f$$

Required plate thickness:

$$t_s = \sqrt{\left(\frac{4M_e}{b_s (0.9) F_y} \right)}$$

Check plate shear:

$$f_v = F_f / 2 b_s t_s$$

The bottom flange to end plate weld size should be identical to the top flange weld, to insure equal strength. The web to end plate weld should be designed as a fillet weld not exceeding ½-in. in size and developing the full bending strength in the web near the flanges.

Design flange weld:

$$\text{Req'd } D = \frac{F_f^*}{1.392 \times \text{flange perimeter}}$$

Design web weld:

$$\text{Req'd minimum length of web weld} = \frac{R}{0.6 F_y \times t_w}$$

---

*Resulting from factored dead plus live load moments.

TABLE A. Values of C_a		
F_y (Ksi)*	A325	A490
36.0	1.13	1.14
42.0	1.11	1.13
45.0	1.10	1.12
50.0	1.09	1.11
55.0	1.08	1.10
60.0	1.07	1.09
65.0	1.06	1.08
90.0	1.03	1.04
*Yield stress of beam.		

TABLE B. Values of A_f/A_w

Section	A_f/A_w	Section	A_f/A_w	Section	A_f/A_w	Section	A_f/A_w
W 36x359	0.899	W 27x217	1.003	W 18x143	1.204	W 12x87	1.748
x328	0.903	x194	0.986	x130	1.186	x79	1.732
x300	0.887	x178	0.909	x119	1.082	x72	1.720
x280	0.882	x161	0.902	x106	1.059	x65	1.706
x260	0.850	x146	0.885	x 97	1.076	x58	1.631
x245	0.835	x129	0.710	x 86	1.056	x53	1.527
x230	0.818	x114	0.646	x 76	1.048	x50	1.281
x256	0.648	x102	0.635	x 71	0.741	x45	1.266
x232	0.644	x 94	0.597	x 65	0.751	x40	1.281
x210	0.588	x 84	0.545	x 60	0.751	x35	0.992
x194	0.587			x 55	0.722	x30	0.963
x182	0.579	W 24x176	1.021	x 50	0.714	x26	0.936
x170	0.573	x162	0.994	x 46	0.604	x22	0.575
x160	0.554	x146	0.959	x 40	0.595	x19	0.520
x150	0.530	x131	0.904	x 35	0.504	x16	0.419
x135	0.463	x117	0.877			x14	0.390
		x104	0.848	W 16x100	1.170		
W 33x354	0.925	x103	0.711	x 89	1.152	W 10x60	1.842
x318	0.926	x 94	0.683	x 77	1.146	x54	1.882
x291	0.913	x 84	0.655	x 67	1.149	x49	1.859
x263	0.909	x 76	0.616	x 57	0.789	x45	1.603
x241	0.853	x 68	0.560	x 50	0.781	x39	1.516
x221	0.829	x 62	0.428	x 45	0.768	x33	1.348
x201	0.807	x 55	0.397	x 40	0.772	x30	1.045
x169	0.667			x 36	0.679	x26	1.033
x152	0.612	W 21x166	1.140	x 31	0.589	x22	0.913
x141	0.583	x147	1.011	x 26	0.506	x19	0.672
x130	0.541	x132	1.002			x17	0.583
x118	0.492	x122	1.003	W 14x120	1.855	x15	0.497
		x111	0.994	x109	1.899	x12	0.463
W 30x235	0.961	x101	0.995	x 99	1.859		
x211	0.905	x 93	0.683	x 90	1.860	W 8x35	1.796
x191	0.887	x 83	0.686	x 82	1.348	x31	1.711
x173	0.861	x 73	0.683	x 74	1.394	x28	1.495
x148	0.672	x 68	0.667	x 68	1.382	x24	1.487
x132	0.606	x 62	0.641	x 61	1.364	x21	1.127
x124	0.590	x 57	0.532	x 53	1.141	x18	1.007
x116	0.558	x 50	0.465	x 48	1.115	x15	0.690
x108	0.516	x 44	0.423	x 43	1.103	x13	0.593
x 99	0.476			x 38	0.861	x10	0.635
				x 34	0.824		
				x 30	0.734	W 6x25	1.580
				x 26	0.633	x20	1.545
				x 22	0.557	x15	1.238
						x16	1.148
						x12	0.890
						x 9	0.911
						W 5x19	1.867
						x16	1.748
						W 4x13	1.442

Req'd weld to develop maximum web bending stress near flanges:

$$D = \frac{0.9F_y \times t_w}{2 \times 1.392}$$

Check adequacy of column size and need for web stiffeners or reinforcement.

EXAMPLE 1

Given:

Design a beam having a factored end reaction of 45 kips and a factored fixed-end negative moment of 180 kip-ft framed to a W14 × 193 column. The beam-to-column connection is to be an end plate connection using ASTM A325 bolts. $F_y = 36$ ksi steel is used for all members. The end plate is to be shop-welded to the beam with E70XX electrodes.

Solution:

A. Beam selection:

Req'd $Z_x = \dfrac{180 \times 12}{32.4} = 66.7$ in.^{3}

Use: W16 × 40 ($Z_x = 72.9$ in.$^3 > 66.7$ in.^{3})

$d = 16.01$ in.; $b_f = 6.995$ in.
$t_f = 0.505$ in.; $t_w = 0.305$ in.

B. Bolt design:

$$F_f = \frac{M}{(d - t_f)} = \frac{180 \times 12}{(16.01 - 0.505)} = 139 \text{ kips}$$

Assume 2 bolts per transverse line.

Req'd nominal bolt area for tension only:

$$A_b = \frac{139}{2 \times 2 \times 67.5} = 0.515 \text{ in.}^2 \text{ (LRFD Specification Table J3.2)}$$

Try ⅞" dia. A325 bolts, $A_b = 0.6013$ in.^{2}

Consider connection to be bearing-type-N.

Try 6 bolts, 4 at tension flange and 2 inside the compression flange.

No. bolts required for reaction = 45/21.1 = 2.13 < 6 **o.k.**

Use: Six ⅞" dia. A325 bolts

C. End plate design:

Maximum design plate width:

6.995 × 1.15 = 8.04 in.

Use: 8 in. plate

Effective span:

Assume ⁷⁄₁₆-in. fillet weld.

$$P_e = \left(\frac{7}{8} + \frac{1}{2}\right) - \left(\frac{1}{4} \times \frac{7}{8}\right) - \left(0.707 \times \frac{7}{16}\right)$$
$$= 0.847 \text{ in.}$$

$C_a = 1.13$ (from Table A)

$C_b = \sqrt{(6.995/8)} = 0.935$

$$\frac{A_f}{A_w} = \frac{6.995 \times 0.505}{[16.01 - (2 \times 0.505)] \times 0.305} = 0.772$$

Note: The above value of A_f/A_w may also be obtained from Table B for W16 × 40.

$$\frac{P_e}{d_b} = \frac{0.847}{0.875} = 0.968$$

$\alpha_m = 1.13 \times 0.935\,(0.772)^{1/3}\,(0.968)^{1/4} = 0.961$

$M_e = 0.961 \times 139 \times (0.847/4) = 28.3 \text{ kip-in.}$

Req'd plate thickness:

$$t_s = \sqrt{\frac{4 \times 28.3}{8 \times 32.4}} = 0.66 \text{ in.}$$

Use: ¾ in. × 8 in. A36 plate

Check plate shear:

$$f_v = \frac{139}{2 \times 8 \times \text{¾}} = 11.6 \text{ ksi} < 26.1 \text{ ksi} \quad \textbf{o.k.}$$

D. Top flange to end plate weld:

Req'd length $= 2(6.995 + 0.505) - 0.305 = 14.7$ in.

For E70XX electrodes:

$$\text{Req'd } D = \frac{139}{14.7 \times 1.392} = 6.8 \text{ sixteenths}$$

Use: ⁷⁄₁₆ in. fillet welds

E. Beam web to end plate weld:

Minimum size fillet weld is ¼ in.

Minimum length of weld required based upon shear capacity of beam web:

$$\frac{45}{0.305 \times 26.1} = 5.65 \text{ in.}$$

Required weld to develop design flexural strength $0.9F_y = 32.4$ ksi in web near flanges.

$$D = \frac{0.305 \times 32.4}{2 \times 1.392} = 3.5 < 4; \text{ ¼ in. weld} \quad \textbf{o.k.}$$

Use: ¼ in. fillet welds continuous on both sides of beam web

F. Column web stiffeners

Stiffener requirements for the column to which the end plate is bolted may be checked as follows.

To prevent column web yielding where the beam flange delivers an axial compressive force, the following formula may be used:

$$F_f < F_{yc} \times t_w \times (t_f + 6k + 2t_s + 2w)$$

An additional check may be required to prevent instability of the column web. In this case, Formula K1-4 may be conservatively used.

To prevent excessive column flange bending where the beam flange delivers an axial tension force, the following "rule of thumb" may be used:

> Stiffeners are not required in the tension region if (a) the column flange thickness equals or exceeds the calculated required bolt diameter, (b) the column flange thickness must be greater than the end plate thickness and, (c) the horizontal distance from the toe of the column web fillet to the center of the hole is no greater than 2 in.

PRELIMINARY END PLATE ESTIMATING TABLE

The Preliminary End Plate Estimating Table provides a means for preliminary sizing of plates, bolts and welds for end plate moment connections. It is emphasized that sizing of plates, bolts and welds for final design should be accomplished using specific parameters of the design connection. The following procedure and example problem illustrate a method by which preliminary selection of plates, bolts and welds may be made.

End Plate Considerations

End plate material is A36 steel regardless of the beam grade of material. Plate width is 1.15 × beam flange width, rounded up to the nearest ½ in. The tabulated plate width values are based on the heaviest beam flange width in each nominal width group, from Part 1 of the LRFD Manual. In some cases, column size considerations indicate a slightly narrower plate to be more suitable. The narrower size can be estimated from:

$$(t_s)_{nar} = \sqrt{\frac{(t_s^2 b_s)_{orig}}{(b_s)_{nar}}}$$

where

t_s = plate thickness
b_s = plate width.

Bolt Considerations

Bolts are ASTM A325 and are designed to develop the flange tensile force resulting from the development of the beam total moment capacity. The four bolt solution is for the tension flange; for reversal of stress, additional bolts are required at the other flange. In addition, transfer of the beam's reaction through fastener shear must be accounted for in the design of the entire connection.

Weld Considerations

F indicates fillet weld size on both sides of beam flange. *R* indicates value of reinforcement of full-penetration groove welds. Welds are designed using E70XX electrodes, with ϕF_v = 31.5 ksi for fillet welds and ϕF_t = 32.4 ksi for full-penetration welds (base metal F_y = 36 ksi), and ϕF_t = 45 ksi for full-penetration welds (base metal F_y = 50 ksi) (Table J2.3). The fillet weld size, for economical reasons, is arbitrarily limited to a maximum size of ½-in. If greater capacity is required, a full-penetration groove weld, with reinforcement as per AWS, is recommended.

EXAMPLE 2

Given:

Estimate the end plate dimensions, bolt sizes and weld for the full moment capacity of a W24 × 84 (F_y = 36 ksi) using Preliminary End Plate Estimating Table.

Solution:

The W24 × 84, by weight, is approximately midway between the maximum W24 × 117 and minimum W24 × 55. From the table, the plate width is indicated as 11 in. The plate thickness and bolt size can be estimated as approximately midway between the table values for the maximum and minimum beam weight:

    Plate thickness: 1⅜ in. max.
                     ⅞ in. min.
                     Estimate 1⅛ in.
    A325 bolt size: 1½ in. max.
                     1 in. min.
                     Estimate 1¼ in.

The weld size would be between a maximum full-penetration weld with ¼ in. reinforcement and a minimum ⁷⁄₁₆-in. fillet weld. Since the maximum fillet weld used in this table is ½ in., the full-penetration weld with ¼-in. reinforcement is conservatively selected.

Estimate:

End plate: 1⅛ in. × 11 in.

A325 bolts: Four 1¼-in. dia.

Weld: Full-penetration weld with ¼-in. reinforcement.

Note: Calculated (exact) results for the same problem are:

End plate: 1⅛ in. × 10½ in.

A325 bolts: Four 1¼-in. dia.

Weld: Full-penetration weld with ¼-in. reinforcement.

PRELIMINARY END PLATE DESIGN

$F_y = 50$ Ksi					$F_y = 36$ Ksi				
Weld Size In.	Bolt Size In.	Plate Thick. In.	Plate Width In.	Shape	Shape	Plate Width In.	Plate Thick. In.	Bolt Size In.	Weld Size In.
					W 36x135	14	1¼	1½	¼(R)
					W 33x130	13½	1¼	1½	¼(R)
					x118	13½	1¼	1½	³⁄₁₆(R)
					W 30x124	12½	1⅜	1½	¼(R)
					x116,108	12½	*	*	*
³⁄₁₆(R)	1½	1⅜	12½	W 30x99	x99	12½	1⅛	1⅜	³⁄₁₆(R)
					W 27x114	12	1⅜	1½	¼(R)
³⁄₁₆(R)	1½	1½	12	W 27x94	x102,94	12	*	*	*
³⁄₁₆(R)	1½	1⅜	12	x84	x84	12	1⅛	1¼	³⁄₁₆(R)
					W 24x117	15	1⅜	1½	¼(R)
¼(R)	1½	1½	11	W 24x84	x104	15	*	*	*
*	*	*	11	x76,68	x94,84,76,68	11	*	*	*
*	*	*	8½	x62	x62	8½	*	*	*
³⁄₁₆(R)	1⅛	1⅛	8½	x55	x55	8½	⅞	1	⁷⁄₁₆(F)
					W 21x111	14½	1⅜	1½	¼(R)
¼(R)	1½	1⅝	10	W 21x93	x101	14½	*	*	*
*	*	*	10	x83,73,68,62	x93,83,73,68,62	10	*	*	*
*	*	*	7½	x57,50	x57,50	7½	*	*	*
½(F)	1	1	7½	x44	x44	7½	¾	⅞	⅜(F)
¼(R)	1½	1½	13	W 18x86	W 18x119	13	1½	1½	⁵⁄₁₆(R)
*	*	*	13	x76	x106,97,86,76	13	*	*	*
*	*	*	9	x71,65,60,55,50	x71,65,60,55,50	9	*	*	*
*	*	*	7	x46,40	x46,40	7	*	*	*
½(F)	1	⅞	7	x35	x35	7	¾	⅞	⅜(F)
¼(R)	1½	1½	12	W 16x77	W 16x100	12	1½	1½	¼(R)
*	*	*	12	x67	x89,77,67	12	*	*	*
*	*	*	8½	x57,50,45,40,36	x57,50,45,40,36	8½	*	*	*
*	*	*	6½	x31	x31	6½	*	*	*
⁷⁄₁₆(F)	⅞	¾	6½	x26	x26	6½	⅝	¾	⁵⁄₁₆(F)

*Interpolate

PRELIMINARY END PLATE DESIGN

					$F_y = 50$ Ksi		$F_y = 36$ Ksi				
Weld Size In.	Bolt Size In.	Plate Thick. In.	Plate Width In.		Shape	Shape		Plate Width In.	Plate Thick. In.	Bolt Size In.	Weld Size In.
						W 14x109		17	1⅜	1½	¼(R)
¼(R)	1½	1⅝	12	W	14x82		x99,90	17	*	*	*
*	*	*	12		x74,68,61		x82,74,68,61	12	*	*	*
*	*	*	9½		x53,48,43		x53,48,43	9½	*	*	*
*	*	*	8		x38,34,30		x38,34,30	8	*	*	*
*	*	*	6		x26		x26	6	*	*	*
⅜(F)	¾	¾	6		x22		x22	6	⅝	¾	5⁄16(F)
3⁄16(R)	1½	1½	14	W	12x79	W	12x87	14	1⅜	1⅜	¼(R)
*	*	*	14		x72,65		x79,72,65	14	*	*	*
*	*	*	11½		x58,53		x58,53	11½	*	*	*
*	*	*	9½		x50,45,40		x50,45,40	9½	*	*	*
*	*	*	7½		x35,30,26		x35,30,26	7½	*	*	*
*	*	*	5		x22,19,16		x22,19,16	5	*	*	*
5⁄16(F)	¾	⅝	5		x14		x14	5	⅝	¾	3⁄16(F)
3⁄16(R)	1⅜	1½	12	W	10x60	W	10x60	12	1⅛	1⅛	3⁄16(R)
*	*	*	12		x54,49		x54,49	12	*	*	*
*	*	*	9½		x45,39,33		x45,39,33	9½	*	*	*
*	*	*	7		x30,26,22		x30,26,22	7	*	*	*
*	*	*	5		x19,17,15		x19,17,15	5	*	*	*
¼(F)	¾	⅝	5		x12		x12	5	⅝	¾	3⁄16(F)
⅛(R)	1	1⅛	9½	W	8x35	W	8x35	9½	⅞	⅞	7⁄16(F)
*	*	*	9½		x31		x31	9½	*	*	*
*	*	*	7½		x28,24		x28,24	7½	*	*	*
*	*	*	6		x21,18		x21,18	6	*	*	*
*	*	*	5		x15,13		x15,13	5	*	*	*
¼(F)	¾	⅝	5		x10		x10	5	⅝	¾	3⁄16(F)

*Interpolate

Notes

SUGGESTED DETAILS
Beam framing

SKEWED AND SLOPED CONNECTIONS

Details on this and succeeding pages are suggested treatments only, and are not intended to limit the use of other connections not illustrated. For additional information on connections, see *Detailing for Steel Construction* and *Engineering for Steel Construction.*

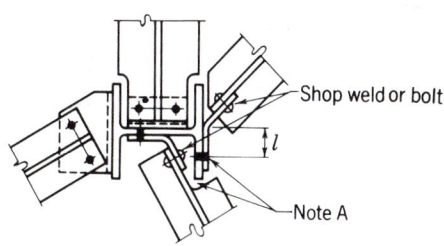

Shop weld or bolt

l

Note A

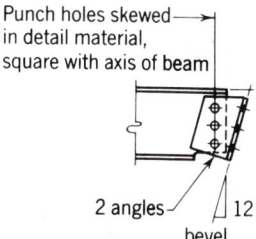

Punch holes skewed in detail material, square with axis of beam

2 angles

12
bevel

Note A: If a combination of several connections occurs at one level, provide field and driving clearance.

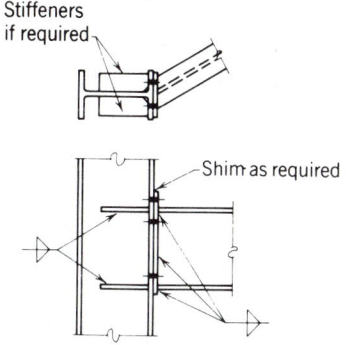

Stiffeners if required

Shim as required

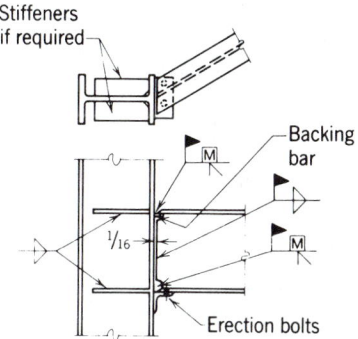

Stiffeners if required

Backing bar

$\frac{1}{16}$

Erection bolts

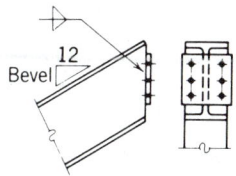

12
Bevel

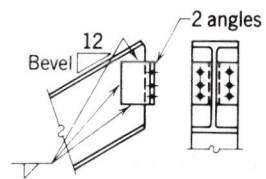

12
Bevel

2 angles

SUGGESTED DETAILS
Beam framing

MOMENT CONNECTIONS

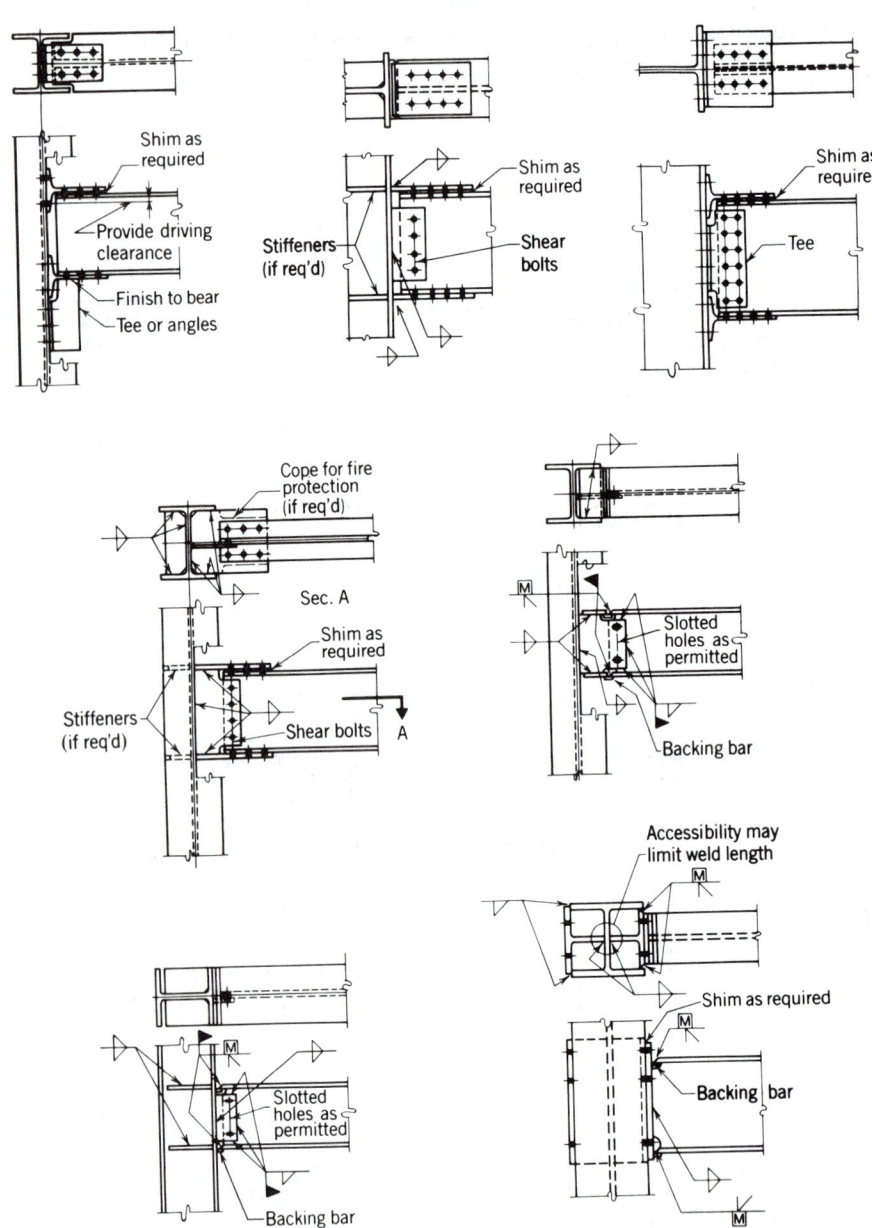

For additional information on moment connections of beams to column webs, refer to AISC, *Engineering for Steel Construction*.

SUGGESTED DETAILS
Beam framing

SHEAR CONNECTIONS

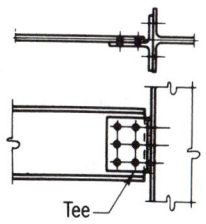

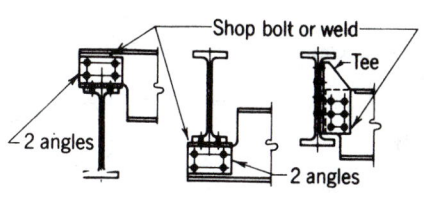

Shop bolt or weld

Tee

2 angles

2 angles

Tee

Note: Check web shear and moment in coped beam.

SHEAR SPLICES

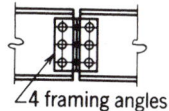

4 framing angles

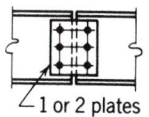

1 or 2 plates

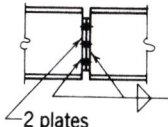

2 plates

Note: Of the above types, 4 framing angles is most flexible.

BOLTED MOMENT SPLICES

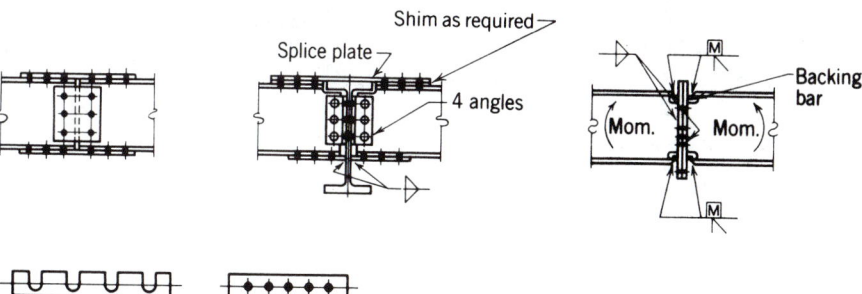

Shim as required

Splice plate

4 angles

M

Backing bar

Mom. Mom.

M

Finger   Strip

TYPICAL SHIMS

SUGGESTED DETAILS
Beam framing

WELDED MOMENT SPLICES

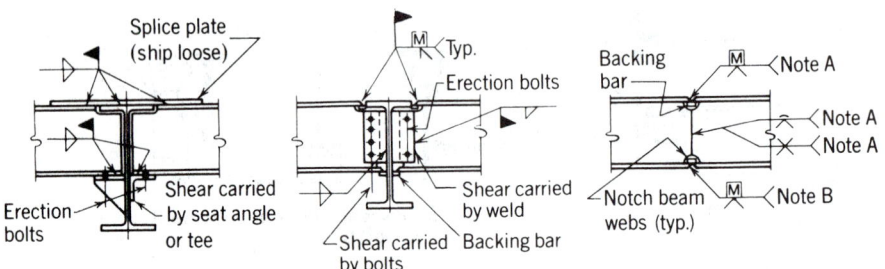

Note A: Joint preparation depends on thickness of material and welding process.
Note B: Invert this joint preparation if beam cannot be turned over.

MOMENT SPLICE AT RIDGE (FIELD BOLTED)

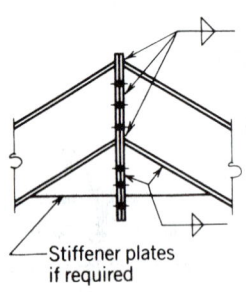

BEAM OVER COLUMN (WITH CONTINUITY)

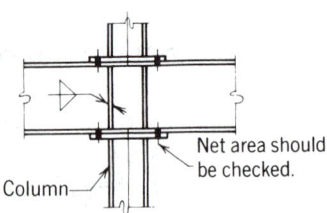

SUGGESTED DETAILS
Column base plates

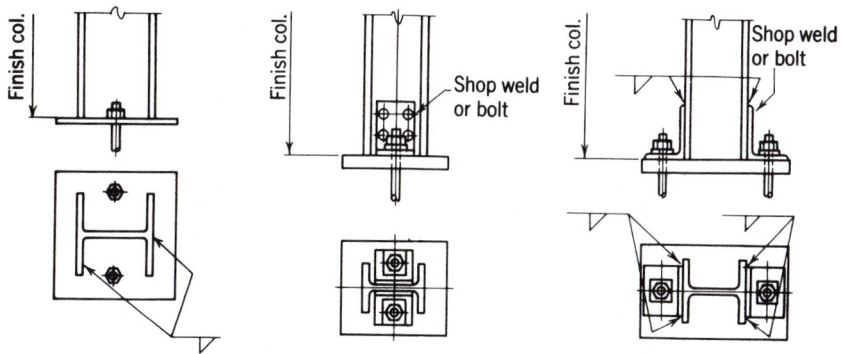

Base plate detailed and shipped loose when required.

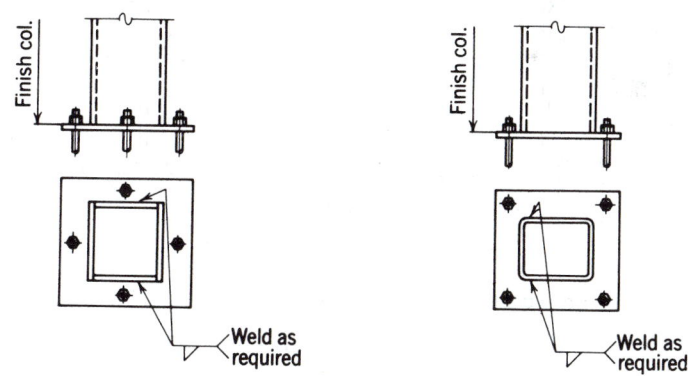

Note: Hole sizes for anchor bolts are normally
made oversize to facilitate erection as follows:
Bolts ¾ to 1"φ — ⁵⁄₁₆" oversize
Bolts 1 to 2"φ — ½" oversize
Bolts over 2"φ — 1" oversize

SUGGESTED DETAILS
Column base plates

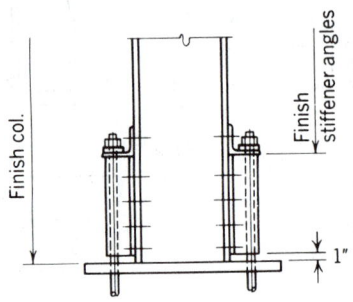 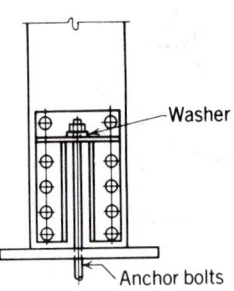

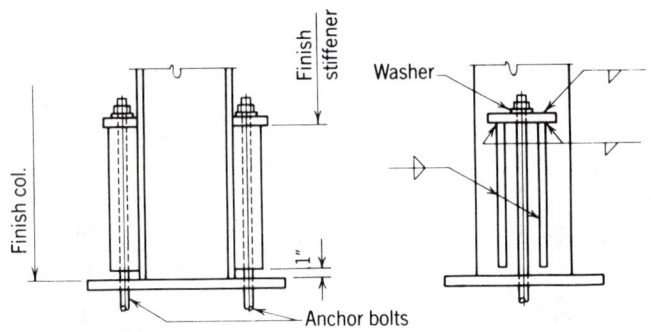

Base plates are normally detailed and shipped loose.

Note: Hole sizes for anchor bolts are normally
made oversize to facilitate erection as follows:
Bolts ¾ to 1"φ — ⁵/₁₆" oversize
Bolts 1 to 2"φ — ½" oversize
Bolts over 2"φ — 1" oversize

SUGGESTED DETAILS
Column splices

RIVETED AND BOLTED

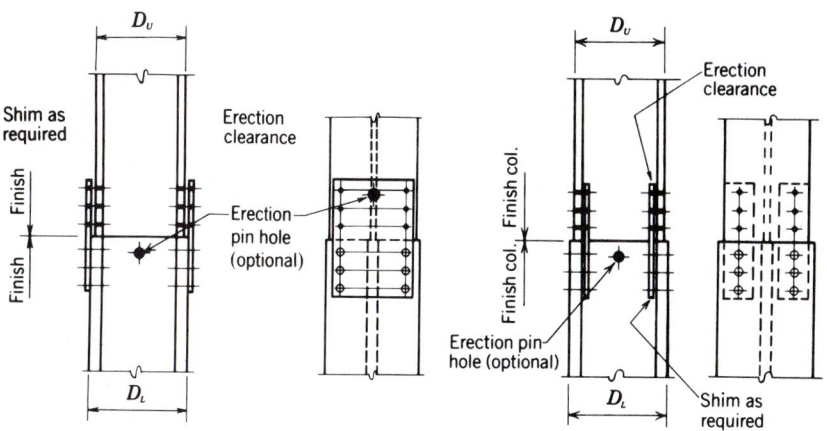

DEPTH OF D_U AND D_L
NOMINALLY THE SAME

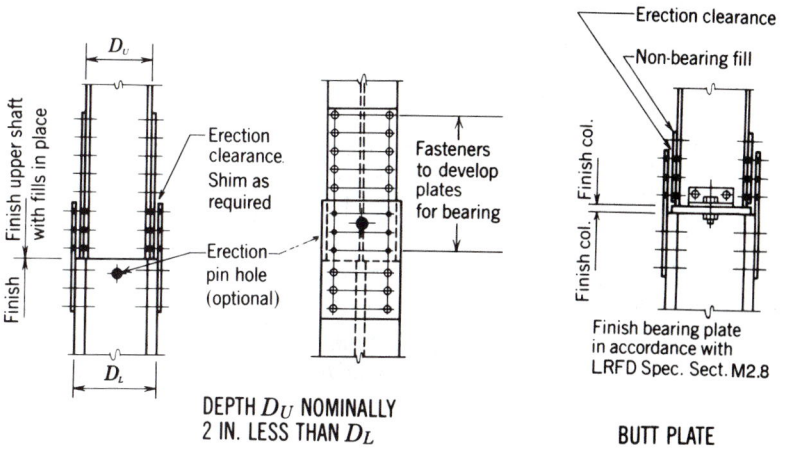

DEPTH D_U NOMINALLY
2 IN. LESS THAN D_L

BUTT PLATE

Note: Erection clearance = 1/8 in.

SUGGESTED DETAILS
Column splices

WELDED

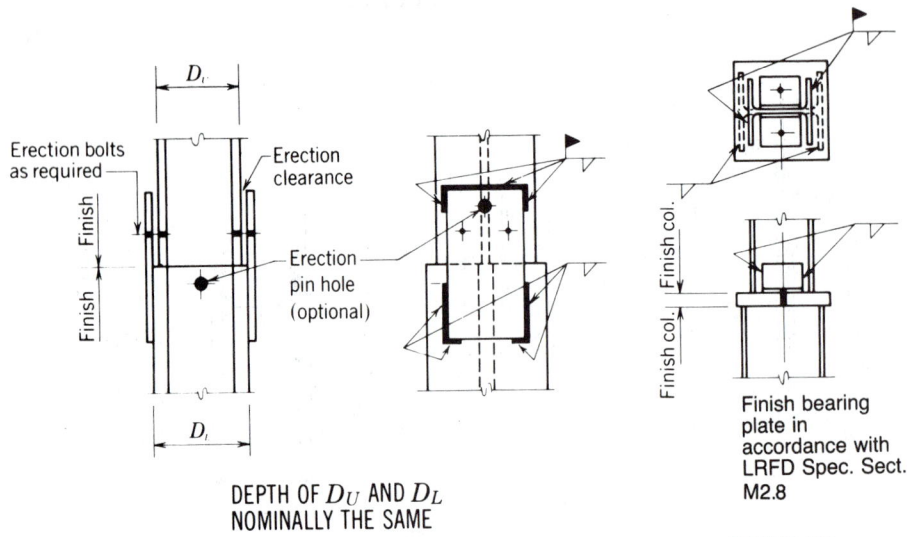

Erection bolts as required

Finish

Finish

D_i

Erection clearance

Erection pin hole (optional)

D_i

DEPTH OF D_U AND D_L
NOMINALLY THE SAME

Finish col.

Finish col.

Finish bearing plate in accordance with LRFD Spec. Sect. M2.8

BUTT PLATE

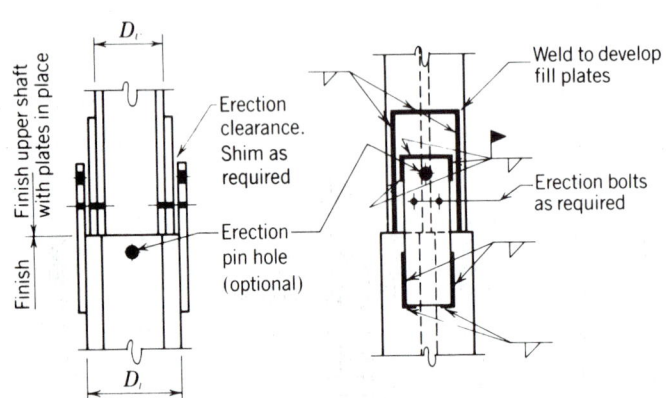

Finish upper shaft with plates in place

Finish

D_i

Erection clearance. Shim as required

Erection pin hole (optional)

D_i

Weld to develop fill plates

Erection bolts as required

DEPTH D_U NOMINALLY
2 IN. LESS THAN D_L

Notes: Erection clearance = $1/16$ in.
When D_U and D_L are nominally the same and thin fills are required, shop may attach splice plate to upper section and provide field clearance over lower section.
Stability of upper shaft, with its loading, should be considered until the final welding is completed.

AMERICAN INSTITUTE OF STEEL CONSTRUCTION

SUGGESTED DETAILS
Column splices

WELDED

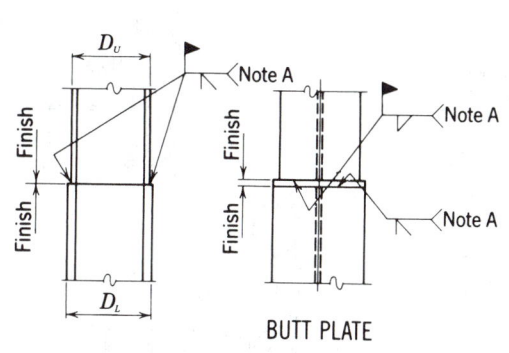

BUTT PLATE

DEPTH OF D_U AND D_L NOMINALLY THE SAME

DEPTH D_U NOMINALLY 2 IN. LESS THAN D_L

ERECTION AID AND STABILITY DEVICE

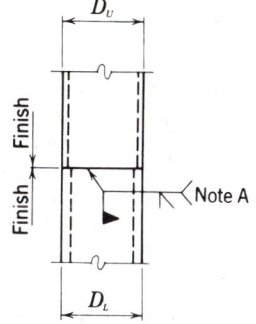

DEPTH OF D_U AND D_L NOMINALLY THE SAME

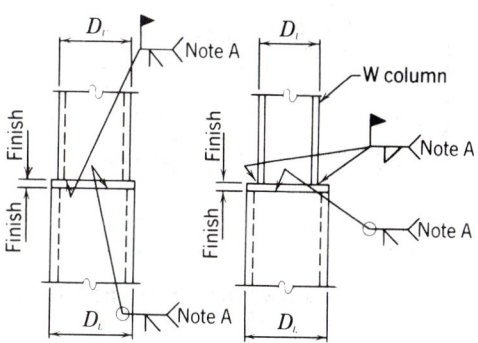

BUTT PLATE

DEPTH D_U NOMINALLY 2 IN. LESS THAN D_L

Note A: Use fillet welds or partial-penetration weld whenever possible.

Finish bearing plates in accordance with LRFD Spec. Sect. M2.8

SUGGESTED DETAILS
Miscellaneous

STRUCTURAL TUBING AND PIPE
BEAM-TO-COLUMN CONNECTIONS

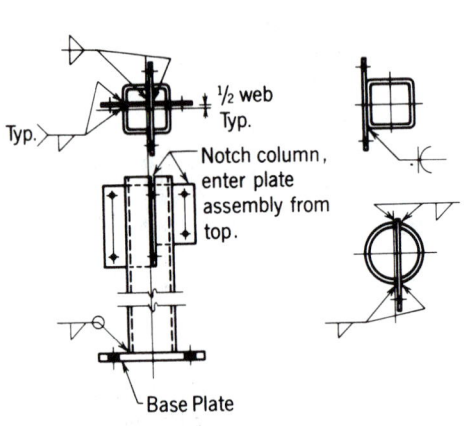

½ web Typ.

Typ.

Notch column, enter plate assembly from top.

Base Plate

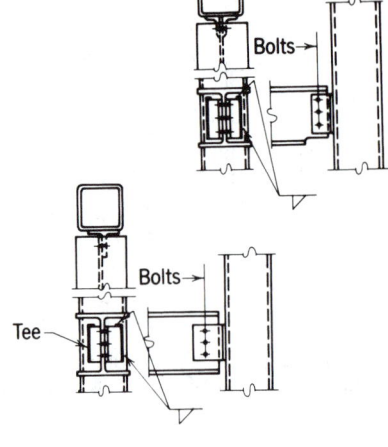

Bolts→

Bolts→

Tee

Note: Details similar for pipe and tubing.

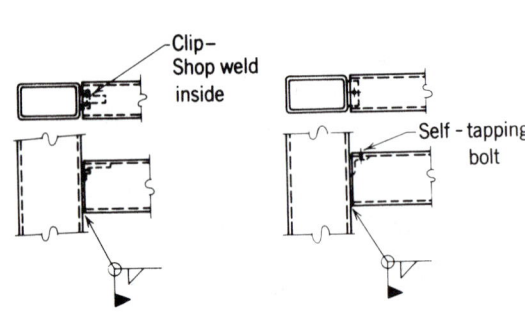

Clip– Shop weld inside

Self – tapping bolt

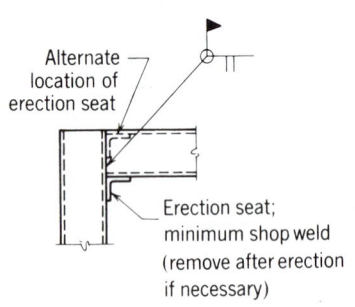

Alternate location of erection seat

Erection seat; minimum shop weld (remove after erection if necessary)

Note: Connections within tubes and pipe may be difficult or impossible to erect.

GIRT CONNECTIONS

PURLIN CONNECTIONS

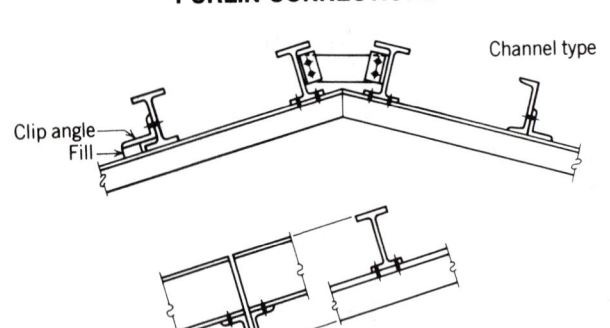

Channel type

Clip angle
Fill

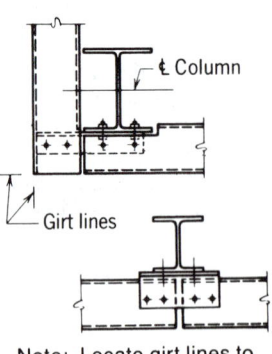

₵ Column

Girt lines

Note: Locate girt lines to avoid blocking girts when possible.

SUGGESTED DETAILS
Miscellaneous

SHELF ANGLES WITH ADJUSTMENT

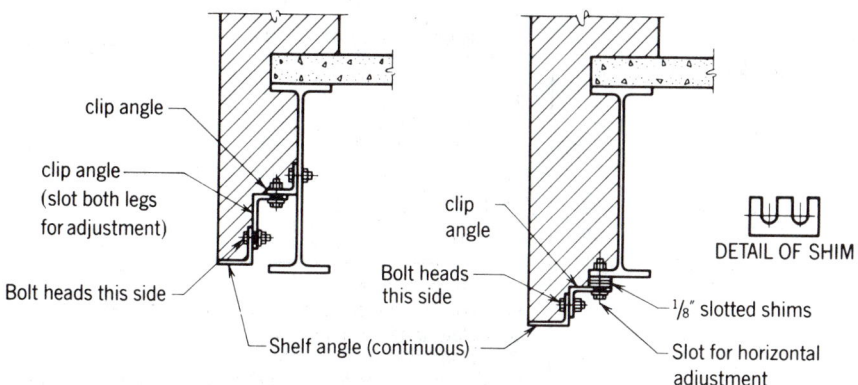

clip angle

clip angle
(slot both legs
for adjustment)

Bolt heads this side

clip
angle

Bolt heads
this side

Shelf angle (continuous)

DETAIL OF SHIM

1/8″ slotted shims

Slot for horizontal
adjustment

Notes: Horizontal adjustment is made by slotted holes; vertical adjustment may be made
by slotted holes or by shims.
For tolerance allowance in alignment, see AISC Code of Standard Practice.

TIE RODS AND ANCHORS

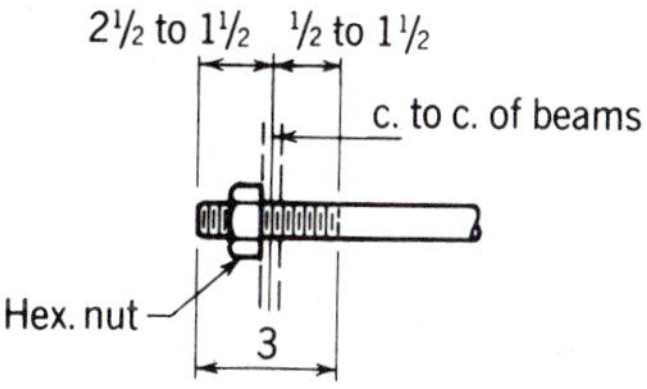

$2^1/2$ to $1^1/2$ $1/2$ to $1^1/2$

c. to c. of beams

Hex. nut

3

Note: Length of rod o. to o.
should be specified in
multiples of 3 in.

Tie Rods

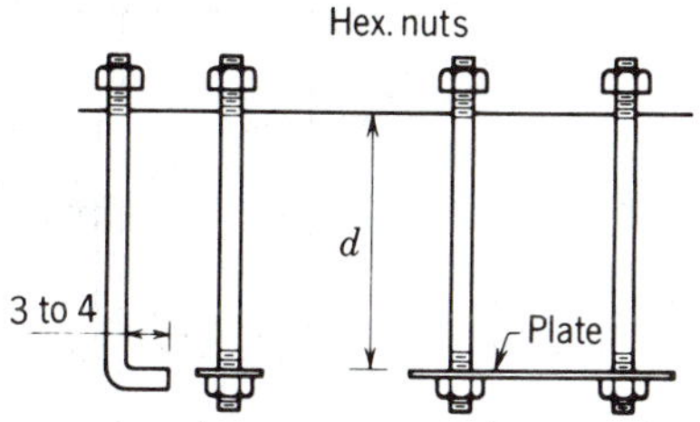

Hex. nuts

d

3 to 4

Plate

Note: Dimension d should
be based on design req't
for uplift

Anchor Bolts

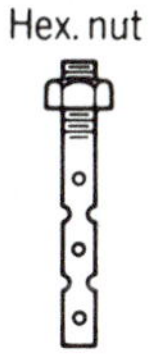

Hex. nut

Swedge Bolts

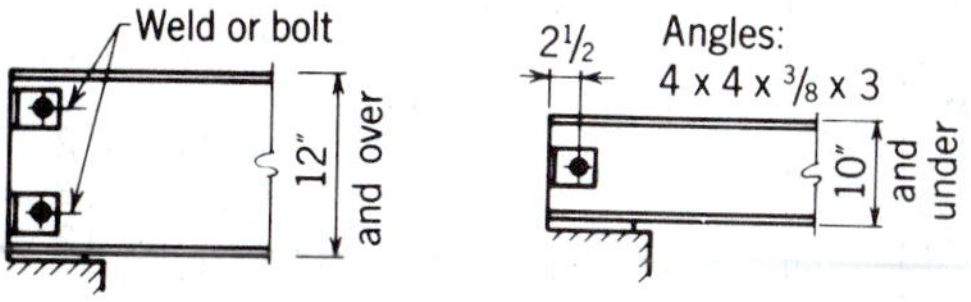

Weld or bolt

12″
and over

Angle Wall Anchors

$2^1/2$ Angles:
4 x 4 x $3/8$ x 3

10″
and
under

2

$3^3/8$

7   $1'-6$

$3/4$″ ⌀ rod

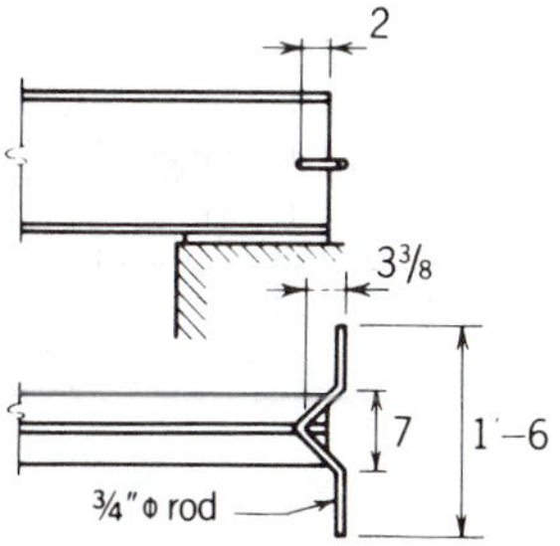

Government Anchor

AMERICAN INSTITUTE OF STEEL CONSTRUCTION

THREADED FASTENERS
Assembling clearances

A325 AND A490 HIGH-STRENGTH BOLTS ENTERING AND TIGHTENING CLEARANCES

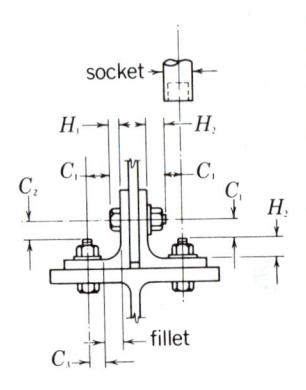

Bolt Dia.	Std. Socket	H_1	H_2	C_1	C_2	C_3 Round	C_3 Clipped
5/8	1 3/4	25/64	1 1/4	1	11/16	11/16	9/16
3/4	2 1/4	15/32	1 3/8	1 1/4	3/4	3/4	11/16
7/8	2 1/2	36/64	1 1/2	1 3/8	7/8	7/8	13/16
1	2 5/8	39/64	1 5/8	1 7/16	15/16	1	7/8
1 1/8	2 7/8	11/16	1 7/8	1 9/16	1 1/16	1 1/8	1
1 1/4	3 1/8	25/32	2	1 11/16	1 1/8	1 1/4	1 1/8
1 3/8	3 1/4	27/32	2 1/8	1 3/4	1 1/4	1 3/8	1 1/4
1 1/2	3 1/2	15/16	2 1/4	1 7/8	1 5/16	1 1/2	1 5/16

H_1 = height of head
H_2 = shank extension, max., based on one flat washer
C_1 = clearance for tightening
C_2 = clearance for entering
C_3 = clearance for fillet, based on std. hardened washer

A325 AND A490 HIGH-STRENGTH BOLTS STAGGER FOR IMPACT WRENCH TIGHTENING

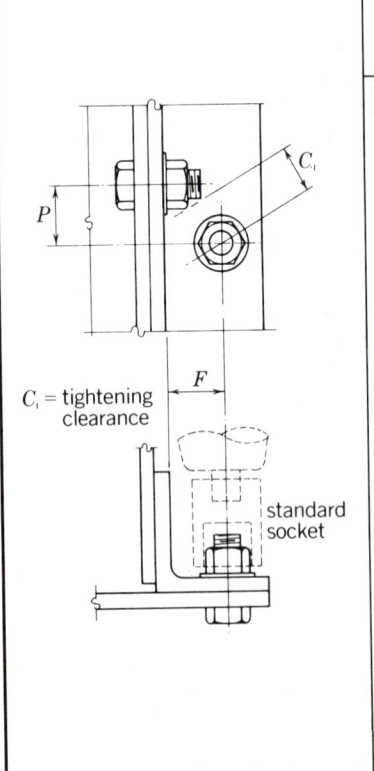

C_1 = tightening clearance

F	Stagger P — High-strength Bolt Diameter							
	5/8	3/4	7/8	1	1 1/8	1 1/4	1 3/8	1 1/2
1	1 5/8							
1 1/8	1 1/2							
1 1/4	1 1/2	1 15/16						
1 3/8	1 7/16	1 7/8	2 3/16					
1 1/2	1 1/4	1 13/16	2 1/8	2 5/16				
1 5/8	1 1/4	1 3/4	2 1/16	2 5/16	2 9/16			
1 3/4	1 3/16	1 11/16	2	2 1/4	2 9/16	2 13/16	3	
1 7/8	1 1/8	1 9/16	1 15/16	2 3/16	2 1/2	2 3/4	3	3 1/4
2	1	1 1/2	1 13/16	2 1/8	2 7/16	2 3/4	2 15/16	3 1/4
2 1/8	13/16	1 3/8	1 11/16	2	2 3/8	2 11/16	2 15/16	3 3/16
2 1/4		1 1/4	1 9/16	1 7/8	2 1/4	2 5/8	2 7/8	3 3/16
2 3/8		1 1/8	1 1/2	1 3/4	2 1/8	2 1/2	2 13/16	3 1/8
2 1/2		7/8	1 3/8	1 5/8	2	2 7/16	2 3/4	3 1/16
2 5/8			1 3/16	1 1/2	1 15/16	2 5/16	2 5/8	3
2 3/4			15/16	1 3/8	1 7/8	2 1/8	2 1/2	2 7/8
2 7/8				1 3/16	1 3/4	2 1/16	2 3/8	2 13/16
3				7/8	1 5/8	2	2 1/4	2 11/16
3 1/8					1 1/2	1 7/8	2 1/8	2 1/2
3 1/4					1 1/4	1 3/4	2	2 3/8
3 3/8					15/16	1 5/8	1 15/16	2 1/4
3 1/2						1 3/8	1 3/4	2 1/8
3 5/8						1 1/16	1 9/16	2
3 3/4							15/16	1 7/8
3 7/8								1 11/16
4								1 3/8

RIVETS AND THREADED FASTENERS
Field erection clearances

RIVET CLEARANCE—W COLUMNS

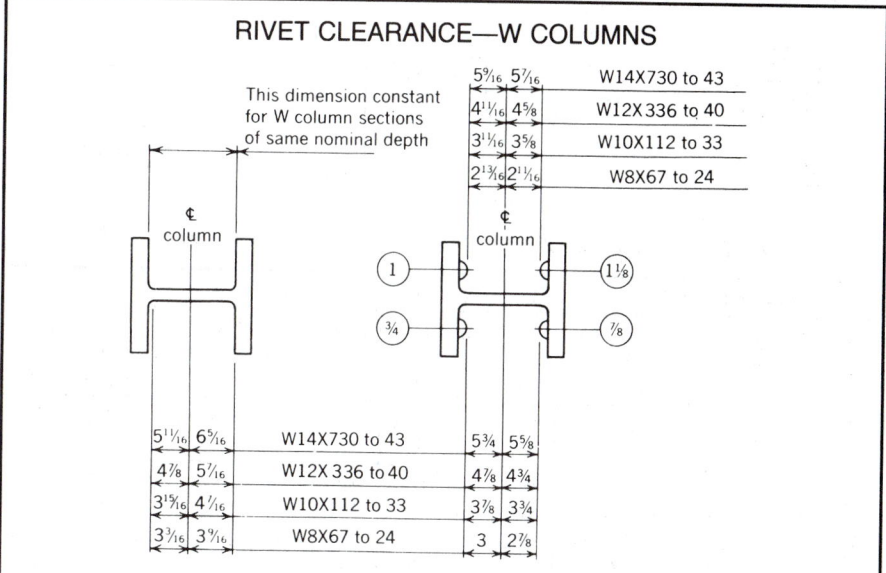

FLANGE CUTS FOR COLUMN WEB CONNECTIONS

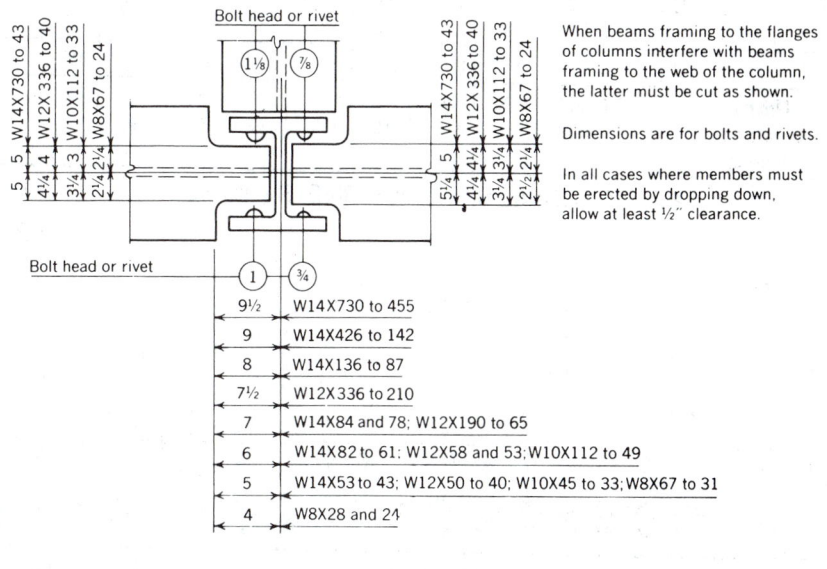

When beams framing to the flanges of columns interfere with beams framing to the web of the column, the latter must be cut as shown.

Dimensions are for bolts and rivets.

In all cases where members must be erected by dropping down, allow at least ½″ clearance.

9½	W14X730 to 455
9	W14X426 to 142
8	W14X136 to 87
7½	W12X336 to 210
7	W14X84 and 78; W12X190 to 65
6	W14X82 to 61; W12X58 and 53; W10X112 to 49
5	W14X53 to 43; W12X50 to 40; W10X45 to 33; W8X67 to 31
4	W8X28 and 24

Notes:
1. Information shown on these clearance diagrams applies to WF and W series. Maximum clearances are shown to accommodate the slight differences in dimensions.
2. Values shown for clearances over rivet heads are applicable when applied to bolt heads but not to the nut and stick-through. See Table of Assembling Clearances for high strength bolt clearance dimensions.
3. Based on Table of Dimensions of Structural Rivets.

RIVETS AND THREADED FASTENERS

DIMENSIONS OF STRUCTURAL RIVETS (HIGH BUTTON OR ACORN HEADS)

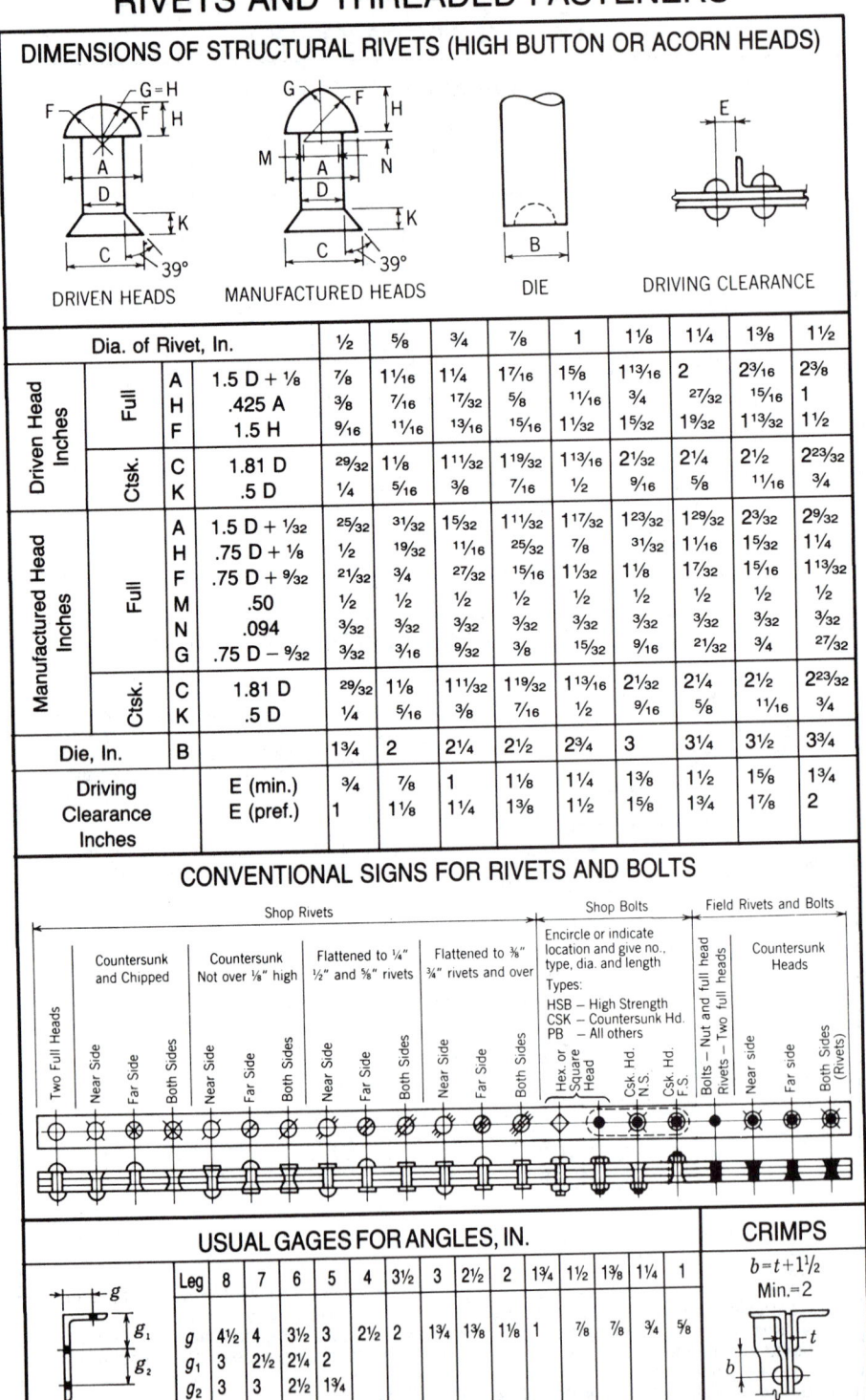

DRIVEN HEADS | MANUFACTURED HEADS | DIE | DRIVING CLEARANCE

			Dia. of Rivet, In.	½	⅝	¾	⅞	1	1⅛	1¼	1⅜	1½
Driven Head Inches	Full	A	1.5 D + ⅛	⅞	1¹⁄₁₆	1¼	1⁷⁄₁₆	1⅝	1¹³⁄₁₆	2	2³⁄₁₆	2⅜
		H	.425 A	⅜	⁷⁄₁₆	1⁷⁄₃₂	⅝	1¹⁄₁₆	¾	²⁷⁄₃₂	¹⁵⁄₁₆	1
		F	1.5 H	⁹⁄₁₆	¹¹⁄₁₆	¹³⁄₁₆	¹⁵⁄₁₆	1¹⁄₃₂	1⁵⁄₃₂	1⁹⁄₃₂	1¹³⁄₃₂	1½
	Ctsk.	C	1.81 D	²⁹⁄₃₂	1⅛	1¹¹⁄₃₂	1¹⁹⁄₃₂	1¹³⁄₁₆	2¹⁄₃₂	2¼	2½	2²³⁄₃₂
		K	.5 D	¼	⁵⁄₁₆	⅜	⁷⁄₁₆	½	⁹⁄₁₆	⅝	¹¹⁄₁₆	¾
Manufactured Head Inches	Full	A	1.5 D + ¹⁄₃₂	²⁵⁄₃₂	³¹⁄₃₂	1⁵⁄₃₂	1¹¹⁄₃₂	1¹⁷⁄₃₂	1²³⁄₃₂	1²⁹⁄₃₂	2³⁄₃₂	2⁹⁄₃₂
		H	.75 D + ⅛	½	¹⁹⁄₃₂	¹¹⁄₁₆	²⁵⁄₃₂	⅞	³¹⁄₃₂	1¹⁄₁₆	1⁵⁄₃₂	1¼
		F	.75 D + ⁹⁄₃₂	²¹⁄₃₂	¾	²⁷⁄₃₂	¹⁵⁄₁₆	1¹⁄₃₂	1⅛	1⁷⁄₃₂	1⁵⁄₁₆	1¹³⁄₃₂
		M	.50	½	½	½	½	½	½	½	½	½
		N	.094	³⁄₃₂	³⁄₃₂	³⁄₃₂	³⁄₃₂	³⁄₃₂	³⁄₃₂	³⁄₃₂	³⁄₃₂	³⁄₃₂
		G	.75 D − ⁹⁄₃₂	³⁄₃₂	³⁄₁₆	⁹⁄₃₂	⅜	¹⁵⁄₃₂	⁹⁄₁₆	²¹⁄₃₂	¾	²⁷⁄₃₂
	Ctsk.	C	1.81 D	²⁹⁄₃₂	1⅛	1¹¹⁄₃₂	1¹⁹⁄₃₂	1¹³⁄₁₆	2¹⁄₃₂	2¼	2½	2²³⁄₃₂
		K	.5 D	¼	⁵⁄₁₆	⅜	⁷⁄₁₆	½	⁹⁄₁₆	⅝	¹¹⁄₁₆	¾
Die, In.		B		1¾	2	2¼	2½	2¾	3	3¼	3½	3¾
Driving Clearance Inches			E (min.)	¾	⅞	1	1⅛	1¼	1⅜	1½	1⅝	1¾
			E (pref.)	1	1⅛	1¼	1⅜	1½	1⅝	1¾	1⅞	2

CONVENTIONAL SIGNS FOR RIVETS AND BOLTS

Shop Rivets				Shop Bolts	Field Rivets and Bolts

(Column subheadings: Countersunk and Chipped | Countersunk Not over ⅛" high | Flattened to ¼" ½" and ⅝" rivets | Flattened to ⅜" ¾" rivets and over | Encircle or indicate location and give no., type, dia. and length — Types: HSB — High Strength, CSK — Countersunk Hd., PB — All others | Countersunk Heads)

Row markers: Two Full Heads | Near Side | Far Side | Both Sides | Near Side | Far Side | Both Sides | Near Side | Far Side | Both Sides | Near Side | Far Side | Both Sides | Hex. or Square Head | Csk. Hd. N.S. | Csk. Hd. F.S. | Bolts — Nut and full head | Rivets — Two full heads | Near side | Far side | Both Sides (Rivets)

USUAL GAGES FOR ANGLES, IN.

Leg	8	7	6	5	4	3½	3	2½	2	1¾	1½	1⅜	1¼	1
g	4½	4	3½	3	2½	2	1¾	1⅜	1⅛	1	⅞	⅞	¾	⅝
g₁	3	2½	2¼	2										
g₂	3	3	2½	1¾										

CRIMPS

$$b = t + 1½$$
$$Min. = 2$$

THREADED FASTENERS
Bolt heads

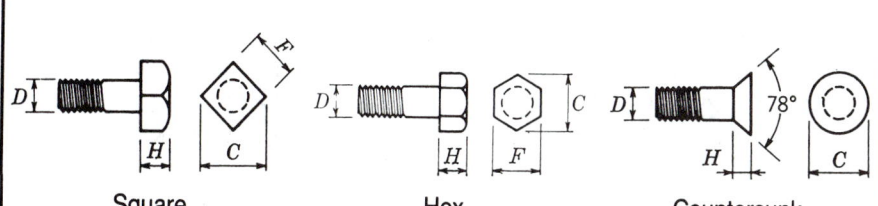

Square   Hex   Countersunk

Bolt head dimensions, rounded to nearest 1/16 in., are in accordance with ANSI B18.2.1-1972 (square and hex) and ANSI 18.5-1971 (countersunk)

Standard Dimensions for Bolt Heads

Dia. of Bolt D	Square			Hex			Heavy Hex			Countersunk	
	Width F	Width C	Height H	Width F	Width C	Height H	Width F	Width C	Height H	Dia. C	Height H
In.	In.	In.	In.	In.	In.	In.	In.	In.	In.	In.	In.
1/4	3/8	1/2	3/16	7/16	1/2	3/16	—	—	—	1/2	1/8
3/8	9/16	13/16	1/4	9/16	5/8	1/4	—	—	—	11/16	3/16
1/2	3/4	1 1/16	5/16	3/4	7/8	3/8	7/8	1	3/8	7/8	1/4
5/8	15/16	1 5/16	7/16	15/16	1 1/16	7/16	1 1/16	1 1/4	7/16	1 1/8	5/16
3/4	1 1/8	1 9/16	1/2	1 1/8	1 5/16	1/2	1 1/4	1 7/16	1/2	1 3/8	3/8
7/8	1 5/16	1 7/8	5/8	1 5/16	1 1/2	9/16	1 7/16	1 11/16	9/16	1 9/16	7/16
1	1 1/2	2 1/8	11/16	1 1/2	1 3/4	11/16	1 5/8	1 7/8	11/16	1 13/16	1/2
1 1/8	1 11/16	2 3/8	3/4	1 11/16	1 15/16	3/4	1 13/16	2 1/16	3/4	2 1/16	9/16
1 1/4	1 7/8	2 5/8	7/8	1 7/8	2 3/16	7/8	2	2 5/16	7/8	2 1/4	5/8
1 3/8	2 1/16	2 15/16	15/16	2 1/16	2 3/8	15/16	2 3/16	2 1/2	15/16	2 1/2	11/16
1 1/2	2 1/4	3 3/16	1	2 1/4	2 5/8	1	2 3/8	2 3/4	1	2 11/16	3/4
1 3/4	—	—	—	2 5/8	3	1 3/16	2 3/4	3 3/16	1 3/16	—	—
2	—	—	—	3	3 7/16	1 3/8	3 1/8	3 5/8	1 3/8	—	—
2 1/4	—	—	—	3 3/8	3 7/8	1 1/2	3 1/2	4 1/16	1 1/2	—	—
2 1/2	—	—	—	3 3/4	4 5/16	1 11/16	3 7/8	4 1/2	1 11/16	—	—
2 3/4	—	—	—	4 1/8	4 3/4	1 13/16	4 1/4	4 15/16	1 13/16	—	—
3	—	—	—	4 1/2	5 3/16	2	4 5/8	5 5/16	2	—	—
3 1/4	—	—	—	4 7/8	5 5/8	2 3/16	—	—	—	—	—
3 1/2	—	—	—	5 1/4	6 1/16	2 5/16	—	—	—	—	—
3 3/4	—	—	—	5 5/8	6 1/2	2 1/2	—	—	—	—	—
4	—	—	—	6	6 15/16	2 11/16	—	—	—	—	—

For dimensions for high-strength bolts, refer to *Load and Resistance Factor Design Specification for Structural Joints Using ASTM A325 or A490 Bolts.*
Countersunk head bolts may be ordered with slotted or socket head.

THREADED FASTENERS
Nuts

Square   Hex

Nut dimensions, rounded to nearest 1/16 in., are in accordance with ANSI B18.2.2-1972

Dimensions for Nuts

Nut Size	Square			Hex			Heavy Square			Heavy Hex		
	Width F	Width C	Height N	Width F	Width C	Height N	Width F	Width C	Height N	Width F	Width C	Height N
In.	In.	In.	In.	In.	In.	In.	In.	In.	In.	In.	In.	In.
1/4	7/16	5/8	1/4	7/16	1/2	1/4	1/2	11/16	1/4	1/2	9/16	1/4
3/8	5/8	7/8	5/16	9/16	5/8	5/16	11/16	1	3/8	11/16	13/16	3/8
1/2	13/16	1 1/8	7/16	3/4	7/8	7/16	7/8	1 1/4	1/2	7/8	1	1/2
5/8	1	1 7/16	9/16	15/16	1 1/16	9/16	1 1/16	1 1/2	5/8	1 1/16	1 1/4	5/8
3/4	1 1/8	1 9/16	11/16	1 1/8	1 5/16	5/8	1 1/4	1 3/4	3/4	1 1/4	1 7/16	3/4
7/8	1 5/16	1 7/8	3/4	1 5/16	1 1/2	3/4	1 7/16	2 1/16	7/8	1 7/16	1 11/16	7/8
1	1 1/2	2 1/8	7/8	1 1/2	1 3/4	7/8	1 5/8	2 5/16	1	1 5/8	1 7/8	1
1 1/8	1 11/16	2 3/8	1	1 11/16	1 15/16	1	1 13/16	2 9/16	1 1/8	1 13/16	2 1/16	1 1/8
1 1/4	1 7/8	2 5/8	1 1/8	1 7/8	2 3/16	1 1/16	2	2 13/16	1 1/4	2	2 5/16	1 1/4
1 3/8	2 1/16	2 15/16	1 1/4	2 1/16	2 3/8	1 3/16	2 3/16	3 1/8	1 3/8	2 3/16	2 1/2	1 3/8
1 1/2	2 1/4	3 3/16	1 5/16	2 1/4	2 5/8	1 5/16	2 3/8	3 3/8	1 1/2	2 3/8	2 3/4	1 1/2
1 3/4	—	—	—	—	—	—	—	—	—	2 3/4	3 3/16	1 3/4
2	—	—	—	—	—	—	—	—	—	3 1/8	3 5/8	2
2 1/4	—	—	—	—	—	—	—	—	—	3 1/2	4 1/16	2 3/16
2 1/2	—	—	—	—	—	—	—	—	—	3 7/8	4 1/2	2 7/16
2 3/4	—	—	—	—	—	—	—	—	—	4 1/4	4 15/16	2 11/16
3	—	—	—	—	—	—	—	—	—	4 5/8	5 5/16	2 15/16
3 1/4	—	—	—	—	—	—	—	—	—	5	5 3/4	3 3/16
3 1/2	—	—	—	—	—	—	—	—	—	5 3/8	6 3/16	3 7/16
3 3/4	—	—	—	—	—	—	—	—	—	5 3/4	6 5/8	3 11/16
4	—	—	—	—	—	—	—	—	—	6 1/8	7 1/16	3 15/16

For dimensions for high-strength bolts, refer to *Load and Resistance Factor Design Specification for Structural Joints Using ASTM A325 or A490 Bolts.*

THREADED FASTENERS
Weight of bolts
With square heads and hexagon nuts in pounds per 100

Length Under Head, In.	Diameter of Bolts, In.								
	¼	⅜	½	⅝	¾	⅞	1	1⅛	1¼
1	2.38	6.11	13.0	24.1	38.9	—	—	—	—
1¼	2.71	6.71	14.0	25.8	41.5	—	—	—	—
1½	3.05	7.47	15.1	27.6	44.0	67.3	95.1	—	—
1¾	3.39	8.23	16.5	29.3	46.5	70.8	99.7	—	—
2	3.73	8.99	17.8	31.4	49.1	74.4	104	143	—
2¼	4.06	9.75	19.1	33.5	52.1	77.9	109	149	—
2½	4.40	10.5	20.5	35.6	55.1	82.0	114	155	206
2¾	4.74	11.3	21.8	37.7	58.2	86.1	119	161	213
3	5.07	12.0	23.2	39.8	61.2	90.2	124	168	221
3¼	5.41	12.8	24.5	41.9	64.2	94.4	129	174	229
3½	5.75	13.5	25.9	44.0	67.2	98.5	135	181	237
3¾	6.09	14.3	27.2	46.1	70.2	103	140	188	246
4	6.42	15.1	28.6	48.2	73.3	107	145	195	254
4¼	6.76	15.8	29.9	50.3	76.3	111	151	202	262
4½	7.10	16.6	31.3	52.3	79.3	115	156	208	271
4¾	7.43	17.3	32.6	54.4	82.3	119	162	215	279
5	7.77	18.1	33.9	56.5	85.3	123	167	222	288
5¼	8.11	18.9	35.3	58.6	88.4	127	172	229	296
5½	8.44	19.6	36.6	60.7	91.4	131	178	236	304
5¾	8.78	20.4	38.0	62.8	94.4	136	183	242	313
6	9.12	21.1	39.3	64.9	97.4	140	188	249	321
6¼	9.37	21.7	40.4	66.7	100	143	193	255	329
6½	9.71	22.5	41.8	68.7	103	147	198	262	337
6¾	10.1	23.3	43.1	70.8	106	151	204	269	345
7	10.4	24.0	44.4	72.9	109	156	209	275	354
7¼	10.7	24.8	45.8	75.0	112	160	214	282	362
7½	11.0	25.5	47.1	77.1	115	164	220	289	371
7¾	11.4	26.3	48.5	79.2	118	168	225	296	379
8	11.7	27.0	49.8	81.3	121	172	231	303	387
8½	—	28.6	52.5	85.5	127	180	241	316	404
9	—	30.1	55.2	89.7	133	189	252	330	421
9½	—	31.6	57.9	93.9	139	197	263	343	438
10	—	33.1	60.6	98.1	145	205	274	357	454
10½	—	34.6	63.3	102	151	213	284	371	471
11	—	36.2	66.0	106	157	221	295	384	488
11½	—	37.7	68.7	110	163	230	306	398	505
12	—	39.2	71.3	115	170	238	316	411	522
12½	—	—	74.0	119	176	246	327	425	538
13	—	—	76.7	123	182	254	338	439	556
13½	—	—	79.4	127	188	263	349	452	572
14	—	—	82.1	131	194	271	359	466	589
14½	—	—	84.8	135	200	279	370	479	605
15	—	—	87.5	140	206	287	381	493	622
15½	—	—	90.2	144	212	296	392	507	639
16	—	—	92.9	148	218	304	402	520	656
Per Inch Additional	1.3	3.0	5.4	8.4	12.1	16.5	21.4	27.2	33.6

Bolt is square bolt, ANSI B18.2.1-72 and nut is hex nut, ANSI B18.2.2-72. This table conforms to weight standards adopted by the Industrial Fasteners Institute.

THREADED FASTENERS
Weight of bolts
Special cases in pounds per 100

VARIATIONS IN BOLT AND NUT TYPES

Weights for combinations of bolt heads and nuts, other than square heads and hex nuts, may be determined by making the appropriate additions and deductions tabulated below from the weight per 100 shown on the previous page.

Combination	Add or Subtract	Diameter of Bolt, In.								
		1/4	3/8	1/2	5/8	3/4	7/8	1	1 1/8	1 1/4
Square bolt with square nut	+	0.1	1.0	2.0	3.4	3.5	5.5	8.0	12.2	16.3
Square bolt with heavy square nut	+	0.6	2.1	4.1	7.0	11.6	17.2	23.2	32.1	41.2
Square bolt with heavy hex nut	+	0.4	1.5	2.8	4.6	7.6	10.7	14.2	18.9	24.3
Hex bolt with square nut	+	0.1	0.6	1.1	1.4	0.2	0.5	−0.2	−0.1	−1.7
Hex bolt with hex nut	−	0.0	0.4	0.9	2.0	3.3	5.0	8.2	12.3	18.0
Hex bolt with heavy square nut	+	0.6	1.7	3.2	5.0	8.3	12.2	15.0	19.8	23.2
Hex bolt with heavy hex nut	+	0.4	1.1	1.9	2.6	4.3	5.7	6.0	6.6	6.3
Heavy hex bolt with heavy square nut	+	—	—	4.7	7.3	11.3	16.5	20.7	27.0	33.6
Heavy hex bolt with heavy hex nut	+	—	—	3.4	4.9	7.3	10.0	11.7	13.8	16.7

LARGE DIAMETER BOLTS

Weights of bolts over 1 1/4 inches in diameter may be calculated from the following data. Standard practice is hex head bolts with heavy hex nut. Square head bolts and square nuts are not standard in sizes over 1 1/2 in.

Weight of 100 each	Diameter of Bolt, In.											
	1 3/8	1 1/2	1 3/4	2	2 1/4	2 1/2	2 3/4	3	3 1/4	3 1/2	3 3/4	4
Square heads	105	130	—	—	—	—	—	—	—	—	—	—
Hex heads	84	112	178	259	369	508	680	900	1120	1390	1730	2130
Heavy hex heads	95	124	195	280	397	541	720	950	—	—	—	—
Square nuts	94.5	122	—	—	—	—	—	—	—	—	—	—
Heavy square nuts	125	161	—	—	—	—	—	—	—	—	—	—
Heavy hex nuts	102	131	204	299	419	564	738	950	1190	1530	1810	2180
Linear inch of threaded shank	35.0	42.5	57.4	75.5	97.4	120	147	178	210	246	284	325
Linear inch of unthreaded shank	42.0	50.0	68.2	89.0	113	139	168	200	235	272	313	356

THREADED FASTENERS
Weight of ASTM A325 or A490 high-strength bolts
Heavy hex structural bolts with heavy hex nuts in pounds per 100

Length Under Head, In.	Diameter of Bolts, In.								
	½	⅝	¾	⅞	1	1⅛	1¼	1⅜	1½
1	16.5	29.4	47.0	—	—	—	—	—	—
1¼	17.8	31.1	49.6	74.4	104	—	—	—	—
1½	19.2	33.1	52.2	78.0	109	148	197	—	—
1¾	20.5	35.3	55.3	81.9	114	154	205	261	333
2	21.9	37.4	58.4	86.1	119	160	212	270	344
2¼	23.3	39.8	61.6	90.3	124	167	220	279	355
2½	24.7	41.7	64.7	94.6	130	174	229	290	366
2¾	26.1	43.9	67.8	98.8	135	181	237	300	379
3	27.4	46.1	70.9	103	141	188	246	310	391
3¼	28.8	48.2	74.0	107	146	195	255	321	403
3½	30.2	50.4	77.1	111	151	202	263	332	416
3¾	31.6	52.5	80.2	116	157	209	272	342	428
4	33.0	54.7	83.3	120	162	216	280	353	441
4¼	34.3	56.9	86.4	124	168	223	289	363	453
4½	35.7	59.0	89.5	128	173	230	298	374	465
4¾	37.1	61.2	92.7	133	179	237	306	384	478
5	38.5	63.3	95.8	137	184	244	315	395	490
5¼	39.9	65.5	98.9	141	190	251	324	405	503
5½	41.2	67.7	102	146	196	258	332	416	515
5¾	42.6	69.8	105	150	201	265	341	426	527
6	44.0	71.9	108	154	207	272	349	437	540
6¼	—	74.1	111	158	212	279	358	447	552
6½	—	76.3	114	163	218	286	367	458	565
6¾	—	78.5	118	167	223	293	375	468	577
7	—	80.6	121	171	229	300	384	479	589
7¼	—	82.8	124	175	234	307	392	489	602
7½	—	84.9	127	179	240	314	401	500	614
7¾	—	87.1	130	183	246	321	410	510	626
8	—	89.2	133	187	251	328	418	521	639
8¼	—	—	—	192	257	335	427	531	651
8½	—	—	—	196	262	342	435	542	664
8¾	—	—	—	—	—	—	444	552	676
9	—	—	—	—	—	—	453	563	689
Per inch additional add	5.5	8.6	12.4	16.9	22.1	28.0	34.4	42.5	49.7
For each 100 plain round washers add	2.1	3.6	4.8	7.0	9.4	11.3	13.8	16.8	20.0
For each 100 beveled square washers add	23.1	22.4	21.0	20.2	19.2	34.0	31.6	—	—

This table conforms to weight standards adopted by the Industrial Fasteners Institute, 1965, updated for washer weights.

THREADED FASTENERS

SCREW THREADS
Unified Standard Series—UNC/UNRC and 4UN/4UNR
ANSI B1.1-1974

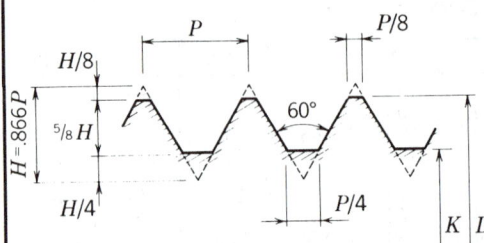

Nominal size (basic major dia.)
No. threads per inch (*n*)
Thread series symbol
Thread class symbol
Left hand thread. No symbol req'd for right hand thread.

$$K \mid D \quad \tfrac{3}{4}\text{--}\mathbf{10\ UNC\ 2A\ LH}$$

Thread Dimensions 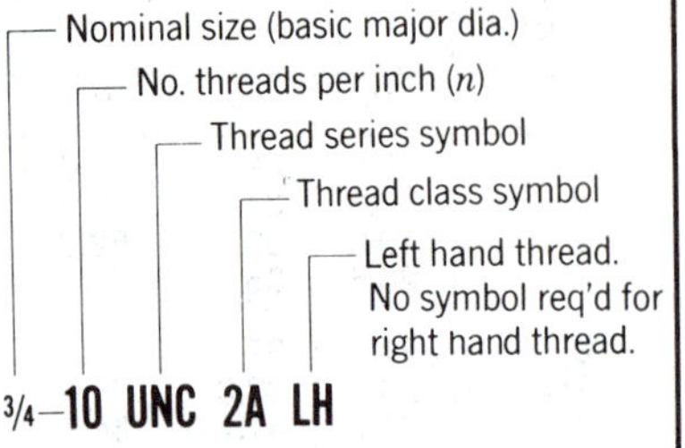   Standard Designations

Dia.		Area			Th'ds[b] per In. *n*	Dia.		Area			Th'ds[b] per In. *n*
Basic Major *D*	Min. Root *K*	Gross A_D	Min. Root A_K	Tensile[a] Stress		Basic Major *D*	Min. Root *K*	Gross A_D	Min. Root A_K	Tensile[a] Stress	
In.	In.	In.²	In.²	In.²		In.	In.	In.²	In.²	In.²	
¼	.189	.049	.028	.032	20	2¾	2.443	5.940	4.69	4.93	4
⅜	.298	.110	.070	.078	16	3	2.693	7.069	5.70	5.97	4
½	.406	.196	.129	.142	13	3¼	2.943	8.296	6.80	7.10	4
⅝	.514	.307	.207	.226	11	3½	3.193	9.621	8.01	8.33	4
¾	.627	.442	.309	.334	10	3¾	3.443	11.045	9.31	9.66	4
⅞	.739	.601	.429	.462	9	4	3.693	12.566	10.71	11.1	4
1	.847	.785	.563	.606	8	4¼	3.943	14.186	12.2	12.6	4
1⅛	.950	.994	.709	.763	7	4½	4.193	15.904	13.8	14.2	4
1¼	1.075	1.227	.908	.969	7	4¾	4.443	17.721	15.5	15.9	4
1⅜	1.171	1.485	1.08	1.16	6	5	4.693	19.635	17.3	17.8	4
1½	1.296	1.767	1.32	1.41	6	5¼	4.943	21.648	19.2	19.7	4
1¾	1.505	2.405	1.78	1.90	5	5½	5.193	23.758	21.2	21.7	4
2	1.727	3.142	2.34	2.50	4½	5¾	5.443	25.967	23.3	23.8	4
2¼	1.977	3.976	3.07	3.25	4½	6	5.693	28.274	25.5	26.0	4
2½	2.193	4.909	3.78	4.00	4						

[a]Tensile stress area $= 0.7854 \left(D - \dfrac{.9743}{n}\right)^2$.

[b]For basic major diameters of ¼ to 4 in. incl., thread series is UNC (coarse); for 4¼ in. diameter and larger, thread series is 4UN.

[c]2A denotes Class 2A fit applicable to external threads, 2B denotes corresponding Class 2B fit for internal threads.

MINIMUM LENGTH OF THREAD ON BOLTS
ANSI B18.2.1-1972

Length of Bolt	Diameter of Bolt *D*, In.																	
	¼	⅜	½	⅝	¾	⅞	1	1⅛	1¼	1⅜	1½	1¾	2	2¼	2½	2¾	3
To 6 in. Incl.	¾	1	1¼	1½	1¾	2	2¼	2½	2¾	3	3¼	3¾	4¼	4¾	5¼	5¾	6¼
Over 6 in.	1	1¼	1½	1¾	2	2¼	2½	2¾	3	3¼	3½	4	4½	5	5½	6	6½

Thread length for bolts up to 6 in. long is 2*D* + ¼. For bolts over 6-in. long, thread length is 2*D* + ½. These proportions may be used to compute thread length for diameters not shown in the table. Bolts which are too short for listed or computed thread lengths are threaded as close to the head as possible.

For thread lengths for high-strength bolts, refer to *Load and Resistance Factor Design Specification for Structural Joints Using ASTM A325 or A490 Bolts.*

CLEVISES

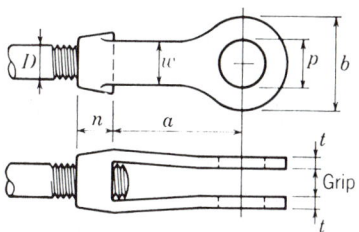

Thread: UNC Class 2B

Grip = thickness
plate + ¼″

Clevis Number	Dimensions, In.							Weight Pounds	Safe Working Load, Kips[a]
	Max. D	Max. p	b	n	a	w	t		
2	5/8	3/4	1 7/16	5/8	3 7/8	1 1/16	5/16(+ 1/32 − 0)	1.0	3.5
2½	7/8	1½	2½	1⅛	4	1¼	5/16(+ 1/32 − 0)	2.0	7.5
3	1⅜	1¾	3	1 5/16	5	1½	½(+ 1/32 − 0)	4.0	15
3½	1½	2	3½	1⅝	6	1¾	½(+ 1/32 − 0)	6.0	18
4	1¾	2¼	4	1¾	6	2	½(+ 1/32 − 0)	8.0	21
5	2	2½	5	2¼	7	2½	⅝(+ 1/16 − 0)	16.0	37.5
6	2½	3	6	2¾	8	3	¾(+ 3/32 − 0)	26.0	54
7	3	3¾	7	3	9	3½	⅞(+ ⅛ − 0)	36.0	68.5
8	4	4	8	4	10	4	1½(+ ⅛ − 0)	80.0	135

[a]Safe working load based on 5:1 safety factor using maximum pin diameter.

CLEVIS NUMBERS FOR VARIOUS RODS AND PINS

Diameter of Tap, In.	Diameter of Pin, In.															
	5/8	3/4	7/8	1	1¼	1½	1¾	2	2¼	2½	2¾	3	3¼	3½	3¾	4
5/8	2	2	2½	2½	2½	2½										
3/4	—	2½	2½	2½	2½	2½										
7/8	—	—	2½	2½	2½	2½										
1	—	—	—	3	3	3	3									
1¼	—	—	—	3	3	3	3	3½								
1⅜	—	—	—	3	3	3	3½	3½	4							
1½	—	—	—	3½	3½	3½	4	4	5							
1¾	—	—	—	—	4	4	5	5	5	5						
2	—	—	—	—	5	5	5	5	5	6	6					
2¼	—	—	—	—	—	—	6	6	6	6	6	7	7			
2½	—	—	—	—	—	—	6	6	6	7	7	7	7	7		
2¾	—	—	—	—	—	—	—	—	7	7	7	7	8	8		
3	—	—	—	—	—	—	—	7	8	8	8	8	8	8	8	8
3¼	—	—	—	—	—	—	—	—	—	8	8	8	8	8	8	8
3½	—	—	—	—	—	—	—	—	—	8	8	8	8	8	8	8
3¾	—	—	—	—	—	—	—	—	—	8	8	8	9	8	8	8
4	—	—	—	—	—	—	—	—	—	8	8	8	8	8	8	8

Above Table of Clevis Sizes is based on the Net Area of Clevis through Pin Hole being equal to or greater than 125% of Net Area of Rod. Table applies to round rods without upset ends. Pins are sufficient for shear but must be investigated for bending. For other combinations of pin and rod or net area ratios, required clevis size can be calculated by reference to the tabulated dimensions.

Weights and dimensions of clevises are typical. Products of all suppliers are similar and essentially the same.

TURNBUCKLES

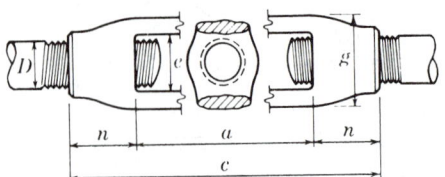

Threads: UNC and 4 UN Class 2B

Dia. D, In.	Standard Turnbuckles					Weight of Turnbuckles, Pounds						Turnbuckle Breaking Strength, Kips[a]
	Dimensions, In.					Length *a*, In.						
	a	n	c	e	g	6	9	12	18	24	36	
3/8	6	9/16	7 1/8	9/16	1 1/32	.41						6.0
1/2	6	3/4	7 1/2	11/16	1 5/16	.75	.80	1.00				11.0
5/8	6	29/32	7 13/16	13/16	1 1/2	1.00	1.38	1.50	2.43			17.5
3/4	6	1 1/16	8 1/8	15/16	1 23/32	1.45	1.63	2.13	3.06	4.25		26.0
7/8	6	1 7/32	8 7/16	1 3/32	1 7/8	1.85		2.83	4.20	5.43		36.0
1	6	1 3/8	8 3/4	1 9/32	2 1/32	2.60		3.20	4.40	6.85	10.0	46.5
1 1/8	6	1 9/16	9 1/8	1 13/32	2 9/32	2.72		4.70	6.10			58.0
1 1/4	6	1 3/4	9 1/2	1 9/16	2 17/32	3.58		4.70	7.13	11.30	13.1	76.0
1 3/8	6	1 15/16	9 7/8	1 11/16	2 3/4	4.50						87.0
1 1/2	6	2 1/8	10 1/4	1 27/32	3 1/32	5.50		8.00	9.13	16.80	19.4	105.0
1 5/8	6	2 1/4	10 1/2	1 31/32	3 9/32	7.50						122.5
1 3/4	6	2 1/2	11	2 1/8	3 9/16	9.50		15.25	16.00	19.50		141.5
1 7/8	6	2 3/4	11 1/2	2 3/8	4	11.50						186.0
2	6	2 3/4	11 1/2	2 3/8	4	11.50		15.25		27.50		186.0
2 1/4	6	3 3/8	12 3/4	2 11/16	4 5/8	18.00		35.25		43.50		240.0
2 1/2	6	3 3/4	13 1/2	3	5	23.25		33.60		42.38		300.0
2 3/4	6	4 1/8	14 1/4	3 1/4	5 5/8	31.50				54.00		375.0
3	6	4 1/2	15	3 5/8	6 1/8	39.50						483.5
3 1/4	6	5 1/4	16 1/2	3 7/8	6 3/4	60.50						611.0
3 1/2	6	5 1/4	16 1/2	3 7/8	6 3/4	60.50						611.0
3 3/4	6	6	18	4 5/8	8 1/2	95.00						839.0
4	6	6	18	4 5/8	8 1/2	95.00						839.0
4 1/4	9	6 3/4	22 1/2	5 1/4	9 3/4		152.0					1169
4 1/2	9	6 3/4	22 1/2	5 1/4	9 3/4		152.0					1169
4 3/4	9	6 3/4	22 1/2	5 1/4	9 3/4		152.0					1169
5	9	7 1/2	24	6	10		200.0					1474

[a]Factor of safety to be determined by user. Usual factor of safety for rigging is 5.
Weights and dimensions of turnbuckles are typical. Products of all suppliers are similar and essentially the same.

SLEEVE NUTS

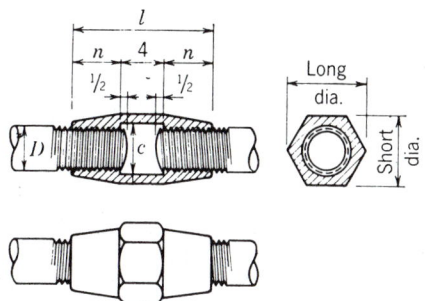

Thread: UNC and 4 UN Class 2B

Diameter of Screw D In.	Dimensions, In.					Weight, Pounds
	Short Dia.	Long Dia.	Length l	Nut n	Clear c	
3/8	11/16	25/32	4	—	—	.27
7/16	25/32	7/8	4	—	—	.34
1/2	7/8	1	4	—	—	.43
9/16	15/16	1 1/16	5	—	—	.64
5/8	1 1/16	1 7/32	5	—	—	.93
3/4	1 1/4	1 7/16	5	—	—	1.12
7/8	1 7/16	1 5/8	7	1 7/16	1	1.75
1	1 5/8	1 13/16	7	1 7/16	1 1/8	2.46
1 1/8	1 13/16	2 1/16	7 1/2	1 5/8	1 1/4	3.10
1 1/4	2	2 1/4	7 1/2	1 5/8	1 3/8	4.04
1 3/8	2 3/16	2 1/2	8	1 7/8	1 1/2	4.97
1 1/2	2 3/8	2 11/16	8	1 7/8	1 5/8	6.16
1 5/8	2 9/16	2 15/16	8 1/2	2 1/16	1 3/4	7.36
1 3/4	2 3/4	3 1/8	8 1/2	2 1/16	1 7/8	8.87
1 7/8	2 15/16	3 5/16	9	2 5/16	2	10.42
2	3 1/8	3 1/2	9	2 5/16	2 1/8	12.24
2 1/4	3 1/2	3 15/16	9 1/2	2 1/2	2 3/8	16.23
2 1/2	3 7/8	4 3/8	10	2 3/4	2 5/8	21.12
2 3/4	4 1/4	4 13/16	10 1/2	2 15/16	2 7/8	26.71
3	4 5/8	5 1/4	11	3 3/16	3 1/8	33.22
3 1/4	5	5 5/8	11 1/2	3 3/8	3 3/8	40.62
3 1/2	5 3/8	6	12	3 5/8	3 5/8	49.07
3 3/4	5 3/4	6 3/8	12 1/2	3 13/16	3 7/8	58.57
4	6 1/8	6 7/8	13	4 1/16	4 1/8	69.22
4 1/4	6 1/2	7 1/2	13 1/2	4 3/4	4 3/8	75.00
4 1/2	6 7/8	7 15/16	14	5	4 3/4	90.00
4 3/4	7 1/4	8 3/8	14 1/2	5 1/4	5	98.00
5	7 5/8	8 7/8	15	5 1/2	5 1/4	110.0
5 1/4	8	9 1/4	15 1/2	5 3/4	5 1/2	122.0
5 1/2	8 3/8	9 3/4	16	6	5 3/4	142.0
5 3/4	8 3/4	10 1/8	16 1/2	6 1/4	6	157.0
6	9 1/8	10 5/8	17	6 1/2	6 1/4	176.0

Strengths are greater than the corresponding connecting rod when same material is used. Weights and dimensions are typical. Products of all suppliers are similar and essentially the same.

RECESSED PIN NUTS AND COTTER PINS

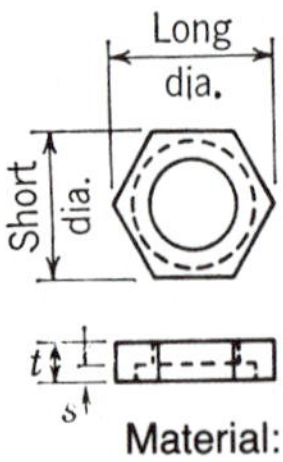
Long
dia.
Short
dia.
t
s

Material: Steel

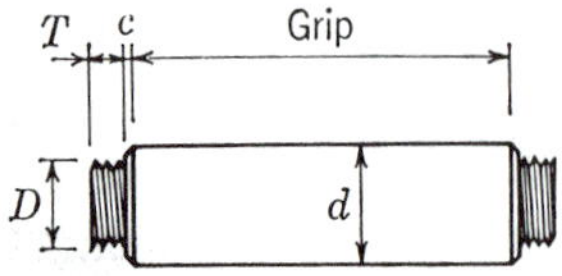
T
c
Grip
D
d

Thread: 6 UN Class 2A/2B

Diameter of Pin *d*			Pin Thread *D*	Pin Thread *T*	Pin *c*	Nut Thickness *t*	Nut Diameter Short Dia.	Nut Diameter Long Dia.	Nut Recess Rough Dia.	Nut Recess *s*	Weight, Pounds
	2	2¼	1½	1	⅛	⅞	3	3⅜	2⅝	¼	1
	2½	2¾	2	1⅛	⅛	1	3⅝	4⅛	3⅛	¼	2
3	3¼	3½	2½	1¼	⅛	1⅛	4⅜	5	3⅞	⅜	3
	3¾	4	3	1⅜	¼	1¼	4⅞	5⅝	4⅜	⅜	4
4¼	4½	4¾	3½	1½	¼	1⅜	5¾	6⅝	5¼	½	5
	5	5¼	4	1⅝	¼	1½	6¼	7¼	5¾	½	6
5½	5¾	6	4½	1¾	¼	1⅝	7	8⅛	6½	⅝	8
	6¼	6½	5	1⅞	⅜	1¾	7⅝	8⅞	7	⅝	10
	6¾	7	5½	2	⅜	1⅞	8⅛	9⅜	7½	¾	12
	7¼	7½	5½	2	⅜	1⅞	8⅝	10	8	¾	14
7¾	8	8¼	6	2¼	⅜	2⅛	9⅜	10⅞	8¾	¾	19
8½	8¾	9	6	2¼	⅜	2⅛	10¼	11⅞	9⅝	¾	24
	9¼	9½	6	2⅜	⅜	2¼	11¼	13	10⅝	¾	32
	9¾	10	6	2⅜	⅜	2¼	11¼	13	10⅝	¾	32

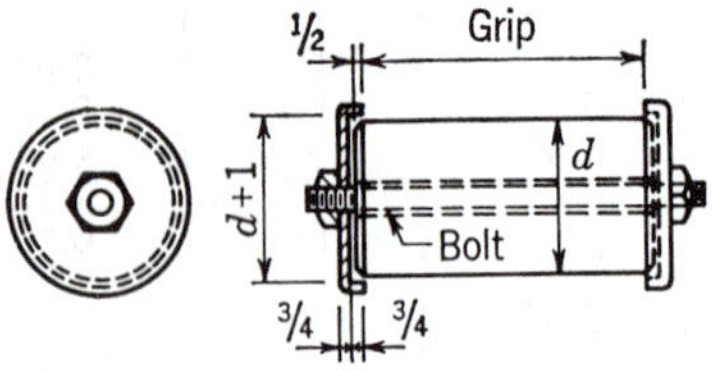
½
Grip
d
d+1
Bolt
¾
¾

Typical Pin Cap Detail for Pins over 10 In. in dia.

Dimensions shown are approximate

Although nuts may be used on all sizes of pins as shown above, for pins over 10″ in diameter the preferred practice is a detail similar to that shown at the left, in which the pin is held in place by a recessed cap at each end and secured by a bolt passing completely through the caps and pin. Suitable provision must be made for attaching pilots and driving nuts.

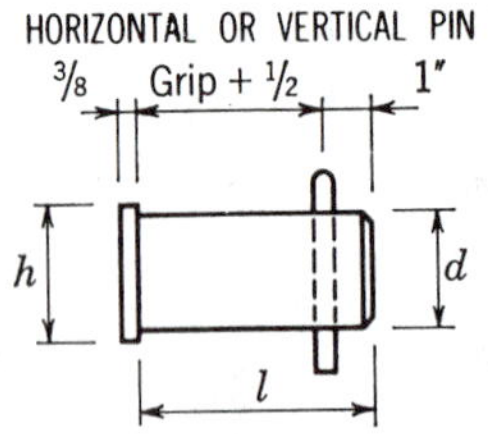
HORIZONTAL OR VERTICAL PIN
⅜
Grip + ½
1″
h
d
l

l = Length of pin, in in.

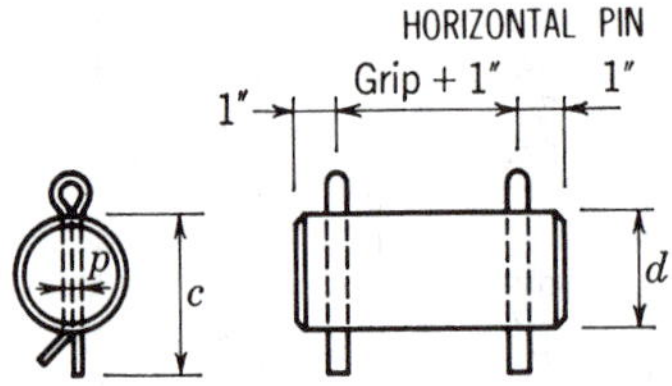
HORIZONTAL PIN
1″
Grip + 1″
1″
p
c
d

Pin Dia. *d*	Pins with Heads Head Dia. *h*	Pins with Heads Weight of One (lb.)	Cotter Length *c*	Cotter Dia. *p*	Cotter Wt. Per 100 (lb.)	Pin Dia. *d*	Pins with Heads Head Dia. *h*	Pins with Heads Weight of One (lb.)	Cotter Length *c*	Cotter Dia. *p*	Cotter Wt. per 100 (lb.)
1¼	1½	$.19 + .35l$	2	¼	2.64	2¾	3⅛	$.82 + 1.68l$	4	⅜	11.4
1½	1¾	$.26 + .50l$	2½	¼	3.10	3	3½	$1.02 + 2.00l$	5	½	28.5
1¾	2	$.33 + .68l$	2¾	¼	3.50	3¼	3¾	$1.17 + 2.35l$	5	½	28.5
2	2⅜	$.47 + .89l$	3	⅜	9.00	3½	4	$1.34 + 2.73l$	6	½	33.8
2¼	2⅝	$.58 + 1.13l$	3¼	⅜	9.40	3¾	4¼	$1.51 + 3.13l$	6	½	33.8
2½	2⅞	$.70 + 1.39l$	3¾	⅜	10.9						

WELDED JOINTS
Requirements

The LRFD Specification and the Structural Welding Code of the American Welding Society exempt from tests and qualification most of the common welded joints used in steel structures. Such exempt joints are designated *prequalified*. AWS prequalification of a weld joint is based upon experience that sound weld metal with appropriate mechanical properties can be deposited, provided work is performed in accordance with all applicable provisions of the Structural Welding Code. Among the applicable provisions are requirements for joint form and geometry, which are reproduced for convenience on the following pages.

Prequalification is intended only to mean that sound weld metal can be deposited and fused to the base metal. Suitability of particular joints for specific applications is not assured merely by the selection of a prequalified joint form. The design and detailing for successful welded construction require consideration of factors which include, but are not limited to, magnitude, type and distribution of forces to be transmitted, accessibility, restraint to weld metal contraction, thickness of connected material, effect of residual welding stresses on connected material and distortion.

In general, all fillet welds are deemed prequalified, whether illustrated or not, provided they conform to requirements of the AWS Code and the LRFD Specification.

These prequalified joints are limited to those made by the shielded metal arc, submerged arc, gas metal arc (except short circuiting transfer) and flux-cored arc welding procedures. Small deviations from dimensions, angles of grooves, and variation in the depth of groove joints are permissible within the tolerances given. Other joint forms and welding procedures may be employed, provided they are tested and qualified in accordance with AWS D1.1-85.

Most prequalified joints illustrated are also applicable for bridge construction. (See notes to Prequalified Welded Joints, immediately preceding the tables, and to prohibited types in Sect. 9 of AWS D1.1-85.)

For information on the subject of highly restrained welded joints, refer to the article "Commentary on Highly Restrained Welded Connections," AISC Engineering Journal, Vol. 10, No. 3, 3rd Quarter 1973, pp. 61–73.

The designations such as **B-L1a, B-U2, B-P3**, which are given on the following pages, are used in the AWS standards. Groove welds are classified using the following convention:

1. *Symbols for Joint Types*

   B—butt joint
   C—corner joint
   T—T-joint
   BC—butt or corner joint
   TC—T- or corner joint
   BTC—butt, T-, or corner joint

2. *Symbols for Base Metal Thickness and Penetration*

   L—limited thickness, complete joint penetration
   U—unlimited thickness, complete joint penetration
   P—partial joint penetration

3. *Symbols for Weld Types*

   1—square groove
   2—single-V groove
   3—double-V groove
   4—single-bevel groove
   5—double-bevel groove
   6—single-U groove
   7—double-U groove
   8—single-J groove
   9—double-J groove

4. *Symbols for Welding Processes*

If not shielded metal arc
(SMAW):
S—submerged arc welding (SAW)
G—gas metal arc welding
(GMAW)
F—flux-cored arc welding (FCAW)

5. *Symbols for Welding Positions*

F—flat
H—horizontal
V—vertical
OH—overhead

6. The lower case letters, e.g., a, b, c, etc., are used to differentiate between joints that would otherwise have the same joint designation.

NOTES TO PREQUALIFIED WELD JOINTS

A: Not prequalified for gas metal arc welding using short circuiting transfer.

B: Joints welded from one side.

Br: Bridge application limits the use of these joints to the horizontal position.

C: Gouge root of joint to sound metal before welding second side.

E: Minimum effective throat (E) as shown in LRFD Specification, Table J2.4; S as specified on drawings.

J: If fillet welds are used in buildings to reinforce groove welds in corner and T-joints, they shall be equal to ¼ T_1 but need not exceed ⅜ in. Groove welds in corner and T-joints in bridges shall be made with fillet welds equal to ¼ T_1, but not more than ⅜ in.

J2: If fillet welds are used in buildings to reinforce groove welds in corner and T-joints, they shall be equal to ¼ T_1, but not more than ⅜ in.

L: Butt and T-joints are not prequalified for bridges.

M: Double-groove welds may have grooves of unequal depth, but the depth of the shallower groove shall be no less than one-fourth of the thickness of the thinner part joined.

Mp: Double-groove welds may have grooves of unequal depth, provided they conform to the limitations of Note E. Also, the effective throat (E), less any reduction, applies individually to each groove.

N: The orientation of the two members in the joints may vary from 135° to 180°, provided the basic joint configuration (groove angle, root face, root opening) remain the same and that design throat thickness is maintained.

Q: For corner and T-joints, the member orientation may be changed, provided the groove angle is maintained as specified.

Q2: The member orientation may be changed, provided the groove dimensions are maintained as specified.

R: The orientation of two members in the joint may vary from 45° to 135° for corner joints and from 45° to 90° for T-joints, provided the basic joint configuration (groove angle, root face, root opening) remain the same and the design throat thickness is maintained.

V: For corner joints, the outside groove preparation may be in either or both members, provided the basic groove configuration is not changed and adequate edge distance is maintained to support the welding operations without excessive edge melting.

Note: Data on welded prequalified joints are reproduced on the following pages by courtesy of the American Welding Society.

WELDED JOINTS
Standard symbols

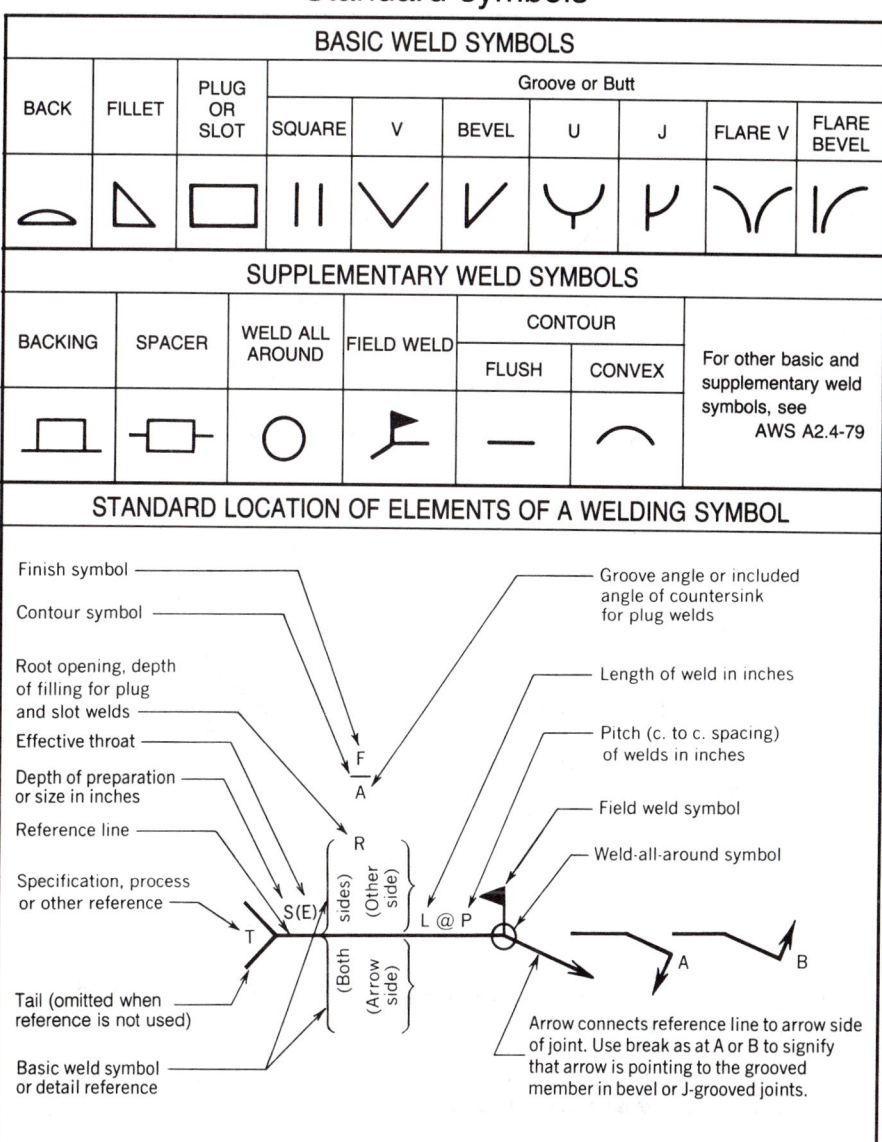

BASIC WELD SYMBOLS									
		PLUG OR SLOT	Groove or Butt						
BACK	FILLET		SQUARE	V	BEVEL	U	J	FLARE V	FLARE BEVEL

SUPPLEMENTARY WELD SYMBOLS						
		WELD ALL AROUND	FIELD WELD	CONTOUR		For other basic and supplementary weld symbols, see AWS A2.4-79
BACKING	SPACER			FLUSH	CONVEX	

STANDARD LOCATION OF ELEMENTS OF A WELDING SYMBOL

Finish symbol

Contour symbol

Root opening, depth of filling for plug and slot welds

Effective throat

Depth of preparation or size in inches

Reference line

Specification, process or other reference

Tail (omitted when reference is not used)

Basic weld symbol or detail reference

Groove angle or included angle of countersink for plug welds

Length of weld in inches

Pitch (c. to c. spacing) of welds in inches

Field weld symbol

Weld-all-around symbol

Arrow connects reference line to arrow side of joint. Use break as at A or B to signify that arrow is pointing to the grooved member in bevel or J-grooved joints.

Note:

Size, weld symbol, length of weld and spacing must read in that order from left to right along the reference line. Neither orientation of reference line nor location of the arrow alters this rule.

The perpendicular leg of ◺, ⊻, ⊬, ⌐ weld symbols must be at left.

Arrow and Other Side welds are of the same size unless otherwise shown. Dimensions of fillet welds must be shown on both the Arrow Side and the Other Side Symbol.

The point of the field weld symbol must point toward the tail.

Symbols apply between abrupt changes in direction of welding unless governed by the "all around" symbol or otherwise dimensioned.

These symbols do not explicitly provide for the case that frequently occurs in structural work, where duplicate material (such as stiffeners) occurs on the far side of a web or gusset plate. The fabricating industry has adopted this convention: that when the billing of the detail material discloses the existence of a member on the far side as well as on the near side, the welding shown for the near side shall be duplicated on the far side.

PREQUALIFIED WELDED JOINTS
Fillet welds

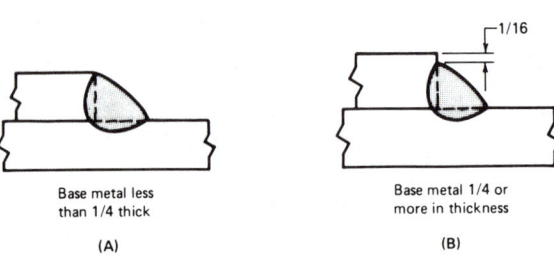

Base metal less
than 1/4 thick

(A)

Base metal 1/4 or
more in thickness

(B)

Maximum detailed size of fillet weld along edges.

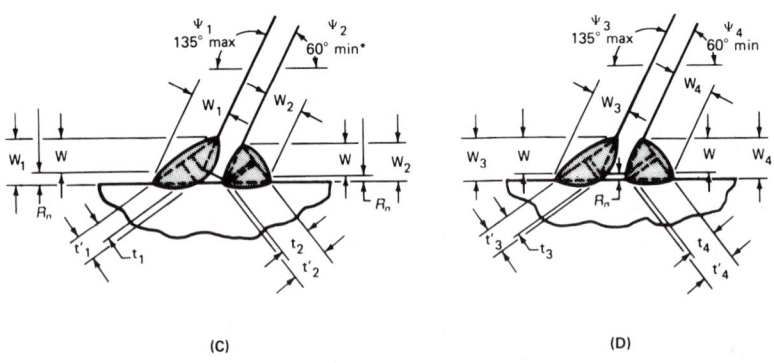

(C)

(D)

Skewed T-joints

Note: E_n, E'_n = effective throats dependent on magnitude of root opening R_n. See AWS 3.3.1. Subscript n represents 1,2,3 or 4.

*Angles smaller than 60 degrees are permitted; however, in such cases, the weld is considered to be a partial joint penetration groove weld.

PREQUALIFIED WELDED JOINTS
Complete penetration groove welds

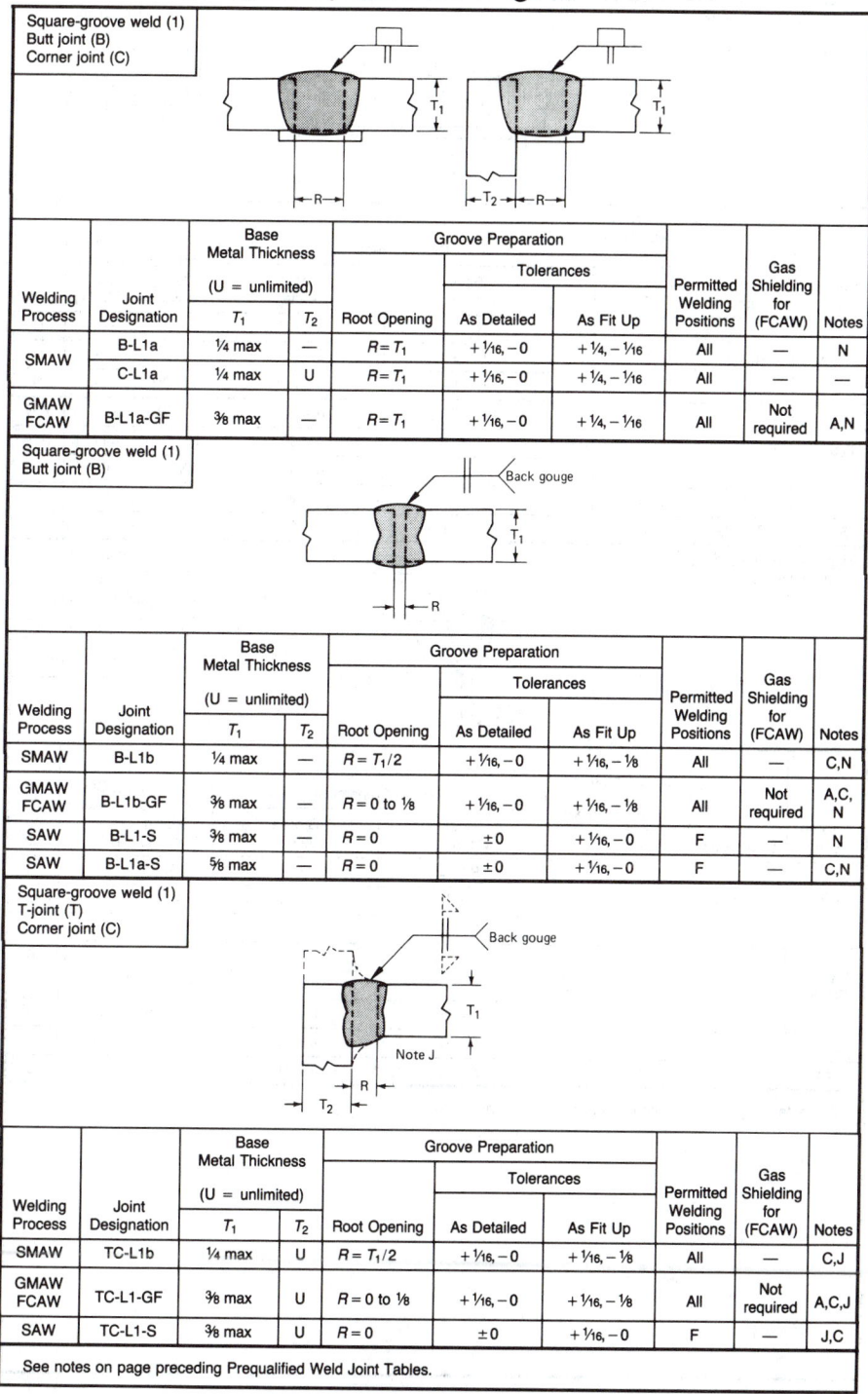

Square-groove weld (1)
Butt joint (B)
Corner joint (C)

Welding Process	Joint Designation	Base Metal Thickness (U = unlimited) T₁	T₂	Root Opening	As Detailed	As Fit Up	Permitted Welding Positions	Gas Shielding for (FCAW)	Notes
SMAW	B-L1a	¼ max	—	$R = T_1$	+1/16, −0	+¼, −1/16	All	—	N
	C-L1a	¼ max	U	$R = T_1$	+1/16, −0	+¼, −1/16	All	—	—
GMAW FCAW	B-L1a-GF	⅜ max	—	$R = T_1$	+1/16, −0	+¼, −1/16	All	Not required	A,N

Square-groove weld (1)
Butt joint (B)

Welding Process	Joint Designation	Base Metal Thickness (U = unlimited) T₁	T₂	Root Opening	As Detailed	As Fit Up	Permitted Welding Positions	Gas Shielding for (FCAW)	Notes
SMAW	B-L1b	¼ max	—	$R = T_1/2$	+1/16, −0	+1/16, −⅛	All	—	C,N
GMAW FCAW	B-L1b-GF	⅜ max	—	$R = 0$ to ⅛	+1/16, −0	+1/16, −⅛	All	Not required	A,C,N
SAW	B-L1-S	⅜ max	—	$R = 0$	±0	+1/16, −0	F	—	N
SAW	B-L1a-S	⅝ max	—	$R = 0$	±0	+1/16, −0	F	—	C,N

Square-groove weld (1)
T-joint (T)
Corner joint (C)

Welding Process	Joint Designation	Base Metal Thickness (U = unlimited) T₁	T₂	Root Opening	As Detailed	As Fit Up	Permitted Welding Positions	Gas Shielding for (FCAW)	Notes
SMAW	TC-L1b	¼ max	U	$R = T_1/2$	+1/16, −0	+1/16, −⅛	All	—	C,J
GMAW FCAW	TC-L1-GF	⅜ max	U	$R = 0$ to ⅛	+1/16, −0	+1/16, −⅛	All	Not required	A,C,J
SAW	TC-L1-S	⅜ max	U	$R = 0$	±0	+1/16, −0	F	—	J,C

See notes on page preceding Prequalified Weld Joint Tables.

PREQUALIFIED WELDED JOINTS
Complete penetration groove welds

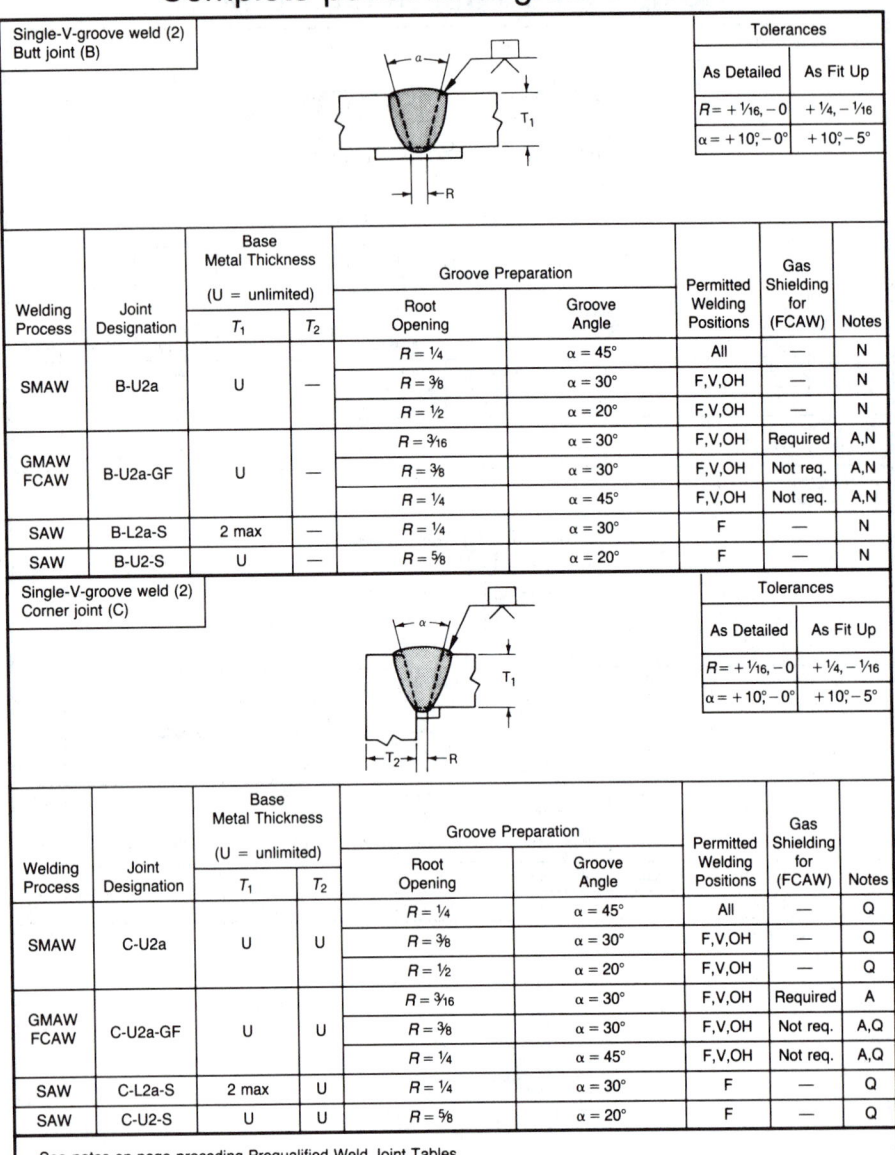

	Single-V-groove weld (2) Butt joint (B)	Tolerances	
As Detailed	As Fit Up
$R = + \frac{1}{16}, - 0$	$+ \frac{1}{4}, - \frac{1}{16}$
$\alpha = + 10°, - 0°$	$+ 10°, - 5°$

Welding Process	Joint Designation	Base Metal Thickness (U = unlimited) T₁	T₂	Groove Preparation Root Opening	Groove Angle	Permitted Welding Positions	Gas Shielding for (FCAW)	Notes
SMAW	B-U2a	U	—	$R = \frac{1}{4}$	$\alpha = 45°$	All	—	N
				$R = \frac{3}{8}$	$\alpha = 30°$	F,V,OH	—	N
				$R = \frac{1}{2}$	$\alpha = 20°$	F,V,OH	—	N
GMAW FCAW	B-U2a-GF	U	—	$R = \frac{3}{16}$	$\alpha = 30°$	F,V,OH	Required	A,N
				$R = \frac{3}{8}$	$\alpha = 30°$	F,V,OH	Not req.	A,N
				$R = \frac{1}{4}$	$\alpha = 45°$	F,V,OH	Not req.	A,N
SAW	B-L2a-S	2 max	—	$R = \frac{1}{4}$	$\alpha = 30°$	F	—	N
SAW	B-U2-S	U	—	$R = \frac{5}{8}$	$\alpha = 20°$	F	—	N

	Single-V-groove weld (2) Corner joint (C)	Tolerances	
As Detailed	As Fit Up
$R = + \frac{1}{16}, - 0$	$+ \frac{1}{4}, - \frac{1}{16}$
$\alpha = + 10°, - 0°$	$+ 10°, - 5°$

Welding Process	Joint Designation	Base Metal Thickness (U = unlimited) T₁	T₂	Groove Preparation Root Opening	Groove Angle	Permitted Welding Positions	Gas Shielding for (FCAW)	Notes
SMAW	C-U2a	U	U	$R = \frac{1}{4}$	$\alpha = 45°$	All	—	Q
				$R = \frac{3}{8}$	$\alpha = 30°$	F,V,OH	—	Q
				$R = \frac{1}{2}$	$\alpha = 20°$	F,V,OH	—	Q
GMAW FCAW	C-U2a-GF	U	U	$R = \frac{3}{16}$	$\alpha = 30°$	F,V,OH	Required	A
				$R = \frac{3}{8}$	$\alpha = 30°$	F,V,OH	Not req.	A,Q
				$R = \frac{1}{4}$	$\alpha = 45°$	F,V,OH	Not req.	A,Q
SAW	C-L2a-S	2 max	U	$R = \frac{1}{4}$	$\alpha = 30°$	F	—	Q
SAW	C-U2-S	U	U	$R = \frac{5}{8}$	$\alpha = 20°$	F	—	Q

See notes on page preceding Prequalified Weld Joint Tables.

PREQUALIFIED WELDED JOINTS
Complete penetration groove welds

Single-V-groove weld (2)
Butt joint (B)

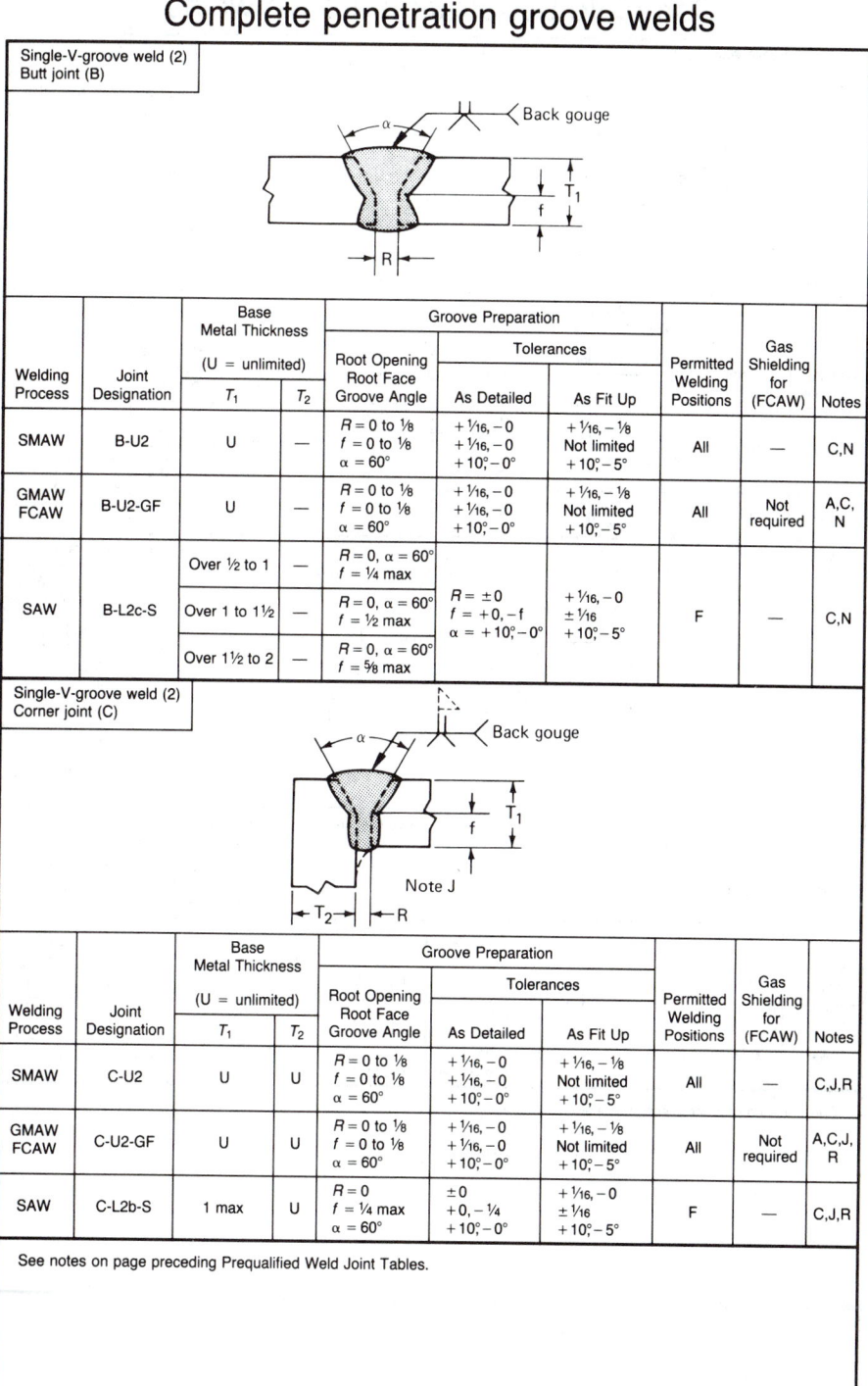

Welding Process	Joint Designation	Base Metal Thickness (U = unlimited)		Groove Preparation			Permitted Welding Positions	Gas Shielding for (FCAW)	Notes
		T_1	T_2	Root Opening Root Face Groove Angle	Tolerances				
					As Detailed	As Fit Up			
SMAW	B-U2	U	—	$R = 0$ to $1/8$ $f = 0$ to $1/8$ $\alpha = 60°$	$+ 1/16, - 0$ $+ 1/16, - 0$ $+ 10°, - 0°$	$+ 1/16, - 1/8$ Not limited $+ 10°, - 5°$	All	—	C,N
GMAW FCAW	B-U2-GF	U	—	$R = 0$ to $1/8$ $f = 0$ to $1/8$ $\alpha = 60°$	$+ 1/16, - 0$ $+ 1/16, - 0$ $+ 10°, - 0°$	$+ 1/16, - 1/8$ Not limited $+ 10°, - 5°$	All	Not required	A,C,N
SAW	B-L2c-S	Over 1/2 to 1	—	$R = 0, \alpha = 60°$ $f = 1/4$ max	$R = \pm 0$ $f = +0, -f$ $\alpha = + 10°, - 0°$	$+ 1/16, - 0$ $\pm 1/16$ $+ 10°, - 5°$	F	—	C,N
		Over 1 to 1 1/2	—	$R = 0, \alpha = 60°$ $f = 1/2$ max					
		Over 1 1/2 to 2	—	$R = 0, \alpha = 60°$ $f = 5/8$ max					

Single-V-groove weld (2)
Corner joint (C)

Back gouge

Note J

Welding Process	Joint Designation	Base Metal Thickness (U = unlimited)		Groove Preparation			Permitted Welding Positions	Gas Shielding for (FCAW)	Notes
		T_1	T_2	Root Opening Root Face Groove Angle	Tolerances				
					As Detailed	As Fit Up			
SMAW	C-U2	U	U	$R = 0$ to $1/8$ $f = 0$ to $1/8$ $\alpha = 60°$	$+ 1/16, - 0$ $+ 1/16, - 0$ $+ 10°, - 0°$	$+ 1/16, - 1/8$ Not limited $+ 10°, - 5°$	All	—	C,J,R
GMAW FCAW	C-U2-GF	U	U	$R = 0$ to $1/8$ $f = 0$ to $1/8$ $\alpha = 60°$	$+ 1/16, - 0$ $+ 1/16, - 0$ $+ 10°, - 0°$	$+ 1/16, - 1/8$ Not limited $+ 10°, - 5°$	All	Not required	A,C,J,R
SAW	C-L2b-S	1 max	U	$R = 0$ $f = 1/4$ max $\alpha = 60°$	± 0 $+ 0, - 1/4$ $+ 10°, - 0°$	$+ 1/16, - 0$ $\pm 1/16$ $+ 10°, - 5°$	F	—	C,J,R

See notes on page preceding Prequalified Weld Joint Tables.

PREQUALIFIED WELDED JOINTS
Complete penetration groove welds

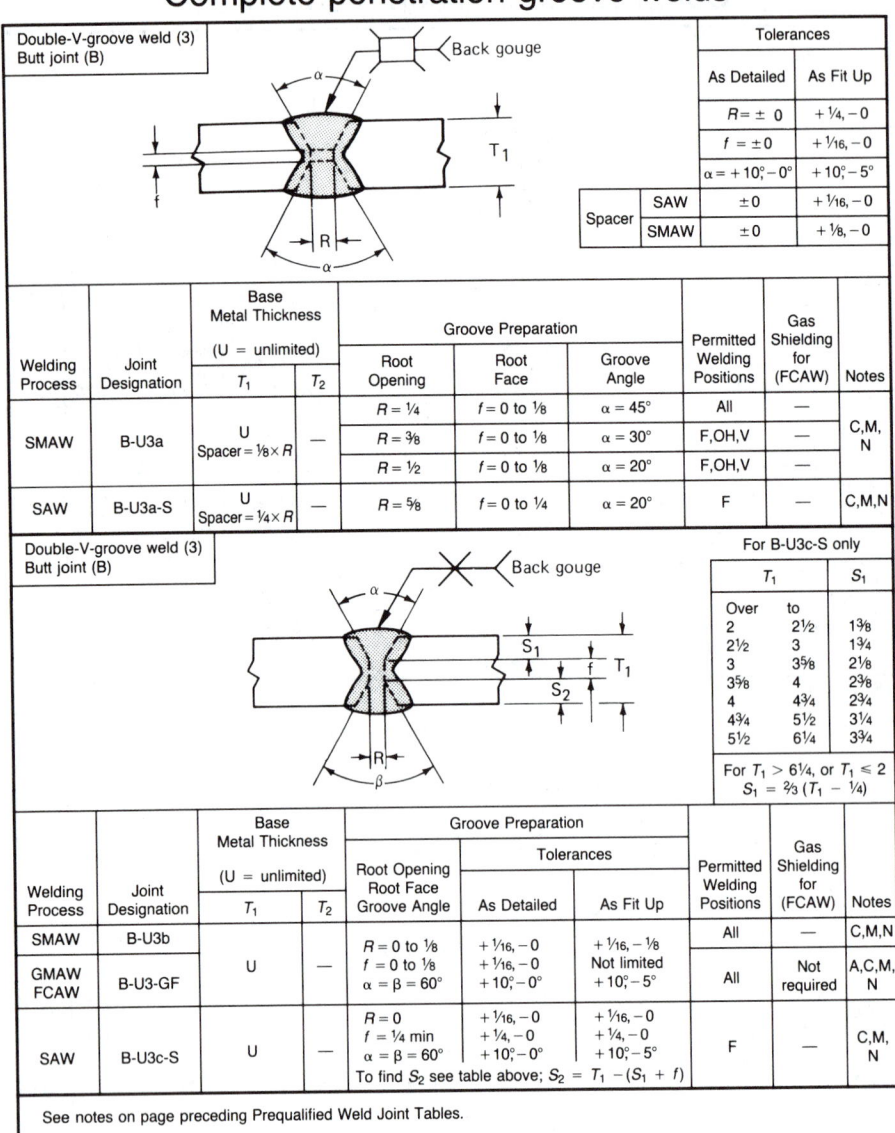

Double-V-groove weld (3)
Butt joint (B)

Tolerances		
	As Detailed	As Fit Up
$R = \pm\ 0$		$+\frac{1}{4}, -0$
$f = \pm 0$		$+\frac{1}{16}, -0$
$\alpha = +10°, -0°$		$+10°, -5°$
Spacer SAW	± 0	$+\frac{1}{16}, -0$
Spacer SMAW	± 0	$+\frac{1}{8}, -0$

Welding Process	Joint Designation	Base Metal Thickness (U = unlimited) T_1	T_2	Groove Preparation Root Opening	Root Face	Groove Angle	Permitted Welding Positions	Gas Shielding for (FCAW)	Notes
SMAW	B-U3a	U Spacer $=\frac{1}{8} \times R$		$R = \frac{1}{4}$	$f = 0$ to $\frac{1}{8}$	$\alpha = 45°$	All	—	C,M, N
				$R = \frac{3}{8}$	$f = 0$ to $\frac{1}{8}$	$\alpha = 30°$	F,OH,V	—	
				$R = \frac{1}{2}$	$f = 0$ to $\frac{1}{8}$	$\alpha = 20°$	F,OH,V	—	
SAW	B-U3a-S	U Spacer $=\frac{1}{4} \times R$	—	$R = \frac{5}{8}$	$f = 0$ to $\frac{1}{4}$	$\alpha = 20°$	F	—	C,M,N

Double-V-groove weld (3)
Butt joint (B)

For B-U3c-S only

T_1		S_1
Over	to	
2	$2\frac{1}{2}$	$1\frac{3}{8}$
$2\frac{1}{2}$	3	$1\frac{3}{4}$
3	$3\frac{5}{8}$	$2\frac{1}{8}$
$3\frac{5}{8}$	4	$2\frac{3}{8}$
4	$4\frac{3}{4}$	$2\frac{3}{4}$
$4\frac{3}{4}$	$5\frac{1}{2}$	$3\frac{1}{4}$
$5\frac{1}{2}$	$6\frac{1}{4}$	$3\frac{3}{4}$

For $T_1 > 6\frac{1}{4}$, or $T_1 \leq 2$
$S_1 = \frac{2}{3}(T_1 - \frac{1}{4})$

Welding Process	Joint Designation	Base Metal Thickness (U = unlimited) T_1	T_2	Groove Preparation Root Opening Root Face Groove Angle	Tolerances As Detailed	As Fit Up	Permitted Welding Positions	Gas Shielding for (FCAW)	Notes
SMAW	B-U3b	U	—	$R = 0$ to $\frac{1}{8}$	$+\frac{1}{16}, -0$	$+\frac{1}{16}, -\frac{1}{8}$	All	—	C,M,N
GMAW FCAW	B-U3-GF			$f = 0$ to $\frac{1}{8}$	$+\frac{1}{16}, -0$	Not limited	All	Not required	A,C,M, N
				$\alpha = \beta = 60°$	$+10°, -0°$	$+10°, -5°$			
SAW	B-U3c-S	U	—	$R = 0$	$+\frac{1}{16}, -0$	$+\frac{1}{16}, -0$	F	—	C,M, N
				$f = \frac{1}{4}$ min	$+\frac{1}{4}, -0$	$+\frac{1}{4}, -0$			
				$\alpha = \beta = 60°$	$+10°, -0°$	$+10°, -5°$			
				To find S_2 see table above; $S_2 = T_1 - (S_1 + f)$					

See notes on page preceding Prequalified Weld Joint Tables.

PREQUALIFIED WELDED JOINTS
Complete penetration groove welds

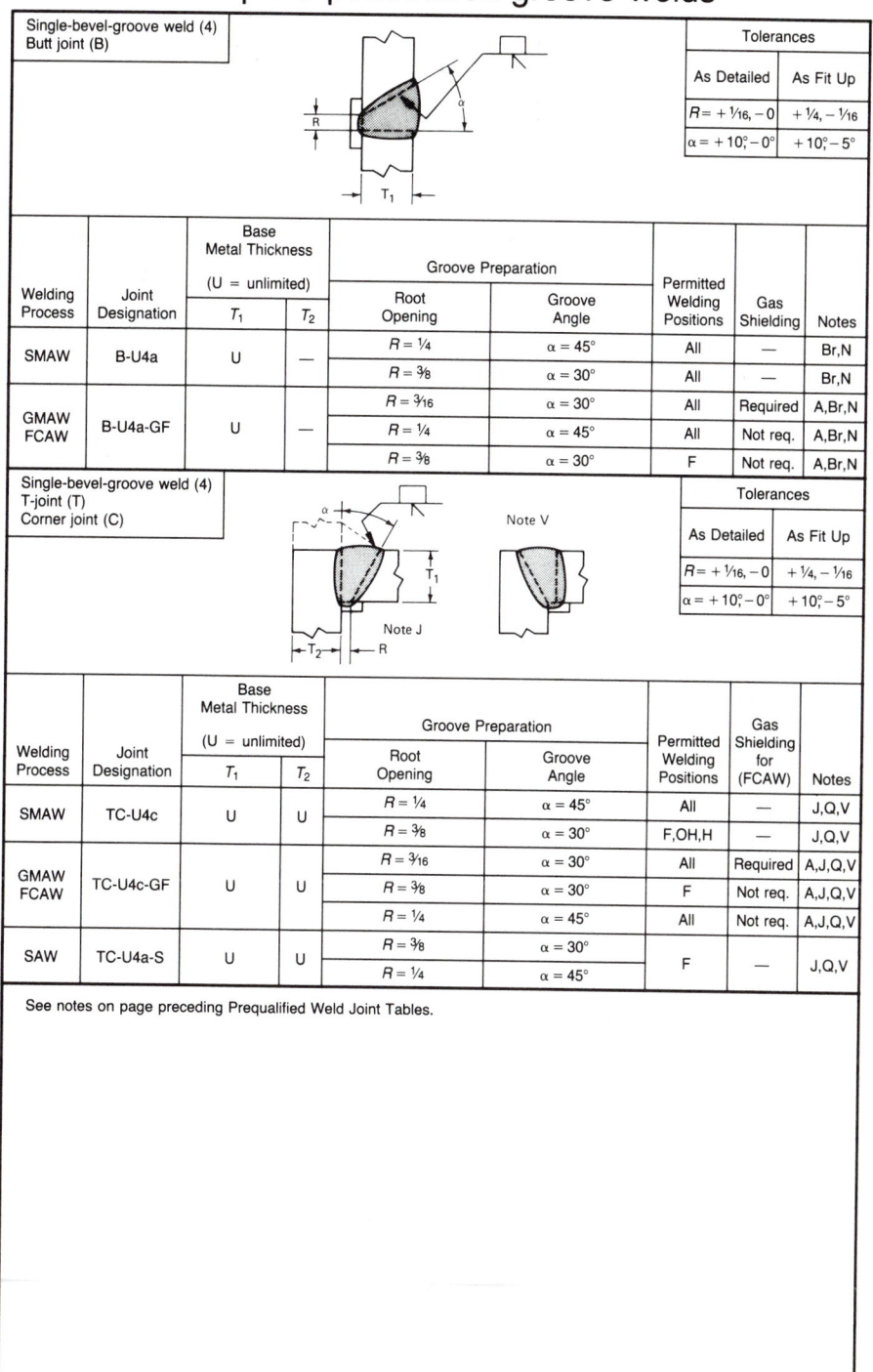

Single-bevel-groove weld (4)
Butt joint (B)

Tolerances		
	As Detailed	As Fit Up
$R = +\frac{1}{16}, -0$		$+\frac{1}{4}, -\frac{1}{16}$
$\alpha = +10°, -0°$		$+10°, -5°$

Welding Process	Joint Designation	Base Metal Thickness (U = unlimited) T_1	T_2	Groove Preparation Root Opening	Groove Angle	Permitted Welding Positions	Gas Shielding	Notes
SMAW	B-U4a	U	—	$R = \frac{1}{4}$	$\alpha = 45°$	All	—	Br,N
				$R = \frac{3}{8}$	$\alpha = 30°$	All	—	Br,N
GMAW FCAW	B-U4a-GF	U	—	$R = \frac{3}{16}$	$\alpha = 30°$	All	Required	A,Br,N
				$R = \frac{1}{4}$	$\alpha = 45°$	All	Not req.	A,Br,N
				$R = \frac{3}{8}$	$\alpha = 30°$	F	Not req.	A,Br,N

Single-bevel-groove weld (4)
T-joint (T)
Corner joint (C)

Note V

Note J

Tolerances		
	As Detailed	As Fit Up
$R = +\frac{1}{16}, -0$		$+\frac{1}{4}, -\frac{1}{16}$
$\alpha = +10°, -0°$		$+10°, -5°$

Welding Process	Joint Designation	Base Metal Thickness (U = unlimited) T_1	T_2	Groove Preparation Root Opening	Groove Angle	Permitted Welding Positions	Gas Shielding for (FCAW)	Notes
SMAW	TC-U4c	U	U	$R = \frac{1}{4}$	$\alpha = 45°$	All	—	J,Q,V
				$R = \frac{3}{8}$	$\alpha = 30°$	F,OH,H	—	J,Q,V
GMAW FCAW	TC-U4c-GF	U	U	$R = \frac{3}{16}$	$\alpha = 30°$	All	Required	A,J,Q,V
				$R = \frac{3}{8}$	$\alpha = 30°$	F	Not req.	A,J,Q,V
				$R = \frac{1}{4}$	$\alpha = 45°$	All	Not req.	A,J,Q,V
SAW	TC-U4a-S	U	U	$R = \frac{3}{8}$	$\alpha = 30°$	F	—	J,Q,V
				$R = \frac{1}{4}$	$\alpha = 45°$			

See notes on page preceding Prequalified Weld Joint Tables.

PREQUALIFIED WELDED JOINTS
Complete penetration groove welds

| Single-bevel-groove weld (4) |
| Butt joint (B) |

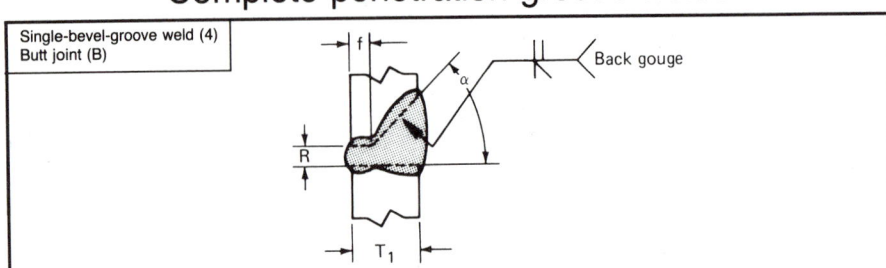

Back gouge

Welding Process	Joint Designation	Base Metal Thickness (U = unlimited)		Groove Preparation			Permitted Welding Positions	Gas Shielding for (FCAW)	Notes
				Root Opening Root Face Groove Angle	Tolerances				
		T_1	T_2		As Detailed	As Fit Up			
SMAW	B-U4	U	—	$R = 0$ to $1/8$	$+1/16, -0$	$+1/16, -1/8$	All	—	Br,C,N
GMAW FCAW	B-U4-GF	U	—	$f = 0$ to $1/8$ $\alpha = 45°$	$+1/16, -0$ $+10°, -0°$	Not limited $+10°, -5°$	All	Not required	A,Br,C, N

See notes on page preceding Prequalified Weld Joint Tables.

PREQUALIFIED WELDED JOINTS
Complete penetration groove welds

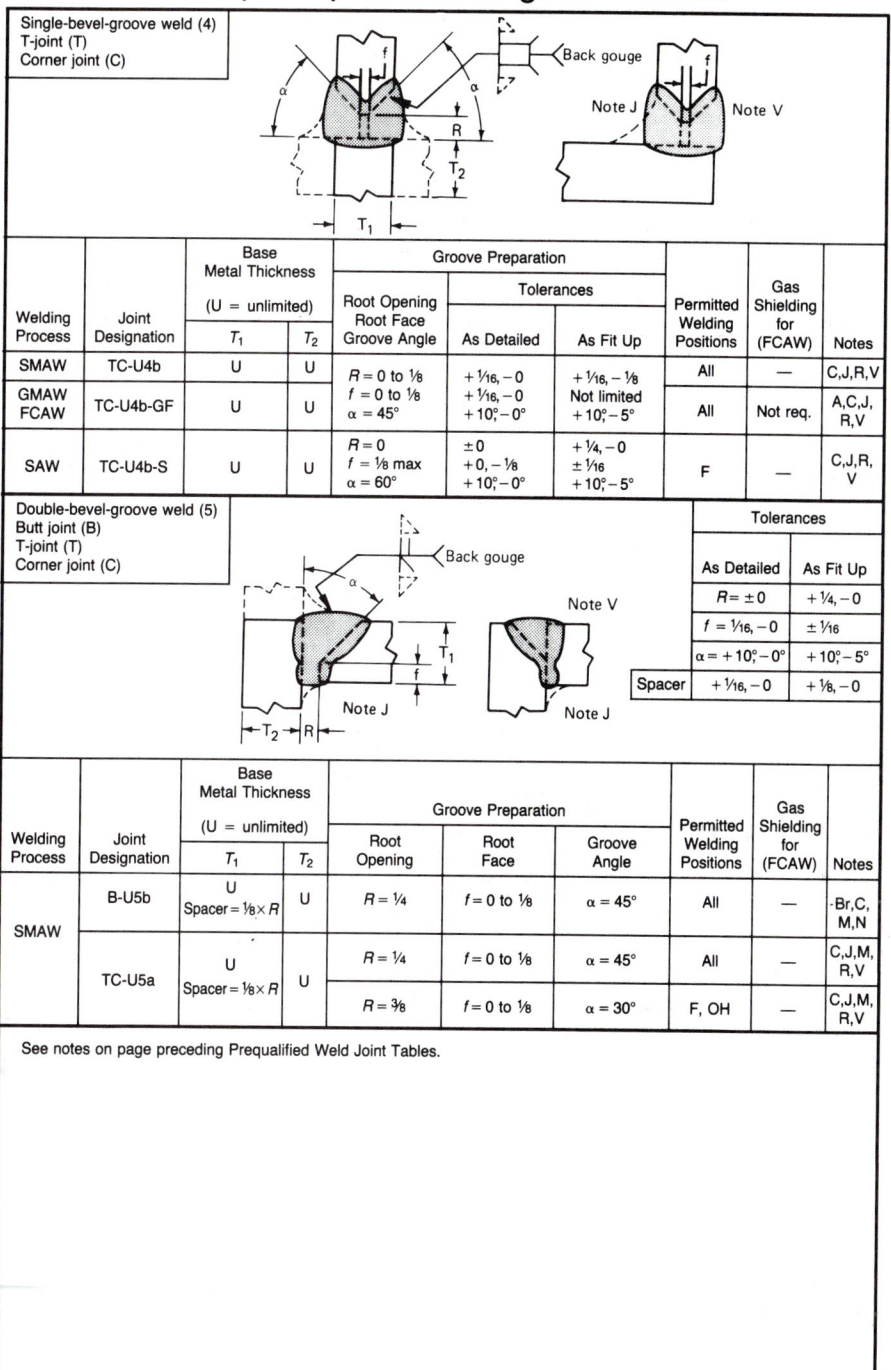

Single-bevel-groove weld (4)
T-joint (T)
Corner joint (C)

Welding Process	Joint Designation	Base Metal Thickness (U = unlimited) T_1	T_2	Groove Preparation — Root Opening Root Face Groove Angle	Tolerances As Detailed	Tolerances As Fit Up	Permitted Welding Positions	Gas Shielding for (FCAW)	Notes
SMAW	TC-U4b	U	U	$R = 0$ to $1/8$ $f = 0$ to $1/8$ $\alpha = 45°$	$+1/16, -0$ $+1/16, -0$ $+10°, -0°$	$+1/16, -1/8$ Not limited $+10°, -5°$	All	—	C,J,R,V
GMAW FCAW	TC-U4b-GF	U	U				All	Not req.	A,C,J, R,V
SAW	TC-U4b-S	U	U	$R = 0$ $f = 1/8$ max $\alpha = 60°$	± 0 $+0, -1/8$ $+10°, -0°$	$+1/4, -0$ $\pm 1/16$ $+10°, -5°$	F	—	C,J,R, V

Double-bevel-groove weld (5)
Butt joint (B)
T-joint (T)
Corner joint (C)

	Tolerances As Detailed	Tolerances As Fit Up
	$R = \pm 0$	$+1/4, -0$
	$f = 1/16, -0$	$\pm 1/16$
	$\alpha = +10°, -0°$	$+10°, -5°$
Spacer	$+1/16, -0$	$+1/8, -0$

Welding Process	Joint Designation	Base Metal Thickness (U = unlimited) T_1	T_2	Groove Preparation — Root Opening	Groove Preparation — Root Face	Groove Preparation — Groove Angle	Permitted Welding Positions	Gas Shielding for (FCAW)	Notes
SMAW	B-U5b	U Spacer $= 1/8 \times R$	U	$R = 1/4$	$f = 0$ to $1/8$	$\alpha = 45°$	All	—	Br,C, M,N
	TC-U5a	U Spacer $= 1/8 \times R$	U	$R = 1/4$	$f = 0$ to $1/8$	$\alpha = 45°$	All	—	C,J,M, R,V
				$R = 3/8$	$f = 0$ to $1/8$	$\alpha = 30°$	F, OH	—	C,J,M, R,V

See notes on page preceding Prequalified Weld Joint Tables.

PREQUALIFIED WELDED JOINTS
Complete penetration groove welds

Double-bevel-groove weld (5) Butt joint (B)

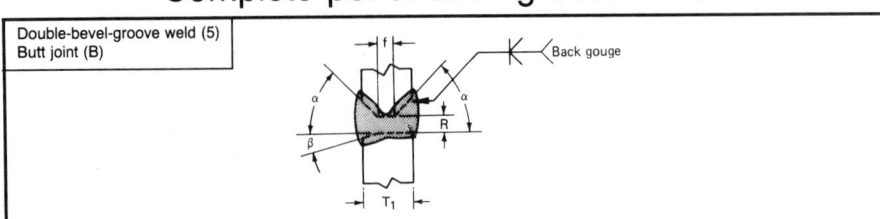

Welding Process	Joint Designation	Base Metal Thickness (U = unlimited)		Groove Preparation			Permitted Welding Positions	Gas Shielding for (FCAW)	Notes
		T_1	T_2	Root Opening Root Face Groove Angle	Tolerances				
					As Detailed	As Fit Up			
SMAW	B-U5a	U	—	$R = 0$ to $\frac{1}{8}$ $f = 0$ to $\frac{1}{8}$ $\alpha = 45°$ $\beta = 0°$ to $15°$	$+\frac{1}{16}, -0$ $+\frac{1}{16}, -0$ $\alpha + \beta, \begin{smallmatrix}+10°\\-0°\end{smallmatrix}$	$+\frac{1}{16}, -\frac{1}{8}$ Not limited $\alpha + \beta, \begin{smallmatrix}+10°\\-5°\end{smallmatrix}$	All	—	Br, C,M,N
GMAW FCAW	B-U5-GF	U	—	$R = 0$ to $\frac{1}{8}$ $f = 0$ to $\frac{1}{8}$ $\alpha = 45°$ $\beta = 0°$ to $15°$	$+\frac{1}{16}, -0$ $+\frac{1}{16}, -0$ $\alpha + \beta = +10°, -0°$	$+\frac{1}{16}, -\frac{1}{8}$ Not limited $\alpha + \beta = +10°, -5°$	All	Not req.	A,Br,C, M,N,

Double-bevel-groove weld (5) T-joint (T) Corner joint (C)

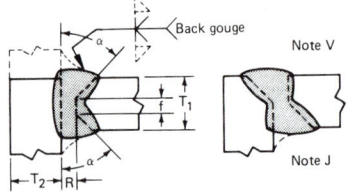

Welding Process	Joint Designation	Base Metal Thickness (U = unlimited)		Groove Preparation			Permitted Welding Positions	Gas Shielding for (FCAW)	Notes
		T_1	T_2	Root Opening Root Face Groove Angle	Tolerances				
					As Detailed	As Fit Up			
SMAW	TC-U5b	U	U	$R = 0$ to $\frac{1}{8}$ $f = 0$ to $\frac{1}{8}$ $\alpha = 45°$	$+\frac{1}{16}, -0$ $+\frac{1}{16}, -0$ $+10°, -0°$	$+\frac{1}{16}, -\frac{1}{8}$ Not limited $+10°, -5°$	All	—	C,J,M R,V
GMAW FCAW	TC-U5-GF	U	U				All	Not required	A,C,J M,R,V
SAW	TC-U5-S	U	U	$R = 0$ $f = \frac{3}{16}$ max $\alpha = 60°$	± 0 $+0, -\frac{3}{16}$ $+10°, -0°$	$+\frac{1}{16}, -0$ $\pm\frac{1}{16}$ $+10°, -5°$	F	—	C,J,M, R,V

See notes on page preceding Prequalified Weld Joint Tables.

PREQUALIFIED WELDED JOINTS
Complete penetration groove welds

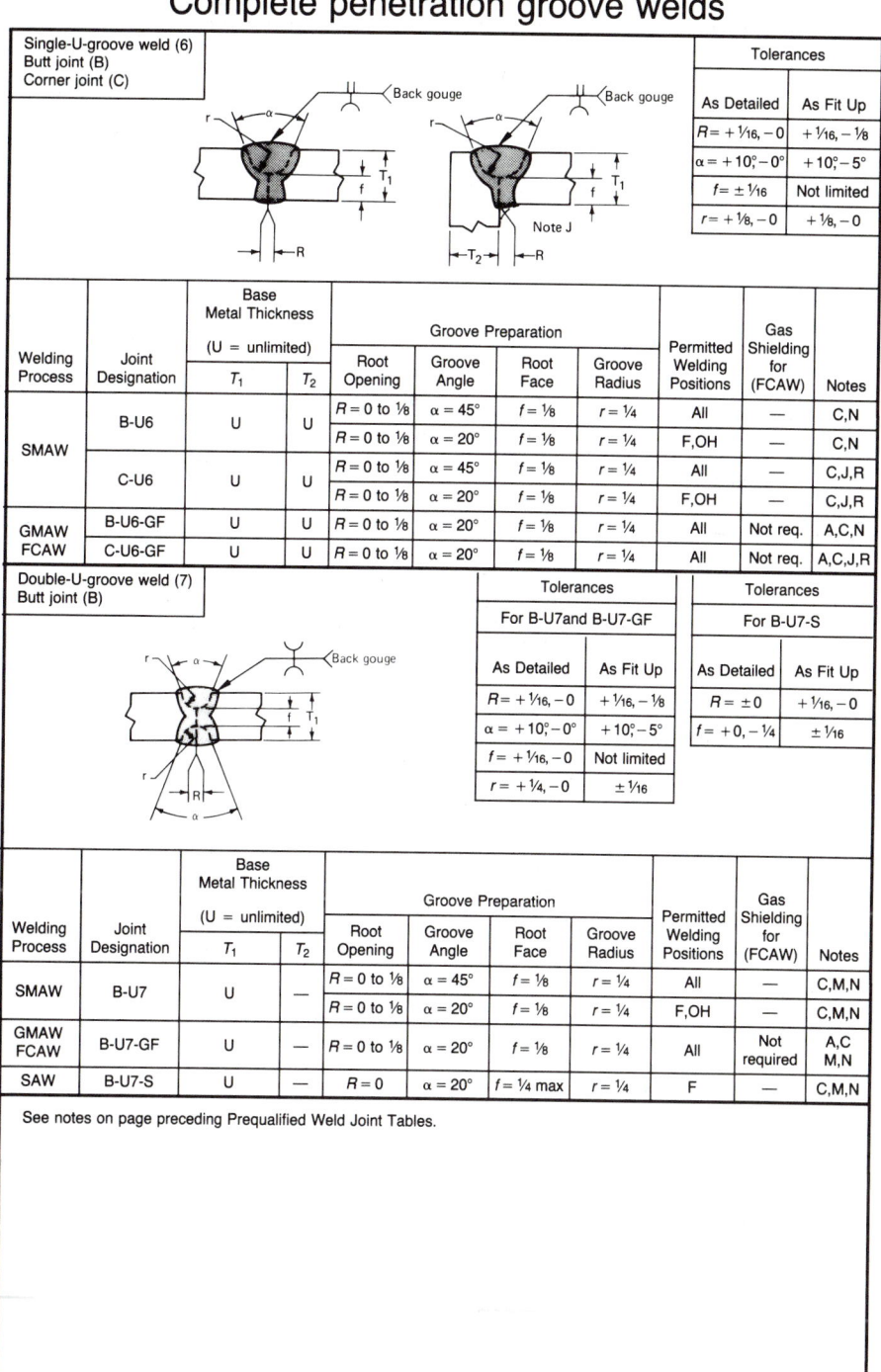

Single-U-groove weld (6)
Butt joint (B)
Corner joint (C)

Tolerances

	As Detailed	As Fit Up
$R = +\frac{1}{16}, -0$		$+\frac{1}{16}, -\frac{1}{8}$
$\alpha = +10°, -0°$		$+10°, -5°$
$f = \pm\frac{1}{16}$		Not limited
$r = +\frac{1}{8}, -0$		$+\frac{1}{8}, -0$

Welding Process	Joint Designation	T_1	T_2	Root Opening	Groove Angle	Root Face	Groove Radius	Permitted Welding Positions	Gas Shielding for (FCAW)	Notes
SMAW	B-U6	U	U	$R = 0$ to $\frac{1}{8}$	$\alpha = 45°$	$f = \frac{1}{8}$	$r = \frac{1}{4}$	All	—	C,N
	B-U6	U	U	$R = 0$ to $\frac{1}{8}$	$\alpha = 20°$	$f = \frac{1}{8}$	$r = \frac{1}{4}$	F,OH	—	C,N
	C-U6	U	U	$R = 0$ to $\frac{1}{8}$	$\alpha = 45°$	$f = \frac{1}{8}$	$r = \frac{1}{4}$	All	—	C,J,R
	C-U6	U	U	$R = 0$ to $\frac{1}{8}$	$\alpha = 20°$	$f = \frac{1}{8}$	$r = \frac{1}{4}$	F,OH	—	C,J,R
GMAW FCAW	B-U6-GF	U	U	$R = 0$ to $\frac{1}{8}$	$\alpha = 20°$	$f = \frac{1}{8}$	$r = \frac{1}{4}$	All	Not req.	A,C,N
	C-U6-GF	U	U	$R = 0$ to $\frac{1}{8}$	$\alpha = 20°$	$f = \frac{1}{8}$	$r = \frac{1}{4}$	All	Not req.	A,C,J,R

Double-U-groove weld (7)
Butt joint (B)

Tolerances For B-U7 and B-U7-GF

	As Detailed	As Fit Up
$R = +\frac{1}{16}, -0$		$+\frac{1}{16}, -\frac{1}{8}$
$\alpha = +10°, -0°$		$+10°, -5°$
$f = +\frac{1}{16}, -0$		Not limited
$r = +\frac{1}{4}, -0$		$\pm\frac{1}{16}$

Tolerances For B-U7-S

	As Detailed	As Fit Up
$R = \pm 0$		$+\frac{1}{16}, -0$
$f = +0, -\frac{1}{4}$		$\pm\frac{1}{16}$

Welding Process	Joint Designation	T_1	T_2	Root Opening	Groove Angle	Root Face	Groove Radius	Permitted Welding Positions	Gas Shielding for (FCAW)	Notes
SMAW	B-U7	U	—	$R = 0$ to $\frac{1}{8}$	$\alpha = 45°$	$f = \frac{1}{8}$	$r = \frac{1}{4}$	All	—	C,M,N
	B-U7	U	—	$R = 0$ to $\frac{1}{8}$	$\alpha = 20°$	$f = \frac{1}{8}$	$r = \frac{1}{4}$	F,OH	—	C,M,N
GMAW FCAW	B-U7-GF	U	—	$R = 0$ to $\frac{1}{8}$	$\alpha = 20°$	$f = \frac{1}{8}$	$r = \frac{1}{4}$	All	Not required	A,C M,N
SAW	B-U7-S	U	—	$R = 0$	$\alpha = 20°$	$f = \frac{1}{4}$ max	$r = \frac{1}{4}$	F	—	C,M,N

See notes on page preceding Prequalified Weld Joint Tables.

PREQUALIFIED WELDED JOINTS
Complete penetration groove welds

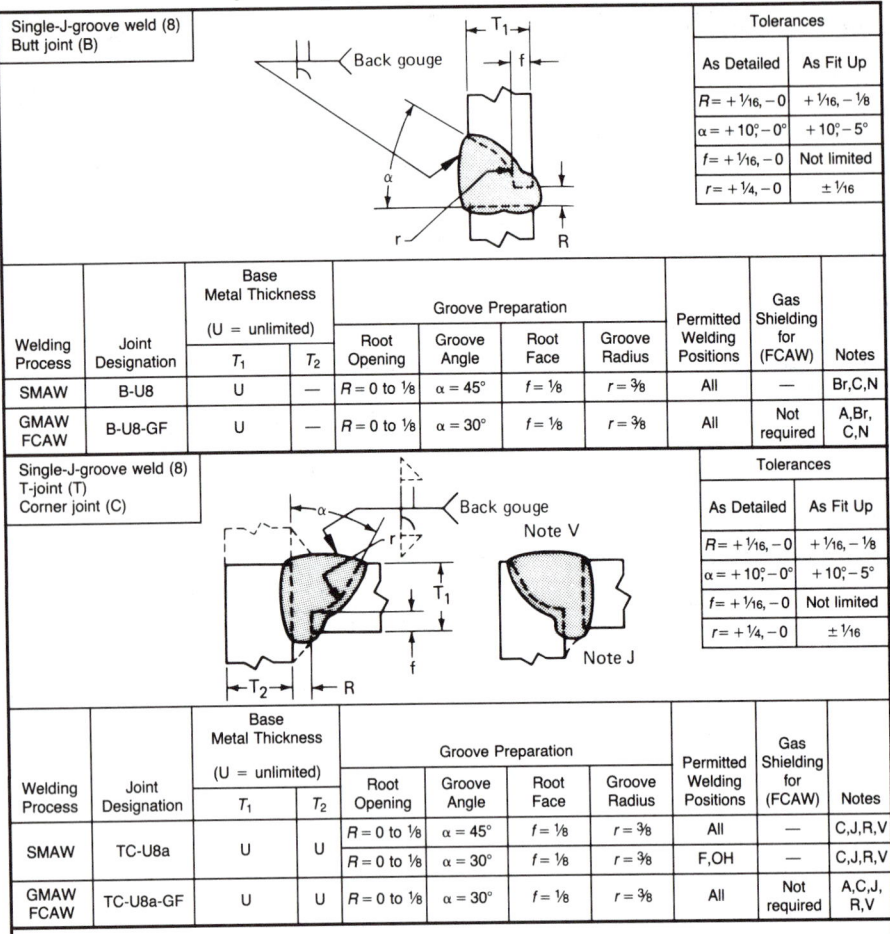

Single-J-groove weld (8) Butt joint (B)	Tolerances		
		As Detailed	As Fit Up
	$R=$	$+1/16, -0$	$+1/16, -1/8$
	$\alpha=$	$+10°, -0°$	$+10°, -5°$
	$f=$	$+1/16, -0$	Not limited
	$r=$	$+1/4, -0$	$\pm 1/16$

Welding Process	Joint Designation	Base Metal Thickness (U = unlimited)		Groove Preparation				Permitted Welding Positions	Gas Shielding for (FCAW)	Notes
		T_1	T_2	Root Opening	Groove Angle	Root Face	Groove Radius			
SMAW	B-U8	U	—	$R=0$ to $1/8$	$\alpha=45°$	$f=1/8$	$r=3/8$	All	—	Br,C,N
GMAW FCAW	B-U8-GF	U	—	$R=0$ to $1/8$	$\alpha=30°$	$f=1/8$	$r=3/8$	All	Not required	A,Br, C,N

Single-J-groove weld (8) T-joint (T) Corner joint (C)	Tolerances		
		As Detailed	As Fit Up
	$R=$	$+1/16, -0$	$+1/16, -1/8$
	$\alpha=$	$+10°, -0°$	$+10°, -5°$
	$f=$	$+1/16, -0$	Not limited
	$r=$	$+1/4, -0$	$\pm 1/16$

Welding Process	Joint Designation	Base Metal Thickness (U = unlimited)		Groove Preparation				Permitted Welding Positions	Gas Shielding for (FCAW)	Notes
		T_1	T_2	Root Opening	Groove Angle	Root Face	Groove Radius			
SMAW	TC-U8a	U	U	$R=0$ to $1/8$	$\alpha=45°$	$f=1/8$	$r=3/8$	All	—	C,J,R,V
				$R=0$ to $1/8$	$\alpha=30°$	$f=1/8$	$r=3/8$	F,OH	—	C,J,R,V
GMAW FCAW	TC-U8a-GF	U	U	$R=0$ to $1/8$	$\alpha=30°$	$f=1/8$	$r=3/8$	All	Not required	A,C,J, R,V

See notes on page preceding Prequalified Weld Joint Tables.

PREQUALIFIED WELDED JOINTS
Complete penetration groove welds

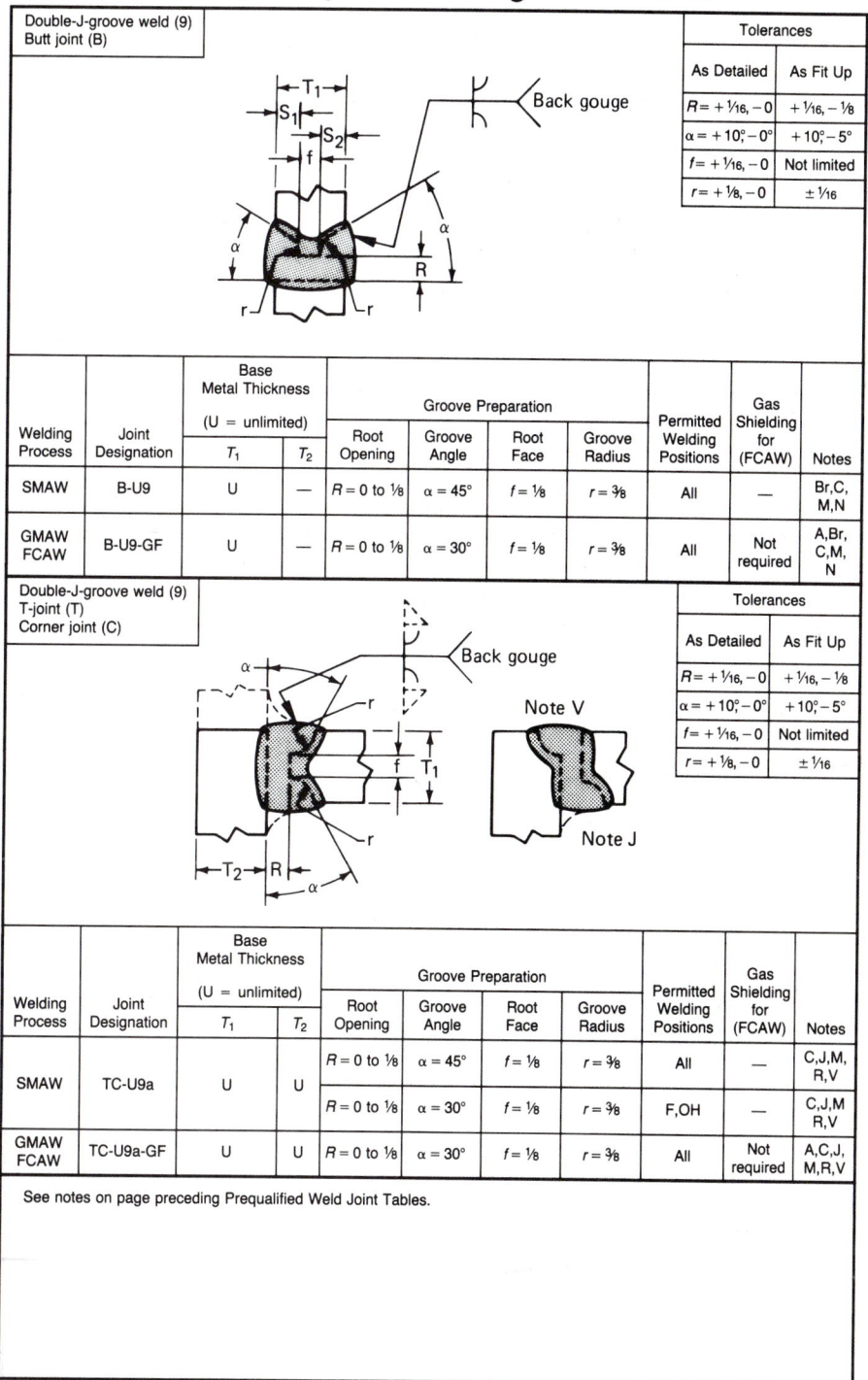

Double-J-groove weld (9)
Butt joint (B)

Back gouge

Tolerances		
	As Detailed	As Fit Up
$R = +\frac{1}{16}, -0$		$+\frac{1}{16}, -\frac{1}{8}$
$\alpha = +10°, -0°$		$+10°, -5°$
$f = +\frac{1}{16}, -0$		Not limited
$r = +\frac{1}{8}, -0$		$\pm\frac{1}{16}$

Welding Process	Joint Designation	Base Metal Thickness (U = unlimited)		Groove Preparation				Permitted Welding Positions	Gas Shielding for (FCAW)	Notes
		T_1	T_2	Root Opening	Groove Angle	Root Face	Groove Radius			
SMAW	B-U9	U	—	$R = 0$ to $\frac{1}{8}$	$\alpha = 45°$	$f = \frac{1}{8}$	$r = \frac{3}{8}$	All	—	Br,C, M,N
GMAW FCAW	B-U9-GF	U	—	$R = 0$ to $\frac{1}{8}$	$\alpha = 30°$	$f = \frac{1}{8}$	$r = \frac{3}{8}$	All	Not required	A,Br, C,M, N

Double-J-groove weld (9)
T-joint (T)
Corner joint (C)

Back gouge

Note V

Note J

Tolerances		
	As Detailed	As Fit Up
$R = +\frac{1}{16}, -0$		$+\frac{1}{16}, -\frac{1}{8}$
$\alpha = +10°, -0°$		$+10°, -5°$
$f = +\frac{1}{16}, -0$		Not limited
$r = +\frac{1}{8}, -0$		$\pm\frac{1}{16}$

Welding Process	Joint Designation	Base Metal Thickness (U = unlimited)		Groove Preparation				Permitted Welding Positions	Gas Shielding for (FCAW)	Notes
		T_1	T_2	Root Opening	Groove Angle	Root Face	Groove Radius			
SMAW	TC-U9a	U	U	$R = 0$ to $\frac{1}{8}$	$\alpha = 45°$	$f = \frac{1}{8}$	$r = \frac{3}{8}$	All	—	C,J,M, R,V
				$R = 0$ to $\frac{1}{8}$	$\alpha = 30°$	$f = \frac{1}{8}$	$r = \frac{3}{8}$	F,OH	—	C,J,M R,V
GMAW FCAW	TC-U9a-GF	U	U	$R = 0$ to $\frac{1}{8}$	$\alpha = 30°$	$f = \frac{1}{8}$	$r = \frac{3}{8}$	All	Not required	A,C,J, M,R,V

See notes on page preceding Prequalified Weld Joint Tables.

PREQUALIFIED WELDED JOINTS
Partial penetration groove welds

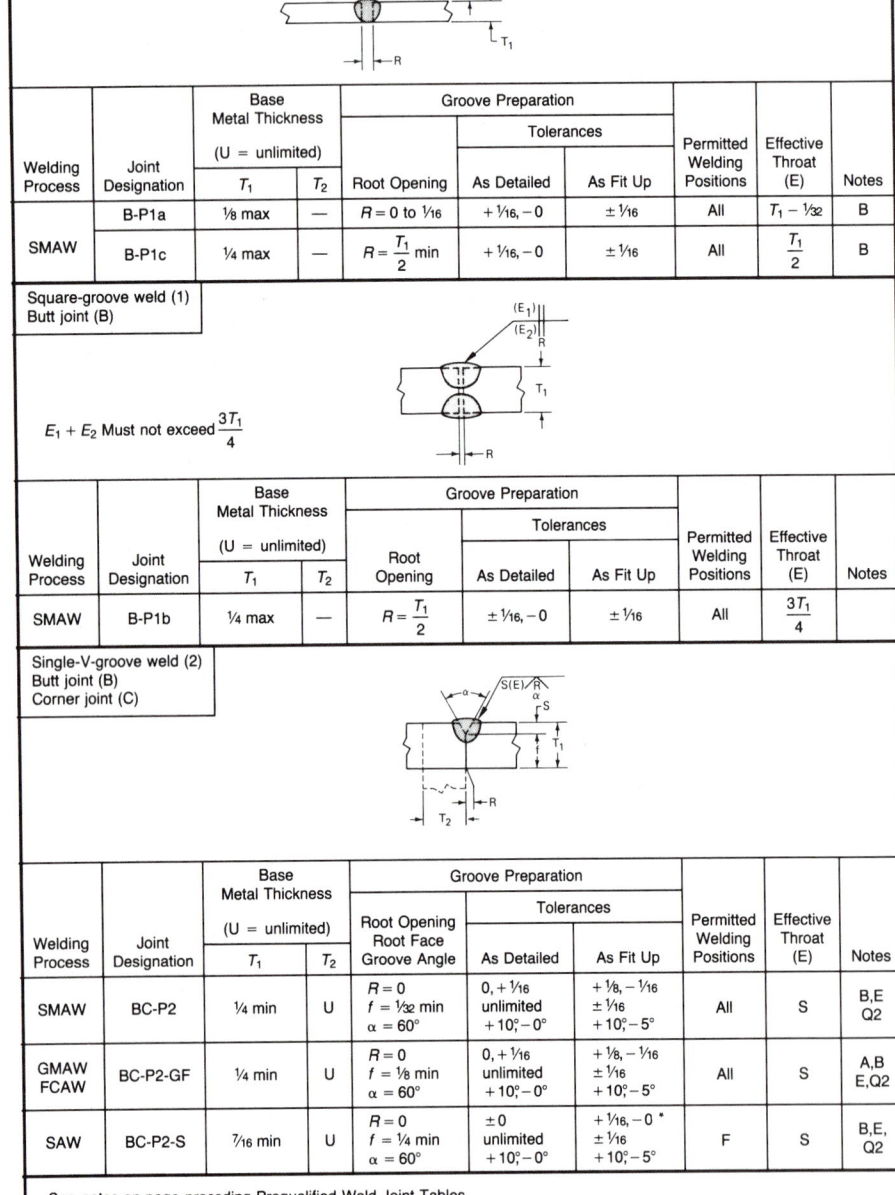

Square-groove weld (1)
Butt joint (B)

Reinforcement 1/32 to 1/8, no tolerance

Welding Process	Joint Designation	Base Metal Thickness (U = unlimited)		Groove Preparation			Permitted Welding Positions	Effective Throat (E)	Notes
		T_1	T_2	Root Opening	Tolerances As Detailed	As Fit Up			
SMAW	B-P1a	1/8 max	—	$R = 0$ to 1/16	$+ 1/16, - 0$	$\pm 1/16$	All	$T_1 - 1/32$	B
	B-P1c	1/4 max	—	$R = \dfrac{T_1}{2}$ min	$+ 1/16, - 0$	$\pm 1/16$	All	$\dfrac{T_1}{2}$	B

Square-groove weld (1)
Butt joint (B)

$E_1 + E_2$ Must not exceed $\dfrac{3T_1}{4}$

Welding Process	Joint Designation	Base Metal Thickness (U = unlimited)		Groove Preparation			Permitted Welding Positions	Effective Throat (E)	Notes
		T_1	T_2	Root Opening	Tolerances As Detailed	As Fit Up			
SMAW	B-P1b	1/4 max	—	$R = \dfrac{T_1}{2}$	$\pm 1/16, - 0$	$\pm 1/16$	All	$\dfrac{3T_1}{4}$	

Single-V-groove weld (2)
Butt joint (B)
Corner joint (C)

Welding Process	Joint Designation	Base Metal Thickness (U = unlimited)		Groove Preparation			Permitted Welding Positions	Effective Throat (E)	Notes
		T_1	T_2	Root Opening Root Face Groove Angle	Tolerances As Detailed	As Fit Up			
SMAW	BC-P2	1/4 min	U	$R = 0$ $f = 1/32$ min $\alpha = 60°$	$0, + 1/16$ unlimited $+10°, -0°$	$+ 1/8, - 1/16$ $\pm 1/16$ $+10°, -5°$	All	S	B,E Q2
GMAW FCAW	BC-P2-GF	1/4 min	U	$R = 0$ $f = 1/8$ min $\alpha = 60°$	$0, + 1/16$ unlimited $+10°, -0°$	$+ 1/8, - 1/16$ $\pm 1/16$ $+10°, -5°$	All	S	A,B E,Q2
SAW	BC-P2-S	7/16 min	U	$R = 0$ $f = 1/4$ min $\alpha = 60°$	± 0 unlimited $+10°, -0°$	$+ 1/16, - 0$ * $\pm 1/16$ $+10°, -5°$	F	S	B,E Q2

See notes on page preceding Prequalified Weld Joint Tables.

*Fit-up tolerance, SAW: for rolled shape R may be 5/16 in. in thick plates if backing is provided.

PREQUALIFIED WELDED JOINTS
Partial penetration groove welds

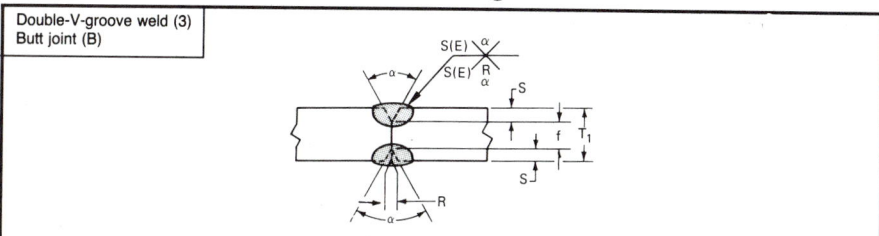

Welding Process	Joint Designation	Base Metal Thickness (U = unlimited) T_1	T_2	Root Opening Root Face Groove Angle	Groove Preparation Tolerances As Detailed	As Fit Up	Permitted Welding Positions	Effective Throat (E)	Notes
SMAW	B-P3	½ min	—	$R=0$ $f = ⅛$ min $α = 60°$	$+ 1/16, -0$ unlimited $+10°, -0°$	$+⅛, -1/16$ $±1/16$ $+10°, -5°$	All	S	E,Mp, Q2
GMAW FCAW	B-P3-GF	½ min	—	$R=0$ $f = ⅛$ min $α = 60°$	$+ 1/16, -0$ unlimited $+10°, -0°$	$+⅛, -1/16$ $±1/16$ $+10°, -5°$	All	S	A,E, Mp, Q2
SAW	B-P3-S	¾ min	—	$R=0$ $f = ¼$ min $α = 60°$	$±0$ unlimited $+10°, -0°$	$+1/16, -0*$ $±1/16$ $+10°, -5°$	F	S	E,Mp, Q2

Double-V-groove weld (3) Butt joint (B)

*Fit-up tolerance, SAW: for rolled shapes R may be 5/16 inches in thick plates if backing is provided.

See notes on page preceding Prequalified Weld Joint Tables.

PREQUALIFIED WELDED JOINTS
Partial penetration groove welds

Single-bevel-groove weld (4)
Butt joint (B)
T-joint (T)
Corner joint (C)

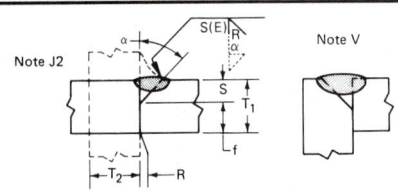

Welding Process	Joint Designation	Base Metal Thickness (U = unlimited) T_1	T_2	Groove Preparation Root Opening Root Face Groove Angle	Tolerances As Detailed	As Fit Up	Permitted Welding Positions	Effective Throat (E)	Notes
SMAW	BTC-P4	U	U	$R = 0$ $f = \frac{1}{8}$ min $\alpha = 45°$	$+\frac{1}{16}, -0$ unlimited $+10°, -0°$	$+\frac{1}{8}, -\frac{1}{16}$ $\pm\frac{1}{16}$ $+10°, -5°$	All	$S - \frac{1}{8}$	B,E,J2, Q2,V
GMAW FCAW	BTC-P4-GF	$\frac{1}{4}$ min	U	$R = 0$ $f = \frac{1}{8}$ min $\alpha = 45°$	$+\frac{1}{16}, -0$ unlimited $+10°, -0°$	$+\frac{1}{8}, -\frac{1}{16}$ $\pm\frac{1}{16}$ $+10°, -5°$	All	$S - \frac{1}{8}$	A,B,E, J2,Q2, V
SAW	TC-P4-S	$\frac{7}{16}$ min	U	$R = 0$ $f = \frac{1}{4}$ min $\alpha = 60°$	± 0 unlimited $+10°, -0°$	$+\frac{1}{16}, -0*$ $\pm\frac{1}{16}$ $+10°, -5°$	F	S	B,E,J2, Q2,V

*Fit-up tolerance, SAW: for rolled shapes R may be $\frac{5}{16}$ inches in thick plates if backing is provided.

Double-bevel-groove weld (5)
Butt joint (B)
T-joint (T)
Corner joint (C)

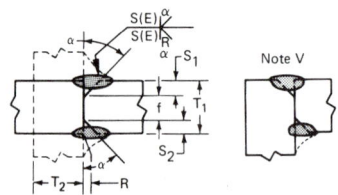

Welding Process	Joint Designation	Base Metal Thickness (U = unlimited) T_1	T_2	Groove Preparation Root Opening Root Face Groove Angle	Tolerances As Detailed	As Fit Up	Permitted Welding Positions	Effective Throat (E)	Notes
SMAW	BTC-P5	$\frac{5}{16}$ min	U	$R = 0$ $f = \frac{1}{8}$ min $\alpha = 45°$	$+\frac{1}{16}, -0$ unlimited $+10°, -0°$	$+\frac{1}{8}, -\frac{1}{16}$ $\pm\frac{1}{16}$ $+10°, -5°$	All	$(S_1 + S_2)$ $- \frac{1}{4}$	E,J2, L,Mp, Q2,V
GMAW FCAW	BTC-P5-GF	$\frac{1}{2}$ min	U	$R = 0$ $f = \frac{1}{8}$ min $\alpha = 45°$	$+\frac{1}{16}, -0$ unlimited $+10°, -0°$	$+\frac{1}{8}, -\frac{1}{16}$ $\pm\frac{1}{16}$ $+10°, -5°$	All	$(S_1 + S_2)$ $- \frac{1}{4}$	A,E,J2, L,Mp, Q2,V
SAW	TC-P5-S	$\frac{3}{4}$ min	U	$R = 0$ $f = \frac{1}{4}$ min $\alpha = 60°$	± 0 unlimited $+10°, -0°$	$+\frac{1}{16}, -0*$ $\pm\frac{1}{16}$ $+10°, -5°$	F	$S_1 + S_2$	E,J2, L,Mp, Q2,V

*Fit-up tolerance, SAW: for rolled shapes R may be $\frac{5}{16}$ inches in thick plates if backing is provided.

See notes on page preceding Prequalified Weld Joint Tables.

PREQUALIFIED WELDED JOINTS
Partial penetration groove welds

Single-U-groove weld (6) Butt joint (B) Corner joint (C)

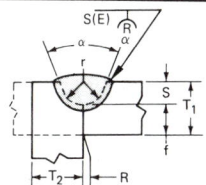

Welding Process	Joint Designation	Base Metal Thickness (U = unlimited)		Groove Preparation			Permitted Welding Positions	Effective Throat (E)	Notes
		T_1	T_2	Root Opening Root Face Groove Radius Groove Angle	Tolerances				
					As Detailed	As Fit Up			
SMAW	BC-P6	¼ min	U	$R = 0$ $f = 1/32$ min $r = ¼$ $α = 45°$	+ 1/16, − 0 unlimited + ¼, − 0 + 10°, − 0°	+ ⅛, − 1/16 ± 1/16 ± 1/16 + 10°, − 5°	All	S	B,E,Q2,
GMAW FCAW	BC-P6-GF	¼ min	U	$R = 0$ $f = ⅛$ min $r = ¼$ $α = 20°$	+ 1/16, − 0 unlimited + ¼, − 0 + 10°, − 0°	+ ⅛, − 1/16 ± 1/16 ± 1/16 + 10°, − 5°	All	S	A,B, E,Q2,
SAW	BC-P6-S	7/16 min	U	$R = 0$ $f = ¼$ min $r = ¼$ $α = 20°$	± 0 unlimited + ¼, − 0 + 10°, − 0°	+ 1/16, − 0* ± 1/16 ± 1/16 + 10°, − 5°	F	S	B,E,Q2,

*Fit-up tolerance, SAW: for rolled shapes R may be 5/16 inches in thick plates if backing is provided.

Double-U-groove weld (7) Butt joint (B)

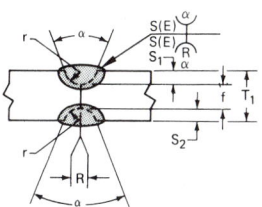

Welding Process	Joint Designation	Base Metal Thickness (U = unlimited)		Groove Preparation			Permitted Welding Positions	Effective Throat (E)	Notes
		T_1	T_2	Root Opening Root Face Groove Radius Groove Angle	Tolerances				
					As Detailed	As Fit Up			
SMAW	B-P7	½ min	—	$R = 0$ $f = ⅛$ min $r = ¼$ $α = 45°$	+ 1/16, − 0 unlimited + ¼, − 0 + 10°, − 0°	+ ⅛, − 1/16 ± 1/16 ± 1/16 + 10°, − 5°	All	$S_1 + S_2$	E, Mp, Q2
GMAW FCAW	B-P7-GF	½ min	—	$R = 0$ $f = ⅛$ min $r = ¼$ $α = 20°$	+ 1/16, − 0 unlimited + ¼, − 0 + 10°, − 0°	+ ⅛, − 1/16 ± 1/16 ± 1/16 + 10°, − 5°	All	$S_1 + S_2$	A,E Mp, Q2
SAW	B-P7-S	¾ min	—	$R = 0$ $f = ¼$ min $r = ¼$ $α = 20°$	± 0 unlimited + ¼, − 0 + 10°, − 0°	+ 1/16, − 0* ± 1/16 ± 1/16 + 10°, − 5°	F	$S_1 + S_2$	E, Mp, Q2

*Fit-up tolerance, SAW: for rolled shapes R may be 5/16 inches in thick plates if backing is provided.

See notes on page preceding Prequalified Weld Joint Tables.

PREQUALIFIED WELDED JOINTS
Partial penetration groove welds

Single-J-groove weld (B)
Butt joint (B)
T-joint (T)
Corner joint (C)

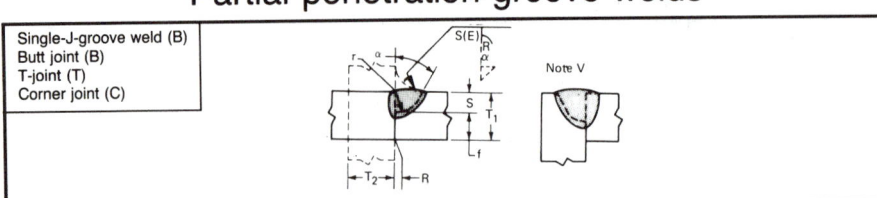

Welding Process	Joint Designation	Base Metal Thickness (U = unlimited)		Groove Preparation			Permitted Welding Positions	Effective Throat (E)	Notes
		T_1	T_2	Root Opening / Root Face / Groove Radius / Groove Angle	Tolerances As Detailed	As Fit Up			
SMAW	TC-P8*	¼ min	U	$R = 0$ $f = $ ⅛ min $r = $ ⅜ $\alpha = 45°$	$+$ ¹⁄₁₆, -0 Not limited $+$ ¼, -0 $+10°, -0°$	$+$ ⅛, $-$ ¹⁄₁₆ $\pm$ ¹⁄₁₆ $\pm$ ¹⁄₁₆ $+10°, -5°$	All	S	E,J2, Q2,V
SMAW	BC-P8**	¼ min	U	$R = 0$ $f = $ ⅛ min $r = $ ⅜ $\alpha = 30°$	$+$ ¹⁄₁₆, -0 Not limited $+$ ¼, -0 $+10°, -0°$	$+$ ⅛, $-$ ¹⁄₁₆ $\pm$ ¹⁄₁₆ $\pm$ ¹⁄₁₆ $+10°, -5°$	All	S	E,J2, Q2,V
GMAW FCAW	TC-P8-GF*	¼ min	U	$R = 0$ $f = $ ⅛ min $r = $ ⅜ $\alpha = 45°$	$+$ ¹⁄₁₆, -0 Not limited $+$ ¼, -0 $+10°, -0°$	$+$ ⅛, $-$ ¹⁄₁₆ $\pm$ ¹⁄₁₆ $\pm$ ¹⁄₁₆ $+10°, -5°$	All	S	A,E, J2,Q2, V
GMAW FCAW	BC-P8-GF**	¼ min	U	$R = 0$ $f = $ ⅛ min $r = $ ⅜ $\alpha = 30°$	$+$ ¹⁄₁₆, -0 Not limited $+$ ¼, -0 $+10°, -0°$	$+$ ⅛, $-$ ¹⁄₁₆ $\pm$ ¹⁄₁₆ $\pm$ ¹⁄₁₆ $+10°, -5°$	All	S	A,E, J2,Q2, V
SAW	TC-P8-S*	⁷⁄₁₆ min	U	$R = 0$ $f = $ ¼ min $r = $ ½ $\alpha = 45°$	± 0 Not limited $+$ ¼, -0 $+10°, -0°$	$+$ ¹⁄₁₆, -0*** $\pm$ ¹⁄₁₆ $\pm$ ¹⁄₁₆ $+10°, -5°$	F	S	E,J2, Q2,V
SAW	C-P8-S**	⁷⁄₁₆ min	U	$R = 0$ $f = $ ¼ min $r = $ ½ $\alpha = 20°$	± 0 Not limited $+$ ¼, -0 $+10°, -0°$	$+$ ¹⁄₁₆, -0*** $\pm$ ¹⁄₁₆ $\pm$ ¹⁄₁₆ $+10°, -5°$	F	S	E,J2, Q2,V

Double-J-groove weld (9)
Butt joint (B)
T-joint (T)
Corner joint (C)

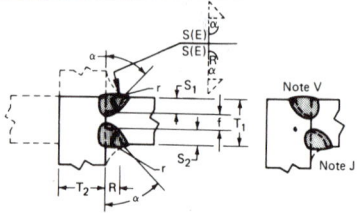

Welding Process	Joint Designation	Base Metal Thickness (U = unlimited)		Groove Preparation			Permitted Welding Positions	Effective Throat (E)	Notes
		T_1	T_2	Root Opening / Root Face / Groove Radius / Groove Angle	Tolerances As Detailed	As Fit Up			
SMAW	BTC-P9*	½ min	U	$R = 0$ $f = $ ⅛ min $r = $ ⅜ $\alpha = 45°$	$+$ ¹⁄₁₆, -0 Not limited $+$ ¼, -0 $+10°, -0°$	$+$ ⅛, $-$ ¹⁄₁₆ $\pm$ ¹⁄₁₆ $\pm$ ¹⁄₁₆ $+10°, -5°$	All	$S_1 + S_2$	E,J2, Mp,Q2, V
GMAW FCAW	BTC-P9-GF**	½ min	U	$R = 0$ $f = $ ⅛ min $r = $ ⅜ $\alpha = 30°$	$+$ ¹⁄₁₆, -0 Not limited $+$ ¼, -0 $+10°, -0°$	$+$ ⅛, $-$ ¹⁄₁₆ $\pm$ ¹⁄₁₆ $\pm$ ¹⁄₁₆ $+10°, -5°$	All	$S_1 + S_2$	A,J2, Mp,Q2, V
SAW	C-P9-S*	¾ min	U	$R = 0$ $f = $ ¼ min $r = $ ½ $\alpha = 45°$	± 0 Not limited $+$ ¼, -0 $+10°, -0°$	$+$ ¹⁄₁₆, -0*** $\pm$ ¹⁄₁₆ $\pm$ ¹⁄₁₆ $+10°, -5°$	F	$S_1 + S_2$	E,J2, Mp,Q2, V
SAW	C-P9-S**	¾ min	U	$R = 0$ $f = $ ¼ min $r = $ ½ $\alpha = 20°$	± 0 Not limited $+$ ¼, -0 $+10°, -0°$	$+$ ¹⁄₁₆, -0*** $\pm$ ¹⁄₁₆ $\pm$ ¹⁄₁₆ $+10°, -5°$	F	$S_1 + S_2$	E,J2, Mp,Q2, V
SAW	T-P9-S	¾ min	U	$R = 0$ $f = $ ¼ min $r = $ ½ $\alpha = 45°$	± 0 Not limited $+$ ¼, -0 $+10°, -0°$	$+$ ¹⁄₁₆, -0*** $\pm$ ¹⁄₁₆ $\pm$ ¹⁄₁₆ $+10°, -5°$	F	$S_1 + S_2$	E,J2, Mp,Q2

*Applies to inside corner joints.
**Applies to outside corner joints.
***Fit-up tolerance, SAW: for rolled shapes R may be ⁵⁄₁₆ in. in thick plates if backing is provided.
See notes on page preceding Prequalified Welded Joint Tables.

BENT PLATES

Minimum radius for cold bending

The following table gives the generally accepted minimum inside radii of bends in terms of thickness t for various steels listed. Values are for bend lines transverse to the direction of final rolling. When bend lines are parallel to the direction of final rolling, the values may have to be approximately doubled. When bend lines are longer than 36 in., all radii may have to be increased if problems in bending are encountered.

Before bending, special attention should be paid to the condition of plate edges transverse to the bend lines. Flame-cut edges of hardenable steels should be machined or softened by heat treatment. Nicks should be ground out. Sharp corners should be rounded.

ASTM Designation		Thickness, in.				
		Up to ¼	Over ¼ to ½	Over ½ to 1	Over 1 to 1½	Over 1½ to 2
A36		1½t	1½t	2t	3t	4t
A242		2t	3t	5t	—[a]	—[a]
A441		2t	3t	5t	—[a]	—[a]
A529		2t	2t	—	—	—
A572[c]	Gr. 42	2t	2t	3t	4t	5t
	Gr. 50	2½t	2½t	4t	—[a]	—[a]
	Gr. 60	3½t	3½t	6t	—[a]	—
	Gr. 65	4t	4t	—[a]	—[a]	—
A588		2t	3t	5t	—[a]	—[a]
A514[b]		2t	2t	2t	3t	3t

[a] It is recommended that steel in this thickness range be bent hot. Hot bending, however, may result in a decrease in the as-rolled mechanical properties.
[b] The mechanical properties of ASTM A514 steel results from a quench-and-temper operation. Hot bending may adversely affect these mechanical properties. If necessary to hot-bend, fabricator should discuss procedure with the steel supplier.
[c] Thickness may be restricted because of columbium content. Consult supplier.

FABRICATING PRACTICES

Maximum efficiency in the fabrication of structural steel by modern shops is entirely dependent upon close cooperation between designing office, drafting room and shop. Designs should be favorable to, the drafting room should recognize and call for, and the shop should adapt its equipment to, the use of recurrent details which have been standardized.

Consideration should be given to duplication of details and multiple punching or drilling. Utilization of standard jigs and machine set-ups eliminates unnecessary handling of material and facilitates the drilling or punching of holes. Gage lines should conform to standard machine set ups. Once determined, they should be duplicated as far as possible throughout any one job. Gages and hole sizes on an individual member should not be varied throughout the length of that member.

Keep gages and longitudinal spacing alike to permit maximum economy in either drilling or punching operations. Longitudinal spacing should preferably be 3 in., or multiples of 3 in., since most shops consider this to be standard.

Copes, blocks and cuts

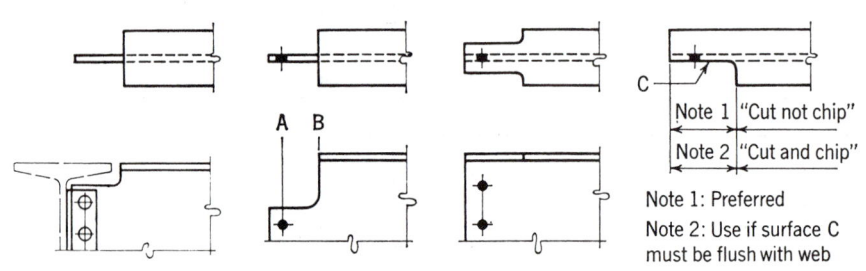

Note 1 | "Cut not chip"
Note 2 | "Cut and chip"

Note 1: Preferred
Note 2: Use if surface C must be flush with web

All re-entrant corners shall be shaped, notch free, to a radius.

The above sketches indicate standard methods of providing clearance for beams connecting to beams or columns. Where possible, a minimum clearance of ½ in. is to be provided. Fabricators may vary in designation and dimensions of copes and blocks. Some fabricators designate all of the operations pictured above by the term "cuts." Note recommended cutting practice in sketch below.

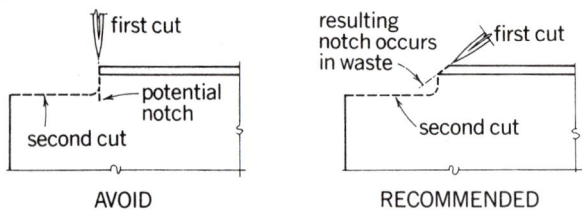

AVOID   RECOMMENDED

For economy, coping or blocking of beams should be avoided if possible. When construction will permit, the elevation of the top of filler beams should be established a sufficient distance below the top of girders to clear the girder fillet. Unusually long or deep copes and blocks, or blocks in beams with thin webs, may materially affect the capacity of the beam. Such beams must be investigated for both shear and moment at lines A and B and, when necessary, adequate reinforcement provided.

PART 6
Specifications and Codes

SPFC

Load and Resistance Factor Design Specification for Structural Steel Buildings

September 1, 1986

AMERICAN INSTITUTE OF STEEL CONSTRUCTION
ONE EAST WACKER DRIVE, SUITE 3100
CHICAGO, ILLINOIS 60601-2001

<u>Notes</u>

PREFACE

The AISC *Load and Resistance Factor Design (LRFD) Specification for Structural Steel Buildings* is the first of a new generation of specifications based on reliability theory. As have all AISC Specifications, this LRFD Specification has been based upon past successful usage, advances in the state of knowledge and changes in design practice. The LRFD Specification has been developed to provide a uniform practice in the design of steel-framed buildings. The intention is to provide design criteria for routine use and not to cover infrequently encountered problems which occur in the full range of structural design. Providing definitive provisions to cover all complex cases would make the LRFD Specification useless for routine designs.

The LRFD Specification is the result of the deliberations of a committee of structural engineers with wide experience and high professional standing, representing a wide geographical distribution throughout the U.S. The committee includes approximately equal numbers of engineers in private practice, engineers involved in research and teaching and engineers employed by steel fabricating companies.

In order to avoid reference to proprietary steels which may have limited availability, only those steels which can be identified by ASTM specifications are approved under this Specification. However, some steels covered by ASTM specifications, but subject to more costly manufacturing and inspection techniques than deemed essential for structures covered by this Specification, are not listed, even though they may provide all of the necessary characteristics of less expensive steels which are listed. Approval of such steels is left to the owner's representative.

The Appendices to this Specification are considered to be an integral part of the Specification.

As used throughout the LRFD Specification, the term structural steel refers exclusively to those items enumerated in Sect. 2 of the AISC *Code of Standard Practice for Steel Buildings and Bridges*; nothing herein contained is intended as a specification for design of items not specifically enumerated in that Code, such as skylights, fire escapes, etc. For the design of cold-formed steel structural members whose profiles contain rounded corners and slender flat elements, the provisions of the American Iron and Steel Institute *Specification for the Design of Cold-Formed Steel Structural Members* are recommended.

A Commentary has been prepared to provide background for the Specification provisions and the user is encouraged to consult it.

The reader is cautioned that professional judgment must be exercised when data or recommendations in this Specification are applied. The publication of the material contained herein is not intended as a representation or warranty on the part of the American Institute of Steel Construction, Inc.—or any other person named herein—that this information is suitable for general or particular use, or freedom from infringement of any patent or patents. Anyone making use of this information assumes all liability arising from such use. The design of structures is within the scope of expertise of a competent licensed structural engineer, architect or other licensed professional for the application of principles to a particular structure.

By the Committee,

A. P. Arndt, Chairman
E. W. Miller,
  Vice Chairman
Horatio Allison
Lynn S. Beedle
Reidar Bjorhovde
Omer W. Blodgett
R. L. Brockenbrough
John H. Busch
Wai-Fah Chen
Duane S. Ellifritt
Bruce Ellingwood
Shu-Jin Fang
Steven J. Fenves
Richard F. Ferguson
James M. Fisher
John W. Fisher
Walter H. Fleischer
Theodore V. Galambos
Geerhard Haaijer
Ira M. Hooper
Jerome S. B. Iffland

A. L. Johnson
Larry A. Kloiber
William J. LeMessurier
Stanley D. Lindsey
William McGuire
William A. Milek
Walter P. Moore, Jr.
William E. Moore, II
Thomas M. Murray
Dale C. Perry
Clarkson W. Pinkham
Egor P. Popov
Donald R. Sherman
Frank Sowokinos
Raymond H. R. Tide
Sophus A. Thompson
William A. Thornton
Ivan M. Viest
Lyle L. Wilson
Joseph A. Yura
Charles Peshek, Jr.,
  Secretary

April 4, 1986

TABLE OF CONTENTS

Commentary

NOMENCLATURE

The section number in parentheses after the definition of a symbol refers to the section where the symbol is first defined.

A Cross-sectional area, in.^{2} (F1.3)

A_B Loaded area of concrete, in.^{2} (I2.4)

A_b Nominal body area of a fastener, in.^{2} (J3.4)

A_b Area of an upset rod based upon the major diameter of its threads (J3.3)

A_c Area of concrete, in.^{2} (I2.2)

A_c Area of concrete slab within effective width (I5.2)

A_e Effective net area, in.^{2} (B3)

A_f Area of flange, in. (Appendix F4)

A_g Gross area, in.^{2} (B1)

A_n Net area, in.^{2} (B2)

A_{ns} Net area subject to shear, in.^{2} (J4)

A_{pb} Projected bearing area, in.^{2} (D3, J8.1)

A_r Area of reinforcing bars, in.^{2} (I2.2)

A_s Area of steel cross section, in.^{2} (I2.2, I5.2)

A_{sc} Cross-sectional area of stud shear connector, in.^{2} (I5.3)

A_{sf} Shear area on the failure path, in.^{2} (D3)

A_w Web area, in.^{2} (F2.1)

A_1 Area of steel bearing concentrically on a concrete support, in.^{2} (J9)

A_2 Total cross-sectional area of a concrete support, in.^{2} (J9)

B Factor for bending stress in web-tapered members, in., defined by Formulas A-F4-7 through A-F4-10 (Appendix F4)

B_1, B_2 Factors used in determining M_u for combined bending and axial forces when first order analysis is employed (H1)

C_{PG} Plate girder coefficient (Appendix G2)

C_b Bending coefficient dependent upon moment gradient (F1.3)

C_m Coefficient applied to bending term in interaction formula for prismatic members and dependent upon column curvature caused by applied moments (H1)

C_m' Coefficient applied to bending term in interaction formula for tapered members and dependent upon axial stress at the small end of the member (Appendix F4)

C_p Ponding flexibility coefficient for primary member in a flat roof (K2)

C_s Ponding flexibility coefficient for secondary member in a flat roof (K2)

C_v Ratio of "critical" web stress, according to linear buckling theory, to the shear yield stress of web material (Appendix G4)

C_w Warping constant, in.^{6} (F1.3)

D Outside diameter of circular hollow section, in. (Appendix B5.3)

D Dead load due to the self-weight of the structural and permanent elements on the structure (A4.1)

D Factor used in Formula A-G4-2, dependent on the type of transverse stiffeners used in a plate girder (Appendix G4)

E Modulus of elasticity of steel (29,000 ksi) (E2)

E Earthquake load (A4.1)

E_c Modulus of elasticity of concrete, ksi (I2.2)

E_m Modified modulus of elasticity, ksi (I2.2)

F_{BM} Nominal strength of the base material to be welded, ksi (J2.4)

F_{EXX} Classification strength of weld metal, ksi (A3.5 and J2.4)

F_a Axial design strength, ksi (A5.1)

$F_{b\gamma}$ Flexural stress for tapered members defined by Formulas A-F4-4 and A-F4-5 (Appendix F4)

F_{cr} Critical stress, ksi (E2)

F_e Elastic buckling stress, ksi (Appendix E3)

F_{ex} Elastic flexural buckling stress about the major axis, ksi (Appendix E3)

F_{ey} Elastic flexural buckling stress about the minor axis, ksi (Appendix E3)

F_{ez} Elastic torsional buckling stress, ksi (Appendix E3)

F_{my} Modified yield stress for composite columns, ksi (I2.2)

F_n Nominal shear rupture strength, ksi (J4)

F_r Compressive residual stress in flange, ksi (F1.3)

$F_{s\gamma}$ Stress for tapered members defined by Formula A-F4-6 (Appendix F4)

F_u Specified minimum tensile strength of the type of steel being used, ksi (D1)

F_w Nominal strength of the weld electrode material, ksi (J2.4)

$F_{w\gamma}$ Stress for tapered members defined by Formula A-F4-7, ksi (Appendix F4)

F_y Specified minimum yield stress of the type of steel being used, ksi. As used in this Specification, "yield stress" denotes either the specified minimum yield point (for those steels that have a yield point) or specified yield strength (for those steels that do not have a yield point)

F_{yf} Specified minimum yield stress of the flange, ksi (B5.1)

F_{ym} Yield stress obtained from mill test reports or from physical tests, ksi (N2)

F_{yr} Specified minimum yield stress of reinforcing bars, ksi (I2.2)

F_{ys} Static yield stress, ksi (N2)

F_{yst} Specified minimum yield stress of the stiffener material, ksi (Appendix G4)

F_{yw} Specified minimum yield stress of the web, ksi (F2.2)

G Shear modulus of elasticity of steel, ksi ($G = 11,200$) (F1.3)

H Horizontal force, kips (H1.2)

H_s Length of stud connector after welding, in. (I3.5)

I Moment of inertia, in.^{4}

I_d Moment of inertia of the steel deck supported on secondary members, in.^{4} (K2)

I_p Moment of inertia of primary members, in.^{4} (K2)

I_s Moment of inertia of secondary members, in.^{4} (K2)

I_{st} Moment of inertia of a transverse stiffener, in.^{4} (Appendix G4)

J Torsional constant for a section, in.^{4} (F1.3)

K Effective length factor for prismatic member (C2)

K_s Slip coefficient (J3)

K_z Effective length factor for torsional buckling (Appendix E3)

K_γ Effective length factor for a tapered member (Appendix F4.3)

L Unbraced length of member measured between the center of gravity of the bracing members, in. (E2)

L Story height, in. (H1.2)

L Distance in line of force from center of a standard or oversized hole or from the center of the end of a slotted hole to an edge of a connected part, in. (J3.6)

L Live load due to occupancy (A4.1)

L_b Laterally unbraced length; length between points which are either braced against lateral displacement of compression flange or braced against twist of the cross section, in. (F1.1)

L_c Length of channel shear connector, in. (I5.4)

L_p Limiting laterally unbraced length for full plastic bending capacity, uniform moment case ($C_b = 1.0$), in. (F1.3)

L_p Column spacing in direction of girder, in. (K2)

L_{pd} Limiting laterally unbraced length for plastic analysis, in. (F1.1)

L_r Limiting laterally unbraced length for inelastic lateral-torsional buckling, in. (F1.3)

L_r Roof live load (A4.1)

L_s Column spacing perpendicular to direction of girder, in. (K2)

M_1 Smaller moment at end of unbraced length of beam or beam-column, kip-in. (F1.1)

M_2 Larger moment at end of unbraced length of beam or beam-column, kip-in. (F1.3)

M_{cr} Elastic buckling moment, kip-in. (F1.4)

$M_{\ell t}$ Required flexural strength in member due to lateral frame translation, kip-in. (H1)

M_n Nominal flexural strength, kip-in. (F1.2)

M'_{nx}, M_{ny} Flexural strength defined in Formulas A-H3-7 and A-H-3-8 for use in alternate interaction equations for combined bending and axial force, kip-in. (Appendix H3)

M_{nt} Required flexural strength in member assuming there is no lateral translation of the frame, kip-in. (H1)

M_p Plastic bending moment, kip-in. (F1.1)

M'_p Moment defined in Formulas A-H3-5 and A-H3-6, for use in alternate interaction equations for combined bending and axial force, kip-in. (Appendix H3)

M_r Limiting buckling moment, M_{cr}, when $\lambda = \lambda_r$ and $C_b = 1.0$, kip-in. (F1.3)

M_u Required flexural strength, kip-in. (H1)

N Length of bearing, in. (K1.3)

N_r Number of stud connectors in one rib at a beam intersection (I3.5)

P Force transmitted by one fastener to the critical connected part, kips (J3.9)

P_e Euler buckling strength, kips (H1)

P_e Elastic buckling load, kips (I4)

P_n Nominal axial strength (tension or compression), kips (D1)

P_p Bearing load on concrete, kips (J9)

P_u Required axial strength (tension or compression), kips (H1)

P_y Yield strength, kips (B5.1)

Q Full reduction factor for slender compression elements (Appendix E3)

Q_a Reduction factor for slender stiffened compression elements (Appendix B5)

Q_n Nominal strength of one stud shear connector, kips (I5)

Q_s Reduction factor for slender unstiffened compression elements (Appendix B5.3)

R Nominal load due to initial rainwater or ice exclusive of the ponding contribution (A4.1)

R_{PG} Plate girder bending strength reduction factor (Appendix G)

R_e Hybrid girder factor (Appendix F1.7)

R_n Nominal resistance (A5.3)

R_v Web shear strength, kips (K1.7)

S Elastic section modulus, in.^{3}

S Spacing of secondary members, in. (K2)

S Snow load (A4.1)

S_x Elastic section modulus about major axis, in.^{3} (F1.3)

S_x' Elastic section modulus of larger end of tapered member about its major axis, in.^{3} (Appendix F4)

$(S_x)_{eff}$ Effective section modulus about major axis, in.^{3} (Appendix F1.7)

S_{xt}, S_{xc} Elastic section modulus referred to tension and compression flanges, respectively, in.^{3} (Appendix G2)

T Factored applied tension force per bolt, kips (J3.5)

T_b Specified pretension load in high-strength bolt, kips (J3.5)

U Reduction coefficient, used in calculating effective net area (B3)

U_p, U_s Ponding stress index for primary and secondary members (Appendix K2)

V Shear force, kips (J10.2)

V_n Nominal shear strength, kips (F2.2)

V_u Required shear strength, kips (Appendix G4)

W Wind load (A4.1)

X_1 Beam buckling factor defined by Formula F1-8 (F1.3)

X_2 Beam buckling factor defined by Formula F1-9 (F1.3)

Z Plastic section modulus, in.^{3}

a Clear distance between transverse stiffeners, in. (F2.2)

a Distance between connectors in a built-up member, in. (E4)

a Shortest distance from edge of pin hole to edge of member measured parallel to direction of force, in. (D3)

a_r Ratio of web area to compression flange area (Appendix G2)

b Compression element width (Table B5.1)

b_e Reduced effective width for slender compression elements, in. (Appendix B5.3)

b_{eff} Effective edge distance, in. (D3)

b_f Flange width, in. (K1.5)

c_1, c_2, c_3 Numerical coefficients (I2.2)

d	Nominal fastener diameter, in. (J3.6)
d	Overall depth of member, in. (F2.1)
d	Pin diameter, in. (D3)
d	Roller diameter, in. (J8.2)
d_L	Depth at larger end of unbraced tapered segment, in. (Appendix F4)
d_c	Web depth clear of fillets, in. (K1.5)
d_h	Diameter of a standard size hole, in. (J3.9)
d_o	Depth at smaller end of unbraced tapered segment, in. (Appendix F4)
e	Base of natural logarithm = 2.71828 . . .
f	Computed compressive stress in the stiffened element, ksi (Appendix B5.3)
f_a	Computed axial stress in column, ksi (A5.1)
f_{b1}	Smallest computed bending stress at one end of a tapered segment, ksi (Appendix F4)
f_{b2}	Largest computed bending stress at one end of a tapered segment, ksi (Appendix F4)
f_c'	Specified compressive strength of concrete, ksi (I2.2)
f_o	Stress due to $1.2D + 1.2R$, ksi (Appendix K2)
f_t	Computed tension stress in bolts or rivets, ksi (J3.4)
f_{un}	Required normal stress, ksi (H2)
f_{uv}	Required shear stress, ksi (H2)
f_v	Computed shear stress in bolts or rivets, ksi (J3.4)
g	Transverse c.-to-c. spacing (gage) between fastener gage lines, in. (B2)
g	Acceleration due to gravity = 32.2 ft/sec.^{2} (A4.3)
h	Clear distance between flanges less the fillet or corner radius for rolled shapes; and for built-up sections, the distance between adjacent lines of fasteners or the clear distance between flanges when welds are used, in. (B5.1)
h_c	Assumed web depth for stability, in. (B5.1)
h_r	Nominal rib height, in. (I3.5)
h_s	Factor used in Formula A-F4-6 for web-tapered members (Appendix F4.3)
h_w	Factor used in Formula A-F4-7 for web-tapered members (Appendix F4.3)
j	Factor defined by Formula F3-1 for minimum moment of inertia for a transverse stiffener (F3)
k	Web plate buckling coefficient (F2.2)
k	Distance from outer face of flange to web toe of fillet, in. (K1.3)
ℓ	Largest laterally unbraced length along either flange at the point of load, in. (K1.5)
ℓ	Length of bearing, in. (J8.2)
m	Ratio of web to flange yield stress or critical stress in hybrid beams (Appendix G2)
r	Governing radius of gyration, in. (E2)
r_{To}	Radius of gyration, r_T, for the smaller end of a tapered member, in. (Appendix F4.3)
r_T	Radius of gyration of compression flange plus one third of the compression portion of the web taken about an axis in the plane of the web, in. (Appendix F1.7)

r_i Minimum radius of gyration of individual component in a built-up member, in. (E4)

r_m Radius of gyration of the steel shape, pipe or tubing in composite columns. For steel shapes it may not be less than 0.3 times the overall thickness of the composite section, in. (I2)

$\overline{r}_o$ Polar radius of gyration about the shear center, in. (Appendix E3)

r_{ox},   Radius of gyration about x and y axes at the smaller end of a tapered member,
$\quad r_{oy}$ respectively, in. (Appendix F4.3)

r_x, r_y Radius of gyration about x and y axes, respectively, in. (F1.1, E3)

s Longitudinal c.-to-c. spacing (pitch) of any two consecutive holes, in. (B2)

t Thickness of connected part, in. (J3.6)

t Thickness of the critical part, in. (J3.9)

t_f Flange thickness, in. (B5.1)

t_f Flange thickness of channel shear connector, in. (I5.4)

t_w Web thickness of channel shear connector, in. (I5.4)

t_w Web thickness, in. (F2.1)

w Plate width; distance between welds, in. (B3)

w Unit weight of concrete, lbs./cu. ft. (I2)

w_r Average width of concrete rib or haunch, in. (I3.5)

x Subscript relating symbol to strong axis bending

x_o, y_o Coordinates of the shear center with respect to the centroid, in. (Appendix E3)

y Subscript relating symbol to weak axis bending

z Distance from the smaller end of tapered member used in Formula A-F4-1 for the variation in depth, in. (Appendix F4)

Δ_{oh} Translation deflection of the story under consideration, in. (H1)

γ Depth tapering ratio (Appendix F4)
       Subscript for tapered members (Appendix F4)

ζ Exponent for alternate beam-column interaction equation (Appendix H3)

η Exponent for alternate beam-column interaction equation (Appendix H3)

λ_c Column slenderness parameter (E2)

λ_e Equivalent slenderness parameter (Appendix E3)

λ_{eff} Effective slenderness ratio defined by Formula A-F4-2 (Appendix F4)

λ_p Limiting slenderness parameter for compact element (B5.1)

λ_r Limiting slenderness parameter for noncompact element (B5.1)

μ Coefficient of friction (J10.2)

ϕ Resistance factor (A5.3)

ϕ_b Resistance factor for flexure (B1)

ϕ_c Resistance factor for compression (E2)

ϕ_c Resistance factor for axially loaded composite columns (I2.2)

ϕ_{sf} Resistance factor for shear on the failure path (D3)

ϕ_t Resistance factor for tension (D1)

ϕ_v Resistance factor for shear (F2.2)

CHAPTER A.
GENERAL PROVISIONS

A1. SCOPE

This *Load and Resistance Factor Design Specification for Structural Steel Buildings* is intended as an alternate to the currently approved *Specification for the Design, Fabrication and Erection of Structural Steel for Buildings* of the American Institute of Steel Construction.

A2. LIMITS OF APPLICABILITY

1. Structural Steel Defined

As used in this Specification, the term *structural steel* refers to the steel elements of the structural steel frame essential to the support of the design loads. Such elements are generally enumerated in Sect. 2.1 of the AISC *Code of Standard Practice for Steel Buildings and Bridges*. For the design of cold-formed steel structural members, whose profiles contain rounded corners and slender flat elements, the provisions of the American Iron and Steel Institute *Specification for the Design of Cold-Formed Steel Structural Members* are recommended.

2. Types of Construction

Two basic types of construction and associated design assumptions are permissible under the conditions stated herein, and each will govern in a specific manner the size of members and the types and strength of their connections. Both types must comply with the stability requirements of Sect. B4.

Type FR (fully restrained), commonly designated as "rigid-frame" (continuous frame), assumes that beam-to-column connections have sufficient rigidity to hold the original angles between intersecting members virtually unchanged.

Type PR (partially restrained) assumes that the connections of beams and girders possess an insufficient rigidity to hold the original angles between intersecting members virtually unchanged.

The design of all connections shall be consistent with assumptions as to type of construction called for on the design drawings.

Type FR construction is unconditionally permitted under the LRFD Specification.

The use of Type PR construction under this Specification depends on the evidence of predictable proportion of full end restraint. Where the connection restraint is ignored, commonly designated "simple framing," it is assumed that under gravity loads the ends of the beams and girders are connected for shear only and are free to rotate. For "simple framing" the following requirements apply:

a. The connections and connected members must be adequate to carry the factored gravity loads as "simple beams."
b. The connections and connected members must be adequate to resist the factored lateral loads.
c. The connections must have sufficient inelastic rotation capacity to avoid overload of fasteners or welds under combined factored gravity and lateral loading.

When the rotational restraint of the connections is used in the design of the connected members or for the stability of the structure as a whole, the capacity of the connection for such restraint must be established by analytical or empirical means.

Type PR construction may necessitate some inelastic, but self-limiting, deformation of a structural steel part.

A3. MATERIAL

1. Structural Steel

Material conforming to one of the following standard specifications is approved for use under this Specification:

Structural Steel, ASTM A36
Welded and Seamless Steel Pipe, ASTM A53, Gr. B
High-strength Low-alloy Structural Steel, ASTM A242
High-strength Low-alloy Structural Manganese-vanadium Steel, ASTM A441
Cold-formed Welded and Seamless Carbon Steel Structural Tubing in Rounds and Shapes, ASTM A500
Hot-formed Welded and Seamless Carbon Steel Structural Tubing, ASTM A501
High Yield-strength Quenched and Tempered Alloy Steel Plate, Suitable for Welding, ASTM A514
Structural Steel with 42,000 psi Minimum Yield Point, ASTM A529
Hot-rolled Carbon Steel Sheets and Strip, Structural Quality, ASTM A570, Gr. 40, 45 and 50
High-strength Low-alloy Columbium-vanadium Steels of Structural Quality, ASTM A572
High-strength Low-alloy Structural Steel with 50,000 psi Minimum Yield Point to 4 in. Thick, ASTM A588

Steel Sheet and Strip, Hot-rolled and Cold-rolled, High-strength, Low-alloy, with Improved Corrosion Resistance, ASTM A606

Steel Sheet and Strip, Hot-rolled and Cold-rolled, High-strength, Low-alloy, Columbium and/or Vanadium, ASTM A607

Hot-formed Welded and Seamless High-strength Low-alloy Structural Tubing, ASTM A618

Certified mill test reports or certified reports of tests made by the fabricator or a testing laboratory in accordance with ASTM A6 or A568, as applicable, shall constitute sufficient evidence of conformity with one of the above ASTM standards. If requested, the fabricator shall provide an affidavit stating that the structural steel furnished meets the requirements of the grade specified.

Unidentified steel, if surface conditions are acceptable according to criteria contained in ASTM A6, may be used for unimportant members or details, where the precise physical properties and weldability of the steel would not affect the strength of the structure.

2. Steel Castings and Forgings

Cast steel shall conform to one of the following standard specifications:

Mild-to-medium-strength Carbon-steel Castings for General Applications, ASTM A27, Gr. 65-35

High-strength Steel Castings for Structural Purposes, ASTM A148, Gr. 80-50

Steel forgings shall conform to the following standard specification:

Steel Forgings Carbon and Alloy for General Industrial Use, ASTM A668

Certified test reports shall constitute sufficient evidence of conformity with the standards.

3. Bolts

Steel bolts shall conform to one of the following standard specifications:

Low-carbon Steel Externally and Internally Threaded Standard Fasteners, ASTM A307

High-strength Bolts for Structural Steel Joints, ASTM A325

Quenched and Tempered Steel Bolts and Studs, ASTM A449

Quenched and Tempered Alloy Steel Bolts for Structural Steel Joints, ASTM A490

Carbon and Alloy Steel Nuts, ASTM A563

Hardened Steel Washers, ASTM F436

A449 bolts may be used only in connections requiring bolt diameters greater than 1½ in. and shall not be used in slip-critical connections. A449 material is acceptable for high-strength anchor bolts and threaded rods of any diameter.

Manufacturer's certification shall constitute sufficient evidence of conformity with the standards.

4. Anchor Bolts and Threaded Rods

Anchor bolt and threaded rod steel shall conform to one of the following standard specifications:

> Structural Steel, ASTM A36
>
> Quenched and Tempered Alloy Steel Bolts, Studs and Other Externally Threaded Fasteners, ASTM A354
>
> High-strength Low-alloy Structural Steel with 50,000 psi Minimum Yield Point to 4 in. Thick, ASTM A588.
>
> High-strength Nonheaded Steel Bolts and Studs, ASTM A687.
>
> Threads on bolts and rods shall conform to Unified Standard Series of latest edition of ANSI B18.1 and shall have Class 2A tolerances.

Steel bolts conforming to other provisions of Sect. A3 may be used as anchor bolts.

Manufacturer's certification shall constitute sufficient evidence of conformity with the standards.

5. Filler Metal and Flux for Welding

Welding electrodes and fluxes shall conform to one of the following specifications of the American Welding Society.*

> Specification for Covered Carbon Steel Arc Welding Electrodes, AWS A5.1
>
> Specification for Low-alloy Steel Covered Arc Welding Electrodes, AWS A5.5 Specification for Carbon Steel Electrodes and Fluxes for Submerged Arc Welding, AWS A5.17
>
> Specification for Carbon Steel Filler Metals for Gas Shielded Arc Welding, AWS A5.18
>
> Specification for Carbon Steel Electrodes for Flux Cored Arc Welding, AWS A5.20
>
> Specification for Low-alloy Steel Electrodes and Fluxes for Submerged-arc Welding, AWS A5.23
>
> Specification for Low-alloy Steel Filler Metals for Gas-shielded Arc Welding, AWS A5.28
>
> Specification for Low-alloy Steel Electrodes for Flux-cored Arc Welding, AWS A5.29

Manufacturer's certification shall constitute sufficient evidence of conformity with the standards.

6. Stud Shear Connectors

Steel stud shear connectors shall conform to the requirements of *Structural Welding Code—Steel*, AWS D1.1.

Manufacturer's certification shall constitute sufficient evidence of conformity with the code.

---

*Approval of these welding electrode specifications is given without regard to weld metal notch toughness requirements, which are generally not critical for building construction.

A4. LOADS AND LOAD COMBINATIONS

The nominal loads shall be the minimum design loads stipulated by the applicable code under which the structure is designed or dictated by the conditions involved. In the absence of a code, the loads and load combinations shall be those stipulated in the American National Standard *Minimum Design Loads for Buildings and Other Structures*, ANSI A58.1. For design purposes, the loads stipulated by the applicable code shall be taken as nominal loads. For ease of reference, the more common ANSI load combinations are listed below.

1. Loads, Load Factors and Load Combinations

The following nominal loads are to be considered:

D : dead load due to the weight of the structural elements and the permanent features on the structure
L : live load due to occupancy and moveable equipment
L_r : roof live load
W : wind load
S : snow load
E : earthquake load
R : load due to initial rainwater or ice exclusive of the ponding contribution

The required strength of the structure and its elements must be determined from the appropriate critical combination of factored loads. The most critical effect may occur when one or more loads are not acting. The following load combinations and the corresponding load factors shall be investigated:

$$1.4 \, D \tag{A4-1}$$
$$1.2 \, D + 1.6 \, L + 0.5 \, (L_r \text{ or } S \text{ or } R) \tag{A4-2}$$
$$1.2 \, D + 1.6 \, (L_r \text{ or } S \text{ or } R) + (0.5 \, L \text{ or } 0.8 \, W) \tag{A4-3}$$
$$1.2 \, D + 1.3 \, W + 0.5 \, L + 0.5 \, (L_r \text{ or } S \text{ or } R) \tag{A4-4}$$
$$1.2 \, D + 1.5 \, E + (0.5 \, L \text{ or } 0.2 \, S) \tag{A4-5}$$
$$0.9 \, D - (1.3 \, W \text{ or } 1.5 \, E) \tag{A4-6}$$

Exception: The load factor on L in combinations A4-3, A4-4 and A4-5 shall equal 1.0 for garages, areas occupied as places of public assembly and all areas where the live load is greater than 100 psf.

2. Impact

For structures carrying live loads which induce impact, the assumed nominal live load shall be increased to provide for this impact in Formulas A4-2 and A4-3.
If not otherwise specified, the increase shall be:

For supports of elevators and elevator machinery 100%
For supports of light machinery, shaft or motor driven, not less than 20%
For supports of reciprocating machinery or power driven units, not less than 50%
For hangers supporting floors and balconies 33%
For cab-operated traveling crane support girders and their connections 25%
For pendant-operated traveling crane support girders and their connections 10%

3. Crane Runway Horizontal Forces

The nominal lateral force on crane runways to provide for the effect of moving crane trolleys shall be a minimum of 20% of the sum of weights of the lifted load and of the crane trolley, but exclusive of other parts of the crane. The force shall be assumed to be applied at the top of the rails, acting in either direction normal to the runway rails, and shall be distributed with due regard for lateral stiffness of the structure supporting the rails.

The longitudinal force shall be a minimum of 10% of the maximum wheel loads of the crane applied at the top of the rail, unless otherwise specified.

A5. DESIGN BASIS

1. Required Strength at Factored Loads

The required strength of structural members and connections shall be determined by structural analysis for the appropriate factored load combinations given in Sect. A4.

Design by either elastic or plastic analysis is permitted, except that plastic analysis is permitted only for steels with yield stress not exceeding 65 ksi and is subject to provisions of Sects. B5.2, C2, E1.2, F1.1, H1, and I1.

Except for hybrid girders and members of A514 steel, beams and girders (including members designed on the basis of composite action) which meet the requirements of the above, and are continuous over supports or are rigidly framed to columns by means of rivets, high-strength bolts, or welds, may be proportioned for 9/10 of the negative moments produced by gravity loading which are maximum at points of support, provided that, for such members, the maximum positive moment shall be increased by 1/10 of the average negative moments. This reduction shall not apply to moments produced by loading on cantilevers. If the negative moment is resisted by a column rigidly framed to the beam or girder, the 1/10 reduction may be used in proportioning the column for the combined axial and bending loading, provided that the stress, f_a, due to any concurrent axial load on the member, does not exceed $0.15F_a$.

2. Limit States

LRFD is a method of proportioning structures so that no applicable limit state is exceeded when the structure is subjected to all appropriate factored load combinations.

Strength limit states are related to safety and concern maximum load carrying capacity. Serviceability limit states are related to performance under normal service conditions. The term "resistance" includes both strength limit states and serviceability limit states.

3. Design for Strength

The design strength of each structural component or assemblage must equal or exceed the required strength based on the factored nominal loads. The design strength ϕR_n is calculated for each applicable limit state as the nominal strength R_n multiplied by a resistance factor ϕ. The required strength is determined for each applicable load combination as stipulated in Section A4.

Nominal strength R_n and resistance factors ϕ are given in the appropriate chapters. Additional strength considerations are given in Chap. K.

4. Design for Serviceability and Other Considerations

The overall structure and the individual members, connections and connectors should be checked for serviceability. Provisions for design for serviceability are given in Chap. L.

A6. REFERENCED CODES AND STANDARDS

The following documents are referenced in this Specification.

American National Standards Institute
    ANSI B18.1
    ANSI A58.1-1982

American Society of Testing and Materials

ASTM A6-84c	ASTM A27-84	ASTM A36-84a
ASTM A53-84a	ASTM A148-84	ASTM A242-84
ASTM A307-84	ASTM A325-84	ASTM A354-84b
ASTM A441-84	ASTM A449-84a	ASTM A490-84
ASTM A500-84	ASTM A501-84	ASTM A514-84a
ASTM A529-84	ASTM A563-84	ASTM A570-84a
ASTM A572-84	ASTM A588-84a	ASTM A606-84
ASTM A607-84a	ASTM A618-84	ASTM A668-83
ASTM A687-84	ASTM C33-85	ASTM C330-85
ASTM F436-84		

American Welding Society

AWS D1.1-85	AWS A5.1-81	AWS A5.5-81
AWS A5.17-80	AWS A5.18-79	AWS A5.20-79
AWS A5.23-80	AWS A5.28-79	AWS A5.29-80

Research Council on Structural Connections
    Specification for Structural Joints Using ASTM A325 or A490 Bolts, 1985

A7. DESIGN DOCUMENTS

1. Plans

The design plans shall show a complete design with sizes, sections and relative locations of the various members. Floor levels, column centers and offsets shall be dimensioned. Drawings shall be drawn to a scale large enough to show the information clearly.

Design documents shall indicate the type or types of construction as defined in Sect. A2.2 and include the nominal loads and design strengths if necessary for preparation of shop drawings.

Where joints are to be assembled with high-strength bolts, the design documents shall indicate the connection type (slip-critical, tension or bearing).

Camber of trusses, beams and girders, if required, shall be called for in the design documents. The requirements for stiffeners and bracing shall be shown on the design documents.

2. Standard Symbols and Nomenclature

Welding and inspection symbols used on plans and shop drawings shall preferably be the American Welding Society symbols. Other adequate welding symbols may be used, provided a complete explanation thereof is shown in the design documents.

3. Notation for Welding

Notes shall be made in the design documents and on the shop drawings of those joints or groups of joints in which the welding sequence and technique of welding should be carefully controlled to minimize distortion.

Weld lengths called for in the design documents and on the shop drawings shall be the net effective lengths.

CHAPTER B.
DESIGN REQUIREMENTS

This chapter contains provisions which are common to the Specification as a whole.

B1. GROSS AREA

The gross area A_g of a member at any point is the sum of the products of the thickness and the gross width of each element measured normal to the axis of the member. For angles, the gross width is the sum of the widths of the legs less the thickness.

Plate girders, coverplated beams and rolled or welded beams shall be proportioned on the basis of the gross section. No deduction shall be made for shop or field bolt holes in either flange unless the reduction of the area of either flange by such holes, calculated in accordance with the provisions of Sect. B2, exceeds 15% of the gross flange area, in which case the area in excess of 15% shall be deducted.

Hybrid girders may be proportioned on the basis of their gross section, provided that they are not required to resist an axial force greater than ϕ_b times $0.15A_gF_{yf}$. No limit is placed on the flexural stress in the web.

Flanges of welded plate girders may be varied in thickness or width by splicing plates or by using cover plates.

B2. NET AREA

The net area A_n of a member is the sum of the products of the thickness and the net width of each element computed as follows.

In computing net area for tension, the width of a bolt hole shall be taken as $\frac{1}{16}$-in. greater than the nominal dimension of the hole normal to the direction of applied stress. For shear the width shall be taken as the nominal dimension of the hole.

For a chain of holes extending across a part in any diagonal or zigzag line, the net width of the part shall be obtained by deducting from the gross width the sum of the diameters or slot dimensions as provided in Sect. J3.7, of all holes in the chain, and adding, for each gage space in the chain, the quantity $s^2/4g$

where

s = longitudinal center-to-center spacing (pitch) of any two consecutive holes, in.
g = transverse center-to-center spacing (gage) between fastener gage lines, in.

For angles, the gage for holes in opposite adjacent legs shall be the sum of the gages from the back of the angles less the thickness.

The critical net area A_n of the part is obtained from that chain which gives the least net width.

In determining the net area across plug or slot welds, the weld metal shall not be considered as adding to the net area.

B3. EFFECTIVE NET AREA

When the load is transmitted directly to each of the cross-sectional elements by connectors, the effective net area A_e is equal to the net area A_n.

When the load is transmitted by bolts or rivets through some but not all of the cross-sectional elements of the member, the effective net area A_e shall be computed as:

$$A_e = U A_n \qquad \text{(B3-1)}$$

where

A_n = net area of the member, in.^{2}
U = reduction coefficient

When the load is transmitted by welds through some but not all of the cross-sectional elements of the member, the effective net area A_e shall be computed as:

$$A_e = U A_g \qquad \text{(B3-2)}$$

where

A_g = gross area of member, in.^{2}

Unless a larger coefficient can be justified by tests or other rational criteria, the following values of U shall be used:

a. W, M or S shapes with flange widths not less than ⅔ the depth, and structural tees cut from these shapes, provided the connection is to the flanges. Bolted or riveted connections shall have no fewer than three fasteners per line in the direction of stress .... $U = 0.90$

b. W, M or S shapes not meeting the conditions of subparagraph a, structural tees cut from these shapes and all other shapes, including built-up cross sections. Bolted or riveted connections shall have no fewer than three fasteners per line in the direction of stress .... $U = 0.85$

c. All members with bolted or riveted connections having only two fasteners per line in the direction of stress .... $U = 0.75$

When load is transmitted by transverse welds to some but not all of the cross sectional elements of W, M or S shapes and structural tees cut from these shapes, A_e shall be taken as the area of the directly connected elements.

When the load is transmitted to a plate by longitudinal welds along both edges at the end of the plate, the length of the welds shall not be less than the width of the plate. The effective net area A_e shall be computed as:

$$A_e = U A_g \qquad \text{(B3-2)}$$

Unless a larger coefficient can be justified by tests or other rational criteria, the following values of U shall be used:

a. When $l > 2w$ .... $U = 1.0$
b. When $2w > l > 1.5w$ .... $U = 0.87$
c. When $1.5w > l > w$ .... $U = 0.75$

where

l = weld length, in.
w = plate width (distance between welds), in.

B4. STABILITY

General stability shall be provided for the structure as a whole and for each compression element.

Consideration shall be given to significant load effects resulting from the deflected shape of the structure or of individual elements of the lateral load resisting system.

B5. LOCAL BUCKLING

1. Classification of Steel Sections

Steel sections are classified as compact, noncompact and slender element sections. For a section to qualify as compact, its flanges must be continuously connected to the web or webs and the width-thickness ratios of its compression elements must not exceed the limiting width-thickness ratios λ_p from Table B5.1. If the width-thickness ratio of one or more compression elements exceeds λ_p but does not exceed λ_r, the section is noncompact. If the width-thickness ratio exceeds λ_r from Table B5.1, the element is referred to as a slender compression element.

For unstiffened elements which are supported along only one edge, parallel to the direction of the compression force, the width shall be taken as follows:

a. For flanges of I-shaped members and tees, the width b is half the full nominal width.

b. For legs of angles and flanges of channels and zees, the width b is the full nominal dimension.

c. For plates, the width b is the distance from the free edge to the first row of fasteners or line of welds.

d. For stems of tees, d is taken as the full nominal depth.

For stiffened elements, i.e., supported along two edges parallel to the direction of the compression force, the width shall be taken as follows:

a. For webs of rolled or formed sections, h is the clear distance between flanges less the fillet or corner radius at each flange; h_c is twice the distance from the neutral axis to the inside face of the compression flange less the fillet or corner radius.

b. For webs of built-up sections, h is the distance between adjacent lines of fasteners or the clear distance between flanges when welds are used, and h_c is twice the distance from the neutral axis to the nearest line of fasteners at the compression flange or the inside face of the compression flange when welds are used.

c. For flange or diaphragm plates in built-up sections, the width b is the distance between adjacent lines of fasteners or lines of welds.

d. For flanges of rectangular hollow structural sections, the width b is the clear distance between webs less the inside corner radius on each side. If the corner radius is not known, the flat width may be taken as the total section width minus three times the thickness.

For tapered flanges of rolled sections, the thickness is the nominal value halfway between the free edge and the corresponding face of the web.

TABLE B5.1
Limiting Width-Thickness Ratios
for Compression Elements

Description of Element	Width-Thick-ness Ratio	Limiting Width-Thickness Ratios	
		λ_p	λ_r
Flanges of I-shaped rolled beams and channels in flexure	b/t	$65/\sqrt{F_y}$ [c]	$141/\sqrt{F_y - 10}$
Flanges of I-shaped hybrid or welded beams in flexure	b/t	$65/\sqrt{F_{yf}}$ [c]	$\dfrac{106}{\sqrt{F_{yw} - 16.5}}$
Flanges of I-shaped sections in pure compression, plates projecting from compression elements; outstanding legs of pairs of angles in continuous contact; flanges of channels in pure compression	b/t	NA	$95/\sqrt{F_y}$
Flanges of square and rectangular box and hollow structural sections of uniform thickness subject to bending or compression; flange cover plates and diaphragm plates between lines of fasteners or welds	b/t	$190/\sqrt{F_y}$	$238/\sqrt{F_y - F_r}$ [e]
Unsupported width of cover plates perforated with a succession of access holes[b]	b/t	NA	$317/\sqrt{F_y - F_r}$ [e]
Legs of single angle struts; legs of double angle struts with separators; unstiffened elements, i.e., supported along one edge	b/t	NA	$76/\sqrt{F_y}$
Stems of tees	d/t	NA	$127/\sqrt{F_y}$
All other uniformly compressed stiffened elements, i.e., supported along two edges	b/t h_c/t_w	NA	$253/\sqrt{F_y}$
Webs in flexural compression[a]	h_c/t_w	$640/\sqrt{F_y}$ [c]	$970/\sqrt{F_y}$
Webs in combined flexural and axial compression	h_c/t_w	for $P_u/\phi_b P_y \le 0.125$ $\dfrac{640}{\sqrt{F_y}}\left(1 - \dfrac{2.75 P_u}{\phi_b P_y}\right)$ [c] for $P_u/\phi_b P_y > 0.125$ $\dfrac{191}{\sqrt{F_y}}\left(2.33 - \dfrac{P_u}{\phi_b P_y}\right) \ge \dfrac{253}{\sqrt{F_y}}$ [c]	$970/\sqrt{F_y}$
Circular hollow sections In axial compression In flexure	D/t	$2{,}070/F_y$ [d] $2{,}070/F_y$ [d]	$3{,}300/F_y$ $8{,}970/F_y$

[a] For hybrid beams, use the yield strength of the flange F_{yf} instead of F_y.
[b] Assumes net area of plate at widest hole.
[c] Assumes an inelastic rotation capacity of 3. For structures in zones of high seismicity, a greater rotation capacity may be required. See Table C-B5.1.
[d] For plastic design use $1{,}300/F_y$.
[e] F_r = compressive residual stress in flange
    = 10 ksi for rolled shapes
    = 16.5 ksi for welded shapes.

2. Sections for Plastic Analysis

Plastic analysis is permitted when flanges subject to compression involving hinge rotation and all webs have a width-thickness ratio less than or equal to the limiting λ_p from Table B5.1. For circular hollow sections see Footnote d of Table B5.1.

Plastic analysis is subject to the limitations as outlined in Sect. A5.1.

3. Slender Compression Elements

For the flexural design of I-shaped sections, channels and rectangular or circular sections with slender compression elements, see Appendix F1.7. For other shapes in flexure or members in axial compression that have slender compression elements, see Appendix B5.3. For plate girders with $h_c/t_w > 970/\sqrt{F_{yf}}$, see Appendix G.

B6. BRACING AT SUPPORTS

At points of support for beams, girders and trusses, restraint against rotation about their longitudinal axis shall be provided unless restraint against rotation is otherwise assured.

B7. LIMITING SLENDERNESS RATIOS

For members whose design is based on compressive force, the slenderness ratio Kl/r preferably should not exceed 200.

For members whose design is based on tensile force, the slenderness ratio L/r preferably should not exceed 300. The above limitation does not apply to rods in tension. Such tension members may be subject to compressive force not exceeding 50% of the design compressive strength when such compressive force is due to wind or earthquake.

Notes

CHAPTER C.
FRAMES AND OTHER STRUCTURES

This chapter specifies general requirements to assure stability of the structure as a whole.

C1. SECOND ORDER EFFECTS

Second order ($P\Delta$) effects shall be considered in the design of frames.

C2. FRAME STABILITY

1. Braced Frames

In trusses and frames where lateral stability is provided by diagonal bracing, shear walls or equivalent means, the effective length factor K for compression members shall be taken as unity, unless structural analysis shows that a smaller value may be used.

The vertical bracing system for a braced multistory frame shall be adequate, as determined by structural analysis, to prevent buckling of the structure and maintain the lateral stability of the structure, including the overturning effects of drift, under the factored loads given in Sect. A4.

The vertical bracing system for a multistory frame may be considered to function together with in-plane shear-resisting exterior and interior walls, floor slabs and roof decks, which are properly secured to the structural frames. The columns, girders, beams and diagonal members, when used as the vertical bracing system, may be considered to comprise a vertical cantilever, simply connected truss in the analyses for frame buckling and lateral instability. Axial deformation of all members in the vertical bracing system shall be included in the lateral stability analysis.

Girders and beams included in the vertical bracing system of a braced multistory frame shall be proportioned for axial force and moment caused by concurrent factored horizontal and gravity loads.

2. Unbraced Frames

In frames where lateral stability depends upon the bending stiffness of rigidly connected beams and columns, the effective length factor K of compression members shall be determined by structural analysis and shall be not less than unity.

Analysis of the required strength of unbraced multistory frames shall include the effects of frame instability and column axial deformation under the factored loads given in Sect. A4.

In plastic design the axial force in the columns caused by factored gravity plus factored horizontal loads shall not exceed $0.75A_gF_y$.

CHAPTER D.
TENSION MEMBERS

This section applies to prismatic members subject to axial tension caused by static forces acting through the centroidal axis. For members subject to combined axial tension and flexure, see Sect. H1. For members subject to fatigue, see Sect. K4. For tapered members, see Appendix F4. For threaded rods see Sect. J3.

D1. DESIGN TENSILE STRENGTH

The design strength of tension members $\phi_t P_n$ shall be the lower value obtained according to the limit states of yielding in the gross section and fracture in the net section.

    a. For yielding in the gross section:

$$\phi_t = 0.90$$
$$P_n = F_y A_g \tag{D1-1}$$

    b. For fracture in the net section:

$$\phi_t = 0.75$$
$$P_n = F_u A_e \tag{D1-2}$$

where

    A_e = effective net area, in.^{2}
    A_g = gross area of member, in.^{2}
    F_y = specified minimum yield stress, ksi
    F_u = specified minimum tensile strength, ksi
    P_n = nominal axial strength, kips

When members without holes are fully connected by welds, the effective net section used in Formula D1-2 shall be computed using the smaller of the gross area of the member or the effective area of the welds as defined in Sect. J2. When holes are present in a welded member between end connections, or at the welded connection in the case of plug or slot welds, the net section through the holes shall be used in Formula D1-2.

D2. BUILT-UP MEMBERS

The longitudinal spacing of connectors between elements in continuous contact consisting of a plate and a shape or two plates shall not exceed:

    24 times the thickness of the thinner plate or 12 in. for painted members or unpainted members not subject to corrosion.
    14 times the thickness of the thinner plate or 7 in. for unpainted members of weathering steel subject to atmospheric corrosion.

The longitudinal spacing of connectors between components should preferably limit the slenderness ratio in any component between the connectors to 300 or less.

Either perforated cover plates or tie plates without lacing may be used on the open sides of built-up tension members. Tie plates shall have a length not less than ⅔ the distance between the lines of welds or fasteners connecting them to the components of the member. The thickness of such tie plates shall not be less than ⅟₅₀ of the distance between these lines. The longitudinal spacing of intermittent welds or fasteners at tie plates shall not exceed 6 in. The spacing of tie plates shall be such that the slenderness ratio of any component in the length between tie plates should preferably not exceed 300.

D3. EYEBARS AND PIN-CONNECTED MEMBERS

The design strength of eyebars shall be determined in accordance with Sect. D1a with A_g taken as the cross-sectional area of the body.

Eyebars shall be of uniform thickness, without reinforcement at the pin holes, and have circular heads whose periphery is concentric with the pin hole.

The radius of transition between the circular head and the eyebar body shall be not less than the head diameter.

The width of the body of the eyebars shall not exceed eight times its thickness.

The thickness can be less than ½-in. only if external nuts are provided to tighten pin plates and filler plates into snug contact. The width b from the hole edge to the plate edge perpendicular to the direction of applied load shall be greater than ⅔ and, for the purpose of calculation, not more than ¾ times the eyebar body width.

The pin diameter shall not be less than ⅞ times the eyebar body width.

The pin-hole diameter shall not be more than ⅟₃₂-in. greater than the pin diameter.

For steels having a yield stress greater than 70 ksi, the hole diameter shall not exceed five times the plate thickness and the width of the eyebar body shall be reduced accordingly.

In pin-connected members the pin hole shall be located midway between the edges of the member in the direction normal to the applied force. For pin-connected members in which the pin is expected to provide for relative movement between connected parts while under full load, the diameter of pin hole shall be not more than ⅟₃₂-in. greater than the diameter of the pin. The width of the plate beyond the pin hole shall be not less than the effective width on either side of the pin hole.

In pin-connected plates other than eyebars, the design strength shall be determined according to Formula D1-2 and the bearing strength of the projected area of the pin shall be determined according to Sect. J8. The minimum net area beyond the bearing end of the pin hole, parallel to the axis of the member, shall not be less than ⅔ of the net area required for strength across the pin hole.

The design strength of a pin-connected member $\phi_t P_n$ shall be the lowest value of the following limit states.

a. Tension on the net effective area:

$$\phi_t = 0.75$$
$$P_n = 2tb_{eff}F_u \tag{D3-1}$$

b. Shear on the effective area:

$$\phi_{sf} = 0.75$$
$$P_n = A_{sf} F_y$$

(D3-2)

c. Bearing on the projected area of the pin:

$$\phi = 1.0$$
$$P_n = A_{pb} F_y$$

(D3-3)

where

a = shortest distance from edge of the pin hole to the edge of the member measured parallel to the direction of the force, in.

A_{pb} = projected bearing area, in.^{2}

A_{sf} = $2t\,(a + d/2)$, in.^{2}

b_{eff} = $2t + 0.63$, but not more than the actual distance from the edge of the hole to the edge of the part measured in the direction normal to the applied force, in.

d = pin diameter, in.

t = thickness of plate, in.

The corners beyond the pin hole may be cut at 45° to the axis of the member, provided the net area beyond the pin hole, on a plane perpendicular to the cut, is not less than that required beyond the pin hole parallel to the axis of the member.

Thickness limitations on both eyebars and pin-connected plates may be waived whenever external nuts are provided so as to tighten pin plates and filler plates into snug contact. When the plates are thus contained, the bearing strength shall be determined according to Sect. J8.

CHAPTER E.
COLUMNS AND OTHER
COMPRESSION MEMBERS

This section applies to prismatic members subject to axial compression through the centroidal axis. For members subject to combined axial compression and flexure, see Chap. H. For tapered members, see Appendix F4.

E1. EFFECTIVE LENGTH AND SLENDERNESS LIMITATIONS

1. Effective Length

The effective length factor K shall be determined in accordance with Sect. C2.

2. Plastic Analysis

Plastic analysis, as limited in Sect. A5.1, is permitted if the column slenderness parameter λ_c defined by Formula E2-4 does not exceed $1.5K$.

E2. DESIGN COMPRESSIVE STRENGTH

The design strength of compression members whose elements have width-thickness ratios less than λ_r of Sect. B5.1 is $\phi_c P_n$

$$\phi_c = 0.85$$
$$P_n = A_g F_{cr} \tag{E2-1}$$

for $\lambda_c \leq 1.5$

$$F_{cr} = (0.658^{\lambda_c^2}) F_y \tag{E2-2}$$

for $\lambda_c > 1.5$

$$F_{cr} = \left[\frac{0.877}{\lambda_c^2}\right] F_y \tag{E2-3}$$

where

$$\lambda_c = \frac{Kl}{r\pi}\sqrt{\frac{F_y}{E}} \tag{E2-4}$$

A_g = gross area of member, in.^{2}
F_y = specified yield stress, ksi
E = modulus of elasticity, ksi
K = effective length factor
l = unbraced length of member, in.
r = governing radius of gyration about plane of buckling, in.

For members whose elements do not meet the requirements of Sect. B5.1, see Appendix B5.3.

E3. FLEXURAL-TORSIONAL BUCKLING

Singly symmetric and unsymmetric columns, such as angle or tee-shaped columns, and doubly symmetric columns such as cruciform or built-up columns with very thin walls, may require consideration of the limit states of flexural-torsional and torsional buckling. See Appendix E3 for the determination of design strength for these limit states.

E4. BUILT-UP MEMBERS

At the ends of built-up compression members bearing on base plates or milled surfaces, all components in contact with one another shall be connected by a weld having a length not less than the maximum width of the member or by bolts spaced longitudinally not more than four diameters apart for a distance equal to 1½ times the maximum width of the member.

Along the length of built-up compression members between the end connections required above, longitudinal spacing for intermittent welds, bolts or rivets shall be adequate to provide for the transfer of calculated stress. However, where a component of a built-up compression member consists of an outside plate, except as provided in the next sentence, the maximum spacing shall not exceed the thickness of the thinner outside plate times $127/\sqrt{F_y}$, nor 12 in., when intermittent welds are provided along the edges of the components or when fasteners are provided on all gage lines at each section. When fasteners are staggered, the maximum spacing on each gage line shall not exceed the thickness of the thinner outside plate times $190/\sqrt{F_y}$, nor 18 in.

For unpainted built-up members made of weathering steel which will be exposed to atmospheric corrosion, the fasteners connecting a plate and a shape or two-plate components in contact with one another shall not exceed 14 times the thickness of the thinnest part nor 7 in. and the maximum edge distance shall not exceed eight times the thickness of the thinnest part, nor 5 in.[91]

Compression members composed of two or more shapes shall be connected to one another at intervals such that the slenderness ratio L/r of either shape, between the fasteners, does not exceed the governing slenderness ratio of the built-up member. The least radius of gyration r shall be used in computing the slenderness ratio of each component part.

The design strength of built-up members composed of two or more shapes shall be determined in accordance with Section E2 or Appendix E3 subject to the following modification. If the buckling mode involves relative deformation that produce shear forces in the connectors between individual shapes, Kl/r is replaced by $(Kl/r)_m$ determined as follows:

a. for snug-tight bolted connectors:

$$\left(\frac{Kl}{r}\right)_m = \sqrt{\left(\frac{Kl}{r}\right)_o^2 + \left(\frac{a}{r_i}\right)^2}$$ (E4-1)

b. for welded connectors and for fully tightened bolted connectors as required for slip-critical joints:

with $\frac{a}{r_i} > 50$

$$\left(\frac{Kl}{r}\right)_m = \sqrt{\left(\frac{Kl}{r}\right)_o^2 + \left(\frac{a}{r_i} - 50\right)^2}$$ (E4-2)

with $\frac{a}{r_i} \le 50$

$$\left(\frac{Kl}{r}\right)_m = \left(\frac{Kl}{r}\right)_o$$ (E4-3)

where

$\left(\frac{Kl}{r}\right)_o$ = column slenderness of built-up member acting as a unit

$\dfrac{a}{r_i}$ = largest column slenderness of individual components

$\left(\dfrac{Kl}{r}\right)_m$ = modified column slenderness of built-up member

a = distance between connectors

r_i = minimum radius of gyration of individual component

Open sides of compression members built up from plates or shapes shall be provided with continuous cover plates perforated with a succession of access holes. The unsupported width of such plates at access holes, as defined in Sect. B5.1, is assumed to contribute to the design strength provided that:

a. The width-thickness ratio conforms to the limitations of Sect. B5.1.
b. The ratio of length (in direction of stress) to width of hole shall not exceed two.
c. The clear distance between holes in the direction of stress shall be not less than the transverse distance between nearest lines of connecting fasteners or welds.
d. The periphery of the holes at all points shall have a minimum radius of 1½ in.

The function of perforated cover plates may be performed by lacing with tie plates at each end and at intermediate points if the lacing is interrupted. Tie plates shall be as near the ends as practicable. In main members providing design strength, the end tie plates shall have a length of not less than the distance between the lines of fasteners or welds connecting them to the components of the member. Intermediate tie plates shall have a length not less than ½ of this distance. The thickness of tie plates shall be not less than ⅟₅₀ of the distance between lines of welds or fasteners connecting them to the segments of the members. In welded construction, the welding on each line connecting a tie plate shall aggregate not less than ⅓ the length of the plate. In bolted and riveted construction, the spacing in the direction of stress in tie plates shall be not more than six diameters and the tie plates shall be connected to each segment by at least three fasteners.

Lacing, including flat bars, angles, channels or other shapes employed as lacing, shall be so spaced that the L/r ratio of the flange included between their connections shall not exceed the governing slenderness ratio for the member as a whole. Lacing shall be proportioned to provide a shearing strength normal to the axis of the member equal to 2% of the compressive design strength of the member. The L/r ratio for lacing bars arranged in single systems shall not exceed 140. For double lacing this ratio shall not exceed 200. Double lacing bars shall be joined at their intersections. For lacing bars in compression, L may be taken as the unsupported length of the lacing bar between welds or fasteners connecting it to the components of the built-up member for single lacing, and 70% of that distance for double lacing. The inclination of lacing bars to the axis of the member shall preferably be not less than 60° for single lacing and 45° for double lacing. When the distance between the lines of welds or fasteners in the flanges is more than 15 in., the lacing shall preferably be double or be made of angles.

E5. PIN-CONNECTED COMPRESSION MEMBERS

Pin-connections of pin-connected compression members shall conform to the requirements of Sect. D3 except Formulas D3-1 and D3-2 do not apply.

CHAPTER F.
BEAMS AND OTHER FLEXURAL MEMBERS

This section applies to singly or doubly symmetric beams including hybrid beams and girders loaded in the plane of symmetry. It also applies to channels loaded in a plane passing through the shear center parallel to the web or restrained against twisting at load points and points of support. For design flexural strength for members not covered in Sect. F1, see Appendix F1.7. For members subject to combined flexural and axial force, see Sect. H1. For unsymmetric beams and beams subject to torsion combined with flexure, see Sect. H2.

F1. DESIGN FOR FLEXURE

1. Unbraced Length for Plastic Analysis

Plastic analysis, as limited in Sect. A5, is permitted when the laterally unbraced length L_b of the compression flange at plastic hinge locations associated with the failure mechanism, for a compact section bent about the major axis, does not exceed L_{pd}, determined as follows:

a. For doubly symmetric and singly symmetric I-shaped members with the compression flange larger than the tension flange (including hybrid members) loaded in the plane of the web

$$L_{pd} = \frac{3,600 + 2,200(M_1/M_p)}{F_y} r_y \qquad \text{(F1-1)}$$

where

F_y = specified minimum yield stress of the compression flange, ksi
M_1 = smaller moment at end of unbraced length of beam, kip-in.
M_p = plastic moment (= $F_y Z$ for homogeneous sections; computed from fully plastic stress distribution for hybrids), kip-in.
r_y = radius of gyration about minor axis, in.
(M_1/M_p) is positive when moments cause reverse curvature

b. For solid rectangular bars and symmetric box beams

$$L_{pd} = \frac{5,000 + 3,000(M_1/M_p)}{F_y} r_y \geq 3,000 \, r_y/F_y \qquad \text{(F1-2)}$$

There is no limit on L_b for members with circular or square cross sections nor for any beam bent about its minor axis.

In the region of the last hinge to form, and in regions not adjacent to a plastic hinge, the flexural design strength shall be determined in accordance with Sect. F1.2.

2. Flexural Design Strength

The flexural design strength, determined by the limit state of lateral-torsional buckling, is $\phi_b M_n$, where the nominal strength M_n shall be determined in accordance with the following sections, and $\phi_b = 0.90$.

3. Compact Section Members with $L_b \leq L_r$

For laterally unsupported compact section members bent about the major axis:

$$M_n = C_b \left[M_p - (M_p - M_r)\left(\frac{L_b - L_p}{L_r - L_p}\right) \right] \leq M_p \qquad \text{(F1-3)}$$

where

$C_b = 1.75 + 1.05(M_1/M_2) + 0.3(M_1/M_2)^2 \leq 2.3$ where M_1 is the smaller and M_2 the larger end moment in the unbraced segment of the beam; M_1/M_2 is positive when the moments cause reverse curvature and negative when bent in single curvature.

$C_b = 1.0$ for unbraced cantilevers and for members where the moment within a significant portion of the unbraced segment is greater than or equal to the larger of the segment end moments.*

L_b = distance between points braced against lateral displacement of the compression flange, or between points braced to prevent twist of the cross section.

For I-shaped members including hybrid sections and channels bent about their major axis:

$$L_p = \frac{300r_y}{\sqrt{F_{yf}}} \qquad \text{(F1-4)}$$

For solid rectangular bars and box beams:

$$L_p = \frac{3,750r_y}{M_p}\sqrt{JA} \qquad \text{(F1-5)}$$

where

A = cross-sectional area, in.^{2}
J = torsional constant, in.^{4}

The limiting laterally unbraced length L_r and the corresponding buckling moment M_r shall be determined as follows:

a. For I-shaped members, doubly symmetric and singly symmetric with the compression flange larger than or equal to the tension flange, and channels loaded in the plane of the web:

$$L_r = \frac{r_y X_1}{(F_{yw} - F_r)}\sqrt{1 + \sqrt{1 + X_2(F_{yw} - F_r)^2}} \qquad \text{(F1-6)}$$

$$M_r = (F_{yw} - F_r)S_x \qquad \text{(F1-7)}$$

where

$$X_1 = \frac{\pi}{S_x}\sqrt{\frac{EGJA}{2}} \qquad \text{(F1-8)}$$

$$X_2 = 4\frac{C_w}{I_y}\left(\frac{S_x}{GJ}\right)^2 \qquad \text{(F1-9)}$$

---

*For the use of larger C_b values, see Structural Stability Research Council *Guide to Stability Design Criteria for Metal Structures*, 3rd Ed., pg. 135.

S_x = section modulus about major axis, in.^{3}
E = modulus of elasticity of steel (29,000 ksi)
G = shear modulus of elasticity of steel (11,200 ksi)
F_{yw} = yield stress of web, ksi
I_y = moment of inertia about y-axis, in.^{4}
C_w = warping constant, in.^{6}
F_r = compressive residual stress in flange; 10 ksi for rolled shapes, 16.5 ksi for welded shapes

b. For singly symmetric, I-shaped members with the compression flange larger than the tension flange, use S_{xc} in place of S_x in Formulas F1-7 through F1-9, or see Table A-F1.1.

c. For symmetric box sections bent about the major axis and loaded in the plane of symmetry, M_r and L_r shall be determined from Formula F1-7 and F1-10 respectively.

d. For solid rectangular bars bent about the major axis:

$$L_r = \frac{57,000 r_y \sqrt{JA}}{M_r} \qquad \text{(F1-10)}$$

$$M_r = F_y S_x \qquad \text{(F1-11)}$$

4. Compact Section Members with $L_b > L_r$

For laterally unsupported members with compact section members bent about the major axis:

$$M_n = M_{cr} \le C_b M_r \qquad \text{(F1-12)}$$

where M_{cr} is the critical elastic moment, determined as follows:

a. For I-shaped members, doubly symmetric and singly symmetric with compression flange larger than the tension flange (including hybrid members) and channels loaded in the plane of the web:

$$M_{cr} = C_b \frac{\pi}{L_b} \sqrt{EI_y\, GJ + \left(\frac{\pi E}{L_b}\right)^2 I_y C_w} \qquad \text{(F1-13)}$$

$$= \frac{C_b S_x X_1 \sqrt{2}}{L_b/r_y} \sqrt{1 + \frac{X_1^2 X_2}{2(L_b/r_y)^2}}$$

b. For solid rectangular bars and symmetric box sections:

$$M_{cr} = \frac{57,000\, C_b \sqrt{JA}}{L_b/r_y} \qquad \text{(F1-14)}$$

5. Tees and Double-angle Beams

The nominal strength of tees and double-angle beams loaded in the plane of symmetry, with flange and web slenderness ratios less than the corresponding values of λ_r in Table B5.1:

$$M_n = M_{cr} = \frac{C_b \pi \sqrt{EI_y\, GJ}}{L_b}[B + \sqrt{1 + B^2}] \le M_y \qquad \text{(F1-15)}$$

where

$$B = \pm 2.3 \, (d/L_b)\sqrt{I_y/J} \tag{F1-16}$$

The plus sign for B applies when the stem is in tension and the minus sign applies when the stem is in compression.

6. Noncompact Plate Girders

The nominal strength of a doubly symmetric, single-web plate girder, including hybrid sections, shall be calculated by the provisions of Appendix F1.7 if $h_c/t_w \leq 970/\sqrt{F_{yf}}$ or by the provisions of Appendix G if $h_c/t_w > 970/\sqrt{F_{yf}}$.

7. Nominal Flexural Strength of Other Sections

There is no lateral-torsional buckling limit state for circular or square shapes nor for any shape bent about its minor axis.

For the nominal strength M_n of other cross section types, including noncompact sections or sections with slender elements, see Appendix F1.7. See Appendix G for design of plate girders with slender webs.

F2. DESIGN FOR SHEAR

This section applies to the web (or webs in the case of multiple web members) of singly or doubly symmetric beams, including hybrid beams, subject to shear in the plane of symmetry, and channels subject to shear in the web. Where failure might occur by shear along a plane through fasteners, refer to Sect. J4. For members subjected to high shear from concentrated loads, see Sect. K1.7.

1. Web Area Determination

The web area A_w shall be taken as the overall depth d times the web thickness t_w.

2. Design Shear Strength

The design shear strength of webs is $\phi_v V_n$, where $\phi_v = 0.90$ and the nominal shear strength V_n is determined as follows:

For $\dfrac{h}{t_w} \leq 187\sqrt{k/F_{yw}}$

$$V_n = 0.6 \, F_{yw} A_w \tag{F2-1}$$

for $187\sqrt{k/F_{yw}} < \dfrac{h}{t_w} \leq 234\sqrt{k/F_{yw}}$

$$V_n = 0.6 \, F_{yw} A_w \, \frac{187\sqrt{k/F_{yw}}}{h/t_w} \tag{F2-2}$$

for $\dfrac{h}{t_w} > 234\sqrt{k/F_{yw}}$

$$V_n = A_w \, \frac{26{,}400k}{(h/t_w)^2} \tag{F2-3}$$

The web plate buckling coefficient k is given by

$$k = 5 + \frac{5}{(a/h)^2} \qquad \text{(F2-4)}$$

Except that k shall be taken as 5 if a/h exceeds 3.0 or $[260/(h/t_w)]^2$. When stiffeners are not required, $k = 5$. In unstiffened girders, h/t shall not exceed 260.

Maximum (h/t_w) limits are given in Appendix G1.

An alternative design method for plate girders utilizing tension field action is given in Appendix G.

F3. TRANSVERSE STIFFENERS

Transverse stiffeners are not required when $h/t_w \leq 418/\sqrt{F_{yw}}$, or when the required shear V_u, as determined by structural analysis for the factored loads, is less than or equal to $\phi_v V_n$ for $k = 5$ given in Sect. F2. Transverse stiffeners used to develop the web design shear strength as provided in Sect. F2 shall have a moment of inertia about an axis in the web center for stiffener pairs or about the face in contact with the web plate for single stiffeners, which shall not be less than $a t_w^3 j$, where

$$j = \frac{2.5}{(a/h)^2} - 2 \geq 0.5 \qquad \text{(F3-1)}$$

Intermediate stiffeners may be stopped short of the tension flange, provided bearing is not needed to transmit a concentrated load or reaction. The weld by which intermediate stiffeners are attached to the web shall be terminated not less than four times nor more than six times the web thickness from the near toe of the web-to-flange weld. When single stiffeners are used, they shall be attached to the compression flange, if it consists of a rectangular plate, to resist any uplift tendency due to torsion in the plate. When lateral bracing is attached to a stiffener, or a pair of stiffeners, these, in turn, shall be connected to the compression flange to transmit one percent of the total flange stress, unless the flange is composed only of angles.

Bolts connecting stiffeners to the girder web shall be spaced not more than 12 in. o. c. If intermittent fillet welds are used, the clear distance between welds shall not be more than 16 times the web thickness nor more than 10 in.

F4. WEB-TAPERED MEMBERS

See Appendix F4.

CHAPTER G.
PLATE GIRDERS

Plate girders shall be distinguished from beams on the basis of the web slenderness ratio h_c/t_w. When this value is greater than $970/\sqrt{F_{yf}}$ the provisions of Appendix G shall apply for design flexural strength, otherwise Appendix F1.7 is applicable.

For design shear strength and transverse stiffener design see appropriate sections in Chap. F or see Appendix G3 and G4 if tension field action is utilized.

CHAPTER H.
MEMBERS UNDER TORSION
AND COMBINED FORCES

This section applies to prismatic members subjected to axial force and flexure about one or both axes of symmetry, with or without torsion, and torsion only. For web-tapered members, see Appendix F4.

H1. SYMMETRIC MEMBERS SUBJECT TO BENDING AND AXIAL FORCE

1. Doubly and Singly Symmetric Members in Flexure and Tension

The interaction of flexure and tension in symmetric shapes shall be limited by Formulas H1-1a and H1-1b

for $\dfrac{P_u}{\phi P_n} \geq 0.2$

$$\frac{P_u}{\phi P_n} + \frac{8}{9}\left(\frac{M_{ux}}{\phi_b M_{nx}} + \frac{M_{uy}}{\phi_b M_{ny}}\right) \leq 1.0 \qquad \text{(H1-1a)}$$

for $\dfrac{P_u}{\phi P_n} < 0.2$

$$\frac{P_u}{2\phi P_n} + \left(\frac{M_{ux}}{\phi_b M_{nx}} + \frac{M_{uy}}{\phi_b M_{ny}}\right) \leq 1.0 \qquad \text{(H1-1b)}$$

where

P_u = required tensile strength, kips
P_n = nominal tensile strength determined in accordance with Sect. D1, kips
M_u = required flexural strength, kip-in.
M_n = nominal flexural strength determined in accordance with Sect. F1, kip-in.
ϕ_t = resistance factor for tension, $\phi_t = 0.90$ (see Sect. D1)
ϕ_b = resistance factor for flexure = 0.90

Second order effects may be considered in the determination of M_u for use in Formulas H1-1a and H1-1b. A more detailed analysis of the interaction of flexure and tension may be made in lieu of using Formulas H1-1a and H1-1b.

2. Doubly and Singly Symmetric Members in Flexure and Compression

The interaction of flexure and compression in symmetric shapes shall be limited by Formulas H1-1a and H1-1b where

P_u = required compressive strength, kips
P_n = nominal compressive strength determined in accordance with Sect. E2, kips
M_u = required flexural strength determined in accordance with subsection a, below, kip-in.
M_n = nominal flexural strength determined in accordance with subsection b, below, kip-in.
ϕ_c = resistance factor for compression, $\phi_c = 0.85$ (see Sect. E2)
ϕ_b = resistance factor for flexure = 0.90

a. Determination of M_u

In structures designed on the basis of elastic analysis, M_u may be determined from a second order elastic analysis using factored loads. In structures designed on the basis of plastic analysis, M_u shall be determined from a plastic analysis that satisfies the requirements of Sects. C1 and C2. In structures designed on the basis of elastic first order analysis the following procedure for the determination of M_u may be used in lieu of a second order analysis:

$$M_u = B_1 M_{nt} + B_2 M_{\ell t} \tag{H1-2}$$

where

M_{nt} = required flexural strength in member assuming there is no lateral translation of the frame, kip-in.
$M_{\ell t}$ = required flexural strength in member as a result of lateral translation of the frame only, kip-in.

$$B_1 = \frac{C_m}{(1 - P_u/P_e)} \geq 1 \tag{H1-3}$$

P_e = $A_g F_y / \lambda_c^2$ where λ_c is defined by Formula E2-4 with $K \leq 1.0$ in the plane of bending.

C_m = a coefficient whose value shall be taken as follows:

i. For restrained compression members in frames braced against joint translation and not subject to transverse loading between their supports in the plane of bending,

$$C_m = 0.6 - 0.4(M_1/M_2) \qquad \text{(H1-4)}$$

where M_1/M_2 is the ratio of the smaller to larger moments at the ends of that portion of the member unbraced in the plane of bending under consideration. M_1/M_2 is positive when the member is bent in reverse curvature, negative when bent in single curvature.

ii. For compression members in frames braced against joint translation in the plane of loading and subjected to transverse loading between their supports, the value of C_m can be determined by rational analysis. In lieu of such analysis, the following values may be used:

for members whose ends are restrained .... $C_m = 0.85$
for members whose ends are unrestrained............. $C_m = 1.0$

$$B_2 = \frac{1}{1 - \Sigma P_u \left(\dfrac{\Delta_{oh}}{\Sigma HL} \right)} \qquad \text{(H1-5)}$$

or $\quad B_2 = \dfrac{1}{1 - \dfrac{\Sigma P_u}{\Sigma P_e}} \qquad \text{(H1-6)}$

ΣP_u = required axial load strength of all columns in a story, kips
Δ_{oh} = translation deflection of the story under consideration, in.
ΣH = sum of all story horizontal forces producing Δ_{oh}, kips
L = story height, in.
P_e = $A_g F_y / \lambda_c^2$, kips, where λ_c is the slenderness parameter defined by Formula E2-4, in which the effective length factor K in the plane of bending shall be determined in accordance with Sect. C2.2, but shall not be less than unity.

b. Determination of M_n

In the use of Formulas H1-1a and H1-1b, M_{nx} shall be determined in accordance with Sect. F1. The actual value of C_b from Sect. F1.3 may be used, provided that the maximum moment M_{ux} occurs at the end of the member or at the end of an unbraced segment of a member. When the maximum moment occurs between the ends, M_{nx} shall be determined with $C_b = 1.0$. When Formula H1-2 is used for determining M_u, the maximum moment for a braced member bent about the strong axis and laterally braced only at its ends will occur at an end whenever the calculated value of B_1 is equal to or less than 1.

H2. UNSYMMETRIC MEMBERS AND MEMBERS UNDER TORSION AND COMBINED TORSION, FLEXURE AND/OR AXIAL FORCE

The design strength ϕF_y of the member shall equal or exceed the required strength expressed in terms of the normal stress f_{un} or the shear stress f_{uv}, determined by elastic analysis for the factored loads:

a. For the limit state of yielding under normal stress:

$$f_{un} \leq \phi F_y \qquad \text{(H2-1)}$$

$$\phi = 0.90$$

b. For the limit state of yielding under shear stress:

$$f_{uv} \leq 0.6\phi F_y \qquad \text{(H2-2)}$$

$$\phi = 0.90$$

c. For the limit state of buckling:

$$f_{un} \text{ or } f_{uv} \leq \phi_c F_{cr}, \text{ as applicable} \qquad \text{(H2-3)}$$

where

$\phi_c = 0.85$ and F_{cr} may be determined from Formula A-E3-2 or A-E3-3, as applicable.

Some constrained local yielding is permitted in areas adjacent to areas which remain elastic.

H3. ALTERNATE INTERACTION EQUATIONS FOR MEMBERS UNDER COMBINED STRESS

See Appendix H3.

CHAPTER I.
COMPOSITE MEMBERS

This chapter applies to composite columns composed of rolled or built-up structural steel shapes, pipe or tubing and structural concrete acting together and to steel beams supporting a reinforced concrete slab so interconnected that the beams and the slab act together to resist bending. Simple and continuous composite beams with shear connectors and concrete-encased beams, constructed with or without temporary shores, are included.

I1. DESIGN ASSUMPTIONS

Force Determination In determining forces in members and connections of a structure that includes composite beams, consideration must be given to the effective sections at the time each increment of load is applied.

Elastic Analysis For an elastic analysis of continuous composite beams without haunched ends, it is acceptable to assume that the stiffness of a beam is uniform throughout the beam length and may be computed using the moment of inertia of the composite transformed section in the positive moment region.

Plastic Analysis When plastic analysis is used, the strength of flexural composite members shall be determined from plastic stress distributions as specified in Sect. I3.

Plastic Stress Distribution for Positive Moment If the slab in the positive moment region is connected to the steel beam with shear connectors, a concrete stress of $0.85 f_c'$ may be assumed uniformly distributed throughout the effective compression zone. Concrete tensile strength shall be neglected. A uniformly distributed steel stress of F_y shall be assumed throughout the tension zone and throughout the compression zone in the structural steel section. The net tensile force in the steel section shall be equal to the compressive force in the concrete slab.

Plastic Stress Distribution for Negative Moment If the slab in the negative moment region is connected to the steel beam with shear connectors, a tensile stress of F_{yr} shall be assumed in all adequately developed longitudinal reinforcing bars within the effective width of the concrete slab. Concrete tensile strength shall be neglected. A uniformly distributed steel stress of F_y shall be assumed throughout the tension zone and throughout the compression zone in the structural steel section. The net compressive force in the steel section shall be equal to the total tensile force in the reinforcing steel.

Elastic Stress Distribution When a determination of elastic stress distribution is required, strains in steel and concrete shall be assumed directly proportional to the distance from the neutral axis. The stress shall equal strain times E or E_c. Concrete tensile strength shall be neglected. Maximum stress in the steel shall not exceed F_y. Maximum compressive stress in the concrete shall not exceed $0.85 f_c'$. In composite hybrid beams, the maximum stress in the steel flange shall not exceed F_{yf} but the strain in the web may exceed the yield strain; the stress shall be taken as F_{yw} at such locations.

Fully Composite Beam Shear connectors are provided in sufficient numbers to develop the maximum flexural strength of the composite beam. For elastic stress distribution it may be assumed that no slip occurs.

Partially Composite Beam   The shear strength of shear connectors governs the flexural strength of the partially composite beam. Elastic computations such as those for deflections, fatigue and vibrations should include the effect of slip.

Concrete-encased Beam   A beam totally encased in concrete cast integrally with the slab may be assumed to be interconnected to the concrete by natural bond, without additional anchorage, provided that: (1) concrete cover over beam sides and soffit is at least 2 in.; (2) the top of the beam is at least 1½ in. below the top and 2 in. above the bottom of the slab; and (3) concrete encasement contains adequate mesh or other reinforcing steel to prevent spalling of concrete.

Composite Column   A steel column fabricated from rolled or built-up steel shapes and encased in structural concrete or fabricated from steel pipe or tubing and filled with structural concrete.

I2. COMPRESSION MEMBERS

1. Limitations

To qualify as a composite column, the following limitations shall be met.

  a. The cross-sectional area of the steel shape, pipe or tubing must comprise at least 4% of the total composite cross section.

  b. Concrete encasement of a steel core shall be reinforced with longitudinal load carrying bars, longitudinal bars to restrain concrete and lateral ties. Longitudinal load carrying bars shall be continuous at framed levels; longitudinal restraining bars may be interrupted at framed levels. The spacing of ties shall be not greater than ⅔ of the least dimension of the composite cross section. The cross-sectional area of the transverse and longitudinal reinforcement shall be at least 0.007 sq. in. per inch of bar spacing. The encasement shall provide at least 1.5 in. of clear cover outside of both transverse and longitudinal reinforcement.

  c. Concrete shall have a specified compressive strength f_c' of not less than 3 ksi nor more than 8 ksi for normal weight concrete and not less than 4 ksi for light weight concrete.

  d. The specified minimum yield stress of structural steel and reinforcing bars used in calculating the strength of a composite column shall not exceed 55 ksi.

  e. The minimum wall thickness of structural steel pipe or tubing filled with concrete shall be equal to $b \sqrt{F_y/3E}$ for each face of width b in rectangular sections and $D \sqrt{F_y/8E}$ for circular sections of outside diameter D.

2. Design Strength

The design strength of axially loaded composite columns is $\phi_c P_n$, where $\phi_c = 0.85$ and the nominal axial compressive strength P_n shall be determined from Formulas E2-1 through E2-4 with the following modifications:

  a. A_s = gross area of steel shape, pipe or tubing, in.^{2} (replaces A_g)
      r_m = radius of gyration of the steel shape, pipe or tubing except that for steel shapes it shall not be less than 0.3 times the overall thickness of the composite cross section in the plane of buckling, in. (replaces r)

  b. Replace F_y with modified yield stress F_{my} from Formula I2-1 and replace E with modified modulus of elasticity E_m from Formula I2-2:

$$F_{my} = F_y + c_1 F_{yr}\,(A_r/A_s) + c_2 f_c'\,(A_c/A_s) \tag{I2-1}$$

$$E_m = E + c_3 E_c\,(A_c/A_s) \tag{I2-2}$$

where

A_c = area of concrete, in.^{2}

A_r = area of longitudinal reinforcing bars, in.^{2}

A_s = area of steel, in.^{2}

E = modulus of elasticity of steel, ksi

E_c = modulus of elasticity of concrete,* ksi

F_y = specified minimum yield stress of steel shape, pipe or tubing, ksi

F_{yr} = specified minimum yield stress of longitudinal reinforcing bars, ksi

f_c' = specified compressive strength of concrete, ksi

c_1, c_2, c_3 = numerical coefficients. For concrete-filled pipe and tubing: $c_1 = 1.0$, $c_2 = 0.85$ and $c_3 = 0.4$; for concrete encased shapes $c_1 = 0.7, c_2 = 0.6$ and $c_3 = 0.2$

3. Columns with Multiple Steel Shapes

If the composite cross section includes two or more steel shapes, the shapes must be interconnected with lacing, tie plates or batten plates to prevent buckling of individual shapes before hardening of concrete.

4. Load Transfer

The portion of the design strength of axially loaded composite columns resisted by concrete shall be developed by direct bearing at connections. When the supporting concrete area is wider than the loaded area on one or more sides and otherwise restrained against lateral expansion on the remaining sides, the maximum design strength of concrete shall be $1.7\phi_c f_c' A_B$, where $\phi_c = 0.60$ is the resistance factor in bearing on concrete and A_B is the loaded area.

I3. FLEXURAL MEMBERS

1. Effective Width

The portion of the effective width of the concrete slab on each side of the beam center-line shall not exceed:

    a. One-eighth of the beam span, center-to-center of supports;

    b. One-half the distance to the centerline of the adjacent beam; or

    c. The distance from the beam centerline to the edge of the slab.

2. Strength of Beams with Shear Connectors

The positive design flexural strength $\phi_b M_n$ shall be determined as follows:

    a. For $h_c/t_w \leq 640/\sqrt{F_{yf}}$:

       $\phi_b = 0.85$; M_n shall be determined from the plastic stress distribution on the composite section.

    b. For $h_c/t_w > 640/\sqrt{F_{yf}}$:

       $\phi_b = 0.90$; M_n shall be determined from the superposition of elastic stresses, considering the effects of shoring.

---

*E_c may be computed from $E_c = w^{1.5} \sqrt{f_c'}$ where w, the unit weight of concrete, is expressed in lbs./cu. ft and f_c' is expressed in ksi.

The negative design flexural strength $\phi_b M_n$ shall be determined for the steel section alone, in accordance with the requirements of Sect. F.

Alternatively, the negative design flexural strength $\phi_b M_n$ may be computed with $\phi_b = 0.85$ and M_n determined from the plastic stress distribution on the composite section, provided that:

a. Steel beam is an adequately braced compact section, as defined in Sect. B5.
b. Shear connectors connect the slab to the steel beam in the negative moment region.
c. Slab reinforcement parallel to the steel beam, within the effective width of the slab, is properly developed.

3. Strength of Concrete-encased Beams

The design flexural strength $\phi_b M_n$ shall be computed with $\phi_b = 0.90$ and M_n determined from the superposition of elastic stresses, considering the effects of shoring.

Alternatively, the design flexural strength $\phi_b M_n$ may be computed with $\phi_b = 0.90$ and M_n determined from the plastic stress distribution on the steel section alone.

4. Strength During Construction

When temporary shores are not used during construction, the steel section alone shall have adequate strength to support all loads applied prior to the concrete attaining 75% of its specified strength f'_c. The design flexural strength of the steel section shall be determined in accordance with the requirements of Sect. F1.

5. Formed Steel Deck

a. General

The design flexural strength $\phi_b M_n$ of composite construction consisting of concrete slabs on formed steel deck connected to steel beams shall be determined by the applicable portions of Sect. I3.2, with the following modifications.

This section is applicable to decks with nominal rib height not greater than 3 in. The average width of concrete rib or haunch w_r shall be not less than 2 in., but shall not be taken in calculations as more than the minimum clear width near the top of the steel deck. See Sect. I3.5c for additional restrictions.

The concrete slab shall be connected to the steel beam with welded stud shear connectors ¾-in. or less in diameter (AWS D1.1). Studs may be welded either through the deck or directly to the steel beam. Stud shear connectors, after installation, shall extend not less than 1½ in. above the top of the steel deck.

The slab thickness above the steel deck shall be not less than 2 in.

b. Deck Ribs Oriented Perpendicular to Steel Beam

Concrete below the top of the steel deck shall be neglected in determining section properties and in calculating A_c for deck ribs oriented perpendicular to the steel beams.

The spacing of stud shear connectors along the length of a supporting beam shall not exceed 32 in.

The nominal strength of a stud shear connector shall be the value stipulated in Sect. I5 multiplied by the following reduction factor:

$$\frac{0.85}{\sqrt{N_r}} (w_r/h_r) [(H_s/h_r) - 1.0] \leq 1.0 \tag{I3-1}$$

where

h_r = nominal rib height, in.

H_s = length of stud connector after welding, in., not to exceed the value $(h_r + 3)$ in computations, although actual length may be greater

N_r = number of stud connectors in one rib at a beam intersection, not to exceed three in computations, although more than three studs may be installed

w_r = average width of concrete rib or haunch (as defined in Sect. I3.5a), in.

To resist uplift, steel deck shall be anchored to all supporting members at a spacing not to exceed 16 in. Such anchorage may be provided by stud connectors, a combination of stud connectors and arc spot (puddle) welds or other devices specified by the designer.

c. Deck Ribs Oriented Parallel to Steel Beam

Concrete below the top of the steel deck may be included in determining section properties and shall be included in calculating A_c for Sect. I5.

Steel deck ribs over supporting beams may be split longitudinally and separated to form a concrete haunch.

When the nominal depth of steel deck is 1½ in. or greater, the average width w_r of the supported haunch or rib shall be not less than 2 in. for the first stud in the transverse row plus 4 stud diameters for each additional stud.

The nominal strength of a stud shear connector shall be the value stipulated in Sect. I5, except that when w_r/h_r is less than 1.5, the value from Sect. I5 shall be multiplied by the following reduction factor:

$$0.6 (w_r/h_r) [(H_s/h_r) - 1.0] \leq 1.0 \tag{I3-2}$$

where h_r and H_s are as defined in Sect. I3.5b and w_r is the average width of concrete rib or haunch as defined in Sect. I3.5a.

6. Design Shear Strength

The design shear strength of composite beams shall be determined by the shear strength of the steel web, in accordance with the requirements of Sect. F2.

I4. COMBINED COMPRESSION AND FLEXURE

The interaction of axial compression and flexure in the plane of symmetry on composite members shall be limited by Formulas H1-1 through H1-6 with the following modifications:

M_n = nominal flexural strength determined from plastic stress distribution on the composite cross section except as provided below, kip-in.

P_e = $A_s F_{my}/\lambda_c^2$, elastic buckling load, kips

F_{my} = modified yield stress, ksi, see Sect. I2

ϕ_b = resistance factor for flexure from Sect. I3
ϕ_c = 0.85
λ_c = column slenderness parameter defined by Formula E2-4 as modified in Sect. I2.2

When the axial term in Formulas H1-1a and H1-1b is less than 0.3, the nominal flexural strength M_n shall be determined by straight line transition between the nominal flexural strength determined from the plastic distribution on the composite cross sections at $(P_u/\phi_b P_n) = 0.3$ and the flexural strength at $P_u = 0$ as determined from Sect. I3. If shear connectors are required at $P_u = 0$, they shall be provided whenever $P_u/\phi_b P_n$ is less than 0.3.

I5. SHEAR CONNECTORS

This section applies to the design of stud and channel shear connectors. For connectors of other types, see Sect. I6.

1. Materials

Shear connectors shall be headed steel studs not less than four stud diameters in length after installation, or hot rolled steel channels. The stud connectors shall conform to the requirements of Sect. A3.6. The channel connectors shall conform to the requirements of Sect. A3. Shear connectors shall be embedded in concrete slabs made with ASTM C33 aggregate or with rotary kiln produced aggregates conforming to ASTM C330, with concrete unit weight not less than 90 pcf.

2. Horizontal Shear Force

Except for concrete-encased beams as defined in Sect. I1, the entire horizontal shear at the interface between the steel beam and the concrete slab shall be assumed to be transferred by shear connectors. For composite action with concrete subject to flexural compression, the total horizontal shear force between the point of maximum positive moment and the point of zero moment shall be taken as the smallest of: (1) $0.85 f'_c A_c$; (2) $A_s F_y^*$; and (3) ΣQ_n;

where

f'_c = specified compressive strength of concrete, ksi
A_c = area of concrete slab within effective width, in.^{2}
A_s = area of steel cross section, in.^{2}
F_y = minimum specified yield stress, ksi
ΣQ_n = sum of nominal strengths of shear connectors between the point of maximum positive moment and the point of zero moment, kips

In continuous composite beams where longitudinal reinforcing steel in the negative moment regions is considered to act compositely with the steel beam, the total horizontal shear force between the point of maximum negative moment and the point of zero moment shall be taken as the smaller of $A_r F_{yr}$ and ΣQ_n;

---

*For hybrid beams, the yield force must be computed separately for each component of the cross section; $A_s F_y$ of the entire cross section is the sum of the component yield forces.

where

A_r = area of adequately developed longitudinal reinforcing steel within the effective width of the concrete slab, in.^{2}

F_{yr} = minimum specified yield stress of the reinforcing steel, ksi

ΣQ_n = sum of nominal strengths of shear connectors between the point of maximum negative moment and the point of zero moment, kips

3. Strength of Stud Shear Connectors

The nominal strength of one stud shear connector embedded in a solid concrete slab is

$$Q_n = 0.5\, A_{sc} \sqrt{f_c' E_c} \le A_{sc} F_u \qquad (I5\text{-}1)$$

where

A_{sc} = cross-sectional area of a stud shear connector, in.^{2}

f_c' = specified compressive strength of concrete, ksi

F_u = minimum specified tensile strength of a stud shear connector, ksi

E_c = modulus of elasticity of concrete,* ksi

For a stud shear connector embedded in a slab on a formed steel deck, refer to Sect. I3 for reduction factors given by Formulas I3-1 and I3-2 as applicable. The reduction factors should be applied only to $0.5 A_{sc} \sqrt{f_c' E_c}$ term in Formula I5-1.

4. Strength of Channel Shear Connectors

The nominal strength of one channel shear connector embedded in a solid concrete slab is

$$Q_n = 0.3\, (t_f + 0.5\, t_w)\, L_c \sqrt{f_c' E_c} \qquad (I5\text{-}2)$$

where

t_f = flange thickness of channel shear connector, in.

t_w = web thickness of channel shear connector, in.

L_c = length of channel shear connector, in.

5. Required Number of Shear Connectors

The number of shear connectors required between the section of maximum bending moment, positive or negative, and the adjacent section of zero moment shall be equal to the horizontal shear force as determined from Sect. I5.2 divided by the nominal strength of one shear connector as determined from Sect. I5.3 or I5.4.

6. Shear Connector Placement and Spacing

Shear connectors required each side of the point of maximum bending moment, positive or negative, may be distributed uniformly between that point and the adjacent points of zero moment. However, the number of shear connectors placed between any

---

*E_c, in ksi, may be computed from $E_c = w^{1.5} \sqrt{f_c'}$ where w, the unit weight of concrete, is expressed in lbs./cu. ft and f_c' is expressed in ksi.

concentrated load and the nearest point of zero moment shall be sufficient to develop the maximum moment required at the concentrated load point.

Except for connectors installed in the ribs of formed steel decks, shear connectors shall have at least 1 in. of lateral concrete cover. Unless located over the web, the diameter of studs shall not be greater than 2.5 times the thickness of the flange to which they are welded. The minimum center-to-center spacing of stud connectors shall be 6 diameters along the longitudinal axis of the supporting composite beam and 4 diameters transverse to the longitudinal axis of the supporting composite beam, except that within the ribs of formed steel decks the center-to-center spacing may be as small as 4 diameters in any direction. The maximum center-to-center spacing of shear connectors shall not exceed 8 times the total slab thickness.

I6. SPECIAL CASES

When composite construction does not conform to the requirements of Sects. I1 through I5, the strength of shear connectors and details of construction shall be established by a suitable test program.

CHAPTER J.
CONNECTIONS, JOINTS AND FASTENERS

J1. GENERAL PROVISIONS

1. Design Basis

Connections consist of connecting elements (e.g., stiffeners, gussets, angles, brackets) and connectors (welds, bolts, rivets). These components shall be proportioned so that their design strength equals or exceeds the required strength determined by (a) structural analysis for factored loads acting on the structure or (b) a specified proportion of the strength of the connected members, whichever is appropriate.

2. Simple Connections

Except as otherwise indicated in the design documents, connections of beams, girders or trusses shall be designed as flexible, and may ordinarily be proportioned for the reaction shears only. Flexible beam connections shall accommodate end rotations of unrestrained (simple) beams. To accomplish this, inelastic deformation in the connection is permitted.

3. Moment Connections

End connections of restrained beams, girders and trusses shall be designed for the combined effect of forces resulting from moment and shear induced by the rigidity of the connections.

4. Compression Members with Bearing Joints

When columns bear on bearing plates or are finished to bear at splices, there shall be sufficient connectors to hold all parts securely in place.

When other compression members are finished to bear, the splice material and its connectors shall be arranged to hold all parts in line and shall be proportioned for 50% of the factored strength of the member.

All compression joints shall be proportioned to resist any tension developed by the factored loads specified by Formula A4-6.

5. Minimum Strength of Connections

Except for lacing, sag rods or girts, connections providing design strength shall be designed to support a factored load not less than 10 kips.

6. Placement of Welds and Bolts

Groups of welds or bolts at the ends of any member which transmit axial force into that member shall be sized so that the center of gravity of the group coincides with the center of gravity of the member, unless provision is made for the eccentricity. The foregoing provision is not applicable to end connections of statically-loaded single angle, double angle and similar members.

7. Bolts in Combination with Welds

In new work, A307 bolts or high-strength bolts proportioned as bearing-type connections shall not be considered as sharing the load in combination with welds. Welds, if used, shall be provided to carry the entire force in the connection. High-strength bolts proportioned for slip-critical connections may be considered as sharing the load with the welds.

In making welded alterations to structures, existing rivets and high-strength bolts tightened to the requirements for slip-critical connections may be utilized for carrying loads resulting from existing dead loads, and the welding need only provide the additional design strength required.

8. High-strength Bolts in Combination with Rivets

In both new work and alterations, in connections designed as slip-critical connections in accordance with the provisions of Sect. J3, high-strength bolts may be considered as sharing the load with rivets.

9. Limitations on Bolted and Welded Connections

Fully tensioned high-strength bolts (see Table J3.1), or welds shall be used for the following connections:

Column splices in all tier structures 200 ft or more in height

Column splices in tier structures 100 to 200 ft in height, if the least horizontal dimension is less than 40% of the height

Column splices in tier structures less than 100 ft in height, if the least horizontal dimension is less than 25% of the height

Connections of all beams and girders to columns and of any other beams and girders on which the bracing of columns is dependent, in structures over 125 ft in height

In all structures carrying cranes of over 5-ton capacity: roof-truss splices and connections of trusses to columns, column splices, column bracing, knee braces and crane supports

Connections for supports of running machinery, or of other live loads which produce impact or reversal of stress

Any other connections stipulated on the design plans.

In all other cases connections may be made with A307 bolts or snug-tight high-strength bolts.

For the purpose of this Section, the height of a tier structure shall be taken as the vertical distance from the curb level to the highest point of the roof beams in the case of flat roofs, or to the mean height of the gable in the case of roofs having a rise of more than 2⅔ in 12. Where the curb level has not been established, or where the structure does not adjoin a street, the mean level of the adjoining land shall be used instead of curb level. Penthouses may be excluded in computing the height of structure.

J2. WELDS

All provisions of the American Welding Society *Structural Welding Code—Steel*, AWS D1.1, except Sects. 2.3.2.4, 2.5, 8.13.1, 9, and 10 as applicable, apply to work performed under this Specification.

1. Groove Welds

a. Effective Area

The effective area of groove welds shall be considered as the effective length of the weld times the effective throat thickness.

The effective length of a groove weld shall be the width of the part joined.

The effective throat thickness of a complete-penetration groove weld shall be the thickness of the thinner part joined.

The effective throat thickness of a partial-penetration groove weld shall be as shown in Table J2.1.

The effective throat thickness of a flare groove weld when flush to the surface of a bar or 90° bend in a formed section shall be as shown in Table J2.2. Random sections of production welds for each welding procedure, or such test sections as may be required by design documents, shall be used to verify that the effective throat is consistently obtained.

Larger effective throat thicknesses than those in Table J2.2 are permitted, provided the fabricator can establish by qualification that he can consistently provide such larger effective throat thicknesses. Qualification shall consist of sectioning the weld normal to its axis, at mid-length and terminal ends. Such sectioning shall be made on a number of combinations of material sizes representative of the range to be used in the fabrication or as required by the designer.

b. Limitations

The minimum effective throat thickness of a partial-penetration groove weld shall be as shown in Table J2.4. Weld size is determined by the thicker of the two parts joined, except that the weld size need not exceed the thickness of the thinnest part joined

TABLE J2.1
Effective Throat Thickness of Partial Penetration Groove Welds

Welding Process	Welding Position	Included Angle at Root of Groove	Effective Throat Thickness
Shielded metal arc Submerged arc Gas metal arc Flux-cored arc	All	J or U joint	Depth of chamfer
		Bevel or V joint $\geq$ 60°	Depth of chamfer
		Bevel or V joint < 60° but $\geq$ 45°	Depth of chamfer minus ⅛-in.

TABLE J2.2
Effective Throat Thickness of Flare Groove Welds

Type of Weld	Radius (R) of Bar or Bend	Effective Throat Thickness
Flare bevel groove	All	⁵⁄₁₆R
Flare V-groove	All	½R[a]

[a] Use ⅜R for Gas Metal Arc Welding (except short circuiting transfer process) when $R \geq$ 1 in.

when a larger size is required by calculated strength. For this exception, particular care shall be taken to provide sufficient preheat for soundness of the weld.

2. Fillet Welds

a. Effective Area

The effective area of fillet welds shall be taken as the effective length times the effective throat thickness.

The effective length of fillet welds, except fillet welds in holes and slots, shall be the overall length of full-size fillets, including returns.

The effective throat thickness of a fillet weld shall be the shortest distance from the root of the joint to the face of the diagrammatic weld, except that for fillet welds made by the submerged arc process, the effective throat thickness shall be taken equal to the leg size for ⅜-in. and smaller fillet welds, and equal to the theoretical throat plus 0.11-in. for fillet welds over ⅜-in.

For fillet welds in holes and slots, the effective length shall be the length of the centerline of the weld along the center of the plane through the throat. In the case of overlapping fillets, the effective area shall not exceed the nominal cross-sectional area of the hole or slot, in the plane of the faying surface.

b. Limitations

The *minimum size of fillet welds* shall be as shown in Table J2.5. Minimum weld size is

TABLE J2.4
Minimum Effective Throat Thickness of
Partial-penetration Groove Welds

Material Thickness of Thicker Part Joined (in.)	Minimum Effective Throat Thickness[a] (in.)
To ¼ inclusive	⅛
Over ¼ to ½	3/16
Over ½ to ¾	¼
Over ¾ to 1½	5/16
Over 1½ to 2¼	⅜
Over 2¼ to 6	½
Over 6	⅝

[a]See Sect. J2.

TABLE J2.5
Minimum Size of Fillet Welds

Material Thickness of Thicker Part Joined (in.)	Minimum Size of Fillet Weld[a] (in.)
To ¼ inclusive	⅛
Over ¼ to ½	3/16
Over ½ to ¾	¼
Over ¾	5/16

[a]Leg dimension of fillet welds.

determined by the thicker of the two parts joined, except that the weld size need not exceed the thickness of the thinner part. For this exception, particular care shall be taken to provide sufficient preheat for soundness of the weld. Weld sizes larger than the thinner part joined are permitted if required by calculated strength. In the as-welded condition, the distance between the edge of the base metal and the toe of the weld may be less than $\frac{1}{16}$-in. provided the weld size is clearly verifiable.

The *maximum size of fillet welds* that may be used along edges of connected parts shall be:

Along edges of material less than $\frac{1}{4}$-in. thick, not greater than the thickness of the material.

Along edges of material $\frac{1}{4}$-in. or more in thickness, not greater than the thickness of the material minus $\frac{1}{16}$-in., unless the weld is especially designated on the drawings to be built out to obtain full-throat thickness.

The *minimum effective length of fillet welds* designed on the basis of strength shall be not less than 4 times the nominal size, or else the size of the weld shall be considered not to exceed $\frac{1}{4}$ of its effective length. If longitudinal fillet welds are used alone in end connections of flat bar tension members, the length of each fillet weld shall be not less than the perpendicular distance between them. The transverse spacing of longitudinal fillet welds used in end connections of tension members shall not exceed 8 in., unless the member is designed on the basis of effective net area in accordance with Sect. B3.

Intermittent fillet welds may be used to transfer calculated stress across a joint or faying surfaces when the strength required is less than that developed by a continuous fillet weld of the smallest permitted size, and to join components of built-up members. The effective length of any segment of intermittent fillet welding shall be not less than 4 times the weld size, with a minimum of $1\frac{1}{2}$ in.

In *lap joints*, the minimum amount of lap shall be 5 times the thickness of the thinner part joined, but not less than 1 in. Lap joints joining plates or bars subjected to axial stress shall be fillet welded along the end of both lapped parts, except where the deflection of the lapped parts is sufficiently restrained to prevent opening of the joint under maximum loading.

Side or end fillet welds terminating at ends or sides, respectively, of parts or members shall, wherever practicable, be returned continuously around the corners for a distance not less than 2 times the nominal size of the weld. This provision shall apply to side and top fillet welds connecting brackets, beam seats and similar connections, on the plane about which bending moments are computed. For framing angles and simple end plate connections which depend upon flexibility of the outstanding legs for connection flexibility, end returns shall not exceed four times the nominal size of the weld. Fillet welds which occur on opposite sides of a common plane shall be interrupted at the corner common to both welds. End returns shall be indicated on the design and detail drawings.

Fillet welds in holes or slots may be used to transmit shear in lap joints or to prevent the buckling or separation of lapped parts and to join components of built-up members. Such fillet welds may overlap, subject to the provisions of Sect. J2. Fillet welds in holes or slots are not to be considered plug or slot welds.

3. Plug and Slot Welds

a. Effective Area

The effective shearing area of plug and slot welds shall be considered as the nominal cross-sectional area of the hole or slot in the plane of the faying surface.

b. Limitations

Plug or slot welds may be used to transmit shear in lap joints or to prevent buckling of lapped parts and to join component parts of built-up members.

The diameter of the holes for a plug weld shall be not less than the thickness of the part containing it plus 5/16-in., rounded to the next larger odd 1/16-in., nor greater than 2¼ times the thickness of the weld metal.

The minimum c.-to-c. spacing of plug welds shall be four times the diameter of the hole.

The length of slot for a slot weld shall not exceed 10 times the thickness of the weld. The width of the slot shall be not less than the thickness of the part containing it plus 5/16-in., rounded to the next larger odd 1/16-in., nor shall it be larger than 2¼ times the thickness of the weld. The ends of the slot shall be semicircular or shall have the corners rounded to a radius not less than the thickness of the part containing it, except those ends which extend to the edge of the part.

The minimum spacing of lines of slot welds in a direction transverse to their length shall be 4 times the width of the slot. The minimum c.-to-c. spacing in a longitudinal direction on any line shall be 2 times the length of the slot.

The thickness of plug or slot welds in material 5/8-in. or less in thickness shall be equal to the thickness of the material. In material over 5/8-in. in thickness, the thickness of the weld shall be at least ½ the thickness of the material but not less than 5/8-in.

4. Design Strength

The design strength of welds shall be the lower value of ϕF_{BM} and ϕF_w, when applicable, where F_{BM} and F_w are the nominal strengths of the base material and the weld electrode material, respectively. The values of ϕ, F_{BM} and F_w and limitations thereon are given in Table J2.3.

5. Combination of Welds

If two or more of the general types of welds (groove, fillet, plug, slot) are combined in a single joint, the design strength of each shall be separately computed with reference to the axis of the group in order to determine the design strength of the combination.

6. Matching Steel

The choice of electrode for use with complete-penetration groove welds subject to tension normal to the effective area is dictated by the requirements for matching steels given in the AWS *Structural Welding Code—Steel* D1.1.

J3. BOLTS, THREADED PARTS AND RIVETS

1. High-strength Bolts

Except as otherwise provided in this Specification, use of high-strength bolts shall conform to the provisions of the *Specification for Structural Joints Using ASTM A325 or A490 Bolts*—1985, as approved by the Research Council on Structural Connections.

If required to be tightened to more than 50% of their minimum specified tensile strength, ASTM A449 bolts in tension and bearing-type shear connections shall have an ASTM F436 hardened washer installed under the bolt head, and the nuts shall meet

TABLE J2.3
Design Strength of Welds

Types of Weld and Stress[a]	Material	Resistance Factor ϕ	Nominal strength F_{BM} or F_w	Required Weld strength level[b,c]
Complete Penetration Groove Weld				
Tension normal to effective area	Base	0.90	F_y	"Matching" weld must be used.
Compression normal to effective area	Base	0.90	F_y	Weld metal with a strength level equal to or less than "matching" may be used.
Tension or compression parallel to axis of weld				
Shear on effective area	Base / Weld electrode	0.90 / 0.80	$0.60F_y$ / $0.60F_{EXX}$	
Partial Penetration Groove Welds				
Compression normal to effective area	Base	0.90	F_y	Weld metal with a strength level equal to or less than "matching" weld metal may be used.
Tension or compression parallel to axis of weld[d]				
Shear parallel to axis of weld	Base[e] / Weld electrode	0.75	$0.60F_{EXX}$	
Tension normal to effective area	Base / Weld Electrode	0.90 / 0.80	F_y / $0.60F_{EXX}$	
Fillet Welds				
Stress on effective area	Base[e] / Weld electrode	0.75	$0.60F_{EXX}$	Weld metal with a strength level equal to or less than "matching" weld metal may be used.
Tension or compression parallel to axis of weld[d]	Base	0.90	F_y	
Plug or Slot Welds				
Shear parallel to faying surfaces (on effective area)	Base[e] / Weld Electrode	0.75	$0.60F_{EXX}$	Weld metal with a strength level equal to or less than "matching" weld metal may be used.

[a]For definition of effective area, see Sect. J2.
[b]For "matching" weld metal, see Table 4.1.1, AWS D1.1.
[c]Weld metal one strength level stronger than "matching" weld metal will be permitted.
[d]Fillet welds and partial-penetration groove welds joining component elements of built-up members, such as flange-to-web connections, may be designed without regard to the tensile or compressive stress in these elements parallel to the axis of the welds.
[e]The design of connected material is governed by Sect. J4.

TABLE J3.1
Minimum Bolt Tension, kips[a]

Bolt Size, in.	A325 Bolts	A490 Bolts
½	12	15
⅝	19	24
¾	28	35
⅞	39	49
1	51	64
1⅛	56	80
1¼	71	102
1⅜	85	121
1½	103	148

[a]Equal to 0.70 of minimum tensile strength of bolts, rounded off to nearest kip, as specified in ASTM specifications for A325 and A490 bolts with UNC threads.

the requirements of ASTM A563. When assembled, all joint surfaces, including those adjacent to the washers, shall be free of scale, except tight mill scale. Except as noted below, all A325 and A490 bolts shall be tightened to a bolt tension not less than that given in Table J3.1. Tightening shall be done by the turn-of-nut method, a direct tension indicator or by calibrated wrench.

Bolts in connections not subject to tension loads, where slip can be permitted and where loosening or fatigue due to vibration or load fluctuations are not design considerations, need only to be tightened to the snug-tight condition. The snug-tight condition is defined as the tightness attained by a few impacts of an impact wrench or the full effort of a worker with an ordinary spud wrench and must bring the connected plies into firm contact. The nominal strength value given in Table J3.2 for bearing-type connections shall be used for bolts tightened to the snug-tight condition. Bolts to be tightened only to the snug-tight condition shall be clearly identified on the design and erection drawings.

2. Effective Bearing Area

The effective bearing area of bolts, threaded parts and rivets shall be the diameter multiplied by the length in bearing, except that for countersunk bolts and rivets ½ the depth of the countersink shall be deducted.

3. Design Tension or Shear Strength

The design strength of bolts and threaded parts shall be taken as the product of the resistance factor ϕ and the nominal strength given in Table J3.2 of the unthreaded nominal body area of bolts and threaded parts other than upset rods (see footnote c, Table J3.2). High-strength bolts required to support the applied load by means of direct tension shall be proportioned so that their average required strength, computed on the basis of nominal bolt area and independent of any initial tightening force, will not exceed the design strength. The applied load shall be the sum of the factored external loads and any tension resulting from prying action produced by deformation of the connected parts.

TABLE J3.2
Design Strength of Fasteners

Description of Fasteners	Tensile Strength		Shear Strength in Bearing-type Connections	
	Resistance Factor ϕ	Nominal Strength, ksi	Resistance Factor ϕ	Nominal Strength ksi
A307 bolts	0.75	45.0[a]	0.60	27.0[b,e]
A325 bolts, when threads are *not* excluded from shear planes		90.0[d]	0.65	54.0[e]
A325 bolts, when threads *are* excluded from shear planes		90.0[d]		72.0[e]
A490 bolts, when threads are *not* excluded from shear planes		112.5[d]		67.5[e]
A490 bolts, when threads *are* excluded from the shear planes		112.5[d]		90.0[e]
Threaded parts meeting the requirements of Sect. A3, when threads are *not* excluded from the shear planes		$0.75F_u$[a,c]		$0.45F_u$
Threaded parts meeting the requirements of Sect. A3, when threads *are* excluded from the shear planes		$0.75F_u$[a,c]		$0.60F_u$
A502, Gr. 1, hot-driven rivets		45.0[a]		36.0[e]
A502, Gr. 2 & 3, hot-driven rivets		60.0[a]		48.0[e]

[a]Static loading only.
[b]Threads permitted in shear planes.
[c]The nominal tensile strength of the threaded portion of an upset rod, based upon the cross-sectional area at its major thread diameter, A_b shall be larger than the nominal body area of the rod before upsetting times F_y.
[d]For A325 and A490 bolts subject to tensile fatigue loading, see Appendix K4.
[e]When bearing-type connections used to splice tension members have a fastener pattern whose length, measured parallel to the line of force, exceeds 50 in., tabulated values shall be reduced by 20%.

TABLE J3.3
Tension Stress Limit (F_t), ksi, for Fasteners
in Bearing-type Connections

Description of Fasteners	Threads Included in the Shear Plane	Threads Excluded from the Shear Plane
A307 bolts	$39 - 1.8f_v \le 30$	
A325 bolts	$85 - 1.8f_v \le 68$	$85 - 1.4f_v \le 68$
A490 bolts	$106 - 1.8f_v \le 84$	$106 - 1.4f_v \le 84$
Threaded parts A449 bolts over 1½-in. diameter	$0.73F_u - 1.8f_v \le 0.56F_u$	$0.73F_u - 1.4f_v \le 0.56F_u$
A502 Gr. 1 rivets	$44 - 1.3f_v \le 34$	
A502 Gr. 2 rivets	$59 - 1.3f_v \le 45$	

4. Combined Tension and Shear in Bearing-type Connections

Bolts and rivets subject to combined tension and shear shall be so proportioned that the tension stress f_t produced by factored loads on the nominal body area A_b does not exceed the values computed from the formulas in Table J3.3. The value of f_v, the shear produced by the same factored loads, shall not exceed the values for shear given in Sect. J3.3.

5. High-strength Bolts in Slip-critical Joints

The design shear resistance of slip-critical joints shall be determined by using the tabulated values from Table J3.4 multiplied by $\phi = 1.0$, except $\phi = 0.85$ for the long-slotted holes when the load is in the direction of the slot. The shear on the bolt due to service loads shall be less than the tabulated values. When the loading combination includes wind or seismic loads in addition to dead and live loads, the total of the combined load effects, at service loads, may be multiplied by 0.75 in accordance with ANSI 58.1.

When specified by the designer, the nominal slip resistance for connections having special faying surface conditions may be increased to the applicable values in RCSC Load and Resistance Factor Design Specification.

When a bolt in a slip-critical connection is subjected to a service tensile force T, the nominal resistance in Table J3.4 shall be multiplied by the reduction factor $(1 - T/T_b)$ where T_b is the minimum pretension load from Table J3.1.

6. Bearing Strength at Bolt Holes

When L is not less than $1\frac{1}{2} d$ and the distance center-to-center of bolts is not less than

TABLE J3.4
Nominal Slip-critical Shear Strength, ksi,
of High-strength Bolts[a]

Type of Bolt	Nominal Shear Strength		
	Standard Size Holes	Oversized and Short-slotted Holes	Long-slotted Holes[b]
A325	17	15	12
A490	21	18	15

[a]Class A (slip coefficient 0.33). Clean mill scale and blast cleaned surfaces with class A coatings. For design strengths with other coatings see RCSC "Load and Resistance Factor Design Specification for Structural Joints Using ASTM A325 or A490 Bolts."
[b]Tabulated values are for the case of load application transverse to the slot. When the load is parallel to the slot multiply tabulated values by 0.85.

$3d$, the design bearing strength on two or more bolts in the line of force is ϕR_n, where $\phi = 0.75$.

In standard or short-slotted holes,

$$R_n = 2.4\ dt\ F_u \tag{J3-1a}$$

In long-slotted holes perpendicular to the load,

$$R_n = 2.0\ dt\ F_u \tag{J3-1b}$$

For the bolt closest to the edge, in all connections not covered by Formulas J3-1a and J3-1b, the design bearing of a single bolt, or two or more bolts in line of force, each with an end distance less than $1\frac{1}{2}\ d$, shall be determined by ϕR_n, where $\phi = 0.75$.

$$R_n = Lt\ F_u \tag{J3-1c}$$

If deformation around the bolt hole is not a design consideration (see Commentary) and adequate spacing and edge distance as required by Sect. J3.9 and J3.10 is provided, the following formula may be used in lieu of J3-1a and J3-1b, where $\phi = 0.75$.

$$R_n = 3.0\ dt\ F_u \tag{J3-1d}$$

where

d = nominal dia. of bolt, in.
t = thickness of connected part, in.
F_u = specified tensile strength of connected part, ksi
L = distance in line of force from the center of a standard or oversized hole or from the center of the end of a slotted hole to an edge of a connected part, in.

7. Size and Use of Holes

a. The *maximum sizes* of holes for rivets and bolts are given in Table J3.5, except that larger holes, required for tolerance on location of anchor bolts in concrete foundations, may be used in column base details.

TABLE J3.5
Nominal Hole Dimensions

Bolt Dia.	Hole Dimensions			
	Standard (Dia.)	Oversize (Dia.)	Short-slot (Width × length)	Long-slot (Width × length)
1/2	9/16	5/8	9/16 × 11/16	9/16 × 1 1/4
5/8	11/16	13/16	11/16 × 7/8	11/16 × 1 9/16
3/4	13/16	15/16	13/16 × 1	13/16 × 1 7/8
7/8	15/16	1 1/16	15/16 × 1 1/8	15/16 × 2 3/16
1	1 1/16	1 1/4	1 1/16 × 1 5/16	1 1/16 × 2 1/2
≥1 1/8	$d + 1/16$	$d + 5/16$	$(d + 1/16) \times (d + 3/8)$	$(d + 1/16) \times (2.5 \times d)$

   b. *Standard holes* shall be provided in member-to-member connections, unless oversized, short-slotted or long-slotted holes in bolted connections are approved by the designer. Finger shims up to 1/4-in. may be introduced into slip-critical connections designed on the basis of standard holes without reducing the nominal shear strength of the fastener to that specified for slotted holes.

   c. *Oversized holes* may be used in any or all plies of slip-critical connections, but they shall not be used in bearing-type connections. Hardened washers shall be installed over oversized holes in an outer ply.

   d. *Short-slotted holes* may be used in any or all plies of slip-critical or bearing-type connections. The slots may be used without regard to direction of loading in slip-critical connections, but the length shall be normal to the direction of the load in bearing-type connections. Washers shall be installed over short-slotted holes in an outer ply; when high-strength bolts are used, such washers shall be hardened.

   e. *Long-slotted holes* may be used in only one of the connected parts of either a slip-critical or bearing-type connection at an individual faying surface. Long-slotted holes may be used without regard to direction of loading in slip-critical connections, but shall be normal to the direction of load in bearing-type connections. Where long-slotted holes are used in an outer ply, plate washers or a continuous bar with standard holes, having a size sufficient to completely cover the slot after installation, shall be provided. In high-strength bolted connections, such plate washers or continuous bars shall be not less than 5/16-in. thick and shall be of structural grade material, but need not be hardened. If hardened washers are required for use of high-strength bolts, the hardened washers shall be placed over the outer surface of the plate washer or bar.

   f. When A490 bolts over 1 in. in diameter are used in slotted or oversize holes in external plies, a single hardened washer conforming to ASTM F436, except with 5/16-in. minimum thickness, shall be used in lieu of the standard washer.

8. Long Grips

A307 bolts providing design strength, and for which the grip exceeds five diameters, shall have their number increased 1% for each additional 1/16-in. in the grip.

TABLE J3.6
Values of Spacing Increment C_1, in.

Nominal Dia. of Fastener	Oversize Holes	Slotted Holes		
		Perpendicular to Line of Force	Parallel to Line of Force	
			Short-slots	Long-slots[a]
$\leq \frac{7}{8}$	$\frac{1}{8}$	0	$\frac{3}{16}$	$1\frac{1}{2}d - \frac{1}{16}$
1	$\frac{3}{16}$	0	$\frac{1}{4}$	$1\frac{7}{16}$
$\geq 1\frac{1}{8}$	$\frac{1}{4}$	0	$\frac{5}{16}$	$1\frac{1}{2}d - \frac{1}{16}$

[a] When length of slot is less than maximum allowed in Table J3.5, C_1 may be reduced by the difference between the maximum and actual slot lengths.

9. Minimum Spacing

The distance between centers of standard, oversized or slotted fastener holes shall not be less than $2\frac{2}{3}$ times the nominal diameter of the fastener* nor less than that required by the following paragraph, if applicable.

Along a line of transmitted forces, the distance between centers of holes shall be not less than $3d$ when R_n is determined by J3-1a and J3-1b. Otherwise the distance between centers of holes shall be not less than the following:

a. For standard holes:

$$\frac{P}{\phi F_u t} + \frac{d_h}{2} \qquad (\text{J3-2})$$

where

$\phi = 0.75$
P = force transmitted by one fastener to the critical connected part, kips
F_u = specified minimum tensile strength of the critical connected part, ksi
t = thickness of the critical connected part, in.
d_h = diameter of standard size hole, in.

b. For oversized and slotted holes, the distance required for standard holes in subparagraph a, above, plus the applicable increment C_1 from Table J3.6, but the clear distance between holes shall not be less than one bolt diameter.

---

*A distance of $3d$ is preferred.

TABLE J3.7
Minimum Edge Distance, in.
(Center of Standard Hole[a] to Edge of Connected Part)

Nominal Rivet or Bolt Diameter (in.)	At Sheared Edges	At Rolled Edges of Plates, Shapes or Bars or Gas Cut Edges[b]
½	⅞	¾
⅝	1⅛	⅞
¾	1¼	1
⅞	1½[c]	1⅛
1	1¾[c]	1¼
1⅛	2	1½
1¼	2¼	1⅝
Over 1¼	1¾ × Diameter	1¼ × Diameter

[a]For oversized or slotted holes, see Table J3.8.
[b]All edge distances in this column may be reduced ⅛-in. when the hole is at a point where stress does not exceed 25% of the maximum design strength in the element.
[c]These may be 1¼ in. at the ends of beam connection angles.

10. Minimum Edge Distance

The distance from the center of a standard hole to an edge of a connected part shall be not less than the applicable value from Table J3.7 nor the value from Formula J3-3, as applicable.

Along a line of transmitted force, in the direction of the force, the distance from the center of a standard hole to the edge of the connected part shall be not less than $1\frac{1}{2}d$ when R_n is determined by J3-1a or J3-1b. Otherwise the edge distance shall be not less than

$$\frac{P}{\phi F_u t} \tag{J3-3}$$

where ϕ, P, F_u, t are defined in Sect. J3.9.

The distance from the center of an oversized or slotted hole to an edge of a connected part shall be not less than that required for a standard hole plus the applicable increment C_2 from Table J3.8.

11. Maximum Edge Distance and Spacing

The maximum distance from the center of any rivet or bolt to the nearest edge of parts in contact shall be 12 times the thickness of the connected part under consideration, but shall not exceed 6 in. Bolted joints in unpainted steel exposed to atmospheric corrosion require special limitations on pitch and edge distance, see Sect. E4.

J4. DESIGN SHEAR RUPTURE STRENGTH

The design strength for the limit state of rupture along a shear failure path in main members shall be taken as $\phi F_n A_{ns}$

TABLE J3.8
Values of Edge Distance Increment C_2, in.

Nominal Diameter of Fastener (in.)	Oversized Holes	Slotted Holes		
		Perpendicular to Edge		Parallel to Edge
		Short Slots	Long Slots[a]	
≤ 7/8	1/16	1/8		
1	1/8	1/8	3/4d	0
≤ 1 1/8	1/8	3/16		

[a]When length of slot is less than maximum allowable (see Table J3.5), C_2 may be reduced by one-half the difference between the maximum and actual slot lengths.

where

$\phi = 0.75$
$F_n = 0.6\,F_u$ (J4-1)
A_{ns} = net area subject to shear.

J5. CONNECTING ELEMENTS

This section applies to the design of connecting elements, such as stiffeners, gussets, angles and brackets and the panel zones of beam-to-column connections.

1. Eccentric Connections

Intersecting axially stressed members shall have their gravity axes intersect at one point, if practicable; if not, provision shall be made for bending and shearing stresses due to the eccentricity.

2. Design Strength of Connecting Elements

The design strength ϕR_n of welded, bolted and riveted connecting elements statically loaded in tension (e.g., splice and gusset plates) shall be the lower value obtained according to the limit states of yielding, fracture of the connecting element and block shear rupture.

a. For yielding of the connecting element:

$$\phi = 0.90$$
$$R_n = A_g F_y$$ (J5-1)

b. For fracture of the connecting element where $A_n \le 0.85A_g$:

$$\phi = 0.75$$
$$R_n = A_n F_u$$ (J5-2)

c. For block shear rupture

Block shear is a failure mode in which the resistance is determined by the sum of the shear strength on a failure path(s) and the tensile strength on a perpendicular segment. When ultimate strength on the net section is used to determine the resistance on one segment, yielding on the gross section shall be used on the perpendicular segment; $\phi = 0.75$. Design strength shall be the larger of the two failure modes.

At beam end connections where the top flange is coped, and in similar situations, failure can occur by shear along a plane through the fasteners, acting in combination with tension along a perpendicular plane. In such cases, the ultimate strength on the net section (shear or tension) shall be used to determine the resistance of one segment and yielding on the gross section (shear or tension) shall be used for the perpendicular segment, with $\phi = 0.75$ for both. By alternating the choice of which segment resistance is based on ultimate strength, two possible values of design strength are obtained. The larger value shall be taken as the design strength.

For all other connecting elements, the design strength ϕR_n shall be determined for the applicable limit state to insure that the design strength is equal to or greater than the required strength, where R_n is the nominal strength appropriate to the geometry and type of loading on the connecting element. The shear limit state is governed by:

$$\phi = 0.80$$
$$R_n = 0.7A_g F_y \tag{J5-3}$$

J6. FILLERS

In welded construction, any filler ¼-in. or more in thickness shall extend beyond the edges of the splice plate and shall be welded to the part on which it is fitted with sufficient weld to transmit the splice plate load, applied at the surface of the filler. The welds joining the splice plate to the filler shall be sufficient to transmit the splice plate load and shall be long enough to avoid overloading the filler along the toe of the weld. Any filler less than ¼-in. thick shall have its edges made flush with the edges of the splice plate and the weld size shall be the sum of the size necessary to carry the splice plus the thickness of the filler plate.

When bolts or rivets carrying loads pass through fillers thicker than ¼-in., except in connections designed as slip-critical connections, the fillers shall be extended beyond the splice material and the filler extension shall be secured by enough bolts or rivets to distribute the total stress in the member uniformly over the combined section of the member and the filler, or an equivalent number of fasteners shall be included in the connection.

J7. SPLICES

Groove-welded splices in plate girders and beams shall develop the full strength of the smaller spliced section. Other types of splices in cross sections of plate girders and beams shall develop the strength required by the forces at the point of splice.

J8. BEARING STRENGTH

The strength of surfaces in bearing is ϕR_n, where $\phi = 0.75$ and R_n is defined below for various types of bearing.

1. Milled or Finished Surfaces

For milled surfaces, pins in reamed, drilled or bored holes, and ends of fitted bearing stiffeners,

$$R_n = 2.0 \, F_y \, A_{pb} \tag{J8-1}$$

where

F_y = specified minimum yield stress, ksi
A_{pb} = projected bearing area, in.^{2}

2. Expansion Rollers and Rockers

For expansion rollers and rockers,

$$R_n = 1.5 \, (F_y - 13)\ell d/20 \tag{J8-2}$$

where

d = diameter, in.
ℓ = length of bearing, in.

J9. COLUMN BASES AND BEARING ON CONCRETE

Proper provision shall be made to transfer the column loads and moments to the footings and foundations.

In the absence of code regulations, design bearing loads on concrete may be taken as $\phi_c P_p$:

On the full area of a concrete support .... $P_p = 0.85 f_c' A_1$
On less than the full area of a concrete
  support .... $P_p = 0.85 f_c' A_1 \sqrt{A_2/A_1}$

where

ϕ_c = 0.60
A_1 = area of steel concentrically bearing on a concrete support, in.
A_2 = maximum area of the portion of the supporting surface that is geometrically similar to and concentric with the loaded area, in.^{2}

$\sqrt{A_2/A_1} \le 2$

J10. ANCHOR BOLTS AND EMBEDMENTS

1. Anchor Bolts

Anchor bolts shall be designed to provide resistance to all design conditions on completed structures of tension and shear at the bases of columns, including the net tensile components of any bending moment which may result from column base restraint.

2. Embedments

The concrete structure shall be designed to safely support the loads from the embedment with an appropriate factor of safety to ensure that the embedment strength is not reduced as a result of local or gross failure of the supporting concrete structure.

The strength and design of the structural steel elements of the embedment shall be in accordance with this Specification. Bolts, studs and bars functioning as embedment anchors resisting tensile loads shall be designed to transfer the design load to the concrete by means of bond, shear, bearing or a combination thereof.

Shear loads shall be considered to be transmitted by the embedment to the concrete by either shear lugs or shear friction.

The friction force V, kips, to resist shear shall be computed as:

$$V = \mu P \tag{J10-1}$$

where

P = normal force, kips
μ = coefficient of friction

The coefficient of friction μ shall be 0.90 for concrete placed against as-rolled steel with contact plane a full plate thickness below the concrete surface; 0.70 for concrete or grout placed against as-rolled steel with contact plane coincidental with the concrete surface; 0.55 for grouted conditions with the contact plane between grout and as-rolled steel above the concrete surface.

3. Prestressed Embedments

Anchorage to concrete structures by means of post-tensioned high-strength steel members is permissible. The material and the design requirements of the high-strength steel members and associated anchorage, as well as the fabrication and installation procedures, shall conform to the appropriate provisions of applicable codes.

CHAPTER K.
STRENGTH DESIGN CONSIDERATIONS

This chapter covers additional member strength design considerations related to introduction of concentrated forces, ponding, torsion, and fatigue.

K1. WEBS AND FLANGES WITH CONCENTRATED FORCES

1. Design Basis

Members with concentrated loads applied normal to *one flange* and symmetric to the web shall have a flange and web design strength sufficient to satisfy the local flange bending, web yielding strength, web crippling and sidesway web buckling criteria of Sects. K1.2, K1.3, K1.4 and K1.5. Members with concentrated loads applied to *both flanges* shall have a web design strength sufficient to satisfy the web yielding, web crippling and column web buckling criteria of Sects. K1.3, K1.4 and K1.6.

Where pairs of stiffeners are provided on opposite sides of the web, at concentrated loads, and extend at least half the depth of the member, Sects K1.2 and K1.3 need not be checked.

For column webs subject to high shears, see Sect. K1.7; for bearing stiffeners, see Sect. K1.8.

2. Local Flange Bending

The flange design strength in bending due to a tensile load shall be ϕR_n, kips

where

$$\phi = 0.90$$
$$R_n = 6.25 t_f^2 F_{yf} \tag{K1-1}$$

F_{yf} = specified minimum yield stress of flange, ksi
t_f = thickness of the loaded flange, in.

If the length of loading measured across the member flange is less than $0.15b$, where b is the member flange width, Formula K1-1 need not be checked.

3. Local Web Yielding

The design strength of the web at the toe of the fillet under concentrated loads shall be ϕR_n, kips, where $\phi = 1.0$ and R_n is determined as follows:

a. When the force to be resisted is a concentrated load producing tension or compression, applied at a distance from the member end that is greater than the depth of the member,

$$R_n = (5k + N)F_{yw}t_w \tag{K1-2}$$

b. When the force to be resisted is a concentrated load applied at or near the end of the member,

$$R_n = (2.5k + N)F_{yw}t_w \tag{K1-3}$$

where

F_{yw} = specified minimum yield stress of the web, ksi
N = length of bearing, in.
k = distance from outer face of flange to web toe of fillet, in.
t_w = web thickness, in.

4. Web Crippling

For unstiffened portions of webs of members under concentrated loads, the design compressive strength shall be ϕR_n, kips, where $\phi = 0.75$ and the nominal strength R_n is determined as follows:

a. When the concentrated load is applied at a distance not less than $d/2$ from the end of the member:

$$R_n = 135t_w^2\left[1 + 3\left(\frac{N}{d}\right)\left(\frac{t_w}{t_f}\right)^{1.5}\right]\sqrt{F_{yw}t_f/t_w} \tag{K1-4}$$

b. When the concentrated load is applied less than a distance $d/2$ from the end of the member:

$$R_n = 68t_w^2\left[1 + 3\left(\frac{N}{d}\right)\left(\frac{t_w}{t_f}\right)^{1.5}\right]\sqrt{F_{yw}t_f/t_w} \tag{K1-5}$$

where

d = overall depth of the member, in.
t_f = flange thickness, in.

If stiffeners are provided and extend at least one-half the web depth, Formulas K1-4 and K1-5 need not be checked.

5. Sidesway Web Buckling

For webs of members with flanges not restrained against relative movement by stiffeners or lateral bracing and subject to concentrated compressive loads, the design compressive strength shall be ϕR_n, kips, where $\phi = 0.85$ and the nominal strength R_n is determined as follows:

a. If the loaded flange is restrained against rotation and $(d_c/t_w)/(\ell/b_f)$ is less than 2.3:

$$R_n = \frac{12{,}000t_w^3}{h}\left[1 + 0.4\left(\frac{d_c/t_w}{\ell/b_f}\right)^3\right] \tag{K1-6}$$

b. If the loaded flange is *not* restrained against rotation and $(d_c/t_w)/(\ell/b_f)$ is less than 1.7:

$$R_n = \frac{12{,}000t_w^3}{h}\left[0.4\left(\frac{d_c/t_w}{\ell/b_f}\right)^3\right] \tag{K1-7}$$

where

ℓ = largest laterally unbraced length along either flange at the point of load, in.
b_f = flange width, in.
t_w = web thickness, in.
$d_c = d - 2k$ = web depth clear of fillets, in.

Formulas K1-6 and K1-7 need not be checked providing $(d_c/t_w)/(\ell/b_f)$ exceeds 2.3 or 1.7, respectively, or for webs subject to distributed load.

If a concentrated load is located at a point where the web flexural stress due to factored load is below yielding, 24,000 may be used in lieu of 12,000 in Formulas K1-6 and K1-7.

6. Compression Buckling of the Web

For unstiffened portions of webs of members under concentrated loads to both flanges, the design compressive strength shall be ϕR_n, kips, where

$$\phi = 0.90$$
$$R_n = \frac{4,100 \, t_w^3 \sqrt{F_{yw}}}{d_c} \tag{K1-8}$$

R_n may be exceeded provided that a transverse stiffener or pair of stiffeners is attached to the web to satisfy Sect. F3.

7. Compression Members with Web Panels Subject to High Shear

For compression members subject to high shear stress in the web, the design web shear strength shall be ϕR_v, kips, where $\phi = 0.90$ and R_v is determined as follows:

    a. For $P_u \leq 0.75 P_n$:
$$R_v = 0.7 \, F_y d_c t_w \tag{K1-9}$$
    b. For $P_u > 0.75 P_n$:
$$R_v = 0.7 \, F_y d_c t_w \, [1.9 - 1.2(P_u/P_n)] \tag{K1-10}$$

where

    P_u = required axial strength, kips
    P_n = nominal axial strength, kips

8. Stiffener Requirements for Concentrated Loads

When required, stiffeners shall be placed in pairs at unframed ends of beams and girders. They shall be placed in pairs at points of concentrated load on the interior of beams, girders or columns if the load exceeds the nominal strength ϕR_n as determined from Sects. K1.2 through K1.6, as applicable.

If the concentrated load, tension or compression exceeds the criteria for ϕR_n of Sects. K1.2 or K1.3, respectively, stiffeners need not extend more than one-half the depth of the web, except as follows:

If concentrated compressive loads are applied to the members and if the load exceeds the compressive strength of the web ϕR_n given in Sects. K1.4 or K1.6, the stiffeners shall be designed as axially compressed members (columns) in accordance with requirements of Sect. E2 with an effective length equal to $0.75h$, a cross section composed of two stiffeners and a strip of the web having a width of $25t_w$ at interior stiffeners and $12t_w$ at the ends of members.

When the load normal to the flange is tensile, the stiffeners shall be welded to the loaded flange. When the load normal to the flange is compressive, the stiffeners shall either bear on or be welded to the loaded flange.

K2. PONDING

The roof system shall be investigated by structural analysis to assure adequate strength and stability under ponding conditions, unless the roof surface is provided with sufficient slope toward points of free drainage or adequate individual drains to prevent the accumulation of rainwater.

The roof system shall be considered stable and no further investigation is needed if:

$$C_p + 0.9C_s \leq 0.25 \qquad \text{(K2-1)}$$

$$\text{and } I_d \geq 25 \ (S^4)10^{-6} \qquad \text{(K2-2)}$$

where

$$C_p = \frac{32L_s L_p^4}{10^7 I_p}$$

$$C_s = \frac{32SL_s^4}{10^7 I_s}$$

L_p = column spacing in direction of girder (length of primary members), ft
L_s = column spacing perpendicular to direction of girder (length of secondary members), ft
S = spacing of secondary members, ft
I_p = moment of inertia of primary members, in.^{4}
I_s = moment of inertia of secondary members, in.^{4}
I_d = moment of inertia of the steel deck supported on secondary members, in.^{4} per ft

For trusses and steel joists, the moment of inertia I_s shall be decreased 15% when used in the above equation. A steel deck shall be considered a secondary member when it is directly supported by the primary members.

See Appendix K2 for an alternate determination of flat roof framing stiffness.

K3. TORSION

For limiting values of normal and shear stress, due to torsion and other loading, see Sect. H2. Some constrained local yielding may be permitted.

K4. FATIGUE

Few members or connections in conventional buildings need to be designed for fatigue, since most load changes in such structures occur only a small number of times or produce only minor stress fluctuations. The occurrence of full design wind or earthquake loads is too infrequent to warrant consideration in fatigue design. However, crane runways and supporting structures for machinery and equipment are often subject to fatigue loading conditions.

Members and their connections subject to fatigue loading shall be proportioned in accordance with the provisions of Appendix K4 for service loads.

CHAPTER L.
SERVICEABILITY DESIGN CONSIDERATIONS

This chapter is intended to provide design guidance for serviceability considerations not covered elsewhere. Serviceability is a state in which the function of a building, its appearance, maintainability, durability and comfort of its occupants are preserved under normal usage.

The general design requirement for serviceability is given in Sect. A5.4. Limiting values of structural behavior to ensure serviceability (e.g., maximum deflections, accelerations, etc.) shall be chosen with due regard to the intended function of the structure.

Where necessary, serviceability shall be checked using realistic loads for the appropriate serviceability limit state.

L1. CAMBER

If any special camber requirements are necessary to bring a loaded member into proper relation with the work of other trades, as for the attachment of runs of sash, the requirements shall be set forth in the design documents.

Beams and trusses detailed without specified camber shall be fabricated so that after erection any camber due to rolling or shop assembly shall be upward. If camber involves the erection of any member under a preload, this shall be noted in the design documents.

L2. EXPANSION AND CONTRACTION

Adequate provision shall be made for expansion and contraction appropriate to the service conditions of the structure.

L3. DEFLECTIONS, VIBRATION AND DRIFT

1. Deflections

Deformations in structural members and structural systems due to service loads shall not impair the serviceability of the structure.

2. Vibration

Vibration shall be considered in designing beams and girders supporting large areas free of partitions or other sources of damping, where vibration due to pedestrian traffic or other sources within the building might not be acceptable.

3. Drift

Lateral deflection or drift of structures due to code-specified wind or seismic loads shall not cause collision with adjacent structures nor exceed the limiting values of such drifts which may be specified or appropriate.

L4. CONNECTION SLIP

For the design of slip-resistant connections see Sect. J3.5.

L5. CORROSION

When appropriate, structural components shall be designed to tolerate corrosion or shall be protected against corrosion that may impair the strength or serviceability of the structure.

CHAPTER M.
FABRICATION, ERECTION AND
QUALITY CONTROL

M1. SHOP DRAWINGS

Shop drawings giving complete information necessary for the fabrication of the component parts of the structure, including the location, type and size of all welds, bolts and rivets, shall be prepared in advance of the actual fabrication. These drawings shall clearly distinguish between shop and field welds and bolts and shall clearly identify slip-critical high-strength bolted connections.

Shop drawings shall be made in conformity with the best practice and with due regard to speed and economy in fabrication and erection.

M2. FABRICATION

1. Cambering, Curving and Straightening

Local application of heat or mechanical means may be used to introduce or correct camber, curvature and straightness. The temperature of heated areas, as measured by approved methods, shall not exceed 1,100°F for A514 steel nor 1,200°F for other steels.

2. Thermal Cutting

Thermal cutting shall preferably be done by machine. Thermally cut edges which will be subjected to substantial stress, or which are to have weld metal deposited on them, shall be reasonably free from notches or gouges; notches or gouges not more than ³⁄₁₆-in. deep will be permitted. Notches or gouges greater than ³⁄₁₆-in. deep that remain from cutting shall be removed by grinding or repaired by welding. All re-entrant corners shall be shaped to provide a smooth transition. If a specific contour is required, it must be shown in the design documents.

3. Planing of Edges

Planing or finishing of sheared or thermally cut edges of plates or shapes will not be required unless specifically called for in the design documents or included in a stipulated edge preparation for welding.

4. Welded Construction

The technique of welding, the workmanship, appearance and quality of welds and the methods used in correcting nonconforming work shall be in accordance with "Sect. 3—Workmanship" and "Sect. 4—Technique" of the AWS *Structural Welding Code— Steel*, D1.1.

5. Bolted Construction

All parts of bolted members shall be pinned or bolted and rigidly held together while assembling. Use of a drift pin in bolt holes during assembling shall not distort the metal or enlarge the holes. Poor matching of holes shall be cause for rejection.

If the thickness of the material is not greater than the nominal diameter of the bolt plus ⅛-in., the holes may be punched. If the thickness of the material is greater than the nominal diameter of the bolt plus ⅛-in., the holes shall be either drilled or sub-punched and reamed. The die for all sub-punched holes, and the drill for all sub-drilled holes, shall be at least 1/16-in. smaller than the nominal diameter of the bolt. Holes in A514 steel plates over ½-in. thick shall be drilled.

Fully inserted finger shims, with a total thickness of not more than ¼-in. within a joint, may be used in joints without changing the design load (based upon hole type) for the design of connections. The orientation of such shims is independent of the direction of application of the load.

The use of high-strength bolts shall conform to the requirements of the RCSC Load and Resistance Factor Design Specification for Structural Joints Using ASTM A325 or A490 Bolts.

6. Compression Joints

Compression joints which depend on contact bearing as part of the splice capacity shall have the bearing surfaces of individual fabricated pieces prepared by milling, sawing or other suitable means.

7. Dimensional Tolerances

Dimensional tolerances shall be as permitted in the *Code of Standard Practice* of the American Institute of Steel Construction, Inc.

8. Finishing of Column Bases

Column bases and base plates shall be finished in accordance with the following requirements:

    a. Steel bearing plates 2 in. or less in thickness may be used without milling, provided a satisfactory contact bearing is obtained. Steel bearing plates over 2 in. but not over 4 in. in thickness may be straightened by pressing or, if presses are not available, by milling for all bearing surfaces (except as noted in subparagraphs b and c of this section), to obtain a satisfactory contact bearing. Steel bearing plates over 4 in. in thickness shall be milled for all bearing surfaces (except as noted in subparagraphs b and c of this section).

    b. Bottom surfaces of bearing plates and column bases which are grouted to insure full bearing contact on foundations need not be milled.

    c. Top surfaces of bearing plates need not be milled when full-penetration welds are provided between the column and the bearing plate.

M3. SHOP PAINTING

1. General Requirements

Shop painting and surface preparation shall be in accordance with the provisions of the *Code of Standard Practice* of the American Institute of Steel Construction.

Unless otherwise specified, steelwork which will be concealed by interior building finish or will be in contact with concrete need not be painted. Unless specifically excluded, all other steelwork shall be given one coat of shop paint.

2. Inaccessible Surfaces

Except for contact surfaces, surfaces inaccessible after shop assembly shall be cleaned and painted prior to assembly, if required by the design documents.

3. Contact Surfaces

Paint is permitted unconditionally in bearing-type connections. For slip-critical connections where the design is based on special faying surface conditions in accordance with Sect. J3.5, shop contact surfaces shall be cleaned prior to assembly in accordance with the provisions of the *Code of Standard Practice* of the American Institute of Steel Construction, but shall not be painted. Field contact surfaces and surfaces meeting the requirements of Sect. J3.5 shall be shop cleaned in accordance with the design documents, except as provided by Sect. M3.5.

4. Finished Surfaces

Machine-finished surfaces shall be protected against corrosion by a rust-inhibiting coating that can be removed prior to erection, or which has characteristics that make removal prior to erection unnecessary.

5. Surfaces Adjacent to Field Welds

Unless otherwise specified in the design documents, surfaces within 2 in. of any field weld location shall be free of materials that would prevent proper welding or produce toxic fumes during welding.

M4. ERECTION

1. Alignment of Column Bases

Column bases shall be set level and to correct elevation with full bearing on concrete or masonry.

2. Bracing

The frame of steel skeleton buildings shall be carried up true and plumb within the limits defined in the *Code of Standard Practice* of the American Institute of Steel Construction. Temporary bracing shall be provided, in accordance with the requirements of the *Code of Standard Practice*, wherever necessary to take care of all loads to which the structure may be subjected, including equipment and the operation of same. Such bracing shall be left in place as long as may be required for safety.

3. Alignment

No permanent bolting or welding shall be performed until as much of the structure as will be stiffened thereby has been properly aligned.

4. Fit of Column Compression Joints

Lack of contact bearing not exceeding a gap of $\frac{1}{16}$-in., regardless of the type of splice used (partial-penetration groove welded or bolted), shall be acceptable. If the gap exceeds $\frac{1}{16}$-in., but is less than $\frac{1}{4}$-in., and if an engineering investigation shows that sufficient contact area does not exist, the gap shall be packed out with non-tapered steel shims. Shims need not be other than mild steel, regardless of the grade of the main material.

5. Field Welding

Any shop paint on surfaces adjacent to joints to be field welded shall be wire brushed to reduce the paint film to a minimum.

Field welding of attachments to installed embedments in contact with concrete shall be done in such a manner as to avoid excessive thermal expansion of the embedment which could result in spalling or cracking of the concrete or excessive stress in the embedment anchors.

6. Field Painting

Responsibility for touch-up painting, cleaning and field painting shall be allocated in accordance with accepted local practices, and this allocation shall be set forth explicitly in the design documents.

7. Field Connections

As erection progresses, the work shall be securely bolted or welded to take care of all dead load, wind and erection stresses.

M5. QUALITY CONTROL

The fabricator shall provide quality control procedures to the extent that he deems necessary to assure that all work is performed in accordance with this Specification. In addition to the fabricator's quality control procedures, material and workmanship at all times may be subject to inspection by qualified inspectors representing the purchaser. If such inspection by representatives of the purchaser will be required, it shall be so stated in the design documents.

1. Cooperation

As far as possible, all inspection by representatives of the purchaser shall be made at the fabricator's plant. The fabricator shall cooperate with the inspector, permitting access for inspection to all places where work is being done. The purchaser's inspector shall schedule his work for minimum interruption to the work of the fabricator.

2. Rejections

Material or workmanship not in reasonable conformance with the provisions of this Specification may be rejected at any time during the progress of the work. The fabricator shall receive copies of all reports furnished to the purchaser by the inspection agency.

3. Inspection of Welding

The inspection of welding shall be performed in accordance with the provisions of Sect. 6 of the AWS *Structural Welding Code—Steel*, D1.1.

When visual inspection is required to be performed by AWS certified welding inspectors, it shall be so specified in the design documents.

When nondestructive testing is required, the process, extent and standards of acceptance shall be clearly defined in the design documents.

4. Inspection of Slip-critical High-strength Bolted Connections

The inspection of slip-critical high-strength bolted connections shall be in accordance with the provisions of the RCSC Load and Resistance Factor Design Specification for Structural Joints Using ASTM A325 or A490 Bolts.

5. Identification of Steel

The fabricator shall be able to demonstrate by a written procedure and by actual practice a method of material application and identification, visible at least through the "fit-up" operation, of the main structural elements of a shipping piece.

The identification method shall be capable of verifying proper material application as it relates to:

1. Material specification designation
2. Heat number, if required
3. Material test reports for special requirements.

APPENDIX B.
DESIGN REQUIREMENTS

B5. LOCAL BUCKLING

3. Slender Compression Elements

Axially loaded members containing elements subject to compression which have a width-thickness ratio in excess of the applicable λ_r as stipulated in Sect. B5.1 shall be proportioned according to this Appendix. Flexural members with slender compression elements shall be designed in accordance with Appendix F1.7. Rolled flexural members with proportions not covered by Appendix F1.7 shall be designed in accordance with this Appendix.

a. Unstiffened Compression Elements

The design strength of unstiffened compression elements whose width-thickness ratio exceeds the applicable limit λ_r as stipulated in Sect. B5.1 shall be subject to a reduction factor Q_s. The value of Q_s shall be determined by Formulas A-B5-1 through A-B5-6, as applicable. When such elements comprise the compression flange of a flexural member, the maximum required bending stress shall not exceed $\phi_b F_y Q_s$, where $\phi_b = 0.90$. The design strength of axially loaded compression members shall be modified by the appropriate reduction factor Q_s, as provided in paragraph c.

For single angles:

When $76.0/\sqrt{F_y} < b/t < 155/\sqrt{F_y}$:

$$Q_s = 1.340 - 0.00447(b/t)\sqrt{F_y} \tag{A-B5-1}$$

When $b/t \geq 155/\sqrt{F_y}$:

$$Q_s = 15{,}500/[F_y(b/t)^2] \tag{A-B5-2}$$

For angles or plates projecting from columns or other compression members, and for projecting elements of compression flanges of girders:

When $95.0/\sqrt{F_y} < b/t < 176/\sqrt{F_y}$:

$$Q_s = 1.415 - 0.00437(b/t)\sqrt{F_y} \qquad \text{(A-B5-3)}$$

When $b/t \geq 176/\sqrt{F_y}$:

$$Q_s = 20,000/[F_y(b/t)^2] \qquad \text{(A-B5-4)}$$

For stems of tees:

When $127/\sqrt{F_y} < b/t < 176/\sqrt{F_y}$:

$$Q_s = 1.908 - 0.00715(b/t)\sqrt{F_y} \qquad \text{(A-B5-5)}$$

When $b/t \geq 176/\sqrt{F_y}$:

$$Q_s = 20,000/[F_y(b/t)^2] \qquad \text{(A-B5-6)}$$

where

b = width of unstiffened compression element as defined in Sect. B5.1, in.
t = thickness of unstiffened element, in.
F_y = specified minimum yield stress, ksi

Unstiffened elements of tees whose proportions exceed the limits of Sect. B5.1 shall conform to the limits given in Table A-B5.1.

b. Stiffened Compression Elements

When the width-thickness ratio of uniformly compressed stiffened elements (except perforated cover plates) exceeds the limit λ_r stipulated in Sect. B5.1, a reduced effective width b_e shall be used in computing the design properties of the section containing the element.

TABLE A-B5.1
Limiting Proportions for Tees

Shape	Ratio of Full Flange Width to Profile Depth	Ratio of Flange Thickness to Web or Stem Thickness
Built-up tees	≥ 0.50	≥ 1.25
Rolled tees	≥ 0.50	≥ 1.10

i. For flanges of square and rectangular sections of uniform thickness:

$$b_e = \frac{326t}{\sqrt{f}}\left[1 - \frac{64.9}{(b/t)\sqrt{f}}\right] \leq b \qquad \text{(A-B5-7)}$$

ii. For other uniformly compressed elements:

$$b_e = \frac{326t}{\sqrt{f}}\left[1 - \frac{57.2}{(b/t)\sqrt{f}}\right] \leq b \qquad \text{(A-B5-8)}$$

where

b = actual width of a stiffened compression element, as defined in Sect. B5.1, in.
b_e = reduced width, in.
t = element thickness, in.
f = computed elastic compressive stress in the stiffened elements, based on the design properties as specified in Sect. C of this Appendix, ksi. If unstiffened elements are included in the total cross section, f for the stiffened element must be such that the maximum compressive stress in the unstiffened element does not exceed $\phi_c F_{cr}$ as defined in Appendix B5.3c with $Q = Q_s$ and $\phi_c = 0.85$, or $\phi_b F_y Q_s$ with $\phi_b = 0.90$, as applicable.

iii. For axially loaded circular sections:

Members with diameter-to-thickness ratios D/t greater than $3,300/F_y$, but having a diameter-to-thickness ratio of less than $13,000/F_y$;

$$Q = \frac{1,100}{F_y(D/t)} + \frac{2}{3} \qquad \text{(A-B5-9)}$$

where

D = outside diameter, in.
t = wall thickness, in.

c. Design Properties

Properties of sections shall be determined using the full cross section, except as follows:

In computing the moment of inertia and elastic section modulus of flexural members, the effective width of uniformly compressed stiffened elements, as determined in Appendix B5.3b, shall be used in determining effective cross-sectional properties.

For unstiffened elements of the cross section, Q_s is determined from Appendix B5.3a. For stiffened elements of the cross section

$$Q_a = \frac{\text{effective area}}{\text{actual area}} \qquad \text{(A-B5-10)}$$

where the effective area is equal to the summation of the effective areas of the cross section.

For axially loaded compression members the gross cross-sectional area and the radius of gyration r shall be computed on the basis of the actual cross section. However, when $\lambda_c \sqrt{Q} \leq 1.5$, the critical stress F_{cr} shall be determined by

where

$$F_{cr} = Q(0.658^{Q\lambda_c^2}) F_y \qquad \text{(A-B5-11)}$$

$$Q = Q_s Q_a \qquad \text{(A-B5-12)}$$

i. Cross sections composed entirely of unstiffened elements, $Q = Q_s (Q_a = 1.0)$
ii. Cross sections composed entirely of stiffened elements, $Q = Q_a (Q_s = 1.0)$
iii. Cross sections composed of both stiffened and unstiffened elements, $Q = Q_s Q_a$

When $\lambda_c \sqrt{Q} > 1.5$, the critical stress F_{cr} shall be determined by

$$F_{cr} = \left[\frac{0.877}{\lambda_c^2}\right] F_y \qquad \text{(A-B5-13)}$$

APPENDIX E.
COLUMNS AND OTHER COMPRESSION MEMBERS

E3. FLEXURAL-TORSIONAL BUCKLING

This section applies to the strength of singly symmetric and unsymmetric columns fo the limiting states of flexural-torsional and torsional buckling.

Torsional buckling of symmetric shapes and flexural-torsional buckling of unsymmetric shapes are modes of buckling usually not considered in the design of hot-rolled columns. (Generally, either these modes do not govern or the critical load differs very little from the weak-axis planar buckling load.) Such buckling modes may, however, control the capacity of columns made from relatively thin plate elements and of unsymmetric columns.

The strength of compression members determined by the limit states of torsional and flexural-torsional buckling is $\phi_c P_n$, where

$\phi_c = 0.85$
P_n = nominal resistance in compression, kips
$\quad = A_g F_{cr}$ (A-E3-1)
A_g = gross area of cross section, in.^{2}
Q = 1.0 for elements meeting the width-thickness ratios λ_r of Sect. B5.1
$\quad = Q_s Q_a$ for elements not meeting the width-thickness ratios λ_r of Sect. B5.1
$\quad\quad$ and determined in accordance with the provisions of Appendix B5.3

The nominal critical stress F_{cr} is determined as follows:

a. For $\lambda_e \sqrt{Q} \le 1.5$:

$$F_{cr} = Q(0.658^{Q\lambda_e^2})F_y \qquad (A\text{-}E3\text{-}2)$$

b. For $\lambda_e \sqrt{Q} > 1.5$:

$$F_{cr} = \left[\frac{0.877}{\lambda_e^2}\right]F_y \qquad (A\text{-}E3\text{-}3)$$

where

$$\lambda_e = \sqrt{F_y/F_e} \qquad \text{(A-E3-4)}$$

F_y = specified minimum yield stress of steel, ksi

The critical torsional or flexural-torsional elastic buckling stress F_e is determined as follows:

a. For doubly symmetric shapes the critical torsional elastic buckling stress is

$$F_e = \left[\frac{\pi^2 E C_w}{(K_z L)^2} + GJ\right]\frac{1}{I_x + I_y} \qquad \text{(A-E3-5)}$$

b. For singly symmetric shapes, where y is the axis of symmetry, the critical flexural-torsional elastic buckling stress is

$$F_e = \frac{F_{ey} + F_{ez}}{2H}\left(1 - \sqrt{1 - \frac{4 F_{ey} F_{ez} H}{(F_{ey} + F_{ez})^2}}\right) \qquad \text{(A-E3-6)}$$

c. For unsymmetric shapes, the critical flexural-torsional elastic buckling stress F_e is the lowest root of the cubic equation

$$(F_e - F_{ex})(F_e - F_{ey})(F_e - F_{ez}) - F_e^2(F_e - F_{ey})(x_o/\overline{r}_o)^2 - F_e^2 (F_e - F_{ex})(y_o/\overline{r}_o)^2 = 0 \qquad \text{(A-E3-7)}$$

where

K_z = effective length factor for torsional buckling
E = modulus of elasticity, ksi
G = shear modulus, ksi
C_w = warping constant, in.[6]
J = torsional constant, in.[4]
I_x, I_y = moment of inertia about the principal axes, in.[4]
x_o, y_o = coordinates of shear center with respect to the centroid, in.

$$\overline{r}_o^2 = x_o^2 + y_o^2 + \frac{I_x + I_y}{A} \qquad \text{(A-E3-8)}$$

$$H = 1 - \left(\frac{x_o^2 + y_o^2}{\overline{r}_o^2}\right) \qquad \text{(A-E3-9)}$$

$$F_{ex} = \frac{\pi^2 E}{(K_x L/r_x)^2} \qquad \text{(A-E3-10)}$$

$$F_{ey} = \frac{\pi^2 E}{(K_y L/r_y)^2} \qquad \text{(A-E3-11)}$$

$$F_{ez} = \left(\frac{\pi^2 E C_w}{(K_z L)^2} + GJ\right)\frac{1}{A\overline{r}_o^2} \qquad \text{(A-E3-12)}$$

where

A = cross-sectional area of member, in.[2]
L = unbraced length, in.
K_x, K_y = effective length factors in x and y directions
r_x, r_y = radii of gyration about the principal axes, in.

APPENDIX F.
BEAMS AND OTHER FLEXURAL
MEMBERS

F1. DESIGN FOR FLEXURE

7. Nominal Flexural Strength

Table A-F1.1 provides a tabular summary of Formulas F1-3 through F1-16 for determining the nominal flexural strength of beams and girders. For slenderness parameters of cross sections not included in Table A-F1.1, see Appendix B5.3.

The nominal flexural strength M_n is the lowest value obtained according to the limit states of: (a) lateral-torsional buckling (LTB); (b) flange local buckling (FLB); and (c) web local buckling (WLB).

The nominal flexural strength M_n shall be determined as follows for each limit state:

For $\lambda \leq \lambda_p$:

$$M_n = M_p \qquad \text{(A-F1-1)}$$

For $\lambda_p < \lambda \leq \lambda_r$:

For the limit state of lateral-torsional buckling:

$$M_n = C_b \left[M_p - (M_p - M_r)\left(\frac{\lambda - \lambda_p}{\lambda_r - \lambda_p} \right) \right] \leq M_p \qquad \text{(A-F1-2)}$$

For the limit states of flange and web local buckling:

$$M_n = M_p - (M_p - M_r)\left(\frac{\lambda - \lambda_p}{\lambda_r - \lambda_p} \right) \qquad \text{(A-F1-3)}$$

For $\lambda > \lambda_r$:
    For the limit state of lateral-torsional buckling and for flange local buckling:

$$M_n = M_{cr} = S \, F_{cr} \qquad \text{(A-F1-4)}$$

For λ of the flange $> \lambda_r$ in shapes not included in Table A-F1.1, see Appendix B5.3.
For λ of the web $> \lambda_r$, see Appendix G.

For all slenderness parameters, the flexural strength of hybrid sections is limited by the elastic buckling strength of the homogeneous sections.

The terms used in the above equations are:

M_n = nominal flexural strength, kip-in.
M_p = plastic moment, kip-in.
M_{cr} = buckling moment, kip-in.
M_r = limiting buckling moment (equal to M_{cr} when $\lambda = \lambda_r$), kip-in.
λ = controlling slenderness parameter
   = minor axis slenderness ratio L_b/r_y for lateral-torsional buckling
   = flange width-thickness ratio b/t for flange local buckling as defined in Sect. B5.1
   = web depth-thickness ratio h/t_w for web local buckling as defined in Sect. B5.1
λ_p = largest value of λ for which $M_n = M_p$
λ_r = largest value of λ for which buckling is inelastic
F_{cr} = critical stress, ksi
C_b = $1.75 + 1.05(M_1/M_2) + 0.3(M_1/M_2)^2 \le 2.3$ where M_1 is the smaller and M_2 the larger end-moment in the unbraced segment of the beam; M_1/M_2 is positive when the moments cause reverse curvature.
S = section modulus, in.^{3}
L_b = laterally unbraced length, in.
r_y = radius of gyration about minor axis, in.

The applicable limit states and equations for M_p, M_r, F_{cr}, λ, λ_p and λ_r are given in Table A-F1.1 for the shapes covered in this Appendix. The terms used in the table are:

A = cross-sectional area, in.^{2}
F_r = compressive residual stress in flange
   = 10 ksi for rolled shapes
   = 16.5 ksi for welded shapes
F_y = specified minimum yield strength, ksi
F_{yf} = yield strength of the flange, ksi
F_{yw} = yield strength of the web, ksi
J = torsional constant, in.^{4}
R_e = see Appendix G2
$(S_x)_{eff}$ = effective section modulus about major axis, in.^{3}
S_{xc} = section modulus of the outside fiber of the compression flange, in.^{3}
S_{xt} = section modulus of the outside fiber of the tension flange, in.^{3}
Z = plastic section modulus, in.^{3}
b = flange width, in.
d = overall depth, in.
h_c = twice the distance from the neutral axis to the inside face of the compression flange less the fillet or corner radius, in.
t_f = flange thickness, in.
t_w = web thickness, in.

TABLE A-F1.1
Nominal Strength Parameters

Shape	Plastic Moment M_p	Limit State of Buckling	Limiting Buckling Moment M_r
Channels and doubly and singly symmetric I-shaped beams (including hybrid beams) bent about major axis	$F_y Z_x$ (e)	LTB Doubly symmetric members and channels	$(F_{yw} - F_r)S_x$
		LTB Singly symmetric members	$(F_{yw} - F_r)S_{xc} \leq F_{yf} S_{xt}$
		FLB	$(F_{yw} - F_r)S_x$
		WLB	$R_e F_{yf} S_x$
Channels and doubly symmetric I-shaped members bent about minor axis	$F_y Z_y$	FLB	$F_y S_y$
Solid symmetric shapes, except rectangular bars, bent about major axis	$F_y Z_x$	Not Applicable	
Solid rectangular bars bent about major axis	$F_y Z_x$	LTB	$F_y S_x$

TABLE A-F1.1 (cont'd)
Nominal Strength Parameters

Critical Stress F_{cr}	Slenderness Parameters			Limitations
	λ	λ_p	λ_r	
$\dfrac{C_b X_1 \sqrt{2}}{\lambda}\sqrt{1 + \dfrac{X_1^2 X_2}{2\lambda^2}}$ (b)	$\dfrac{L_b}{r_y}$	$\dfrac{300}{\sqrt{F_{yf}}}$	(a,b)	1. Applicable for I-shaped members if $h_c/t_w \le 970 / \sqrt{F_{yf}}$ when $h_c/t_w > 970 / \sqrt{F_{yf}}$ See Appendix G
(c)	$\dfrac{L_b}{r_y}$	$\dfrac{300}{\sqrt{F_y}}$	Value of λ for which $M_{cr} = S_{xc}(F_{yw} - F_r)$	2. When $\lambda > \lambda_r$ a) Applicable for built-up channels if $b/d \le 0.25$ and $t_f/t_w \le 3.0$
(g)	$\dfrac{b}{t}$	$\dfrac{65}{\sqrt{F_{yf}}}$	(h)	b) Applicable for rolled channels if $b/d \le 0.25$ and $t_f/t_w \le 2.0$
N.A.	$\dfrac{h_c}{t_w}$	$\dfrac{640}{\sqrt{F_{yf}}}$	$\dfrac{970}{\sqrt{F_{yf}}}$	
	Same as for major axis			
	Not Applicable			
$\dfrac{57,000 C_b \sqrt{JA}}{S_x}$	$\dfrac{L_b}{r_y}$	$\dfrac{3,750\sqrt{JA}}{M_p}$	$\dfrac{57,000\sqrt{JA}}{M_r}$	

TABLE A-F1.1 (cont'd)
Nominal Strength Parameters

Shape	Plastic Moment M_p	Limit State of Buckling	Limiting Buckling Moment M_r
Symmetric box sections loaded in a plane of symmetry	$F_y Z_x$	LTB	$(F_{yf} - F_r)S_x$
		FLB	$F_y S_x$
		WLB	Same as for I-shape
Circular Tubes	$F_y Z$	LTB	Not Applicable
		FLB	$M_n = \left(\dfrac{600}{D/t} + F_y\right)S$ [f]
		WLB	Not Applicable

a $\lambda_r = \dfrac{X_1}{(F_{yw} - F_r)}\sqrt{1 + \sqrt{1 + X_2(F_{yw} - F_r)^2}}$

b $X_1 = \dfrac{\pi}{S_x}\sqrt{\dfrac{EGJA}{2}}$ $X_2 = 4\dfrac{C_w}{I_y}\left(\dfrac{S_x}{GJ}\right)^2$

c $M_{cr} = \dfrac{57,000 C_b}{L_b}\sqrt{I_y J}\,(B_1 + \sqrt{1 + B_2 + B_1^2}) \leq S_{xt} F_y$

where
$B_1 = 2.25\,[2(I_{yc}/I_y) - 1](h/L_b)\sqrt{(I_y/J)}$ $B_2 = 25(1 - I_{yc}/I_y)(I_{yc}/J)(h/L_b)^2$

If the compression flange is larger than the tension flange, M_{cr} may be evaluated conservatively by the formula for doubly symmetric members, using for X_1 and X_2 the values for a symmetric section both of whose flanges are the same as the compression flange of the singly symmetric section.

TABLE A-F1.1 (cont'd)
Nominal Strength Parameters

Critical Stress F_{cr}	Slenderness Parameters			Limitations
	λ	λ_p	λ_r	
$\dfrac{57{,}000C_b\sqrt{JA}}{\lambda S_x}$	$\dfrac{L_b}{r_y}$	$\dfrac{3{,}750\sqrt{JA}}{M_p}$	$\dfrac{57{,}000\sqrt{JA}}{M_r}$	Applicable if $h_c/t_w \leq 970/\sqrt{F_{yf}}$
$\dfrac{(S_x)_{eff}}{S_x}F_y{}^{(d)}$	$\dfrac{b}{t}$	$\dfrac{190}{F_y}$	$\dfrac{238}{\sqrt{(F_y - F_r)}}$	LTB applies only if $d > b$
Same as for I-shape				
Not Applicable				
$\dfrac{9{,}750}{D/t}$	D/t	$\dfrac{2{,}070}{F_y}$	$\dfrac{8{,}970}{F_y}$	$D/t < \dfrac{13{,}000}{F_y}$
Not Applicable				

[d]$(S_x)_{eff}$ is the section modulus for a section with a compression-flange width b_e given by

$$b_e = \frac{326\,t}{\sqrt{f}}\left(1 - \frac{64.9}{(b/t)\sqrt{f}}\right) \text{ for flanges of square and rectangular sections of uniform thickness}$$

[e]Computed from fully plastic stress distribution for hybrid sections.

[f]This equation is to be used in place of Formula A-F1-3.

[g]$F_{cr} = \dfrac{20{,}000}{\lambda^2}$ for rolled shapes

$F_{cr} = \dfrac{11{,}200}{\lambda^2}$ for welded shapes

[h]$\lambda_r = \dfrac{141}{\sqrt{F_{yw} - 10}}$ for rolled shapes

$\lambda_r = \dfrac{106}{\sqrt{F_{yw} - 16.5}}$ for welded shapes

F4. WEB-TAPERED MEMBERS

The design of tapered members meeting the requirements of this section shall be governed by the provisions of Chap. F, except as modified by this Appendix.

1. General Requirements

In order to qualify under this Specification, a tapered member must meet the following requirements:

    a. It shall possess at least one axis of symmetry which shall be perpendicular to the plane of bending if moments are present.

    b. The flanges shall be of equal and constant area.

    c. The depth shall vary linearly as

$$d = d_o\left(1 + \gamma\frac{z}{L}\right) \qquad \text{(A-F4-1)}$$

where

d_o = depth at smaller end of member, in.
d_L = depth at larger end of member, in.
γ = $(d_L - d_o)/d_o \le$ the smaller of $0.268(L/d_o)$ or 6.0
z = distance from the smaller end of member, in.
L = unbraced length of member measured between the center of gravity of the bracing members, in.

2. Design Tensile Strength

The design strength of tapered tension members shall be determined in accordance with Sect. D1.

3. Design Compressive Strength

The design strength of tapered compression members shall be determined in accordance with Sect. E2, using an effective slenderness parameter λ_{eff} computed as follows:

$$\lambda_{eff} = \frac{S}{\pi}\sqrt{\frac{QF_y}{E}} \qquad \text{(A-F4-2)}$$

where

S = KL/r_{oy} for weak axis bending and $K_\gamma L/r_{ox}$ for strong axis bending
K = effective length factor for a prismatic member
K_γ = effective length factor for a tapered member as determined by a rational analysis
r_{ox} = strong axis radius of gyration at the smaller end of a tapered member, in.
r_{oy} = weak axis radius of gyration at the smaller end of a tapered member, in.
F_y = specified minimum yield stress, ksi
Q = reduction factor
    = 1.0 if all elements meet the limiting width-thickness ratios λ_r of Sect. B5.1
    = $Q_s Q_a$, determined in accordance with Appendix B, if any stiffened and/or unstiffened elements exceed the ratios λ_r of Sect. B5.1
E = modulus of elasticity for steel, ksi

The smallest area of the tapered member shall be used for A_g in Formula E2-1.

4. Design Flexural Strength

The design flexural strength of tapered flexural members for the limit state of lateral-torsional buckling is $\phi_b M_n$, where $\phi_b = 0.90$ and the nominal strength is

$$M_n = (5/3)S'_x F_{b\gamma} \qquad (\text{A-F4-3})$$

where

S'_x = the section modulus of the critical section of the unbraced beam length under consideration

$$F_{b\gamma} = \frac{2}{3}\left[1.0 - \frac{F_y}{6B\sqrt{F_{s\gamma}^2 + F_{w\gamma}^2}}\right] F_y \leq 0.60 F_y \qquad (\text{A-F4-4})$$

unless $F_{b\gamma} \leq F_y/3$, in which case

$$F_{b\gamma} = B \sqrt{F_{s\gamma}^2 + F_{w\gamma}^2} \qquad (\text{A-F4-5})$$

In the above equations,

$$F_{s\gamma} = \frac{12 \times 10^3}{h_s L d_o / A_f} \qquad (\text{A-F4-6})$$

$$F_{w\gamma} = \frac{170 \times 10^3}{(h_w L / r_{To})^2} \qquad (\text{A-F4-7})$$

where

h_s = factor equal to $1.0 + 0.0230\gamma\sqrt{L d_o / A_f}$
h_w = factor equal to $1.0 + 0.00385\gamma\sqrt{L / r_{To}}$
r_{To} = radius of gyration of a section at the smaller end, considering only the compression flange plus ⅓ of the compression web area, taken about an axis in the plane of the web, in.
A_f = area of the compression flange, in.^{2}

and where B is determined as follows:

a. When the maximum moment M_2 in three adjacent segments of approximately equal unbraced length is located within the central segment and M_1 is the larger moment at one end of the three-segment portion of a member:*

$$B = 1.0 + 0.37\left(1.0 + \frac{M_1}{M_2}\right) + 0.50\gamma\left(1.0 + \frac{M_1}{M_2}\right) \geq 1.0 \qquad (\text{A-F4-8})$$

b. When the largest computed bending stress f_{b2} occurs at the larger end of two adjacent segments of approximately equal unbraced lengths and f_{b1} is the computed bending stress at the smaller end of the two-segment portion of a member:**

$$B = 1.0 + 0.58\left(1.0 + \frac{f_{b1}}{f_{b2}}\right) - 0.70\gamma\left(1.0 + \frac{f_{b1}}{f_{b2}}\right) \geq 1.0 \qquad (\text{A-F4-9})$$

---

*M_1/M_2 is considered as negative when producing single curvature. In the rare case where M_1/M_2 is positive, it is recommended that it be taken as zero.
**f_{b1}/f_{b2} is considered as negative when producing single curvature. If a point of contraflexure occurs in one of two adjacent unbraced segments, f_{b1}/f_{b2} is considered as positive. The ratio $f_{b1}/f_{b2} \neq 0$.

c. When the largest computed bending stress f_{b2} occurs at the smaller end of two adjacent segments of approximately equal unbraced length and f_{b1} is the computed bending stress at the larger end of the two-segment portion of a member:[**]

$$B = 1.0 + 0.55\left(1.0 + \frac{f_{b1}}{f_{b2}}\right) + 2.20\gamma\left(1.0 + \frac{f_{b1}}{f_{b2}}\right) \geq 1.0 \qquad \text{(A-F4-10)}$$

In the foregoing, $\gamma = (d_L - d_o)/d_o$ is calculated for the unbraced length that contains the maximum computed bending stress.

d. When the computed bending stress at the smaller end of a tapered member or segment thereof is equal to zero:

$$B = \frac{1.75}{1.0 + 0.25 \sqrt{\gamma}} \qquad \text{(A-F4-11)}$$

where $\gamma = (d_L - d_o)/d_o$, calculated for the unbraced length adjacent to the point of zero bending stress.

5. Design Shear Strength

The design shear strength of tapered flexural members shall be determined in accordance with Sect. F2.

6. Combined Flexure and Axial Force

For tapered members with a single web taper subject to compression and bending about the major axis, Formulas H1-1 through H1-3 apply, with the following modifications: P_n and P_{ex} shall be determined for the properties of the smaller end, using appropriate effective length factors. M_{nx}, M_u and M_{px} shall be determined for the larger end; $M_{nx} = (5/3)S'_x F_{b\gamma}$, where S'_x is the elastic section modulus of the larger end, and $F_{b\gamma}$ is the design flexural stress of tapered members. C_{mx} is replaced by C'_m, determined as follows:

a. When the member is subjected to end moments which cause single curvature bending and approximately equal computed moments at the ends:

$$C'_m = 1.0 + 0.1\left(\frac{P_u}{\phi_b P_{ex}}\right) + 0.3\left(\frac{P_u}{\phi_b P_{ex}}\right)^2 \qquad \text{(A-F4-12)}$$

b. When the computed bending moment at the smaller end of the unbraced length is equal to zero:

$$C'_m = 1.0 + 0.9\left(\frac{P_u}{\phi_b P_{ex}}\right) + 0.6\left(\frac{P_u}{\phi_b P_{ex}}\right)^2 \qquad \text{(A-F4-13)}$$

When the effective slenderness parameter $\lambda_{eff} \geq 1.0$ and combined stress is checked incrementally along the length, the actual area and the actual section modulus at the section under investigation may be used.

[**]f_{b1}/f_{b2} is considered as negative when producing single curvature. If a point of contraflexure occurs in one of two adjacent unbraced segments, f_{b1}/f_{b2} is considered as positive. The ratio $f_{b1}/f_{b2} \neq 0$.

APPENDIX G
PLATE GIRDERS

G1. LIMITATIONS

Doubly and singly symmetric single-web non-hybrid and hybrid plate girders loaded in the plane of the web may be proportioned according to the provisions of this Appendix or Sect. F2, provided that these limits are satisfied.

a. For $\frac{a}{h} \leq 1.5$:

$$\frac{h}{t_w}\max = \frac{2{,}000}{\sqrt{F_{yf}}} \tag{A-G1-1}$$

b. For $\frac{a}{h} > 1.5$

$$\frac{h}{t_w}\max = \frac{14{,}000}{\sqrt{F_{yf}(F_{yf} + 16.5)}} \tag{A-G1-2}$$

where

a = clear distance between transverse stiffeners, in.
h = clear distance between flanges less the fillet or corner radius for rolled shapes; and for built-up sections, the distance between adjacent lines of fasteners or the clear distance between flanges when welds are used, in.
t_w = web thickness, in.
F_{yf} = specified minimum yield stress of the flange, ksi

In unstiffened girders h/t_w must be less than 260.

G2. DESIGN FLEXURAL STRENGTH

The design flexural strength for plate girders with slender webs ($h_c/t_w > 970/\sqrt{F_{yf}}$) shall be $\phi_b M_n$, where $\phi_b = 0.90$ and M_n is the lower value obtained according to the limit states of tension-flange yield and compression-flange buckling. When $h_c/t_w \leq 970/\sqrt{F_{yf}}$, see Appendix F1.7.

For tension-flange yield:

$$M_n = S_{xt} R_{PG} R_e F_{yt} \tag{A-G2-1}$$

For compression flange buckling:

$$M_n = S_{xc} R_{PG} R_e F_{cr} \tag{A-G2-2}$$

where

$$R_{PG} = 1 - 0.0005 a_r \left(\frac{h_c}{t_w} - \frac{970}{\sqrt{F_{cr}}} \right) \leq 1.0 \tag{A-G2-3}$$

R_e = hybrid girder factor
  = $1.0 - 0.1(1.3 + a_r)(0.81 - m) \leq 1.0$ (for non-hybrid girders, $R_e = 1.0$)
a_r = ratio of web area to compression flange area
m = ratio of web yield stress to flange yield stress or F_{cr}
F_{cr} = critical compression flange stress, ksi
F_{yt} = yield stress of tension flange, ksi
S_{xc} = section modulus referred to compression flange, in.^{3}
S_{xt} = section modulus referred to tension flange, in.^{3}
h_c = twice the distance from the neutral axis to the inside face of the compression flange less the fillet or corner radius, in.

The critical stress F_{cr} to be used is dependent upon the slenderness parameters $\lambda, \lambda_p, \lambda_r$ and C_{PG} as follows:

a. For $\lambda \leq \lambda_p$:

$$F_{cr} = F_{yf} \tag{A-G2-4}$$

b. For $\lambda_p < \lambda \leq \lambda_r$:

$$F_{cr} = C_b F_{yf} \left[1 - \frac{1}{2} \left(\frac{\lambda - \lambda_p}{\lambda_r - \lambda_p} \right) \right] \leq F_{yf} \tag{A-G2-5}$$

c. For $\lambda > \lambda_r$:

$$F_{cr} = \frac{C_{PG}}{\lambda^2} \tag{A-G2-6}$$

In the foregoing, the slenderness parameter shall be determined for both the limit state of lateral-torsional buckling and the limit state of flange local buckling; the slenderness parameter which results in the lowest value of F_{cr} governs.

For the limit state of lateral-torsional buckling:

$$\lambda = \frac{L_b}{r_T} \tag{A-G2-7}$$

$$\lambda_p = \frac{300}{\sqrt{F_{yf}}} \tag{A-G2-8}$$

$$\lambda_r = \frac{756}{\sqrt{F_{yf}}} \tag{A-G2-9}$$

$$C_{PG} = 286,000 \, C_b \tag{A-G2-10}$$

where

$$C_b = 1.75 + 1.05(M_1/M_2) + 0.3(M_1/M_2)^2 \le 2.3$$

r_T = radius of gyration of compression flange plus one-sixth the web, in.

For the limit state of flange local buckling:

$$\lambda = \frac{b_f}{2t_f} \qquad \text{(A-G2-11)}$$

$$\lambda_p = \frac{65}{\sqrt{F_{yf}}} \qquad \text{(A-G2-12)}$$

$$\lambda_r = \frac{150}{\sqrt{F_{yf}}} \qquad \text{(A-G2-13)}$$

$$C_{PG} = 11,200 \qquad \text{(A-G2-14)}$$

$$C_b = 1$$

The limit state of flexural web local buckling is not applicable.

G3. DESIGN SHEAR STRENGTH WITH TENSION FIELD ACTION

The design shear strength shall be $\phi_v V_n$, kips, where $\phi_v = 0.90$ and V_n is determined as follows:

a. For $h/t_w \le 187\sqrt{k/F_{yw}}$:

$$V_n = 0.6\, A_w F_{yw} \qquad \text{(A-G3-1)}$$

b. For $h/t_w > 187\sqrt{k/F_{yw}}$:

where

$$V_n = 0.6\, A_w F_{yw}\left(C_v + \frac{1-C_v}{1.15\sqrt{1+(a/h)^2}}\right) \qquad \text{(A-G3-2)}$$

C_v = ratio of "critical" web stress, according to linear buckling theory, to the shear yield stress of web material

except for end-panels in non-hybrid plate girders, for all panels in hybrid and web-tapered plate girders and when a/h exceeds 3.0 or $[260/(h/t_w)]^2$. In these cases, tension field action is not permitted and

$$V_n = 0.6\, A_w F_{yw} C_v \qquad \text{(A-G3-3)}$$

The web plate buckling coefficient k is given as

$$k = 5 + \frac{5}{(a/h)^2} \qquad \text{(A-G3-4)}$$

except that k shall be taken as 5.0 if a/h exceeds 3.0 or $[260/(h/t_w)]^2$. The shear coefficient C_v is determined as follows:

For $187\sqrt{\dfrac{k}{F_{yw}}} \le \dfrac{h}{t_w} \le 234\sqrt{\dfrac{k}{F_{yw}}}$:

$$C_v = \frac{187\sqrt{k/F_{yw}}}{h/t_w} \qquad \text{(A-G3-5)}$$

For $\dfrac{h}{t_w} > 234 \sqrt{\dfrac{k}{F_{yw}}}$:

$$C_v = \frac{44{,}000\,k}{(h/t_w)^2 F_{yw}} \tag{A-G3-6}$$

G4. TRANSVERSE STIFFENERS

Transverse stiffeners are not required in plate girders when $h/t_w \le 418/\sqrt{F_{yw}}$, or when the required shear V_u, as determined by structural analysis for the factored loads, is less than or equal to $0.6\phi A_w F_{yw} C_v$, where C_v is determined for $k = 5$ and $\phi = 0.90$. Stiffeners may be required in certain portions of a plate girder to develop the required shear or to satisfy the limitations given in Appendix G1.

The moment of inertia I_{st} of a transverse stiffener about an axis in the web center for stiffener pairs or about the face in contact with the web plate for single stiffeners shall not be less than $a t_w^3 j$, where

$$j = \frac{2.5}{(a/h)^2} - 2 \ge 0.5 \tag{A-G4-1}$$

and the stiffener area A_{st} when designing for tension field action shall not be less than

$$\frac{F_{yw}}{F_{yst}}\left[0.15\,Dht_w(1 - C_v)\frac{V_u}{\phi_v V_n} - 18\,t_w^2\right] \ge 0 \tag{A-G4-2}$$

where

F_{yst} = specified yield stress of the stiffener material, ksi
D = 1 for stiffeners in pairs
  = 1.8 for single angle stiffeners
  = 2.4 for single plate stiffeners

C_v and V_n are defined in Appendix G3, and V_u is the required shear at the location of the stiffener.

G5. FLEXURE-SHEAR INTERACTION

Plate girders with webs that depend on tension field action must satisfy flexure-shear interaction criteria.

When stiffeners are required and

$$\frac{0.6V_n}{M_n} \le \frac{V_u}{M_u} \le \frac{V_n}{0.75\,M_n}$$

the following interaction equation must be satisfied:

$$\frac{M_u}{M_n} + .625\frac{V_u}{V_n} \le 1.375\,\phi \tag{A-G5-1}$$

where M_n is the nominal flexural strength of plate girders from Appendix G2, $\phi = 0.90$ and V_n is the nominal shear strength from Appendix G3, except that M_u may not exceed ϕM_n ($\phi = 0.90$) and V_u may not exceed ϕV_n ($\phi = 0.90$).

APPENDIX H.
MEMBERS UNDER TORSION
AND COMBINED FORCES

H3. ALTERNATE INTERACTION EQUATIONS FOR MEMBERS UNDER COMBINED STRESS

For biaxially loaded I-shaped members used in braced frames only, the following interaction equations may be used in lieu of Formulas H1-1a and H1-1b.

$$\left(\frac{M_{ux}}{\phi_b M_{px}'}\right)^{\zeta} + \left(\frac{M_{uy}}{\phi_b M_{py}'}\right)^{\zeta} \leq 1.0 \qquad \text{(A-H3-1)}$$

$$\left(\frac{C_{mx} M_{ux}}{\phi_b M_{nx}'}\right)^{\eta} + \left(\frac{C_{my} M_{uy}}{\phi_b M_{ny}'}\right)^{\eta} \leq 1.0 \qquad \text{(A-H3-2)}$$

For $0.5 \leq b_f/d \leq 1.0$:

$$\zeta = 1.6 - \frac{P_u/P_y}{2[\ln(P_u/P_y)]} \qquad \text{(A-H3-3)}$$

For $b_f/d \geq 0.3$:

$$\eta = 0.4 + \frac{P_u}{P_y} + \frac{b_f}{d} \geq 1.0 \qquad \text{(A-H3-4)}$$

$$= 1.0 \text{ for } b_f/d < 0.3$$

where

b_f = flange width, in.
d = member depth, in.

$$M_{px}' = 1.2 M_{px}[1 - (P_u/P_y)] \leq M_{px} \qquad \text{(A-H3-5)}$$

$$M_{py}' = 1.2 M_{py}[1 - (P_u/P_y)^2] \leq M_{py} \qquad \text{(A-H3-6)}$$

$$M_{nx}' = M_{nx}\left(1 - \frac{P_u}{\phi_c P_n}\right)\left(1 - \frac{P_u}{P_{ex}}\right) \qquad \text{(A-H3-7)}$$

$$M_{ny}' = M_{ny}\left(1 - \frac{P_u}{\phi_c P_n}\right)\left(1 - \frac{P_u}{P_{ey}}\right) \qquad \text{(A-H3-8)}$$

where

P_n = nominal compressive strength determined in accordance with Sect. E2, kips
P_u = required axial strength, kips
P_y = compressive yield strength $A_g F_y$, kips
ϕ_b = resistance factor for flexure = 0.90
ϕ_c = resistance factor for compression = 0.85
P_e = Euler buckling strength $A_g F_y/\lambda_c^2$, where λ_c is the column slenderness parameter defined by Formula E2-4, kips
M_u = required flexural strength, kip-in.
M_n = nominal flexural strength, determined in accordance with Sect. F1, kip-in.
M_p = plastic moment, kip-in.
C_m = coefficient defined in Sect. H1.

Notes

APPENDIX K.
STRENGTH DESIGN CONSIDERATIONS

K2. PONDING

The provisions of this Appendix may be used when a more exact determination of flat roof framing stiffness is needed than that given by the provision of Sect. K2 that $C_p + 0.9C_s \leq 0.25$.

For any combination of primary and secondary framing, the stress index is computed as

$$U_p = \left(\frac{\phi_b F_y - f_o}{f_o}\right)_p \text{ for the primary member} \qquad \text{(A-K2-3)}$$

$$U_s = \left(\frac{\phi_b F_y - f_o}{f_o}\right)_s \text{ for the secondary member} \qquad \text{(A-K2-4)}$$

where

f_o = the stress due to $1.2D + 1.2R$ (D = nominal dead load, R = nominal load due to rain water or ice exclusive of the ponding contribution)*

ϕ_b = resistance factor for flexure = 0.90 (Sect. F)

---

*Depending upon geographic location, this loading should include such amount of snow as might also be present, although ponding failures have occurred more frequently during torrential summer rains when the rate of precipitation exceeded the rate of drainage runoff and the resulting hydraulic gradient over large roof areas caused substantial accumulation of water some distance from the eaves.

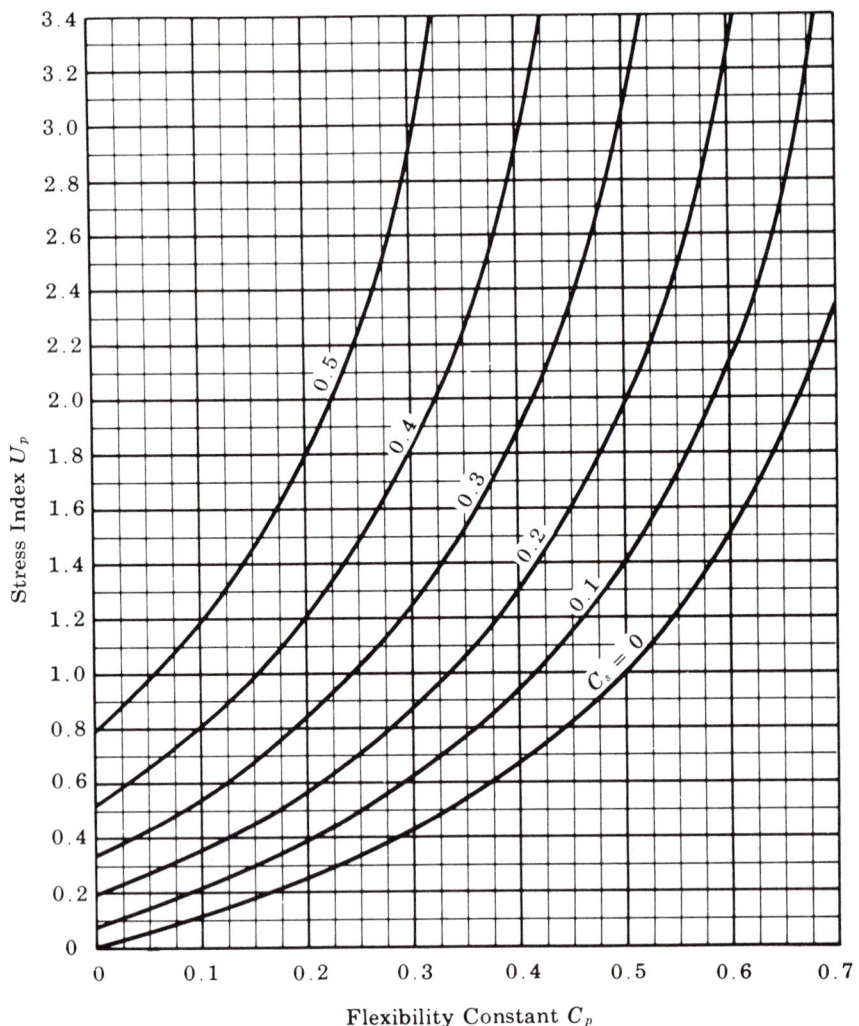

Stress Index U_p

Flexibility Constant C_p

Fig. A-K2.1.   Flexibility coefficients for combined primary and secondary systems

Enter Fig. A-K2.1 at the level of the computed stress index U_p determined for the primary beam; move horizontally to the computed C_s-value of the secondary beams and then downward to the abscissa scale. The combined stiffness of the primary and secondary framing is sufficient to prevent ponding if the flexibility constant read from this latter scale is more than the value of C_p computed for the given primary member; if not, a stiffer primary or secondary beam, or combination of both, is required. In the above,

$$C_p = \frac{32L_s L_p^4}{10^7 I_p}$$

$$C_s = \frac{32S L_s^4}{10^7 I_s}$$

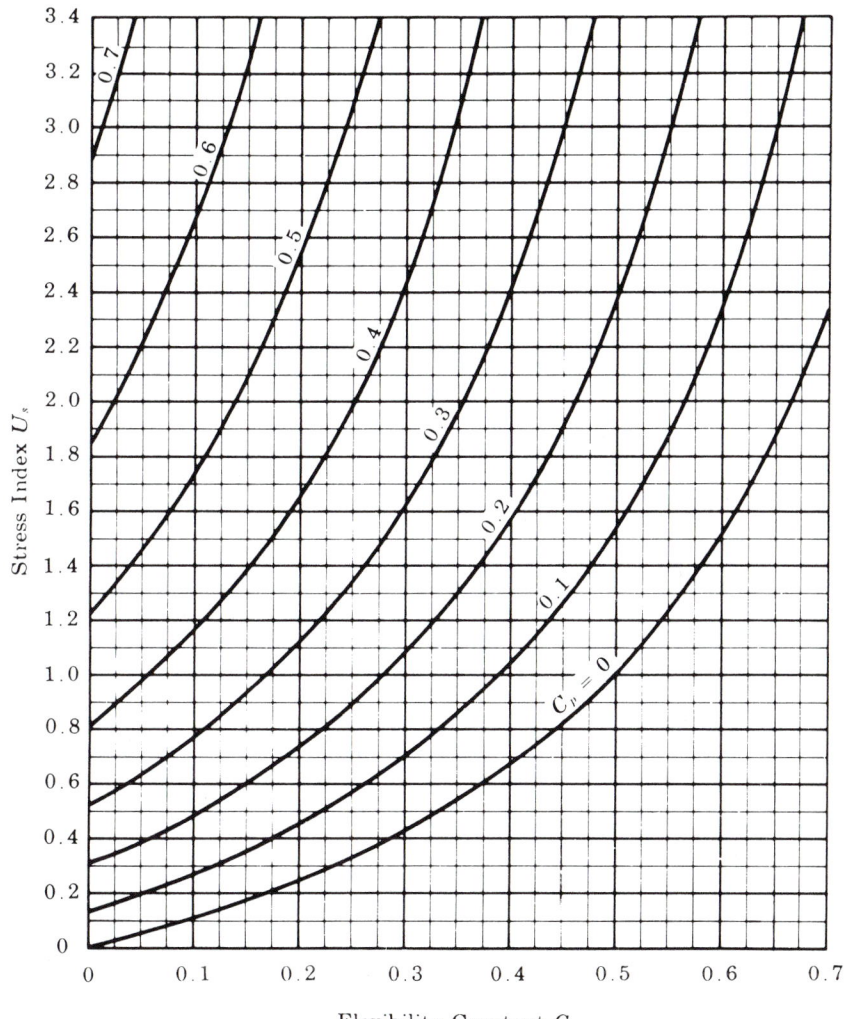

Fig. A-K2.2. *Flexibility coefficients for secondary beams alone*

where

L_p = column spacing in direction of girder (length of primary members), ft
L_s = column spacing perpendicular to direction of girder (length of secondary members), ft
S = spacing of secondary members, ft
I_p = moment of inertia of primary members, in.[4]
I_s = moment of inertia of secondary members, in.[4]

Roof framing consisting of a series of equally spaced wall-bearing beams is considered as consisting of secondary members supported on an infinitely stiff primary member. For this case, enter Fig. A-K2.2 with the computed stress index U_s. The limiting value of C_s is determined by the intercept of a horizontal line representing the U_s-value and the curve for $C_p = 0$.

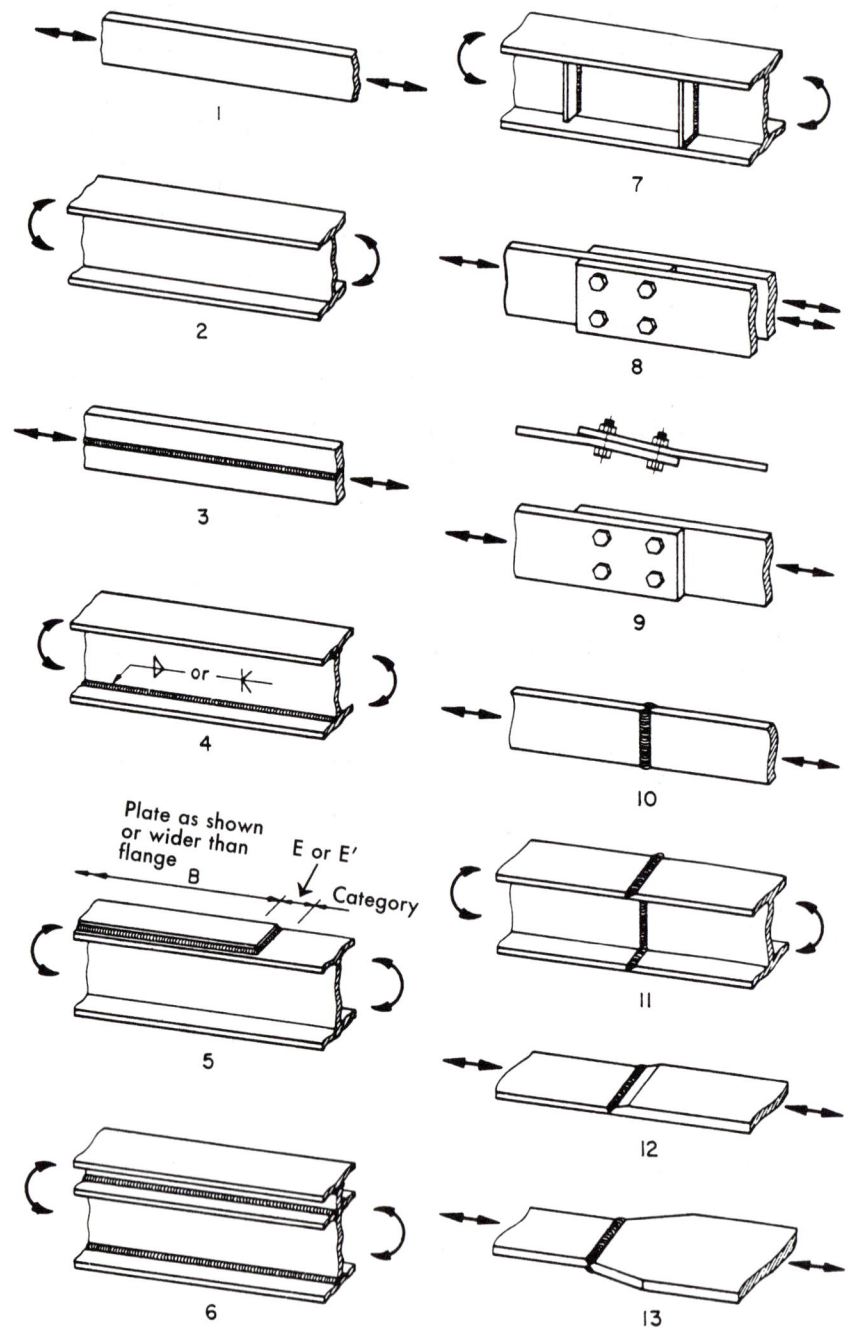

Fig. A-K4.1. Illustrative examples

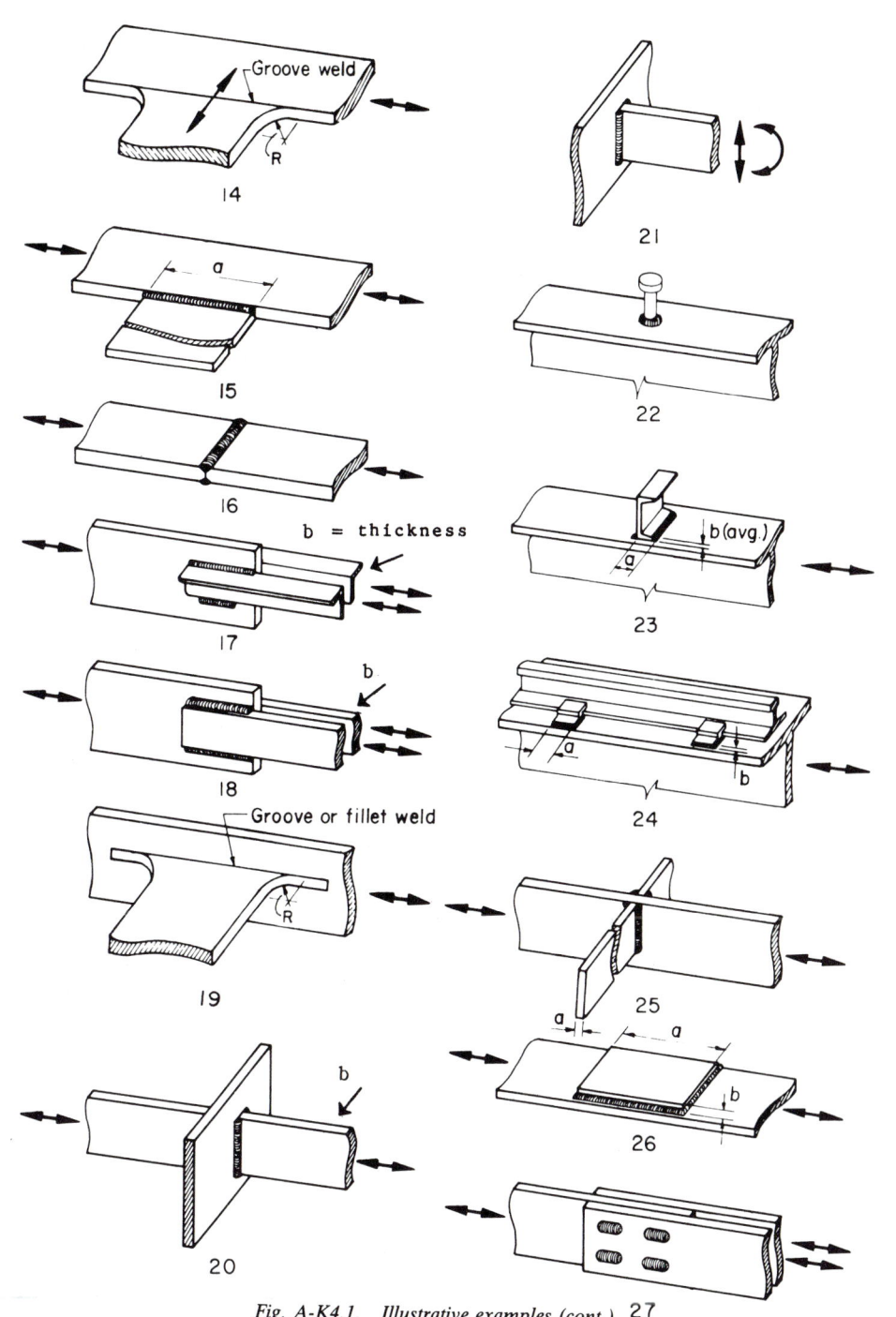

Fig. A-K4.1. Illustrative examples (cont.)

The ponding deflection contributed by a metal deck is usually such a small part of the total ponding deflection of a roof panel that it is sufficient merely to limit its moment of inertia (per foot of width normal to its span) to 0.000025 times the fourth power of its span length. However, the stability against ponding of a roof consisting of a metal roof deck of relatively slender depth-span ratio, spanning between beams supported directly on columns, may need to be checked. This can be done using Fig. A-K2.1 or A-K2.2 using as C_s the flexibility constant for one foot width of the roof deck ($S = 1.0$).

Since the shear rigidity of the web system of steel joists and trusses is less than that of a solid plate, their moment of inertia should be taken as 85% of their chords.

K4. FATIGUE

Members and connections subject to fatigue loading shall be proportioned in accordance with the provisions of this Appendix.

Fatigue, as used in this Specification, is defined as the damage that may result in fracture after a sufficient number of fluctuations of stress. Stress range is defined as the magnitude of these fluctuations. In the case of a stress reversal, the stress range shall be computed as the numerical sum of maximum repeated tensile and compressive stresses or the sum of maximum shearing stresses of opposite direction at a given point, resulting from differing arrangement of live load.

1. Loading Conditions; Type and Location of Material

In the design of members and connections subject to repeated variation of live load, consideration shall be given to the number of stress cycles, the expected range of stress and the type and location of member or detail.

Loading conditions shall be classified according to Table A-K4.1.

The type and location of material shall be categorized according to Table A-K4.2.

2. Design Stress Range

The maximum range of stress at service loads shall not exceed the design stress range specified in Table A-K4.3.

3. Design Strength of Bolts in Tension

When subject to tensile fatigue loading, properly tightened A325 or A490 bolts shall be designed for the combined tensile design strength due to external and prying forces within limits given in Table A-K4.4.

TABLE A-K4.1
Number of Loading Cycles

Loading Condition	From	To
1	20,000[a]	100,000[b]
2	100,000	500,000[c]
3	500,000	2,000,000[d]
4	Over 2,000,000	

[a]Approximately equivalent to two applications every day for 25 years.
[b]Approximately equivalent to 10 applications every day for 25 years.
[c]Approximately equivalent to 50 applications every day for 25 years.
[d]Approximately equivalent to 200 applications every day for 25 years.

TABLE A-K4.2
Type and Location of Material

General Condition	Situation	Kind of Stress[a]	Stress Category (see Table A-K4.3)	Illustrative Example Nos. (see Fig. A-K4.1)[b]
Plain Material	Base metal with rolled or cleaned surface. Flame-cut edges with ANSI smoothness of 1,000 or less	T or Rev.	A	1,2
Built-up Members	Base metal and weld metal in members without attachments, built-up plates or shapes connected by continuous full-penetration groove welds or by continuous fillet welds parallel to the direction of applied stress	T or Rev.	B	3,4,5,6
	Base metal and weld metal in members without attachments, built-up plates, or shapes connected by full-penetration groove welds with backing bars not removed, or by partial-penetration groove welds parallel to the direction of applied stress	T or Rev.	B′	3,4,5,6
	Base metal at toe of welds on girder webs or flanges adjacent to welded transverse stiffeners	T or Rev.	C	7
	Base metal at ends of partial length welded coverplates narrower than the flange having square or tapered ends, with or without welds across the ends or wider than flange with welds across the ends			
	Flange thickness ≤ 0.8 in.	T or Rev.	E	5
	Flange thickness > 0.8 in.	T or Rev.	E′	5
	Base metal at end of partial length welded coverplates wider than the flange without welds across the ends		E′	5

[a] "T" signifies range in tensile stress only; "Rev." signifies a range involving reversal of tensile or compressive stress; "S" signifies a range in shear, including shear stress reversal.

[b] These examples are provided as guidelines and are not intended to exclude other reasonably similar situations.

[c] Allowable fatigue stress range for transverse partial-penetration and transverse fillet welds is a function of the effective throat, depth of penetration and plate thickness. See Frank and Fisher, *Journal of the Structural Division*, Vol. 105 No. ST9, Sept. 1979.

TABLE A-K4.2 (cont'd)
Type and Location of Material

General Condition	Situation	Kind of Stress[a]	Stress Category (see Table A-K4.3)	Illus-trative Example Nos. (see Fig. A-K4.1)[b]
Groove Welds	Base metal and weld metal at full-penetration groove welded splices of parts of similar cross section ground flush, with grinding in the direction of applied stress and with weld soundness established by radiographic or ultrasonic inspection in accordance with the requirements of 9.25.2 or 9.25.3 of AWS D1.1-85	T or Rev.	B	10,11
	Base metal and weld metal at full-penetration groove welded splices at transitions in width or thickness, with welds ground to provide slopes no steeper than 1 to 2½ with grinding in the direction of applied stress, and with weld soundness established by radiographic or ultrasonic inspection in accordance with the requirements of 9.25.2 or 9.25.3 of AWS D1.1-85			
	A514 base metal	T or Rev.	B′	12,13
	Other base metals	T or Rev.	B	12,13
	Base metal and weld metal at full-penetration groove welded splices, with or without transitions having slopes no greater than 1 to 2½ when reinforcement is not removed but weld soundness is established by radiographic or ultrasonic inspection in accordance with requirements of 9.25.2 or 9.25.3 of AWS D1.1-85	T or Rev.	C	10,11,12, 13
Partial-Penetration Groove Welds	Weld metal of partial-penetration transverse groove welds, based on effective throat area of the weld or welds	T or Rev.	F[c]	16

TABLE A-K4.2 (cont'd)
Type and Location of Material

General Condition	Situation	Kind of Stress[a]	Stress Category (see Table A-K4.3)	Illustrative Example Nos. (see Fig. A-K4.1)[b]
Fillet-welded Connections	Base metal at intermittent fillet welds	T or Rev.	E	
	Base metal at junction of axially loaded members with fillet-welded end connections. Welds shall be disposed about the axis of the member so as to balance weld stresses $b \leq 1$ in. $b > 1$ in.	T or Rev. T or Rev.	E E′	17,18 17,18
	Base metal at members connected with transverse fillet welds $b \leq \frac{1}{2}$ in. $b > \frac{1}{2}$ in.	T or Rev.	C See Note c	20,21
Fillet Welds	Weld metal of continuous or intermittent longitudinal or transverse fillet welds	S	F[c]	15,17,18 20,21
Plug or Slot Welds	Base metal at plug or slot welds	T or Rev.	E	27
	Shear on plug or slot welds	S	F	27
Mechanically Fastened Connections	Base metal at gross section of high-strength bolted slip-critical connections, except axially loaded joints which induce out-of-plane bending in connected material	T or Rev.	B	8
	Base metal at net section of other mechanically fastened joints	T or Rev.	D	8,9
	Base metal at net section of fully tensioned high-strength, bolted-bearing connections	T or Rev.	B	8,9

TABLE A-K4.2 (cont'd)
Type and Location of Material

General Condition	Situation	Kind of Stress[a]	Stress Category (see Table A-K4.3)	Illustrative Example Nos. (see Fig. A-K4.1)[b]
Attachments	Base metal at details attached by full-penetration groove welds subject to longitudinal and/or transverse loading when the detail embodies a transition radius *R* with the weld termination ground smooth and for transverse loading, the weld soundness established by radiographic or ultrasonic inspection in accordance with 9.25.2 or 9.25.3 of AWS D1.1-85			
	Longitudinal loading			
	$R > 24$ in.	T or Rev.	B	14
	24 in. $> R > 6$ in.	T or Rev.	C	14
	6 in. $> R > 2$ in.	T or Rev.	D	14
	2 in. $> R$	T or Rev.	E	14
	Detail base metal for transverse loading: equal thickness and reinforcement removed			
	$R > 24$ in.	T or Rev.	B	14
	24 in. $> R > 6$ in.	T or Rev.	C	14
	6 in. $> R > 2$ in.	T or Rev.	D	14
	2 in. $> R$	T or Rev.	E	14,15
	Detail base metal for transverse loading: equal thickness and reinforcement not removed			
	$R > 24$ in.	T or Rev.	C	14
	24 in. $> R > 6$ in.	T or Rev.	C	14
	6 in. $> R > 2$ in.	T or Rev.	D	14
	2 in. $> R$	T or Rev.	E	14,15

TABLE A-K4.2 (cont'd)
Type and Location of Material

General Condition	Situation	Kind of Stress[a]	Stress Category (see Table A-K4.3)	Illustrative Example Nos. (see Fig. A-K4.1)[b]
Attachments (cont'd)	Detail base metal for transverse loading: unequal thickness and reinforcement removed			
	$R > 2$ in.	T or Rev.	D	14
	2 in. $> R$	T or Rev.	E	14,15
	Detail base metal for transverse loading: unequal thickness and reinforcement not removed			
	all R	T or Rev.	E	14,15
	Detail base metal for transverse loading			
	$R > 6$ in.	T or Rev.	C	19
	6 in. $> R > 2$ in.	T or Rev.	D	19
	2 in. $> R$	T or Rev.	E	19
	Base metal at detail attached by full-penetration groove welds subject to longitudinal loading			
	$2 < a < 12b$ or 4 in.	T or Rev.	D	15
	$a > 12b$ or 4 in. when $b \le 1$ in.	T or Rev.	E	15
	$a > 12b$ or 4 in. when $b > 1$ in.	T or Rev.	E′	15
	Base metal at detail attached by fillet welds or partial-penetration groove welds subject to longitudinal loading			
	$a < 2$ in.	T or Rev.	C	15,23,24, 25,26
	2 in. $< a < 12b$ or 4 in.	T or Rev.	D	15,23, 24,26
	$a > 12b$ or 4 in. when $b \le 1$ in.	T or Rev.	E	15,23, 24,26
	$a > 12b$ or 4 in. when $b > 1$ in.	T or Rev.	E′	15,23, 24,26

TABLE A-K4.2 (cont'd)
Type and Location of Material

General Condition	Situation	Kind of Stress[a]	Stress Category (see Table A-K4.3)	Illus-trative Example Nos. (see Fig. A-K4.1)[b]
Attachments (cont'd)	Base metal attached by fillet welds or partial-penetration groove welds subjected to longitudinal loading when the weld termination embodies a transition radius with the weld termination ground smooth:			
	$R > 2$ in.	T or Rev.	D	19
	$R \leq 2$ in.	T or Rev.	E	19
	Fillet-welded attachments where the weld termination embodies a transition radius, weld termination ground smooth, and main material subject to longitudinal loading:			
	Detail base metal for transverse loading:			
	$R > 2$ in.	T or Rev.	D	19
	$R < 2$ in.	T or Rev.	E	19
	Base metal at stud-type shear connector attached by fillet weld or automatic end weld	T or Rev.	C	22
	Shear stress on nominal area of stud-type shear connectors	S	F	

TABLE A-K4.3
Allowable Stress Range, ksi

Category (From Table A-K4.2)	Loading Condition 1	Loading Condition 2	Loading Condition 3	Loading Condition 4
A	63	37	24	24
B	49	29	18	16
B′	39	23	15	12
C	35	21	13	10[a]
D	28	16	10	7
E	22	13	8	4.5
E′	16	9.2	5.8	2.6
F	15	12	9	8

[a]Flexural stress range of 12 ksi permitted at toe of stiffener welds or flanges.

TABLE A-K4.4
Design Strength of A325 or A490 Bolts Subject to Tension

Number of cycles	Design strength
Not more than 20,000	As specified in Section J3
From 20,000 to 500,000	$0.30\,A_b F_u$[a]
More than 500,000	$0.25\,A_b F_u$[a]

[a]At service loads.

Notes

NUMERICAL VALUES

TABLE 1
Design Strength as a Function of F_y

F_y (ksi)	Design Stress (ksi)				
	$0.54F_y$[a]	$0.56F_y$[b]	$0.63F_y$[c]	$0.85F_y$[d]	$0.90F_y$[e]
33	17.8	18.5	20.8	28.1	29.7
35	18.9	19.6	22.1	29.8	31.5
36	19.4	20.2	22.7	30.6	32.4
40	21.6	22.4	25.2	34.0	36.0
42	22.7	23.5	26.5	35.7	37.8
45	24.3	25.2	28.4	38.3	40.5
46	24.8	25.8	29.0	39.1	41.4
50	27.0	28.0	31.5	42.5	45.0
55	29.7	30.8	34.7	46.8	49.5
60	32.4	33.6	37.8	51.0	54.0
65	35.1	36.4	41.0	55.3	58.5
70	37.8	39.2	44.1	59.5	63.0
90	48.6	50.4	56.7	76.5	81.0
100	54.0	56.0	63.0	85.0	90.0

[a]See Sect. F2, Formulas F2-1 and F2-2
[b]See Sect. J5.2, Formula J5-3
[c]See Sect. K1.7, Formulas K1-9 and K1-10
[d]See Sect. E2, Formula E2-1
[e]See Sect. D1, Formula D1-1

Notes

TABLE 2
Design Strength as a Function of F_u

Item	ASTM Designation	F_y (ksi)	F_u (ksi)	Connected Part of Designated Steel — Tension $0.75 \times F_u{}^a$	Bearing $0.75 \times 2.4F_u{}^b$	Bolt or Threaded Part of Designated Steel — Tension $0.75 \times 0.75F_u{}^c$	Shear $0.65 \times 0.45F_u{}^d$	Shear $0.65 \times 0.6F_u{}^e$
Shapes, Plates, Bars, Sheet and Tubing or Threaded Parts	A36	36	58–80	43.5	104	32.6	17.0	22.6
	A53	35	60	45.0	108	—	—	—
	A242 / A441 / A588	50	70	52.5	126	39.4	20.5	27.3
		46	67	50.3	121	37.7	19.6	26.1
		42	63	47.3	113	35.4	18.4	24.6
		40[f]	60	45.0	108	33.8	17.6	23.4
	A500	33/39[g]	45	33.8	81	—	—	—
		42/46[g]	58	43.5	104	—	—	—
		46/50[g]	62	46.5	112	—	—	—
	A501	36	58	43.5	104	—	—	—
	A529	42	60–85	45.0	108	33.8	17.6	23.4
	A570	40	55	41.3	99	—	—	—
		42	58	43.5	104	—	—	—
	A572	42	60	45.0	108	33.8	17.6	23.4
		50	65	48.8	117	36.6	19.0	25.4
		60	75	56.3	135	42.2	21.9	29.3
		65	80	60.0	144	45.0	23.4	31.2
	A514	100	110–130	82.5	198	61.9	32.2	42.9
		90	100–130	75.0	180	56.3	29.3	39.0
	A606	45	65	48.8	117	—	—	—
		50	70	52.5	126	—	—	—
	A607	45	60	45.0	108	—	—	—
		50	65	48.8	117	—	—	—
		55	70	52.5	126	—	—	—
		60	75	56.3	135	—	—	—
		65	80	60.0	144	—	—	—
		70	85	63.8	153	—	—	—
	A618	50	70	52.5	126	—	—	—
		50	65	48.8	117	—	—	—
Bolts	A449	92	120	—	—	67.5	35.1	46.8
		81	105	—	—	59.1	30.7	41.0
		58	90	—	—	50.6	26.3	35.1

[a] On effective net area, see Sects. D1, J5.2.
[b] Produced by fastener in shear, see Sect. J3.6. Note that smaller maximum design bearing stresses, as a function of hole spacing, may be required by Sects. J3.9 and J3.10.
[c] On nominal body area, see Table J3.2.
[d] Threads not excluded from shear plane, see Table J3.2.
[e] Threads excluded from shear plane, see Table J3.2.
[f] For A441 material only.
[g] Smaller value for circular shapes, larger for square or rectangular shapes.
Note: For dimensional and size limitations, see the appropriate ASTM Specification.

TABLE 3-36
Design Stress for Compression Members of 36 ksi Specified Yield stress Steel, $\phi_c = 0.85$[a]

$\frac{Kl}{r}$	$\phi_c F_{cr}$ (ksi)	$\frac{Kl}{r}$	$\phi_c F_{cr}$ (ksi)	$\frac{Kl}{r}$	$\phi_c F_{cr}$ (ksi)	$\frac{Kl}{r}$	$\phi_c F_{cr}$ (ksi)	$\frac{Kl}{r}$	$\phi_c F_{cr}$ (ksi)
1	30.60	41	28.01	81	21.66	121	14.16	161	8.23
2	30.59	42	27.89	82	21.48	122	13.98	162	8.13
3	30.59	43	27.76	83	21.29	123	13.80	163	8.03
4	30.57	44	27.64	84	21.11	124	13.62	164	7.93
5	30.56	45	27.51	85	20.92	125	13.44	165	7.84
6	30.54	46	27.37	86	20.73	126	13.27	166	7.74
7	30.52	47	27.24	87	20.54	127	13.09	167	7.65
8	30.50	48	27.11	88	20.36	128	12.92	168	7.56
9	30.47	49	26.97	89	20.17	129	12.74	169	7.47
10	30.44	50	26.83	90	19.98	130	12.57	170	7.38
11	30.41	51	26.68	91	19.79	131	12.40	171	7.30
12	30.37	52	26.54	92	19.60	132	12.23	172	7.21
13	30.33	53	26.39	93	19.41	133	12.06	173	7.13
14	30.29	54	26.25	94	19.22	134	11.88	174	7.05
15	30.24	55	26.10	95	19.03	135	11.71	175	6.97
16	30.19	56	25.94	96	18.84	136	11.54	176	6.89
17	30.14	57	25.79	97	18.65	137	11.37	177	6.81
18	30.08	58	25.63	98	18.46	138	11.20	178	6.73
19	30.02	59	25.48	99	18.27	139	11.04	179	6.66
20	29.96	60	25.32	100	18.08	140	10.89	180	6.59
21	29.90	61	25.16	101	17.89	141	10.73	181	6.51
22	29.83	62	24.99	102	17.70	142	10.58	182	6.44
23	29.76	63	24.83	103	17.51	143	10.43	183	6.37
24	29.69	64	24.67	104	17.32	144	10.29	184	6.30
25	29.61	65	24.50	105	17.13	145	10.15	185	6.23
26	29.53	66	24.33	106	16.94	146	10.01	186	6.17
27	29.45	67	24.16	107	16.75	147	9.87	187	6.10
28	29.36	68	23.99	108	16.56	148	9.74	188	6.04
29	29.28	69	23.82	109	16.37	149	9.61	189	5.97
30	29.18	70	23.64	110	16.19	150	9.48	190	5.91
31	29.09	71	23.47	111	16.00	151	9.36	191	5.85
32	28.99	72	23.29	112	15.81	152	9.23	192	5.79
33	28.90	73	23.12	113	15.63	153	9.11	193	5.73
34	28.79	74	22.94	114	15.44	154	9.00	194	5.67
35	28.69	75	22.76	115	15.26	155	8.88	195	5.61
36	28.58	76	22.58	116	15.07	156	8.77	196	5.55
37	28.47	77	22.40	117	14.89	157	8.66	197	5.50
38	28.36	78	22.22	118	14.70	158	8.55	198	5.44
39	28.25	79	22.03	119	14.52	159	8.44	199	5.39
40	28.13	80	21.85	120	14.34	160	8.33	200	5.33

[a]When element width-to-thickness ratio exceeds λ_r, see Appendix B5.3.

TABLE 3-50
Design Stress for Compression Members of 50 ksi Specified Yield stress Steel, $\phi_c = 0.85^a$

$\frac{Kl}{r}$	$\phi_c F_{cr}$ (ksi)	$\frac{Kl}{r}$	$\phi_c F_{cr}$ (ksi)	$\frac{Kl}{r}$	$\phi_c F_{cr}$ (ksi)	$\frac{Kl}{r}$	$\phi_c F_{cr}$ (ksi)	$\frac{Kl}{r}$	$\phi_c F_{cr}$ (ksi)
1	42.50	41	37.59	81	26.31	121	14.57	161	8.23
2	42.49	42	37.36	82	26.00	122	14.33	162	8.13
3	42.47	43	37.13	83	25.68	123	14.10	163	8.03
4	42.45	44	36.89	84	25.37	124	13.88	164	7.93
5	42.42	45	36.65	85	25.06	125	13.66	165	7.84
6	42.39	46	36.41	86	24.75	126	13.44	166	7.74
7	42.35	47	36.16	87	24.44	127	13.23	167	7.65
8	42.30	48	35.91	88	24.13	128	13.02	168	7.56
9	42.25	49	35.66	89	23.82	129	12.82	169	7.47
10	42.19	50	35.40	90	23.51	130	12.62	170	7.38
11	42.13	51	35.14	91	23.20	131	12.43	171	7.30
12	42.05	52	34.88	92	22.89	132	12.25	172	7.21
13	41.98	53	34.61	93	22.58	133	12.06	173	7.13
14	41.90	54	34.34	94	22.28	134	11.88	174	7.05
15	41.81	55	34.07	95	21.97	135	11.71	175	6.97
16	41.71	56	33.79	96	21.67	136	11.54	176	6.89
17	41.61	57	33.51	97	21.36	137	11.37	177	6.81
18	41.51	58	33.23	98	21.06	138	11.20	178	6.73
19	41.39	59	32.95	99	20.76	139	11.04	179	6.66
20	41.28	60	32.67	100	20.46	140	10.89	180	6.59
21	41.15	61	32.38	101	20.16	141	10.73	181	6.51
22	41.02	62	32.09	102	19.86	142	10.58	182	6.44
23	40.89	63	31.80	103	19.57	143	10.43	183	6.37
24	40.75	64	31.50	104	19.28	144	10.29	184	6.30
25	40.60	65	31.21	105	18.98	145	10.15	185	6.23
26	40.45	66	30.91	106	18.69	146	10.01	186	6.17
27	40.29	67	30.61	107	18.40	147	9.87	187	6.10
28	40.13	68	30.31	108	18.12	148	9.74	188	6.04
29	39.97	69	30.01	109	17.83	149	9.61	189	5.97
30	39.79	70	29.70	110	17.55	150	9.48	190	5.91
31	39.62	71	29.40	111	17.27	151	9.36	191	5.85
32	39.43	72	29.09	112	16.99	152	9.23	192	5.79
33	39.25	73	28.79	113	16.71	153	9.11	193	5.73
34	39.06	74	28.48	114	16.42	154	9.00	194	5.67
35	38.86	75	28.17	115	16.13	155	8.88	195	5.61
36	38.66	76	27.86	116	15.86	156	8.77	196	5.55
37	38.45	77	27.55	117	15.59	157	8.66	197	5.50
38	38.24	78	27.24	118	15.32	158	8.55	198	5.44
39	38.03	79	26.93	119	15.07	159	8.44	199	5.39
40	37.81	80	26.62	120	14.82	160	8.33	200	5.33

^{a}When element width-to-thickness ratio exceeds λ_r, see Appendix B5.3.

TABLE 4
Values of $\phi_c F_{cr}/F_y$, $\phi_c = 0.85$
For Determining Design Stress for Compression Members for Steel of Any Yield Stress[a]

<table>
<tr><td>λ_c</td><td>$\phi_c F_{cr}/F_y$</td><td>λ_c</td><td>$\phi_c F_{cr}/F_y$</td><td>λ_c</td><td>$\phi_c F_{cr}/F_y$</td><td>λ_c</td><td>$\phi_c F_{cr}/F_y$</td></tr>
<tr><td>0.02</td><td>0.850</td><td>0.82</td><td>0.641</td><td>1.62</td><td>0.284</td><td>2.42</td><td>0.127</td></tr>
<tr><td>0.04</td><td>0.849</td><td>0.84</td><td>0.632</td><td>1.64</td><td>0.277</td><td>2.44</td><td>0.125</td></tr>
<tr><td>0.06</td><td>0.849</td><td>0.86</td><td>0.623</td><td>1.66</td><td>0.271</td><td>2.46</td><td>0.123</td></tr>
<tr><td>0.08</td><td>0.848</td><td>0.88</td><td>0.614</td><td>1.68</td><td>0.264</td><td>2.48</td><td>0.121</td></tr>
<tr><td>0.10</td><td>0.846</td><td>0.90</td><td>0.605</td><td>1.70</td><td>0.258</td><td>2.50</td><td>0.119</td></tr>
<tr><td>0.12</td><td>0.845</td><td>0.92</td><td>0.596</td><td>1.72</td><td>0.252</td><td>2.52</td><td>0.117</td></tr>
<tr><td>0.14</td><td>0.843</td><td>0.94</td><td>0.587</td><td>1.74</td><td>0.246</td><td>2.54</td><td>0.116</td></tr>
<tr><td>0.16</td><td>0.841</td><td>0.96</td><td>0.578</td><td>1.76</td><td>0.241</td><td>2.56</td><td>0.114</td></tr>
<tr><td>0.18</td><td>0.839</td><td>0.98</td><td>0.568</td><td>1.78</td><td>0.235</td><td>2.58</td><td>0.112</td></tr>
<tr><td>0.20</td><td>0.836</td><td>1.00</td><td>0.559</td><td>1.80</td><td>0.230</td><td>2.60</td><td>0.110</td></tr>
<tr><td>0.22</td><td>0.833</td><td>1.02</td><td>0.550</td><td>1.82</td><td>0.225</td><td>2.62</td><td>0.109</td></tr>
<tr><td>0.24</td><td>0.830</td><td>1.04</td><td>0.540</td><td>1.84</td><td>0.220</td><td>2.64</td><td>0.107</td></tr>
<tr><td>0.26</td><td>0.826</td><td>1.06</td><td>0.531</td><td>1.86</td><td>0.215</td><td>2.66</td><td>0.105</td></tr>
<tr><td>0.28</td><td>0.823</td><td>1.08</td><td>0.521</td><td>1.88</td><td>0.211</td><td>2.68</td><td>0.104</td></tr>
<tr><td>0.30</td><td>0.819</td><td>1.10</td><td>0.512</td><td>1.90</td><td>0.206</td><td>2.70</td><td>0.102</td></tr>
<tr><td>0.32</td><td>0.814</td><td>1.12</td><td>0.503</td><td>1.92</td><td>0.202</td><td>2.72</td><td>0.101</td></tr>
<tr><td>0.34</td><td>0.810</td><td>1.14</td><td>0.493</td><td>1.94</td><td>0.198</td><td>2.74</td><td>0.099</td></tr>
<tr><td>0.36</td><td>0.805</td><td>1.16</td><td>0.484</td><td>1.96</td><td>0.194</td><td>2.76</td><td>0.098</td></tr>
<tr><td>0.38</td><td>0.800</td><td>1.18</td><td>0.474</td><td>1.98</td><td>0.190</td><td>2.78</td><td>0.096</td></tr>
<tr><td>0.40</td><td>0.795</td><td>1.20</td><td>0.465</td><td>2.00</td><td>0.186</td><td>2.80</td><td>0.095</td></tr>
<tr><td>0.42</td><td>0.789</td><td>1.22</td><td>0.456</td><td>2.02</td><td>0.183</td><td>2.82</td><td>0.094</td></tr>
<tr><td>0.44</td><td>0.784</td><td>1.24</td><td>0.446</td><td>2.04</td><td>0.179</td><td>2.84</td><td>0.092</td></tr>
<tr><td>0.46</td><td>0.778</td><td>1.26</td><td>0.437</td><td>2.06</td><td>0.176</td><td>2.86</td><td>0.091</td></tr>
<tr><td>0.48</td><td>0.772</td><td>1.28</td><td>0.428</td><td>2.08</td><td>0.172</td><td>2.88</td><td>0.090</td></tr>
<tr><td>0.50</td><td>0.765</td><td>1.30</td><td>0.419</td><td>2.10</td><td>0.169</td><td>2.90</td><td>0.089</td></tr>
<tr><td>0.52</td><td>0.759</td><td>1.32</td><td>0.410</td><td>2.12</td><td>0.166</td><td>2.92</td><td>0.087</td></tr>
<tr><td>0.54</td><td>0.752</td><td>1.34</td><td>0.401</td><td>2.14</td><td>0.163</td><td>2.94</td><td>0.086</td></tr>
<tr><td>0.56</td><td>0.745</td><td>1.36</td><td>0.392</td><td>2.16</td><td>0.160</td><td>2.96</td><td>0.085</td></tr>
<tr><td>0.58</td><td>0.738</td><td>1.38</td><td>0.383</td><td>2.18</td><td>0.157</td><td>2.98</td><td>0.084</td></tr>
<tr><td>0.60</td><td>0.731</td><td>1.40</td><td>0.374</td><td>2.20</td><td>0.154</td><td>3.00</td><td>0.083</td></tr>
<tr><td>0.62</td><td>0.724</td><td>1.42</td><td>0.365</td><td>2.22</td><td>0.151</td><td>3.02</td><td>0.082</td></tr>
<tr><td>0.64</td><td>0.716</td><td>1.44</td><td>0.357</td><td>2.24</td><td>0.149</td><td>3.04</td><td>0.081</td></tr>
<tr><td>0.66</td><td>0.708</td><td>1.46</td><td>0.348</td><td>2.26</td><td>0.146</td><td>3.06</td><td>0.080</td></tr>
<tr><td>0.68</td><td>0.700</td><td>1.48</td><td>0.339</td><td>2.28</td><td>0.143</td><td>3.08</td><td>0.079</td></tr>
<tr><td>0.70</td><td>0.692</td><td>1.50</td><td>0.331</td><td>2.30</td><td>0.141</td><td>3.10</td><td>0.078</td></tr>
<tr><td>0.72</td><td>0.684</td><td>1.52</td><td>0.323</td><td>2.32</td><td>0.138</td><td>3.12</td><td>0.077</td></tr>
<tr><td>0.74</td><td>0.676</td><td>1.54</td><td>0.314</td><td>2.34</td><td>0.136</td><td>3.14</td><td>0.076</td></tr>
<tr><td>0.76</td><td>0.667</td><td>1.56</td><td>0.306</td><td>2.36</td><td>0.134</td><td>3.16</td><td>0.075</td></tr>
<tr><td>0.78</td><td>0.659</td><td>1.58</td><td>0.299</td><td>2.38</td><td>0.132</td><td>3.18</td><td>0.074</td></tr>
<tr><td>0.80</td><td>0.650</td><td>1.60</td><td>0.291</td><td>2.40</td><td>0.129</td><td>3.20</td><td>0.073</td></tr>
</table>

(Left margin, rotated:) **All grades of steel**

[a]When element width-to-thickness ratios exceed λ_r, see Appendix B5.3.
Values of $\lambda_c > 2.24$ exceed Kl/r of 200 for $F_y = 36$.
Values of $\lambda_c > 2.64$ exceed Kl/r of 200 for $F_y = 50$.

TABLE 5
Values of Kl/r for $F_y = 36$ and 50 ksi

| λ_c | Kl/r | | λ_c | Kl/r | |
	$F_y = 36$	$F_y = 50$		$F_y = 36$	$F_y = 50$
0.02	1.8	1.5	0.82	73.1	62.0
0.04	3.6	3.0	0.84	74.9	63.6
0.06	5.3	4.5	0.86	76.7	65.1
0.08	7.1	6.1	0.88	78.5	66.6
0.10	8.9	7.6	0.90	80.2	68.1
0.12	10.7	9.1	0.92	82.0	69.6
0.14	12.5	10.6	0.94	83.8	71.1
0.16	14.3	12.1	0.96	85.6	72.6
0.18	16.0	13.6	0.98	87.4	74.1
0.20	17.8	15.1	1.00	89.2	75.7
0.22	19.6	16.6	1.02	90.9	77.2
0.24	21.4	18.2	1.04	92.7	78.7
0.26	23.2	19.7	1.06	94.5	80.2
0.28	25.0	21.2	1.08	96.3	81.7
0.30	26.7	22.7	1.10	98.1	83.2
0.32	28.5	24.2	1.12	99.9	84.7
0.34	30.3	25.7	1.14	101.6	86.3
0.36	32.1	27.2	1.16	103.4	87.8
0.38	33.9	28.8	1.18	105.2	89.3
0.40	35.7	30.3	1.20	107.0	90.8
0.42	37.4	31.8	1.22	108.8	92.3
0.44	39.2	33.3	1.24	110.6	93.8
0.46	41.0	34.8	1.26	112.3	95.3
0.48	42.8	36.3	1.28	114.1	96.8
0.50	44.6	37.8	1.30	115.9	98.4
0.52	46.4	39.3	1.32	117.7	99.9
0.54	48.1	40.9	1.34	119.5	101.4
0.56	49.9	42.4	1.36	121.3	102.9
0.58	51.7	43.9	1.38	123.0	104.4
0.60	53.5	45.4	1.40	124.8	105.9
0.62	55.3	46.9	1.42	126.6	107.4
0.64	57.1	48.4	1.44	128.4	108.9
0.66	58.8	49.9	1.46	130.2	110.5
0.68	60.6	51.4	1.48	132.0	112.0
0.70	62.4	53.0	1.50	133.7	113.5
0.72	64.2	54.5	1.52	135.5	115.0
0.74	66.0	56.0	1.54	137.3	116.5
0.76	67.8	57.5	1.56	139.1	118.0
0.78	69.5	59.0	1.58	140.9	119.5
0.80	71.3	60.5	1.60	142.7	121.1

TABLE 5 (cont'd)
Values of Kl/r for $F_y = 36$ and 50 ksi

	Kl/r			Kl/r
λ_c	$F_y = 36$	$F_y = 50$	λ_c	$F_y = 50$
1.62	144.4	122.6	2.42	183.1
1.64	146.2	124.1	2.44	184.6
1.66	148.0	125.6	2.46	186.1
1.68	149.8	127.1	2.48	187.6
1.70	151.6	128.6	2.50	189.1
1.72	153.4	130.1	2.52	190.7
1.74	155.1	131.6	2.54	192.2
1.76	156.9	133.2	2.56	193.7
1.78	158.7	134.7	2.58	195.2
1.80	160.5	136.2	2.60	196.7
1.82	162.3	137.7	2.62	198.2
1.84	164.1	139.2	2.64	199.7
1.86	165.8	140.7		
1.88	167.6	142.2		
1.90	169.4	143.8		
1.92	171.2	145.3		
1.94	173.0	146.8		
1.96	174.8	148.3		
1.98	176.5	149.8		
2.00	178.3	151.3		
2.02	180.1	152.8		
2.04	181.9	154.3		
2.06	183.7	155.9		
2.08	185.5	157.4		
2.10	187.2	158.9		
2.12	189.0	160.4		
2.14	190.8	161.9		
2.16	192.6	163.4		
2.18	194.4	164.9		
2.20	196.2	166.5		
2.22	197.9	168.0		
2.24	199.7	169.5		
2.26		171.0		
2.28		172.5		
2.30		174.0		
2.32		175.5		
2.34		177.0		
2.36		178.6		
2.38		180.1		
2.40		181.6		

Heavy line indicates Kl/r of 200.

TABLE 6
Slenderness Ratios of Elements as a Function of F_y
From Table B5.1

Ratio	F_y (ksi)					
	36	42	46	50	60	65
$65/\sqrt{F_y}$	10.8	10.0	9.6	9.2	8.4	8.1
$76/\sqrt{F_y}$	12.7	11.7	11.2	10.7	9.8	9.4
$95/\sqrt{F_y}$	15.8	14.7	14.0	13.4	12.3	11.8
$106/\sqrt{F_y-16.5}$	24.0	21.0	19.5	18.3	16.1	15.2
$127/\sqrt{F_y}$	21.2	19.6	18.7	18.0	16.4	15.8
$141/\sqrt{F_y-10}$	27.7	24.9	23.5	22.3	19.9	19.0
$190/\sqrt{F_y}$	31.7	29.3	28.0	26.9	24.5	23.6
$238/\sqrt{F_y-10}$	46.7	42.1	39.7	37.6	33.7	32.1
$238/\sqrt{F_y-16.5}$	53.9	47.1	43.8	41.1	36.1	34.2
$253/\sqrt{F_y}$	42.2	39.0	37.3	35.8	32.7	31.4
$317/\sqrt{F_y-10}$	62.2	56.0	52.8	50.1	44.8	42.7
$640/\sqrt{F_y}$	107	98.8	94.4	90.5	82.6	79.4
$970/\sqrt{F_y}$	162	150	143	137	125	120
$1,300/F_y$	36.1	31.0	28.3	26.0	21.7	20.0
$2,070/F_y$	57.5	49.3	45.0	41.4	34.5	31.8
$3,300/F_y$	91.7	78.6	71.7	66.0	55.0	50.8
$8,970/F_y$	249	214	195	179	150	138

TABLE 7
Values of C_b
For Use in Chapters F and G

$\dfrac{M_1}{M_2}$	C_b	$\dfrac{M_1}{M_2}$	C_b	$\dfrac{M_1}{M_2}$	C_b
− 1.00	1.00	− 0.45	1.34	0.10	1.86
− 0.95	1.02	− 0.40	1.38	0.15	1.91
− 0.90	1.05	− 0.35	1.42	0.20	1.97
− 0.85	1.07	− 0.30	1.46	0.25	2.03
− 0.80	1.10	− 0.25	1.51	0.30	2.09
− 0.75	1.13	− 0.20	1.55	0.35	2.15
− 0.70	1.16	− 0.15	1.60	0.40	2.22
− 0.65	1.19	− 0.10	1.65	0.45	2.28
− 0.60	1.23	− 0.05	1.70	≥ 0.47	2.30
− 0.55	1.26	0	1.75		
− 0.50	1.30	0.05	1.80		

Note 1: $C_b = 1.75 + 1.05(M_1/M_2) + 0.3\,(M_1/M_2)^2 \le 2.3$
Note 2: M_1/M_2 positive for reverse curvature and negative for single curvature.

TABLE 8
Values of C_m
For Use in Section H1

$\dfrac{M_1}{M_2}$	C_m	$\dfrac{M_1}{M_2}$	C_m	$\dfrac{M_1}{M_2}$	C_m
− 1.00	1.00	− 0.45	0.78	0.10	0.56
− 0.95	0.98	− 0.40	0.76	0.15	0.54
− 0.90	0.96	− 0.35	0.74	0.20	0.52
− 0.85	0.94	− 0.30	0.72	0.25	0.50
− 0.80	0.92	− 0.25	0.70	0.30	0.48
− 0.75	0.90	− 0.20	0.68	0.35	0.46
− 0.70	0.88	− 0.15	0.66	0.40	0.44
− 0.65	0.86	− 0.10	0.64	0.45	0.42
− 0.60	0.84	− 0.05	0.62	0.50	0.40
				0.60	0.36
− 0.55	0.82	0	0.60	0.80	0.28
− 0.50	0.80	0.05	0.58	1.00	0.20

Note 1: $C_m = 0.6 - 0.4(M_1/M_2)$.
Note 2: M_1/M_2 is positive for reverse curvature and negative for single curvature.

TABLE 9
Values of P_e/A_g
For Use in Section H1 for Steel of Any Yield Stress

$\frac{Kl}{r}$	P_e/A_g (ksi)	$\frac{Kl}{r}$	P_e/A_g (ksi)	$\frac{Kl}{r}$	P_e/A_g (ksi)	$\frac{Kl}{r}$	P_e/A_g (ksi)	$\frac{Kl}{r}$	P_e/A_g (ksi)	$\frac{Kl}{r}$	P_e/A_g (ksi)
21	649.02	51	110.04	81	43.62	111	23.23	141	14.40	171	9.79
22	591.36	52	105.85	82	42.57	112	22.82	142	14.19	172	9.67
23	541.06	53	101.89	83	41.55	113	22.42	143	14.00	173	9.56
24	496.91	54	98.15	84	40.56	114	22.02	144	13.80	174	9.45
25	457.95	55	94.62	85	39.62	115	21.64	145	13.61	175	9.35
26	423.40	56	91.27	86	38.70	116	21.27	146	13.43	176	9.24
27	392.62	57	88.09	87	37.81	117	20.91	147	13.25	177	9.14
28	365.07	58	85.08	88	36.96	118	20.56	148	13.07	178	9.03
29	340.33	59	82.22	89	36.13	119	20.21	149	12.89	179	8.93
30	318.02	60	79.51	90	35.34	120	19.88	150	12.72	180	8.83
31	297.83	61	76.92	91	34.56	121	19.55	151	12.55	181	8.74
32	279.51	62	74.46	92	33.82	122	19.23	152	12.39	182	8.64
33	262.83	63	72.11	93	33.09	123	18.92	153	12.23	183	8.55
34	247.59	64	69.88	94	32.39	124	18.61	154	12.07	184	8.45
35	233.65	65	67.74	95	31.71	125	18.32	155	11.91	185	8.36
36	220.85	66	65.71	96	31.06	126	18.03	156	11.76	186	8.27
37	209.07	67	63.76	97	30.42	127	17.75	157	11.61	187	8.18
38	198.21	68	61.90	98	29.80	128	17.47	158	11.47	188	8.10
39	188.18	69	60.12	99	29.20	129	17.20	159	11.32	189	8.01
40	178.89	70	58.41	100	28.62	130	16.94	160	11.18	190	7.93
41	170.27	71	56.78	101	28.06	131	16.68	161	11.04	191	7.85
42	162.26	72	55.21	102	27.51	132	16.43	162	10.91	192	7.76
43	154.80	73	53.71	103	26.98	133	16.18	163	10.77	193	7.68
44	147.84	74	52.57	104	26.46	134	15.94	164	10.64	194	7.60
45	141.34	75	50.88	105	25.96	135	15.70	165	10.51	195	7.53
46	135.26	76	49.55	106	25.47	136	15.47	166	10.39	196	7.45
47	129.57	77	48.27	107	25.00	137	15.25	167	10.26	197	7.38
48	124.23	78	47.04	108	24.54	138	15.03	168	10.14	198	7.30
49	119.21	79	45.86	109	24.09	139	14.81	169	10.02	199	7.23
50	114.49	80	44.72	110	23.65	140	14.60	170	9.90	200	7.16

Note: $P_e/A_g = \dfrac{\pi^2 E}{(Kl/r)^2}$

TABLE 10-36

$$\frac{\phi_v V_n}{A_w} \text{ (ksi) for Plate Girders by Section F2}$$

For 36 ksi Yield stress Steel, Tension Field Action Not Included

$\dfrac{h}{t_w}$	Aspect Ratio a/h: Stiffener Spacing to Web Depth													
	0.5	0.6	0.7	0.8	0.9	1.0	1.2	1.4	1.6	1.8	2.0	2.5	3.0	Over 3.0
60	19.4	19.4	19.4	19.4	19.4	19.4	19.4	19.4	19.4	19.4	19.4	19.4	19.4	19.4
70	19.4	19.4	19.4	19.4	19.4	19.4	19.4	19.4	19.4	19.4	19.4	19.4	19.4	19.4
80	19.4	19.4	19.4	19.4	19.4	19.4	19.4	19.4	19.4	19.4	18.9	18.2	17.9	16.9
90	19.4	19.4	19.4	19.4	19.4	19.4	19.4	18.5	17.8	17.2	16.8	16.2	15.9	14.7
100	19.4	19.4	19.4	19.4	19.4	19.2	17.6	16.6	16.0	15.5	14.9	13.8	13.2	11.9
110	19.4	19.4	19.4	19.4	18.4	17.4	16.0	14.8	13.7	12.8	12.3	11.4	10.9	9.8
120	19.4	19.4	19.4	18.1	16.9	16.0	14.0	12.5	11.5	10.8	10.3	9.6	9.2	8.3
130	19.4	19.4	18.2	16.7	15.6	14.1	11.9	10.6	9.8	9.2	8.8	8.2	7.8	7.0
140	19.4	18.8	16.9	15.5	13.5	12.1	10.3	9.2	8.4	7.9	7.6	7.0	6.7	6.1
150	19.4	17.6	15.7	13.5	11.8	10.6	8.9	8.0	7.3	6.9	6.6	6.1	5.9	5.3
160	18.9	16.5	14.1	11.9	10.4	9.3	7.9	7.0	6.5	6.1	5.8	5.4		4.6
170	17.8	15.5	12.5	10.5	9.2	8.2	7.0	6.2	5.7	5.4	5.1			4.1
180	16.8	13.9	11.1	9.4	8.2	7.3	6.2	5.5	5.1	4.8	4.6			3.7
200	14.9	11.2	9.0	7.6	6.6	5.9	5.0	4.5	4.1					3.0
220	12.3	9.3	7.5	6.3	5.5	4.9	4.2							2.5
240	10.3	7.8	6.3	5.3	4.6	4.1								2.1
260	8.8	6.6	5.3	4.5	3.9	3.5								1.8
280	7.6	5.7	4.6	3.9										
300	6.6	5.0	4.0											
320	5.8	4.4												

TABLE 10-50

$\dfrac{\phi_v V_n}{A_w}$ (ksi) for Plate Girders by Section F2

For 50 ksi Yield stress Steel, Tension Field Action Not Included

$\dfrac{h}{t_w}$	Aspect Ratio a/h: Stiffener Spacing to Web Depth													
	0.5	0.6	0.7	0.8	0.9	1.0	1.2	1.4	1.6	1.8	2.0	2.5	3.0	Over 3.0
60	27.0	27.0	27.0	27.0	27.0	27.0	27.0	27.0	27.0	27.0	27.0	27.0	27.0	26.6
70	27.0	27.0	27.0	27.0	27.0	27.0	27.0	27.0	26.9	26.1	25.5	24.6	24.0	22.8
80	27.0	27.0	27.0	27.0	27.0	27.0	26.0	24.5	23.5	22.8	22.3	21.5	20.6	18.6
90	27.0	27.0	27.0	27.0	26.5	25.1	23.1	21.8	20.4	19.2	18.3	17.0	16.3	14.7
100	27.0	27.0	27.0	25.6	23.9	22.6	20.1	17.9	16.5	15.5	14.9	13.8	13.2	11.9
110	27.0	27.0	25.3	23.2	21.7	19.6	16.6	14.8	13.7	12.8	12.3	11.4	10.9	9.8
120	27.0	25.9	23.2	21.1	18.4	16.5	14.0	12.5	11.5	10.8	10.3	9.6	9.2	8.3
130	27.0	23.9	21.4	18.0	15.7	14.1	11.9	10.6	9.8	9.2	8.8	8.2	7.8	7.0
140	25.5	22.2	18.4	15.5	13.5	12.1	10.3	9.2	8.4	7.9	7.6	7.0	6.7	6.1
150	23.8	19.9	16.1	13.5	11.8	10.6	8.9	8.0	7.3	6.9	6.6	6.1	5.9	5.3
160	22.3	17.5	14.1	11.9	10.4	9.3	7.9	7.0	6.5	6.1	5.8	5.4		4.6
170	20.6	15.5	12.5	10.5	9.2	8.2	7.0	6.2	5.7	5.4	5.1			4.1
180	18.3	13.9	11.1	9.4	8.2	7.3	6.2	5.5	5.1	4.8	4.6			3.7
200	14.9	11.2	9.0	7.6	6.6	5.9	5.0	4.5	4.1					3.0
220	12.3	9.3	7.5	6.3	5.5	4.9	4.2							2.5
240	10.3	7.8	6.3	5.3	4.6	4.1								2.1
260	8.8	6.6	5.3	4.5	3.9	3.5								
280	7.6	5.7	4.6	3.9										

TABLE 11-36

$$\frac{\phi_v V_n}{A_w}$$ (ksi) for Plate Girders by Appendix G

For 36 ksi Yield stress Steel, Tension Field Action Included[b]
(*Italic* values indicate gross area, as percent of $(h \times t_w)$ required for
pairs of intermediate stiffeners of 36 ksi yield stress steel with $V_u/\phi V_n = 1.0$.)[a]

$\dfrac{h}{t_w}$	Aspect Ratio a/h: Stiffener Spacing to Web Depth													
	0.5	0.6	0.7	0.8	0.9	1.0	1.2	1.4	1.6	1.8	2.0	2.5	3.0	Over 3.0[c]
60	19.4	19.4	19.4	19.4	19.4	19.4	19.4	19.4	19.4	19.4	19.4	19.4	19.4	19.4
70	19.4	19.4	19.4	19.4	19.4	19.4	19.4	19.4	19.4	19.4	19.4	19.4	19.4	19.4
80	19.4	19.4	19.4	19.4	19.4	19.4	19.4	19.4	19.4	19.4	19.1	18.6	18.3	16.9
90	19.4	19.4	19.4	19.4	19.4	19.4	19.4	19.0	18.5	18.2	17.8	17.3	16.8	14.7
100	19.4	19.4	19.4	19.4	19.4	19.3	18.6	18.1	17.6	17.2	16.6	15.6	14.9	11.9
110	19.4	19.4	19.4	19.4	19.1	18.7	17.9	17.2	16.3	15.6	15.1	14.0	13.3	9.8
120	19.4	19.4	19.4	19.0	18.5	18.1	17.0	16.0	15.1	14.4	13.9	12.8	12.0	8.3
130	19.4	19.4	19.1	18.6	18.1	17.4	16.1	15.1	14.2	13.5	12.9	11.8	11.0	7.0
140	19.4	19.3	18.7	18.2	17.4	16.6	15.4	14.4	13.5	12.8	12.2	11.0	10.2	6.1
150	19.4	19.0	18.4	17.5	16.7	16.0	14.8	13.8	12.9	12.2	11.6	10.4	9.6	5.3
160	19.3	18.7	17.9	17.0	16.2	15.5	14.3	13.3	12.4	11.7	11.1	9.9		4.6
170	19.1	18.4	17.4	16.6	15.8	15.1	13.9	12.9	12.0	11.3 *0.3*	10.7 *0.4*			4.1
180	18.9	18.0	17.1	16.2	15.5	14.8	13.6 *0.2*	12.6 *0.7*	11.7 *1.1*	11.0 *1.3*	10.4 *1.5*			3.7
200	18.4	17.3	16.4	15.6 *0.1*	14.9 *0.9*	14.2 *1.4*	13.1 *2.1*	12.0 *2.5*	11.2 *2.8*					3.0
220	17.8	16.9	16.0 *1.1*	15.2 *2.0*	14.5 *2.6*	13.8 *3.0*	12.7 *3.6*							2.5
240	17.4	16.5 *1.5*	15.7 *2.7*	14.9 *3.4*	14.2 *3.9*	13.5 *4.3*								2.1
260	17.1 *1.3*	16.2 *3.0*	15.4 *4.0*	14.6 *4.6*	14.0 *5.0*	13.3 *5.4*								1.8
280	16.8 *2.7*	16.0 *4.2*	15.2 *5.0*	14.4 *5.6*										
300	16.6 *3.9*	15.8 *5.2*	15.0 *5.9*											
320	16.4 *4.9*	15.6 *6.0*												

[a]For area of single-angle and single-plate stiffeners, or when $V_u/\phi V_n < 1.0$, see Formula A-G4-2.
[b]For end-panels and all panels in hybrid and web-tapered plate girders use Table 10-36.
[c]Same as for Table 10-36.
Note: Girders so proportioned that the computed shear is less than that given in right-hand column do not require intermediate stiffeners.

TABLE 11-50

$$\frac{\phi_v V_n}{A_w} \text{ (ksi) for Plate Girders by Appendix G}$$

For 50 ksi Yield stress Steel, Tension Field Action Included[b]
(*Italic* values indicate gross area, as percent of ($h \times t_w$) required for
pairs of intermediate stiffeners of 50 ksi yield stress steel with $V_u/\phi V_n = 1.0$.)[a]

$\frac{h}{t_w}$	Aspect Ratio a/h: Stiffener Spacing to Web Depth													
	0.5	0.6	0.7	0.8	0.9	1.0	1.2	1.4	1.6	1.8	2.0	2.5	3.0	Over 3.0[c]
60	27.0	27.0	27.0	27.0	27.0	27.0	27.0	27.0	27.0	27.0	27.0	27.0	27.0	26.6
70	27.0	27.0	27.0	27.0	27.0	27.0	27.0	27.0	26.9	26.5	26.1	25.4	24.9	22.8
80	27.0	27.0	27.0	27.0	27.0	27.0	26.5	25.8	25.1	24.6	24.1	23.3	22.4	18.6
90	27.0	27.0	27.0	27.0	26.8	26.3	25.3	24.4	23.4	22.5	21.7	20.2	19.2	14.7
100	27.0	27.0	27.0	26.5	25.9	25.3	24.0	22.5	21.4	20.4	19.6	18.0	17.0	11.9
110	27.0	27.0	26.5	25.8	25.1	24.2	22.4	21.0	19.8	18.8	18.0	16.4	15.3	9.8
120	27.0	26.7	25.9	25.1	24.0	23.0	21.2	19.8	18.6	17.6	16.8	15.2	14.1	8.3
130	27.0	26.2	25.4	24.1	23.0	22.0	20.3	18.9	17.7	16.7	15.9	14.2	13.1	7.0
140	26.7	25.8	24.5	23.3	22.2	21.3	19.6	18.2	17.0	16.0	15.1	13.5	12.3	6.1
150	26.3	25.2	23.9	22.7	21.6	20.7	19.0	17.6	16.4	15.4	14.5	12.9	11.7	5.3
160	26.0	24.6	23.3	22.2	21.1	20.2	18.5	17.1	15.9	14.9	14.0	12.4		4.6
									0.2	*0.4*	*0.5*	*0.8*		
170	25.6	24.1	22.8	21.7	20.7	19.8	18.1	16.7	15.5	14.5	13.6			4.1
							0.5	*1.0*	*1.2*	*1.4*	*1.6*			
180	25.1	23.7	22.4	21.3	20.3	19.4	17.8	16.4	15.2	14.2	13.3			3.7
					0.4	*0.9*	*1.5*	*1.9*	*2.2*	*2.3*	*2.5*			
200	24.3	23.0	21.8	20.8	19.8	18.9	17.3	15.9	14.7					3.0
		1.0	*1.8*	*2.3*	*2.7*	*3.2*	*3.5*	*3.7*						
220	23.7	22.5	21.4	20.4	19.4	18.5	16.9							2.5
		1.7	*2.7*	*3.3*	*3.8*	*4.1*	*4.5*							
240	23.2	22.1	21.0	20.0	19.1	18.2								2.1
	1.8	*3.2*	*4.0*	*4.6*	*4.9*	*5.2*								
260	23.0	21.8	20.8	19.8	18.8	18.0								
	3.2	*4.4*	*5.1*	*5.6*	*5.9*	*6.1*								
280	22.7	21.6	20.6	19.6										
	4.4	*5.4*	*6.0*	*6.4*										

[a]For area of single-angle and single-plate stiffeners, or when $V_u/\phi V_n < 1.0$, see Formula A-G4-2.
[b]For end-panels and all panels in hybrid and web-tapered plate girders use Table 10-50.
[c]Same as for Table 10-50.
Note: Girders so proportioned that the computed shear is less than that given in right-hand column do not require intermediate stiffeners.

TABLE 12
Nominal Horizontal Shear Load for One Connector Q_n, kips[a]
From Formulas I5-1 and I5-2

Connector[b]	Specified Compressive Strength of Concrete, f'_c, ksi[d]		
	3.0	3.5	4.0
½-in. dia. × 2-in. hooked or headed stud	9.4	10.5	11.6
⅝-in. dia. × 2½-in. hooked or headed stud	14.6	16.4	18.1
¾-in. dia. × 3-in. hooked or headed stud	21.0	23.6	26.1
⅞-in. dia. × 3½-in. hooked or headed stud	28.6	32.1	35.5
Channel C3 × 4.1	$10.2\ L_c$[c]	$11.5\ L_c$[c]	$12.7\ L_c$[c]
Channel C4 × 5.4	$11.1\ L_c$[c]	$12.4\ L_c$[c]	$13.8\ L_c$[c]
Channel C5 × 6.7	$11.9\ L_c$[c]	$13.3\ L_c$[c]	$14.7\ L_c$[c]

[a]Applicable only to concrete made with ASTM C33 aggregates.
[b]The design horizontal loads tabulated may also be used for studs longer than shown.
[c]L_c = length of channel, inches.
[d]$F_u > 0.5(f'_c w)^{.75}$, w = 145 lbs./cu.ft.

TABLE 13
Coefficients for Use with Concrete Made with C330 Aggregates to Adjust Values from Table 12 for Lightweight Concrete

Specified Compressive Strength of Concrete (f'_c)	Air Dry Unit Weight of Concrete, pcf						
	90	95	100	105	110	115	120
≤ 4.0 ksi	0.73	0.76	0.78	0.81	0.83	0.86	0.88
≥ 5.0 ksi	0.82	0.85	0.87	0.91	0.93	0.96	0.99

Commentary

on the
Load and Resistance Factor Design
Specification for
Structural Steel Buildings
(September 1, 1986)

INTRODUCTION

This Commentary provides information on the basis and limitations of various provisions of the LRFD Specification, so that designers, fabricators and erectors (users) can make more efficient use of the Specification. The Commentary and Specification documents do not attempt to anticipate and/or set forth all the questions or possible problems that may be encountered, or situations in which special consideration and engineering judgment should be exercised in using the documents. Such a recitation would make the documents unduly lengthy and cumbersome. WARNING is given that AISC assumes that the users of its documents are competent in their fields of endeavor and are informed on current developments and findings related to their fields.

CHAPTER A.
GENERAL PROVISIONS

A1. SCOPE

Load and Resistance Factor Design (LRFD) is an improved approach to the design of structural steel for buildings. It involves explicit consideration of limit states, multiple load factors and resistance factors, and implicit probabilistic determination of reliability. The designation LRFD reflects the concept of factoring both loads and resistance. This type of factoring differs from Part 1 of the allowable stress design (ASD) 1978 AISC Specification,[1] where only the resistance is divided by a factor of safety (to obtain allowable stress) and from Part 2, where only the loads are multiplied by a common load factor. The LRFD method was devised to offer the designer greater flexibility, more rationality and possible overall economy.

The format of using resistance factors and multiple load factors is not new, as several such design codes are in effect (the ACI-318 Strength Design for Reinforced Concrete[2] and the AASHTO Load Factor Design for Bridges).[3] Nor should the new LRFD method give designs radically different from the older methods, since it was tuned, or "calibrated," to typical representative designs of the earlier methods. The principal new ingredient is the use of a probabilistic mathematical model in the development of the load and resistance factors, which made it possible to give proper weight to the degree of accuracy with which the various loads and resistances can be determined. Also, it provides a rational methodology for transference of test results into design provisions. A more rational design procedure leading to more uniform reliability is the practical result.

A2. LIMITS OF APPLICABILITY

The AISC *Specification for the Design, Fabrication and Erection of Structural Steel for Buildings* has evolved through numerous versions from the 1st Edition, published June 1, 1923. Each succeeding edition has been based upon past successful usage, advances in the state of knowledge and changes in engineering design practice. The data included has been developed to provide a uniform practice in the design of steel-framed buildings.

The intention of the new LRFD Specification, as with the allowable stress design AISC Specification, is to cover the many everyday design criteria required for routine design office usage. It is not intended to cover the infrequently encountered problems within the full range of structural design practice, because to provide such definitive provisions covering all possible cases and their complexities would diminish the Specification's usefulness for routine design office use.

The LRFD Specification is the result of the deliberations of a committee of structural engineers with wide experience and high professional standing, representing a wide geographical distribution throughout the United States. The membership of the committee is made up of approximately equal numbers representing design engineers in private practice, engineers involved in research and teaching and engineers employed by steel fabricating companies.

In order to avoid reference to proprietary steels which may have limited availability, only those steels which can be identified by ASTM specifications are listed as approved under the Specification. However, some steels covered by ASTM specification, but subject to more costly manufacturing and inspection techniques than deemed essential for structures covered by this Specification, are not listed, even though they may provide all of the necessary characteristics of less expensive steels which are listed. Approval of such steels is left to the owner's representative.

As used throughout the Specification, the term *structural steel* refers exclusively to those items enumerated in Sect. 2 of the AISC *Code of Standard Practice for Steel Buildings and Bridges*, and nothing contained herein is intended as a specification for design of items not specifically enumerated in that code, such as skylights, fire escapes, etc. For the design of cold-formed steel structural members, whose profiles contain rounded corners and slender flat elements, the provisions of the American Iron and Steel Institute *Specification for the Design of Cold-Formed Steel Structural Members* are recommended.

The reader is cautioned that independent professional judgment must be exercised when data or recommendations set forth in the Specification and Commentary are applied. The publication of the material contained herein is not intended as a representation or warranty on the part of the American Institute of Steel Construction, Inc.—or any other person named herein—that this information is suitable for general or particular use, or freedom from infringement of any patent or patents. Anyone making use of this information assumes all liability arising from such use. The design of structures is within the scope of expertise of a competent licensed professional for the application of principles to a particular structure.

2. Types of Construction

The provisions for these types of construction have been revised to provide for a truer recognition of the actual degree of connection restraint in the structural design. All connections possess an amount of restraint. Depending on the amount of restraint offered by the connection they are classified as either Type FR or PR. This new classification renames the Type I connection of the 1978 AISC Specification to Type FR and includes both Type II and Type III of that Specification under a new, more general classification of Type PR.

Just as in the allowable stress design (ASD) provisions, construction utilizing Type FR connections may be designed in LRFD using either elastic or plastic analysis provided the appropriate Specification provisions are satisfied.

For Type PR construction which uses the "simple framing" approach, the restraint of the connection is ignored, provided the given conditions are met. This is no change from the ASD provisions. Where there is evidence of the actual moment rotation capability of a given type of connection, the use of designs incorporating the connection restraint is permitted just as in ASD. The designer should, when incorporating connection restraint into the design, take into account the reduced connection stiffness on the stability of the structure and its effect on the magnitude of second order effects.

A3. MATERIAL

The grades of structural steel approved for use under the LRFD Specification, covered by ASTM standard specifications, extend to a yield stress of 100 ksi. Some of these ASTM standards specify a minimum yield point, while others specify a minimum yield strength. The term "yield stress" is used in the Specification as a generic term to denote either the yield point or the yield strength.

It is important to be aware of limitations of availability that may exist for some combinations of strength and size. Not all structural section sizes are included in the various material specifications. For example, the 60 ksi yield strength steel in the A572 specification includes plate only up to 1¼ in. in thickness. Another limitation on availability is that even when a product is included in the specifications, it may be only infrequently produced by the mills. Specifying these products may result in procurement delays or require ordering large quantities directly from the producing mills. Consequently, it is prudent to check availability before completing the details of a design.

The direction parallel to the direction of rolling is the direction of principal interest in the design of steel structures. Hence, yield stress as determined by standard tensile test is the principal mechanical property recognized in the selection of the steels approved for use under the Specification. It must be recognized that other mechanical and physical properties of rolled steel, such as anisotropy, ductility, notch toughness, formability, corrosion resistance, etc., may also be important to the satisfactory performance of a structure.

It is not possible to incorporate in the Commentary adequate information to impart full understanding of all factors which might merit consideration in the selection and specification of materials for unique or especially demanding applications. In such a situation the user of the Specification is advised to make use of reference material contained in the literature on the specific properties of concern and to specify supplementary material production or quality requirements as provided for in ASTM material specifications. One such case is the design of highly restrained welded connections.[90] Rolled steel is anisotropic, especially insofar as ductility is concerned; therefore, weld contraction strains in the region of highly restrained welded connections may exceed the capabilities of the material if special attention is not given to material selection, details, workmanship and inspection.

Another special situation is that of fracture control design for certain types of service conditions.[3] The relatively warm temperatures of steel in buildings, the essentially static strain rates, the stress intensity and the number of cycles of full design stress make the probability of fracture in building structures extremely remote. Good design details which incorporate good workmanship and joint geometry that avoids severe stress concentrations are generally the most effective means of providing fracture-resistant construction. However, for especially demanding service conditions such as low temperatures with impact loading, the specification of steels with superior notch toughness may be warranted.

The ASTM standard for A307 bolts covers two grades of fasteners. Either grade may be used under the LRFD Specification; however, it should be noted that Gr. B is intended for pipe flange bolting and Gr. A is the quality long in use for structural applications.

When specifying filler metal and/or flux by AWS designation, the applicable standard specifications should be carefully reviewed to assure a complete understanding of the designation reference. This is necessary because the AWS designation systems are not consistent. For example, in the case of electrodes for shielded metal arc welding (AWS A5.1), the first two or three digits indicate the nominal tensile strength classification, in ksi, of the weld metal and the final two digits indicate the type of coating; however, in the case of mild steel electrodes for submerged arc welding (AWS A5.17), the first one or two digits times 10 indicate the nominal tensile strength classification, while the final digit or digits times 10 indicate the testing temperature in degrees F, for weld metal impact tests. In the case of low-alloy steel covered arc welding electrodes (AWS A5.5), certain portions of the designation indicate a requirement for stress relief, while others indicate no stress relief requirement.

4. Anchor Bolts and Threaded Rods

New criteria on anchor bolts and threaded rods have been included in the LRFD Specification. Since there is a limit on the maximum available length of A325 and A490, the use of these bolts for anchor bolts with design lengths longer than the maximum available lengths has presented problems in the past. The inclusion of A687 material in this Specification allows the use of higher strength material for bolts longer than A325 and A490 bolts. The designer should be aware that pretensioning anchor bolts is not recommended due to relaxation and stress corrosion after pretensioning.

A new provision for threads has been included. The designer should specify the appropriate thread and SAE fit for threaded rods used as load-carrying members.

A4. LOADS AND LOAD COMBINATIONS

1. Loads, Load Factors and Load Combinations

The load factors and load combinations given in Sect. A4.1 were developed to be used with the recommended minimum loads given in ANSI A58.1 *Minimum Design Loads for Buildings and Other Structures.*[6] The load factors and load combinations are developed in Ref. 5. The target reliability indices underlying the load factors are $\beta = 3.0$ for combinations with gravity loads only (dead, snow and live loads), 2.5 for combinations with wind included and 1.75 for combinations with earthquake loads. See Commentary A5.3 for definition of β.

The load factors and load combinations recognize that when several loads act in combination with the dead load, e.g., dead plus live plus wind loads, only one of these takes on its maximum lifetime value, while the other load is at its "arbitrary point-in-time value," i.e., at a value which can be expected to be on the structure at any time. For example, under dead, live and wind loads the following combinations are appropriate:

$$\gamma_D D + \gamma_L L \qquad \text{(C-A4-1)}$$

$$\gamma_D D + \gamma_{L_a} L_a + \gamma_W W \qquad \text{(C-A4-2)}$$

$$\gamma_D D + \gamma_L L + \gamma_{W_a} W_a \qquad \text{(C-A4-3)}$$

where γ is the appropriate load factor as designated by the subscript symbol. Subscript *a* refers to an "arbitrary point-in-time" value.

The mean value of arbitrary point-in-time live load L_a is on the order of 0.24 to 0.4 times the mean maximum lifetime live load L for many occupancies, but its dispersion is far greater. The arbitrary point-in-time wind load W_a, acting in conjunction with the maximum lifetime live load, is the maximum daily wind. It turns out that $\gamma_{W_a} W_a$ is a negligible quantity so only two load combinations remain:

$$1.2\,D + 1.6\,L \qquad \text{(C-A4-4)}$$

$$1.2\,D + 0.5\,L + 1.3\,W \qquad \text{(C-A4-5)}$$

The load factor 0.5 assigned to L in the second equation reflects the statistical properties of L_a, but to avoid having to calculate yet another load, it is reduced so it can be combined with the maximum lifetime wind load.

The nominal loads D, L, W, E and S are the code loads or the loads given in ANSI A58.1-1982.[6] The dead and live loads in Ref. 6 are essentially identical to ANSI A58.1-1972, but there have been substantial changes in the manner of determining wind, snow and earthquake loads. These changes are discussed in the appendix to Ref. 6. Only one item will be mentioned. The treatment of the live load reduction has changed to the form

$$L = L_o (0.25 + 15/\sqrt{A_I}) \le L_o \qquad \text{(C-A4-6)}$$

where L is the nominal live load, L_o is the basic live load assigned to the type of occupancy and A_I is an influence area which is equal to the tributary area for a two-way floor system, twice the tributary area for beams and four times the tributary area for columns.

2. Impact

A mass of the total moving load (wheel load) is used as the basis for impact loads on crane runway girders, because maximum impact load results when cranes travel while supporting lifted loads.

The increase in load, in recognition of random impacts, is not required to be applied to supporting columns because the impact load effects (increase in eccentricities or increases in out-of-straightness) will not develop or will be negligible during the short duration of impact. For additional information on crane girder design criteria see AISE Technical Report No. 13.

A5. DESIGN BASIS

1. Required Strength at Factored Loads

LRFD permits the use of both elastic and plastic structural analyses. LRFD provisions result in essentially the same methodology for, and end product of, plastic design as Part 2 of the 1978 AISC Specification,[1] except that the LRFD provisions tend to be more liberal reflecting added experience and the results of further research. The 10% redistribution permitted is carried forward from Part 1 of the 1978 AISC Specification.[1]

2. Limit States

A limit state is a condition which represents the limit of structural usefulness. Limit states may be dictated by functional requirements, such as maximum deflections or drift; they may be conceptual, such as plastic hinge or mechanism formation; or they may represent the actual collapse of the whole or part of the structure, such as fracture or instability. Design criteria insure that a limit state is violated only with an acceptably small probability by selecting the load and resistance factors and nominal load and resistance values which will never be exceeded under the design assumptions.

Two kinds of limit states apply for structures: limit states of strength which define safety against the extreme loads during the intended life of the structure, and limit states of serviceability which define the functional requirements. The LRFD Specification, like other structural codes, focuses on the limit states of strength because of the overriding considerations of public safety for the life, limb and property of human beings. This does not mean that limit states of serviceability are not important to the designer, who must equally insure functional performance and economy of design.

However, these latter considerations permit more exercise of judgment on the part of designers. Minimum considerations of public safety, on the other hand, are not matters of individual judgment and, therefore, specifications dwell more on the limit states of strength than on the limit states of serviceability.

Limit states of strength vary from member to member, and several limit states may apply to a given member. The following limit states of strength are the most common: onset of yielding, formation of a plastic hinge, formation of a plastic mechanism, overall frame or member instability, lateral-torsional buckling, local buckling, tensile fracture, development of fatigue cracks, deflection instability, alternating plasticity and excessive deformation.

The most common serviceability limit states include unacceptable elastic deflections and drift, unacceptable vibrations and permanent deformations.

3. Design for Strength

The general format of the LRFD Specification is given by the formula:

$$\Sigma \gamma_i Q_i \leq \phi R_n \qquad \text{(C-A5-1)}$$

where

Σ	= summation
i	= type of load, i.e., dead load, live load, wind, etc.
Q_i	= nominal load effect
γ_i	= load factor corresponding to Q_i
$\Sigma \gamma_i Q_i$	= required resistance
R_n	= nominal resistance
ϕ	= resistance factor corresponding to R_n
ϕR_n	= design strength

The left side of Formula C-A5-1 represents the required resistance which is computed by structural analysis based upon assumed loads and the right side of Formula C-A5-1 represents a limiting structural capacity provided by the selected members. In LRFD, the designer compares the effect of factored loads to the strength actually provided. The term design strength refers to the resistance or strength ϕR_n that must be provided by the selected member. The load factors γ and the resistance factor ϕ reflect the fact that loads, load effects (the computed forces and moments in the structural elements) and the resistances can be determined only to imperfect degrees of accuracy. The resistance factor ϕ is equal to or less than 1.0 because there is always a chance for the actual resistance to be less than the nominal value R_n computed by the formulas given in Chaps. D through K. Similarly, the load factors γ reflect the fact that the actual load effects may deviate from the nominal values of Q_i computed from the specified nominal loads. These factors account for unavoidable inaccuracies in the theory, variations in the material properties and dimensions and uncertainties in the determination of loads. They provide a margin of reliability to account for unexpected loads. They do not account for gross error or negligence.

The LRFD Specification is based on (1) a probabilistic model,[4,5] (2) a calibration of the new criteria to the 1978 AISC Specification,[1] and (3) the evaluation of the resulting criteria by judgment and past experience aided by comparative design office studies of representative structures.

The following is a brief probabilistic basis for LRFD.[4,5] The load effects Q and the resistance factor R are assumed to be statistically independent random variables. In

FREQUENCY

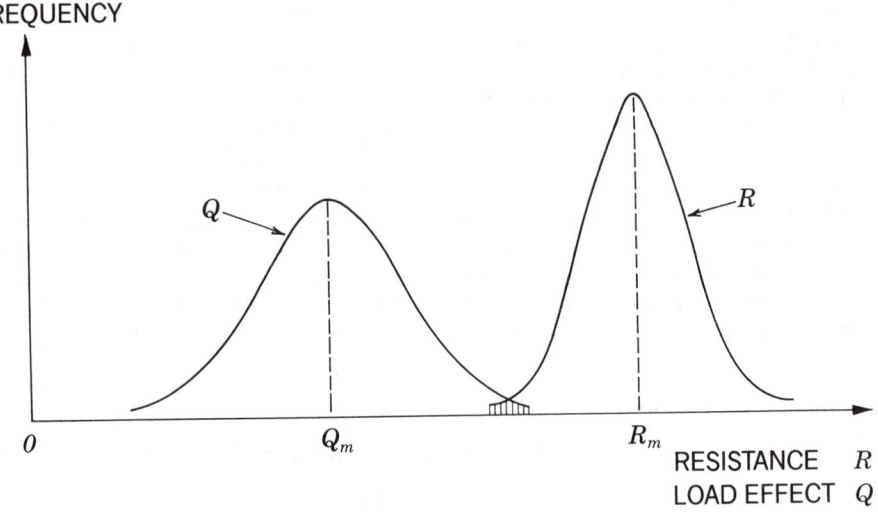

Fig. C-A5.1. Frequency distribution of load effect Q *and resistance* R

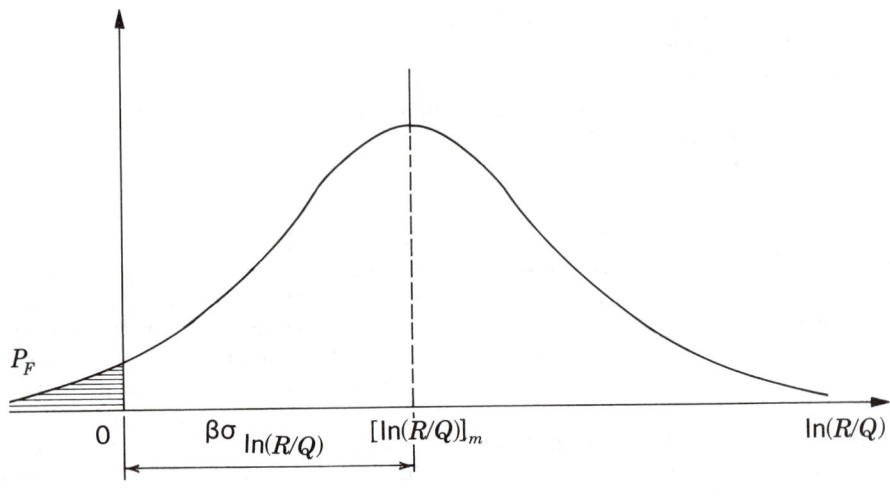

Fig. C-A5.2. Definition of reliability index

Fig. C-A5.1, frequency distributions for Q and R are portrayed as separate curves on a common plot for a hypothetical case. As long as the resistance R is greater than (to the right of) the effects of the loads Q, a margin of safety for the particular limit state exists. However, because Q and R are random variables, there is some small probability that R may be less than Q, $(R < Q)$. This is portrayed by the small shaded area where the distribution curves cross.

An equivalent situation may be represented as in Fig. C-A5.2. If the expression $R < Q$ is divided by Q and the result expressed logarithmically, the result will be a single frequency distribution curve combining the uncertainties of both R and Q. The probability of attaining a limit state ($R < Q$) is equal to the probability that $\ln(R/Q) < 0$ and is represented by the shaded area in the diagram.

The shaded area may be reduced and thus reliability increased in either of two ways: (1) by moving the mean of $\ln(R/Q)$ to the right, or (2) by reducing the spread of the curve for a given position of the mean relative to the origin. A convenient way of combining these two approaches is by defining the position of the mean using the standard deviation of the curve, $\ln(R/Q)$, as the unit of measure. Thus, the distance from the origin to the mean is measured as the number of standard deviations of the function $\ln(R/Q)$. As shown in Fig. C-A5.2, this is stated as β times $\sigma_{\ln R/Q}$, the standard deviation of $\ln R/Q$. The factor β therefore is called the "reliability index."

If the actual distribution shape of $\ln(R/Q)$ were known, and if an acceptable value of the probability of reaching the limit state could be agreed upon, one could establish a completely probability-based set of design criteria. Unfortunately, this much information frequently is not known. The distribution shape of each of the many variables (material, loads, etc.) has an influence on the shape of the distribution of $\ln(R/Q)$. Often only the means and the standard deviations of the many variables involved in the makeup of the resistance and the load effect can be estimated. However, this information is enough to build an approximate design criterion which is independent of the knowledge of the distribution, by stipulating the following design condition:

$$\beta\sigma_{\ln(R/Q)} \approx \beta\sqrt{V_R^2 + V_Q^2} \le \lambda\nu(R/Q)_m \approx \ln(R_m/Q_m) \qquad \text{(C-A5-2)}$$

In this formula, the standard deviation has been replaced by the approximation $\sqrt{V_R^2 + V_Q^2}$, where $V_R = \sigma_R/R_m$ and $V_Q = \sigma_Q/Q_m$ (σ_R and σ_Q are the standard deviations, R_m and Q_m are the mean values, V_R and V_Q are the coefficients of variation, respectively, of the resistance R and the load effect Q). For structural elements and the usual loadings R_m, Q_m, and the coefficients of variation, V_R and V_Q, can be estimated, so a calculation of

$$\beta = \frac{\ln(R_m/Q_m)}{\sqrt{V_R^2 + V_Q^2}} \qquad \text{(C-A5-3)}$$

will give a comparative value of the measure of reliability of a structure or component.

The description of the determination of β as given above is a simple way of defining the probabilistic method used in the development of LRFD. A more refined method, which can accommodate more complex design situations (such as the beam-column interaction equation) and include probabilistic distributions other than the lognormal distribution used to derive Formula C-A5-3, has been developed since the publication of Ref. 4 and is fully described in Ref. 5. This latter method has been used in the development of the recommended load factors (see Sect. A4). The two methods give essentially the same β values for most steel structural members and connections.

Statistical properties (mean values and coefficients of variations) are presented for the basic material properties and for steel beams, columns, composite beams, plate girders, beam-columns and connection elements in a series of eight articles in the September 1978 issue of the *Journal of the Structural Division of ASCE* (Vol. 104, ST9). The corresponding load statistics are given in Ref. 5. Based on these statistics, the values of β inherent in the 1978 AISC Specification[1] were evaluated under different load combinations (live/dead, wind/dead, etc.), and for various tributary areas for typical members (beams, columns, beam-columns, structural components, etc.). As can be expected, there was a considerable variation in the range of β values.

Examination of the many β values which were computed revealed certain trends. For example, typical beams have β values on the order of 3, and for typical connectors β was on the order of 4 to 5. The reliability index β for load combinations involving wind and earthquake tended to be lower.

One of the features of the probability-based method used in the development of LRFD is that the variations of β can be reduced by specifying several "target" β values, and selecting multiple load and resistance factors to meet these targets.

The following targets were selected: (1) under dead plus live and/or snow loading, β = 3.0 for members and β = 4.5 for connections; (2) under dead plus live plus wind loading, β = 2.5 for members; and (3) under dead plus live plus earthquake loading, β = 1.75 for members. The larger value of β = 4.5 for connectors reflects the fact that connections are expected to be stronger than the members they connect. The lower β-values for load combinations involving wind or earthquake loading are consistent with previous specifications.

Based on the target values of β given above, common load factors for various structural materials (steel, reinforced concrete, etc.) were developed in Ref. 5. Computer methods as well as charts are also given in Ref. 5 for the use of specification writers to determine the resistance factors ϕ. These factors can also be approximately determined by the following:

$$\phi = (R_m/R_n) \exp(- 0.55 \beta V_R) \qquad \text{(C-A5-4)}^*$$

where

R_m = mean resistance
R_n = nominal resistance according to the formulas in Chaps. D through K
V_R = coefficient of variation of the resistance

4. Design for Serviceability and Other Considerations

Nominally, serviceability should be checked at the unfactored loads. For combinations of gravity and wind or seismic loads some additional reduction factor may be warranted.

---

*Note that exp(x) is identical to the more familiar e^x.

CHAPTER B.
DESIGN REQUIREMENTS

B3. EFFECTIVE NET AREA

Section B3 deals with the effect of shear lag. The inclusion of welded members acknowledges that shear lag is also a factor in determining the effective area of welded connections where the welds are so distributed as to directly connect some, but not all, of the elements of a tension member. However, since welds are applied to the unreduced cross-sectional area, the reduction coefficient U is applied to the gross area A_g. With this modification the values of U are the same as for similar shapes connected by bolts and rivets except that: (1) the provisions for members having only two fasteners per line in the direction of stress have no application to welded connections; and (2) tests[7] have shown that flat plates, or bars axially loaded in tension and connected only by longitudinal fillet welds, may fail prematurely by shear lag at their corners if the welds are separated by too great a distance. Therefore, the values of U are specified unless the member is designed on the basis of effective net area as discussed below.

As the length of a connection ℓ is increased the intensity of shear lag is diminished. The concept can be expressed empirically as:

$$U = 1 - \overline{x}/\ell$$

Munse and Chesson have shown,[43,92,93] using this expression to compute an effective net area, that with few exceptions, the estimated strength of some 1,000 test specimens correlated with observed test results within a scatterband of ± 10%. For any given profile and connected elements, $\overline{x}$ is a fixed geometric property. Length ℓ, however, is dependent upon the number of fasteners or length of weld required to develop the given tensile force, and this in turn is dependent upon the mechanical properties of the member and the capacity of the fasteners or weld used. The values of U, given as reduction coefficients in Sect. B3, are reasonably lower bounds for the profile types and connection means described, based upon the use of the above expression.

B4. STABILITY

The stability of structures must be considered from the standpoint of the structures as a whole, including not only the compression members, but also the beams, bracing system and connections. The stability of individual elements must also be provided. Considerable attention has been given to this subject in the technical literature, and various methods of analysis are available to assure stability. The SSRC *Guide to Design Criteria for Metal Compression Members*[11] devotes several chapters to the stability of different types of members considered as individual elements, and then considers the effects of individual elements on the stability of the structure as a whole.

B5. LOCAL BUCKLING

For the purposes of this Specification, steel sections are divided into compact sections, noncompact sections and sections with slender compression elements. Compact sections are capable of developing a *fully plastic* stress distribution and possess rotation capacity of approximately 3 before the onset of local buckling.[14] Noncompact sections can develop the yield stress in compression elements before local buckling occurs, but will not resist inelastic local buckling at the strain levels required for a fully plastic stress distribution. Slender compression elements buckle elastically before the yield stress is achieved.

The dividing line between compact and noncompact sections is the limiting width-thickness ratio λ_p. For a section to be compact, all of its compression elements must have width-thickness ratios smaller than the limiting λ_p.

A greater inelastic rotation capacity than provided by the limiting values λ_p given in Table B5.1 may be required for some structures in areas of high seismicity. It has been suggested that in order to develop a ductility of from 3 to 5 in a structural member, ductility factors for elements would have to lie in the range of 5 to 15. Thus, in this case, it is prudent to provide for an inelastic rotation of 7 to 9 times the elastic rotation.[8] In order to provide for this rotation capacity the limits λ_p for local flange and web buckling would be as shown in Table C-B5.1.[9]

Another limiting width-thickness ratio is λ_r, representing the distinction between noncompact sections and sections with slender compression elements. As long as the width-thickness ratio of a compression element does not exceed the limiting value λ_r, local elastic buckling will not govern its strength. However, for those cases where the width-thickness ratios exceed λ_r, elastic buckling strength must be considered. A design procedure for such slender compression elements, based on elastic buckling of plates, is given in Appendix B5.3. An exception is plate girders with slender webs. Such plate girders are capable of developing postbuckling strength in excess of the elastic buckling load. A design procedure for plate girders including tension field action is given in Appendix G.

The values of the limiting ratios λ_p and λ_r specified in Table B5.1 were obtained from Sects. 1.9 and 2.7 of Ref. 1 and from Table 2.3.3.3 of Ref. 9, except that: (1) $\lambda_p = 65/\sqrt{F_y}$, limited in Ref. 9 to indeterminate beams when moments are determined by elastic analysis and to determinate beams, was adopted for all conditions on the basis of Ref. 14; and (2) $\lambda_p = 1{,}300/F_y$ for circular hollow sections was obtained from Ref. 10.

The high shape factor for circular hollow sections makes it impractical to use the same slenderness limits to define the regions of behavior for different types of loading. In Table B5.1, the values of λ_p for a compact shape that can achieve the plastic moment, and λ_r for bending, are based on an analysis of test data from several projects involving the bending of pipes in a region of constant moment.[79,81] The same analysis produced the equation for the inelastic moment capacity in Table A-F1.1 in Appendix F1.7. However, a more restrictive value of λ_p is required to prevent inelastic local buckling from limiting the plastic hinge rotation capacity needed to develop a mechanism in a circular hollow beam section.[10]

The values of λ_r for axial compression and for bending are both based on test data. The former value has been used in building specifications since 1968.[80] Appendices B5.3 and F1.7 also limit the diameter-to-thickness ratio for any circular section to $13{,}000/F_y$. Beyond this, the local buckling strength decreases rapidly, making it impractical to use these sections in building construction.

Following the SSRC recommendations[11] and the approach used for other shapes

TABLE C-B5.1
Limiting Width-thickness Ratios for Compression Elements

Description of Element	Width-thick-ness Ratio	Limiting Width-thickness Ratios λ_p	
		Non-seismic	Seismic
Flanges of I-shaped sections (including hybrid sections) and channels in flexure[a]	b/t	$65/\sqrt{F_y}$	$52/\sqrt{F_y}$
Webs in combined flexural and axial compression	h_c/t_w	For $P_n/\phi_b P_y \leqq 0.125$	
		$\dfrac{640}{\sqrt{F_y}}\left(1 - \dfrac{2.75P}{\phi_b P_y}\right)$	$\dfrac{520}{\sqrt{F_y}}\left(1 - \dfrac{1.54P}{\phi_b P_y}\right)$
		For $P_n/\phi_b P_y > 0.125$	
		$\dfrac{191}{\sqrt{F_u}}\left(2.33 - \dfrac{P}{\phi_b P_y}\right) \geq \dfrac{253}{\sqrt{F_y}}$	$\dfrac{152}{\sqrt{F_y}}\left(2.89 - \dfrac{P}{\phi_b P_y}\right)$

[a]For hybrid beams use F_{yf} in place of F_y.

with slender compression elements, a Q-factor is used for circular sections to account for an interaction between local and column buckling. The Q-factor is the ratio between the local buckling stress and the yield stress. The local buckling stress for the circular section is taken from the inelastic AISI criteria[80] and is based on tests conducted on fabricated and manufactured cylinders. More recent tests on fabricated cylinders[81] confirm that this equation is conservative.

The definitions of the width and thickness of compression elements were taken from Sect. 1.9 of the 1978 AISC Specification[1] with minor modifications extending their applicability to sections formed by bending and to unsymmetrical and hybrid sections.

B7. LIMITING SLENDERNESS RATIOS

Chapters D and E provide reliable criteria for resistance of axially loaded members based on theory and confirmed by test for all significant parameters including slenderness. The advisory upper limits on slenderness contained in B7 are based on professional judgement and practical considerations of economics, ease of handling, and care required to minimize inadvertent damage during fabrication, transport and erection. Out-of-straightness within reasonable tolerances does not affect the strength of tension members, and the effect of out-of-straightness within specified tolerances on the strength of compression members is accounted for in formulas for resistance. Applied tension tends to reduce whereas compression tends to amplify out-of-straightness. Therefore, more liberal criteria are suggested for tension members, including those subject to small compressive force resulting from transient loads such as earthquake and wind. For members with slenderness ratios greater than 200, these compressive forces correspond to stresses less than 2.6 ksi.

CHAPTER C.
FRAMES AND OTHER STRUCTURES

C1. SECOND ORDER EFFECTS

While resistance to wind and seismic loading can be provided in certain buildings by means of concrete or masonry shear walls, which also provide for overall frame stability at factored gravity loading, other building frames must provide this resistance by acting alone. This resistance can be achieved in several ways, e.g., by a system of bracing, by a moment-resisting frame or by any combination of lateral force-resisting elements.

For frames under combined gravity and lateral loads, drift (horizontal deflection caused by applied loads) occurs at the start of loading. At a given value of the applied loads, the frame has a definite amount of drift Δ. In unbraced frames, significant additional secondary bending moments, known as the $P\Delta$ moments, may be developed in the columns and beams of the lateral load-resisting systems in each story. P is the total gravity load above the story and Δ is the story drift. As the applied load increases, the $P\Delta$ moments also increase. Therefore, the $P\Delta$ effect often should be accounted for in frame design. Similarly, in braced frames, increases in axial forces occur in the members of the bracing systems; however, such effects are usually less significant. The designer should consider these effects for all types of frames and determine if they are significant for his particular case. When considering frame instability effects ($P\Delta$) in the design of these frames, the equations in Chap. H offer the designer a direct method for including such considerations in calculations of required strength moments. The designer also has the option to compute second order effects directly. Since $P\Delta$ effects can cause frame drifts to be larger than those calculated by ignoring them, $P\Delta$ effects should also be included in the drift analysis when they are significant.

In plastically designed unbraced frames the limit of $0.75\,P_y$ on column axial loads has also been retained to help insure proper stability.

C2. FRAME STABILITY

The stability of structures must be considered from the standpoint of the structure as a whole, including not only the compression members, but also the beams, bracing system and connections. The stability of individual elements must also be provided. Considerable attention has been given in the technical literature to this subject, and various methods of analysis are available to assure stability. The SSRC *Guide to Design Criteria for Metal Compression Members*[11] devotes several chapters to the stability of different types of members considered as individual elements, and then considers the effects of individual elements on the stability of the structure as a whole.

The effective length concept is one method for estimating the interaction effects of the total frame on a compression element being considered. This concept uses K-factors to equate the strength of a framed compression element of length L to an equivalent pin-ended member of length KL subject to axial load only. Other rational methods are available for evaluating the stability of frames subject to gravity and side loading and individual compression members subject to axial load and moments. However, the effective length concept is the only tool currently available for handling several cases which occur in practically all structures, and it is an essential part of many analysis procedures. Although the concept is completely valid for ideal structures, its practical implementation involves several assumptions of idealized conditions which will be mentioned later.

Table C-C2.1

	(a)	(b)	(c)	(d)	(e)	(f)
Buckled shape of column is shown by dashed line	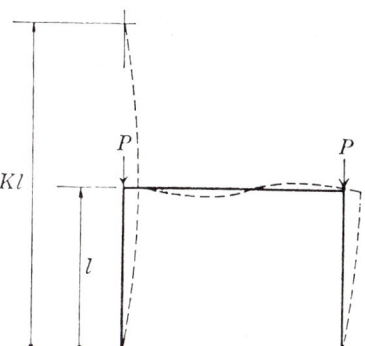					
Theoretical K value	0.5	0.7	1.0	1.0	2.0	2.0
Recommended design value when ideal conditions are approximated	0.65	0.80	1.2	1.0	2.10	2.0
End condition code		Rotation fixed and translation fixed				
		Rotation free and translation fixed				
		Rotation fixed and translation free				
		Rotation free and translation free				

Figure C-C2.1

Two conditions, opposite in their effect upon column strength under axial loading, must be considered. If enough axial load is applied to the columns in an unbraced frame dependent entirely on its own bending stiffness for resistance to lateral deflection of the tops of the columns with respect to their bases (see Fig. C-C2.1), the effective length of these columns will exceed the actual length. On the other hand, if the same frame were braced to resist such lateral movement, the effective length would be less than the actual length, due to the restraint (resistance to joint rotation) provided by the bracing or other lateral support. The ratio K, effective column length to actual unbraced length, may be greater or less than 1.0.

The theoretical K-values for six idealized conditions in which joint rotation and translation are either fully realized or nonexistent are tabulated in Table C-C2.1. Also shown are suggested design values recommended by the Structural Stability Research Council (formerly the Column Research Council) for use when these conditions are approximated in actual design. In general, these suggested values are slightly higher than their theoretical equivalents, since joint fixity is seldom fully realized.

If the column base in Case f of Table C-C2.1 were truly pinned, K would actually exceed 2.0 for a frame such as that pictured in Fig. C-C2.1, because the flexibility of the horizontal member would prevent realization of full fixity at the top of the column. On the other hand, it has been shown[51] that the restraining influence of foundations, even where these footings are designed only for vertical load, can be very substantial in the case of flat-ended column base details with ordinary anchorage. For this condition, a design K-value of 1.5 would generally be conservative in Case f.

While in some cases the existence of masonry walls provides enough lateral support for their building frames to control lateral deflection, the increasing use of light curtain wall construction and wide column spacing for high-rise structures not provided with a positive system of diagonal bracing can create a situation where only the bending stiffness of the frame itself provides this support.

In this case the effective length factor K for an unbraced length of column L is dependent upon the amount of bending stiffness provided by the other inplane members entering the joint at each end of the unbraced segment. If the combined stiffness provided by the beams is sufficiently small, relative to that of the unbraced column segments, KL could exceed two or more story heights.*

Several rational methods are available to estimate the effective length of the columns in an unbraced frame with sufficient accuracy. These range from simple interpolation between the idealized cases shown in Table C-C2.1 to very complex analytical procedures. Once a trial selection of framing members has been made, the use of the alignment chart in Fig. C-C2.2 affords a fairly rapid method for determining adequate K-values.

However, it should be noted that this alignment chart is based upon assumptions of idealized conditions which seldom exist in real structures.[11] These assumptions are as follows:

1. Behavior is purely elastic.
2. All members have constant cross section.
3. All joints are rigid.
4. For braced frames, rotations at opposite ends of beams are equal in magnitude, producing single curvature bending.
5. For unbraced frames, rotations at opposite ends of the restraining beams are equal in magnitude, producing reverse curvature bending.
6. The stiffness parameters $L\sqrt{P/EI}$ of all columns are equal.
7. Joint restraint is distributed to the column above and below the joint in proportion to I/L of the two columns.
8. All columns buckle simultaneously.

Where the actual conditions differ from these assumptions, unrealistic designs may result. There are design procedures available[52,53] which may be used in the calculation of G for use in Fig. C-C2.2 to give results more truly representative of conditions in real structures.

It is worth mentioning here that frames which use PR connections violate the condition that all joints are rigid, and special attention should be paid to calculation of a proper G for these instances.

In frames which depend upon their own bending stiffness for stability, the amplified moments are accounted for in the design of columns by means of the

---

*Ref. 50, pp. 260–265.

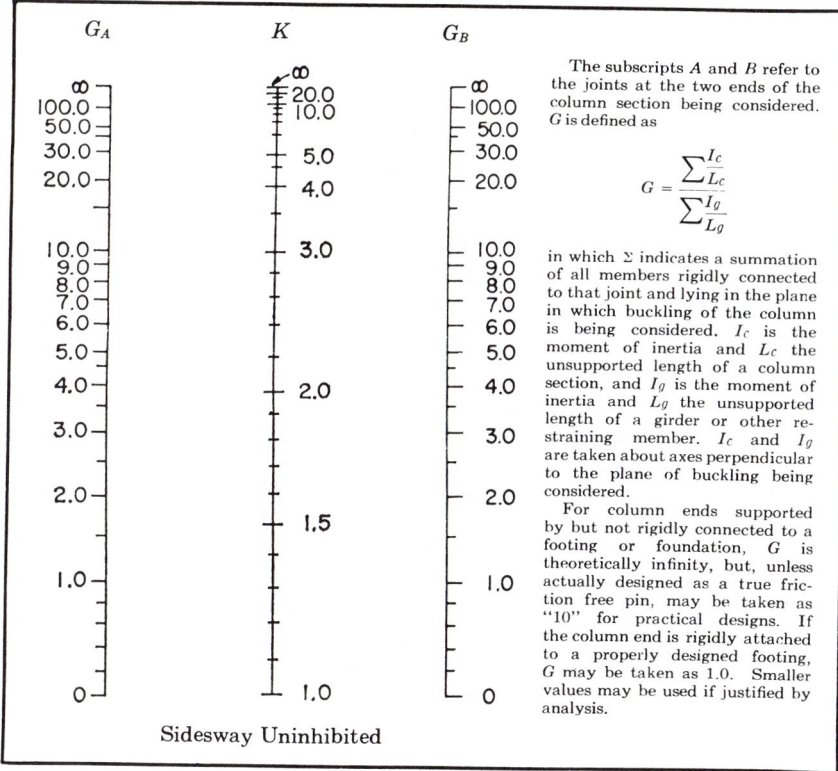

Alignment Chart for Effective Length of Columns in Continuous Frames

Figure C-C2.2

interaction formulas of Sect. H1. However, moments are also induced in the beams which restrain the columns; thus, consideration must be given to the amplification of those portions of the beam moments that are increased when the frame drifts. The effect may be particularly important in frames in which the contribution to individual beam moments from story shears becomes small as a result of distribution to many bays, but in which the $P\Delta$ moment in individual columns and beams is not diminished and becomes dominant.

If roof decks or floor slabs, anchored to shear walls or vertical plane bracing systems, are counted upon to provide lateral support for individual columns in a building frame, due consideration must be given to their stiffness when functioning as a horizontal diaphragm.[59]

While translation of the joints in the plane of a truss is inhibited and, due to end restraint, the effective length of compression members might therefore be assumed to be less than the distance between panel points, it is usual practice to take K as equal to 1.0,[11] since, if all members of the truss reached their ultimate load capacity simultaneously, the restraints at the ends of the compression members would disappear or, at least, be greatly reduced.

CHAPTER D.
TENSION MEMBERS

D1. DESIGN TENSILE STRENGTH

Due to strain hardening, a ductile steel bar loaded in axial tension can resist, without fracture, a force greater than the product of its gross area and its coupon yield stress. However, excessive elongation of a tension member due to uncontrolled yielding of its gross area not only marks the limit of its usefulness, but can precipitate failure of the structural system of which it is a part. On the other hand, depending upon the scale of reduction of gross area and the mechanical properties of the steel, the member can fail by fracture of the net area at a load smaller than required to yield the gross area. Hence, general yielding of the gross area and fracture of the net area both constitute failure limit states. The relative values of ϕ_t given for yielding and fracture reflect the same basic difference in factor of safety as between design of members and design of connections in the 1978 AISC Specification, Part 1.[1]

The part of the member occupied by the net area at fastener holes has a negligible length relative to the total length of the member. As a result, the strain hardening condition is quickly reached and yielding of the net area at fastener holes does not constitute a limit state of practical significance.

D2. BUILT-UP MEMBERS

The slenderness ratio L/r of tension members other than rods, tubes or straps should preferably not exceed the limiting values of 240. This slenderness limit recommended for tension members is not essential to the structural integrity of such members; they merely assure a degree of stiffness such that undesirable lateral movement ("slapping" or vibration) will be unlikely.

See Commentary Sect. E4.

D3. EYEBARS AND PIN-CONNECTED MEMBERS

Forged eyebars have generally been replaced by pin-connected plates or eyebars thermally cut from plates. Provisions for the proportioning of eyebars contained in the LRFD Specification are based upon standards evolved from long experience with forged eyebars. Through extensive destructive testing, eyebars have been found to provide balanced designs when they are thermally cut instead of forged. The somewhat more conservative rules for pin-connected members of nonuniform cross section and those not having enlarged "circular" heads are likewise based on the results of experimental research.[4] For greater clarity, the provisions relating to pin plate design have been extensively reworded and updated.

Somewhat stockier proportions are provided for eyebars and pin-connected members fabricated from steel having a yield stress greater than 70 ksi, in order to eliminate any possibility of their "dishing" under the higher working stress for which they may be designed.

CHAPTER E.
COLUMNS AND OTHER
COMPRESSION MEMBERS

E1. EFFECTIVE LENGTH AND SLENDERNESS LIMITATIONS

1. Effective Length

The Commentary on Sect. C2 regarding frame stability and effective length factors applies here. Further analytic methods, formulas, charts and references for the determination of effective length are provided in Chap. 15 of the SSRC *Guide to Design Criteria for Metal Compression Members.*[11]

2. Plastic Analysis

The limitation on λ_c is essentially the same as that for L/r in Sect. 2.4 of the 1978 AISC Specification.[1]

E2. DESIGN COMPRESSIVE STRENGTH*

Formulas E2-2 and E2-3 are based on a reasonable conversion of research data into design equations. Conversion of the allowable stress design (ASD) equations based on the Structural Stability Research Council (SSRC) (formerly the Column Research Council)[11] was found to be cumbersome for two reasons. The first was the nature of the ASD variable safety factor. Secondly, the difference in philosophical origins of the two design procedures requires an assumption of a live load-to-dead load ratio (L/D).

Since all L/D ratios could not be considered, a value of approximately 1.1 at λ equal to 1.0 was used to calibrate the exponential equation for columns with the lower range of λ against the appropriate ASD provision. The coefficient with the Euler equation was obtained by equating the two equations at the common λ of 1.5.

Formulas E2-2 and E2-3 are essentially the same as column-strength curve 2P of the 4th edition SSRC *Guide* for an out-of-straightness criterion of $L/1500$.[81,99]

It should be noted that this set of column equations has a range of reliability (β)-values. At low and high column slenderness, β-values exceeding 3.0 and 3.3 respectively are obtained compared to β of 2.60 at L/D of 1.1. This is considered satisfactory, since the limits of out-of-straightness combined with residual stress have not been clearly established. Furthermore, there has been no history of unacceptable behavior of columns designed using the ASD procedure. This includes cases with L/D ratios greater than 1.1.

Formulas E2-2 and E2-3 can be restated in terms of the more familiar slenderness $K\ell/r$. First, Formula E2-2 is expressed in exponential form,

$$F_{cr} = [\exp(-0.419\lambda_c^2)]F_y \qquad \text{(C-E2-1)}$$

---

*For tapered members, also see Commentary Sect. D2.

Note that exp(x) is identical to e^x. Substitution of λ_c according to Formula E2-4 gives,

for $\dfrac{K\ell}{r} \leq 4.71 \sqrt{\dfrac{E}{F_y}}$

$$F_{cr} = \left\{ \exp\left[-0.0424 \frac{F_y}{E} \left(\frac{K\ell}{r} \right)^2 \right] \right\} F_y \qquad \text{(C-E2-2)}$$

for $\dfrac{K\ell}{r} > 4.71 \sqrt{\dfrac{E}{F_y}}$

$$F_{cr} = \frac{0.877\pi^2 E}{\left(\dfrac{K\ell}{r} \right)^2} \qquad \text{(C-E2-3)}$$

The design strength of columns and other compression members of 36 and 50 ksi structural steels is included in Tables 3-36 and 3-50 of the LRFD Specification for the convenience of the designer.

E3. FLEXURAL-TORSIONAL BUCKLING

Torsional buckling of symmetric shapes and flexural-torsional buckling of unsymmetric shapes are failure modes usually not considered in the design of hot-rolled columns. They generally do not govern or the critical load differs very little from the weak axis planar buckling load. Such buckling loads may, however, control the capacity of symmetric columns made from relatively thin plate elements and of unsymmetric columns. Design equations for determining the strength of such columns are given in Appendix E3.

E4. BUILT-UP MEMBERS

Requirements for detailing and design of built-up members, which cannot be stated in terms of calculated stress, are based upon judgment and experience.

The longitudinal spacing of connectors connecting components of built-up compression members must be such that the slenderness ratio L/r of individual shapes does not exceed the slenderness ratio of the entire member. Additional requirements are imposed for built-up members consisting of angles. However, these minimum requirements do not necessarily ensure that the effective slenderness ratio of the built-up member is equal to that for the built-up member acting as a single unit. Appendix E4 gives formulas for modified slenderness ratios that are based on research and take into account the effect of shear deformation in the connectors.[94] The connectors must be designed to resist the shear forces which develop in the buckled member. The shear stresses are highest where the slope of the buckled shape is maximum.[50]

Maximum fastener spacing less than that required for strength may be needed to ensure a close fit over the entire faying surface of components in continuous contact. Specific requirements are given for weathering steel members exposed to atmospheric corrosion.[91]

The provisions governing the proportioning of perforated cover plates are based upon extensive experimental research.[95]

CHAPTER F.
BEAMS AND OTHER FLEXURAL MEMBERS

F1. DESIGN FOR FLEXURE

1. Unbraced Length for Plastic Analysis

In the 1978 AISC Specification,[1] Part 2, the unbraced length of a beam which permits the attainment of plastic moments and sufficient rotation capacity to redistribute moments is given by two formulas which depend on the moment ratio at the ends of the unbraced length. One length was permitted for $M_1/M_p < -0.5$ (almost uniform moment), and a substantially larger length for $M_1/M_p > -0.5$. These two formulas are replaced by (F1-1) which provides for a continuous function between unbraced length and end moment ratio so there is no abrupt change for a slight change in moment ratio near -0.5. At $M_1/M_p = -1.0$ (uniform moment) the maximum unbraced length is almost the same as the 1978 Specification. There is a substantial increase in unbraced length for positive moment ratios (reverse curvature) because the yielding is confined to zones close to the brace points.[14] Formulas F1-1 and F1-2 assume that the moment diagram within the unbraced length next to plastic hinge locations is reasonably linear. For nonlinear diagrams between braces, judgement should be used in choosing a representative ratio.

Formulas F1-1 and F1-2 were developed to provide rotations capacities of at least 3.0, which are sufficient for most applications.[14] When inelastic rotations of 7 to 9 are deemed appropriate in areas of high seismicity, as discussed in Commentary Sect. B5, Formula F1-1 would become:[9]

$$L_{pd} = \frac{146 \, r_y}{\sqrt{F_y}} \qquad \text{(C-F1-1)}$$

3. Compact Section Members with $L_b \le L_r$

The basic relationship between nominal moment M_n and unbraced length L_b is shown in Fig. C-F1.1 for a compact section with $C_b = 1.0$. There are four principal zones defined on the basic curve by L_{pd}, L_p and L_r. Formula F1-4 defines the maximum unbraced length L_p to reach M_p with uniform moment. Elastic lateral-torsional buckling will occur when the unbraced length is greater than L_r given by Formula F1-6. Formula F1-3 defines the inelastic lateral-torsional buckling as a straight line between the defined limits L_p and L_r. Buckling strength in the elastic region $L_b > L_r$ is given by Formula F1-13 for I-shaped members.

For other moment diagrams, the lateral buckling strength is obtained by multiplying the basic strength by C_b as shown in Fig. C-F1.1. The maximum M_n, however, is limited to M_p. Note that L_p given by Formula F1-4 is merely a definition which has physical meaning when $C_b = 1.0$. For C_b greater than 1.0, larger unbraced lengths are permitted to reach M_p as shown by the curve for $C_b = 2.3$. For design, this length could be calculated by setting Formula F1-3 equal to M_p and solving this equation for L_b using the desired C_b value.

For nonlinear moment diagrams between brace points, especially where the largest moment is not at the ends, C_b larger than 1.0 may be warranted. See Ref. 11 for a summary of such cases.

The elastic strength of hybrid beams is identical to homogeneous beams. The strength advantage of hybrid sections becomes evident only in the inelastic and plastic slenderness ranges.

4. Compact Section Members with $L_b > L_r$

The equation given in the Specification assumes that the loading is applied along the beam centroidal axis. If the load is placed on the top flange and is not braced, there is a tipping effect that reduces the critical moment; conversely, if the load is suspended from the bottom flange and is not braced, there is a stabilizing effect which increases the critical moment.[81] For unbraced top flange loading, the reduced critical moment may be conservatively approximated by setting the warping buckling factor X_2 to zero.

5. Tees and Double-angle Beams

The lateral-torsional buckling strength of singly symmetric tee beams is given by a fairly complex formula.[11] This buckling formula has been simplified considerably in Ref. 15 and Formula F1-15 is based on this work.

7. Nominal Flexural Strength of Other Sections

Formulae for the nominal strength of various types of sections are given in Appendix F1.

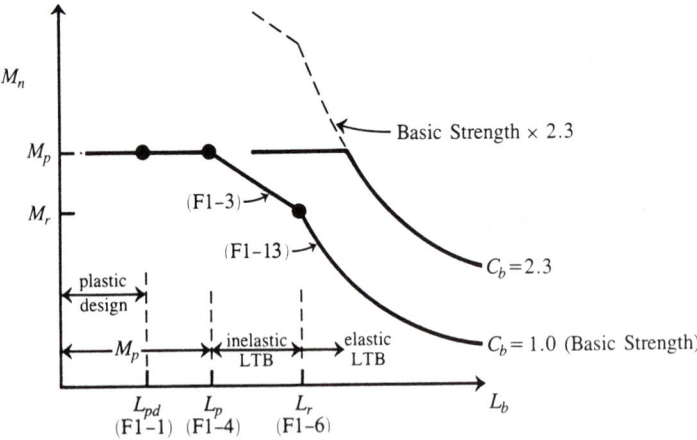

Fig. C-F1.1. *Nominal moment as a function of unbraced length and moment gradient.*

F2. DESIGN FOR SHEAR

For webs with $h/t_w \leq 187\sqrt{k/F_{yw}}$, the nominal shear strength V_n is based on shear yielding of the web, Formula F2-1. This h/t_w limit was determined by setting the critical stress causing shear buckling F_{cr} equal to the yield point of the web F_{yw} in Formula 35 of Ref. 16. When $h/t_w > 187 \sqrt{k/F_{yw}}$, the web shear strength is based on buckling. Basler[17] suggested taking the proportional limit as 80% of the yield stress of the web. This corresponds to $h/t_w = (187/0.8)(\sqrt{k/F_{yw}})$. Thus, when $h/t_w > 234(\sqrt{k/F_{yw}})$, the web strength is determined from the elastic buckling stress given by Formula 6 of Ref. 16:

$$F_{cr} = \frac{\pi^2 Ek}{12(1 - v^2)(h/t_w)^2} \qquad \text{(C-F2-1)}$$

The nominal shear strength, given by Formula F2-3, was obtained by multiplying F_{cr} by the web area and using $E = 29{,}000$ ksi and $v = 0.3$. A straight line transition, Formula F2-2, is used between the limits $187(\sqrt{k/F_{yw}})$ and $234(\sqrt{k/F_{yw}})$.

The shear strength of flexural members follows the approach used in the 1978 AISC Specification,[1] except for two simplifications. First, the expression for the plate buckling coefficient k has been simplified; it corresponds to that given in the AASHTO *Standard Specification for Highway Bridges*.[3] The earlier expression for k was a curve fit to the exact expression; the new recommendation is just as accurate. Second, the alternate method (tension field action) for web shear strength which utilized post buckling strength has been placed in Appendix G because the AISC Specification Advisory Committee recommends that only one method appear in the main body of the Specification with alternate methods given in the Appendix. When designing plate girders, thicker unstiffened webs will frequently be less costly than lighter stiffened web designs because of the additional fabrication. If a stiffened girder design has economic advantages, the tension field method in Appendix G will require fewer stiffeners.

The formulas in this section were established assuming monotonically increasing loads. If a flexural member is subjected to load reversals causing cyclic yielding over large portions of a web, such as may occur during a major earthquake, special design considerations may apply.[18]

CHAPTER H.
MEMBERS UNDER TORSION
AND COMBINED FORCES

H1. SYMMETRIC MEMBERS SUBJECT TO BENDING AND AXIAL FORCE

The new Formulas H1-1a and H1-1b are simplifications and clarifications of Formulas 1.6-1a and 1.6-1b used in the AISC Specification[1] since 1961. Previously, both equations had to be checked, Formula 1.6-1a being the check for stability and Formula 1.6-1b being the check for strength. In the new formulation the applicable equation is governed by the value of the first term, $P_u/\phi P_n$. For bending about one axis only, the equations have the form shown in Fig. C-H1.1.

The first term $P_u/\phi P_n$ has the same significance as the axial load term f_a/F_a in Formula 1.6-1a of the 1978 AISC Specification. This means that P_n must be based on the largest effective slenderness ratio $K\ell/r$. In the development of Formulas H1-1a and H1-1b, a number of alternative formulations were compared to the exact inelastic solutions of 82 sidesway cases reported in Ref. 61. In particular, the possibility of using $K\ell/r$ as the actual column length ($K = 1$) in determining P_n, combined with an elastic second order moment M_u, was studied. In those cases where the true P_n based on $K\ell/r$, with $K = 1.0$, was in the inelastic range, the errors proved to be unacceptably large without the additional check that $P_u \le \phi_c P_n$, P_n being based on effective length. Although deviations from exact solutions were reduced they still remained high.

In summary, it is not possible to formulate a safe general interaction equation for compression without considering effective length directly (or indirectly by a second equation). Therefore, the requirement that the nominal compressive strength P_n be

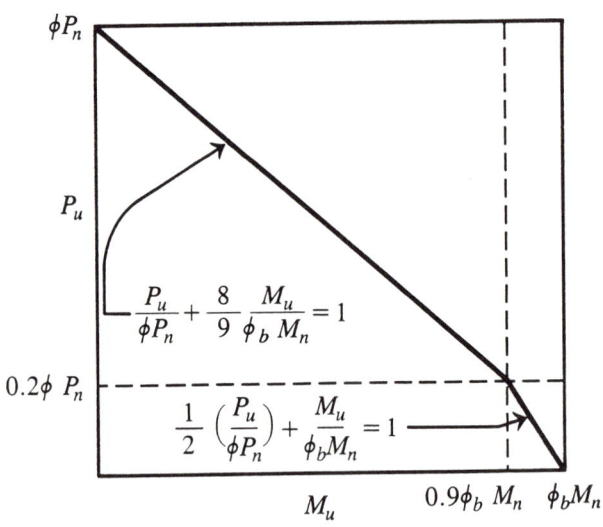

Figure C-H1.1

based on the effective length KL in the general interaction equation is continued in the LRFD Specification as it has been in the AISC Specification since 1961. It is not intended these provisions be applicable to limit nonlinear secondary flexure that might be encountered in large amplitude earthquake stability design.[28]

An important benefit which follows from including the correct value of P_n in Formulas H1-1a and H1-1b is that the factor B_2 may be based on first order sidesway deflection. Alternatively, the product $B_2 M_{lt}$ may be directly determined by a $P\Delta$ analysis as outlined in Ref. 62.

The newly defined term M_u is the maximum moment in a member, including the magnifying effect of compressive axial loads. In the general case, a member may have first order moments not associated with sidesway which are multiplied by B_1, and first order moments produced by forces causing sidesway which are multiplied by B_2. Unlike the 1978 AISC Specification,[1] B_1 is never less than one in the LRFD Specification.

In frames subject to sidesway, earlier Specifications required that all moments be multiplied by the factor $0.85/(1 - f_a/F'_e)$. This often unnecessarily increased gravity moments not associated with sidesway. The new factor B_2 applies only to moments caused by forces producing sidesway and is calculated for an entire story. In building frames designed to limit Δ_{oh}/L to a predetermined value, the factor B_2 may be found in advance of designing indidual members. Drift limits may also be set for design of various categories of buildings so that the effect of secondary bending can be insignificant.[27,28] Alternative formulations of $B_2 M_{lt}$ may be made at the designer's option.[24,25,26,27] It is conservative to use the B_2 factor with the total of the sway and no sway moments, as in the 1978 AISC Specification.

The two kinds of first order moment M_{nt} and M_{lt} may both occur in sidesway frames from gravity loads. M_{nt} may be defined as a moment developed in a member with frame sidesway prevented. If a significant restraining force is necessary to prevent sidesway of an unsymmetrical structure (or an unsymmetrically loaded symmetrical structure), the moments induced by releasing the restraining force will be M_{lt} moments, to be multiplied by B_2. In most reasonably symmetric frames, this effect will be trivial. If such a moment $B_2 M_{lt}$ is added algebraically to the $B_1 M_{nt}$ moment developed with sidesway prevented, a fairly accurate value of M_{lt} will result. Of course end moments produced in sidesway frames by lateral loads from wind or earthquake will always be M_{lt} moments to be multiplied by B_2.

For compression members in braced frames, Formulas H1-1a and H1-1b (which is a continuous function) are similar in application to the previous Formula 1.61-1a. B_1 is determined from C_m values which are unchanged from the 1978 AISC Specification except that the 0.4 limit has been removed in Formula H1-4. A significant change, however, is that B_1 is never less than 1. When $C_m = 1$ for a compression member loaded between its supports the factors of ⅜ and ½ make the new equations more liberal than Formula 1.6-1a. For $C_m \leq 1$ (for members with unequal end moments) the new equations will be slightly more conservative than the 1978 AISC Specification for a very slender member with low C_m. For the entire range of l/r and C_m, the equations compare very closely to exact inelastic solutions of braced members.

For braced frames, $\phi_c P_n$ will always be based on $K \leq 1$, and B_1 will always be determined from P_e based on $K \leq 1$. The actual length of members is used in the structural analysis. In braced and unbraced frames, P_n is governed by the maximum slenderness ratio regardless of the plane of bending. P_e on the other hand, is always governed by the slenderness ratio in the plane of bending. Thus, when flexure is about the strong axis only, two different values of slenderness ratio may be involved in solving a given problem.

When bending occurs about both the x- and y-axes, the required flexural strength calculated about each axis is adjusted by the value of C_m and P_e corresponding to the distribution of moment and the slenderness ratio in its plane of bending, and is then taken as a fraction of the design bending strength, $\phi_b M_n$, about that axis, with due regard to the unbraced length of compression flange where this is a factor.

Formulas H1-3 and H1-4 approximate the maximum second order moments in compression members with no relative joint translation and no transverse loads between the ends of the member. This approximation is compared to an exact solution[97] in Fig. C-H1.2. For single curvature, Formula H1-4 is slightly unconservative, for a zero end moment it is almost exact; and for double curvature it is conservative. The 1978 AISC Specification limits $C_m \geq 0.4$ which corresponds to a M_1/M_2 ratio of 0.5. However, Fig. C-H1.2 shows that if, for example, $M_1/M_2 = 0.8$, the $C_m = 0.28$ is already very conservative, so the limit has been removed. The limit was originally adopted from Ref. 29, which was intended to apply to lateral-torsional buckling not second order in plane bending strength. The AISC Specifications, both in the 1978 and LRFD, use a modification factor C_b as given in Formula F1-3 for lateral-torsional buckling. C_b, which is limited to 2.3, is approximately the inverse of C_m as presented in Ref. 29 with a 0.4 limit. In Ref. 94 it was pointed out that Formula H1-4 could be used for in plane second order moments if the 0.4 limit was eliminated. Unfortunately, Ref. 29 was misinterpreted and a lateral-torsional buckling solution was used for an in plane second order analysis. This oversight has now been corrected.

For beam columns with transverse loadings, the second order moment can be approximated by using the following equation

$$C_m = 1 + \psi P_u/P_e$$

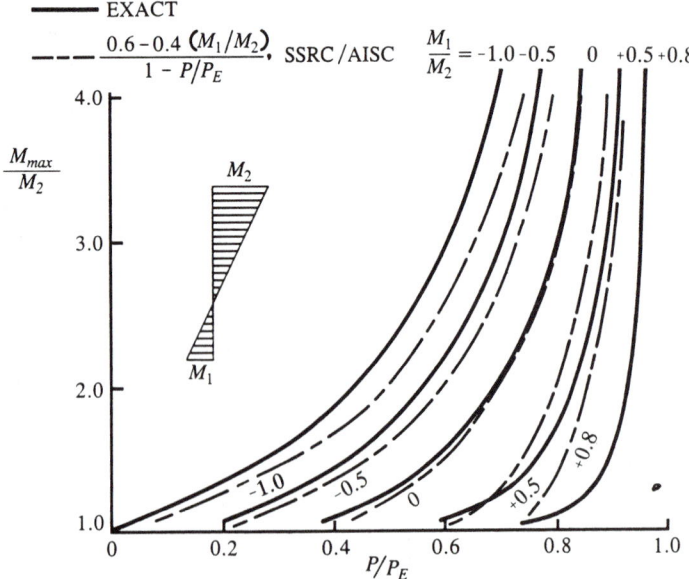

Figure C-H1.2

TABLE C-H1.1
Amplification Factors ψ and C_m

Case	ψ	C_m
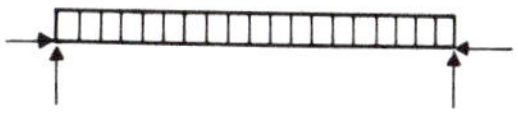	0	1.0
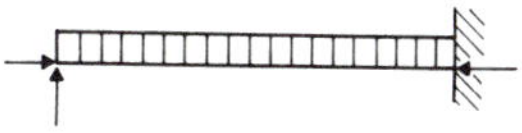	-0.4	$1 - 0.4\dfrac{f_a}{F'_e}$
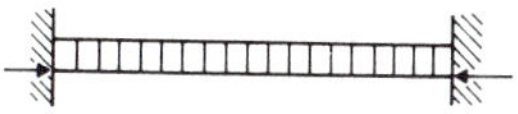	-0.4	$1 - 0.4\dfrac{f_a}{F'_e}$
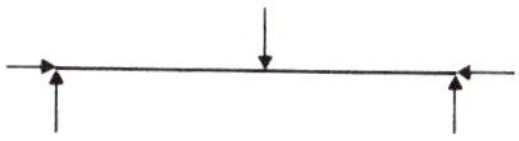	-0.2	$1 - 0.2\dfrac{f_a}{F'_e}$
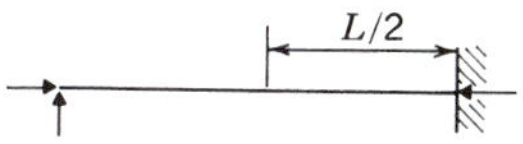	-0.3	$1 - 0.3\dfrac{f_a}{F'_e}$
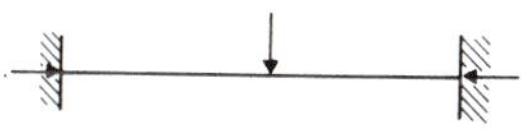	-0.2	$1 - 0.2\dfrac{f_a}{F'_e}$

For simply supported members

$$\psi = \frac{\pi^2 \delta_o EI}{M_o L^2} - 1$$

where

δ_o = maximum deflection due to transverse loading, in.
M_o = maximum factored design moment between supports due to transverse loading, kip-in.

For restrained ends some limiting cases[98] are given in Table C-H1.1 together with two cases of simply supported beam-columns. These values of C_m are always used with the maximum moment in the member. For the restrained end cases, the values of B_1 will be most accurate if values of $K < 1.0$ corresponding to the end boundary conditions are used in calculating P_e. In lieu of using the equations above, $C_m = 1.0$

can be used conservatively for transversely loaded members with unrestrained ends and 0.85 for restrained ends.

If, as in the case of a derrick boom, such a beam-column is subject to transverse (gravity) load and a calculable amount of end moment, the value δ_o should include the deflection between supports produced by this moment.

The interaction equations in Appendix H3 have been recommended for biaxially loaded H and wide flange shapes in Refs. 11 and 23. These equations which can be used only in braced frames represent a considerable liberalization over the provisions given in Sect. H1; it is, therefore, also necessary to check yielding under service loads, using the appropriate load and resistance factors for the serviceability limit state in Formula H1-1a or H1-1b with $M_{ux} = S_x F_y$ and $M_{uy} = S_y F_y$.

H2. UNSYMMETRIC MEMBERS AND MEMBERS UNDER TORSION AND COMBINED TORSION, FLEXURE AND/OR AXIAL FORCE

This section deals with types of cross sections and loadings not covered in Sect. H1, especially where torsion is a consideration. For such cases it is recommended to perform an elastic analysis based on the theoretical numerical methods available from the literature for the determination of the maximum normal and shear stresses, or for the elastic buckling stresses. In the buckling calculations an equivalent slenderness parameter is determined for use in Formulas E2-2 or E3-3, as follows:

$$\lambda_e = \sqrt{F_y/F_e}$$

where F_e is the elastic buckling stress determined from a stability analysis. This procedure is similar to that of Appendix E3.

CHAPTER I.
COMPOSITE MEMBERS

I1. DESIGN ASSUMPTIONS

Force Determination

Loads applied to an unshored beam before the concrete has hardened are resisted by the steel section alone, and only loads applied after the concrete has hardened are considered as resisted by the composite section. It is usually assumed for design purposes that concrete has hardened when it attains 75% of its design strength. In beams properly shored during construction, all loads may be assumed as resisted by the composite cross-section. Loads applied to a continuous composite beam with shear connectors throughout its length, after the slab was cracked in the negative moment region, are resisted in that region by the steel section and by properly anchored longitudinal slab reinforcement.

For purposes of plastic analysis all loads are considered resisted by the composite cross section, since a fully plastic strength is reached only after considerable yielding at the locations of plastic hinges.

Elastic Analysis

The use of constant stiffness in elastic analyses of continuous beams is analogous to the practice in reinforced concrete design.

Plastic Analysis

The requirement of the plastic strength of cross section for plastically analyzed beams means that, for composite beams with shear connectors, plastic analysis may be used only when the steel section in the positive moment region has a compact web, i.e., $h_c/t_w \leq 640/\sqrt{F_{yf}}$, and in the negative moment region the steel section is compact, as required for steel beams alone. No compactness limitations are placed on encased beams, but plastic analysis is permitted only if the direct contribution of concrete to the strength of sections is neglected; the concrete is relied upon only to prevent buckling.

Plastic Stress Distribution for Positive Moment

Plastic stress distributions are described in the Commentary in Sect. I3, along with a discussion of the composite participation of slab reinforcement.

Plastic Stress Distribution for Negative Moment

Plastic stress distributions are described in the Commentary in Sect. I3.

Elastic Stress Distribution

The strain distribution at any cross section of a composite beam is related to slip between the structural steel and concrete elements. Prior to slip, strain in both steel and concrete is proportional to the distance from the neutral axis for the elastic transformed section. After slip, the strain distribution is discontinuous, with a jump at the top of the steel shape. The strains in steel and concrete are proportional to distance from separate neutral axes, one for steel and the other for concrete.

Fully Composite Beam

Either the tensile yield strength of the steel section or the compressive strength of the concrete slab governs the maximum flexural strength of a fully composite beam subjected to a positive moment. The tensile yield strength of the longitudinal reinforcing bars in the slab governs the maximum flexural strength of a fully composite beam subjected to a negative moment. When shear connectors are provided in sufficient numbers to fully develop this maximum flexural strength, any slip that occurs prior to yielding is minor and has negligible influence both on stresses and stiffness.

Partially Composite Beam

The effects of slip on elastic properties of a partially composite beam can be significant and should be accounted for in calculations of deflections and stresses at service loads. Approximate elastic properties of partially composite beams are given in the Commentary in Sect. I3.*

Concrete-encased Beam

When the dimensions of a concrete slab supported on steel beams are such that the slab can effectively serve as the flange of a composite T-beam, and the concrete and steel are adequately tied together so as to act as a unit, the beam can be proportioned on the assumption of composite action.

    Two cases are recognized: fully encased steel beams, which depend upon natural bond for interaction with the concrete, and those with mechanical anchorage to the slab (shear connectors), which do not have to be encased.

I2. COMPRESSION MEMBERS

1. Limitations

    a. The lower limit of 4% on the cross-sectional area of structural steel differentiates between composite and reinforced concrete columns. If the area is less than 4%, a column with a structural steel core should be designed as a reinforced concrete column.

    b. The specified minimum quantity of transverse and longitudinal reinforcement in the encasement should be adequate to prevent severe spalling of the surface concrete during fires.

    c. Very little of the supporting test data involved concrete strengths in excess of 6 ksi, even though the cylinder strength for one group of four columns was 9.6

---

*For simplified design methods, see Ref. 37.

ksi. Normal weight concrete is believed to have been used in all tests. Thus, the upper limit of concrete strength is specified as 8 ksi for normal weight concrete. A lower limit of 3 ksi is specified for normal weight concrete and 4 ksi for lightweight concrete to encourage the use of good quality, yet readily available, grades of structural concrete.

d. Encased steel shapes and longitudinal reinforcing bars are restrained from buckling as long as the concrete remains sound. A limit strain of 0.0018, at which unconfined concrete remains unspalled and stable, serves analytically to define a failure condition for composite cross sections under uniform axial strain. The limit strain of 0.0018 corresponds approximately to 55 ksi.

e. The specified minimum wall thicknesses are identical to those in the 1983 ACI Building Code.[2] Their purpose is to prevent buckling of the steel pipe or tubing before yielding.

2. Design Strength

The procedure adopted for the design of axially loaded composite columns is described in detail in Ref. 34, which utilized the equation for the strength of a short column, derived in Ref. 35, and the same reductions for slenderness as those specified for steel columns in Sect. E2. The design follows the same path as the design of steel columns, except that the yield stress of structural steel, the modulus of elasticity of steel and the radius of gyration of the steel section are modified to account for the effect of concrete and of longitudinal reinforcing bars. A detailed explanation of the origin of these modifications may be found in Ref. 35. Reference 34 includes comparisons of the design procedure with tests of 48 axially loaded stub columns, 96 tests of concrete-filled pipes or tubing, and 26 tests of concrete-encased steel shapes. The mean ratio of the test failure loads to the predicted strengths is 1.18 for all 170 tests, and the corresponding coefficient of variation is 0.19.

3. Columns with Multiple Steel Shapes

This limitation is based on Australian research reported in Ref. 76, which demonstrated that after hardening of concrete the composite column will respond to loading as a unit even without lacing, tie plates or batten plates connecting the individual steel sections.

4. Load Transfer

To avoid overstressing either the structural steel section or the concrete at connections, a transfer of load to concrete by direct bearing is required.

When a supporting concrete area is wider on all sides than the loaded area, the maximum design strength of concrete is specified by Ref. 2 as $1.7\phi_B f'_c A_b$ where $\phi_B = 0.7$ is the strength reduction factor in bearing on concrete and A_B is the loaded area. Because the AISC LRFD Specification is based on the lower ANSI A58 load factors, $\phi_B = 0.60$ in the AISC LRFD Specification. The portion of the design load of an axially loaded column P_n resisted by the concrete may be expressed as $(c_2 f'_c A_c / A_s F_{my})\phi_B P_n$.

Accordingly,

$$A_B \geq \frac{\phi_B}{\phi_B} \frac{c_2}{1.7} \frac{A_c}{A_s} \frac{P_n}{F_{my}} = \frac{c_2}{1.7} \frac{A_c}{A_s} \frac{P_n}{F_{my}}$$

(C-I2-1)

I3. FLEXURAL MEMBERS

1. Effective Width

LRFD provisions for effective width omit any limit based on slab thickness, in accord with both theoretical and experimental studies, as well as current composite beam codes in other countries.[36] The same effective width rules apply to composite beams with a slab on either one side or both sides of the beam. To simplify design, effective width is based on the full span, c.-to-c. of supports, for both simple and continuous beams.

2. Strength of Beams with Shear Connectors

This section applies to simple and continuous composite beams with shear connectors, constructed with or without temporary shores.

Positive Flexural Design Strength

Flexural strength of a composite beam in the positive moment region may be limited by the plastic strength of the steel section, the concrete slab or shear connectors. In addition, web buckling may limit flexural strength if the web is slender and a significantly large portion of the web is in compression.

According to Table B5.1, local web buckling does not reduce the plastic strength of a bare steel beam if the beam depth-to-web thickness ratio is not larger than $640/\sqrt{F_y}$. In the absence of web buckling research on composite beams, the same ratio is conservatively applied to composite beams. Furthermore, for more slender webs, the LRFD Specification conservatively adopts first yield as the flexural strength limit. In this case, stresses on the steel section from permanent loads applied to unshored beams before the concrete has hardened must be superimposed on stresses on the composite section from loads applied to the beams after hardening of concrete. In this superposition, all permanent loads should be multiplied by the dead load factor and all live loads should be multiplied by the live load factor. For shored beams, all loads may be assumed as resisted by the composite section.

When first yield is the flexural strength limit, the elastic transformed section is used to calculate stresses on the composite section. The modular ratio $n = E/E_c$ used to determine the transformed section depends on the specified unit weight and strength of concrete. Note that this procedure for compact beams differs from the requirements of Sect. 1.11.2.2 of the 1978 AISC Specification.[1]

Plastic Stress Distribution for Positive Moment

When flexural strength is determined from the plastic stress distribution shown in Fig. C-I3.1, compression force C in the concrete slab is the smallest of:

$$C = A_{sw}F_{yw} + 2A_{sf}F_{yf} \qquad \text{(C-I3-1)}$$

$$C = 0.85 f_c' A_c \qquad \text{(C-I3-2)}$$

$$C = \Sigma Q_n \qquad \text{(C-I3-3)}$$

For a non-hybrid steel section, Formula C-I3-1 becomes $C = A_s F_y$

where

f'_c = specified compressive strength of concrete, ksi
A_c = area of concrete slab within effective width, in.²
A_s = area of steel cross section, in.²
A_{sw} = area of steel web, in.²
A_{sf} = area of steel flange, in.²
F_y = minimum specified yield stress of steel, ksi
F_{yw} = minimum specified yield stress of web steel, ksi
F_{yf} = minimum specified yield stress of flange steel, ksi
ΣQ_n = sum of nominal strengths of shear connectors between the point of maximum positive moment and the point of zero moment to either side, kips

Longitudinal slab reinforcement makes a negligible contribution to the compression force, except when Formula C-I3-2 governs. In this case, the area of longitudinal reinforcement within the effective width of the concrete slab times the yield stress of the reinforcement may be added in determining C.

The depth of the compression block is

$$a = \frac{C}{0.85 \, f'_c b} \qquad \text{(C-I3-4)}$$

where

b = effective width of concrete slab, in.

A fully composite beam corresponds to the case of C governed by the yield strength of the steel beam or the compressive strength of the concrete slab, as in Formulas C-I3-1 or C-I3-2. The number and strength of shear connectors govern C for a partially composite beam as in Formula C-I3-3.

The plastic stress distribution may have the plastic neutral axis (PNA) in the web, in the top flange of the steel section or in the slab, depending on the value of C. These alternatives are shown in Fig. C-I3.1.

For convenient reference, plastic moments for composite beams using a symmetrical non-hybrid steel shape are presented below. The more general case using an unsymmetrical hybrid steel shape is treated in the *Washington University Report No. 45*.[9]

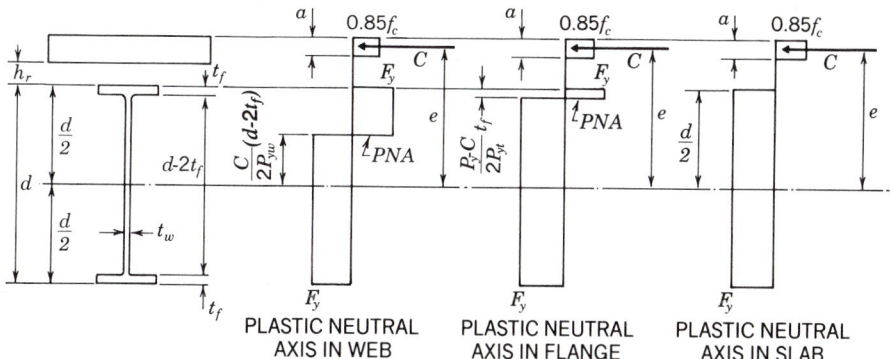

Fig. C-I3.1. Plastic stress distribution for positive moment in composite beams

It is convenient to introduce the following notation for symmetrical steel shapes:

Let

$P_{yw} = (d - 2t_f) t_w F_y$ = web yield force, kips
$P_{yf} = 0.5 (A_s F_y - P_{yw})$ = flange yield force, kips
$P_y = P_{yw} + 2P_{yf}$ = steel section yield force, kips
$M_{pw} = 0.25 P_{yw} (d - 2t_f)$ = web plastic moment, kip-in.
$M_{pf} = P_{yf} (d - t_f)$ = flange plastic moment, kip-in.
$M_p = M_{pw} + M_{pf}$ = steel section plastic moment, kip-in.

If the compression force C is less than the web yield force, i.e., $C \le P_{yw}$, the PNA is in the web and the nominal plastic moment of the composite beam is

$$M_n = M_p - (C/P_{yw})^2 M_{pw} + Ce \qquad \text{(C-I3-5)}$$

where e = distance from center of steel section to the center of the compression stress block in the slab.
  $= 0.5d + h_r + t_c - 0.5a$

The upper limit is reached when $C = P_{yw}$. Then, the PNA is at the top of the web and the nominal plastic moment of the composite beam is

$$M_n = M_{pf} + P_{yw}e \qquad \text{(C-I3-6)}$$

If the compression force C is larger than the web yield force, but smaller than the steel section yield force, i.e., $P_{yw} \le C \le P_y$, the PNA is in the steel flange and the nominal plastic moment of the composite beam is

$$M_n = 0.5(P_y - C)\left[d - \left(\frac{P_y - C}{2P_{yf}}\right)t_f\right] + Ce \qquad \text{(C-I3-7)}$$

At the lower limit $C = P_{yw}$, Formula C-I3-7 reduces to Formula C-I3-6 for the PNA at the top of the web. At the upper limit $C = P_y$, the entire steel section is at tension yield and the nominal plastic moment of the composite beam is

$$M_n = P_y e \qquad \text{(C-I3-8)}$$

Formula C-I3-8 also applies when compression force is less than $0.85f_c'A_c$, i.e., when the neutral axis is in the slab.*

Approximate Elastic Properties of Partially Composite Beams

Elastic calculations for stress and deflection of partially composite beams should include the effects of slip.
  The effective moment of inertia I_{eff} for a partially composite beam is approximated by

$$I_{eff} = I_s + \sqrt{(\Sigma Q_n/C_f)} (I_{tr} - I_s) \qquad \text{(C-I3-9)}$$

where

I_s = moment of inertia for the structural steel section, in.[4]
I_{tr} = moment of inertia for the fully composite uncracked transformed section, in.[4]

---

*For a simplified design method, see Ref. 37.

ΣQ_n = strength of shear connectors between points of maximum and zero moment, kips

C_f = compression force in concrete slab for fully composite beam; smaller of Formulas C-I3-1 and C-I3-2, kips

The effective section modulus S_{eff}, referred to the tension flange of the steel section for a partially composite beam, is approximated by

$$S_{eff} = S_s + \sqrt{(\Sigma Q_n / C_f)}(S_{tr} - S_s) \qquad \text{(C-I3-10)}$$

where

S_s = section modulus for the structural steel section, referred to the tension flange, in.^{3}

S_{tr} = section modulus for the fully composite uncracked transformed section, referred to the tension flange of the steel section, in.^{3}

Formulas C-I3-9 and C-I3-10 should not be used for ratios $\Sigma Q_n / C_f$ less than 0.25. This restriction is to prevent excessive slip, as well as substantial loss in beam stiffness. Studies indicate that Formulas C-I3-9 and C-I3-10 adequately reflect the reduction in strength and beam stiffness, respectively, when fewer connectors are used than required for full composite action.[38]

Negative Flexural Design Strength

The flexural strength in the negative moment region is the strength of the steel beam alone or the plastic strength of the composite section made up of the longitudinal slab reinforcement and the steel section.

Plastic Stress Distribution for Negative Moment

When an adequately braced compact steel section and adequately developed longitudinal reinforcing bars act compositely in the negative moment region, the nominal flexural strength is determined from the plastic stress distributions as shown in Fig. C-I3.2. The tensile force T in the reinforcing bars is the smaller of:

$$T = A_r F_{yr} \qquad \text{(C-I3-11)}$$

$$T = \Sigma Q_n \qquad \text{(C-I3-12)}$$

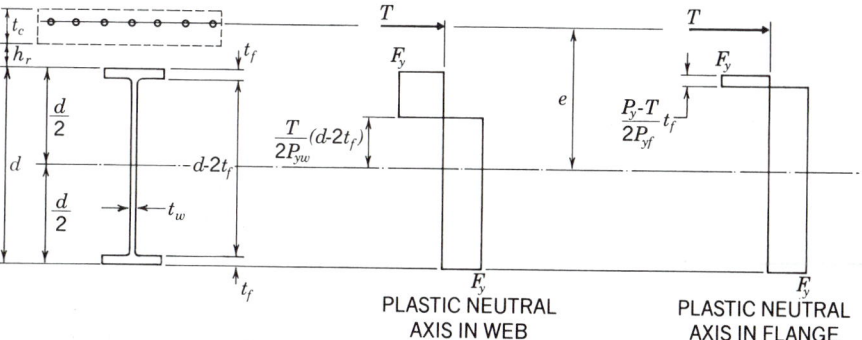

Fig. C-I3.2. Plastic stress distribution for negative moment

where

A_r = area of properly developed slab reinforcement parallel to the steel beam and within the effective width of the slab, in.^{2}

F_{yr} = specified yield stress of the slab reinforcement, ksi

ΣQ_n = sum of the nominal strengths of shear connectors between the points of maximum negative moment and zero moment to either side, kips

A third theoretical limit on T is the product of the area and yield stress of the steel section. However, this limit is redundant in view of practical limitations on slab reinforcement and shear connectors.

The plastic bending strength formulas for positive moment, Formulas C-I3-5 and C-I3-7, can be applied to the negative moment case by substituting T for C and by defining e as the distance from the center of steel section to the centroid of longitudinal slab reinforcement.

In determining web slenderness ratio λ_p, P is replaced with axial tensile force T obtained as the smaller of values given by Formulas C-I3-11 and C-I3-12.

3. Strength of Concrete-encased Beams

Tests of concrete-encased beams demonstrated that (1) the encasement drastically reduces the possibility of lateral-torsional instability and prevents local buckling of the encased steel, (2) the restrictions imposed on the encasement practically prevent bond failure prior to first yielding of the steel section, and (3) bond failure does not necessarily limit the moment capacity of an encased steel beam.[36] Accordingly, LRFD Specification permits two alternate design methods: one based on the first yield in the tension flange of the composite section and the other based on the plastic moment capacity of the steel beam alone. No limitations are placed on the slenderness of either the composite beam or the elements of the steel section, since the encasement effectively inhibits both local and lateral buckling.

In the method based on first yield, stresses on the steel section from permanent loads applied to unshored beams before the concrete has hardened must be superimposed on stresses on the composite section from loads applied to the beams after hardening of concrete. In this superposition, all permanent loads should be multiplied by the dead load factor and all live loads should be multiplied by the live load factor. For shored beams, all loads may be assumed as resisted by the composite section. Complete interaction (no slip) between the concrete and steel is assumed.

The contribution of concrete to the strength of the composite section is ordinarily larger in the positive moment regions than in the negative moment regions. Accordingly, the design based on composite section is more advantageous in the regions of positive moments.

4. Strength During Construction

When temporary shores are not used during construction, the steel beam alone must resist all loads applied before the concrete has hardened enough to provide composite action. Unshored beam deflection caused by wet concrete tends to increase slab thickness and dead load. For longer spans this may lead to instability analogous to roof ponding. An excessive increase of slab thickness may be avoided by beam camber.

When forms are not attached to the top flange, lateral bracing of the steel beam during construction may not be continuous and the unbraced length may control flexural strength, as defined in Sect. F1.

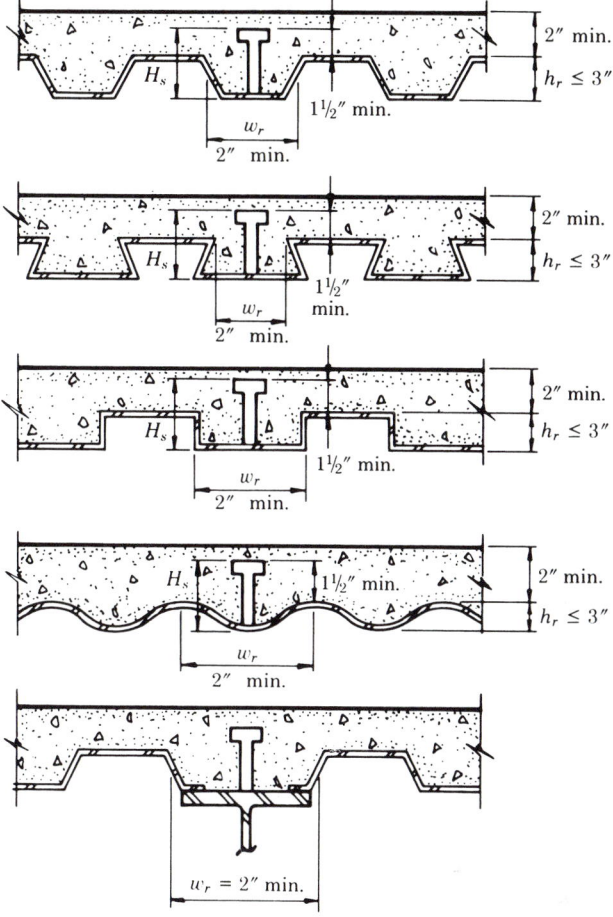

Figure C-I3.3

The LRFD Specification does not include special requirements for a margin against yield during construction. According to Sect. F1, maximum factored moment during construction is $0.90\,F_y Z$, where $F_y Z$ is the plastic moment. $(0.90\,F_y Z \approx 0.90 \times 1.1\,F_y S)$ This is equivalent to approximately the yield moment, $F_y S$. Hence, required flexural strength during construction prevents moment in excess of the yield moment.

Load factors for construction loads should be determined for individual projects according to local conditions, with the factors listed in Sect. A4 as a guide. Once the concrete has hardened, slab weight becomes a permanent dead load and the dead load factor applies to any load combinations.

5. Formed Steel Deck

Figure C-I3.3 is a graphic presentation of the terminology used in Sect. I3.5.

When studs are used on beams with formed steel deck, they may be welded directly through the deck or through prepunched or cut-in-place holes in the deck. The usual procedure is to install studs by welding directly through the deck; however, when the deck thickness is greater than 16 gage for single thickness, or 18 gage for each sheet of double thickness, or when the total thickness of galvanized coating is greater than 1.25 ounces/sq. ft, special precautions and procedures recommended by the stud manufacturer should be followed.

The design rules for composite construction with formed steel deck are based upon a study[38] at Lehigh University of the then available test results. The limiting parameters listed in Sect. I3.5 were established to keep composite construction with formed steel deck within the available research data.

Seventeen full size composite beams with concrete slab on formed steel deck were tested at Lehigh University and the results supplemented by the results of 58 tests performed elsewhere. The range of stud and steel deck dimensions encompassed by the 75 tests were limited to:

1. Stud dimensions: ¾ in.-dia. × 3.00 to 7.00 in.
2. Rib width: 1.94 in. to 7.25 in.
3. Rib height: 0.88 in. to 3.00 in.
4. Ratio w_r/h_r: 1.30 to 3.33
5. Ratio H_s/h_r: 1.50 to 3.41
6. Number of studs
   in any one rib: 1, 2 or 3

The strength of stud connectors installed in the ribs of concrete slabs on formed steel deck with the ribs oriented perpendicular to the steel beam is reasonably estimated by the strength of stud connectors in flat soffit composite slabs multiplied by values computed from Formula I3-1.

For the case where ribs run parallel to the beam, limited testing[38] has shown that shear connection is not significantly affected by the ribs. However, for narrow ribs, where the ratio w_r/h_r is less than 1.5, a shear stud reduction factor, Formula I3-2, has been suggested in view of lack of test data.

The Lehigh study[38] also indicated that Formula C-I3-10 for effective section modulus and Formula C-I3-9 for effective moment of inertia were valid for composite construction with formed steel deck.

When metal deck includes units for carrying electrical wiring, crossover headers are commonly installed over the cellular deck perpendicular to the ribs. They create trenches which completely or partially replace sections of the concrete slab above the deck. These trenches, running parallel to or transverse to a composite beam, may reduce the effectiveness of the concrete flange. Without special provisions to replace the concrete displaced by the trench, the trench should be considered as a complete structural discontinuity in the concrete flange.

When trenches are parallel to the composite beam, the effective flange width should be determined from the known position of the trench.

Trenches oriented transverse to composite beams should, if possible, be located in areas of low bending moment and the full required number of studs should be placed between the trench and the point of maximum positive moment. Where the trench cannot be located in an area of low moment, the beam should be designed as non-composite.

6. Design Shear Strength

A conservative approach to vertical shear provisions for composite beams is adopted by assigning all shear to the steel section web. This neglects any concrete slab contribution and serves to simplify the design.

I4. COMBINED COMPRESSION AND FLEXURE

The procedure adopted for the design of beam-columns is described and supported by comparisons with test data in Ref. 34. The basic approach is identical to that specified for steel columns in Sect. H1.

The nominal axial strength of a beam-column is obtained from Sect. I2.2, while the nominal flexural strength is determined from the plastic stress distribution on the composite section. An approximate formula for this plastic moment resistance of a composite column is given in Ref. 34:

$$M_n = M_p = ZF_y + \frac{1}{3}(h_2 - 2c_r)A_rF_{yr} + \left(\frac{h_2}{2} - \frac{A_wF_y}{1.7f_c'h_1}\right)A_wF_y \qquad \text{(C-I4-1)}$$

where

A_w = web area of encased steel shape; for concrete-filled tubes, $A_w = 0$, in.^{2}

Z = plastic section modulus of the steel section, in.^{3}

c_r = average of distance from compression face to longitudinal reinforcement in that face and distance from tension face to longitudinal reinforcement in that face, in.

h_1 = width of composite cross section perpendicular to the plane of bending, in.

h_2 = width of composite cross section parallel to the plane of bending, in.

The supporting comparisons with beam-column tests included 48 concrete-filled pipes or tubing and 44 concrete-encased steel shapes.[34] The overall mean test-to-prediction ratio was 1.23 and the coefficient of variation 0.21.

The last paragraph in Sect. I4 provides a transition from beam-columns to beams. It involves bond between the steel section and concrete. Section I3 for beams requires either shear connectors or full, properly reinforced encasement of the steel section. Furthermore, even with full encasement, it is assumed that bond is capable of developing only the moment at first yielding in the steel of the composite section. No test data are available on the loss of bond in composite beam-columns. However, consideration of tensile cracking of concrete suggests $P_u/\phi_b P_n = 0.3$ as a conservative limit. It is assumed that when $P_u/\phi_b P_n$ is less than 0.3, the nominal flexural strength is reduced below that indicated by plastic stress distribution on the composite cross section unless the transfer of shear from the concrete to the steel is provided for by shear connectors.

I5. SHEAR CONNECTORS

1. Materials

Tests[39] have shown that fully composite beams with concrete meeting the requirements of Part 3, Chapter 4, "Concrete Quality," of Ref. 2, made with ASTM C33 or rotary-kiln produced C330 aggregates, develop their full flexural capacity.

2. Horizontal Shear Force

Composite beams in which the longitudinal spacing of shear connectors has been varied according to the intensity of statical shear, and duplicate beams where the number of connectors were uniformly spaced, have exhibited the same ultimate strength and the same amount of deflection at normal working loads. Only a slight deformation in the concrete and the more heavily stressed connectors is needed to redistribute the horizontal shear to other less heavily stressed connectors. The important consideration is that the total number of connectors be sufficient to develop the shear V_h on either side of the point of maximum moment. The provisions of the LRFD Specification are based upon this concept of composite action.

In computing the design flexural strength at points of maximum negative bending, reinforcement parallel to the steel beam and lying within the effective width of slab may be included, provided such reinforcement is properly anchored beyond the region of negative moment. However, enough shear connectors are required to transfer, from the slab to the steel beam, the ultimate tensile force in the reinforcement.

3. Strength of Stud Shear Connectors

Studies have defined stud shear connector strength in terms of normal weight and lightweight aggregate concretes as a function of both concrete modulus of elasticity and concrete strength as given by Formula I5-1.

Formula I5-1, obtained from Ref. 39, corresponds to Tables 1.11.4 and 4A in Sect. 1.11.4 of the 1978 AISC Specification.[1] Note that an upper bound on stud shear strength is the product of the cross-sectional area of the stud times its ultimate tensile strength.

The LRFD Specification does not specify a resistance factor for shear connector strength. The resistance factor for the flexural strength of a composite beam accounts for all sources of variability, including those associated with the shear connectors.

4. Strength of Channel Shear Connectors

Formula I5-2 is a modified form of the formula for the strength of channel connectors developed by Slutter and Driscoll.[40] The modification has extended its use to lightweight concrete.

6. Shear Connector Placement and Spacing

As in Part 1 of the 1978 AISC Specification,[1] uniform spacing of shear connectors is permitted except in the presence of heavy concentrated loads. The second sentence in the first paragraph of Sect. I5.6 serves the same purpose as Formula 1.11-7 in Sect. 1.11.4 of Ref. 1.

When stud shear connectors are installed on beams with formed steel deck, concrete cover at the sides of studs adjacent to sides of steel ribs is not critical. Tests have shown that studs installed as close as is permitted to accomplish welding of studs does not reduce the composite beam capacity.

Studs not located directly over the web of a beam tend to tear out of a thin flange before attaining their full shear-resisting capacity. To guard against this contingency, the size of a stud not located over the beam web is limited to 2½ times the flange thickness.[87]

The minimum spacing of connectors along the length of the beam, in both flat soffit concrete slabs and when ribs of formed steel deck are parallel to the beam, is 6 diameters; this spacing reflects development of shear planes in the concrete slab.[39] Since most test data are based on the minimum transverse spacing of 4 diameters, this transverse spacing was set as the minimum permitted. If the steel beam flange is narrow, this spacing requirement may be achieved by staggering the studs with a minimum transverse spacing of 3 diameters between the staggered row of studs. The reduction in connector capacity in the ribs of formed steel decks is provided by the factor $0.85/\sqrt{N_r}$, which accounts for the reduced capacity of multiple connectors, including the effect of spacing. When deck ribs are parallel to the beam and the design

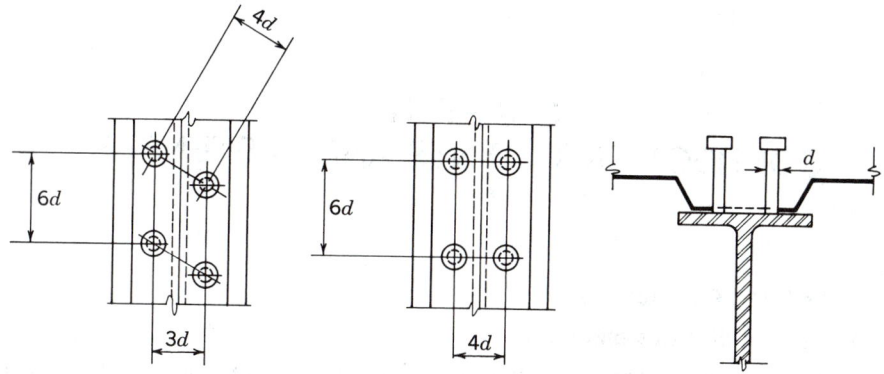

Fig. C-I5.1. Connector arrangements

requires more studs than can be placed in the rib, the deck may be split so that adequate spacing is available for stud installation. Figure C-I5.1 shows possible connector arrangements.

I6. SPECIAL CASES

This section is a modified version of Sect. 1.11.6 of the 1978 AISC Specification.[1] The phrase "and details of construction" was added in recognition of the fact that different types of shear connectors may require different spacing and other detailing than stud and channel connectors.

CHAPTER J.
CONNECTIONS, JOINTS AND FASTENERS

J1. GENERAL PROVISIONS

6. Placement of Welds and Bolts

Slight eccentricities between the gravity axis of single and double angle members and the center of gravity of their connecting rivets or bolts have long been ignored as having negligible effect on the static strength of such members. Tests[41] have shown that similar practice is warranted in the case of welded members in statically loaded structures.

However, the fatigue life of eccentrically loaded welded angles has been shown to be very short.[42] Notches at the roots of fillet welds are harmful when reverse stresses are normal to the axis of the weld, as could occur when axial cyclic loading is applied to angles with end welds not balanced about the neutral axis. Accordingly, balanced welds are indicated when such members are subjected to cyclic loading (see Fig. C-J1.1).

7. Bolts in Combination with Welds

Welds will not share the load equally with mechanical fasteners in bearing-type connections. Before ultimate loading occurs, the fastener will slip and the weld will carry an indeterminately larger share of the load.

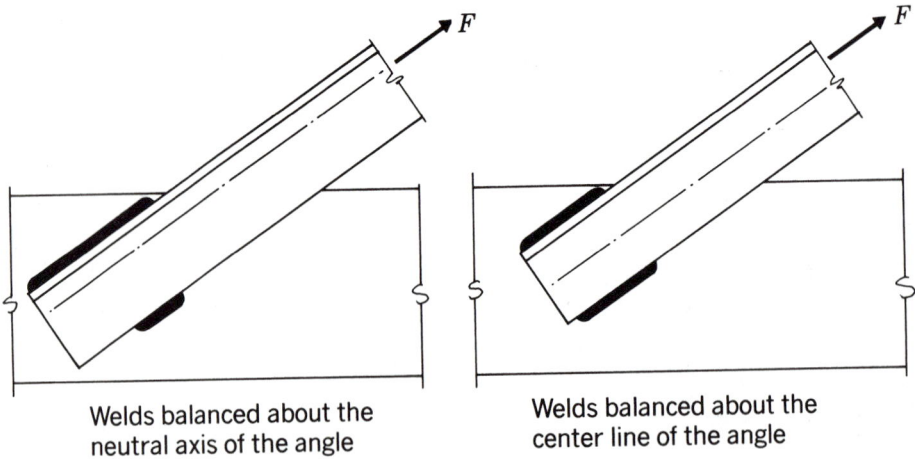

Welds balanced about the
neutral axis of the angle

Welds balanced about the
center line of the angle

Fig. C-J1.1. Welds at end connection

Accordingly, the sharing of load between welds and A307 bolts or high-strength bolts in a bearing-type connection is not recommended. For similar reasons, A307 bolts and rivets should not be assumed to share loads in a single group of fasteners.

For high-strength bolts in slip-resistant connections to share the load with welds it is advisable to properly torque the bolt before the weld is made. Were the weld to be placed first, angular distortion from the heat of the weld might prevent the faying action required for development of the slip-resistant force. When the bolts are properly torqued before the weld is made, the slip-resistant bolts and the weld may be assumed to share the load on a common shear plane.[43] The heat of welding near bolts will not alter the mechanical properties of the bolt.

In making alterations to existing structures, it is assumed that whatever slip is likely to occur in high-strength bolted bearing-type connections or riveted connections will have already taken place. Hence, in such cases the use of welding to resist all stresses, other than those produced by existing dead load present at the time of making the alteration, is permitted.

It should be noted that combination of fasteners as defined herein does not refer to connections such as shear plates for beam-to-column connections which are welded to the column and bolted to the beam flange or web[43] and other comparable connections.

8. High-strength Bolts in Combination with Rivets

When high-strength bolts are used in combination with rivets, the ductility of the rivets permits the direct addition of the capacity of both fastener groups.

J2. WELDS

1. Groove Welds

The engineer preparing contract design drawings cannot specify the depth of groove without knowing the welding process and the position of welding. Accordingly, only the effective throat for partial-penetration groove welds should be specified on design drawings, allowing the fabricator to produce this effective throat with his own choice of welding process and position.

The weld reinforcement is not used in determining the effective throat thickness of a groove weld (see Table J2.1).

2. Fillet Welds

a. Effective Area

The effective throat of a fillet weld is based upon the root of the joint and the face of the diagrammatic weld, hence this definition gives no credit for weld penetration or reinforcement at the weld face. If the fillet weld is made by the submerged arc welding process, some credit for penetration is made. If the leg size of the resulting fillet weld does not exceed ⅜-in., then 0.11 in. is added to the theoretical throat. This increased weld throat is allowed because the submerged arc process produces deep penetration welds of consistent quality. However, it is necessary to run a short length of fillet weld to be assured that this increased penetration is obtained.

In practice, this is usually done initially by cross-sectioning the runoff plates of the joint. Once this is done, no further testing is required, as long as the welding procedure is not changed.

b. Limitations

Table J2.5 provides a minimum size of fillet weld for a given thickness of the thicker part joined.

The requirements are not based upon strength considerations, but upon the quench effect of thick material on small welds. Very rapid cooling of weld metal may result in a loss of ductility. Further, the restraint to weld metal shrinkage provided by thick material may result in weld cracking. Because a 5/16-in. fillet weld is the largest that can be deposited in a single pass by manual process, 5/16-in. applies to all material 3/4-in. and greater in thickness, but minimum preheat and interpass temperature are required by AWS D1.1.* Both the design engineer and the shop welder must be governed by the requirements.

Table J2.4 gives the minimum effective throat of a partial-penetration groove weld. Notice that Table J2.4 for partial-penetration groove welds goes up to a plate thickness of over 6 in. and a minimum weld throat of 5/8-in., whereas, for fillet welds, Table J2.5 goes up to a plate thickness of over 3/4-in. and a minimum leg size of fillet weld of only 5/16-in. The additional thickness for partial-penetration welds is to provide for reasonable proportionality between weld and material thickness.

For plates of 1/4-in. or more in thickness, it is necessary that the inspector be able to identify the edge of the plate to position the weld gage. This is assured if the weld is kept back at least 1/16-in. from this edge, as shown in Fig. C-J2.1.

Where longitudinal fillet welds are used alone in a connection (see Fig. C-J2.4), Sect. J2.2b requires the length of each weld to be at least equal to the width of the connecting material, because of shear lag.[44] (See Fig. C-J2.2.) By providing a minimum lap of five times the thickness of the thinner part of a lap joint, the resulting rotation of the joint when pulled will not be excessive, as shown in Fig. C-J2.3.

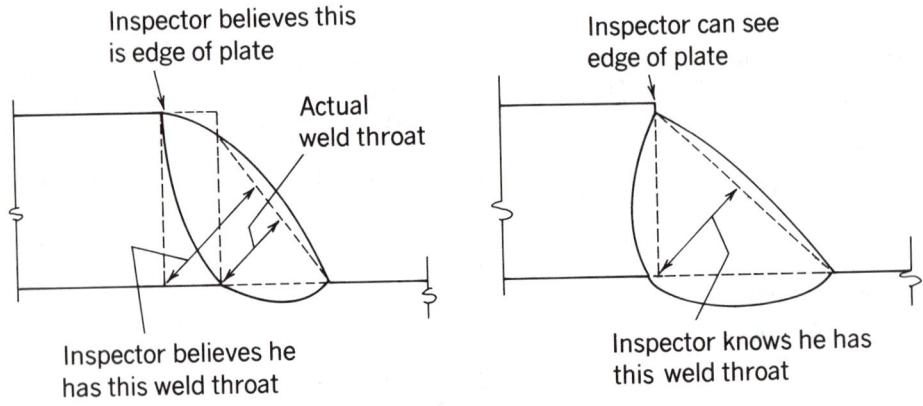

Fig. C-J2.1. Identification of plate edge

---

*See Table J2.5.

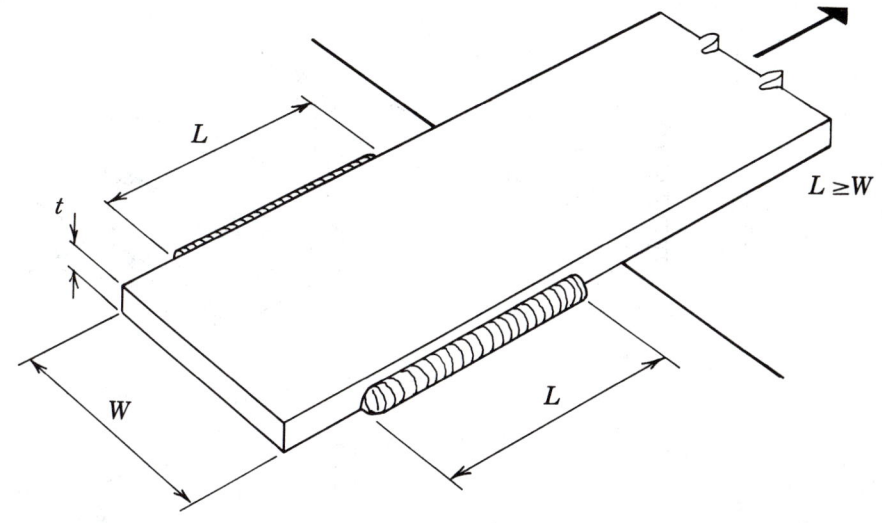

Fig. C-J2.2.   Longitudinal fillet welds

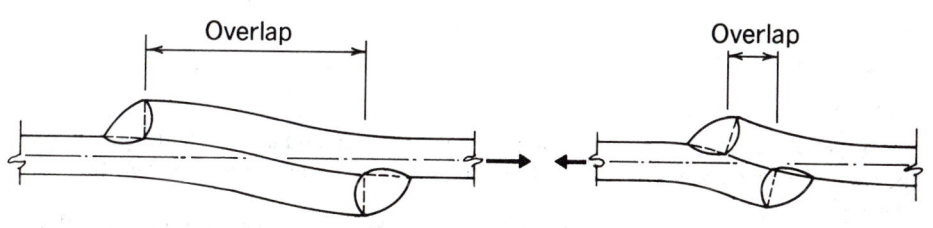

Fig. C-J2.3.   Minimum lap

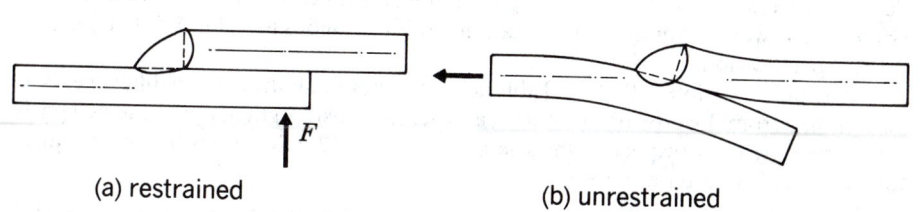

(a) restrained   (b) unrestrained

Fig. C-J2.4.   Restraint of lap joints

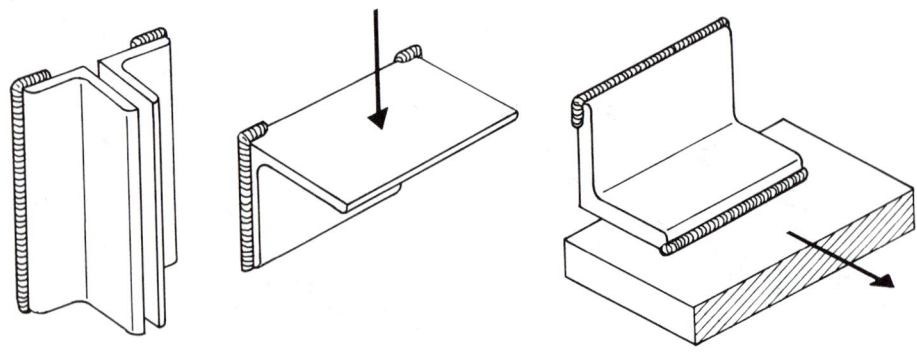

Fig. C-J2.5.   Return welds

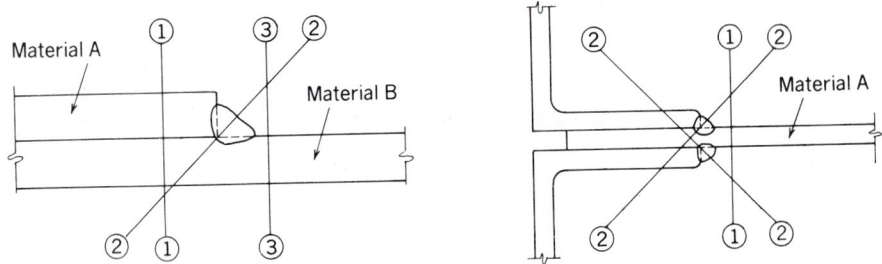

Fig. C-J2.6.   Shear planes for fillet welds loaded in longitudinal shear

Fillet welded lap joints under tension tend to open and apply a tearing action at the root of the weld as shown in Fig. C-J2.4b, unless restrained by a force F as shown in Fig. C-J2.4a.

While end returns (see Fig. C-J2.5) do not substantially increase the strength of the connection, they do increase the carrying capacity of the structure by providing more ultimate plastic movement of the connection before failure. The end return delays the initial tearing of the weld in the connection.

4. Design Strength

The strength of welds is governed by the strength of either the base material or the deposited weld metal. Table J2.3 contains the resistance factors and nominal weld strengths, as well as a number of limitations. It is analogous to Table 1.5.3 in the 1978 AISC Specification.[1]

It should be noted that in Table J2.3 the nominal strength of fillet welds is determined from the effective throat area, whereas the strength of the connected parts is governed by their respective thicknesses. Figure C-J2.6 illustrates the shear planes for fillet welds and base material:

   a. Plane 1-1, in which the resistance is governed by the shear strength for material A.

b. Plane 2-2, in which the resistance is governed by the shear strength of the weld metal

c. Plane 3-3 in which the resistance is governed by the shear strength of material B

The resistance of the welded joint is the lowest of the resistances calculated in each plane of shear transfer. Note that planes 1-1 and 3-3 are positioned away from the fusion areas between the weld and the base material. Tests have demonstrated that the stress on this fusion area is not critical in determining the shear strength of fillet welds.*

5. Combination of Welds

This method of adding weld capacities does not apply to a welded joint using a partial-penetration single bevel groove weld with a superimposed fillet weld. In this case, the effective throat of the combined joint must be determined and the design capacity based upon this throat area.

J3. BOLTS, THREADED PARTS AND RIVETS

1. High-strength Bolts

In general, the use of high-strength bolts is required to conform to the provisions of the *Specification for Structural Joints Using ASTM A325 or A490 Bolts* [68] as approved by the Research Council on Structural Connections.

Occasionally the need arises for the use of high-strength bolts of diameters in excess of those available for A325 and A490 bolts, as for example, anchor bolts for fastening machine castings. For this situation Sect. A3.3 permits the use of A449 bolts.

3. Design Tension or Shear Strength

Tension loading of fasteners is usually accompanied by some bending due to the deformation of the connected parts. Hence, the resistance factor ϕ, by which R_n is multiplied to obtain the design tensile strength of fasteners, is relatively low. The nominal tensile strength values in Table J3.2 were obtained from the formula

$$R_n = 0.75\, A_b F_u \qquad \text{(C-J3-1)}$$

While the formula was developed for bolted connections (Ref. 43, p. 68), it was also conservatively applied to threaded parts and to rivets. The nominal strength of A307 bolts was discounted by 5 ksi.

In connections consisting of only a few fasteners, the effects of strain on the shear in bearing fasteners is negligible.[43,45] In longer joints, the differential strain produces an uneven distribution between fasteners (those near the end taking a disproportionate part of the total load), so that the maximum strength per fastener is reduced. The 1978 AISC Specification[1] permits connections up to 50 in. in length without a reduction in maximum shear stress. With this in mind the resistance factor ϕ for shear in bearing-type connections has been adjusted to accommodate the same range of connections.

The values of nominal shear strength in Table J3.2 were obtained from the formula

$$R_n/mA_b = 0.60 F_u \qquad \text{(C-J3-2)}$$

*F. R. Preece, "AWS-AISC Fillet Weld Study—Longitudinal and Transverse Shear Tests," Testing Engineers, Inc., Los Angeles, May 31, 1968.

when threads are excluded from the shear planes and

$$R_n/mA_b = 0.45F_u \qquad\qquad \text{(C-J3-3)}$$

when threads are not excluded from the shear plane (Ref. 43, p. 68), where m is the number of shear planes. While developed for bolted connections,[43] the formulas were also conservatively applied to threaded parts and rivets. The value given for A307 bolts was obtained from Formula C-J3-3 but is specified for all cases regardless of the position of threads. For A325 bolts, no distinction is made between small and large diameters, even though the minimum tensile strength F_u is lower for bolts with diameters in excess of 1 in. It was felt that such a refinement of design was not justified, particularly in view of the low resistance factor ϕ and other compensating factors.

4. Combined Tension and Shear in Bearing-type Connections

Tests have shown that the strength of bearing fasteners subject to combined shear and tension resulting from externally applied forces can be closely defined by an ellipse.[43] Such a curve can be replaced, with only minor deviations, by three straight lines as shown in Fig. C-J3.1. This latter representation offers the advantage that no modification of either type stress is required in the presence of fairly large magnitudes of other types. This linear representation was adopted for Table J3.3, giving a limiting tensile stress F_t as a function of the shearing stress f_v for bearing-type connections.

5. High-strength Bolts in Slip-critical Joints

The onset of slipping in a high-strength bolted, slip-critical connection is not an indication that maximum capacity of the connection has been reached. Its occurrence may be only a serviceability limit state. However, poor fatigue performance due to joint slippage may occur in slip-critical connections subjected to repeated loads producing large stress reversal. If slip occurs in slip-critical connections with oversized or slotted holes, the rigid body movement could introduce second order effects that would reduce the maximum load capacity of the structure. Slip of slip-critical connections is likely to occur at approximately 1.4 to 1.5 times the service loads.

While the possibility of a slip-critical connection slipping into bearing under anticipated service conditions is small, such connections must comply with the provisions of Sect. J3.6 in order to prevent connection failure at the maximum load condition.

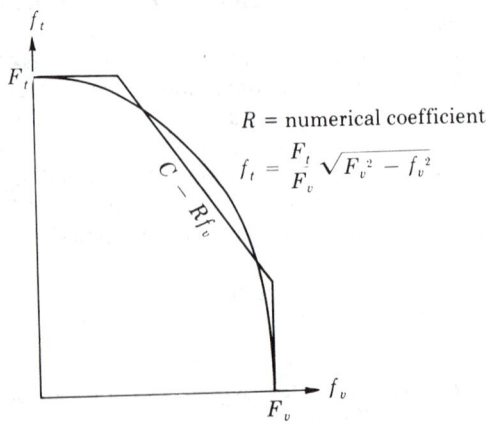

R = numerical coefficient

$$f_t = \frac{F_t}{F_v}\sqrt{F_v^2 - f_v^2}$$

Figure C-J3.1

6. Bearing Strength at Bolt Holes

The recommended bearing stress on pins is not the same as for bolts.

Bearing values are not provided as a protection to the fastener, because it needs no such protection. Therefore, the same bearing value applies to joints assembled by bolts, regardless of fastener shear strength or the presence or absence of threads in the bearing area.

Recent tests[96] have demonstrated that hole elongation greater than 0.25 in. will begin to develop as the bearing stress is increased beyond the values given in Formulas J3-1a and J3-1b, especially if it is combined with high tensile stress on the net section, even though rupture does not occur. Formula J3-1d considers the effect of hole ovalization.

7. Size and Use of Holes

To provide some latitude for adjustment in plumbing up a frame during erection, three types of enlarged size holes are permitted, subject to the approval of the designer. The nominal maximum sizes of these holes are given in Table J3.5.

The use of these oversize holes is restricted to connections assembled with bolts and is subject to the provisions of Sects. J3.9 and J3.10.

8. Long Grips

Provisions requiring a decrease in calculated stress for A307 bolts having long grips (by arbitrarily increasing the required number in proportion to the grip length) are not required for high-strength bolts. Tests[46] have demonstrated that the ultimate shearing strength of high-strength bolts having a grip of 8 or 9 diameters is no less than that of similar bolts with much shorter grips.

9. Minimum Spacing

The *maximum* factored strength R_n at a bolt or rivet hole in bearing requires that the distance between the centerline of the first fastener and the edge of a plate toward which the force is directed should not be less than $1\frac{1}{2}d$, where d is the fastener diameter.[43] By similar reasoning the distance measured in the line of force, from the centerline of any fastener to the nearest edge of an adjacent hole, should not be less than $3d$, to insure maximum design strength in bearing. Plotting of numerous test results indicates that the critical bearing strength is directly proportional to the above defined distances up to a maximum value of $3d$, above which no additional bearing strength is achieved.[43] By adding $d/2$, Formula J3-2 defines the maximum pitch in terms of bearing capacity for standard holes. Table J3.6 lists the increments that must be added to adjust the spacing upward to compensate for an increase in hole dimension parallel to the line of force.

10. Minimum Edge Distance

Critical bearing stress is a function of the material tensile strength, the spacing of fasteners and the distance from the edge of the part to the center line of the nearest fastener. Tests have shown* that a linear relationship exists between the ratio of

---

*Ref. 43, pp. 109–112, 137.

critical bearing stress to tensile strength (of the connected material) and the ratio of fastener spacing (in the line of force) to fastener diameter. The following equation affords a good lower bound to published test data for single-fastener connections with standard holes, and is conservative for adequately spaced multi-fastener connections:

$$\frac{F_{pcr}}{F_u} = \frac{\ell_e}{d} \qquad \text{(C-J3-4)}$$

where

F_{pcr} = critical bearing stress, ksi
F_u = tensile strength of the connected material, ksi
ℓ_e = distance, along a line of transmitted force, from the center of a fastener to the nearest edge of an adjacent fastener or to the free edge of a connected part (in the direction of stress), in.
d = diameter of a fastener, in.

The above equation, modified by a resistance factor $\phi = 0.75$, is the basis for Formulas J3-2 and J3-3 in the LRFD Specification.

The provisions of Sect. J3.9 are concerned with ℓ_e as hole spacing, whereas Sect. J3.10 is concerned with ℓ_e as edge distance in the direction of stress. Section J3.6 establishes a maximum bearing strength. Spacing and/or edge distance may be increased to provide for a required bearing strength, or bearing force may be reduced to satisfy a spacing and/or edge distance limitation.

It has long been known that the critical bearing stress of a single fastener connection is more dependent upon a given edge distance than multi-fastener connections.[63] For this reason, longer edge distances (in the direction of force) are required for connections with one fastener in the line of transmitted force than required for those having two or more.

11. Maximum Edge Distance and Spacing

Limiting the edge distance to not more than 12 times the thickness of an outside connected part, but not more than 6 in., is intended to provide for the exclusion of moisture in the event of paint failure, thus preventing corrosion between the parts which might accumulate and force these parts to separate. More restrictive limitations are required for connected parts of unpainted steel exposed to atmospheric corrosion.

J4. DESIGN SHEAR RUPTURE STRENGTH

Tests[47] on coped beams indicated that a tearing failure mode can occur along the perimeter of the bolt holes as shown in Fig. C-J5.1. This block shear mode combines tensile strength on one plane and shear strength on a perpendicular plane. The 1978 AISC Specification[1] adopted an analytical model which sums the tensile net area strength and shear net area fracture strength to predict the block shear strength. The shear fracture stress is taken as $0.6F_u$. The failure path is defined by the center lines of the bolt holes. The block shear failure mode is not limited to the coped ends of beams. Other examples are shown in Fig. C-J5.2. The block shear failure mode should also be checked around the periphery of welded connections.

Based on more recent tests,[101,102] the LRFD Specification has adopted a more conservative model to predict block shear strength. The previous model[1] summed the fracture strengths on two perpendicular planes which implies that ultimate fracture strength on both planes occur simultaneously. If fracture occurs first on one plane, the

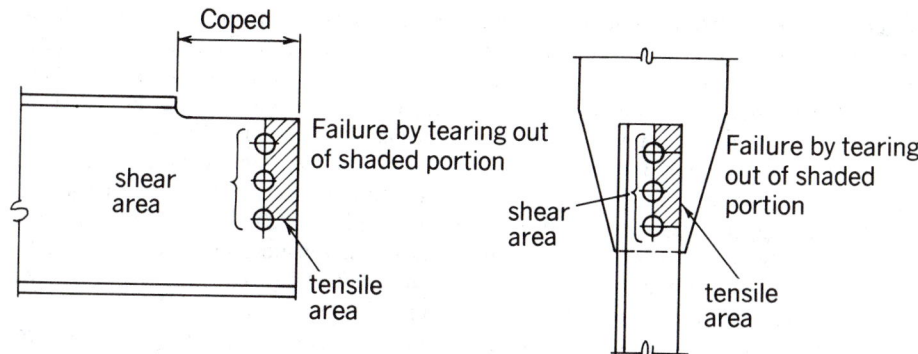

Fig. C-J5.1. Failure surface for tearing mode

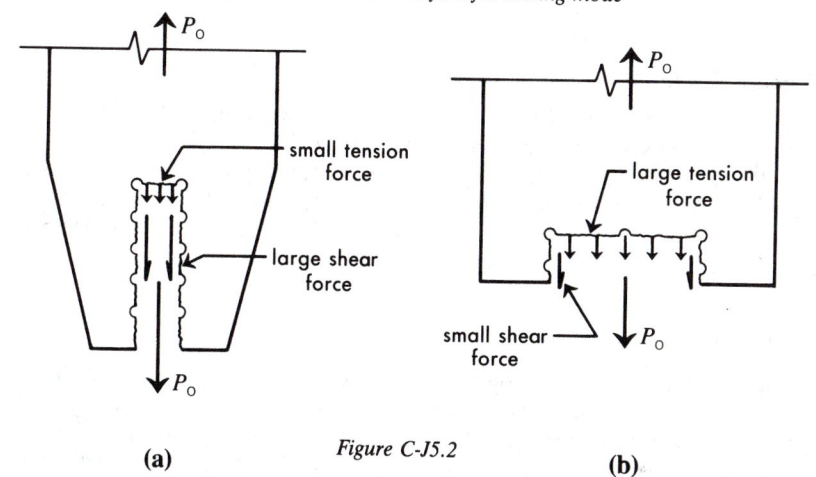

(a)   *Figure C-J5.2*   **(b)**

strength is lost and the total force must be supported by the perpendicular plane. Test results suggest that it is reasonable to add the yield strength on one plane to the fracture strength of the perpendicular plane. Therefore, two possible block shear strengths can be calculated; fracture strength F_u on the net tensile section along with shear yielding $0.6\,F_y$ on the gross section on the shear plane(s) or fracture $0.6F_u$ on the net shear area(s) combined with yielding F_y on the gross tensile area as given by the following formulas:

$$\phi[0.6\,F_y A_{vg} + F_u A_{nt}] \qquad\qquad \text{(C-J4-1)}$$

$$\phi[0.6\,F_u A_{ns} + F_y A_{tg}] \qquad\qquad \text{(C-J4-2)}$$

where

$\phi\ = 0.75$
$A_{vg} =$ gross area subjected to shear, in.^{2}
$A_{tg} =$ gross area subjected to tension, in.^{2}
$A_{ns} =$ net area subjected to shear, in.^{2}
$A_{nt} =$ net area subjected to tension, in.^{2}

These formulas are consistent with the philosophy in Chap. D for tension members, where gross area is used for the limit state of yielding and net area is used for fracture. The controlling equation is one that produces the *larger* force. This can be explained by the two extreme examples given in Fig. C-J5.2. In Case a, the total force is resisted primarily by shear, therefore shear fracture, not shear yielding, should control the block shear tearing mode; therefore, use Formula C-J4-2. For Case b, block shear cannot occur until the tension area fractures as given by Formula C-J4-1. If Formula C-J4-2 (shear fracture on the small area and yielding on the large tension area) is checked for Case b, a smaller P_o will result. In fact, as the shear area gets smaller and approaches zero, the use of Formula C-J4-2 for Case b would give a block shear strength based totally on *yielding* of the gross tensile area. Block shear is a fracture or tearing phenomenon not a yielding limit state. Therefore, the proper formula to use is the one in which the fracture term is larger than the yield term. When it is not obvious which failure plane fractures, it is easier just to use the larger of the two formulas.

J5. CONNECTING ELEMENTS

2. Design Strength of Connecting Elements

Tests have shown that yield will occur on the gross section area before the tensile capacity of the net section is reached, if the ratio $A_n/A_g \leq 0.85$.[43] Since the length of connecting elements is small compared to the member length, inelastic deformation of the gross section is limited. Hence, the effective net area A_n of the connecting element is limited to $0.85 A_g$ in recognition of the limited inelastic deformation and to provide a reverse capacity.

J6. FILLERS

The practice of securing fillers by means of additional fasteners, so that they are in effect an integral part of a shear-connected component, is not required where a connection is designed to be a slip-critical connection using high-strength bolts. In such connections, the resistance to slip between filler and either connected part is comparable to that which would exist between the connected parts if no fill were present.

Filler plates may be used in lap joints of welded connections that splice parts of different thickness, or where there may be an offset in the joint.

J8. BEARING STRENGTH

The LRFD Specification provisions for bearing on milled surfaces, Sect. J8.1, and on expansion rollers and rockers, Sect. J8.2, are equivalent to those provisions in the 1978 AISC Specification,[1] Sects. 1.5.1.5.1 and 1.5.1.5.2, respectively. The specific formulas were transformed from allowable stress to LRFD format by multiplying the numerical constants 0.9 and 0.66 by 1.67, resulting in 1.5 and 1.1 for their LRFD equivalents.

As used throughout the LRFD Specification, the terms "milled surface," "milled" or "milling" are intended to include surfaces which have been accurately sawed or finished to a true plane by any suitable means.

J9. COLUMN BASES AND BEARING ON CONCRETE

The formulas for resistance of concrete in bearing are the same as ACI 318-83, except that AISC formulas are used with $\phi = 0.60$ while ACI uses $\phi = 0.70$. The difference is because ACI specifies larger load factors than the ANSI load factors specified by AISC.

CHAPTER K.
STRENGTH DESIGN CONSIDERATIONS

K1. WEBS AND FLANGES WITH CONCENTRATED FORCES

The LRFD Specification separates web strength requirements into distinct categories representing different limit state criteria, i.e., local flange bending (Sect. K1.2), local web yielding (Sect. K1.3), web crippling (Sect. K1.4), sidesway web buckling (Sect. K1.5) and column buckling of the web (Sect. K1.6). The following is a discussion of each criterion with respect to a concentrated load applied at an interior location; the same argument would be valid for an exterior reaction load except as modified by the appropriate equations. An interior location is defined as being at least a distance equal to the member's depth from the end of the member.

2. Local Flange Bending

Where a tension force is applied through a plate welded to a flange, that flange must be sufficiently rigid not to deform and cause an area of high stress concentration in the weld in line with the web. This is handled in the 1978 AISC Specification[1] by Formula 1.15-3 which has been retained in the LRFD Specification, in a modified form, as Formula K1-1.

3. Local Web Yielding

The web strength criteria have been established to limit the stress in the web of a member into which a force is being transmitted. It should matter little whether the member receiving the force is a beam or a column; however, the reference material[9,1] upon which the LRFD Specification is based did make such a distinction. For beams, a 2:1 stress gradient through the flange was used, whereas the gradient was 2½:1 through column flanges. In Sect. K1.3, the 2½:1 gradient is used for both.

4. Web Crippling

The expression for resistance to web crippling at a concentrated load is a departure from previous specifications.[30,31,32,64] Formulas K1-4 and K1-5 are based on research by Roberts.[88]

5. Sidesway Web Buckling

The sidesway web buckling criteria were developed after observing several unexpected failures in beams tested at the University of Texas-Austin.[33] In these tests the compression flanges were braced at the concentrated load, the web was squeezed into compression and the tension flange buckled (see Fig. C-K1.1).

Sidesway web buckling will not occur in the following cases:

For flanges restrained against rotation:

$$\frac{d_c/t_w}{\ell/b_f} > 2.3$$

For flange rotation not restrained:

$$\frac{d_c/t_w}{\ell/b_f} > 1.7$$

Sidesway web buckling can also be prevented by the proper design of lateral bracing or stiffeners at the load point. It is suggested that local bracing at both flanges be designed for 1% of the concentrated load applied to that point. Stiffeners must extend from the load point through at least one-half the girder depth. In addition, the pair of stiffeners should be designed to carry the full load. If flange rotation is permitted at the loaded flange, stiffeners will not be effective.

7. Compression Members with Web Panels Subject to High Shear

In rigid beam-to-column connections, the column web stresses are often quite high and may require a pair of diagonal stiffeners or a web doubler plate. Referring to Fig. C-K1.2, a minimum web thickness not requiring reinforcement is computed as follows:

The factored column web strength is ϕR_v, where

$$\phi = 0.90$$
$$R_v = 0.7\, F_{yc} d_c t_w$$

The maximum beam forces are assumed to act through the flanges as a couple with a moment arm of $0.95d_b$. Noting that the factored resistance must be greater than the factored load, the following inequality results:

$$(0.9)\,(0.7)\, F_{yc} d_c t_w \geq \frac{12}{0.95d_b}\,(M_1 + M_2) - V_u$$

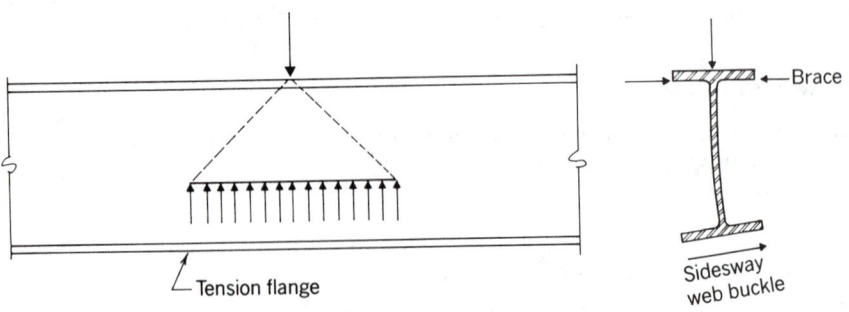

Fig. C-K1.1. Sidesway web buckling

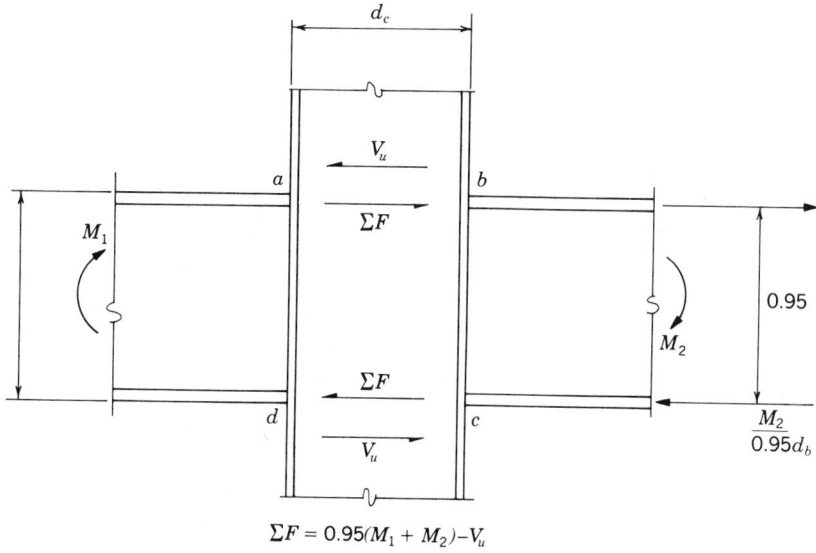

$$\Sigma F = 0.95(M_1 + M_2) - V_u$$

Fig. C-K1.2. Equilibrium of forces on web

Solving the above equation for t_w gives:

$$t_w = \frac{20.1(M_1 + M_2) - 1.6V_u d_b}{F_{yc} A_{bc}} = t_{min}$$

where

M_1, M_2 = factored beam moments, kip-in.
V_u = factored column shear, kips
F_{yc} = yield strength of the column web, ksi
A_{bc} = planar area bounded by the points *abcd* in Fig. C-K1.2, in.[2]

For webs subject to high shear in combination with axial loads exceeding $0.75P_n$, Formula K1-10 was developed to give a straight line transition from maximum shear at 0.75 axial load to maximum axial load at 0.70 shear; see Fig. C-K1.3.

K2. PONDING

As used in the LRFD Specification, *ponding* refers to the retention of water due solely to the deflection of flat roof framing. The amount of this water is dependent upon the flexibility of the framing. Lacking sufficient framing stiffness, its accumulated weight can result in collapse of the roof if a strength evaluation is not made (see ANSI A58.1).

Representing the deflected shape of the primary and critical secondary member as a half-sine wave, the weight and distribution of the ponded water can be estimated

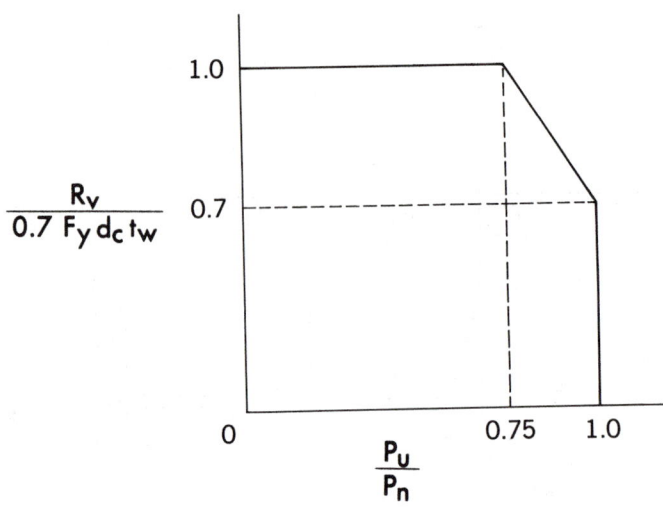

Fig. C-K1.3. Interaction of shear and axial force

and, from this, the contribution that the deflection each of these members makes to the total ponding deflection can be expressed:[65]

For the primary member:

$$\Delta_w = \frac{\alpha_p \Delta_o \, 1 + 0.25\pi\alpha_s + 0.25\pi \, \rho(1 + \alpha_s)}{1 - 0.25\pi\alpha_p\alpha_s}$$

For the secondary member:

$$\delta_w = \frac{\alpha_s \delta_o \, 1 + \dfrac{\pi^3}{32}\alpha_p + \dfrac{\pi^2}{8\rho}(1 + \alpha_p) + 0.185\alpha_s\alpha_p}{1 - 0.25\pi\alpha_p\alpha_s}$$

In these expressions Δ_o and δ_o are, respectively, the primary and secondary beam deflections due to loading present at the initiation of ponding, $\alpha_p = C_p/(1 - C_p)$, $\alpha_s = C_s/(1 - C_s)$ and $\rho = \delta_o/\Delta_o = C_s/C_p$.

Using the above expressions for Δ_w and δ_w, the ratios Δ_w/Δ_o and δ_w/δ_o can be computed for any given combination of primary and secondary beam framing using, respectively, the computed value of parameters C_p and C_s defined in the LRFD Specification.

Even on the basis of unlimited elastic behavior, it is seen that the ponding deflections would become infinitely large unless

$$\left(\frac{C_p}{1 - C_p}\right)\left(\frac{C_s}{1 - C_s}\right) < \frac{4}{\pi}$$

Since elastic behavior is not unlimited, the effective bending strength available in each member to resist the stress caused by ponding action is restricted to the difference between the yield stress of the member and the stress f_o produced by the total load supported by it before consideration of ponding is included.

Note that elastic deflection is directly proportional to stress. The admissible amount of ponding in either the primary or critical (midspan) secondary member, in terms of the applicable ratio Δ_w/Δ_o or δ_w/δ_o, can be represented as $(0.8F_y - f_o)/f_o$. Substituting this expression for Δ_w/Δ_o and δ_w/δ_o, and combining with the foregoing expressions for Δ_w and δ_w, the relationship between critical values for C_p and C_s and the available elastic bending strength to resist ponding is obtained. The curves presented in Figs. C-K2.1 and C-K2.2 are based upon this relationship. They constitute a design aid for use when a more exact determination of required flat roof framing stiffness is needed than given by the LRFD Specification provision that $C_p + 0.9C_s \leq 0.25$.

Given any combination of primary and secondary framing, the stress index is computed as

$$U_p = \left(\frac{0.8F_y - f_o}{f_o}\right)_p \text{ for the primary member}$$

$$U_s = \left(\frac{0.8F_y - f_o}{f_o}\right)_s \text{ for the secondary member}$$

where f_o, in each case, is the computed bending stress, ksi, in the member due to the supported loading, neglecting ponding effect. Depending upon geographic location, this loading should include such amount of snow as might also be present, although ponding failures have occurred more frequently during torrential summer rains when the rate of precipitation exceeded the rate of drainage runoff and the resulting hydraulic gradient over large roof areas caused substantial accumulation of water some distance from the eaves.

Given the size, spacing and span of a tentatively selected combination of primary and secondary beams, for example, one may enter Fig. C-K2.1 at the level of the computed stress index U_p, determined for the primary beam; move horizontally to the computed C_s value of the secondary beams; then move downward to the abscissa scale. The combined stiffness of the primary and secondary framing is sufficient to prevent ponding if the flexibility constant read from this latter scale is more than the value of C_p computed for the given primary member; if not, a stiffer primary or secondary beam, or combination of both, is required.

If the roof framing consists of a series of equally-spaced wall-bearing beams, they would be considered as secondary members, supported on an infinitely stiff primary member. For this case, one would use Fig. C-K2.2. The limiting value of C_s would be determined by the intercept of a horizontal line representing the U_s value and the curve for $C_p = 0$.

The ponding deflection contributed by a metal deck is usually such a small part of the total ponding deflection of a roof panel that it is sufficient merely to limit its moment of inertia (per foot of width normal to its span) to 0.000025 times the fourth power of its span length, as provided in the LRFD Specification. However, the stability against ponding of a roof consisting of a metal roof deck of relatively slender depth-span ratio, spanning between beams supported directly on columns, may need

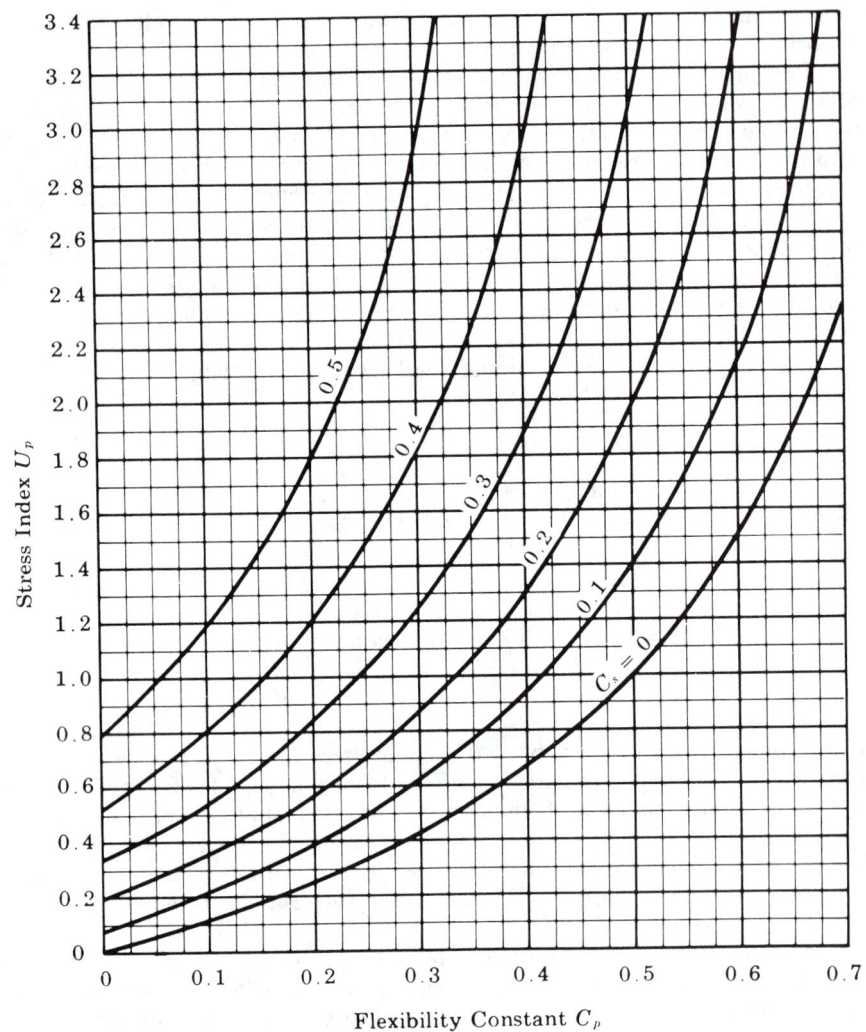

Figure C-K2.1

to be checked. This can be done using Fig. C-K2.1 or C-K2.2 with the following computed values:

U_p = stress index for the supporting beam
U_s = stress index for the roof deck
C_p = flexibility constant for the supporting beams
C_s = flexibility constant for one foot width of the roof deck ($S = 1.0$)

Since the shear rigidity of their web system is less than that of a solid plate, the moment of inertia of steel joists and trusses should be taken as somewhat less than that of their chords.

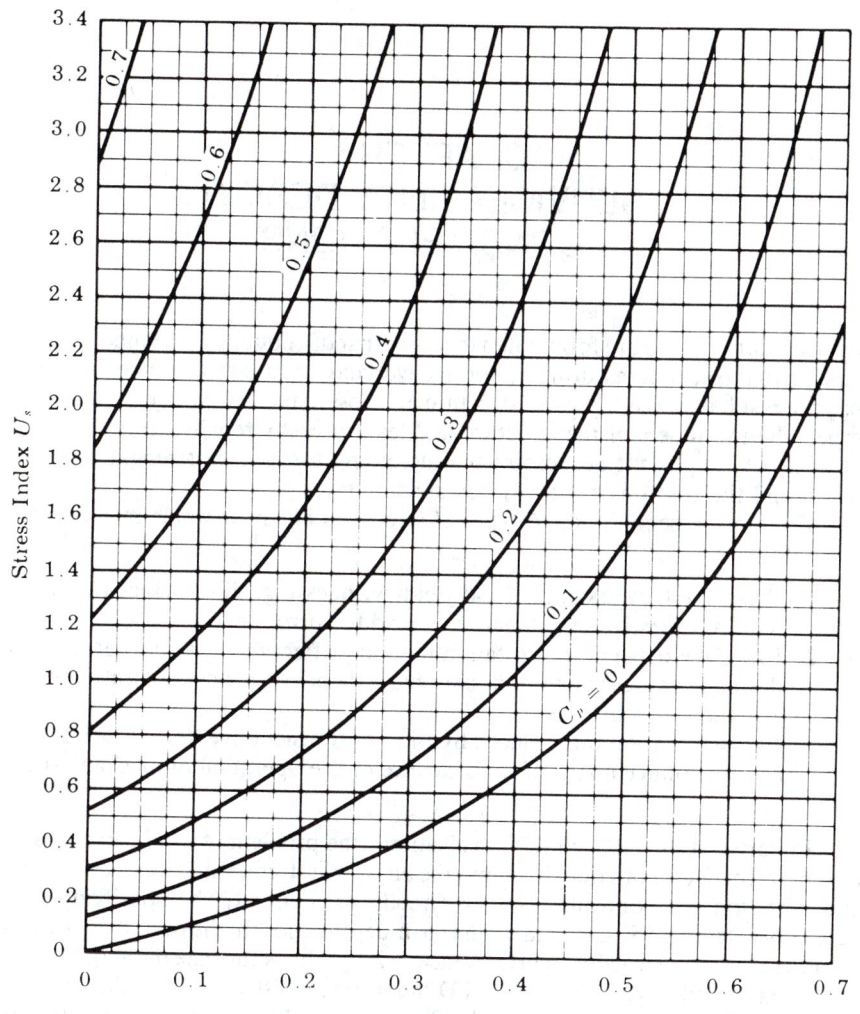

Figure C-K2.2

CHAPTER L.
SERVICEABILITY DESIGN
CONSIDERATIONS

Serviceability criteria are formulated to insure that disruptions of the functional use or damage to the structure during its normal everyday use are rare. While malfunctions may not result in the collapse of a structure or in loss of life or injury, they can seriously impair the usefulness of the structure and lead to costly repairs. Concern with serviceability is important because of the increasing use of high-strength materials in design and the resulting relatively flexible structures.

There are essentially three types of structural behavior which may impair serviceability:

1. Excessive local damage (local yielding, buckling, slip or cracking) that may require excessive maintenance or lead to corrosion.
2. Excessive deflection or rotation that may affect the appearance, function or drainage of the structure, or may cause damage to nonstructural components and their attachments.
3. Excessive vibrations induced by wind or transient live loads which affect the comfort of occupants of the structure or the operation of mechanical equipment.

In Part 1 of the 1978 AISC Specification,[1] the problem of local damage is taken care of by the factor of safety built into the allowable stress, while excessive deflection and vibration are controlled, either directly or indirectly, by specifying limiting deflections, lateral drifts and maximum span-depth ratios. In the past, these rules have led to satisfactory structural performance, with perhaps the exception of large open floor areas without partitions. In LRFD the serviceability checks should consider the appropriate loads, the response of the structure and the reaction of the occupants to the structural response.

Examples of loads that may require consideration in serviceability checking include permanent live loads, wind and earthquake; effects of human activities such as walking, dancing, etc.; temperature fluctuations; and vibrations induced by traffic near the building or by the operation of mechanical equipment within the building.

Serviceability checks are concerned with adequate performance under the appropriate load conditions. The response of the structure can usually be assumed to be elastic. However, some structural elements may have to be examined with respect to their long-term behavior under load.

It is difficult to specify limiting values of structural performance based on serviceability considerations because these depend to a great extent on the type of structure, its intended use and subjective physiological reaction. For example, acceptable structural motion in a hospital clearly would be much less than in an ordinary industrial building. It should be noted that humans perceive levels of structural motion that are far less than motions that would cause any structural damage. Serviceability limits must be determined through careful consideration by the designer and client.

L1. CAMBER

The engineer should consider camber when deflections at the appropriate load level present a serviceability problem.

L2. EXPANSION AND CONTRACTION

As in the case of deflections, the satisfactory control of expansion cannot be reduced to a few simple rules, but must depend largely upon the good judgment of qualified engineers.

The problem is more serious in buildings with masonry walls than with prefabricated units. Complete divorcement of the framing, at widely spaced expansion joints, is generally more satisfactory than more frequently located devices dependent upon the sliding of parts in bearing, and usually less expensive than rocker or roller expansion bearings.

Creep and shrinkage of concrete and yielding of steel are among the causes, other than temperature, for dimensional changes.

L3. DEFLECTIONS, VIBRATION AND DRIFT

1. Deflections

Excessive transverse deflections or lateral drift may lead to permanent damage to building elements, separation of cladding or loss of weathertightness, damaging transfer of load to non-load-supporting elements, disruption of operation of building service systems, objectionable changes in appearance of portions of the buildings and discomfort to occupants.

The LRFD Specification does not provide specific limiting deflections for individual members or structural assemblies. Such limits would depend on the function of the structure.[36, 77, 78] Provisions that limit deflections to a percentage of span may not be adequate for certain long-span floor systems; a limit on maximum deflection that is independent of span length may also be necessary to minimize the possibility of damage to adjoining or connecting nonstructural elements.

2. Vibration

The increasing use of high-strength materials and efficient structural schemes leads to longer spans and more flexible floor systems. Even though the use of a deflection limit related to span length generally precluded vibration problems in the past, some floor systems may require explicit consideration of the dynamic, as well as the static, characteristics of the floor system.

The dynamic response of structures or structural assemblies may be difficult to analyze because of difficulties in defining the actual mass, stiffness and damping characteristics. Moreover, different load sources cause varying responses. For example, a steel beam-concrete slab floor system may respond to live loading as a non-composite system, but to transient excitation from human activity as an orthotropic composite plate. Nonstructural partitions, cladding and built-in furniture significantly increase the stiffness and damping of the structure and frequently eliminate potential vibration problems. The damping can also depend on the amplitude of excitation.

The general objective in minimizing problems associated with excessive structural motion is to limit accelerations, velocities and displacements to levels that would not be disturbing to the building occupants. Generally, occupants of a building find sustained vibrations more objectionable than transient vibrations.

The levels of peak acceleration that people find annoying depend on frequency of response. Thresholds of annoyance for transient vibrations are somewhat higher and depend on the amount of damping in the floor system. These levels depend on the individual and his activity at the time of excitation.[36, 48, 77, 78, 103, 104]

The most effective way to reduce effects of continuous vibrations is through vibration isolation devices. Care should be taken to avoid resonance, where the frequency of steady-state excitation is close to the fundamental frequency of the system. Transient vibrations are reduced most effectively by increasing the damping in the structural assembly. Mechanical equipment which can produce objectionable vibrations in any portion of a structure should be adequately isolated to reduce the transmission of such vibrations to critical elements of the structure.

3. Drift

The LRFD Specification does not provide specific limiting values for lateral drift. If a drift analysis is desired, the stiffening effect of non-load-supporting elements such as partitions and infilled walls may be included in the analysis of drift if substantiating information regarding their effect on response is available.

Some irrecoverable inelastic deformations may occur at given load levels in certain types of construction, e.g., flexural members where the shape factor Z/S is in excess of 1.5. The effect of such deformations may be negligible or serious, depending on the function of the structure, and should be considered by the designer on a case by case basis.

The deformation limits should apply to structural assemblies as a whole. Reasonable tolerance should also be provided for creep. Where load cycling occurs, consideration should be given to the possibility of increases in residual deformation that may lead to incremental failure.

L5. CORROSION

Steel members may deteriorate in particular service environments. This deterioration may appear either in external corrosion, which would be visible upon inspection, or in undetected changes in the material that would reduce its load-carrying capacity. The designer should recognize these problems by either factoring a specific amount of damage tolerance into his design or providing adequate protection systems (e.g., coatings, cathodic protection) and/or planned maintenance programs so that such problems do not occur.

CHAPTER M.
FABRICATION, ERECTION
AND QUALITY CONTROL

M2. FABRICATION

1. Cambering, Curving and Straightening

The use of heat for straightening or cambering members is permitted for A514 steel, as it is for other steels. However, the maximum temperature permitted is 1,100°F for A514 steel, as contrasted with 1,200°F for other steels.

The cambering of flexural members when required by contract documents, is accomplished in various ways. In the case of trusses and girders, the desired curvature can be built in during assembly of the component parts. Within limits, rolled beams can be cold-cambered at the producing mill.

The local application of heat has come into common use as a means of straightening or cambering beams and girders. The method depends upon an ultimate shortening of the heat-affected zones. A number of such zones, on the side of the member that would be subject to compression during cold-cambering or "gagging," are heated enough to be "upset" by the restraint provided by surrounding unheated areas. Shortening takes place upon cooling.

While the final curvature or camber can be controlled by these methods, it must be realized that some deviation, due to workmanship error and permanent change due to handling, is inevitable.

5. Bolted Construction

In the past, it has been required to tighten to a specified tension all ASTM A325 and A490 bolts in both slip-critical and bearing-type connections. The requirement was changed in 1985 to permit some bearing-type connections to be only tightened to a snug-tight condition.

To qualify as a snug-tight bearing connection, the bolts cannot be subject to tension loads, slip can be permitted and loosening or fatigue due to vibration or load fluctuations are not design considerations.

It is suggested that snug-tight bearing-type connections be used in applications when A307 bolts would be permitted. The 1978 Specification Sect. 1.15.12 serves as a guide to these applications.

This section provides rules for the use of oversized and slotted holes paralleling the provisions which have been in the RCSC *Specification for Structural Joints Using ASTM A325 or A490 Bolts*[68] since 1972, extended to include A307 bolts which are outside the scope of the high-strength bolt specifications.

M3. SHOP PAINTING

The surface condition of steel framing disclosed by the demolition of long-standing buildings has been found to be unchanged from the time of its erection, except at isolated spots where leakage may have occurred. Even in the presence of leakage, the shop coat is of minor influence.[69]

The LRFD Specification does not define the type of paint to be used when a shop coat is required. Conditions of exposure and individual preference with regard to finish paint are factors which bear on the selection of the proper primer. Hence, a single formulation would not suffice. For a comprehensive treatment of the subject, see Ref. 70.

M4. ERECTION

4. Fit of Column Compression Joints

Tests at the University of California-Berkeley[71] on spliced full-size columns with joints that had been intentionally milled out-of-square, relative to either strong or weak axis, demonstrated that their load-carrying capacity was the same as that for a similar unspliced column. In the tests, gaps of $\frac{1}{16}$-in. were not shimmed; gaps of $\frac{1}{4}$-in. were shimmed with non-tapered mild steel shims. Minimum size partial-penetration welds were used in all tests. No tests were performed on specimens with gaps greater than $\frac{1}{4}$-in.

APPENDIX E.
COLUMNS AND OTHER
COMPRESSION MEMBERS

E3. FLEXURAL-TORSIONAL BUCKLING

A possible mode of buckling of columns is torsional buckling for symmetric shapes and flexural-torsional buckling for unsymmetric shapes. These modes are usually not considered in design for the hot-rolled columns because they generally do not govern, or the critical load differs very little from the weak-axis planar buckling load. Such a buckling mode may, however, control the capacity of columns made from plate elements which are relatively thin, and for unsymmetric columns. Formulas for determining the flexural-torsional elastic buckling loads of such columns are derived in texts on structural stability (Refs. 49, 50, 72, for example). Since these equations for flexural-torsional buckling apply only to elastic buckling, they must be modified for inelastic buckling when $F_{cr} > 0.5 \, F_y$. This is accomplished through the use of the equivalent slenderness factor $\lambda_e = \sqrt{F_y/F_e}$.

APPENDIX F.
BEAMS AND OTHER
FLEXURAL MEMBERS

F1. DESIGN FOR FLEXURE

7. Nominal Flexural Strength of Other Sections

Three limit states must be investigated to determine the moment capacity of flexural members: lateral-torsional buckling (LTB), local buckling of the compression flange (FLB) and local buckling of the web (WLB). These limit states depend, respectively, on the beam slenderness ratio L_b/r_y, the width-thickness ratio b/t of the compression flange and the width-thickness ratio h_c/t_w of the web. For convenience, all three measures of slenderness are denoted by λ.

Variations in M_n with L_b are shown in Fig. C-F1.1. The discussion of plastic, inelastic and elastic buckling in Commentary Sect. F1 with reference to lateral-torsional buckling applies here except for an important difference in the significance of λ_p for lateral-torsional buckling and local buckling. Values of λ_p for FLB and WLB produce a compact section with a rotation capacity of about four (after reaching M_p) before the onset of local buckling, and therefore meet the requirements for plastic analysis of load effects (Commentary Sect. B5). On the other hand, values of λ_p for LTB do not allow plastic analysis because they do not provide rotation capacity beyond that needed to develop M_p. Instead $L_b \leq L_{pd}$ (Sect. F1.1) must be satisfied.

Analyses to include restraint effects of adjoining elements are discussed in Ref. 11. Analysis of the lateral stability of members with shapes not covered in this appendix must be performed according to the available literature.[11]

See the Commentary for Sect. B5 for the discussion of the equation regarding the bending capacity of circular sections.

F4. WEB-TAPERED MEMBERS

1. General Requirements

The provision contained in Appendix F4 covers only those aspects of the design of tapered members that are unique to tapered members. For other criteria of design not specifically covered in Appendix F4, see the appropriate portions of this Specification and Commentary.

The design of wide-flange columns with a single web taper and constant flanges follows the same procedure as for uniform columns according to Sect. E2, except the column slenderness parameter λ_c for major axis buckling is determined for a slenderness ratio $K_\gamma L/r_{ox}$, and for minor axis buckling for KL/r_{oy}, where K_γ is an effective length factor for tapered members, K is the effective length factor for prismatic members and r_{ox} and r_{oy} are the radii of gyration about the x and the y axes, respectively, taken at the smaller end of the tapered member.

For stepped columns or columns with other than a single web taper, the elastic critical stress is determined by analysis or from data in reference texts or research reports (Chaps. 11 and 13 in Ref. 49 and Refs. 50 and 15), and then the same procedure of using λ_{eff} is utilized in calculating the factored resistance.

This same approach is recommended for open section built-up columns (columns with perforated cover plates, lacing and battens) where the elastic critical buckling stress determination must include a reduction for the effect of shear. Methods for calculating the elastic buckling strength of such columns are given in Chap. 12 of the SSRC *Guide to Design Criteria for Metal Compression Members*[11] and in Refs. 49 and 50.

3. Design Compressive Strength

The approach in formulating $F_{a\gamma}$ of tapered columns is based on the concept that the critical stress for an axially loaded tapered column is equal to that of a prismatic column of different length, but of the same cross section as the smaller end of the tapered column. This has resulted in an equivalent effective length factor K_γ for a tapered member subjected to axial compression.[73] This factor, which is used to determine the value of S in Formulas A-F4-2 and E2-3, can be determined accurately for a symmetrical rectangular rigid frame composed of prismatic beams and tapered columns.

With modifying assumptions, such a frame can be used as a mathematical model to determine with sufficient accuracy the influence of the stiffness $\Sigma(I/b)_g$ of beams and rafters which afford restraint at the ends of a tapered column in other cases such as those shown in Fig. C-A-F4.1. From Formulas A-F4-2 and E2-3, the critical load P_{cr} can be expressed as $\pi^2 EI_o/(K\gamma l)^2$. The value of K_γ can be obtained by interpolation, using the appropriate chart from Ref. 73 and restraint modifiers G_T and G_B. In each of these modifiers the tapered column, treated as a prismatic member having a moment of inertia I_o, computed at the smaller end, and its actual length ℓ, is assigned the stiffness I_o/ℓ, which is then divided by the stiffness of the restraining members at the end of the tapered column under consideration.

4. Design Flexural Strength

The development of the design bending stress for tapered beams follows closely with that for prismatic beams. The basic concept is to replace a tapered beam by an equivalent prismatic beam with a different length, but with a cross section identical to

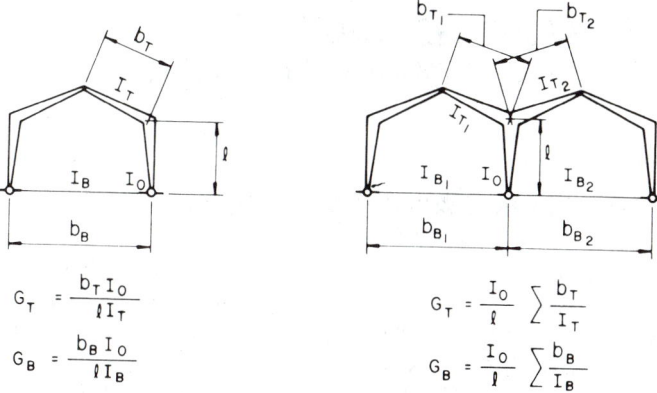

Figure C-A-F4.1

that of the smaller end of the tapered beam.[73] This has led to the modified length factors h_s and h_w in Formulas A-F4-6 and A-F4-7.

Formulas A-F4-6 and A-F4-7 are based on total resistance to lateral buckling, using both St. Venant and warping resistance. The factor B modifies the basic $F_{b\gamma}$ to account for moment gradient and lateral restraint offered by adjacent segments. For members which are continuous past lateral supports, categories 1, 2 and 3 of Sect. D3 usually apply; however, it is to be noted that they apply only when the axial force is small and adjacent unbraced segments are approximately equal in length. For a single member, or segments which do not fall into category 1, 2, 3 or 4, the recommended value of B is unity. The value of B should also be taken as unity when computing the value of $F_{b\gamma}$ to obtain M_n to be used in Formulas H1-1 through H1-3, since the effect of moment gradient is provided for by the factor C_m. The background material is given in WRC Bulletin No. 192.[74]

APPENDIX G.
PLATE GIRDERS

Appendix G is taken from AISI Bulletin 27.[9] Comparable provisions are included in Sect. 1.10 of the 1978 AISC Specification.[1] The provisions have been moved to an appendix as they are seldom used and produce designs which are often less economical than plate girders designed without tension field action.

The web slenderness ratio $h_c/t_w = 970/\sqrt{F_{yf}}$ that distinguishes plate girders from beams is written in terms of the flange yield stress, because for hybrid girders inelastic buckling of the web due to bending depends on the flange strain.

APPENDIX H.
MEMBERS UNDER TORSION
AND COMBINED FORCES

H3. ALTERNATE INTERACTION EQUATIONS FOR MEMBERS UNDER COMBINED STRESS

In the case of members not subject to local column or flexural buckling, i.e., $L_b < L_{pd}$, the use of somewhat more liberal interaction Formulas A-H3-5 and A-H3-6 are acceptable as an alternate when the flexure is about one axis only. Formula A-H3-5 is retained from Part 2 of the 1978 AISC Specification.[1]

The alternate interaction Formulas A-H3-1 and A-H3-2 for biaxially loaded H and wide flange column shapes were taken from Refs. 11 and 23.

APPENDIX K.
STRENGTH DESIGN
CONSIDERATIONS

K4. FATIGUE

Because most members in building frames are not subject to a large enough number of cycles of full design stress application to require design for fatigue, the provisions covering such designs have been placed in Appendix K4.

When fatigue is a design consideration, its severity is most significantly affected by the number of load applications, the magnitude of the stress range and the severity of the stress concentrations associated with the particular details. These factors are not encountered in normal building designs; however, when encountered and when fatigue is of concern, all provisions of Appendix K4 must be satisfied.

Members or connections subject to less than 20,000 cycles of loading will not involve a fatigue condition, except in the case of repeated loading involving large ranges of stress. For such conditions, the admissible range of stress can conservatively be taken as one and one-half times the applicable value given in Table A-K4.3 for "Loading Condition 1."

Fluctuation in stress which does not involve tensile stress does not cause crack propagation and is not considered to be a fatigue situation. On the other hand, in elements of members subject solely to calculated compression stress, fatigue cracks may initiate in regions of high tensile residual stress. In such situations, the cracks generally do not propagate beyond the region of the residual tensile stress, because the residual stress is relieved by the crack. For this reason stress ranges that are completely in compression are not included in the column headed by "Kind of Stress" in Table A-K4.2 of Appendix K4. This is also true of comparable tables of the current AASHTO and AREA specifications.

When fabrication details involving more than one category occur at the same location in a member, the stress range at that location must be limited to that of the most restrictive category. By locating notch-producing fabrication details in regions subject to a small range of stress, the need for a member larger than required by static loading will often be eliminated.

Extensive test programs[66, 67] using full size specimens, substantiated by theoretical stress analysis, have confirmed the following general conclusions:

1. Stress range and notch severity are the dominant stress variables for welded details and beams.
2. Other variables such as minimum stress, mean stress and maximum stress are not significant for design purposes.
3. Structural steels with yield points of 36 to 100 ksi do not exhibit significantly different fatigue strength for given welded details fabricated in the same manner.

Allowable stress ranges can be read directly from Table A-K4.3 for a particular category and loading condition. The values are based on recent research.[100]

Provisions for bolts subjected to tension are given in Table A-K4.4. Tests have uncovered dramatic differences in fatigue life, not completely predictable from the various published formulas for estimating the actual magnitude of prying force.[43] To limit the uncertainties regarding prying action on the fatigue behavior of these bolts, the tensile stresses given in Table J3.2 are approved for use under extended cyclic loading only if the prying force, included in the design tensile force, is small. When this cannot be assured, the design tensile stress is drastically reduced to cover any conceivable prying effect.

The use of other types of mechanical fasteners to resist applied cyclic loading in tension is not recommended. Lacking a high degree of assured pretension, the range of stress is generally too great to resist such loading for long.

However, all types of mechanical fasteners survive unharmed when subject to cyclic shear stresses sufficient to fracture the connected parts, which is provided for elsewhere in Appendix K4.

REFERENCES

1. *American Institute of Steel Construction, Inc.* Specification for the Design, Fabrication and Erection of Structural Steel for Buildings *1978, Chicago, Ill.*
2. *American Concrete Institute* Building Code Requirements for Reinforced Concrete *ACI 318-83, 1983.*
3. *American Association of State Highway and Transportation Officials* Standard Specification for Highway Bridges *1977.*
4. *Ravindra, M. K. and T. V. Galambos* Load and Resistance Factor Design for Steel *ASCE Journal of the Structural Division, Vol. 104, No. ST9, September 1978.*
5. *Ellingwood, B., et al* Development of a Probability Based Load Criterion for American National Standard A58 Building Code Requirements for Minimum Design Loads in Buildings and Other Structures *Special Publication 577, National Bureau of Standards, June 1980.*
6. *American National Standards Institute* Minimum Design Loads for Buildings and Other Structures *ANSI A58.1-82.*
7. *American Welding Bureau* Report of Structural Welding Committee *1931.*
8. *Chopra, A. K. and N. M. Newmark* Design of Earthquake Resistant Structures *E. Rosenblueth, Ed., 1980. John Wiley and Sons, Inc., New York, N.Y.*
9. *Galambos, T. V.* Proposed Criteria for Load Resistance Factor Design of Steel Building Structures *Research Report No. 45, Civil Engineering Dept., Washington Univ., St. Louis, Mo., May 1976. Also, American Iron and Steel Institute, Bulletin No. 27, January 1978.*
10. *Sherman, D. R.* Tentative Criteria for Structural Applications of Steel Tubing and Pipe *August 1976, American Iron and Steel Institute, Washingon, D.C.*
11. *Johnston, B. G., Ed.* Guide to Design Criteria for Metal Compression Members *3rd. Ed., 1976, Structural Stability Research Council, John Wiley and Sons, New York, N.Y.*
12. *Galambos, T. V.* Reliability of Axially Loaded Columns *December 1980. Washington Univ., Dept. of Civil Engineering, St. Louis, Mo.*
13. *Hall, Dann H.* Proposed Steel Column Strength Criteria *ASCE Journal of the Structural Division, Vol. 107, No. ST4, April 1981.*
14. *Yura, J. A., et al* The Bending Resistance of Steel Beams *ASCE Journal of the Structural Division, Vol. 104, No. ST9, September 1978.*
15. *Kitipornchai, S. and N. S. Trahair* Buckling Properties of Monosymmetric I-Beams *ASCE Journal of the Structural Division, Vol. 109, No. ST5, May 1980.*
16. *Cooper, P. B., et al* LRFD Criteria for Plate Girders *ASCE Journal of the Structural Division, Vol. 104, No. ST9, September 1978.*
17. *Basler, Konrad* Strength of Plate Girders in Shear *ASCE Journal of the Structural Division, Vol. 104, No. ST9, October 1961.*
18. *Popov, E. P.* An Update on Eccentric Seismic Bracing *AISC Engineering Journal, 3rd Qtr., 1980.*
19. *Galambos, T. V. and M. K. Ravindra* Tentative Load and Resistance Factor Design Criteria for Steel Buildings *Research Report No. 18, Washington Univ., Dept. of Civil Engineering, St. Louis, Mo., September 1973.*
20. *American Society of Civil Engineers* Plastic Design in Steel—A Guide and a Commentary *ASCE Manual 41, 2nd Ed., 1971.*
21. *Lim, L. C. and L. W. Lu* The Strength and Behavior of Laterally Unsupported Columns *Fritz Engineering Laboratory Report No. 329.5, Lehigh Univ., Bethlehem, Pa., June 1970.*
22. *Ross, D. A. and W. F. Chen* Design Criteria for Steel I-Columns Under Axial Load and Biaxial Bending *Fritz Engineering Laboratory Report No. 389.6/393.3A, Lehigh Univ., Bethlehem, Pa., August 1975.*
23. *Springfield, J.* Design of Columns Subject to Biaxial Bending *AISC Engineering Journal, 3rd Qtr., 1975.*
24. *LeMessurier, W. J., R. J. McNamara and J. C. Scrivener* Approximate Analytical Model for Multi-story Frames *AISC Engineering Journal, 4th Qtr., 1974.*
25. *LeMessurier, W. J.* A Practical Method of Second Order Analysis, Part 1—Pin-jointed Frames *AISC Engineering Journal, 4th Qtr., 1976.*
26. *LeMessurier, W. J.* A Practical Method of Second Order Analysis, Part 2—Rigid Frames *AISC Engineering Journal, 2nd Qtr., 1977.*

27. *Kanchanalai, T. and L. W. Lu* Analysis and Design of Framed Columns Under Minor Axis Bending *AISC Engineering Journal, 2nd Qtr., 1979.*

28. Tentative Provisions for the Development of Seismic Regulations for Buildings *ATC Publication 3-06, June 1978.*

29. *Austin, W. J.* Strength and Design of Metal Beam-Columns *ASCE Journal of the Structural Division, Vol. 87, No. ST4, April 1961.*

30. *International Association of Bridge and Structural Engineering* Final Report of the Eighth Congress *September 1968.*

31. *Bergfelt, A.* Studies and Tests on Slender Plate Girders Without Stiffeners *March 1971.*

32. *Hoglund, T.* Simply Supported Long Thin Plate I-Girders Without Web Stiffeners, Subjected to Distributed Transverse Load *Dept. of Building Statics and Structural Engineering of the Royal Institute of Technology, Stockholm, Sweden.*

33. *Yura, J. A.* Web Behavior at Points of Concentrated Loads *Univ. of Texas-Austin.*

34. *Galambos, T. V. and J. Chapuis* LRFD Criteria for Composite Columns and Beam Columns *Revised Draft, December 1980. Washington Univ., Dept. of Civil Engineering, St. Louis, Mo.*

35. *SSRC Task Group 20* A Specification for the Design of Steel-Concrete Composite Columns *AISC Engineering Journal, 4th Qtr., 1979.*

36. *American Society of Civil Engineers* Structural Design of Tall Steel Buildings *1979.*

37. *Hansell, W. C., T. V. Galambos, M. K. Ravindra and I. M. Viest* Composite Beam Criteria in LRFD *ASCE Journal of the Structural Division, Vol. 104, No. ST9, September 1978.*

38. *Grant, J. A., Jr., J. W. Fisher and R. G. Slutter* Composite Beams with Formed Steel Deck *AISC Engineering Journal, 1st Qtr., 1977.*

39. *Ollgaard, J. G., R. G. Slutter and J. W. Fisher* Shear Strength of Stud Connectors in Light Weight and Normal Weight Concrete *AISC Engineering Journal, April 1971.*

40. *Slutter, R. G. and G. C. Driscoll, Jr.* Flexural Strength of Steel-concrete Composite Beams *ASCE Journal of the Structural Division, Vol. 91, No. ST2, April 1965.*

41. *Gibson, G. T. and B. T. Wake* An Investigation of Welded Connections for Angle Tension Members *AWS, The Welding Journal, January 1942.*

42. *Kloppel, K. and T. Seeger* Dauerversuche Mit Einsohnittigen Hv-Verbindurgen Aus ST37 *Der Stahlbau, 33(8):225–245, August 1964 and Vol. 33, No. 11, November 1964.*

43. *Fisher, J. W. and J. H. A. Struik* Guide to Design Criteria for Bolted and Riveted Joints *John Wiley and Sons, Inc., New York, N.Y., 1974.*

44. *Freeman, F. R.* The Strength of Arc-welded Joints *Proc. Inst. Civil Engineers 231, London, England, 1930.*

45. *Fisher, J. W., T. V. Galambos, G. L. Kulak and M. K. Ravindra* Load and Resistance Factor Design Criteria for Connectors *ASCE Journal of the Structural Division, Vol. 104, No. ST9, September 1978.*

46. *Bendigo, R. A., R. H. Hansen and J. L. Rumpf* Long-bolted Joints *ASCE Journal of the Structural Division, Vol. 89, No. ST6, December 1963.*

47. *Birkemoe, P. C. and M. I. Gilmor* Behavior of Bearing Critical Double-angle Beam Connections *AISC Engineering Journal, 4th Qtr., 1978.*

48. *International Organization for Standardization* Guide for the Evaluation of Human Exposure to Whole-Body Vibration *Document ISO 2631, September 1974.*

49. *Timoshenko, S. P. and J. M. Gere* Theory of Elastic Stability *McGraw-Hill Book Co., 1961.*

50. *Bleich, F.* Buckling Strength of Metal Structures *McGraw-Hill Book Co., 1952.*

51. *Galambos, T. V.* Influence of Partial Base Fixity on Frame Stability *ASCE Journal of the Structural Division, Vol. 86, No. ST5, May 1960.*

52. *Yura, J. A.* The Effective Length of Columns in Unbraced Frames *AISC Engineering Journal, April 1971.*

53. *Disque, R. O.* Inelastic K-Factor in Design *AISC Engineering Journal, 2nd Qtr., 1973.*

54. *Lu, L. W., E. Ozer, J. H. Daniels, O. S. Okten, S. Morino* Strength and Drift Characteristics of Steel Frames *ASCE Journal of the Structural Division Vol. 103, No. ST11, November 1977.*

55. *Cheong-Siat Moy, F., E. Ozer and L. W. Lu* Strength of Steel Frames under Gravity Loads *ASCE Journal of the Structural Division, Vol. 103, No. ST6, June 1977.*

56. *Springfield, J. and P. F. Adams* Aspects of Column Design in Tall Steel Buildings *ASCE Journal of the Structural Division, Vol. 98, No. ST5, May 1972.*

57. *Liapunow, S.* Ultimate Load Studies of Plane Multi-story Steel Rigid Frames *ASCE Journal of the Structural Division, Vol. 100, No. ST8, Proc. Paper 10750, August 1974.*

58. *Daniels, J. H. and L. W. Lu* Plastic Subassemblage Analysis for Unbraced Frames *ASCE Journal of the Structural Division, Vol. 98, No. ST8, August 1972.*

59. *Winter, G.* Lateral Bracing of Columns and Beams *ASCE Journal of the Structural Division, Vol. 84, No. ST2, March 1958.*
60. *Lu, Le-Wu* Design of Braced Multi-story Frames by the Plastic Method *AISC Engineering Journal, January 1967.*
61. *Kanchanalai, T.* The Design and Behavior of Beam-columns in Unbraced Steel Frames *AISI Project No. 189, Report No. 2, Civil Engineering/Structures Research Lab, Univ. of Texas-Austin, October 1977.*
62. *Wood, B. R., D. Beaulieu and P. F. Adams* Column Design by P-Delta Method *ASCE Journal of the Structural Division, Vol. 102, No. ST2, February 1976.*
63. *Jones, J.* Static Tests on Riveted Joints *Civil Engineering, May 1940.*
64. *Elgaaly, M.* Web Design under Compressive Edge Loads *AISC Engineering Journal, 4th Qtr., 1983.*
65. *Marino, F. J.* Ponding of Two-way Roof Systems *AISC Engineering Journal, July 1966.*
66. *Fisher, J. W., K. H. Frank, M. A. Hirt and B. M. McNamee* Effect of Weldments on the Fatigue Strength of Beams *National Cooperative Highway Research Program, Report 102, 1970.*
67. *Fisher, J. W., P. A. Albrecht, B. T. Yen, D. J. Klingerman and B. M. McNamee* Fatigue Strength of Steel Beams with Welded Stiffeners and Attachments *National Cooperative Highway Research Program, Report 147, 1974.*
68. *Research Council on Riveted and Bolted Structural Joints* Specification for Structural Joints Using ASTM A325 or A490 Bolts *1980.*
69. *Bigos, J., G. W. Smith, E. F. Ball and P. J. Foehl* Shop Paint and Painting Practice *AISC National Engineering Conference, Proceedings, 1954.*
70. *Steel Structures Painting Council* Steel Structures Painting Manual, Vol. 2, Systems and Specifications *Pittsburgh, Pa.*
71. *Popov, E. P. and R. M. Stephen* Capacity of Columns with Splice Imperfections *AISC Engineering Journal, 1st Qtr., 1977.*
72. *Galambos, T. V.* Structural Members and Frames *Prentice-Hall, Englewood Cliffs, N.J., 1968.*
73. *Lee, G. C., M. L. Morrell and R. L. Ketter* Design of Tapered Members *WRC Bulletin No. 173, June 1972.*
74. *Morrell, M. L. and G. C. Lee* Allowable Stress for Web-tapered Beams with Lateral Restraints *WRC Bulletin No. 192, February 1974.*
75. *Johnston, B. G. and L. F. Green* Flexible Welded Angle Connections *The Welding Journal, October 1940.*
76. *Bridge, P. Q. and J. W. Roderick* Behavior of Built-up Composite Columns *ASCE Journal of the Structural Division, Vol. 104, No. ST7, July 1978, pp. 1,141–1,165.*
77. *Galambos, T. V., et al* Structural Deflections: A Literature and State of the Art Survey *National Bureau of Standards Building Science Series 47, Washington, D.C.*
78. *Canadian Standards Association* Steel Structures for Buildings, Appendices G, H and I *CSA S16.1–1974, Rexdale, Ontario, Canada, 1974.*
79. *Sherman, D. R. and A. S. Tanavde* Comparative Study of Flexural Capacity of Pipes *Civil Engineering Department Report, Univ. of Wisconsin-Milwaukee, March 1984.*
80. *Winter, G.* Commentary on the 1968 Edition of Light Gage Cold-formed Steel Design Manual *American Iron and Steel Institute, 1970.*
81. *Johnston, B. G., Ed.* Guide to Stability Design Criteria for Metal Structures *3rd Ed., Structural Stability Research Council.*
82. *Galambos, T. V. and M. K. Ravindra* Load and Resistance Factor Design Criteria for Steel Beams *Research Report No. 27, Washington Univ., Dept. of Civil Engineering, St. Louis, Mo., February 1976.*
83. *Rao, N. R. N., M. Lohrmann and L. Tall* Effect of Strain Rate on the Yield Stress of Structural Steels *Journal of Materials, Vol. 1, No. 1, ASTM, March 1966.*
84. *Galambos, T. V. and M. K. Ravindra* Properties of Steel for Use in LRFD *ASCE Journal of the Structural Division, Vol. 104, No. ST9, September 1978.*
85. *Beedle, L. S. and L. Tall* Basic Column Strength *ASCE Journal of the Structural Division, Vol. 86, No. ST7, July 1960.*
86. *Galambos, T. V., B. Ellingwood, J. G. MacGregor and C. A. Cornell* Probability Based Load Criteria: Assessment of Current Design Practice *ASCE Journal of the Structural Division, Vol. 108, No. ST5, May 1982.*
87. *Goble, G. G.* Shear Strength of Thin Flange Composite Specimens *AISC Engineering Journal, April 1968, Chicago, Ill.*

88. *Roberts, T. M.* Slender Plate Girders Subjected to Edge Loading *Proceedings of Institute of Civil Engineers, Part 2, 71, September 1981.*

89. *Research Council on Riveted and Bolted Structural Joints* Specification for Structural Joints Using ASTM A325 or A490 Bolts Load and Resistance Factor Design.

90. *American Institute of Steel Construction, Inc.* Commentary on Highly Restrained Welded Connections *AISC Engineering Journal, 2nd Qtr., 1973.*

91. *Brockenbrough, R. L.* Considerations in the Design of Bolted Joints for Weathering Steel *AISC Engineering Journal, 1st Qtr., 1983, Chicago, Ill. (p. 40).*

92. *Munse, W. H. and E. Chesson, Jr.* Riveted and Bolted Joints: Net Section Design *ASCE Journal of the Structural Division, Vol. 89, No. ST1, February 1963.*

93. *Gaylord, E. H. and C. N. Gaylord* Design of Steel Structures *2nd Ed., McGraw-Hill Book Co., 1972, New York, N.Y.*

94. *Zandonini, R.* Stability of Compact Built-up Struts: Experimental Investigation and Numerical Simulation *Contruzione Metalliche, November 4, 1985.*

95. *Stang, A. H. and B. S. Jaffe* Perforated Cover Plates for Steel Columns *Research Paper RP 1861, National Bureau of Standards, 1984.*

96. *Frank, K. H. and J. A. Yura* An Experimental Study of Bolted Shear Connections *FHWA/ RD-81/148, December 1981.*

97. *Ketter, R. L.* Further Studies of the Strength of Beam Columns *ASCE Journal of the Structural Division, Vol. 87, No. ST6, August 1961.*

98. *Iwankiw, N.* Note on Beam-Column Moment Amplification Factor *AISC Engineering Journal, 1st Qtr., 1984, Chicago, Ill. (p. 21).*

99. *Tide, R. H. R.* Reasonable Column Design Equations *Annual Technical Session of Structural Stability Research Council, April 16–17, 1985.*

100. *Keating, P. B. and J. W. Fisher* Review of Fatigue Tests and Design Criteria on Welded Details *NCHRP Project 12-15(50), October 1985.*

101. *Ricles, J. M. and J. A. Yura* Strength of Double-row Bolted Web Connections *ASCE Journal of the Structural Division, Vol. 109, No. ST1, January 1983.*

102. *Hardash, S. and R. Bjorhovde* New Design Criteria for Gusset Plates in Tension *AISC Engineering Journal, 2nd Qtr., 1985, Chicago, Ill. (p. 77).*

103. *Murray, T. M.* Design to Prevent Floor Vibration *AISC Engineering Journal, 3rd Qtr., 1975, Chicago, Ill. (p. 82).*

104. *Murray, T. M.* Acceptability Criterion for Occupant-induced Floor Vibrations *AISC Engineering Journal, 2nd Qtr., 1981, Chicago, Ill. (p. 62).*

Notes

GLOSSARY

Alignment chart for columns.   A nomograph for determining the effective length factor K for some types of columns

Amplification factor.   A multiplier of the value of moment or deflection in the unbraced length of an axially loaded member to reflect the secondary values generated by the eccentricity of the applied axial load within the member

Aspect ratio.   In any rectangular configuration, the ratio of the lengths of the sides

Batten plate.   A plate element used to join two parallel components of a built-up column, girder or strut rigidly connected to the parallel components and designed to transmit shear between them

Beam.   A structural member whose primary function is to carry loads transverse to its longitudinal axis

Beam-column.   A structural member whose primary function is to carry loads both transverse and parallel to its longitudinal axis

Bent.   A plane framework of beam or truss members which support loads and the columns which support these members

Biaxial bending.   Simultaneous bending of a member about two perpendicular axes

Bifurcation.   The phenomenon whereby a perfectly straight member under compression may either assume a deflected position or may remain undeflected, or a beam under flexure may either deflect and twist out of plane or remain in its in-plane deflected position

Braced frame.   A frame in which the resistance to lateral load or frame instability is primarily provided by a diagonal, a K-brace or other auxiliary system of bracing

Brittle fracture.   Abrupt cleavage with little or no prior ductile deformation

Buckling load.   The load at which a perfectly straight member under compression assumes a deflected position

Built-up member.   A member made of structural metal elements that are welded, bolted or riveted together

Cladding.   The exterior covering of the structural components of a building

Cold-formed members.   Structural members formed from steel without the application of heat

Column.   A structural member whose primary function is to carry loads parallel to its longitudinal axis

Column curve.   A curve expressing the relationship between axial column strength and slenderness ratio

Combined mechanism.   A mechanism determined by plastic analysis procedure which combines elementary beam, panel and joint mechanisms

Compact section.   Compact sections are capable of developing a fully plastic stress distribution and possess rotation capacity of approximately 3 before the onset of local buckling

Composite beam.   A steel beam structurally connected to a concrete slab so that the beam and slab respond to loads as a unit. See also concrete-encased beam

Composite column.   A steel column fabricated from rolled or built-up steel shapes and encased in structural concrete or fabricated from steel pipe or tubing and filled with structural concrete

Concrete-encased beam.   A beam totally encased in concrete cast integrally with the slab

Connection. Combination of joints used to transmit forces between two or more members. Categorized by the type and amount of force transferred (moment, shear, end reaction). See also splices

Critical load. The load at which bifurcation occurs as determined by a theoretical stability analysis

Curvature. The rotation per unit length due to bending

Design documents. See structural design documents

Design strength. Resistance (force, moment, stress, as appropriate) provided by element or connection; the product of the nominal strength and the resistance factor

Diagonal bracing. Inclined structural members carrying primarily axial load employed to enable a structural frame to act as a truss to resist horizontal loads

Diaphragm. Floor slab, metal wall or roof panel possessing a large in-plane shear stiffness and strength adequate to transmit horizontal forces to resisting systems

Diaphragm action. The in-plane action of a floor system (also roofs and walls) such that all columns framing into the floor from above and below are maintained in their same position relative to each other

Double curvature. A bending condition in which end moments on a member cause the member to assume an S-shape

Drift. Lateral deflection of a building

Drift index. The ratio of lateral deflection to the height of the building

Ductility factor. The ratio of the total deformation at maximum load to the elastic-limit deformation

Effective length. The equivalent length KL used in compression formulas and determined by a bifurcation analysis

Effective length factor K. The ratio between the effective length and the unbraced length of the member measured between the centers of gravity of the bracing members

Effective moment of inertia. The moment of inertia of the cross section of a member that remains elastic when partial plastification of the cross section takes place, usually under the combination of residual stress and applied stress. Also, the moment of inertia based on effective widths of elements that buckle locally. Also, the moment of inertia used in the design of partially composite members

Effective stiffness. The stiffness of a member computed using the effective moment of inertia of its cross section

Effective width. The reduced width of a plate or slab which, with an assumed uniform stress distribution, produces the same effect on the behavior of a structural member as the actual plate width with its nonuniform stress distribution

Elastic analysis. Determination of load effects (force, moment, stress, as appropriate) on members and connections based on the assumption that material deformation disappears on removal of the force that produced it

Elastic-perfectly plastic. A material which has an idealized stress-strain curve that varies linearly from the point of zero strain and zero stress up to the yield point of the material, and then increases in strain at the value of the yield stress without any further increases in stress

Embedment.   A steel component cast in a concrete structure which is used to transmit externally applied loads to the concrete structure by means of bearing, shear, bond, friction or any combination thereof. The embedment may be fabricated of structural-steel plates, shapes, bars, bolts, pipe, studs, concrete reinforcing bars, shear connectors or any combination thereof

Encased steel structure.   A steel-framed structure in which all of the individual frame members are completely encased in cast-in-place concrete

Euler formula.   The mathematical relationship expressing the value of the Euler load in terms of the modulus of elasticity, the moment of inertia of the cross section and the length of a column

Euler load.   The critical load of a perfectly straight centrally loaded pin-ended column

Eyebar.   A particular type of pin-connected tension member of uniform thickness with forged or flame cut head of greater width than the body proportioned to provide approximately equal strength in the head and body

Factored load.   The product of the nominal load and a load factor

Fastener.   Generic term for welds, bolts, rivets or other connecting device

Fatigue.   A fracture phenomenon resulting from a fluctuating stress cycle

First-order analysis.   Analysis based on first-order deformations in which equilibrium conditions are formulated on the undeformed structure

Flame-cut plate.   A plate in which the longitudinal edges have been prepared by oxygen cutting from a larger plate

Flat width.   For a rectangular tube, the nominal width minus twice the outside corner radius. In absence of knowledge of the corner radius, the flat width may be taken as the total section width minus three times the thickness

Flexible connection.   A connection permitting a portion, but not all, of the simple beam rotation of a member end

Floor system.   The system of structural components separating the stories of a building

Force.   Resultant of distribution of stress over a prescribed area. A reaction that develops in a member as a result of load (formerly called total stress or stress). Generic term signifying axial loads, bending moment, torques and shears

Fracture toughness.   Measurement of the ability to absorb energy without fracture. Generally determined by impact loading of specimens containing a notch having a prescribed geometry

Frame buckling.   A condition under which bifurcation may occur in a frame

Frame instability.   A condition under which a frame deforms with increasing lateral deflection under a system of increasing applied monotonic loads until a maximum value of the load called the stability limit is reached, after which the frame will continue to deflect without further increase in load

Fully composite beam.   A composite beam with sufficient shear connectors to develop the full flexural strength of the composite section

High-cycle fatigue.   Failure resulting from more than 20,000 applications of cyclic stress

Hybrid beam.   A fabricated steel beam composed of flanges with a greater yield strength than that of the web. Whenever the maximum flange stress is less than or equal to the web yield stress the girder is considered homogeneous

Hysteresis loop.   A plot of force versus displacement of a structure or member subjected to reversed, repeated load into the inelastic range, in which the path followed during release and removal of load is different from the path for the addition of load over the same range of displacement

Inclusions.   Nonmetallic material entrapped in otherwise sound metal

Incomplete fusion.   Lack of union by melting of filler and base metal over entire prescribed area

Inelastic action.   Material deformation that does not disappear on removal of the force that produced it

Instability.   A condition reached in the loading of an element or structure in which continued deformation results in a decrease of load-resisting capacity

Joint.   Area where two or more ends, surfaces, or edges are attached. Categorized by type of fastener or weld used and method of force transfer

K-bracing.   A system of struts used in a braced frame in which the pattern of the struts resembles the letter *K*, either normal or on its side

Lamellar tearing.   Separation in highly restrained base metal caused by through-thickness strains induced by shrinkage of adjacent weld metal

Lateral bracing member.   A member utilized individually or as a component of a lateral bracing system to prevent buckling of members or elements and/or to resist lateral loads

Lateral (or lateral-torsional) buckling.   Buckling of a member involving lateral deflection and twist

Limit state.   A condition in which a structure or component becomes unfit for service and is judged either to be no longer useful for its intended function (*serviceability limit state*) or to be unsafe (*strength limit state*)

Limit states.   Limits of structural usefulness, such as brittle fracture, plastic collapse, excessive deformation, durability, fatigue, instability and serviceability

Load factor.   A factor that accounts for unavoidable deviations of the actual load from the nominal value and for uncertainties in the analysis that transforms the load into a load effect

Loads.   Forces or other actions that arise on structural systems from the weight of all permanent construction, occupants and their possessions, environmental effects, differential settlement and restrained dimensional changes. *Permanent* loads are those loads in which variations in time are rare or of small magnitude. All other loads are *variable* loads. See *nominal loads.*

LRFD (Load and Resistance Factor Design).   A method of proportioning structural components (members, connectors, connecting elements and assemblages) such that no applicable limit state is exceeded when the structure is subjected to all appropriate load combinations

Local buckling.   The buckling of a compression element which may precipitate the failure of the whole member

Low-cycle fatigue.   Fracture resulting from a relatively high stress range resulting in a relatively small number of cycles to failure

Lower bound load.   A load computed on the basis of an assumed equilibrium moment diagram in which the moments are not greater then M_p that is less than or at best equal to the true ultimate load

Mechanism.   An articulated system able to deform without an increase in load, used in the special sense that the linkage may include real hinges or plastic hinges, or both

Mechanism method.   A method of plastic analysis in which equilibrium between external forces and internal plastic hinges is calculated on the basis of an assumed mechanism. The failure load so determined is an upper bound

Nominal loads.   The magnitudes of the loads specified by the applicable code

Nominal strength.   The capacity of a structure or component to resist the effects of loads, as determined by computations using specified material strengths and dimensions and formulas derived from accepted principles of structural mechanics or by field tests or laboratory tests of scaled models, allowing for modeling effects and differences between laboratory and field conditions

Noncompact section.   Noncompact sections can develop the yield stress in compression elements before local buckling occurs, but will not resist inelastic local buckling at strain levels required for a fully plastic stress distribution

P-Delta effect.   Secondary effect of column axial loads and lateral deflection on the moments in members

Panel zone.   The zone in a beam-to-column connection that transmits moment by a shear panel

Partially composite beam.   A composite beam for which the shear strength of shear connectors governs the flexural strength

Plane frame.   A structural system assumed for the purpose of analysis and design to be two-dimensional

Plastic analysis.   Determination of load effects (force, moment, stress, as appropriate) on members and connections based on the assumption of rigid-plastic behavior, i.e., that equilibrium is satisfied throughout the structure and yield is not exceeded anywhere. Second order effects may need to be considered

Plastic design section.   The cross section of a member which can maintain a full plastic moment through large rotations so that a mechanism can develop; the section suitable for plastic design

Plastic hinge.   A yielded zone which forms in a structural member when the plastic moment is attained. The beam is assumed to rotate as if hinged, except that it is restrained by the plastic moment M_p

Plastic-limit load.   The maximum load that is attained when a sufficient number of yield zones have formed to permit the structure to deform plastically without further increase in load. It is the largest load a structure will support, when perfect plasticity is assumed and when such factors as instability, second-order effects, strain hardening and fracture are neglected

Plastic mechanism.   See mechanism

Plastic modulus.   The section modulus of resistance to bending of a completely yielded cross section. It is the combined static moment about the neutral axis of the cross-sectional areas above and below that axis

Plastic moment.   The resisting moment of a fully yielded cross section

Plastic strain.   The difference between total strain and elastic strain

Plastic zone.   The yielded region of a member

Plastification.   The process of successive yielding of fibers in the cross section of a member as bending moment is increased

Plate girder.   A built-up structural beam

Post-buckling strength.   The load that can be carried by an element, member or frame after buckling

Redistribution of moment.   A process which results in the successive formation of plastic hinges so that less highly stressed portions of a structure may carry increased moments

Required strength.   Load effect (force, moment, stress, as appropriate) acting on element or connection determined by structural analysis from the factored loads (using most appropriate critical load combinations)

Residual stress.   The stresses that remain in an unloaded member after it has been formed into a finished product. (Examples of such stresses include, but are not limited to, those induced by cold bending, cooling after rolling, or welding.)

Resistance.   The capacity of a structure or component to resist the effects of loads. It is determined by computations using specified material strengths, dimensions and formulas derived from accepted principles of structural mechanics, or by field tests or laboratory tests of scaled models, allowing for modeling effects and differences between laboratory and field conditions. Resistance is a generic term that includes both strength and serviceability limit states

Resistance factor.   A factor that accounts for unavoidable deviations of the actual strength from the nominal value and the manner and consequences of failure

Rigid frame.   A structure in which connections maintain the angular relationship between beam and column members under load

Root of the flange.   Location on the web of the corner radius termination point or the toe of the flange-to-web weld. Measured as the k distance from the far side of the flange

Rotation capacity.   The incremental angular rotation that a given shape can accept prior to local failure defined as $R = (\theta_u / \theta_p) - 1$ where θ_u is the overall rotation attained at the factored load state and θ_p is the idealized rotation corresponding to elastic theory applied to the case of $M = M_p$

St. Venant torsion.   That portion of the torsion in a member that induces only shear stresses in the member

Second-order analysis.   Analysis based on second-order deformations, in which equilibrium conditions are formulated on the deformed structure

Service load.   Load expected to be supported by the structure under normal usage; often taken as the nominal load

Serviceability limit state.   Limiting condition affecting the ability of a structure to preserve its appearance, maintainability, durability or the comfort of its occupants or function of machinery under normal usage

Shape factor.   The ratio of the plastic moment to the yield moment, or the ratio of the plastic modulus to the section modulus for a cross section

Shear-friction.   Friction between the embedment and the concrete that transmits shear loads. The relative displacement in the plane of the shear load is considered to be resisted by shear-friction anchors located perpendicular to the plane of the shear load.

Shear lugs.   Plates, welded studs, bolts and other steel shapes that are embedded in the concrete and located transverse to the direction of the shear force and that transmit shear loads, introduced into the concrete by local bearing at the shear lug-concrete interface

Shear wall.   A wall that in its own plane resists shear forces resulting from applied wind, earthquake or other transverse loads or provides frame stability. Also called a structural wall

Sidesway.   The lateral movement of a structure under the action of lateral loads, unsymmetrical vertical loads or unsymmetrical properties of the structure

Sidesway buckling.   The buckling mode of a multistory frame precipitated by the relative lateral displacements of joints, leading to failure by sidesway of the frame

Simple plastic theory.   See plastic design

Single curvature.   A deformed shape of a member having one smooth continuous arc, as opposed to double curvature which contains a reversal

Slender section.   The cross section of a member which will experience local buckling in the elastic range

Slenderness ratio.   The ratio of the effective length of a column to the radius of gyration of the column, both with respect to the same axis of bending

Slip-critical joint.   A bolted joint in which the slip resistance of the connection is required

Space frame.   A three-dimensional structural framework (as contrasted to a plane frame)

Splice.   The connection between two structural elements joined at their ends to form a single, longer element

Stability-limit load.   Maximum (theoretical) load a structure can support when second-order instability effects are included

Stepped column.   A column with changes from one cross section to another occurring at abrupt points within the length of the column

Stiffener.   A member, usually an angle or plate, attached to a plate or web of a beam or girder to distribute load, to transfer shear or to prevent buckling of the member to which it is attached

Stiffness.   The resistance to deformation of a member or structure measured by the ratio of the applied force to the corresponding displacement

Story drift.   The difference in horizontal deflection at the top and bottom of a story

Strain hardening.   Phenomenon wherein ductile steel, after undergoing considerable deformation at or just above yield point, exhibits the capacity to resist substantially higher loading than that which caused initial yielding

Strain-hardening strain.   For structural steels that have a flat (plastic) region in the stress-strain relationship, the value of the strain at the onset of strain hardening

Strength design.   A method of proportioning structural members using load factors and resistance factors such that no applicable limit state is exceeded (also called load and resistance factor design)

Strength limit state.   Limiting condition affecting the safety of the structure, in which the ultimate load-carrying capacity is reached

Stress.   Force per unit area

Stress concentration.   Localized stress considerably higher than average (even in uniformly loaded cross sections of uniform thickness) due to abrupt changes in geometry or localized loading

Strong axis.   The major principal axis of a cross section

Structural design documents.   Documents prepared by the designer (plans, design details and job specifications)

Structural system.   An assemblage of load-carrying components which are joined together to provide regular interaction or interdependence

Stub column.   A short compression-test specimen, long enough for use in measuring the stress-strain relationship for the complete cross section, but short enough to avoid buckling as a column in the elastic and plastic ranges

Subassemblage.   A truncated portion of a structural frame

Supported frame.   A frame which depends upon adjacent braced or unbraced frames for resistance to lateral load or frame instability. (This transfer of load is frequently provided by the floor or roof system through diaphragm action or by horizontal cross bracing in the roof.)

Tangent modulus.   At any given stress level, the slope of the stress-strain curve of a material in the inelastic range as determined by the compression test of a small specimen under controlled conditions

Temporary structure. A general term for anything that is built or constructed (usually to carry construction loads) that will eventually be removed before or after completion of construction and does not become part of the permanent structural system

Tensile strength. The maximum tensile stress that a material is capable of sustaining

Tension field action. The behavior of a plate girder panel under shear force in which diagonal tensile stresses develop in the web and compressive forces develop in the transverse stiffeners in a manner analogous to a Pratt truss

Toe of the fillet. Termination point of fillet weld or of rolled section fillet

Torque-tension relationship. Term applied to the wrench torque required to produce specified pre-tension in high-strength bolts

Turn-of-nut method. Procedure whereby the specified pre-tension in high-strength bolts is controlled by rotation of the wrench a predetermined amount after the nut has been tightened to a snug fit

Unbraced frame. A frame in which the resistance to lateral load is provided by the bending resistance of frame members and their connections

Unbraced length. The distance between braced points of a member, measured between the centers of gravity of the bracing members

Undercut. A notch resulting from the melting and removal of base metal at the edge of a weld

Universal-mill plate. A plate in which the longitudinal edges have been formed by a rolling process during manufacture. Often abbreviated as UM plate

Upper bound load. A load computed on the basis of an assumed mechanism which will always be at best equal to or greater than the true ultimate load

Vertical bracing system. A system of shear walls, braced frames or both, extending through one or more floors of a building

Von Mises yield criterion. A theory which states that inelastic action at any point in a body under any combination of stresses begins only when the strain energy of distortion per unit volume absorbed at the point is equal to the strain energy of distortion absorbed per unit volume at any point in a simple tensile bar stressed to the elastic limit under a state of uniaxial stress. It is often called the maximum strain-energy-of-distortion theory. Accordingly, shear yield occurs at 0.58 times yield strength

Warping torsion. That portion of the total resistance to torsion that is provided by resistance to warping of the cross section

Weak axis. The minor principal axis of a cross section

Weathering steel. A type of high-strength, low-alloy steel which can be used in normal environments (not marine) and outdoor exposures without protective paint covering. This steel develops a tight adherent rust at a decreasing rate with respect to time

Web buckling. The buckling of a web plate

Web crippling. The local failure of a web plate in the immediate vicinity of a concentrated load or reaction

Working load. Also called service load. The actual load assumed to be acting on the structure

Yield moment. In a member subjected to bending, the moment at which an outer fiber first attains the yield stress

Yield plateau. The portion of the stress-strain curve for uniaxial tension or compression in which the stress remains essentially constant during a period of substantially increased strain

Yield point. The first stress in a material at which an increase in strain occurs without an increase in stress, the yield point less than the maximum attainable stress

Yield strength. The stress at which a material exhibits a specified limiting deviation from the proportionality of stress to strain. Deviation expressed in terms of strain

Yield stress. Yield point, yield strength or yield stress level as defined

Yield-stress level. The average stress during yielding in the plastic range, the stress determined in a tension test when the strain reaches 0.005 in. per in.

Code of Standard Practice

FOR STEEL BUILDINGS AND BRIDGES

Adopted Effective September 1, 1986
American Institute of Steel Construction, Inc.

**AMERICAN INSTITUTE OF STEEL CONSTRUCTION
ONE EAST WACKER DRIVE, SUITE 3100
CHICAGO, ILLINOIS 60601-2001**

PREFACE

When contractual documents do not contain specific provisions to the contrary, existing trade practices are considered to be incorporated into the relationships between the parties to a contract. As in any industry, trade practices have developed among those involved in the purchase, design, fabrication and erection of structural steel. The American Institute of Steel Construction has continuously surveyed the structural steel fabrication industry to determine standard practices and, commencing in 1924, published its *Code of Standard Practice*. Since that date, the Code has been periodically updated to reflect new and changing technology and practices of the industry.

It is the Institute's intention to provide to owners, architects, engineers, contractors and others associated with construction, a useful framework for a common understanding of acceptable standards when contracting for structural steel construction.

This edition is the third complete revision of the Code since it was first published. It includes a number of new sections covering new subjects not included in the previous Code, but which are an integral part of the relationship of the parties to a contract.

The Institute acknowledges the valuable information and suggestions provided by trade associations and other organizations associated with construction and the fabricating industry in developing this current *Code of Standard Practice*.

While every precaution has been taken to insure that all data and information presented is as accurate as possible, the Institute cannot assume responsibility for errors or oversights in the information published herein, or the use of the information published or incorporation of such information in the preparation of detailed engineering plans. The Code should not replace the judgment of an experienced architect or engineer who has the responsibility of design for a specific structure.

Code of Standard Practice
for Steel Buildings and Bridges

Adopted Effective September 1, 1986
American Institute of Steel Construction, Inc.

SECTION 1. GENERAL PROVISIONS

1.1. Scope

The practices defined herein have been adopted by the AISC as the commonly accepted standards of the structural steel fabricating industry. In the absence of other instructions in the contract documents, the trade practices defined in this *Code of Standard Practice*, as revised to date, govern the fabrication and erection of structural steel.

1.2. Definitions

AISC Specification—The *Specification for the Design, Fabrication and Erection of Structural Steel for Buildings* as adopted by the American Institute of Steel Construction.

ANSI—American National Standards Institute.

Architect/Engineer—The owner's designated representative with full responsibility for the design and integrity of the structure.

ASTM—The material standard of the American Society for Testing and Materials.

AWS Code—The *Structural Welding Code* of the American Welding Society.

Code—The *Code of Standard Practice* as adopted by the American Institute of Steel Construction.

Contract Documents—The documents which define the responsibilities of the parties involved in bidding, purchasing, supplying and erecting structural steel. Such documents normally consist of a contract, plans and specifications.

Drawings—Shop and field erection drawings prepared by the fabricator and erector for the performance of the work.

Erector—The party responsible for the erection of the structural steel.

Fabricator—The party responsible for furnishing fabricated structural steel.

General Contractor—The owner's designated representative with full responsibility for the construction of the structure.

MBMA—Metal Building Manufacturers Association.

Mill Material—Steel mill products ordered expressly for the requirements of a specific project.

Owner—The owner of the proposed structure or his designated representatives, who may be the architect, engineer, general contractor, public authority or others.

Plans—Design drawings furnished by the party responsible for the design of the structure.

Release for Construction—The release by the owner permitting the fabricator to commence work under the contract, including ordering material and the preparation of shop drawings.

SSPC—The Steel Structures Painting Council, publishers of the *Steel Structures Painting Manual*, Vol. 2, "Systems and Specifications."

Tier—The word Tier used in Sect. 7.11 is defined as a column shipping piece.

1.3. Design Criteria for Buildings and Similar Type Structures

In the absence of other instructions, the provisions of the AISC Specification govern the design of the structural steel.

1.4. Design for Bridges

In the absence of other instructions, the following provisions govern, as applicable:

Standard Specifications for Highway Bridges of American Association of State Highway and Transportation Officials

Specifications for Steel Railway Bridges of American Railway Engineering Association

Structural Welding Code of American Welding Society

1.5. Responsibility for Design

1.5.1. When the owner provides the design, plans and specifications, the fabricator and erector are not responsible for the suitability, adequacy or legality of the design. The fabricator is not responsible for the practicability or safety of erection if the structure is erected by others.

1.5.2. If the owner desires the fabricator or erector to prepare the design, plans and specifications, or to assume any responsibility for the suitability, adequacy or legality of the design, he clearly states his requirements in the contract documents.

1.6. Patented Devices

Except when the contract documents call for the design to be furnished by the fabricator or erector, the fabricator and erector assume that all necessary patent rights have been obtained by the owner and that the fabricator or erector will be fully protected in the use of patented designs, devices or parts required by the contract documents.

SECTION 2.0. CLASSIFICATION OF MATERIALS

2.1. Definition of Structural Steel

"Structural Steel," as used to define the scope of work in the contract documents, consists of the steel elements of the structural steel frame essential to support the design loads. Unless otherwise specified in the contract documents, these elements consist of material as shown on the structural steel plans and described as:

Anchor bolts for structural steel
Base or bearing plates
Beams, Girders, Purlins and Girts
Bearings of steel for girders, trusses or bridges
Bracing
Columns, posts
Connecting materials for framing structural steel to structural steel
Crane rails, splices, stops, bolts and clamps
Door frames constituting part of the steel frame
Expansion joints connected to steel frame
Fasteners for connecting structural steel items:
    Shop rivets
    Permanent shop bolts
    Shop bolts for shipment
    Field rivets for permanent connections
    Field bolts for permanent connections
    Permanent pins
Floor Plates (checkered or plain) attached to steel frame
Grillage beams and girders
Hangers essential to the structural steel frame
Leveling plates, wedges, shims & leveling screws
Lintels, if attached to the structural steel frame
Marquee or canopy framing
Machinery foundations of rolled steel sections and/or plate attached to the structural frame
Monorail elements of standard structural shapes when attached to the structural frame
Roof frames of standard structural shapes
Shear connectors—if specified shop attached
Struts, tie rods and sag rods forming part of the structural frame
Trusses

2.2. Other Steel or Metal Items

The classification "Structural Steel," does not include steel, iron or other metal items not generally described in Paragraph 2.1, even when such items are shown on the structural steel plans or are attached to the structural frame. These items include but are not limited to:

Cables for permanent bracing or suspension systems
Chutes and hoppers
Cold-formed steel products
Door and corner guards
Embedded steel parts in precast or poured concrete
Flagpole support steel
Floor plates (checkered or plain) not attached to the steel frame
Grating and metal deck
Items required for the assembly or erection of materials supplied by trades other than structural steel fabricators or erectors
Ladders and safety cages
Lintels over wall recesses
Miscellaneous metal

Non-steel bearings
Open-web, long-span joists and joist girders
Ornamental metal framing
Shear connectors field installed
Stacks, tanks and pressure vessels
Stairs, catwalks, handrail and toeplates
Trench or pit covers.

SECTION 3. PLANS AND SPECIFICATIONS

3.1. Structural Steel

In order to insure adequate and complete bids, the contract documents provide complete structural steel design plans clearly showing the work to be performed and giving the size, section, material grade and the location of all members, floor levels, column centers and offsets, camber of members, with sufficient dimensions to convey accurately the quantity and nature of the structural steel to be furnished. Structural steel specifications include any special requirements controlling the fabrication and erection of the structural steel.

3.1.1. Wind bracing, connections, column stiffeners, bearing stiffeners on beams and girders, web reinforcement, openings for other trades, and other special details where required are shown in sufficient detail so that they may be readily understood.

3.1.2. Plans include sufficient data concerning assumed loads, shears, moments and axial forces to be resisted by members and their connections, as may be required for the development of connection details on the shop drawings and the erection of the structure.

3.1.3. Where connections are not shown, the connections are to be in accordance with the requirements of the AISC Specification.

3.1.4. When loose lintels and leveling plates are required to be furnished as part of the contract requirements, the plans and specifications show the size, section and location of all pieces.

3.2. Architectural, Electrical and Mechanical

Architectural, electrical and mechanical plans may be used as a supplement to the structural steel plans to define detail configurations and construction information, provided all requirements for the structural steel are noted on the structural steel plans.

3.3. Discrepancies

In case of discrepancies between plans and specifications for buildings, the specifications govern. In case of discrepancies between plans and specifications for bridges, the plans govern. In case of discrepancies between scale dimensions on the plans and figures written on them, the figures govern. In case of discrepancies between the structural steel plans and plans for other trades, the structural steel plans govern.

3.4. Legibility of Plans

Plans are clearly legible and made to a scale not less then ⅛-inch to the foot. More complex information is furnished to an adequate scale to convey the information clearly.

3.5. Special Conditions

When it is required that a project be advertised for bidding before the requirements of Article 3.1 can be met, the owner must provide sufficient information in form of scope, drawings, weights, outline specifications, and other descriptive data to enable the fabricator and erector to prepare a knowledgeable bid.

SECTION 4. SHOP AND ERECTION DRAWINGS

4.1. Owner Responsibility

To enable the fabricator and erector to properly and expeditiously proceed with the work, the owner furnishes, in a timely manner and in accordance with the contract documents, complete structural steel plans and specifications released for construction. "Released for construction" plans and specifications are required by the fabricator for ordering the mill material and for the preparation and completion of shop and erection drawings.

4.2. Approval

When shop drawings are made by the fabricator, prints thereof are submitted to the owner for his examination and approval. The fabricator includes a maximum allowance of fourteen (14) calendar days in his schedule for the return of shop drawings. Return of shop drawings is noted with the owner's approval, or approval subject to corrections as noted. The fabricator makes the corrections, furnishes corrected prints to the owner, and is released by the owner to start fabrication.

4.2.1. Approval by the owner of shop drawings prepared by the fabricator indicates that the fabricator has correctly interpreted the contract requirements, and is released by the owner to start fabrication. This approval constitutes the owner's acceptance of all responsibility for the design adequacy of any detail configuration of connections developed by the fabricator as part of his preparation of these shop drawings. Approval does not relieve the fabricator of the responsibility for accuracy of detail dimensions on shop drawings, nor the general fit-up of parts to be assembled in the field.

4.2.2. Unless specifically stated to the contrary, any additions, deletions or changes indicated on the approval of shop and erection drawings are authorizations by the owner to release the additions, deletions or revisions for construction.

4.3. Drawings Furnished by Owner

When the shop drawings are furnished by the owner, he must deliver them to the fabricator in time to permit material procurement and fabrication to proceed in an orderly manner in accordance with the prescribed time schedule. The owner prepares these shop drawings, insofar as practicable, in accordance with the shop and drafting room standards of the fabricator. The owner is responsible for the completeness and accuracy of shop drawings so furnished.

SECTION 5. MATERIALS

5.1. Mill Materials

5.1.1. Mill tests are performed to demonstrate material conformance to ASTM specifications in accordance with the contract requirements. Unless special requirements are included in the contract documents, mill testing is limited to those tests required by the applicable ASTM material specifications. Mill test reports are furnished by the fabricator only if requested by the owner, either in the contract documents or in separate written instructions prior to the time the fabricator places his material orders with the mill.

5.1.2. When material received from the mill does not satisfy ASTM A6 tolerances for camber, profile, flatness or sweep, the fabricator is permitted to perform corrective work by the use of controlled heating and mechanical straightening, subject to the limitations of the AISC Specification.

5.1.3. Corrective procedures described in ASTM A6 for reconditioning the surface of structural steel plates and shapes before shipment from the producing mill may also be performed by the fabricator, at his option, when variations described in ASTM A6 are discovered or occur after receipt of the steel from the producing mill.

5.1.4. When special requirements demand tolerances more restrictive than allowed by ASTM A6, such requirements are defined in the contract documents and the fabricator has the option of corrective measures as described above.

5.2. Stock Materials

5.2.1. Many fabricators maintain stocks of steel products for use in their fabricating operations. Materials taken from stock by the fabricator for use for structural purposes must be of a quality at least equal to that required by the ASTM specifications applicable to the classification covering the intended use.

5.2.2. Mill test reports are accepted as sufficient record of the quality of materials carried in stock by the fabricator. The fabricator reviews and retains the mill test reports covering the materials he purchases for stock, but he does not maintain records that identify individual pieces of stock material against individual mill test reports. Such records are not required if the fabricator purchases for stock under established specifications as to grade and quality.

5.2.3. Stock materials purchased under no particular specifications or under specifications less rigid than those mentioned above, or stock materials which have not been subject to mill or other recognized test reports, are not used without the express approval of the owner, except where the quality of the material could not affect the integrity of the structure.

SECTION 6. FABRICATION AND DELIVERY

6.1. Identification of Material

6.1.1. High strength steel and steel ordered to special requirements is marked by the supplier, in accordance with ASTM A6 requirements, prior to delivery to the fabricator's shop or other point of use.

6.1.2. High strength steel and steel ordered to special requirements that has not been marked by the supplier in accordance with Sect. 6.1.1 is not used until its identification is established by means of tests as specified in Sect. 1.4.1.1 of the AISC Specification, and until a fabricator's identification mark, as described in Sect. 6.1.3, has been applied.

6.1.3. During fabrication, up to the point of assembling members, each piece of high strength steel and steel ordered to special requirements carries a fabricator's identification mark or an original supplier's identification mark. The fabricator's identification mark is in accordance with the fabricator's established identification system, which is on record and available for the information of the owner or his representative, the building commissioner and the inspector, prior to the start of fabrication.

6.1.4. Members made of high strength steel and steel ordered to special requirements are not given the same assembling or erecting mark as members made of other steel, even though they are of identical dimensions and detail.

6.2. Preparation of Material

6.2.1. Thermal cutting of structural steel may be by hand or mechanically guided means.

6.2.2. Surfaces noted as "finished" on the drawings are defined as having a maximum ANSI roughness height value of 500. Any fabricating technique, such as friction sawing, cold sawing, milling, etc., that produces such a finish may be used.

6.3. Fitting and Fastening

6.3.1. Projecting elements of connection attachments need not be straightened in the connecting plane if it can be demonstrated that installation of the connectors or fitting aids will provide reasonable contact between faying surfaces.

6.3.2. Runoff tabs are often required to produce sound welds. The fabricator or erector does not remove them unless specified in the contract documents. When their removal is required, they may be hand flame-cut close to the edge of the finished member with no further finishing required, unless other finishing is specifically called for in the contract documents.

6.3.3. All high-strength bolts for shop attached connection material are to be installed in the shop in accordance with *Specification for Structural Joints Using A325 or A490 Bolts*, unless otherwise noted on the shop drawings.

6.4. Dimensional Tolerances

6.4.1. A variation of $\frac{1}{32}$-inch is permissible in the overall length of members with both ends finished for contact bearing as defined in Sect. 6.2.2.

6.4.2. Members without ends finished for contact bearing, which are to be framed to other steel parts of the structure, may have a variation from the detailed length not greater than ⅟₁₆-inch for members 30 ft or less in length, and not greater than ⅛-inch for members over 30 ft in length.

6.4.3. Unless otherwise specified, structural members, whether of a single-rolled shape or built-up, may vary from straightness within the tolerances allowed for wide-flange shapes by ASTM Specification A6, except that the tolerance on deviation from straightness of compression members is ⅟₁₀₀₀ of the axial length between points which are to be laterally supported.

Completed members should be free from twists, bends and open joints. Sharp kinks or bends are cause for rejection of material.

6.4.4. Beams and trusses detailed without specified camber are fabricated so that after erection any camber due to rolling or shop fabrication is upward.

6.4.5. Any permissible deviation in depths of girders may result in abrupt changes in depth at splices. Any such difference in depth at a bolted joint, within the prescribed tolerances, is taken up by fill plates. At welded joints the weld profile may be adjusted to conform to the variation in depth, provided that the minimum cross section of required weld is furnished and that the slope of the weld surface meets AWS Code requirements.

6.5. Shop Painting

6.5.1. The contract documents specify all the painting requirements, including members to be painted, surface preparation, paint specifications, manufacturer's product identification and the required dry film thickness, in mils, of the shop coat.

6.5.2. The shop coat of paint is the prime coat of the protective system. It protects the steel for only a short period of exposure in ordinary atmospheric conditions, and is considered a temporary and provisional coating. The fabricator does not assume responsibility for deterioration of the prime coat that may result from extended exposure to ordinary atmospheric conditions, nor from exposure to corrosive conditions more severe than ordinary atmospheric conditions.

6.5.3. In the absence of other requirements in the contract documents, the fabricator hand cleans the steel of loose rust, loose mill scale, dirt and other foreign matter, prior to painting, by means of wire brushing or by other methods elected by the fabricator, to meet the requirements of SSPC-SP2. The fabricator's workmanship on surface preparation is considered accepted by the owner unless specifically disapproved prior to paint application.

6.5.4. Unless specifically excluded, paint is applied by brush, spray, roller coating, flow coating or dipping, at the election of the fabricator. When the term "shop coat" or "shop paint" is used with no paint system specified, the fabricator's standard paint shall be applied to a minimum dry film thickness of one mil.

6.5.5 Steel not requiring shop paint is cleaned of oil or grease by solvent cleaners and cleaned of dirt and other foreign material by sweeping with a fiber brush or other suitable means.

6.5.6. Abrasions caused by handling after painting are to be expected. Touch-up of these blemished areas is the responsibility of the contractor performing field touch-up or field painting.

6.6. Marking and Shipping of Materials

6.6.1. Erection marks are applied to the structural steel members by painting or other suitable means, unless otherwise specified in the contract documents.

6.6.2. Rivets and bolts are commonly shipped in separate containers according to length and diameter; loose nuts and washers are shipped in separate containers according to sizes. Pins and other small parts, and packages of rivets, bolts, nuts and washers are usually shipped in boxes, crates, kegs or barrels. A list and description of the material usually appears on the outside of each closed container.

6.7. Delivery of Materials

6.7.1 Fabricated structural steel is delivered in such sequence as will permit the most efficient and economical performance of both shop fabrication and erection. If the owner wishes to prescribe or control the sequence of delivery of materials, he reserves such right and defines the requirements in the contract documents. If the owner contracts separately for delivery and erection, he must coordinate planning between contractors.

6.7.2. Anchor bolts, washers and other anchorage or grillage materials to be built into masonry should be shipped so that they will be on hand when needed. The owner must give the fabricator sufficient time to fabricate and ship such materials before they are needed.

6.7.3. The quantities of material shown by the shipping statement are customarily accepted by the owner, fabricator and erector as correct. If any shortage is claimed, the owner or erector should immediately notify the carrier and the fabricator in order that the claim may be investigated.

6.7.4. The size and weight of structural steel assemblies may be limited by shop capabilities, the permissible weight and clearance dimensions of available transportation and the job site conditions. The fabricator limits the number of field splices to those consistent with minimum project cost.

6.7.5. If material arrives at its destination in damaged condition, it is the responsibility of the receiving party to promptly notify the fabricator and carrier prior to unloading the material, or immediately upon discovery.

SECTION 7. ERECTION

7.1. Method of Erection

When the owner wishes to control the method and sequence of erection, or when certain members cannot be erected in their normal sequence, the owner so specifies in the contract documents. In the absence of such restrictions, the erector will proceed using the most efficient and economical method and sequence available to him consistent with the contract documents. When the owner contracts separately for fabrication and erection services, the owner is responsible for coordinating planning between contractors.

7.2. Site Conditions

The owner provides and maintains adequate access roads into and through the site for the safe delivery of derricks, cranes, other necessary equipment, and the material to be erected. The owner affords the erector a firm, properly graded, drained, convenient and adequate space at the site for the operation of his equipment, and removes all overhead obstructions such as power lines, telephone lines, etc., in order to provide a safe working area for erection of the steelwork. The erector provides and installs the safety protection required for his own work. Any protection for other trades not essential to the steel erection activity is the responsibility of the owner. When the structure does not occupy the full available site, the owner provides adequate storage space to enable the fabricator and erector to operate at maximum practicable speed.

7.3. Foundations, Piers and Abutments

The accurate location, strength, suitability and access to all foundations, piers and abutments is the sole responsibility of the owner.

7.4. Building Lines and Bench Marks

The owner is responsible for accurate location of building lines and bench marks at the site of the structure, and for furnishing the erector with a plan containing all such information.

7.5. Installation of Anchor Bolts and Embedded Items

7.5.1. Anchor bolts and foundation bolts are set by the owner in accordance with an approved drawing. They must not vary from the dimensions shown on the erection drawings by more than the following:

(a) ⅛-inch center to center of any two bolts within an anchor bolt group, where an anchor bolt group is defined as the set of anchor bolts which receive a single fabricated steel shipping piece.

(b) ¼-inch center to center of adjacent anchor bolt groups.

(c) Elevation of the top of anchor bolts ± ½-inch

(d) Maximum accumulation of ¼-inch per hundred feet along the established column line of multiple anchor bolt groups, but not to exceed a total of 1 in., where the established column line is the actual field line most representative of the centers of the as-built anchor bolt groups along a line of columns.

(e) ¼-inch from the center of any anchor bolt group to the established column line through that group.

(f) The tolerances of paragraphs b, c and d apply to offset dimensions shown on the plans, measured parallel and perpendicular to the nearest established column line for individual columns shown on the plans to be offset from established column lines.

7.5.2. Unless shown otherwise, anchor bolts are set perpendicular to the theoretical bearing surface.

7.5.3. Other embedded items or connection materials between the structural steel and the work of other trades are located and set by the owner in accordance with approved location or erection drawings. Accuracy of these items must satisfy the erection tolerance requirements of Sect. 7.11.3.

7.5.4. All work performed by the owner is completed so as not to delay or interfere with the erection of the structural steel.

7.6. Bearing Devices

The owner sets to line and grade all leveling plates and loose bearing plates which can be handled without a derrick or crane. All other bearing devices supporting structural steel are set and wedged, shimmed or adjusted with leveling screws by the erector to lines and grades established by the owner. The fabricator provides the wedges, shims or leveling screws that are required, and clearly scribes the bearing devices with working lines to facilitate proper alignment. Promptly after the setting of any bearing devices, the owner checks lines and grades, and grouts as required. The final location and proper grouting of bearing devices are the responsibility of the owner. Tolerance on elevation relative to established grades of bearing devices, whether set by the owner or by the erector, is plus or minus ⅛-inch.

7.7 Field Connection Material

7.7.1 The fabricator provides field connection details consistent with the requirements of the contract documents which will, in his opinion, result in the most economical fabrication and erection cost.

7.7.2. When the fabricator erects the structural steel, the fabricator supplies all materials required for temporary and permanent connection of the component parts of the structural steel.

7.7.3 When the erection of the structural steel is performed by someone other than the fabricator, the fabricator furnishes the following field connection material:

(a) Bolts of required size and in sufficient quantity for all field connections of steel to steel which are to be permanently bolted. Unless high-strength bolts or other special types of bolts and washers are specified, common bolts are furnished. An extra 2 percent of each bolt size (diameter and length) are furnished.

(b) Rivets of required size and in sufficient quantity for all field connections of steel to steel which are to be riveted field connections. An extra 10 percent of each rivet size are furnished.

(c) Shims shown as necessary for make-up of permanent connections of steel to steel.

(d) Back-up bars or run-off tabs that may be required for field welding.

7.7.4. When the erection of the structural steel is performed by someone other than the fabricator, the erector furnishes all welding electrodes, fit-up bolts and drift pins used for erection of the structural steel.

7.7.5. Field-installed shear connectors are supplied by the shear connector applicator.

7.7.6. Metal deck support angles are the responsibility of the metal deck supplier.

7.8. Loose Material

Loose items of structural steel not connected to the structural frame are set by the owner without assistance from the erector, unless otherwise specified in the contract documents.

7.9. Temporary Support of Structural Steel Frames.

7.9.1. General

Temporary supports, such as temporary guys, braces, falsework, cribbing or other elements required for the erection operation will be determined and furnished and installed by the erector. These temporary supports will secure the steel framing, or any partly assembled steel framing, against loads comparable in intensity to those for which the structure was designed, resulting from wind, seismic forces and erection operations, but not the loads resulting from the performance of work by or the acts of others, nor such unpredictable loads as those due to tornado, explosion or collision.

7.9.2. Self-supporting Steel Frames

A self-supporting steel frame is one that provides the required stability and resistance to gravity loads and design wind and seismic forces without interaction with other elements of the structure. The erector furnishes and installs only those temporary supports that are necessary to secure any element or elements of the steel framing until they are made stable without external support.

7.9.3. Non-Self-supporting Steel Frames

A non-self-supporting steel frame is one that requires interaction with other elements not classified as Structural Steel to provide the required stability or resistance to wind and seismic forces. Such frames shall be clearly identified in the contract documents. The contract documents specify the sequence and schedule of placement of such elements and the effect of loads imposed by these partially or completely installed interacting elements on the bare steel frame. The erector determines the need and furnishes and installs the temporary supports in accordance with this information. The owner is responsible for the installation and timely completion of all elements that are required for stability of the frame.

7.9.4. Special Erection Conditions

When the design concept of a structure is dependent upon use of shores, jacks or loads which must be adjusted as erection progresses to set or maintain camber or prestress, such requirement is specifically stated in the contract documents.

7.9.5. Removal of Temporary Supports

The temporary guys, braces, falsework, cribbing and other elements required for the erection operation, which are furnished and installed by the erector, are not the property of the owner.

In *self-supporting structures*, temporary supports are not required after the structural steel for a self-supporting element is located and finally fastened within the required tolerances. After such final fastening, the erector is no longer responsible for temporary support of the self-supporting element and may remove the temporary supports.

In *non-self-supporting structures*, the erector may remove temporary supports when the necessary non-structural steel elements are complete. Temporary supports are not to be removed without the consent of the erector. At completion of steel erection, any temporary supports that are required to be left in place are removed by the owner and returned to the erector in good condition.

7.9.6. Temporary Supports for Other Work

Should temporary supports beyond those defined as the responsibility of the erector in Sects. 7.9.1, 7.9.2 and 7.9.3 be required, either during or after the erection of the structural steel, responsibility for the supply and installation of such supports rests with the owner.

7.10. Temporary Floors and Handrails for Buildings

The erector provides floor coverings, handrails and walkways as required by law and applicable safety regulations for protection of his own personnel. As work progresses, the erector removes such facilities from units where the erection operations are completed, unless other arrangements are included in the contract documents. The owner is responsible for all protection necessary for the work of other trades. When permanent steel decking is used for protective flooring and is installed by the owner, all such work is performed so as not to delay or interfere with erection progress and is scheduled by the owner and installed in a sequence adequate to meet all safety regulations.

7.11. Frame Tolerances

7.11.1. Overall Dimensions

Some variation is to be expected in the finished overall dimensions of structural steel frames. Such variations are deemed to be within the limits of good practice when they do not exceed the cumulative effect of rolling tolerances, fabricating tolerances and erection tolerances.

7.11.2. Working Points and Working Lines

Erection tolerances are defined relative to member working points and working lines as follows:

(a) For members other than horizontal members, the member work point is the actual center of the member at each end of the shipping piece.

(b) For horizontal members, the working point is the actual center line of the top flange or top surface at each end.

(c) Other working points may be substituted for ease of reference, providing they are based upon these definitions.

(d) The member working line is a straight line connecting the member working points.

7.11.3. Position and Alignment

The tolerances on position and alignment of member working points and working lines are as follows:

7.11.3.1. Columns

Individual column shipping pieces are considered plumb if the deviation of the working line from a plumb line does not exceed 1:500, subject to the following limitations:

(a) The member working points of column shipping pieces adjacent to elevator shafts may be displaced no more than 1 in. from the established column line in the first 20 stories; above this level, the displacement may be increased ⅟₃₂-inch for each additional story up to a maximum of 2 in.

(b) The member working points of exterior column shipping pieces may be displaced from the established column line no more than 1 in. toward nor 2 in. away from the building line in the first 20 stories; above the 20th story, the displacement may be increased ⅟₁₆-inch for each additional story, but may not exceed a total displacement of 2 in. toward nor 3 in. away from the building line.

(c) The member working points of exterior column shipping pieces at any splice level for multi-tier buildings and at the tops of columns for single tier buildings may not fall outside a horizontal envelope, parallel to the building line, 1½-inch wide for buildings up to 300 ft in length. The width of the envelope may be increased by ½-inch for each additional 100 ft in length, but may not exceed 3 in.

(d) The member working points of exterior column shipping pieces may be displaced from the established column line, in a direction parallel to the building line, no more than 2 in. in the first 20 stories; above the 20th story, the displacement may be increased ⅟₁₆-inch for each additional story, but may not exceed a total displacement of 3 in. parallel to the building line.

7.11.3.2. Members Other Than Columns

(a) Alignment of members which consist of a single straight shipping piece containing no field splices, except cantilever members, is considered acceptable if the variation in alignment is caused solely by the variation of column alignment and/or primary supporting member alignment within the permissible limits for fabrication and erection of such members.

(b) The elevation of members connecting to columns is considered acceptable if the distance from the member working point to the upper milled splice line of the column does not deviate more than plus ³⁄₁₆-inch or minus ⁵⁄₁₆-inch from the distance specified on the drawings.

(c) The elevation of members which consist of a single shipping piece, other than members connected to columns, is considered acceptable if the variation in actual elevation is caused solely by the variation in elevation of the supporting members which are within permissible limits for fabrication and erection of such members.

(d) Individual shipping pieces which are segments of field assembled units containing field splices between points of support are considered plumb, level and aligned if the angular variation of the working line of each shipping piece relative to the plan alignment does not exceed 1:500.

(e) The elevation and alignment of cantilever members shall be considered plumb, level and aligned if the angular variation of the working line from a straight line extended in the plan direction from the working point at its supported end does not exceed 1:500.

(f) The elevation and alignment of members which are of irregular shape shall be considered plumb, level and aligned if the fabricated member is within its tolerance and its supporting member or members are within the tolerances specified in this Code.

7.11.3.3. Adjustable Items

The alignment of lintels, wall supports, curb angles, mullions and similar supporting members for the use of other trades, requiring limits closer than the foregoing tolerances, cannot be assured unless the owner's plans call for adjustable connections of these members to the supporting structural frame. When adjustable connections are specified, the owner's plans must provide for the total adjustment required to accommodate the tolerances on the steel frame for the proper alignment of these supports for other trades. The tolerances on position and alignment of such adjustable items are as follows:

(a) Adjustable items are considered to be properly located in their vertical position when their location is within ⅜-inch of the location established from the upper milled splice line of the nearest column to the support location as specified on the drawings.

(b) Adjustable items are considered to be properly located in their horizontal position when their location is within ⅜-inch of the proper location relative to the established finish line at any particular floor.

7.11.4. Responsibility for Clearances

In the design of steel structures, the owner is responsible for providing clearances and adjustments of material furnished by other trades to accommodate all of the foregoing tolerances of the structural steel frame.

7.11.5. Acceptance of Position and Alignment

Prior to placing or applying any other materials, the owner is responsible for determining that the location of the structural steel is acceptable for plumbness, level and alignment within tolerances. The erector is given timely notice of acceptance by the owner or a listing of specific items to be corrected in order to obtain acceptance. Such notice is rendered immediately upon completion of any part of the work and prior to the start of work by other trades that may be supported, attached or applied to the structural steelwork.

7.12. Correction of Errors

Normal erection operations include the correction of minor misfits by moderate amounts of reaming, chipping, welding or cutting, and the drawing of elements into line through the use of drift pins. Errors which cannot be corrected by the foregoing means or which require major changes in member configuration are reported immediately to the owner and fabricator by the erector, to enable whoever is responsible either to correct the error or to approve the most efficient and economic method of correction to be used by others.

7.13. Cuts, Alterations and Holes for Other Trades

Neither the fabricator nor the erector will cut, drill or otherwise alter his work, or the work of other trades, to accommodate other trades, unless such work is clearly specified in the contract documents. Whenever such work is specified, the owner is responsible for furnishing complete information as to materials, size, location and number of alterations prior to preparation of shop drawings.

7.14. Handling and Storage

The erector takes reasonable care in the proper handling and storage of steel during erection operations to avoid accumulation of unnecessary dirt and foreign matter. The erector is not responsible for removal from the steel of dust, dirt or other foreign matter which accumulates during the erection period as the result of exposure to the elements.

7.15. Field Painting

The erector does not paint field bolt heads and nuts, field rivet heads and field welds, nor touch up abrasions of the shop coat, nor perform any other field painting.

7.16 Final Cleaning Up

Upon completion of erection and before final acceptance, the erector removes all of his falsework, rubbish and temporary buildings.

SECTION 8.   QUALITY CONTROL

8.1.   General

8.1.1.   The fabricator maintains a quality control program to the extent deemed necessary so that the work is performed in accordance with this Code, the AISC Specification and contract documents. The fabricator has the option to use the AISC Quality Certification Program in establishing and administering the quality control program.

8.1.2.   The erector maintains a quality control program to the extent the erector deems necessary so that all of the work is performed in accordance with this Code, the AISC Specification and the contract documents. The erector shall be capable of performing the erection of the structural steel, and shall provide the equipment, personnel and management for the scope, magnitude and required quality of each project.

8.1.3.   When the owner requires more extensive quality control or independent inspection by qualified personnel, or requires the fabricator to be certified by the AISC Quality Certification Program, this shall be clearly stated in the contract documents, including a definition of the scope of such inspection.

8.2.   Mill Material Inspection

The fabricator customarily makes a visual inspection, but does not perform any material tests, depending upon mill reports to signify that the mill product satisfies material order requirements. The owner relies on mill tests required by contract and on such additional tests as he orders the fabricator to have made at the owner's expense. If mill inspection operations are to be monitored, or if tests other than mill tests are desired, the owner so specifies in the contract documents and should arrange for such testing through the fabricator to assure coordination.

8.3. Non-destructive Testing

When non-destructive testing is required, the process, extent, technique and standards of acceptance are clearly defined in the contract documents.

8.4. Surface Preparation and Shop Painting Inspection

Surface preparation and shop painting inspection must be planned for acceptance of each operation as completed by the fabricator. Inspection of the paint system, including material and thickness, is made promptly upon completion of the paint application. When wet film thickness is inspected, it must be measured immediately after application.

8.5. Independent Inspection

When contract documents specify inspection by other than the fabricator's and erector's own personnel, both parties to the contract incur obligations relative to the performance of the inspection.

8.5.1.   The fabricator and erector provide the inspector with access to all places where work is being done. A minimum of 24 hours notification is given prior to commencement of work.

8.5.2.   Inspection of shop work by the owner or his representative is performed in the fabricator's shop to the fullest extent possible. Such inspections should be in sequence, timely, and performed in such a manner as to minimize disruptions in operations and to permit the repair of all non-conforming work while the material is in process in the fabricating shop.

8.5.3.   Inspection of field work is completed promptly, so that corrections can be made without delaying the progress of the work.

8.5.4.   Rejection of material or workmanship not in conformance with the contract documents may be made at any time during the progress of the work. However, this provision does not relieve the owner of his obligation for timely, in-sequence inspections.

8.5.5.   The fabricator and erector receive copies of all reports prepared by the owner's inspection representative.

SECTION 9.   CONTRACTS

9.1.   Types of Contracts

9.1.1.   For contracts stipulating a lump sum price, the work required to be performed by the fabricator and erector is completely defined by the contract documents.

9.1.2.   For contracts stipulating a price per pound, the scope of work, type of materials, character of fabrication, and conditions of erection are based upon the contract documents which must be representative of the work to be performed.

9.1.3. For contracts stipulating a price per item, the work required to be performed by the fabricator and erector is based upon the quantity and the character of items described in the contract documents.

9.1.4. For contracts stipulating unit prices for various categories of structural steel, the scope of the work required to be performed by the fabricator and erector is based upon the quantity, character and complexity of the items in each category as described in the contract documents. The contract documents must be representative of the work to be done in each category.

9.2. Calculation of Weights

Unless otherwise set forth in the contract, on contracts stipulating a price per pound for fabricated structural steel delivered and/or erected, the quantities of materials for payment are determined by the calculation of gross weight of materials as shown on the shop drawings.

9.2.1. The unit weight of steel is assumed to be 490 pounds per cubic foot. The unit weight of other materials is in accordance with the manufacturer's published data for the specific product.

9.2.2. The weights of shapes, plates, bars, steel pipe and structural tubing are calculated on the basis of shop drawings showing actual quantities and dimensions of material furnished, as follows:

(a) The weight of all structural shapes, steel pipe and structural tubing is calculated using the nominal weight per foot and the detailed overall length.

(b) The weight of plates and bars is calculated using the detailed overall rectangular dimensions.

(c) When parts can be economically cut in multiples from material of larger dimensions, the weight is calculated on the basis of the theoretical rectangular dimensions of the material from which the parts are cut.

(d) When parts are cut from structural shapes, leaving a non-standard section not useable on the same contract, the weight is calculated on the basis of the nominal unit weight of the section from which the parts are cut.

(e) No deductions are to be made for material removed by cuts, copes, clips, blocks, drilling, punching, boring, slot milling, planing or weld joint preparation.

9.2.3. The calculated weights of castings are determined from the shop drawings of the pieces. An allowance of 10 percent is added for fillets and overrun. Scale weights of rough castings may be used if available.

9.2.4. The items for which weights are shown in tables in the AISC *Manual of Steel Construction* are calculated on the basis of tabulated unit weights.

9.2.5. The weight of items not included in the tables in the AISC *Manual of Steel Construction* shall be taken from the manufacturers' catalog and the manufacturers' shipping weight shall be used.

9.2.6. The weight of shop or field weld metal and protective coatings is not included in the calculated weight for pay purposes.

9.3. Revisions to Contract Documents

9.3.1. Revisions to the contract are made by the issuance of new documents or the reissuance of existing documents. In either case, all revisions are clearly indicated and the documents are dated.

9.3.2. A revision to the requirements of the contract documents are made by change orders, extra work orders, or notations on the shop and erection drawings when returned upon approval.

9.3.3. Unless specifically stated to the contrary, the issuance of a revision is authorization by the owner to release these documents for construction.

9.4. Contract Price Adjustment

9.4.1. When the scope of work and responsibilities of the fabricator and erector are changed from those previously established by the contract documents, an appropriate modification of the contract price is made. In computing the contract price adjustment, the fabricator and erector consider the quantity of work added or deleted, modifications in the character of the work, and the timeliness of the change with respect to the status of material ordering, detailing, fabrication and erection operations.

9.4.2. Requests for contract price adjustments are presented by the fabricator and erector in a timely manner and are accompanied by a description of the change in sufficient detail to permit evaluation and timely approval by the owner.

9.4.3. Price per pound and price per item contracts generally provide for additions or deletions to the quantity of work prior to the time work is released for construction. Changes to the character of the work, at any time, or additions and/or deletions to the quantity of the work after it is released for construction, may require a contract price adjustment.

9.5. Scheduling

9.5.1. The contract documents specify the schedule for the performance of the work. This schedule states when the "released for construction" plans will be issued and when the job site, foundations, piers and abutments will be ready, free from obstructions and accessible to the erector, so that erection can start at the designated time and continue without interference or delay caused by the owner or other trades.

9.5.2. The fabricator and erector have the responsibility to advise the owner, in a timely manner, of the effect any revision has on the contract schedule.

9.5.3. If the fabrication or erection is significantly delayed due to design revisions, or for other reasons which are the owner's responsibility, the fabricator and erector are compensated for additional costs incurred.

9.6 Terms of Payment. The terms of payment for the contract shall be outlined in the contract documents.

SECTION 10. ARCHITECTURALLY EXPOSED STRUCTURAL STEEL

10.1. Scope

This section of the Code defines additional requirements which apply only to members specifically designated by the contract documents as "Architecturally Exposed Structural Steel" (AESS). All provisions of Sects. 1 through 9 of the Code apply unless specifically modified in this section. AESS members or components are fabricated and erected with the care and dimensional tolerances indicated in this section.

10.2. Additional Information Required in Contract Documents

(a) Specific identification of members or components which are to be AESS.

(b) Fabrication and erection tolerances which are more restrictive than provided for in this section.

(c) Requirements, if any, of a test panel or components for inspection and acceptance standards prior to the start of fabrication.

10.3. Fabrication

10.3.1. Rolled Shapes

Permissible tolerances for out-of-square or out-of-parallel, depth, width and symmetry of rolled shapes are as specified in ASTM Specification A6. No attempt to match abutting cross-sectional configurations is made unless specifically required by the contract documents. The as-fabricated straightness tolerances of members are one-half of the standard camber and sweep tolerances in ASTM A6.

10.3.2. Built-up Members

The tolerances on overall profile dimensions of members made up from a series of plates, bars and shapes by welding are limited to the accumulation of permissible tolerances of the component parts as provided by ASTM Specification A6. The as-fabricated straightness tolerances for the member as a whole are one-half the standard camber and sweep tolerances for rolled shapes in ASTM A6.

10.3.3. Weld Show-through

It is recognized that the degree of weld show-through, which is any visual indication of the presence of a weld or welds on the side away from the viewer, is a function of weld size and material thickness. The members or components will be acceptable as produced unless specific acceptance criteria for weld show-through are included in the contract documents.

10.3.4. Joints

All copes, miters and butt cuts in surfaces exposed to view are made with uniform gaps of ⅛-inch if shown to be open joints, or in reasonable contact if shown without gap.

10.3.5. Welding

Reasonably smooth and uniform as-welded surfaces are acceptable on all welds exposed to view. Butt and plug welds do not project more than ¹⁄₁₆-inch above the exposed surface. No finishing or grinding is required except where clearances or fit of other components may necessitate, or when specifically required by the contract documents.

10.3.6. Weathering Steel

Members fabricated of weathering steel which are to be AESS shall not have erection marks or other painted marks on surfaces that are to be exposed in the completed structure. If cleaning other than SSPC-SP6 is required, these requirements shall be defined in the contract documents.

10.4. Delivery of Materials

The fabricator uses special care to avoid bending, twisting or otherwise distorting individual members.

10.5 Erection

10.5.1. General

The erector uses special care in unloading, handling and erecting the steel to avoid marking or distorting the steel members. Care is also taken to minimize damage to any shop paint. If temporary braces or erection clips are used, care is taken to avoid unsightly surfaces upon removal. Tack welds are ground smooth and holes are filled with weld metal or body solder and smoothed by grinding or filing. The erector plans and executes all operations in such a manner that the close fit and neat appearance of the structure will not be impaired.

10.5.2. Erection Tolerances

Unless otherwise specifically designated in the contract documents, members and components are plumbed, leveled and aligned to a tolerance not to exceed one-half the amount permitted for structural steel. These erection tolerances for AESS require that the owner's plans specify adjustable connections between AESS and the structural steel frame or the masonry or concrete supports, in order to provide the erector with means for adjustment.

10.5.3. Components with Concrete Backing

When AESS is backed with concrete, it is the general contractor's responsibility to provide sufficient shores, ties and strongbacks to assure against sagging, bulging, etc., of the AESS resulting from the weight and pressure of the wet concrete.

Commentary

on the

Code of Standard Practice

PREFACE

This Commentary has been prepared to assist those who use the *Code of Standard Practice* in understanding the background, basis and intent of its provisions.

Each section in the Commentary is referenced to the corresponding section or subsection in the Code. Not all sections of the Code are discussed; sections are covered only if it is believed that additional explanation may be helpful.

While every precaution has been taken to insure that all data and information presented is as accurate as possible, the Institute cannot assume responsibility for errors or oversights in the information published herein or the use of the information published or incorporating such information in the preparation of detailed engineering plans. The figures are for illustrative purposes only and are not intended to be applicable to any actual design. The information should not replace the judgment of an experienced architect or engineer who has the responsibility of design for a specific structure.

Commentary

ON THE CODE OF STANDARD PRACTICE
FOR STEEL BUILDINGS AND BRIDGES

(Adopted Effective September 1, 1986)

SECTION 1. GENERAL PROVISIONS

1.1. Scope

This Code is not applicable to metal building systems, which are the subject of standards published by the Metal Building Manufacturers Association in their *Metal Building Systems Manual*. AISC has not participated in the development of the MBMA code and, therefore, takes no position and is not responsible for any of its provisions.

This Code is not applicable to standard steel joists, which are the subject of *Recommended Code of Standard Practice for Steel Joists*, published by the Steel Joist Institute. AISC has not participated in the development of the SJI code and, therefore, takes no position and is not responsible for any of its provisions.

SECTION 2. CLASSIFICATION OF MATERIALS

2.2. Other Steel or Metal Items

These items include materials which may be supplied by the steel fabricator which require coordination between other material suppliers and trades. If they are to be supplied by the fabricator, they must be specifically called for and detailed in contract documents.

SECTION 3. PLANS AND SPECIFICATIONS

3.1. Structural Steel

Project specifications vary greatly in complexity and completeness. There is a benefit to the owner if the specifications leave the contractor reasonable latitude in performing his work. However, critical requirements affecting the integrity of the structure or necessary to protect the owner's interest must be covered in the contract documents. The following checklist is included for reference:

Standard codes and specifications governing structural steelwork
Material specifications
Mill test reports
Welded joint configuration
Weld procedure qualification
Bolting specifications
Special requirements for work of other trades
Runoff tabs
Surface preparation and shop painting
Shop inspection
Field inspection
Non-destructive testing, including acceptance criteria
Special requirements on delivery
Special erection limitations
Temporary bracing for non-self-supporting structures
Special fabrication and erection tolerances for AESS
Special pay weight provisions

SECTION 4. SHOP AND ERECTION DRAWINGS

4.1. Owner's Responsibility

The owner's responsibility for the proper planning of the work and the communication of all facts of his particular project is a requirement of the Code, not only at the time of bidding, but also throughout the term of any project. The contract documents, including the plans and specification, are for the purpose of communication. It is the owner's responsibility to properly define the scope of work, and to define information or items required and outlined in the plans and specifications. When the owner releases plans and specifications for construction, the fabricator and erector rely on the fact that these are the owner's requirements for his project.

The Code defines the owner as including a designated representative such as the architect, engineer or project manager, and when these representatives direct specific action to be taken, they are acting as and for the owner.

On phased construction projects, to insure the orderly flow of material procurement, detailing, fabrication and erection activities, it is essential that designs are not continuously revised after progressive releases for construction are made. In essence, once a portion of a design is released for construction, the essential elements of that design should be "frozen" to assure adherence to the construction schedule or all parties should reach an understanding on the effects of future changes as they affect scheduled deliveries and added costs, if any.

4.2. Approval

4.2.1. In those instances where a fabricator develops the detail configuration of connections during the preparation of shop drawings, he does not thereby become responsible for the design of that part of the overall structure. The Engineer-of-Record has the final and total responsibility for the adequacy and safety of a structure, and is the only individual who has all the information necessary to evaluate the total impact of the connection details on the structural design. The structural steel fabricator is in no position to accept such design responsibility, for two practical reasons:

(a) The structural steel plans may be released for construction with incomplete or preliminary member reaction data, forcing a review by the Engineer at the time of approval.

(b) Few fabricators have engineers registered in all of the states in which they do business.

In practice, the fabricator develops connection details which satisfy two basic criteria:

(a) The connections must be of suitable strength and rigidity to meet the requirements of the design information provided by the engineer-of-record.

(b) The detail configuration accommodates the fabricator's shop equipment and procedures.

Since each shop has different equipment and skills, the fabricator is best suited to develop connection details which satisfy the second requirement. However, the overriding first requirement necessitates acceptance of responsibility and approval by the engineer.

SECTION 5. MATERIALS

5.1. Mill Materials

5.1.2. Mill dimensional tolerances are completely set forth as part of ASTM Specification A6. Variation in cross section geometry of rolled members must be recognized by the designer, the fabricator and erector (see Fig. 1). Such tolerances are

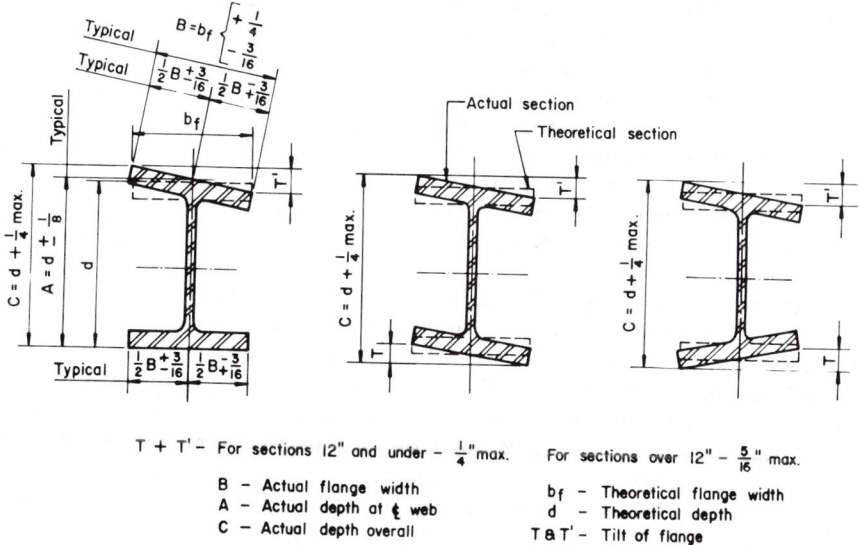

T + T' – For sections 12" and under – $\frac{1}{4}$" max.   For sections over 12" – $\frac{5}{16}$" max.

B – Actual flange width   bf – Theoretical flange width
A – Actual depth at ₵ web   d – Theoretical depth
C – Actual depth overall   T & T' – Tilt of flange

Fig. 1. Mill tolerances on cross-section dimensions

mandatory because roll wear, thermal distortions of the hot cross section immediately after leaving the forming rolls, and differential cooling distortions that take place on the cooling beds are economically beyond precise control. Absolute perfection of cross section geometry is not of structural significance and, if the tolerances are recognized and provided for, also not of architectural significance. ASTM A6 also stipulates straightness and camber tolerances which are adequate for most conventional construction; however, these characteristics may be controlled or corrected to closer tolerances during the fabrication process when the unique demands of a particular project justify the added cost.

SECTION 6. FABRICATION AND DELIVERY

6.4. Dimensional Tolerances

Fabrication tolerances are stipulated in several specification documents, each applicable to a special area of construction. Basic fabrication tolerances are stipulated in Sects. 6.4 and 10 of the Code and Sect. 1.23.8.1 of the AISC Specification. Other specifications and codes frequently incorporated by reference in the contract documents are the AWS *Structural Welding Code* and AASHTO *Standard Specifications for Highway Bridges*.

6.5. Shop Painting

6.5.2., 6.5.3. The selection of a paint system is a design decision involving many factors, including owner's preference, service life of the structure, severity of environmental exposure, the cost of both initial application and future renewals, and the compatibility of the various components comprising the paint system, i.e., surface preparation, prime coat and subsequent coats.

Because inspection of shop painting needs to be concerned with workmanship at each stage of the operation, the fabricator provides notice of the schedule of operations and affords access to the work site to inspectors. Inspection must be coordinated with that schedule in such a way as to avoid delay of the scheduled operations.

Acceptance of the prepared surface must be made prior to application of the prime coat, because the degree of surface preparation cannot be readily verified after painting. Also, time delay between surface preparation and application of the prime coat can, especially with blast-cleaned surfaces, result in unacceptable deterioration of a properly prepared surface, necessitating a repetition of surface preparation. Therefore, to avoid potential deterioration of the surface it is assumed that surface preparation is accepted unless it is inspected and rejected prior to the scheduled application of the prime coat.

The prime coat in any paint system is designed to maximize the wetting and adherence characteristics of the paint, usually at the expense of its weathering capabilities. Consequently, extended exposure of the prime coat to weather or to a corrosive atmosphere will lead to its deterioration and may necessitate repair, possibly including repetition of surface preparation and primer application in limited areas. With the introduction of high performance paint systems in the recent past, delay in the application of the prime coat has become more critical. High performance paint systems generally require a greater degree of surface preparation, as well as early application of weathering protection for the prime coat.

Since the fabricator does not control the selection of the paint system, the compatibility of the various components of the total paint system, nor the length of exposure of the prime coat, he cannot guarantee the performance of the prime coat or any other part of the system. Rather, he performs specific operations to the requirements established in the contract documents.

Section 6.5.3 stipulates cleaning the steel to the requirements of SSPC-SP2. This is not meant as an exclusive cleaning level, but rather that level of surface preparation which will be furnished if the steel is to be painted and if the job specifications are silent or do not require more stringent surface preparation requirements.

Further information regarding shop painting is available in *A Guide to Shop Painting of Structural Steel*, published jointly by the Steel Structures Painting Council and the American Institute of Steel Construction.

6.5.5. Extended exposure of unpainted steel which has been cleaned for subsequent fire protection material application can be detrimental to the fabricated product. Most levels of cleaning require the removal of all loose mill scale, but permit some amount of "tightly adhering mill scale." When a piece of structural steel which has been cleaned to an acceptable level is left exposed to a normal environment, moisture can penetrate behind the scale, and some "lifting" of the scale by the oxidation products is to be expected. Cleanup of "lifted" mill scale is not the responsibility of the fabricator, but is assigned by contract requirement to an appropriate contractor.

Section 6.5.5 of the Code is not applicable to weathering steel, for which special cleaning specifications are always required in the contract documents.

SECTION 7.   ERECTION

7.5.   Installation of Anchor Bolts and Embedded Items

7.5.1. While the general contractor must make every effort to set anchor bolts accurately to theoretical drawing dimensions, minor errors may occur. The tolerances set forth in this section were compiled from data collected from general contractors and erectors. They can be attained by using reasonable care and will ordinarily allow the steel to be erected and plumbed to required tolerances. If special conditions require closer tolerances, the contractor responsible for setting the anchor bolts should be so informed by the contract documents. When anchor bolts are set in sleeves, the adjustment provided may be used to satisfy the required anchor bolt setting tolerances.

The tolerances established in this section of the Code have been selected to be compatible with oversize holes in base plates, as recommended in the AISC textbook *Detailing for Steel Construction*.

An *anchor bolt group* is the set of anchor bolts which receive a single fabricated steel shipping piece.

The *established column line* is the actual field line most representative of the centers of the as-built anchor bolt groups along a line of columns. It must be straight or curved as shown on the plans.

7.6.   Bearing Devices

The ⅛-inch tolerance on elevation of bearing devices relative to established grades is provided to permit some variation in setting bearing devices and to account for attainable accuracy with standard surveying instruments.

The use of leveling plates larger than 12 in. × 12 in. is discouraged and grouting is recommended with larger sizes.

7.9.3. Non Self-supporting Steel Frames

To rationally provide temporary supports and/or bracing, the erector must be informed by the owner of the sequence of installation and the effect of loads imposed by such elements at various stages during the sequence until they become effective. For example, precast tilt-up slabs or channel slab facia elements which depend upon attachment to the steel frame for stability against overturning due to eccentricity of their gravity load, may induce significant unbalanced lateral forces on the bare steel frame when partially installed.

7.11. Framing Tolerances

The erection tolerances defined in this section of the Code have been developed through long-standing usage as practical criteria for the erection of structural steel. Erection tolerances were first defined by AISC in its *Code of Standard Practice* of October, 1924 in Paragraph 7 (f), "Plumbing Up." With the changes that took place in the types and use of materials in building construction after World War II, and the increasing demand by architects and owners for more specific tolerances, AISC adopted new standards for erection tolerances in Paragraph 7 (h) of the March 15, 1959 edition of the Code. Experience has proven that those tolerances can be economically obtained.

The current requirements were first published in the October 1, 1972 edition of the Code. They provide an expanded set of criteria over earlier Code editions. The basic premise that the final accuracy of location of any specific point in a structural steel frame results from the combined mill, fabrication and erection tolerances, rather than from the erection tolerances alone, remains unchanged in this edition of the Code. However, to improve clarity, pertinent standard fabrication tolerances are now stipulated in Sect. 7.11, rather than by reference to the AISC Specification as in previous editions. Additionally, expanded coverage has been given to definition of working points and working lines governing measurements of the actual steel location. Illustrations for defining and applying the applicable Code tolerances are provided in this Commentary.

The recent trend in building work is away from built-in-place construction wherein compatibility of the frame and the facade or other collateral materials is automatically provided for by the routine procedures of the crafts. Building construction today frequently incorporates prefabricated components wherein large units are developed with machine-like precision to dimensions that are theoretically correct for a perfectly aligned steel frame with ideal member cross sections. This type of construction has made the magnitude of the tolerances allowed for structural steel building frames increasingly of concern to owners, architects and engineers. This has led to the inclusion in job specifications of unrealistically small tolerances, which indicate a general lack of recognition of the effects of the accumulation of dead load, temperature effects and mill, fabrication and erection tolerances. Such tolerances are not economically feasible and do not measurably increase the structure's functional value. This edition of the Code incorporates tolerances previously found to be practical and presents them in a precise and clear manner. Actual application methods have been considered and the application of the tolerance limitations to the actual structure defined.

7.11.3. Position and Alignment

The limitations described in Sect. 7.11.3.1 and illustrated in Figs. 2 and 3 make it possible to maintain built-in-place or prefabricated facades in a true vertical plane up to the 20th story, if connections which provide for 3-in. adjustment are used. Above the 20th story, the facade may be maintained within $\frac{1}{16}$-inch per story with a maximum total deviation of 1 in. from a true vertical plane, if the 3-in. adjustment is provided.

Section 7.11.3.1(c) limits the position of exterior column working points at any given splice elevation to a narrow horizontal envelope parallel to the building line (see Fig. 4). This envelope is limited to a width of $1\frac{1}{2}$ in., normal to the building line, in up to 300 ft of building length. The horizontal location of this envelope is not necessarily directly above or below the corresponding envelope at the adjacent splice elevations, but should be within the limitation of the 1:500 allowable tolerance in plumbness of the controlling columns (see Fig. 3).

Connections permitting adjustments of plus 2 in. to minus 3 in. (5 in. total) will be necessary in cases where the architect or owner insists upon attempting to construct the facade to a true vertical plane above the 20th story.

Usually there is a differential shortening of the internal versus the external columns during construction, due to non-uniform rate of accumulation of dead load stresses (see Fig. 5). The amount of such differential shortening is indeterminate because it varies dependent upon construction sequence from day to day as the construction progresses, and does not reach its maximum shortening until the building is in service. When floor concrete is placed while columns are supporting different percentages of their full design loads, the floor must be finished to slopes established by measurements from the tops of beams at column connections. The effects of differential shortening, plus mill camber and deflections, all become very important when there is little cover over the steel, when there are electrical fittings mounted on

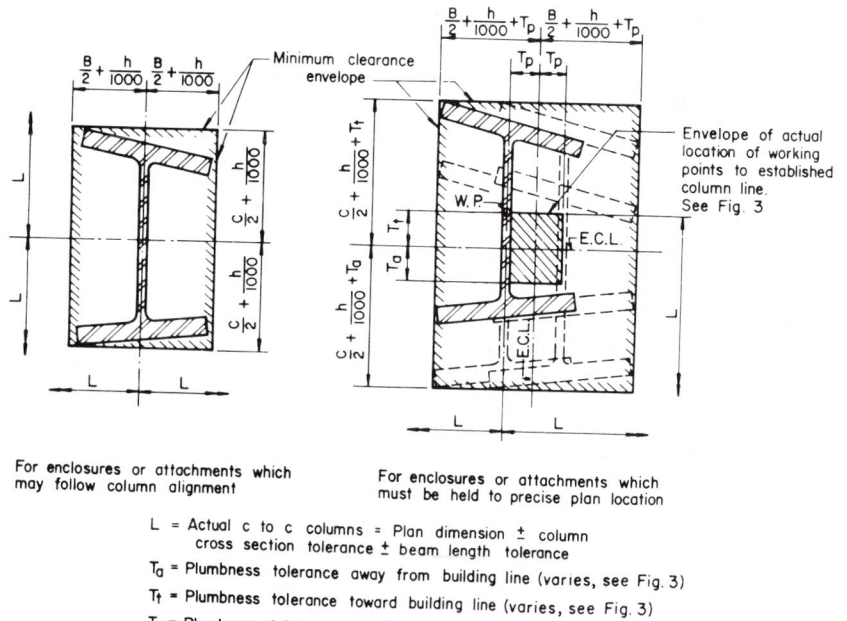

Fig. 2. Clearance required to accommodate accumulated column tolerances

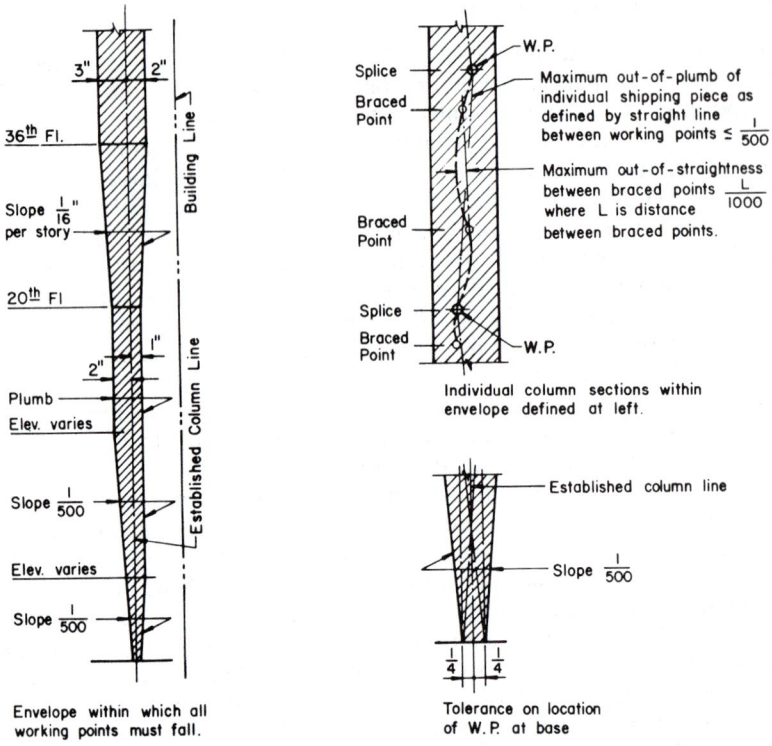

Fig. 3. *Exterior column plumbness tolerances normal to building line*

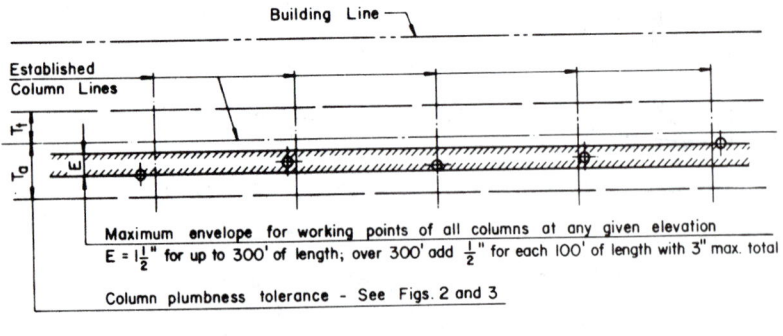

Fig. 4. *Tolerances in plan at any splice elevation of exterior columns*

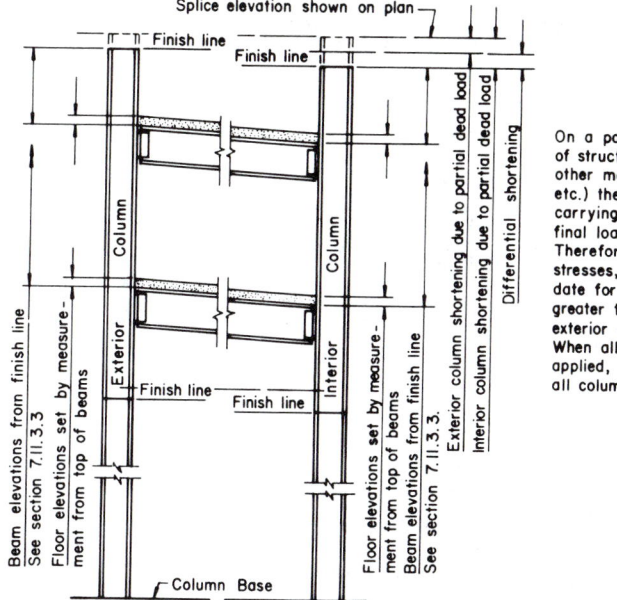

On a particular date during the erection of structural steel and placement of other material, (floor concrete, facade etc.) the interior columns will be carrying a higher percentage of their final loads than the exterior columns Therefore, for equal design unit stresses, the actual stress on that date for interior columns will be greater than the actual stresses on exterior columns.
When all dead loads have been applied, stresses and shortening in all columns will be approximately equal.

Fig. 5. Effect of differential column shortening

the steel flooring whose tops are supposed to be flush with the finished floor, when there is small clearance between bottom of beams and top of door frames, etc., and when there is little clearance around ductwork. To finish floors to precise level plane, for example by the use of laser leveling techniques, can result in significant differential floor thicknesses, different increases above design dead loads for individual columns and, thus, permanent differential column shortening and out-of-level completed floors.

Similar considerations make it infeasible to attempt to set the elevation of a given floor in a multistory building by reference to a bench mark at the base of the structure. Columns are fabricated to a length tolerance of plus or minus ⅟₃₂-inch while under a zero state of stress. As dead loads accumulate, the column shortening which takes place is negligible within individual stories and in low buildings, but will accumulate to significant magnitude in tall buildings; thus, the upper floors of tall buildings will be excessively thick and the lower floors will be below the initial finish elevation if floor elevations are established relative to a ground level bench mark.

If foundations and base plates are accurately set to grade and the lengths of individual column sections are checked for accuracy prior to erection, and if floor elevations are established by reference to the elevation of the top of beams, the effect of column shortening due to dead load will be minimized.

Since a long unencased steel frame will expand or contract ⅛-in. per 100 feet for each change of 15°F in temperature, and since the change in length can be assumed to act about the center of rigidity, the end columns anchored to foundations will be plumb only when the steel is at normal temperature (see Fig. 6). It is, therefore, necessary to correct field measurements of offsets to the structure from established baselines for the expansion or contraction of the exposed steel frame. For example, a building 200-ft long that is plumbed up at 100°F should have working points at the tops of end columns positioned ½-inch out from the working point at the base in order for the column to be

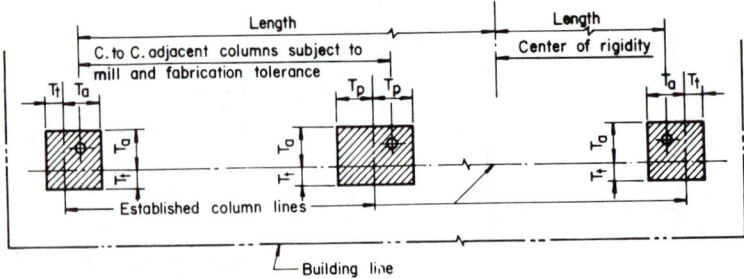

When plumbing end columns, apply temperature adjustment at rate $\frac{1}{8}$" per 100' of length from center of rigidity per each 15°F of difference between erection and working temperatures.

Fig. 6.   Tolerances in plan location of columns

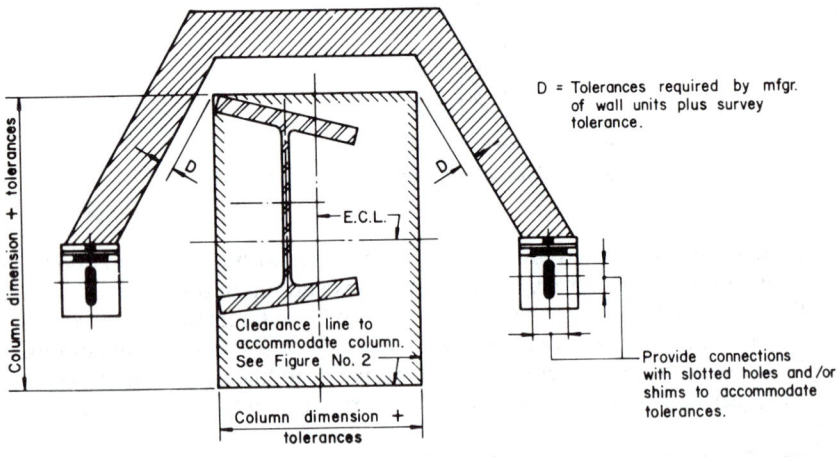

D = Tolerances required by mfgr. of wall units plus survey tolerance.

Provide connections with slotted holes and/or shims to accommodate tolerances.

If fascia joints are set from nearest column finish line, allow $\pm\frac{5}{8}$" for vertical adjustment. Owners plans for fascia details must allow for progressive shortening of steel columns.

Fig. 7.   Clearance required to accommodate fascia

plumb at 60°F. Differential temperature effects on column length should also be taken into account in plumbing surveys when tall steel frames are subject to strong sun exposure on one side.

The alignment of lintels, spandrels, wall supports and similar members used to connect other building construction units to the steel frame should have an adjustment of sufficient magnitude to allow for the accumulative effect of mill, fabrication and erection tolerances on the erected steel frame (see Fig. 7).

7.11.3.2.   Alignment Tolerance for Members with Field Splices

The angular misalignment of the working line of all fabricated shipping pieces relative to the line between support points of the member as a whole in erected position must not exceed 1 in 500. Note that the tolerance is not stated in terms of a

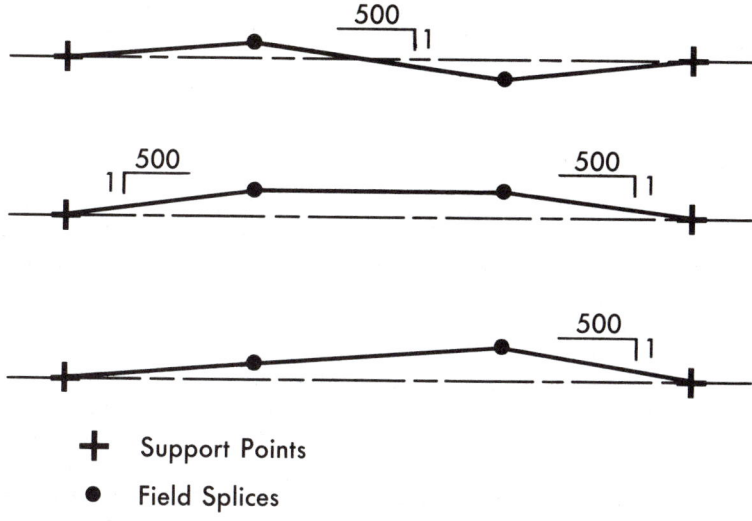

+ Support Points

● Field Splices

Fig. 8. Alignment tolerances for members with field splices

linear displacement at any point and is not to be taken as the overall length between supports divided by 500. Typical examples are shown in Fig. 8. Numerous conditions within tolerance for these and other cases are possible. This condition applies to both plan and elevation tolerances.

7.11.4. Responsibility for Clearances

In spite of all efforts to minimize inaccuracies, deviations will still exist; therefore, in addition, the designs of prefabricated wall panels, partition panels, fenestrations, floor-to-ceiling door frames and similar elements must provide for clearance and details for adjustment, as described in Sect. 7.11.4. Designs must provide for adjustment in the vertical dimension of prefabricated facade panels supported by the steel frame, because the accumulation of shortening of stressed steel columns will result in the unstressed facade supported at each floor level being higher than the steel frame connections to which it must attach. Observations in the field have shown that where a heavy facade is erected to a greater height on one side of a multistory building than on the other, the steel framing will be pulled out of alignment. Facades should be erected at a relatively uniform rate around the perimeter of the structure.

SECTION 8. QUALITY CONTROL

8.1.1. The AISC Quality Certification Program confirms to the construction industry that a certified structural steel fabricating plant has the capability by reason of commitment, personnel, organization, experience, procedures, knowledge and equipment to produce fabricated structural steel of the required quality for a given category of structural steelwork. The AISC Quality Certification Program is not intended to involve inspection and/or judgement of product quality on individual projects. Neither is it intended to guarantee the quality of specific fabricated steel products.

9.2. Calculation of Weights

The standard procedure for calculation of weights that is described in the Code meets the need for a universally acceptable system for defining "pay weights" in contracts based on the weight of delivered and/or erected materials. This procedure permits owners to easily and accurately evaluate price per pound proposals from potential suppliers and enables both parties to a contract to have a clear understanding of the basis for payment.

The Code procedure affords a simple, readily understood method of calculation which will produce pay weights which are consistent throughout the industry and which may be easily verified by the owner. While this procedure does not produce actual weights, it can be used by purchasers and suppliers to define a widely accepted basis for bidding and contracting for structural steel. However, any other system can be used as the basis for a contractual agreement. When other systems are used, both supplier and purchaser should clearly understand how the alternate procedure is handled.

9.3. Revisions to Contract Documents

9.3.1. Revisions to the contract are proposed by the issuance of new documents or the re-issuance of existing documents. Individual revisions are noted where they occur and documents are dated with latest issue date and reasons for issuing are identified.

9.3.2. Revisions to the contract are also proposed by change order, extra work order or notations on the shop and erection drawings when returned from approval. However, revisions proposed in this manner are incorporated subsequently as revisions to the plans and/or specifications and re-issued in accordance with Article 9.3.1.

9.3.3. Unless specifically stated to the contrary in the contract documents, the issuance of revisions authorizes the fabricator and erector to incorporate the revisions in the work. This authorization obligates the owner to pay the fabricator and erector for costs associated with changed and/or additional work.

When authorization for revisions is not granted to the fabricator and erector by issuance of new or revised documents, revisions affecting contract price and/or schedule are only incorporated by issuance of (1) change order, (2) extra work order, or (3) other documents expressing the agreement of all contract parties to such revisions. The fabricator and erector must promptly notify the owner of the effect and cost of proposed revisions to contract price and schedule, enabling orderly progress of the work.

9.6. Terms of Payment

These terms include such items as progress payments for material, fabrication, erection, retainage, performance and payment bonds and final payment. If a performance or payment bond, paid for by the owner, is required by contract, then no retainage shall be required.

SECTION 10. ARCHITECTURALLY EXPOSED STRUCTURAL STEEL

The rapidly increasing use of exposed structural steel as a medium of architectural expression has given rise to a demand for closer dimensional tolerances and smoother finished surfaces than required for ordinary structural steel framing.

This section of the Code establishes standards for these requirements which take into account both the desired finished appearance and the abilities of the fabrication shop to produce the desired product. These requirements were previously contained in the AISC *Specification for Architecturally Exposed Structural Steel* which Architects and Engineers have specified in the past. It should be pointed out that the term "Architecturally Exposed Structural Steel" (AESS), as covered in this section, must be specified in the contract documents if the fabricator is required to meet the fabricating standards of Sect. 10, and applies only to that portion of the structural steel so identified. In order to avoid misunderstandings and to hold costs to a minimum, only those steel surfaces and connections which will remain exposed and subject to normal view by pedestrians or occupants of the completed structure should be designated AESS.

LOAD AND RESISTANCE FACTOR DESIGN

Specification for Structural Joints Using ASTM A325 or A490 Bolts

September 1, 1986

At the time of preparation of the AISC Load and Resistance Factor Design Specification for Structural Steel Buildings, the Research Council on Structural Connections (RCSC) had not completed the LRFD version of its design specification. Basic provisions for the design of bolted structural connections are included in the AISC LRFD Specification. When reference is made to the RCSC Specification, the Allowable Stress Design (ASD) Specification may be used except for Sects. 4, 5 and 6, which should be replaced by the following corresponding sections stated in LRFD format.

LRFD Provisions Corresponding to RCSC Allowable Stress Design Specification:

4. Design of Bolted Connections

   (a) Tension and Shear Strength Limit States. See AISC LRFD Specification Sect. J3.3.

   (b) Bearing Strength Limit State. See AISC LRFD Specification Sect. J3.6.

   (c) Prying Action. The force in bolts required to support loads by means of direct tension shall be calculated considering the effects of the external load and any tension resulting from prying action produced by deformation of the connected parts.

   (d) Tensile Fatigue. See AISC Specification Appendix Sect. K4.3.

5. Design Check for Slip Resistance

(a) Slip-critical Joints. Slip-critical joints are defined as joints in which slip would be detrimental to the serviceability of the structure. They include:

(1) Joints subject to fatigue loading.

(2) Joints with bolts installed in oversized holes.

(3) Except where the engineer intends otherwise, and so indicates in the contract documents, joints with bolts installed in slotted holes where the force on the joint is in a direction other than normal (between approximately 80 and 100 degrees) to the axis of the slot.

(4) Joints subject to significant load reversal.

(5) Joints in which welds and bolts share in transmitting load at a common faying surface. See Commentary.

(6) Joints in which, in the judgement of the engineer, any slip would be critical to the performance of the joint or the structure and so designated on the contract plans and specifications.

(b) Slip Resistance. In addition to the requirements of Sect. 4, the force due to the service loads acting on a slip-critical joint shall not exceed the design slip resistance (ϕP_s) of the connection (see Commentary) according to:

$$\phi P_s = \phi F_s A_b N_b N_s$$

where

F_s = nominal slip resistance per unit area of bolt from Table 3 of the RCSC Allowable Stress Design Specification, ksi.
A_b = area corresponding to the nominal body area of the bolt sq. in.
N_b = number of bolts in the joint
N_s = number of slip planes
ϕ = 1.0

Class A, B or C surface conditions of the bolted parts as defined in Table 3 (ASD) shall be used in joints designated as slip-critical, except as permitted in ASD 3 (b) (3).

6. Increase in Design Strengths

The reduced probabilities of loads acting in combinations are accounted for by the use of lower load factors provided by the authority having jurisdiction or the Load and Resistance Factor Design Specification with which this Specification is being used; therefore the resistances given in this Specification shall not be increased.

ALLOWABLE STRESS DESIGN

Specification for Structural Joints Using ASTM A325 or A490 Bolts

Approved by the Research Council on Structural Connections
of the Engineering Foundation, November 13, 1985
Endorsed by American Institute of Steel Construction
Endorsed by Industrial Fasteners Institute

AMERICAN INSTITUTE OF STEEL CONSTRUCTION
ONE EAST WACKER DRIVE, SUITE 3100
CHICAGO, ILLINOIS 60601-2001

PREFACE

The purpose of the Research Council on Structural Connections is to stimulate and support such investigation as may be deemed necessary and valuable to determine the suitability and capacity of various types of structural connections, to promote the knowledge of economical and efficient practices relating to such structural connections, and to prepare and publish related standards and such other documents as necessary to achieving its purpose.

The Council membership consists of qualified structural engineers from the academic and research institutions, practicing design engineers, suppliers and manufacturers of threaded fasteners, fabricators and erectors and code writing authorities. Each version of the Specification is based upon deliberations and letter ballot of the full Council membership.

The first *Specification for Assembly of Structural Joints Using High Tensile Steel Bolts* approved by the Council was published in January 1951. Since that time the Council has published 11 succeeding editions each based upon past successful usage, advances in the state of knowledge and changes in engineering design practice. This twelfth version of the Council's *Allowable Stress Design Specification* is significantly reorganized and revised from earlier versions.

The intention of the Specifications is to cover the design criteria and normal usage and practices involved in the everyday use of high-strength bolts in steel-to-steel structural connections. It is not intended to cover the full range of structural connections using threaded fasteners nor the use of high-strength bolts other than those included in ASTM A325 or ASTM A490 Specifications nor the use of ASTM A325 or A490 Bolts in connections with material other than steel within the grip.

A Commentary has been prepared to accompany these Specifications to provide background and aid the user to better understand and apply the provisions.

The user is cautioned that independent professional judgement must be exercised when data or recommendations set forth in these Specifications are applied. The design and the proper installation and inspection of bolts in structural connections is within the scope of expertise of a competent licensed architect, structural engineer or other licensed professional for the application of the principles to a particular case.

ALLOWABLE STRESS DESIGN

Specification for Structural Joints Using ASTM A325 or A490 Bolts

Approved by Research Council on Structural Connections of the
Engineering Foundation, November 13, 1985.
Endorsed by American Institute of Steel Construction
Endorsed by Industrial Fasteners Institute

1. Scope

This Specification relates to the allowable stress design for strength and slip resistance of structural joints using ASTM A325 high-strength bolts, ASTM A490 heat treated high-strength bolts or equivalent fasteners, and for the installation of such bolts in connections of structural steel members. The Specification relates only to those aspects of the connected materials that bear upon the performance of the fasteners.

Construction shall conform to an applicable code or specification for structures of carbon, high strength low alloy steel or quenched and tempered structural steel.

The attached Commentary provides background information in order that the user may better understand the provisions of the Specification.

2. Bolts, Nuts, Washers and Paint

(a) **Bolt Specifications.** Bolts shall conform to the requirements of the current edition of the Specifications of the American Society for Testing and Materials for High-strength Bolts for Structural Steel Joints, ASTM A325, or Heat Treated Steel Structural Bolts, 150 ksi Minimum Tensile Strength, ASTM A490, except as provided in paragraph (d) of this section. The designer shall specify the type of bolts to be used.

(b) **Bolt Geometry.** Bolt dimensions shall conform to the current requirements of the American National Standards Institute for Heavy Hex Structural Bolts, ANSI Standard B18.2.1, except as provided in paragraph (d) of this section. The length of bolts shall be such that the end of the bolt will be flush with or outside the face of the nut when properly installed.

(c) **Nut Specifications.** Nuts shall conform to the current chemical and mechanical requirements of the American Society for Testing and Materials Standard Specification for Carbon and Alloy Steel Nuts, ASTM A563 or Standard Specification for Carbon and Alloy Steel Nuts for Bolts for High Pressure and High Temperatures Service, ASTM A194. The grade and surface finish of nuts for each type shall be as follows:

A325 Bolt Type	Nut Specification, Grade and Finish
1 and 2, plain (uncoated)	A563 C, C3, D, DH, and D3 or A194 2 and 2H; plain
1 and 2, galvanized	A563 DH or A194 2H; galvanized
3 plain	A563 C3 and DH3

A490 Bolt Type	Nut Specification, Grade and Finish
1 and 2, plain	A563 DH and DH3 or A194 2H; plain
3 plain	A563 DH3

Nut dimensions shall conform to the current requirements of the American National Standards Institute for Heavy Hex Nuts, ANSI Standard B18.2.2., except as provided in paragraph (d) of this section.

(d) **Alternate Fastener Designs.** Other fasteners or fastener assemblies which meet the materials, manufacturing and chemical composition requirements of ASTM Specification A325 or A490 and which meet the mechanical property requirements of the same specifications in full-size tests, and which have a body diameter and bearing areas under the head and nut not less than those provided by a bolt and nut of the same nominal dimensions prescribed by paragraphs 2(b) and 2(c), may be used subject to the approval of the responsible Engineer. Such alternate fasteners may differ in other dimensions from those of the specified bolts and nuts. Their installation procedure and inspection may differ from procedures specified for regular high-strength bolts in Sections 8 and 9. When a different installation procedure or inspection is used, it shall be detailed in a supplemental specification applying to the alternate fastener and that specification must be approved by the engineer responsible for the design of the structure.

(e) **Washers.** Flat circular washers and square or rectangular beveled washers shall conform to the current requirements of the American Society for Testing and Materials Standard Specification for Hardened Steel Washers, ASTM F436.

(f) **Load Indicating Devices.** Load indicating devices may be used in conjunction with bolts, nuts and washers specified in 2(a) through 2(e) provided they satisfy the requirements of 8(d)(4). Their installation procedure and inspection shall be detailed in supplemental specifications provided by the manufacturer and subject to the approval of the engineer responsible for the design of the structure.

(g) **Faying Surface Coatings.** Paint, if used on faying surfaces of connections which are not specified to be slip critical, may be of any formulation. Paint, used on the faying surfaces of connections specified to be slip critical, shall be qualified by test in accordance with *Test Method to Determine the Slip Coefficient for Coatings Used in Bolted Joints* as adopted by the Research Council on Structural Connections, see Appendix A. Manufacturer's certification shall include a certified copy of the test report.

3. Bolted Parts

(a) **Connected Material.** All material within the grip of the bolt shall be steel. There shall be no compressible material such as gaskets or insulation within the grip. Bolted steel parts shall fit solidly together after the bolts are tightened, and may be coated or noncoated. The slope of the surfaces of parts in contact with the bolt head or nut shall not exceed 1:20 with respect to a plane normal to the bolt axis.

(b) **Surface Conditions.** When assembled, all joint surfaces, including surfaces adjacent to the bolt head and nut, shall be free of scale, except tight mill scale, and shall be free of dirt or other foreign material. Burrs that would prevent solid seating of the connected parts in the snug tight condition shall be removed.

Paint is permitted on the faying surfaces unconditionally in connections except in slip critical connections as defined in Section 5(a).

The faying surfaces of slip critical connections shall meet the requirements of the following paragraphs, as applicable.

(1) In noncoated joints, paint, including any inadvertent overspray, shall be excluded from areas closer than one bolt diameter but not less than one inch from the edge of any hole and all areas within bolt pattern.

(2) Joints specified to have painted faying surfaces shall be blast cleaned and coated with a paint which has been qualified as class A or B in accordance with the requirements of paragraph 2(g), except as provided in 3(b)(3).

(3) Subject to the approval of the Engineer, coatings providing a slip coefficient less than 0.33 may be used provided the mean slip coefficient is established by test in accordance with the requirements of paragraph 2(g), and the allowable slip load per unit area established. The allowable slip load per unit area shall be taken as equal to the allowable slip load per unit area from Table 3 for Class A coatings as appropriate for the hole type and bolt type times the slip coefficient determined by test divided by 0.33.

(4) Coated joints shall not be assembled before the coatings have cured for the minimum time used in the qualifying test.

(5) Galvanized faying surfaces shall be hot dip galvanized in accordance with ASTM Specification A123 and shall be roughened by means of hand wire brushing. Power wire brushing is not permitted.

(c) **Hole Types.** Hole types recognized under this specification are standard holes, oversize holes, short slotted holes and long slotted holes. The nominal dimensions for each type hole shall be not greater than those shown in Table 1. Holes not more than $1/32$ inch larger in diameter than the true decimal equivalent of the nominal diameter that may result from a drill or reamer of the nominal diameter are considered acceptable. The slightly conical hole that naturally results from punching operations is considered acceptable. The width of slotted holes which are produced by flame cutting or a combination of drilling or punching and flame cutting shall generally be not more than $1/32$ inch greater than the nominal width except that gouges not more than $1/16$ inch deep shall be permitted. For statically loaded connections, the flame cut surface need not be ground. For dynamically loaded connections, the flame cut surface shall be ground smooth.

Table 1. Nominal Hole Dimensions

Bolt Dia.	Hole Dimensions			
	Standard (Dia.)	Oversize (Dia.)	Short Slot (Width x Length)	Long Slot (Width x Length)
$1/2$	$9/16$	$5/8$	$9/16 \times 11/16$	$9/16 \times 1 1/4$
$5/8$	$11/16$	$13/16$	$11/15 \times 7/8$	$11/16 \times 19/16$
$3/4$	$13/16$	$15/16$	$13/16 \times 1$	$13/16 \times 17/8$
$7/8$	$15/16$	$1 1/16$	$15/16 \times 1 1/8$	$15/16 \times 23/16$
1	$1 1/16$	$1 1/4$	$1 1/16 \times 1 5/16$	$1 1/16 \times 2 1/2$
$\geq 1 1/8$	$d + 1/16$	$d + 5/16$	$(d + 1/16) \times (d + 3/8)$	$(d + 1/16) \times (2.5 \times d)$

4. Design for Strength of Bolted Connections

(a) **Allowable Strength.** The allowable working stress in shear and bearing, independent of the method of tightening, for A325 and A490 bolts is given in Table 2. Also given in Table 2 is the allowable working stress in axial tension for A325 and A490 bolts which are tightened to the minimum fastener tension specified in Table 4. The allowable working stresses in Table 2 are to be used in conjunction with the cross sectional area of the bolt corresponding to the nominal diameter.

(b) **Bearing Force.** The computed bearing force shall be assumed to be distributed over an area equal to the nominal bolt diameter times the thickness of the connected part.

A value of allowable bearing pressure on the connected material at a bolt greater than permitted by Table 2 can be justified provided deformation around the bolt hole is not a design consideration and adequate pitch and end distance L is provided according to:

$$F_p = LF_u/2d \leq 1.5F_u$$

Table 2. Allowable Working Stress[a] on Fasteners or Connected Material (ksi)

Load Condition	A325	A490
Applied Static Tension [b,c]. Shear on bolt with threads in shear plane. Shear on bolt without threads in shear plane.	44 21[d] 30[d]	54 28[d] 40[d]
Bearing on connected material with single bolt in line of force in a standard or short slotted hole.	$1.0F_u$[e,f,g,h]	
Bearing on connected material with 2 or more bolts in line of force in standard or short slotted holes.	$1.2F_u$[e,f,g,h]	
Bearing on connected material in long slotted holes.	$1.0F_u$[e,f,g,h]	

[a]Ultimate failure load divided by factor of safety.

[b]Bolts must be tensioned to requirements of Table 4.

[c]See 4 (d) for bolts subject to tensile fatigue.

[d]In connections transmitting axial force whose length between extreme fasteners measured parallel to the line of force exceeds 50 inches, tabulated values shall be reduced 20 percent.

[e]F_u = specified minimum tensile strength of connected part.

[f]Connections using high strength bolts in slotted holes with the load applied in a direction other than approximately normal (between 80 and 100 degrees) to the axis of the hole and connections with bolts in oversize holes shall be designed for resistance against slip at working load in accordance with Section 5.

[g]Tabulated values apply when the distance L parallel to the line of force from the center of the bolt to the edge of the connected part is not less than $1\,^1/_2 d$ and the distance from the center of a bolt to the center of an adjacent bolt is not less than $3d$. When either of these conditions is not satisfied, the distance L requirement of 4(b) determines allowable bearing stress. See Commentary.

[h]Except as may be justified under provision 4(b).

Where

d = bolt diameter

F_p = allowable bearing pressure at a bolt

F_u = specified minimum tensile strength of connected part

(c) **Prying Action.** The force in bolts required to support loads by means of direct tension shall be calculated considering the effects of the external load and any tension resulting from prying action produced by deformation of the connected parts.

(d) **Tensile Fatigue.** When subject to tensile fatigue loading, the tensile stress in the bolt due to the combined applied load and prying forces shall not exceed the following values dependent upon the bolt grade and number of cycles, and the prying force shall not exceed 60 percent of the externally applied load.

Number of Cycles	A325	A490
Not more than 20,000	44	54
From 20,000 to 500,000	40	49
More than 500,000	31	38

Bolts must be tensioned to requirements of Table 4.

5. Design Check for Slip Resistance

(a) **Slip-Critical Joints.** Slip-critical joints are defined as joints in which slip would be detrimental to the serviceability of the structure. They include:

(1) Joints subject to fatigue loading.

(2) Joints with bolts installed in oversized holes.

(3) Except where the Engineer intends otherwise and so indicates in the contract documents, joints with bolts installed in slotted holes where the force on the joint is in a direction other than normal (between approximately 80 and 100 degrees) to the axis of the slot.

(4) Joints subject to significant load reversal.

(5) Joints in which welds and bolts share in transmitting load at a common faying surface. See Commentary.

(6) Joints in which, in the judgement of the Engineer, any slip would be critical to the performance of the joint or the structure and so designated on the contract plans and specifications.

(b) **Allowable Slip Load.** In addition to the requirements of Section 4, the force on a slip-critical joint shall not exceed the allowable resistance (P_s) of the connection (See Commentary) according to:

$$P_s = F_s A_b N_b N_s$$

Where

F_s = allowable slip load per unit area of bolt from Table 3
A_b = area corresponding to the nominal body area of the bolt
N_b = number of bolts in the joint
N_s = number of slip planes

Class A, B or C surface conditions of the bolted parts as defined in Table 3 shall be used in joints designated as slip-critical except as permitted in 3(b)(3).

6. Increase in Allowable Stresses

When the applicable code or specification for design of connected members permits an increase in working stress for loads in combination with wind or seismic forces, the permitted increases in working stresses may be applied with wind or seismic forces, the permitted increases in working stresses may be applied to the allowable stresses in Sections 4 and 5. When the effect of loads in combination with wind or seismic forces are accounted for by reduction in the load factors, the allowable stresses in Sections 4 and 5 may not be increased.

7. Design Details of Bolted Connections

(a) **Standard Holes.** In the absence of approval by the engineer for use of other hole types, standard holes shall be used in high strength bolted connections.

Table 3. Allowable Load for Slip-critical Connections
(Slip Load per Unit of Bolt Area, ksi)

Contact Surface of Bolted Parts	Hole Type and Direction of Load Application							
	Any Direction				Transverse		Parallel	
	Standard		Oversize & Short Slot		Long Slots		Long Slots	
	A325	A490	A325	A490	A325	A490	A325	A490
Class A (Slip Coefficient 0.33) Clean mill scale and blast-cleaned surfaces with Class A coatings[a]	17	21	15	18	12	15	10	13
Class B (Slip Coefficient 0.50) Blast-cleaned surfaces and blast-cleaned surfaces with Class B coatings[a]	28	34	24	29	20	24	17	20
Class C (Slip Coefficient 0.40) Hot dip Galvanized and rough-ened surfaces	22	27	19	23	16	19	14	16

[a]Coatings classified as Class A or Class B includes those coatings which provide a mean slip coefficient not less than 0.33 or 0.50, respectively, as determined by Testing Method to Determine the Slip Coefficient for Coatings Used in Bolted Joints, see Appendix A.

(b) Oversize and Slotted Holes. When approved by the Engineer, oversize, short slotted holes or long slotted holes may be used subject to the following joint detail requirements:

(1) Oversize holes may be used in all plies of connections in which the allowable slip resistance of the connection is greater than the applied load.

(2) Short slotted holes may be used in any or all plies of connections designed on the basis of allowable stress on the fasteners in Table 2 provided the load is applied approximately normal (between 80 and 100 degrees) to the axis of the slot. Short slotted holes may be used without regard for the direction of applied load in any or all plies of connections in which the allowable slip resistance is greater than the applied force.

(3) Long slotted holes may be used in one of the connected parts at any individual faying surface in connections designed on the basis of allowable stress on the fasteners in Table 2 provided the load is applied approximately normal (between 80 and 100 degrees) to the axis of the slot. Long slotted holes may be used in one of the connected parts at any individual faying surface without regard for the direction of applied load on connections in which the allowable slip resistance is greater than the applied force.

(4) Fully inserted finger shims between the faying surfaces of load transmitting elements of connections are not to be considered a long slot element of a connection.

(c) **Washer Requirements.** Design details shall provide for washers in high strength bolted connections as follows:

(1) Where the outer face of the bolted parts has a slope greater than 1:20 with respect to a plane normal to the bolt axis, a hardened beveled washer shall be used to compensate for the lack of parallelism.

(2) Hardened washers are not required for connections using A325 and A490 bolts except as required in paragraphs 7(c)(3) through 7(c)(7) for slip-critical connections and connections subject to direct tension or as required by paragraph 8(c) for shear/bearing connections.

(3) Hardened washers shall be used under the element turned in tightening when the tightening is to be performed by calibrated wrench method.

(4) Irrespective of the tightening method, hardened washers shall be used under both the head and the nut when A490 bolts are to be installed and tightened to the tension specified in Table 4 in material having a specified yield point less than 40 ksi.

(5) Where A325 bolts of any diameter or A490 bolts equal to or less than 1 inch in diameter are to be installed and tightened in an oversize or short slotted hole in an outer ply, a hardened washer conforming to ASTM F436 shall be used.

(6) When A490 bolts over 1 inch in diameter are to be installed and tightened in an oversize or short slotted hole in an outer ply, hardened washers conforming to ASTM F436 except with $5/16$ inch minimum thickness shall be used under both the head and the nut in lieu of standard thickness hardened washers. Multiple hardened washers with combined thickness equal to or greater than $5/16$ inch do not satisfy this requirement.

(7) Where A325 bolts of any diameter or A490 bolts equal to or less than 1 inch in diameter are to be installed and tightened in a long slotted hole in an outer ply, a plate washer or continuous bar of at least $5/16$ inch thickness with standard holes shall be provided. These washers or bars shall have a size sufficient to completely cover the slot after installation and shall be of structural grade material, but need not be hardened except as follows. When A490 bolts over 1 inch in diameter are to be used in long slotted holes in external plies, a single hardened washer conforming to ASTM F436 but with $5/16$ inch minimum thickness shall be used in lieu of washers or bars of structural grade material. Multiple hardened washers with combined thickness equal to or greater than $5/16$ inch do not satisfy this requirement.

(8) Alternate design fasteners meeting the requirements of 2(d) with a geometry which provides a bearing circle on the head or nut with a diameter equal to or greater than the diameter of hardened washers meeting the requirements ASTM F436 satisfy the requirements for washers specified in paragraphs 7(c)(4) and 7(c)(5).

8. Installation and Tightening

(a) **Handling and Storage of Fasteners.** Fasteners shall be protected from dirt and moisture at the job site. Only as many fasteners as are anticipated to be installed and tightened during a work shift shall be taken from protected storage. Fasteners not used shall be returned to protected storage at the end of the shift. Fasteners shall not be cleaned of lubricant that is present in as-delivered condition. Fasteners for slip critical connections which must be cleaned of accumulated rust or dirt resulting from job site conditions, shall be cleaned and relubricated prior to installation.

(b) Tension Calibrator. A tension measuring device shall be required at all job sites where bolts in slip-critical joints or connections subject to direct tension are being installed and tightened. The tension measuring device shall be used to confirm: (1) the suitability to satisfy the requirements of Table 4 of the complete fastener assembly, including lubrication if required to be used in the work, (2) calibration of wrenches, if applicable, and (3) the understanding and proper use by the bolting crew of the method to be used. The frequency of confirmation testing, the number of tests to be performed and the test procedure shall be as specified in 8(d), as applicable. The accuracy of the tension measuring device shall be confirmed through calibration by an approved testing agency at least annually.

(c) Joint Assembly and Tightening of Shear/Bearing Connections. Bolts in connections not within the slip-critical category as defined in Section 5(a) nor subject to tension loads nor required to be fully tensioned bearing-type connections shall be installed in properly aligned holes, but need only be tightened to the snug tight condition. The snug tight condition is defined as the tightness that exists when all plies in a joint are in firm contact. This may be attained by a few impacts of an impact wrench or the full effort of a man using an ordinary spud wrench. See Commentary. If a slotted hole occurs in an outer ply, a flat hardened washer or common plate washer shall be installed over the slot. Bolts which may be tightened only to a snug tight condition shall be clearly identified on the drawings.

(d) Joint Assembly and Tightening of Connections Requiring Full Pre-tensioning. In slip-critical connections, connections subject to direct tension, and fully pre-tensioned bearing connections, fasteners, together with washers of size and quality specified, located as required by Section 7(c), shall be installed in properly aligned holes and tightened by one of the methods described in Subsections 8(d)(1) through 8(d)(4) to at least the minimum tension specified in Table 4 when all the fasteners are tight. Tightening may be done by turning the bolt while the nut is prevented from rotating when it is impractical to turn the nut. Impact wrenches, if used, shall be of adequate capacity and sufficiently supplied with air to perform the required tightening of each bolt in approximately 10 seconds.

(1) Turn-of-nut Tightening. When turn-of-nut tightening is used, hardened washers are not required except as may be specified in 7(c).

A representative sample of not less than three bolts and nuts of each diameter, length and grade to be used in the work shall be checked at the start of work in a device capable of indicating bolt tension. The test shall demonstrate that the method of estimating the snug-tight condition and controlling turns from snug tight to be used by the bolting crews develops a tension not less than five percent greater than the tension required by Table 4.

Bolts shall be installed in all holes of the connection and brought to a snug-tight condition. Snug tight is defined as the tightness that exist when the plies of the joint are in firm contact. This may be attained by a few

Table 4. Fastener Tension Required for Slip-critical Connections and Connections Subject to Direct Tension

Nominal Bolt Size, Inches	Minimum Tension[a] in 1000's of Pounds (kips)	
	A325 Bolts	A490 Bolts
1/2	12	15
5/8	19	24
3/4	28	35
7/8	39	49
1	51	64
1 1/8	56	80
1 1/4	71	102
1 3/8	85	121
1 1/2	103	148

[a]Equal to 70 percent of specified minimum tensile strengths of bolts (as specified in ASTM Specifications for tests of full size A325 and A490 bolts with UNC threads loaded in axial tension) rounded to the nearest kip.

impacts of an impact wrench or the full effort of a man using an ordinary spud wrench. Snug tightening shall progress systematically from the most rigid part of the connection to the free edges, and then the bolts of the connection shall be retightened in a similar systematic manner as neccessary until all bolts are simultaneously snug tight and the connection is fully compacted. Following this initial operation all bolts in the connection shall be tightened further by the applicable amount of rotation specified in Table 5. During the tightening operation there shall be no rotation of the part not turned by the wrench. Tightening shall progress systematically from the most rigid part of the joint to its free edges.

(2) Calibrated Wrench Tightening. Calibrated wrench tightening may be used only when installation procedures are calibrated on a daily basis and when a hardened washer is used under the element turned in tightening. See the Commentary to this Section. This specification does not recognize standard torques determined from tables or from formulas which are assumed to relate torque to tension.

When calibrated wrenches are used for installation, they shall be set to provide a tension not less than 5 percent in excess of the minimum tension specified in Table 4. The installation procedures shall be calibrated at least once each working day for each bolt diameter, length and grade using fastener assemblies that are being installed in the work. Calibration shall be accomplished in a device capable of indicating actual bolt tension by tightening three typical bolts of each diameter, length and grade from the bolts being installed and with a hardened washer from the washers being used in the work under the element turned in tightening. Wrenches shall be recalibrated when significant difference is noted in the surface condition of the bolts threads, nuts or washers. It shall be verified during actual installation in the assembled steelwork that the wrench adjustment selected by the calibration does not produce a nut or bolt head rotation from snug tight greater than that permitted in Table 5. If manual torque wrenches are used, nuts shall be turned in the tightening direction when torque is measured.

Table 5. Nut Rotation from Snug Tight Condition[a,b]

Bolt length (Under side of head to end of bolt)	Disposition of Outer Face of Bolted Parts		
	Both faces normal to bolt axis	One face normal to bolt axis and other sloped not more than 1:20 (beveled washer not used)	Both faces sloped not more than 1:20 from normal to the bolt axis (beveled washer not used)
Up to and including 4 diameters	1/3 turn	1/2 turn	2/3 turn
Over 4 diameters but not exceeding 8 dia.	1/2 turn	2/3 turn	5/6 turn
Over 8 diameters but not exceeding 12 dia.[c]	2/3 turn	5/6 turn	1 turn

[a]Nut rotation is relative to bolt regardless of the element (nut or bolt) being turned. For bolts installed by 1/2 turn and less, the tolerance should be plus or minus 30 degrees; for bolts installed by 2/3 turn and more, the tolerance should be plus or minus 45 degrees.

[b]Applicable only to connections in which all material within the grip of the bolt is steel.

[c]No research has been performed by the Council to establish the turn-of-nut procedure for bolt lengths exceeding 12 diameters. Therefore, the required rotation must be determined by actual test in a suitable tension measuring device which simulates conditions of solidly fitted steel.

When calibrated wrenches are used to install and tension bolts in a connection, bolts shall be installed with hardened washers under the element turned in tightening bolts in all holes of the connection and brought to a snug tight condition. Following this initial tightening operation, the connection shall be tightened using the calibrated wrench. Tightening shall progress systematically from the most rigid part of the joint to its free edges. The wrench shall be returned to "touch up" previously tightened bolts which may have been relaxed as a result of the subsequent tightening of adjacent bolts until all bolts are tightened to the prescribed amount.

(3)   Installation of Alternate Design Bolts.   When fasteners which incorporate a design feature intended to indirectly indicate the bolt tension or to automatically provide the tension required by Table 4 and which have been qualified under Section 2(d) are to be installed, a representative sample of not less than three bolts of each diameter, length and grade shall be checked at the job site in a device capable of indicating bolt tension. The test assembly shall include flat hardened washers, if required in the actual connection, arranged as in the actual connections to be tensioned. The calibration test shall demonstrate that each bolt develops a tension not less than five percent greater than the tension required by Table 4. Manufacturer's installation procedure as required by Section 2(d) shall be followed for installation of bolts in the calibration device and in all connections.

When alternate design features of the fasteners involve an irreversible mechanism such as yield or twist-off of an element, bolts shall be installed in all holes of the connection and initially brought to a snug tight condition. All fasteners shall then be tightened, progressing systematically from the most rigid part of the connection to the free edges in a manner that will minimize relaxation of previously tightened fasteners prior to final twist-off or yielding of the control or indicator element of the individual fasteners. In some cases, proper tensioning of the bolts may require more than a single cycle of systematic tightening.

(4) Direct Tension Indicator Tightening. Tightening of bolts using direct tension indicator devices is permitted provided the suitability of the device can be demonstrated by testing a representative sample of not less than three devices for each diameter and grade of fastener in a calibration device capable of indicating bolt tension. The test assembly shall include flat hardened washers, if required in the actual connection, arranged as those in the actual connections to be tensioned. The calibration test shall demonstrate that the device indicates a tension not less than five percent greater than that required by Table 4. Manufacturer's installation procedure as required by Section 2(d) shall be followed for installation of bolts in the calibration device and in all connections. Special attention shall be given to proper installation of flat hardened washers when load indicating devices are used with bolts installed in oversize or slotted holes and when the load indicating devices are used under the turned element.

When the direct tension indicator involves an irreversible mechanism such as yielding or fracture of an element, bolts shall be installed in all holes of the connection and brought to snug tight condition. All fasteners shall then be tightened, progressing systematically from the most rigid part of the connection to the free edges in a manner that will minimize relaxation of previously tightened fasteners prior to final twist-off or yielding of the control or indicator element of the individual devices. In some cases, proper tensioning of the bolts may require more than a single cycle of systematic tightening.

(e) **Reuse of Bolts.** A490 bolts and galvanized A325 bolts shall not be reused. Other A325 bolts may be reused if approved by the Engineer responsible. Touching up or retightening previously tightened bolts which may have been loosened by the tightening of adjacent bolts shall not be considered as reuse provided the snugging up continues from the initial position and does not require greater rotation, including the tolerance, than that required by Table 5.

9. Inspection

(a) **Inspector Responsibility.** While the work is in progress, the Inspector shall determine that the requirements of Sections 2, 3 and 8 of this Specification are met in the work. The Inspector shall observe the calibration procedures when such procedures are required by contract documents and shall monitor the installation of bolts to determine that all plies of connected material have been drawn together and that the selected procedure is properly used to tighten all bolts.

In addition to the requirement of the foregoing paragraph, for all connections specified to be slip critical or subject to axial tension, the Inspector shall assure that the specified procedure was followed to achieve the pretension specified in Table 4. Bolts installed by procedures in Section 8(d) may reach tensions substantially greater than values given in Table 4, but this shall not be cause for rejection.

Bolts in connections identified as not being slip-critical nor subject to direct tension need not be inspected for bolt tension other than to ensure that the plies of the connected elements have been brought into snug contact.

(b) **Arbitration Inspection.** When high strength bolts in slip-critical connections and connections subject to direct tension have been installed by any of the tightening methods in Section 8(d) and inspected in accordance with Section 9(a) and a disagreement exists as to the minimum tension of the installed bolts, the following arbitration procedure may be used. Other methods for arbitration inspection may be used if approved by the engineer.

(1) The Inspector shall use a manual torque wrench which indicates torque by means of a dial or which may be adjusted to give an indication that the job inspecting torque has been reached.

(2) This Specifcation does not recognize standard torques determined from tables or from formulas which are assumed to relate torque to tension. Testing using such standard torques shall not be considered valid.

(3) A representative sample of five bolts from the diameter, length and grade of the bolts used in the work shall be tightened in the tension measuring device by any convenient means to an initial condition equal to approximately 15 percent of the required fastener tension and then to the minimum tension specified in Table 4. Tightening beyond the initial condition must not produce greater nut rotation than 1 1/2 times that permitted in Table 5. The job inspecting torque shall be taken as the average of three values thus determined after rejecting the high and low values. The inspecting wrench shall then be applied to the tightened bolts in the work and the torque necessary to turn the nut or head 5 degrees (approximately 1 inch at 12 inch radius) in the tightening direction shall be determined.

(4) Bolts represented by the sample in the foregoing paragraph which have been tightened in the structure shall be inspected by applying, in the tightening direction, the inspecting wrench and its job torque to 10 percent of the bolts, but not less than 2 bolts, selected at random in each connection in question. If no nut or bolt head is turned by application of the job inspecting torque, the connection shall be accepted as properly tightened. If any nut or bolt is turned by the application of the job inspecting torque, all bolts in the connection shall be tested, and all bolts whose nut or head is turned by the job inspecting torque shall be tightened and reinspected. Alternatively, the fabricator or erector, at his option, may retighten all of the bolts in the connection and then resubmit the connection for the specified inspection.

(c) **Delayed Verification Inspection.** The procedure specified in Sections 9(a) and (b) are intended for inspection of bolted connections and verification of pretension at the time of tensioning the joint. If verification of bolt tension is required after a passage of a period of time and exposure of the completed joints, the procedures of Section 9(b) will provide indication of bolt tension which is of questionable accuracy. Procedures appropriate to the specific situation should be used for verification of bolt tension. This might involve use of the arbitration inspection procedure contained herein, or might require the development and use of alternate procedures.

See Commentary.

APPENDIX A

Testing Method To Determine the Slip Coefficient for Coatings Used in Bolted Joints

Reprinted from *Engineering Journal*
American Institute of Steel Construction, Third Quarter, 1985.

JOSEPH A. YURA and KARL H. FRANK

In 1975, the Steel Structures Painting Council (SSPC) contacted the Research Council In Riveted and Bolted Structural Joints (RCRBSJ), now the Research Council on Structural Connections (RCSC), regarding the difficulties and costs which steel fabricators encounter with restrictions on coatings of contact surfaces for friction-type structural joints. The SSPC also expressed the need for a "standardized test which can be conducted by any certified testing agency at the initiative and expense of any interested party, including the paint manufacturer." And finally, the RCSC was requested to "prepare and promulgate a specification for the conduct of such a standard test for slip coefficients."

The following Testing Method is the answer of Research Council on Structural Connections to the SSPC request. The test method was developed by Professors Joseph A. Yura and Karl H. Frank of The University of Texas at Austin under a grant from the Federal Highway Administration. The Testing Method was approved by the RCSC on June 14, 1984.

1.0 GENERAL PROVISIONS

1.1 Purpose and Scope

The purpose of the testing procedure is to determine the slip coefficient of a coating for use in high-strength bolted connections. The testing specification ensures that the creep deformation of the coating due to both the clamping force of the bolt and the service load joint shear are such that the coating will provide satisfactory performance under sustained loading.

Joseph A. Yura, M. ASCE, is Warren S. Bellows Centennial Professor in Civil Engineering, University of Texas at Austin, Austin, Texas.

Karl H. Frank, A.M. ASCE, is Associate Professor, Department of Civil Engineering, University of Texas at Austin, Austin, Texas.

1.2 Definition of Essential Variables

Essential variables mean those variables which, if changed, will require retesting of the coating to determine its slip coefficient. The essential variables are given below. The relationship of these variables to the limitation of application of the coating for structural joints is also given.

The *time interval* between application of the coating and the time of testing is an essential variable. The time interval must be recorded in hours and any special curing procedures detailed. Curing according to published manufacturer's recommendations would not be considered a special curing procedure. The coatings are qualified for use in structural connections which are assembled after coating for a time equal to or greater than the interval used in the test specimens. Special curing conditions used in the test specimens will also apply to the use of the coating in the structural connections.

The *coating thickness* is an essential variable. The maximum average coating thickness allowed on the bolted structure will be the average thickness, rounded to the nearest whole mil, of the coating used on the creep test specimens minus 2 mils.

The *composition of the coating,* including the thinners used, and its method of manufacture are essential variables. Any change will require retesting of the coating.

1.3 Retesting

A coating which fails to meet the creep or the post-creep slip test requirements given in Sect. 4 may be retested in accordance with methods in Sect. 4 at a lower slip coefficient, without repeating the static short-term tests specified in Sect. 3. Essential variables must remain unchanged in the retest.

2.0 TEST PLATES AND COATING OF THE SPECIMENS

2.1 Test Plates

The test specimen plates for the short-term static tests are shown in Fig. 1. The plates are 4×4 in. (102×102 mm) plates, 5/8-in. (16 mm) thick, with a 1-in. (25 mm) dia. hole drilled 1 1/2 in. ± 1/16 in. (38 mm ± 1.6 mm) from one edge. The specimen plates for the creep specimen are shown in Fig. 2. The plates are 4×7 in. (102×178 mm), 5/8-in. (16 mm) thick, with two 1-in. (25 mm) holes, 1 1/2 in. ± 1/16 in. (38 mm ± 1.6 mm) from each end. The edges of the plates may be milled, as rolled or saw cut.

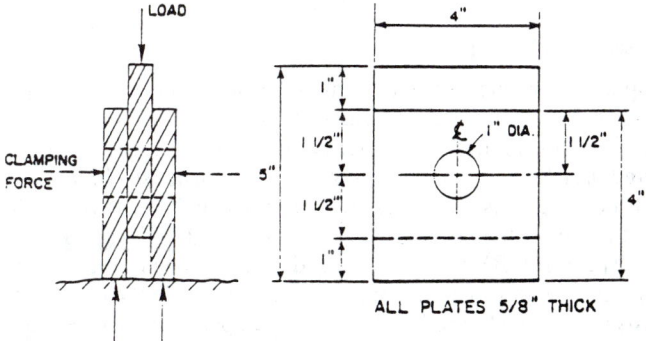

Fig. 1. Compression test specimen

Flame cut edges are not permitted. The plates should be flat enough to ensure they will be in reasonably full contact over the faying surface. Any burrs, lips or rough edges should be filed or milled flat. The arrangement of the specimen plates for the testing is shown in Figs. 2 and 3. The plates are to be fabricated from a steel with a minimum yield strength between 36 to 50 ksi (250 to 350 MPa).

If specimens with more than one bolt are desired, the contact surface per bolt should be 4×3 in. (102×76.5 mm) as shown for the single bolt specimen in Fig. 1.

2.2 Specimen Coating

The coatings are to be applied to the specimens in a manner consistent with the actual intended structural application. The method of applying the coating and the surface preparation should be given in the test report. The specimens are to be coated to an average thickness 2 mils (0.05 mm) greater than average thickness to be used in the structure. The thickness of the total coating and the primer, if used, shall be measured on the contact surface of the specimens. The thickness should be measured in accordance with the Steel Structures Painting Council specification SSPC-PA2, Measurement of Dry Paint Thickness with Magnetic Gages.[1] Two spot readings (six gage readings) should be made for each contact surface. The overall average thickness from the three plates comprising a specimen is the average thickness for the specimen. This value should be reported for each specimen. The average coating thickness of the three creep specimens will be calculated and reported. The average thickness of the creep specimen minus two mils rounded to the nearest whole mil is the maximum average thickness of the coating to be used in the faying surface of a structure.

The time between painting and specimen assembly is to be the same for all specimens within ± 4 hours. The average time is to be calculated and reported. The two coating applications required in Sect. 3 are to use the same equipment and procedures.

3.0 SLIP TESTS

The methods and procedures described herein are used to determine experimentally the slip coefficient (sometimes called the coefficient of friction) under short-term static loading for high-strength bolted connections. The slip coefficient will be determined by testing two sets of five specimens. The two sets are to be coated at different times at least one week apart.

3.1 Compression Test Setup

The test setup shown in Fig. 3 has two major loading components, one to apply a clamping force to the specimen plates and another to apply a compressive load to the specimen so that the load is transferred across the faying surfaces by friction.

Clamping Force System. The clamping force system consists of a ⁷/₈-in. (22 mm) dia. threaded rod which passes through the specimen and a centerhole compression ram. A 2H nut is used at both ends of the rod, and a hardened washer is used at each side of the test specimen. Between the ram and the specimen is a specially fabricated ⁷/₈-in. (22 mm) 2H nut in which the threads have been drilled out so that it will slide with little resistance along the rod. When oil is pumped into the centerhole ram, the piston rod extends, thus forcing the special nut against one of the outside plates of the specimen. This action puts tension in the threaded rod and applies a clamping force to

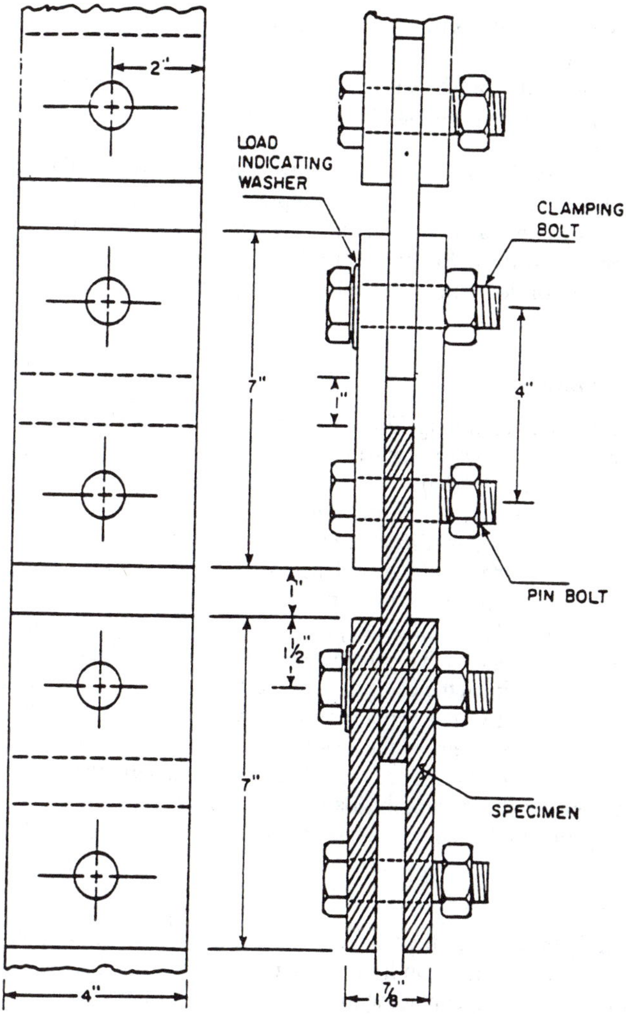

Fig. 2. *Creep test specimens*

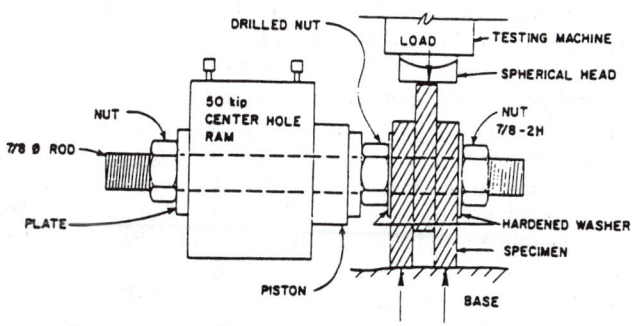

Fig. 3. *Test setup*

the specimen which simulates the effect of a tightened bolt. If the diameter of the centerhole ram is greater than 1 in. (25 mm), additional plate washers will be necessary at the ends of the ram. The clamping force system must have a capability to apply a load of at least 49 kips (219 kN) and maintain this load during the test with an accuracy of ±1%.

Compressive Load System. A compressive load is applied to the specimen until slip occurs. This compressive load can be applied by a compression test machine or compression ram. The machine, ram and the necessary supporting elements should be able to support a force of 90 kips (400 kN).

The compression loading system should have an accuracy of 1.0% of the slip load.

3.2 Instrumentation

Clamping Force. The clamping force must be measured within 0.5 kips (2.2 kN). This may be accomplished by measuring the pressure in the calibrated ram or placing a load cell in series with the ram.

Compression Load. The compression load must be measured during the test. This may be accomplished by direct reading from a compression testing machine, a load cell in series with the specimen and the compression loading device, or pressure readings on a calibrated compression ram.

Slip Deformation. The relative displacement of the center plate and the two outside plates must be measured. This displacement, called slip for simplicity, should be the average which occurs at the centerline of the specimen. This can be accomplished by using the average of two gages placed on the two exposed edges of the specimen or by monitoring the movement of the loading head relative to the base. If the latter method is used, due regard must be taken for any slack that may be present in the loading system prior to application of the load. Deflections can be measured by dial gages or any other calibrated device which has an accuracy of 0.001 in. (0.20 mm).

3.3 Test Procedure

The specimen is installed in the test setup as shown in Fig. 3. Before the hydraulic clamping force is applied, the individual plates should be positioned so that they are in, or are close to, bearing contact with the 7/8-in. (22 mm) threaded rod in a direction opposite to the planned compressive loading to ensure obvious slip deformation. Care should be taken in positioning the two outside plates so that the specimen will be straight and both plates are in contact with the base.

After the plates are positioned, the centerhole ram is engaged to produce a clamping force of 49 kips (219 kN). The applied clamping force should be maintained within ±0.5 kips (2.2 kN) during the test until slip occurs.

The spherical head of the compression loading machine should be brought in contact with the center plate of the specimen after the clamping force is applied. The spherical head or other appropriate device ensures uniform contact along the edge of the plate, thus eliminating eccentric loading. When 1 kip (4.45 kN) or less of compressive load is applied, the slip gages should be engaged or attached. The purpose of engaging the deflection gage(s), after a slight load is applied, is to eliminate initial specimen settling deformation from the slip readings.

When the slip gages are in place, the compression load is applied at a rate not exceeding 25 kips (109 kN) per minute, or 0.003 in. (0.07 mm) of slip displacement per minute until the slip load is reached. The test should be terminated when a slip of

0.05 in. (1.3 mm) or greater is recorded. The load-slip relationship should preferably be monitored continuously on an *X-Y* plotter throughout the test, but in lieu of continuous data, sufficient load-slip data must be recorded to evaluate the slip load defined below.

3.4 Slip Load

Typical load-slip response is shown in Fig. 4. Three types of curves are usually observed and the slip load associated with each type is defined as follows:

Curve (a).   Slip load is the maximum load, provided this maximum occurs before a slip of 0.02 in. (0.5 mm) is recorded.

Curve (b).   Slip load is the load at which the slip rate increases suddenly.

Curve (c).   Slip load is the load corresponding to a deformation of 0.02 in. (0.5 mm). This definition applies when the load vs. slip curves show a gradual change in response.

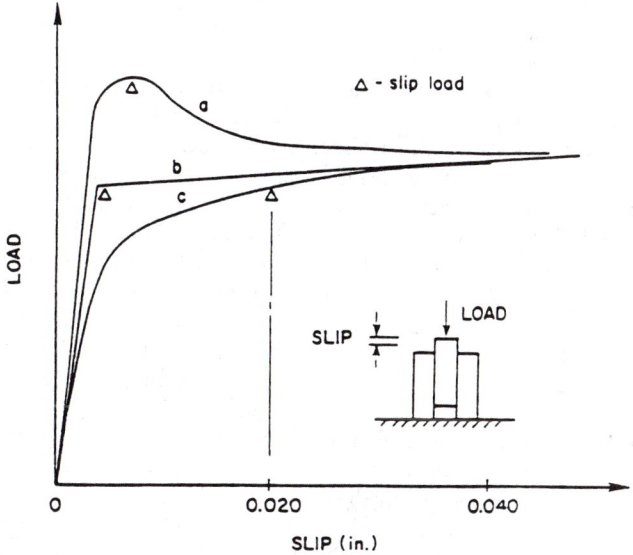

Fig. 4.   Definition of slip load

3.5 Coefficient of Slip

The slip coefficient k_s for an individual specimen is calculated as follows:

$$k_s = \frac{\text{slip load}}{2 \times \text{clamping force}}$$

The mean slip coefficient for both sets of five specimens must be compared. If the two means differ by more than 25%, using the smaller mean as the base, a third five-specimen set must be tested. The mean and standard deviation of the data from all specimens tested define the slip coefficient of the coating.

3.6 Alternate Test Methods

Other test meethods to determine slip may be used provided the accuracy of load measurement and clamping satisfies the conditions presented in the previous sections. For example, the slip load may be determined from a tension-type test setup rather than the compression-type as long as the contact surface area per fastener of the test specimen is the same as shown in Fig. 1. The clamping force of at least 49 kips (219 kN) may be applied by any means provided the force can be established within ±1%. Strain-gaged bolts can usually provide the desired accuracy. However, bolts installed by turn-of-nut method, tension indicating fasteners and load indicator washers usually show too much variation to be used in the slip test.

4.0 TENSION CREEP TESTS

The test method outlined is intended to ensure the coating will not undergo significant creep deformation under service loading. The test also determines the loss in clamping force in the fastener due to the compression or creep of the paint. Three replicate specimens are to be tested.

4.1 Test Setup

Tension-type specimens, as shown in Fig. 2, are to be used. The replicate specimens are to be linked togther in a single chain-like arrangement, using loose pin bolts, so the same load is applied to all specimens. The specimens shall be assembled so the specimen plates are bearing against the bolt in a direction opposite to the applied tension loading. Care should be taken in the assembly of the specimens to ensure the centerline of the holes used to accept the pin bolts is in line with the bolts used to assemble the joint. The load level, specified in Sect. 4.2, shall be maintained constant within ±1% by springs, load maintainers, servo controllers, dead weights or other suitable equipment. The bolts used to clamp the specimens together shall be 7/8-in. (22 mm) dia. A490 bolts. All bolts should come from the same lot.

The clamping force in the bolts should be a minimum of 49 kips (219 kN). The clamping force is to be determined by calibrating the bolt force with bolt elongation, if standard bolts are used. Special fasteners which control the clamping force by other means such as bolt torque or strain gages may be used. A minimum of three bolt calibrations must be performed using the technique selected for bolt force determination. The average of the three-bolt calibration is to be calculated and reported. The method of measuring bolt force must ensure the clamping force is within ±2 kips (9 kN) of the average value.

The relative slip between the outside plates and the center plates shall be measured to an accuracy of 0.001 in. (0.02 mm). This is to be measured on both sides of each specimen.

4.2 Test Procedure

The load to be placed on the creep specimens is the service load permitted for 7/8-in. A490 bolts in slip-critical connections by the latest edition of the *Specification for Structural Joints Using ASTM A325 or A490 Bolts*[2] for the particular slip coefficient category under consideration. The load is to be placed on the specimen and held for 1,000 hours. The creep deformation of a specimen is calculated using the average reading of the two dispacements on each side of the specimen. The difference between the average after 1,000 hours and the initial average reading taken within one-half hour after loading the specimens is defined as the creep deformation of the specimen. This value is to be reported for each specimen. If the creep deformation of any specimen exceeds 0.005 in. (0.12 mm), the coating has failed the test for the slip coefficient used. The coating may be retested using new specimens in accordance with this section at a load corresponding to a lower value of slip coefficient.

If the value of creep deformation is less than 0.005 in. (0.12 mm) for all specimens, the specimens are to be loaded in tension to a load calculated as

$$P_u = \text{average clamping force} \times \text{design slip coefficient} \times 2$$

since there are two slip planes. The average slip deformation which occurs at this load must be less than 0.015 in. (0.38 mm) for the three specimens. If the deformation is greater than this value, the coating is considered to have failed to meet the requirements for the particular slip coefficient used. The value of deformation for each specimen is to be reported.

COMMENTARY

The slip coefficient under short-term static loading has been found to be independent of clamping force, paint thickness and hole diameter.[3] The slip coefficient can be easily determined using the hydraulic bolt test setup included in this specification. The sip load measured in this setup yields the slip coefficient directly since the clamping force is controlled. The slip coefficient k_s is given by

$$k_s = \frac{\text{slip load}}{2 \times \text{clamping force}}$$

The resulting slip coefficient has been found to correlate with both tension and compression tests of bolted specimens. However, tests of bolted specimens revealed that the clamping force may not be constant but decreases with time due to the compressive creep of the coating on the faying surfaces and under the nut and bolt head. The reduction of the clamping force can be considerable for joints with high clamping force and thick coatings, as much as a 20% loss. This reduction in clamping force causes a corresponding reduction in the slip load. The resulting reduction in slip load must be considered in the procedure used to determine the design allowable slip loads for the coating.

The loss in clamping force is a characteristic of the coating. Consequently, it cannot be accounted for by an increase in the factor of safety or a reduction in the clamping force used for design without unduly penalizing coatings which do not exhibit this behavior.

The creep deformation of the bolted joint under the applied shear loading is also an important characteristic and a function of the coating applied. Thicker coatings tend to creep more than thinner coatings. Rate of creep deformation increases as the applied load approaches the slip load. Extensive testing has shown the rate of creep is not constant with time, rather it decreases with time. After 1,000 hours of loading, the additional creep deformation is negligible.

The proposed test methods are designed to provide the necessary information to evaluate the suitability of a coating for slip critical bolted connections and to determine the slip coefficient to be used in the design of the connections. The initial testing of the compression specimens provides a measure of the scatter of the slip coefficient. In order to get better statistical information, a third set of specimens must be tested whenever the means of the initial two sets differ by more than 25%.

The creep tests are designed to measure the paint's creep behavior under the service loads determined by the paint's slip coefficient based on the compression test results. The slip test conducted at the conclusion of the creep test is to ensure the loss of clamping force in the bolt does not reduce the slip load below that associated with the design slip coefficient. A490 bolts are specified, since the loss of clamping force is larger for these bolts than A325 bolts. Qualifying of the paint for use in a structure at an averge thickness of 2 mils less than the test specimen is to ensure that a casual buildup of paint due to overspray, etc., does not jeopardize the coating's performance.

The use of 1-in. (25 mm) holes in the specimens is to ensure that adequate clearance is available for slip. Fabrication tolerances, coating buildup on the holes and assembly tolerances reduce the apparent clearances.

REFERENCES

1. *Steel Structures Painting Council* Steel Structures Painting Manual *Vols. 1 and 2. Pittsburgh, Pa., 1982.*
2. *Research Council on Structural Connections* Specification for Structural Joints Using ASTM A325 or A490 Bolts *American Institute of Steel Construction., Inc., Chicago, Ill., November 1985.*
3. *Frank, K.H., and J. A. Yura* An Experimental Study of Bolted Shear Connections *FHWA/RD-81-148, Federal Highway Administration, Washington, D.C., December 1981.*

---

*Presently identified as A502, Grade 1.

Commentary on Specifications for Structural Joints Using ASTM A325 or A490 Bolts

November 13, 1985

PREFACE

This Commentary to *Specifications for Structural Joints Using ASTM A325 or A490 Bolts* is equally applicable to the *Allowable Stress Design* version or the *Load and Resistance Factor Design* version. It is provided as an aid to the user of either Specification. By providing background information, references to source information and discussion relative to questions raised over the years by users of earlier versions of the Council Specifications, it is intended to provide a record of the reasoning behind the requirements and understanding of the intent of the Specification provisions.

Historical Notes

When first approved by the Research Council on Structural Connections of the Engineering Foundation, January 1951, the *Specification for Assembly of Structural Joints Using High-Strength Bolts* merely permitted the substitution of a like number of A325 high-strength bolts for hot driven ASTM A141* steel rivets of the same nominal diameter. It was required that all contact surfaces be free of paint. As revised in 1954, the omission of paint was required to apply only to "joints subject to stress reversal, impact or vibration, or to cases where stress redistribution due to joint slippage would be undesirable." This relaxation of the earlier provision recognized the fact that, in a great many cases, movement of the connected parts that brings the bolts into bearing against the sides of their holes is in no way detrimental.

In the first edition of the Specification published in 1951, a table of torque to tension relationships for bolts of various diameters was included. It was soon demonstrated in research that a variation in the torque to tension relationship of as high as plus or minus 40 percent must be anticipated unless the relationship is established individually for each bolt lot, diameter and fastener condition. Hence, by the 1954 edition of the Specification, recognition of standard torque to tension relationships in the form of tabulated values or formulas was withdrawn. Recognition of the calibrated wrench method of tightening was retained however until 1980, but with the requirement that the torque required for installation or inspection be determined specifically for the bolts being installed on a daily basis. Recognition of the method was withdrawn in 1980 because of continuing controversy resulting from failure of users to adhere to the detailed requirements for valid use of the method both during installation and inspection. With this version of the Specification, the calibrated wrench method has been reinstated, but with more detailed requirements which should be carefully followed.

The increasing use of high-strength steels created the need for bolts substantially stronger than A325, in order to resist the much greater forces they support without resort to very large connections. To meet this need, a new ASTM specification, A490, was developed. When provisions for the use of these bolts were included in this Specification in 1964, it was required that they be tightened to their specified proof load, as was required for the installation of A325 bolts. However, the ratio of proof load to specified minimum tensile strength is approximately 0.7 for A325 bolts, whereas it is 0.8 for A490 bolts. Calibration studies have shown that high-strength bolts have ultimate load capacities in torqued tension which vary from about 80 to 90 percent of the pure-tension tensile strength.[1] Hence, if minimum strength A490 bolts were supplied and they experienced the maximum reduction due to torque required to induce the tension, there is a possibility that these bolts could not be tightened to proof load by any method of installation. Also, statistical studies have shown that tightening to the 0.8 times tensile strength under calibrated wrench control may result in some "twist-off" bolt failures during installation or in some cases a slight amount of under tightening.[2] Therefore, the required installed tension for A490 bolts was reduced to 70 percent of the specified minimum tensile strength. For consistency, but with only minor change, the initial tension required for A325 bolts was also set at 70 percent of their specified minimum tensile strength and at the same time the values for minimum required pretension were rounded off to the nearest kip.

C1 Scope

This Specification deals only with two types of high-strength bolts, namely, ASTM A325 and A490, and to their installation in structural steel joints. The provisions may not be relied upon for high-strength fasteners of other chemical composition or mechanical properties or size. The provisions do not apply to ASTM A325 or A490 fasteners when material other than steel is included in the grip. The provisions do not apply to high-strength anchor bolts.

The Specification relates only to the performance of fasteners in structural steel connections and those few aspects of the connected material that affect the performance of the fasteners in connections. Many other aspects of connection design and fabrication are of equal importance and must not be overlooked. For information on questions of design of connected material, not covered herein, the user is directed to standard textbooks on design of structural steel and also to "Fisher, J.W. and J.H.A. Struik," Guide to Design Criteria for Bolted and Riveted Joints, John Wiley & Sons, New York, 1974. (Hereinafter referred to as the Guide.)

C2 Bolts, Nuts, Washers and Paint

Complete familiarity with the referenced ASTM Specification requirements is necessary for the proper application of this Specification. Discussion of referenced specifications in this Commentary is limited to only a few frequently overlooked or little understood items.

In this Specification a single style of fastener (heavy hex structural bolts with heavy hex nuts), available in two strength grades (A325 and A490) is specified as a principal style, but conditions for acceptance of other types of fasteners are provided.

Bolt Specifications ASTM A325 and A490 bolts are manufactured to dimensions specified in ANSI Standard B18.2.1 for Heavy Hex Structural Bolts. The basic dimensions as defined in Figure C1 are shown in Table C1. The principal geometric features of heavy hex structural bolts that distinguish them from bolts for general application are the size of the head and the body length. The head of the heavy hex

---

[1]Christopher, R.J., G. L. Kulak, and J. W. Fisher, "Calibration of Alloy Steel Bolts," *Journal of the Structural Division*, ASCE, Vol. 92, No. ST2, Proc. Paper 4768, April, 1966, pp. 19–40.

[2]Gill, P.J., "Specifications of Minimum Preloads for Structural Bolts," Memorandum 30, G. K. N. Group Research Laboratory, England, 1966 (unpublished report).

structural bolt is specified to be the same size as a heavy hex nut of the same nominal diameter in order that the ironworker may use a single size wrench or socket on both the head and the nut. Heavy hex structural bolts have shorter thread length than bolts for general application. By making the body length of the bolt the control dimension, it has been possible to exclude the thread from all shear planes, except in the case of thin outside parts adjacent to the nut. Depending upon the amount of bolt length added to adjust for incremental stock lengths, the full thread may extend into the grip by as much as ³/₈ inch for ¹/₂ in., ⁵/₈ in., ³/₄ in., ⁷/₈ in., 1¹/₄ in., and 1¹/₂ in. diameter bolts and as much as ¹/₂ inch for 1 in., 1¹/₈ in. and 1³/₈ in. diameter bolts. Inclusion of some thread run-out in the plane of shear is permissible. Of equal or even greater importance is exercise of care to provide sufficient thread for nut tightening to keep the nut threads from jamming into the thread run-out. When the thickness of an outside part is less than the amount the threads may extend into the grip tabulated above, it may be necessary to call for the next increment of bolt length together with sufficient flat washers to insure full tightening of the nut without jamming nut threads on the thread run-out.

Table C1

Nominal Bolt Size, Inches D	Bolt Dimensions, Inches Heavy Hex Structural Bolts			Nut Dimensions, Inches Heavy Hex nuts	
	Width across flats, F	Height, H	Thread length	Width across flats, W	Height, H
¹/₂	⁷/₈	⁵/₁₆	1	⁷/₈	³¹/₆₄
⁵/₈	1¹/₁₆	²⁵/₆₄	1¹/₄	1¹/₁₆	³⁹/₆₄
³/₄	1¹/₄	¹⁵/₃₂	1³/₈	1¹/₄	⁴⁷/₆₄
⁷/₈	1 ⁷/₁₆	³⁵/₆₄	1¹/₂	1⁷/₁₆	⁵⁵/₆₄
1	1⁵/₈	³⁹/₆₄	1³/₄	1⁵/₈	⁶³/₆₄
1¹/₈	1¹³/₁₆	¹¹/₁₆	2	1¹³/₁₆	1⁷/₆₄
1¹/₄	2	²⁵/₃₂	2	2	1⁷/₃₂
1³/₈	2³/₁₆	²⁷/₃₂	2¹/₄	2³/₁₆	1¹¹/₃₂
¹/₂	2³/₈	¹⁵/₁₆	2¹/₄	2³/₈	1¹⁵/₃₂

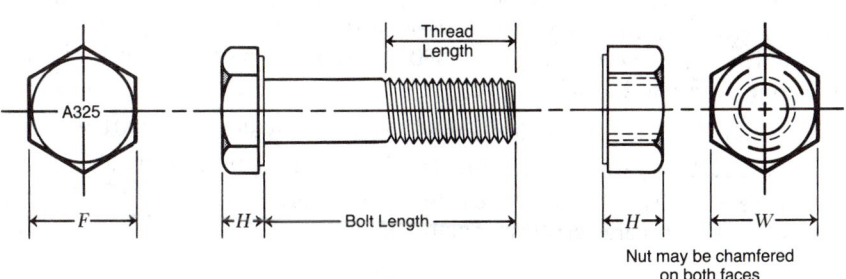

Fig. C1.   Heavy hex structural bolt and heavy hex nut

There is an exception to the short thread length requirements for ASTM A325 bolts discussed in the foregoing. Beginning with ASTM A325-83, supplementary requirements have been added to the ASTM A325 Specification which permit the purchaser, when the bolt length is equal to or shorter than four times the nominal diameter, to specify that the bolt be threaded for the full length of the shank. This exception to the requirements for thread length of heavy hex structural bolts was provided in the Specification in order to increase economy through simplified ordering and inventory control in the fabrication and erection of structures using relatively thin materials where strength of the connection is not dependent upon shear strength of the bolt, whether threads are in the shear plane or not. The Specification requires that bolts ordered to such supplementary requirements be marked with the symbol A325T.

In order to determine the required bolt length, the value shown in Table C2 should be added to the grip (i.e., the total thickness of all connected material, exclusive of washers). For each hardened flat washer that is used, add $5/32$ inch, and for each beveled washer add $5/16$ inch. The tabulated values provide appropriate allowances for manufacturing tolerances, and also provide for full thread engagement* with an installed heavy hex nut. The length determined by the use of Table C2 should be adjusted to the next longer $1/4$ inch length.

Table C2

Nominal Bolt Size, Inches	To Determine Required Bolt Length, Add to Grip, in Inches
$1/2$	$11/16$
$5/8$	$7/8$
$3/4$	1
$7/8$	$1 1/8$
1	$1 1/4$
$1 1/8$	$1 1/2$
$1 1/4$	$1 5/8$
$1 3/8$	$1 3/4$
$1 1/2$	$1 7/8$

ASTM A325 and ASTM A490 currently provide for three types (according to metallurgical classification) of high-strength structural bolts, supplied in sizes $1/2$ inch to $1 1/2$ inch inclusive except for A490 Type 2 bolts which are available in diameters from $1/2$ inch to 1 inch inclusive:

Type 1. Medium carbon steel for A325 bolts, alloy steel for A490 bolts.

Type 2. Low carbon martensitic steel for both A325 and A490 bolts.

Type 3. Bolts having improved atmospheric corrosion resistance and weathering characteristics for both A325 and A490 bolts.

---

*Defined as: Having the end of the bolt at least flush with the face of the nut.

When the bolt type is not specified, either Type 1, Type 2 or Type 3 may be supplied at the option of the manufacturer. Special attention is called to the requirement in ASTM A325 that where elevated temperature applications are involved, Type 1 bolts shall be specified by the purchaser. This is because the chemistry of Type 2 bolts permits heat treatment at sufficiently low temperatures that subsequent heating to elevated temperatures may affect the mechanical properties.

Heavy Hex Nuts. Heavy hex nuts for use with A325 bolts may be manufactured to the requirements of ASTM A194 for grades 2 or 2H or the requirements of ASTM A563 for grades C, C3, D, DH or DH3 except that nuts to be galvanized for use with galvanized bolts must be hardened nuts meeting the requirements for 2H, DH or DH3.

The heavy hex nuts for use with A490 bolts may be manufactured to the requirements of ASTM A194 for grade 2H or the requirements of ASTM A563 for grade DH or DH3.

Galvanized High-strength Bolts. Galvanized high-strength bolts and nuts must be considered as a manufactured matched assembly; hence, comments relative to them have not been included in the foregoing paragraphs where bolts and nuts were considered separately. Insofar as the galvanized bolt and nut assembly, per se, is concerned, four principal factors need be discussed in order that the provisions of the Specification may be understood and properly applied. They are (1) the effect of the galvanizing process on the mechanical properties of high-strength steels, (2) the effect of galvanized coatings on the nut stripping strength, (3) the effect of galvanizing upon the torque involved in the tightening operation and (4) shipping requirements.

Effect of Galvanizing on the Strength of Steels. Steels in the 200 ksi and higher tensile strength range are subject to embrittlement if hydrogen is permitted to remain in the steel and the steel is subjected to high tensile stress. The minimum tensile strength of A325 bolts is 105 or 120 ksi, depending upon the size, comfortably below the critical range. The required minimum tensile strength for A490 bolts was set at 170 ksi in order to provide a little more than a ten percent margin below 200 ksi; however, because manufacturers must target their production slightly higher than the required minimum, A490 bolts close to the critical range of tensile strength must be anticipated. For black bolts this is not a cause for concern, but, if the bolt is galvanized, a hazard of delayed brittle fracture in service exists because of the real possibility of introduction of hydrogen into the steel during the pickling operation of the galvanizing process and the subsequent "sealing-in" of the hydrogen by the zinc coating. ASTM specifications provide for the galvanizing of A325 bolts but not A490 bolts. Galvanizing of A490 bolts is not permitted.

The heat treatment temperatures for Type 2 ASTM A325 bolts is in the range of the molten zinc temperatures for hot-dip galvanizing; therefore there is a potential for diminishing the heat treated mechanical properties of Type 2 A325 bolts by the galvanizing process. For this reason, the Specifications require that such fasteners be tension tested after galvanizing to check the mechanical properties. Special attention should be given to specifying only Type 1 bolts for hot-dip galvanizing or to assuring that this requirement has not been overlooked if galvanized A325 bolts with Type 2 head markings are supplied. Because it is recommended that A490 bolts not be hot-dip galvanized, a similar requirement is not part of the ASTM Specification for Type 2 A490 bolts.

Nut Stripping Strength. Hot-dip galvanizing affects the stripping strength of the nut-bolt assembly primarily because to accommodate the relatively thick zinc coatings on bolt threads, it is usual practice to galvanize the blank nut and then to tap the nut oversize after galvanizing. This overtapping results in a reduction in the amount of engagement between the steel portions of the male and female threads with a consequent approximately 25 percent reduction in the stripping strength. Only the stronger hardened nuts have adequate strength to meet specification requirements even with the reduction due to overtapping; therefore, ASTM A325 specifies that only Grades DH and 2H be used for galvanized nuts. This requirement should not be overlooked if non-galvanized nuts are purchased and then sent to a local galvanizer for hot-dip galvanizing.

Effect of Galvanizing upon Torque Involved in Tightening. Research[3] has shown that, as galvanized, hot-dip galvanizing both increases the friction between the bolt and nut threads and also makes the torque induced tension much more variable. Lower torque and more consistent results are provided if the nuts are lubricated; thus, ASTM A325 requires that a galvanized bolt and a tapped oversize lubricated galvanized nut intended to be used with the bolt shall be assembled in a steel joint with a galvanized washer and tested in accordance with ASTM A536 by the manufacturer prior to shipment to assure that the galvanized nut with the lubricant provided may be rotated from the snug tight condition well in excess of the rotation required for full tensioning of the bolts without stripping.

Shipping Requirements for Galvanized Bolts and Nuts. The above requirements clearly indicate that galvanized bolts and nuts are to be treated as an assembly and shipped together. Purchase of galvanized bolts and galvanized nuts from separate sources is not recommended because the amount of over tapping appropriate for the bolt and the testing and application of lubricant would cease to be under the control of a single supplier and the responsibility for proper performance of the nut/bolt assembly would become obscure. Because some of the lubricants used to meet the requirements of ASTM Specification are water soluble, it is advisable that galvanized bolts and nuts be shipped and stored in plastic bags in wood or metal containers.

Washers. The primary function of washers is to provide a hardened non-galling surface under the element turned in tightening for those installation procedures which depend upon torque for control. Circular hardened washers meeting the requirements of ASTM A436 provide an increase in bearing area of 45 to 55 percent over the area provided by a heavy hex bolt head or nut; however, tests have shown that standard thickness washers play only a minor role in distributing the pressure induced by the bolt pretension, except where oversize or short slotted holes are used. Hence, consideration is given to this function only in the case of oversize and short slotted holes. The requirement for standard thickness hardened washers, when such washers are specified as an aid in the distribution of pressure, is waived for alternate design fasteners which incorporate a bearing surface under the head of the same diameter as the hardened washer; however, the requirements for hardened washers to satisfy the principal requirement of providing a non-galling surface under the element turned in tightening is not waived. The maximum thickness is the same for all standard washers up to and including 1 1/2 inch bolt diameter in order that washers may be produced from a single stock of material.

---

[3]Birkemoe, P. C., and D. C. Herrschaft, "Bolted Galvanized Bridges—Engineering Acceptance Near," *Civil Engineering*, ASCE, April 1970.

The requirement that heat-treated washers not less than $5/16$ inch thick be used to cover oversize and slotted holes in external plies, when A490 bolts of $1\,1/8$ inch or larger diameter are used, was found necessary to distribute the high clamping pressure so as to prevent collapse of the hole perimeter and enable development of the desired clamping force. Preliminary investigation has shown that a similar but less severe deformation occurs when oversize or slotted holes are in the interior plies. The reduction in clamping force may be offset by "keying" which tends to increase the resistance to slip. These effects are accentuated in joints of thin plies.

Marking. Heavy hex structural bolts and heavy hex nuts are required by ASTM Specifications to be distinctively marked. Certain markings are mandatory. In addition to the mandatory markings the manufacturer may apply other distinguishing markings. The mandatory and optional markings are shown in Figure C2.

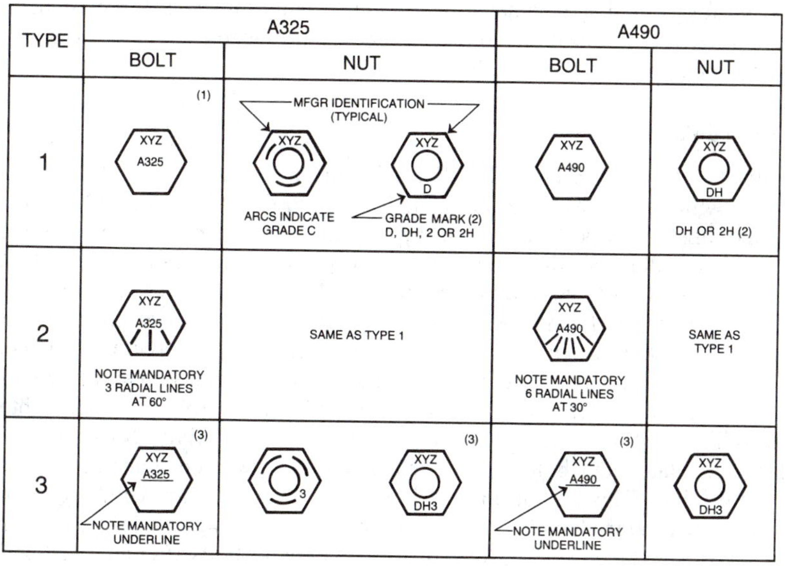

(1) ADDITIONAL OPTIONAL 3 RADIAL LINES AT 120° MAY BE ADDED.
(2) TYPE 3 ALSO ACCEPTABLE.
(3) ADDITIONAL OPTIONAL MARK INDICATING WEATHERING GRADE MAY BE ADDED.

Fig. C2. Required marking for acceptable bolt and nut assemblies

Paint. In the previous edition of the Specification, generic names for paints applied to faying surfaces was the basis for categories of allowable working stresses in "friction" type connections. Research[4] completed since the adoption of the 1980 Specification has demonstrated that the slip coefficients for paints described by a generic type are not single values but depend also upon the type vehicle used. Small differences in formulation from manufacturer to manufacturer or from lot to lot with a single manufacturer, if certain essential variables within a generic type were changed, significantly affected slip coefficients; hence it is unrealistic to assign paints to categories with relatively small incremental differences between categories based solely upon a generic description. As a result of the research, a test method was developed

---

[4]Frank, Karl H. and J. A. Yura, "An Experimental Study of Bolted Shear Connections," FHWA/RD-81/148, Dec. 1981.

and adopted by the Council titled "Test Method to Determine the Slip Coefficient for Coatings Used in Bolted Joints." A copy of this document is appended to this Specification as Appendix A. The method, which requires requalification if an essential variable is changed, is the sole basis for qualification of any paint to be used under this Specification. Further, normally only 2 categories of slip coefficient for paints to be used in slip critical joints are recognized, Class A for coatings which do not reduce the slip coefficient below that provided by clean mill scale, and Class B for paints which do not reduce the slip coefficient below that of blast-cleaned steel surfaces.

The research cited in the preceding paragraph also investigated the effect of varying the time from coating the faying surfaces to assembly of the connection and tightening the bolts. The purpose was to ascertain if partially cured paint continued to cure within the assembled joint over a period of time. It was learned that all curing ceased at the time the joint was assembled and tightened and that paint coatings that were not fully cured acted much as a lubricant would; thus, the slip resistance of the joint was severely reduced from that which was provided by faying surfaces which were fully cured prior to assembly.

C3 Bolted Parts

Material within the Grip. The Specification is intended to apply to structural joints in which all of the material within the grip of the bolt is steel because predictable and satisfactory performance of slip critical joints is dependent upon predictable and stable installed tension in the bolts. The Test Method to Determine the Slip Coefficient for Coatings Used in Bolted Joints includes long term creep test requirements to assure reliable performance for qualified paint coatings. However, it must be recognized that in the case of hot dip galvanized coatings, especially if the joint consists of many plies of thickly coated material, relaxation of bolt tension may be significant and may require retensioning of the bolts subsequent to the initial tightening. Research[5] has shown that a loss of pretension of approximately 6.5 percent for galvanized plates and bolts due to relaxation as compared with 2.5 percent for uncoated joints. This loss of bolt tension occurred in five days with negligible loss recorded thereafter. This loss can be allowed for in design or pretension may be brought back to the prescribed level by retightening the bolts after an initial period of "settling-in."

This Specification has permitted the use of bolt holes $1/16$ inch larger than the bolts installed in them since it was first published. Research[6] has shown that, where greater latitude is needed in meeting dimensional tolerances during erection, somewhat larger holes can be permitted for bolts $5/8$ inch diameter and larger without adversely affecting the performance of shear connections assembled with high-strength bolts. The oversize and slotted hole provisions of this Specification are based upon these findings. Because an increase in hole size generally reduces the net area of a connected part, the use of oversize holes is subject to approval by the Engineer.

---

[5]Munse, W. H., "Structural Behavior of Hot Galvanized Bolted Connections," 8th International Conference on Hot-dip Galvanizing," London, England, June 1967.

[6]Allen, R. N., and J. W. Fisher, "Bolted Joints With Oversize or Slotted Holes," *Journal of the Structural Division*, ASCE, Vol. 94, No. ST9, September 1968.

[7]Polyzois, D. and J. A. Yura, "Effect of Burrs on Bolted Friction Connections," AISC *Engineering Journal*, Vol. 22, No. 3, Third Quarter 1985.

Burrs. Based upon tests,[7] which demonstrated that the slip resistance of joints was unchanged or slightly improved by the presence of burrs, burrs which do not prevent solid seating of the connected parts in the snug tight condition need not be removed. On the other hand, parallel tests in the same program demonstrated that large burrs can cause a small increase in the required turns from snug tight condition to achieve specified pretension with turn-of-nut method of tightening.

Unqualified Paint on Faying Surfaces. An extension to the research on the slip resistance of shear connections cited in footnote 4 investigated the effect of ordinary paint coatings on limited portions of the contact area within joints and the effect of overspray over the total contact area. The tests demonstrated that the effective area for transfer of shear by friction between contact surfaces was concentrated in an annular ring around and close to the bolts. Paint on the contact surfaces approximately one inch but not less then the bolt diameter away from the edge of the hole did not reduce the slip resistance. Because in connections of thick material involving a number of bolts on multiple gage lines, bolt pretension might not be adequate to completely flatten and pull thick material into tight contact around every bolt, the Specification requires that all areas between bolts also be free of paint. See Figure C3. The new requirements have a potential for increased economy because the paint free area may easily be protected using masking tape located relative to the hole pattern, and further, the narrow paint strip around the perimeter of the faying surface will minimize uncoated material outside the connection requiring field touch up.

This research also investigated the effect of various degrees of inadvertent overspray on slip resistance. It was found that even the smallest amount of overspray of ordinary paint (that is, not qualified as Class A) within the specified paint free area on clean mill scale reduced the slip resistance significantly. On blast cleaned surfaces, the presence of a small amount of overspray was not as detrimental. For simplicity, the Specification prohibits any overspray from areas required to be free of paint in slip-critical joints regardless of whether the surface is clean mill scale or blast cleaned.

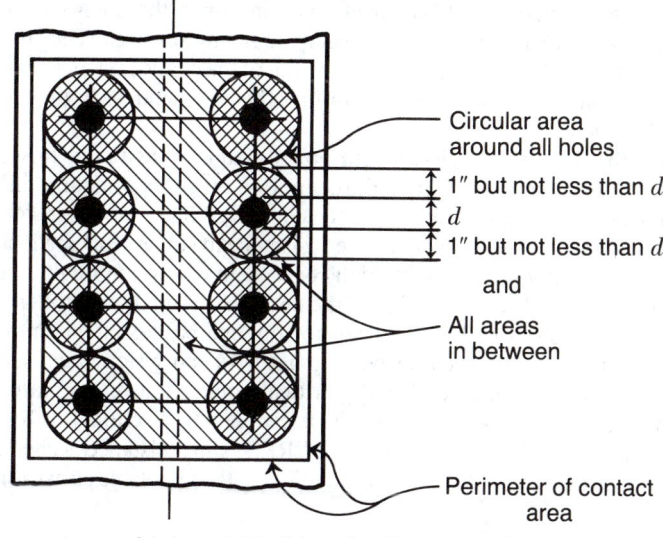

Circular area
around all holes

1″ but not less than d
d
1″ but not less than d

and

All areas
in between

Perimeter of contact
area

Areas outside the defined area need not
be free of paint.

Figure C3

Galvanized Faying Surfaces. The slip factor for initial slip with clean hot-dip galvanized surfaces is of the order of 0.19 as compared with a factor of about 0.35 for clean mill scale. However, research[3] has shown that the slip factor of galvanized surfaces is significantly improved by treatments such as hand wire brushing or light "brush-off" grit blasting. In either case, the treatment must be controlled in order to achieve the necessary roughening or scoring. Power wire brushing is unsatisfactory because it tends to polish rather than roughen the surface.

Field experience and test results have indicated that galvanized members may have a tendency to continue to slip under sustained loading.[8] Tests of hot-dip galvanized joints subject to sustained loading show a creep type behavior. Treatments to the galvanized faying surfaces prior to assembly of the joint which caused an increase in the slip resistance under short duration loads did not significantly improve the slip behavior under sustained loading.

C4 Design for Strength of Bolted Connections

Background for Design Stresses. With this edition of the Specification, the arbitrary designations "friction type" and "bearing type" connections used in former editions, and which were frequently misinterpreted as implying an actual difference in the manner of performance or strength of the two types of connection, are discontinued in order to focus attention more upon the real manner of performance of bolted connections.

In bolted connections subject to shear type loading, the load is transferred between the connected parts by friction up to a certain level of force which is dependent upon the total clamping force on the faying surfaces and the coefficient of friction of the faying surfaces. The connectors are not subject to shear nor is the connected material subject to bearing stress. As loading is increased to a level in excess of the frictional resistance between the faying surfaces, slip occurs, but failure in the sense of rupture does not occur. As even higher levels of load are applied, the load is resisted by shear upon the fastener and bearing upon the connected material plus some uncertain amount of friction between the faying surfaces. The final failure will be by shear failure of the connectors or tear out of the connected material or unacceptable ovalization of the holes. Final failure load is independent of the clamping force provided by the bolts.[9]

Thus, the design of high-strength bolted connections under this Specification begins with consideration of strength required to prevent premature failure by shear of the connectors or bearing failure of the connected material. Next, for connections which are defined as "slip-critical" the resistance to slip at working load is checked. Because high clamping force with high coefficient of friction might mathematically exceed the ultimate shear or bearing of the fasteners, even though the fasteners would not be subject to shear or bearing prior to slip, and because the combined effect of frictional resistance with shear or bearing has not been systematically studied and is uncertain, the allowable force for slip critical connections conservatively must not exceed the lesser of the allowable loads determined by Section 4 or Section 5 of the allowable stress design specification. For LRFD, slip resistance is checked at service load. The resistance at service load is identical to the allowable stress for ASD.

---

[8]Fisher, J. W. and J. H. A. Struik, *Guide to Design Criteria for Bolted and Riveted Joints*, John Wiley & Sons, New York, 1974, Pg. 205 and 206. (Hereinafter referred to as the Guide.)

[9]Ibid., pp. 49–52.

Connection Slip. There are practical cases in the design of structures where slip of the connection is desirable in order to permit rotation in a joint or to minimize the transfer of moment. Additionally there are cases where, because of the number of fasteners in a joint, the probability of slip is extremely small or where, if slip did occur, it would not be detrimental to the serviceability of the structure. In order to provide for such cases while at the same time making use of the higher shear strength of high-strength bolts, as contrasted to ASTM A307 bolts, the Specification now permits joints tightened only to the snug tight condition.

The maximum amount of slip that can occur in connections that are not classified as slip-critical under the Specification rules is limited theoretically to $^1/_{16}$ inch. In most practical cases, however, the real magnitude of slip would probably be much less because the acceptable inaccuracies in the location of holes within a pattern of bolts would usually cause one or more bolts to be in bearing in the initial unloaded condition. Further, in statically loaded structures, even with perfectly positioned holes, the usual method of erection would cause the weight of the connected elements to put the bolts into direct bearing at the time the member is supported on loose bolts and the lifting crane is unhooked. Subsequent additional gravity loading could not cause additional connection slip.

Connections classified as slip-critical include those cases where slip could theoretically exceed $^1/_{16}$ inch, and thus, possibly affect the suitability for service of the structure by excessive distortion or reduction in strength or stability even though the resistance to fracture of the connection, per se, may be adequate. Also included are those cases where slip of any magnitude must be prevented, for example, joints subject to fatigue loading.

Shear and Bearing on Fasteners. Several interrelated parameters influence the shear and bearing strength of connections. These include such geometric parameters as the net-to-gross-area ratio of the connected parts, the ratio of the net area of the connected parts to the total shear-resisting area of the fasteners, and the ratio of transverse fastener spacing to fastener diameter and the ratio of transverse fastener spacing to connected part thickness. In addition, the ratio of yield strength to tensile strength of the steel comprising the connected parts, as well as the total distance between extreme fasteners, measured parallel to the line of direct tensile force, play a part.

In the past, a balanced design concept had been sought in developing criteria for mechanically fastened joints to resist shear between connected parts by means of bearing of the fasteners against the sides of the holes. This philosophy resulted in wide variations in the factor of safety for the fasteners, because the ratio of yield to tensile strength increases significantly with increasingly stronger grades of steel. It had no application at all in the case of very long joints used to transfer direct tension, because the end fasteners "unbutton" before the plate can attain its full strength or before the interior fasteners can be loaded to their rated shear capacity.

By means of a mathematical model it was possible to study the interrelationship of the previously mentioned parameters.[10,11] It has been shown that the factor of safety against shear failure ranged from 3.3 for compact (short) joints to approximately 2.0 for joints with an overall length in excess of 50 inches. It is of interest to

---

[10]Fisher, J.W., and L. S. Beedle, "Analysis of Bolted Butt Joints," *Journal of the Structural Division,* ASCE, Vol. 91, No. ST5, October 1965.

[11]Guide, pp. 84–107 and 123–127.

note that the longest (and often the most important) joints had the lowest factor, indicating that a factor of safety of 2.0 has proven satisfactory in service.

The absence of any working stress or design strength provisions for the case where a bolt in double shear has a nonthreaded shank in one shear plane and a threaded section in the other shear plane recognizes that knowledge as to the bolt placement (which might leave both shear planes in the threaded section) is not ordinarily available to the detailer.

The allowable working stresses and corresponding design strengths for fasteners subject to applied tension or shear given in Table 2 are unchanged from the 1980 edition of the Specification. The values are based upon the research and recommendations reported in the Guide. With the wealth of data available, it was possible, through statistical analyses, to adjust allowable working stresses to provide uniform reliability for all loading and joint types. The design of connections is more conservative than that of the connected members of buildings and bridges by a substantial margin, in the sense that the failure load of the fasteners is substantially in excess of the maximum serviceability limit (yield) of the connected material.

Design for Tension. The allowable working stresses and design strengths specified for applied tension[12] are intended to apply to the external bolt load plus any tension resulting from prying action produced by deformation of the connected parts. When stressed in tension to the recommended working value (approximately equal to two-thirds of the initial tightening force), high-strength bolts tensioned to the requirements of Table 4 will experience little if any actual change in stress. For this reason, bolts in connections in which the applied loads subject the bolts to axial tension are required to be fully tensioned even though the connection may not be subject to fatigue loading nor classified as slip-critical.

Properly tightened A325 and A490 bolts are not adversely affected by repeated application of the recommended working tensile stress, provided the fitting material is sufficiently stiff, so that the prying force is a relatively small part of the applied tension.[13] The provisions covering bolt tensile fatigue are based upon study of test reports of bolts that were subjected to repeated tensile load to failure.

Design for Shear. The strength in shear is based upon the assumption of a ratio of shear strength to tensile strength of 0.6.[14] In the allowable stress design specification, the allowable shear is based upon a factor of safety of approximately 2.35. Load and Resistance Factor Design uses nominal shear strength with a resistance factor of 0.65 or 0.75, depending on the position of the shear plane relative to the bolt threads to establish the design shear strength.

Design for Bearing Bearing stress produced by a high-strength bolt pressing against the side of the hole in a connected part is important only as an index to behavior of the connected part. It is of no significance to the bolt. The critical value can be derived from the case of a single bolt at the end of a tension member.

---

[12]Ibid., pp. 257–276.

[13]Ibid., pg. 266.

[14]Ibid., pp. 46–53.

It has been shown,[15] using finger-tight bolts, that a connected plate will not fail by tearing through the free edge of the material if the distance L, measured parallel to the line of applied force from a single bolt to the free edge of the member toward which the force is directed, is not less than the diameter of the bolt multiplied by the ratio of the bearing stress to the tensile strength of the connected part.

Providing a factor of safety of 2.0, the working stress design criterion given in 4(b) is

$$L/d \geq 2(F_p/F_u).$$

where

F_p = allowable working stress in bearing.
F_u = specified minimum tensile strength of the connected part.

When using factored loads, the load and resistance factor criterion given in 4(b) is

$$L/d \geq R_n/F_u$$

where

R_n = nominal bearing pressure.
F_u = specified minimum tensile strength of the connected part.

As a practical consideration, a lower limit of 1.5 is placed on the ratio L/d and an upper limit of 1.5 on the ratio F_p/F_u and an upper limit of 3.0 on the ratio R_n/F_u.

The foregoing leads to the rules governing bearing strength in both versions of the specification. The permitted bearing pressure in the 1980 Specification and the current provisions are fully justifiable from the standpoint of strength of the connected material. However, recent tests have demonstrated that severe ovalization of the hole will begin to develop, even though rupture does not occur, as bearing stress is increased beyond the previously permitted stess, especially if it is combined with high-tensile stress on the net section. Thus, the rules for bearing strength have been revised to provide increased conservatism, unless special consideration is given to the effect of possible hole ovalization.

For connections with more than a single bolt in the direction of force, the resistance may be taken as the sum of the resistances of the individual bolts.

C5 Design Check for Slip Resistance

The Specification recognizes that, for a significant number of cases, slip of the joint would be undesirable or must be precluded. Such joints are termed "slip-critical" joints. This is somewhat different from the previous term "friction type" connection because it recognizes that all tensioned high-strength bolted joints resist load by friction between faying surfaces up to the slip load and subsequently are able to resist even greater loads by shear and bearing. The Specification requires that, in addition to assuring that the connection has adequate strength, the design of slip-critical joints be checked to assure that slip will not occur at working load.

It must be recognized that the formula for P_s in 5(b) is for connections subject to a linear load. For cases in which the load tends to rotate the connection in the plane of the faying surface, a modified formula accounting for the placement of bolts relative to the center of rotation should be used.

The safety index for serviceability considerations has traditionally been less than that required for strength considerations. In the consideration of the consequences of slip of bolts into bearing, a single criterion cannot apply. In the case of bolts in holes

---

[15]Ibid., pg. 137.

affording only small clearance as in standard holes, oversize holes, slotted holes loaded transverse to the axis of the slot and short slotted holes loaded parallel to the axis of the slot, the consequences of slip are trivial except for fatigue applications. In the case of slip critical connections in which load is applied to bolts parallel to the axis of a long slot in which they are installed, it is conceivable that failure by slip could lead to critical geometric distortions of the frame, endangering the strength of the structure, even though danger of strength failure of the connection does not exist.

Extensive test data developed through research sponsored by the Council and others have made possible a statistical analysis of the slip probability of connections tensioned to the requirements of Table 4. The frequency distribution and mean value of clamping force for bolts tightened by the calibrated wrench method[16] is due to variation in the torque-tension ratio from bolt to bolt, the tolerance on wrench performance, hole type and human error. Both the variation in the slip coefficient (or degree of surface roughness) and the frequency distribution of the magnitude of clamping force provided by A325 and A490 bolts in the connection were considered.[17]

In the 1978 edition of this Specification, nine classes of faying surface conditions were introduced and significant increases were made in the recommended working stresses for proportioning connections which function by transfer of shear between connected parts by friction. These classes and the stresses were adopted on the basis of statistical evaluation of the information then available.

The allowable stresses or nominal resistances for bolts in standard holes in Table 3 of these Specifications were developed for a 10 percent probability of slip considering only faying surface treatment and bolt clamping force. This is analogous to the product of the resistance factor and the nominal strength (ϕR) used in reliability concepts.[18] The bolt clamping force was based upon *calibrated wrench* method of installation using the clamping force variation shown in Figure 5.8 of the Guide. An examination of the slip coefficient for a wide range of surface conditions, including additional data developed during the past ten years,[4] indicate that the variability for each surface class was about the same ($\sigma = 0.007$ to 0.09). Rather than providing a separate class for each of the individual surface conditions, the surface conditions with approximately the same mean values were grouped into three classes, with mean slip coefficients of 0.33, 0.40 and 0.50. The revised edition of the Guide is scheduled for printing in 1986. Table 5.5 of the revised edition provides the equivalent shear stress values for mean slip coefficients between 0.2 and 0.6, with calibrated wrench installation. Mean values for turn-of-nut method of installation are provided in Table 5.4.

Because of the effects of oversize and slotted holes on the induced tension in bolts using any of the specified installation methods, lower loads are provided in Table 3 for bolts in these hole types. In the case of bolts in long slotted holes, even though the slip load is the same for bolts loaded transverse or parallel to the axis of the slot, the load for bolts loaded parallel to the axis has been further reduced in recognition of the greater consequences of slip.

---

[16]Guide, pg. 82, Fig. 5.8.

[17]The probability distribution function of the product of two independent random variables can be determined using standard statistical techniques as outlined, for example, in "Introductory Probability and Statistical Applications," by P. L. Meyer, Addison Publishing Company, 1965.

[18]Fisher, J. W., etal "Load and Resistance Design Criteria for Connectors," *Journal of the Structural Division*, ASCE, Vol. 104, No. ST9, September 1978.

The frequency distribution and mean value of clamping force for bolts tightened by turn-of-nut method are higher, due to the elimination of variables which affect torque-tension ratios and to higher-than-specified minimum strength of production bolts. Because properly applied turn-of-nut installation induces yield point strain in the bolt, the higher-than-specified yield strength of production bolts will be mobilized and result in higher clamping force by the method. On the other hand, with the calibrated wrench method, which is dependent upon the calibration of wrenches in a tension indicating device independent of the actual bolt properties, any additional strength of production bolts will not be mobilized. High clamping force might be achieved by the calibrated wrench method if the wrench was set to a higher torque value. However, this would require more attention to the degrees of rotation (because the limitation on the maximum rotation of the nut specified in the second paragraph of 8(d)(2) to prevent excessive deformation of the bolt would control more often), otherwise a torsional bolt failure might result. Because of the increased clamping force, connections having bolts installed by turn-of-nut method provide a greater resistance to slip (lower probability of slip).

Connections of the type shown in Figure C4(a), in which some of the bolts (A) lose a part of their clamping force due to applied tension, suffer no overall loss of frictional resistance. The bolt tension produced by the moment is coupled with a compensating compressive force (C) on the other side of the axis of bending. In a connection of the type shown in Figure C4(b), however, all fasteners (B) receive applied tension which reduces the initial compression force at the contact surface. If slip under load cannot be tolerated, the design slip load value of the bolts in shear should be reduced in proportion to the ratio of residual axial force to initial tension. If slip of the joint can be tolerated, the bolt shear stress should be reduced according to the tension-shear interaction as outlined in the Guide page 69. Because the bolts are subject to applied axial tension, they are required to be pretensioned in either case.

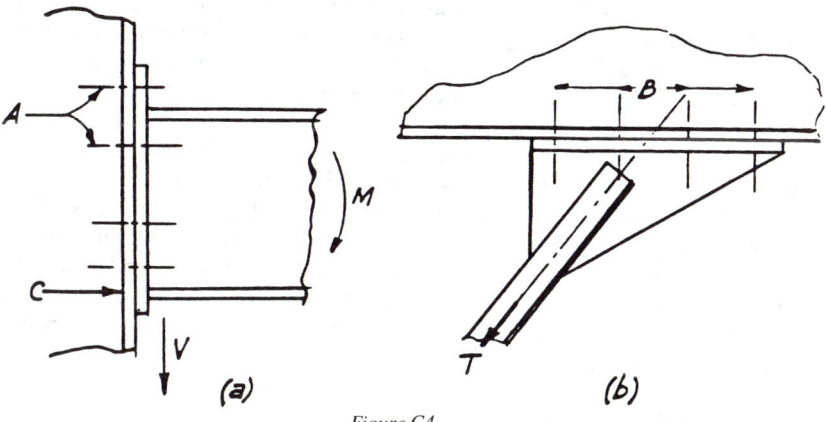

Figure C4

While connections with bolts pretensioned to the levels specified in Table 4 do not ordinarily slip into bearing against the sides of the hole when subjected to the allowable loads of Table 3, it is required that they meet the requirements for allowable stress in Table 2 in order to maintain the factor of safety of 2 against fracture in the event that the bolts do slip into bearing as a result of large unforeseen loads.

To cover those cases where a coefficient of friction less than 0.33 might be adequate for a given situation the Specification provides that, subject to the approval of the Engineer, and provided the mean slip coefficient is determined by the specified test procedure allowable slip loads or nominal resistances less than those provided by Class A faying surface coating may be used.

It should be noted that both Class A and Class B coatings are required to be applied to blast cleaned steel.

High-strength Bolts in Combination with Weld or Rivets. For high-strength bolts in combination with welds in statically loaded conditions, the allowable load or nominal strength may be taken as the sum of two contributions.[19] One results from the slip resistance of the bolted parts and may be determined in accordance with Section 5(b). The second results from the resistance of the welds and may be determined on the basis of the allowable stresses for welds given in the applicable specifications.

For high-strength bolts in combination with welds in fatigue loaded applications, data available are not sufficient to develop general design recommendations at this time. High-strength bolts in combination with rivets are rarely encountered in modern practice. If need arises, guidance may be found in the Guide.

C7 Design Details of Bolted Connections

A new section has been added with this edition of the Specification in order to bring together a number of requirements for proper design and detailing of high-strength bolted connections. The material covered in the Specification, and Section 7 in particular, is not intended to provide comprehensive coverage of the design of high-strength bolted connections. For example, other design considerations of importance to the satisfactory performance of the connected material such as block shear, shear lag, prying action, connection stiffness, effect on the performance of the structure and others are beyond the scope of this Specification and Commentary.

Proper location of hardened washers is as important as other elements of a detail to the peformance of the fasteners. Drawings and details should clearly reflect the number and disposition of washers, especially the thick hardened washers that are required for several slotted hole applications. Location of washers is a design consideration which should not be left to the experience of the iron worker.

Finger shims are a necessary device or tool of the trade to permit adjusting alignment and plumbing of structures. When these devices are fully and properly inserted, they do not have the same effect on bolt tension relaxation or the connection performance as do long slotted holes in an outer ply which is the basis for allowable design stresses. When fully inserted, the shim provides support around approximately 75 percent of the perimeter of the bolt in contrast to the greatly reduced area that exists with a bolt centered in a long slot. Further, finger shims would always be enclosed on either side by the connected material which would be fully effective in bridging the space between the fingers.

C8 Installation and Tightening

Several methods for installation and tensioning of high-strength bolts, when tensioning is required, are provided without preference in the Specification. Each method recognized in Section 8, when properly used as specified, may be relied upon to provide satisfactory results. All methods may be misused or abused.

At the expense of redundancy, the provisions stipulating the manner in which each method is intended to be used is set forth in complete detail in order that the rules

---

[19]Guide, pp. 238 to 240.

for each method may stand alone without need for footnotes or reference to other sections. If the methods are conscientiously implemented, good results should be routinely achieved.

Connections not Requiring Full Tensioning. In the Commentary, Section C6 of the previous edition of the Specification it was pointed out that "bearing" type connections need not be tested to assure that the specified pretension in the bolts had been provided, but specific provision permitting relaxation of the tensioning requirement was not contained in the body of the Specification. In this edition of the Specification, separate installation procedures are provided for bolts that are not within the slip-critical or direct tension category. The intent in making this change is to improve the quality of bolted steel construction and reduce the frequency of costly controversies by focusing attention, both during the installation and tensioning phase and during inspection, on the true slip-critical connections rather than diluting the effort through the requirement for costly tensioning and tension testing of the great many connections where such effort serves no useful purpose. The requirement for identification of connections on the drawings may be satisfied either by identifying the slip-critical and direct tension connections which must be fully tightened and inspected or by identifying the connections which need be tightened only to the snug tight condition.

In the Specification, snug tight is defined as the tightness that exists when all plies are in firm contact. This may usually be attained by a few impacts of an impact wrench or the full effort of a man using an ordinary spud wrench. In actuality, snug tight is a degree of tightness which will vary from joint to joint depending upon the thickness and degree of parallelism of the connected material. In most joints the plies will pull together; however, in some joints, it may not be possible at snug tight to have contact throughout the faying surface area.

Tension Calibrating Devices. At the present time, there is no device or economical means for determining the actual tension in a bolt that is installed in a connection. However, the actual tension in a bolt installed in a tension calibrator (hydraulic tension indicating device) is directly indicated by the dial of the device. Thus, such a device is an economical and valuable tool that should be readily available whenever high-strength bolts in slip critical joints or bolts subject to applied axial tension are to be tensioned to the pretension specified in Table 4. Although each element of a fastener assembly may conform to the minimum requirements of their separate ASTM specifications, their compatability in an assembly or the need for lubrication can only be assured by testing the assembly. Therefore, such devices are important for testing the complete fastener assembly as it will be used with any method of tightening to assure the suitability of bolts and nuts (probably produced by different manufacturers), as well as other elements, and the ability of the assembly to provide the specified tension using the selected method. Testing before starting to install fasteners in the work will also identify potential sources of problems, such as the need for lubrication, weakened nuts due to excessive over tapping in the case of galvanized nuts, or failure of bolts subject to combined torque and tension due to over strength load indicators, and to clarify, for the bolting crews and inspectors, the proper implementation of the selected installation method to be used. Such devices are essential to the specified procedure for the calibrated wrench method of installation and for specified procedure for arbitration inspection when such inspection is required.

Experience on many projects has shown that bolts and/or nuts not meeting the requirements of the applicable ASTM specification, but which were intended for installation by turn-of-nut method, would have been identified prior to installation; thus saving the controversy and great expense of replacing bolts installed in the structure when difficulties were discovered at a later date.

Hydraulic tension calibrating devices capable of indicating bolt tension undergo a slight deformation under load. Hence the nut rotation corresponding to a given tension reading may be somewhat larger than it would be if the same bolt were tightened against a solid steel abutment. Stated differently, the reading of the calibrating device tends to underestimate the tension which a given rotation of the turned element would induce in a bolt in an actual joint. This should be borne in mind when using such devices to establish a tension-rotation relationship.

Slip-critical Connections and Connections Subject to Direct Tension. Four methods for joint assembly and tightening are provided for slip-critical and direct tension connections. Regardless of the method used, it should be demonstrated prior to the commencement of work that the procedure to be used with the fasteners to be used and by the crews who will be doing the work that the specified pretension is achieved. For this reason, it is a requirement that a tension measuring device be provided at the job site.

With any of the four described tensioning methods, it is important to install bolts in all holes of the connection and bring them to an intermediate level of tension generally corresponding to snug tight in order to compact the joint. Even after being fully tightened, some thick parts with uneven surfaces may not be in contact over the entire faying surface. This is not detrimental to the performance of the joint. As long as the specified bolt tension is present in all bolts of the completed connection, the clamping force equal to the total of the tensions in all bolts will be transferred at the locations that are in contact and be fully effective in resisting slip through friction. If however, bolts are not installed in all holes and brought to an intermediate level of tension to compact the joint, bolts which are tightened first will be subsequently relaxed by the tightening of the adjacent bolts. Thus the total of the forces in all bolts will be reduced which will reduce the slip load whether there is uninterrupted contact between the surfaces or not.

With all methods, tightening should begin at the most rigidly fixed or stiffest point and progress toward the free edges, both in the initial snugging up and in the final tightening.

Turn-of-Nut-Tightening. When properly implemented, turn-of-nut method provides more uniform tension in the bolts than does torque controlled tensioning methods because it is primarily dependent upon bolt elongation slightly into the inelastic range.

Consistency and reliability is dependent upon assuring that the joint is well compacted and all bolts at a snug tight condition prior to application of the final required partial turn. Reliability is also dependent upon assuring that the turn that is applied is relative between the bolt and nut; thus the element not turned in tightening should be prevented from rotating while the required degree of turn is applied to the turned element. Reliability and inspectability of the method may be improved by having the outer face of the nut match-marked to the protruding end of the bolt after the joint has been snug tightened but prior to final tightening. Such marks may be applied by the wrench operator using a crayon or dab of paint. Such marks in their relatively displaced position after tightening will afford the inspector a means for noting the rotation that was applied.

Some problems with turn-of-nut tightening encountered with galvanized bolts have been attributed to especially effective lubricant applied by the manufacturer to meet ASTM Specification requirements. Job site tests in the tension indicating device demonstrated the lubricant reduced the coefficient of friction between the bolt and nut to the degree that "the full effort of a man using an ordinary spud wrench" to snug tighten the joint actually induced the full required tension. Because the nuts could be removed by an ordinary spud wrench they were erroneously judged improperly tightened by the inspector. Research[5] confirms that lubricated high-strength bolts may require only one-half as much torque to induce the specified tension. For such situations, use of a tension indicating device and the fasteners being installed may be helpful in establishing alternate criteria for snug tight at about one-half the tension required by Table 4.

Because reliability of the method is independent of the presence or absence of washers, washers are not required except for oversize and slotted holes in an outer ply. Thus, in the absence of washers, testing after the fact using a torque wrench method is highly unreliable. That is, the turn-of-nut method of installation, properly applied, is more reliable and consistent than the testing method. The best method for inspection of the method is for the inspector to observe any job site confirmation testing of the fasteners and the method to be used followed by monitoring of the work in progress to assure that the method is routinely properly applied.

Calibrated Wrench Method.   Research has demonstrated that scatter in induced tension is to be expected when torque is used as an indirect indicator of tension. Numerous variables, which are not related to tension, affect torque. For example, the finish and tolerance on bolt threads, the finish and tolerance on the nut threads, the fact that the bolt and nut may not be produced by the same manufacturer, the degree of lubrication, the job site conditions contributing to dust and dirt or corrosion on the threads, the friction that exists to varying degree between the turned element and the supporting surface, the variability of the air pressure on the torque wrenches due to length of air lines or number of wrenches operating from the same source, the condition and lubrication of the wrench which may change within a work shift and other factors all bear upon the relationship between torque and induced tension.

Recognition of the calibrated wrench method of tightening was removed from the Specification with the 1980 edition. This action was taken because it is the least reliable of all methods of installation and many costly controversies had occured. It is to be suspected that short cut procedures in the use of the calibrated wrench method of installation, not in accordance with the Specification provisions, were being used. Further, torque controlled inspection procedures based upon "standard" or calculated inspection torques rather than torques determined as required by the Specification were being routinely used. These incorrect procedures plus others had a compounding effect upon the uncertainty of the installed bolt tension, and were responsible for many of the controversies.

It is recognized, however, that if the calibrated wrench method is implemented without short cuts as intended by the Specification, that there will be a ninety percent assurance that the tensions specified in Table 4 will be equaled or exceeded. Because the Specification should not prohibit any method which will give acceptable results when used as specified, the calibrated wrench method of installation is reinstated in this edition of the Council Specification. However, to improve upon the previous situation, the 1985 version of the Specification has been modified to require better control. Wrenches must be calibrated daily. Hardened washers must be used. Fasteners must be protected from dirt and moisture at the job site. Additionally, to achieve reliable results attention should be given to the control, insofar as it is practical, of those controllable factors which contribute to variability. For example, bolts and nuts should be purchased from single sources, insofar as practical, to minimize the variability of the fit. Bolts and nuts should be adequately and uniformly lubricated. Water soluble lubricants should be avoided.

Installation of Alternate Design Fasteners.   It is the policy of the Council to recognize only fasteners covered by ASTM Specifications, however, it cannot be denied that a general type of alternate design fastener, produced by several manufacturers, are used on a significant number of projects as permitted by Section 2(d). The bolts referenced involve a splined end extending beyond the threaded portion of the bolt which is gripped by a special design wrench chuck providing a means for turning the nut relative to the bolt until the splined end is sheared off. While such bolts are subject to many of the variables affecting torque mentioned in the preceding section, they are produced and shipped by the manufacturers as a nut-bolt assembly under good quality control which apparently minimizes some of the negative aspects of the torque controlled process.

While these alternate design fasteners have been demonstrated to consistently provide tension in the fastener meeting the requirements of Table 5 in controlled tests in tension indicating devices, it must be recognized that the fastener may be misused and provide results as unreliable as those with other methods. The requirements of this Specification and the installation requirements of the manufacturer's specification required by Section 2(d) must be adhered to.

As with other methods, a representative sample of the bolts to be used should be tested to assure that they do, in fact, when used in accordance with the manufacturer's instructions, provide tension as specified in Table 5. In actual joints, bolts must be installed in all holes of a connection and all fasteners tightened to an intermediate level of tension adequate to pull all material into contact. Only after this has been accomplished should the fasteners be fully tensioned in a systematic manner and the splined end sheared off. The sheared off splined end merely signifies that at some time the bolt has been subjected to a torque adequate to cause the shearing. If the fasteners are installed and tensioned in a single continuous operation, they will give a misleading indication to the inspector that the bolts are properly tightened. Therefore, the only way to inspect these fasteners with assurance is to observe the job site testing of the fasteners and installation procedure and then monitor the work while in progress to assure that the specified procedure is routinely followed.

Direct Tension Indicator Tightening.   Proprietary load indicating devices, not yet covered by an ASTM Specification, but recognized under this Specification in Section 2(f) are being specified and used in a significant number of projects. The referenced device is a hardened washer incorporating several small formed arches which are designed to deform in a controlled manner when subjected to load. These load indicator washers are the single device known which is directly dependent upon the tension load in the bolt, rather than upon some indirect parameter, to indicate the tension in a bolt.

As with the alternate design load indicating bolts, load indicating washers are dependent upon the quality control of the producer and proper use in accordance with the manufacturer's installation procedures and these Specifications. Load indicator washers delivered for use in a specific application should be tested at the job site to demonstrate that they do, in fact, provide a proper indication of bolt tension, and that they are properly used by the bolting crews. Because the washers depend upon an irreversible mechanism (inelastic deformation of the formed arches) bolts together with the load indicator washer plus any other washers required by Specification should be installed in all holes of the connection and the bolts tightened to approximately one-half the specified tension. Only after this initial tightening operation should the bolts be fully tensioned in a systematic manner. If the bolts are installed and tensioned in a single continuous operation, the load indicator washers will give the inspector a misleading indication that bolts are properly tightened. Therefore, the only way to inspect fasteners with which load indicator washers are used with assurance is to observe the job site testing of the devices and installation procedure and then routinely monitor the work in progress to assure that the specified procedure is followed.

During installation care must be taken to assure that the indicator nubs are oriented to bear against the hardened bearing surface of the bolt head or against an extra hardened flat washer if used under the nut.

C9 Inspection

It is apparent from the commentary on installation procedures that the inspection procedures giving the best assurance that bolts are properly installed and tensioned is provided by inspector observation of the calibration testing of the fasteners using the selected installation procedure followed by monitoring of the work in progress to assure that the procedure which was demonstrated to provide the specified tension is routinely adhered to. When such a program is followed, no further evidence of proper bolt tension is required.

However, if testing for bolt tension using torque wrenches is conducted subsequent to the time the work of installation and tightening of bolts is performed, the test procedure is subject to all of the uncertainties of torque controlled calibrated wrench installation. Additionally, the absence of many of the controls necessary to minimize variablity of the torque to tension relationship, which are unnecessary for the other methods of bolt installation, such as, use of hardened washers, careful attention to lubrication and the uncertainty of the effect of passage of time and exposure in the installed condition all reduce the reliability of the arbitration inspection results. The fact that it may, of necessity, have to be based upon a job test torque determined by bolts only assumed to be representative of the bolts in the actual job or bolts removed from completed joints, in many cases, makes the test procedure less reliable than a properly implemented installation procedure it is used to verify. Other verification inspection procedures available at this time are more accurate but too costly and time consuming for all but the most critical structural applications. The arbitration inspection procedure contained in the Specification is provided, in spite of its limitations, as the most feasible available at this time.

AISC
Quality Certification
Program

AMERICAN INSTITUTE OF STEEL CONSTRUCTION
ONE EAST WACKER DRIVE, SUITE 3100
CHICAGO, ILLINOIS 60601-2001

Notes

AISC
Quality Certification
Program

In recent years, the quality of construction methods and materials has become the subject of increasing concern to building officials, highway officials, and designers. One result of this concern has been the enactment of ever more demanding inspection requirements intended to ensure product quality. In many cases, however, these more demanding inspection requirements have not been based upon demonstrated unsatisfactory performance of structures in service. Rather, they have been based upon the capacity of sophisticated test equipment, or upon standards developed for nuclear construction rather than conventional construction. Adding to the problem, arbitrary interpretation of specifications by inspectors has too often been made without rational consideration of the type of construction involved. The result has been spiraling increases in the costs of fabrication of structural steel and of inspection, which must be paid by owners without necessarily assuring that the product quality required has been improved.

Product inspection, although it has a valid place in the construction process, is not the most logical or practical way to assure that structural steelwork will conform to the requirements of contract documents and satisfy the intended use. A better solution can be found in the exercise of good quality control and quality assurance by the fabricator *throughout the entire production process.*

Recognizing this fact, and seeking some valid, objective method whereby a fabricator's capability for assuring a quality product could be evaluated, a number of code authorities have, in recent years, instituted steps to establish fabricator registration programs. However, these independent efforts resulted in extremely inconsistent criteria. They were developed primarily by inspectors or inspection agencies who were experienced in testing, but were not familiar with the complexities of the many steps, procedures, techniques, and controls required to assure quality throughout the fabricating process. Neither were these inspection agencies qualified to determine the various levels of quality required to assure satisfactory performance in meeting the service requirements of the many different types of steel structures.

Recognizing the need for a comprehensive national standard for fabricator certification, and concerned by the trend toward costly inspection requirements that could not be justified by rational quality standards, the American Institute of Steel Construction has developed and implemented a voluntary Quality Certification Program, whereby any structural steel fabricating plant—whether a member of AISC or not—can have its capability for assuring quality production evaluated on a fair and impartial basis.

THE AISC PROGRAM

The AISC Quality Certification Program does not involve inspection and/or judgment of product quality on individual projects. Neither does it guarantee the quality of specific fabricated steel products. Rather, the purpose of the AISC Quality Certification Program is to confirm to the construction industry that a Certified structural steel fabricating plant *has the personnel, organization, experience, procedures, knowledge, equipment, capability and commitment to produce fabricated steel of the required quality for a given category of structural steelwork.*

The AISC Quality Certification Program was developed by a group of highly qualified shop operation personnel from large, medium, and small structural steel fabricating plants throughout the United States. These individuals all had extensive experience and were fully aware of where and how problems can arise during the production process and of the steps and procedures that must be followed during fabrication to assure that the finished product meets the quality requirements of the contract.

The program was reviewed and strongly endorsed by an Independent Board of Review comprised of 17 prominent structural engineers from throughout the United States, who were not associated with the steel fabricating industry, but were well qualified in matters of quality requirements for reliable service of all types of steel structures.

CATEGORIES OF CERTIFICATION

A fabricator may apply for certification of a plant in one of the following categories of structural steelwork:

I: **Conventional Steel Structures** — Small Public Service and Institutional Buildings, (Schools, etc.), Shopping Centers, Light Manufacturing Plants, Miscellaneous and Ornamental Iron Work, Warehouses, Sign Structures, Low Rise, Truss Beam/Column Structures, Simple Rolled Beam Bridges.

II: **Complex Steel Building Structures** — Large Public Service and Institutional Buildings, Heavy Manufacturing Plants, Powerhouses (fossil, non-nuclear), Metal Producing/Rolling Facilities, Crane Bridge Girders, Bunkers and Bins, Stadia, Auditoriums, High Rise Buildings, Chemical Processing Plants, Petroleum Processing Plants.

III: **Major Steel Bridges** — All bridge structures other than simple rolled beam bridges.

MB: **Metal Building Systems** — Pre-engineered Metal Building Structures.

Supplement: **Auxiliary and Support Structures for Nuclear Power Plants** — This supplement, applicable to nuclear plant structures designed under the AISC Specification, but not to pressure-retaining structures, offers utility companies and designers of nuclear power plants a certification program that will eliminate the need for many of the more costly, conflicting programs now in use. A fabricator must hold certification in either Category I, II or III prior to application for certification in this category.

Certification in Category II automatically includes Category I. Certification in Category III automatically includes Categories I and II. Certification in Category MB is not transferable to any other Category.

INSPECTION-EVALUATION PROCEDURE

An outside, experienced, professional organization, ABS Worldwide Technical Services, Inc. (a subsidiary of American Bureau of Shipping) has been retained by AISC to perform the plant Inspection-Evaluation in accordance with a standard check list and rating procedure established by AISC for each certification category in the program. Upon completion of this Inspection-Evaluation, ABS Worldwide Technical Services (commonly known as ABSTECH) will recommend to AISC that a fabricator be approved or disapproved for certification. ABSTECH's Inspection-Evaluation is totally independent of the fabricator's and AISC's influence, and their evaluation is not subject to review by AISC.

At a time mutually agreed upon by the fabricator, AISC, and ABSTECH, the Inspection-Evaluation team visits the plant to investigate and rate the following basic plant functions directly and indirectly affecting quality assurance: General Management, Engineering and Drafting, Procurement, Shop Operations, and Quality Control. The Inspection-Evaluation team will perform the following:

1. Confirm data submitted with the Application for Certification.
2. Interview key supervisory personnel and subordinate employees.
3. Observe and rate the organization in operation, including procedures used in functions affecting quality assurance.
4. Inspect and rate equipment and facilities.
5. At an "exit interview," review with plant management the completed check list observations and evaluation scoring, including discussions of deficiencies and omissions, if any.

The number of days required for Inspection-Evaluation varies according to the size and complexity of the plant, but usually requires two to five days.

CERTIFICATION

Following recommendation for Certification by the Inspection-Evaluation team, AISC will issue a certificate identifying the fabricator, the plant, and the Category of Certification. The certificate is valid for a three year period, subject to annual review in the form of unannounced inspections early in the second and third year periods. The certificate is endorsed annually, provided there is successful completion of the unannounced second and third year inspection.

An annual self-audit, based on the standard check list, must be made by plant management during the 11th and 23rd months after initial Certification. This self-audit must be retained at the plant and made available to the Inspection-Evaluation team during the unannounced second and third year inspections.

At the end of the third year, the cycle begins again with a complete prescheduled Inspection-Evaluation and the issuance of a new certificate.

PRESENT STATUS

Two of the major Building Code bodies in the country have recognized that the AISC Quality Certification Program assures uniform minimum standards of quality in structural steel fabrication. AISC has been named a Quality Assurance Agency by Southern Building Code Congress International, Inc. in their report number Q.A. 7801-78 and by Building Officials and Code Administrators International, Inc. in their report RR 77-61.

For additional information on this program, write to: AISC Quality Certification Administrator, 400 North Michigan Ave., Chicago, IL 60611-4185.

PART 7
Miscellaneous Data and
Mathematical Tables

MISC.

Notes

WIRE AND SHEET METAL GAGES
Equivalent thickness in decimals of an inch

Gage No.	U.S. Standard Gage for Uncoated Hot & Cold-Rolled Sheets[b]	Galvanized Sheet Gage for Hot-Dipped Zinc Coated Sheets[b]	USA Steel Wire Gage	Gage No.	U.S. Standard Gage for Uncoated Hot & Cold-Rolled Sheets[b]	Galvanized Sheet Gage for Hot-Dipped Zinc Coated Sheets[b]	USA Steel Wire Gage
7/0	—	—	.490	13	.0897	.0934	.092[a]
6/0	—	—	.462[a]	14	.0747	.0785	.080
5/0	—	—	.430[a]	15	.0673	.0710	.072
4/0	—	—	.394[a]	16	.0598	.0635	.062[a]
3/0	—	—	.362[a]	17	.0538	.0575	.054
2/0	—	—	.331	18	.0478	.0516	.048[a]
1/0	—	—	.306	19	.0418	.0456	.041
1	—	—	.283	20	.0359	.0396	.035[a]
2	—	—	.262[a]	21	.0329	.0366	—
3	.2391	—	.244[a]	22	.0299	.0336	—
4	.2242	—	.225[a]	23	.0269	.0306	—
5	.2092	—	.207	24	.0239	.0276	—
6	.1943	—	.192	25	.0209	.0247	—
7	.1793	—	.177	26	.0179	.0217	—
8	.1644	.1681	.162	27	.0164	.0202	—
9	.1495	.1532	.148[a]	28	.0149	.0187	—
10	.1345	.1382	.135	29	—	.0172	—
11	.1196	.1233	.120[a]	30	—	.0157	—
12	.1046	.1084	.106[a]				

[a]Rounded value. The steel wire gage has been taken from ASTM A510 "General Requirements for Wire Rods and Coarse Round Wire, Carbon Steel." Sizes originally quoted to 4 decimal equivalent places have been rounded to 3 decimal places in accordance with rounding procedures of ASTM "Recommended Practice" E29.
[b]The equivalent thicknesses are for information only. The product is commonly specified to decimal thickness, not to gage number.

AISI STANDARD NOMENCLATURE FOR FLAT CARBON STEEL

Thickness (Inches)	Width (Inches)					
	To 3½ incl.	Over 3½ To 6	Over 6 To 8	Over 8 To 12	Over 12 To 48	Over 48
0.2300 & thicker	Bar	Bar	Bar	Plate	Plate	Plate
0.2299 to 0.2031	Bar	Bar	Strip	Strip	Sheet	Plate
0.2030 to 0.1800	Strip	Strip	Strip	Strip	Sheet	Plate
0.1799 to 0.0499	Strip	Strip	Strip	Strip	Sheet	Sheet
0.0448 to 0.0344	Strip	Strip				
0.0343 to 0.0255	Strip		Hot-rolled sheet and strip not generally produced in these widths and thicknesses			
0.0254 & thinner						

EFFECT OF HEAT ON STRUCTURAL STEEL

Short-time elevated-temperature tensile tests on the constructional steels permitted by the AISC Specification indicate that the ratios of the elevated-temperature yield and tensile strengths to their respective room-temperature strength values are reasonably similar at any particular temperature for the various steels in the 300 to 700° F range, except for variations due to strain aging. (The tensile strength ratio may increase to a value greater than unity in the 300 to 700° F range when strain aging occurs.) Above this range, the ratio of elevated-temperature to room-temperature strength decreases as the temperature increases.

The composition of the steels is usually such that the carbon steels exhibit strain aging with attendant reduced notch toughness. The high-strength low-alloy and heat-treated constructional alloy steels exhibit less-pronounced or little strain aging.

As examples of the decreased ratio levels obtained at elevated temperature, the yield strength ratios for carbon and high-strength low-alloy steels are approximately 0.77 at 800° F, 0.63 at 1,000° F, and 0.37 at 1,200° F.

FIRE-RESISTANT CONSTRUCTION

ASTM Specification E119, *Standard Methods of Fire Tests of Building Construction and Materials*, outlines the procedures of fire testing of structural elements located inside a building and exposed to fire within the compartment or room in which they are located. The temperature criterion used requires that the average of the temperature readings not exceed 1,000° F for columns and 1,100° F for beams. An individual temperature reading may not exceed 1,100° F for columns and 1,200° F for beams.

Steel buildings whose condition of exterior exposure and whose combustible contents under fire hazards will not produce a steel temperature greater than the foregoing criteria may therefore be considered fire-resistive without the provision of insulating protection for the steel.

A fire exposure of severity and duration sufficient to raise the temperature of the steel much above the fire test criteria temperature will seriously impair its ability to sustain loads at the unit stresses or plasticity load factors permitted by the AISC Specification. In such cases, the members upon which the stability of the structure depends should be insulated by fire-resistive materials or constructions capable of holding the average temperature of the steel to not more than that specified for the fire test standard.

Under the E119 specification, each tested assembly is subjected to a standard fire of controlled extent and severity. The fire resistance rating is expressed as the time, in hours, that the assembly is able to withstand the fire exposure before the first critical point in its behavior is reached. These tests indicate the minimum period of time during which structural members, such as columns and beams, are capable of maintaining their strength and rigidity when subjected to the standard fire. They also establish the minimum period of time during which floors, roofs, walls, or partitions will prevent fire spread by protecting against the passage of flame, hot gases, and excessive heat.

Tables of fire resistance ratings for various insulating materials and constructions applied to structural elements are published in the AISI booklets *Fire Resistant Steel Frame Construction, Designing Fire Protection for Steel Columns*, and *Designing Fire Protection for Steel Trusses*. Ratings may also be found in publications of the Underwriters' Laboratories, Inc.

A new rational fire-protection design procedure for exposed columns and beams at building exteriors has been developed by the American Iron and Steel Institute, and is described in AISI publication No. FS3, *Fire-Safe Structural Steel—A Design Guide*. The Design Guide provides a step by step procedure which enables building designers to estimate the maximum steel temperature that would occur during a fire at any location on a structural member located outside a building. The design procedure is accepted by some building codes and is under study for adoption by others.

To judge the effect of a fire on structural steel, it is necessary to consider what happens in such an exposure. Peculiarities of this exposure are: (1) temperature attained by the steel can only be estimated, (2) time of exposure at any given temperature is unknown, (3) heating is uneven, (4) cooling rates vary and can only be estimated, and (5) the steel is usually under load, and is sometimes restrained from normal expansion.

Carbon and high-strength low-alloy steels that show no evidence of gross damage from exposure to high temperatures, or from sudden cooling from high temperatures, can usually be straightened as necessary and be reused without reduction of working stress. Quenched and tempered alloy steels should not be heated to temperatures within 50° F of the tempering temperature used in heat treatment. Thus, for the quenched and tempered constructional alloy steels approved by the AISC Specification, i.e., ASTM A514, for which the tempering temperature is 1,150° F, the maximum steel temperature should be 1,100° F.

Steel that has been exposed to very high temperatures can be identified by very heavy scale, pitting, and surface erosion. Such temperatures may not only cause a loss of cross section, but may also result in metallurgical changes. Normally these conditions will be accompanied by such severe deformation that the cost and difficulty of straightening such members, as compared to replacement, dictates that they be discarded.

Steel members that have suffered rapid cooling will usually be so severely distorted that straightening for reuse will seldom be considered practicable.

In some cases, there may be some deformation in members whose normal thermal expansion is inhibited or prevented by the nature of the construction. Such members may usually be straightened and reused.

Connections require special attention to make sure that the stresses induced by a fire, and by subsequent cooling after the fire, have not sheared or loosened bolts or rivets, or cracked welds.

COEFFICIENT OF EXPANSION

The average coefficient of expansion for structural steel between 70° F and 100° F is 0.0000065 for each degree. For temperatures of 100° F to 1,200° F the coefficient is given by the approximate formula:

$$\epsilon = (6.1 + 0.0019t) \times 10^{-6}$$

in which ϵ is the coefficient of expansion for each degree Fahrenheit and t is the temperature in degrees Fahrenheit.

The modulus of elasticity of structural steel is approximately 29,000 ksi at 70° F. It decreases linearly to about 25,000 ksi at 900° F, and then begins to drop at an increasing rate at higher temperatures.

EFFECT OF HEAT DUE TO WELDING

Application of heat by welding produces residual stresses, which are generally accompanied by distortion of various amounts. Both the stresses and distortions are minimized by controlled welding procedures and fabrication methods. In normal structural practice, it has not been found necessary or desirable to use heat treatment (stress-relieving) as a means of reducing residual stresses. Procedures normally followed include: (1) proper positioning of the components of joints before welding, (2) selection of welding sequences determined by experience, (3) deposition of a minimum volume of weld metal with a minimum number of passes for the design condition, and (4) preheating as determined by experience (usually above the specified minimums).

USE OF HEAT TO STRAIGHTEN, CAMBER
OR CURVE MEMBERS

With modern fabrication techniques, a controlled application of heat can be effectively used to either straighten or to intentionally curve structural members. By this process, the member is rapidly heated in selected areas; the heated areas tend to expand, but are restrained by adjacent cooler areas. This action causes a permanent plastic deformation or "upset" of the heated areas and, thus, a change of shape is developed in the cooled member.

"Heat straightening" is used in both normal shop fabrication operations and in the field to remove relatively severe accidental bends in members. Conversely, "heat cambering" and "heat curving" of either rolled beams or welded girders are examples of the use of heat to effect a desired curvature.

As with many other fabrication operations, the use of heat to straighten or curve will cause residual stresses in the member as a result of plastic deformations. These stresses are similar to those that develop in rolled structural shapes as they cool from the rolling temperature; in this case, the stresses arise because all parts of the shape do not cool at the same rate. In like manner, welded members develop residual stresses from the localized heat of welding.

In general, the residual stresses from heating operations do not affect the ultimate strength of structural members. Any reduction in column strength due to residual stresses is incorporated in the present design provisions.

The mechanical properties of steels are largely unaffected by heating operations, provided that the maximum temperature does not exceed 1,100° F for quenched and tempered alloy steels, and 1,300° F for other steels. The temperature should be carefully checked by temperature-indicating crayons or other suitable means during the heating process.

COEFFICIENTS OF EXPANSION

The coefficient of linear expansion (ϵ) is the change in length, per unit of length, for a change of one degree of temperature. The coefficient of surface expansion is approximately two times the linear coefficient, and the coefficient of volume expansion, for solids, is approximately three times the linear coefficient.

A bar, free to move, will increase in length with an increase in temperature and will decrease in length with a decrease in temperature. The change in length will be ϵtl, where ϵ is the coefficient of linear expansion, t the change in temperature, and l the length. If the ends of a bar are fixed, a change in temperature (t) will cause a change in the unit stress of $E\epsilon t$, and in the total stress of $AE\epsilon t$, where A is the cross sectional area of the bar and E the modulus of elasticity.

The following table gives the coefficient of linear expansion for 100°, or 100 times the value indicated above.

Example: A piece of medium steel is exactly 40 ft long at 60° F. Find the length at 90° F assuming the ends free to move.

$$\text{Change of length} = \epsilon tl = \frac{.00065 \times 30 \times 40}{100} = .0078 \text{ ft}$$

The length at 90° is 40.0078 ft

Example: A piece of medium steel is exactly 40 ft long and the ends are fixed. If the temperature increases 30° F, what is the resulting change in the unit stress?

$$\text{Change in unit stress} = E\epsilon t = \frac{29,000,000 \times .00065 \times 30}{100} = 5,655 \text{ lbs. per sq. in.}$$

COEFFICIENTS OF EXPANSION FOR 100 DEGREES = 100ϵ

Materials	Linear Expansion Centigrade	Linear Expansion Fahrenheit	Materials	Linear Expansion Centigrade	Linear Expansion Fahrenheit
METALS AND ALLOYS			**STONE AND MASONRY**		
Aluminum, wrought	.00231	.00128	Ashlar masonry	.00063	.00035
Brass	.00188	.00104	Brick masonry	.00061	.00034
Bronze	.00181	.00101	Cement, portland	.00126	.00070
Copper	.00168	.00093	Concrete	.00099	.00055
Iron, cast, gray	.00106	.00059	Granite	.00080	.00044
Iron, wrought	.00120	.00067	Limestone	.00076	.00042
Iron, wire	.00124	.00069	Marble	.00081	.00045
Lead	.00286	.00159	Plaster	.00166	.00092
Magnesium, various alloys	.0029	.0016	Rubble masonry	.00063	.00035
Nickel	.00126	.00070	Sandstone	.00097	.00054
Steel, mild	.00117	.00065	Slate	.00080	.00044
Steel, stainless, 18-8	.00178	.00099			
Zinc, rolled	.00311	.00173			
TIMBER			**TIMBER**		
Fir	.00037	.00021	Fir	.0058	.0032
Maple } parallel to fiber	.00064	.00036	Maple } perpendicular to fiber	.0048	.0027
Oak	.00049	.00027	Oak	.0054	.0030
Pine	.00054	.00030	Pine	.0034	.0019

EXPANSION OF WATER
Maximum Density = 1

C°	Volume	C°	Volume	C°	Volume	C°	Volume	C°	Volume	C°	Volume
0	1.000126	10	1.000257	30	1.004234	50	1.011877	70	1.022384	90	1.035829
4	1.000000	20	1.001732	40	1.007627	60	1.016954	80	1.029003	100	1.043116

WEIGHTS AND SPECIFIC GRAVITIES

Substance	Weight Lb. per Cu Ft	Specific Gravity	Substance	Weight Lb. per Cu Ft	Specific Gravity
ASHLAR, MASONRY			**MINERALS**		
Granite, syenite, gneiss	165	2.3-3.0	Asbestos	153	2.1-2.8
Limestone, marble	160	2.3-2.8	Barytes	281	4.50
Sandstone, bluestone	140	2.1-2.4	Basalt	184	2.7-3.2
			Bauxite	159	2.55
MORTAR RUBBLE			Borax	109	1.7-1.8
MASONRY			Chalk	137	1.8-2.6
Granite, syenite, gneiss	155	2.2-2.8	Clay, marl	137	1.8-2.6
Limestone, marble	150	2.2-2.6	Dolomite	181	2.9
Sandstone, bluestone	130	2.0-2.2	Feldspar, orthoclase	159	2.5-2.6
			Gneiss, serpentine	159	2.4-2.7
DRY RUBBLE MASONRY			Granite, syenite	175	2.5-3.1
Granite, syenite, gneiss	130	1.9-2.3	Greenstone, trap	187	2.8-3.2
Limestone, marble	125	1.9-2.1	Gypsum, alabaster	159	2.3-2.8
Sandstone, bluestone	110	1.8-1.9	Hornblende	187	3.0
			Limestone, marble	165	2.5-2.8
BRICK MASONRY			Magnesite	187	3.0
Pressed brick	140	2.2-2.3	Phosphate rock, apatite	200	3.2
Common brick	120	1.8-2.0	Porphyry	172	2.6-2.9
Soft brick	100	1.5-1.7	Pumice, natural	40	0.37-0.90
			Quartz, flint	165	2.5-2.8
CONCRETE MASONRY			Sandstone, bluestone	147	2.2-2.5
Cement, stone, sand	144	2.2-2.4	Shale, slate	175	2.7-2.9
Cement, slag, etc.	130	1.9-2.3	Soapstone, talc	169	2.6-2.8
Cement, cinder, etc.	100	1.5-1.7			
VARIOUS BUILDING					
MATERIALS			**STONE, QUARRIED, PILED**		
Ashes, cinders	40-45	—	Basalt, granite, gneiss	96	—
Cement, portland, loose	90	—	Limestone, marble, quartz	95	—
Cement, portland, set	183	2.7-3.2	Sandstone	82	—
Lime, gypsum, loose	53-64	—	Shale	92	—
Mortar, set	103	1.4-1.9	Greenstone, hornblende	107	—
Slags, bank slag	67-72	—			
Slags, bank screenings	98-117	—			
Slags, machine slag	96	—			
Slags, slag sand	49-55	—	**BITUMINOUS SUBSTANCES**		
			Asphaltum	81	1.1-1.5
EARTH, ETC., EXCAVATED			Coal, anthracite	97	1.4-1.7
Clay, dry	63	—	Coal, bituminous	84	1.2-1.5
Clay, damp, plastic	110	—	Coal, lignite	78	1.1-1.4
Clay and gravel, dry	100	—	Coal, peat, turf, dry	47	0.65-0.85
Earth, dry, loose	76	—	Coal, charcoal, pine	23	0.28-0.44
Earth, dry, packed	95	—	Coal, charcoal, oak	33	0.47-0.57
Earth, moist, loose	78	—	Coal, coke	75	1.0-1.4
Earth, moist, packed	96	—	Graphite	131	1.9-2.3
Earth, mud, flowing	108	—	Paraffine	56	0.87-0.91
Earth, mud, packed	115	—	Petroleum	54	0.87
Riprap, limestone	80-85	—	Petroleum, refined	50	0.79-0.82
Riprap, sandstone	90	—	Petroleum, benzine	46	0.73-0.75
Riprap, shale	105	—	Petroleum, gasoline	42	0.66-0.69
Sand, gravel, dry, loose	90-105	—	Pitch	69	1.07-1.15
Sand, gravel, dry, packed	100-120	—	Tar, bituminous	75	1.20
Sand, gravel, wet	118-120	—			
EXCAVATIONS IN WATER					
Sand or gravel	60	—	**COAL AND COKE, PILED**		
Sand or gravel and clay	65	—	Coal, anthracite	47-58	—
Clay	80	—	Coal, bituminous, lignite	40-54	—
River mud	90	—	Coal, peat, turf	20-26	—
Soil	70	—	Coal, charcoal	10-14	—
Stone riprap	65	—	Coal, coke	23-32	—

The specific gravities of solids and liquids refer to water at 4°C, those of gases to air at 0°C and 760 mm pressure. The weights per cubic foot are derived from average specific gravities, except where stated that weights are for bulk, heaped or loose material, etc.

WEIGHTS AND SPECIFIC GRAVITIES

Substance	Weight Lb. per Cu Ft	Specific Gravity	Substance	Weight Lb. per Cu Ft	Specific Gravity
METALS, ALLOYS, ORES			**TIMBER, U.S. SEASONED**		
Aluminum, cast,			Moisture Content by		
hammered	165	2.55-2.75	Weight:		
Brass, cast, rolled	534	8.4-8.7	Seasoned timber 15 to 20%		
Bronze, 7.9 to 14% Sn	509	7.4-8.9	Green timber up to 50%		
Bronze, aluminum	481	7.7	Ash, white, red	40	0.62-0.65
Copper, cast, rolled	556	8.8-9.0	Cedar, white, red	22	0.32-0.38
Copper ore, pyrites	262	4.1-4.3	Chestnut	41	0.66
Gold, cast, hammered	1205	19.25-19.3	Cypress	30	0.48
Iron, cast, pig	450	7.2	Fir, Douglas spruce	32	0.51
Iron, wrought	485	7.6-7.9	Fir, eastern	25	0.40
Iron, speigel-eisen	468	7.5	Elm, white	45	0.72
Iron, ferro-silicon	437	6.7-7.3	Hemlock	29	0.42-0.52
Iron ore, hematite	325	5.2	Hickory	49	0.74-0.84
Iron ore, hematite in bank ..	160-180	—	Locust	46	0.73
Iron ore, hematite loose	130-160	—	Maple, hard	43	0.68
Iron ore, limonite	237	3.6-4.0	Maple, white	33	0.53
Iron ore, magnetite	315	4.9-5.2	Oak, chestnut	54	0.86
Iron slag	172	2.5-3.0	Oak, live	59	0.95
Lead	710	11.37	Oak, red, black	41	0.65
Lead ore, galena	465	7.3-7.6	Oak, white	46	0.74
Magnesium, alloys	112	1.74-1.83	Pine, Oregon	32	0.51
Manganese	475	7.2-8.0	Pine, red	30	0.48
Manganese ore, pyrolusite ...	259	3.7-4.6	Pine, white	26	0.41
Mercury	849	13.6	Pine, yellow, long-leaf	44	0.70
Monel Metal	556	8.8-9.0	Pine, yellow, short-leaf ...	38	0.61
Nickel	565	8.9-9.2	Poplar	30	0.48
Platinum, cast, hammered ...	1330	21.1-21.5	Redwood, California	26	0.42
Silver, cast, hammered	656	10.4-10.6	Spruce, white, black	27	0.40-0.46
Steel, rolled	490	7.85	Walnut, black	38	0.61
Tin, cast, hammered	459	7.2-7.5	Walnut, white	26	0.41
Tin ore, cassiterite	418	6.4-7.0			
Zinc, cast, rolled	440	6.9-7.2			
Zinc ore, blende	253	3.9-4.2			
			VARIOUS LIQUIDS		
			Alcohol, 100%		
			Acids, muriatic 40%.....	49	0.79
			Acids, nitric 91%.....	75	1.20
VARIOUS SOLIDS			Acids, sulphuric 87%.....	94	1.50
Cereals, oats .... bulk	32	—	Lye, soda 66%.....	112	1.80
Cereals, barley ...... bulk	39	—	Oils, vegetable	106	1.70
Cereals, corn, rye ..... bulk	48	—	Oils, mineral, lubricants ..	58	0.91-0.94
Cereals, wheat ...... bulk	48	—	Water, 4°C max. density ..	57	0.90-0.93
Hay and Straw .... bales	20	—	Water, 4°C max. density ..	62.428	1.0
Cotton, Flax, Hemp	93	1.47-1.50	Water, 100°C	59.830	0.9584
Fats	58	0.90-0.97	Water, ice	56	0.88-0.92
Flour, loose	28	0.40-0.50	Water, snow, fresh fallen	8	.125
Flour, pressed	47	0.70-0.80	Water, sea water	64	1.02-1.03
Glass, common	156	2.40-2.60			
Glass, plate or crown	161	2.45-2.72			
Glass, crystal	184	2.90-3.00			
Leather	59	0.86-1.02	**GASES**		
Paper	58	0.70-1.15	Air, 0°C 760 mm	.08071	1.0
Potatoes, piled	42	—	Ammonia	.0478	0.5920
Rubber, caoutchouc	59	0.92-0.96	Carbon dioxide	.1234	1.5291
Rubber goods	94	1.0-2.0	Carbon monoxide	.0781	0.9673
Salt, granulated, piled	48	—	Gas, illuminating	.028-.036	0.35-0.45
Saltpeter	67	—	Gas, natural	.038-.039	0.47-0.48
Starch	96	1.53	Hydrogen	.00559	0.0693
Sulphur	125	1.93-2.07	Nitrogen	.0784	0.9714
Wool	82	1.32	Oxygen	.0892	1.1056

The specific gravities of solids and liquids refer to water at 4°C, those of gases to air at 0°C and 760 mm pressure. The weights per cubic foot are derived from average specific gravities, except where stated that weights are for bulk, heaped or loose material, etc.

WEIGHTS OF BUILDING MATERIALS

Materials	Weight Lb. per Sq Ft	Materials	Weight Lb. per Sq Ft
CEILINGS		**PARTITIONS**	
Channel suspended		Clay Tile	
system	1	3 in.	17
Lathing and plastering	See Partitions	4 in.	18
Acoustical fiber tile	1	6 in.	28
		8 in.	34
		10 in.	40
FLOORS	See	Gypsum Block	
Steel Deck	Manufacturer	2 in.	9½
		3 in.	10½
Concrete-Reinforced 1 in.		4 in.	12½
Stone	12½	5 in.	14
Slag	11½	6 in.	18½
Lightweight	6 to 10	Wood Studs 2 × 4	
		12-16 in. o.c.	2
Concrete-Plain 1 in.		Steel partitions	4
Stone	12	Plaster 1 inch	
Slag	11	Cement	10
Lightweight	3 to 9	Gypsum	5
		Lathing	
Fills 1 inch		Metal	½
Gypsum	6	Gypsum Board ½ in.	2
Sand	8		
Cinders	4		
		WALLS	
Finishes		Brick	
Terrazzo 1 in.	13	4 in.	40
Ceramic or Quarry Tile ¾		8 in.	80
in.	10	12 in.	120
Linoleum ¼ in.	1	Hollow Concrete Block	
Mastic ¾ in.	9	(Heavy Aggregate)	
Hardwood ⅞ in.	4	4 in.	30
Softwood ¾ in.	2½	6 in.	43
		8 in.	55
		12½ in.	80
ROOFS		Hollow Concrete Block	
Copper or tin	1	(Light Aggregate)	
Corrugated steel	See Manufacturer	4 in.	21
3-ply ready roofing	1	6 in.	30
3-ply felt and gravel	5½	8 in.	38
5-ply felt and gravel	6	12 in.	55
		Clay tile	
Shingles		(Load Bearing)	
Wood	2	4 in.	25
Asphalt	3	6 in.	30
Clay tile	9 to 14	8 in.	33
Slate ¼	10	12 in.	45
		Stone 4 in.	55
Sheathing		Glass Block 4 in.	18
Wood ¾ in.	3	Window, Glass, Frame	8
Gypsum 1 in.	4	& Sash	
		Curtain Walls	See Manufacturer
Insulation 1 in.		Structural Glass 1 in.	15
Loose	½	Corrugated Cement	
Poured	2	Asbestos ¼ in.	3
Rigid	1½		

For weights of other materials used in building construction, see pages 7-8 and 7-9

WEIGHTS AND MEASURES
International System of Units (SI)[a]
(Metric practice)

BASE UNITS

Quantity	Unit	Symbol
length	metre	m
mass	kilogram	kg
time	second	s
electric current	ampere	A
thermodynamic temperature	kelvin	K
amount of substance	mole	mol
luminous intensity	condela	cd

SUPPLEMENTARY UNITS

Symbol	Unit	Symbol
plane angle	radian	rad
solid angle	steradian	sr

DERIVED UNITS (WITH SPECIAL NAMES)

Quantity	Unit	Symbol	Formula
force	newton	N	$kg\text{-}m/s^2$
pressure, stress	pascal	Pa	N/m^2
energy, work, quantity of heat	joule	J	N-m
power	watt	W	J/s

DERIVED UNITS (WITHOUT SPECIAL NAMES)

Quantity	Unit	Formula
area	square metre	m^2
volume	cubic metre	m^3
velocity	metre per second	m/s
acceleration	metre per second squared	m/s^2
specific volume	cubic metre per kilogram	m^3/kg
density	kilogram per cubic metre	kg/m^3

SI PREFIXES

Multiplication Factor	Prefix	Symbol
$1\ 000\ 000\ 000\ 000\ 000\ 000 = 10^{18}$	exa	E
$1\ 000\ 000\ 000\ 000\ 000 = 10^{15}$	peta	P
$1\ 000\ 000\ 000\ 000 = 10^{12}$	tera	T
$1\ 000\ 000\ 000 = 10^{9}$	giga	G
$1\ 000\ 000 = 10^{6}$	mega	M
$1\ 000 = 10^{3}$	kilo	k
$100 = 10^{2}$	hecto[b]	h
$10 = 10^{1}$	deka[b]	da
$0.1 = 10^{-1}$	deci[b]	d
$0.01 = 10^{-2}$	centi[b]	c
$0.001 = 10^{-3}$	milli	m
$0.000\ 001 = 10^{-6}$	micro	μ
$0.000\ 000\ 001 = 10^{-9}$	nano	n
$0.000\ 000\ 000\ 001 = 10^{-12}$	pico	p
$0.000\ 000\ 000\ 000\ 001 = 10^{-15}$	femto	f
$0.000\ 000\ 000\ 000\ 000\ 001 = 10^{-18}$	atto	a

[a]Refer to ASTM E380-79 for more complete information on SI.
[b]Use is not recommended.

WEIGHTS AND MEASURES
United States System

LINEAR MEASURE

Inches	Feet	Yards	Rods	Furlongs	Miles
1.0 =	.08333 =	.02778 =	.0050505 =	.00012626 =	.00001578
12.0 =	1.0 =	.33333 =	.0606061 =	.00151515 =	.00018939
36.0 =	3.0 =	1.0 =	.1818182 =	.00454545 =	.00056818
198.0 =	16.5 =	5.5 =	1.0 =	.025 =	.003125
7920.0 =	660.0 =	220.0 =	40.0 =	1.0 =	.125
63360.0 =	5280.0 =	1760.0 =	320.0 =	8.0 =	1.0

SQUARE AND LAND MEASURE

Sq. Inches	Square Feet	Square Yards	Square Rods	Acres	Sq. Miles
1.00 =	.006944 =	.000772			
144.0 =	1.0 =	.111111			
1296.0 =	9.0 =	1.0 =	.03306 =	.000207	
39204.0 =	272.25 =	30.25 =	1.0 =	.00625 =	.0000098
	43560.0 =	4840.0 =	160.0 =	1.0 =	.0015625
		3097600.0 =	102400.0 =	640.0 =	1.0

AVOIRDUPOIS WEIGHTS

Grains	Drams	Ounces	Pounds	Tons
1.0 =	.03657 =	.002286 =	.000143 =	.0000000714
27.34375 =	1.0 =	.0625 =	.003906 =	.00000195
437.5 =	16.0 =	1.0 =	.0625 =	.00003125
7000.0 =	256.0 =	16.0 =	1.0 =	.0005
14000000.0 =	512000.0 =	32000.0 =	2000.0 =	1.0

DRY MEASURE

Pints	Quarts	Pecks	Cubic Feet	Bushels
1.0 =	.5 =	.0625 =	.01945 =	.01563
2.0 =	1.0 =	.125 =	.03891 =	.03125
16.0 =	8.0 =	1.0 =	.31112 =	.25
51.42627 =	25.71314 =	3.21414 =	1.0 =	.80354
64.0 =	32.0 =	4.0 =	1.2445 =	1.0

LIQUID MEASURE

Gills	Pints	Quarts	U.S. Gallons	Cubic Ft.
1.0 =	.25 =	.125 =	.03125 =	.00418
4.0 =	1.0 =	.5 =	.125 =	.01671
8.0 =	2.0 =	1.0 =	.250 =	.03342
32.0 =	8.0 =	4.0 =	1.0 =	.1337
			7.48052 =	1.0

SI CONVERSION FACTORS[a]

Quantity	Multiply	by	to obtain	
Length	inch	[b]25.400	millimetre	mm
	foot	[b] 0.304 800	metre	m
	yard	[b] 0.914 400	metre	m
	mile (U.S. Statute)	1.609 347	kilometre	km
	millimetre	$39.370\ 079 \times 10^{-3}$	inch	in
	metre	3.280 840	foot	ft
	metre	1.093 613	yard	yd
	kilometre	0.621 370	mile	mi
Area	square inch	[b] $0.645\ 160 \times 10^{3}$	square millimetre	mm^2
	square foot	[b] 0.092 903	square metre	m^2
	square yard	0.836 127	square metre	m^2
	square mile (U.S. Statute)	2.589 998	square kilometre	km^2
	acre	$4.046\ 873 \times 10^{3}$	square metre	m^2
	acre	0.404 687	hectare	
	square millimetre	$1.550\ 003 \times 10^{-3}$	square inch	in^2
	square metre	10.763 910	square foot	ft^2
	square metre	1.195 990	square yard	yd^2
	square kilometre	0.386 101	square mile	mi^2
	square metre	$0.247\ 104 \times 10^{-3}$	acre	
	hectare	2.471 044	acre	
Volume	cubic inch	[b]$16.387\ 06 \times 10^{3}$	cubic millimetre	mm^3
	cubic foot	$28.316\ 85 \times 10^{-3}$	cubic metre	m^3
	cubic yard	0.764 555	cubic metre	m^3
	gallon (U.S. liquid)	3.785 412	litre	l
	quart (U.S. liquid)	0.946 353	litre	l
	cubic millimetre	$61.023\ 759 \times 10^{-6}$	cubic inch	in^3
	cubic metre	35.314 662	cubic foot	ft^3
	cubic metre	1.307 951	cubic yard	yd^3
	litre	0.264 172	gallon (U.S. liquid)	gal
	litre	1.056 688	quart (U.S. liquid)	qt
Mass	ounce (avoirdupois)	28.349 52	gram	g
	pound (avoirdupois)	0.453 592	kilogram	kg
	short ton	$0.907\ 185 \times 10^{3}$	kilogram	kg
	gram	$35.273\ 996 \times 10^{-3}$	ounce (avoirdupois)	oz av
	kilogram	2.204 622	pound (avoirdupois)	lb av
	kilogram	$1.102\ 311 \times 10^{-3}$	short ton	

[a]Refer to ASTM E380-79 for more complete information on SI.
[b]Indicates exact value.

SI CONVERSION FACTORS[a]

Quantity	Multiply	by	to obtain	
Force	ounce-force	0.278 014	newton	N
	pound-force	4.448 222	newton	N
	newton	3.596 942	ounce-force	
	newton	0.224 809	pound-force	lbf
Bending Moment	pound-force-inch	0.112 985	newton-metre	N-m
	pound-force-foot	1.355 818	newton-metre	N-m
	newton-metre	8.850 748	pound-force-inch	lbf-in
	newton-metre	0.737 562	pound-force-foot	lbf-ft
Pressure, Stress	pound-force per square inch	6.894 757	kilopascal	kPa
	foot of water (39.2 F)	2.988 98	kilopascal	kPa
	inch of mercury (32 F)	3.386 38	kilopascal	kPa
	kilopascal	0.145 038	pound-force per square inch	lbf/in^2
	kilopascal	0.334 562	foot of water (39.2 F)	
	kilopascal	0.295 301	inch of mercury (32 F)	
Energy, Work, Heat	foot-pound-force	1.355 818	joule	J
	[c]British thermal unit	$1.055\ 056 \times 10^3$	joule	J
	[c]calorie	[b] 4.186 800	joule	J
	kilowatt hour	[b] $3.600\ 000 \times 10^6$	joule	J
	joule	0.737 562	foot-pound-force	ft-lbf
	joule	$0.947\ 817 \times 10^{-3}$	[c]British thermal unit	Btu
	joule	0.238 846	[c]calorie	
	joule	$0.277\ 778 \times 10^{-6}$	kilowatt hour	kW-h
Power	foot-pound-force/second	1.355 818	watt	W
	[c]British thermal unit per hour	0.293 071	watt	W
	horsepower (550 ft lbf/s)	0.745 700	kilowatt	kW
	watt	0.737 562	foot-pound-force/ second	ft-lbf/s
	watt	3.412 141	[c]British thermal unit per hour	Btu/h
	kilowatt	1.341 022	horsepower (550 ft-lbf/s)	hp
Angle	degree	$17.453\ 29 \times 10^{-3}$	radian	rad
	radian	57.295 788	degree	
Temperature	degree Fahrenheit	$t°C = (t°F - 32)/1.8$	degree Celsius	
	degree Celsius	$t°F = 1.8 \times t°C + 32$	degree Fahrenheit	

[a]Refer to ASTM E380-79 for more complete information on SI.
[b]Indicates exact value.
[c]International Table.

BRACING FORMULAS

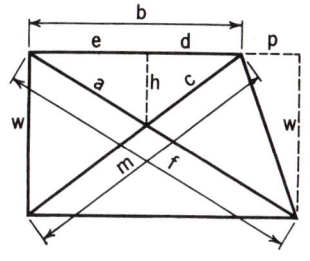
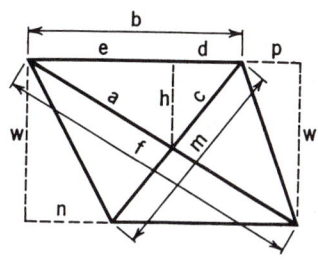

Given	To Find	Formula	Given	To Find	Formula
bpw	f	$\sqrt{(b+p)^2 + w^2}$	bpw	f	$\sqrt{(b+p)^2 + w^2}$
bw	m	$\sqrt{b^2 + w^2}$	bnw	m	$\sqrt{(b-n)^2 + w^2}$
bp	d	$b^2 \div (2b+p)$	bnp	d	$b(b-n) \div (2b+p-n)$
bp	e	$b(b+p) \div (2b+p)$	bnp	e	$b(b+p) \div (2b+p-n)$
bfp	a	$bf \div (2b+p)$	bfnp	a	$bf \div (2b+p-n)$
bmp	c	$bm \div (2b+p)$	bmnp	c	$bm \div (2b+p-n)$
bpw	h	$bw \div (2b+p)$	bnpw	h	$bw \div (2b+p-n)$
afw	h	$aw \div f$	afw	h	$aw \div f$
cmw	h	$cw \div m$	cmw	h	$cw \div m$

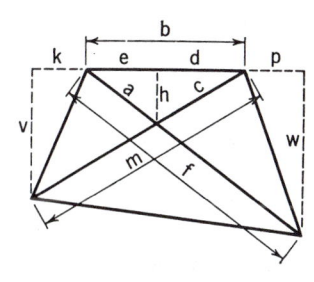

Given	To Find	Formula
bpw	f	$\sqrt{(b+p)^2 + w^2}$
bkv	m	$\sqrt{(b+k)^2 + v^2}$
bkpvw	d	$bw(b+k) \div [v(b+p) + w(b+k)]$
bkpvw	e	$bv(b+p) \div [v(b+p) + w(b+k)]$
bfkpvw	a	$fbv \div [v(b+p) + w(b+k)]$
bkmpvw	c	$bmw \div [v(b+p) + w(b+k)]$
bkpvw	h	$bvw \div [v(b+p) + w(b+k)]$
afw	h	$aw \div f$
cmv	h	$cv \div m$

PARALLEL BRACING

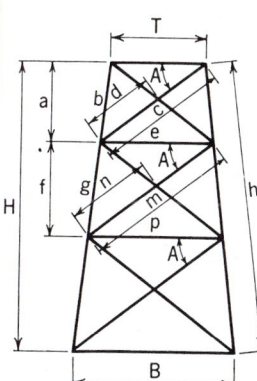

$k = (\log B - \log T) \div$ no. of panels. Constant k plus the logarithm of any line equals the log of the corresponding line in the next panel below.

$a = TH \div (T + e + p)$
$b = Th \div (T + e + p)$
$c = \sqrt{(\frac{1}{2}T + \frac{1}{2}e)^2 + a^2}$
$d = ce \div (T + e)$
$\log e = k + \log T$
$\log f = k + \log a$
$\log g = k + \log b$
$\log m = k + \log c$
$\log n = k + \log d$
$\log p = k + \log e$

The above method can be used for any number of panels.
In the formulas for "a" and "b" the sum in parenthesis, which in the case shown is $(T + e + p)$, is always composed of all the horizontal distances except the base.

PROPERTIES OF PARABOLA AND ELLIPSE

PARABOLA	ELLIPSE

PARABOLA

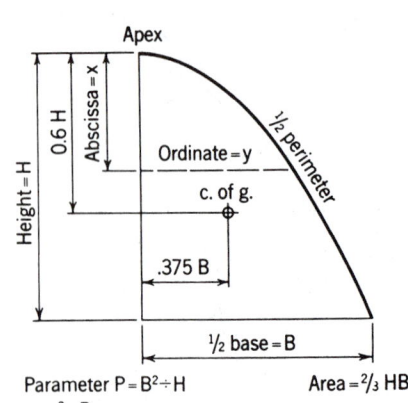

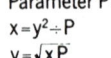

Parameter $P = B^2 \div H$ Area $= \frac{2}{3} HB$

$x = y^2 \div P$

$y = \sqrt{xP}$

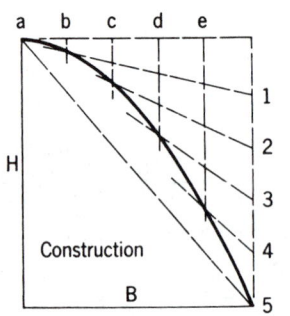

Construction

ELLIPSE

$(x^2 \div H^2) + (y^2 \div B^2) = 1$

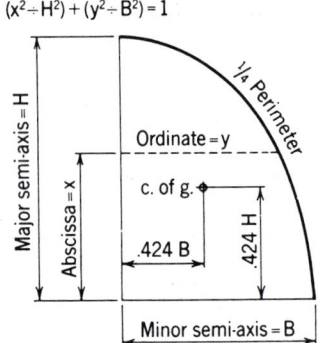

Area $= .7854\ Dd$

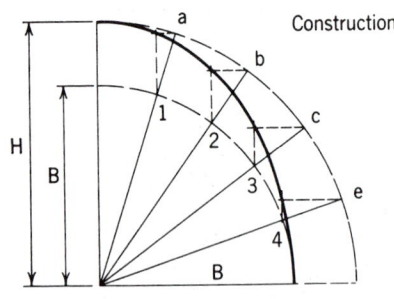

Construction

AREA BETWEEN PARABOLIC CURVE AND SECANT

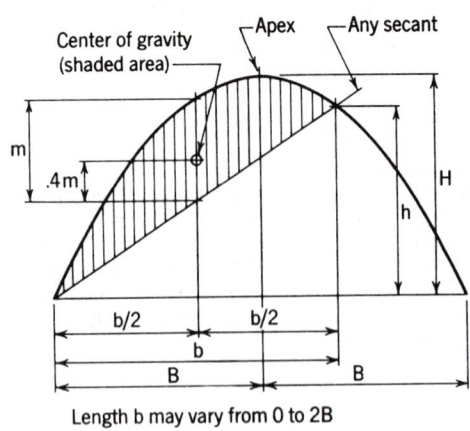

Length b may vary from 0 to 2B

PROPERTIES OF THE CIRCLE

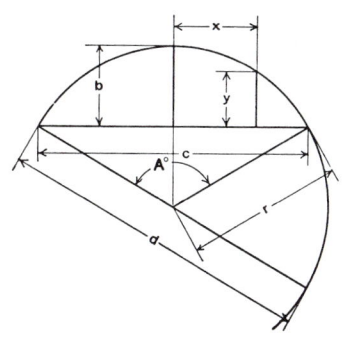

Circumference $= 6.28318\ r = 3.14159\ d$

Diameter $= 0.31831$ circumference

Area $= 3.14159\ r^2$

Arc $\quad a = \dfrac{\pi r\ A^\circ}{180^\circ} = 0.017453\ r\ A^\circ$

Angle $A^\circ = \dfrac{180^\circ\ a}{\pi r} = 57.29578\ \dfrac{a}{r}$

Radius $r = \dfrac{4\ b^2 + c^2}{8\ b}$

Chord $c = 2\ \sqrt{2\ br - b^2} = 2\ r \sin \dfrac{A}{2}$

Rise $\quad b = r - \tfrac{1}{2}\ \sqrt{4\ r^2 - c^2} = \dfrac{c}{2}\ \tan \dfrac{A}{4}$

$\qquad = 2\ r \sin^2 \dfrac{A}{4} = r + y - \sqrt{r^2 - x^2}$

$\quad y = b - r + \sqrt{r^2 - x^2}$

$\quad x = \sqrt{r^2 - (r + y - b)^2}$

Diameter of circle of equal periphery as square = 1.27324 side of square
Side of square of equal periphery as circle = 0.78540 diameter of circle
Diameter of circle circumscribed about square = 1.41421 side of square
Side of square inscribed in circle = 0.70711 diameter of circle

CIRCULAR SECTOR

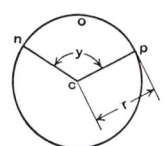

r = radius of circle $\quad y$ = angle ncp in degrees

Area of Sector ncpo = ½ (length of arc nop $\times$ r)

$\qquad$ = Area of Circle $\times \dfrac{y}{360}$

$\qquad = 0.0087266 \times r^2 \times y$

CIRCULAR SEGMENT

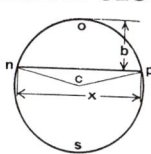

r = radius of circle $\quad x$ = chord $\quad b$ = rise

Area of Segment nop = Area of Sector ncpo — Area of triangle ncp

$\qquad = \dfrac{(\text{Length of arc nop} \times r) - x\ (r - b)}{2}$

Area of Segment nsp = Area of Circle — Area of Segment nop

VALUES FOR FUNCTIONS OF π

$\pi = 3.14159265359, \quad \log = 0.4971499$

$\pi^2 = 9.8696044,\ \log = 0.9942997 \quad \dfrac{1}{\pi} = 0.3183099,\ \log = \overline{1}.5028501 \quad \sqrt{\dfrac{1}{\pi}} = 0.5641896,\ \log = \overline{1}.7514251$

$\pi^3 = 31.0062767,\ \log = 1.4914496 \quad \dfrac{1}{\pi^2} = 0.1013212,\ \log = \overline{1}.0057003 \quad \dfrac{\pi}{180} = 0.0174533,\ \log = \overline{2}.2418774$

$\sqrt{\pi} = 1.7724539,\ \log = 0.2485749 \quad \dfrac{1}{\pi^3} = 0.0322515,\ \log = \overline{2}.5085504 \quad \dfrac{180}{\pi} = 57.2957795,\ \log = 1.7581226$

Note: Logs of fractions such as $\overline{1}.5028501$ and $\overline{2}.5085500$ may also be written 9.5028501 − 10 and 8.5085500 − 10 respectively.

PROPERTIES OF GEOMETRIC SECTIONS

SQUARE

Axis of moments through center

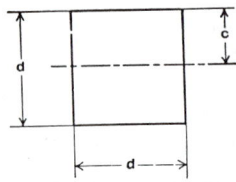

$$A = d^2$$

$$c = \frac{d}{2}$$

$$I = \frac{d^4}{12}$$

$$S = \frac{d^3}{6}$$

$$r = \frac{d}{\sqrt{12}} = .288675\ d$$

$$Z = \frac{d^3}{4}$$

SQUARE

Axis of moments on base

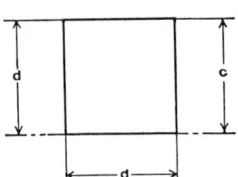

$$A = d^2$$

$$c = d$$

$$I = \frac{d^4}{3}$$

$$S = \frac{d^3}{3}$$

$$r = \frac{d}{\sqrt{3}} = .577350\ d$$

SQUARE

Axis of moments on diagonal

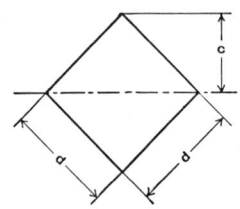

$$A = d^2$$

$$c = \frac{d}{\sqrt{2}} = .707107\ d$$

$$I = \frac{d^4}{12}$$

$$S = \frac{d^3}{6\sqrt{2}} = .117851\ d^3$$

$$r = \frac{d}{\sqrt{12}} = .288675\ d$$

$$Z = \frac{2c^3}{3} = \frac{d^3}{3\sqrt{2}} = .235702 d^3$$

RECTANGLE

Axis of moments through center

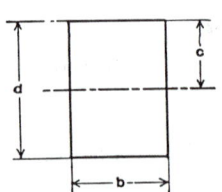

$$A = bd$$

$$c = \frac{d}{2}$$

$$I = \frac{bd^3}{12}$$

$$S = \frac{bd^2}{6}$$

$$r = \frac{d}{\sqrt{12}} = .288675\ d$$

$$Z = \frac{bd^2}{4}$$

PROPERTIES OF GEOMETRIC SECTIONS

RECTANGLE
Axis of moments on base

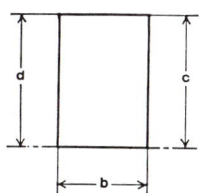

$A = bd$

$c = d$

$I = \dfrac{bd^3}{3}$

$S = \dfrac{bd^2}{3}$

$r = \dfrac{d}{\sqrt{3}} = .577350\,d$

RECTANGLE
Axis of moments on diagonal

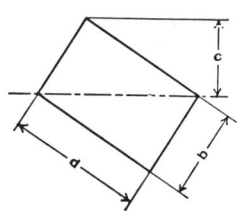

$A = bd$

$c = \dfrac{bd}{\sqrt{b^2 + d^2}}$

$I = \dfrac{b^3 d^3}{6(b^2 + d^2)}$

$S = \dfrac{b^2 d^2}{6\sqrt{b^2 + d^2}}$

$= \dfrac{bd}{\sqrt{6(b^2 + d^2)}}$

RECTANGLE
Axis of moments any line
through center of gravity

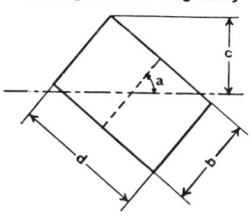

$A = bd$

$c = \dfrac{b \sin a + d \cos a}{2}$

$I = \dfrac{bd(b^2 \sin^2 a + d^2 \cos^2 a)}{12}$

$S = \dfrac{bd(b^2 \sin^2 a + d^2 \cos^2 a)}{6(b \sin a + d \cos a)}$

$r = \sqrt{\dfrac{b^2 \sin^2 a + d^2 \cos^2 a}{12}}$

HOLLOW RECTANGLE
Axis of moments through center

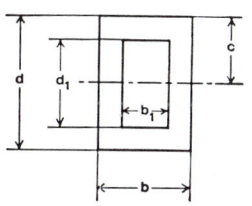

$A = bd - b_1 d_1$

$c = \dfrac{d}{2}$

$I = \dfrac{bd^3 - b_1 d_1^3}{12}$

$S = \dfrac{bd^3 - b_1 d_1^3}{6d}$

$r = \sqrt{\dfrac{bd^3 - b_1 d_1^3}{12\,A}}$

$Z = \dfrac{bd^2}{4} - \dfrac{b_1 d_1^2}{4}$

PROPERTIES OF GEOMETRIC SECTIONS

EQUAL RECTANGLES

Axis of moments through
center of gravity

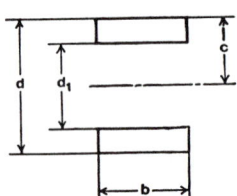

$$A = b(d - d_1)$$

$$c = \frac{d}{2}$$

$$I = \frac{b(d^3 - d_1^3)}{12}$$

$$S = \frac{b(d^3 - d_1^3)}{6d}$$

$$r = \sqrt{\frac{d^3 - d_1^3}{12(d - d_1)}}$$

$$Z = \frac{b}{4}(d^2 - d_1^2)$$

UNEQUAL RECTANGLES

Axis of moments through
center of gravity

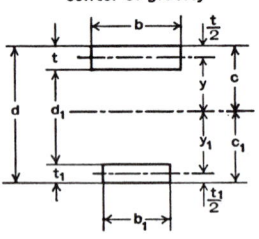

$$A = bt + b_1 t_1$$

$$c = \frac{\frac{1}{2}bt^2 + b_1 t_1(d - \frac{1}{2}t_1)}{A}$$

$$I = \frac{bt^3}{12} + bty^2 + \frac{b_1 t_1^3}{12} + b_1 t_1 y_1^2$$

$$S = \frac{I}{c} \qquad S_1 = \frac{I}{c_1}$$

$$r = \sqrt{\frac{I}{A}}$$

$$Z = \frac{A}{2}\left[d - \left(\frac{t + t_1}{2}\right)\right]$$

TRIANGLE

Axis of moments through
center of gravity

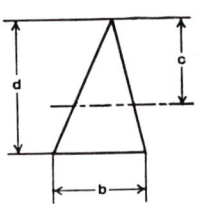

$$A = \frac{bd}{2}$$

$$c = \frac{2d}{3}$$

$$I = \frac{bd^3}{36}$$

$$S = \frac{bd^2}{24}$$

$$r = \frac{d}{\sqrt{18}} = .235702\,d$$

TRIANGLE

Axis of moments on base

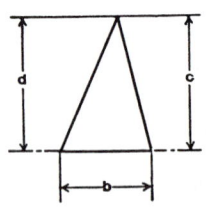

$$A = \frac{bd}{2}$$

$$c = d$$

$$I = \frac{bd^3}{12}$$

$$S = \frac{bd^2}{12}$$

$$r = \frac{d}{\sqrt{6}} = .408248\,d$$

PROPERTIES OF GEOMETRIC SECTIONS

TRAPEZOID

Axis of moments through
center of gravity

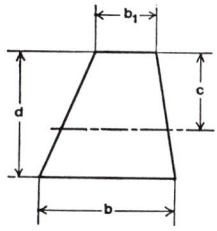

$$A = \frac{d(b + b_1)}{2}$$

$$c = \frac{d(2b + b_1)}{3(b + b_1)}$$

$$I = \frac{d^3(b^2 + 4bb_1 + b_1^2)}{36(b + b_1)}$$

$$S = \frac{d^2(b^2 + 4bb_1 + b_1^2)}{12(2b + b_1)}$$

$$r = \frac{d}{6(b + b_1)} \sqrt{2(b^2 + 4bb_1 + b_1^2)}$$

CIRCLE

Axis of moments
through center

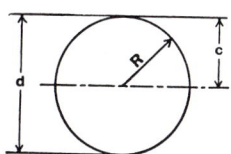

$$A = \frac{\pi d^2}{4} = \pi R^2 = .785398\, d^2 = 3.141593\, R^2$$

$$c = \frac{d}{2} = R$$

$$I = \frac{\pi d^4}{64} = \frac{\pi R^4}{4} = .049087\, d^4 = .785398\, R^4$$

$$S = \frac{\pi d^3}{32} = \frac{\pi R^3}{4} = .098175\, d^3 = .785398\, R^3$$

$$r = \frac{d}{4} = \frac{R}{2}$$

$$Z = \frac{d^3}{6}$$

HOLLOW CIRCLE

Axis of moments
through center

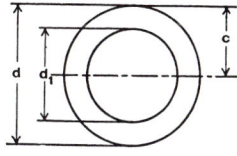

$$A = \frac{\pi(d^2 - d_1^2)}{4} = .785398\,(d^2 - d_1^2)$$

$$c = \frac{d}{2}$$

$$I = \frac{\pi(d^4 - d_1^4)}{64} = .049087\,(d^4 - d_1^4)$$

$$S = \frac{\pi(d^4 - d_1^4)}{32d} = .098175\,\frac{d^4 - d_1^4}{d}$$

$$r = \frac{\sqrt{d^2 + d_1^2}}{4}$$

$$Z = \frac{d^3}{6} - \frac{d_1^3}{6}$$

HALF CIRCLE

Axis of moments through
center of gravity

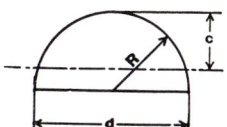

$$A = \frac{\pi R^2}{2} = 1.570796\, R^2$$

$$c = R\left(1 - \frac{4}{3\pi}\right) = .575587\, R$$

$$I = R^4\left(\frac{\pi}{8} - \frac{8}{9\pi}\right) = .109757\, R^4$$

$$S = \frac{R^3}{24}\frac{(9\pi^2 - 64)}{(3\pi - 4)} = .190687\, R^3$$

$$r = R\frac{\sqrt{9\pi^2 - 64}}{6\pi} = .264336\, R$$

PROPERTIES OF GEOMETRIC SECTIONS

PARABOLA

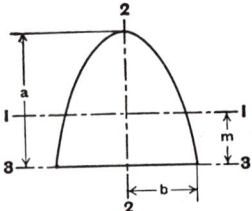

$$A = \frac{4}{3} ab$$

$$m = \frac{2}{5} a$$

$$I_1 = \frac{16}{175} a^3b$$

$$I_2 = \frac{4}{15} ab^3$$

$$I_3 = \frac{32}{105} a^3b$$

HALF PARABOLA

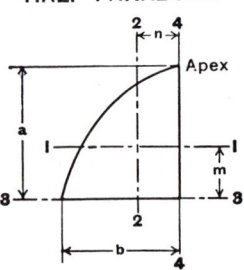

$$A = \frac{2}{3} ab$$

$$m = \frac{2}{5} a$$

$$n = \frac{3}{8} b$$

$$I_1 = \frac{8}{175} a^3b$$

$$I_2 = \frac{19}{480} ab^3$$

$$I_3 = \frac{16}{105} a^3b$$

$$I_4 = \frac{2}{15} ab^3$$

COMPLEMENT OF HALF PARABOLA

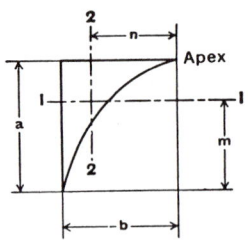

$$A = \frac{1}{3} ab$$

$$m = \frac{7}{10} a$$

$$n = \frac{3}{4} b$$

$$I_1 = \frac{37}{2100} a^3b$$

$$I_2 = \frac{1}{80} ab^3$$

PARABOLIC FILLET IN RIGHT ANGLE

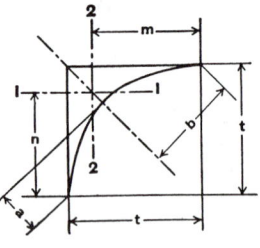

$$a = \frac{t}{2\sqrt{2}}$$

$$b = \frac{t}{\sqrt{2}}$$

$$A = \frac{1}{6} t^2$$

$$m = n = \frac{4}{5} t$$

$$I_1 = I_2 = \frac{11}{2100} t^4$$

PROPERTIES OF GEOMETRIC SECTIONS

* HALF ELLIPSE

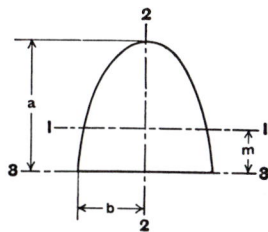

$$A = \frac{1}{2}\pi ab$$

$$m = \frac{4a}{3\pi}$$

$$I_1 = a^3b\left(\frac{\pi}{8} - \frac{8}{9\pi}\right)$$

$$I_2 = \frac{1}{8}\pi ab^3$$

$$I_3 = \frac{1}{8}\pi a^3 b$$

* QUARTER ELLIPSE

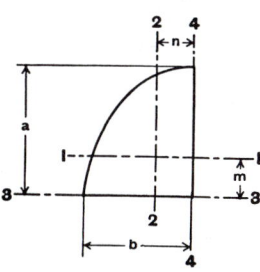

$$A = \frac{1}{4}\pi ab$$

$$m = \frac{4a}{3\pi}$$

$$n = \frac{4b}{3\pi}$$

$$I_1 = a^3b\left(\frac{\pi}{16} - \frac{4}{9\pi}\right)$$

$$I_2 = ab^3\left(\frac{\pi}{16} - \frac{4}{9\pi}\right)$$

$$I_3 = \frac{1}{16}\pi a^3 b$$

$$I_4 = \frac{1}{16}\pi ab^3$$

* ELLIPTIC COMPLEMENT

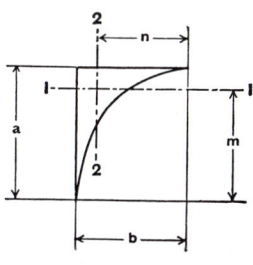

$$A = ab\left(1 - \frac{\pi}{4}\right)$$

$$m = \frac{a}{6\left(1 - \frac{\pi}{4}\right)}$$

$$n = \frac{b}{6\left(1 - \frac{\pi}{4}\right)}$$

$$I_1 = a^3b\left(\frac{1}{3} - \frac{\pi}{16} - \frac{1}{36\left(1 - \frac{\pi}{4}\right)}\right)$$

$$I_2 = ab^3\left(\frac{1}{3} - \frac{\pi}{16} - \frac{1}{36\left(1 - \frac{\pi}{4}\right)}\right)$$

* To obtain properties of half circle, quarter circle and circular complement substitute a = b = R.

PROPERTIES OF GEOMETRIC SECTIONS AND STRUCTURAL SHAPES

REGULAR POLYGON

Axis of moments through center

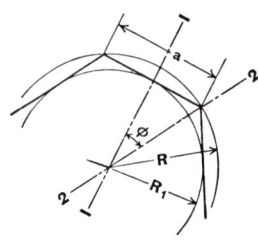

n = Number of sides

$\phi = \dfrac{180°}{n}$

$a = 2\sqrt{R^2 - R_1{}^2}$

$R = \dfrac{a}{2 \sin \phi}$

$R_1 = \dfrac{a}{2 \tan \phi}$

$A = \dfrac{1}{4} na^2 \cot \phi = \dfrac{1}{2} nR^2 \sin 2\phi = nR_1{}^2 \tan \phi$

$I_1 = I_2 = \dfrac{A(6R^2 - a^2)}{24} = \dfrac{A(12R_1{}^2 + a^2)}{48}$

$r_1 = r_2 = \sqrt{\dfrac{6R^2 - a^2}{24}} = \sqrt{\dfrac{12R_1{}^2 + a^2}{48}}$

ANGLE

Axis of moments through center of gravity

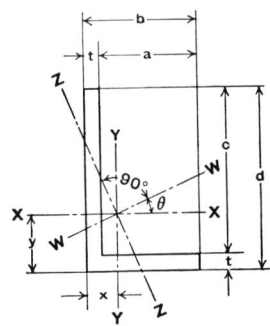

Z-Z is axis of minimum I

$\tan 2\theta = \dfrac{2K}{I_Y - I_X}$

$A = t(b + c) \quad x = \dfrac{b^2 + ct}{2(b + c)} \quad y = \dfrac{d^2 + at}{2(b + c)}$

K = Product of Inertia about X-X & Y-Y

$\quad = + \dfrac{abcdt}{4(b + c)}$

$I_x = \dfrac{1}{3}\left(t(d - y)^3 + by^3 - a(y - t)^3 \right)$

$I_Y = \dfrac{1}{3}\left(t(b - x)^3 + dx^3 - c(x - t)^3 \right)$

$I_z = I_x \sin^2\theta + I_Y \cos^2\theta + K \sin 2\theta$

$I_w = I_x \cos^2\theta + I_Y \sin^2\theta - K \sin 2\theta$

K is negative when heel of angle, with respect to c. g., is in 1st or 3rd quadrant, positive when in 2nd or 4th quadrant.

BEAMS AND CHANNELS

Transverse force oblique through center of gravity

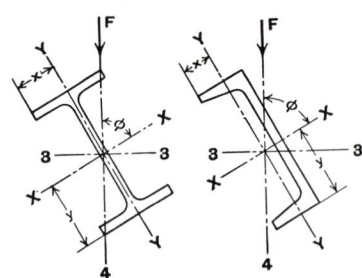

$I_3 = I_x \sin^2\phi + I_Y \cos^2\phi$

$I_4 = I_x \cos^2\phi + I_Y \sin^2\phi$

$f_b = M\left(\dfrac{y}{I_x} \sin\phi + \dfrac{x}{I_Y} \cos\phi \right)$

where M is bending moment due to force F.

TRIGONOMETRIC FORMULAS

TRIGONOMETRIC FUNCTIONS

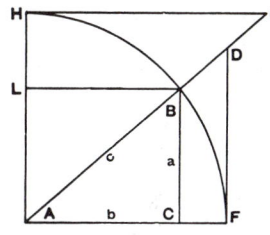

Radius AF $= 1$
$= \sin^2 A + \cos^2 A = \sin A \cosec A$
$= \cos A \sec A = \tan A \cot A$

Sine A $= \dfrac{\cos A}{\cot A} = \dfrac{1}{\cosec A} = \cos A \tan A = \sqrt{1 - \cos^2 A} = BC$

Cosine A $= \dfrac{\sin A}{\tan A} = \dfrac{1}{\sec A} = \sin A \cot A = \sqrt{1 - \sin^2 A} = AC$

Tangent A $= \dfrac{\sin A}{\cos A} = \dfrac{1}{\cot A} = \sin A \sec A$ $= FD$

Cotangent A $= \dfrac{\cos A}{\sin A} = \dfrac{1}{\tan A} = \cos A \cosec A$ $= HG$

Secant A $= \dfrac{\tan A}{\sin A} = \dfrac{1}{\cos A}$ $= AD$

Cosecant A $= \dfrac{\cot A}{\cos A} = \dfrac{1}{\sin A}$ $= AG$

RIGHT ANGLED TRIANGLES

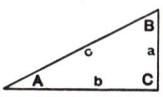

$a^2 = c^2 - b^2$
$b^2 = c^2 - a^2$
$c^2 = a^2 + b^2$

Known	Required					
	A	B	a	b	c	Area
a, b	$\tan A = \dfrac{a}{b}$	$\tan B = \dfrac{b}{a}$			$\sqrt{a^2 + b^2}$	$\dfrac{ab}{2}$
a, c	$\sin A = \dfrac{a}{c}$	$\cos B = \dfrac{a}{c}$		$\sqrt{c^2 - a^2}$		$\dfrac{a\sqrt{c^2 - a^2}}{2}$
A, a		$90° - A$		$a \cot A$	$\dfrac{a}{\sin A}$	$\dfrac{a^2 \cot A}{2}$
A, b		$90° - A$	$b \tan A$		$\dfrac{b}{\cos A}$	$\dfrac{b^2 \tan A}{2}$
A, c		$90° - A$	$c \sin A$	$c \cos A$		$\dfrac{c^2 \sin 2A}{4}$

OBLIQUE ANGLED TRIANGLES

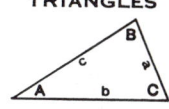

$s = \dfrac{a + b + c}{2}$

$K = \sqrt{\dfrac{(s - a)(s - b)(s - c)}{s}}$

$a^2 = b^2 + c^2 - 2bc \cos A$
$b^2 = a^2 + c^2 - 2ac \cos B$
$c^2 = a^2 + b^2 - 2ab \cos C$

Known	Required					
	A	B	C	b	c	Area
a, b, c	$\tan \frac{1}{2}A = \dfrac{K}{s - a}$	$\tan \frac{1}{2}B = \dfrac{K}{s - b}$	$\tan \frac{1}{2}C = \dfrac{K}{s - c}$			$\sqrt{s(s-a)(s-b)(s-c)}$
a, A, B			$180° - (A + B)$	$\dfrac{a \sin B}{\sin A}$	$\dfrac{a \sin C}{\sin A}$	
a, b, A		$\sin B = \dfrac{b \sin A}{a}$			$\dfrac{b \sin C}{\sin B}$	
a, b, C	$\tan A = \dfrac{a \sin C}{b - a \cos C}$				$\sqrt{a^2 + b^2 - 2ab \cos C}$	$\dfrac{ab \sin C}{2}$

DECIMALS OF AN INCH
For each 64th of an inch
With millimeter equivalents

Fraction	⁄64ths	Decimal	Millimeters (Approx.)	Fraction	⁄64ths	Decimal	Millimeters (Approx.)
—	1	.015625	0.397	—	33	.515625	13.097
1/32	2	.03125	0.794	17/32	34	.53125	13.494
—	3	.046875	1.191	—	35	.546875	13.891
1/16	4	.0625	1.588	9/16	36	.5625	14.288
—	5	.078125	1.984	—	37	.578125	14.684
3/32	6	.09375	2.381	19/32	38	.59375	15.081
—	7	.109375	2.778	—	39	.609375	15.478
1/8	8	.125	3.175	5/8	40	.625	15.875
—	9	.140625	3.572	—	41	.640625	16.272
5/32	10	.15625	3.969	21/32	42	.65625	16.669
—	11	.171875	4.366	—	43	.671875	17.066
3/16	12	.1875	4.763	11/16	44	.6875	17.463
—	13	.203125	5.159	—	45	.703125	17.859
7/32	14	.21875	5.556	23/32	46	.71875	18.256
—	15	.234375	5.953	—	47	.734375	18.653
1/4	16	.250	6.350	3/4	48	.750	19.050
—	17	.265625	6.747	—	49	.765625	19.447
9/32	18	.28125	7.144	25/32	50	.78125	19.844
—	19	.296875	7.541	—	51	.796875	20.241
5/16	20	.3125	7.938	13/16	52	.8125	20.638
—	21	.328125	8.334	—	53	.828125	21.034
11/32	22	.34375	8.731	27/32	54	.84375	21.431
—	23	.359375	9.128	—	55	.859375	21.828
3/8	24	.375	9.525	7/8	56	.875	22.225
—	25	.390625	9.922	—	57	.890625	22.622
13/32	26	.40625	10.319	29/32	58	.90625	23.019
—	27	.421875	10.716	—	59	.921875	23.416
7/16	28	.4375	11.113	15/16	60	.9375	23.813
—	29	.453125	11.509	—	61	.953125	24.209
15/32	30	.46875	11.906	31/32	62	.96875	24.606
—	31	.484375	12.303	—	63	.984375	25.003
1/2	32	.500	12.700	1	64	1.000	25.400

DECIMALS OF A FOOT
For each 32nd of an inch

Inch	0	1	2	3	4	5
0	0	.0833	.1667	.2500	.3333	.4167
1/32	.0026	.0859	.1693	.2526	.3359	.4193
1/16	.0052	.0885	.1719	.2552	.3385	.4219
3/32	.0078	.0911	.1745	.2578	.3411	.4245
1/8	.0104	.0938	.1771	.2604	.3438	.4271
5/32	.0130	.0964	.1797	.2630	.3464	.4297
3/16	.0156	.0990	.1823	.2656	.3490	.4323
7/32	.0182	.1016	.1849	.2682	.3516	.4349
1/4	.0208	.1042	.1875	.2708	.3542	.4375
9/32	.0234	.1068	.1901	.2734	.3568	.4401
5/16	.0260	.1094	.1927	.2760	.3594	.4427
11/32	.0286	.1120	.1953	.2786	.3620	.4453
3/8	.0313	.1146	.1979	.2812	.3646	.4479
13/32	.0339	.1172	.2005	.2839	.3672	.4505
7/16	.0365	.1198	.2031	.2865	.3698	.4531
15/32	.0391	.1224	.2057	.2891	.3724	.4557
1/2	.0417	.1250	.2083	.2917	.3750	.4583
17/32	.0443	.1276	.2109	.2943	.3776	.4609
9/16	.0469	.1302	.2135	.2969	.3802	.4635
19/32	.0495	.1328	.2161	.2995	.3828	.4661
5/8	.0521	.1354	.2188	.3021	.3854	.4688
21/32	.0547	.1380	.2214	.3047	.3880	.4714
11/16	.0573	.1406	.2240	.3073	.3906	.4740
23/32	.0599	.1432	.2266	.3099	.3932	.4766
3/4	.0625	.1458	.2292	.3125	.3958	.4792
25/32	.0651	.1484	.2318	.3151	.3984	.4818
13/16	.0677	.1510	.2344	.3177	.4010	.4844
27/32	.0703	.1536	.2370	.3203	.4036	.4870
7/8	.0729	.1563	.2396	.3229	.4063	.4896
29/32	.0755	.1589	.2422	.3255	.4089	.4922
15/16	.0781	.1615	.2448	.3281	.4115	.4948
31/32	.0807	.1641	.2472	.3307	.4141	.4974

DECIMALS OF A FOOT
For each 32nd of an inch

Inch	6	7	8	9	10	11
0	.5000	.5833	.6667	.7500	.8333	.9167
1/32	.5026	.5859	.6693	.7526	.8359	.9193
1/16	.5052	.5885	.6719	.7552	.8385	.9219
3/32	.5078	.5911	.6745	.7578	.8411	.9245
1/8	.5104	.5938	.6771	.7604	.8438	.9271
5/32	.5130	.5964	.6797	.7630	.8464	.9297
3/16	.5156	.5990	.6823	.7656	.8490	.9323
7/32	.5182	.6016	.6849	.7682	.8516	.9349
1/4	.5208	.6042	.6875	.7708	.8542	.9375
9/32	.5234	.6068	.6901	.7734	.8568	.9401
5/16	.5260	.6094	.6927	.7760	.8594	.9427
11/32	.5286	.6120	.6953	.7786	.8620	.9453
3/8	.5313	.6146	.6979	.7813	.8646	.9479
13/32	.5339	.6172	.7005	.7839	.8672	.9505
7/16	.5365	.6198	.7031	.7865	.8698	.9531
15/32	.5391	.6224	.7057	.7891	.8724	.9557
1/2	.5417	.6250	.7083	.7917	.8750	.9583
17/32	.5443	.6276	.7109	.7943	.8776	.9609
9/16	.5469	.6302	.7135	.7969	.8802	.9635
19/32	.5495	.6328	.7161	.7995	.8828	.9661
5/8	.5521	.6354	.7188	.8021	.8854	.9688
21/32	.5547	.6380	.7214	.8047	.8880	.9714
11/16	.5573	.6406	.7240	.8073	.8906	.9740
23/32	.5599	.6432	.7266	.8099	.8932	.9766
3/4	.5625	.6458	.7292	.8125	.8958	.9792
25/32	.5651	.6484	.7318	.8151	.8984	.9818
13/16	.5677	.6510	.7344	.8177	.9010	.9844
27/32	.5703	.6536	.7370	.8203	.9036	.9870
7/8	.5729	.6563	.7396	.8229	.9063	.9896
29/32	.5755	.6589	.7422	.8255	.9089	.9922
15/16	.5781	.6615	.7448	.8281	.9115	.9948
31/32	.5807	.6641	.7474	.8307	.9141	.9974

PART 8
INDEX

General Nomenclature

A	Cross-sectional area, in.^{2}
A_B	Loaded area of concrete, in.^{2}
A_H	Area of H-shaped portion of base plate in light columns, in.^{2}
A_b	Nominal body area of a fastener, in.^{2}
A_b	Area of an upset rod based upon the major diameter of its threads, in.^{2}
A_{bc}	Planar area of web at beam-to-column connection, in.^{2}
A_c	Area of concrete in a composite column, in.^{2}
A_c	Area of concrete slab within effective width, in.^{2}
A_e	Effective net area, in.^{2}
A_f	Area of flange, in.^{2}
A_g	Gross area, in.^{2}
A_n	Net area, in.^{2}
A_{ns}	Net area subject to shear, in.^{2}
A_{nt}	Net area subject to tension, in.^{2}
A_p	Area of top or bottom plate in a moment connection, in.^{2}
A_{pb}	Projected bearing area, in.^{2}
A_r	Area of reinforcing bars, in.^{2}
A_s	Area of steel cross section, in.^{2}
A_{sc}	Cross-sectional area of stud shear connector, in.^{2}
A_{sf}	Shear area on the failure path, in.^{2}
A_{st}	Cross-sectional area of stiffener or pair of stiffeners, in.^{2}
A_{tg}	Gross area subject to tension, in.^{2}
A_{vg}	Gross area subject to shear, in.^{2}
A_w	Web area, in.^{2}
A_1	Area of steel bearing concentrically on a concrete support, in.^{2}
A_2	Total cross-sectional area of a concrete support, in.^{2}
B	Design tensile strength of bolt, kips
B	Factor for bending stress in web-tapered members; defined by Formulas A-F4-7 through A-F4-10, in.
B	Base plate width, in.
B_c	Factored load per bolt including prying action, kips

B_1, B_2 Factors used in determining M_u for combined bending and axial forces when elastic, first order analysis is employed

BF A factor that can be used to calculate the design nominal moment capacity for unbraced length L_b, between L_p and L_r, defined in Part 3

C Coefficient for determining permissible loads in kips for eccentrically loaded welded connections

C Coefficient used to determine the number of fasteners required in a one-sided framed beam connection or other eccentrically loaded bolted connections

C_{PG} Plate girder coefficient

C_{Tot} Sum of compressive forces in a composite beam, kips

C_a Constant given in Table A, Part 5, used in calculating the design moment M_e for end plate connections

C_b Coefficient used in calculating the design moment M_e for end plate connections

C_b Bending coefficient dependent upon moment gradient

$$= 1.75 + 1.05 \left(\frac{M_1}{M_2} \right) + 0.3 \left(\frac{M_1}{M_2} \right)^2$$

C_{conc} Effective concrete flange force for a composite beam, kips

C_m Coefficient applied to bending term in interaction formula for prismatic members and dependent upon column curvature caused by applied moments

C_m' Coefficient applied to bending term in interaction formula for tapered members and dependent upon axial stress at the small end of the member

C_p Ponding flexibility coefficient for primary member in a flat roof

C_s Ponding flexibility coefficient for secondary member in a flat roof

C_{stl} Compressive force in steel in a composite beam, kips

C_v Ratio of "critical" web stress, according to linear buckling theory, to the shear yield stress of web material

C_w Warping constant, in.[6]

C_1 Loading constant, Fig. 3.3, Part 3, used in deflection calculations

C_1 Coefficient used to determine the capacity of an eccentrically loaded weld group based on electrode strength

C_1, C_2 Coefficients for web tear-out (block shear)

D Outside diameter of circular hollow section, in.

D Dead load due to the self-weight of the structural and permanent elements on the structure

D Factor used in Formula A-G4-2, dependent on the type of transverse stiffeners used in a plate girder

D Weld size in sixteenths of an inch

E Modulus of elasticity of steel (29,000 ksi)

E Earthquake load

E_c Modulus of elasticity of concrete, ksi

E_m	Modified modulus of elasticity, ksi

E_m — Modified modulus of elasticity, ksi

F_{BM} — Nominal strength of the base material to be welded, ksi

F_{EXX} — Classification strength of weld metal, ksi

F_a — Axial design stress, psi

$F_{b\gamma}$ — Flexural stress for tapered members defined by Formulas A-F4-4 and A-F4-5, ksi

F_{cr} — Critical stress, ksi

F_e — Elastic buckling stress, ksi

F_{ex} — Elastic flexural buckling stress about the major axis, ksi

F_{ey} — Elastic flexural buckling stress about the minor axis, ksi

F_{ez} — Elastic torsional buckling stress, ksi

F_f — Nominal flange force resulting from factored live plus dead load moments, kips

F_{my} — Modified yield stress for composite columns, ksi

F_n — Nominal shear rupture strength, ksi

F_p — Nominal bearing stress on fastener, ksi

F_r — Compressive residual stress in flange, ksi

$F_{s\gamma}$ — Stress for tapered members defined by Formula A-F4-6

F_t — Nominal tensile stress in fastener or connecting element, ksi

F_u — Specified minimum tensile strength of the type of steel being used, ksi

F_v — Nominal shear stress in fastener, ksi

F_{vp} — Nominal shear stress in plate, ksi

F_w — Nominal strength of the weld electrode material, ksi

$F_{w\gamma}$ — Stress for tapered members defined by Formula A-F4-7, ksi

F_y — Specified minimum yield stress of the type of steel being used, ksi. As used in this Specification, "yield stress" denotes either the specified minimum yield point (for steels that have a yield point) or specified yield strength (for steels that do not have a yield point)

F_y''' — The theoretical maximum yield stress (ksi) based on the web depth-thickness ratio (h_c/t_w) below which a particular shape may be considered "compact" for any condition of combined bending and axial force (see LRFD Specification Table B5.1)

$$= \left[\frac{253}{h_c/t_w}\right]^2$$

F_{yc} — Yield strength of column web, ksi

F_{yf} — Specified minimum yield stress of the flange, ksi

F_{ym} — Yield stress obtained from mill test reports or from physical tests, ksi

F_{yr} — Specified minimum yield stress of reinforcing bars, ksi

F_{ys} — Static yield stress, ksi

F_{yst} — Specified minimum yield stress of the stiffener material, ksi

F_{yw} — Specified minimum yield stress of the web, ksi

G — Shear modulus of elasticity of steel (11,200 ksi)

G — Ratio of the total column stiffness framing into a joint to that of the stiffening members framing into the same joint

H	Horizontal force, kips
H	Flexural constant
H_s	Length of stud connector after welding, in.
I	Moment of inertia, in.⁴
I_{LB}	Lower bound moment of inertia for composite section, in.⁴
I_c	Moment of inertia of column section about axis perpendicular to plane of buckling, in.⁴
I_d	Moment of inertia of the steel deck supported on secondary members, in.⁴
I_g	Moment of inertia of girder about axis perpendicular to plane of buckling, in.⁴
I_p	Moment of inertia of primary member in flat roof framing, in.⁴
I_s	Moment of inertia of secondary member in flat roof framing, in.⁴
I_{st}	Moment of inertia of a transverse stiffener, in.⁴
J	Torsional constant for a section, in.⁴
K	Effective length factor for a prismatic member
K_{area}	An idealized area representing the contribution of the fillet to the steel beam area, as defined in the composite beam model of Part 4, in.² $= [A_s - 2A_f - (d-2k)t_w]/2$
K_{dep}	Weld depth, $(k-t_f)$, in.
K_s	Slip coefficient
K_z	Effective length factor for torsional buckling
K_γ	Effective length factor for a tapered member
L	Unbraced length of member measured between the center of gravity of the bracing members, in. or ft as indicated
L	Story height, in. or ft as indicated
L	Live load due to occupancy
L	Connection angle or plate length, in.
L	Distance in the line of force from center of a standard or oversized hole, or from the center of the end of a slotted hole to an edge of a connected part, in.
L'	Connection angle length for staggered holes, in.
L_b	Laterally unbraced length; length between points which are either braced against lateral displacement of compression flange or braced against twist of the cross section, in. or ft as indicated
L_c	Length of channel shear connector, in.
L_c	Unsupported length of a column section, ft
L_g	Unsupported length of a girder or other restraining member, ft
L_m	Limiting laterally unbraced length for full plastic bending capacity, in. or ft as indicated
L'_m	Limiting laterally unbraced length for the maximum design flexural strength for noncompact shapes, in. or ft as indicated
L_p	Column spacing in direction of girder, in. or ft as indicated
L_p	Limiting laterally unbraced length for full plastic bending capacity, uniform moment case ($C_b = 1.0$), in. or ft as indicated

L'_p Limiting laterally unbraced length for the maximum design flexural strength for noncompact shapes, uniform moment case ($C_b = 1.0$), in. or ft as indicated

L_{pd} Limiting laterally unbraced length for plastic analysis, in. or ft as indicated

L_r Limiting laterally unbraced length for inelastic lateral-torsional buckling, in. or ft as indicated

L_r Roof live load

L_s Column spacing perpendicular to direction of girder, in. or ft as indicated

L_v Span length below which shear, $\phi_v V_n$, in the beam web governs, in. or ft as indicated

M Beam bending moment, kip-in. or kip-ft as indicated

M Factored connection moment, kip-ft

M' Critical moment at fastener line, face of tee stem or angle leg, kip-in.

M_{LL} Beam moment due to live load, kip-in. or kip-ft as indicated

M_{cr} Elastic buckling moment, kip-in. or kip-ft as indicated

M_e Design moment for end plate connections, kip-in.

M_{lt} Required flexural strength in member due to lateral frame translation, kip-in.

M_n Nominal flexural strength, kip-in. or kip-ft as indicated

M'_n Maximum design flexural strength for noncompact shapes, when $L_b \leq L'_m$, kip-in. or kip-ft as indicated

M'_{nx}, M'_{ny} Flexural strength defined in Formulas A-H3-7 and A-H3-8, for use in alternate interaction equations for combined bending and axial force, kip-in. or kip-ft as indicated

M_{nt} Required flexural strength in member assuming there is no lateral translation of the frame, kip-in.

M_p Plastic bending moment, kip-in. or kip-ft as indicated

M'_p Moment defined in Formulas A-H3-5 and A-H3-6, for use in alternate interaction formulas for combined bending and axial force, kip-in. or kip-ft as indicated.

M_r Limiting buckling moment, M_{cr}, when $\lambda = \lambda_r$, kip-in. or kip-ft as indicated

M_u Required flexural strength, kip-in. or kip-ft as indicated

M_1 Smaller moment at end of unbraced length of beam or beam-column, kip-in.

M_2 Larger moment at end of unbraced length of beam-column, kip-in.

N Length of base plate, in.

N Length of bearing, in.

N_r Number of stud connectors in one rib at a beam intersection, not to exceed 3 in calculations

P Force transmitted by one fastener to the critical connected part, kips

P_{bf} Applied factored beam flange force in moment connections, kips

P_e Equivalent buckling strength, kips

P_e Effective span used to compute bending moment in end plate, in.

P_e Elastic buckling load, kips

P_f Distance from center line of bolt to nearer surface of tension flange in end plate connections, in.

P_{fb} Maximum column web resisting force at beam tension flange, kips

P_n Nominal axial strength (tension or compression), kips

P_o Load contributary to base plate area enclosed by the column, kips

P_p Bearing load on concrete, kips

P_u Applied factored load on eccentrically loaded bolt group or weld configuration, kips

P_u Factored concentrated beam load, kips

P_u Required axial strength (tension or compression), kips

P_{wb} Maximum column web resisting force at beam compression flange, kips

P_{wi} A factor consisting of combined terms from LRFD Specification Formula K1-2, used in a column web stiffener check for local web yielding, kips/in.
   $= F_{yw}t_w$

P_{wo} A factor consisting of combined terms from LRFD Specification Formula K1-2, used in a column web stiffener check for local web yielding, kips
   $= 5F_{yw}t_w k$

P_y Yield strength, kips

Q Factored prying force per bolt at design load, kips

Q Full reduction factor for slender compression elements

Q_a Reduction factor for slender stiffened compression elements

Q_f Statical moment for a point in the flange directly above the vertical edge of the web, in.^{3}

Q_n Nominal strength of one stud shear connector, kips

Q_s Reduction factor for slender unstiffened compression elements

Q_w Statical moment at mid-depth of the section, in.^{3}

R Nominal load due to initial rainwater or ice exclusive of the ponding contribution

R Design reaction, kips

R Shear force in a single fastener or weld element at any given deformation, kips or kips/in./weld size, respectively

R ($N=$ 3½ in.) Maximum permitted reaction for a bearing length of 3½ in., kips

R_{BS} Resistance to web tear-out (block shear), kips

R_{PG} Plate girder bending strength reduction factor

R_e Hybrid girder factor

R_n Nominal resistance, kips

R_{ult} Ultimate shear load of a single fastener or weld element, kips or kips/in./weld size, respectively

R_v Web shear strength, kips

R_1 An expression consisting of the first portion of LRFD Specification Formula K1-3, kips

R_2	An expression consisting of the second portion of LRFD Specification Formula K1-3, kips
R_3	An expression consisting of the first portion of LRFD Specification Formula K1-5, kips
R_4	An expression consisting of the second portion of LRFD Specification Formula K1-5, kips
S	Elastic section modulus, in.^{3}
S	Spacing of secondary members in a flat roof, ft
S	Snow load
S'	Additional section modulus corresponding to $\frac{1}{16}$-in. increase in web thickness for built-up wide flange sections, in.^{3}
S_w	Warping statical moment at a point at the flange edge, in.^{2}
S_x	Elastic section modulus about major axis, in.^{3}
S_x'	Elastic section modulus of larger end of tapered member about its major axis, in.^{3}
S_{xt}, S_{xc}	Elastic section modulus referred to tension and compression flanges, respectively, in.^{3}
$(S_x)_{eff}$	Effective section modulus about major axis, in.^{3}
T	Distance between web toes of fillet at top and at bottom of web, in. $= d - 2k$
T	Factored applied tension per bolt (exclusive of initial tightening), kips
T	Horizontal force in flanges of a beam to form a couple equal to beam end moment, kips
T'	Half of the factored load on two angles or on a structural tee, in a hanger-type connection, using max. bending strength, kips/in.
T_b	Specified pretension load in high-strength bolt, kips
T_{stl}	Tensile force in steel in a composite beam, kips
T_{Tot}	Sum of tensile forces in a composite beam, kips
U	Approximate beam-column design coefficient
U	Reduction coefficient, used in calculating effective net area
U_p, U_s	Ponding stress index for primary and secondary members
V	Nominal shear strength per bolt, kips
V	Shear force, kips
V_n	Nominal shear strength, kips
V_u	Required shear strength, kips
W	Wind load
W	Uniformly distributed design load, kips
W_c	Uniform load constant for beams, kip-ft
W_{no}	Normalized warping function at a point at the flange edge, in.^{2}
W_u	Total factored uniformly distributed load, kips
X_1	Beam buckling factor defined by Formula F1-8
X_2	Beam buckling factor defined by Formula F1-9
Y_{ENA}	Distance from bottom of beam to elastic neutral axis, in.
Y_{con}	Distance from top of steel beam to top of concrete, in.

$Y1$ Distance from the plastic neutral axis to beam top flange, in.

$Y2$ Distance from concrete flange force to beam top flange in a composite beam, in.

Z Plastic section modulus, in.

Z' Additional plastic section modulus corresponding to $\frac{1}{16}$-inch increase in web thickness for built-up wide flange sections, in.[3]

a Clear distance between transverse stiffeners, in.

a Distance between connectors in a built-up member, in.

a Distance from bolt line to application of prying force Q, in.

a Eccentricity coefficient for weld groups

a Effective concrete flange thickness of a composite beam, in.

a Shortest distance from edge of pinhole to edge of member measured parallel to direction of force, in.

a_r Ratio of web area to compression flange area

b Compression element width, in.

b Distance from bolt centerline to the face of angle leg or tee stem in determining prying action, in.

b Effective concrete flange width in a composite beam, in.

b Vertical fastener spacing, in.

b Width of composite column section, in.

b_e Reduced effective width for slender compression elements, in.

b_{eff} Effective edge distance, in.

b_f Flange width of rolled beam or plate girder, in.

b_s End plate width, in.

c Length of cantilever measured from the centerline of column web or flange to edge of H-shaped area in base plate design, in.

c_1, c_2, c_3 Numerical coefficients used in the calculation of the modified yield stress and modulus of elasticity for composite columns

d Nominal fastener diameter, in.

d Overall depth of member, in.

d Pin diameter, in.

d Roller diameter, in.

d' Width of bolt hole parallel to tee stem or angle leg, in.

d_L Depth at larger end of unbraced tapered segment, in.

d_c Web depth clear of fillets, in.

d_h Diameter of a standard size hole, in.

d_o Depth at smaller end of unbraced tapered segment, in.

e Base of natural logarithm $= 2.71828\ldots$

e_o Horizontal distance from the outer edge of channel web to its shear center, in.

f Computed compressive stress in the stiffened element, ksi

f_a Computed axial stress in column, ksi

f_{b1} Smallest computed bending stress at one end of a tapered segment, ksi

f_{b2} Largest computed bending stress at one end of a tapered segment, ksi

f'_c Specified compressive strength of concrete at 28 days, ksi
f_o Stress due to $1.2D + 1.2R$, ksi
f_t Computed tension stress in bolts or rivets, ksi
f_{un} Required normal stress, ksi
f_{uv} Required shear stress, ksi
f_v Computed shear stress in bolts or rivets, ksi
g Transverse center-to-center spacing (gage) between fastener gage lines, in.
g Acceleration due to gravity $= 32.2$ ft/sec.^{2}
h Clear distance between flanges less the fillet or corner radius for rolled shapes; and for built-up sections, the distance between adjacent lines of fasteners or the clear distance between flanges when welds are used, in.
h Depth of composite column section, in.
h_c Assumed web depth for stability, as defined in Sect. B5.1, in.
h_r Nominal rib height, in.
h_s Factor used in Formula A-F4-6 for web-tapered members
h_w Factor used in Formula A-F4-7 for web-tapered members
j Factor defined by Formula F3-1, for minimum moment of inertia for a transverse stiffener
k Web plate buckling coefficient
k Distance from outer face of flange to web toe of fillet, in.
k_1 Distance from web center line to flange toe of fillet, in.
k_1, k_2 Regression coefficients for determining ultimate shear strength of a weld group
l Largest laterally unbraced length along either flange at the point of load, in.
l Length of bearing, in.
l Length of vertical weld, in.
l Load eccentricity for eccentrically loaded bolt group, measured perpendicular to the applied force line of action, in.
l_h Distance from center of fastener hole to end of beam web, in.
l_v Distance from center of fastener hole to free edge of connected part in direction of force, in.
m Cantilever dimension of base plate as defined in Part 2, in.
m Ratio of web to flange yield stress or to critical stress in hybrid beams
m Ratio of average root area to body area of structural bolts (taken as 0.75)
m Coefficient for converting bending to an approximate equivalent axial load in columns subjected to combined loading conditions
n Cantilever dimension of base plate as defined in Part 2, in.
n Number of shear connectors between point of maximum positive moment and the point of zero moment to each side
n Number of fasteners in a vertical row
p Length of flange that the hanger is attached to, parallel to stem or leg of hanger, tributary to each bolt, in determining prying action, in.
r Governing radius of gyration, in.

r	Distance from instantaneous center of rotation to individual fastener or weld element force, in.
r_T	Radius of gyration of compression flange plus one third of the compression portion of the web taken about an axis in the plane of the web, in.
r_{To}	Radius of gyration, r_T, for the smaller end of a tapered member, in.
r_i	Minimum radius of gyration of individual component in a built-up member, in.
r_m	Radius of gyration of steel shape, pipe or tubing in composite columns. For steel shapes it may not be less than 0.3 times the overall thickness the composite section, in.
r_o	Distance from center of gravity of bolt group to the instantaneous center of rotation, in.
$\bar{r}_o$	Polar radius of gyration about the shear center, in.
r_{ox}, r_{oy}	Radius of gyration about x and y axes at the smaller end of a tapered member respectively, in.
r_v	Nominal shear or bearing value for one fastener, kips
r_x, r_y	Radius of gyration about x and y axes respectively, in.
s	Longitudinal center-to-center spacing (pitch) of any two consecutive holes, in.
t	Thickness of connected part, in.
t	Thickness of the critical part, in.
t	Thickness of bearing plate, in.
t	Weld throat dimension, in.
t_b	Thickness of beam flange or connection plate delivering concentrated force, in.
t_c	Flange or angle thickness required to develop the design tensile strength in the bolts with no prying action, in.
t_f	Flange thickness, in.
t_f	Flange thickness of channel shear connector, in.
t_p	Thickness of base plate, in.
t_s	End plate thickness, in.
t_w	Web thickness of channel shear connector, in.
t_w	Web thickness, in.
w	Fillet weld size, in.
w	Plate width; distance between welds, in.
w	Subscript relating symbol to strong principal axis of angle
w	Unit weight of concrete, lbs./cu. ft
w_r	Average width of concrete rib or haunch, in.
x	Subscript relating symbol to strong axis bending
x_o	Horizontal distance from centroid of bolt group to eccentrically applied factored load, in.
x_o, y_o	Coordinates of the shear center with respect to the centroid, in.
x_p	Horizontal distance from designated edge of member to plastic neutral axis, in.

x_1	Horizontal distance from designated edge of member to center of gravity, in.
y	Moment arm between centroid of tensile forces and compressive forces, in.
y	Subscript relating symbol to weak axis bending
y_p	Vertical distance from designated edge of member to plastic neutral axis, in.
y_1, y_2	Vertical distance from designated edge of member to center of gravity, in.
z	Distance from the smaller end of tapered member used in Formula A-F4-1 for the variation in depth, in.
z	Subscript relating symbol to weak principal axis of angle
Δ	Beam deflection, in.
Δ	Total deformation of fastener or weld element, in.
Δ_{oh}	Translation deflection of the story under consideration, in.
Ω_o	Shear center coordinate, in.
α	Ratio of moment at bolt line to moment at stem line for determining prying action in hanger-type connection
α_m	Coefficient used in calculating the design moment, M_e, for end plate connections
γ	Depth tapering ratio
γ	Subscript relating symbol to tapered members
δ	Ratio of net area (at bolt line) and gross area (at face of stem or angle leg) used in determining prying action for hanger-type connections
ζ	Exponent for alternate beam-column interaction equation
η	Exponent for alternate beam-column interaction equation
λ_c	Column slenderness parameter
λ_e	Equivalent slenderness parameter
λ_{eff}	Effective slenderness ratio defined by Formula A-F4-2
λ_p	Limiting slenderness parameter for compact element
λ_r	Limiting slenderness parameter for noncompact element
μ	Coefficient of friction
ϕ	Resistance factor
ϕ_b	Residence factor for flexure
ϕ_c	Resistance factor for bearing on concrete
ϕ_c	Resistance factor for compression
ϕ_c	Resistance factor for axially loaded composite columns
ϕ_p	Resistance factor for base plate
ϕ_r	Resistance factor for compression, used in web crippling equations
ϕ_{sf}	Resistance factor for shear on the failure path
ϕ_t	Resistance factor for tension
ϕ_v	Resistance factor for shear
kip	1,000 pounds
ksi	Expression for stress in kips per sq. in.

INDEX